Teacher's Edition

Technology & Engineering

Sixth Edition

R. Thomas Wright
Professor Emeritus, Industry and Technology
Ball State University
Muncie, Indiana

Publisher
The Goodheart-Willcox Company, Inc.
Tinley Park, Illinois
www.g-w.com

Manufactured in the United States of America.

ISBN 978-1-60525-414-2

1 2 3 4 5 6 7 8 9 – 12 – 16 15 14 13 12 11 10

TSA Modular Activities: The Technology Student Association (TSA) is a nonprofit national student organization devoted to teaching technology education to young people. TSA's mission is to inspire its student members to prepare for careers in a technology-driven economy and culture. The demand for technological expertise is escalating in American industry. Therefore, TSA's teachers strive to promote technological literacy, leadership, and problem solving to their student membership.

TSA Modular Activities are based on the Technology Student Association competitive events current at the time of writing. Please refer to the *Official TSA Competitive Events Guide,* which is periodically updated, for actual regulations for current TSA competitive events. TSA publishes two *Official TSA Competitive Events Guides*: one for middle school events and one for high school events.

To obtain additional information about starting a TSA chapter at your school, to order the *Official TSA Competitive Events Guide,* or to learn more about the Technology Student Association and technology education, contact TSA: Technology Student Association, 1914 Association Drive, Reston, VA 20191-1540, www.tsaweb.org.

Teacher's Edition Contents

Technology & Engineering

A teaching package to help students learn how technology affects people and the world in which we live.

- Fully correlated with ITEEA's new national Standards for Technological Literacy.
- Covers the scope of technology, technological system components, tools of technology, problem solving and design, technological contexts, technological enterprises, and technology and society.
- An activity-based program.
- Full-color illustrations, referenced within the text, help students associate visual images with written material.
- Logical organization, colorful headings, and readable typeface facilitate reading comprehension.

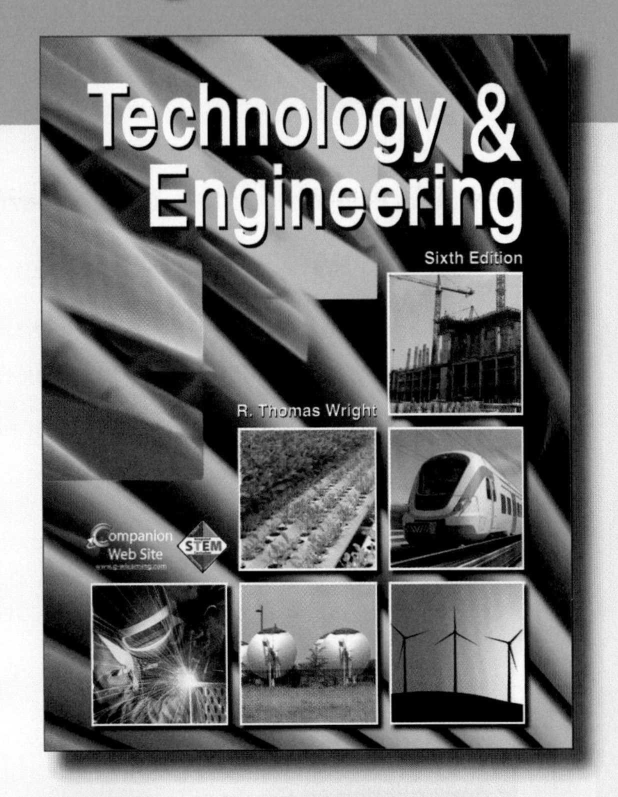

Essential Program Components

A variety of support materials are available to help you teach *Technology & Engineering.*

Student Text

Numerous illustrations, easy-to-read text, and student-friendly features enable understanding of how people use technology and why technological systems work the way they do. (See page 6–9 for a summary of text features.)

Teacher's Edition

Each chapter opens with a chapter outline and a list of the Standards for Technological Literacy addressed. Page margins include teaching aids to guide your review and reinforcement of key topics. Answer keys appear with the chapter review questions.

Tech Lab Workbook

Includes activities related to textbook activities to help students review and apply chapter concepts.

Teacher's Manual

Includes teaching strategies, answer keys, chapter tests, and reproducible masters.

Teacher's Resource CD

Gives you easy access to the content of the *Teacher's Manual* and the *Teacher's Edition* of the text, both in PDF format. Also includes lesson slides and a lesson planning feature for customizing daily lesson plans.

EXAMVIEW® Assessment Suite

Allows you to quickly and easily create tests from a bank of questions relating to the textbook. Lets you choose specific questions, add your own, and create different versions of a test.

Instructor's Presentations for PowerPoint®

Includes colorful presentations with a focus topic for each chapter.

Companion Web Site

Online content, such as Animations, Crossword Puzzles, and Matching Games help you motivate and engage students beyond the classroom.

Introduction

Technology and Technology Education—A Background

People live in a human-built world. We eat food that was raised and processed using technology. We live in constructed homes and work in constructed offices, stores, and factories. We use energy through thousands of technological devices. We use thousands of manufactured products, which range from the simple wooden pencil to complex computer systems. We communicate using technological devices, systems, and products such as printed and electronic media. We maintain our health and treat illnesses using technology. We move freely on transportation vehicles, and we receive goods through transportation systems. Our life is enriched because of technology.

Not all the results of technology are positive, however. People using technology have polluted our water and air. Technological devices such as wind generators and cellular telephone towers mar the beauty of the landscape. Communication media shape our attitudes. Neighborhoods are impacted by transportation systems. Construction projects can change some people's entire way of life.

Without an adequate knowledge of technology, people are afraid of the results of technological change. The hope for the future lies in a populace that can understand and direct the path of technological development. The technology education discipline is accepting the challenge of teaching this understanding. Technology education interprets the practices of designing, operating, and using technology and the impacts of these actions on people, society, and the environment.

Through technology education experiences, students discover that the word *technology* has different meanings in different contexts and that different people see it differently. Specifically, students learn that technology can be described as products (computers, lasers, fiber-optic communication systems), as processes (computer-aided design [CAD], nuclear power generation, robotic assembly), or as organizations (industrial organizations, company structures), **Figure IN-1.**

Furthermore, through the study of technology, students are able to differentiate between technology and other areas of human knowledge and practice, such

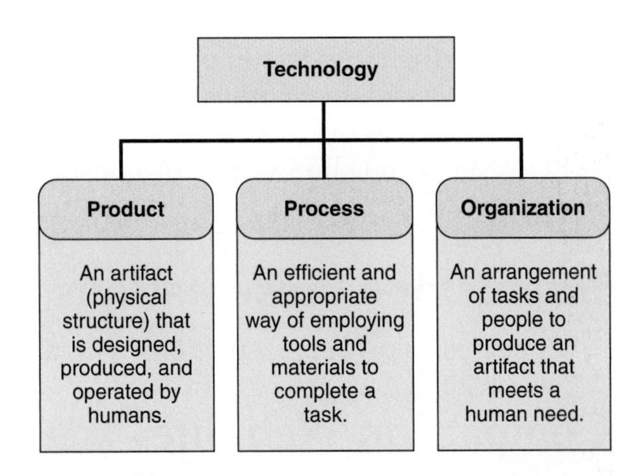

Figure IN-1.

as science. Students discover that science searches for laws and principles that explain the natural world. Technology, in contrast, explores how people develop and use tools, materials, and systems to control and live in harmony with natural and human-made environments, **Figure IN-2.**

Technology education presents technology as a systematic human activity. It shows that all technologies have inputs, processes, and outputs. Technology education demonstrates how technological systems use feedback to ensure a system operates within established parameters, **Figure IN-3.** Finally, technology is presented as purposeful human action. Technology is the result of human volition. Technology is developed to meet an identified human want or need.

Technology . . .

... is human knowledge about using tools, materials, and systems to produce artifacts (outputs) that modify or control the environment.

Figure IN-2.

Technology can be viewed in its entirety. Questions such as "What is technology?" "How does technology impact human life?" and "What is a technological system?" can be explored. This is useful information. This broad approach, however, does not allow students to explore specific technologies. For this reason, most technology educators will first introduce a broad view of technology and then explore several contexts in which technology is applied. These contexts allow students to explore a number of technological systems. One common grouping of technologies views them as human productive activities. This grouping explores those technological actions that people have done, are doing, and most likely will be doing in the future. These groupings are agriculture, construction, information and communication, energy-conversion, manufacturing, medical, and transportation technologies, **Figure IN-4.** This approach proposes that people have used, are using, and will use technology to grow and process foods, communicate and process information, construct structures, convert and apply energy, promote wellness, treat illnesses, manufacture products, and transport cargo.

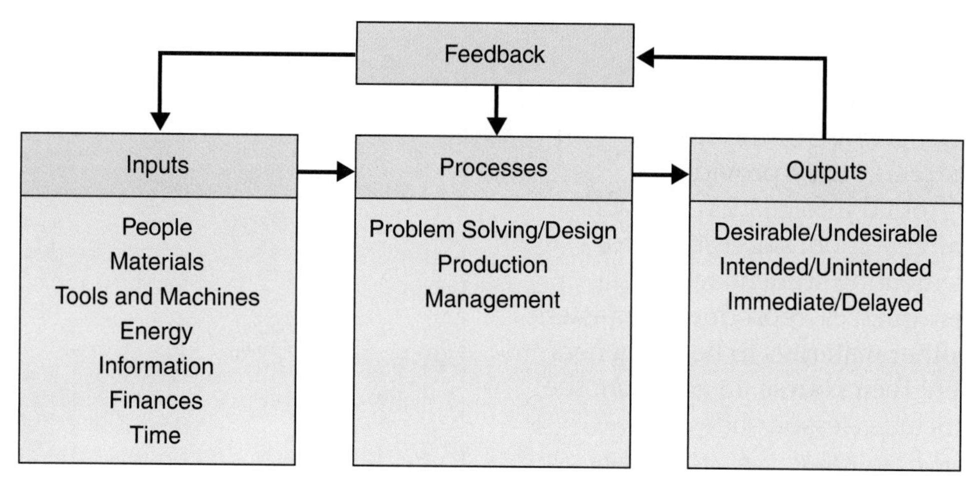

Figure IN-3.

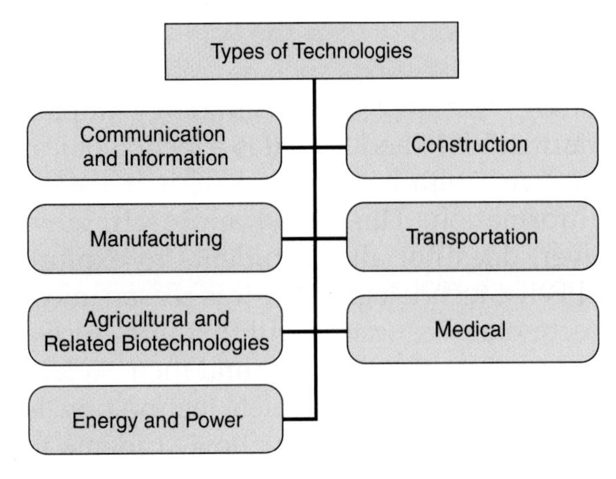

Figure IN-4.

Technology Systems—An Instructional Approach

This instructional package has been developed to allow students systematically to study technology systems as phenomena that directly impact their lives and to study technology as something they have the power to develop, use, and control. The package includes three important teaching resources:

- *Technology & Engineering*, which includes information about technology and its various contexts, lists of key words, review questions, and laboratory activities.
- The teacher's resource, either the *Teacher's Manual* or the *Teacher's Resource CD*. Each provides background information for each teaching section, suggestions for implementing student activities, review quizzes, reproducible masters, and other materials to help teachers present their course. In addition, the CD includes lesson slides.
- *Tech Lab Workbook*, which includes forms that support the activities in the textbook.

These materials have been organized into teaching sections that correspond to the sections of the student textbook. Each section covers several chapters in the textbook. Each section has a specific theme and includes an estimate of the time needed to cover the material. This structure is shown in **Figure IN-5.**

The instructional system for the student of technology has been developed with specific roles for the teacher, students, textbook, and activities in mind. These roles are summarized in the following.

Role of the Student

School programs are designed to help students prepare for meaningful and rewarding participation in society. The roles students play in their various classes during a typical school day are markedly different. Technology education has some

36-Week (Two-semester) Schedule

	Section	Focus	Text Chapters	Activities	Weeks
Semester I	1	Technology	1–3	1A & B	1
	2	Systems	4–6	2A & B	1
	3	Tools	7–8	3A & B	3
	4	Design	9–12	4A & B	5
	5	Production	13–18	5A–D	6
		Unassigned time			2
Semester II	6	Communication	19–23	6A & B	4
	7	Transportation	24–26	7A & B	3
	8	Energy	27–28	8A & B	3
	9	Agriculture and Medicine	29–31	9A & B	3
	10	Management	32–34	10A & B	3
	11	Society	35–36	11A & B	1
		Unassigned time			1

Figure IN-5.

features that make it a unique subject, **Figure IN-6.** One of these features relates to the role of students in approaching the course content. Students are considered active participants. Students are not viewed as passive receptors of wisdom. Students in technology education are included in the decision-making process. The teacher sets the boundaries of the study, but the students are involved in developing the structure and the activities that exist within these boundaries. For example, students do not have a choice on whether to study construction technology. However, they should be directly involved in deciding which structures they will model during the study. This approach requires students to be accountable for their learning and directed in their actions.

Characteristic	Description
Mission	Technology education is general education in the same way science, mathematics, and language arts are.
Focus	Technology education focuses on technological systems and their social/cultural impacts. Technical means are used to help students develop deeper understandings of the essential concepts of technology.
Emphasis	Technology education is an *activity-centered* study that helps students understand the technological portion of society by emphasizing the *why* behind the *how.*
Goal	Technology education strives to enable students to become technologically literate (able to find, interpret, apply, and evaluate information about technology).
Instructional Practices	Technology education is designed to encourage students to become problem solvers, flexible thinkers, and group participants through the use of a combination of individual and group learning activities that students help plan and carry out. Teacher-developed procedures have limited use in this program.
Thinking Mode	Technology education encourages divergent thinking (What are the WAYS we could do this task?), followed by convergent thinking (Which of their ways will work best at what point in time?).
Teacher Role	The technology teacher is a facilitator of learning, as opposed to the person who knows THE answer. He or she manages the learning environment.
Student Role	The students in technology education are encouraged to be creative and active participants. This requires teachers to have faith in people—a *can do* philosophy. Technology education has high expectations for all students.
Safety	Technology education places strong emphasis on safe laboratory practices. It views safety as both an attitude and a practice developed through teacher-student activities. Therefore, the teacher is expected to work with the students to develop essential safety rules and procedures.

Figure IN-6.

Role of the Teacher

Teachers of technology education must have three areas of competence. First, they must possess a broad understanding of technology and industry and the impacts they have on society, individuals, and the environment. Second, the teachers must possess a level of technical skill in the designing and production of products and structures, the use of tools and machines, and the maintenance of technological devices and systems. Third, technology teachers must be capable of selecting and using a wide range of teaching methods and strategies to facilitate student learning.

Technology teachers must also have a special attitude about the subject matter and students. They must believe that students can, and are willing to, learn. This belief must be communicated through an excitement about the subject matter—"Technology is fun and interesting to learn about!"—and a concern for students—"I am truly interested in you and your educational achievements!" Finally, technology teachers must realize that there are not many "right" answers. There are usually a group of possible answers from which an appropriate answer may be selected.

The attributes and attitudes are reflected in the classroom where the technology teacher:

- Structures the overall class and interacts with students to structure specific learning activities.
- Guides students toward appropriate solutions.
- Presents short, exciting content presentations that include student interactions.
- Uses student input to select products, structures, and systems to be fabricated and tested.
- Develops and demonstrates, with student interaction, appropriate and safe production procedures that are used to fabricate products, structures, and systems. (Student safety is always important and is the responsibility of the teacher. No textbook or set of rules can teach safety. They can only support teacher actions. The proper, safe operation of tools and machines should be presented and reinforced in the laboratory setting.)
- Serves as a technical adviser for students. It is better to react to student solutions than to give them the one "right" way to do a task. This approach requires the student to ask, "Do you think the design our group developed can be produced?" rather than "What are we going to make?" This approach may cause the class to cover less content, but it helps students learn the content more thoroughly. The goal of technology education is to develop key concepts about technology, not to cover large volumes of specific information that may become obsolete before it is useful to the student. This goal allows a course to sacrifice a certain amount of depth in order to cover broad content in a more meaningful way.
- Develops appropriate evaluation instruments and procedures. The Teacher's Resource and the textbook contain review questions and sample quizzes. Do not hesitate to develop your own instruments that more closely match your instruction methods and your students.
- Adjusts content for individual student abilities and learning styles. The problem-solving and open-ended nature of technology education activities makes this task much easier. Each instructional section of this manual provides hints for adjusting the activities to accommodate varying student abilities.

People learn at differing rates, using different approaches. This is coupled with people possessing varying degrees of eye-hand coordination (physical skill) and ability to see spatial relationships (visualization skill). These factors present the technology educator with a unique challenge: structuring the course so every student has the opportunity to succeed.

This requires that various learning strategies be selected to reach course objectives and that these strategies be selected with student abilities in mind. The chart in **Figure IN-7** presents some approaches and types of activities that can be used to meet instructional objectives.

Helping Students Gain Basic Information	
Textbook Reading	Films
Filmstrips	Field Trips
Slide Series	Guest Speakers
Videotapes	
Helping Students Develop Creativity	
Product Design	Brainstorming
Structure Design	Advertising Design
Helping Students Develop Problem-Solving Skills	
Product Design	Financial Planning
Process Planning	Structure Design
Helping Students Apply Technology	
Product Design Structure	Material Processing
Product Production	Managing
System Testing	Structure Fabrication
Helping Students Develop Communication Skills	
Class Discussions	Market Surveys
Engineering Reports	Product Presentations
Laboratory Summaries	Product Sales
Helping Students Develop Leadership Skills	
Decision Making	Managing
Cooperating	

Figure IN-7.

These activities should be modified to meet the needs of the gifted, the average, and students with special needs.

Gifted students present a challenge because they may complete assigned work quickly or become bored with routine assignments. These students often need to be challenged with open-ended activities that focus on societal issues and require the use of problem-solving skills and creativity.

Special needs students may require more structure, short-range goals, and activities that build on their strengths. Class activities in technology education often involve group work, allowing these students to complete simpler, yet essential, tasks that contribute to group success and develop self-esteem. At no time should special needs students be relegated to insignificant or nonessential tasks. The laboratories have many tasks that special needs students can complete that are still important for the success of the class activity.

Each section of the Teacher's Resource provides suggestions for adjusting course content to accommodate the gifted and the special needs student.

Role of the Textbook

The textbook is a vital part of an instructional system. It is designed to give students the background to understand the course content. Technology education does not accept skill development and mastery of technical knowledge as a central mission. Therefore, *Technology & Engineering* does not contain specific process or machine operation procedures, technical data (such as screw sizes or abrasive grits), or specific sets of safety rules to be memorized.

Technology & Engineering should be used in delivering the basic content for the course and as a reference in completing laboratory activities. Students should be given reading assignments for their study periods or as homework. They should bring the book to class each day, so they can use it as a source of specific information for completing laboratory assignments.

Role of the Teacher's Resource

The Teacher's Resource is a resource to aid the teacher. The Teacher's Resource is *not* the final answer for presenting a course about technology systems. The final answer resides in the teacher. The teacher is ultimately responsible for selecting, delivering, and evaluating student learning activities.

To assist the teacher, the Teacher's Resource serves several purposes. This resource:

- Offers a rationale for teaching technology systems and for structuring the program elements.
- Presents approaches and procedures for selecting, organizing, delivering, and evaluating student learning activities.
- Lists references for teacher information and teacher development.
- Presents possible laboratory activities for the various sections in the course.
- Indicates the conceptual structure for the various topics to be covered.
- Suggests ways to accommodate students with varying abilities and interests.
- Provides reproducible masters and lesson slides (lesson slides are on the CD only) to aid in student interest and understanding. These materials can be used to introduce fundamental concepts, to generate discussion about the topic, and to facilitate your presentation to the class. Many of the lesson slides can be reproduced as a handout, as well.

The Teacher's Resource should be read and studied well in advance of implementing a course in technology systems. At no time does this resource attempt to be a "cookbook" containing set directions, laboratory procedures, or lecture outlines for teaching the course. The Teacher's Resource contains suggestions to aid teachers in designing and presenting an exciting course.

The Teacher's Resource is not designed to fit the concept of an "ideal" laboratory, an "ideal" class, an "ideal" course length, or the illusive "average" student. It is designed with enough flexibility to facilitate a program in a variety of traditional and contemporary laboratories with students of varying abilities and interests.

Reading the Teacher's Resource from cover to cover prior to implementing the course provides you with:

- An overview of the course and how its parts are logically integrated.
- An opportunity to prepare long-range plans for:
  A. Lesson and activity planning.
  B. Supplemental resources.
  C. Support material development (transparencies and slides, for example).
  D. Laboratory modification and improvement.

The Teacher's Resource contains features that include:

- Clearly defined instructional objectives.
- Instructional strategies that can be used in teaching situations for students of varying ability levels.
- Suggestions and directions to reinforce, extend, and enrich the learning of students.
- Instructional strategies and extensive activities that require the use of higher-order thinking skills.

- Suggestions for appropriate questions to guide student discovery and exploration.
- Answers to questions and solutions to problems.
- A variety of evaluative procedures for assessing student achievement and for documenting mastery of content. These procedures include written quizzes and laboratory experiences.
- Background information for teaching content areas in the textbook that have not traditionally been in a subject or course.
- Problems and activities that require the application of technology concepts to life skills.

Evaluation instruments (quizzes and rating scales) and procedures help the teacher address three important aspects of the program:

- Student progress and achievement.
- Teacher and teaching method effectiveness.
- Program adequacy.

Evaluating for the purpose of assigning student grades is only a portion of the value of the continuous assessment of student progress. Data gathered during evaluation allow the teacher to identify student learning difficulties, determine concepts that have been miscommunicated or misunderstood, and measure student progress toward course objectives.

Role of the Tech Lab Workbook

The *Tech Lab Workbook* for *Technology & Engineering* is used up during the course. It contains many of the forms and activity sheets needed to conduct the laboratory activities. Most of the forms and activity sheets are designed to be flexible. They are structured to fit more than one product or process.

Individual teachers may want to supplement sheets in the *Tech Lab Workbook* with forms tailored to the specific laboratory activities they select for the course.

A significant feature of the *Tech Lab Workbook* is the professional-looking forms that can save the teacher many hours of preparation. These forms have been designed to allow an open program or approach.

Laboratory Activities

The goal of technology education is *technological literacy*, which contributes to the students' ability to participate in society. These contributions are made through citizenship (voting and community service involvement, for example), careers (selection and contribution), and consumer roles (purchasing and using, for example).

The laboratory activities are designed to make abstract concepts easier for the student to understand. The activities are *not* designed to develop tool skills or specific technical knowledge. That task is the purview of occupational education at vocational schools, career centers, community colleges, and universities.

The laboratory activities allow students to gain experience in solving problems. Students participate in group and individual activities involving the design, construction, operation, and maintenance of technological structures and systems. They also give students practice in communicating technological information.

Each section in this course has two types of laboratory activities—design problems and fabrication problems. Design problems use a *design brief* to communicate a situation, state a problem, or describe an opportunity. They then challenge the student to action. Students use their own creativity, problem-solving abilities, and management skills to address the design brief. Design problems do not have procedures developed and presented by the author. Rather, they have general guidelines, which are given to help teachers work with students in identifying processing and safety procedures.

Fabrication problems are more structured. The students are given a set of plans or a procedure to use investigating a specific technology. An effective way to present fabrication problems to the students is to have them record the laboratory procedures as they are demonstrated. This method allows the teacher to interact with the class, as opposed to simply dictating "the way" to do the task. Also, by having students complete procedure sheets while the demonstration is presented, they are encouraged to give close attention to the task being presented.

A final suggestion for laboratory activities is to use a combination of the design and the fabrication problems. This keeps the students challenged, and it keeps the laboratory experiences fresh and challenging for the teacher. Using the same type of activity over an extended period of time may cause students and parents to inappropriately view the course as stagnant.

The laboratory and laboratory activities are vital parts of a strong technology education program. This program has been designed with few laboratory restrictions. It can be conducted in both traditional material-centered (woodworking or metalworking) laboratories and general technology laboratories. This characteristic has been achieved by not specifying activities that "must" be conducted in order to complete the mission of the course. Each section presents a number of activities. Thus, the program is relatively easy to implement in any broadly equipped laboratory that supports material processing.

"Taking Your Technology Knowledge Home" Activities

Sections 3 through 10 have special activities called "Taking Your Technology Knowledge Home." These activities allow students to work with parents and guardians to apply what they learned in their technology education class to everyday challenges around their home and neighborhood. The activities can also be adapted to be performed in school.

Section 1
Technology

Chapter 1—Technology: A Dynamic, Human-Created System
Chapter 2—Technology as a System
Chapter 3—Types of Technological Systems

Teaching Materials

Text

Chapter 1, pages 22–37
  Test Your Knowledge, pages 36–37
  STEM Applications, page 37
Chapter 2, pages 38–53
  Test Your Knowledge, pages 52–53
  STEM Applications, page 53
Chapter 3, pages 54–65
  Test Your Knowledge, pages 64–65
  STEM Applications, page 65
Section Activities
  Activity 1A—Design Problem, page 66
  Activity 1B—Fabrication Problem, pages 66–67

Tech Lab Workbook

  Activity 1A—Design Problem, pages 7–8
  Activity 1B—Fabrication Problem, pages 9–10

Teacher's Resource

  Chapter 1 Quiz, pages 56–57
  Chapter 2 Quiz, pages 61–62
  Chapter 3 Quiz, pages 70–71
Reproducible Masters
  1-1: Technology Can Help or Harm, page 58
  2-1: Technology as a System, page 63
  2-2: Technological Processes, page 64
  2-3: Technological Processes, page 65
  2-4: Technological Processes, page 66
  2-5: Technological Processes, page 67
  2-6: Technological Processes, page 68
  3-1: Types of Technological Systems, page 72
Lesson Slides (CD only)
  1-1: Technology
  2-1: System Components
  3-1: How to Classify Technology

Organization of Section 1

In any educational program, students should understand the scope and sequence of the course they are taking. It is also a good practice to introduce the students to your operational procedures and practices early in the course. Therefore, Section 1 is designed to address three major topics:

- The structure (scope and sequence) of the course.
- The hands-on nature of the study.
- Technology as an intellectual discipline and as a mode of addressing human problems and needs.

Specifically, this section allows students to develop an understanding of the role of technology in their lives and society. As shown in **Figure 1-1,** they will be introduced to the ideas that:

- Technology is human knowledge.
- Technology involves the use of tools, materials, and systems.
- Technology results in artifacts (physical objects) and other outputs (pollution and scrap, for example).
- Technology is developed by people to modify or control the environment.

Then, technology is contrasted with other types of human knowledge, **Figure 1-2.** Students will learn that people must use three types of knowledge to understand the world in which they live and to solve problems that arise. These types of knowledge are:

- **Science.** The knowledge of the natural world.
- **Technology.** The knowledge of the human-made world.

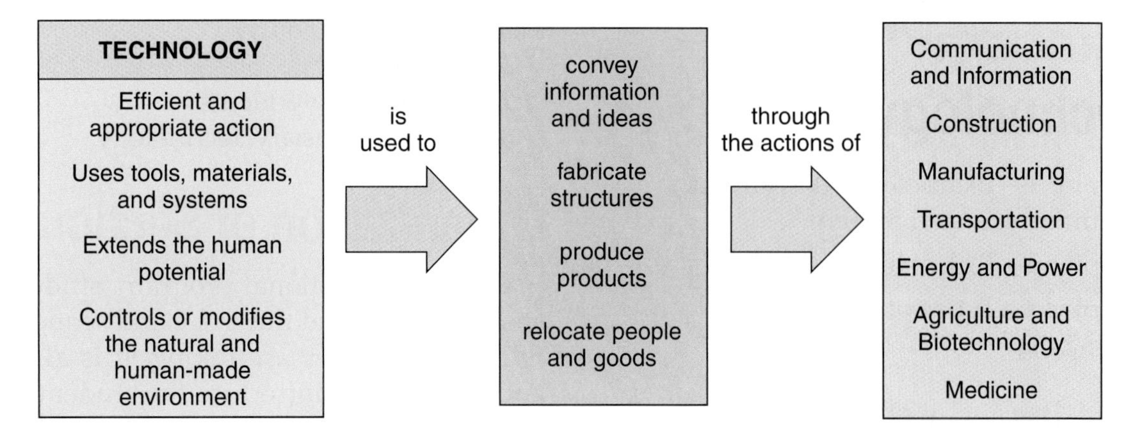

Figure 1-1.

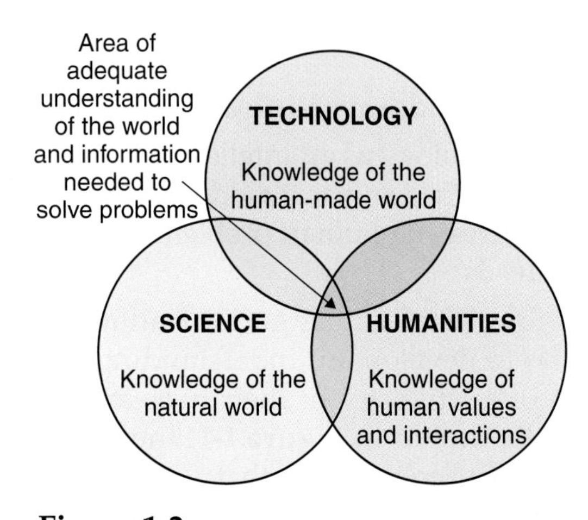

Figure 1-2.

- **Humanities.** The knowledge of human values and interactions.

From there, the students explore technology as a human-designed and human-operated system. The students examine the major parts of all systems and see how they interrelate, **Figure 1-3.** This study includes:

Inputs—The elements that flow into the system and are consumed or processed by the system:

- People.
- Materials.
- Tools and machines.
- Energy.
- Information.

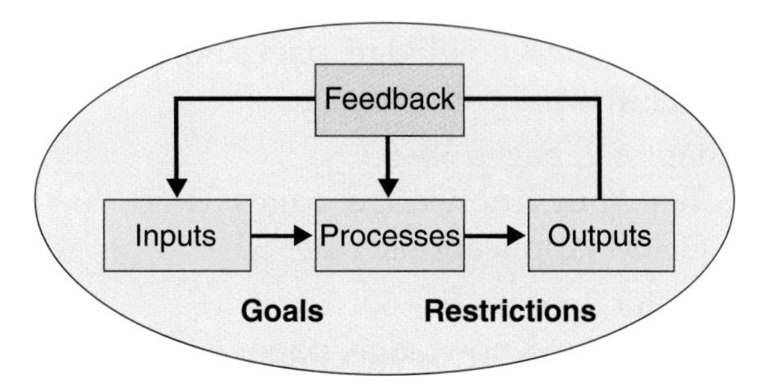

Figure 1-3.

- Finances.
- Time.

Processes—Actions used to design, produce, and deliver goods and services:

- Problem-solving/design processes.
- Production processes.
- Management processes.

Outputs—The results of operating the system:

- Intended and unintended.
- Desirable and undesirable.
- Immediate and delayed.

Feedback and Control—Using information about the outputs of a process or system to regulate the system.

Finally, the students see how technology is applied to economic activity. They are introduced to the major actions that change an idea into a product or service that is sold for a profit, **Figure 1-4.**

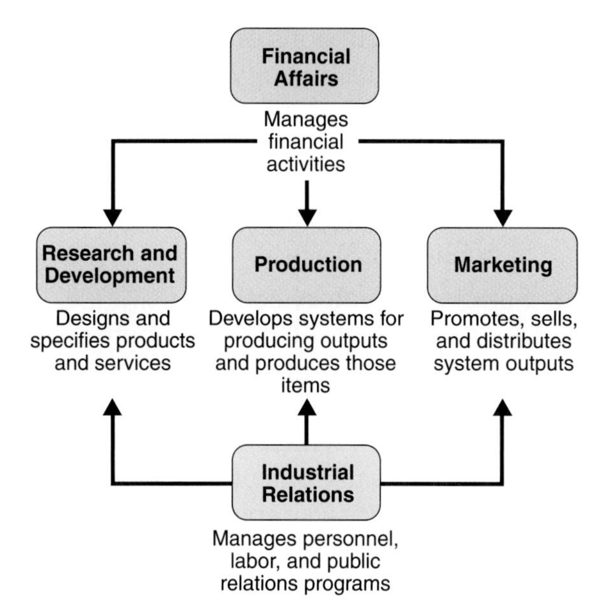

Figure 1-4.

- **Research and development.** Designing, developing, and specifying the characteristics of a product, structure, or service.
- **Production.** Developing and operating systems for producing the product, structure, or service.
- **Marketing.** Promoting, selling, and delivering a product, structure, or service.
- **Industrial relations.** Developing and administrating programs to ensure the relations between the company and its workers and the public are positive.
- **Financial affairs.** Obtaining the capital and physical resources and maintaining the financial records to manage the system properly.

This section includes three chapters and one laboratory activity. The chapters are Chapter 1—Technology: A Dynamic, Human-Created System; Chapter 2—Technology as a System; and Chapter 3—Types of Technological Systems.

As with each section, two different types of activities are provided. Those teachers wanting students to address the content from a problem-solving approach should select Activity 1A—Design Problem. A more structured laboratory activity is presented by Activity 1B—Fabrication Problem.

Presenting Section 1

Estimated length—1 week

Section 1 is designed to help students understand technology as:

- A body of knowledge of efficient and appropriate action.
- A system with inputs, processes, outputs, feedback and control, and goals.
- A group of practices that includes agricultural and related biotechnologies, communication and information technologies, construction technologies, energy and power technologies, manufacturing technologies, medical technologies, and transportation technologies.

The content of Section 1 lends itself to lecture/discussion strategies. It is more important, however, to use presentations that are reinforced with practical laboratory activities. This approach is appropriate for two reasons. First, students learn abstract concepts best when they are integrated with problem solving, critical thinking, and laboratory practice. Second, it is important to send the students a message that says, "This class is a fun, hands-on class where I can learn a great deal!"

Thus, it is important, starting with the first or second day, that all lectures/presentations be limited to only 10 to 15 minutes. All class periods need to have a large portion dedicated to laboratory activity. For this course, a laboratory activity is any experience in which students are actively involved in seeking solutions to a problem or any activity in which the students are engaged with tools, materials, or technological systems. This could involve designing products, brainstorming ideas, making sketches and drawings, completing activity sheets, building prototypes, or evaluating procedures.

A five-day schedule, shown in **Figure 1-5,** has been developed for this section. During this period of time, you will notice that three important topics are introduced through presentations and discussions. In addition, the students use simple tools to complete either a design or a fabrication problem.

Note
It is important that the students use tools and machines safely. Demonstrate appropriate safety procedures to your class prior to any laboratory activity.

To prepare to present this section, you should:

1. Read Chapters 1, 2, and 3 in *Technology & Engineering.*

Section 1: Technology	
Day	Activity
1	Complete "Start of the School Year" administrative details Discussion: The Technology Systems Course (Scope and sequence)
2	Discussion: "What Is Technology?" (Chapter 1) Introduce laboratory activity
3	Discussion: "What Is a Technological System?" (Chapter 2) Students work on laboratory activity
4	Students work on laboratory activity
5	Discussion: "What Are the Types of Technological Systems?" (Chapter 3) Students complete laboratory activity

Figure 1-5.

2. Prepare discussions on:
   A. What is technology? (Chapter 1).
   B. What are technological systems and their parts? (Chapter 2).
   C. What are the seven main types of technological systems? (Chapter 3).
3. Select a laboratory activity.
   A. Activity 1A—This is an open-ended problem-solving activity.
   B. Activity 1B—This is a structured activity in which groups of students use one of two techniques to produce a simple product. They then compare the process without technology to the process that uses technology.
4. Practice the selected laboratory activity.
   A. Prepare teaching materials (transparencies or bulletin boards, for example) to support the content.
   B. Determine which study questions and test questions will be used.

Reinforcing and Enriching

The following activities can be used to supplement and enrich the laboratory activities at the end of each textbook section and the STEM Applications included at the end of each textbook chapter.

1. Research early pottery making in Pueblo Indian cultures. Prepare either a report or a display to communicate the information gained from the research.
2. Select a simple device (tool or machine, for example). Make a bulletin board that shows the device as a technological system with inputs, processes, and outputs.

Adjustments for Exceptional Students

Both laboratory activities are designed to be done rather quickly. The level of sophistication can accommodate students of all ability levels. Gifted students should have more creative answers and may produce a more complex solution. Lower achievers can still solve the problem but with solutions within their ability level. In all cases, the amount of structure and teacher guidance should be adjusted to meet students' needs. Guard against providing too much structure, however, because this action often thwarts creativity and is less appealing to gifted students. In addition, careful grouping of students allows each student to bring his or her own special abilities to the group. Some students have more artistic abilities, others may have more dexterity, and still other students have more abstract reasoning ability. Throughout the course, group work allows students to develop cooperative work habits and to contribute their own unique abilities to the team.

Evaluation Procedures

The understanding that students have of technology and technological systems should be measured in several ways including:

- Observation and critical listening during class presentations and discussions.
- Worksheets and quizzes covering the cognitive material presented.
- Open-ended questions about the learned material and how technology is developed.
- Assessment of individual participation and contributions to group work.
- Assessment of the quality and creativeness of the solutions to the assigned design/fabrication problems.

Section 2
Technological-System Components

Chapter 4—Inputs to Technological Systems

Chapter 5—Technological Processes

Chapter 6—Outputs and Feedback and Control

Teaching Materials

Text

Chapter 4, pages 70–91
- Test Your Knowledge, pages 90–91
- STEM Applications, page 91

Chapter 5, pages 92–115
- Test Your Knowledge, pages 114–115
- STEM Applications, page 115

Chapter 6, pages 116–133
- Test Your Knowledge, pages 132–133
- STEM Applications, page 133

Section Activities
- Activity 2A—Design Problem, page 134
- Activity 2B—Fabrication Problem, pages 134–135

Tech Lab Workbook
- Activity 2A—Design Problem, pages 13–14
- Activity 2B—Fabrication Problem, pages 15–16

Teacher's Resource
- Chapter 4 Quiz, pages 81–82
- Chapter 5 Quiz, pages 88–89
- Chapter 6 Quiz, pages 94–95

Reproducible Masters
- 4-1: Basic Mechanisms, page 83
- 4-2: Classification of Materials, page 84
- 4-3: Solid Materials, page 85
- 4-4: Properties of Materials, page 86
- 5-1: Problem-Solving/Design Processes, page 90
- 5-2: Functions of Management, page 91
- 6-1: Types of Results from Technological Systems, page 96

Lesson Slides (CD only)
- 4-1: Inputs to Technological Systems
- 4-2: Types of Tools
- 5-1: Problem-Solving and Design Process

Organization of Section 2

Section 2 is designed to allow students to explore the important components of all systems: inputs, processes, outputs, and feedback and control. First, students will explore the inputs common to all technology systems, **Figure 2-1.** These inputs are:

- People.
- Tools and machines.
- Materials.
- Information.
- Energy.
- Finances.
- Time.

Next, students will investigate the three major types of processes used in technological systems in further depth than they did in Section 1, **Figure 2-2.** These processes are:

- **Problem-solving design processes.** Used to design and engineer the outputs of systems and the systems themselves.

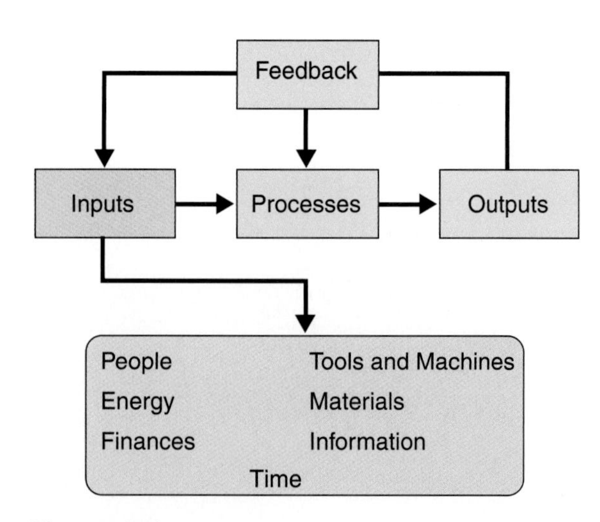

Figure 2-1.

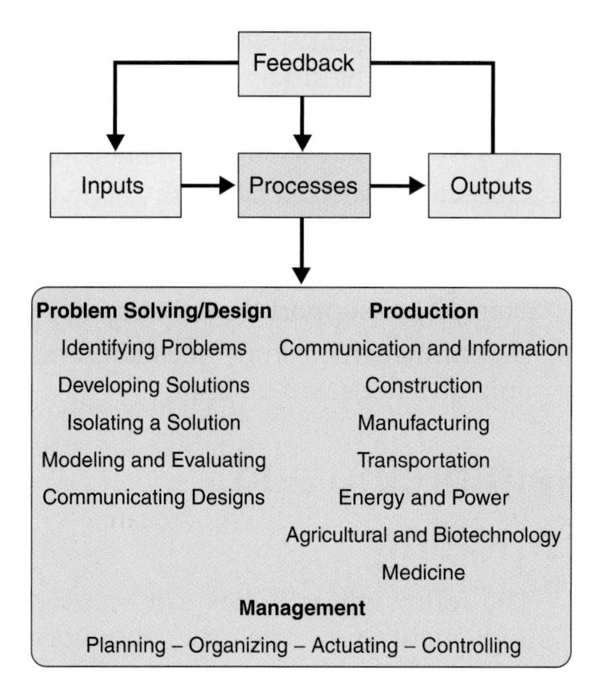

Figure 2-2.

- **Production processes.** Used to grow and process crops, produce products, convert and use energy, structures, produce media messages, treat illnesses, and transport people and cargo.
- **Management processes.** Used to plan, organize, actuate, and control technological systems.

Finally, students will explore the outputs of technological systems and feedback and control systems, **Figure 2-3.**

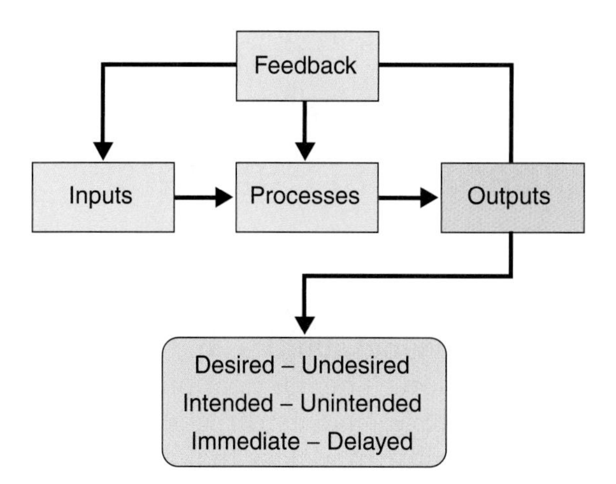

Figure 2-3.

For this section, the students spend at least five days covering the content of the three chapters and in completing a laboratory activity. These chapters are Chapter 4—Inputs to Technological Systems; Chapter 5—Technological Processes; and Chapter 6—Outputs and Feedback and Control.

This section, like all others in this course, has two different types of activities provided. The design problem allows students to engage in a creative, problem-solving approach, whereas the fabrication problem provides a more structured approach to the laboratory experience.

Presenting Section 2

Estimated length—1 week

Section 2 helps students understand technology as a system that:

- Has materials, tools and machines, people, information, energy, finances, and time as resources.
- Uses problem-solving/design, production, and management processes.
- Has desired and undesired outputs, intended and unintended outputs, and immediate and delayed outputs.
- Uses feedback to control the outputs.

The content of Section 2 can be presented by using just lecture/discussion strategies. Note, however, that technology education is based on the concept that presentations should be reinforced with realistic laboratory activities. Thus, about 75 percent of the time assigned to this section is dedicated to laboratory activity. The five-day schedule developed for this section reflects this emphasis, **Figure 2-4.**

Three important topics are introduced through presentations and discussions. These are (1) inputs to technology, (2) processes of technology, and (3) outputs and feedback and controls of technology. The students will be using simple tools to complete either a design or a fabrication problem. It is important that the students use

Section 2: Technological Systems	
Day	Activity
1	Discussion: "Inputs to Technological Systems" (Chapter 4) Introduce laboratory activity
2	Students work on laboratory activity
3	Discussion: "Technological Processes" (Chapter 5) Students continue work on laboratory activity
4	Discussion: "Outputs and Feedback and Control" (Chapter 6) Complete construction part of laboratory activity
5	Students demonstrate results of laboratory activity

Figure 2-4.

tools and machines *safely*. Demonstrate appropriate safety procedures prior to any laboratory activity.

To prepare to present this section, you should:

1. Read Chapters 4, 5, and 6 in *Technology & Engineering*.
2. Prepare discussions on:
   A. Inputs to technology systems (Chapter 4).
   B. Processes of technology (Chapter 5).
   C. Outputs and feedback and control of technological systems (Chapter 6).
3. Select a laboratory activity.
   A. Activity 2A—This activity is a creative, problem-solving activity.
   B. Activity 2B—This activity is a structured activity. Groups of students construct a simple technological system to lift a weight. The students then analyze the system in terms of inputs, processes, and outputs.
4. Practice the selected laboratory activity.
5. Prepare teaching materials (transparencies or bulletin boards, for example) to support the content.
6. Determine which study questions and quiz questions will be used.

Reinforcing and Enriching

The following activities can be used to supplement and enrich the laboratory activities at the end of each textbook section and the STEM Applications included at the end of each textbook chapter.

1. Have the students use a block of wood and a tongue depressor (craft stick) to make a lever. Have them experiment with first-class, second-class, and third-class levers. They should explain the levers as force or distance multipliers. The students should then give examples of where each type of lever could be used.
2. Prepare a bulletin board or three-dimensional display that shows a simple product as a technological system with inputs, processes, and outputs.

Adjustments for Exceptional Students

Both laboratory activities are designed to be done in a short period of time. The design problem is open ended to accommodate varying abilities and levels of sophistication. The fabrication problem is fairly simple, and its directions can be followed easily. In a class that has a wide range of abilities, students needing more direction and structure should be assigned the fabrication problem, and gifted students should complete the design problem. When this approach is used, both groups should report their results to the entire class.

Evaluation Procedures

The understanding that students have of technology and technological systems should be measured in several ways, including:

- Observation and critical listening during class presentations and discussions.
- Worksheets and quizzes covering the cognitive material presented.
- Open-ended questions about the learned material and how technology is developed.
- Assessment of individual participation and contributions to group work.
- Assessment of the quality and creativeness of the solutions to the assigned design/fabrication activities.

Section 3
Tools of Technology

Chapter 7—Production Tools and Their Safe Use

Chapter 8—Measurement Systems and Tools and Their Role in Technology

Teaching Materials

Text

Chapter 7, pages 138–167
- Test Your Knowledge, page 166
- STEM Applications, page 167

Chapter 8, pages 168–185
- Test Your Knowledge, page 184
- STEM Applications, page 185

Section Activities
- Activity 3A—Design Problem, page 186
- Activity 3B—Fabrication Problem, pages 186–187

Tech Lab Workbook
- Activity 3A—Design Problem, pages 19–21
- Activity 3B—Fabrication Problem, pages 23–25
- Taking Your Technology Knowledge Home, pages 27–30

Teacher's Resource
- Chapter 7 Quiz, pages 107–108
- Chapter 8 Quiz, pages 115–116

Reproducible Masters
- 7-1: Types of Motion, page 109
- 7-2: Sawing Machines, page 110
- 7-3: Electric Motor, page 111
- 8-1: Measurement, page 117

Lesson Slides (CD only)
- 7-1: Internal Combustion Engine
- 8-1: Standard- and Precision-Measurement Tools

Organization of Section 3

Section 3 has been developed to introduce the common tools and machines of technology and to give students an opportunity to use these tools and machines. The students will be introduced to tools and machines used in three major types of processing, **Figure 3-1:**

- Material processing.
- Energy processing.
- Information processing.

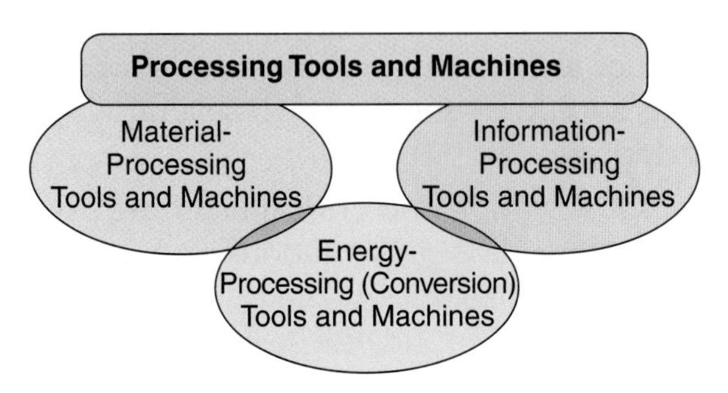

Figure 3-1.

When the students study material-processing tools and machines, they will explore the three essential elements that make machine tools what they are, **Figure 3-2:**

- **The cutting element or tool.** A device (tool) that removes excess material to produce the desired size and shape.
- **Motion.** The movement between the work and the tool that produces the cutting force (cutting motion) and brings new material into the cutting arena (feed motion).

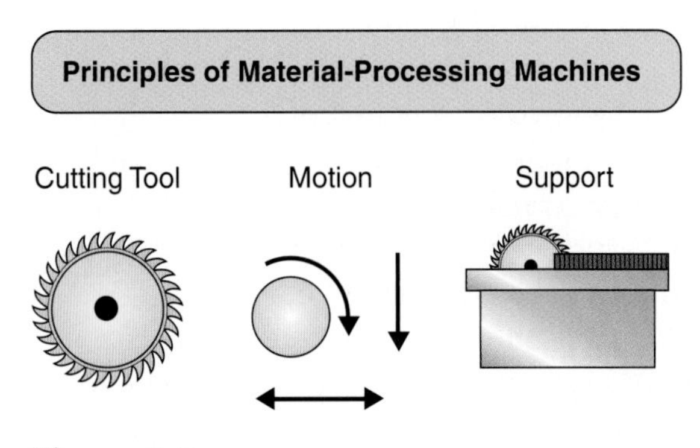

Figure 3-2.

- **Support.** Devices such as tables, arbors, or chucks that support and/or rotate the work or the tool.

The students will apply their knowledge of these three essential elements as they look at the major types of machine tools: turning, sawing, drilling, shaping, planing, and grinding. Following an introduction to material-processing tools and machines, the students will study three common energy-processing converters that have provided the foundation for the development of industry and advanced technology. These are the electric generator, the internal combustion engine, and the electric motor. This study will be followed by an opportunity for the students to study some information-processing developments. The specific focus will be on improving our ability to communicate information and ideas.

The students will then learn the guidelines for using technological devices safely. They will learn general safety rules and those specific to working safely with people, materials, and tools and machines.

The students will then be introduced to measurement systems and tools. They will explore the two major types of measuring systems: the U.S. customary system and the International System of Units (abbreviated SI from the French name *Système International d'Unités*) or metric system. They will learn that people have developed systems to measure size and space, mass, time, temperature, number of particles, electric current, and light intensity. They will also learn about standard and precision measurement and how to apply measurement to production and quality control activities. For this section, the students will spend approximately three weeks exploring material-, energy-, and information-processing tools and machines; general and specific safety rules; and various measurement systems and tools. They will spend their time covering and applying the content of the two chapters assigned to this section. These chapters are:

- Chapter 7—Production Tools and Their Safe Use.
- Chapter 8—Measurement Systems and Tools and Their Role in Technology.

As with most of the sections in this program, two different types of activities are provided. The design problem lets students engage in a creative, problem-solving approach to using information-processing technology. The fabrication problem provides a structured approach to using material-processing technology. Measurement should be stressed in both activities. Also, if time permits, you can use both activities to give students an opportunity to experience the two different types of processing technology and both design and fabrication practices.

Presenting Section 3

Estimated length—3 weeks

Section 3 is designed to help the students understand that all technology processes use inputs to make more useful outputs. The basic inputs are materials, energy, and information. Each type of processing relies on measurement, as does all technological activity. Section 3 lends itself to laboratory activities. These activities include information processing, which is presented in Activity 3A—Design Problem; material processing, which is presented in Activity 3B—Fabrication Problem; and energy processing, which can be presented in an alternate activity developed by you. Also, each process will be practiced in activities included in later sections. The 15-day schedule shown in **Figure 3-3** has been developed for this section. During this period of time, you will notice that four major topics are introduced through presentations and discussions. In addition, the students will use simple processing techniques to complete either a design or fabrication problem.

Note
It is important to explain how to use the tools and machines safely. Demonstrate appropriate safety procedures prior to any laboratory activity.

Section 3: Tools of Technology	
Day	Activity
1	Discussion: "Tools and Technology" Introduce laboratory activity
2	Discussion: "Material-Processing Tools" (Chapter 7) Demonstrate laboratory activity procedures
3–4	Students work on laboratory activity
5	Discussion: "Energy-Converting Tools" (Chapter 7) Students work on laboratory activity
6	Discussion: "Information-Processing Tools" (Chapter 7) Students work on laboratory activity
7	Students work on laboratory activity
8	Discussion: "Tools and Safety" (Chapter 7) Students work on laboratory activity
9	Discussion: "Measurement Systems and Tools" (Chapter 8) Students work on laboratory activity
10	Students work on laboratory activity or Introduce second laboratory activity
11–14	Students complete laboratory activity
15	Students present results of activity Summarize section

Figure 3-3.

To prepare to present this section, you should:

1. Read Chapters 7 and 8 in *Technology & Engineering.*
   A. Material-processing tools and machines (Chapter 7).
   B. Energy-converting tools and machines (Chapter 7).
   C Information-processing tools and machines (Chapter 7).

Note
More detailed experiences in material processing, energy processing, and information processing will be presented in Sections 5, 6, and 8.

   D. Safety rules when working with other people, materials, and tools and machines (Chapter 7).
   E. Measurement systems and tools (Chapter 8).
2. Select a laboratory activity.
   A. Activity 3A—This activity is an open-ended problem-solving activity that uses information-processing tools.
   B. Activity 3B—This activity is a structured activity where groups of students use material-processing tools to produce a simple product.
3. Practice the selected laboratory activity.
4. Prepare teaching materials (transparencies or bulletin boards, for example) to support the content.
5. Determine which study questions and quiz questions will be used.

Reinforcing and Enriching

The following activities can be used to supplement and enrich the laboratory activities at the end of each textbook section and the STEM Applications included at the end of each textbook chapter.

1. Have the students select an electrically powered product and analyze it in terms of the:
   A. Material-processing tools used to make it.
   B. Energy-conversion technology it uses.
   C. Information-processing activities used to produce the package.
2. Have the students who completed Activity 3B—Fabrication Problem develop a package for the product using information-processing practices.

Taking Your Technology Knowledge Home

This take-home activity allows students to apply their knowledge of tools to design a special tool for work around their home. They will:

1. Identify a task for which they could design a tool.
2. Research similar tools in magazines and catalogs.
3. List the features that the researched tools have.
4. List the features they will include in their tool.
5. Sketch four possible designs for the tool.
6. Select and refine the best design.
7. Build a prototype of the tool.
8. Ask one or more people to try out the tool.
9. List suggestions for improving the tool that the users made.
10. Prepare a sketch for an improved design for the tool.

Adjustments for Exceptional Students

Both laboratory activities are designed to accommodate students of all ability levels. Activity 3A is good for gifted students who can develop a wide range of creative answers and produce more complex solutions. Lower achievers can still solve the design problem but with solutions within their ability level.

Activity 3B is simple enough to accommodate students of several dexterity levels. You can build the tooling to allow less-skilled students to produce an acceptable product.

In all cases, the amount of structure and teacher guidance should be adjusted to meet the students' needs. Students should be given the minimum amount of structure they need to be successful. Practices that hamper creativity should be avoided.

Careful assignment of groups will allow students to contribute their own special skills and abilities to the activity. Using student teams also allows students to develop cooperative work habits.

Evaluation Procedures

The understanding that students have of technology and technological systems should be measured in several ways, including:

- Observation and critical listening during class presentations and discussions.
- Worksheets and quizzes covering the cognitive material presented.
- Open-ended questions about the learned material and how technology is developed.
- Assessment of individual participation and contributions to group work.
- Assessment of the quality and creativeness of the solutions to the assigned design and fabrication problems.

Section 4
Problem Solving and Design in Technology

Teaching Materials

Text

Tech Lab Workbook

Teacher's Resource

Organization of Section 4

Technology has two major components: a way to develop new knowledge or artifacts and an accumulated body of knowledge about technological practices. We can categorize the latter as knowledge about the following systems: agriculture and related biotechnologies, communication and information, construction, energy and power, manufacturing, medicine, and transportation. The students will explore these types of knowledge in later sections. This section will allow them to investigate the first component, which is often called the problem-solving or design method. It involves four major phases, as shown in **Figure 4-1:**

- Identifying or recognizing a technological problem.
- Developing a solution.
- Evaluating the solution.
- Communicating the solution.

During the study of these major steps, the students will first identify and define a problem. They will then use various sketching techniques to record, evaluate, and modify product or structure ideas. As shown in **Figure 4-2,** they will:

- Use rough sketching techniques to generate multiple product ideas.

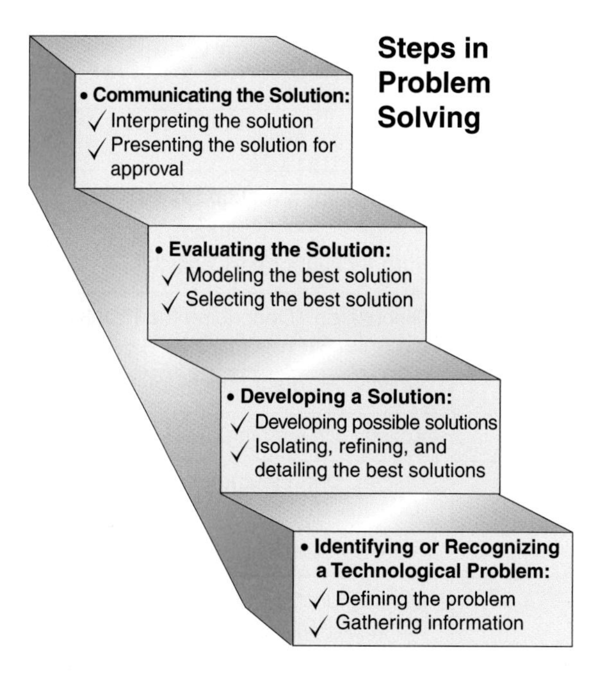

Figure 4-1.

- Develop refined sketches to modify and improve promising ideas.
- Generate dimensioned sketches to communicate the best ideas that have been developed.

Students will improve and evaluate the ideas using various models. The techniques that the students will be introduced to are:

- Graphic models.
- Mathematical models.
- Physical models.

In addition, the students will use brainstorming techniques to gather suggestions for improving any product ideas they are developing. Section 4 is designed to address the major topics involved in designing technological products. It will include content from four chapters and one or more laboratory activities. The chapters are:

- Chapter 9—The Problem-Solving and Design Process.
- Chapter 10—Developing Design Solutions.
- Chapter 11—Evaluating Design Solutions.
- Chapter 12—Communicating Design Solutions.

As with other sections in this program, two different types of activities are provided. Teachers wanting students to address the content from a problem-solving approach should select Activity 4A—Design Problem, in which the students will use a design brief to direct their activities. A more structured laboratory activity in which students develop a bookend is presented by Activity 4B—Fabrication Problem.

Presenting Section 4

Estimated length—5 weeks

Section 4 is designed to help students use the problem-solving method as their new technological devices and systems are developed. They will explore the design processes in four major stages.

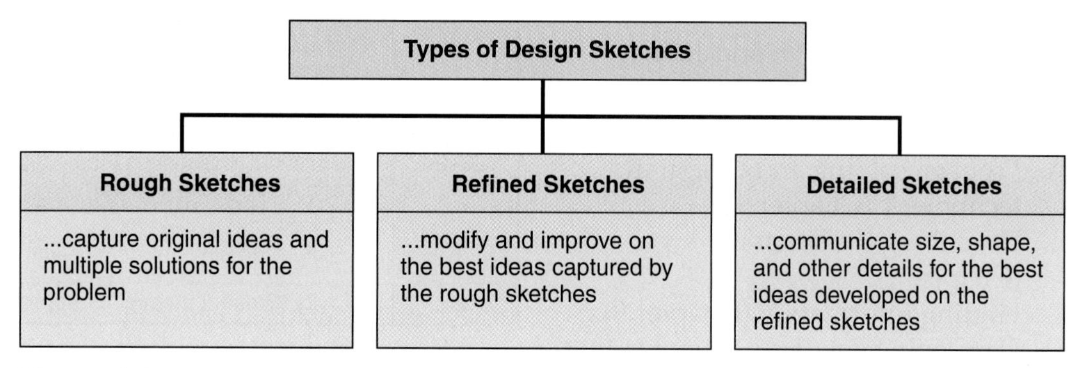

Figure 4-2.

First, the students will develop a design statement or problem definition. Second, they will develop several product ideas that meet the problem or opportunity statement. This activity will include rough and refined sketching and brainstorming activities.

The third phase will involve the students developing specifications for the products. This phase will primarily include producing engineering drawings and bills of materials. Finally, the students will present their product for approval.

The content of Section 4 is best presented through short and frequent discussions and demonstrations followed by significant laboratory activities. At each step, the students will be moving a product idea toward a specified, approved product ready for production.

The 25-day schedule, shown in **Figure 4-3,** has been developed for this section. During this time you will notice that a number of major topics are introduced through presentations and discussions. In addition, the students will use simple sketching and drawing techniques to capture and develop ideas for a new product.

Note
It is important that the students use appropriate sketching and drawing procedures. Developing advanced drawing skills is beyond the scope of this section, however.

In preparation for presenting this section, you should:

1. Read Chapters 9, 10, 11, and 12 in *Technology & Engineering.*
2. Prepare discussions on:
   A. Problem solving and design in technology (Chapter 9).
   B. Describing the situation (Chapter 9).
   C. Finding information (Chapter 9).
   D. Developing solutions (Chapter 10).
   E. Brainstorming (Chapter 10).

Section 4: Problem Solving and Design

Day	Activity
1	Discussion: "Problem Solving and Design in Technology" Introduce laboratory activity
2	Discussion: "Describing the Problem" (Chapter 9) Activity: Problem definition
3	Activity: Developing product criteria
4–5	Discussion: "Finding Information" (Chapter 9) Activity: Researching information
6	Discussion: "Developing Solutions" (Chapter 10) Demonstration: Sketching
7–8	Activity: Rough and refined sketches
9	Discussion: "Brainstorming" (Chapter 10) Activity: Brainstorming
10	Activity: Detailed (dimensioned) sketches Demonstration: Dimensioning
11	Activity: Detailed sketches
12	Discussion: "Modeling Solutions" (Chapter 11) Demonstration: Using machines for producing models
13–17	Activity: Producing a physical model
18	Discussion: "Specifying Products" (Chapter 12) Demonstration: Multiview drawings or sketches
19–21	Activity: Engineering drawing
22	Discussion: "Bills of Materials" (Chapter 12) Activity: Bill of materials
23–24	Discussion: "Presenting Final Product Plans for Approval" (Chapter 12) Activity: Complete products plans
25	Activity: Present product for approval

Figure 4-3.

   F. Modeling solutions (Chapter 11).
   G. Specifying products (Chapter 12).
   H. Bills of materials (Chapter 12).
   I. Presenting products for approval (Chapter 12).
3. Prepare demonstrations on:
   A. Sketching (rough and refined—pictorial).
   B. Dimensioning.
   C. Model building.
   D. Multiview sketching or drawing.
4. Select a laboratory activity.
   A. Activity 4A—This is an open-ended problem-solving activity, in which students design a travel game.
   B. Activity 4B—This is a structured activity in which groups of students use design techniques to develop a bookend.
5. Prepare teaching materials (transparencies or bulletin boards, for example) to support the content.
6. Determine the study questions and quiz questions that will be used.

Reinforcing and Enriching

The following activities can be used to supplement and enrich the laboratory activities at the end of each textbook section and the STEM Applications included at the end of each textbook chapter.

1. Invite a product designer or architect to speak to the class about the process he or she uses to develop a new product or structure.
2. Take the class on a field trip to a company that uses CAD (computer-aided design) to design products and structures.

Taking Your Technology Knowledge Home

This take-home activity allows students to apply their knowledge of problem solving and design to devise a means of sending coded messages to a friend. The code would be transmitted by either visual methods (such as flags or flashes of light) or electronic means (such as telegraph or radio). To complete the activity, the students will:

1. Decide on the method to be used for transmitting the message.
2. Devise a substitution code (letter, number, or symbol) for each letter of the alphabet and the numbers 0–9.
3. Provide a copy of the code to the friend who will receive the message so the message can be decoded.
4. Sketch the technical system to be used for transmitting the code. Examples of such sketches might be the electrical circuit for a telegraph or flashing light system or the construction details for flags used in a semaphore system.
5. Construct and test the sketched transmission system.
6. Encode the test phrase, "Can you read this?"
7. Transmit the encoded phrase to a friend.
8. Receive and translate (decode) the friend's encoded response.

Adjustments for Exceptional Students

The two laboratory activities included in the textbook are designed to accommodate students of all ability levels. The design problem will be challenging for gifted students, who may be able to develop a broader range of creative answers. Lower achievers can still solve the design problem but with solutions within their ability level. The fabrication problem is appropriate for students who need more structure and guidance. Whichever activity is chosen, the amount of structure and teacher guidance should be adjusted to meet students' needs. Students should be given only as much structure as they require for a successful experience, however. The fabrication problem includes considerable drawing and sketching. It should be used as a communication tool and not as a skill in its own right. The course is not designed to produce drafters but to acquaint students with the role of drawing in designing products and structures and how these designs are communicated. Throughout the course, group work allows students to develop cooperative work habits and to contribute their own unique abilities to the team.

Evaluation Procedures

The understanding students have of technology and technological systems should be measured in several ways, including:

- Observation and critical listening during class presentations and discussions.
- Worksheets and quizzes covering the cognitive material presented.
- Open-ended questions about the learned material and how technology is developed.
- Assessment of individual participation and contributions to group work.
- Assessment of the quality and creativeness of the solutions to the assigned design/fabrication activities.

Section 5
Applying Technology: Producing Products and Structures

Chapter 13—Using Technology to Produce Artifacts
Chapter 14—The Types of Material Resources and How They Are Obtained
Chapter 15—Processing Resources
Chapter 16—Manufacturing Products
Chapter 17—Constructing Structures
Chapter 18—Using and Servicing Products and Structures

Teaching Materials

Text

Chapter 13, pages 260–269
- Test Your Knowledge, page 268
- STEM Applications, page 269

Chapter 14, pages 270–287
- Test Your Knowledge, page 286
- STEM Applications, page 287

Chapter 15, pages 288–307
- Test Your Knowledge, page 306
- STEM Applications, page 307

Chapter 16, pages 307–334
- Test Your Knowledge, pages 333–334
- STEM Applications, page 334

Chapter 17, pages 336–361
- Test Your Knowledge, pages 359–360
- STEM Applications, page 361

Chapter 18, pages 364–375
- Test Your Knowledge, page 375
- STEM Applications, page 375

Section Activities
- Activity 5A—Design Problem: Manufacturing Technology, page 376
- Activity 5B—Fabrication Problem: Manufacturing Technology, pages 376–377
- Activity 5C—Design Problem: Construction Technology, page 378
- Activity 5D—Fabrication Problem: Construction Technology, pages 378–379

Tech Lab Workbook
- Activity 5A/5B—Manufacturing Design/Fabrication Problem, pages 45–51
- Activity 5C—Construction Design Problem, pages 53–54
- Activity 5D—Construction Design Problem, pages 55–56
- Activity 5E—Product Servicing Activity, pages 57–59
- Activity 5F—Papermaking/Paper Recycling Activity, pages 61–62
- Taking Your Technology Knowledge Home, pages 63–66

Teacher's Resource
- Chapter 13 Quiz, pages 160–161
- Chapter 14 Quiz, pages 165–166
- Chapter 15 Quiz, pages 170–171
- Chapter 16 Quiz, pages 175–176
- Chapter 17 Quiz, pages 185–186
- Chapter 18 Quiz, pages 192–193

Reproducible Masters
- 14-1: Manufacturing, page 167
- 15-1: Types of Primary Processes, page 172
- 16-1: Casting and Molding, page 177
- 16-2: Forming, page 178
- 16-3: Separating, page 179
- 16-4: Conditioning, page 180
- 16-5: Assembly, page 181
- 16-6: Finishing, page 182
- 17-1: Construction, page 187
- 17-2: Steps in Constructing a Structure, page 188
- 17-3: Types of Civil Structures, page 189
- 18-1: Steps in Using a Technological Product, page 194

Lesson Slides (CD only)
- 13-1: Production Activities
- 14-1: Types of Natural Resources
- 15-1: Production Process
- 16-1: Types of Manufacturing
- 17-1: Types of Construction
- 17-2: Types of Buildings
- 18-1: Servicing

Organization of Section 5

Section 5 is structured to allow students to investigate the major actions that are used to obtain and convert natural resources into manufactured products and constructed structures. This study involves four major topics:

- Locating, extracting, and converting natural resources into industrial materials.
- Manufacturing products from industrial materials.
- Constructing structures from industrial materials and manufactured products.
- Using and maintaining products and structures.

The section is organized so the students will explore the various actions that are used to locate and change raw materials into products and structures. This process is shown in **Figure 5-1.** Next, they will investigate the types of raw materials and the techniques used to locate and obtain them, **Figure 5-2.** Specifically, they will learn about:

- **Genetic materials.** Organic materials obtained during the normal life cycle of plants or animals.
- **Fossil fuel materials.** Organic materials made up predominantly of carbon and hydrogen, such as petroleum, natural gas, and coal.
- **Minerals.** Naturally occurring substances with a specific chemical composition, such as ores, nonmetallic minerals, clay, and gems.

In addition, the students will explore ways these materials are located and obtained, including harvesting, drilling, and mining.

Next, the students will learn how materials are converted into industrial materials, called standard stock. They will learn about:

- Mechanical processes.
- Thermal processes.
- Chemical and electrochemical processes.

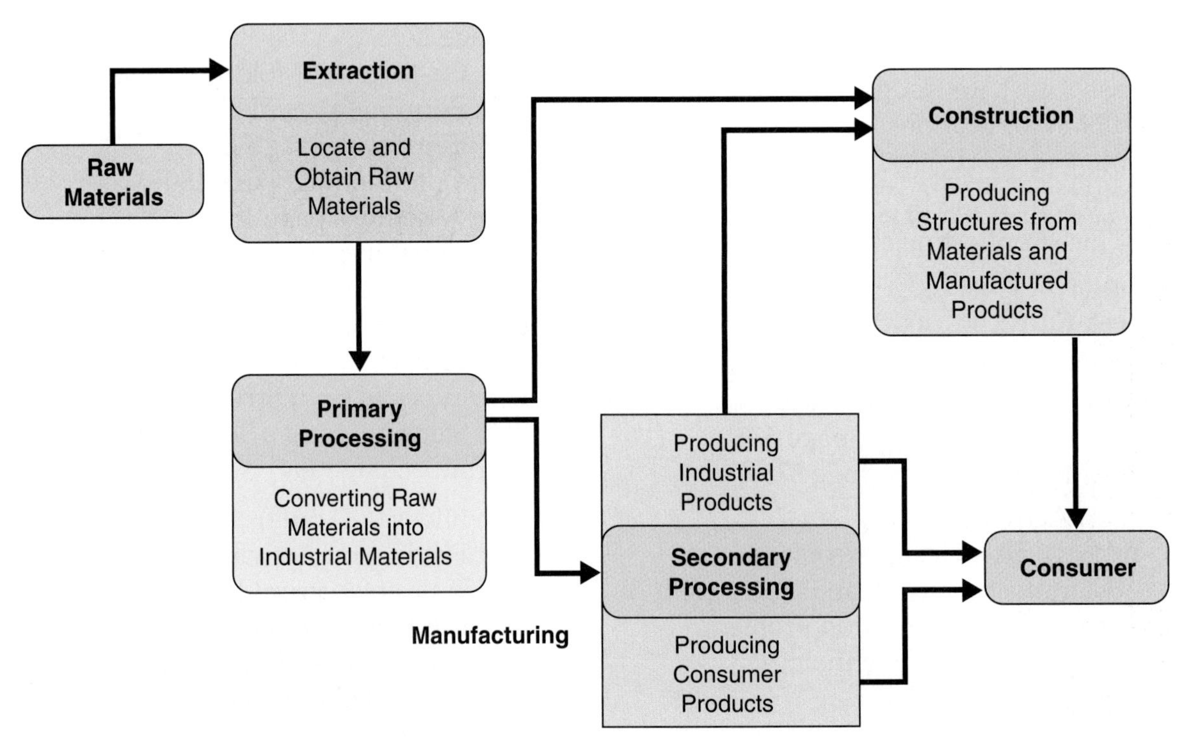

Figure 5-1.

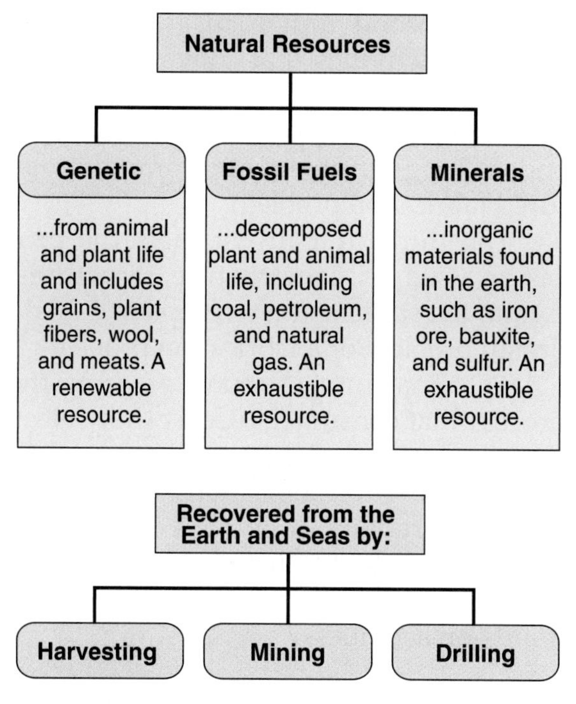

Figure 5-2.

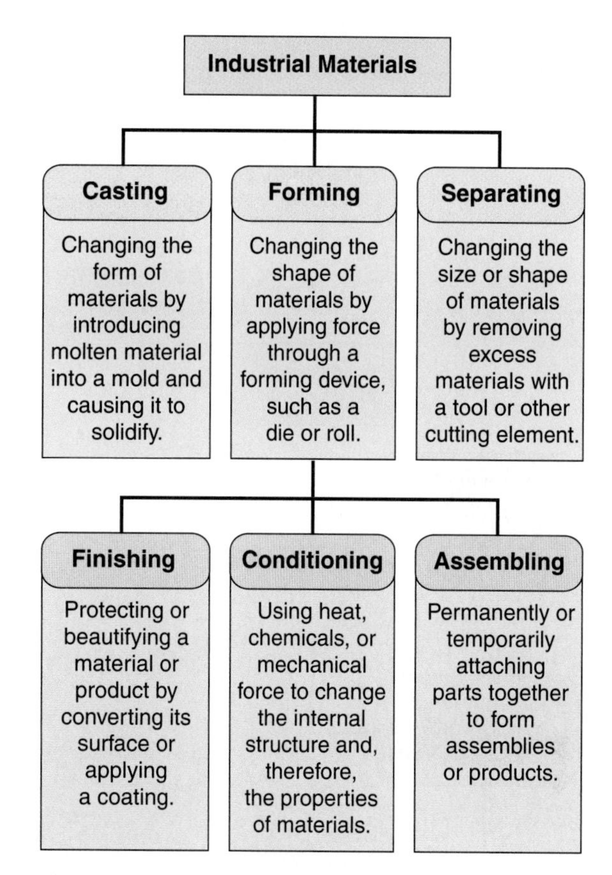

Figure 5-3.

Following this study, the students will explore secondary manufacturing, where industrial materials are changed into industrial and consumer products. This type of manufacturing, as shown in **Figure 5-3,** involves six major types of processes:

- **Casting and molding.** Processes that change the form of materials by introducing a liquid material into a mold where it is allowed to solidify before being removed.
- **Forming.** Processes that change the form of material by applying a force above the material's yield point and below its fracture point, causing the material to change shape.
- **Separating.** Processes that change a product's size and shape by removing excess material through machining or shearing.
- **Conditioning.** Processes that change the internal structure of a material and, therefore, its properties, through thermal, mechanical, and chemical means.
- **Assembling.** Processes that temporarily or permanently hold parts together using mechanical or bonding techniques.
- **Finishing.** Processes that protect or improve the appearance of a product by changing its surface or by applying a coating.

Next, the students will study the practices used to construct a structure. The typical steps, as shown in **Figure 5-4,** are:

- Preparing the site.
- Setting foundations.
- Building the framework.
- Enclosing the structure.
- Installing utilities.
- Finishing the interior and exterior.
- Completing the site.

Finally, the students will learn how to select, operate, service, and dispose

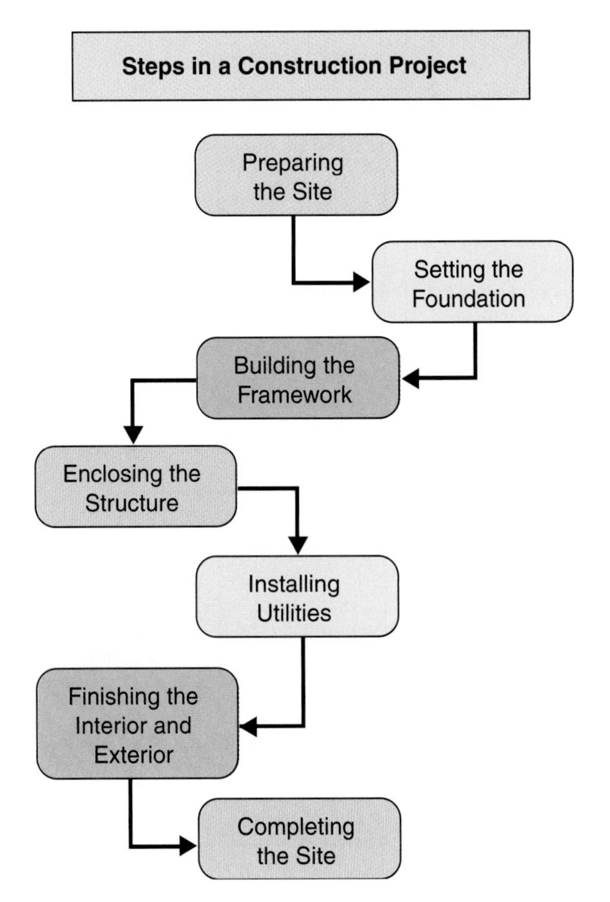

Figure 5-4.

of technological products. The students will investigate these technological topics through six chapters:

- Chapter 13—Using Technology to Produce Artifacts.
- Chapter 14—The Types of Material Resources and How They Are Obtained.
- Chapter 15—Processing Resources.
- Chapter 16—Manufacturing Products.
- Chapter 17—Constructing Structures.
- Chapter 18—Using and Servicing Products and Structures.

These chapters have been grouped into four units of study. Each area has at least one activity associated with it. The first unit includes Chapters 13–15 and focuses on locating resources, extracting resources, and converting them into industrial materials. During this unit, the students will make paper or plywood.

The second unit focuses on manufacturing activities that convert standard stock (industrial materials) into products. This activity is supported with an open-ended design problem and a more structured fabrication problem.

The third unit investigates the processes used in constructing structures. Again, the students may complete either a design problem or a fabrication problem.

The final unit of study presents the process that consumers use as they select, use, service, and dispose of products and structures. This short activity has the students prepare an owner's manual for a simple product. During the study, the students will engage in four hands-on/minds-on activities.

Presenting Section 5

Estimated length—7 weeks

Section 5 is designed to help students understand technology as it is applied to:

- Locating and extracting material resources.
- Converting raw materials into industrial materials (standard stock).
- Manufacturing products from standard stock.
- Constructing structures from standard stock (i.e., lumber, plywood, and PVC pipe) and manufactured products (i.e., windows, doors, and lavatories).
- Selecting, operating, maintaining, and disposing of (recycling or scrapping) products and structures.

The content of Section 5 lends itself to being introduced through lecture/discussion strategies and reinforced by laboratory activities.

The 35-day schedule, shown in **Figure 5-5,** has been developed as a suggestion for scheduling both the presentations and activities for this section.

During this period of time, you will notice that four units of study (major topics) are introduced through a series of presentations and discussions. In addition, the students use common manufacturing and construction tools to complete either a design or a fabrication problem.

Note
It is important that the students use tools and machines safely. Demonstrate appropriate safety procedures to your class prior to any laboratory activity.

To prepare to present this section, the teacher should:

1. Read Chapters 13 through 18 in *Technology & Engineering.*
2. Prepare discussions on:
   A. Production and technology (Chapter 13).
   B. Obtaining material resources (Chapter 14).
   C. Producing industrial materials (Chapter 15).
   D. Manufacturing (secondary) processes (Chapter 16).

Note
The discussion of secondary manufacturing may be divided into three separate presentations, which are given as they are needed during Unit 2. These presentations are (1) sizing and shaping materials through casting, forming, and machining processes, (2) changing internal properties through conditioning processes, and (3) assembling and finishing products.

   E. Constructing processes (Chapter 17).

Section 5: Production Technology	
Day	Activity
Unit 1—Obtaining and Converting Resources	
1	Discussion: "Production and Technology" (Chapter 13) Introduce primary processing laboratory activity
2–3	Discussion: "Obtaining Resources" (Chapter 14) Students work on laboratory activity
4–7	Discussion: "Primary Processing" (Chapter 15) Primary Processing Film Students complete laboratory activity
Unit 2—Manufacturing Products	
8	Discussion: "Manufacturing Processes" (Chapter 16) Introduce first processing activity
9–19	Demonstrate manufacturing processes Students work on laboratory activity
Unit 3—Constructing Structures	
20	Discussion: "Construction Processes" (Chapter 17) Introduce construction laboratory activity
21–31	Demonstrate construction practices Students work on laboratory activity
Unit 4—Maintaining Products and Structures	
32–34	Discussion: "Maintaining Artifacts" (Chapter 18) Servicing activity
35	Summarize section

Figure 5-5.

Note
This presentation may be divided into three separate presentations, which are given as they are needed during Unit3. These presentations are (1) preparing to build—site work, (2) erecting the structure, and (3) finishing the structure and site.

F. Selecting and using products and structures (Chapter 18).

Note
Be sure to stress proper product selection, use, and maintenance and the proper and safe disposal of obsolete products.

3. Select a laboratory activity for each unit of study.

   Unit 1—Use either Activity 5F (Papermaking), included in the *Tech Lab Workbook*, or Activity 5G (Making Plywood), included in the Teacher's Resource.

   Unit 2—Use either Activity 5A or Activity 5B. Activity 5A is an open-ended problem-solving manufacturing activity. Activity 5B is a structured activity in which groups of students use manufacturing technology to produce a simple product.

   Unit 3—Use either Activity 5C or Activity 5D. Activity 5C is a construction activity that involves an open-ended problem-solving approach to design and fabricate a nontraditional structure. Activity 5D is a structured construction activity in which groups of students construct and test a wall section for its sound absorption.

   Unit 4—Use Activity 5E in the *Tech Lab Workbook*, which has students troubleshoot and repair a technological device.
4. Practice the selected laboratory activities.
5. Prepare teaching materials (transparencies or bulletin boards, for example) to support the content.
6. Choose the study questions and test that will be used for the section.

Additional Activities

Activity 5E—Product Servicing Activity

A common technological action is servicing. This activity will allow students to troubleshoot and repair a technological device—a flashlight. To conduct this activity:

1. Obtain a flashlight for each two students in the class.
2. Obtain an inexpensive VOM or continuity checker for each group.
3. Create a problem with the flashlight by using any one of the following actions:
   A. Placing a burned-out bulb in it.
   B. Placing a dead battery in it.
   C. Coating the switch contacts with fingernail polish.
4. Divide the class into groups of two.
5. Discuss troubleshooting as a systematic elimination of possible problems.
6. Distribute the following items to each group:
   A. "Faulty" flashlight.
   B. Tester.
7. Have the students:
   A. Read the "Service Manual" in the *Tech Lab Workbook.*
   B. Conduct the troubleshooting procedure.
   C. Complete the "Job Ticket" and the "Customer's Bill" in the *Tech Lab Workbook.*

Activities 5F and 5G—Producing Industrial Materials

Students should produce an industrial material from raw materials. This type of processing is difficult to show in a technology education laboratory. Activities such as smelting iron, converting petroleum to plastics, and refining aluminum are complex and hazardous processes.

Still, students can experience the production of a common industrial material by one or both of the following activities.

Activity 5F—Papermaking

Students will produce a sheet of paper from reconstituted pulp. To complete the activity, students need the following materials and supplies:

Paper towels
Toilet tissue or newspaper
Hand mixer
Ink roller (brayer)
16-ounce cup
Electric iron
Production screen
8″ × 10″ tray

Divide the class into groups of two to four students for this activity. Have each group produce a sheet of paper using the procedures in the *Tech Lab Workbook.* If newspaper is used, the paper the students make will be gray in color.

Activity 5G—Making Plywood

1. Provide each student with five pieces of 1/8″ × 3 1/2″ × 3 1/2″ veneer cut from a construction grade 2 × 4.
2. Have the students:
   A. Apply aliphatic resin or polyvinyl glue to one side of four pieces of veneer.
   B. Lay up a plywood sheet from the veneer. See **Figure 5-4.**
   C. Place a sheet of waxed paper on each side of the plywood.
   D. Place the plywood in a vise and clamp the sheet for at least 45 minutes.
   E. Remove the plywood from the vise and let it cure for 24 hours.
3. The students should:
   A. Trim the sheet to 3″ × 3″.
   B. Sand the faces.
4. Each student can produce a peg game from the sheet of plywood. Some typical peg and board games are included in Section 3.

Taking Your Technology Knowledge Home

This take-home activity allows students to apply their knowledge of manufacturing and design and build a product they can sell or give to a charity. Typical simple products would be toy cars and trucks, CD holders, and pencil holders, for example. To complete the activity, the students will:

1. Decide whether they want to make a product to sell or give away.
2. Identify a type of product on which they will work.
3. Describe the specific function of the product.
4. List the features the product will have.
5. Identify the constraints within which they must work.
6. Sketch several possible designs for the product.
7. Select and refine the best design.
8. Build a prototype of the product.
9. Develop a procedure for making several products.
10. Design a package for the product.

Adjustments for Exceptional Students

The laboratory activities are designed to be done quickly. The level of sophistication can accommodate students of all ability levels. Gifted students should have more creative answers and possibly produce a more complex solution. Lower achievers can still solve the problem but with solutions within their ability level. In all cases, the amount of structure and teacher guidance should be adjusted to meet students' needs. Guard against providing too much structure, however, because this action often thwarts creativity and is less appealing to gifted students.

Also, careful grouping of students will allow each student to bring his or her own special abilities to the group. Some students will have more artistic ability, others might have more dexterity, and others might have more abstract reasoning skills. Throughout the course, group work allows students to develop cooperative work habits and to contribute their own unique abilities to the team.

Evaluation Procedures

The understanding students have of technology and technological systems should be measured in several ways, including:

- Observation and critical listening during class presentations and discussions.
- Worksheets and quizzes covering the cognitive material presented.
- Open-ended questions about the learned material and how technology is developed.
- Assessment of individual participation and contributions to group work.
- Assessment of the quality and creativeness of the solutions to the assigned design/fabrication activities.

Section 6
Applying Technology: Communicating Information and Ideas

Chapter 19—Using Technology to Communicate

Chapter 20—Printed Graphic Communication

Chapter 21—Photographic Communication

Chapter 22—Telecommunication

Chapter 23—Computer and Internet Communication

Teaching Materials

Text

Chapter 19, pages 382–395
- Test Your Knowledge, page 394
- STEM Applications, page 395

Chapter 20, pages 396–418
- Test Your Knowledge, pages 417–418
- STEM Applications, page 418

Chapter 21, pages 422–439
- Test Your Knowledge, page 438
- STEM Applications, page 439

Chapter 22, pages 442–463
- Test Your Knowledge, page 462
- STEM Applications, page 463

Chapter 23, pages 468–487
- Test Your Knowledge, pages 486–487
- STEM Applications, page 487

Section Activities
- Activity 6A—Design Problem, page 490
- Activity 6B—Production Problem, pages 490–491

Tech Lab Workbook
- Activity 6A—Design Problem, pages 69–71
- Activity 6B—Production Problem, pages 73–75
- Activity 6C—Telecommunication Activity, pages 77–78
- Taking Your Technology Knowledge Home, pages 79–80

Teacher's Resource
- Chapter 19 Quiz, pages 205–206
- Chapter 20 Quiz, pages 213–214
- Chapter 21 Quiz, pages 222–223
- Chapter 22 Quiz, pages 226–227
- Chapter 23 Quiz, pages 230–231

Reproducible Masters
- 19-1: Functions of Communication Media, page 207
- 19-2: Communication Process, page 208
- 19-3: Types of Communication Systems, page 209
- 20-1: Creating Media Messages, page 215
- 20-2: Designing the Message, page 216
- 20-3: Preparing to Produce the Message, page 217
- 20-4: Producing the Message, page 218
- 20-5: Delivering the Message, page 219
- 22-1: Telecommunication Systems, page 228

Lesson Slides (CD only)
- 19-1: Communication
- 19-2: Types of Communication
- 20-1: Ways to Communicate
- 22-1: Radio Broadcast Systems
- 23-1: Information and Communication Technology

Organization of Section 6

Section 6 is designed to present communication technology as a technological system that connects a sender with a receiver, **Figure 6-1.** Communication technology is different from language in that communication technology must have technical means involved.

The students will first explore the scope of communication technology. They will learn that five major types of communication technology exist: printed graphic communication, photographic communication,

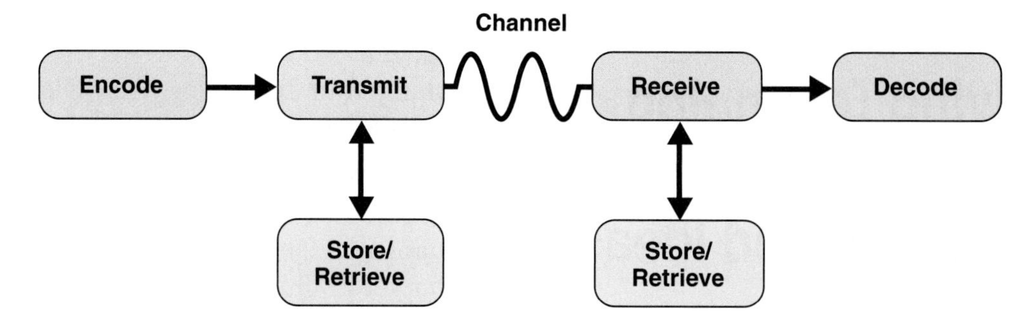

Figure 6-1.

telecommunication (electronic communication), technical graphic communications (which was presented in Section 4—Problem Solving and Design in Technology), and computer and Internet communication. See **Figure 6-2.**

Then they will address printed graphic, photographic, telecommunication, and computer and Internet communication independently. During the study of printed graphic communication they will learn about the six major types of printing, **Figure 6-3:**

- **Relief.** The oldest of all printing processes; prints an image that is raised above a surface.
- **Lithography.** The most widely used method of printing; carries the image on a flat surface.
- **Gravure.** Process in which the image carrier has the message chemically etched or scribed into its surface.

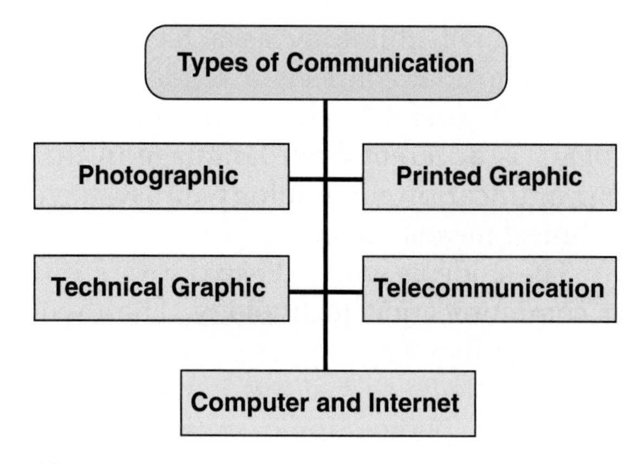

Figure 6-2.

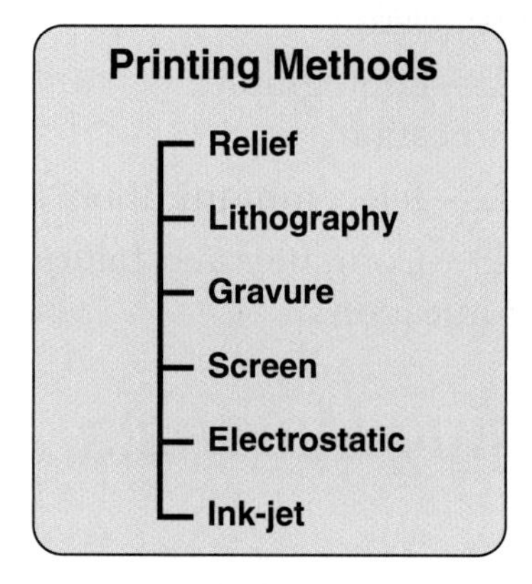

Figure 6-3.

- **Screen.** Process that uses a stencil with openings identical to the message.
- **Electrostatic.** Process that uses a machine with a special drum and a strong light to create an electrically charged image that is transferred to paper.
- **Ink-jet.** Process in which the message generated by a computer directs a special printer that sprays jets of ink onto paper to produce the message.

Following the introduction to the printing processes, the students will investigate the planning and production activities used to produce a printed product. These are:

- Designing the message.
- Preparing to produce the message.
- Producing the message.

The second area of study will be photographic communication. In this area, the students will learn about light waves, cameras, and film. They will then apply this knowledge as they design, produce, and reproduce photographic messages.

The third area of study in this section involves the students with telecommunication. They will explore two major types of telecommunication systems:

- **Hard-wired systems.** Systems that use electrical codes that are conducted over a permanent waveguide between the sender and the receiver.
- **Broadcast systems.** Systems that use radio waves to carry the signal from the sender to the receiver.

The students will use their knowledge of telecommunication systems as they complete two tasks: designing messages and producing and broadcasting messages. Specifically, they will complete the following steps under these two categories:

Designing Communication Messages

- Identifying the audience.
- Selecting an approach.
- Developing the message.

Producing and Broadcasting Communication Messages

- Casting and rehearsing.
- Performing and recording.
- Broadcasting and receiving.

During the final area of study in this section, the students will explore the newest area of communication: computer and Internet communication. They will review the parts of a computer system they studied in Chapter 7. They will then learn about computer networks and how they are united to form the World Wide Web and the Internet.

During this section, the students will study from five chapters and complete two laboratory activities. The chapters are:

- Chapter 19—Using Technology to Communicate.
- Chapter 20—Printed Graphic Communication.
- Chapter 21—Photographic Communication.
- Chapter 22—Telecommunication.
- Chapter 23—Computer and Internet Communication.

Both activities should be used as the students complete this section. The design problem engages the students in printed graphic communications through a creative, problem-solving approach. The fabrication problem uses a more structured approach to studying photographic communication. In addition, two other activities that are included in this guide should be completed. These are:

- A telecommunication activity that includes the study of the printed graphic, photographic, and telecommunication systems.
- An Internet activity.

Presenting Section 6

Estimated length—5 weeks

Section 6 is designed to help students understand communication technology as a means of using technical means (equipment and machines) to deliver information, feelings, and ideas. The students will explore communication technology through four basic media in this section:

- Printed graphic communication using printing processes.
- Photographic communication using photography or digital imaging.
- Telecommunication using electronically generated carrier signals.
- Computer and Internet communication using computer systems to send and receive messages.

The content of Section 6 should be delivered through presentations, discussions, and textbook readings. These approaches should be reinforced by at least two or three realistic laboratory activities.

The 25-day schedule, shown in **Figure 6-4,** has been developed for this section. During this period of time, you will notice that five major topics are introduced through presentations and discussions. In addition, the students will receive equipment demonstrations and complete assigned laboratory activities.

To prepare to present this section, you should:

1. Read Chapters 19–23 in *Technology & Engineering.*
2. Prepare discussions on:
   A. Using technology to communicate (Chapter 19).
   B. Printed graphic communication (Chapter 20).
   C. Photographic communication (Chapter 21).
   D. Telecommunication (Chapter 22).
   E. Computer and Internet communication (Chapter 23).
3. Prepare four laboratory activities.
   A. Activity 6A—Design and produce a printed advertisement to promote technological literacy.
   B. Activity 6B—Design and produce a picture set.
   C. Electronic communication—Develop and produce a 30-second commercial to promote recycling.
   D. Internet communication—Prepare a directory of Internet sites for a specific topic.
4. Practice the selected laboratory activities.
5. Prepare teaching materials (transparencies or bulletin boards, for example) to support the content.
6. Determine the study questions and quiz that will be used.

Section 6: Communication Technology	
Day	Activity
1	Discussion: "Using Technology to Communicate" (Chapter 19) Introduce laboratory activity 6A
2	Discussion: "Printed Graphic Communication" (Chapter 20) Review sketching—rough and refined
3–4	Students work on laboratory activity—Planning phase
5–9	Demonstrate printing techniques Students work on laboratory activity—Production phase
10	Discussion: "Photographic Communication" (Chapter 21) Introduce laboratory activity 6B
11	Students work on laboratory activity—Planning phase
12–13	Demonstrate using a camera Students work on laboratory activity—Production phase
14	Demonstrate developing film and making a print Students work on laboratory activity
15	Discussion: "Telecommunications" Introduce telecommunications activity
16–17	Students work on laboratory activity—Planning
18–20	Demonstrate recording messages Students complete laboratory activity
21	Discussion: "Computers and the Internet" Students work on laboratory activity
22–24	Demonstrate Internet and World Wide Web Students complete laboratory activity
25	Summarize section

Figure 6-4.

Additional Activities

Activity 6C—Telecommunication Activity

This activity is presented in this resource as the third communication activity. The students are challenged to produce a 30-second radio or television commercial to promote an issue of societal concern (for example, environmental protection, pollution, drug use, recycling).

This activity uses teams of three to five students to complete two major tasks, program development and program production. During the program development phase:

1. Assign each team a topic for their advertisement. (The same one can be assigned to each team, or each team can develop a separate one.)
2. Have each team member:
   A. Develop three themes for the assigned topic.
   B. Present the themes to his/her group.
3. Have each team select an appropriate theme.
4. Have each team member:
   A. Write a script and storyboard (for TV commercial) for the program.
   B. Present it to the team.
5. Have each team select the best script.

For the production phase of the commercial, each team should:

1. Develop production directions for camera location (for TV) and sound effects, for example.
2. Assign actor roles and production duties (recording engineer and camera operator, for example) to members of their group.
3. Rehearse the production.
4. Record and edit the production.
5. Present their production to the class.

Activity 6D—Internet Communication

This activity is presented in this resource as the fourth communication activity. The students are challenged to develop an annotated bibliography of Internet sites that relate to a specific topic.

This activity uses teams of three to four students to complete the assignment.

1. Have the teams select a topic from a list you provide. A good source of topics is elementary teachers who are teaching technology sections.
2. Have each member of each team:
   A. Use search engines to locate sites for the topic.
   B. Select five appropriate sites.
   C. Record the name and URL address for the site on the laboratory sheet.
   D. Write a brief description of the site on the laboratory sheet.
3. Have each team select the best sites and prepare a master bibliography.

Reinforcing and Enriching

The following activities can be used to supplement and enrich the laboratory activities at the end of each textbook section and the STEM Applications included at the end of each textbook chapter.

1. Have all assignments focus on the same theme, such as stopping littering. Then the class should integrate these materials and present them as a package to such people as the school administration or city officials, for example.
2. During the telecommunication activity, have some groups produce radio commercials while other groups produce television commercials on the same theme. Have the class judge the effectiveness of these two media in communicating information and ideas.

Taking Your Technology Knowledge Home

This take-home activity allows students to apply their knowledge of communication systems to develop a sign for use in their neighborhood. Challenge the students to:

1. Work with their parents, guardians, neighbors, or youth group leader to identify a poster that could be used to promote an event such as a meeting or a garage sale, for example.
2. Establish the purpose for the sign.
3. Identify the information it will communicate.
4. Develop layout for the sign.
5. Present the layout to the proper adult for approval.
6. Produce the sign.
7. Give the sign to the appropriate adult.

Adjustments for Exceptional Students

The laboratory activities in this section are designed to be done quickly. Individual differences among students in a class can be accommodated by accepting varying levels of sophistication in the material presented. Also, communication media use a wide range of talents in their development. Students with artistic abilities, intellectual talents, writing skills, and production capabilities can all find a place to contribute to group work and feel needed. For this reason, all the activities in this section are designed for group work.

Evaluation Procedures

The understanding that students have of technology and technological systems should be measured in several ways, including:

- Observation and critical listening during class presentations and discussions.
- Worksheets and quizzes covering the cognitive material presented.
- Open-ended questions about the learned material and how technology is developed.
- Assessment of individual participation and contributions to group work.
- Assessment of the quality and creativeness of the solutions to the assigned design/fabrication activities.

Section 7
Applying Technology: Transporting People and Cargo

Chapter 24—Using Technology to Transport
Chapter 25—Transportation Vehicles
Chapter 26—Operating Transportation Systems

Teaching Materials

Text

Chapter 24, pages 494–505
Test Your Knowledge, page 504
STEM Applications, pages 504–505
Chapter 25, pages 506–539
Test Your Knowledge, pages 538–539
STEM Applications, page 539
Chapter 26, pages 540–555
Test Your Knowledge, page 555
STEM Applications, page 555
Section 7 Activities
Activity 7A—Design Problem, page 558
Activity 7B—Fabrication Problem, pages 558–559

Tech Lab Workbook

Activity 7A—Design Problem, pages 83–84
Activity 7B—Fabrication Problem, pages 85–87
Taking Your Technology Knowledge Home, pages 89–92

Teacher's Resource

Chapter 24 Quiz, page 240
Chapter 25 Quiz, pages 245–246
Chapter 26 Quiz, pages 254–255
Reproducible Masters
24-1: Transportation, page 241
24-2: Transportation Systems, page 242
25-1: Parts of a Transportation System, page 247
25-2: Vehicles, page 248
25-3: Vehicle Systems: Propulsion, page 249
25-4: Vehicle Systems: Suspension, page 250
25-5: Vehicle Systems: Guidance, page 251
25-6: Vehicle Systems: Structure, page 252
26-1: Types of Transportation Systems, page 256
26-2: Transportation Processes, page 257
Lesson Slides (CD only)
24-1: Transportation Systems: Environment
24-2: Transportation Systems: Physical Elements
25-1: Vehicle Systems
26-1: Transportation System

Organization of Section 7

This teaching section is designed to introduce students to transportation technology: its types, components, and vehicles, **Figure 7-1.** This study approaches

Transportation Technology		
Types of Transportation Systems	**Types of Transportation Systems Components**	**Transportation Vehicle Systems**
Land Water Air Space	Pathways Vehicles Support Structures	Structure Propulsion Suspension Guidance Control

Figure 7-1.

transportation as systems that are used to move people and cargo from an origin to a destination. This action involves an environmental medium (land, water, air, and space), a vehicle that has five basic systems (structure, suspension, propulsion, guidance, and control), and support structures (terminals or pathways, for example).

- Transportation as technological action.
- Transportation vehicles and their systems.
- Operating transportation systems, including the integration of vehicles, pathways, and support structures into a dynamic system.

During this study the students will first take a general look at transportation. They will study transportation from a systems point of view. Specifically, they will study:

- **Land transportation.** Highway systems, rail systems, and continuous flow systems.
- **Water transportation.** Inland waterway and ocean-going systems.
- **Air transportation.** Commercial aviation and general aviation.
- **Space transportation.** Manned and unmanned space flights.

This study will be followed by an exploration of the three components common to all transportation systems:

- Pathways or guideways.
- Vehicles or carriers.
- Support structures.

Then the students will concentrate on the five systems that are commonly found in all transportation vehicles. These systems, shown in **Figure 7-2,** are structure, propulsion, suspension, guidance, and control.

Transportation Vehicle Systems	
Structure	The physical frame that protects cargo and people.
Propulsion	Moves the vehicle along its pathway.
Suspension	Supports the vehicle and its load as it moves along its path.
Control	Adjusts the speed and direction of the vehicle's travel.
Guidance	Provides information needed to direct the vehicle's travel.

Figure 7-2.

Finally, the students will investigate the procedures used to operate a transportation system. They will look at both personal transportation systems (travel in a vehicle that is owned and operated by an individual) and commercial transportation systems (enterprises that move people and goods for money).

They will then learn that all commercial transportation systems have vehicles to contain and transport people and cargo, routes the vehicles travel from the origin to the destination, established schedules for the transportation of people and goods, and terminals at the origin and destination points of the system. While completing this section, students will read three chapters and complete one laboratory activity. The chapters are:

- Chapter 24—Using Technology to Transport.
- Chapter 25—Transportation Vehicles.
- Chapter 26—Operating Transportation Systems.

Two different types of activities are provided for this section. If you want students to address the content from a problem-solving approach, select Activity 7A—Design Problem. If you want a more structured approach, select Activity 7B—Fabrication Problem.

Presenting Section 7

Estimated length—3 weeks

Section 7 is designed to help students understand how technology is applied to transportation. They will learn that transportation:

- Is a system that uses technology.
- Generally uses vehicles to move people and cargo.
- Is a combination of vehicles along with support structures and activities.

The content of Section 7 is presented through two major media: the student textbook and lecture/discussion sessions. It is critically important to reinforce these presentations with realistic laboratory activities. The 15-day schedule shown in **Figure 7-3** has been developed for this section. During this time, you will notice that three major topics are introduced through presentations and discussions. In addition, the students use simple tools to complete either a design or a fabrication problem.

Section 7: Transportation Technology	
Day	Activity
1	Discussion: "Using Technology to Transport" (Chapter 24) Introduce laboratory activity
2–4	Demonstrate laboratory activity procedures Students work on laboratory activity
5–8	Discussion: "Transportation Vehicles" (Chapter 25) Students work on laboratory activity
9–14	Discussion: "Operating Transportation Systems" (Chapter 26) Students work on laboratory activity
15	Students present results of activity Summarize section

Figure 7-3.

Note
It is important that the students use tools and machines safely.

To prepare for presenting this section, you should:

1. Read Chapters 24, 25, and 26 in *Technology & Engineering.*
2. Prepare discussions on:
   A. Using technology to transport people and cargo (Chapter 24).
   B. Transportation vehicles (Chapter 25).
   C. Operating transportation systems (Chapter 26).
3. Select a laboratory activity.
   A. Activity 7A—This is an open-ended problem-solving activity in which the students design and test a transportation vehicle kit.
   B. Activity 7B—This is a structured activity in which groups of students construct and test wind-powered vehicles.

Note
Mousetrap or balloon-powered vehicles can be substituted for electric motors and propellers.

4. Practice the selected laboratory activity.
5. Prepare teaching materials (transparencies or bulletin boards, for example) to support the content.
6. Determine which study questions and quiz questions will be used.

Reinforcing and Enriching

The following activities can be used to supplement and enrich the laboratory activities at the end of each textbook section and the STEM Applications included at the end of each textbook chapter.

1. Activity 7A—Design Problem can be enlarged to include a directional control system so the vehicle can be steered along a pathway.
2. Activity 7B—Fabrication Problem can be enlarged by having the students design auxiliary power systems using balloons, rubber bands, or mousetraps.

Taking Your Technology Knowledge Home

This take-home activity allows students to apply their knowledge of transportation systems to plan a trip. They will:

1. Work with their parents or guardians to establish a possible trip.
2. Establish the purpose of the trip.
3. Identify things to see or do along the way.
4. Develop a route for the trip.
5. Identify places to stay using Internet sites or motel/hotel guides.
6. Estimate the cost of each day of travel.
7. Summarize the trip and its costs.

Adjustments for Exceptional Students

Both laboratory activities are designed to be done in a relatively short time—10 to 12 class periods. The level of expectation for the final product can be adjusted to accommodate students of different ability levels. Gifted students can be challenged to develop more creative answers and possibly produce a more complex solution. Lower achievers can still solve the problem but with solutions within their ability level.

With all students, the amount of structure and guidance should be adjusted to meet the students' needs. Avoid providing too much structure because this often stifles the creativity and has less appeal to gifted students.

Finally, careful grouping of students will allow each student to contribute special talents to the group. Some students may have more dexterity, others will have more artistic abilities; and still others may have greater abstract reasoning ability. Throughout this course, group work allows students to develop cooperative work habits while contributing unique abilities to the team.

Evaluation Procedures

The understanding that students have of technology and technological systems should be measured in several ways, including:

- Observation and critical listening during class presentations and discussions.
- Worksheets and quizzes covering the cognitive material presented.
- Open-ended questions about the learned material and how technology is developed.
- Assessment of individual participation and contributions to group work.
- Assessment of the quality and creativeness of the solutions to the assigned design/fabrication activities.

Section 8
Applying Technology: Using Energy

Chapter 27—Energy: The Foundation of Technology
Chapter 28—Energy-Conversion Systems

Teaching Materials

Text

Chapter 27, pages 562–575
- Test Your Knowledge, pages 574–575
- STEM Applications, page 575

Chapter 28, pages 576–603
- Test Your Knowledge, page 602
- STEM Applications, page 603

Section 8 Activities
- Activity 8A—Design Problem, page 604
- Activity 8B—Fabrication Problem, pages 604–605

Tech Lab Workbook

- Activity 8A—Design Problem, pages 95–98
- Activity 8B—Fabrication Problem, pages 99–100
- Taking Your Technology Knowledge Home, pages 101–104

Teacher's Resource

- Chapter 27 Quiz, pages 256–266
- Chapter 28 Quiz, pages 271–272

Reproducible Masters
- 27-1: Types of Energy page 267
- 27-2: Forms of Energy, page 268
- 28-1: Energy Converters, page 273
- 28-2: Energy Conversion, page 273
- 28-3: Creating Motion, page 275

Lesson Slides (CD only)
- 27-1: Energy
- 27-2: Energy Resources

Organization of Section 8

Energy is fundamental to all technological activities. It involves converting one of six basic types of energy into heat, light, sound, or motion, **Figure 8-1.** This section is designed to introduce students to:

- The concepts of energy and work.
- The types of energy and energy sources.
- Common energy converters.

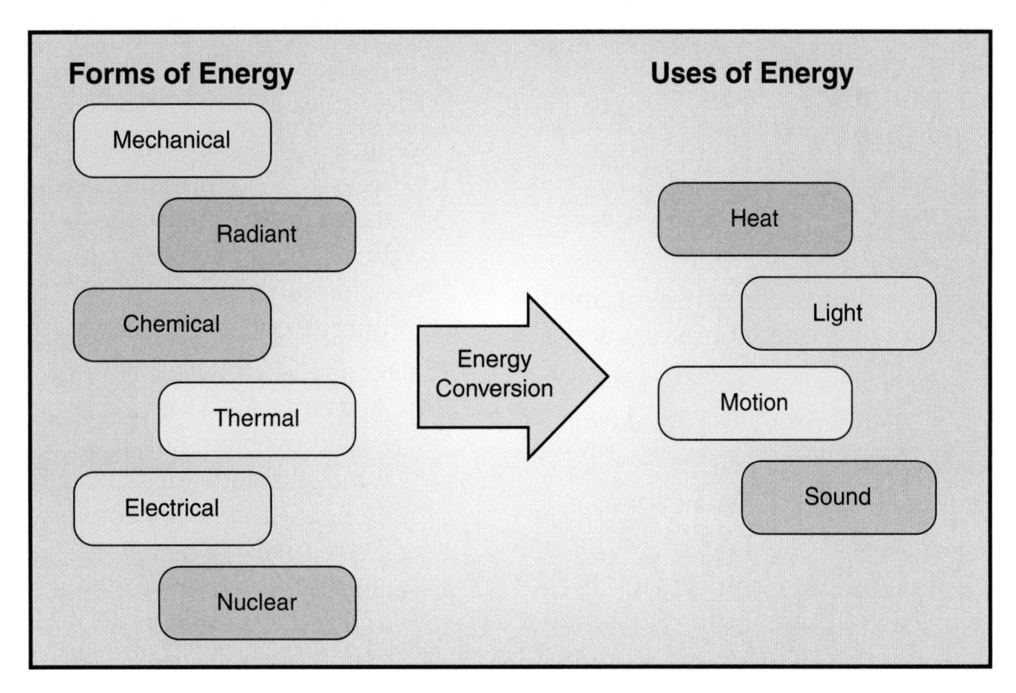

Figure 8-1.

As the students complete Section 8, they will first learn about the types of energy. They will be able to describe energy as either the force doing the work or a force that has the ability of doing work. This will introduce them to both potential energy (stored) and kinetic energy (created by motion). Then they will learn that work means applying a force to move a mass a certain distance.

Using this foundation, the students will explore the six major forms of energy:

- **Mechanical.** Energy produced by motion of technological devices and natural forces, such as wind and falling water.
- **Radiant.** Energy given off by hot objects.
- **Chemical.** Energy stored with a chemical compound or complex substance.
- **Thermal.** Energy related to heat (rapid movement of atoms or molecules in a substance).
- **Electrical.** Energy associated with electrons moving along a conductor.
- **Nuclear.** Energy associated with the internal bonds of atoms.

Finally, the students will investigate four major types of energy-conversion systems:

- Inexhaustible energy converters.
- Renewable energy converters.
- Thermal energy converters.
- Electrical energy converters.

Section 8 includes content contained in two chapters and one laboratory activity. The chapters are:

- Chapter 27—Energy: The Foundation of Technology.
- Chapter 28—Energy-Conversion Systems.

As with the other learning sections in this program, two different types of activities are provided. If you want students to address the content from a problem-solving approach, select Activity 8A—Design Problem. A more structured approach is provided in Activity 8B—Fabrication Problem, which allows students to build and test a specific energy converter.

Presenting Section 8

Estimated length—3 weeks

Section 8 is designed to help students understand that:

- Energy is fundamental to all technologies.
- Energy sources may be exhaustible, renewable, or inexhaustible.
- Energy is constantly being converted in form as it is applied to agriculture, communication and information processing, construction, manufacturing, medicine, and transportation systems.
- Conservation of energy is important.

The teaching strategies for this section include lectures and discussions to present content and realistic laboratory activities to reinforce these presentations. The fifteen-day schedule, shown in **Figure 8-2,** has been developed for this section. To prepare for presenting this section, you should:

1. Read Chapters 27 and 28 in *Technology & Engineering.*
2. Prepare discussions on:
   A. Energy: The foundation of technology (Chapter 27).
   B. Energy-conversion systems (Chapter 28).
   C. Energy conservation (Chapter 28).
3. Select a laboratory activity.
   A. Activity 8A—This is an open-ended problem-solving activity in which students design, build, and evaluate an energy converter.
   B. Activity 8B—This is a structured activity in which students build and test a wind-powered electric generation system.

Section 8: Energy Conversion Technology	
Day	Activity
1	Discussion: "Energy: The Foundation of Technology" (Chapter 27) Introduce laboratory activity
2–4	Demonstrate laboratory activity procedures Students work on laboratory activity
5–8	Discussion: "Energy-Conversion Systems" (Chapter 28) Students work on laboratory activity
9–13	Discussion: "Energy-Conservation Systems" (Chapter 28) Students work on laboratory activity
14	Students present results of laboratory activity
15	Summarize section

Figure 8-2.

4. Practice the selected laboratory activity.
5. Prepare teaching materials (transparencies or bulletin boards, for example) to support the content.
6. Determine which study questions and quiz questions will be used.
7. Review the Taking Your Technology Knowledge Home activity in the *Tech Lab Workbook.*

Reinforcing and Enriching

The following activities can be used to supplement and enrich the laboratory activities at the end of each textbook section and the STEM Applications included at the end of each textbook chapter.

1. If time permits, the section can be enhanced by having the students complete both activities. Another approach is to have half of the class complete the design problem and have the other half complete the fabrication problem. At the end of the section, each group could share its experiences with the other group.
2. The design problem could be based on an energy conversion problem developed by the students. This approach would build on students' interests.

Taking Your Technology Knowledge Home

This take-home activity allows students to apply their knowledge of energy and energy conservation to daily life. They will:

1. Survey several rooms in their home and identify the type of lighting used.
2. Record the wattage of each bulb used.
3. Estimate the number of hours each bulb is used in a month.
4. Calculate the cost of operating each bulb for one month.
5. Identify bulbs that can be replaced with compact fluorescent bulbs.
6. Recalculate the cost of operating the light for the month.
7. Determine the possible monthly savings that could be achieved using compact fluorescent bulbs.
8. Determine the cost of exchanging the bulbs.
9. Determine the number of months that it would take for the savings to pay for the bulbs.

Adjustments for Exceptional Students

Either activity in this section can be completed in relatively short time. Adjustments for students with special needs can be made in several ways. One approach would be to adjust the level of sophistication for the design and fabrication problems. Gifted students could be challenged to provide more creative answers and possibly produce a more complex solution. Lower achievers can still complete either activity but with solutions reflecting their ability levels.

Also, the level of structure provided for the activity can be adjusted. Often, lower-ability students need more structure and shorter, more frequent demonstrations and presentations. They may require reinforcement almost daily. Guard against providing too much structure, however, because this often stifles creativity and is less appealing to gifted students.

Finally, having groups complete the activities allows students of all abilities to contribute their unique talents. This approach requires careful grouping so individual abilities complement one another. Some students will possess more abstract reasoning ability, others will have more artistic abilities, and still others might have more manual dexterity. Group work allows students to develop cooperative work habits and to contribute their own unique abilities to the team.

Evaluation Procedures

The understanding that students have of technology and technological systems should be measured in several ways, including:

- Observation and critical listening during class presentations and discussions.
- Worksheets and quizzes covering the cognitive material presented.
- Open-ended questions about the learned material and how technology is developed.
- Assessment of individual participation and contributions to group work.
- Assessment of the quality and creativeness of the solutions to the assigned design/fabrication activities.

Section 9
Applying Technology: Meeting Needs through Biorelated Technologies

Chapter 29—Agricultural and Related Biotechnologies
Chapter 30—Food-Processing Technologies
Chapter 31—Medical and Health Technologies

Teaching Materials

Text

Chapter 29, pages 608–631
Test Your Knowledge, pages 630–631
STEM Applications, page 631

Chapter 30, pages 634–657
Test Your Knowledge, pages 656–657
STEM Applications, page 657

Chapter 31, pages 658–679
Test Your Knowledge, page 678
STEM Applications, page 679

Section 9 Activities
Activity 9A—Design Problem, page 682
Activity 9B—Fabrication Problem, pages 682–683

Tech Lab Workbook

Activity 9A—Design Problem, pages 107–109
Activity 9B—Fabrication Problem, page 111
Taking Your Technology Knowledge Home, pages 113–116

Teacher's Resource

Chapter 29 Quiz, pages 283–284
Chapter 30 Quiz, pages 291–292
Chapter 31 Quiz, pages 297–298

Reproducible Masters
29-1: Major Farm Crops, page 285
29-2: Types of Farm Equipment, page 286
29-3: Parts of a Tractor, page 287
30-1: Types of Food Processing, page 293
30-2: Methods of Preserving Food, page 294
31-1: Factors in Wellness, page 299
31-2: Technology in Sports, page 300
31-3: Goals of Medicine, page 301

Lesson Slides (CD only)
29-1: Types of Agriculture
30-1: Food Processing
31-1: Focus of Health and Medicine

Organization of Section 9

Section 9 is designed to allow students to explore three major applications of technology as it applies to living things. These, as shown in **Figure 9-1,** are:

- Growing and harvesting food.
- Processing food.
- Promoting human health and treating illnesses.

The students will first investigate the field of agriculture. They will explore the two types of agriculture, **Figure 9-2.** These are crop production, which grows crops, and animal husbandry, which involves breeding, raising, and training animals.

The second area of study in Section 9 is food processing. It will allow students to investigate the technology used in preparing and processing foods. It will focus on

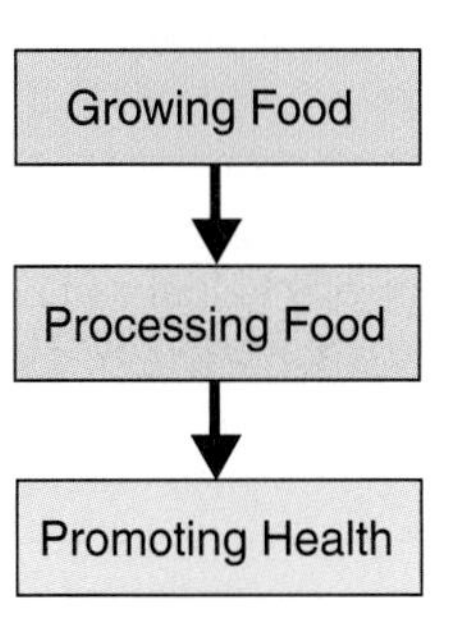

Figure 9-1.

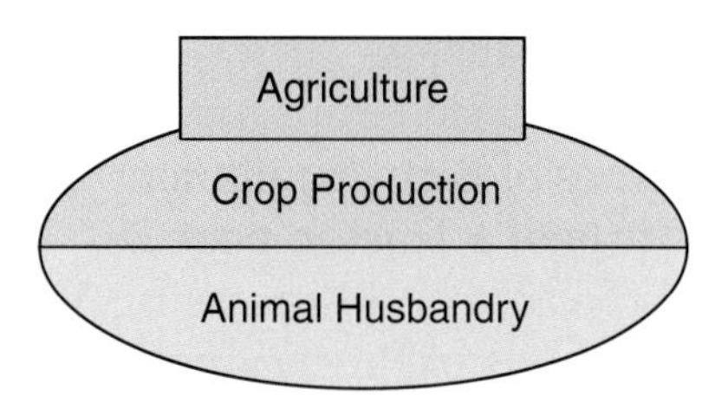

Figure 9-2.

the two types of food processing shown in **Figure 9-3:**

- Primary processing.
- Secondary processing.

They will discover that food is processed by material conversion and food preservation techniques. Finally, they will learn about secondary food processing and how a new food product is developed.

The last area of study in Section 9 is medical and health technologies. In this area, the students will learn about two areas of this topic:

- **Wellness.** A state of physical well-being.
- **Illness.** A state of poor health.

Students will then explore how exercise and sports have been developed to promote wellness. Next, they will explore the types of technology that have been developed to diagnose and treat illnesses and medical conditions. Finally, they will explore how new drugs and vaccines are developed and tested.

Raw agricultural materials are converted by

Primary Processing

into food commodities or ingredients that can be converted by

Secondary Processing

into edible products.

Figure 9-3.

For this section, the students spend at least ten days covering the content of three chapters and in completing two laboratory activities. These chapters are:

- Chapter 29—Agricultural and Related Biotechnologies.
- Chapter 30—Food-Processing Technologies.
- Chapter 31—Medical and Health Technologies.

This section, like all others in this course, has two different types of activities provided. The design problem allows students to engage in a creative, problem-solving approach, and the fabrication problem provides a more structured approach to the laboratory experience.

Presenting Section 9

Estimated length—2 weeks

Section 9 helps students understand several areas where technology is applied to living things. Specifically, this section deals with:

- Agriculture.
- Food processing.
- Medicine and health care.

The content of Section 9 could be presented solely using lecture/discussion strategies. Technology education, however, is based on the concept that presentations should be reinforced with realistic laboratory activities. Thus, about 75 percent of the time assigned to this section is dedicated to laboratory activity. The ten-day schedule developed for this section reflects this emphasis, **Figure 9-4.**

Three important topics are introduced through presentations and discussions. These are: (1) agricultural and related biotechnologies, (2) food-processing technologies, and (3) medical and health technologies. The students will be using simple tools to complete both a design problem and a fabrication problem. It is important

Section 9: Applying Technology: Meeting Needs through Biorelated Technologies	
Day	Activity
1–3	Discussion: "Agricultural and Related Technologies" (Chapter 29) Complete laboratory activity 9B
4–7	Discussion: "Food-Processing Technologies" (Chapter 30) Introduce laboratory activity 9A
8–9	Discussion: "Medical and Health Technologies" (Chapter 31) Complete laboratory activity 9A
10	Summarize section

Figure 9-4.

that the students use tools and machines safely. Demonstrate appropriate safety procedures prior to any laboratory activity.

To prepare to present this section, you should:

1. Read Chapters 29, 30, and 31 in *Technology & Engineering.*
2. Prepare discussions on:
   A. Agricultural and Related Biotechnologies (Chapter 29).
   B. Food-Processing Technologies (Chapter 30).
   C. Medical and Health Technologies (Chapter 31).
3. Review the laboratory activities.
   A. Activity 9A—This activity is a creative, problem-solving activity.
   B. Activity 9B—This activity is a structured activity. Groups of students will use a food dehydrator to preserve an agricultural product.
4. Practice the selected laboratory activity.
5. Prepare teaching materials (transparencies or bulletin boards, for example) to support the content.
6. Determine which study questions and quiz questions will be used.

Reinforcing and Enriching

The following activities can be used to supplement and enrich the laboratory activities at the end of each textbook section and the STEM Applications included at the end of each textbook chapter.

1. Have the students develop a package for their dehydrated product. The package should include appropriate graphics, nutritional information, and other texts.
2. Prepare a user's manual for their biomechanical product. It should tell product users how to use, care for, and maintain their products.

Taking Your Technology Knowledge Home

This take-home activity allows students to apply their knowledge of food processing at home. They will:

1. Use primary food-processing techniques to make peanut butter.
2. Use secondary food-processing techniques to make biscuits.
3. Prepare a snack using the products of their efforts.

Adjustments for Exceptional Students

Both laboratory activities are designed to be done in a short period of time. The design problem is open-ended to accommodate varying abilities and levels of sophistication. The fabrication problem is fairly simple, and its directions can be followed easily. By using carefully structured groups for each activity, individual differences in student abilities can be accommodated.

Evaluation Procedures

The understanding that students have of technology and technological systems should be measured in several ways, including:

- Observation and critical listening during class presentations and discussions.
- Worksheets and quizzes covering the cognitive material presented.
- Open-ended questions about the learned material and how technology is developed.
- Assessment of individual participation and contributions to group work.
- Assessment of the quality and creativeness of the solutions to the assigned design/fabrication activities.

Section 10
Managing a Technological Enterprise

Chapter 32—Organizing a Technological Enterprise
Chapter 33—Operating Technological Enterprises
Chapter 34—Using and Assessing Technology

Teaching Materials

Text

Chapter 32, pages 686–699
  Test Your Knowledge, pages 698–699
  STEM Applications, page 699
Chapter 33, pages 700–723
  Test Your Knowledge, pages 722–723
  STEM Applications, page 723
Chapter 34, pages 724–735
  Test Your Knowledge, page 734
  STEM Applications, page 735
Section 10 Activities
  Activity 10A—Forming the Company, pages 736–737
  Activity 10B—Operating the Company, pages 737–738

Tech Lab Workbook

  Activity 10A—Forming the Company, page 119
  Activity 10B—Operating the Company, pages 121–129
  Activity 10C—Using and Assessing Activity, page 131
  Taking Your Technology Knowledge Home, pages 133–134

Teacher's Resource

  Chapter 32 Quiz, pages 313–314
  Chapter 33 Quiz, pages 321–322
  Chapter 34 Quiz, page 330

Reproducible Masters
  32-1: Management, page 315
  32-2: Functions of Management, page 316
  32-3: Levels of Authority and Responsibility, page 317
  33-1: Research and Development, page 323
  33-2: Production, page 324
  33-3: Marketing, page 325
  33-4: Industrial Relations, page 326
  33-5: Financial Affairs, page 327
  34-1: Using Technological Products and Services, page 331
  34-2: Assessing the Impacts of Technology, page 332
Lesson Slide (CD only)
  33-1: Manufacturing Systems

Organization of Section 10

Up to this point in the course, the students have been studying technological practices that are used to design and use agricultural, construction, energy-conversion, information and communication, manufacturing, medical, and transportation systems. These practices can be applied to all types of activities. Individuals can use them in the home workshop, and service clubs can use them to meet their goals. Most technology, however, is developed by one societal institution: industry. This institution is created and organized to provide economic value in growing and processing crops, communicating information, converting and applying energy, treating illnesses, transporting people and cargo, and producing products and structures.

In this section, the students will explore the concepts associated with organizing and operating an enterprise. They will address a number of topics, including:

- What is an enterprise?
- What is entrepreneurship?
- How are enterprises organized and financed?

- What is the role of management in an enterprise?
- What are the managed areas of activity in an enterprise?
- How do enterprises develop, produce, and market their products and services?
- How should people use and assess technology?

As the students address these concepts, they will study management and learn that it involves planning, organizing, actuating, and controlling activities, **Figure 10-1.** They will also learn that owners/managers organize a company by selecting the type of ownership, establishing the enterprise, and securing financing.

Next, the students will learn that society has five major institutions:

- **Family.** The basic unit of society that provides the foundation for social and economic actions.
- **Religion.** The institution that develops and communicates values and beliefs about life and appropriate living.
- **Education.** The institution that communicates information, ideas, and skills from one person to another and from one generation to another.
- **Political/Legal.** The institution that establishes and enforces the rules of behavior and conduct for the society.
- **Economic.** The institution that designs, produces, and delivers the basic goods and services required by society.

Students then discover that industry is part of the economic institution and that industrial enterprises engage in five important managed activities, **Figure 10-2.** They are:

- **Research and development.** The managed activities that develop new or improved products and processes.
- **Production.** The managed activities that develop methods for producing products or services and that produce the desired outputs.
- **Marketing.** The managed activities that encourage the flow of goods and services from the producer to the consumer.

Functions of Management	
Planning	...setting goals and a course of action.
Organizing	...structuring the tasks and assigning work to people.
Actuating	...starting work and supervising employees.
Directing	...comparing results to the plan.

Figure 10-1.

Managed Areas of Corporate Activity

Research and Development	Production	Marketing
The managed activities that develop new or improved products and processes.	The managed activities that develop methods for producing products or services and that produce the desired outputs.	The managed activities that encourage the flow of goods and services from the producer to the consumer.

Industrial Relations	Financial Affairs
The managed activities that develop an efficient work force and maintain positive relations with workers and the public.	The managed activities that obtain, account for, and disburse funds.

Figure 10-2.

- **Industrial relations.** The managed activities that develop an efficient workforce and maintain positive relations with the workers and the public.
- **Financial affairs.** The managed activities that obtain, account for, and disburse funds.

As the students develop a conceptual understanding of industry, they organize model enterprises that let them apply their newly acquired knowledge. Three chapters and a multistage laboratory activity deliver this educational experience, **Figure 10-3.** The chapters are:

- Chapter 32—Organizing a Technological Enterprise.
- Chapter 33—Operating Technological Enterprises.
- Chapter 34—Using and Assessing Technology.

The activities in this section consist of three phases for the enterprise and one using-and-assessing activity. The first phase is included in Activity 10A. Here, students are to organize an enterprise that will design, produce, and market a personalized calendar. They will sell space on any given date to promote a social or sporting event, a person's birthday, or other similar entries.

Activity 10B includes the remaining two phases, **Figure 10-4.** First, the students develop the financial plans, develop the product format, and promote and sell calendar entries. Then, they control financial activities, produce the calendar, and promote and sell the product.

The using-and-assessing activity is contained in this guide. It allows the students to study how to use and assess properly the technological product they developed.

Presenting Section 10

Estimated length—4 weeks

Section 10 is designed to guide the students through the typical phases a new enterprise moves through. First, they will organize a company. Then, they will determine their financial needs and design a product. After that, they will produce and sell the product. Finally, they will determine how to use and assess the product. These activities help the students understand an enterprise as:

- A human-designed activity.
- The primary source of new technology.

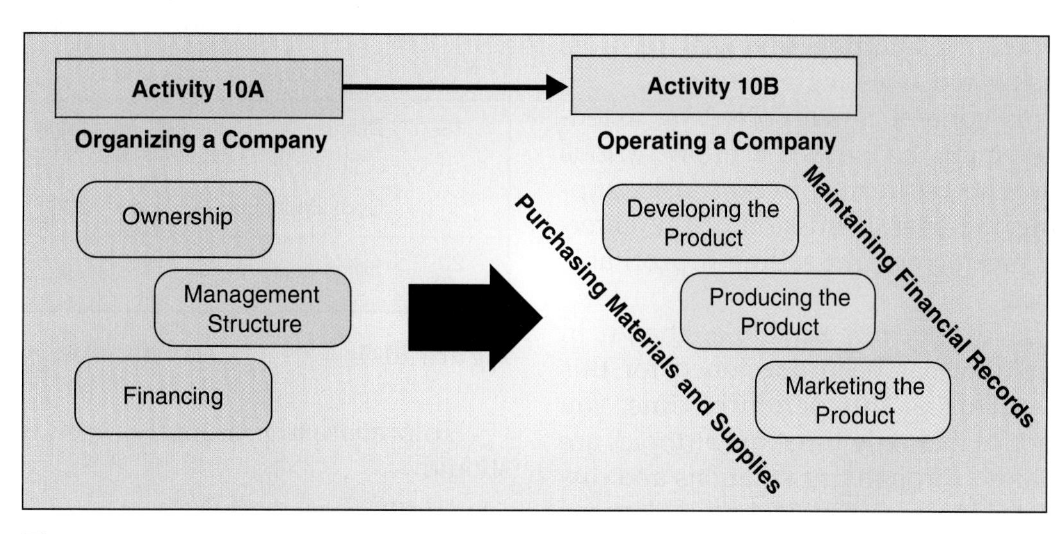

Figure 10-3.

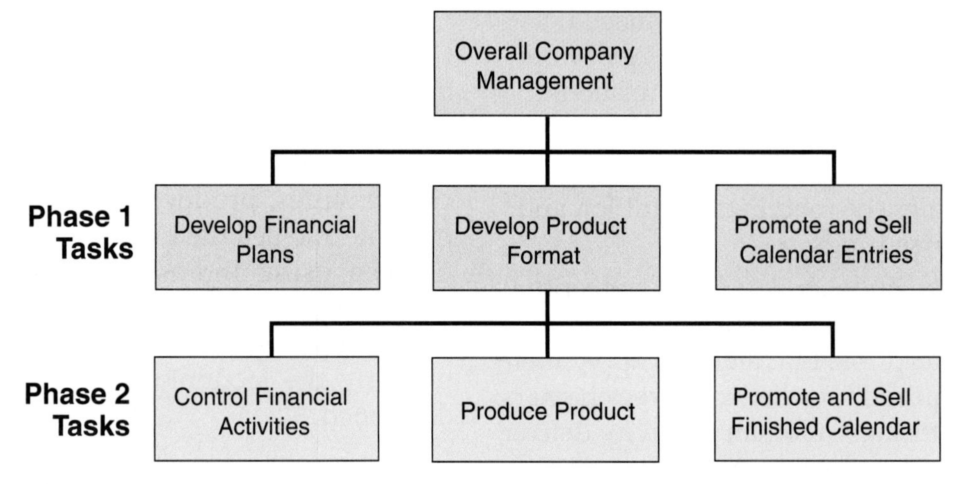

Figure 10-4.

- A dynamic organization in which many tasks are addressed simultaneously.
- An organization that includes design, production, and marketing activities.
- A major employer of managerial and technological talents.
- A source of products people use and assess.

The content of Section 10 is fairly general. Complete courses in organizing and operating an enterprise are often offered. Therefore, this four-week study, by its very nature, can only highlight selected enterprise activities. These activities include designing, producing, and selling products; purchasing materials to support production activities; maintaining financial records; and using and assessing products.

This content is supported by extensive laboratory activities that allow various students to perform different tasks supporting the overall mission of the enterprise: producing and selling a profitable product.

The twenty-day schedule, shown in **Figure 10-5,** has been developed for this section. During this period of time, you will notice that only three major topics are introduced through presentations and discussions. The majority of time is devoted to the students operating their enterprise.

Section 10: Managing a Technological Enterprise	
Day	Activity
1	Discussion: "Organizing a Technological Enterprise" (Chapter 32) Introduce laboratory activity 10A
2–3	Discussion: Financing Enterprises (Chapter 32) Complete laboratory activity 10A
4	Discussion: "Operating Technological Enterprises" (Chapter 33) Introduce laboratory activity 10B
5–18	Students operate their enterprise Demonstrate as needed: Calendar software Product duplication (printing) Producing advertisements
19	Discussion: "Using and Assessing Technology" (Chapter 34) Complete laboratory activity 10B
20	Summarize section

Figure 10-5.

To prepare to present this section, you should:

1. Read Chapters 32, 33, and 34 in *Technology & Engineering.*

2. Prepare discussions on:
   A. Organizing a technological enterprise (Chapter 32).
   B. Financing enterprises (Chapter 32).
   C. Operating an enterprise (Chapter 33).
   D. Using and assessing technology (Chapter 34).

Note
This presentation can be divided into three short presentations: (1) research and development activities, (2) production activities, and (3) marketing activities.

3. Practice the selected laboratory activity.
4. Prepare teaching materials (transparencies or bulletin boards, for example) to support the content.
5. Determine which study questions and quiz questions will be used.

Additional Activities

Activity 10C—Using and Assessing Activity

During this activity, the students will work in groups to determine how the product should be used and assess any impacts it could have on people, society, and the environment. They will complete this activity by following the directions on the appropriate sheet in the *Tech Lab Workbook.*

Alternate Activity

This activity will have the students use a managed production system to produce a kite that can be sold to elementary and middle school students. Before the activity is introduced, collect a number of commercial kites and kite plans. Present the activity's three phases, **Figure 10-6.** Then the class should complete these phases.

Phase 1—Establishing the Enterprise

During this phase, have the students work in groups to:

1. Organize the company.
2. Develop financial plans (budgets, break-even charts).
3. Develop product drawings (dimensioned sketches) by analyzing several existing kites or kite designs.
4. Identify the major tasks that will need to be completed to produce the kite.

Note
As you review Phases 2 and 3, these tasks will become apparent.

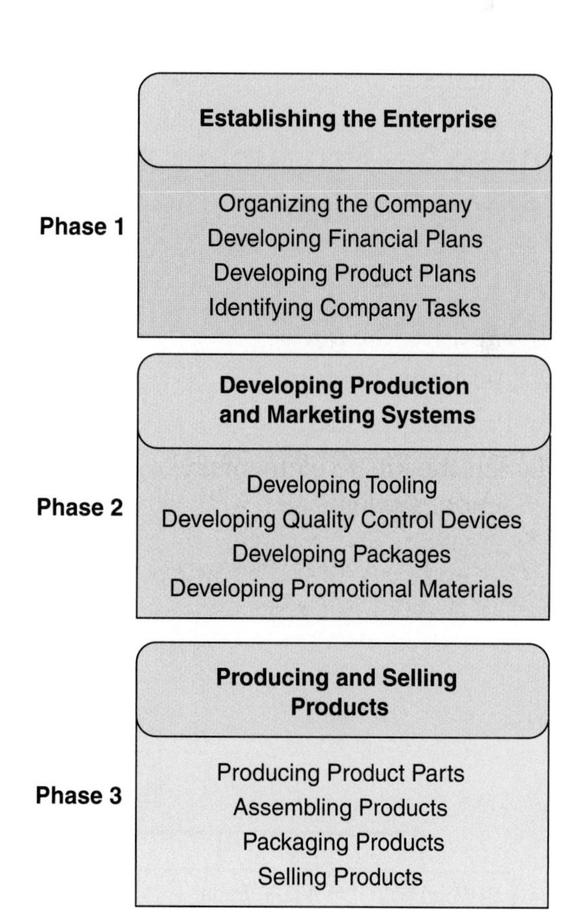

Figure 10-6.

Phase 2—Developing Production and Marketing Systems

During this phase, the class is divided into five groups as shown in the organizational chart in **Figure 10-7.** Each group is assigned one of the following tasks:

1. Develop a system to produce the sticks and paper cover for the kite (flowcharts or tooling, for example).
2. Develop a system to control the quality of the product (inspection gauges or quality control posters, for example).
3. Develop a product package (container or graphics, for example).
4. Develop product promotion (advertising) materials (flyers, radio spots for the school's public address system, or video commercials, for example).

Phase 3—Producing and Selling Products

During this phase of the activity, the students will complete the following tasks:

1. Produce kite parts.
2. Assemble the kits.
3. Package the kits.
4. Sell the kits to elementary and middle school students.

Reinforcing and Enriching

The following activities can be used to supplement and enrich the laboratory activities at the end of each textbook section and the STEM Applications included at the end of each textbook chapter.

1. A package could be developed to increase the product's marketing appeal.
2. A group of students could use their communication technology knowledge to develop a video or radio commercial promoting the product.

Taking Your Technology Knowledge Home

This take-home activity allows students to apply their knowledge of management systems to organize a social or civic event. Challenge the students to:

1. Work with their parents, guardians, neighbors, or youth group leader on an event for which they can prepare an organization and management system.
2. Determine the management functions that must be organized, such as production (day of the event management), marketing, financial affairs, and industrial relations (personnel assignment and training).

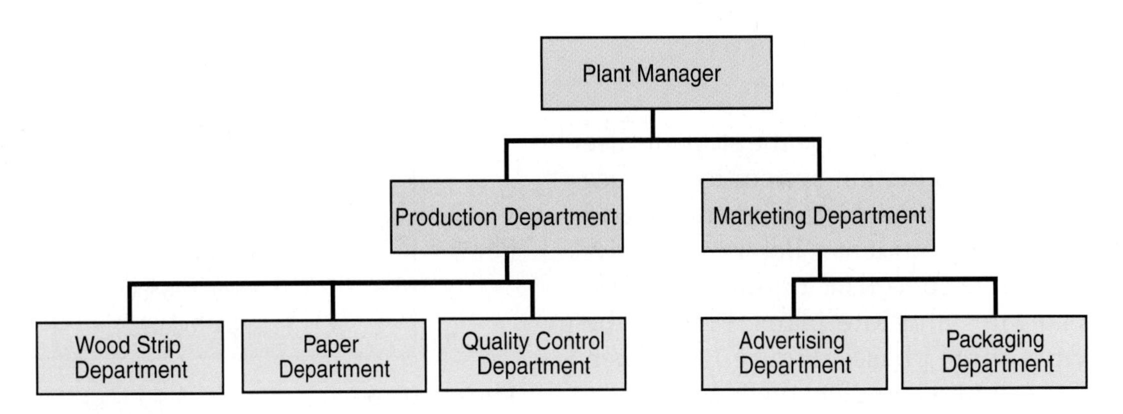

Figure 10-7.

3. Prepare a management plan.
4. Present the plan to the event organizers.

Adjustments for Exceptional Students

The laboratory activities in this section, by their very nature, accommodate a wide range of student abilities and interests. The various activities include needs for artistic, mechanical, design, engineering, and managerial skills. Also, some of the tasks are routine, whereas others require more self-direction. Adjusting for individual differences requires that you carefully assign the tasks in terms of student abilities. This purposeful grouping of students will allow each student to bring his or her own special abilities to the group.

Evaluation Procedures

The understanding that students have of technology and technological systems should be measured in several ways, including:

- Observation and critical listening during class presentations and discussions.
- Worksheets and quizzes covering the cognitive material presented.
- Open-ended questions about the learned material and how technology is developed.
- Assessment of individual participation and contributions to group work.
- Assessment of the quality and creativeness of the solutions to the assigned design/fabrication activities.

Section 11
Technological Systems in Modern Society

Chapter 35—Technology: A Societal View
Chapter 36—Technology: A Personal View

Teaching Materials

Text

Chapter 35, pages 742–757
  Test Your Knowledge, page 756
  STEM Applications, page 757

Chapter 36, pages 758–773
  Test Your Knowledge, page 772
  STEM Applications, page 773

Section 11 Activities
  Activity 11A—Design Problem, page 774
  Activity 11B—Production Problem, pages 774–775

Tech Lab Workbook

  Activity 11A—Design Problem, pages 137–138
  Activity 11B—Production Problem, pages 139–140

Teacher's Resource

  Chapter 35 Quiz, page 341
  Chapter 36 Quiz, page 346

Reproducible Masters
  35-1: Features of Futuring, page 342
  35-2: Some Challenges to the Environment, page 343
  36-1: Some Technology-Based Jobs, page 347
  36-2: Job Requirements, page 348

Lesson Slides (CD only)
  35-1: Switching Resources
  35-2: Pollution

Organization of Section 11

This last section studies technology and its design, operation, and impacts. It is not, however, to be considered less important than the other sections. In fact, in the final analysis, it might be the most important because it focuses on technology as it impacts people, society, and the environment.

It involves a dual view of technology. The first view is an outward, encompassing view of technology. It looks at technology and society. It also explores its impacts on people as a group and their environment.

The second view is an inwardly looking personal view, dealing with technology and the individual. It explores how technology makes life different for each person. It also introduces the student to the demands and hierarchy of technological careers.

During Section 11, the students will learn about futuring, which is a process that helps people choose wisely among alternate, and sometimes conflicting, courses of action. They will learn that futuring has five distinct features, **Figure 11-1:**

- Alternate avenues—possible answers rather than the answer.
- Different futures—seeking an entirely new future.

Features of Futuring
Alternate Avenues Different Futures Rational Decision Making Designing the Future Interrelationships

Figure 11-1.

- Rational decision making—using logical thinking and considering the consequences to make decisions.
- Designing the future—predicting a possible future that can be created.
- Interrelationships—seeing alternatives, cross-impacts, and leaps forward.

The students will use some of these ways of thinking to address common technological problems, such as energy conservation, pollution, and global warming.

Later in the section, the students will take personal looks at technology as they explore a wide range of careers, such as those of the production worker, technician, engineer, technologist, and manager. Then they will learn how to select a career that meets their abilities and desired lifestyles.

To complete the study, the students will complete two chapters and two laboratory activities. The chapters are:

- Chapter 35—Technology: A Societal View.
- Chapter 36—Technology: A Personal View.

As with most of the sections, two different types of activities are provided. If you want the students to address the content from a problem-solving approach, select Activity 11A—Design Problem. If you want a more structured approach, select Activity 11B—Production Problem.

Presenting Section 11

Estimated length—1 week

Section 11 is designed to help the students understand technology as a neutral force that can be made positive or negative when applied by people. For example, nuclear reactions are neutral. However, people can apply them to generate electricity, which is a positive force, or to create a nuclear winter through weapons, which is a negative force.

The content of Section 11 lends itself to a lecture/discussion strategy, where the students could simply discuss the impacts of technology on themselves and their environments. More meaningful learning, however, is accomplished by using realistic laboratory activities to reinforce the discussions.

The five-day schedule, shown in **Figure 11-2,** has been developed for this section. During this period of time, you will notice that two major focuses are used for the study—the individual and society/environment. To prepare to present this section, you should:

1. Read Chapters 35 and 36 in *Technology & Engineering.*
2. Prepare discussions on:
   A. Technology, society, and the environment (Chapter 35).
   B. You and technology (Chapter 36).
3. Select a laboratory activity.
   A. Activity 11A—This is an open-ended activity where students select a technological issue, analyze it, and suggest an action plan to improve it.

Section 11: Technological Systems in Modern Society	
Day	Activity
1–2	Discussion: "Technology: A Societal View" (Chapter 35) Introduce laboratory activity
3	Students work on laboratory activity
4	Discussion: "Technology: A Personal View" (Chapter 36) Students work on laboratory activity
5	Students complete laboratory activity Summarize section and course

Figure 11-2.

B. Activity 11B—This is a structured activity where students develop 60-second television commercials that promote a positive technological action.

4. Practice the selected laboratory activity.
5. Prepare teaching materials (transparencies or bulletin boards, for example) to support the content.
6. Determine which study questions and quiz questions will be used.

Reinforcing and Enriching

The following activity can be used to supplement and enrich the laboratory activities at the end of each textbook section and the STEM Applications included at the end of each textbook chapter. Activity 11B—Production Problem could be reinforced by working with the social studies and language arts teachers so the research and writing activities could be coordinated with those classes. Your class could handle the production activities.

Adjustments for Exceptional Students

Both laboratory activities are designed to be done rather quickly. The level of sophistication can be adjusted to accommodate students of all ability levels. Gifted students should have more creative answers and possibly produce a more complex solution. Lower achievers can still solve the problem but with solutions that fit within their ability level. Also, by grouping students of various interests and abilities, each student can bring his or her own special talents to the group. Some students will have more artistic abilities, others might have more dexterity, and still others might have more abstract reasoning ability.

Evaluation Procedures

The understanding that students have of technology and technological systems should be measured in several ways, including:

- Observation and critical listening during class presentations and discussions.
- Worksheets and quizzes covering the cognitive material presented.
- Open-ended questions about the learned material and how technology is developed.
- Assessment of individual participation and contributions to group work.
- Assessment of the quality and creativeness of the solutions to the assigned design/production activities.

Standards for Technological Literacy Correlation Chart

The International Technology and Engineering Educators Association (ITEEA) and its Technology for All Americans Project developed *Standards for Technological Literacy: Content for the Study of Technology* to identify the essential core of technological knowledge and skills for students in grades K–12. This work defined twenty separate standards, divided into five broad categories. Within each standard, benchmark topics are defined for four different grade levels:

- Grades K–2
- Grades 3–5
- Grades 6–8
- Grades 9–12

The following chart lists the standards and the benchmark topics for grades 9–12. Adjacent to each standard and benchmark topic are the chapter and page references identifying material in *Technology & Engineering* relating to the item.

Standards for Technological Literacy Correlation Chart	
Standard 1. Students will develop an understanding of the characteristics and scope of technology.	Chapter 1: 23–26 Chapter 3: 57–63 Chapter 4 Chapter 5: 94–95, 97–112 Chapter 9 Chapter 32: 687
Benchmark Topics:	
The nature and development of technological knowledge and processes are functions of the setting.	Chapter 3: 60–63 Chapter 5: 97–112
The rate of technological development and diffusion is increasing rapidly.	Chapter 1: 26 Chapter 4
Inventions and innovations are the results of specific, goal-directed research.	Chapter 4 Chapter 9
Most development of technologies these days is driven by the profit motive and the market.	Chapter 1: 25 Chapter 3: 57–59 Chapter 5: 94–95 Chapter 32: 687
Standard 2. Students will develop an understanding of the core concepts of technology.	Chapter 1: 23–25 Chapter 2: 40–50 Chapter 5: 93–97, 112–113 Chapter 6: 121–125 Chapter 8: 182–183 Chapter 9 Chapter 10: 205–207 Chapter 11: 221–230, 233 Chapter 32: 688–691 Chapter 33: 710–711

Standards for Technological Literacy Correlation Chart	
Benchmark Topics:	
Systems thinking applies logic and creativity with appropriate compromises in complex real-life problems.	Chapter 10: 205–207 Chapter 11: 221–230
Systems, which are the building blocks of technology, are embedded within larger technological, social, and environmental systems.	Chapter 10: 205–207
The stability of a technological system is influenced by all of the components in the system, especially those in the feedback loop.	Chapter 2: 40–50
Selecting resources involves trade-offs between competing values, such as availability, cost, desirability, and waste.	Chapter 10: 207
Requirements involve the identification of the criteria and constraints of a product or system and the determination of how they affect the final design and development.	Chapter 5: 93–97 Chapter 9 Chapter 10: 207
Optimization is an ongoing process or methodology of designing or making a product and is dependent on criteria and constraints.	Chapter 11: 226, 233
New technologies create new processes.	Chapter 1: 26
Quality control is a planned process to ensure that a product, service, or system meets established criteria.	Chapter 6: 121–124 Chapter 8: 182–183 Chapter 33: 710–711
Management is the process of planning, organizing, and controlling work.	Chapter 2: 48–49 Chapter 5: 112–113 Chapter 32: 688–691
Complex systems have many layers of controls and feedback loops to provide information.	Chapter 2: 50 Chapter 6: 120–131
Standard 3. Students will develop an understanding of the relationships among technologies and the connections between technology and other fields of study.	Chapter 1: 26–29 Chapter 2: 43 Chapter 3: 62 Chapter 4 Chapter 5: 104 Chapter 6: 122, 126 Chapter 7: 146 Chapter 8: 176 Chapter 9: 195 Chapter 10: 210 Chapter 12: 245 Chapter 13: 266 Chapter 14: 278 Chapter 15: 294 Chapter 17: 348 Chapter 18: 371 Chapter 19: 386 Chapter 20: 402 Chapter 21: 429 Chapter 22: 443–447, 449–450

Standards for Technological Literacy Correlation Chart	
Standard 3. Students will develop an understanding of the relationships among technologies and the connections between technology and other fields of study.	Chapter 23: 469, 476 Chapter 24: 501 Chapter 25: 516 Chapter 26: 550 Chapter 27: 567 Chapter 28: 592 Chapter 29: 618 Chapter 30: 646 Chapter 31: 666–674 Chapter 33: 707
Benchmark Topics:	
Technology transfer occurs when a new user applies an existing innovation developed for one purpose in a different function.	Chapter 4
Technological innovation often results when ideas, knowledge, or skills are shared within a technology, among technologies, or across other fields.	Chapter 1: 28–29 Chapter 4
Technological ideas are sometimes protected through the process of patenting.	Chapter 4: 71–74
Technological progress promotes the advancement of science and mathematics.	Chapter 2: 43 Chapter 3: 62 Chapter 4: 77 Chapter 6: 122, 126 Chapter 8: 176 Chapter 10: 210 Chapter 13: 266 Chapter 14: 278 Chapter 15: 294 Chapter 18: 371 Chapter 19: 386 Chapter 20: 402 Chapter 21: 429 Chapter 22: 449–450 Chapter 23: 476 Chapter 24: 501 Chapter 25: 516 Chapter 26: 550 Chapter 28: 592 Chapter 30: 646 Chapter 31: 666–674 Chapter 33: 707
Standard 4. Students will develop an understanding of the cultural, social, economic, and political effects of technology.	Chapter 1: 23–28 Chapter 3: 55–57 Chapter 6: 117–120 Chapter 9: 199–200 Chapter 32: 691–696 Chapter 33: 702–704, 715–720 Chapter 35 Chapter 36: 769–770

Standards for Technological Literacy Correlation Chart	
Benchmark Topics:	
Changes caused by the use of technology can range from gradual to rapid and from subtle to obvious.	Chapter 3: 55–57 Chapter 6: 119–120 Chapter 35: 745–747 Chapter 36: 769–770
Making decisions about the use of technology involves weighing the trade-offs between the positive and negative effects.	Chapter 1: 26–29 Chapter 6: 117–119 Chapter 32: 691–692 Chapter 35 Chapter 36: 769–770
Ethical considerations are important in the development, selection, and use of technologies.	Chapter 9: 199–200
The transfer of a technology from one society to another can cause cultural, social, economic, and political changes affecting both societies to varying degrees.	Chapter 1: 23–25
Standard 5. Students will develop an understanding of the effects of technology on the environment.	Chapter 1: 26–29 Chapter 6: 125–127 Chapter 18: 372–374 Chapter 25: 530–533 Chapter 28: 581–584 Chapter 35: 744, 750–753 Chapter 36: 769–770
Benchmark Topics:	
Humans can devise technologies to conserve water, soil, and energy through such techniques as reusing, reducing, and recycling.	Chapter 18: 372–374
When new technologies are developed to reduce the use of resources, considerations of trade-offs are important.	Chapter 35: 750–753
With the aid of technology, various aspects of the environment can be monitored to provide information for decision making.	Chapter 6: 125–127 Chapter 25: 530–533 Chapter 35: 750–753
The alignment of technological processes with natural processes maximizes performance and reduces negative impacts on the environment.	Chapter 1: 26–29 Chapter 18: 372–374 Chapter 28: 581–582
Humans devise technologies to reduce the negative consequences of other technologies.	Chapter 18: 372–374
Decisions regarding the implementation of technologies involve the weighing of trade-offs between predicted positive and negative effects on the environment.	Chapter 1: 26–29 Chapter 35: 752–753 Chapter 36: 769–770
Standard 6. Students will develop an understanding of the role of society in the development and use of technology.	Chapter 2 Chapter 4: 71–74, 87–89 Chapter 6: 117 Chapter 22: 454 Chapter 33: 702–703, 713–714, 720 Chapter 35: 744–745, 753–755 Chapter 36: 759–762

Standards for Technological Literacy Correlation Chart

Benchmark Topics:	
Different cultures develop their own technologies to satisfy their individual and shared needs, wants, and values.	Chapter 6: 117 Chapter 36: 759–762
The decision whether to develop a technology is influenced by societal opinions and demands, in addition to corporate cultures.	Chapter 2 Chapter 4: 71–74
A number of different factors, such as advertising, the strength of the economy, the goals of a company, and the latest fads, contribute to shaping the design of and demand for various technologies.	Chapter 4: 71–74, 87–89 Chapter 22: 454 Chapter 33: 713–714, 720 Chapter 35: 744–745, 753–755
Standard 7. Students will develop an understanding of the influence of technology on history.	Chapter 1: 23–26, 29–35 Chapter 2: 41–43 Chapter 4: 71–84, 87–89 Chapter 5: 104 Chapter 7: 139–154, 156–158 Chapter 8: 169–172, 178–181 Chapter 9: 195 Chapter 12: 239 Chapter 21: 429 Chapter 22: 449–450 Chapter 23: 476 Chapter 25: 525–526 Chapter 27: 567 Chapter 28: 578–581 Chapter 29: 618 Chapter 33: 705–706 Chapter 35: 745–747 Chapter 36: 759–762
Benchmark Topics:	
Most technological development has been evolutionary, the result of a series of refinements to a basic invention.	Chapter 1: 23–25, 29–35 Chapter 4: 71 Chapter 8: 169–172 Chapter 33: 705–706
The evolution of civilization has been directly affected by, and has in turn affected, the development and use of tools and materials.	Chapter 1: 25, 29–35 Chapter 2: 41–43 Chapter 4: 74–84 Chapter 7: 139–151 Chapter 8: 178–181
Throughout history, technology has been a powerful force in reshaping the social, cultural, political, and economic landscape.	Chapter 1: 23–25, 29–35 Chapter 4: 74–84 Chapter 35: 745–747
Early in the history of technology, the development of many tools and machines was based not on scientific knowledge but on technological know-how.	Chapter 1: 30–33 Chapter 22: 449–450
The Iron Age was defined by the use of iron and steel as the primary materials for tools.	Chapter 1: 33 Chapter 28: 578–580

***Standards for Technological Literacy* Correlation Chart**	
The Middle Ages saw the development of many technological devices that produced long-lasting effects on technology and society.	Chapter 1: 33 Chapter 7: 157–158 Chapter 28: 580–581
The Renaissance, a time of rebirth of the arts and humanities, was also an important development in the history of technology.	Chapter 1: 33
The Industrial Revolution saw the development of continuous manufacturing, sophisticated transportation and communication systems, advanced construction practices, and improved education and leisure time.	Chapter 1: 33–35 Chapter 7: 143, 152–154 Chapter 12: 239 Chapter 22: 449–450 Chapter 25: 525–526 Chapter 36: 760–761
The Information Age places emphasis on the processing and exchange of information.	Chapter 1: 35 Chapter 36: 759-762
Standard 8. Students will develop an understanding of the attributes of design.	Chapter 2: 47–48 Chapter 5: 94–97 Chapter 9: 191–200 Chapter 10: 207–217 Chapter 11 Chapter 12
Benchmark Topics:	
The design process includes defining a problem, brainstorming, researching and generating ideas, identifying criteria and specifying constraints, exploring possibilities, selecting an approach, developing a design proposal, making a model or prototype, testing and evaluating the design using specifications, refining the design, creating or making it, and communicating processes and results.	Chapter 2: 47–48 Chapter 5: 94–97 Chapter 9: 191–195 Chapter 10: 207–217 Chapter 11: 221–230 Chapter 12
Design problems are seldom presented in a clearly defined form.	Chapter 5: 94–96 Chapter 9: 197–200
The design needs to be continually checked and critiqued, and the ideas of the design must be redefined and improved.	Chapter 5: 96 Chapter 10: 212 Chapter 11: 230–233
Requirements of a design, such as criteria, constraints, and efficiency, sometimes compete with each other.	Chapter 5: 94–96 Chapter 10: 208
Standard 9. Students will develop an understanding of engineering design.	Chapter 5: 94–97 Chapter 9 Chapter 10: 208 Chapter 11: 228–233 Chapter 12 Chapter 20

Standards for Technological Literacy Correlation Chart

Benchmark Topics:	
Established design principles are used to evaluate existing designs, to collect data, and to guide the design process.	Chapter 12: 249 Chapter 20
Engineering design is influenced by personal characteristics, such as creativity, resourcefulness, and the ability to visualize and think abstractly.	Chapter 5: 94–97 Chapter 9 Chapter 12
A prototype is a working model used to test a design concept by making actual observations and necessary adjustments.	Chapter 5: 96 Chapter 11: 228
The process of engineering design takes into account a number of factors.	Chapter 11: 228–233
Standard 10. Students will develop an understanding of the role of troubleshooting, research and development, invention and innovation, and experimentation in problem solving.	Chapter 9 Chapter 33: 703–707
Benchmark Topics:	
Research and development is a specific problem-solving approach that is used intensively in business and industry to prepare devices and systems for the marketplace.	Chapter 33: 703–707
Technological problems must be researched before they can be solved.	Chapter 9: 191–192, 196–197, 199–200
Not all problems are technological, and not every problem can be solved using technology.	Chapter 9: 191–192
Many technological problems require a multidisciplinary approach.	Chapter 9
Standard 11. Students will develop abilities to apply the design process.	Chapter 5: 94–97 Chapter 9: 196–200 Chapter 10 Chapter 12 Chapter 20: 419–421 Chapter 21: 440–441 Chapter 22: 464–466 Chapter 23: 488–489
Benchmark Topics:	
Identify the design problem to solve and decide whether or not to address it.	Chapter 5: 95–96 Chapter 9: 197–200 Chapter 10: 209
Identify criteria and constraints and determine how these will affect the design process.	Chapter 5: 95–96 Chapter 9: 196–199
Refine a design by using prototypes and modeling to ensure quality, efficiency, and productivity of the final product.	Chapter 5: 96–97

Standards for Technological Literacy Correlation Chart

Evaluate the design solution using conceptual, physical, and mathematical models at various intervals of the design process in order to check for proper design and to note areas where improvements are needed.	Chapter 5: 96–97
Develop and produce a product or system using a design process.	Chapter 10 Chapter 20: 419–421 Chapter 21: 440–441 Chapter 22: 464–466 Chapter 23: 488–489
Evaluate final solutions and communicate observation, processes, and results of the entire design process, using verbal, graphic, quantitative, virtual, and written means, in addition to three-dimensional models.	Chapter 5: 96–97 Chapter 12
Standard 12. Students will develop the abilities to use and maintain technological products and systems.	Chapter 7 Chapter 11: 236–237 Chapter 12: 254–255 Chapter 18: 367–370 Chapter 20: 420–421 Chapter 23: 488–489 Chapter 26: 552–553 Chapter 31: 680–681 Chapter 34: 725–732
Benchmark Topics:	
Document processes and procedures and communicate them to different audiences using appropriate oral and written techniques.	Chapter 12: 254–255
Diagnose a system that is malfunctioning and use tools, materials, machines, and knowledge to repair it.	Chapter 18: 367–370 Chapter 34: 729
Troubleshoot, analyze, and maintain systems to ensure safe and proper function and precision.	Chapter 7 Chapter 18: 367–370 Chapter 26: 552–553
Operate systems so that they function in the way they were designed.	Chapter 7: 162–165
Use computers and calculators to access, retrieve, organize, process, maintain, interpret, and evaluate data and information in order to communicate.	Chapter 11: 236–237 Chapter 12: 254–255 Chapter 20: 420–421 Chapter 23: 488–489 Chapter 31: 680–681
Standard 13. Students will develop the abilities to assess the impact of products and systems.	Chapter 18 Chapter 31: 680–681 Chapter 34 Chapter 35: 745–747

Standards for Technological Literacy Correlation Chart

Benchmark Topics:	
Collect information and evaluate its quality.	Chapter 31: 680–681
Synthesize data, analyze trends, and draw conclusions regarding the effect of technology on the individual, society, and the environment.	Chapter 31: 680–681
Use assessment techniques, such as trend analysis and experimentation to make decisions about the future development of technology.	Chapter 34 Chapter 35: 745–747
Design forecasting techniques to evaluate the results of altering natural systems.	Chapter 18 Chapter 34
Standard 14. Students will develop an understanding of and be able to select and use medical technologies.	Section 1: 21 Chapter 3: 62 Chapter 5: 109–110 Chapter 31
Benchmark Topics:	
Medical technologies include prevention and rehabilitation, vaccines and pharmaceuticals, medical and surgical procedures, genetic engineering, and the systems within which health is protected and maintained.	Chapter 5: 109–110 Chapter 31
Telemedicine reflects the convergence of technological advances in a number of fields, including medicine, telecommunications, virtual presence, computer engineering, informatics, artificial intelligence, robotics, materials science, and perceptual psychology.	Chapter 31
The sciences of biochemistry and molecular biology have made it possible to manipulate the genetic information found in living creatures.	Section 1: 21 Chapter 3: 62
Standard 15. Students will develop an understanding of and be able to select and use agricultural and related biotechnologies.	Section 1: 21 Chapter 1 Chapter 3: 62 Chapter 5: 109–110 Section 9: 607 Chapter 29 Chapter 30 Chapter 31: 665–677 Section 11: 741 Chapter 35
Benchmark Topics:	
Agriculture includes a combination of businesses that use a wide array of products and systems to produce, process, and distribute food, fiber, fuel, chemical, and other useful products.	Chapter 29: 609–610

***Standards for Technological Literacy* Correlation Chart**	
Biotechnology has applications in such areas as agriculture, pharmaceuticals, food and beverages, medicine, energy, the environment, and genetic engineering.	Section 1: 15 Chapter 3: 62 Section 9: 607 Chapter 29: 609–610 Chapter 30 Chapter 31: 665–677
Conservation is the process of controlling soil erosion, reducing sediment in waterways, conserving water, and improving water quality.	Chapter 1 Chapter 35
The engineering design and management of agricultural systems require knowledge of artificial ecosystems and the effects of technological development on flora and fauna.	Section 11: 741
Standard 16. Students will develop an understanding of and be able to select and use energy and power technologies.	Chapter 4: 83, 86–87 Chapter 5: 104–106 Chapter 7: 151–157 Chapter 15: 303 Chapter 27 Chapter 28
Benchmark Topics:	
Energy cannot be created nor destroyed; however, it can be converted from one form to another.	Chapter 5: 105 Chapter 7: 151–157 Chapter 27: 569 Chapter 28: 578–595
Energy can be grouped into major forms: thermal, radiant, electrical, mechanical, chemical, nuclear, and others.	Chapter 4: 86–87 Chapter 5: 105 Chapter 15: 303 Chapter 27: 566–569
It is impossible to build an engine to perform work that does not exhaust thermal energy to the surroundings.	Chapter 27 Chapter 28
Energy resources can be renewable or nonrenewable.	Chapter 4: 83 Chapter 27: 570–572 Chapter 28
Power systems must have a source of energy, a process, and loads.	Chapter 5: 104–106 Chapter 28: 596–600
Standard 17. Students will develop an understanding of and be able to select and use information and communication technologies.	Chapter 5: 99–101 Chapter 7: 157–162 Chapter 19 Chapter 20 Chapter 21 Chapter 22 Chapter 23
Benchmark Topics:	
Information and communication technologies include the inputs, processes, and outputs associated with sending and receiving information.	Chapter 5: 99–101 Chapter 7: 157–162

Standards for Technological Literacy Correlation Chart	
Information and communication systems allow information to be transferred from human to human, human to machine, machine to human, and machine to machine.	Chapter 19: 387–388
Information and communication systems can be used to inform, persuade, entertain, control, manage, and educate.	Chapter 19: 384–385 Chapter 22
Communication systems are made up of source, encoder, transmitter, receiver, decoder, storage, retrieval, and destination.	Chapter 5: 99–101 Chapter 19: 385–387 Chapter 22
There are many ways to communicate information, such as graphic and electronic means.	Chapter 5: 99–101 Chapter 19: 388–392 Chapter 20 Chapter 21: 426–436 Chapter 22: 444–461
Technological knowledge and processes are communicated using symbols, measurement, conventions, icons, graphic images, and languages that incorporate a variety of visual, auditory, and tactile stimuli.	Chapter 23
Standard 18. Students will develop an understanding of and be able to select and use transportation technologies.	Chapter 5: 110–112 Chapter 24 Chapter 25 Chapter 26
Benchmark Topics:	
Transportation plays a vital role in the operation of other technologies, such as manufacturing, construction, communication, health and safety, and agriculture.	Chapter 24: 496 Chapter 26
Intermodalism is the use of different modes of transportation, such as highways, railways, and waterways as part of an interconnected system that can move people and goods easily from one mode to another.	Chapter 24
Transportation services and methods have led to a population that is regularly on the move.	Chapter 24 Chapter 26: 548–550
The design of intelligent and nonintelligent transportation systems depends on many processes and innovative techniques.	Chapter 24: 496–502 Chapter 25: 507–509 Chapter 26: 542–548

***Standards for Technological Literacy* Correlation Chart**	
Standard 19. Students will develop an understanding of and be able to select and use manufacturing technologies.	Chapter 1 Chapter 4 Chapter 5: 106–109 Chapter 7: 143 Chapter 12: 239 Chapter 13 Chapter 14 Chapter 15 Chapter 16 Chapter 18: 365–370 Chapter 33: 707–715 Chapter 36: 780–781
Benchmark Topics:	
Servicing keeps products in good operating condition.	Chapter 13: 266–267 Chapter 18: 367–370
Materials have different qualities and may be classified as natural, synthetic, or mixed.	Chapter 4 Chapter 14
Durable goods are designed to operate for a long period of time, while nondurable goods are designed to operate for a short period of time.	Chapter 13
Manufacturing systems may be classified into types, such as customized production, batch production, and continuous production.	Chapter 7: 143 Chapter 33: 707–715
The interchangeability of parts increases the effectiveness of manufacturing processes.	Chapter 1 Chapter 12: 239 Chapter 36: 780–781
Chemical technologies provide a means for humans to alter or modify materials and to produce chemical products.	Chapter 15
Marketing involves establishing a product's identity, conducting research on its potential, advertising it, distributing it, and selling it.	Chapter 33: 713–715
Standard 20. Students will develop an understanding of and be able to select and use construction technologies.	Chapter 5: 101–104 Chapter 13 Chapter 17 Chapter 18: 367–372 Chapter 34: 731
Benchmark Topics:	
Infrastructure is the underlying base or basic framework of a system.	Chapter 17
Structures are constructed using a variety of processes and procedures.	Chapter 5: 101–104 Chapter 34: 731
The design of structures includes a number of requirements.	Chapter 17
Structures require maintenance, alteration, or renovation periodically to improve them or to alter their intended use.	Chapter 5: 104 Chapter 13 Chapter 18: 367–372
Structures can include prefabricated materials.	Chapter 17

Basic Skills Chart

The Basic Skills Chart has been designed to identify those activities in the *Technology & Engineering* and the *Technology & Engineering Tech Lab Workbook* that specifically encourage the development of basic skills.

Academic areas addressed in the chart include reading, writing, verbal (other than reading and writing), math, science, and analytical.

- **Reading** activities include assignments designed to improve comprehension of information presented in the chapter. Some are designed to improve understanding of vocabulary terms.
- **Writing** activities allow students to practice composition skills, such as letter writing, informative writing, and creative writing.
- **Verbal** activities encourage students to organize ideas, develop interpersonal and group speaking skills, and respond appropriately to verbal messages. Activities include oral reports and interviews.
- **Math** activities require students to use basic principles of math, as well as computation skills, to solve typical problems.
- **Science** activities call for students to use fundamental principles of science to solve typical problems.
- **Analytical** activities involve the higher-order skills needed for thinking creatively, making decisions, solving problems, visualizing information, reasoning, and knowing how to learn.

Activities are broken down by chapter, and a page number is given to locate the activity.

Basic Skills Chart	
Section 1	
Reading	**Text:** 1: *Technology* Defined (23); Technology as a Dynamic Process (26); Positive and Negative Aspects of Technology (26); The Technology and Types of Knowledge (28); The Evolution of Technology (29); Presidential Election of 1960 (30); Smart Houses (32) 2: Inputs (41); Processes (45); Solar Collectors (46); Outputs (49); Feedback and Control (50) 3: Level of Development (55); Economic Structure (57); Number of People Involved (59); Type of Technology Developed and Used (60)
Writing	**Text:** 2: Activity 2 (52)
Verbal	**Text:** 3: Activity 3 (65)
Math	
Science	**Text:** 1: Activity 2 (37); Activity 3 (37) 2: Newton's Third Law of Motion (43) 3: Genetic Engineering (62)

Basic Skills Chart	
Section 1 *(Continued)*	
Analytical	**Text:** 1: Activity 1 (37); Activity 2 (37); Activity 3 (37) 2: Activity 1 (52) 3: Activity 1 (65); Activity 2 (65); Activity 3 (65) Activity 1A: Design Problem (66) Activity 1B: Fabrication Problem (66) **Tech Lab Workbook:** Activity 1A: Design Problem (7) Activity 1B: Fabrication Problem (9)
Section 2	
Reading	**Text:** 4: People (71); Tools and Machines (74); Materials (82); Information (85); Energy (86); Finances (87); Time (88) 5: Problem-Solving or Design Processes (94); Production Processes (97); The Tennessee Valley Authority (104); Hybrid Vehicles (107); Management Processes (112) 6: Outputs (117); Feedback and Control (120)
Writing	
Verbal	
Math	**Text:** 4: The Law of Equilibrium (77)
Science	**Text:** 6: Chlorofluorocarbons (122); Integrated Circuits (126)
Analytical	**Text:** 4: Activity 1 (91); Activity 2 (91); Activity 3 (91) 5: Activity 1 (115); Activity 2 (115); Activity 3 (115) 6: Activity 1 (133); Activity 2 (133); Activity 3 (133) Activity 2A: Design Problem (134) Activity 2B: Fabrication Problem (134) **Tech Lab Workbook:** Activity 2A: Design Problem (13) Activity 2B: Fabrication Problem (15)
Section 3	
Reading	**Text:** 7: Material-Processing Tools and Machines (139); Types of Machine Tools (142); Computer Numerical Control (143); Computer Bugs (146); Energy-Processing Converters (157); Information-Processing Machines (157); Using Technology Safely (162) 8: Measurement Systems: Past and Present (169); Qualities Measured (172); Measurement Tools (178); Measurement and Control (182)

Basic Skills Chart	
Section 3 *(Continued)*	
Writing	**Text:** 7: Activity 2 (162); Activity 4 (162); Activity 5 (162) Activity 3A: Design Problem (186) **Tech Lab Workbook:** Activity 3A: Design Problem (19)
Verbal	
Math	**Text:** 8: Measuring Area (176); Types of Measurement (177); Activity 1 (185); Activity 2 (185)
Science	
Analytical	**Text:** 7: Activity 1 (162); Activity 3 (162); Activity 4 (162); Activity 6 (162) 8: Activity 3 (185) Activity 3A: Design Problem (186) Activity 3B: Fabrication Problem (186) **Tech Lab Workbook:** Activity 3A: Design Problem (19) Activity 3B: Fabrication Problem (23) Taking Your Technology Knowledge Home: Designing a Tool (27)
Section 4	
Reading	**Text:** 9: The Problem-Solving and Design Process in General (191); The Problem-Solving and Design Process in Technology (192); The Origin of Radar (195); Steps in Solving Technological Problems and Meeting Opportunities (196); Identifying a Technological Problem or Opportunity (197) 10: System Design (205); Product Design (207); Steps for Developing Design Solutions (208) 11: Modeling Design Solutions (221); GPS (227); Redesigning Products and Structures (233) 12: Product Documents and Reports (239); Principles of Design (245); Computer-Aided Design (248); Approval Documents and Reports (251)
Writing	**Text:** 9: Activity 1 (203) 11: Activity 2 (235) **Tech Lab Workbook:** Activity 4A/4B: Design/Fabrication Problem (33)
Verbal	
Math	**Text:** 10: Solid Geometry (210)
Science	

<table>
<tr><th colspan="2">Basic Skills Chart</th></tr>
<tr><th colspan="2">Section 4 (Continued)</th></tr>
<tr><td>Analytical</td><td>Text:
9: Activity 1 (203); Activity 2 (203)
10: Activity 1 (219); Activity 2 (219); Activity 3 (219); Activity 4 (219)
11: Analyzing the Design (230); Activity 1 (235); Activity 2 (235); Computer-Aided Design, Engineering with Animation (236)
12: Activity 1 (253); Activity 2 (253); Computer-Aided Design, Architecture with Animation (254)
Activity 4A: Design Problem (256)
Activity 4B: Fabrication Problem (256)
Tech Lab Workbook:
Activity 4A/4B: Design/Fabrication Problem (33)
Taking Your Technology Knowledge Home: Designing a Code (39)</td></tr>
<tr><th colspan="2">Section 5</th></tr>
<tr><td>Reading</td><td>Text:
13: Production Activities (261); Servicing and Repairing Products and Structures (266)
14: Types of Natural Material Resources (271); Locating and Obtaining Natural Resources (275); Activity 3 (287)
15: Mechanical Processes (290); Thermal Processes (294); Chemical and Electrochemical Processes (304)
16: Types of Manufacturing Processes (310); Robots (316)
17: Buildings (337); Word Origins (348); Heavy Engineering Structures (353)
18: Selecting Technological Products (365); Installing Technological Products (366); Maintaining Technological Products (367); Repairing Technological Products (368); Altering Technological Products (370); Disposing of Technological Products (372)</td></tr>
<tr><td>Writing</td><td>Text:
14: Activity 3 (287)</td></tr>
<tr><td>Verbal</td><td></td></tr>
<tr><td>Math</td><td>Text:
15: Calculating Board Footage (294)</td></tr>
<tr><td>Science</td><td>Text:
13: The Principles of Expansion and Contraction (266)
14: Synthetic Fuels (278)
15: Nuclear Energy (303)
18: Materials Science (371)</td></tr>
<tr><td>Analytical</td><td>Text:
13: Activity 1 (269)
14: Activity 1 (287); Activity 2 (287)
15: Activity 1 (307); Activity 2 (307); Activity 3 (307)
16: Activity 1 (335); Activity 2 (335)
17: Activity 1 (361); Activity 2 (361); Structural Engineering (362)
18: Activity 1 (375); Activity 2 (375)</td></tr>
</table>

Basic Skills Chart

Section 5 *(Continued)*

Analytical	**Text:** Activity 5A: Design Problem—Manufacturing Technology (376) Activity 5B: Fabrication Problem—Manufacturing Technology (376) Activity 5C: Design Problem—Construction Problem (378) Activity 5D: Fabrication Problem—Construction Problem (378) **Tech Lab Workbook:** Activity 5A/5B: Manufacturing Design/Fabrication Problem (45) Activity 5C: Construction Design Problem (53) Activity 5D: Construction Design Problem (55) Activity 5E: Product Servicing Activity (57) Activity 5F: Papermaking/Paper Recycling Activity (61) Taking Your Technology Knowledge Home: Designing a Manufactured Product (63)

Section 6

Reading	**Text:** 19: Communicated Items (384); Goals of Communication (384); The Communication Model (385); Types of Communication (387); Communication Systems (388); Digital Video Disc (389) 20: Printing Methods (397); Steps in Producing Printed Graphic Messages (401); Fax Machines (408); Computer-Based Publishing (414) 21: Fundamentals of Photographic Communication (426); The Beginnings of Photojournalism (429); Digital Theaters (432); Types of Photographic Communication (435) 22: Types of Telecommunication Systems (447); Advertising (450); Communicating with Telecommunication Systems (452); Fiber Optics (454); Other Communication Technologies (461) 23: Computer Systems (469); Networks (471); The Internet (472); Virtual Reality (481)
Writing	**Text:** 20: Desktop Publishing (420) 22: Activity 2 (463); Film (464) Activity 6B: Production Problem (490) **Tech Lab Workbook:** Activity 6C: Telecommunication Activity (77) Taking Your Technology Knowledge Home: Promoting a Neighborhood Event (79)
Verbal	**Text:** 19: The Power of Radio (386) 22: Activity 3 (463); Film (464) Activity 6B: Production Problem (490) **Tech Lab Workbook:** Activity 6C: Telecommunication Activity (77) Taking Your Technology Knowledge Home: Promoting a Neighborhood Event (79)

Basic Skills Chart	
Section 6 *(Continued)*	
Math	**Text:** 20: Measuring Type (402)
Science	**Text:** 21: Light and Photography (423) 22: The Physics of Telecommunication (443); Film (464)
Analytical	**Text:** 19: Activity 1 (395); Activity 2 (395) 20: Activity 1 (418); Activity 2 (418); Promotional Graphics (419); Desktop Publishing (420) 21: Activity 1 (439); Activity 2 (439); Imaging Technology (440) 22: Activity 1 (463); Activity 2 (463); Activity 3 (463); Film (464) 23: Activity 1 (487); Activity 2 (487); Cyberspace Pursuit (488) Activity 6A: Design Problem (490) Activity 6B: Production Problem (490) **Tech Lab Workbook:** Activity 6A: Design Problem (69) Activity 6B: Production Problem (73) Activity 6C: Telecommunication Activity (77) Taking Your Technology Knowledge Home: Promoting a Neighborhood Event (79)
Section 7	
Reading	**Text:** 24: *Transportation:* A Definition (496); Transportation as a System (496); Types of Transportation Systems (496); Transportation-System Components (498) 25: Vehicular Systems (507); Land-Transportation Vehicles (509); Water-Transportation Vehicles (517); Maglev Trains (520); Air-Transportation Vehicles (525); Space-Transportation Vehicles (533) 26: Types of Transportation (541); Components of a Transportation System (542); Transporting People and Cargo (548); Maintaining Transportation Systems (550); Regulating Transportation Systems (552)
Writing	**Text:** 26: System Control Technology (556) Activity 7A: Design Problem (558) **Tech Lab Workbook:** Taking Your Technology Knowledge Home: Planning a Trip (89)
Verbal	
Math	**Text:** 25: Calculating Buoyant Force (516) 26: Relating Speed, Time, and Distance (550) **Tech Lab Workbook:** Taking Your Technology Knowledge Home: Planning a Trip (89)

Basic Skills Chart	
Section 7 *(Continued)*	
Science	**Text:** 24: Newton's First Law of Motion (501)
Analytical	**Text:** 24: Activity 1 (504); Activity 2 (505) 25: Activity 1 (539); Activity 2 (539); Activity 3 (539) 26: Activity 1 (555); Activity 2 (555); Activity 3 (555); Activity 4 (555); System Control Technology (556) Activity 7A: Design Problem (558) Activity 7B: Fabrication Problem (558) **Tech Lab Workbook:** Activity 7A: Design Problem (83) Activity 7B: Fabrication Problem (85) Taking Your Technology Knowledge Home: Planning a Trip (89)
Section 8	
Reading	**Text:** 27: Types of Energy (563); Energy, Work, and Power (564); Forms of Energy (566); The Origin of Horsepower (567); Sources of Energy (569); Effects of Energy Technology (572) 28: Inexhaustible-Energy Converters (578); Renewable-Energy Converters (586); Thermal-Energy Converters (588); Electrical-Energy Converters (593); Applying Energy to Do Work (596)
Writing	**Text:** 27: Activity 2 (575) 28: Activity 1 (603); Activity 2 (603)
Verbal	
Math	**Text:** 27: Energy, Work, and Power (564); The Origin of Horsepower (567) **Tech Lab Workbook:** Taking Your Technology Knowledge Home: Investigating Lighting Costs and Savings (101)
Science	**Text:** 27: Types of Energy (563); Sources of Energy (569); Activity 1 (575) 28: Inexhaustible-Energy Converters (578); Renewable-Energy Converters (586); Thermal-Energy Converters (588); Laws of Gases (592); Electrical-Energy Converters (593); Activity 1 (603) Activity 8A: Design Problem (604) Activity 8B: Fabrication Problem (604)

Basic Skills Chart

Section 8 *(Continued)*

Analytical	**Text:** 27: Activity 2 (575) 28: Activity 2 (603) **Tech Lab Workbook:** Activity 8A: Design Problem (95) Activity 8B: Fabrication Problem (99) Taking Your Technology Knowledge Home: Investigating Lighting Costs and Savings (101)

Section 9

Reading	**Text:** 29: Types of Agriculture (610); The Homestead Act and the Morrill Act (618); Agriculture and Biotechnology (628) 30: Primary Food Processing (636); Secondary Food Processing (649) 31: Technology and Wellness (659); Technology and Illness (665); Dialysis Machines (673); Activity 1 (679)
Writing	**Text:** 29: Activity 1 (631); Activity 2 (631) 30: Activity 1 (657); Activity 2 (657) 31: Activity 1 (679); Activity 2 (679); Medical Technology (680) **Tech Lab Workbook:** Activity 9A: Design Problem (107)
Verbal	**Text:** 31: Medical Technology (680) **Tech Lab Workbook:** Activity 9A: Design Problem (107) Activity 9B: Fabrication Problem (111) Taking Your Technology Knowledge Home: Practicing Primary and Secondary Food Processing (113)
Math	
Science	**Text:** 29: Agriculture and Biotechnology (628) 30: Irradiation (646); Activity 1 (657); Activity 2 (657) 31: Aerodynamics (670) Activity 9B: Fabrication Problem (682) **Tech Lab Workbook:** Taking Your Technology Knowledge Home: Practicing Primary and Secondary Food Processing (113)

Basic Skills Chart	
Section 9 *(Continued)*	
Analytical	**Text:** 29: Activity 2 (631) 31: Activity 1 (679) Activity 9A: Design Problem (682) **Tech Lab Workbook:** Activity 9A: Design Problem (107)
Section 10	
Reading	**Text:** 32: Technology and the Entrepreneur (687); Technology and Management (688); Risks and Rewards (691); Forming a Company (692) 33: Societal Institutions (701); Economic Enterprises (702); Industry (703); Areas of Industrial Activity (703); Industry-Consumer Product Cycle (720) 34: Using Technology (725); Assessing Technology (732)
Writing	**Text:** 32: Activity 2 (699) Activity 10A: Forming the Company (736) Activity 10B: Operating the Company (737) **Tech Lab Workbook:** Taking Your Technology Knowledge Home: Developing an Organization and Management System (133)
Verbal	**Text:** 32: Activity 2 (699) **Tech Lab Workbook:** Taking Your Technology Knowledge Home: Developing an Organization and Management System (133)
Math	**Text:** 33: Calculating Bids (707) Activity 10B: Operating the Company (737) **Tech Lab Workbook:** Activity 10B: Operating the Company (121)
Science	**Text:** 33: Wind Tunnels (712) 34: Earth-Sheltered Buildings (731)

Basic Skills Chart	
Section 10 *(Continued)*	
Analytical	**Text:** 32: Activity 1 (699) 33: Activity 1 (723); Activity 2 (723) 34: Activity 1 (735); Activity 2 (735); Activity 3 (735) Activity 10A: Forming the Company (736) Activity 10B: Operating the Company (737) **Tech Lab Workbook:** Activity 10A: Forming the Company (119) Activity 10B: Operating the Company (121) Activity 10C: Using and Assessing Activity (131) Taking Your Technology Knowledge Home: Developing an Organization and Management System (133)
Section 11	
Reading	**Text:** 35: Technology and Natural Forces (744); Technology's Global Impacts (744); Technology and the Future (745); Technology's Challenges and Promises (747) 36: Technology and Lifestyle (759); Technology and Employment (762); Technology and Individual Control (767); Technology and Major Concerns (769); Technology and New Horizons (770); Activity 2 (773)
Writing	**Text:** 35: Activity 1 (757) 36: Activity 1 (773); Activity 2 (773) Activity 11A: Design Problem (774) **Tech Lab Workbook:** Activity 11B: Production Problem (139)
Verbal	**Text:** Activity 11A: Design Problem (774)
Math	
Science	**Text:** 35: Technology and Natural Forces (744)
Analytical	**Text:** 35: Activity 1 (757); Activity 2 (757) **Tech Lab Workbook:** Activity 11A: Design Problem (137) Activity 11B: Production Problem (139)

Scope and Sequence Chart

In planning your program, you may want to use this Scope and Sequence Chart to identify the major concepts presented in each chapter of the *Technology & Engineering* text. Refer to the chart to select topics that meet your curriculum needs. The chart is divided into five sections. Within these sections, bold numbers indicate the chapters in which the concepts are found. Topics and their corresponding page numbers follow the chapter numbers for easy reference to the text.

Scope and Sequence Chart

Section 1

Core Concepts of Technology

1: *Technology* Defined (23); Technology as a Dynamic Process (26); Technology and Types of Knowledge (28);

Connections to Other Fields of Study

1: The Evolution of Technology (29); The Stone Age (30); The Bronze Age (31); The Iron Age (33); The Middle Ages (33); The Renaissance (33); The Industrial Revolution (33); The Information Age (35);

Technology and Its Impacts

1: Positive and Negative Aspects of Technology (26); The Evolution of Technology (29); The Stone Age (30); The Bronze Age (31); The Iron Age (33); The Middle Ages (33); The Renaissance (33); The Industrial Revolution (33); The Information Age (35)

Technological Systems and Processes

2: Goals (40); Inputs (41); Processes (45); Outputs (49); Feedback and Control (50)
3: Level of Development (55); Obsolete Technologies (56); Current Technologies (56); Emerging Technologies (56); Economic Structure (57); Number of People Involved (59); Type of Technology Developed and Used (60)

Section 2 *(Continued)*

Technology and Its Impacts

6: Outputs (117)

Technological Tools and Materials

4: Tools and Machines (74); Hand Tools (75); Machines (76); Materials (82); Types of Materials (82); Properties of Materials (84)

Technological Systems and Processes

4: People (71); Tools and Machines (74); Information (85); Energy (86); Types of Energy (87); Sources of Energy (87); Finances (87); Time (88)
5: Problem-Solving or Design Processes (94); Identifying the Problem or Opportunity (95); Developing Multiple Solutions (96); Isolating, Refining, and Detailing the Best Solution (96); Modeling and Evaluating the Selected Solution (96); Communicating the Final Solution (96); Production Processes (97); Agricultural and Related Biotechnical Processes (97); Communication and Information Processes (99); Construction Processes (101); Energy and Power Processes (104); Manufacturing Processes (106);

Scope and Sequence Chart

Section 2

Technological Systems and Processes

5: Medical Processes (109);
Transportation Processes (110);
Management Processes (112)

6: Outputs (117);
Feedback and Control (120);
Internal Controls (121);
External Controls (130)

Applications of Technology

5: Agricultural and Related Biotechnical Processes (97)

Technological Enterprises

4: Finances (87)

5: Management Processes (112)

Section 3

Design and Problem Solving in Technology

8: Measurement Systems: Past and Present (169);
Qualities Measured (172);
Size and Space (172);
Mass (175);
Temperature (176);
Time (177);
Types of Measurement (177);
Standard Measurement (177);
Precision Measurement (178);
Measurement and Control (182);
Measurement and Production Processes (182);
Measurement and Quality Control (182)

Technological Tools and Materials

7: Material-Processing Tools and Machines (139);
Characteristics of Machine Tools (139);
The Cutting Tool (140);
Motion (142);
Support (142);
Types of Machine Tools (142);
Turning Machines (144);
Sawing Machines (145);
Drilling Machines (148);
Shaping and Planing Machines (150);
Grinding Machines (150);
Energy-Processing Converters (151);
The Electric Motor (152);
The Internal Combustion Engine (155);
Information-Processing Machines (157);
Computers (159);
Radios (161);
Using Technology Safely (162);
Safety with People (163);
Safety with Materials (163);
Safety with Tools and Machines (163)

8: Measurement Tools (178);
Direct-Reading Measurement Tools (178);
Indirect-Reading Measurement Tools (181)

Technological Systems and Processes

7: Printing (157)

8: Measurement and Production Processes (182)

Section 4 *(Continued)*

Design and Problem Solving in Technology

9: The Problem-Solving and Design Process in General (191);
The Problem-Solving and Design Process in Technology (192);
Technological Problems (192);
Technological Opportunities (194);
Steps in Solving Technological Problems and Meeting Opportunities (196);
Identifying a Technological Problem or Opportunity (197);
Defining the Problem or Opportunity (197);
Gathering Information (199)

10: System Design (205); Product Design (207);
Steps for Developing Design Solutions (208);
Developing Preliminary Solutions (209);
Isolating and Refining Design Solutions (212);
Detailing Design Solutions (212)

11: Modeling Design Solutions (221);
Graphic Models (223);
Mathematical Models (226);
Physical Models (226);
Computer Models (228);
Analyzing the Design (230);
Functional Analysis (230);
Specification Analysis (231);
Human Factors Analysis (231);
Market Analysis (232); Economic Analysis (232);
Redesigning Products and Structures (233)

Scope and Sequence Chart

Section 4

Design and Problem Solving in Technology

12: Product Documents and Reports (239); Engineering Drawings (240); Bills of Materials (249); Specification Sheets (250); Approval Documents and Reports (251)

Section 5

Design and Problem Solving in Technology

14: Genetic Materials (271); Fossil Fuel Materials (272); Minerals (274)

Technological Tools and Materials

14: Types of Natural Material Resources (271); Genetic Materials (271); Fossil Fuel Materials (272); Minerals (274); Locating and Obtaining Natural Resources (275); Locating and Obtaining Genetic Materials (275); Locating and Obtaining Fossil Fuel Resources (277); Locating and Obtaining Minerals (284)
15: Producing Lumber (290); Producing Plywood (293); Producing Steel (296); Producing Glass (299); Refining Petroleum (300)

Technological Systems and Processes

13: Production Activities (261); Resource-Processing Systems (262); Production-Manufacturing Systems (263); Structure-Construction Systems (264)
15: Mechanical Processes (290); Producing Lumber (290); Producing Plywood (293); Thermal Processes (294); Producing Steel (296); Producing Glass (299); Refining Petroleum (300); Chemical and Electrochemical Processes (304)
16: Types of Manufacturing Processes (310); Casting and Molding Processes (310); Forming Processes (315); Separating Processes (319); Conditioning Processes (323); Assembling Processes (325); Finishing Processes (328)

Applications of Technology

13: Production Activities (261); Servicing and Repairing Products and Structures (266)
14: Locating and Obtaining Natural Resources (275)
15: Producing Lumber (290); Producing Plywood (293); Producing Steel (296); Producing Glass (299); Refining Petroleum (300)
17: Buildings (337); Heavy Engineering Structures (353)
18: Selecting Technological Products (365); Installing Technological Products (366); Maintaining Technological Products (367); Repairing Technological Products (368); Altering Technological Products (370); Disposing of Technological Products (372)

Section 6 *(Continued)*

Connections to Other Fields of Study

21: Light and Photography (423)
22: The Physics of Telecommunication (443); Electrical Principles (443); Electromagnetic Waves (444)

Technology and Its Impacts

20: Designing the Message (401)
22: Designing the Message (453)
23: The World Wide Web (476)

Design and Problem Solving in Technology

21: Designing a Photographic Message (426)
22: Designing the Message (453)

Scope and Sequence Chart

Section 6

Technological Systems and Processes

19: Communicated Items (384); Goals of Communication (384); The Communication Model (385); Types of Communication (387); Communication Systems (388); Printed Graphic Communication (388); Photographic Communication (390); Telecommunication (390); Technical Graphic Communication (390); Computer and Internet Communication (392)
20: Computer-Based Publishing (414)
22: Types of Telecommunication Systems (447); Hard-Wired Systems (447); Broadcast Systems (449)
23: Computer Systems (469)

Applications of Technology

20: Printing Methods (397); Relief Printing (398); Lithographic Printing (398); Gravure Printing (399); Screen Printing (400); Electrostatic Printing (400); Ink-Jet Printing (401); Steps in Producing Printed Graphic Messages (401); Designing the Message (401); Preparing to Produce the Message (405); Producing the Message (410)
21: Fundamentals of Photographic Communication (426); Designing a Photographic Message (426); Types of Photographic Communication (435)
22: Communicating with Telecommunication Systems (452); Designing the Message (453); Producing Message (456); Delivering the Message (458); Other Communication Technologies (461); Mobile Communication Systems (461); Sound Recording Systems (461)
23: Networks (471); The Internet (472); Internet Access (473); Internet Domains (474); The World Wide Web (476); Web Browsers (477); Web Servers (477); Web Pages (477); Links (478); Search Engines (478); E-Mail (479); E-Mail System (482); E-Mail Messages (482); E-Mail Mailing Lists (483)

Technological Enterprises

23: Electronic Commerce (484)

Section 7

Core Concepts of Technology

24: *Transportation:* A Definition (496)

Technological Systems and Processes

24: Transportation as a System (496); Transportation-System Components (498); Pathways (498); Vehicles (500); Support Structures (502)
25: Vehicular Systems (507); The Structural System (508); The Propulsion System (508); The Suspension System (508); The Guidance System (508); The Control System (509)
26: Components of a Transportation System (542); Transportation Routes (543); Transportation Schedules (545); Maintaining Transportation Systems (550); Regulating Transportation Systems (552)

Applications of Technology

24: Types of Transportation Systems (496); Land (496); Water (497); Air (497); Space (498)
25: Land-Transportation Vehicles (509); Water-Transportation Vehicles (517); Air-Transportation Vehicles (525); Space-Transportation Vehicles (533); Types of Space Travel (533); Areas of Operation (537)
26: Types of Transportation (541); Components of a Transportation System (542); Vehicles (543); Transportation Terminals (547); Transporting People and Cargo (548); Loading (548);

Scope and Sequence Chart

Section 7 *(Continued)*

Applications of Technology

26: Moving (549); Unloading (549); Structure Repairs (551); Vehicle Repairs (551)

Section 8

Core Concepts of Technology

27: Types of Energy (563); Energy, Work, and Power (564); Forms of Energy (566)

Connections to Other Fields of Study

27: Renewable Energy Resources (570); Inexhaustible Energy Resources (570)

Technology and Its Impacts

27: Effects of Energy Technology (572)
28: Renewable-Energy Converters (586); Thermal-Energy Converters (588); Electrical-Energy Converters (593)

Technological Systems and Processes

28: Inexhaustible-Energy Converters (578); Renewable-Energy Converters (586); Electrical-Energy Converters (593); Mechanical-Power Systems (597); Fluid-Power Systems (599)

Applications of Technology

27: Measuring Work (564); Measuring Power (565); Mechanical Energy (566); Radiant Energy (567); Chemical Energy (568); Thermal Energy (568); Electrical Energy (568); Nuclear Energy (568); Sources of Energy (569); Exhaustible Energy Resources (570); Renewable Energy Resources (570); Inexhaustible Energy Resources (570)
28: Inexhaustible-Energy Converters (578); Wind-Energy Conversion (578); Water-Energy Conversion (580); Solar-Energy Conversion (581); Geothermal-Energy Conversion (584); Ocean-Energy Conversion (584); Renewable-Energy Converters (586); Thermochemical Conversion (586); Biochemical Conversion (587); Thermal-Energy Converters (588); Heat Engines (588); Space Heating (589); Heat Production (590); Electrical-Energy Converters (593); Applying Energy to Do Work (596)

Section 9

Connections to Other Fields of Study

29: Agriculture and Biotechnology (628)
31: Goals of Medicine (666); Technology in Medicine (666)

Technology and Its Impacts

29: Agriculture and Biotechnology (628)
30: Food Preservation (643)
31: Technology and Wellness (659); Technology and Exercise (661); Technology and Sports (662); Technology and Illness (665); Goals of Medicine (666); Technology in Medicine (666)

Design and Problem Solving in Technology

29: Animal Waste–Disposal Facilities (626)

Technological Tools and Materials

29: Power or Pulling Equipment (613); Planting Equipment (614); Pest-Control Equipment (615); Irrigation Equipment (615); Harvesting Equipment (618); Transportation Equipment (621); Storage Equipment (622); Livestock Buildings (624); Fences and Fencing (625); Buildings and Machines for Feeding (626); Animal Waste–Disposal Facilities (626)
31: Technology in Medicine (666)

Scope and Sequence Chart

Section 9 *(Continued)*

Technological Systems and Processes

29: Types of Agriculture (610); Agriculture and Biotechnology (628)

30: Primary Food Processing (636); Secondary Food Processing (649)

Applications of Technology

29: Crop Production (610); Technology in Crop Production (612); A Special Type of Crop Production (623); Animal Waste–Disposal Facilities (626); A Special Type of Animal Husbandry (627)

30: Material Conversion (637); Food Preservation (643); Food-Product Development (650); Food-Product Manufacture (651); An Example of Secondary Food Processing: Pasta Making (652)

31: Technology and Illness (665); Goals of Medicine (666); Technology in Medicine (666)

Section 10

Technology and Its Impacts

33: Research and Development (705)

34: Using Technology (725); Selecting Appropriate Devices and Systems (726); Using Technological Services (730); Assessing Technology (732)

Applications of Technology

33: Production (707)

34: Selecting Appropriate Devices and Systems (726); Operating Products and Systems (727); Servicing Devices and Systems (728); Disposing of Devices and Systems (729)

Technological Enterprises

32: Technology and the Entrepreneur (687); Technology and Management (688); Functions of Management (689); Authority and Responsibility (689); Risks and Rewards (691); Forming a Company (692); Selecting a Type of Ownership (692); Establishing the Enterprise (694); Electing a Board of Directors (695); Equity Financing (696); Debt Financing (696)

33: Societal Institutions (701); Economic Enterprises (702); Industry (703); Areas of Industrial Activity (703); Research and Development (705); Research (705); Development (705); Engineering (706); Production (707); Planning (709); Producing (710); Maintaining Quality (710); Marketing (713); Industrial Relations (715); Employee Relations (716); Labor Relations (718); Public Relations (718); Financial Affairs (718); Industry-Consumer Product Cycle (720)

Section 11

Technology and Its Impacts

35: Technology and Natural Forces (744); Technology's Global Impacts (744); Technology and the Future (745); Technology's Challenges and Promises (747); Energy Use (747); Environment Protection (750); Global Economic Competition (753)

36: Technology and Lifestyle (759); Colonial Life and Technology (759); The Industrial Revolution and Technology (760); The Information Age and Technology (761); Technology and Individual Control (767); Technology and Major Concerns (769); Technology and New Horizons (770)

Technological Enterprises

35: Global Economic Competition (753)

36: The Industrial Revolution and Technology (759); The Information Age and Technology (761); Technology and Employment (762); Types of Technical Jobs (763); Selecting a Job (764)

Technology & Engineering

Sixth Edition

R. Thomas Wright
Professor Emeritus, Industry and Technology
Ball State University
Muncie, Indiana

Publisher
The Goodheart-Willcox Company, Inc.
Tinley Park, Illinois
www.g-w.com

Manufactured in the United States of America.

Library of Congress Catalog Card Number 2010033179

ISBN 978-1-60525-412-8

1 2 3 4 5 6 7 8 9 – 12 – 16 15 14 13 12 11 10

Library of Congress Cataloging-in-Publication Data

Wright, R. Thomas
Technology & engineering / by R. Thomas Wright
p. cm.
Rev. ed. of Technology / R. Thomas Wright, 2004.
Includes index.
ISBN 978-1-60525-412-8
1. Technology and engineering. 2. Technology. 3. Engineering. I. Title.

T47.W74 2012
600--dc22 2010033179

Introduction

Technology & Engineering will help you to understand the following:

- How people use technology to make our world work.
- Why technological systems work the way they do.
- In what ways technology affects both people and our planet.

This book covers the seven areas of technological activity:

- Communication and information.
- Transportation.
- Construction.
- Manufacturing.
- Medicine.
- Agriculture and biotechnology.
- Energy and power.

In this book, you will learn that technology is a reaction to problems and opportunities—a human adaptive system. You will learn that technological systems are made up of many parts requiring tools. Also, you will learn about the problem-solving and design process, especially the testing, evaluating, and communicating of design solutions.

Sections explore, in depth, the production of products and structures, communication and information, agricultural and biorelated technologies, transportation, and the use of energy. The management of technological systems is covered because every system must have direction. The examination of societal and personal views of technology rounds out the book.

Technology & Engineering is illustrated with photographs, drawings, diagrams, and original artwork to help explain the concepts in the text. Most of these illustrations are in color. This material has been carefully selected to make technology easy to understand. Each chapter begins with objectives so you know what is covered. Key words are in bold italics to help make you aware of them. Review questions and activities will improve your understanding. The activities between sections provide you with valuable hands-on experience.

Impacts, both positive and negative, accompany the use of technology. The only way people in the modern world can choose and apply technology responsibly is to understand how technology develops and how the various technological systems interact. A Student Activity Manual has activities and exercises giving you important experience, while fully enriching the concepts developed in the text.

A sound understanding of technology is vital for making wise choices. As you study, you will see the effects of your choices. These choices control how technology is used. Each person can make a difference to be sure that technology is used responsibly. With a solid understanding of technology, you can understand and take an active part in our human-built world.

About the Author

Dr. R. Thomas Wright is one of the leading figures in technology-education curriculum development in the United States. He is the author or coauthor of many Goodheart-Willcox technology textbooks. Dr. Wright is the author of *Manufacturing and Automation Technology*, *Processes of Manufacturing*, and *Technology*. He is the coauthor of *Exploring Design, Technology, & Engineering* with Dr. Ryan A. Brown.

Dr. Wright has served the profession through many professional offices. These offices include President of the International Technology and Engineering Educators Association (ITEEA) and President of the Council on Technology Teacher Education (CTTE). His work has been recognized through the ITEEA Academy of Fellows award and Award of Distinction, the CTTE Technology Teacher Educator of the Year, the Epsilon Pi Tau Laureate Citation and Distinguished Service Citation, the Sagamore of the Wabash Award from the Governor of Indiana, the Bellringer Award from the Indiana Superintendent of Public Instruction, the Ball State University Faculty of the Year Award and George and Frances Ball Distinguished Professorship, and the EEA-Ship Citation.

Dr. Wright's educational background includes a bachelor's degree from Stout State University, a master of science degree from Ball State University, and a doctoral degree from the University of Maryland. His teaching experience consists of 3 years as a junior high instructor in California and 37 years as a university instructor at Ball State University. In addition, he has also been a visiting professor at Colorado State University; Oregon State University; and Edith Cowan University in Perth, Australia.

Technology Student Association (TSA) Modular Activities

The Technology Student Association (TSA) is a nonprofit, national student organization devoted to teaching technology education to young people. TSA's mission is to inspire the organization's student members to prepare for careers in a technology-driven economy and culture. The demand for technological expertise is escalating in American industry. Therefore, TSA's teachers strive to promote technological literacy, leadership, and problem solving to their student membership.

TSA Modular Activities are based on the TSA competitive events current at the time of writing. Please refer to the *Official TSA Competitive Events Guide* for actual regulations for current TSA competitive events. This guide is periodically updated. TSA publishes two *Official TSA Competitive Events Guides.* One guide is for middle school events. The other guide is for high school events. To obtain additional information about starting a TSA chapter at your school, to order the *Official TSA Competitive Events Guide,* or to learn more about TSA and technology education, contact TSA:

TSA
1914 Association Drive
Reston, VA 20191-1540
www.tsaweb.org

The Career Clusters

The Career Clusters are 16 groups of different types of occupational and career specialties, which are further divided into pathways. Looking ahead at these pathways will help determine the course of study for your chosen career. The Career Cluster icons are being used with permission of the:

States' Career Clusters Initiative, 2010, www.careerclusters.org.

Tomorrow's Technology Today

Terraforming

What comes to mind when you think of an extraterrestrial? You may imagine the small, green beings you see on the covers of science fiction novels. Today, scientists are beginning to envision a different picture of life in space. Through the process of terraforming, they hope to someday sustain human life on another moon or planet.

Terraforming is the alteration of a planet or moon's surface to make it suitable for human life. In other words, it is an attempt to make it like Earth. Because the planets and moons are so diverse, terraforming would be a big challenge. It could take centuries, or even longer, to successfully complete a terraforming project and begin human habitation in a place other than Earth.

The process of terraforming is currently hypothetical. Many questions need to be answ[illegible]ore we can even consider such a project. Will a country, several countries [illegible]private organization be responsible for the terraforming process? How [illegible] cost and who would provide funding? And most importantly, what [illegible] would be the best candidate for terraforming? Determining which [illegible] would ensure an easier and more successful conversion.

[illegible]c composition, distance from the Sun, and the presence of water are [illegible] factors to consider when evaluating a planet or moon's potential for [illegible]th a rotation rate and axial tilt similar to Earth, Mars has emerged [illegible]ntender. Water is frozen at Mars' polar caps while carbon, oxygen, [illegible] are all present in its atmosphere. However, the atmosphere is very thin, oxygen levels low, and temperatures significantly cooler than on Earth. Scientists have suggested several methods to heat the planet, including placing large, sun-reflecting mirrors on its surface or building factories that would produce greenhouse gases. Greenhouse gases have proven to raise the temperature on Earth.

Terraforming an entire planet would be extremely costly and time-consuming. As a result, scientists are now exploring the option of paraterraforming. Paraterraforming would transform only a portion of the surface within an enclosed structure. The structure could then be expanded over time as financial capabilities and knowledge of the terraforming process increase.

Should the Earth ever suffer a major disaster or crippling overpopulation, terraforming would offer humans an opportunity to escape and thrive elsewhere in the solar system. Many questions still need to be answered, but scientists move one step closer every day to establishing human life on other moons and planets.

Tomorrow's Technology Today features in each section highlight an emerging technology.

Learning Objectives

After studying this chapter, you will be able to do the following:

- Understand the term *technology.*
- Recall the basic features of technology.
- Explain technology as a dynamic process.
- Give examples of positive and negative aspects of technology.
- Compare science and technology.
- Recall the major divisions in the evolution of technology.
- Recall technological developments in each period of technologica[illegible]
- Recall characteristics of the information age.

Key Terms

artifact
Bronze Age
civilized conditions
development
dynamic process
humanities
Industrial Revolution
information age
Iron Age
Middle Ages
primitive conditions
profit
Renaissance
research
science
Stone Age
technologically literate
technology

Strategic Reading

Before you read the first chapter, write down what you think the term technolog[illegible] As you read the chapter, see how your perception of the term matches or differs [illegible] description in the text.

Learning Objectives identify the topics covered and goals to be achieved by students.

Key Terms lists new vocabulary covered in the chapter, enhancing student recognition of important concepts.

Strategic Reading features provide students with questions to think about while reading the chapter.

New Terms appear in bold italics where they are defined.

Academic Connections provide information on a topic relevant to the chapter material that connects the content to communication and history.

Alaska Airlines, United Airlines

Figure 24-7. Vehicles such as these aircraft are designed differently for cargo and people.

STEM Connections provide information on a topic relevant to the chapter material that connects the content to math or science.

Career Corner features identify and explain different careers related to the chapter material.
Safety notes identify activities that can result in personal injury, if proper procedures or safety measures are not followed.
Think Green features briefly explain environmental concepts related to technology.
Summary provides the student a review of major concepts covered in the chapter.
Test Your Knowledge questions help students review the topics and the material covered in the chapter.
STEM Applications at the end of each chapter encourage students to apply concepts to real-life situations and develop skills related to chapter content.

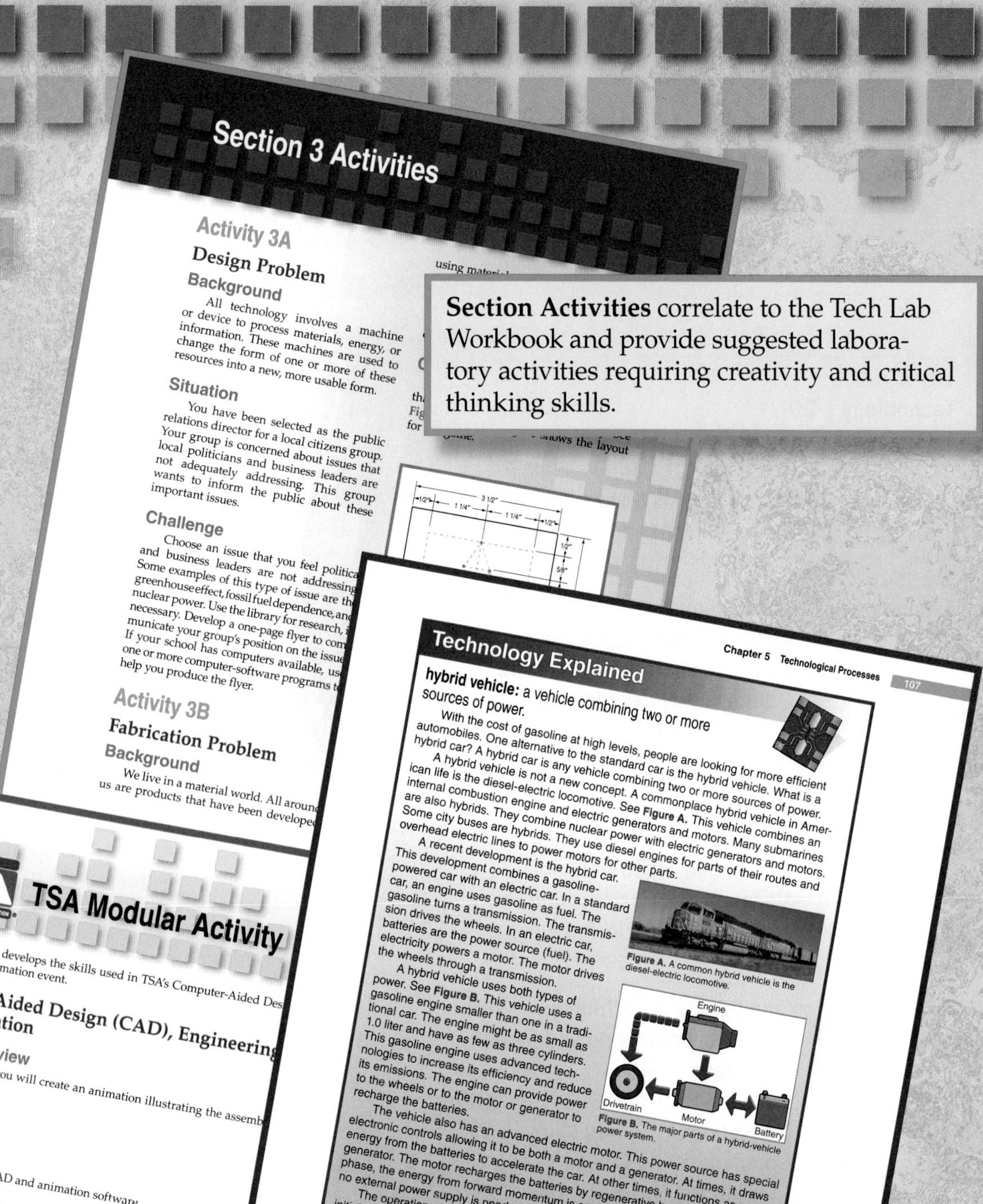

Section 3 Activities

Activity 3A

Design Problem

Background

All technology involves a machine or device to process materials, energy, or information. These machines are used to change the form of one or more of these resources into a new, more usable form.

Situation

You have been selected as the public relations director for a local citizens group. Your group is concerned about issues that local politicians and business leaders are not adequately addressing. This group wants to inform the public about these important issues.

Challenge

Activity 3B

Fabrication Problem

Background

Technology Explained

hybrid vehicle: a vehicle combining two or more sources of power.

With the cost of gasoline at high levels, people are looking for more efficient automobiles. One alternative to the standard car is the hybrid vehicle. What is a hybrid car? A hybrid car is any vehicle combining two or more sources of power.

A hybrid vehicle is not a new concept. A commonplace hybrid vehicle in American life is the diesel-electric locomotive. See Figure A. This vehicle combines an internal combustion engine and electric generators and motors. Many submarines are also hybrids. They combine nuclear power with electric generators and motors. Some city buses are hybrids. They use diesel engines for parts of their routes and overhead electric lines to power motors for other parts.

Figure A. A common hybrid vehicle is the diesel-electric locomotive.

Figure B. The major parts of a hybrid-vehicle power system.

TSA Modular Activity

This activity develops the skills used in TSA's Computer-Aided Design, Engineering with Animation event.

Computer-Aided Design (CAD), Engineering with Animation

Activity Overview

Materials

- Paper.
- A pencil.
- A computer with CAD and animation software.

Background Information

Section Activities correlate to the Tech Lab Workbook and provide suggested laboratory activities requiring creativity and critical thinking skills.

Technology Explained features briefly explain how technological devices and systems work.

TSA Modular Activities features are additional activities intended to develop skills used in TSA's competitive events.

Brief Contents

Contents

Features

Tomorrow's Technology Today

Career Corners

Academic Connections

History

Communication

STEM Connections

Mathematics

Science

Technology Explained

Think Green

TSA Modular Activities

Section 1

Technology

Tomorrow's Technology Today

Cloning

Identical twins often have stories of switching places with one another to avoid trouble—or to cause it. If you have ever wished for a twin to sit in your place on an exam day, your wish might not be altogether unrealistic. Modern science is now making it possible to create a human genetically identical to you, or a clone. A clone is, for all practical purposes, an identical twin of the original because every single gene is an exact copy of the original.

Recently, scientists have successfully cloned several types of plants and animals. These types include some endangered species. This is important because it opens up the possibility of repopulating species of plants and animals at risk of becoming extinct and maybe even species that already are! The main purpose of cloning, though, is to mass-produce plants and animals with specific genetic qualities, such as the best tasting fruit or the ability to produce human insulin.

Cloning plants and animals has some undeniable benefits for humans and the quality of human life. Although there has been some debate about the ethical implications of plant and animal cloning, it is the idea of human cloning that has really started controversies among scientists, the government, and the public. Until recently, the idea of cloning humans was science fiction. Now, it is a real possibility.

Researchers are close to being able to successfully clone humans. Most cloning would not, however, result in entirely new humans. Scientists are studying the possibility of using cloning to produce repair and replacement kits for people with severe medical problems. It might be possible to clone healthy cells and to fix diseased cells. This type of cloning is currently being used in stem cell research. A person's DNA is used to grow an embryonic clone. This clone is then used to grow stem cells. These cells can grow replacement organs and neurons to cure certain diseases.

Cloning, especially human cloning, creates many ethical dilemmas. The success rate of animal cloning is currently less than 3%. The process is very risky. Surviving clones often suffer from problematic genetic abnormalities. Some are born with defective lungs, hearts, blood vessels, or immune systems. It will be a more serious problem if human clones are created and born with these same problems. Cloning holds much promise. This solution, however, also holds many risks. As interest in human cloning continues to grow, the ethical debate certainly will as well.

Discussion Starters

Explain the processes of artificial embryo twinning and cell nuclear transfer. Ask your students if they can think of instances when each process might be used.

Group Activity

Divide the class into pairs. Have one member of the pair develop an argument in favor of human cloning research. Have the other member develop an argument against human cloning. Have the students debate their cases in front of the class.

Writing Assignment

Have your students select and research a single high-profile case study of successful cloning. Have the students write a brief report describing the successes and failures of that particular cloning exercise.

This chapter covers the benchmark topics for the following Standards for Technological Literacy:

1 2 3 4 5 7

Learning Objectives

After studying this chapter, you will be able to do the following:

- Understand the term *technology.*
- Recall the basic features of technology.
- Explain technology as a dynamic process.
- Give examples of positive and negative aspects of technology.
- Compare science and technology.
- Recall the major divisions in the evolution of technology.
- Recall technological developments in each period of technological history.
- Recall characteristics of the information age.

Key Terms

artifact
Bronze Age
civilized conditions
development
dynamic process
humanities
Industrial Revolution
information age
Iron Age
Middle Ages
primitive conditions
profit
Renaissance
research
science
Stone Age
technologically literate
technology

Strategic Reading

Before you read the first chapter, write down what you think the term technology means. As you read the chapter, see how your perception of the term matches or differs from the description in the text.

Do you know what technology is? Is it mainframe and desktop computers, industrial robots, laser scanners, fiber-optic communications, space shuttles, and satellites circling Earth? Does technology consist of people using tools and machines to make their work easier and better? Is it an organization devoted to operating agricultural and related biotechnology, communication and information, energy and power, medical, manufacturing, construction, or transportation systems? See **Figure 1-1.**

Chances are that you are not sure if technology is one, two, or all three of these things. You are not alone in this lack of understanding. Throughout the world, people use technology every day. They understand, however, little about it. We go to school for years to learn how to read and write and to study mathematics, science, history, foreign languages, and other subjects. Few people, however, spend time learning about technology and technology's impacts on everyday life.

This book has been developed to aid in understanding this vast area of human activity. *Technology* will help you become more technologically literate. After this study of technology, you will be able to find, select, and use knowledge about tools and materials to solve problems. This ability leads to an increased understanding of technology as it affects your life as a citizen, consumer, and worker. See **Figure 1-2.**

Technology Defined

Almost everyone uses the word *technology*. What, however, does it really mean? To some people, it means complicated electronic devices and hard-to-understand equipment. To others, it means the source of the radical changes that are happening in all phases of life. Some people fear it. Others see it as the source of longer and more complete lives. Some people believe it to be a development of the twentieth century. Each of these views is partly correct.

Figure 1-1. We encounter technology in many areas of our lives.

Figure 1-2. Technology affects our daily lives.

Standards for Technological Literacy

Research

Have your students find three definitions of *technology* in books or on the Internet.

Extend

Give examples of very simple technology (paper clip, Band-Aid®, etc.) and very complex technology (spacecraft, automobile, etc.). Have each student list three simple technological devices and three complex ones.

TechnoFact

In order to meet the demands of a technological world, a technologically literate person applies his or her knowledge and abilities.

TechnoFact

The government agency dedicated to developing and promoting measurement, standards, and technology to enhance productivity, facilitate trade, and improve the quality of life is the National Institute of Standards and Technology.

Technology is not necessarily a complex or space-age phenomenon. This phenomenon can be primitive and crude. On the other hand, technology can be complex and sophisticated. This phenomenon has been here as long as humans have been on Earth. ***Technology*** is humans using objects (tools, machines, systems, and materials) to change the natural and human-made (built) environments. This phenomenon is conscious, purposeful actions people design to extend human ability or potential to do work.

These actions have four basic features. As shown in **Figure 1-3**, these are the following:

- Technology is human knowledge.
- Technology uses tools, materials, and systems.
- The application of technology results in ***artifacts*** (human-made things) and other outputs (pollution and scrap, for example).
- People develop technology to modify or control the environment.

Figure 1-3. Technology has four characteristics.

These four characteristics suggest that the products and services available in society are the result of technology. The products and services are designed through technological innovation (***development***). They are produced by technological means (processes). These means are integrated with people, machines, and materials to meet an identified need (management). The products are distributed to customers by technology. They are maintained and serviced through technological actions. When the products fulfill their function or become obsolete, they are (or should be) recycled by technological means. As you can see, without technology, the world as we know it today would not exist.

Profit-centered businesses develop and produce nearly all technology. People organize these businesses to make a product or perform a service. Through this action, the owners have a product to sell. They hope to pay their business expenses (wages, material costs, taxes, and so on) and have money left over—a ***profit***. To stay competitive, these businesses change and improve their products and services. See **Figure 1-4.** For example, computer companies develop new models that they hope will capture a large share of the market. Likewise, cereal companies bring out new products to increase their sales. A competitive edge is gained through employing technology. These technologies include standardizing products, using interchangeable parts on machines, and adopting laborsaving practices.

Competition also exists between nations. Each country's economy is directly dependent on the country's ability to produce products and services that can compete in the world market. The country that produces the best products will have a healthy economy and a high number of citizens working. Countries that fail to develop competitive products will experience high levels of unemployment and growing poverty.

Standards for Technological Literacy

1 2 4 7

Extend

Discuss the role of profit in causing technological change (advances). Encourage class participation in the discussion.

TechnoFact

Profit sharing is a practice in which a portion of profits is given to workers as a bonus. The goal of the program is to increase worker output.

Extend

Discuss the personal computer in terms of (1) how it is changing and (2) the changes it has caused. Encourage class participation in the discussion.

Figure Discussion

Have your students look at Figure 1-4. Ask the students what changes have been made to automobiles since the vehicle on the left was manufactured.

©iStockphoto.com/sakakawea7, ©iStockphoto.com/InStock

Figure 1-4. Products must change over time to meet competitive forces. The threshing machine was used to harvest grain in the early 1900s. The combine replaced it and is in use today.

Standards for Technological Literacy

1 2 3 4 5 7

Extend

Discuss how we use technology, on a daily basis, to extend our potential. Use simple, real-life examples. Ask students to come up with their own examples.

TechnoFact

In 1867, Christopher Sholes, Carlos Glidden, and Samuel Soulé invented the first practical typewriter in the United States. As manufacturers of rifles and sewing machines, the Remington Company began to produce Christopher Sholes's patented typewriter in 1873.

Technology as a Dynamic Process

All these points suggest that, by its very nature, technology is a ***dynamic process***. Technology is changing. This process causes change. Technology almost always seems to improve on existing technology. See **Figure 1-5.** The typewriter is an improvement over writing by hand. The electric typewriter is faster than the manual typewriter. The word processor and laser printer outperform the electric typewriter. Now, voice-recognition software enables humans to talk to computers, providing even faster communication. This is not the end. New devices will improve on this system.

Past technology improved productivity. Future technology will increase productivity even more. There is no turning back. We cannot feed the world's population using the walking plow and mules. Commerce cannot be maintained using covered wagons. People cannot build hand-hewn log cabins fast enough to provide housing for a rapidly expanding population. Technology is necessary for survival and the hope for a better future.

Keytec, Inc.

Figure 1-5. Technology is a dynamic process. People are always trying to improve existing artifacts and systems.

Positive and Negative Aspects of Technology

Technology increases human capabilities. Through technology, we can see more clearly and further by using microscopes and telescopes. Technology helps us to lift heavy loads by using hoists, pulleys, and cranes. This process allows us to communicate better and faster by radio, television, and telephone. Technology makes distant places close at hand when traveling by automobiles, trains, and aircraft. This process makes life more enjoyable with video games, motion pictures, and compact disc recordings. Technology brings us new products and materials, such as computers, optical fibers, and artificial human organs.

This process, however, also has negative aspects. Poorly designed and used technology pollutes the air we breathe and the water we drink and can cause soil erosion. See **Figure 1-6.** Inappropriately used technology threatens calm sunsets and unspoiled wilderness areas.

Technology can cause unemployment and radical changes in the ways people live. New technological devices might displace workers from traditional jobs. Families might have to move long distances to find employment. Their new homes might be in areas that have vastly different values and lifestyles than the areas where they used to live. The consequences of technology are now more feared than natural events. Many people are more concerned about nuclear winter, acid rain, and air pollution than they are of earthquakes and tornadoes.

U.S. Department of Agriculture, ©iStockphoto.com/Anutik

Figure 1-6. Technology has positive and negative aspects. Poorly used technology can cause such problems as soil erosion and air pollution.

Career Corner

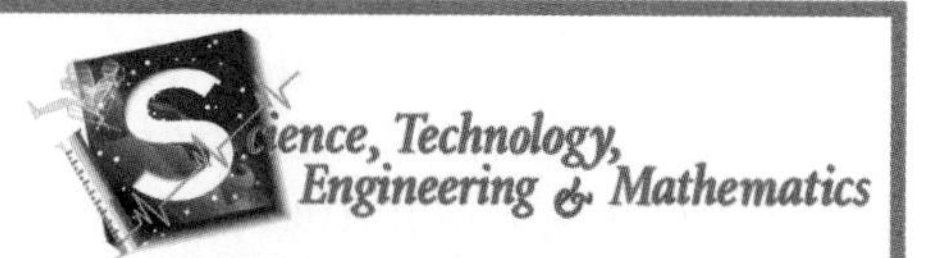

Industrial Engineering Technicians

Industrial engineering technicians study the efficient use of personnel, materials, and machines in production facilities. They prepare layouts of machinery and equipment, plan the flow of work, conduct statistical studies, and analyze production costs. Their work is more practical and limited in scope than the work of industrial engineers. Almost all industrial engineering technicians have completed high school with an emphasis in mathematics and science. Many also complete a postsecondary program in engineering technology, earning either an associate's degree or a bachelor's degree.

Standards for Technological Literacy

3 4 5 7

Brainstorm

Ask your students to list some positive and negative impacts of robotics on employment.

TechnoFact

The period in British history from the middle 1700s to the middle 1800s was a time of rapid and radical technological change. The Industrial Revolution is the term used to describe the social and economic changes caused by this transition from a simple agricultural and commercial society relying on common tools to a modern industrial society relying on complex technology.

TechnoFact

Countries that are generally poor and are technologically less developed are called third world countries. The economies of these nations depend on the export of raw materials to the developed countries while importing most of their finished products.

Standards for Technological Literacy

3 4 5

Extend

List some of the situations or events that scientists research. Have your students expand the list.

TechnoFact

Research is the process of seeking knowledge, data, or the truth about a specific topic.

Quotes

"Technological change defines the horizon of our material world as it shapes the limiting conditions of what is possible and what is barely imaginable."
—Shoshana Zuboff, *In the Age of the Smart Machine: The Future of Work and Power.*

Quotes

"Television thus illustrates the mixed blessings of technological change in American society. It is a new medium, promising extraordinary benefits: great educational potential, a broadening of experience, enrichment of daily life, entertainment for all. But it teaches children the uses of violence, offers material consumption as the answer to life's problems, sells harmful products, habituates viewers to constant stimulation, and undermines family interaction and other forms of learning such as play and reading."—Kenneth Keniston, *All Our Children.*

This good-news and bad-news view of technology requires a new type of citizen. Technology must be developed and used wisely and appropriately. This challenge requires individuals who understand and can direct new technology. People who have this understanding and ability are called ***technologically literate***.

Technology and Types of Knowledge

Some people think technology is applied science. They are mistaken! Technology and science are closely related. They are, however, different. To see this difference, let us use your imagination. Assume that you plan to see your first glacier in Rocky Mountain National Park. You drive to the parking lot and leave your car. As you climb up the trail, the trees are getting shorter. After a while, there are no trees. The type of plant life is changing. Lichen is growing on the north side of the rocks. Finally, you reach your destination. You notice that the small lake in front of the glacier appears to be green. A cold wind is blowing over the continental divide and across the glacier. It is midsummer. Snow, however, still covers the ground. If you want to know why these phenomena happen, you must consult ***science***, the knowledge of the natural world. See **Figure 1-7.** Scientific knowledge is gathered from detached observation. Scientists distance themselves from the phenomena as they try to develop their explanations. They try to explain why something exists or happens in a certain way. Their work can be described as ***research***.

Figure 1-7. Scientific knowledge is different from technological knowledge. Scientists observe phenomena from a distance in an effort to understand how or why something exists.

Let's return to our make-believe hike. The long hike has made you tired. Now you must climb down the mountain. As you come around a curve in the trail you see a welcome sight—your car! You get into it, start the engine, drive along a paved road, and pass a small hydroelectric dam with electric power lines leading in all directions. Along the road, there are homes with lights and heating systems. Finally, you reach town, with its stores, gas stations, and pizza shop. Food at last! All these things are a result of technology. They are part of the human-made world, or the built environment.

Innovators have developed these parts of the world using technological knowledge. These people work with materials and use machines and tools to make things happen. Their work can be described as ***development***. They design and build products and structures to make our lives better. Those who work in technology are not observers. They become directly involved with the processes they develop and use.

Science and technology are two major types of knowledge. A third type is the ***humanities***. This type of knowledge describes the relationships between and among groups of people. Those in the humanities study how people behave individually (psychology), how they work in groups (sociology), what they value (religion and ethics), and how they express themselves (art, music, and literature). People in the humanities also look at human behavior over time (history and anthropology).

You can begin to understand the world by using these three types of knowledge: the knowledge of the natural world (science), the knowledge of the human-made world (technology), and the knowledge of human actions within these environments (the humanities). They all interact with one another. Knowledge from one area directly affects development in other areas. See **Figure 1-8.** Leave out any one of them, and you will make errors in judgment and do things incorrectly.

The Evolution of Technology

Many people see technology as new and dramatic. Technology is not new or dramatic. This knowledge is as old as humanity. The world is said to be about 5 billion years old. Humans have been on Earth for at least 2.5 million years. During this time, two major factors have distinguished humans from other species. These factors are humans' ability to make tools and humans' ability to use those tools. The level of this technology determines the type of life available.

Standards for Technological Literacy

3

Figure Discussion

Ask your students why the use of the three types of knowledge shown in Figure 1-8 is important. Ask the students what they think would happen if they developed a product without all three.

TechnoFact

Research and Development (R&D) is a process used by businesses to design and specify a product before its launch. The research part of the activity is usually scientific research, while development is the technological part that develops and specifies the product.

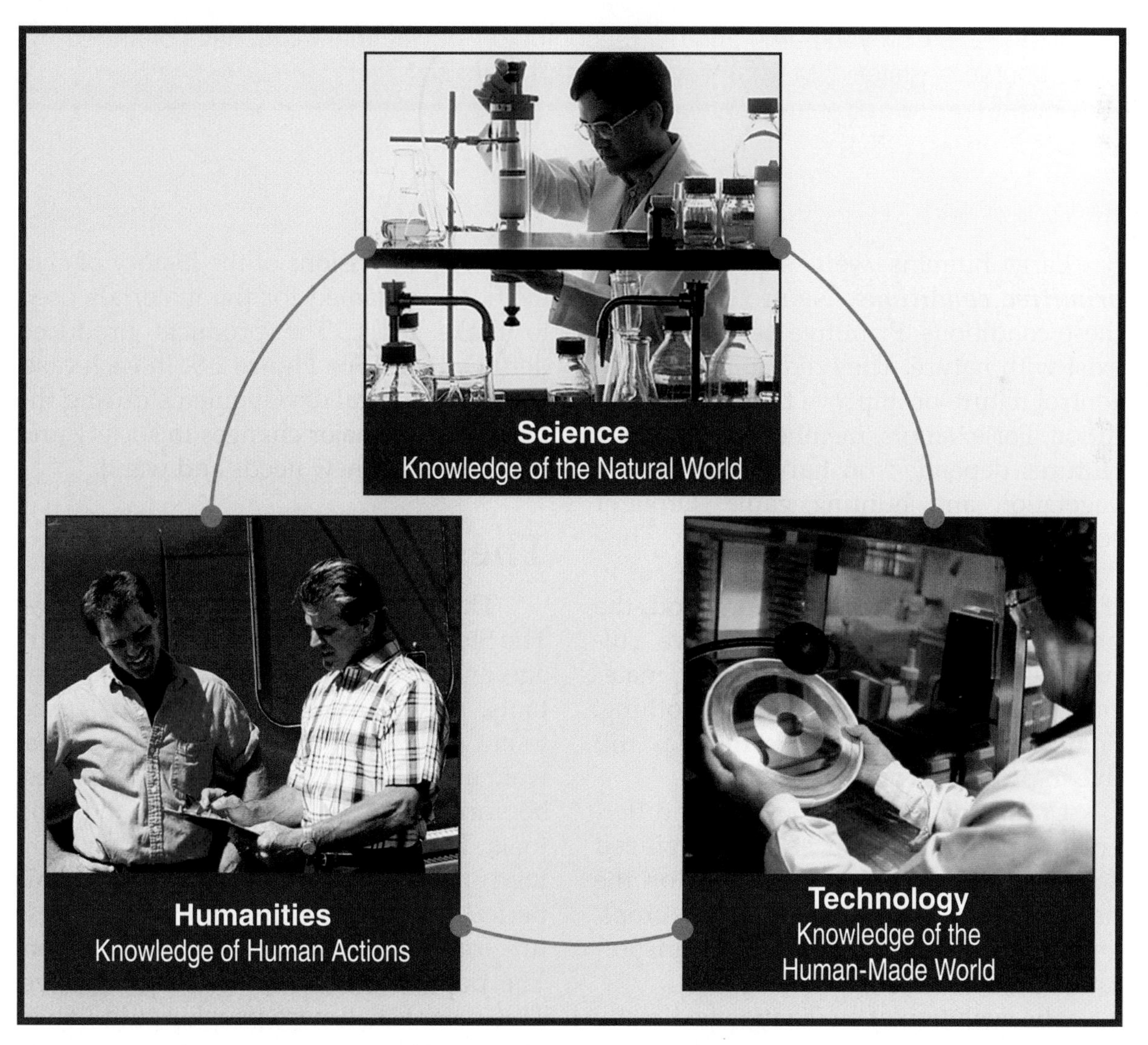

Figure 1-8. The various types of knowledge impact each other.

Standards for Technological Literacy

7

Demonstrate

Show how a tool can be made by flaking a piece of stone.

TechnoFact

Because early artifacts demonstrate a unique trait of invention, the names of the major archaeological ages (Stone Age, the Bronze Age, and the Iron Age) relate to the materials used for making tools and weapons during that age.

Academic Connections: History

The Presidential Election of 1960

As vice president during the eight-year term of a popular president, most observers expected Richard Nixon to win the presidential election of 1960. He was also running against a young and somewhat inexperienced senator, John Kennedy of Massachusetts. This further improved his chances.

Still, in large part because of a relatively new technological device called the *television*, the perceptions of the two candidates changed dramatically as election day grew closer. In a series of debates that were televised for the first time in history, Kennedy came across as confident and mature. Nixon was seen as nervous and self-conscious. Interestingly enough, according to most people who listened to the debates on the radio, Nixon had won the contests. According to most of those who watched on television, however, Kennedy had won.

In the end, Kennedy won the election. Television had forever changed the political process. Can you name another technological device that might have changed the course of history? In what way or ways did it do so?

Early humans were said to live in ***primitive conditions.*** Nature determines these conditions. Primitive people tried to exist with nature. They did not attempt to control nature or improve the natural condition. For example, members of primitive cultures depended on harvesting natural vegetation and hunting game. Drought severely affected their food supply. They could not, however, do anything about it. If nature did not provide ample food, the people starved. Likewise, primitive cultures used only naturally occurring materials to build shelters and make clothing. Some groups of people on Earth today still live in primitive conditions.

Civilized conditions are much different from primitive existence. In civilized societies, humans exert their wills on the natural scene. They make tools, grow crops, engineer materials, and develop transportation systems.

The evolution of civilization is directly related to the tools of the time. Many of the early major divisions of the history of civilization are named for the materials used to make tools. The products produced define others. See **Figure 1-9.** In each case, the technological developments during the period led to major changes in society and the creation of new needs and wants.

The Stone Age

The earliest period was the ***Stone Age.*** The Stone Age began about 2 million years ago. In this period, humans used rocks as tools. These simple stone tools were used to cut and pound vegetables and cut meat from animal carcasses. From these modest beginnings came pointed stone hunting tools. These tools allowed people to obtain food more efficiently. Also during this period, humans learned how to harness fire for heating, cooking, and protection. The population became more productive. This meant that more people could live in a given area.

Stone Age (1,000,000–3000 BC)	Bronze Age (3000–1200 BC)	Iron Age (1200 BC–1300 AD)	Middle Ages (500 AD–1500 AD)	Renaissance (1300 AD–1600 AD)	Industrial Revolution (1750 AD–ongoing)	Information Age (Present)
Stone tools, fire, cave paintings, pottery	Copper and bronze tools, smelting, frescoes, writing, paper, ink	Iron tools, furnaces, aqueducts, body armor, ox-drawn plows, spinning wheels, windmills	Printing press, magnetic compass, paper money, waterwheel	Improved magnetic compass, telescope, hydraulic press, calculating machine, modern architecture	Steam engine, cotton gin, power looms, factories, electricity, automobile, airplane	Desktop computers, robots, lasers, solar energy, cell research, satellites

MACHINE FOR HURLING STONES.

Iron Age

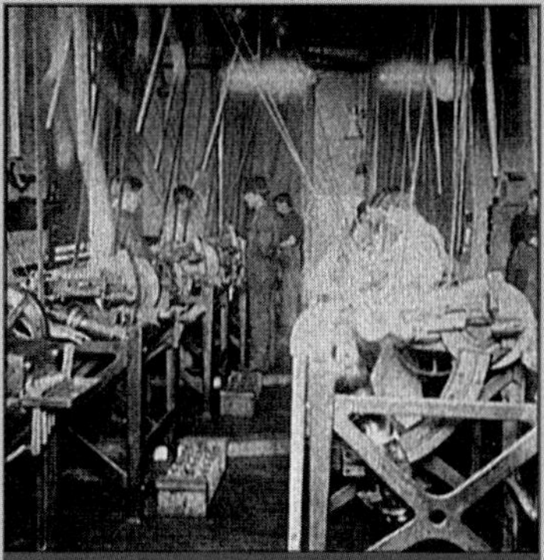

Industrial Revolution

Information Age

www.PDImages.com, Library of Congress, NOAA

Figure 1-9. The major divisions in the history of civilization are identified by the tools and products produced and used.

Standards for Technological Literacy

7

Research

Have your students research what the earliest Bronze Age tools were and where they were developed.

TechnoFact

The earliest bronze found in the New World was cast in Bolivia about 1000 years ago.

The Bronze Age

As populations grew, new technology was needed to support the demand for food and shelter. This need was coupled with the discovery of copper. People also learned that copper could be heated and then melted with other ores to produce bronze, a stronger metal than copper. These events allowed humans to enter the second major historical period, the ***Bronze Age***. The Bronze Age began around 3000 BC. During this period, humans used copper and copper-based metals as the primary materials for tools because of the hardness and durability of these materials. Through such developments as large-scale irrigation systems, humans transformed agriculture to the extent that they did not have to depend on native vegetation and animal life for survival. They also created better ways for storing food and developed writing, navigation, and other basic technologies.

Technology Explained

smart house: a house that allows computers to control appliances and energy use.

Computer technology has been used to create an entirely new type of house called a *smart house*. The smart-house system makes home automation possible. This system allows the electrical, telephone, gas, and television systems and appliances in a house to be interactive. The smart-house system is being built into new homes across the United States. See **Figure A.**

Figure A. In the years to come, many homes will use smart-house systems to control energy use, security, communications, lighting, and entertainment.

The heart of a smart house is found in the system controllers. See **Figure B.** These units serve as the hubs for messages between products within the house. The system controllers constantly check the system for problems. Electrical power, telephone service, and cable television are brought into the house at a service center. A system controller is installed there.

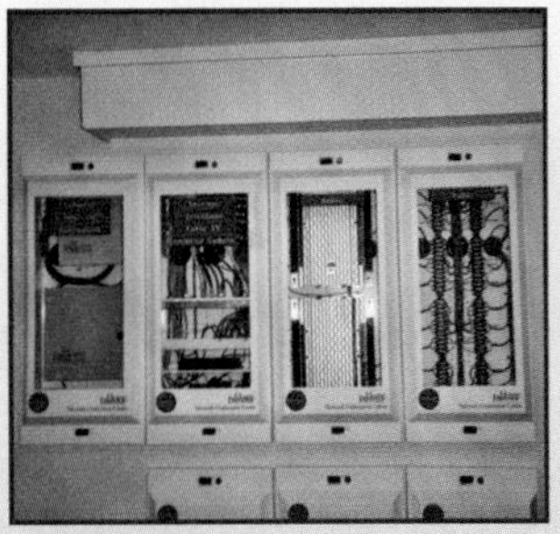

Figure B. A series of smart-house controllers in a recently built home.

The smart-house system uses microprocessors and a unique wiring system to allow the electrical appliances and products in a house to interact. Participating companies are developing products with built-in smart-house chips. These products can then interact with the system controllers and the rest of the smart-house system.

Smart houses allow home owners to change how their houses work. This is done by programming the system controllers. Reprogramming a switch changes what the switch controls. The heating system can be programmed to vary the home's temperature during the day. There can be cool bedrooms at night and warm bathrooms in the morning. The water heater can be set to turn on, so hot water will be available for a morning shower. The electrical range can be turned on to cook a meal in the oven. The dishwasher can be set to start at night, in order to take advantage of lower utility rates.

The wiring system in a smart house is unique. Smart houses use three types of cables. See **Figure C.** Hybrid branch cables carry 120-volt alternating current (AC) power and control signals. These cables provide power for standard appliances, such as coffeemakers, digital videodisc (DVD) players, and lamps. Communications cables carry audio, video, telephone, and computer data signals. Applications cables carry 12-volt direct current (DC) power and control signals. These cables provide power for low-voltage devices, such as smoke detectors, security systems, and switches. Large appliances, such as ovens and clothes dryers, use conventional high-amperage wiring. Radio frequency and infrared devices can be used to make a smart house wireless.

Cables

A Convenience Center

Figure C. The wiring used in a smart house is special. The three types of cables (shown from top to bottom) are applications, communications, and hybrid branch. Convenience centers are special outlets that can be changed to meet the needs of the home owner.

The Iron Age

The next historical period is called the ***Iron Age***. The Iron Age began around 1200 BC. Iron and steel became the primary materials for tools during this period because they were more plentiful and cheaper than copper and bronze. Now more people could afford tools. This affordability created more opportunities for sustained technological advancement. Progress continued even during the Dark Ages (about 500 AD to 1000 AD). Many cultural and governmental institutions stagnated or declined in importance during this time. During the Iron Age, the alphabet came into general use, as did coins. Trade, transportation, and communication all improved. Civilization expanded.

The Middle Ages

The ***Middle Ages*** began about 400 AD. Germanic tribes invaded and conquered the Roman Empire around this time. This period is known for its various upheavals. Tribes continually fought each other for territory. Technology, however, still progressed. For example, one of the major inventions of this time was the printing press, developed by Johann Gutenberg in 1445.

This device changed people's lives forever. Previously, to make books, people had to copy words and illustrations onto parchment by hand, a laborious process that made each book very expensive. With the printing press, books could be produced quickly and with less labor, thus making books less costly and available to more people.

Other items discovered and improved on during this time include the magnetic compass and the waterwheel. The waterwheel changed the power of water into mechanized energy. The adoption and use of paper money increased the amount of goods bought and sold.

The Renaissance

The ***Renaissance***, the period that began in the early 1300s in Italy and lasted until 1600, was a time of great cultural advancement. This period is known as a period of new ideas in art, literature, history, and political science. Technological developments, however, also occurred. For example, because of improvements in ships, European voyagers traveled to America. They brought with them new crops and farming tools and learned new methods of farming from Native Americans, thus improving agriculture for both groups.

In Italy, Leonardo da Vinci, the famous artist, kept a notebook in which he drew plans for many inventions. These inventions included a flying machine and a movable bridge. The drawings greatly influenced future inventors. Other technologies developed during this period include the calculator and the telescope. Gunsmiths invented another new artifact, the screwdriver, because they needed an instrument to adjust gun mechanisms.

The Industrial Revolution

During the last 250 years, technology has dramatically changed the world. For example, humans applied technology to agriculture and went from cultivating food by hand to using a horse and plow to adopting the mechanical reaper. A smaller percentage of the population could now grow the food and clothing fiber needed to sustain the population because of these developments. This allowed large numbers of people to leave the farms and migrate to growing towns and cities. In these places, a new phenomenon was developing—the marvel of manufacturing.

This era is called the ***Industrial Revolution***. The Industrial Revolution started in England about 1750 and moved to America in the late 1700s. Tremendous

Standards for Technological Literacy

7

Research

Have your students research the major technological advancements and inventions of the Middle Ages.

TechnoFact

Because there are no exact dates for the beginning and conclusion of the Middle Ages, historians speculate the period began with the Council of Chalcedon (451 AD) or the downfall of the Roman Empire (476 AD) and ended with the Italian expeditions of Charles VIII of France (1494–1496) or with the revival of literature in the 15th century.

Standards for Technological Literacy

7

Extend

Discuss the work of Fredrick Taylor and scientific management. Encourage class participation in the discussion.

Brainstorm

Ask your students what positive and negative impacts industrialization (the Industrial Revolution) had on people and society.

TechnoFact

The Industrial Revolution started in England during the late 18th century. It brought major social and economic changes as well as the automation of production systems, which resulted in changing production from a home craft system for making goods to large-scale factory production.

changes in technology occurred during this period. For example, Edmund Cartwright changed weaving from a manual process to a mechanical one with his power loom. Joseph Jacquard further revolutionized weaving by creating a series of punched cards with recorded instructions. These cards controlled the intricate patterns the weaver designed, thus freeing up the weaver to spend more time on managing. (These cards later became the inspiration for computer programming.) James Watt improved the steam engine and made it available for the driving of machinery. This development revolutionized both the manufacturing process and methods of transportation. Eli Whitney, inventor of the cotton gin, developed the system of manufacturing standardized, interchangeable parts. These parts dramatically expanded manufacturing capabilities. As has been true throughout the history of technology, many of these developments were refinements of existing inventions. Here, however, scientific knowledge began to be applied to technological knowledge, greatly accelerating technical progress.

As manufacturing changed, so did the character of the workforce. Employees were divided into production workers and managers. Each division was given specific tasks. Efficiency of production became an area of serious study. For example, in the early twentieth century, Frederick Winslow Taylor developed his "four principles of scientific management" to reduce waste and increase productivity. Revolutionary for the time, Taylor's ideas included studying production workers' motions and rearranging equipment to decrease the time the workers spent not producing goods.

The various developments in industry led to continuous manufacture. See **Figure 1-10.** Continuous manufacture is characterized by the following:

- Improved machine life because of interchangeable parts.
- Division of the job into parts assigned to separate production workers.

©iStockphoto.com/epixx

Figure 1-10. Some workers are involved with the continuous-manufacturing process.

- Creation of material-handling devices that bring the work to the production workers.
- Classification of management as a professional group.

Later during the Industrial Revolution, enterprising people developed sophisticated transportation and communication systems to support the growing industrial activities. Dirt and gravel roads became paved highways. The diesel-electric locomotive replaced the steam locomotive. The motortruck and the airplane successfully challenged the dominance of transportation by the railroad. The telegraph, telephone, radio, and television replaced pony express letter carriers.

Construction practices advanced to provide the factories, stores, homes, and other buildings needed to meet a growing demand for shelter. Mass-produced dwellings replaced log cabins. Metal buildings became the factory of choice. Shopping centers and shopping malls made downtown shopping areas less important.

Efficient production was coupled with rising consumer demand. Extended free time was available for the first time, with the 40-hour workweek and annual vacations. Children could stay in school longer because they were not needed on the farm and in the factories. Universal literacy became a possibility, although it has yet to be reached.

The Information Age

The Industrial Revolution moved us from an agrarian era into the industrial era. We moved from a period when most people worked raising food and fiber to a period in which people worked in manufacturing. Technology is now moving many nations into a new period. This stage of development is called many different names. These names include the ***information age***.

During the industrial age, the most successful companies processed material better than their competitors. The information age changes this emphasis. This age places more importance on information processing and cooperative working relations between production workers and managers. The information age has several characteristics. These characteristics include the following:

- Wide use of automatic machines and information-processing equipment.
- High demand for trained technicians, technologists, and engineers.
- Blurring of the previously sharp line between production workers and managers.
- Constant need for job-related training and retraining of production workers.

These factors promise change. This is not, however, something new. Change takes place with every generation. One generation traveled west in covered wagons during their youth, and their children saw an astronaut circle the globe in their later years. What will you see in your lifetime?

Standards for Technological Literacy

5 7

Extend

Discuss the changes that the Information Age has brought about for individuals, families, and businesses. Have your students contribute their own thoughts to the discussion.

TechnoFact

The nations of the world became acutely aware of the progress made by the United States during the Industrial Revolution during the first World's Fair in London (1851). Because the Industrial Revolution developed more quickly in the United States than in Britain, the United States was the world leader in manufacturing by the end of the 19th century. The American economy emerged as the largest and most productive on the globe during this time.

TechnoFact

Computers have become important tools for everyday life and are now used for many different purposes, from helping children do homework to encoding information.

Think Green

Overview

You have probably heard a great deal about the ongoing efforts people are making to conserve resources, or to be more environmentally friendly. There are many ways you can help by making your habits more "green." For example, you may recycle plastic, aluminum, and paper. You can use reusable shopping bags rather than paper or plastic bags. You can use compact fluorescent lights (CFLs) instead of incandescent lights to help save energy. Throughout this text, we will be looking at several ways that people can make personal and group choices to help the environment.

Answers to Test Your Knowledge Questions

1. Technology is humans using objects (tools, machines, systems, and materials) to change the natural and human-made (built) environment.
2. C. Technology is found in nature.
3. True
4. True
5. Evaluate individually.
6. Science
7. Science, technology, and the humanities
8. F. Industrial Revolution
9. D. Middle Ages
10. A. Stone Age
11. G. Information Age
12. B. Bronze Age
13. E. Renaissance
14. C. Iron Age
15. D. Middle Ages
16. A. Stone Age
17. E. Renaissance
18. F. Industrial Revolution
19. steam engine
20. True

Summary

Technology has always been a part of human life. This part changes life and is changed as life progresses. Technology can be described as the use of tools, materials, and systems to extend the human potential for controlling and modifying the environment. This use makes life better for many people. If it is improperly used, however, it can cause serious damage to people, society, and the environment. Throughout history, technology has had a powerful influence on the economic, political, social, and cultural characteristics of societies. Early technological development was based on tools, not scientific knowledge. Later inventors, however, applied scientific principles to refine and advance existing technology. Only technologically literate people can properly develop, select, and responsibly use technology. This is the challenge for today's youth.

Test Your Knowledge

Write your answers on a separate piece of paper. Please do not write in this book.

1. What is technology?
2. Which one of the following is *not* a feature of technology?
   A. Technology uses tools, materials, and systems.
   B. Technology results in artifacts and other outputs.
   C. Technology is found in nature.
   D. People develop technology to control their environment.
3. Competition causes businesses to develop new products. True or false?
4. Technology almost always improves on existing technology. True or false?
5. Name one positive aspect and one negative aspect of technology.
6. ______ is the knowledge of the natural world.
7. List the three major types of knowledge.

Matching questions: For Questions 8 through 18, match each technological development on the left with the correct historical period on the right. Some letters will be used more than once.

Technological Development

8. ______ Airplane.
9. ______ Calculating machine.
10. ______ Copper tools.
11. ______ Electricity.
12. ______ Fire.
13. ______ Ox-drawn plows.
14. ______ Paper money.
15. ______ Pottery.
16. ______ Printing press.
17. ______ Solar energy.
18. ______ Telescope.

Historical Period

A. Stone Age.
B. Bronze Age.
C. Iron Age.
D. Middle Ages.
E. Renaissance.
F. Industrial Revolution.
G. Information age.

19. James Watt's improvements to the ______ revolutionized both the manufacturing process and the transportation system.
20. The information age is characterized by the blurring of the line between production workers and managers. True or false?

STEM Applications

1. Design a simple technological device (a tool to make a job easier) that can be used in your daily activities.
2. Develop a chart similar to the one shown below. For each problem on the left, list the scientific and technological knowledge that would help solve it.

Problem	Scientific Knowledge	Technological Knowledge
Depletion of the ozone layer		
Rapid depletion of petroleum		
Rising cost of electricity		
Injuries on the football field		

3. Select three major problems in your community. Develop a chart similar to the one shown above. For each problem listed on the left, list the scientific and technological knowledge that would help solve it.

This chapter covers the benchmark topics for the following Standards for Technological Literacy:

2 3 5 6 7 8

Learning Objectives

After studying this chapter, you will be able to do the following:

- Explain how technology is a system.
- Recall the major components of a technological system.
- Recall the inputs to a technological system.
- Recall the steps in the problem-solving and design process.
- Recall the major activities involved in production processes.
- Recall the major activities involved in management processes.
- Give examples of positive and negative technological outputs.
- Summarize feedback and control.

Key Terms

actuate	input	problem-solving and design process
control	machine	process
energy	management process	production process
feedback	material	system
finance	organize	time
goal	output	tool
information	plan	

Strategic Reading

As you read this chapter, make a list of the steps that are necessary for producing a technological product. Think of an everyday technological product and see if you can imagine its creation from start to finish.

Technology involves performing a task by using an object that is not part of the human body. For example, suppose you want to crack the shell of a walnut. If you put it in your mouth and bite down, you are not using technology because the human body is doing the work. Also, you might find that your jaw is not strong enough to do the job or that you value your teeth and do not want to damage them. You need another way to open the nutshell. Seeing some rocks, you realize their value in this situation. You place the shell on one rock and strike it with another rock. Now you have employed technology. The rocks extended your potential, or ability, to do a task.

This example shows that technology is the development and application of knowledge, tools, and human skills to solve problems and extend human potential. See **Figure 2-1.** Technology arises and moves forward out of human wants and needs. This development is designed and evaluated by people and, in time, is modified or abandoned. Moreover, every technology, whatever the force behind it, is developed through a ***system***. This means that each technology has parts and that each part has a relationship with all other parts and to the whole. The parts work together in a predictable way. They are designed to achieve a goal. See **Figure 2-2.**

A technological system has several major components. See **Figure 2-3.** These components are the following:

- Goals.
- Inputs.
- Processes.
- Outputs.
- Feedback and control.

Mazda Motor Corp.

Figure 2-2. The parts making up this automobile engine are designed to work together in a predictable way.

Standards for Technological Literacy

2 6

Demonstrate

Have your students try to open a walnut with their hands (without technology). Then, have them use a hammer to crack the shell (with technology).

Brainstorm

Ask your students what tools they used today.

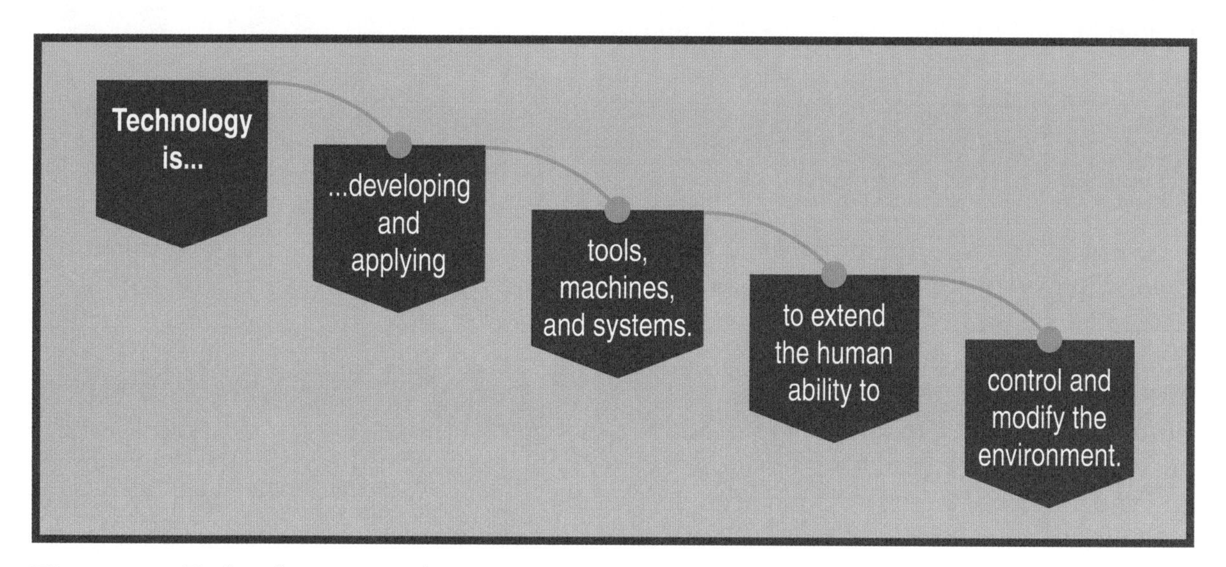

Figure 2-1. Technology extends one's ability to do a certain task.

Standards for Technological Literacy

2 6

Extend

Have your students list three products and write a clear goal statement for each, explaining for what purpose(s) the product was developed.

TechnoFact

In order to direct their attention to the task at hand, designers and engineers make a goal statement.

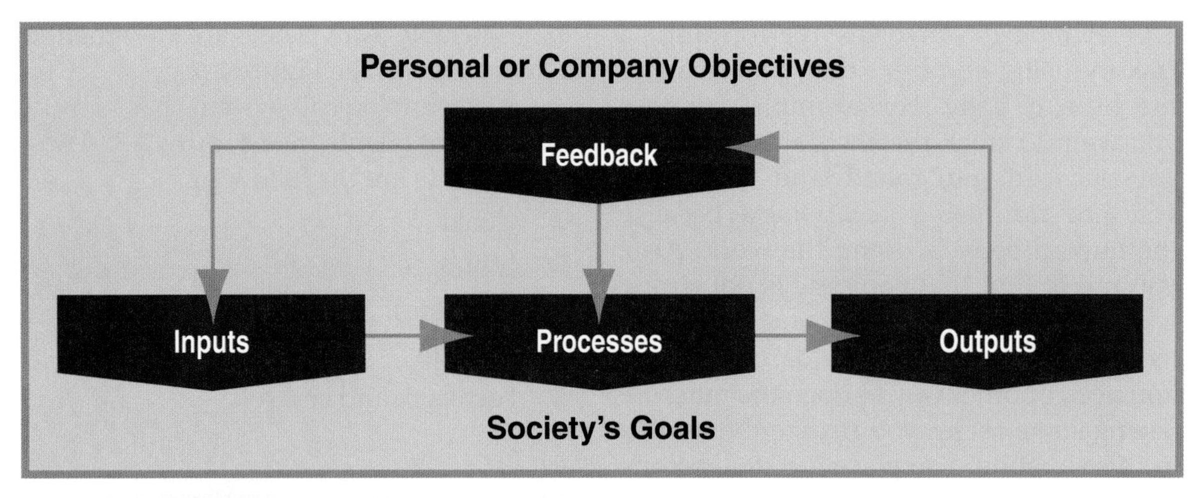

Figure 2-3. Technological systems have several components.

Career Corner

Purchasing Agents

Purchasing agents find quality materials, supplies, and services at the lowest costs possible. They determine which items are best, select a supplier for the material or service, negotiate a low price, establish delivery dates, and release purchase orders. Most purchasing agents work in offices and often work more than 40 hours a week.

Many companies prefer to hire purchasing agents who have a college degree and some experience. Newly hired purchasing employees often attend specialized training programs. They generally spend time learning about their company's purchasing practices, while working with experienced purchasers.

Goals

As we said above, humans develop technology to meet needs. This means humans have a ***goal*** in mind when developing the technology. Each artifact, however, generally meets more than one goal. For example, humans developed radar (*ra*dio *d*etecting *a*nd *r*anging) to determine the position and speed of aircraft. Radar had a major impact on military strategies and actions during World War II. Today, this technological innovation contributes to safe, reliable air transportation. We also use it to help predict the weather and to detect surface features on planets.

To examine this point a little further, let us look at another technology. Suppose a company is developing a new coal-powered automobile. It is not enough to say that the company is creating the automobile to transport people. This is certainly the primary goal. Developing the automobile might also be an opportunity for the innovators to make money. Local government leaders might look at its potential impact on economic growth in their city. The federal government might examine the automobile in terms of whether or not it will help reaffirm national technological leadership. The general public might consider whether or not this automobile is a good alternative to petroleum-powered vehicles, thus reducing its dependence on foreign oil. Workers might look at the pros and cons in terms of whether or not developing the automobile will improve job security. Coal-mining companies and oil companies might examine it with regard to whether it will increase or decrease their markets. Environmental groups might comment on how the car affects the environment. Note the number of different goals and concerns highlighted here. As you can see, a technological development can meet a number of different goals and concerns that are important to different groups.

Inputs

All natural and human-made systems have inputs. ***Inputs*** are the resources that go into the system and are used by the system. Technological systems have at least seven inputs. See **Figure 2-4.**

- People.
- Materials.
- Tools and machines.
- Energy.
- Information.
- Finances.
- Time.

Standards for Technological Literacy

Research

Have your students use the Internet to find three goal statements for technological endeavors.

TechnoFact

Technology that was developed before World War II is used today in a modified and improved capacity for modern consumer use. Such devices include jets, radar, and computers.

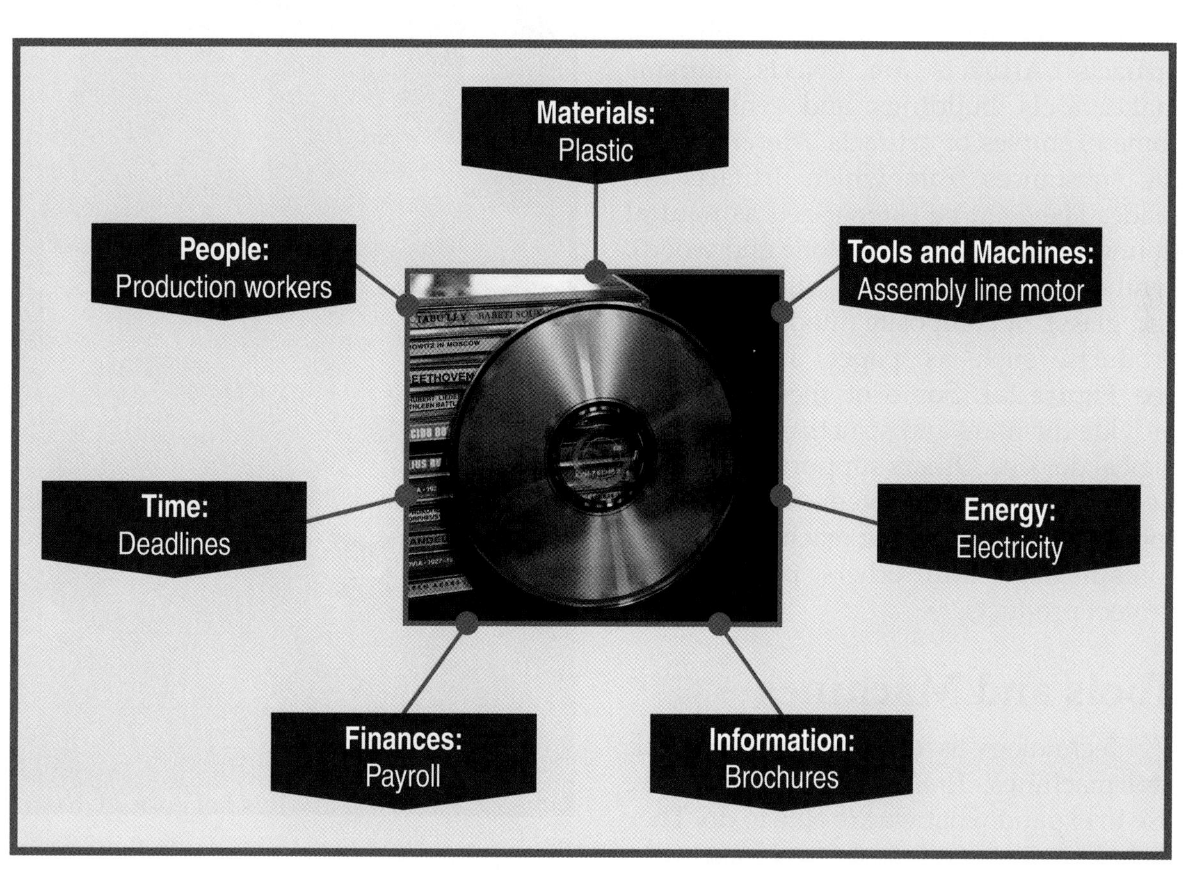

Figure 2-4. Technological systems have at least seven inputs.

Standards for Technological Literacy

2 6 7

Brainstorm

Ask your students to describe the ways that they contribute to technological systems.

Extend

Show a product and have the students give examples for each of the seven inputs used in its development, manufacture, and sale.

Extend

Discuss the simple tools used in everyday life, such as pencils, knives and forks, doorknobs and latches.

TechnoFact

Water, air, fossil fuels, soils, plants, animals, and mineral deposits are common natural resources that are used or impacted by technological systems.

People

People are an essential input to technological systems. See **Figure 2-5.** Human needs and wants give rise to the systems. Human will and purposes decide the types of systems that will be developed. People bring to the systems specific knowledge, attitudes, and skills. They provide the management and technical know-how to design and direct the systems. Their labors make the systems function. Human ethics and values control and direct the systems. Through their local, state, and national governments, people make policies that promote or hinder technological systems. Finally, people are the consumers of technological outputs. They use the products and services that the systems provide.

Figure 2-5. People bring knowledge and skills to each technological system. For example, they operate machines, design devices, and provide management.

Materials

All technology involves physical artifacts. Artifacts are objects humans make. Tools, buildings, and vehicles are some examples of artifacts. ***Materials*** are the substances from which artifacts are made. They can be categorized as natural (found in nature, such as stone and wood), synthetic (made by humans, such as plastics and glass), or composite (modified natural products, such as paper and leather). See **Figure 2-6.** Some of these materials provide the mass and structure for technological devices. Others support the productive actions of the system. For example, some materials lubricate machines. Others contain data. Still others package and protect products.

Tools and Machines

Technology is characterized by tools and machines. ***Tools*** are the artifacts we use to expand what we are able to do. They include both hand tools and machines. Hand tools expand muscle power and include such artifacts as hammers and screwdrivers. ***Machines*** amplify the speed,

Figure 2-6. Materials can be found in nature. Here, for example, we see water that can be converted into energy, rocks that can be made into hand tools, and trees that can be made into paper.

amount, or direction of a force. Early humans relied on power from animals and water to affect force. Now, however, we use energy from such sources as electricity to power machines. We use tools and machines to do many things, including building structures, communicating information and ideas, converting energy, improving our health, growing crops, diagnosing illnesses, and transporting people and goods.

Energy

The development and application of knowledge, tools, and skills involve doing something. All technological activities require energy. ***Energy*** is the ability to do work. The form the energy takes, however, varies. This energy form can range from human muscle power to nuclear power and from heat energy to sound energy.

Technological systems require energy to be converted, transmitted, and applied. For example, a hydroelectric generator might convert the energy of falling water into electrical energy. See Figure 2-7. The electricity might then be carried to a factory over transmission lines. At the factory, motors might convert the electrical energy into mechanical motion. Halogen and fluorescent bulbs

Standards for Technological Literacy

2 3 6 7

TechnoFact

Labor is a term that can be used to describe the efforts expended in completing a task and also the workers engaged in doing the task.

TechnoFact

The human potential to do work is extended by machines, which can alter the amount of force and distance needed to move objects.

STEM Connections: Science

Integrated STEM Curriculum — Science, Technology, Engineering, Mathematics

Newton's Third Law of Motion

Newton's third law of motion states that, for every action by a force, there is an equal and opposite reaction by another force. We can see this law at work with a jet engine. The engine is designed to shoot out a mass of hot gas. The force of this gas in one direction creates an equally strong force in the opposite direction. Thus, when the hot gas shoots backward, the jet plane shoots forward.

We can also see this law in other areas. For example, a book lying on a table is actually exerting a force on the table because of the pull of gravity. In turn, the table exerts an equal and opposite reaction on the book. What might be another example of this law at work?

Standards for Technological Literacy

2 6

Brainstorm

Ask your students to name three examples of data, information, and knowledge.

TechnoFact

The Information Age is founded on Information Technology that processes and distributes data using electronic means, including digital electronics, computer software and hardware, and telecommunications.

Figure 2-7. The energy of falling water can be converted into electrical energy.

might convert it into light energy. In another case, radiant heaters might convert it into heat energy.

Information

Everywhere we turn, we encounter facts and figures called *data*. When we organize this data and group it according to its type, we create ***information***. Information is essential for operating all technological systems.

An example shows the difference between data and information. You might measure the size and weight of everything you can find. This is data because it is random and assorted. If you sort the data so the height and weight of all people are grouped together, you have information. Using this information, people can see relationships and draw conclusions. For example, it becomes obvious that adults are generally taller than children and that men are generally taller than women. With this final step, you have developed knowledge. Knowledge is people using information to understand, interpret, or describe a specific situation or series of events.

Finances

We have seen that technology-system inputs include people, materials, tools and machines, energy, and information. These resources have value and, therefore, must be purchased. Also, the outputs of technology systems have value and can be sold. Thus, ***finances***, the money and credit necessary for the economic system to operate, are another technology input.

For example, a construction company might determine that a family needs a new home. The company must purchase the land and materials to build the structure. Members of the company must then obtain plans for the dwelling. They rent or buy equipment. Members of the company hire workers to build the house. All these actions require money. The members of the company then sell the finished home to cover these costs. See **Figure 2-8.** Generally,

Standards for Technological Literacy

2 6

Figure 2-8. The construction company is selling this newly built home to cover the costs of creating the home and to earn a profit.

money is left after all expenses are paid. As noted in the last chapter, this amount of money is called *profit*. Profit is the goal of most economic activities.

Time

All jobs and activities take ***time***. Each person has only 60 minutes in an hour and 24 hours in a day. This time is allotted to the various tasks that need to be done. Time is set aside for working, eating, sleeping, and recreation. Likewise, time must be allotted to all technological endeavors during a workday. The most important ones are most likely completed. If time is not available, less critical tasks are left undone or postponed to a later date. Therefore, not all needed technology can be developed immediately. Some activities have to wait until time and other resources are available.

Processes

All technological systems are characterized by action in which a series of tasks must be completed. The steps needed to complete these tasks are called ***processes***. Technological systems use three major types of processes. These types are problem-solving and design processes, production processes, and management processes. See **Figure 2-9.**

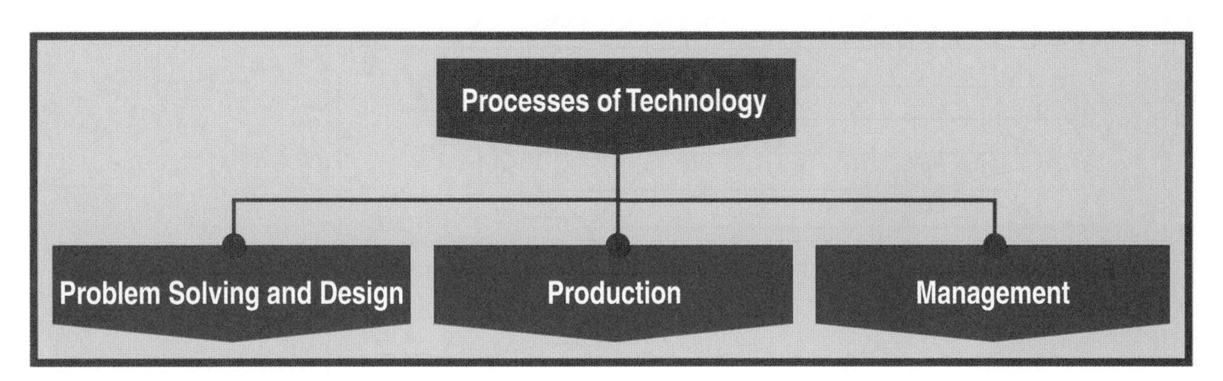

Figure 2-9. Technology uses problem-solving and design, production, and management processes to complete tasks.

Standards for Technological Literacy

2 6

Technology Explained

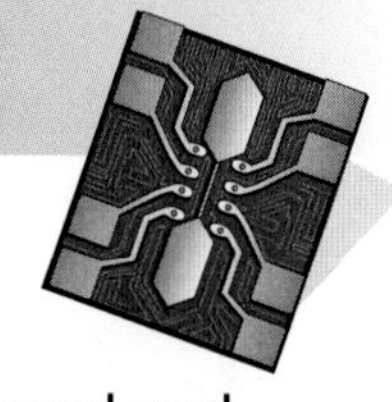

solar collector: a high-temperature device heated by concentrated solar energy.

Solar collectors can be of two types. These types are solar thermal and solar electric. Solar thermal collectors capture heat from sunlight by moving air or another fluid through the collector. The fluid absorbs the heat and carries it away for storage or use. Solar electric collectors use photoelectric materials. When light strikes a photoelectric material, electrons move and create an electric current.

Both solar thermal and solar electric collectors can be placed together in a solar orchard. See **Figure A.** Solar thermal collectors are used to produce steam. This steam is fed to a generating plant to make electricity. Solar electric collectors can be connected directly to the electric utility grid. Both types of collectors use tracking systems to follow the Sun across the sky.

Another method of collecting and using solar energy is the solar furnace. The solar furnace uses the reflected light from mirrors called *heliostats*. Each device tracks the Sun and reflects the solar energy onto a parabolic mirror. The mirror, in turn, reflects the light from all the heliostats onto a single point. Temperatures can reach 5400°F (3000°C). **Figure B** shows a drawing of a solar furnace built for research in France.

Figure A. This solar orchard is in Phoenix, Arizona. These collectors use solar electric cells with concentrating lenses to increase the power output.

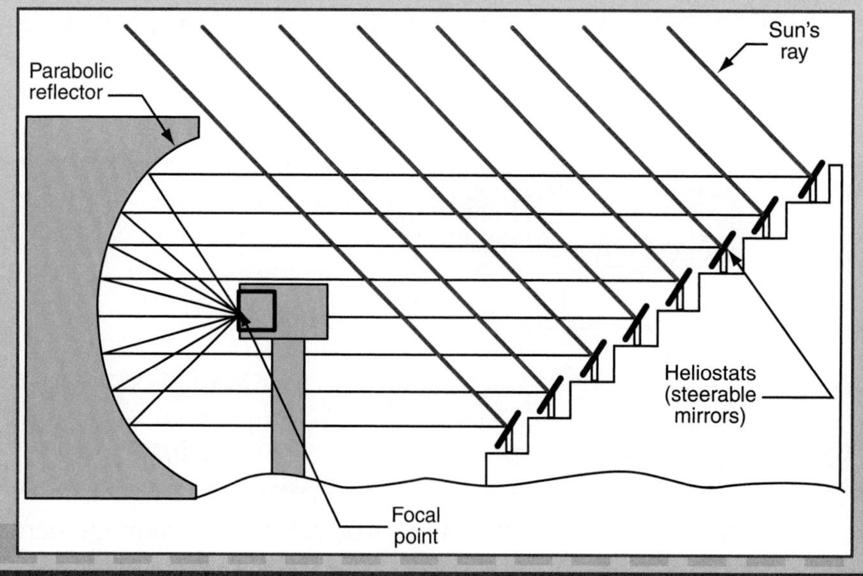

Figure B. The solar furnace in France uses 63 flat mirrors to collect and reflect the Sun's energy.

The Problem-Solving and Design Process

In Chapter 1, we learned that science involves activities that study and interpret the natural world. Scientists use research to develop their descriptions. They carry out their work through a set of procedures called the *scientific method.* This method structures the research so valid results can be obtained. Through the scientific method, scientists attempt to understand the world that now exists.

On the other hand, technology, as described earlier, develops and uses tools and machines, systems, and materials to extend the human ability to control and modify the environment. One key word in this definition is *develop.* This part of technology requires creative action. The procedure used to develop technology is called the ***problem-solving and design process.*** This process involves five major steps. See Figure 2-10.

1. Identifying the problem. A person or group develops basic information about the problem and the design limitations.
2. Developing solutions to the problem. A person or group develops and refines several possible solutions.
3. Isolating, refining, and detailing the best solution. A person or group selects and refines the most promising ideas and then details (describes) the best solution.
4. Modeling and evaluating the solution. A person or group produces and tests physical, graphic, or mathematical models of the selected solution.
5. Communicating the final solution. A person or group selects a final solution and prepares documents needed to produce and use the device or system.

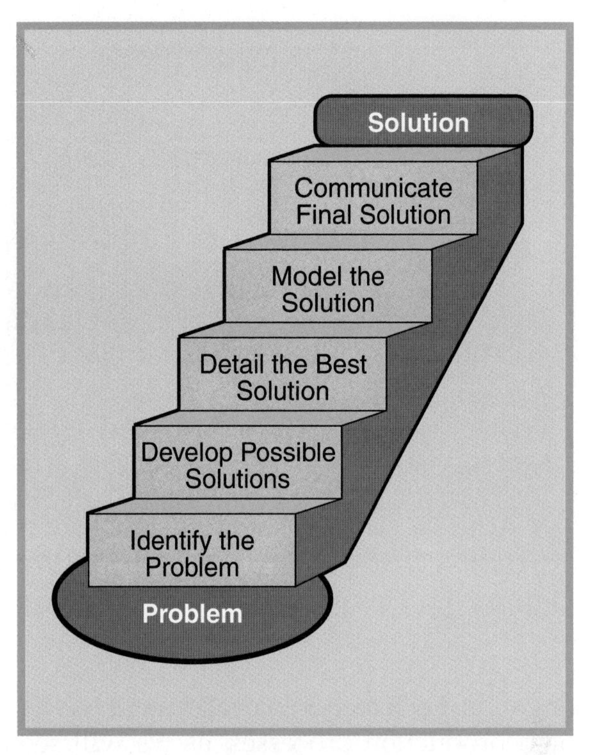

Figure 2-10. The problem-solving process involves five major steps to reach a solution.

Production Processes

Production processes are actions completed to perform the function of the technological system. For example, a company might use a series of production processes to produce an informational booklet. See Figure 2-11. The company writes and edits the message or copy to be communicated. Members of the company produce photographs to illustrate the document. They put the photographs and copy together into a page layout. The members of the company convert the layout into printing plates. Copies of the booklet are printed from these plates. Production processes are used to grow and harvest crops, change natural resources into industrial materials, prevent and treat illnesses, convert materials into products or structures, transform information into media messages, convert a form of energy, and use energy to power transportation vehicles to relocate people

Standards for Technological Literacy

2 6 8

Research

Ask your students to find some common definitions for problem solving.

TechnoFact

According to recent studies, a widespread idea is that design plays a key role in many business operations. Small- and medium-sized businesses often adopt this attitude toward design's function in the process of technology.

Standards for Technological Literacy

2 6 8

Brainstorm

Have your students write down the production processes they think are used to make a common wooden pencil. One answer can be found on the following web site: www.pencils.com.

TechnoFact

The following information is correlated with the Brainstorm challenge above. A typical pencil can write 45,000 words or draw a line 35 miles long. More than 2 billion pencils are used in the United States every year, and most of them have erasers. However, most pencils sold in Europe do not have erasers.

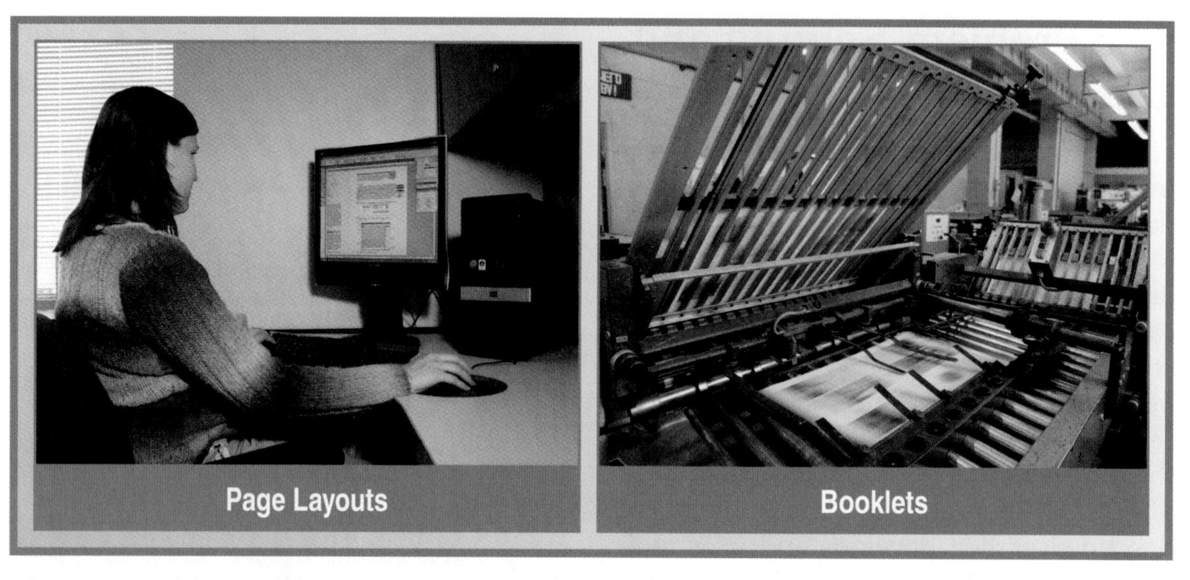

Figure 2-11. This production process converts page layouts into booklets.

or goods. Each technological system has its own production processes, as we will see later in this book.

Management Processes

Management processes are all the actions people use to ensure that the production processes operate efficiently and appropriately. People use these processes to direct the design, development, production, and marketing of the technological device, service, or system. Management activities involve four functions. See Figure 2-12.

- ***Planning.*** Setting goals and developing courses of action to reach the goals.
- ***Organizing.*** Dividing the tasks into major segments so the goals can be met and resources can be assigned to complete each task.
- ***Actuating.*** Starting the system to operate by assigning and supervising work.
- ***Controlling.*** Comparing system output to the goal.

Individuals and groups use management processes to organize and direct their activities. For example, you might have a

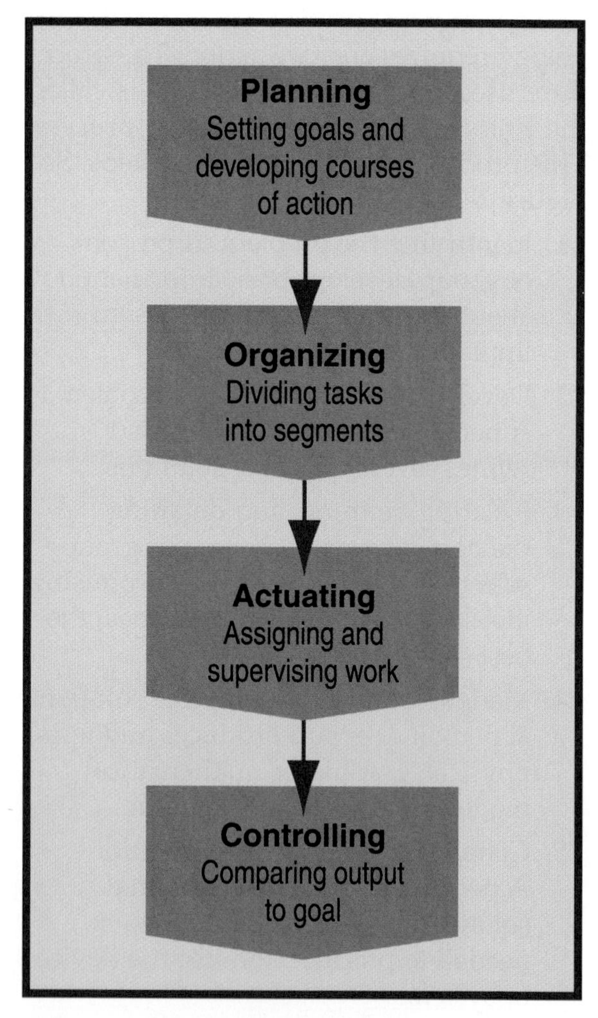

Figure 2-12. Management activities direct the operation of the production process.

Think Green

Sustainability

The word *sustainability* is used to describe the idea of the world's ongoing ability to produce natural resources. The resources in question may be anything from trees to petroleum. Several natural resources are known as *exhaustible resources*. This means that once we have used up these resources, they will be gone forever and will not grow back. In several instances, these resources aren't being depleted due to consumption, but merely by the way humans interact with the environment. For example, the use of pesticides may lead to water contamination.

Individuals, companies, and nonprofit and government organizations are working to encourage sustainability and to prevent the loss of our natural resources. They are looking for ways to reduce the use of exhaustible resources while finding alternative renewable resources. With a common goal, people are making an effort to help preserve our natural resources.

Standards for Technological Literacy

2 5 6

Extend

Discuss how management processes can be used for everyday activities, such as selling Girl Scout cookies, putting on a play, or publishing a school newspaper.

Brainstorm

Ask your students how they would use problem solving/design, production, and management processes to make 100 CD holders.

TechnoFact

People primarily used hand tools to manufacture simple objects for themselves and for sale in small shops before the Industrial Revolution. Managers were needed in the late 19th century, when people were brought together with steam-powered machines in factories.

task to complete, such as writing a term paper. First, you must plan this activity by selecting a topic, establishing major steps to be completed, and setting deadlines for each task. You then must obtain and organize resources and secure reference materials, writing or word processing equipment, and time. Next, you actuate the work by reading and viewing reference material, taking notes, preparing a draft of the paper, and editing the draft into final form. Finally, your instructor compares your paper to established standards and gives you the results. This step is control. Control includes evaluation, feedback, and corrective action.

Outputs

All technological systems are designed to produce specific outputs. These ***outputs*** might be manufactured products, constructed structures, communicated messages, or transported people or goods. These primary outputs are not, however, the only ones that come from technology.

In operating technological systems, we can produce other less direct and unwanted outputs. For example, our manufacturing and construction activities generate scrap and waste. Manufacturing operations also create chemical by-products, fumes, noise, and other types of pollution. Cars, airplanes, trains, and trucks often produce noise, air pollution, and congestion. Poorly designed housing developments, industrial parks, and shopping centers can contribute to soil erosion, as can careless farming practices.

Technological systems also have social and personal impacts. The products of technology shape society and are shaped by society. For example, the automobile was a novelty in the early 1900s. The wide acceptance of the automobile, however, has greatly changed where and how we live. We travel more and live farther from our places of work. Sprawling cities have replaced the small compact towns. Shopping is no longer close to the neighborhood. On the other hand, we have shaped the automobile. Rising gasoline prices and government fuel-economy standards have resulted in improved engine fuel efficiency with fewer exhaust products. Personal buying habits also dictate the types of cars that are built.

Standards for Technological Literacy

2 6

Research

Have your students use the Internet to find out how the Environmental Protection Agency monitors and controls air, soil, and water pollution.

TechnoFact

A governor used to control the speed on an engine is an example of simple feedback. The governor reduces the fuel supply when the speed of the engine exceeds an established limit. The engine's speed decreases because it has less fuel to burn. All living creatures possess feedback control systems.

TechnoFact

Automation is the automatic operation and control of machines and systems by using self-correcting control systems that employ feedback.

Feedback and Control

All systems are characterized by ***feedback*** and control. Feedback and control involves using information about the outputs of a process or system to regulate the system. The process of feedback and control is used in many common situations. For example, many homes have heating systems that are controlled automatically with a thermostat. See **Figure 2-13.** The occupant sets the thermostat at a desired temperature. The thermostat measures the room temperature. If the temperature is too low, the thermostat closes a switch that turns on the furnace. As the room is heated, the thermostat monitors the temperature. The thermostat is using the output of the furnace to determine the necessary adjustments for the system. When this output (heat) warms the room to the proper temperature, the thermostat changes the system's operation. The thermostat opens the switch and stops the furnace. The cycle is repeated when the room cools to a specific temperature. Thus, through feedback and control (in this case, automatic control), the thermostat regulates the heating system.

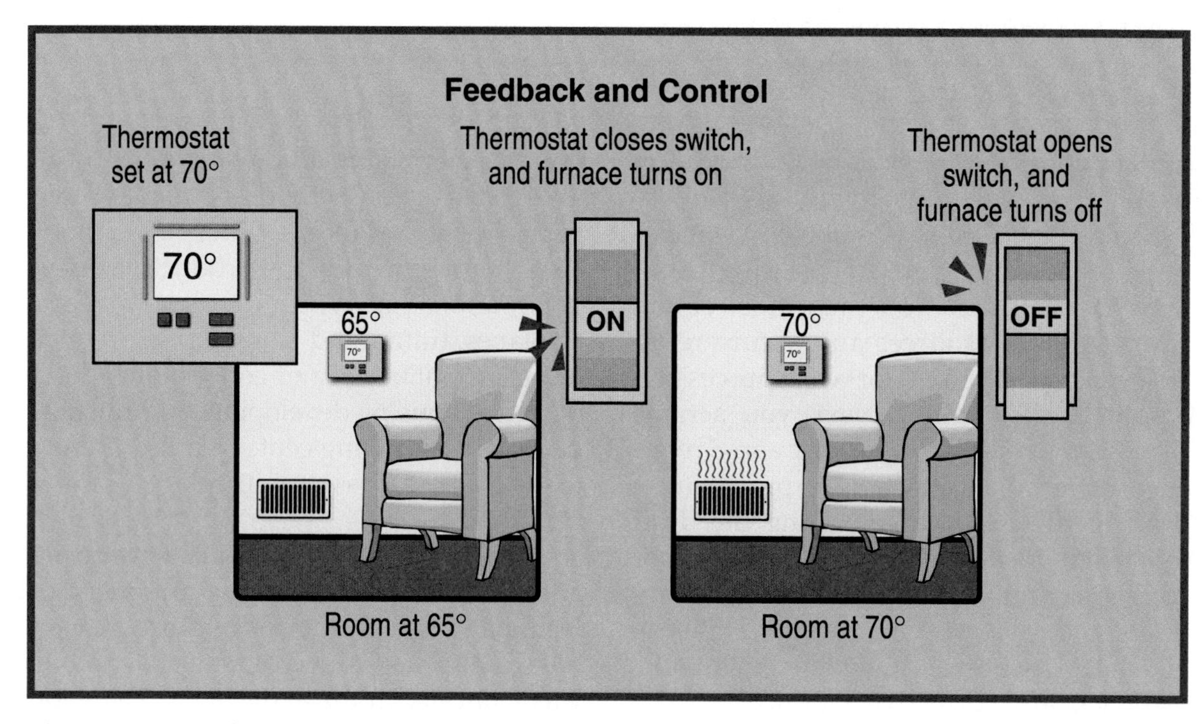

Figure 2-13. A thermostat regulates a heating system through feedback and control.

This GPS navigator was developed using the major components of the technology system.

Answers to Test Your Knowledge Questions

1. Each technology has parts, and each part has a relationship with all other parts and to the whole. The parts work together in a predictable way, and they are designed to achieve a goal.
2. Goals, inputs, processes, outputs, feedback and control
3. False
4. People, materials, tools and machines, energy, information, finances, and time
5. True
6. True
7. Information is data that has been organized and grouped according to its type.
8. False
9. Identifying the problem. Developing solutions to the problem. Isolating, refining, and detailing the best solution. Modeling and evaluating the solution. Communicating the final solution.
10. production processes
11. Planning, organizing, actuating, and controlling
12. Evaluate individually.
13. Evaluate individually.
14. Evaluate individually.
15. feedback and control

Summary

Technology involves human-made systems that use resources to produce desired outputs. The systems use seven major types of resources. The types of resources are people, materials, tools and machines, energy, information, finances, and time. These resources are used during the operation of the technological systems. The operation stage is called the *process*. This stage involves three major types of processes. These processes are problem solving and design, production, and management. The result of each system includes desired and unwanted outputs. We use technological systems to produce products, construct structures, communicate information, and transport people and goods. These things improve our way of life. They are the desired outputs. We also get pollution, scrap and waste, altered lifestyles, and increased health risks, however. These are the negative impacts or outputs of technology. We use feedback and control to regulate the system. The challenge for each person developing or using technology is to use the feedback and control to maximize the desired outputs and minimize the negative impacts at the same time.

Test Your Knowledge

Write your answers on a separate piece of paper. Please do not write in this book.

1. In what way or ways is technology a system?
2. List the five major components of a technological system.
3. Each artifact is developed to meet one goal. True or false?
4. List the seven inputs to a technological system.
5. Materials can be classified as natural, synthetic, or composite. True or false?
6. Energy needs to be converted before it can be used in a technological system. True or false?
7. Explain the difference between data and information.
8. Technology primarily uses the scientific method to develop new artifacts and services. True or false?
9. List the five steps in the problem-solving and design process.
10. Actions completed to perform the function of a technological system are called ______.
11. What are the four functions of management activities?
12. List three desired outputs of a technological system.
13. List three undesired outputs of a technological system.

14. Give one example of how a technological development has affected society.
15. The process of using information about the outputs of a system to regulate the system is called _____.

STEM Applications

1. Select a technological device you use. List its inputs, processes, and desired and undesired outputs. Organize your answer in a form similar to the one shown below.
   Device:
   Major inputs:
   Production processes used to make the device:
   Desired outputs:
   Undesired outputs:
2. Select an early technological advancement (invention). Prepare a short report including the inventor's name, the nation from which the invention came, events that led up to the invention, the invention's impact on life at the time, and the invention's later refinements. Include a sketch of the item.

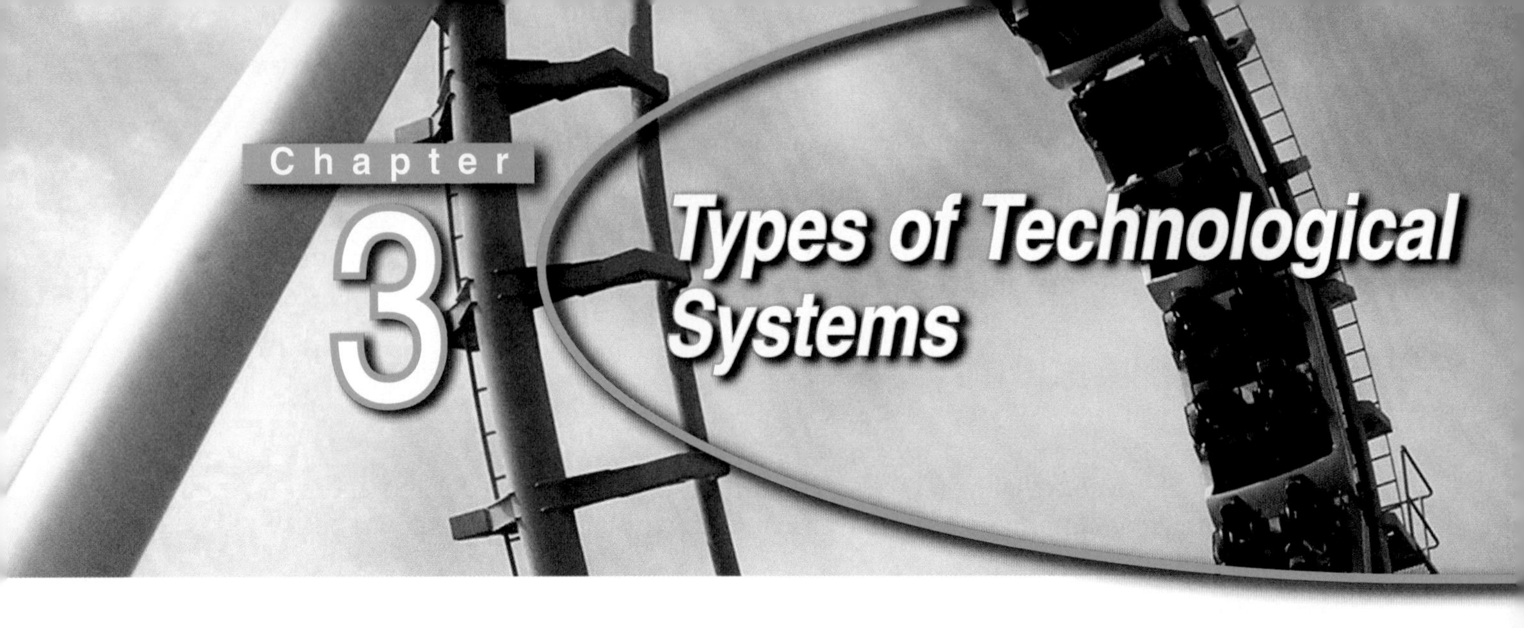

Chapter Outline

This chapter covers the benchmark topics for the following Standards for Technological Literacy:

1 3 4 7 14 15

Learning Objectives

After studying this chapter, you will be able to do the following:

- Recall the four general ways technology systems can be categorized.
- Give examples of obsolete, current, and emerging technologies.
- Explain how technology evolves in a private enterprise system.
- Give examples of government-sponsored technology.
- Summarize why cooperation is important in technological development.
- Recall tasks of each major area of technology.
- Explain how technological systems are related.

Key Terms

agricultural and related biotechnology
communication and information technology
construction technology
corporate participation
current technology
emerging technology
energy and power technology
financial affairs
high technology (high tech)
industrial relations
manufacturing technology
marketing
medical technology
obsolete technology
production
research and development
transportation technology

Strategic Reading

As you read the chapter, you will learn about the three stages of changing technology. Think of the seven areas of technology and list examples for them for each of the three stages.

Technological systems are everywhere. They support our daily lives, produce the artifacts constituting the human-made world, and are vital to the economic and political health of every nation. Technological systems, as shown in **Figure 3-1**, can be grouped in four general ways:

- Level of development.
- Economic structure.
- Number of people involved.
- Type of technology developed and used.

Level of Development

Technology is a constantly changing phenomenon. Sophisticated technologies are constantly replacing simpler ones. For example, the sail was used on boats as early as 5000 BC in Mesopotamia. This device made water transportation more efficient and took less human effort than rowing a boat. In 1807, Robert Fulton used the steam engine to power a ship down the Hudson River. This advancement soon replaced the sail as the primary power source for commercial shipping. The steam engine did not last as long as the sail, however. The diesel engine replaced it in the 1900s. In 1954, nuclear power was first used in the submarine USS *Nautilus*. Nuclear power is still in use in naval vessels around the world. At any stage of history, then, there are obsolete, current, and emerging technologies. See **Figure 3-2.**

Standards for Technological Literacy

4 7

Extend

Explain that most histories are concerned with determining why certain things happened as they did and how these actions are related to earlier events.

TechnoFact

Sources of historical information include books, personal and government records, newspapers, artifacts, and oral accounts.

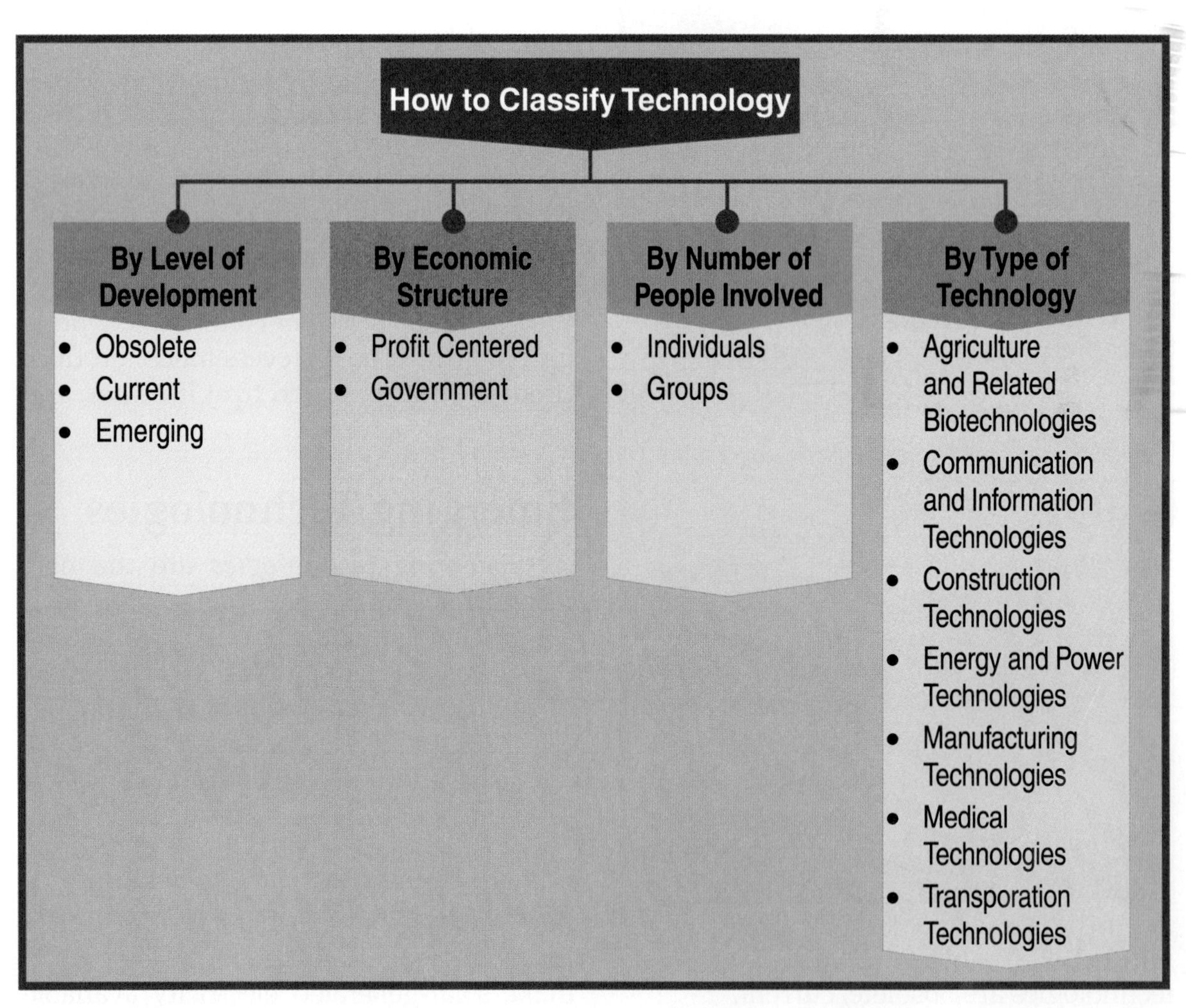

Figure 3-1. Technology systems can be grouped in four general ways.

Standards for Technological Literacy

4 7

Extend

Present your students with examples of how, over time, devices evolve from emerging technology to current technology to obsolete technology: fountain pens, typewriters, audio tape recordings, etc.

TechnoFact

With the developments of the personal computer, the typewriter has become an example of a high-tech machine that is now almost obsolete.

Library of Congress, Dave Sizer

Figure 3-2. At any stage of history, technologies are obsolete, current, or emerging.

Obsolete Technologies

Obsolete technologies are those that can no longer efficiently meet human needs for products and services. For example, hand spinning and weaving of cloth are obsolete. Some artists and home hobbyists still use these technologies to produce works of art and personally designed items. Likewise, artists use the skills of the old-time blacksmith to produce decorative, forged steel items. This technology is seldom used to produce products for the mass market, however. Instead, machines are used to produce these items more quickly and efficiently.

Current Technologies

Current technologies include the range of techniques used to produce most of the products and services today. You can see these technologies everywhere you look. Trucks, trains, and aircraft are the vehicles we usually use to transport the majority of goods in this country. The print and electronic media deliver the majority of the information to the general public. We also use a series of common technologies to grow crops, process foods, produce products, diagnose and treat illnesses, and construct structures.

Emerging Technologies

Emerging technologies are the new technologies that are not widely employed today. They might, however, be commonly used in a later period of time. For example, one emerging technology combines paper with tools to diagnose human diseases and medical conditions. This technology can provide a low-cost medical diagnostic tool for use in poor countries. Another example is a new low-cost battery made with all-liquid active materials that can store energy from solar collectors. This technology can make solar-generated electricity available during both day and night.

We often call these technologies ***high technology (high tech)***. This term is somewhat misleading because many of yesterday's high techs are now current technologies. See **Figure 3-3**. Today's high-tech processes might be tomorrow's current technology. For example, most Americans do not remember a world without television. Yet television was considered high tech in the early 1950s. Fiber-optic communication was unknown to most people until the 1980s. Today, it is used in most telephone systems in the United States. Therefore, we should think of high tech as a moving focus. High tech is new and strange today, current and common tomorrow, and obsolete and seldom used at a later date.

Figure 3-3. Many people remember a time without cellular telephones. Today, cell phones can be seen everywhere.

Economic Structure

People have developed technology to serve humankind. Technology has also become an integral part of the economic system. From this point of view, we can see

Standards for Technological Literacy

1 4 7

Extend

Give your students examples of private technology (automobiles, TV, etc.) and public technology (space shuttle, early atomic reactors, etc.).

TechnoFact

In order to increase profit, many technological advancements were developed to make items less expensively. This action resulted in many timesaving and labor-saving devices.

Career Corner

Finance

Controllers

Controllers manage workers who prepare financial reports that summarize and forecast a company's financial position. Common reports these workers prepare include income statements and balance sheets. The workers also help predict future earnings or expenses for a company. Controllers generally provide overall management for the accounting, audit, and budget departments. Many controllers oversee the preparation of reports that governmental regulatory agencies require.

For most controller positions, a bachelor's degree in finance, accounting, economics, or business administration is required. Often, preference is given to applicants who have a master's degree in one of these areas. Controllers work in offices close to the departments that develop the financial data for top management. They typically have direct access to advanced computer systems. Many controllers work up to 50 or 60 hours per week.

Standards for Technological Literacy

1

Brainstorm

Ask your students why the three stages and two activities of profit-centered companies are important. Ask the students if any of these can be left out by successful enterprises.

TechnoFact

The snap mousetrap and the paper clip were both patented in 1899. Many attempts have been made to improve on the early inventions of these everyday products. The original inventions, however, have proven to be effective and long-lasting solutions.

technology as either private (profit centered) or public (government). Most technology is developed and put into use by private enterprises. These businesses produce the goods and services people want. They plan to make a profit by meeting these human needs.

Private Technologies

Profit-centered companies generally move technology through three stages as they meet human needs and wants. These stages, as shown in **Figure 3-4,** are the following:

1. ***Research and development.*** Designing, developing, and specifying the characteristics of the product, structure, or service.
2. ***Production.*** Developing and operating systems for producing the product, structure, or service.
3. ***Marketing.*** Promoting, selling, and delivering the product, structure, or service.

Two major activities that provide the resources needed to operate the systems support these stages. These are the following:

- ***Industrial relations.*** Setting up and managing programs to ensure that the relations among the company, its workers, and the public are positive.
- ***Financial affairs.*** Obtaining the money and physical resources and keeping the financial records needed to manage the system.

Public Technologies

Some technology that is important to the general public is too expensive or risky to be developed by private enterprise. In this case, the government might address what it sees as the general welfare of the citizens by developing a specific technology. The goal is not to make a profit from the undertaking. The space program, with

Figure 3-4. Technology generally progresses through three stages. These stages are research and development, production, and marketing. Here, people created and sold books by researching content, printing and binding the pages into books, and displaying the books in a manner to attract customers.

its specific hardware, is an example of this type of public-funded technological development. See **Figure 3-5.** Out of this type of program come many products and systems that individuals and companies can later use. Examples of the National Aeronautics and Space Administration's (NASA's) space-program, technological innovations that have become commonplace include the following:

- Mylar® polyester film from the early Echo space program is now used for clothing, packaging tape, and decorative balloons.
- High-temperature composite materials from the Lewis Research Center are used in aircraft engines and aircraft structures.
- Satellite communications systems from many space programs now make it possible to communicate around the globe instantly.
- Film with a v-groove design reduces water and wind resistance. This design was developed for increasing fuel efficiency of jet aircraft. The film was used on the America's Cup–winning yacht *Stars and Stripes.*
- A corrosion-resistant coating developed for launchpads was applied to the refurbished Statue of Liberty.

Figure 3-5. The government sometimes develops technology that benefits all people.

Number of People Involved

We can also categorize technology systems by the number of people involved and how these people work with the technological system. In isolated cases, individuals might develop technology for their own uses. A cattle rancher might create an automatic feeding system for his livestock. A home gardener might develop a unique hoe that meets her personal needs. See **Figure 3-6.** A home craftsperson might design and build a desk for his own use. This type of personal technology is not common, however, and accounts for only a small portion of all the technology in use today.

Most technology is developed and applied by groups or teams of people. The key to these activities is cooperative attitudes. People often say we live in a competitive society. This belief is partly true. We

Figure 3-6. A gardener developed this home-built garden hoe for her own use.

Standards for Technological Literacy

1

Research

Identify and explore a recent development by a government agency, such as NASA.

Extend

Explain how Thomas Edison moved from an individual inventor to a person who ran a research laboratory employing a number of people.

TechnoFact

At age 15, Thomas Edison, the "Father of the Research Laboratory," was a self-educated telegraph operator. Edison filed for his first patent for an electric vote-recording machine at age 21. He set up his own company in Menlo Park, NJ at age 29. In the next 10 years, with his small group of engineers, scientists, and technicians, Edison filed for 500 patents. He had received over 1,300 U.S. and foreign patents by the end of his life.

Standards for Technological Literacy

1

Reinforce

Explain why each of the categories fits the definitions of technology given in earlier chapters.

Figure Discussion

Have your students discuss the reasons cooperation is necessary for technological activities.

TechnoFact

The Natufians lived in the region of the Middle East near the Dead Sea about 10,000 years ago. They were a hunting-gathering tribe who are credited with being the first people to start farming.

TechnoFact

The fourth most common element on Earth is iron. The widespread commercial use of cast iron originated with Abraham Darby, who used coke to smelt iron in his Coalbrookdale Ironworks, central England in 1709. Cast iron was used by the English and many advances in construction, transportation, and manufacturing came from using iron.

compete to get on a sports team or to get a job. Companies compete to sell products and gain market share. Nations compete for prestige. Within each of these groups, however, is cooperation. The winning sports team is a study in cooperative attitudes. The most profitable companies exhibit high levels of management and worker cooperation. See **Figure 3-7.** Nations cooperate in organizations such as the United Nations. Once a person gets a job, cooperation helps to ensure success.

This spirit of cooperation in developing and operating a technological system can be described as ***corporate participation.*** Corporate participation does not mean that the corporate form of ownership must be present. This participation means that people are united and combined into one body. They share a common vision, work toward similar goals, and share in the success of the enterprise.

Figure 3-7. Most technology requires a high level of cooperation among people for its development and efficient operation.

Type of Technology Developed and Used

The three ways of looking at technology discussed so far are useful in a general way. They give only part of the picture, however. A more meaningful approach to understanding technology categorizes the area by the type of technology developed and used. This approach looks at technology as human actions and groups them accordingly. This grouping includes seven types of technology. See **Figure 3-8.** Humans have used all these types of technology throughout history to modify and control their environment:

- ***Agricultural and related biotechnologies.*** Used in growing food and producing natural fibers.
- ***Communication and information technologies.*** Used in processing data into information and in communicating ideas and information.
- ***Construction technologies.*** Used in building structures for housing, business, transportation, and energy transmission.
- ***Energy and power technologies.*** Used in converting and applying energy to power devices and systems.
- ***Manufacturing technologies.*** Used in converting materials into industrial and consumer products.
- ***Medical technologies.*** Used in maintaining health and curing illnesses.
- ***Transportation technologies.*** Used in moving people and cargo from one place to another.

In each technological system, inputs are processed and transformed into outputs. For example, in agricultural technology, materials (seeds and fertilizers, for instance) are changed into food and fibers. In medical technology, information and materials are

Types of Technology

Communication

Construction

Manufacturing

Transportation

Agriculture

Medicine

Energy

Figure 3-8. Technology can be categorized by the type of product developed.

Standards for Technological Literacy

1

Research

Have your students select a technology and research a major advancement that has occurred in that technology in the past 20 years.

Figure Discussion

Give your students an example of each technology shown in Figure 3-8.

TechnoFact

To celebrate our centennial, America hosted a World's Fair in Philadelphia in 1876. We had changed from a colony of England into a great industrial leader during the previous century. Technology developed by many inventors and innovators drove America to this transformation.

Standards for Technological Literacy

1 3 14 15

Extend

Give your students examples of the evolution of a product from raw resources to final product. For example, you may discuss how trees are processed into lumber and ultimately made into furniture, or how silica sand is converted into glass and then made into bottles.

TechnoFact

A range of substances is involved in genetically engineered products. Since these products are in widespread use, approval from agencies such as the Department of Agriculture, Food and Drug Administration, or Environmental Protection Agency is usually required for production and distribution.

changed into devices that aid in maintaining health. In transportation technology, energy is changed into a form of power that moves vehicles containing goods.

These systems can be looked at individually. Such a focus, however, gives an inaccurate view. The systems are all closely related and are part of a single effort. This effort is to help humans live better. These systems work together to support one another. In one case, manufacturing might be the focus, with the other systems in supporting roles. In another situation, manufacturing might be the supporting technological system. To see this relationship, look at **Figure 3-9.** Let us follow one material from its natural state to a finished product that is designed to make life better for us. We will follow iron ore on its journey to becoming a stainless steel cookie sheet.

Our story starts with manufacturing technology. With regard to the cookie sheet, this technology system involves three major activities:

1. Locating and extracting the raw materials (iron ore, limestone, and coal) to make steel.
2. Producing strips of steel from the raw materials.
3. Making (stamping) the cookie sheet from the steel strips.

STEM Connections: Science

Integrated STEM Curriculum — Science, Technology, Engineering, Mathematics

Genetic Engineering

The technology behind today's genetically engineered food products can be traced to scientific experiments conducted a century and a half ago. In the mid-1800s, Gregor Mendel, an Austrian botanist and monk, began experimenting with the breeding of garden peas. He studied the traits of various pea plants. Mendel then started breeding and crossbreeding these plants to see what characteristics would be reproduced in succeeding generations. His findings on dominant and recessive genes led to the development of the science of genetics.

We now use the principles Mendel developed to produce better and more cost-effective foods. For example, because of genetic engineering, dairy cows injected with a growth hormone developed from genetically engineered bacteria produce greater amounts of milk. Beef cattle injected with the hormone have leaner meat. Genetically altered bacteria have also been used to make crops more insect resistant.

Genetic engineering has been used to great benefit. Some people have expressed concerns, however, about the practice. Can you think of a problem that might result from genetic engineering? If one does not immediately come to mind, you might want to check your local newspaper. The pros and cons of various types of genetic engineering are in the news frequently.

©iStockphoto.com/AndreasReh

©iStockphoto.com/thegoodphoto

Figure 3-9. The production of each manufacturing product, such as these automotive wheels, involves a variety of steps. Each step depends on agricultural and related biotechnologies and communication and information, construction, energy and power, manufacturing, medical, and transportation technologies.

These manufacturing activities could not exist without the other technological systems. For example, communication and information technologies play a role at every point in the product's development. The need for the new product is communicated through sales orders. The specifications for the steel and the cookie sheet are communicated through engineering drawings and specification sheets. The availability of the product is communicated to potential customers through advertising. Sales reports communicate product success to the company's management.

Energy and power technologies help to power the melting furnaces and stamping presses, light the work areas, and heat the offices and control rooms. Agricultural and related biotechnologies facilitate the growth of trees. The trees are used to make the pallets containing the boxes of finished products. Medical technologies help to ensure that the workers remain in good health.

Likewise, construction technologies are essential. Constructed roads create access to the iron-ore and coal mines and to the limestone quarries. Workers travel to the steel mill and the production factory on constructed roads and work in constructed buildings. Constructed power lines bring electricity to the various manufacturing sites. Constructed dams and pipelines make water available.

Transportation technologies also play a major role. They move the raw materials to the steel mill and the steel strips from the mill to the product manufacturer. These technologies deliver finished products to stores. Customers use private cars or public transportation systems to visit a store to purchase the product.

Throughout the rest of this book, we will focus on these seven major technological systems. We will examine the components for each of these systems. Also, we will explore the productive processes in greater depth.

Standards for Technological Literacy

Brainstorm

Ask your students how all the technological areas were used in making the pencils that they use in class.

Reinforce

Review how the phrases "technology" and "technology as a system" that were discussed in Chapters 1 and 2 are related to the types of technology presented in this chapter.

TechnoFact

An example of a technological device using many technologies is a space probe. It is a vehicle that carries complex instrumentation but has no crew. Factories, which were built using construction techniques, manufacture space probes. A rocket that used chemical energy transports it in space, and using electronic signals, it communicates information back to earth.

TechnoFact

Power technology is an example of technology that is developed before its time, becomes obsolete, and then comes back into use. All power came from renewable sources of wind, water, and wood until the 1200s. Coal was then used, and steam power became the central energy source. With such energy sources as fossil fuels and nuclear power, many people now advocate the return to renewable energy sources.

Answers to Test Your Knowledge Questions

1. Level of development, economic structure, number of people involved, and the type of technology developed and used
2. Evaluate individually.
3. True
4. Research and development, production, and marketing
5. financial affairs
6. Evaluate individually.
7. False
8. People are united and combined into one body. They share a common vision, work toward similar goals, and share in the success of the enterprise.
9. F. Medical
10. C. Construction
11. E. Manufacturing
12. A. Agricultural and related biotechnologies
13. G. Transportation
14. B. Communication and information
15. D. Energy and power
16. A. Agricultural and related biotechnologies
17. B. Communication and information
18. F. Medical
19. B. Communication and information
20. Evaluate individually.

Summary

Technology has always been a part of human existence. As civilization has become more advanced, technology has evolved. Older technologies are often replaced with newer, more efficient technologies. The high tech of today often becomes the current technologies of tomorrow.

Most current technology is developed and applied by industrial and business enterprises. These companies hire research and development personnel to design and develop these technologies. The technologies are then put into use through production activities and sold through marketing efforts. A desire to profit from the development and application of technologies drives the system. Some technology is also supported by government agencies. Most technology is developed by groups of people working together in a cooperative manner. Technologies are commonly developed to help people grow food, communicate information, construct structures, convert energy, manufacture products, improve health, and transport people and cargo. The systems are closely related and work together to provide better lives for everyone.

Test Your Knowledge

Write your answers on a separate piece of paper. Please do not write in this book.

1. List the four general ways technology systems can be categorized.
2. Give one example each of an obsolete technology, a current technology, and an emerging technology.
3. Emerging technology is also referred to as *high tech.* True or false?
4. Name the three stages a technological development goes through in a profit-centered company.
5. The two resources needed to support the stages of technological development are industrial relations and _____.
6. Give an example of a government-sponsored technology.
7. Most technology is developed by individuals attempting to improve their personal situations. True or false?
8. Define the term *corporate participation.*

Matching questions: For Questions 9 through 19, match each area of use on the left with the correct technological system on the right. Some letters will be used more than once.

9. _____ Used to maintain health.
10. _____ Used to build structures.
11. _____ Used to convert materials into products.
12. _____ Used to grow food.
13. _____ Used to move people.
14. _____ Used to process data.
15. _____ Used to apply energy.
16. _____ Used to produce natural fibers.
17. _____ Used to communicate ideas.
18. _____ Used to cure illness.
19. _____ Used to transmit information.

A. Agricultural and related biotechnologies.
B. Communication and information technologies.
C. Construction technologies.
D. Energy and power technologies.
E. Manufacturing technologies.
F. Medical technologies.
G. Transportation technologies.

20. Give a specific example of how one technological system is related to another.

STEM Applications

1. Develop a chart similar to the following one. List five technological devices from each type of technology that you use daily.

Technology	Technological Device
Agricultural and related biotechnologies	
Communication and information technologies	
Construction technologies	
Energy and power technologies	
Manufacturing technologies	
Medical technologies	
Transportation technologies	

2. Identify a new technological device you can use. Sketch what it would look like and how it would work.
3. Ask older people (grandparents or neighbors) to tell you about five technological devices they used in their lifetimes that are no longer around. List these devices (obsolete technologies), the devices that have replaced them (current technologies), and the devices that you think will replace the current ones (emerging technologies).

Section 1 Activities

Activity 1A

Design Problem

Background

Technology is the application of knowledge to create machines, materials, and systems to help us make work easier, make life more comfortable, or control the natural or human-made environment.

Situation

The Easy-Play game company has developed a new board game for two players. The game uses five red marbles and five green marbles as the playing pieces. The company has found that counting out the marbles one by one is too costly.

Challenge

Design a technological device (machine) that counts five marbles at a time from a box containing a large number of marbles.

Activity 1B

Fabrication Problem

Background

All products of technology have been developed to meet needs or opportunities. Most early technological products answered functional needs. Some of the earliest products of technology were clay containers for transporting liquids and storing food.

Challenge

Divide your class into two groups. Each group will use different clay-forming techniques to produce a clay pot and lid. Group 1 will form the two parts without the aid of any external devices, and Group 2 will use a form to aid in the production process. The form Group 2 will use is a frozen orange-juice concentrate can. See **Figure 1B-1.** The product should be the diameter of the orange-juice can and 1 1/2″ tall. The lid should be 1/2″ tall.

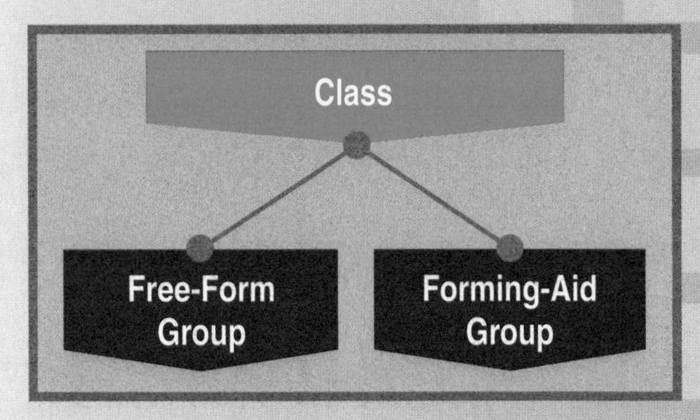

Figure 1B-1. The pot- and lid-forming groups.

Materials and Equipment

- Potter's clay.
- Frozen orange-juice concentrate cans.
- A sponge.
- A table knife.

Procedure

Produce a clay pot and lid using the procedure that has been assigned to your group.

Free-form group

1. Obtain the supplies.
2. Study **Figure 1B-2** to see the procedure that will be used to make the pot and lid.
3. Separate a portion of clay.

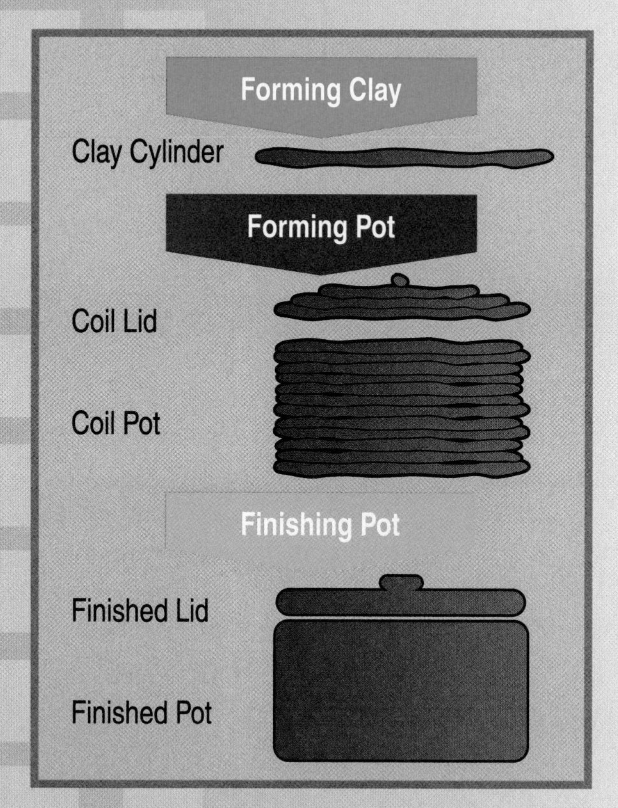

Figure 1B-2. The major stages in free-forming a pot and lid.

4. Roll several 3/8″-diameter strips of clay.
5. Coil a strip of clay flat on the tabletop until a disk the diameter of the juice can is produced. This disk is the bottom of the pot.
6. Build the wall of the pot by layering strips of clay until the 1 1/2″ height is reached.
7. Use a damp sponge and your fingers to smooth the clay and form the finished pot.
8. Coil a strip of clay flat on the tabletop until a disk the diameter of the juice can is produced. This disk is the lid for the pot.
9. Gently shape the disk into a concave shape 1/2″ tall.
10. Form a handle (1/2″ ball). Attach it to the center of the lid.
11. Use a damp sponge and your fingers to smooth the clay and form the finished lid.
12. Allow the pot to air-dry.
13. Fire the pot in a kiln. (Optional)

Forming-aid group

1. Obtain the supplies.
2. Separate a portion of clay.
3. Roll several 3/8″-diameter strips of clay.
4. Coil a strip of clay flat on the tabletop until a disk just a bit larger than the diameter of the juice can is produced. This disk is the bottom of the pot.
5. Use the can as a cookie cutter to cut a perfectly round disk.
6. Place the can on the tabletop with the open end down.
7. Build the wall of the pot by loosely coiling strips of clay around the can until the 1 1/2″ height is reached. Do not push the clay against the can because it will stick.
8. Carefully slide a table knife between the can and pot to separate them.
9. Remove the can.
10. Place the pot wall on top of the bottom.
11. Use a damp sponge and your fingers to smooth the clay and form the finished pot.
12. Coil a strip of clay flat on the tabletop until a disk the diameter of the juice can is produced. This disk is the lid for the pot.
13. Gently shape the disk into a concave shape 1/2″ tall.
14. Form a handle (1/2″ ball). Attach it to the center of the lid.
15. Use a damp sponge and your fingers to smooth the clay to form the finished lid.
16. Allow the pot to air-dry.
17. Fire the pot in a kiln. (Optional)

Analysis

Meet as a class. Analyze the two clay-forming techniques. Answer the following questions:

1. Do you think the free-form method is a common technique for producing modern ceramic products? Why or why not?
2. Which method is easier to use?
3. Which method produces a higher-quality pot?

Section 2
Technological-System Components

Tomorrow's Technology Today

Home Fuel Cells

Have you ever thought about where electric power comes from? It is transmitted into homes, schools, and a great deal of other buildings. Our society would be very different without it. It provides light in the dark, and it powers everything from refrigerators to computers. Nearly half of the energy consumed in the United States is used for electricity.

Steam-powered electrical plants are commonly used to generate electricity. Fossil fuels are frequently used. The fuels used in this type of energy conversion are coal, natural gas, or fuel oils. These resources are burned to produce thermal energy. The thermal energy must then be converted into mechanical energy in order to provide power to a generator.

This method of electricity generation consumes great amounts of fossil fuels while emitting harmful carbon dioxide into the atmosphere. One alternative to this conventional method of electrical energy production is the use of home fuel cells. A fuel cell uses a chemical reaction to produce electricity. Although home fuel cells have been in the works for some time, they are now becoming a reality.

There are several companies attempting to create home fuel cells. One type of fuel cell uses solid oxide fuel cells to generate electricity. The fuel cells use the reaction between oxygen and natural gas to generate power. Biogas could also be used instead of natural gas, and that would completely eliminate any carbon dioxide emissions from using fuel cells. Some larger businesses in the United States are already using fuel cells to generate their electricity.

The companies attempting to create home fuel cells are building them using solid oxide fuel cells. Solid oxide fuel cells have a lower cost and are more efficient than proton exchange membrane fuel cells. Solid oxide fuel cells use a ceramic electrolyte. Such companies working to produce home fuel cells have begun working on more efficient alternatives to this electrolyte, however. Companies have been looking for lower-cost electrolytes to make home fuel cells more affordable for the general public.

Attempts have been made in the past to manufacture inexpensive home fuel cells for residential use. However, the companies producing fuel cells typically spend great amounts of money to create them. With companies working to find less expensive ways to create home fuel cells, however, we may see them in residential use soon.

Discussion Starters

Discuss the use of fuel cells for residential use.

Group Activities

Divide the class into groups of three or four students. Have each group research solid oxide fuel cells and proton exchange membrane fuel cells. Have each group compare and contrast the two types of fuel cell technology.

Writing Assignment

Have your students research how solid oxide fuel cells have been used to create home fuel cells by various companies. Have them write a short paper discussing the use.

Chapter Outline
People
Tools and Machines
Materials
Information
Energy
Finances
Time

This chapter covers the benchmark topics for the following Standards for Technological Literacy:

Learning Objectives

After studying this chapter, you will be able to do the following:

- Recall the inputs to technological systems.
- Summarize the types of skills and knowledge various groups of people bring to technological systems.
- Recall the types of tools and machines used as inputs to technological systems.
- Summarize the types and properties of materials that are inputs to technological systems.
- Recall the types of information that are inputs to technological systems.
- Recall the major types and sources of energy used as inputs to technological systems.
- Recall the sources of finances used as inputs to technological systems.
- Summarize the importance of time, with regard to technological systems.

Key Terms

composite material
consumer
corporation
creativity
data
debt financing
distance multiplier
engineer
entrepreneur
equity financing
exhaust
exhaustible material
first-class lever
force multiplier
fulcrum
gas
genetic material
hand tool
inclined plane
inexhaustible
inorganic material
knowledge
lever
lever arm
liquid
manager
mechanic
natural material
organic material
partnership
production worker
pulley
renewable
scientist
screw
second-class lever
sole proprietor
solid
support staff
synthetic material
technician
third-class lever
wedge
wheel and axle

Strategic Reading

As you read the chapter, outline the hierarchy of the different groups of people involved in technological systems. Where do you fit in?

Humans have lived on the Earth about 2.5 million years. This might seem to be a long time, but it is relatively short, considering the Earth is about 5 billion years old. In the short span of human history, people have developed a special ability. They have learned how to build and use tools, machines, systems, and materials to change their environment. You learned in Chapter 1 that this process is called *technology*. This ability to develop technology has led to many kinds of technological systems. You learned in Chapter 2 that all systems include several components. These components are inputs, processes, outputs, feedback and control, and goals. This chapter explores in more depth the inputs that are common to all technological systems.

As you might recall from Chapter 2, all inputs can be grouped into seven major categories. See **Figure 4-1.** These categories are the following:

- People.
- Tools and machines.
- Materials.
- Information.
- Energy.
- Finances.
- Time.

The inputs are the resources used to make a system operate. They are the elements that are changed by technological processes or are used by technology to change other inputs. Let us look at each one separately.

People

As stated in Chapter 2, people are a key input to technological systems. People produce these systems. Their minds create and design the systems and the systems' outputs. People use their skills to make and operate the systems. Their management abilities make systems operate efficiently. The systems satisfy people's needs and wants. Therefore, it is easy to see why people are fundamental to technology.

Various groups of people bring different knowledge, skills, and abilities to technological systems. For example, ***scientists*** generally develop the basic knowledge needed to help create products and processes. See **Figure 4-2.** Fields of knowledge scientists use include physics, materials science, geology, and chemistry.

Standards for Technological Literacy

1 3 6 7

Reinforce

Discuss technology as described in Chapter 1. Explain what a system is, as it is described in Chapter 2.

Brainstorm

Have your students list examples of each of the seven types of inputs that were used to produce their textbooks.

TechnoFact

A combination of parts forming a complex whole used in many daily activities is a system. The measurement system is a common system. Distance and other factors are divided into separate parts. In this way, the system allows us to measure whole objects and events or their parts.

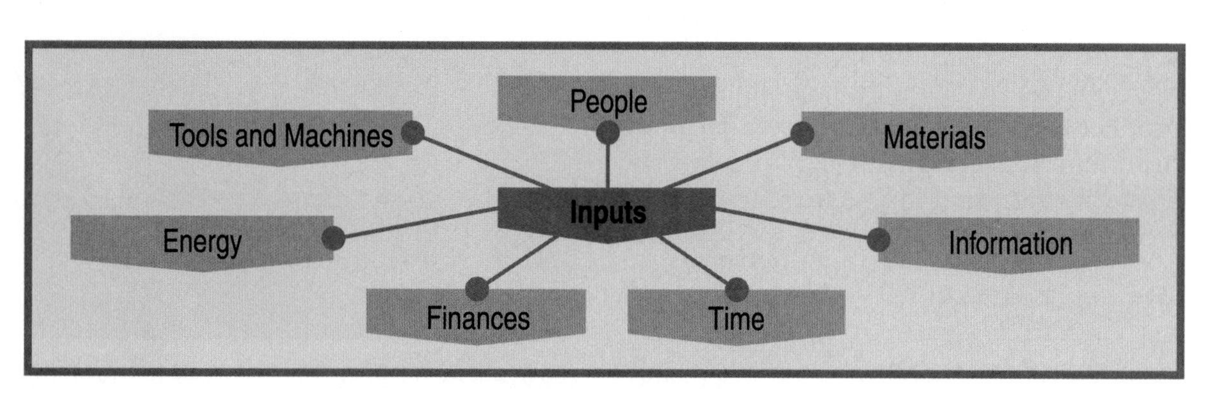

Figure 4-1. Inputs to technological systems can be grouped into seven major categories.

Standards for Technological Literacy

1 3 6 7 19

Extend

Discuss how scientists and engineers work together to solve technological problems.

Example

Discuss Henry Ford as an example of an entrepreneur industrialist. Ask the students to name others.

TechnoFact

University training is generally required for the engineering profession. Machines, structures, and other devices that are used in industry and everyday life are designed by engineers.

Northern Telecom

Figure 4-2. Scientists are involved in various fields of knowledge when creating technological products and processes.

Engineers apply scientific and technological knowledge in designing products, structures, and systems. See **Figure 4-3.** They determine appropriate materials and processes needed to produce products or perform services. For example, civil engineers determine the correct structure for a bridge to carry vehicles across a river. Electrical engineers design circuits for computers.

Still other people work at such tasks as harvesting and processing foods, building products and structures, producing communication messages, treating illnesses, and transporting people and cargo. These people can be referred to as ***production workers***. Production workers in manufacturing and construction are often categorized as unskilled, semiskilled, and skilled workers. The specific designation depends on one's training and work experience. Skilled workers in laboratories and product-testing facilities are usually called ***technicians***. See **Figure 4-4.** Skilled workers in service operations are often called ***mechanics***.

Other groups of people establish or help direct businesses. Those who create businesses are called ***entrepreneurs***. They have a vision of what can be done and are willing to take risks to see it happen.

©iStockphoto.com/lisafx

Figure 4-3. This engineer is marking up plans for a new building.

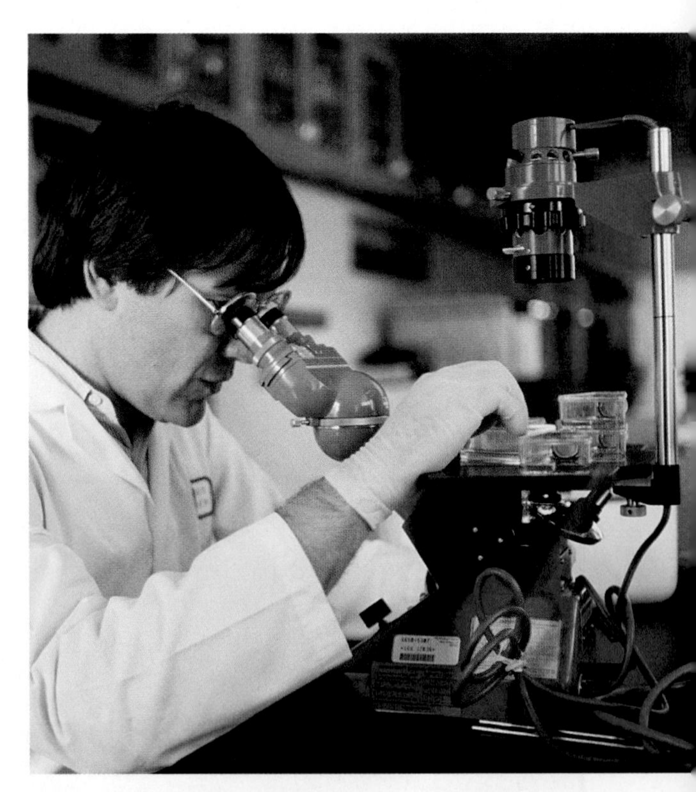

Figure 4-4. This skilled worker is involved with medical technology.

Career Corner

Construction Laborers

Construction laborers perform a wide range of tasks involving building and maintaining constructed works. These works can include buildings, highways, tunnels, and dams. Construction laborers prepare construction sites, dig holes and trenches, mix concrete, unload buildings, and tend machines. They often help concrete workers, carpenters, plasterers, and masons.

Most laborers do physically demanding work that requires lifting and carrying heavy objects. They often work outdoors in all weather conditions. Construction laborers generally work eight-hour shifts, although longer shifts are common. Also, their work is often seasonal, working when the weather permits construction activity. Many construction-laborer jobs do not require experience or training related to the occupation. Most laborers learn skills by observing and learning from experienced workers.

Standards for Technological Literacy

1 3 6 7 19

Research

Have your students select a type of job and then use the U.S. Department of Labor Web site to research the qualifications for the job.

TechnoFact

In the 19th century, the term entrepreneur came about to describe employers who assumed the management and risk of businesses, as opposed to capitalists, who did not take part in the day-to-day operations of the enterprises they owned.

Managers organize and direct the work of others in these businesses. They set goals, structure tasks to be completed, assign work, and monitor results. Nonmanagerial ***support staff*** bring other knowledge to the businesses. They carry out such tasks as keeping financial records, maintaining sales documents, and developing personnel systems.

The final group of people involved in technological systems is ***consumers***. See **Figure 4-5.** They are the reason for the system in the first place. Their attitudes about styling, price, and service directly affect the systems' outputs. Consumers' money is spent on the products or services. Therefore, consumers financially support the system.

Figure 4-5. Consumers are fundamental to all technological systems.

One element common to the development of technology within all these groups of people is ***creativity***. Creative people have the ability to see a need or a way of making life easier. They design systems and products to meet the need or desire. In some instances, they develop ideas for a product

Standards for Technological Literacy

Research

Have your students research the tools that Stone Age people used.

TechnoFact

The earliest tools were made by Stone Age people. In various times in different places, they abandoned stone for metal tools.

and decide what size, shape, and color the product will be. See **Figure 4-6.** Their decisions can add beauty to the world through well-designed products and structures. Creative people's innovations can lead to improvements in existing technology.

One of the more famous examples of a creative person is Art Fry, a chemist who is now retired from the 3M Company. Fry was aware that another scientist at 3M was attempting to invent a stronger adhesive for some of the company's tapes by experimenting with various molecules. This scientist, instead, discovered a new adhesive—one that adheres to paper, but is not as sticky as tape. He was not sure what to do with this discovery, but he did mention it to other employees at 3M. A short time later, Mr. Fry, who was a member of his church's choir, was in church and, as usual, was trying not to get frustrated as he marked his place in the songbook with little pieces of paper that invariably fell out of the book. As the service continued, he suddenly thought—he could make a bookmark with the new adhesive—one that would stick to the page, but could then be removed.

American Greetings

Figure 4-6. This person used her creative abilities to design a greeting card.

He prepared samples of his idea and distributed them to people at 3M to use and react to. In that process, he and his manager realized they had something more than a bookmark. They had a new way of communication and organizing information: self-attaching notes, which we now call *Post-it® notes.* At this point, other workers at 3M, with the support of management, joined the effort to design and produce this new item. After testing various forms and solving a series of manufacturing problems, they came up with a product that many people cannot imagine being without.

We can see from this example that various groups of people contributed to this invention in different ways. Some products create their own demand. In this example, we can see that some changes in society lead to even more needs and demands. In this case, 3M has expanded its product line in the Post-it area to include such items as flip charts, tape flags, and room-decorator kits. More items are sure to come.

Tools and Machines

Humans are the only species on Earth that can develop and use technology because, as we said earlier, humans are tool builders and users. Tools are the artifacts that expand what humans are able to do. From early humans on, people have used tools to increase their ability to do work. See **Figure 4-7.** Today, tools include such diverse artifacts as milking machines used on dairy farms, machine tools used in factories, hammers and saws used on construction sites, medical equipment used in hospitals, and automobiles used to transport people and cargo.

One way to look at tools is by their area of activity. See **Figure 4-8.** For example, microscopes and telescopes are used in scientific activities. Tennis rackets and pitching machines are used in recreational activities.

Figure 4-7. Humans have long used tools to increase their ability to do work.

Another way to group tools is by the technological system in which they are used. For example, some people in agriculture use plows, some people in communication use radio transmitters, and some workers in the field of transportation use trucks. As we noted in Chapter 2, however, the most common way we classify tools is to divide them into the two categories of hand tools and machines. Each category is described in more depth below.

Standards for Technological Literacy

1 3 6 7 19

Brainstorm

Ask your students what tools they would need to make a birdhouse. Have the students identify the category of each tool.

Hand Tools

Almost every technology uses a common set of hand tools to produce, maintain, and service products and equipment. ***Hand tools*** are those simple hand-held artifacts requiring human-muscle power, air, or electric power to make them work. These tools can be classified by their

Figure 4-8. Tools are used in various kinds of activities.

Standards for Technological Literacy

TechnoFact

Though the saws, planes, chisels, and clamps of today may look different from those of earlier times, most woodworking tools have not changed in character since their development in the 17th century. They are easy to recognize for what they were made to do.

purpose. See **Figure 4-9.** These categories are the following:

- **Measuring tools.** These tools are used to determine the size and shape of materials and parts.
- **Cutting tools.** These tools are used to separate materials into two or more pieces.
- **Drilling tools.** These tools are used to produce holes in materials.
- **Gripping tools.** These tools are used to grasp and, in many cases, turn parts and fasteners.
- **Pounding tools.** These tools are used to strike materials, parts, and fasteners.
- **Polishing tools.** These tools are used to abrade and smooth surfaces.

Machines

Machines are artifacts that transmit or change the application of power, force, or motion. They can be simple or complex, with complex machines being made up of more than one simple machine. Simple machines work on two basic principles—the principle of the lever and the principle of the inclined plane—and can be grouped under six categories. See **Figure 4-10.** The categories are the following:

- Lever.
- Wheel and axle.
- Pulley.
- Inclined plane.
- Wedge.
- Screw.

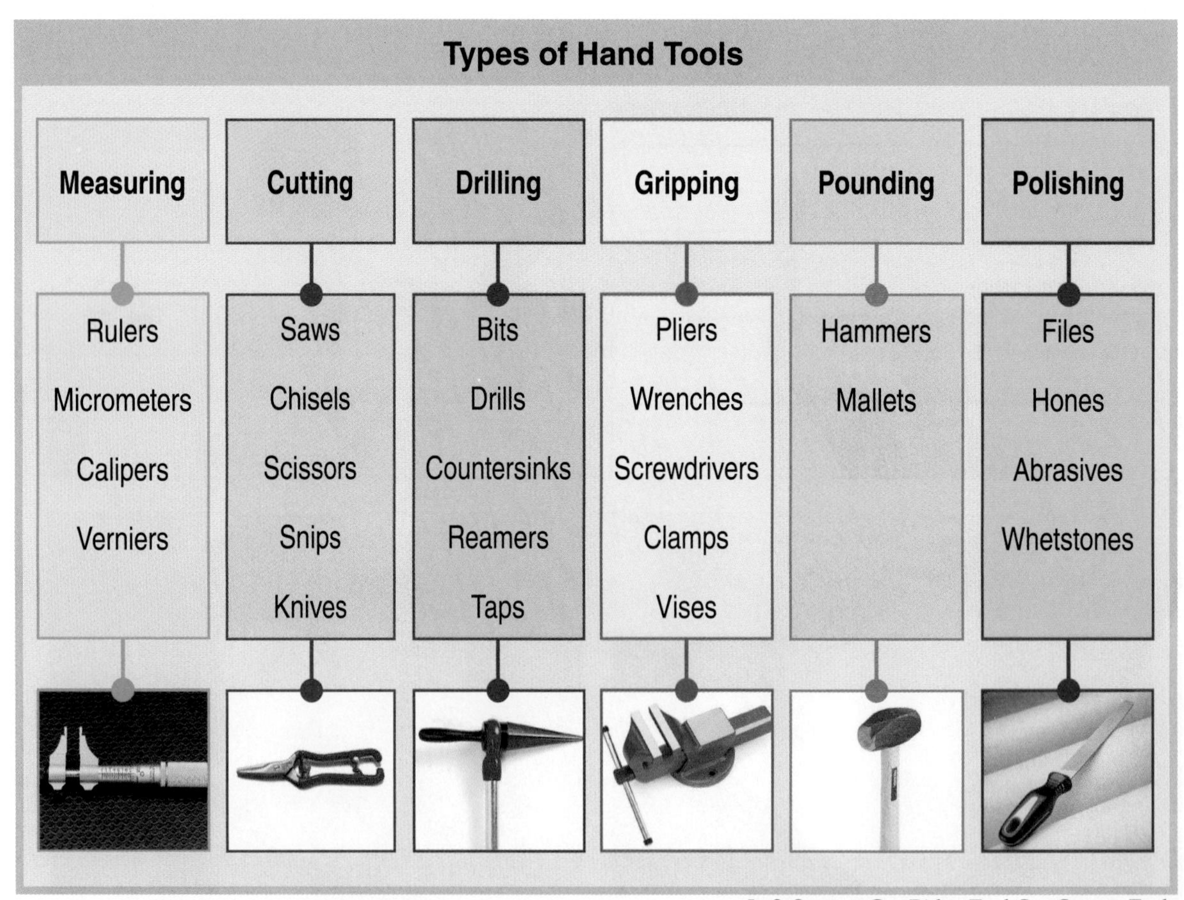

L. S. Starrett Co., Ridge Tool Co., Cooper Tools

Figure 4-9. Hand tools can be grouped by the actions they perform.

Simple Mechanisms	Principle
Lever Wheel and Axle Pulley	Lever 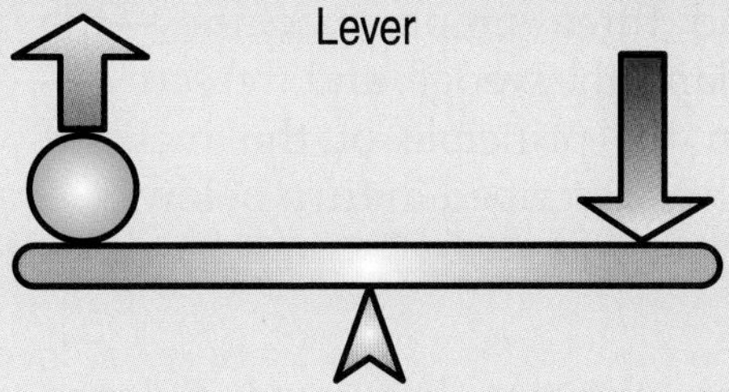
Inclined Plane Wedge Screw	Inclined Plane 

Figure 4-10. The six types of mechanisms work on two basic principles.

Standards for Technological Literacy

1 3 7 19

Extend

Draw a picture of a shovel and explain why it is a lever.

STEM Connections: Mathematics

The Law of Equilibrium

A Greek mathematician who lived more than 2000 years ago actually proved the law behind the workings of the lever. Archimedes had observed how a small force can move a great weight. From this observation, the following law of equilibrium was created: A lever is in equilibrium when the product of the weight (w_1) and distance (d_1) on one side of the fulcrum (the center of gravity) is equal to the product of the weight (w_2) and distance (d_2) on the other side of the fulcrum. The mathematical formula is: $w_1 \times d_1 = w_2 \times d_2$.

Thus, if Roberto, who weighs 150 pounds, is 2′ from the fulcrum of a seesaw, how far from the fulcrum would Becky, who is 60 pounds, have to sit to achieve balance? Using the above formula, where w_1 is 150, d_1 is 2, and w_2 is 60, we can calculate the distance (d_2) as follows:

$60 \times d_2 = 150 \times 2$

$60 \times d_2 = 300$

$d_2 = 300 \div 60$

$d_2 = 5$

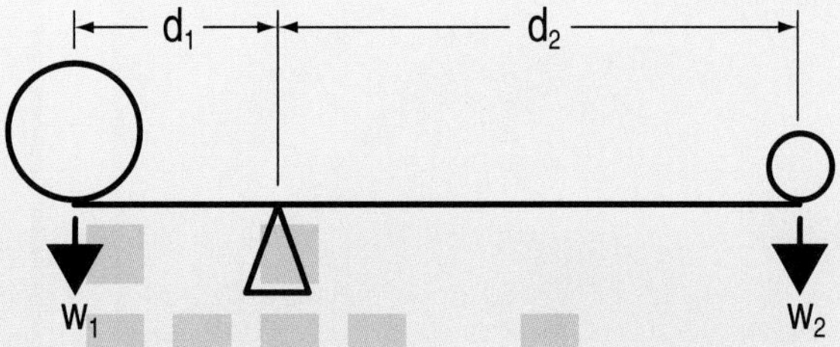

Thus, Becky would have to sit 5′ from the fulcrum. If Isaac, who weighs 120 pounds, is 3′ from the fulcrum, how far from the fulcrum would Josie, who is 90 pounds, have to sit to achieve balance?

Standards for Technological Literacy

Research

Have your students research how first-class, second-class, and third-class levers are used in technological devices.

TechnoFact

A machine that combines the lever and the wheel is the wheelbarrow. The operator lifts only a small part of the load because it is centered behind a single wheel. The person easily controls its movement with the two handles.

The first three simple machines—the lever, the wheel and axle, and the pulley—operate on the basic principle of the lever. The second three simple machines—the inclined plane, the wedge, and the screw—operate on the principle of the inclined plane. Each is described in turn below.

Levers

Almost everyone has used a ***lever***. If you have ever pried open a crate with a crowbar or pulled a nail with a claw hammer, you have used a lever. A lever has a rod, or bar, (the ***lever arm***) that rests and turns on a support. This support is called a ***fulcrum***. See **Figure 4-11.** You apply a force at one end of the rod, or bar, to lift a load at the other end. The purpose is to help lift weight more easily.

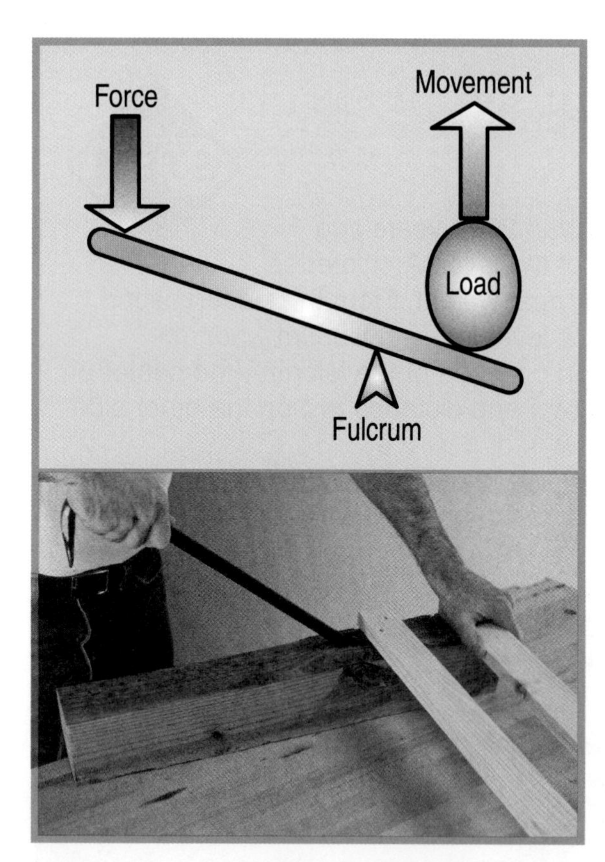

Figure 4-11. Levers use a lever arm and a fulcrum, to which a force is applied and a load is moved. The crowbar uses the principle of the lever.

Levers are grouped into three categories: first class, second class, and third class. See **Figure 4-12.** Each class of lever applies force differently to move the load. In ***first-class levers***, the fulcrum is between the load and the effort. A pry bar is an example

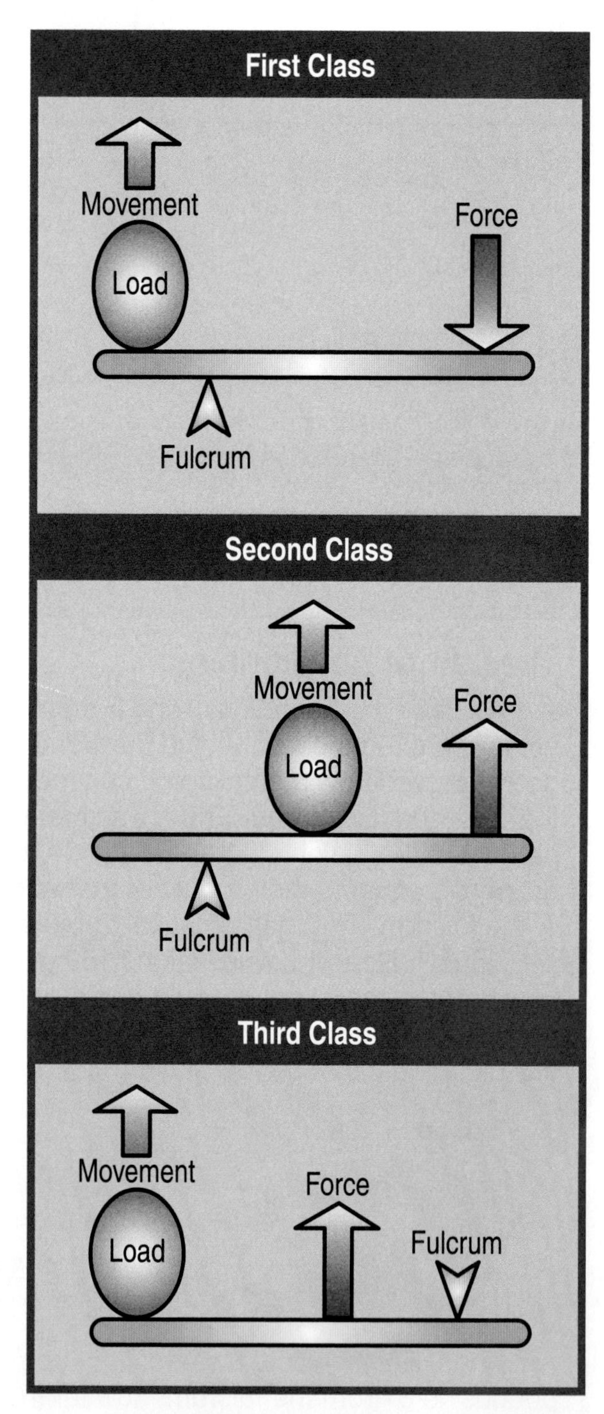

Figure 4-12. The three classes of levers are first class, second class, and third class.

of a first-class lever. In ***second-class levers,*** the load is between the effort and the fulcrum. The wheelbarrow uses the principle of a second-class lever. In ***third-class levers,*** the effort is placed between the load and the fulcrum. A person moving dirt with a shovel applies the principle of a third-class lever.

These simple machines can be either force multipliers or distance multipliers. When levers increase the force applied to the work at hand, they are ***force multipliers.*** The fulcrum is close to the load, and the force is applied at the other end. We, thus, can move a heavy load with a light force.

A distance-multiplier lever is just the opposite. When a lever is a ***distance multiplier,*** the fulcrum is close to the force, and the load is at the other end. The load moves a greater distance than the force, but a large force is required to move a light load. These two applications of levers are shown in **Figure 4-13.**

Wheel and Axles

A ***wheel and axle*** is a shaft attached to a disk. This simple machine acts as a second-class lever. The shaft, or axle, acts as the fulcrum. The circumference of the disk acts as the lever arm. If the load is applied to the shaft, the wheel and axle becomes a force multiplier. See **Figure 4-14.** Automotive steering wheels use this principle. A 15″ wheel attached to a 1/4″ shaft multiplies the force by 60 times.

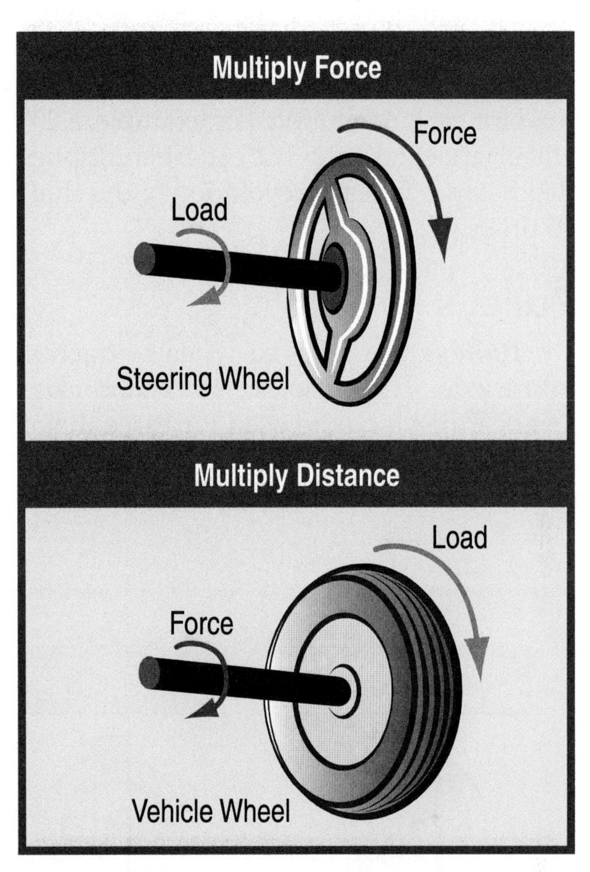

Figure 4-14. A wheel and axle can multiply force or distance.

Standards for Technological Literacy

1 3 7 19

Extend

Describe ways levers are used as force multipliers and distance multipliers in technological devices.

Reinforce

Ask your students why rubber tires, rough surfaces, or gears are added to wheels that are used to transmit power.

TechnoFact

According to archaeological evidence, the first wheel was invented in an area that is now Iraq and Iran about 5500 years ago.

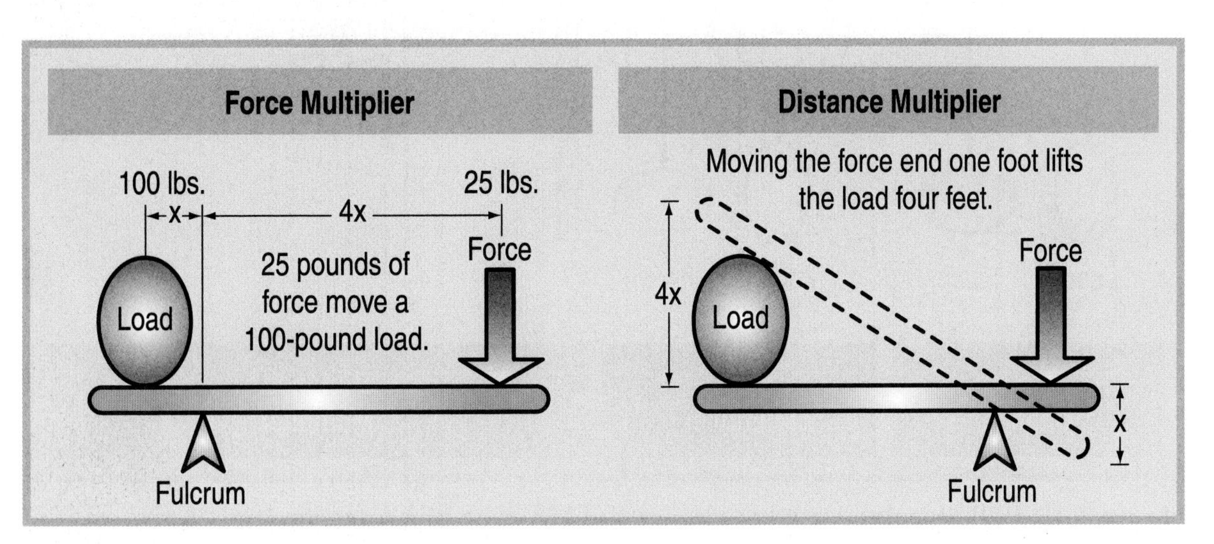

Figure 4-13. Levers can be used to multiply force or distance.

Standards for Technological Literacy

1 3 7 19

Extend

Describe how a pulley uses the principle of a wheel and axle.

Research

Have your students identify the ways wheel and axle mechanisms are used as force multipliers and distance multipliers in technological devices.

Figure Discussion

Show how pulleys can be used to change direction, multiply force, or multiply distance as shown in Figure 4-15.

TechnoFact

A steam engine or water turbine might have run all the machines in a factory before the use of electric motors and power machines. The power source was connected to a system of pulleys, shafts, and belts. The drive shaft consisted of the central shaft and pulleys. Flat belts and pulleys connected the drive shaft to each machine. To keep the belt centered, the pulleys were crowned (high in the center). To provide a variety of speeds, the size of the drive and machine pulleys varied.

TechnoFact

The principle of the inclined plane is used in woodworking planes, drill bits, and chisels. Switchbacks on steep hiking trails and mountain roads, though increasing travel distance, reduce the effort that must be expended at any given time to reach the top.

If the load is applied to the disk, the wheel and axle becomes a distance multiplier. Automobile transaxles use this type of wheel and axle. One revolution of the axle causes the wheel to revolve one time. The circumference of the wheel is many times that of an axle. Therefore, the vehicle moves a considerable distance down the road for each revolution. For example, a 20″ wheel attached to a 1/2″ shaft multiplies the distance for each revolution of the shaft 40 times.

Pulleys

Pulleys are grooved wheels attached to an axle. They also act as second-class levers. Pulleys can be used for three major purposes. See **Figure 4-15.** A single pulley can be used to change the direction of a force. Two or more pulleys can be used to multiply force or to multiply distance. The number and diameters of the pulleys used determine the mechanical advantage (force multiplication) of a pulley system.

Inclined Planes

Inclined planes are sloped surfaces used to make a job easier to do. The principle of the three simple machines in this category (inclined plane, wedge, and screw) is that it is easier to move up a slope than up a vertical surface. See **Figure 4-16.** The simplest application of this principle is the inclined plane. The inclined plane is used to roll or drag a load from one elevation to another. Common examples of inclined planes are roadways in mountains and ramps to load trucks.

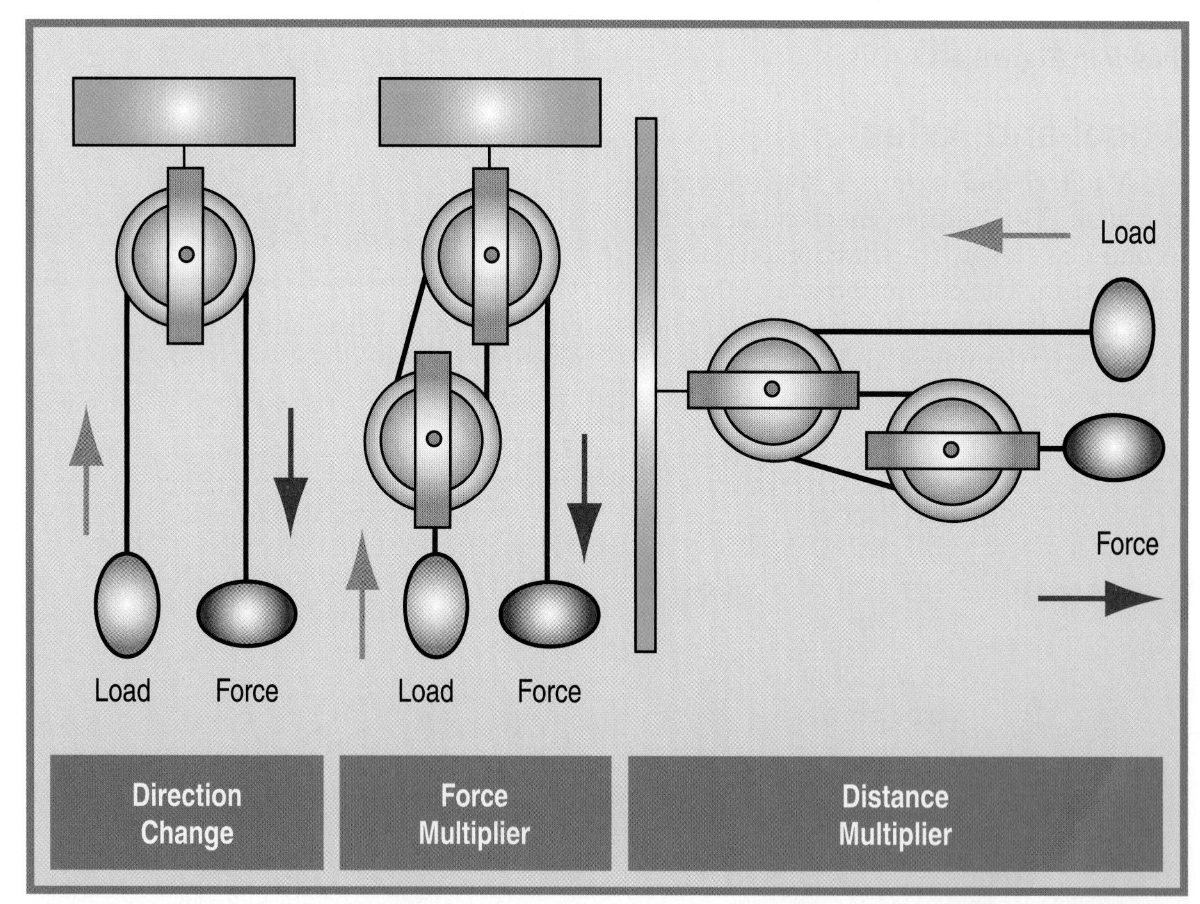

Figure 4-15. Pulleys can be used to change the direction of a force, multiply force, or multiply distance.

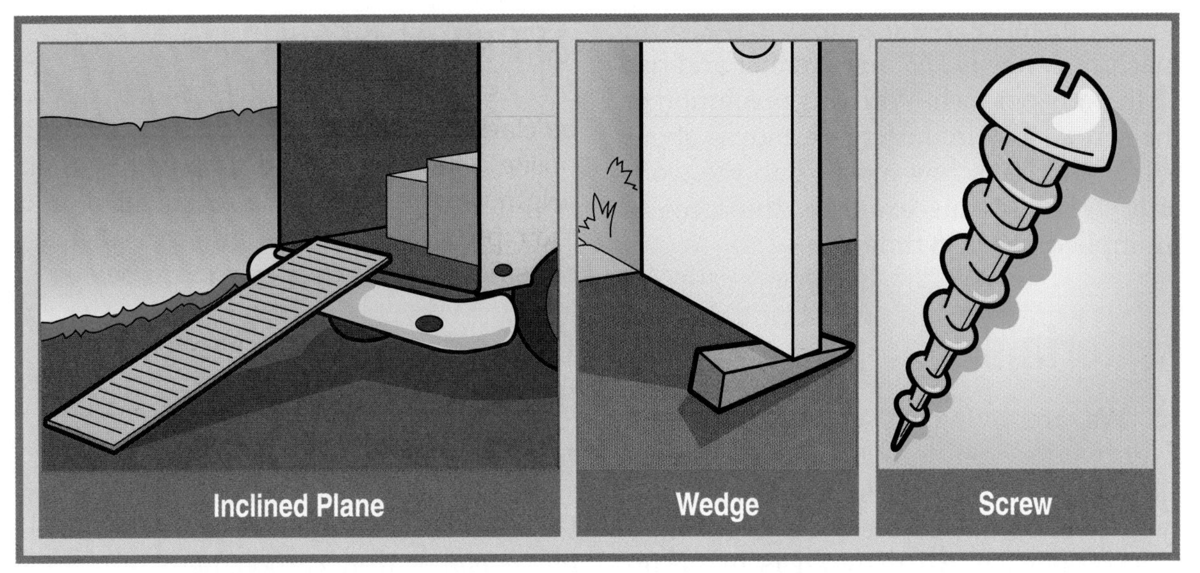

Figure 4-16. The three types of inclined plane mechanisms are the inclined plane, the wedge, and the screw.

Standards for Technological Literacy

1 3 5 7 19

Research

Have your students research the origin of the screw.

TechnoFact

For over 2200 years, screw threads have been used in machines. Ancient screws were threads wound around a straight cylinder. The development of tapered screw threads, which are common today, was a major advancement.

Think Green

Organic Cotton

Traditional farming of cotton uses more pesticides than farming of other types of crops. Farming cotton this way can cause harm to the environment, and could potentially hurt the people who work in these fields or who use products made from that cotton. The chemicals in the pesticides get into the ground, the water, and the air, causing contamination. Even wildlife near these cotton fields may be affected. Cotton is a common material used in making several everyday items, including clothing, towels, and sheets.

Organic cotton is grown without the use of pesticides or fertilizer. The farmers of these crops do not take the health risks other cotton farmers take. This approach is also better for the land, the water supply, and the surrounding wildlife. In recent years, the number of organic cotton crops have increased. As a result, major manufacturers have started using organic cotton in their products, and the products can be found in several stores.

Wedges

A second application of the inclined plane principle is the ***wedge***. This device is used to split and separate materials and to grip parts. A wood chisel, a firewood-splitting wedge, and a doorstop are examples of this simple machine.

Screws

The ***screw*** is the third simple machine using an inclined plane. This simple machine is actually an inclined plane wrapped around a shaft. The screw is a force multiplier. Each revolution of the screw moves the screw into the work only

Standards for Technological Literacy

1 3 7 16 19

Research

Have your students identify the natural materials produced in your state and rank them by economic importance.

TechnoFact

As a natural material, a spider's web is only 1/30th the diameter of a human hair. Cobwebs were used by the ancient Greeks as an application to wounds. Spider silk was used by astronomers in the 19th century for crosshairs in telescopes.

TechnoFact

Spider silk has some interesting properties. It can be stretched to 140% of its length. However, it isn't elastic. Instead, it absorbs the energy used to stretch it and doesn't bounce back. This property is called inelasticity. Spider silk is stronger than steel. Also, it retains its properties in low temperatures.

a short distance. For example, a 1/2″ × 12 machine screw is 1/2″ in diameter and has 12 threads per inch. With one revolution of the screw, the circumference moves about 1 1/2″, but the screw moves into the work only 1/12″. In this example, the force is multiplied about 18 times.

Materials

We are living in a material world. Everywhere you look, you see materials. They come in all sizes, shapes, and types. Materials possess a number of specific properties. All material is made up of one or more of the elements occurring naturally on Earth. These elements, which number less than 100, combine to produce literally thousands of compounds. To understand materials and to use them effectively, a person must know about the types of materials and the properties materials exhibit.

Types of Materials

As we noted in Chapter 2, materials can be classified as natural, synthetic, or composite. They can also be classified by their origin, their ability to be regenerated, and their physical state. See **Figure 4-17.** Each classification is described in turn below.

Classification of Materials as Natural, Synthetic, or Composite

Many materials occur naturally on Earth. These materials are often called *natural resources.* Iron, carbon, petroleum, and silica are examples of ***natural materials.*** They can be refined and combined to make products.

Other materials are human-made, or synthetic, materials. The most common ***synthetic materials*** are plastics. They are developed and produced from cellulose (vegetable fibers), natural gas, and petroleum.

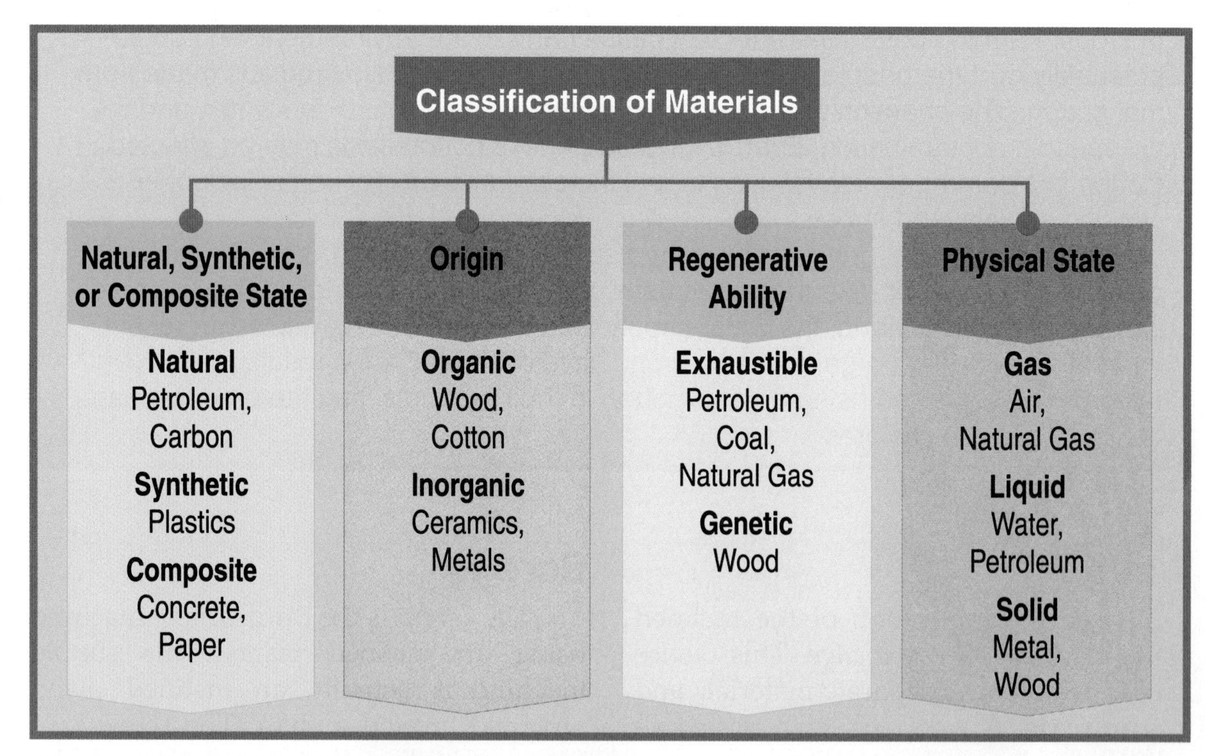

Figure 4-17. Materials can be grouped as being natural, synthetic, or composite and by their origin, ability to be regenerated, and physical state.

Composite materials are combinations of natural and synthetic materials that are mixed to create items with other desirable properties. For example, concrete is a mixture of water, cement, sand, and gravel. Concrete molds to almost any shape and then hardens into a long-lasting material requiring little care.

Classification of Materials by Origin

Materials that come from living organisms are called ***organic materials***. For example, wood, cotton, and flax are products of plant fibers. Wool and leather are products of animals. Petroleum, coal, and natural gas are the products of decayed and fossilized organic materials. All materials that do not come from living organisms are called ***inorganic materials***. For example, metals and ceramic materials are inorganic.

Classification of Materials by Ability to Be Regenerated

Some materials naturally occur on Earth in a specific amount. Human action or nature cannot replace these materials. The quantity of these materials on Earth is finite (limited). Once they are used up, there will be no more. They are called ***exhaustible materials***. Metal ores, coal, petroleum, and natural gas fall into this category of materials.

Other materials have a life cycle and can be regenerated. Human action or nature can produce them. They are called ***genetic materials*** because living things produce them. These materials are the results of farming, forestry, and fishing activities, among others. Wood, meat, wool, cotton, and leather are all genetic materials. The technology associated with producing these materials is called *biorelated technology*.

Still other materials are the products of natural reactions. For example, carbon dioxide and oxygen are the by-products of natural processes. Water is purified through natural processes in lakes, wetlands, and rivers. Natural evaporation produces salt around saltwater.

Classification of Materials by Physical State

Another important way of grouping materials is by their physical states—that is, as gases, liquids, or solids. ***Gases*** are materials that easily disperse and expand to fill any space. They have no physical shape, but they do occupy space and have volume. Gases can be compressed and put into containers. Examples of gases are the air we breathe, the fuel we use for rockets, the carbonation we add to beverages, and the compressed air we use to inflate tires.

Liquids are visible, fluid materials that will not normally hold their sizes and shapes. They cannot be easily compressed. Common liquids include drinking water, fuels used for transportation vehicles, and coolants for industrial processes.

Solids are materials that hold their sizes and shapes. They have internal structures that cause them to be rigid. These materials can support loads without losing their shapes. Solids can be divided into four categories. See **Figure 4-18:**

- **Metallic materials (metals).** Metals are inorganic substances that have crystalline structures—that is, their molecules are arranged in boxlike frameworks called *crystals*. They are the most widely used of all engineering materials. Metals are generally used as alloys—that is, mixtures of a base metal and other metals or nonmetallic materials. For example, steel is primarily an iron-carbon alloy, and brass is a copper-zinc alloy.

Standards for Technological Literacy

1 3 7 19

TechnoFact

The Stone Age started to draw to a close about 5500 years ago when humans started to learn how to work metals. Their first attempts were in hammering shapes out of natural metals such as gold nuggets, chunks of pure copper, or pieces of meteoric iron. These acts were followed by learning how to smelt metal from ores. Smelting copper and making bronze launched civilization into the Bronze Age. People learned to extract pure metals from ores about 3500 years ago and to process iron about 3200 years ago. This new material ushered in a new age, the Iron Age.

Standards for Technological Literacy

1 3 7 19

Brainstorm

Have your students decide which material properties are important to a surfboard, a pencil, and a ceiling tile.

TechnoFact

One of the most unusual uses for composites was in the railroad car wheels developed by Richard Allen shortly after the Civil War. To replace the noisy cast-iron railway wheels that were in use at that time, he developed laminated paper wheels. For about 25 years, George Pullman used the wheels on his luxury railway cars. Steel wheels replaced the paper wheels, however, as cars got heavier and longer.

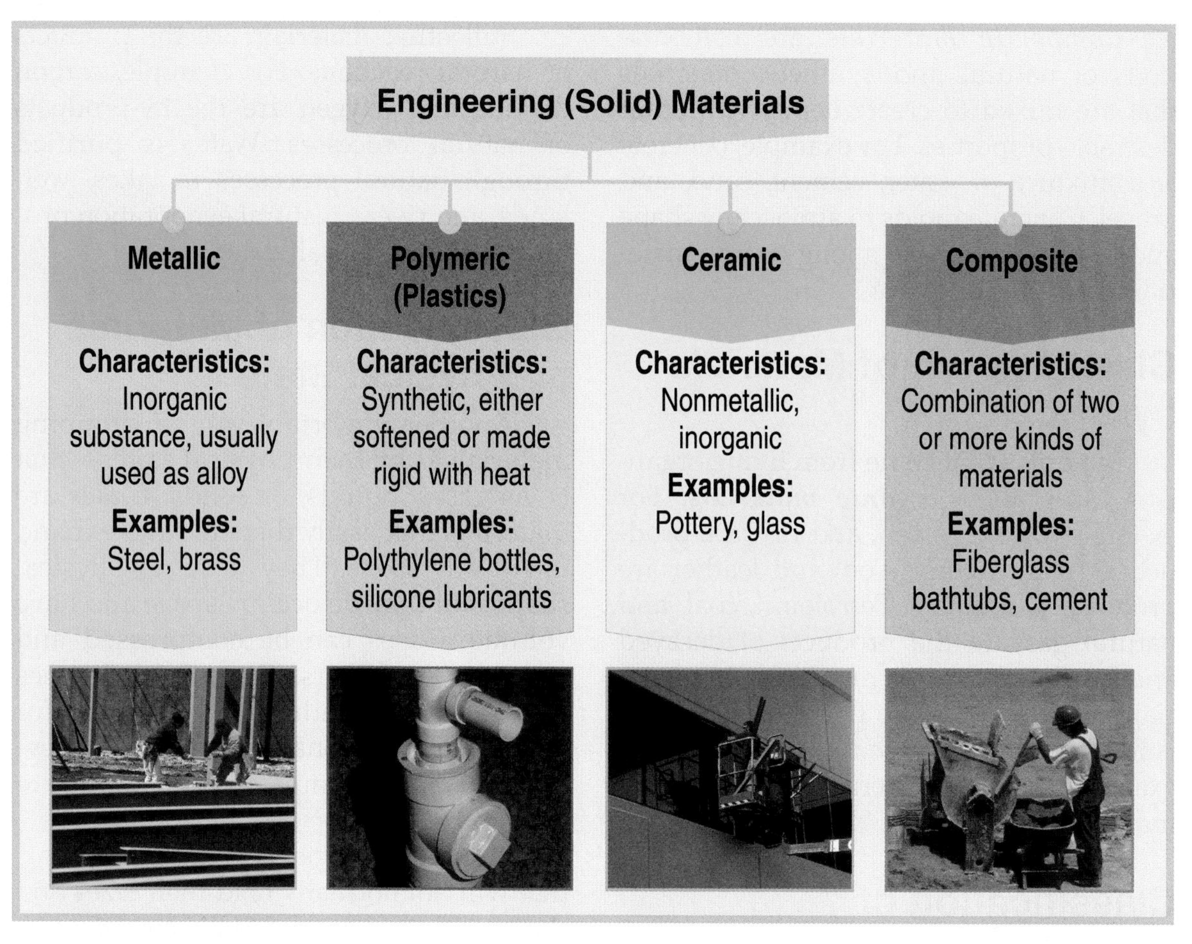

Figure 4-18. Solids can be grouped under the four categories of metallic, polymeric, ceramic, and composite.

- **Polymeric materials (plastics).** Plastics are synthetic materials containing complex chains of hydrogen-carbon (hydrocarbon) molecules. These materials are either thermoplastic (softened when heated) or thermosetting (made rigid by heat).
- **Ceramics.** These solids are nonmetallic, mostly inorganic crystalline materials, such as clay, cement, plaster, glass, abrasives, or refractory material.
- **Composites.** These solids are a combination of two or more kinds of materials. One material forms the matrix or structure. The other material fills the structure. Fiberglass is a composite with a glass fiber structure filled with a plastic resin. Wood is a natural composite with a cellulose-fiber structure filled with lignin, a natural glue, which bonds the structure together.

Properties of Materials

All materials exhibit a specific set of properties. For example, the properties of iron are different from those of oak. These properties are considered when materials are selected for specific uses. The common properties can be grouped under seven categories. See **Figure 4-19:**

- **Physical properties.** These properties are the characteristics due to the

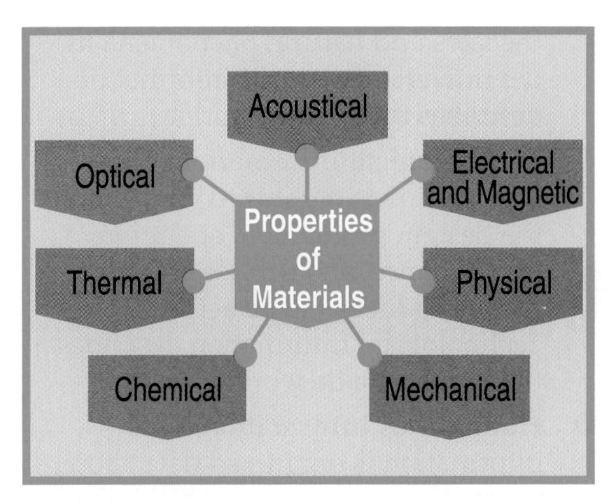

Figure 4-19. Properties of materials can be grouped under seven categories.

structure of a material, including size, shape, density, moisture content, and porosity.

- **Mechanical properties.** These properties are the reaction of a material to a force or load. This property affects the material's strength (ability to withstand stress), plasticity (ability to flow under pressure), elasticity (ability to stretch and return to the original shape), ductility (ability to be bent), and hardness (ability to withstand scratching and denting).
- **Chemical properties.** These properties are the reaction of a material to one or more chemicals in the outside environment. This property is often described in terms of chemical activity (the degree to which the material enters into a chemical action) and corrosion resistance (the ability to resist attack from other chemicals).
- **Thermal properties.** These properties are the reaction of a material to heating and cooling. This property is expressed as thermal conductivity (the ability to conduct heat), thermal shock resistance (the ability to withstand fracture from rapid changes in temperature), and thermal expansion (the change in size due to temperature change).
- **Electrical and magnetic properties.** These properties are the reaction of a material to electrical and magnetic forces. This property is described in terms of electrical conductivity (the ability to conduct electrical current) and magnetic permeability (the ability to retain magnetic forces).
- **Acoustical properties.** These properties are the reaction of a material to sound waves. Acoustical transmission (the ability to conduct sound) and acoustical reflectivity (the ability to reflect sound) are measures of acoustical properties.
- **Optical properties.** These properties are the reaction to visible light waves. Optical properties include color (reflected waves), optical transmission (the ability to pass light waves), and optical reflectivity (the ability to reflect light waves).

Information

In contrast to other living species, humans have the ability to think, reason, and enter into articulate speech. They, therefore, can observe what is happening around them, make judgments about those observations, and explain those judgments to other people. This unique human ability requires ***knowledge***, which is derived from data and information. See **Figure 4-20.**

Data are all the raw facts and figures people and machines collect. Information is data that has been sorted and categorized for human use. Data processing involves collecting, categorizing, and presenting data so humans can interpret it. We can group information into three areas:

- **Scientific information.** This information is organized data about

Standards for Technological Literacy

1 3 17 19

Extend

Discuss why all four types of knowledge are needed to develop, operate, use, and assess technological systems.

TechnoFact

Various means of communication have been used as a result of the continuous search for ways to share information over a distance, including systems of drum beating, smoke signals, and flags.

Standards for Technological Literacy

1 3 16 17 19

Brainstorm

Ask your students to list all types of energy they have used today.

TechnoFact

The law of conservation of energy states that energy can be transformed into another sort of energy, but it cannot be created or destroyed. Therefore, while energy has always existed, technological devices can change it from one form to another.

AC-Rochester

Figure 4-20. Data are processed into information so humans can gain knowledge.

the laws and natural phenomena in the universe. Scientific information describes the natural world.

- **Technological information.** This information is organized data about the design, production, operation, maintenance, and service of human-made products and structures. Technological information describes the human-made world.
- **Humanities information.** This information is organized data about the values and actions of individuals and society. Humanities information describes how people interact with society and the values individuals and groups of people hold.

Information that people learn and apply is called *knowledge*. Knowledge is the result of reasoned human action. This result guides people as they determine which course of action to take. Although knowledge can be described in terms of being derived from scientific, technological, or humanities information, in reality, all these types of knowledge must be brought to bear on any problem needing a solution. Knowledge derived from science might provide a theoretical base for the solution. Knowledge derived from technology is used to implement the solution. Knowledge derived from the humanities tells us if the solution is acceptable to society.

Energy

Energy is key to our survival, but it cannot be created or destroyed. This category of input can only be converted from one form to another. After it is converted, energy powers our factories, heats and lights our homes, cooks our foods, propels our vehicles, drives our communication systems, powers our farm machinery, and supports our construction activities.

Various types of energy exist. For example, everyone uses human energy to complete tasks. Human energy falls short of meeting all our needs, however. We have a limited supply, and we do not want to use it all on work. Also, some tasks cannot be done with human energy alone. For example, we cannot heat a house with human energy because the heat radiated from the human body is not enough. Therefore, people throughout history have used other sources of energy. We can look at energy from two vantage points: type and source. See **Figure 4-21.**

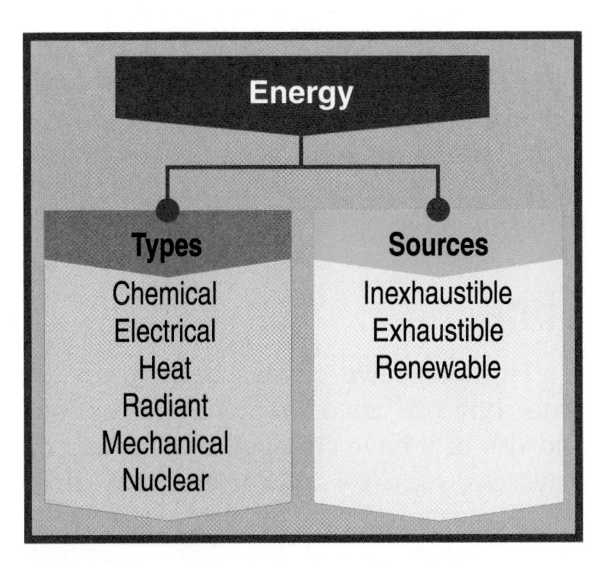

Figure 4-21. The two main factors in understanding energy are the type of energy and the source from which it comes.

Types of Energy

Energy can be grouped into six major types. These are the following:

- **Chemical energy.** This energy is stored in a substance and released by chemical reactions.
- **Electrical energy.** Moving electrons create electrical energy.
- **Thermal (heat) energy.** Heat energy comes from the increased molecular action heat causes.
- **Radiant (light) energy.** The Sun, fire, and other matter, which includes light, radio waves, X rays, and UV and infrared waves, produce radiant energy.
- **Mechanical energy.** Moving water, animals, people, and machines produce mechanical energy.
- **Nuclear energy.** Splitting atoms or uniting atomic matter produces nuclear energy.

Sources of Energy

Energy is available from three major sources. These sources are grouped in terms of the supply. The first source of energy is the Sun. The Sun is the fundamental source of energy for our solar system. This source's actions warm the Earth, cause the wind to blow, create weather, generate lightning, and indirectly with our Moon, create the ocean tides. Solar energy has always been with us and will continue to be so. This energy is said to be ***inexhaustible***.

A second source of energy comes from living matter. Human- and animal-muscle power can provide valuable sources of energy. Wood and other plant matter can be burned as a fuel. These sources are called ***renewable***. They are used up, but they can be replaced with the normal life cycle of the energy source.

The third source of energy is termed ***exhaustible***. The quantity found on Earth limits these sources. Neither the Sun nor human action can create additional supplies. Coal, petroleum, and natural gas are exhaustible energy sources. The goal of wise energy utilization is to maximize the use of inexhaustible sources, recycle the renewable sources, and use a minimum of exhaustible energy.

Finances

As we have seen, technological systems require people, tools and machines, materials, and energy. These resources are generally purchased. For example, we pay

Standards for Technological Literacy

1 3 7 16 19

Extend

Give an example of each type of energy.

Research

Have your students select an energy resource and find out what its advantages and disadvantages are.

TechnoFact

While renewable energy sources have less impact on the environment and are able to be replenished, the production of these sources is not constant.

Standards for Technological Literacy

1 3 6 7 19

Extend

Identify examples of sole-proprietorships, partnerships, and corporations in your community.

Brainstorm

Have your students list the advantages and disadvantages of debt and equity financing.

people wages, or salaries, for their labor and knowledge. We buy or lease machines and purchase materials and energy. All these actions require money, which provides the financial foundation of the technological activity.

Money to develop and operate technological systems can be obtained in two ways—equity financing and debt financing. See **Figure 4-22.** In ***equity financing***, a person or group of people buys a technological system or company. If one person owns the company, the operation is called a ***sole proprietorship***. If two or more people own it, they might form a ***partnership***, in which each person owns a portion of the company. In another circumstance, people might form a legal entity called a ***corporation*** to own the operation. They then sell shares, which are certificates of ownership in the corporation. In all three cases, the person or group of people raises the money by selling equity—a portion of the company. Therefore, raising money in this way is called *equity financing*.

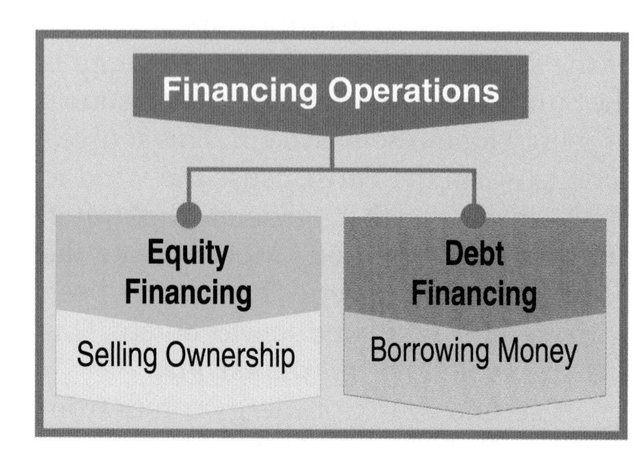

Figure 4-22. Financing the operations of technological systems and companies is achieved through equity and debt financing.

People also raise money by borrowing from other sources. Banks, insurance companies, or investment groups might loan money to support the activities of a company. This loan constitutes a debt that must be repaid. This type of financing is called ***debt financing***.

Time

Humans have always been aware of time, but our measurement, perception, and use of it have changed in a variety of ways. See **Figure 4-23.** For example, early people measured time by way of the rising and setting of the Sun. They knew they had

OutdoorDecor.com, National Institute of Standards and Technology Boulder Laboratories, U.S. Department of Commerce

Figure 4-23. Time has been measured in different ways throughout human history. Early Egyptians used a sundial, which measured the time of day by the Sun's shadow. We currently use various forms of electric and battery-operated clocks and watches. The new NIST-F1 at the National Institute of Standards and Technology keeps time with an uncertainty of about one second in 80 million years.

only so much daylight in which to hunt and gather food, and they acted accordingly. Later, farmers were very aware of particular months of the year because certain periods were more conducive to growing healthy crops.

Now, although time in and of itself has not changed, technology has accelerated its use and changed our standards of measurement. We allocate machine time, computer time, and sales response time, for example. At one point, we measured time only in years, months, and days. Hours and seconds were then observed. Now, engineers worry about nanoseconds (billionths of a second) in computer processing. Time is becoming an even more valuable resource for technological systems as the rate of development increases rapidly. New technologies build on old ones at a rapid rate, and the resulting products are made and disseminated quickly.

Standards for Technological Literacy

1 3 6 7 19

Extend

Discuss the saying "time is money."

Extend

Expand on the statement in the summary: The quality and quantity of inputs will often determine the type of technology a society can use.

TechnoFact

In the way many early technological devices were developed as labor-saving devices, technology is now also developing timesaving devices.

Answers to Test Your Knowledge Questions

1. inputs
2. Their attitudes about styling, price, and service directly affect the systems' outputs. Their money is spent on the products or services.
3. False
4. pulley
5. wedge
6. In first-class levers, the fulcrum is between the load and the effort. In second-class levers, the load is between the effort and the fulcrum. In third-class levers, the effort is placed between the load and the fulcrum.
7. Evaluate individually.
8. liquids
9. True
10. scientific
11. True
12. Evaluate individually.
13. debt financing
14. False

Summary

All technological systems involve inputs that are processed into outputs. People are the most important input, and they contribute to technological systems through their various individual knowledge and skills. They use tools and machines to increase their ability to do work. The various materials in the world make up the forms that different technologies take, and our understanding of their types and properties is key to continuing advancement. Different kinds of information help create knowledge, which is needed to solve problems and implement solutions. The various forms of energy support the resources necessary to complete technical tasks. We use finances to purchase other inputs. Finally, time affects the amount and kinds of technology we produce.

The quality and quantity of these inputs often determine the type of technology a society can use. Many of these inputs are in limited supply. Therefore, everyone designing, building, or using technology should use inputs appropriately and recycle as many of them as possible.

Test Your Knowledge

Write your answers on a separate piece of paper. Please do not write in this book.

1. The resources used to make a technological system operate are called ______.
2. Explain the statement, "Consumers are the reason for technological systems."
3. A drilling tool is used to abrade and smooth surfaces. True or false?
4. The three types of simple machines that use the principle of the lever are the lever, the wheel and axle, and the ______.
5. The three types of simple machines that use the principle of the inclined plane are the inclined plane, the screw, and the ______.
6. Explain the difference between first-class, second-class, and third-class levers.
7. Give an example of a genetic material.
8. In their physical state, materials can be classified as solids, gases, and ______.
9. The mechanical properties of a material deal with the material's reaction to a force or load. True or false?
10. Information can be classified into three areas: technological, humanities, and ______.
11. Energy that comes from the increased molecular action heat causes is called *thermal energy*. True or false?

12. Give an example of an exhaustible source of energy.
13. The two ways to raise money to develop and operate technological systems are equity financing and _____.
14. Our perception of time has remained constant throughout history. True or false?

STEM Applications

1. Look around the room you are in. List the human abilities that were used to design, construct, and decorate it and the products within it and three ways each ability was used.
2. Design and sketch a simple device that uses at least three of the six simple machines (mechanisms) to do a job.
3. Select a product made from exhaustible materials or a task that uses an exhaustible energy source. Describe how that product could be made or how the task could be completed using renewable materials or renewable energy sources.

Chapter Outline

This chapter covers the benchmark topics for the following Standards for Technological Literacy:

1 2 3 7 8 11 14 15 16 17 18 19 20

Learning Objectives

After studying this chapter, you will be able to do the following:

- Summarize the three major types of technological processes and their relationship to one another.
- Recall the steps used in the problem-solving and design process.
- Explain the major processes or activities used in the agricultural and related biotechnological, communication and information, construction, energy and power, manufacturing, medical, and transportation processes.
- Recall the steps involved in management processes.

Key Terms

assemble
building
casting and molding
civil engineering structure
commercial structure
condition
constraint
conversion and processing
criterion
decode
diagnosis
drilling
encode
finish
form
foundation
graphic communication
growth
guidance
harvest
heavy engineering structure
ideation
industrial structure
mining
pathway
planning
prevention
primary process
propagation
propulsion
receive
residential structure
retrieve
secondary process
separate
service
site preparation
store
structure
superstructure
support system
suspension
telecommunication
telecommunications technology
terminal
transmit
treatment
utility
vehicular system

Strategic Reading

Before you read the chapter, think of an everyday piece of technology. As you read, think of how the five steps of the problem-solving and design process were used in creating that technology.

A technological system has inputs, processes, and outputs. Inputs are the resources that go into the system. Processes are what happen within a system. They change inputs into outputs. Processes are used to design products, structures, energy-conversion systems, communication messages, and transportation systems. They are also used to do the following:

- Grow and process crops.
- Produce products and structures.
- Treat medical conditions and illnesses.
- Convert and apply energy.
- Communicate information and ideas.
- Transport people and cargo.

In addition, managerial processes see that the technological system runs efficiently and produces quality products and few unwanted outputs.

All of the above processes can be classified under three major headings. See **Figure 5-1.** These headings are problem-solving and design processes, production processes, and management processes. Both the problem-solving and design process and the management process can be seen as generic in that the same basic process can

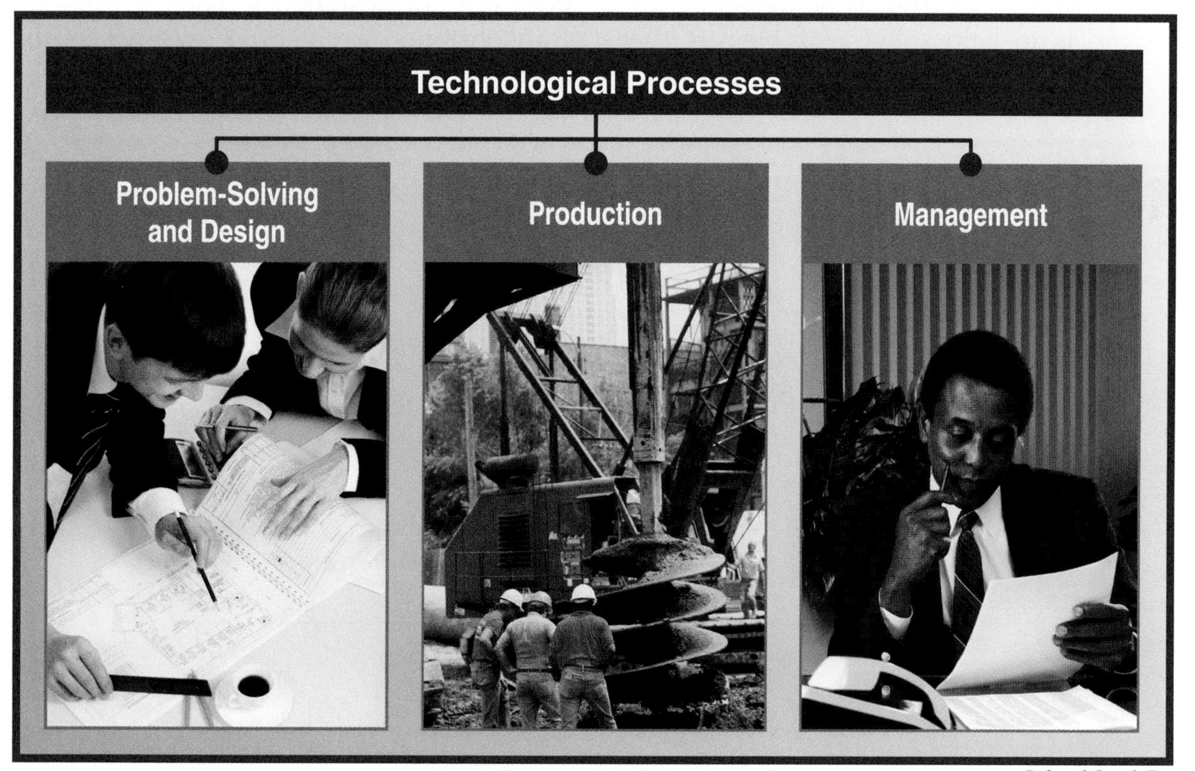

Inland Steel Co.

Figure 5-1. The problem-solving and design, production, and management processes are used to develop and operate technological systems.

Standards for Technological Literacy

1 2 8 11

Reinforce

Review technology (Chapter 1), technology systems (Chapter 2), and technological contexts (Chapter 3).

Figure Discussion

Ask your students why all three types of technologies shown in Figure 5-1 are important if a technological activity is to be successful.

Research

Have your students search books and the Internet to find three definitions for problem solving.

Standards for Technological Literacy

1 2 8 11

TechnoFact

The management of the Industrial Revolution is being replaced with a new type of shared management. Before, the organization relied on one person managing several below him or her, and all workers were expected to obey orders, whether or not the ultimate purpose of their activities was understood. With shared management, all levels of workers collaborate to set goals and carry out work.

TechnoFact

One of the first processes humans had to master was the use of stone tools. Humans first used stone tools by imitating animals. For example, digging sticks were reproductions of an aardvark's digging feet. Copying various animals with their inventions brought humans to developing their own answers to processing challenges.

be used, no matter what the technology. See **Figure 5-2.** The steps in the production process, however, vary according to the technological system under consideration. We will now examine each process separately to understand the processes in more depth.

Problem-Solving and Design Processes

All technology has been developed to meet human needs and wants. Each device or system is designed to solve a problem or to meet an opportunity. See **Figure 5-3.** Early technology was almost always problem oriented. Tools were designed to make work easier. New housing technology made living more comfortable. New equipment made farming easier. New transportation devices helped humans move loads from place to place. New communication methods made information exchange easier.

Many modern technological devices are designed for the same reason. This reason is to help solve a problem. Other technologies, though, are developed to meet an opportunity. For example, spacecraft are not designed to solve a pressing transportation problem. They give us the

Figure 5-3. Technology is designed to solve a problem or meet an opportunity. For example, new medical scanning devices have led to improved diagnoses.

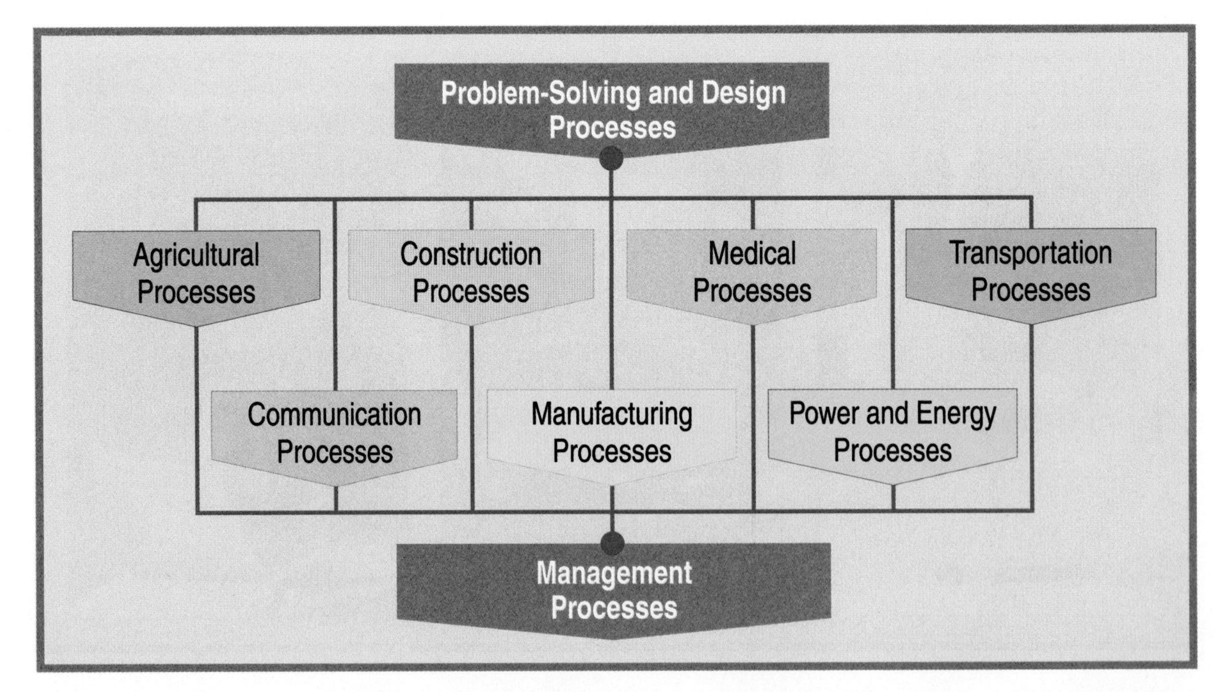

Figure 5-2. The same problem-solving and design processes and management processes are used for all seven technological areas.

opportunity to study the Earth from orbit. Likewise, many products are designed to make money. How many people really need an electric toothbrush? Do standard toothbrushes fail to meet our needs? The electric model was most likely designed to give the inventors and manufacturers an opportunity to profit from their idea. There is nothing wrong with this motive. In fact, the drive to make money has brought us many things we now take for granted. The change simply suggests that our civilization has progressed a great deal. Now, we have time to apply technology to luxuries. We are beyond the survival mode of our early ancestors.

No matter what the motive, however, designers and problem solvers follow a common procedure in designing and developing technology. The procedure is the same for solving problems or meeting opportunities. The five most important steps in the problem-solving and design process are shown in **Figure 5-4.** They are the following:

1. Identifying the problem or opportunity.
2. Developing multiple solutions to the problem or opportunity.
3. Isolating, refining, and detailing the best solution.
4. Modeling and evaluating the selected solution.
5. Communicating the final solution.

Each of these steps leads to the next one. Remember, however, that all technological processes have feedback. The results from one step might cause the designers to retrace their steps. For example, a prototype built in the modeling step might show major problems with a design. The solution might have to be modified by changing it and selecting a new "best" solution. Now, let us look at each of these steps separately.

Identifying the Problem or Opportunity

As we said, technology starts with a problem or an opportunity. These problems and opportunities, however, are seldom clearly seen or felt. The first step in the design and problem-solving process identifies and describes the task being undertaken.

Standards for Technological Literacy

Brainstorm

Ask your students how they would use the design processes shown in Figure 5-4 to develop a poster for a school play.

TechnoFact

Most of today's inventions and discoveries are credited to large research organizations supported by various schools, groups, industries, or foundations.

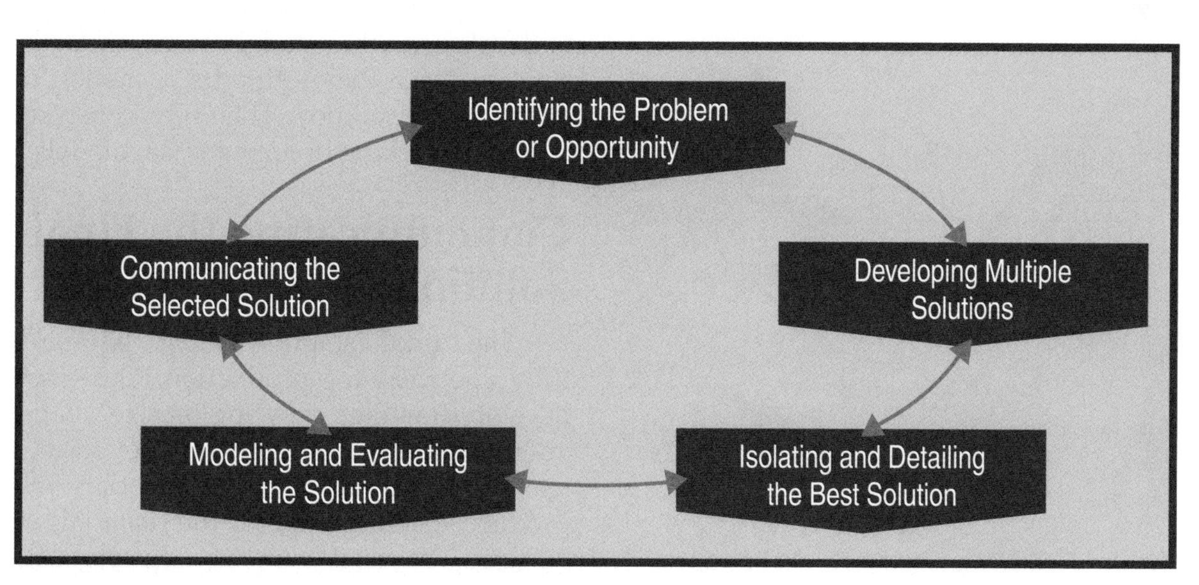

Figure 5-4. The problem-solving and design process has five steps. Each step affects the others.

Standards for Technological Literacy

2 8 11

Figure Discussion

Ask your students what they think the drawings shown in Figure 5-5 would be used for.

TechnoFact

Ideation can find unique solutions to everyday problems. Until the turn of the 20th century, we buttoned or laced clothing. Then in 1913, Gideon Sundback designed a closing device very similar to the zipper. It swept the nations and is now used widely in many products.

This process involves defining the problem or opportunity and listing limitations or requirements for a solution. These requirements include the ***criteria*** and ***constraints*** regarding such characteristics as function, appearance, size, operations, manufacturing, marketing, and finances. Problems that might result from various environmental, political, ethical, and social issues might also be identified. These criteria and constraints communicate the expectations for the solution.

Developing Multiple Solutions

Once the problem or opportunity is defined, a designer seeks solutions. Designers create many possible answers by letting their minds create solutions (a process called ***ideation***) and by brainstorming with others. They continually check and critique their answers, redefining their solutions and improving the designs. Their sketches show ways to meet the challenge. Designers first make rough sketches to capture thoughts. They refine the sketches to mold thoughts into more specific solutions. See **Figure 5-5.**

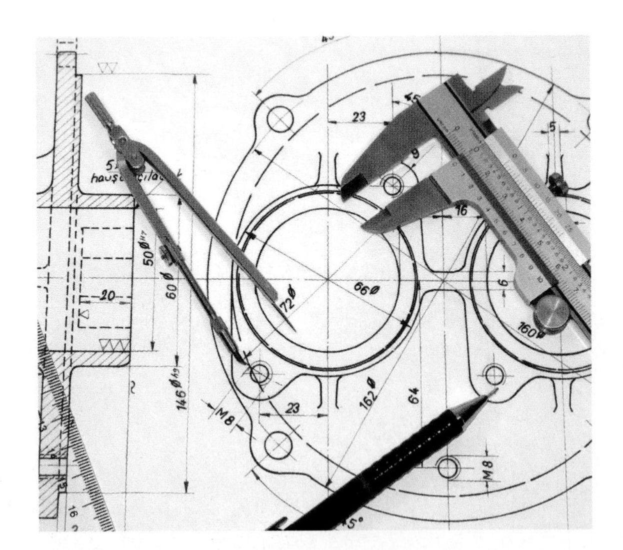

Figure 5-5. Designers use sketches as they explore possible solutions for design problems or opportunities.

Isolating, Refining, and Detailing the Best Solution

Designers then isolate the most promising ideas and refine their work. They analyze ideas to select the most promising, realizing that criteria and constraints sometimes compete with one another and that trade-offs must occur. Designers know that this best solution might not meet all the criteria and constraints perfectly. They do understand, however, that the solution must be safe, be efficient, and be functional (work properly) and that it must be one that can be produced and marketed within cost limits. Once the solution is chosen, it must be detailed. Specifications and general characteristics of the final solution must be established.

Modeling and Evaluating the Selected Solution

Frequently, designers produce a model of the expected solution. See **Figure 5-6.** A model allows designers and managers to review the solution's performance. Models might be physical, graphic, or mathematical. Often, designers first use a computer to develop graphic models based on mathematics. Next, they refine these models to optimize the solution. They then construct physical (working or appearance) models.

Communicating the Final Solution

The final solution must be carefully specified for production. Designers develop drawings showing its size, shape, and component arrangement. They formulate a material list or bill of materials and produce specifications for the materials to be used. These documents communicate the characteristics of the product, structure, media, or system. Designers also prepare

Figure 5-6. Models are used to test product and structure designs. They can be created on computers or made physically, as is this model house.

written and oral reports to gain approval for the solution from decision makers.

The design and problem-solving process ends with a specified solution. Section 4 of this book contains a more detailed discussion of this process. For now, we should note that the solution the design and problem-solving process indicates moves to the next step. This step is the production process.

Production Processes

Production processes are actions that create the physical solution to the problem or opportunity. For example, they grow and harvest crops, construct structures, and generate communication messages. As noted earlier, the problem-solving and design process is used across system lines. This process generates the description of the solution. The same processes or sequence of events can be used to design agricultural, food-processing, manufacturing, construction, energy and power, medical, communication, information-processing, and transportation systems.

Each of the technological systems has its own unique production processes, however. Manufacturing processes are different from medical processes. Agricultural and related biotechnological processes are different from construction processes. Communication and information processes are different from transportation processes. These processes move the solutions the design process describes into tangible solutions. Let us now examine each technological area, with regard to its unique production processes.

Agricultural and Related Biotechnological Processes

Since our early history, people have raised crops and domesticated animals. This activity is called *farming*, or *agriculture*. Agriculture is the art of cultivating the soil, growing crops, and raising livestock. Broadly speaking, this area includes farming, fishing, and forestry. Each of these areas involves a crop with a biological cycle including birth, growth, maturity, and death.

Farming is growing plants and animals for commercial use. See **Figure 5-7.** Typical crops are fruits, vegetables, grains, and forage for animals. Individuals apply

Standards for Technological Literacy

1 2 8 11 15

Brainstorm

Ask your students how people apply agricultural processes to their personal activities.

TechnoFact

Because pictures can be considered a language understood by almost everyone, pictographs and hieroglyphics were an early form of writing. Illustrating important events was both a means of preserving history and of communication.

Standards for Technological Literacy

1 15

TechnoFact

The survival of early people depended on their skills in hunting, fishing, and food gathering. Most of these hunter/gatherers were nomadic until the cultivation of wild plants and domestication of wild animals began about 10,000 years ago.

Figure 5-7. These trees are a crop that will be sold as Christmas trees.

farming processes as they grow plants for landscapes and home gardens.

Fishing is harvesting fish from lakes and oceans for commercial use. Fish can also be raised and harvested in controlled areas called *fish farms*. Individuals also use fishing processes for recreational purposes.

Forestry is growing trees for commercial use. The trees might be used for lumber and veneer (thin sheets of wood), paper, or other products. Agricultural practices involve the following major steps. See **Figure 5-8.**

1. Propagation.
2. Growth.
3. Harvesting.
4. Conversion and processing.

Agriculture produces food and fiber for people to use. This activity starts with the birth of a crop. Types of births include planting a seed, rooting a cutting from a plant, and allowing animals to breed. This step is called ***propagation***. Propagation allows a biological organism to reproduce.

After the new animal or plant appears, the next step is ***growth***. Agriculture provides a proper environment for this to happen. Growth can involve providing feed and water for animals or cultivating and watering (irrigating) crops. When the animal or plant has reached maturity, ***harvesting*** occurs. Harvesting involves such actions as removing edible parts of plants from trees and stocks and butchering animals to produce meat and other products for consumption.

Finally, the food product undergoes ***conversion and processing*** to create a foodstuff. For example, wheat is ground into flour. Meat is cut into steaks and roasts. Trees are cut into lumber or chipped for papermaking.

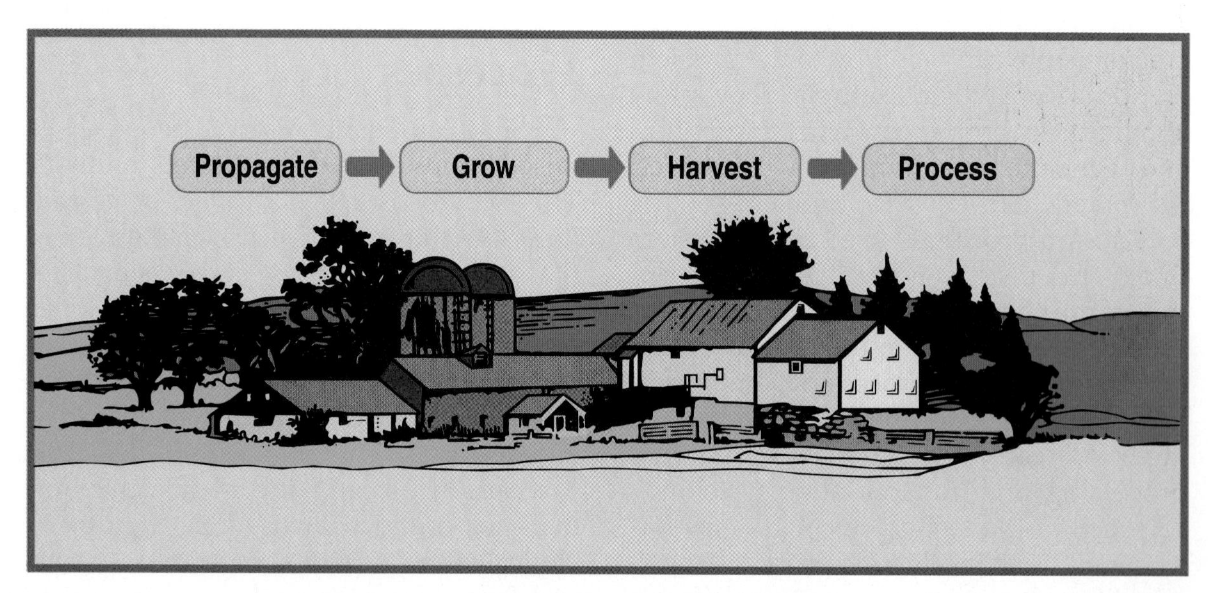

Figure 5-8. Agricultural practices using technology involve four major steps.

We have enhanced agricultural practices by the application of biotechnological processes. For example, through combining biological processes with physical technology, we have genetically altered bacteria. These bacteria are used to protect crops from insects. More complete discussions of agricultural and related biotechnological processes are contained in Chapters 29 and 30.

Communication and Information Processes

Humans have always exchanged information and ideas. This process is called *communication*. The simplest communication processes involve spoken language. Spoken language is not technology, however, because no technical means are used in the process. Technology became involved as civilization grew and people developed additional techniques to help them communicate better. For example, we now use technology to produce printed, graphic, and photographic media to impart information and express ideas. We group these techniques under the general heading called ***graphic communications***. Also, we have developed ***telecommunications technology***. This technology allows us to communicate using electromagnetic waves. Recently, a special type of communication, called *information processing*, has been developed. Each process is described in more detail below.

Graphic Communications

In graphic-communications processes, messages are visual and have two dimensions. This category includes the printed messages commonly found in books, magazines, owners' and service manuals, and promotional flyers. Graphic-communications media also include technical graphic messages, such as technical illustrations and engineering drawings. The final graphic-communications medium is photographic communication. This group includes the film and print media coming from photographic processes.

Standards for Technological Literacy

1 15 17

Extend

Discuss how changes in communication have changed our view of the world.

TechnoFact

The production processes of communication and transportation are similar in that communication can be viewed as having an origin point and destination. Instead of transferring goods and people, however, communication is the transfer of information, such as messages.

Career Corner

Production Managers

Production managers coordinate the people, materials, equipment, and processes needed to manufacture industrial and consumer products. Their major responsibilities include scheduling production activities, requisitioning and maintaining equipment, staffing operations, and maintaining quality and inventory control. Production managers work closely with sales and purchasing managers.

Most industrial production managers work more than 40 hours a week. In plants that have more than one shift, managers often work late shifts. Many companies seek candidates with college degrees in industrial management or business administration. Companies also want managers who have good interpersonal and communication skills.

Standards for Technological Literacy

1 17

Figure Discussion

Describe the communication model in Figure 5-9 using examples of graphic and electronic media.

TechnoFact

Because sending and receiving messages was critically important to ruling all parts of the Roman Empire, the Romans developed an elaborate postal system.

Telecommunications

Telecommunication is the transmission of information over a distance for the purpose of communication. Telecommunications processes depend on electromagnetic waves to carry their messages. Telecommunications include audio, video, and Internet communications services. Telecommunications techniques include broadcast (television and radio), hardwired (telephone and telegraph), and surveillance (radar and sonar) systems.

Information Processing

Information processing can be described as manipulating data to produce useful information. This processing involves changing (processing) information so the information is in a useful form. Information processing includes the actions of obtaining, recording, organizing, storing, retrieving, displaying, and sharing information. This processing generally involves the use of computer systems to locate and obtain the information, manipulate it into a useful form, and output it to a client or customer. The outputting and sharing of the information is the communication part of the process.

Steps in the Communication Process

All communication technologies involve five major steps. These technologies are discussed in depth in Section 6 of this book. See **Figure 5-9.**

1. Encoding.
2. Transmitting.
3. Receiving.
4. Decoding.
5. Storing and retrieving.

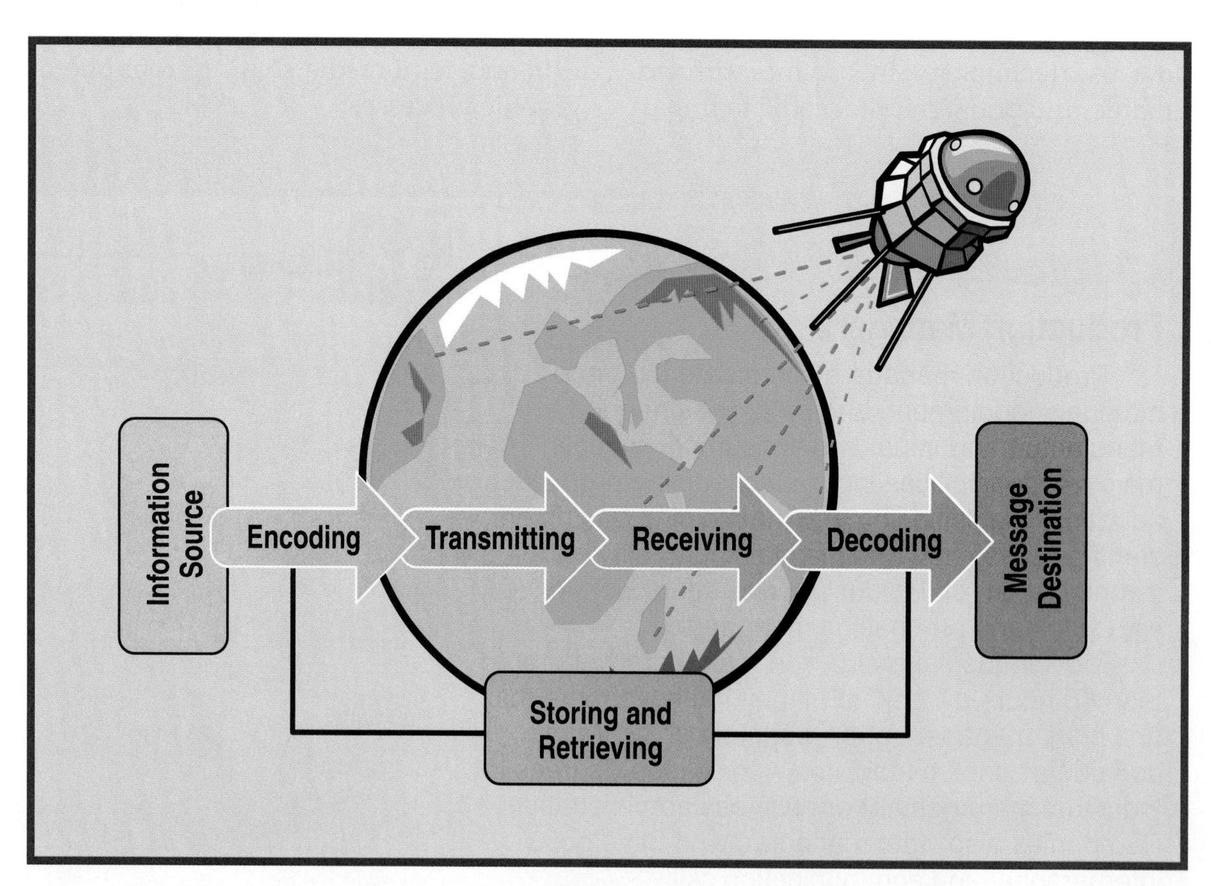

Figure 5-9. All communication technologies involve five major steps.

Communication technology organizes information so people can receive the information. See Figure 5-10. The communication process starts with an information source. Most often, this source is the human brain. In some cases, machines are used. The first step in the communication process is ***encoding*** the information. Encoding involves placing the information in a format or pattern the communication technology can use. This process might involve producing images on film, pulses on a light wave, electrical charges on a tape, or graphics on paper.

Transmitting the coded message from the sender to the receiver is the next step. Transmission can include moving printed materials, broadcasting radio and television programs, transmitting telephone messages along wires or fiber-optic strands, and using pulses of light to send messages between ships at sea. Transmitting a message is not enough. Someone or something must ***receive*** the message. Receiving the message requires recognizing and accepting the information. A radio must recognize the coded radio waves. The telephone receiver must recognize the pulses of electricity. The human eye must see the semaphore flags (ship-to-ship communication).

©iStockphoto.com/webphotographeer

Figure 5-10. Communication technology allows us to send, receive, and store data and information.

The received message must then undergo ***decoding***. Decoding means the coded information must be changed back into a recognizable form. The decoded information is presented in an audio format or a visual format humans can understand. The radio receiver changes radio waves into sound waves. The human brain places meaning relative to the position of the semaphore flags. The telephone changes electrical impulses to sound.

Throughout the communication process, information undergoes storing and retrieving. Storage processes allow the information to be retained for later use. Books can be shelved in libraries. Recorded music can be stored on tapes or compact discs. Television programs can be stored on discs. Data can be stored on computer discs or tapes. Pictures can be stored in file folders. Later, this information can be ***retrieved*** (brought back). The information can be selected and delivered back into the communication process. Storage and retrieval can happen at any time in the communication system. Information is commonly ***stored*** at the source or at the destination.

Construction Processes

Humans first used construction technology to produce shelter. This effort allowed early humans to move out of caves and into crude homes. These humble beginnings took the form of huts and tents. We now construct many types of ***buildings***. Construction technology does not stop with buildings. We also construct many types of structures to support other activities. These include the following:

- Roads, canals, and runways for transportation systems.
- Factories and warehouses for manufacturing activities.

Standards for Technological Literacy

1 17 19

Brainstorm

Ask your students what technology they would need to transmit a secret message.

TechnoFact

Some of the oldest surviving constructed structures had religious significance, including the Parthenon, Egyptian Pyramids, and Stonehenge.

Standards for Technological Literacy

1 20

Extension

Give local examples of building and civil structures.

TechnoFact

Rockefeller Center covers a number of blocks in New York City. The center is a famous complex of buildings in Manhattan. The first 14 buildings were built between 1931 and 1939. These commercial structures provide space for offices, shops, restaurants, exhibition rooms, broadcasting studios, and Radio City Music Hall.

- Studios, transmitter towers, and telephone lines for communication.
- Dams and power lines as a part of our electricity-generation and -distribution systems.

Constructed works are everywhere.

These works can be grouped into two major categories. The categories are buildings and civil or heavy engineering structures. See **Figure 5-11.** Buildings are all the structures erected to protect people and machines from the outside environment. They can be used for three major purposes. First, they are used as ***residential structures***, that is, the places where people live. These structures can be homes, town houses, condominiums, and apartment buildings. Buildings are also used as ***commercial structures***. These structures are the stores and offices used to conduct business. Government buildings, such as schools, city halls, and state capitols, are also placed in this category. Finally, buildings can be ***industrial structures***. These structures are the power plants, factories, transportation terminals, and communication studios major companies use.

Figure 5-11. Construction technology is used to build buildings and civil structures.

The second major type of construction produces ***civil engineering structures***, which are also called ***heavy engineering structures***. These structures are primarily designed with the knowledge of the civil engineer. Common civil structures are roads, dams, communication towers, railroad tracks, pipelines, airport runways, irrigation systems, canals, aqueducts, and electricity-transmission lines.

Most construction projects include several steps. These steps are presented in more depth in Chapters 17 and 18. See **Figure 5-12.** They include the following:

1. Preparing the site.
2. Setting the foundation.
3. Building the superstructure.
4. Installing the utilities.
5. Enclosing the superstructure.
6. Finishing the structure.
7. Completing the site.
8. Servicing the structure.

Preparing the Site

Not all construction projects use all the steps just listed. Most projects, however, start with ***site preparation***. This task includes removing existing buildings, structures, brush, and trees that interfere with locating the new structure. Workers roughly grade the site and establish the desired slope. Surveyors then locate the exact spot for the new structure.

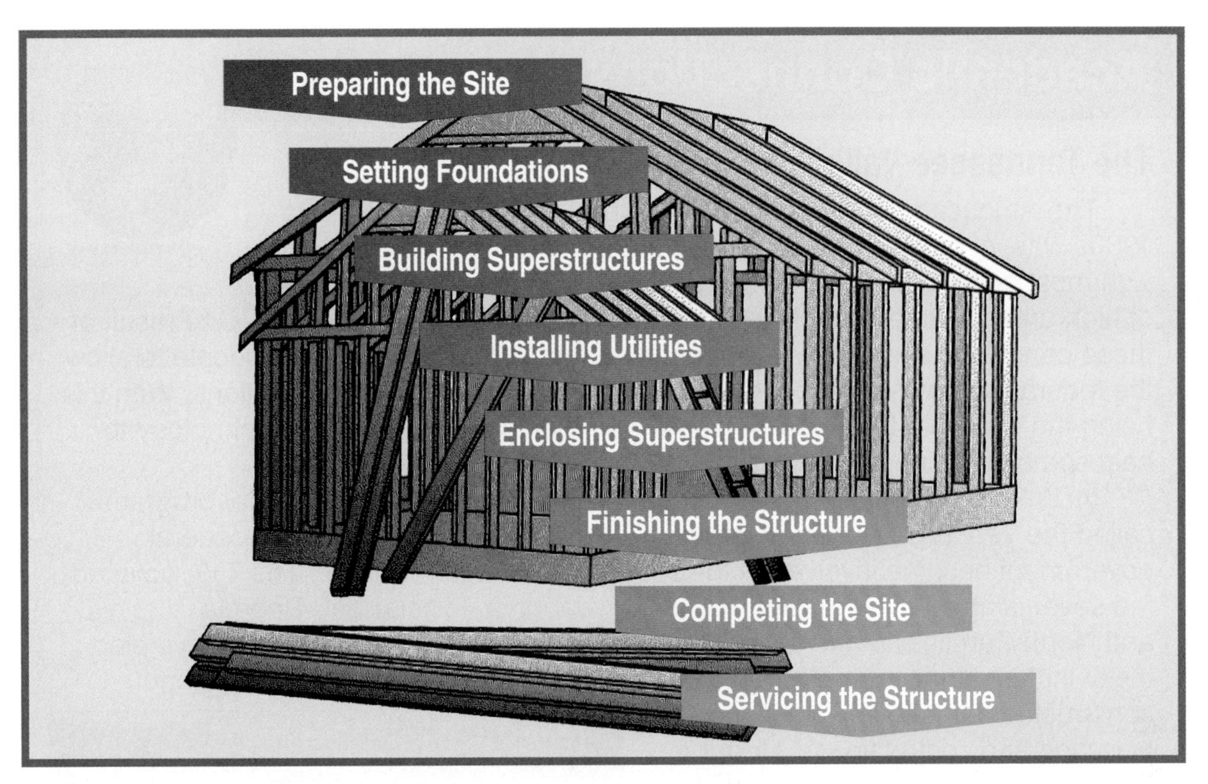

Figure 5-12. Most construction projects move through the eight steps shown here.

Standards for Technological Literacy

1 20

Reinforce

Compare the foundations and superstructures for a building, a road, and a communication tower.

Setting the Foundation

The base, or ***foundation***, for the structure is constructed. This step generally involves digging pits or trenches. Concrete or rock is then placed in the holes. If the hole extends to solid rock, additional foundations might not be needed. Foundations provide a stable surface onto which the building can be constructed.

Building the Superstructure

The ***superstructure*** of the project is constructed on the foundation. Superstructures include the framework of the building or tower. They also include the pipes for pipelines, surfaces for roads and airport runways, and tracks for railroads.

Installing the Utilities

In many cases, the constructed structure includes utilities. ***Utilities*** are such items as water, gas, and waste pipes; electrical and communication wire; and heating and cooling ducts. Many runways, railroads, and highways also need lighting and communication systems installed.

Enclosing the Superstructure

After the utilities are installed, the superstructure must be enclosed. Walls need interior and exterior skins. Roofs, ceilings, and floors must be covered. Doors and windows must be installed to close openings.

Finishing the Structure

Once all this work is done, the structure must be completed. Walls, ceilings, doors, windows, and building trim must be painted. Likewise, communication towers must be coated to protect them from the natural elements. Runways and roadways must have traffic stripes painted on them and signs installed.

Standards for Technological Literacy

1 3 16 20

TechnoFact

The various structures and appearances of buildings often reflect such elements as the time period constructed, region of the world, and religious influences. For example, the style and structure of a cathedral built in Medieval Europe differs greatly from an Indian Temple built in the 17th century.

Academic Connections: History

The Tennessee Valley Authority (TVA)

The American experience with the Great Depression of the 1930s was, of course, not a pleasant one. The United States, along with most of the other countries in the world, went through a period of high unemployment, declining sales, falling prices, and rising debt. One result of these conditions was a new willingness on the part of the American people to allow the federal government to become involved in improving social conditions. With this being the situation, the Roosevelt administration saw a way to use technology to help some of the poorest areas of the country.

In 1933, Congress authorized the administration to create a federal program called the *Tennessee Valley Authority (TVA)*. Under this program, the federal government became involved in the production of electric power. The TVA covered the seven states of Tennessee, Kentucky, Alabama, Mississippi, Georgia, Virginia, and North Carolina. Some of the residents of these states did not even have electricity in their homes yet. The TVA built and operated hydroelectric plants and constructed dams to improve transportation and control floods. Under the TVA, the residents of this region were able to purchase cheap power. Thus, although the Depression created distressing conditions, some people's lives actually improved in some ways because of the application of technology. The TVA is still in effect today. In what areas is it now involved?

Completing the Site

The area around the structure must, also, be completed. The site must be graded to reduce erosion and increase its beauty. Shrubs, grass, and trees are added to landscape the site and keep the soil in place. Hardscape, such as driveways, sidewalks, and retaining walls, is completed.

Servicing the Structure

The result of this sequence of steps is a new constructed structure. During its life, the structure will require ***servicing***. Servicing means the structure will undergo maintenance, repair, and reconditioning. For example, various people will periodically paint the surfaces, repair the roofs, and replace utility-systems components. Servicing attempts to keep the structure in good working order.

Energy and Power Processes

Energy is the ability to do work. All technological systems require energy. Power, which is sometimes confused with energy, is actually the rate at which the work is accomplished or the energy is changed from one form to another. This rate is calculated by dividing the work done by the time taken to do it.

We explore energy and power processes in more detail in Chapter 27. For now, we should note that six basic forms of energy exist and that technology aids us in transforming and applying energy. Each form of energy is discussed in turn.

The six basic forms of energy are as follows:

- **Mechanical energy.** This is the energy found in moving objects.
- **Chemical energy.** This is the energy contained in molecules of a substance.
- **Radiant (light) energy.** This is energy in the form of electromagnetic waves.
- **Thermal (heat) energy.** This is the energy associated with the movement of atoms and molecules.
- **Electrical energy.** This is the energy associated with moving electrons.
- **Nuclear energy.** This is the energy released when atoms are split or united.

The process of transforming and applying energy involves gathering or collecting the resources, converting the energy into a new form, transmitting the energy, and applying the energy to do work. See **Figure 5-13.**

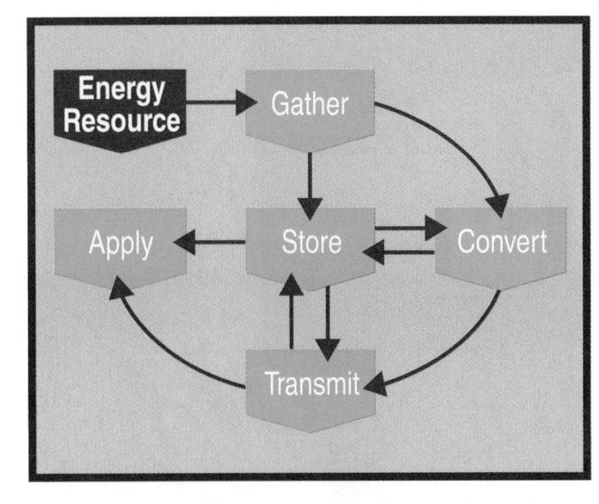

Figure 5-13. The process of transforming energy involves four major steps. These steps are gathering the resources, converting the energy, transmitting the energy, and applying the energy.

Gathering Resources

Energy resources include such items as petroleum, coal, wind, sunlight, and moving water. The type of gathering techniques used depends on the resource to be used. We obtain coal through mining. Natural gas and petroleum are obtained through drilling. Sunlight is captured with collectors. We collect water through the use of dams. Wind is captured with turbines.

Converting Energy

Energy resources are converted from one form to another through a variety of means. Internal combustion engines convert fuels (gasoline, diesel fuel, and natural gas, for example) into mechanical energy. Hydroelectric projects use turbines and generators to change the mechanical energy in flowing water into electricity. Home furnaces change the chemical energy in fuels such as coal, fuel oil, and natural gas into heat (thermal energy). Hundreds of energy converters exist that are designed to perform specific tasks.

Transmitting Energy

Energy must often be moved from one place to another. For example, the energy from an engine must be transmitted to the wheels driving a car. Heat energy must be moved from a furnace to the appropriate rooms in a house. Electrical energy must be moved from generating stations to homes, businesses, and factories.

Applying Energy

Energy, in itself, is of little use to people until it is applied to a task. The energy must power a device or power a process. For example, energy is useful when it causes a vehicle to move, a machine to perform a task, or a device to light a space. See **Figure 5-14.**

Standards for Technological Literacy

1 16

Extend

Give examples of each of the six forms of energy.

Figure Discussion

Using the model in Figure 5-13, explain the operation of an electric light. Start with the generating plant and end with the light.

TechnoFact

The application of available energy resources has greatly impacted the development of technology. Once gathered, these resources may be converted, transmitted, and finally applied to various products. Energy is used in society in mechanical, chemical, radiant (light), thermal (heat), electrical, and nuclear forms. Before these forms of energy are applied, however, they may be stored for later use.

Standards for Technological Literacy

1 16 19

Figure Discussion

Trace the materials in a chair (wood, metals, and plastics) through the three steps shown in Figure 5-15.

TechnoFact

Before the Industrial Revolution, most people lived simple lives. During this time when most people lived in the countryside, the only manufacturing was done by blacksmiths using simple hand tools. The only "industries" were cloth making and ironworks, but this changed when factories began to rise in the late 18th century.

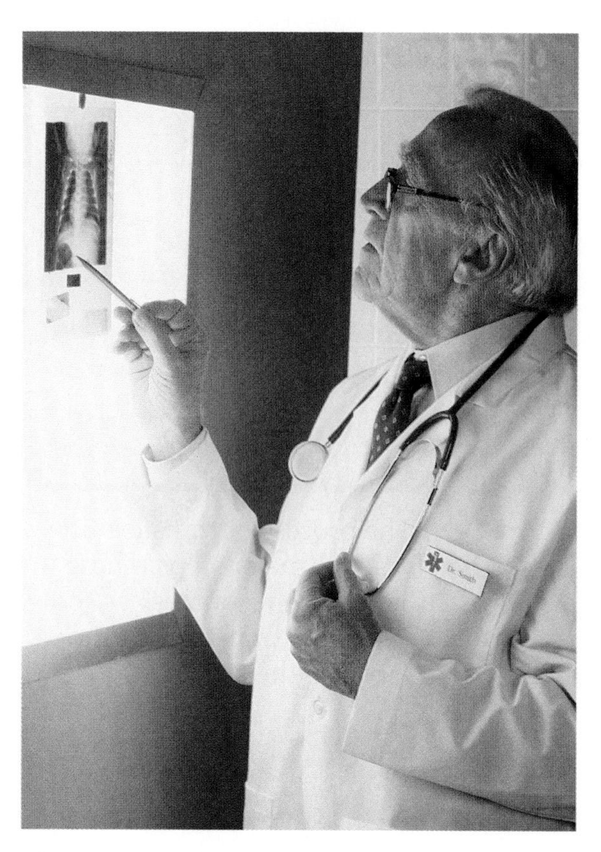

Figure 5-14. Energy is used to do work. Here, light enables a physician to examine X-ray film.

Not all energy is used immediately after it is gathered. We often want to store it for later use. The reservoir behind a dam stores the energy in moving water. A battery stores the chemical energy of its parts. Tank farms store petroleum for later energy applications.

Manufacturing Processes

In everything you do, you use manufactured products. You ride to school in a manufactured car or bus. You look out of the building through a manufactured window. You wear manufactured clothing. You are reading a manufactured book and sitting on a manufactured chair in a room lighted with manufactured fixtures. It is difficult to imagine a world without manufactured products. No matter what the product, however, all manufacturing activities involve three stages. See **Figure 5-15.** These stages are obtaining resources, producing industrial materials through ***primary processing***, and creating finished products through secondary processing.

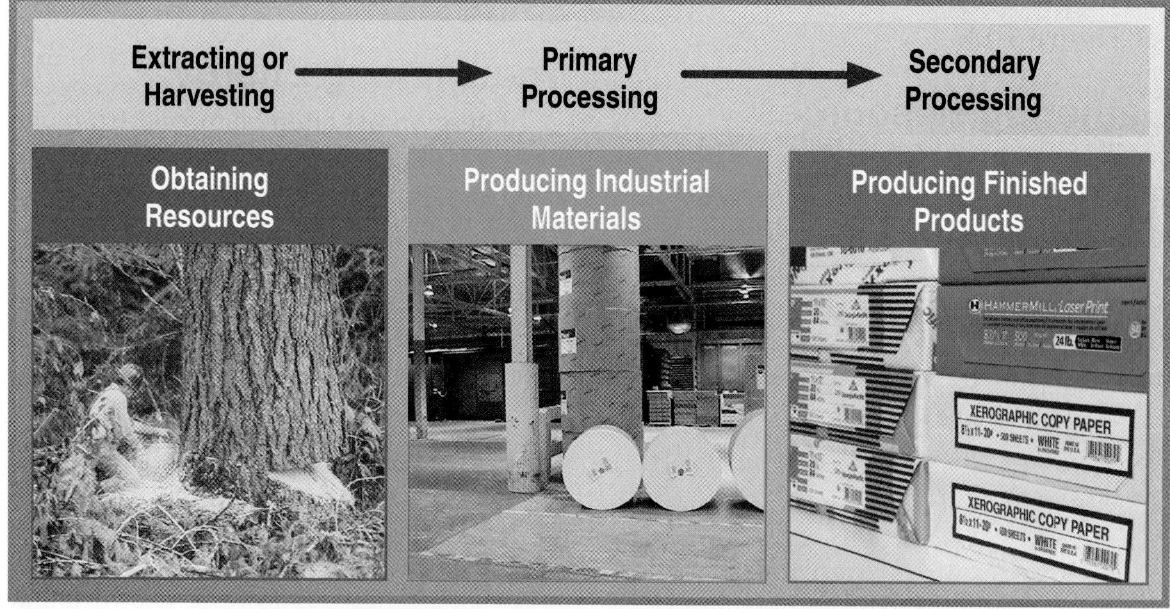

Weyerhauser Co., MeadWestvaco

Figure 5-15. Manufacturing obtains material resources. These resources are changed into industrial materials. The industrial materials are used to make products.

Standards for Technological Literacy

1 16 18

Technology Explained

hybrid vehicle: a vehicle combining two or more sources of power.

With the cost of gasoline at high levels, people are looking for more efficient automobiles. One alternative to the standard car is the hybrid vehicle. What is a hybrid car? A hybrid car is any vehicle combining two or more sources of power.

A hybrid vehicle is not a new concept. A commonplace hybrid vehicle in American life is the diesel-electric locomotive. See **Figure A.** This vehicle combines an internal combustion engine and electric generators and motors. Many submarines are also hybrids. They combine nuclear power with electric generators and motors. Some city buses are hybrids. They use diesel engines for parts of their routes and overhead electric lines to power motors for other parts.

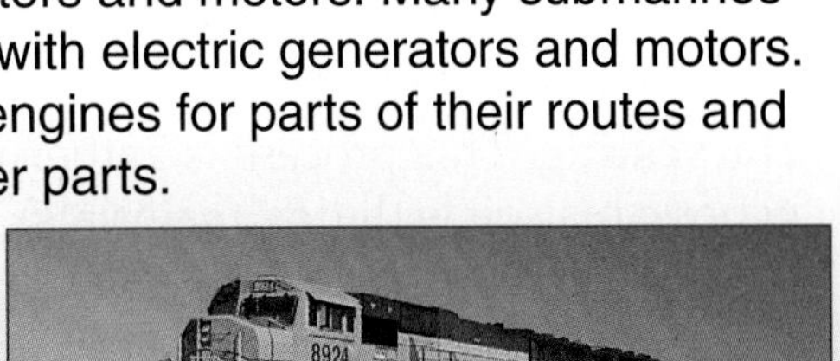

Figure A. A common hybrid vehicle is the diesel-electric locomotive.

A recent development is the hybrid car. This development combines a gasoline-powered car with an electric car. In a standard car, an engine uses gasoline as fuel. The gasoline turns a transmission. The transmission drives the wheels. In an electric car, batteries are the power source (fuel). The electricity powers a motor. The motor drives the wheels through a transmission.

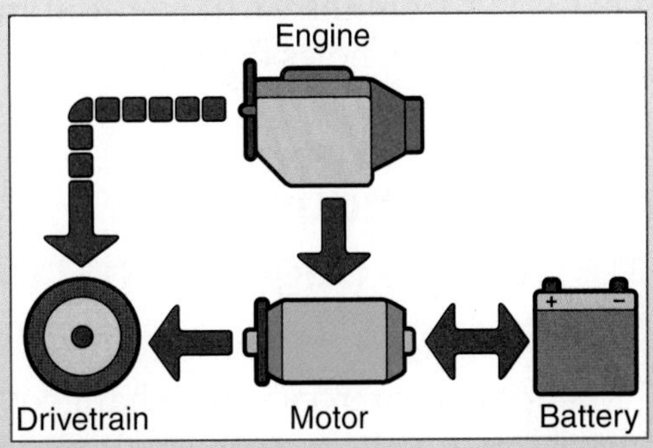

Figure B. The major parts of a hybrid-vehicle power system.

A hybrid vehicle uses both types of power. See **Figure B.** This vehicle uses a gasoline engine smaller than one in a traditional car. The engine might be as small as 1.0 liter and have as few as three cylinders. This gasoline engine uses advanced technologies to increase its efficiency and reduce its emissions. The engine can provide power to the wheels or to the motor or generator to recharge the batteries.

The vehicle also has an advanced electric motor. This power source has special electronic controls allowing it to be both a motor and a generator. At times, it draws energy from the batteries to accelerate the car. At other times, it functions as a generator. The motor recharges the batteries by regenerative braking. During this phase, the energy from forward momentum is captured during braking. Therefore, no external power supply is needed for recharging the batteries.

The operation of the vehicle changes under different driving conditions. During initial acceleration, the electric motor is the primary source of power. The gas engine starts up under heavy acceleration or to turn the generator. The generator, in turn, charges the battery.

During city driving, the electronic control system controls both power sources. The gas engine and electric motor are used equally. The engine starts and stops, depending on the situation. At high speeds, the gas engine is the primary source of power. The electric motor provides some power.

Standards for Technological Literacy

1 19

Extend

Differentiate between raw materials and standard stock.

TechnoFact

Primary processing had existed for thousands of years before the Industrial Revolution because people had been able to shape metals. However, in order to make the machines and transportation systems, advancements in iron and steel were needed. The Industrial Revolution brought about secondary processing in this way.

Obtaining Resources

Manufacturing activities start when we obtain material resources. This effort first involves searching for or growing materials that can be harvested or extracted from the earth. For example, farmers and foresters grow the trees, plants, and animals needed to support the manufacturing processes. Exploration companies locate petroleum, coal, and metal-ore reserves. Once we grow or find the resources, we must gather or obtain them. These activities use one of three major processes:

- **Harvesting.** This process is gathering genetic materials (living materials) from the earth or bodies of water at the proper stage of their life cycles.
- ***Mining.*** This process is obtaining materials from the earth through shafts or pits.
- ***Drilling.*** This process is obtaining materials from the earth by pumping them through holes drilled into the earth.

Producing Industrial Materials through Primary Processing

Material resources are changed to industrial materials in primary-processing plants. For example, steel mills change iron ore, limestone, coke, and other materials into steel sheets and bars. Forest-products plants change trees into paper, lumber, plywood, particleboard, and hardboard. Smelters change aluminum and copper ore into usable metal sheets, bars, and rods.

Typically, these materials are refined using heat (smelting and melting, for example), mechanical action (crushing, screening, sawing, and slicing, for example), and chemical processes (oxidation, reduction, and polymerization, for example). The result is an industrial material or a standard stock. Examples include a sheet of plywood, glass or steel, a plastic pellet, and metal rods and bars.

Creating Finished Products through Secondary Processing

Most industrial materials have little value to the consumer. What would you do with a sheet of plywood, a bar of copper, or a bag of plastic pellets? These materials become valuable when they are made into products. They are the material inputs to the ***secondary processes*** of manufacturing. These processes change industrial materials into industrial equipment and consumer products. See **Figure 5-16.**

Figure 5-16. Secondary manufacturing produces industrial equipment and consumer products.

Secondary processing changes materials in six basic ways. Casting and molding processes, forming processes, and separating processes give materials size and shape. Materials also undergo conditioning, finishing, and assembling.

- *Casting and molding* introduce a liquid material into a mold cavity. There, the material solidifies into the proper size and shape.
- *Forming* uses force applied from a die or roll for reshaping materials.
- *Separating* uses tools to shear or machine away unwanted material. This activity shapes and sizes parts and products.
- *Conditioning* uses heat, chemicals, or mechanical forces to change the internal structure of the material. The result is a material with new, desirable properties.
- *Finishing* coats or modifies the surfaces of parts or products to protect them, make them more appealing to the consumer, or both.
- *Assembling* brings materials and parts together to make a finished product. They are bonded or fastened together to make a functional device.

The results of obtaining resources, producing industrial materials, and manufacturing products are the devices we use daily. More information about these manufacturing activities is in Chapters 14, 15, and 16.

Medical Processes

People in the industrialized nations are living longer. This longer life expectancy is partly the result of better medical care. Technological and scientific advances have given health-care professionals new tools to treat diseases and the aging process. These tools are used in three major areas. See Figure 5-17:

- ***Prevention.*** This area is using knowledge, technological devices, and other means to help people maintain healthy bodies. Prevention focuses on such health aspects as proper nutrition, exercise, immunizations against diseases, and proper actions at work and recreation.
- ***Diagnosis.*** This area is using knowledge, technological devices, and other means to determine the causes of abnormal body conditions. Diagnosis is a type of problem solving that matches symptoms to possible causes. For example, a person who experiences back pains might have vertebrae damage, muscle strains, or cancer of the spine. By using various tests, health-care professionals can determine the actual cause of the pain.

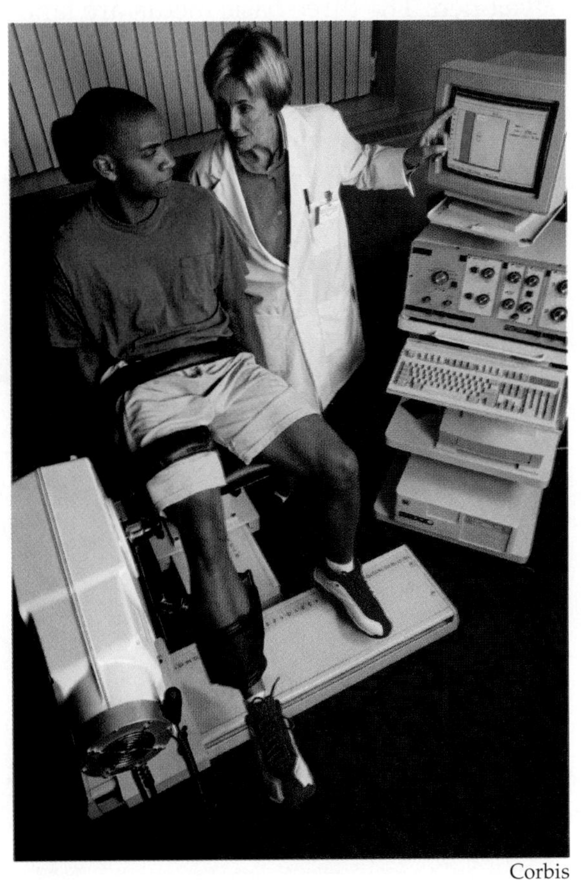

Corbis

Figure 5-17. Medical technologies help people stay well through disease prevention, diagnosis, and treatment.

Standards for Technological Literacy

1 14

Extend

Discuss why many people are putting emphasis on wellness (prevention).

TechnoFact

Humans have been using various medical treatments throughout time. One of the first implements of medical technology came from ancient Egypt, where several potent drugs were developed. Although many medical advancements have been made since that time, various Egyptian remedies are still being used today to treat less serious illnesses.

Standards for Technological Literacy

1 3 7 14 18

Figure Discussion

Ask your students for examples of vehicles used in each transportation media shown in Figure 5-18.

TechnoFact

Transportation is the moving of people or cargo from one destination to another by means of various transportation media. These methods of transportation include land, water, air, and space. Vehicles play a key role in travel. Until the invention of the hot-air balloon, the only means of travel were land and water.

- ***Treatment.*** This area is using knowledge, technological devices, and other means to fight diseases, correct body malfunctions, or reduce the impact of a body condition. Treatment includes such things as surgery, chemicals (drugs), physical therapy, and other medical actions.

You have the opportunity to learn more about medical technology in Chapter 31.

Transportation Processes

Humans have always had the desire to move around. They frequently move themselves and their possessions from one place to another. This movement is called *transportation*. One way to understand transportation is to look at the media in which the various transportation systems have been developed. The three basic media are land, water, and air. A fourth one, space, is on the horizon. See **Figure 5-18.** Another way to understand transportation is to look at its subsystems. The subsystems are vehicular systems and support systems. Both the media and the subsystems are discussed as follows.

Transportation Media

The earliest movement occurred on land. People first walked the land. They then tamed animals to help them move from one place to another. Finally, they developed various land vehicles. These vehicles range from the horse-drawn wagons of old to modern magnetic-levitation trains.

The second transportation medium is water. Again, water transportation has a long history. Means of transportation in this area have ranged from dugout log canoes to nuclear submarines.

The third major transportation medium is air. This medium is of recent vintage.

Figure 5-18. Transportation can be conducted on land, over water or underwater, and through the air. Later, we will probably use space transportation systems more.

Air can be traced back to early balloon travel and now extends to supersonic airplanes.

We are starting to develop a fourth transportation medium—space. Today, we explore only the near reaches of space. See **Figure 5-19.** The future holds the promise of space transportation systems, however.

Figure 5-19. Space travel is presently used for experimental work and study.

Transportation Subsystems

Most transportation systems contain two major subsystems: vehicular systems and support systems. ***Vehicular systems*** are the onboard technical systems that make a vehicle work. Most vehicles are made up of five systems. See **Figure 5-20:**

- ***Structure.*** This system provides spaces for devices. The structure, or framework, includes passenger, cargo, and power-system compartments.

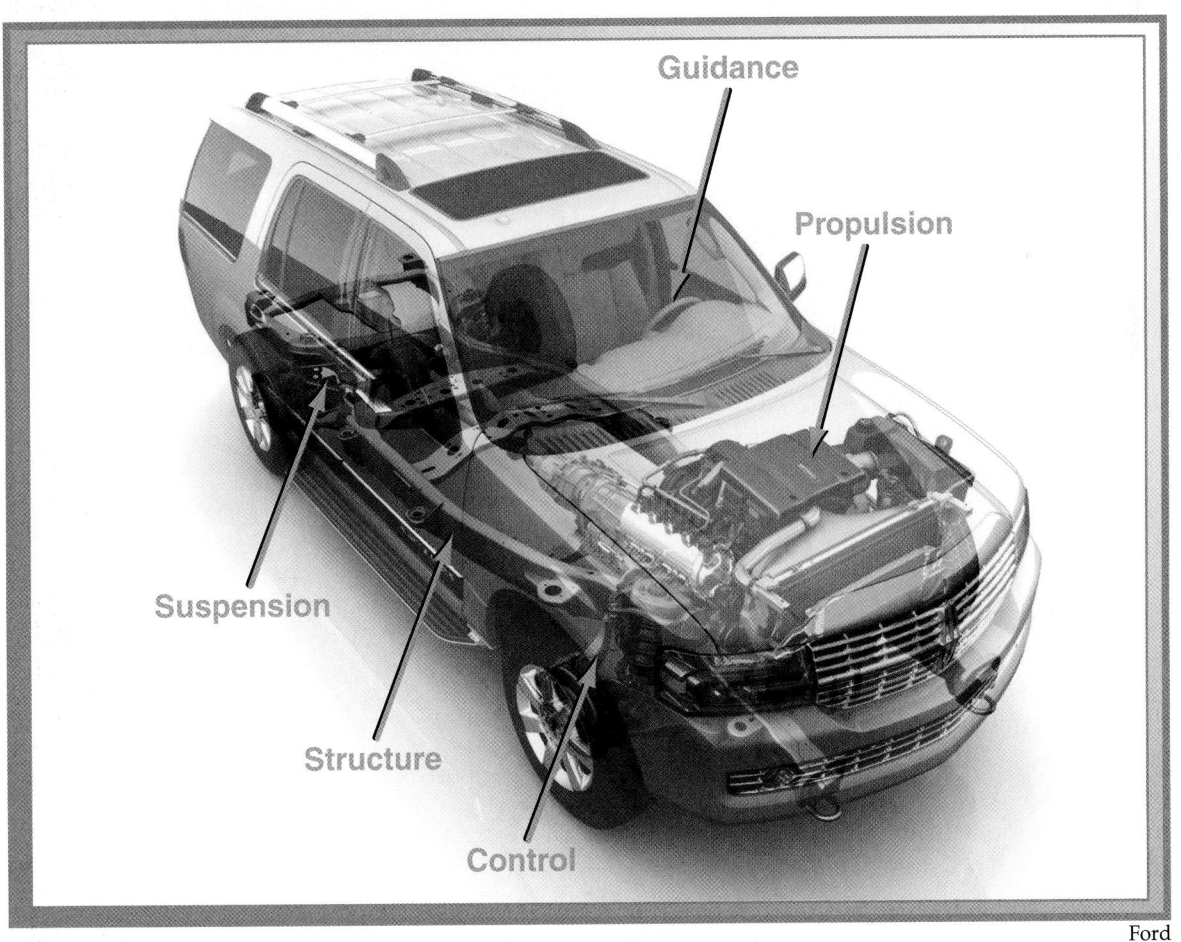

Ford

Figure 5-20. Most vehicles are made up of five systems.

Standards for Technological Literacy

Brainstorm

Ask your students what they think lies ahead in space travel.

Figure Discussion

Contrast the subsystems in Figure 5-20 with those of a bicycle.

TechnoFact

Although air transportation started in 1783 with the invention of the hot-air balloon, the idea of commercial air travel didn't occur until the beginning of the 20th century when the Wright brothers successfully completed their flight at Kitty Hawk, North Carolina in 1903.

Standards for Technological Literacy

1 2 18

Extend

Discuss the difference between a fixed pathway (rail line, highway) and a flexible pathway (seaway, flight path).

Reinforce

Discuss management as a technology that uses knowledge and machines to extend the human potential.

TechnoFact

Pathways are an essential part of all transportation systems, but visible pathways are more central to travel on land. About 4000 years ago, workers in China built the world's first permanent road system. The Romans built 53,000 miles (85,000 km) of roads throughout their vast empire.

- ***Propulsion.*** This system generates motion through energy conversion and transmission.
- ***Guidance.*** This system gathers and displays information so the vehicle can be kept on course.
- **Control.** This system makes changes in speed and direction of the vehicle possible.
- ***Suspension.*** This system keeps the vehicle held in or onto the medium being used (land, water, air, or space).

Support systems include pathways and terminals. See **Figure 5-21:**

- ***Pathways.*** This system is the structures along which the vehicles travel. Pathways include roads, railways, waterways, and flight paths.
- ***Terminals.*** This system is the structures that house passenger and cargo storage and loading facilities.

A complete discussion of transportation technology is included in Section 7 of this book.

©iStockphoto.com/DNY59, Amtrak

Figure 5-21. Transportation support systems include pathways such as roads and railway terminals.

Management Processes

The third major type of process used in technology is management. Management processes are those designed to guide and direct the other processes. Management provides the vision for the activity. Specifically, management includes four steps. See **Figure 5-22.** The first step is ***planning.*** Planning involves developing goals and objectives. The goals can be broad or specific and focus on such areas as production, finance, and marketing.

Once plans are developed, the activity must be structured. Procedures to reach the goal must be established. Lines and levels of authority within the group or enterprise must also be drawn. These actions are called *organizing.*

After the activity is organized, actual work must be started. The starting of the work is called *actuating.* Workers must be assigned to tasks. They must be motivated to complete the tasks accurately and efficiently.

Finally, the outputs must be checked against the plan. This checking is termed

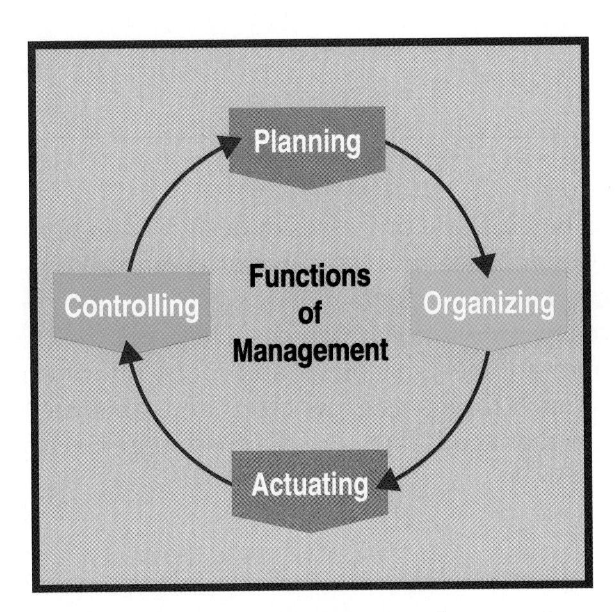

Figure 5-22. Management involves setting goals (planning), establishing a structure to meet them (organizing), assigning and supervising work (actuating), and monitoring the results (controlling).

controlling. Control is the feedback loop that causes management activities to be adjusted.

Sometimes, we think of management only in terms of companies. In reality, every human activity is managed. Goals are set. A procedure for finishing the task is established. Work is started and completed. Results are evaluated. For example, a coach plans, directs, and evaluates the plays a football team makes. A Girl Scout troop leader manages cookie sales. You develop a plan to mow a lawn.

We manage not only personal, but also group, activities. Some group activities take place within social, religious, and educational organizations. Others occur in technological enterprises. The management of technological activities and companies is explored in more detail in Section 10 of this book.

Standards for Technological Literacy

2

TechnoFact

The need for the management process arose with the dawn of the Industrial Revolution in the late 18th century. Instead of a person managing his or her own activities by creating and selling objects alone, several people began to work together in factories to manufacture numbers of products simultaneously. In order to ensure the work was done properly, managers supervised the factory workers.

TechnoFact

With the development of automation and computerized information processing, a second industrial revolution is taking place. Today, emphasis is put on improving the productivity and product quality in manufacturing. New approaches to management and teamwork are being sought as a result.

Answers to Test Your Knowledge Questions

1. management
2. False
3. Evaluate individually.
4. False
5. conversion/processing
6. True
7. Evaluate individually.
8. foundation
9. C. Mechanical
10. Harvesting, mining, and drilling
11. True
12. treatment
13. land
14. Structure, propulsion, guidance, control, and suspension
15. Planning, organizing, actuating, and controlling

Summary

All technological systems have processes. They include processes to design and engineer the outputs and the systems themselves. Often, these processes use problem-solving methods. These methods guide the creative activities of the field.

Once we design the system and outputs, we use production processes. These processes transform inputs into outputs. Materials, human abilities, machines and tools, information, energy, finances, and time are used to do such things as grow crops and construct structures. Finally, all technological systems are managed. They are planned, organized, actuated, and controlled.

Test Your Knowledge

Write your answers on a separate piece of paper. Please do not write in this book.

1. The three major types of technological processes are problem solving and design, production, and ______.
2. The three major technological processes can be applied generically across technological systems. True or false?
3. Give three examples of criteria that might need to be considered when identifying a problem in the problem-solving and design process.
4. The best solution to a problem resolves all the criteria and constraint requirements. True or false?
5. The agricultural production process includes propagation, growth, harvesting, and ______.
6. The first step in the communication process is encoding the information. True or false?
7. Give one example of a heavy engineering structure.
8. The part of a constructed work supporting the superstructure is called a(n) ______.
9. ______ energy is energy found in moving objects.
   A. Thermal
   B. Chemical
   C. Mechanical
   D. Electrical
10. List the three major processes used to gather or obtain manufacturing materials.
11. Secondary processing changes industrial materials into equipment and products. True or false?
12. The three areas in which health-care professionals use technological tools are prevention, diagnosis, and ______.

13. Transportation media include water, air, space, and ______.
14. List the five vehicular systems common to most vehicles.
15. List the four steps in the management process.

STEM Applications

1. Select a product or structure about which you are familiar. Develop a chart similar to the one below. List five considerations used in each of the technological processes used to produce the item.

Product or structure name:	
Process	**Factors Considered**
Design and problem solving	
Production	
Management	

2. Select a problem existing in the world that can be partially solved through technology. List the role each technology would play in solving the challenge.
3. List and describe how you have used the production processes in one of the technological areas (agriculture and related biotechnology, communication and information, construction, energy and power, manufacturing, medicine, or transportation) to meet a need. Use a chart similar to the one below.

Product or structure name:	
Technology	**What You Did and the Need That Was Met**
Agriculture and related biotechnology	
Communication and information	
Construction	
Energy and power	
Manufacturing	
Medicine	
Transportation	

Chapter Outline
Outputs
Feedback and Control

This chapter covers the benchmark topics for the following Standards for Technological Literacy:

2 3 4 5 6

Learning Objectives

After studying this chapter, you will be able to do the following:

- Recall the major types of outputs of various technology systems.
- Explain the relationship between outputs and feedback and control in technology systems.
- Compare desirable and undesirable outputs of technological systems.
- Compare intended and unintended outputs of technological systems.
- Compare immediate and delayed outputs of technological systems.
- Recall the definition of *feedback,* as used in technology systems.
- Recall the two major types of internal control systems and their functions.
- Recall the major components of internal control systems and their functions.
- Compare manual and automatic control in internal control systems.
- Give examples of external controls in technology systems.

Key Terms

adjusting device
analytical system
automatic control system
closed loop control
data-comparing device
delayed output
desirable output
electrical and electronic controllers
electrical or electronic sensor
electromechanical controller
fluidic controller
immediate output
intended output
judgmental system
magnetic (electromagnetic) sensor
manual control system
mechanical controller
mechanical sensor
monitoring device
open loop control
optical sensor
thermal sensor
undesirable output
unintended output

Strategic Reading

As you read this chapter, think of an example technology, either current or no longer in use. Use the three categories given in this chapter to describe this technology's outputs.

All technological systems are designed to meet specific needs and wants of people. Therefore, a direct relationship exists between the need (the reason for the system) and the output (the satisfying of the need). These outputs can be categorized in three ways. See **Figure 6-1:**

- Desirable or undesirable.
- Intended or unintended.
- Immediate or delayed.

The people designing and using technology should strive to maximize the desirable results. The undesirable outputs can never be totally eliminated. They need to be held, however, to a minimum. To achieve this goal, people have created control systems that compare the results of the outputs with the goals. As seen in Chapter 2, these systems use the feedback to regulate the inputs. We now examine in more depth both the outputs and the feedback and control systems essential to understanding and improving technological systems.

Outputs

From the systems model you studied in Chapter 2, remember that resources (inputs) are transformed (processed) into things people want or need (outputs). These outputs can be seen as positive (desirable) or negative (undesirable). They can also be described as planned for (intended) or unplanned (unintentional). Finally, outputs can happen right now (immediately) or appear at a later date (delayed). In this section, you will explore each of these types of outputs.

Desirable and Undesirable Outputs

The needs for and outputs of technology are countless. Each of the technological systems we have already classified (agriculture and related biotechnology, communication and information technology, construction technology, energy and power technology, manufacturing technology, medical technology, and transportation technology) has a general type of output. See **Figure 6-2.** For example, agricultural and related biotechnology helps people satisfy their needs for nourishment. Biotechnology's outputs are plants and animals that can be processed into food. Communication technology helps satisfy the need for information. This technology's outputs are media messages. Construction

Standards for Technological Literacy

4 6

Reinforce

Review outputs and feedback as part of a technological system (Chapter 2).

Brainstorm

Have your students select technologies and list the desirable and negative outputs of those technologies.

TechnoFact

The Colorado River, a valuable natural resource, has a flow of 200,000 cubic feet of water per second and travels over 1800 miles through dry Western states. Its reservoir, called Lake Mead, backs up 140 miles in Nevada and Arizona. The dam and its reservoir provide irrigation and municipal water to several states, control flooding, regulate river levels through the Grand Canyon, and can produce 2000 megawatts of electric power.

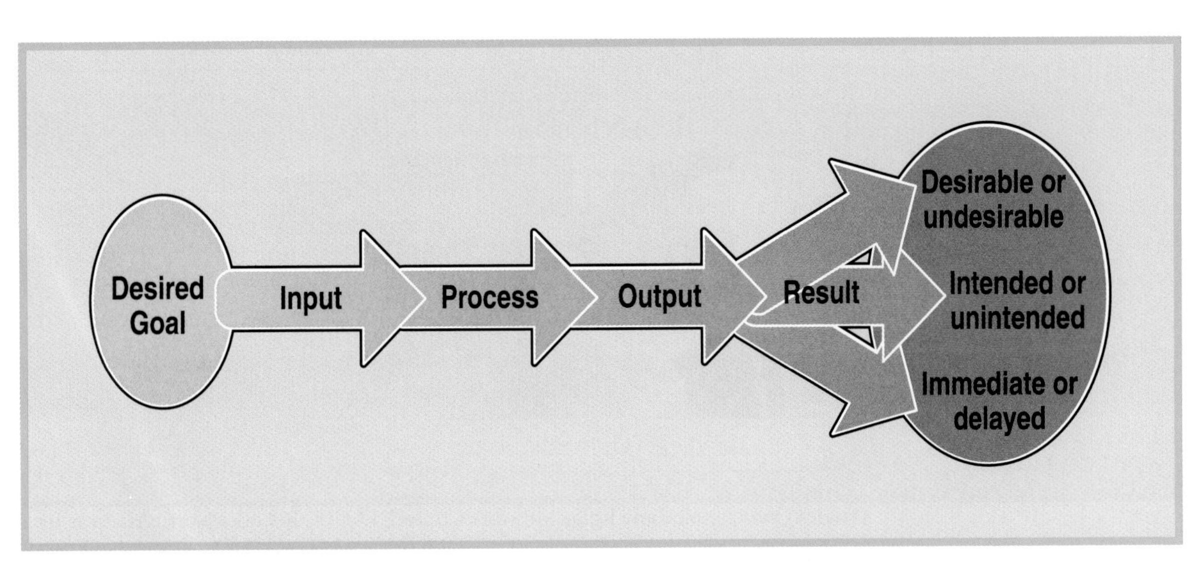

Figure 6-1. The outputs of a technological system can be categorized in three ways.

Standards for Technological Literacy

4

Figure Discussion

Discuss the desirable and undesirable impacts of each of the seven technologies shown in Figure 6-2.

TechnoFact

Even before the Industrial Revolution, the desire for exploration of the world gave way to the advancement of water transportation. This made travel to other continents more available to people, and economies grew stronger because of trade among various nations.

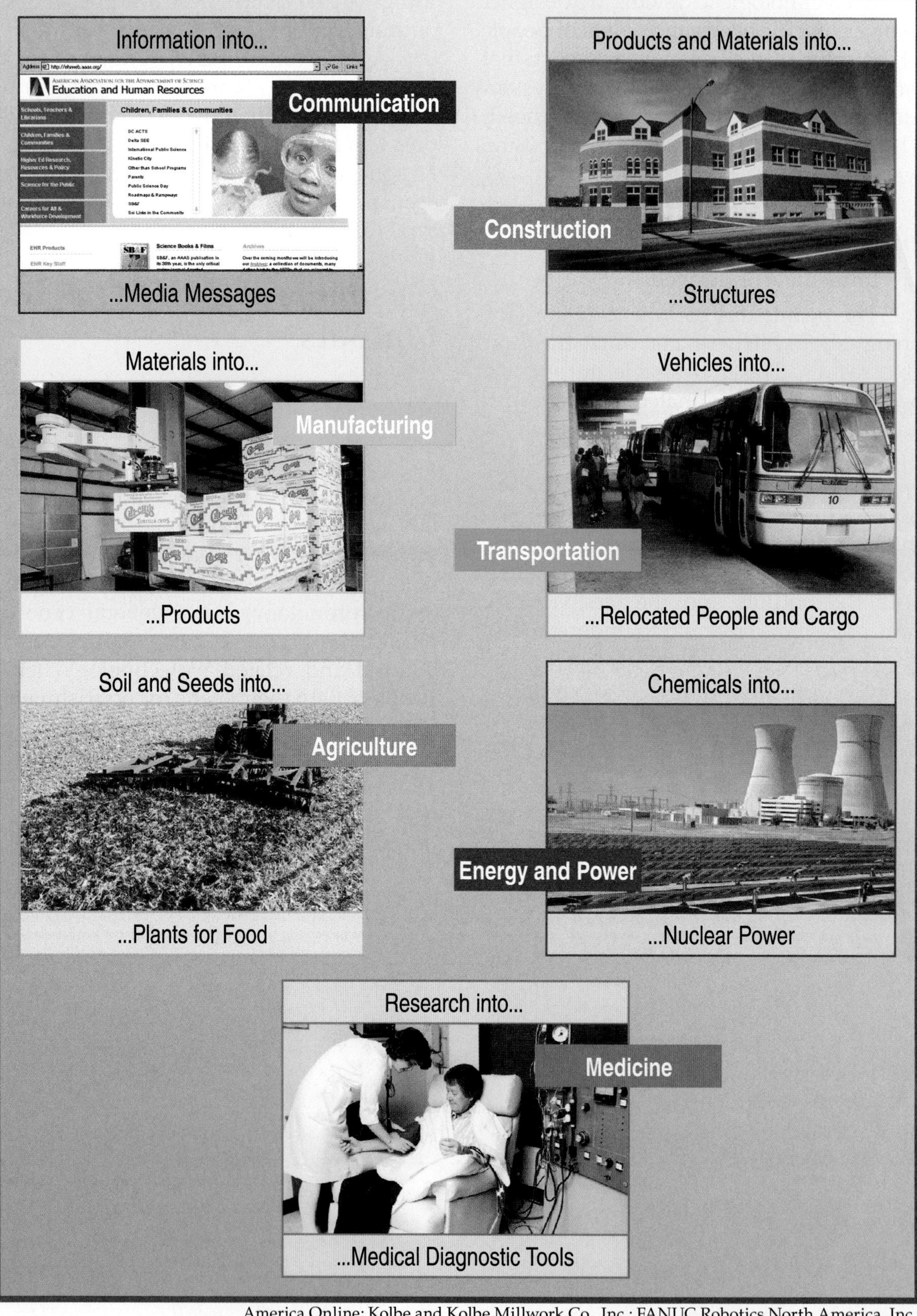

America Online; Kolbe and Kolbe Millwork Co., Inc.; FANUC Robotics North America, Inc.; John Deere and Co.; U.S. Department of Energy

Figure 6-2. Each major type of technological system has a general type of output.

technology satisfies the need for shelter and support structures for other activities. This technology's outputs are buildings and civil structures. Energy and power technology helps satisfy the needs for heat, light, and motion. The output is mechanical, thermal, radiant, chemical, electrical, or nuclear energy. Manufacturing technology satisfies the need for tangible goods. This technology's outputs are consumer and industrial products. Medical technology helps satisfy our need for health and well-being. This technology's outputs are devices and treatments for medical ailments and for promotion of healthy bodies. Transportation technology satisfies a desire to move humans and things. This technology's output is the movement of people and goods (cargo). These are the ***desirable outputs*** of the systems.

In addition to desirable outputs, however, ***undesirable outputs*** also exist. See Figure 6-3. For example, some manufacturing activities produce fumes and toxic chemical by-products. If improperly treated, these outputs poison the air and water around us. Poorly planned agricultural and construction projects can cause soil erosion. The result can be the loss of valuable topsoil and increased flooding. Communication technology can use billboards. Some people find billboards unsightly. Also, people differ on the value of some communication messages. Transportation systems create outputs we do not want. Exhaust fumes from cars and trucks pollute the air. Airport and highway noise impinge on residential areas. New rail lines and roads can separate neighborhoods or divide farms. All these are examples of the undesirable outputs of technology.

John Deere and Co.

Figure 6-3. Outputs of technological systems can be seen as desirable or undesirable.

Intended and Unintended Outputs

As we said, technological systems are designed to produce specific outputs. These outputs include such varied items as refrigerators, weekly magazines, high-rise apartments, frozen foods, municipal bus transportation, and thousands of other products and services. They are the ***intended outputs***.

Depending on your point of view, the intended outputs might be desirable or undesirable. For example, suppose you plan to make 10″-diameter plywood disks. You have to cut them from rectangular sheets of plywood. Cutting circles from rectangular material produces waste material. You can plan for the waste, consider it in pricing your product, and reduce it to a minimum. The waste, however, cannot be totally eliminated. Both the disks and the

Standards for Technological Literacy

4 6

Research

Have your students select a technology and research its impacts on people, society, and the environment.

Research

Have your students select a technological system and explore its intended and unintended outputs.

TechnoFact

While technology has its advantages, it also has its drawbacks. It fulfills our needs and wants, but it affects the natural environment. For example, we benefit from what is produced in steel mills, but at the same time, smokestacks pollute the air.

Standards for Technological Literacy

4 6

Brainstorm

Ask your students to list the immediate and delayed outputs of a nuclear power generation.

TechnoFact

The Industrial Revolution in America created a demand for steel, which gave rise to great steel mills in Pittsburgh. This area was so important because it was near coal fields and cheap water transportation to move iron ore. Though the city grew with much success, it now has a major air-pollution problem.

TechnoFact

All changes impact the natural environment, whether they are technological or natural. An undesirable output from the developing technology around us is the effect of that technology on the environment. Although this output may not be entirely prevented, it is possible to take the environment into consideration when developing new technology.

waste are intended. They are in the plan for the system. The disks are the intended, desirable output. The waste is the intended, undesirable output.

Sometimes, technological systems produce ***unintended outputs***, that is, outputs that were not considered when the system was designed. For example, some heating systems produced in the 1940s through the 1960s used asbestos to insulate pipes. The material was considered an excellent insulator. Later, people who worked with asbestos developed lung cancer. This was an undesirable, unintended output of the technological system.

Immediate and Delayed Outputs

Technological systems are commonly designed to produce a product or service for use now. Thus, we have ***immediate outputs***. For example, we do not produce steel for use in the year 2051. We produce it for use this year. This practice reduces inventory costs and loss due to theft and spoilage. The production of steel is an example of an immediate, intended, and desirable output.

We also create ***delayed outputs***, however. The chemicals used as propellants in aerosol (spray) cans and as refrigerants are now affecting the ozone layer around the Earth. The accumulation of this matter has taken decades to reach a dangerous level. The same is true of the sulfur dioxide coal-fired power plants produce. The resulting acid rain is killing forests in the United States and Canada. These examples show some delayed, unintended, and undesirable outputs of technology. Thus, although we design technological systems to produce specific outputs, many kinds of outputs can occur. The outputs of various types of technological systems are shown in **Figure 6-4.**

Feedback and Control

As we noted earlier, humans have created control systems so they can reach their desired goals. Most of these systems use

Career Corner

Dental Hygienists

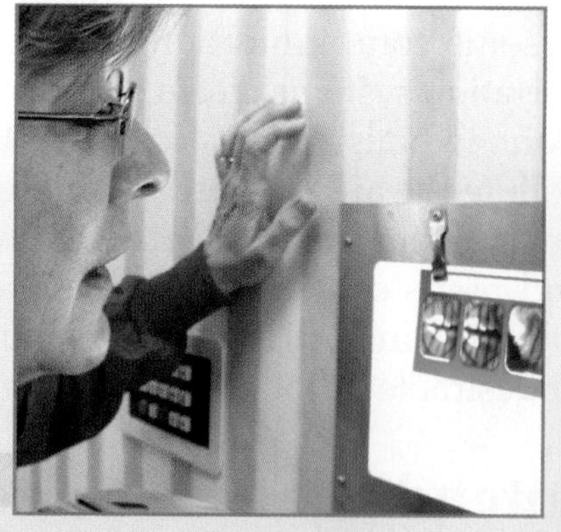

Dental hygienists remove deposits from teeth, teach good oral-hygiene practices, and provide preventive dental care. They generally work in dentists' offices. Dental hygienists examine patients' teeth and gums and record the presence of diseases in these offices. Hygienists remove plaque from teeth, take dental X rays, and apply cavity-preventive agents. They use a variety of tools and machines. These tools and machines include instruments and ultrasonics to clean teeth and X-ray machines to take dental pictures.

Dentists frequently hire hygienists to work only part of each week. Many hygienists have part-time jobs or work in more than one dental office. Dental hygienists must be licensed by the State and, therefore, must graduate from an accredited dental-hygiene school. Also, they must pass a written and clinical examination.

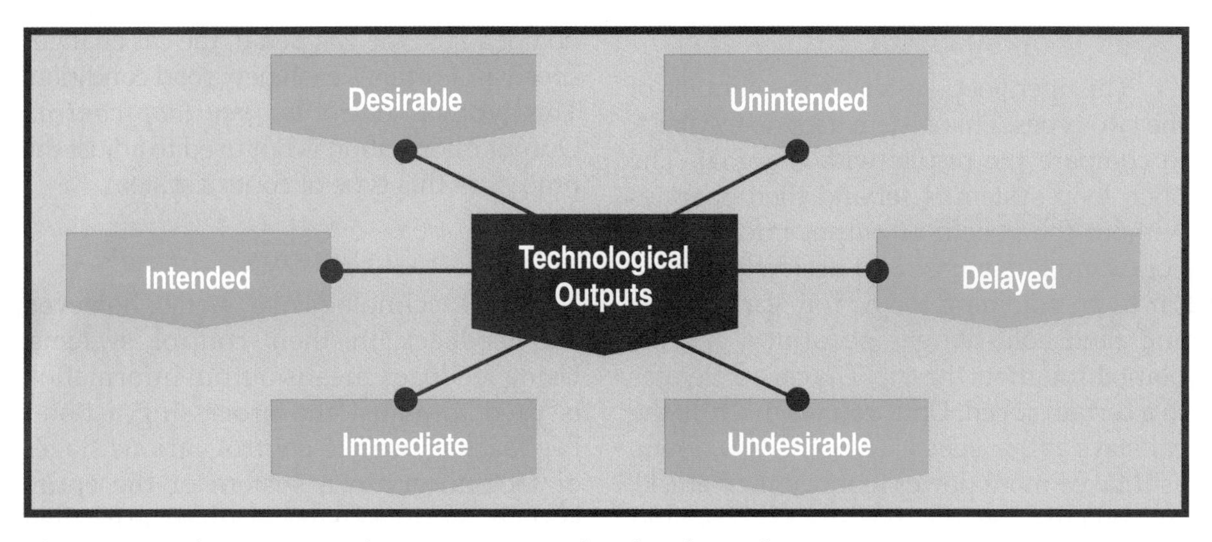

Figure 6-4. The outputs of various types of technological systems.

Standards for Technological Literacy

2 4

Figure Discussion

Discuss the impacts that a technological device has, as shown in Figure 6-4.

Brainstorm

Ask your students how they think the control device in the TechnoFact on this page worked.

TechnoFact

Control systems have been in existence for over 2000 years. This is proven by the feedback control device created by Hero of Alexandria. In order for his wine bowl to refill itself, he invented a float valve that sensed the level of wine in a bowl.

feedback. *Feedback* can be defined as using information from the output of a system to control, or regulate, the system. We often view control systems as internal. That is, we look at how operating processes are controlled so outputs meet specifications. To understand technological system control completely, we also need to consider controls that are external to the system. For example, political actions and personal values also exercise control. We explore both internal and external areas of control in the rest of this chapter.

Internal Controls

We can view internal control systems from three different angles. First, we can explore the systems according to their type: open loop or closed loop. Second, we can explore the components found in many control systems—monitoring, data-comparing, and adjusting devices. Finally, we can explore the operation of control systems—manual versus automatic.

Types of Control Systems

In technology, we use two major types of control systems. These types are the open loop system and the closed loop system. See Figure 6-5.

FMC Corp., American Electric Power

Figure 6-5. Technology uses open loop and closed loop control systems. This food-processing equipment runs at a set speed (open loop). These power generators automatically adjust to increased demand for electricity (closed loop).

Standards for Technological Literacy

2 3 5

Extend

Describe steering a bicycle as an open-loop control system.

Research

Have your students research the ways control systems are used in an automobile.

TechnoFact

An automobile is controlled by a number of systems. One type is speed and direction controls. These controls are made possible by the steering, braking, and fuel rate control systems. These systems are open-loop systems that the driver controls through personal decisions. He or she may turn the steering wheel, step on the brakes, or let up on the accelerator without information from the vehicle.

Open loop systems

The open loop system is the simpler of the two types. This system uses no feedback to compare the results with the goal. The open loop system is set and then operates without the benefit of output information. For example, suppose you decide to drive a car to an adjoining town. You start the car and then hold the gas pedal at a specific point throughout the trip. The car accelerates to a certain speed. Until you reach a hill, the car stays at that speed. The car slows down, as it takes more power to move the vehicle. As you crest the hill, the car speeds up and coasts down the other side. As long as you do not adjust the gas pedal, the car changes speed as it reaches each new road condition. This type of control is ***open loop control***. Output information is not used to adjust the process in this type of control system.

Closed loop systems

Most technological systems, however, use feedback in their control systems. Using feedback means output information is used to adjust the processing actions. Feedback is used to control various stages or factors within a system or the entire system. Systems using feedback are called *closed loop control systems*.

STEM Connections: Science

Science Technology Integrated STEM Curriculum Engineering Mathematics

Chlorofluorocarbons (CFCs)

As this chapter indicates, we sometimes produce new technologies that turn out to have undesirable consequences. Such was the case with a group of synthetic organic compounds called *chlorofluorocarbons (CFCs)*. CFCs are a group of compounds that contain carbon, chlorine, and fluorine and often include hydrogen. They change easily from liquid to gas or from gas to liquid and are used as refrigerants.

The development of a group of CFCs, in the late 1920s, was originally considered to be a miracle find. Until that time, refrigerators contained toxic gases. Sometimes, fatalities occurred when the gases leaked from the refrigerators. Freon refrigerant seemed to solve the problem. This refrigerant is odorless, colorless, nonflammable, and noncorrosive. Within a few years, Freon refrigerant became the standard refrigerant for refrigerators and air conditioners.

As time went on, however, scientists discovered that CFCs damage the environment because these compounds break down ozone molecules in the upper atmosphere of the Earth. In 1978, the United States started to ban the use of some CFCs. An international treaty, called the *Montréal Protocol*, followed this action in 1987. All but one of the United Nations–member nations ratified this treaty. The treaty banned the production of ozone-depleting substances, such as CFCs, in industrial countries by 1996, while developing countries were allowed to continue to produce these substances until 2010. What do we now use as a refrigerant in most air conditioners and refrigerators?

Many early air conditioners used Freon refrigerant (a CFC).

Let us look at ***closed loop control*** in one technology system—manufacturing. See Figure 6-6. After using market research to gather and analyze customers' positive and negative reactions, manufacturers evaluate present and future product designs. Using a process called *quality control*, manufacturers compare parts, assemblies, and finished products with engineering standards. See Figure 6-7. Quality control ensures that the product performs within an acceptable range. Using a process termed *inventory control*, manufacturers compare finished goods in warehouses and products in process with sales projections. See Figure 6-8. The goal is to match as closely as possible the product production with the product demand. With process control, manufacturers closely monitor the operation of machines and equipment. The goal is to see that each manufacturing process produces the proper outputs. Using material-resource control (or planning), manufacturers monitor the need for materials and the quantity on order. The goal of this activity is to reduce the raw material inventory. Finally, using a process called *wage control*, manufacturers monitor the hours employees producing products work. The goal of this activity is to keep the labor content of a product within planned limits.

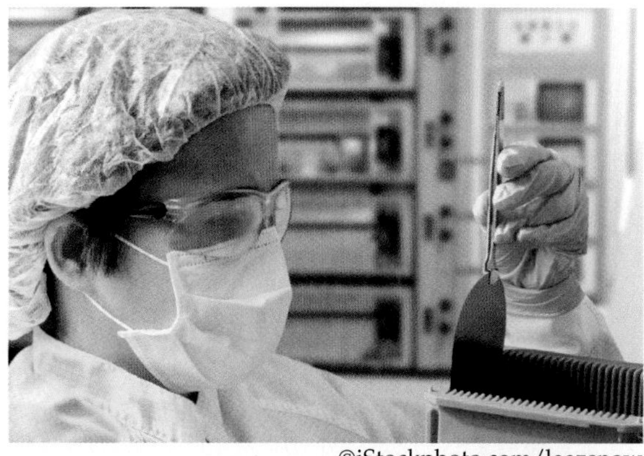

©iStockphoto.com/leezsnow

Figure 6-7. This inspector is gathering quality control data on a silicon-chip manufacturing line.

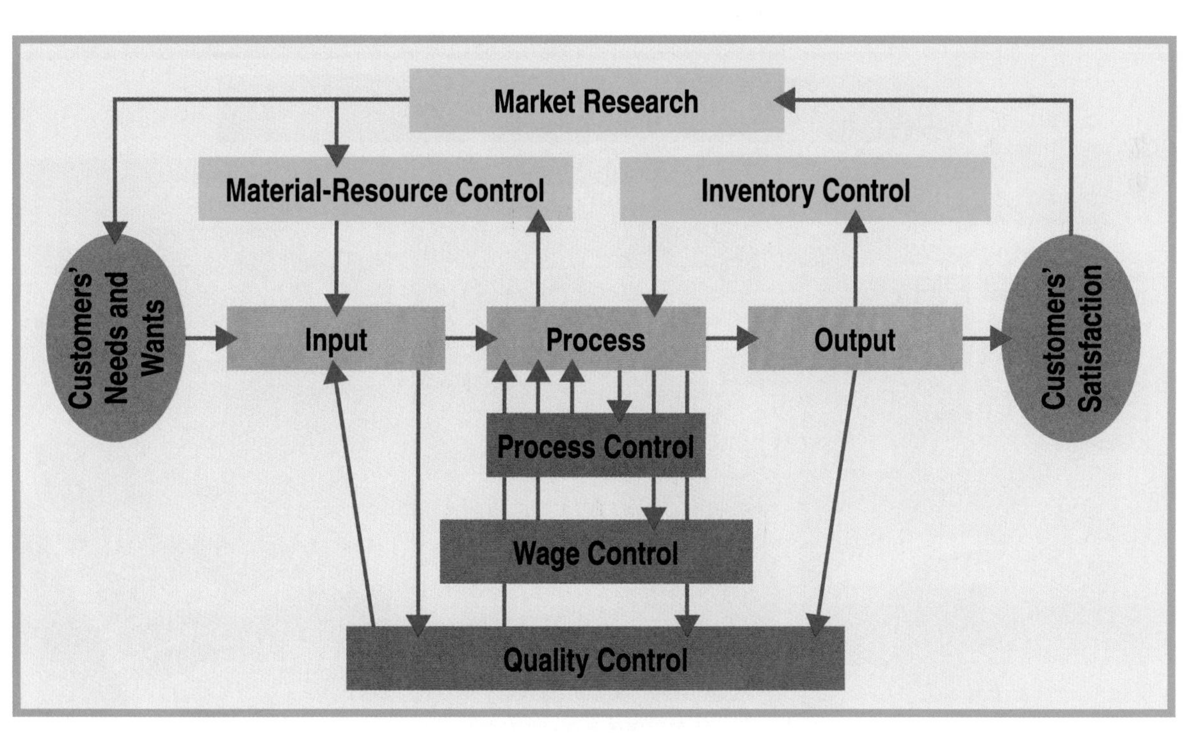

Figure 6-6. With feedback from market research, manufacturers use various types of control to meet customers' needs and wants. Information enters the control system, is processed, and is then fed back into the system.

Standards for Technological Literacy

2

Figure Discussion

Have your students discuss the advantages of using the control systems shown in Figure 6-6.

TechnoFact

During World War II, a control system to guide Allied torpedoes was developed. It consisted of using random frequencies to send signals to radio-controlled torpedoes. This system helped avoid radio-jamming equipment.

Standards for Technological Literacy

2 5

Figure Discussion

Referring to Figure 6-9, discuss the meaning of monitor, compare, and adjust.

Reynolds Metals Co.

Figure 6-8. This worker is checking the inventory of tread plates used in making truck bodies.

Control is not limited to manufacturing, of course. Automobiles have emission control systems. Engines have governors to control their operating speeds. Control systems monitor the ink being applied to paper during the printing process. Thermostats control the temperatures in buildings, ovens, and kilns. Fuses and circuits breakers control the maximum amount of electric current in a circuit. In other words, control is everywhere.

Components of Control Systems

We want technological services, products, information, and structures to meet our needs. To ensure that this happens, we design control systems with three major components. See **Figure 6-9.** These components are the following:

- Monitoring devices.
- Data-comparing devices.
- Adjusting devices.

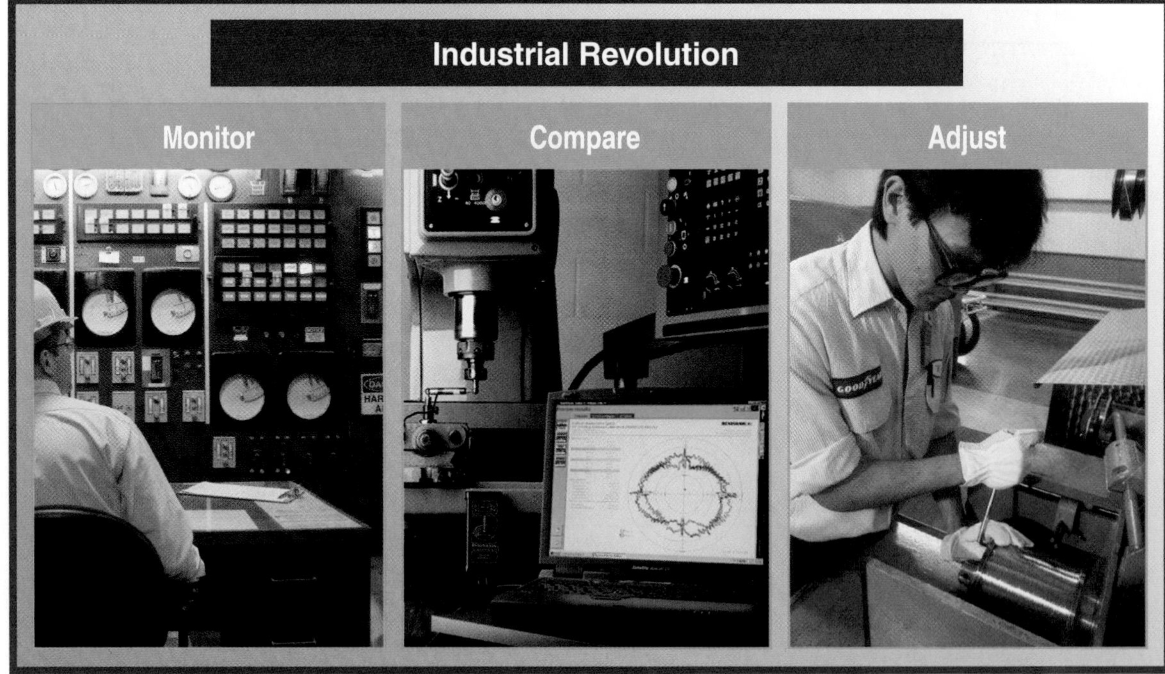

©iStockphoto.com/kickstand; Renishaw, Inc.; Goodyear Tire and Rubber Co.

Figure 6-9. Control systems monitor performance, compare it against standards, and adjust the system to ensure the output meets the goal.

Monitoring devices

The first component of a control system, the ***monitoring device***, gathers information about the action being controlled. This component monitors inputs, processes, outputs, or reactions. The information gathered can be of many types. For example, information can include such varied data as material size and shape, temperatures of enclosures, speed of vehicles, and impacts of advertising. The information can also vary, in that it can be very specific and mathematical or more general. In addition, information can differ, in that it might be gathered and recorded in short intervals or over long periods of time.

The type of information gathered during the monitoring step depends on how the data will be used. Process-performance data is very narrow and focused. Data about the impacts of technological systems is much more general. For example, the data needed to determine the emission levels of a motorcycle engine are more specific than the data needed to determine the impact of the air-transportation system on business. Process-operating information can be gathered using several types of monitoring devices. These include the following:

- ***Mechanical sensors.*** These sensors can be used to determine the position of components, force applied, or movement of parts.
- ***Thermal sensors.*** These sensors can be used to determine changes in temperature.
- ***Optical sensors.*** These sensors can be used to determine the level of light or changes in the intensity of light. See **Figure 6-10.**
- ***Electrical or electronic sensors.*** These sensors can be used to determine the frequency of or changes in electric current or electromagnetic waves. See **Figure 6-11.**

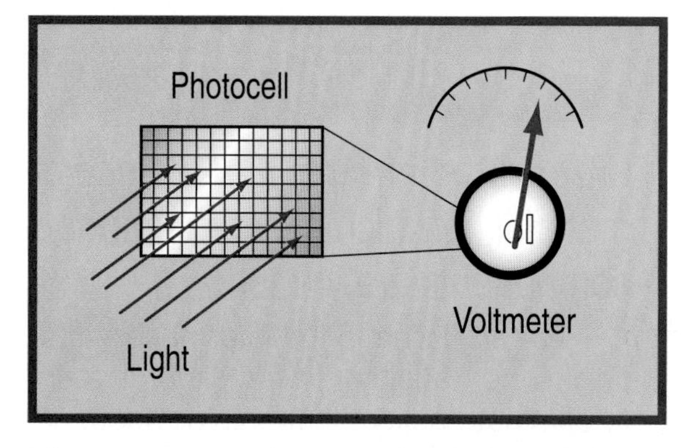

Figure 6-10. A photographic light meter is a common optical sensor. Light striking the photocell generates electricity. A voltmeter measures the electricity. Strong light produces more voltage than weak light.

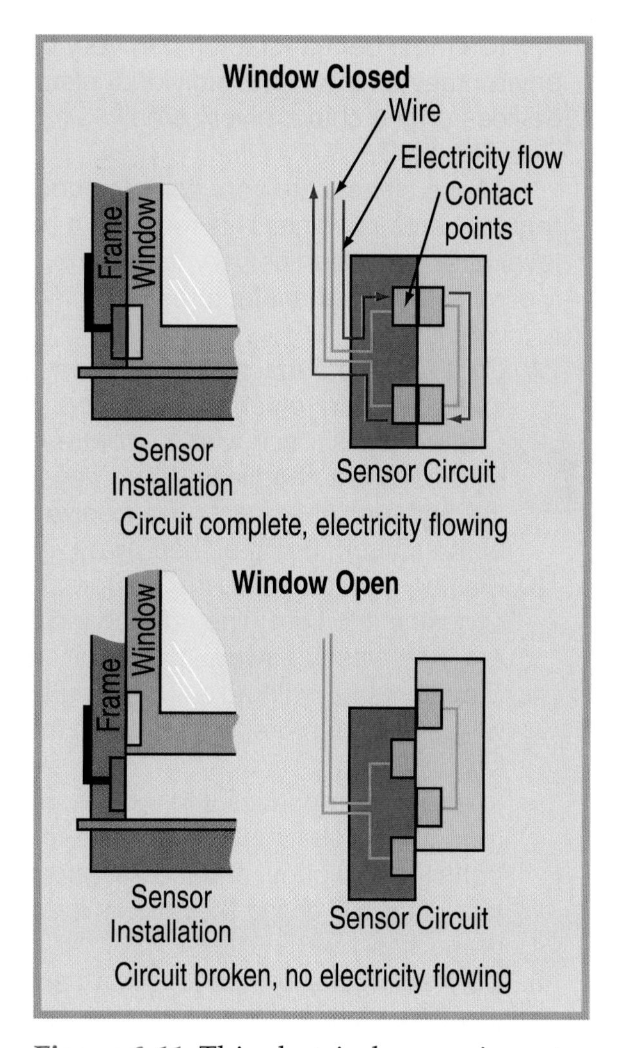

Figure 6-11. This electrical sensor is part of a home security system. The sensor senses when windows are open.

Standards for Technological Literacy

2

Research

Have your students select a type of sensor and research the way it works.

TechnoFact

As a component of internal control systems, monitoring devices are used in various activities. While these are mainly focused on controlling machines and systems, government agencies might monitor broadcasts or conversations in order to discover illegal activities.

Standards for Technological Literacy

3 4

Technology Explained

intregrated circuit (IC): a piece of semiconducting material, in which a large number of electronic components are formed.

We are living in the information age. The information age began with the invention of the transistor. A transistor is a tiny electronic device that allowed engineers to eliminate bulky electronic devices, such as the vacuum tube. Technology has made even greater advancements in the field of electronics, expanding transistor technology to create the integrated circuit (IC), or microchip.

An IC is a thin piece of pure semiconductor material, usually silicon. See **Figure A.** Tens of thousands of electronic devices and their interconnections can be produced in one IC. See **Figure B.**

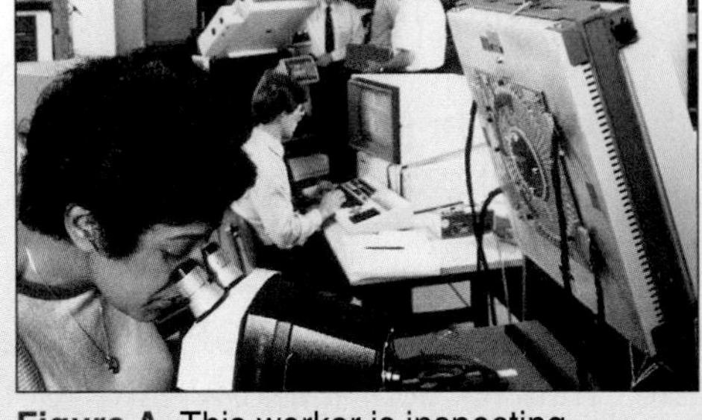

Figure A. This worker is inspecting an IC.

ICs must be manufactured in very clean environments. Often, the individual electronic devices on the chip are very small. See **Figure C.**

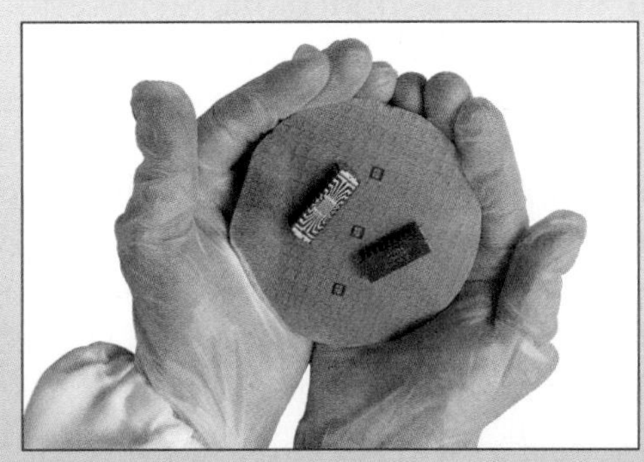

Figure B. Microchips are contained inside these cases. The electrical connections extend out of the case.

These circuits are very complex and require careful design. This requires a high level of skill and can be very time-consuming. Methods used to manufacture ICs include the following steps:

1. Oxidizing the surface. This prevents impurities from entering the silicon.
2. Coating the surface with a photoresist, a light-sensitive chemical. When developed, the photoresist prevents the oxidized layer of the IC from being etched away.
3. Placing a pattern over the photoresist and exposing it to light. The unexposed areas are then washed away.
4. Etching the IC with acid. This creates windows of unprotected areas on the silicon.
5. Introducing impurities into the silicon (doping). These impurities create the tiny transistors, diodes, capacitors, and resistors in the IC, giving the chip its electronic properties.
6. Connecting the components with small aluminum leads.
7. Placing the IC in a protective casing.

BAE Systems

Figure C. A close-up of a silicon-sensor chip.

The IC is responsible for the enormous growth of the computer industry in the twentieth century.

- *Magnetic (electromagnetic) sensors.* These sensors can be used to determine whether or not changes are occurring in the amount of current flowing in a circuit.

Data-comparing devices

The information gathered by monitoring devices is matched against expectations. The results are compared to the intent of the system through the use of ***data-comparing devices***. This process is very similar to a balance scale, in that the results should balance with the goal.

Comparisons can be analytical or judgmental. ***Analytical systems*** mathematically or scientifically make the comparison. Statistics might be used to see whether or not the output meets the goal. For example, the temperature of an oven can be compared with the setting on the knob. The size of a part can be checked with lasers and compared with data in a computer file. The position of a ship can be compared with the signals from a satellite. The analytical type of comparison is commonly used in closed loop control systems.

Analytical systems can also be used for managerial control systems. For example, the number of products produced might be directly compared with the production schedule. The number of passengers buying bus tickets might be compared with sales projections. The number of people listening to a radio station might be compared with the total listeners for all stations. Analytical systems try to remove human opinion from the evaluation process. The information is reduced to percentages, averages, or deviations from expectations.

Judgmental systems allow human opinions and values to enter into the control process. Open loop control systems require human judgment. For example, the speed at which you drive a car or the movement you use to cut on a line is based on your judgment. You do not apply mathematical formulas to decide when to accelerate and when to brake as you drive on a mountain road.

Likewise, human opinion is used in determining the value of system outputs. Judgments are used in the design phase of technological systems. It is almost impossible to select a design by analytical means. For example, people decide the styling, color, and shape of most products and buildings. They select the content of magazines and telecommunications programs. After the output is available, analytical means might be used to compare the results with expectations. After a television program is produced, the number of viewers watching the program can be measured. The success of the program can be subjected to mathematical analysis. Likewise, a furniture manufacturer can mathematically compare the sales of Early American tables with the sales of French Provincial tables. The decision to produce the show or manufacture the particular style of table, however, is a judgment.

Adjusting devices

A comparison of outputs to expectations might indicate that the system is not operating correctly. The output might not meet specifications. Thus, the system might need modifying. Controllers do this modifying through the use of ***adjusting devices***. These devices might speed up motors, close valves, increase the volume of burning gases, or perform a host of other actions. The goal is to cause the system to change and, therefore, produce better outputs. Several types of adjusting devices, or system controllers, exist. These, as shown in **Figure 6-12**, include the following:

- ***Mechanical controllers.*** These controllers use cams, levers, and other types of linkages to adjust machines or other devices.
- ***Electromechanical controllers.*** These controllers use electromagnetic coils

Standards for Technological Literacy

2

Research

Have your students research the way a complete home security system works.

TechnoFact

Virtual reality is a computer-based technology that allows people to have the sensation of being in another world. Sensors are found in helmets and gloves, which allow users to change the field of vision and manipulate objects in the virtual setting.

Standards for Technological Literacy

2

Demonstrate

Show your class how a rheostat can control the speed of a small motor.

TechnoFact

By the mid-1700s, windmills contained three sophisticated automatic adjusting devices. These devices controlled the speed of rotation, adjusted the pitch of the fan blades in order to produce maximum power at any wind speed, and oriented the fan so it faced directly into the wind.

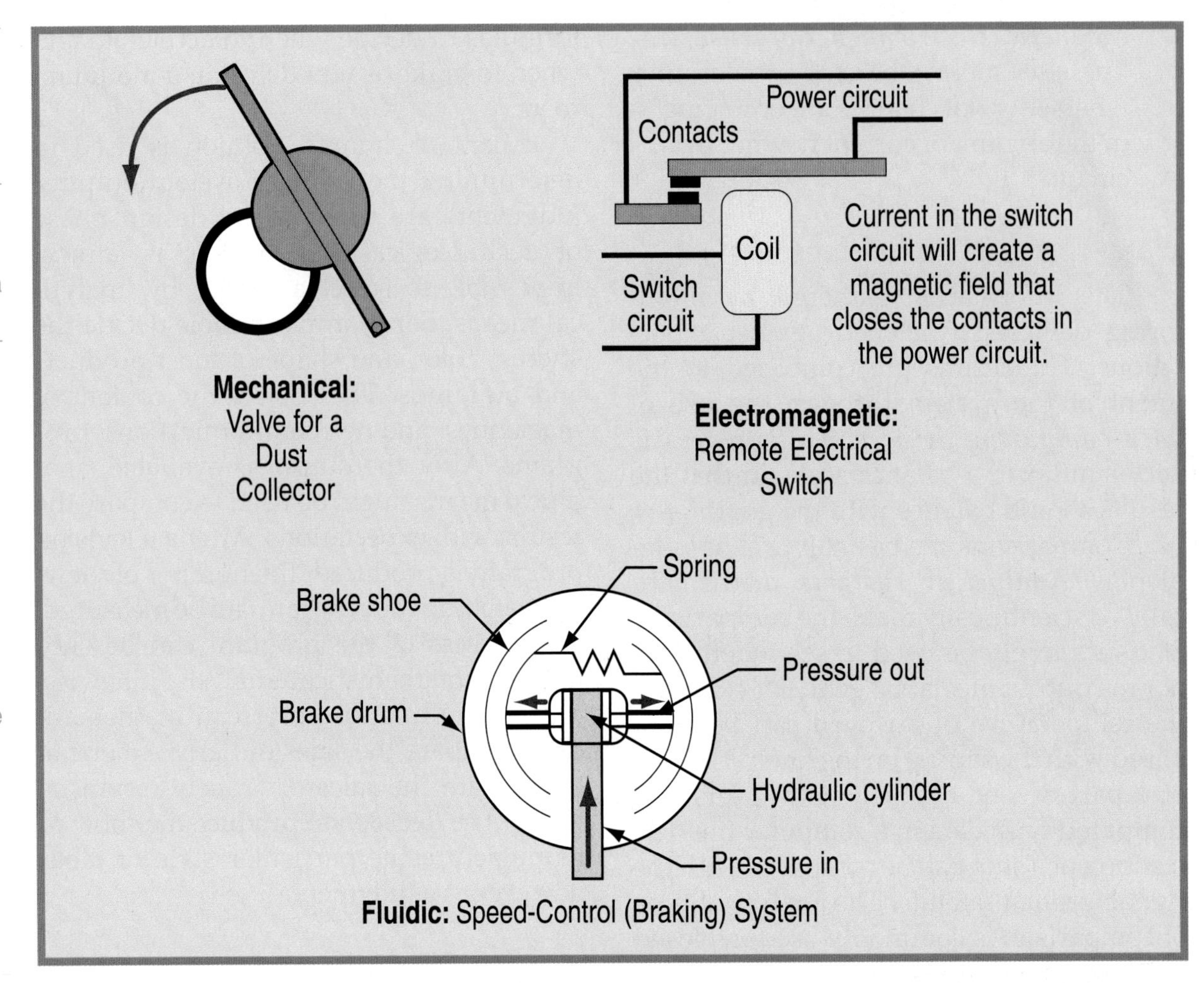

Figure 6-12. System controllers, or adjusting devices, modify the system to produce better outputs.

and forces to move control linkages and operate switches to adjust machines or other devices.

- *Electrical and electronic controllers.* These controllers use electrical devices (switches, relays, and motors, for example) and electronic devices (diodes, transistors, and ICs, for example) to adjust machines or other devices.
- *Fluidic controllers.* These controllers use fluids to adjust machines or other devices. The fluids include oil (in hydraulic controllers) and air (in pneumatic controllers).

The Operation (Manual versus Automatic) of Control Systems

Technological control systems can also be classified as manual and automatic. ***Manual control systems*** require humans to adjust the processes. ***Automatic control systems*** can monitor, compare, and adjust the system without human interference.

Manual control

A good example of a manual control system is the automobile. The driver uses various speed and direction controls to guide the vehicle to the destination (goal). As the vehicle enters a city, the driver eases

United Airlines

Figure 6-13. This pilot is using a manual control system to land an airplane. His eyes monitor the approach to the runway. His mind compares the actual flight path with the desired one. His hands and feet change control surfaces and speed to land the plane safely.

up on the accelerator to reduce the speed to the new speed limit. Brakes might be applied to meet the new conditions. The steering wheel is turned as the road curves. In this example, the human eye gathers information. The eye sees a new speed limit sign. To determine the speed of the vehicle, the eye also scans the speedometer. The brain compares the posted speed limit with the speedometer reading. To allow the speed to be adjusted until it matches the speed limit, the brain then commands the foot to change its position. In this example, the human eye monitors the system. The brain compares the system's actual and desired performances. The foot physically adjusts the system. Manual control systems are everywhere. We use them in such simple actions as riding a bicycle, adjusting the temperature of a gas stove when cooking, setting the focus on a camera, and checking a fence post for plumb. Also, we use them in performing more complicated actions, such as landing an airplane. See **Figure 6-13.**

Automatic control

Many technological systems have automatic control. A simple example of automatic

Standards for Technological Literacy

2

Extend

Compare the control systems of a bicycle and an automobile.

TechnoFact

To control the speed of an electric vehicle, a controller must be connected to the accelerator pedal to adjust the flow of electricity to the motor. A similar controller is needed to change the vehicle's direction.

TechnoFact

An example of manual control is the use of the tiller in a boat. Turning the tiller a certain way steers the boat in that direction.

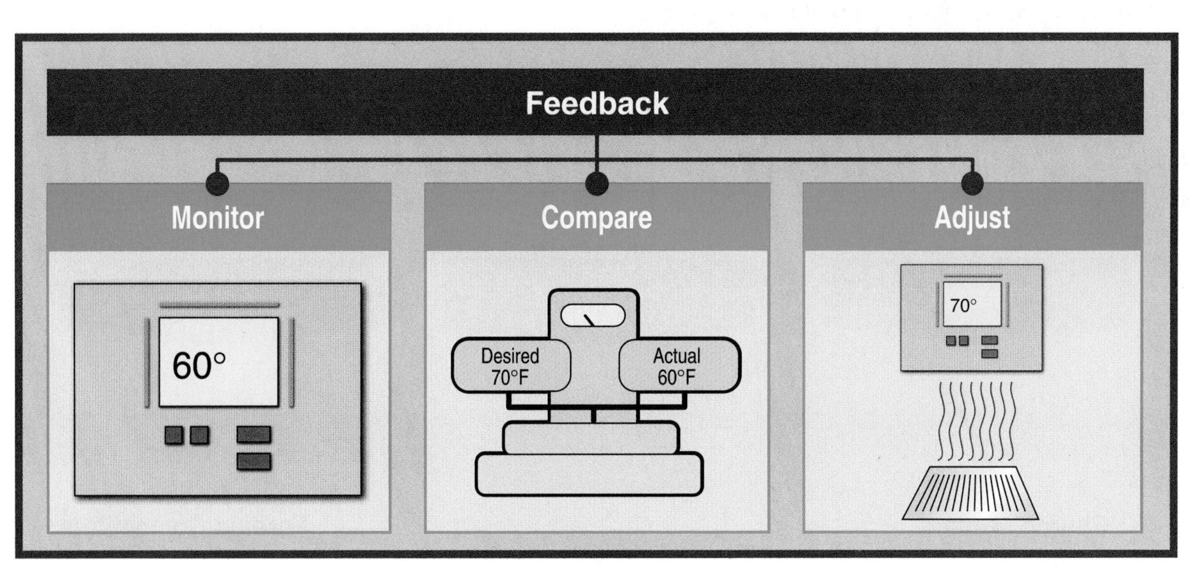

Figure 6-14. A thermostat is a common automatic control system found in many homes and buildings.

Standards for Technological Literacy

Extend

Discuss why external controls are necessary and how they direct technology activities.

Brainstorm

Have your students select a technological advancement and list its advantages and disadvantages for the public.

TechnoFact

An airplane can be moved along three axes by making use of airflow on control surfaces. Depending on how the control surfaces are moved, the airplane will turn left or right and move up or down.

TechnoFact

An example of external control on technology is the Pure Food and Drug Act of 1906. It requires drug companies to state the content, strength, and purity of each drug they manufacture. Until this time, it was not necessary for people to support their claims with evidence of what their home-brewed remedies did.

control is the thermostat in a heating system. See **Figure 6-14.** The device measures the room temperature. The thermostat compares the temperature of the room with the desired temperature. As long as the temperatures are within a preset range, no action is taken. When the room temperature drops below the set temperature range, a switch is activated. This switch turns on the heating unit. The unit operates until the upper limit of the temperature range is reached. At this point, the thermostat turns off the unit. Automatic doors in stores, electronic fuel-injection systems in vehicles, and Digital Video Recorders (DVRs) are other examples of automatic control systems you see every day. More sophisticated automatic control systems help pilots land the new generation of jet airliners, assist workers in controlling the output of electric generating plants, and help contractors lay flat concrete roads.

External Controls

We have looked at control primarily as an internal system component. As we said earlier, control also comes from outside the system. See **Figure 6-15.** For example, environmental responsibility is a societal goal that technological systems must address. Systems, therefore, are designed to reduce air, water, and noise pollution. Likewise, working conditions in factories and offices, on construction sites, and at transportation terminals and warehouses are controlled through political and labor union actions.

Public opinion is also a control factor. The opinions of civic groups, churches, labor unions, and various associations affect public policies. Our news media constantly report on issues controlling technological systems. The debate between environmentalists and forest-products companies over timber cutting is

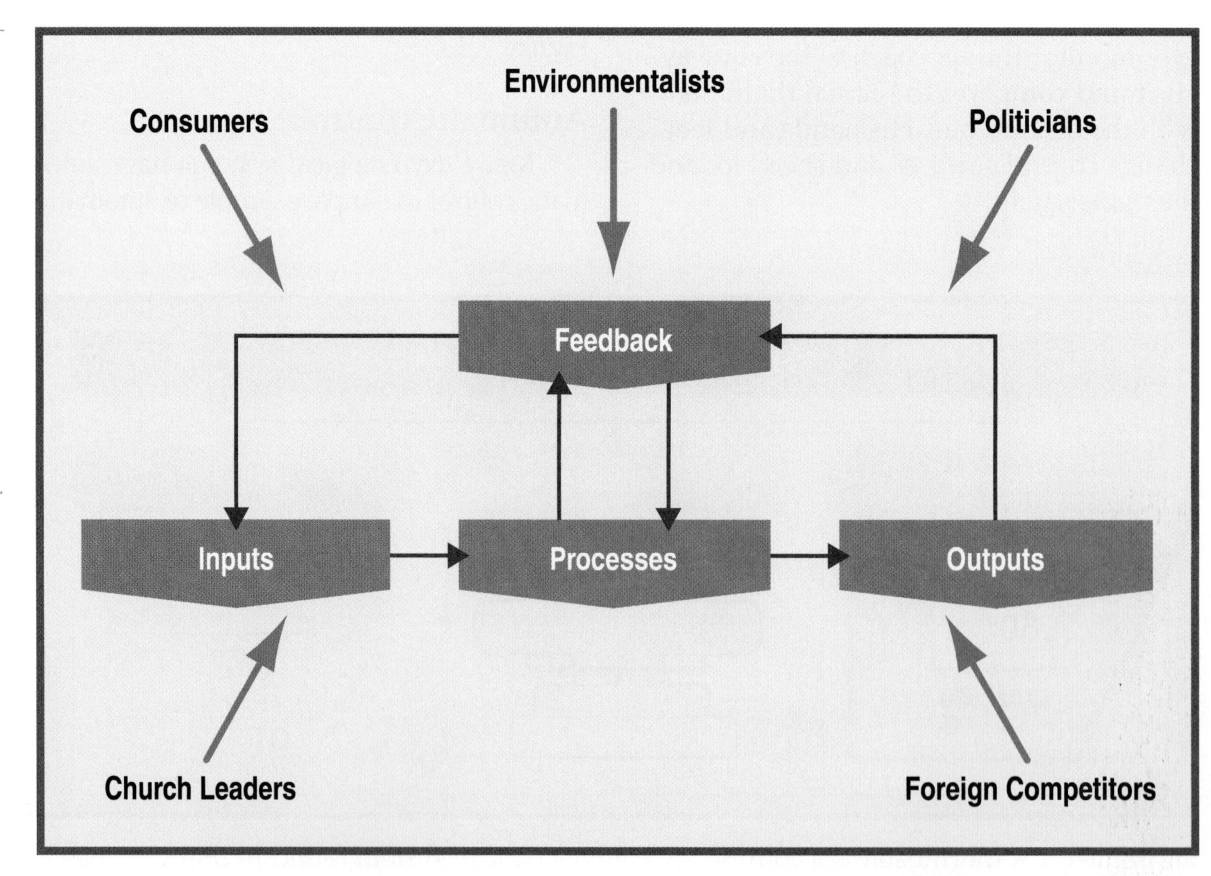

Figure 6-15. External forces exercise control on technological systems.

one example. Another example is the dialogue between the pronuclear and antinuclear power groups. The controversy over the types of nets that should be used in fishing is still another example of external control.

This discussion is not designed to list all the important public issues exercising control on technology. Looking only at the physical controls built into technological systems, however, gives an incomplete view. We need to see that, primarily through feedback, both internal and external control systems affect technology by creating new inputs, processes, and outputs. See **Figure 6-16.**

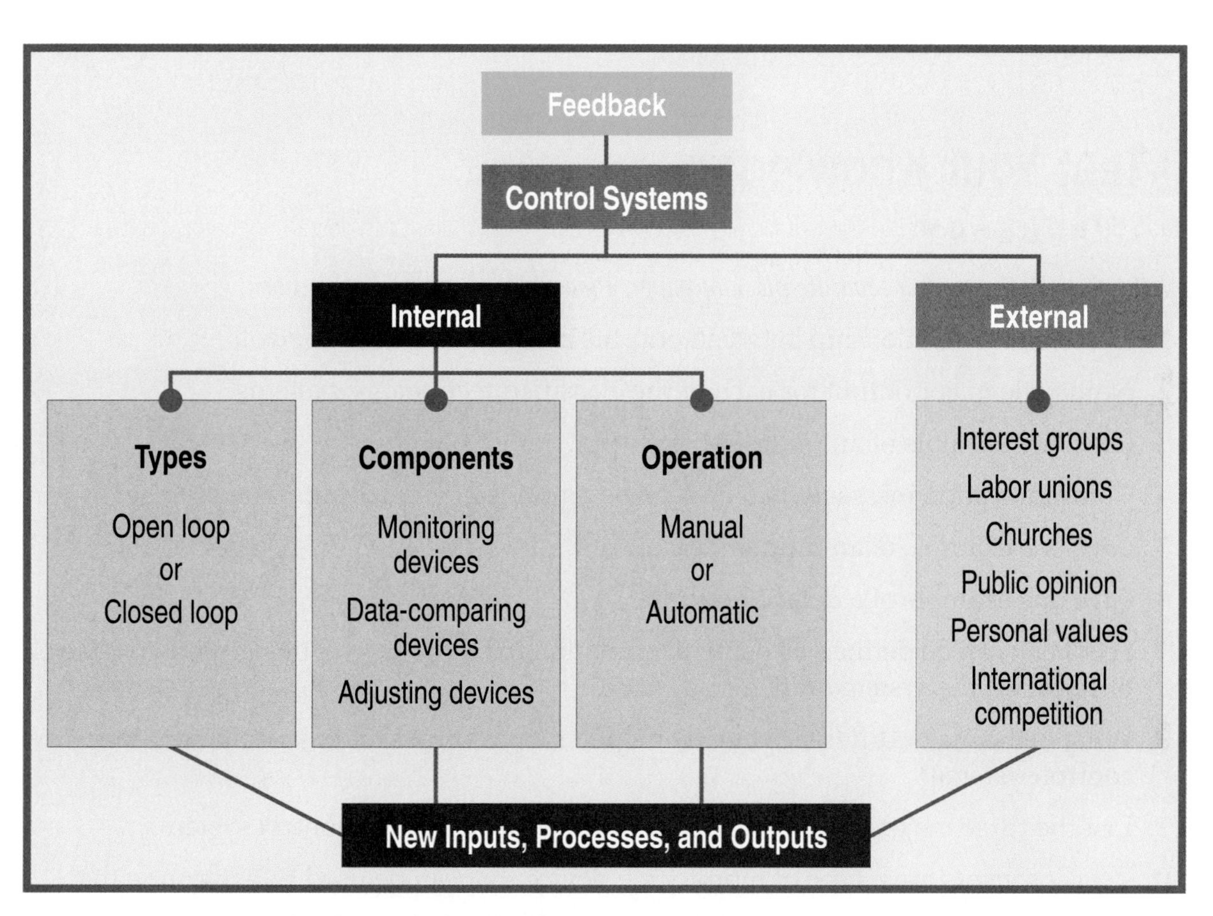

Figure 6-16. Primarily through feedback, both internal and external control systems lead to new inputs, processes, and outputs.

Standards for Technological Literacy

2

Extend

Discuss the way public opinion and laws govern technological actions.

Brainstorm

Have your students select a technological action, such as clear-cut logging, and list the groups that would be for and against it.

TechnoFact

The need for new systems of control often accompanies the development of new technology. For example, traffic regulation came from the invention of the automobile.

TechnoFact

Public utilities are groups of companies granted a monopoly status. Under government regulations, they control power, water, telephone, and sewage removal services.

Answers to Test Your Knowledge Questions

1. Closed-loop control systems use feedback to regulate the system or its inputs to affect the output.
2. Evaluate individually.
3. Evaluate individually.
4. True
5. Evaluate individually.
6. Evaluate individually.
7. Output
8. Output information is used to adjust the processing actions in closed-loop control systems. Output information is not used to adjust the process in open-loop control systems.
9. Monitoring devices, data-comparing devices, and adjusting devices.
10. C. Optical
11. B. Mechanical
12. Evaluate individually.
13. True

Summary

Technology is a series of complex human-made systems that change resources into useful items and services. The outputs of the systems can be desirable or undesirable, intended or unintended, and immediate or delayed. Humans have created control systems. Most of these systems use feedback to increase the probability that the outputs will be desirable. These control systems range from being parts of the internal operating process to outside forces influencing the system.

Test Your Knowledge

Write your answers on a separate piece of paper. Please do not write in this book.

1. Describe the relationship between outputs and feedback and control.
2. Name one major output for each of the classified technology systems.
3. Give one example of an undesirable output.
4. Communication messages can be seen as an undesirable output. True or false?
5. Give one example of an unintended output.
6. Give one example of a delayed output.
7. Feedback can be defined as using information from the _____ of a system to control, or regulate, the system or its inputs.
8. What is the major difference between open loop control systems and closed loop control systems?
9. List the three major types of devices found in most internal-control systems.
10. _____ sensors are a type of monitoring device that can be used to determine the level of light or changes in the intensity of light.
    A. Mechanical
    B. Thermal
    C. Optical
    D. Electrical
11. _____ controllers are a type of adjusting device using cams, levers, and other types of linkages to adjust machines.
    A. Electromechanical
    B. Mechanical
    C. Electrical
    D. Fluidic

12. Give one example of an automatic control system.
13. Public opinion is an external-control factor. True or false?

STEM Applications

1. Select a technology you use every day. Prepare and complete a chart similar to the following.

Outputs	
Desirable	Undesirable
Intended	Unintended
Immediate	Delayed

2. For the same technology, diagram the control system used. List the ways information is gathered, comparisons are made, and adjustment techniques are used.
3. For the same technology, list the outside influences that helped determine its design and help control its operation.
4. With two or more of your classmates, complete the following challenge:
   A. Identify a technological system or device you all use, such as bicycles or desk lighting.
   B. List a control system needed for the device or system.
   C. Brainstorm ways to provide this control.
   D. Sketch your solution.
5. A commonly used technological device is a motion-activated light or electrical switch. Design and conduct an experiment that determines the slowest motion this device will detect and the furthest angle to each side that the device can recognize motion. Present a brief summary of your research, with appropriate sketches, or drawings.

Section 2 Activities

Activity 2A

Design Problem

Background

Technology uses tools to process materials to create products. The tools often use one or more of the six basic mechanisms. These mechanisms are the lever, the wheel and axle, the pulley, the inclined plane, the wedge, and the screw.

Situation

The Science Experiments Company markets instructional kits for use in elementary and middle school science programs. This company has hired you as a designer for its physics kits.

Challenge

Design a kit containing common materials, such as tongue depressors, mousetraps, rubber bands, thread spools, small wooden blocks, and string. Using these materials, develop a technological device (machine) containing a power source and using three or more of the basic mechanisms to lift a tennis ball at least 6″ off a table.

Activity 2B

Fabrication Problem

Background

Technological devices are often built to increase human ability. Some devices multiply our ability to lift loads, while others help us see better. Still other devices improve our ability to hear, move people and products from place to place, or communicate over long distances.

Challenge

Construct the device shown in the drawings on the next page. Test the device to determine how well it helps you lift a load. Analyze how well the device works.

Materials and Equipment

- A test stand.
- A 1/2″-diameter plastic syringe.
- A 3/4″- or 1″-diameter plastic syringe.
- 12″ of clear tubing.
- A baseball or another load.

Procedure

1. Obtain the supplies.
2. Assemble the hydraulic system. See **Figure 2B-1.**

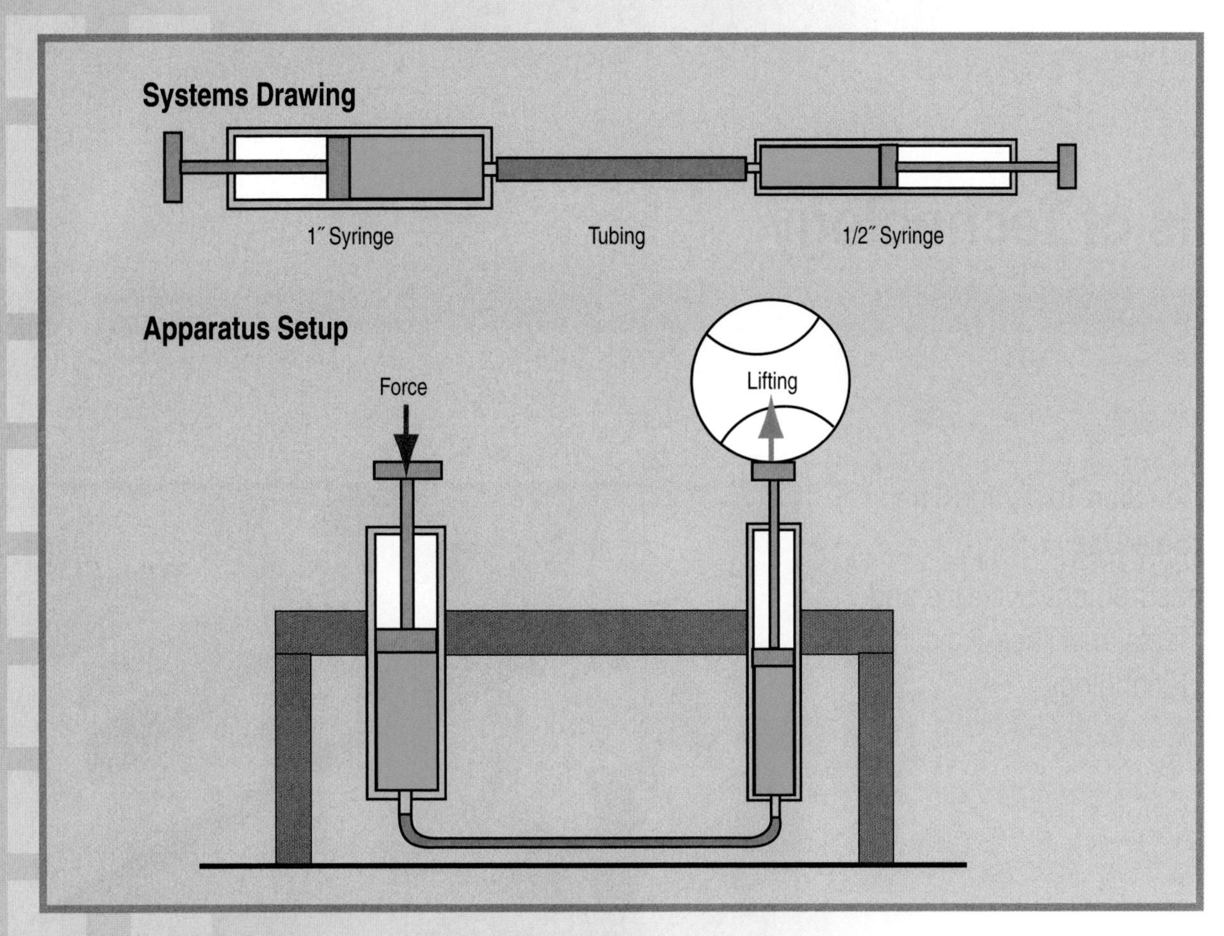

Figure 2B-1. The experiment-apparatus setup.

3. Place the load on the small syringe.
4. Press the large-syringe plunger to cause the load to rise.
5. Reverse the load, placing it on the large-diameter plunger.
6. Press the small-syringe plunger to cause the load to rise.
7. Complete the following analysis of the activity.

Analysis

Analyze the two circuits by answering the following questions:

1. Which arrangement of cylinders lifts the load with the least effort?
   A. The force on the large cylinder and the load on the small cylinder.
   B. The force on the small cylinder and the load on the large cylinder.
2. Which arrangement of cylinders moves the load the greatest distance when the force cylinder is depressed?
   A. The force on the large cylinder and the load on the small cylinder.
   B. The force on the small cylinder and the load on the large cylinder.
3. Which arrangement is a force multiplier?
4. Which arrangement is a distance multiplier?
5. Which type of multiplier would you use as an automobile lift in a service station?
6. Which type of multiplier would you use for a hydraulic elevator in a four-story building?

Section 3

Tools of Technology

Tomorrow's Technology Today

Cryonics

The phrase "life is too short" rings true with many. Adventurers, scientists, and futurists alike have sought the elusive dream of long life for centuries, but to no avail. Today, scientific advancements in the field of cryonics have brought us one step closer to making that seemingly impossible dream a reality.

Cryonics is the process of preserving humans and animals after death at an extremely low temperature. Scientists have yet to reanimate a cryopreserved human. Therefore, the study is classified as a protoscience, or a field of study that is still theoretical because adequate research has yet to be done. People who elect for cryopreservation upon their death do so in the hopes that future medical advancements will allow for their resuscitation. There is no guarantee that technological advances will make that a reality. However, cryopreserved cells, tissues, and small organs have been successfully revived. The cryopreservation of human embryos is frequently reversed, allowing the embryo to be fertilized and grow into a healthy fetus.

Legally, cryopreservation cannot begin until the patient has suffered cardiac arrest and been declared clinically dead. Supporters of cryonics believe that because the cells of the body continue to live for minutes, sometimes hours, after the heart stops beating, their preservation maintains the individual's identity and memories stored within each cell. Once the patient has been declared clinically dead, a form of cardiopulmonary resuscitation (CPR) is administered to maintain blood flow and slow the death of cells until cryonics technicians can begin their work.

In a process called *vitrification*, the body is cooled without freezing. Blood and excess water is flushed from the patient and replaced with a cryoprotectant, which will reduce damaging ice formation within cells. The body is cooled on a bed of dry ice until it reaches –130°C (–202°F). Patients are then placed in a cylindrical container and submerged in a tank of liquid nitrogen maintained at –196°C (–320°F).

Patients are stored at cryonics facilities around the world. Membership at these facilities provides indefinite storage and maintenance until medical advancements make resuscitation feasible.

The developing science of nanotechnology—the manipulation and placement of individual atoms may provide a method of reversing cryopreservation. However, scientists have many obstacles to overcome: unknown damage caused by toxic cryoprotectants, treatment of the fatal ailment, repair of the body and brain, and finally rejuvenation. As the science of cryonics advances, we may one day see solutions to these problems and find a way to increase our life span.

Discussion Starters

As a class, discuss the ways cryonics may change the world.

Group Assignment

Divide the class into two groups. Have one group brainstorm the positive effects cryonics may have. Have the second group brainstorm the negative effects cryonics may have.

Writing Assignment

Nanotechnology may one day help reverse the effects of cryopreservation. Have your students write a brief paper speculating what other technologies might be required in order to successfully revive patients.

Chapter 7 Production Tools and Their Safe Use

Chapter Outline

- Material-Processing Tools and Machines
- Energy-Processing Converters
- Information-Processing Machines
- Using Technology Safely

This chapter covers the benchmark topics for the following Standards for Technological Literacy:

3 5 7 12 16 17 19

Learning Objectives

After studying this chapter, you will be able to do the following:

- Recall the categories of production tools used in technology.
- Recall the characteristics of machine tools.
- Recall the major types of machine tools.
- Summarize the major types of energy-processing tools and machines.
- Summarize the major types of information-processing tools and machines.
- Recall guidelines for using tools and machines properly and safely.

Key Terms

arbor
armature
band saw
chop saw
chuck
circular saw
computer
cutting motion
cutting tool
cylindrical grinder
drilling machine
energy conversion
energy-processing converter
feed motion
field magnet
Forstner bit
grinding machine
information processing
lathe
linear motion
machine tool
material processing
multiple-point tool
planing machine
radial saw
reciprocating motion
rotating motion
sawing machine
scroll saw
shaping machine
single-point tool
spade bit
stroke
surface grinder
table saw
turning machine
twist drill

Strategic Reading

As you read this chapter, make an outline of the various types of tools discussed.

The human ability to design and use tools provides the foundation for technology. You might recall from Chapter 2 that tools range in kind from simple hand tools to complex machines. We have such varied items as buildings, vehicles, communications media, and energy-conversion machines because of tools. See **Figure 7-1.**

Tools can make our lives better, but they can also threaten our existence. For example, we travel in comfort in modern automobiles, but these automobiles pollute the atmosphere. High-rise buildings allow us to live and work in cities, yet social problems and crime in many of these cities frighten people. Television allows events around the globe to instantaneously touch us, but it threatens family interaction.

Our challenge is to design and use the tools of technology wisely. To do this, we must first understand the tools and machines around us. Today, we use thousands of different tools and machines. We cannot discuss them all here because of space limitations. Instead, this chapter introduces you to selected tools and machines used in three major types of processing:

- Material processing.
- Energy processing.
- Information processing.

Understanding the tools and machines around us, however, involves knowing, not only their particular features, but also how they should be handled. The final section of this chapter covers the safety rules for tools and machines.

©iStockphoto.com/thelinke

Figure 7-1. Manufacturing most products requires many machines, such as these packaging machines.

Material-Processing Tools and Machines

The world around us is full of artifacts. Each was made using material-processing tools and machines that change the form of materials. Tools and machines are used to cast, form, and machine materials into specific shapes. They also help assemble products and apply protective coatings.

Fundamental to all ***material processing*** is a group of tools called *machine tools.* These are the machines used to make other machines. To understand them better, we will now look at their characteristics and types.

Characteristics of Machine Tools

Machine tools have some common characteristics. See **Figure 7-2.** These characteristics include the following:

- A method of cutting materials to produce the desired size and shape. The new size and shape are achieved with what is called a ***cutting tool.***
- A series of motions between the material and the tool. This movement causes the tool to cut the material.
- Support of the tool and the workpiece (the material being machined).

Standards for Technological Literacy

7

Reinforce

Review the difference between tools and machines (Chapter 2).

Extend

Discuss the difference between machine tools and other material processing machines.

TechnoFact

Because the Industrial Revolution impacted England and America in different ways, it is considered to consist of two stages. Steam engines began the Industrial Revolution in England, and the widespread use of interchangeable parts and assembly-line techniques started in America.

TechnoFact

John H. Lienhard, M.D. Anderson Professor of Mechanical Engineering and History at the University of Houston, wrote in his The Engines of Our Ingenuity series: "There's a strong view that we shouldn't call humans homo sapiens, or man-the-wise, but rather homo technologicus, or man-the-user-of-technology."

Standards for Technological Literacy

7

Research

Have your students research the features of a cutting tool.

TechnoFact

In 1790, Honoré Blanc made his muskets the first products to be made using interchangeable parts.

The Cutting Tool

Most cutting actions require a cutting tool. Cutting tools come in many sizes and shapes, but they must meet certain requirements. First, they must be harder than the material they are cutting. For example, a diamond cuts steel, steel cuts wood, and wood cuts butter. Second, the cutting tool must have the proper shape. That is, it needs a sharpened edge and relief angles. See **Figure 7-3.** The sharpened edge allows the tool to cut into the material. The relief angles keep the sides of the tool from rubbing against the material as the material is cut. The rake angle also helps to create a chip as the material is cut. This action allows waste material to be carried away efficiently. Two basic types of cutting tools are used in all hand tools and machines. These are single-point and multiple-point tools.

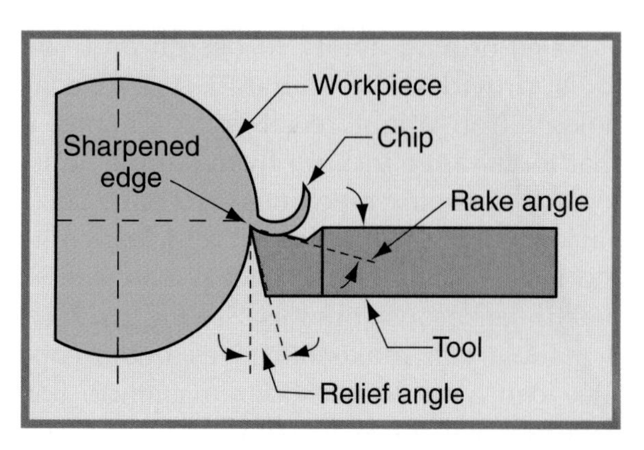

Figure 7-3. Cutting tools have a sharpened edge, a relief angle, and a rake angle.

Single-point tools

The ***single-point tool*** is the simplest cutting device available. This tool has a cutting edge along the edge or end of a rod, bar, or strip. Common hand tools using single points are knives, chisels, and wood

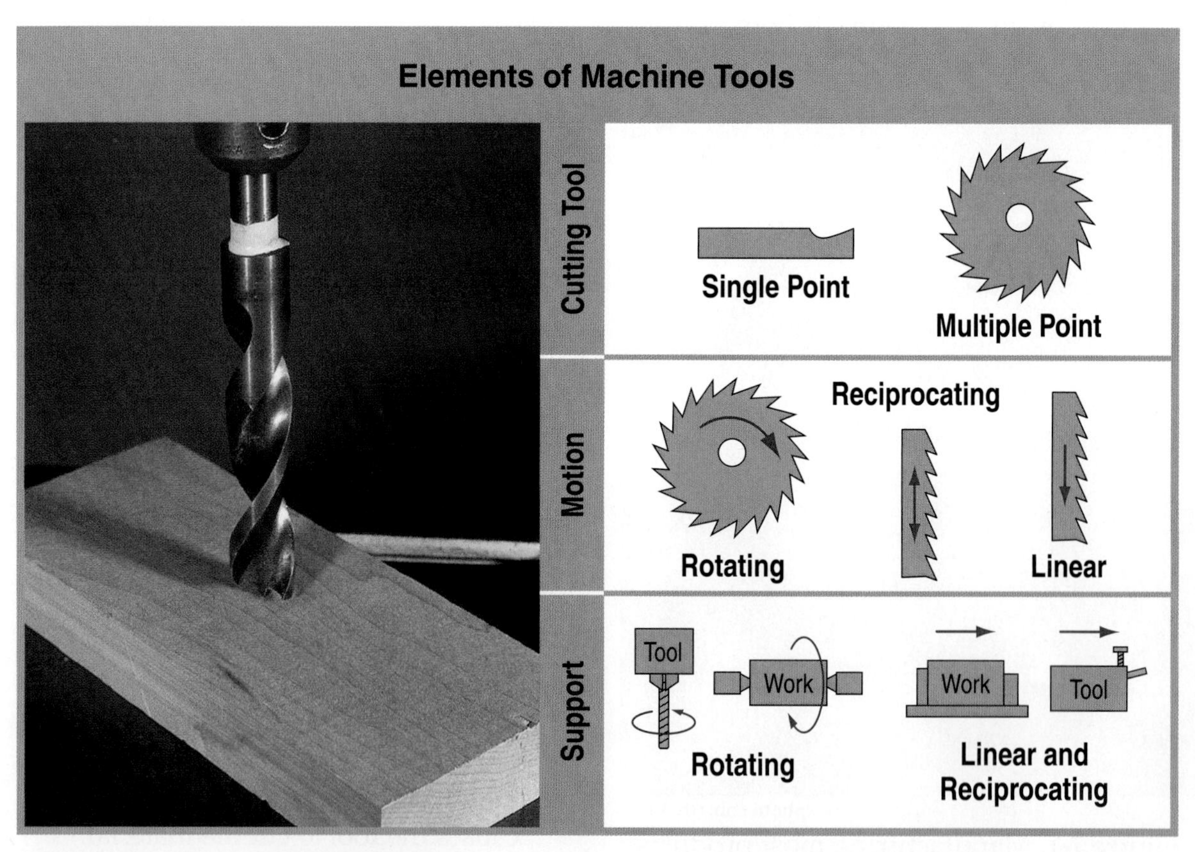

Figure 7-2. All machine tools have three basic elements: a cutting tool, a method of moving the tool or workpiece, and a method of supporting the tool and the workpiece.

planes. See **Figure 7-4.** Most lathe-turning processes (where work is turned on a horizontal axis) use single-point tools. See "Turning Machines" later in this chapter.

Multiple-point tools

With a ***multiple-point tool***, a series of single-point tools is arranged on a cutting device. See **Figure 7-5.** Most often, these single points are arranged in a set pattern. For example, the teeth of a circular saw are evenly spaced around the circumference of the blade. Likewise, the teeth on band saw and scroll saw blades are spaced along the edge of a metal strip. In some cases, there is no pattern for the arrangement of cutting points. The cutting edges, or points, have a random arrangement. Abrasive papers and grinding wheels are good examples of this type of cutting tool.

Standards for Technological Literacy

Demonstrate

Show and use common hand tools that fit the two classifications of cutting tools.

TechnoFact

Although Honoré Blanc used the idea of interchangeable parts in musket, Eli Whitney is credited as the first to use the parts in manufacturing. His use of interchangeable parts was much like Blanc's, producing muskets that could be repaired with more standard parts.

Figure 7-4. Single-point tools have a cutting edge on the end or along the edge. Among the tools pictured here are a hand plane, a utility knife, and various types of chisels.

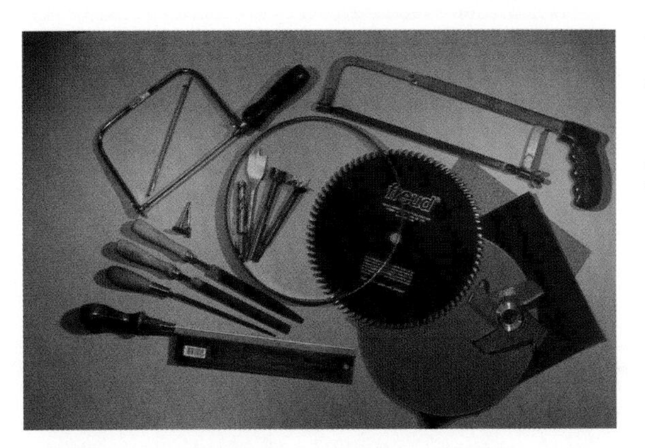

Figure 7-5. Multiple-point tools have a series of single-point tools arranged on a cutting device. Among the tools pictured here are circular, band, and scroll saw blades; grinding wheels; milling cutters; hand files; abrasive paper; and drill bits.

Career Corner

Manufacturing

Machinists

Machinists interpret blueprints or specifications, set up machines, and produce precision parts. They use many different machine tools, including lathes, milling machines, grinders, and machining centers. Machinists must be able to calculate cutting and feed speeds, select proper cutting tools, plan the sequence of operations, monitor machine operations, and use measuring tools to check the quality of the parts. Many machinists must be able to write and modify computer numerical control (CNC) programs. After a machinist has developed and tested operations and setups, less-skilled machine operators generally do repetitive operations. Most machinists are trained through apprenticeship programs, on-the-job training, vocational schools, or community colleges.

Standards for Technological Literacy

7

Extend

Give examples of machines that use each type of cutting motion.

TechnoFact

In many instances, machines are named according to their principle function. Examples include specific machines, such as a surface grinder, but also include the six general categories of machines.

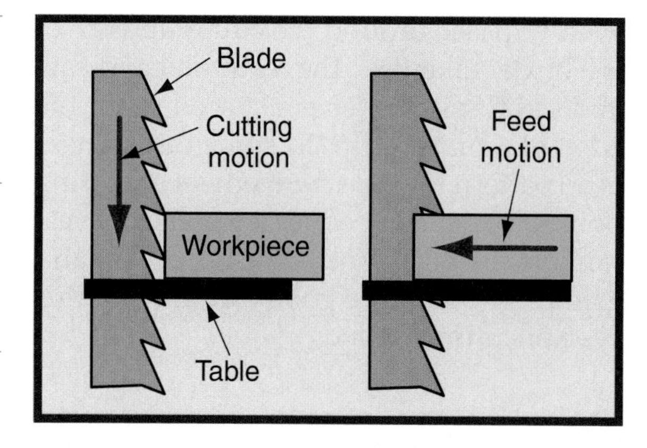

Figure 7-6. All cutting operations include cutting motion and feed motion.

Motion

Movement between the tool and the workpiece must occur before cutting can take place. All cutting operations have two basic motions. These motions are cutting motion and feed motion.

Cutting motion is the action causing material to be removed from the work. This motion causes the excess material to be cut away. ***Feed motion*** is the action bringing new material into the cutter. This motion allows the cutting action to be continuous.

To understand these movements, consider a ***band saw***. See Figure 7-6. Imagine a piece of wood placed against the blade. When the machine is turned on, the first tooth cuts the material. This is the cutting motion. The next tooth follows the path of the first tooth, but does not produce a chip. The first tooth has already removed the material in the kerf (the slot a blade cut). The cutting action continues only if the wood is pushed into the blade. This makes new material available for the next tooth to cut. Moving the wood is the feed motion.

Cutting and feed motions can be rotating, linear, or reciprocating. ***Rotating motion*** uses round cutters or spins the work around an axis. ***Linear motion*** moves a cutter or work in one direction along a straight line. ***Reciprocating motion*** moves the tool or the work back and forth. Look back to Figure 7-2 for a diagram of these three types of motion.

Support

The final element present in all machine tools is support for the tool and the workpiece. The types of cutting and feed motions used determine the support system needed. Rotating motion requires a holder revolving around an axis. ***Chucks*** are attachments used to hold and rotate drills and router bits. ***Arbors*** are spindles, or shafts, used to hold table-saw blades and milling cutters. Parts are placed between centers on lathes and cylindrical grinders.

Linear motion is produced in several ways. A tool can be clamped or held on a rest and then be moved in a straight line. This practice is common for wood and metal lathes and hand wood planes. Band saw blades travel around two or three wheels to produce a linear cutting motion. Material can be pushed through a saw blade while the machine table supports it.

Reciprocating motion is common with scroll saws and hacksaws. The blade is clamped at both ends into the machine and then moved back and forth. Likewise, a workpiece can be clamped to a table reciprocating under or across a cutter. This action is used in a bench grinder. The operator moves a workpiece back and forth across the face of a grinding wheel. This action is used to sharpen hand tools, lawn mower blades, and kitchen knives.

Types of Machine Tools

Hundreds of different machine tools exist. All machine tools can be grouped into six categories, however. See Figure 7-7. These categories are the following:

- Turning machines.
- Sawing machines.
- Drilling machines.
- Shaping machines.
- Planing machines.
- Grinding machines.

Standards for Technological Literacy

7 19

Technology Explained

computer numerical control: using a computer and sequenced instructions to control a machine tool.

Computer numerical control (CNC) has greatly changed how manufacturing systems work. The Industrial Revolution was based on producing large numbers of uniform products using a continuous manufacturing system. Any change in the design of a product necessitates changes in the machines that make the product. These changes are very expensive to make. Numerical control changed how objects were made. This process used to be done manually with punch tapes until the computer was used to enter information into the machine.

The advent of computers has changed how products are designed. Products can be designed, tested, and refined on a computer. The design for a product is expressed in a computer language. This language can be used to control machines and inspect product quality. See **Figure A.** Computers are also used to control how products move through a sequence of operations. The computers are used to feed codes into the machine tools, which then cut the objects.

These systems use a number of technologies, including computer-aided design, to develop and specify a product's size and shape. Robots are used to load and unload machines and move products from machine to machine. CNC is used to guide machine operation. Electronic measurement and computer data processing help to improve inspection and quality control. See **Figure B.**

Hitachi Seiki

Figure A. CNC manufacturing is based on computer control.

An Assembly Machine

A Material-Handling System

An Inspection System

Figure B. Computers are used to direct the operation of machine tools and assembly machines, material-handling systems, and inspection systems.

Standards for Technological Literacy

7

Extend

Show your students some common products or parts made on a metalworking or woodworking lathe.

TechnoFact

The lathe, a machine whose operation is based on a potter's wheel, is the primary example of a turning machine. A painting at an Egyptian tomb suggests the lathe has been in use for over 2300 years.

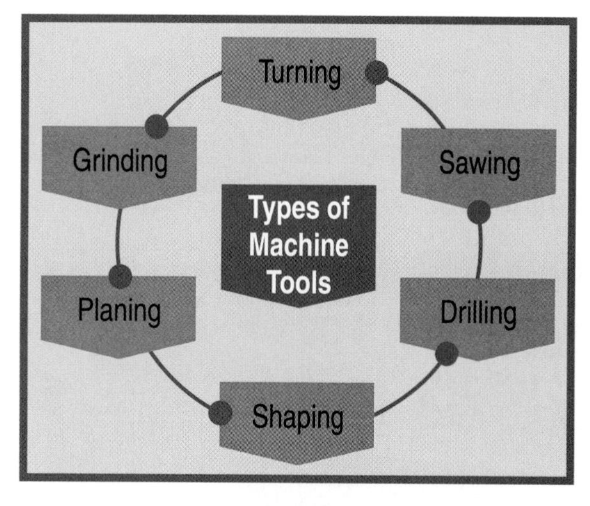

Figure 7-7. The six categories of machine tools are turning, sawing, drilling, shaping, planing, and grinding.

Turning Machines

Turning machines use a process in which a workpiece is held and rotated on an axis. As noted earlier, this action is produced on machines called ***lathes***. See **Figure 7-8.** All lathes produce their cutting motion by rotating the workpiece. Linear movement of the tool generates the feed motion.

Lathes are primarily used to machine wood and metal. Plastics can be machined on lathes as well. Wood and metal lathes contain four main parts or a system. A headstock, which contains the machine's power unit, is the heart of the lathe. This unit has systems to rotate the workpiece and to adjust the speed of rotation. At the opposite end of the machine is a tailstock. This unit supports the opposite end of a part that is gripped at the headstock. The headstock and tailstock are attached to the bed of the lathe. Finally, a tool rest or toolholder is provided to support the tool. Tool rests on metal lathes clamp the tool in position and feed it into or along the work. Wood lathes commonly have a flat tool rest, along which the operator moves the tool by hand.

The work can be held in a lathe in two basic ways. Works can be placed between centers. One center is in the headstock, and the other is in the tailstock. These centers support the workpiece and can be of two types. Live centers rotate with the workpiece, whereas dead centers are fixed as the work rotates around them.

Both wood lathes and metal lathes have centers, but they differ in the way the turning force is applied to the workpiece. Wood lathes apply the force through the headstock using a spur center (a center with cross-shaped extensions that seat into the wood). Metal lathes use a device called a *dog* to rotate the workpiece. The dog clamps onto the work and has a finger that engages the faceplate attached to the drive spindle.

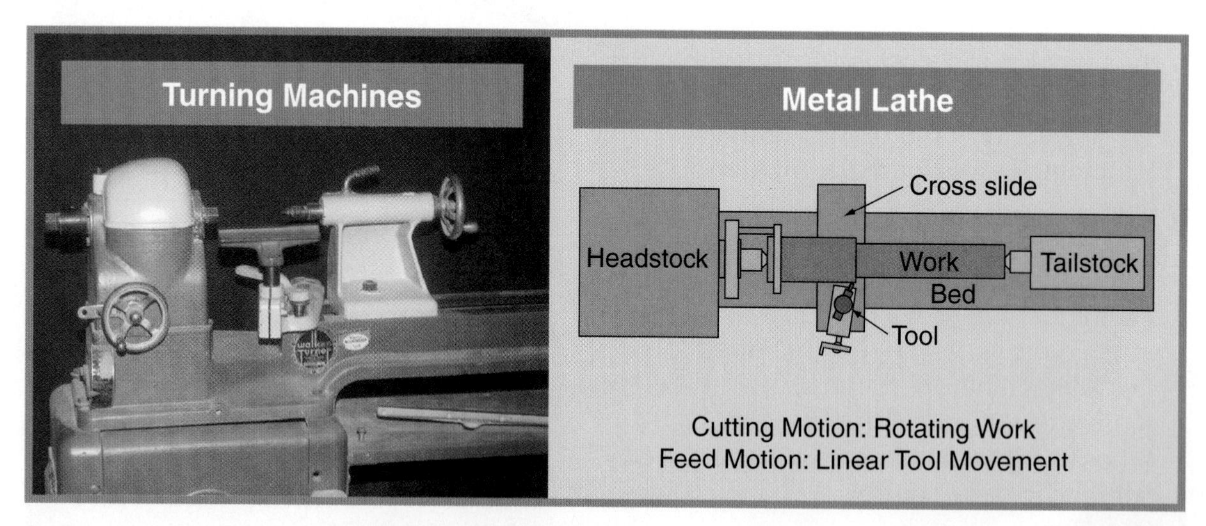

Figure 7-8. Lathes are used for turning operations.

A small workpiece can be rotated using only the headstock in the device called the *chuck*. Three-jaw, four-jaw, or collet chucks are commonly used to grip and rotate the part. Chucks are found on metal lathes. These chucks have jaws that squeeze and hold the part.

Lathes are used to perform many operations. See **Figure 7-9.** The most common are the following:

- **Turning.** This operation is cutting along the length of a workpiece. Turning produces a cylinder of uniform diameter.
- **Tapering.** This operation is cutting along the length of a cylinder at a slight angle to produce a cylindrical shape with a uniformly decreasing diameter.
- **Facing.** This operation is cutting across the end of a rotating workpiece. Facing produces a true (or square) end on the workpiece.
- **Grooving (shouldering).** This operation is cutting into a workpiece to produce a channel with a diameter less than the main diameter of the workpiece.
- **Chamfering.** This operation is cutting an angled surface between two diameters on the workpiece.
- **Parting.** This operation is cutting off a part from the main workpiece.
- **Threading.** This operation is cutting threads along the outside diameter or inside a hole in the workpiece.
- **Knurling.** This operation is producing a diamond pattern of grooves on the outside diameter of a portion of the workpiece. Knurling produces a surface that is easier to grip and turn.

In addition, drills and reamers can be placed in the tailstock to produce and finish holes in the workpiece.

Sawing Machines

Sawing machines use teeth on a blade to cut material to a desired size and shape. These machines are designed to perform a number of different cutting actions. The actions, as shown in **Figure 7-10,** include the following:

- **Crosscutting (or cutoff).** This action is reducing the length of a material.
- **Ripping (or edging).** This action is reducing the width of a material.
- **Resawing.** This action is reducing the thickness of a material.
- **Grooving, dadoing, and notching.** These actions are cutting rectangular slots in or across a part.

Standards for Technological Literacy

7

Brainstorm

Ask your students what kind of products can be made on a lathe. Have the students give examples.

TechnoFact

Early lathes were powered by pulling a cord wrapped around a spindle. A continuous crank-driven belt drive was not invented until the early 15th century. In the 1770s, the first industrial lathe was developed.

TechnoFact

In the late 18th century, the French toy maker Jacques de Vaucanson designed the first modern industrial lathe. This lathe was used to cut threads while inventing his toy mechanical duck.

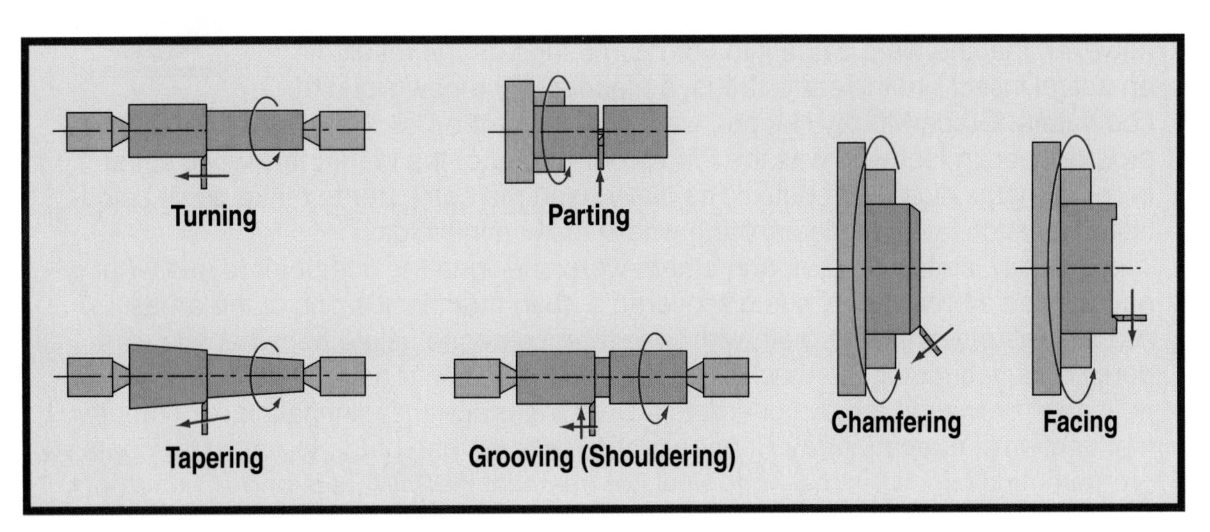

Figure 7-9. Common lathe operations

Standards for Technological Literacy

7

TechnoFact

Before waterpower made it less difficult to saw lumber, log structures were common. The amount of labor needed to cut boards from logs by hand was too great. However, in the Middle Ages, the use of waterpower was turned from grinding grain to sawing boards using sawmills.

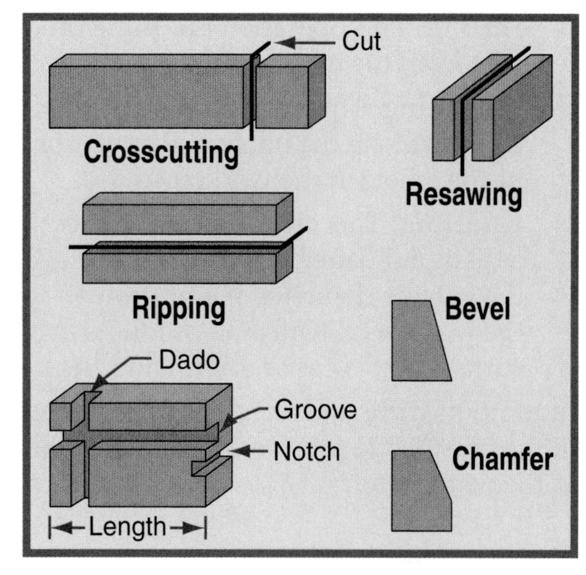

Figure 7-10. Typical sawing operations are crosscutting; ripping; resawing; grooving, dadoing, and notching; and chamfering and beveling.

- **Chamfering and beveling.** These actions are cutting an angled surface between two primary surfaces of a material.

These machines can be grouped according to the type of blade they use and the methods used to produce the cutting action. This grouping identifies three basic types of saws. See Figure 7-11.

- Circular saws.
- Band saws.
- Scroll saws.

Circular saws

Circular saws use a blade in the shape of a disk with teeth arranged around the edge. These teeth vary in shape and arrangement, depending on the operation to be performed. Common blades are available for crosscutting, ripping, and combination cutting (crosscutting or ripping).

The three basic types of circular saws are the table saw, radial saw, and chop saw. See Figure 7-12. All three of these machines generate the cutting motion by rotating the blade. Their feed motions are different, however. The ***table saw*** uses a linear feed of the material. The workpiece is pushed

Academic Connections: Communication

Computer Bugs

We have all probably heard or used the term *debug* to describe the repair of a computer problem. How many of us are aware, however, that the word came into common usage as the result of an actual insect? In the early 1940s, a pioneer in the new world of computers, Grace Murray Hopper, was working for the U.S. Navy as a computer programmer. In fact, she was the third programmer of the Mark I, the world's first programmable digital computer. The Navy used the computer to make quick calculations of such tasks as determining where to lay minefields.

One day, as Ms. Hopper and others were checking the computer to see what had caused a breakdown, she discovered a dead moth inside one of the areas. As she was removing the moth with tweezers, someone asked her what she was doing. "I'm debugging the machine," she reportedly said. The term stuck and is now even defined in the dictionary as a word meaning "the elimination of computer malfunctions." Grace Murray Hopper continued to be one of the foremost computer-programming experts in the world until her death in 1992. What is one of her other major accomplishments?

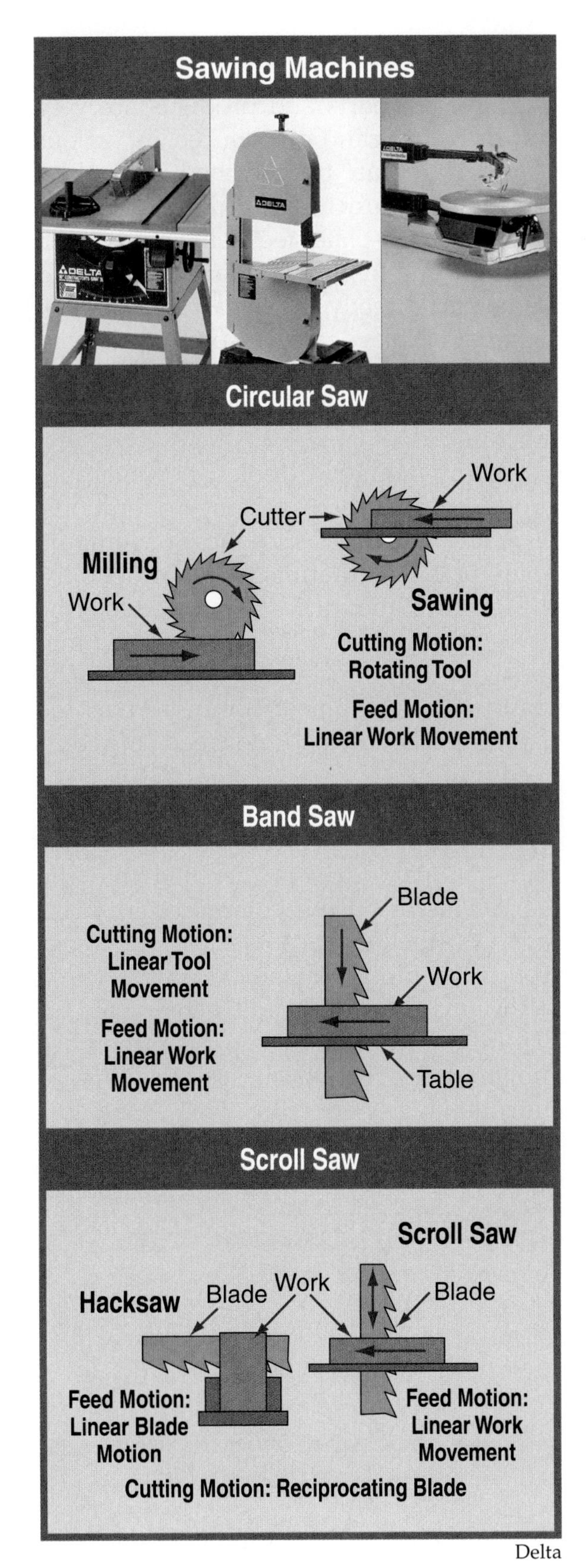

Delta

Figure 7-11. The three basic types of saws are circular saws, band saws, and scroll saws.

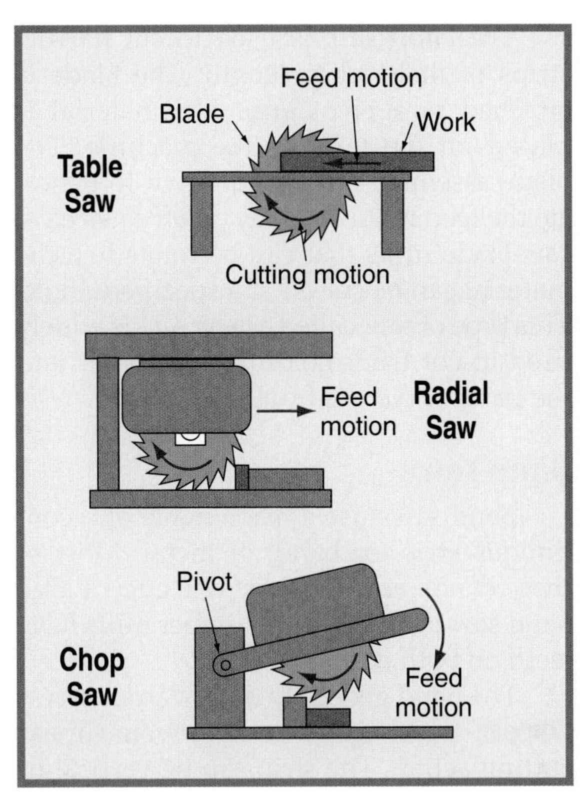

Figure 7-12. The three types of circular saws are the table saw, the radial saw, and the chop saw.

into the rotating blade to generate the cut. The machine operator manually feeds the material in low-volume production settings. Automatic feeding devices are used to increase the speed, accuracy, and safety of many high-volume sawing operations.

Many other machines use cutting and feed motions similar to the circular saw. These include metal milling machines and wood shapers, routers, jointers, and planers. The machines feed the workpiece into a rotating cutter.

The ***radial saw*** moves the rotating blade across the workpiece. The workpiece is positioned on a stationary table. The rotating blade produces the cutting motion. This blade is manually drawn across the material for the feed motion. The position of the arm of the machine determines the direction of the blade travel. Generally, the machine can cut a square angle and any angle up to 45° on each side of square.

Standards for Technological Literacy

Figure Discussion

Discuss the advantages and disadvantages of the three types of saws shown in Figure 7-11.

TechnoFact

Saws were invented as straight blades that cut in one direction, but did nothing on the backward stroke. Although circular saws were patented in England in 1777, the first circular saw wasn't used in America until 1814. Some people credit this to a Shaker woman who got the idea from watching a spinning wheel work.

Standards for Technological Literacy

7

Research

Have your students explore the types of drilling machines in use today.

TechnoFact

The drill was one of the earliest tools invented. Drills were designed with the range of motion of the arm and wrist in mind. Even before they were perfected by adding bowstrings to turn the drills more quickly, they were easy to use.

The ***chop saw*** is used to cut narrow strips of material to length. The blade is attached to a pivot arm. The material is placed on the table of the machine. The blade assembly is pivoted down to generate the feed motion. Many cutoff saws have saw-blade units that can be rotated so the material can be cut off at a specified angle. This type of saw, called a *miter saw,* is widely used to cut trim moldings for homes and the parts of picture frames.

Band saws

Band saws use a blade made of a continuous strip, or band, of metal. Most of these bands have teeth on one edge. Large band saw blades used in lumber mills have teeth on both edges, however.

The band generally travels around two wheels, which gives a continuous linear cutting action. The strip can be vertical or horizontal. Horizontal band saws are usually used as cutoff saws for metal rods and bars. They have replaced hacksaws as the primary method of cutting metal stock to length. Horizontal machines hold the material stationary in vises or clamps. They produce the feed motion by allowing the blade unit to pivot into the work.

Vertical band saws are widely used to cut irregular shapes from wood or metal sheets. The material is placed on the machine table. These sheets are then fed into the blade to produce the cut. Most often, the operator manually feeds the material to produce the desired cut.

Scroll saws

Scroll saws use a straight blade that is a strip of metal with teeth on one edge. The blade is clamped into the machine. The machine then moves the blade up and down to produce a reciprocating cutting motion. The material is placed on the table and manually fed into the blade. Portable scroll saws grip only one end of the blade. The saw is then fed into the work to produce the feed motion.

Drilling Machines

Holes in parts and products are very common and can be produced in various ways. They can be punched into sheet metal using punches and dies. Holes can be burned into plate steel using an oxyacetylene cutting torch. They can be produced in the part directly as it is cast from molten material. Holes can be produced with powerful beams of light (laser machining) or electrical sparks (electro-discharge machining).

Many cylindrical holes, however, are produced through drilling operations. ***Drilling machines*** produce or enlarge holes using a rotating cutter. Generally, the cutting motion is produced as the drill, or bit, rotates. The drill is moved into the work to produce the feed motion. The drill press is the most common machine using these cutting and feed motions. See **Figure 7-13.**

Drilling can also be done by rotating the work to produce the cutting motion. The stationary drill is then fed into the work. This practice is common with drilling on a metal lathe or a computer-controlled machining center.

Many operations can be completed on drilling machines. The most common are shown in **Figure 7-14.** These operations are the following:

- **Drilling.** This operation is producing straight, cylindrical holes in a material. These holes can be used to accommodate bolts, screws, shafts, and pins for assembly. They can also be a functional feature of a product. For example, holes are essential for the functioning of furnace burners, automobile carburetors, and compact disc recordings.
- **Counterboring.** This operation is producing two holes of different diameters and depths around the same center point. Counterbores are

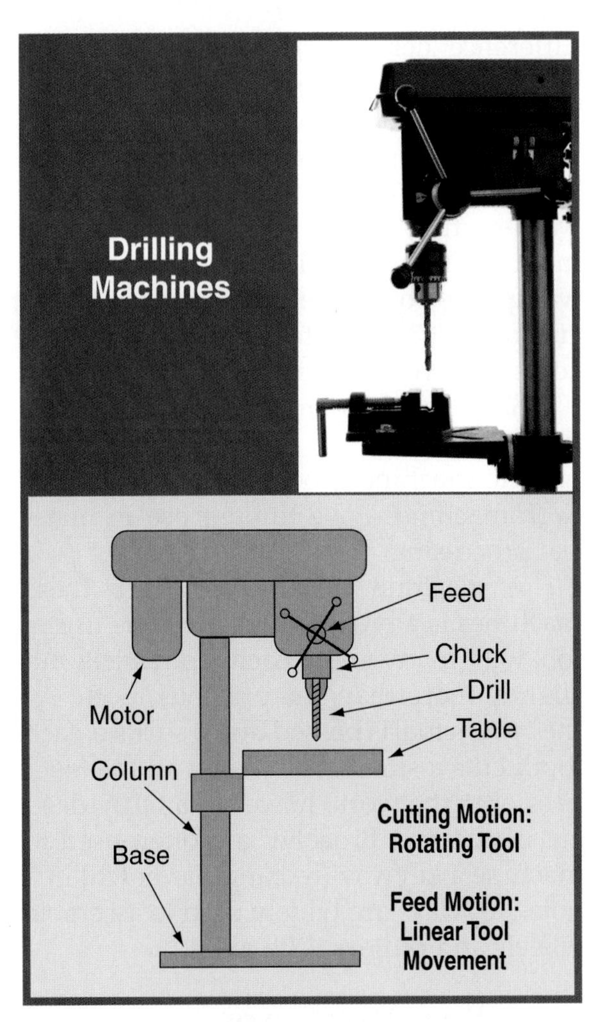

Figure 7-13. A drill press is the most common drilling machine.

used to position shafts and recess heads of fasteners, among other purposes.

- **Countersinking.** This operation is producing a beveled outer portion of a hole. Most often, countersinking is used with flathead wood and metal screws. Countersinking holes allows screw heads to be flush with the surface of the part.
- **Reaming.** This operation is enlarging the diameter of a hole. The action is generally performed to produce an accurate diameter for a bolt-hole.

Drilling operations use various drilling tools. These tools include twist drills, spade bits, and Forstner bits. ***Twist drills*** are shafts of steel with points on the end to produce a chip. These chips are carried from the work on helical flutes circling the shaft. ***Spade bits*** are flat cutters on the end of a shaft. The bottoms of the cutters are shaped to produce the cut. ***Forstner bits*** are two-lipped woodcutters that produce a flat-bottomed hole. Hole saws, which use sawing machine action, and fly cutters, which use lathe-type tools, can also be used to produce holes. A few of these bits and drills are shown in Figure 7-15.

Standards for Technological Literacy

7

Demonstrate

Show your students the techniques used to drill, countersink, and counterbore a part.

TechnoFact

When canals and railways became important methods of transportation during the 19th century, problems with constructing tunnels occurred. These problems led to the development of special-purpose drilling machines, such as the pneumatic drill. This drill was invented when problems arose from drilling through mountains.

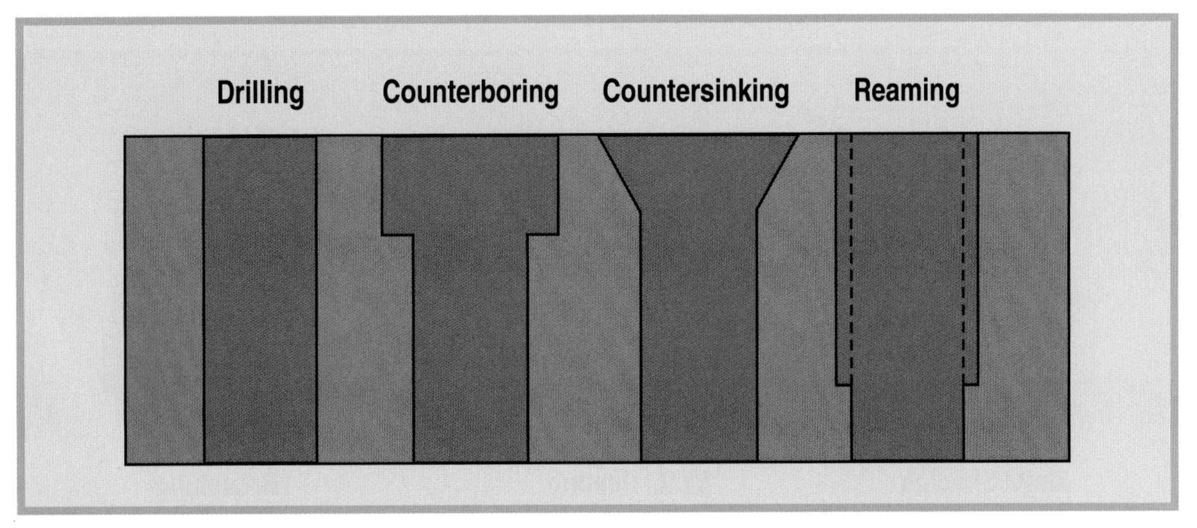

Figure 7-14. The most common operations performed on drilling machines are drilling, counterboring, countersinking, and reaming.

Standards for Technological Literacy

7

Figure Discussion

Describe, explain, and show the uses for the three drilling tools shown in Figure 7-15.

TechnoFact

In order to reduce human error and increase accuracy, automatic computer-controlled machines can be used to shape and plane metal.

Figure 7-15. Common drilling tools are the twist drill, spade bit, and Forstner bit.

Shaping and Planing Machines

Shaping machines and *planing machines* are two metalworking machine tools that produce flat surfaces. See Figure 7-16. These machines should not be confused with the woodworking shaping and planing machines, which operate on the same principles as sawing machines. Wood shapers and planers use a rotating cutter, into which the wood is fed.

Both the metal shaper and the metal planer use single-point tools and reciprocating motion to produce the cut. Their difference lies in the movements of the tool and the workpiece. The metal shaper moves the tool back and forth over the workpiece to produce the cutting motion. The work is moved over after each forward-cutting stroke to produce the feed motion. The metal planer reciprocates the workpiece under the tool to generate the cutting motion. The tool is moved one step across the work for each cutting stroke.

Shapers and planers both can cut on the face or side of the part. They can also be used to machine grooves into the surface. Both machines have limited use in material processing.

A machine closely related to these machines is a broach. This machine uses a tool with many teeth. Each tooth sticks out slightly more than the previous tooth. As the broach tool is passed over a surface, each tooth cuts a small chip. When all the teeth pass over the work, however, a fairly deep cut is possible. Broaches are often used to machine a keyway (rectangular notch) in a hole. Keyways are widely used to assemble wheels and pulleys with axles.

Grinding Machines

Grinding machines use bonded abrasives (grinding wheels) to cut the material.

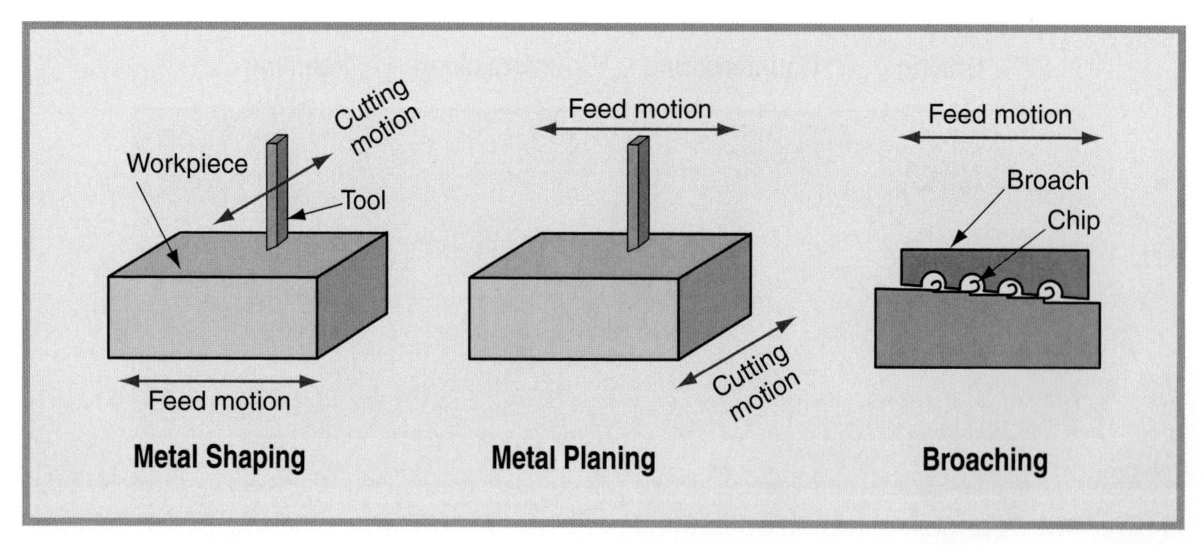

Figure 7-16. Shaping and planing machines use different feed and cutting motions.

These wheels have random cutting surfaces that remove the material in the form of very small chips. Grinders are basically adaptations of other machine tools. The two most common are cylindrical grinders and surface grinders. See **Figure 7-17.** ***Cylindrical grinders*** use the lathe principle to machine the material. The workpiece is held in a chuck or between centers and rotated. A grinding wheel is rotated in the opposite direction. The opposing rotating forces produce the cutting motion. The grinding wheel is fed into the work to produce the feed motion.

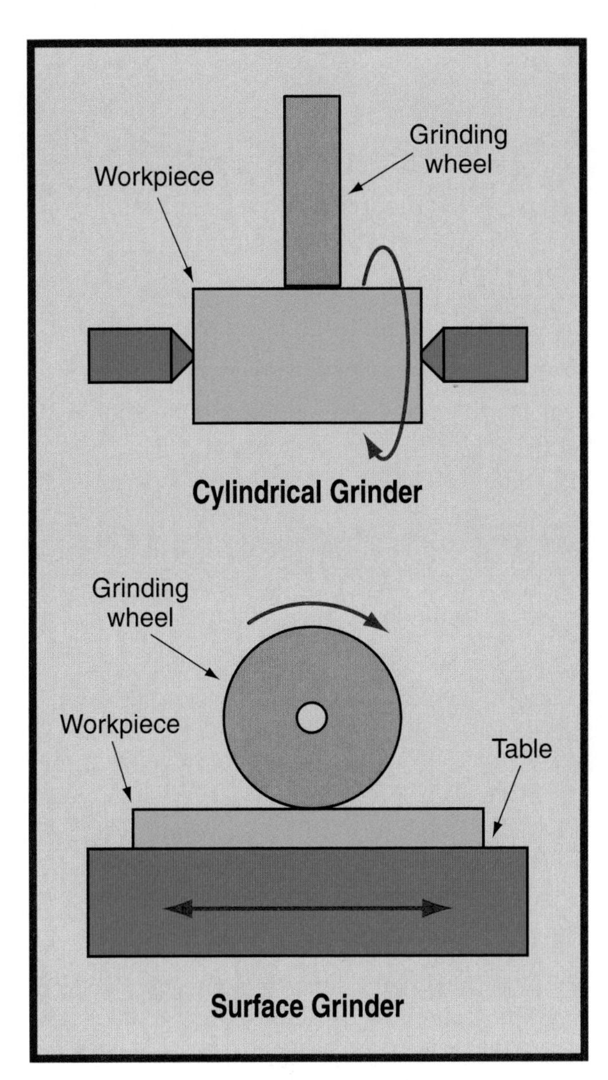

Figure 7-17. Two common grinding machines are the cylindrical grinder and the surface grinder.

Surface grinders work on the metal-planer principle. A rotating grinding wheel is suspended above the workpiece. The work is moved back and forth under the wheel to produce the cutting motion and moved slightly, or indexed, after each grinding pass to produce the feed motion.

A third type of grinder used is the pedestal grinder. The rotating grinding wheel produces the cutting motion. The operator manually moves the workpiece across the face of the wheel to produce the feed motion.

Energy-Processing Converters

At the heart of all technological systems is energy. Energy powers most devices that process materials and information. As we learned earlier, energy can be neither made nor destroyed, but it can be converted and applied to do work. This is the task of energy conversion or energy-processing converters. Literally hundreds of these converters exist, and they process energy in various ways. For example, energy-processing converters do the following:

- Convert mechanical energy into electrical energy (example: electric generator).
- Convert electrical energy into mechanical energy (example: electric motor).
- Convert radiant energy into thermal energy (example: solar heater).
- Convert heat energy into mechanical energy (example: internal combustion engine).

Through a study of human history, we can quickly identify three major ***energy-processing converters*** that have helped to shape life as we know it. They are the electric generator, the electric motor, and the internal combustion engine. We will focus on these converters in this chapter to gain

Standards for Technological Literacy

7 16

Extend

Differentiate between grinding and polishing.

Extend

Give additional examples of each type of energy converter.

Reinforce

Review energy as an input to technological systems (Chapter 2).

TechnoFact

Grinding wheels are used as cutting elements in most grinding machines. They remove very small chips from the material being cut. Abrasive grains are generally used to create grinding wheels.

Standards for Technological Literacy

7 16

Research

Have your students research the way steam engines were adapted for transportation vehicles.

TechnoFact

Before steam was harnessed and used, inventors had already contributed ideas about using steam as a power source. Almost 2000 years before the steam engine, for example, Hero of Alexandria created a steam-powered mechanism.

TechnoFact

In the development of modern technology, we often see a product's size become more compact as time goes on. This was no different over 200 years ago when the steam engine was first patented by Nicholas Cugnot. These early steam engines reached almost two stories tall, but they developed into progressively smaller and more efficient devices.

an understanding of the importance of ***energy conversion***. Other energy converters are discussed in Chapter 27.

Existing energy can be converted to be made more useful. Some energy conversions require special equipment to change one form of energy into another. An early example of energy-conversion technology is Watt's steam engine. The steam engine was the first device to convert energy effectively from one form to another. A later example is the electric generator. The electric generator is a device that converts mechanical energy to electrical energy.

The Electric Motor

The first energy-processing device discussed in this section is the electric motor. This device is probably the most universally used source of power. The average home has more than 40 motors in it. You can find motors in clocks, refrigerators and freezers, video recorders, compact disc players, furnaces and air conditioners, clothes washers and dryers, shavers, hair dryers, and many other appliances. They are also on construction sites and in factories. See **Figure 7-18.** They play a vital part in most agricultural, transportation, and communication systems.

The electric motor is based on the laws of magnetism and electromagnetism. See **Figure 7-19.** These laws state the following:

- Like poles of a magnet repel one another, and unlike poles attract one another.
- Current flowing in a wire creates an electromagnetic field around the conductor.

Look at **Figure 7-20** to see how these laws are applied. There are two magnets. The outer magnet is stationary and is called the ***field magnet***. The inner magnet is an electromagnet that can rotate and is called the ***armature***.

Minster Machine Co.

Figure 7-18. Electric motors power all the equipment in this metal-stamping line.

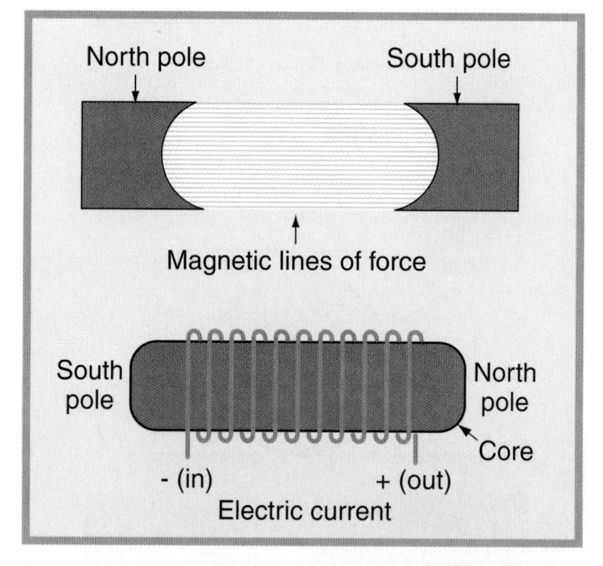

Figure 7-19. Motors take advantage of two principles of physics. The first principle is that magnetic lines of force travel between opposite poles of a magnet. The second principle is that current flowing through wires wrapped around an iron core produces magnetic poles at the ends of the core.

Electrical current is allowed to pass through the armature. The current induces magnetism in the armature core. A north pole is on one end, and a south pole is on the other. The direction the wire wraps around the core and the direction of the current determine which end is the north pole.

The north field pole and the north armature pole are next to each other. These are like poles, and they repel each other. Therefore, the armature spins one-fourth turn. Now, the unlike poles attract each other, and the armature continues rotating another one-fourth turn. The direction of the current is reversed. This action occurs

Standards for Technological Literacy

5 7 16

Extend

Discuss why the fractional horsepower electric motor revolutionized American life.

TechnoFact

In the early 20th century, Iowa bicycle makers August and Fred Duesenberg began working with gasoline engines. They attached a gasoline engine to a bicycle to make a motorcycle before manufacturing engines for automobiles.

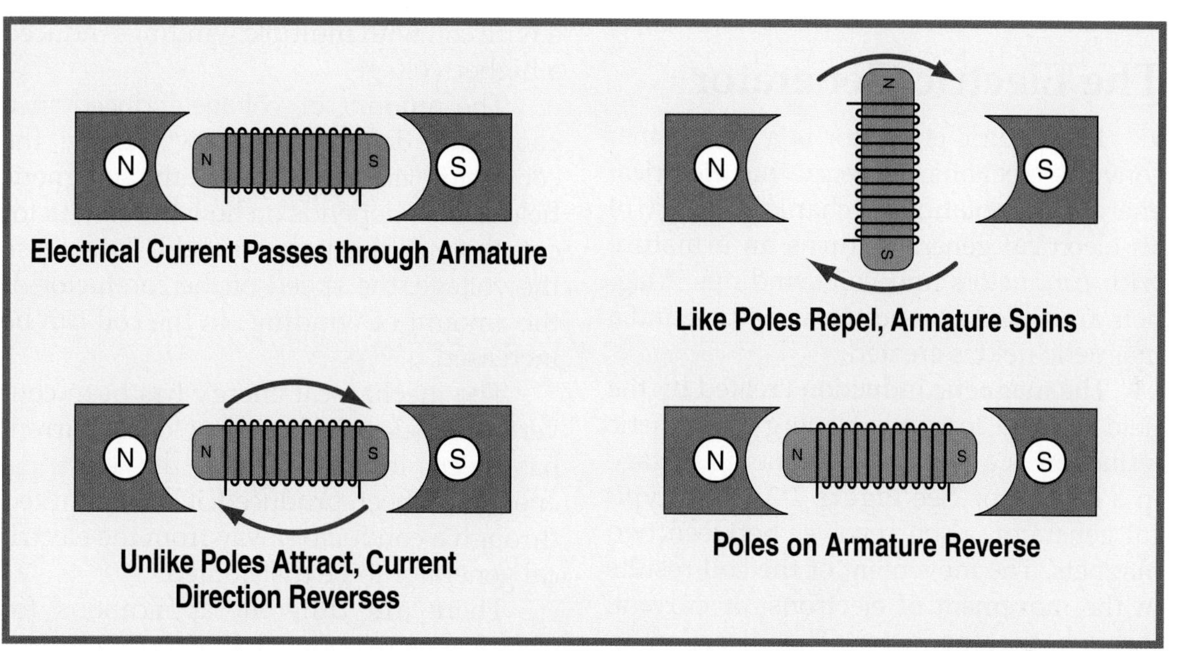

Figure 7-20. Electric motors operate by using a stationary magnet and an armature that switches its magnetic poles.

Think Green

Recycling

Recycling is the reprocessing of useful parts of used materials to help create new objects. Recycling is beneficial to the environment in two ways: it reduces the use of raw materials, and it reduces the amount of waste collecting in landfills. Raw materials take more energy to process. Also, using recycled materials helps reduce the consumption of exhaustible resources. Some of the materials that can be recycled (such as glass bottles, aluminum cans, and plastic packaging) does not decompose well, causing landfills to grow.

Several types of materials have been considered recyclable for a long time, and many communities participate in recycling programs. When you go tot the store, look at the packaging of the products you intend to buy. Many types of packaging on new products contains a percentage of recycled material.

Standards for Technological Literacy

7 16

TechnoFact

The Otto-cycle engine created by Nikolaus Otto in the 1880s is the modern standard internal combustion engine. His machine was the first to use gasoline as a power source. About twenty years later, Rudolf Diesel patented the diesel internal combustion engine.

mechanically in direct current motors. Alternating current motors use the existing direction changes in the line current. The current-direction change reverses the poles on the armatures. Now, the entire action is repeated. Like poles repel and turn the armature another one-fourth of a turn. Unlike poles then attract, turning the armature the final one-fourth turn to complete a full revolution of the armature.

The Electric Generator

The electric generator is a device that converts mechanical energy into electrical energy. The rotating mechanical energy of an electrical generator turns an armature with conductors and coil windings. Magnets are added around the armature, and a magnetic field is created.

The magnetic induction created by the field is used to convert energy. Magnetic induction is a method of inducing voltage in a conductor. See **Figure 7-21.** In a typical generator, a coil revolves between two magnets. The movement of the coil results in the movement of electrons, or current, through the conductor. The direction in which the current will move depends on the direction the coil moves through the magnetic field. The magnetic field around the conductor is created from this current.

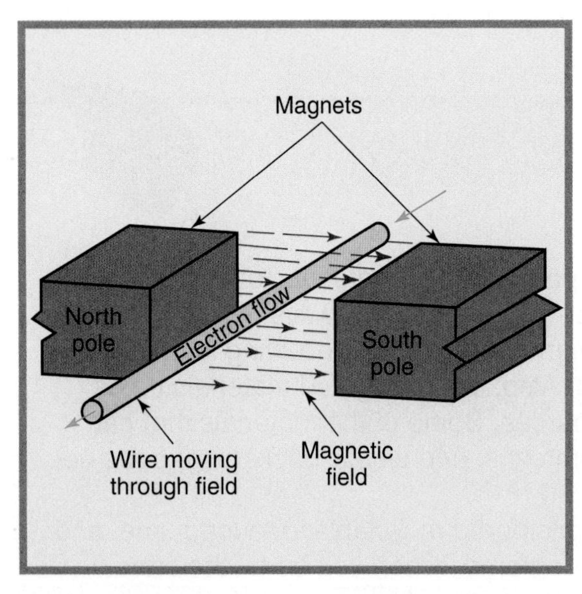

Figure 7-21. Magnetic induction induces voltage in a conductor.

Michael Faraday developed the first electric generator in the 1830s. Faraday moved a copper disk between magnets to create a small voltage. The voltage produced using the disc, however, was too low to be practical. It was then discovered that a wire coil with multiple windings induced a higher voltage.

The amount of voltage induced in a conductor depends on how quickly the conductor moves through the magnetic field. It also depends on how the conductor cuts through the magnetic field. To increase the voltage, the speed of the conductor or the amount of windings in the coil can be increased.

The mechanical energy has been converted to electrical energy once current has been induced. After the electrical energy has been produced, it is transmitted through a conductor away from the electrical generator to be distributed.

There are different applications for electrical generators. Large power-generating facilities are needed to power buildings. Smaller electrical generators may be portable or used for automobiles. Therefore, there are different types of electrical generators. See **Figure 7-22.** Electricity may be generated by engines, wind, fossil fuels (such as coal), and nuclear power.

Alternators are a type of electrical generator that are used in automobiles. An alternator's energy comes from a belt being turned on a pulley. This belt is turned by the engine.

Portable electric generators may be used to temporarily power a house. These generators are similar to alternators, in that a gas-powered engine is used to turn the turbine.

Water is commonly used to produce electricity. Dams are used to store water. This water may flow through turbines. The

Figure 7-22. Electrical energy may be generated by different types of devices.

Standards for Technological Literacy

16

Brainstorm

Have your students list all of the uses for internal combustion engines that they can think of.

TechnoFact

Mechanical energy develops from the chemical energy that is released when fuel is burned in a chamber in an internal combustion engine. The most common of the various types of internal combustion engines are gasoline and diesel engines.

turbine, which is a series of blades, turns, creating the mechanical energy needed in an electrical generator. This type of electricity is called *hydroelectricity*. Similarly, wind can be used to turn turbines. One wind turbine does not generate enough power for large-scale power distribution. Wind turbines are typically grouped into wind farms.

Nuclear power stations and fossil fuel power stations both use thermal energy to create the mechanical energy used to produce electricity. These methods of generating electricity are among the most common used to power cities.

The Internal Combustion Engine

Other energy converters have almost totally replaced the steam engine because of technological advances. The internal combustion engine was developed in the late 1800s. This engine now powers most land-transportation vehicles.

The common internal combustion engines are gasoline and diesel engines. Both of these engines change heat energy into mechanical energy. They drive a reciprocating piston by igniting a fuel. A crankshaft changes the piston's

Standards for Technological Literacy

7 16

TechnoFact

The first electric motors were developed in the 1830s by several inventors, including Michael Faraday, Orange Smalley, and Thomas Davenport. Their devices all centered around a magnetic force. While the first industrial electric motors were large and used to replace steam power, individual motors were in more widespread distribution before long.

reciprocating motion into rotary motion. See Figure 7-23. Gasoline engines can be either two stroke–cycle engines or four stroke–cycle engines. The four stroke–cycle engine is the most common engine and is the focus of this discussion.

A ***stroke*** is the movement of a piston from one end of a cylinder to another. A cycle is a complete set of motions needed to produce a surge of power. A two stroke–cycle engine moves the piston up and back once (two strokes: one up and one down) to produce a power stroke. A four-stroke engine moves the piston up and back twice (four strokes) to produce a power stroke.

Look at Figure 7-24 to see these four strokes. These strokes are intake, compression, power, and exhaust. During these strokes, the following actions take place:

- **Intake stroke.** The piston moves downward to create a partial vacuum. The intake valve opens. Atmospheric pressure forces a fuel-and-air mixture into the cylinder.

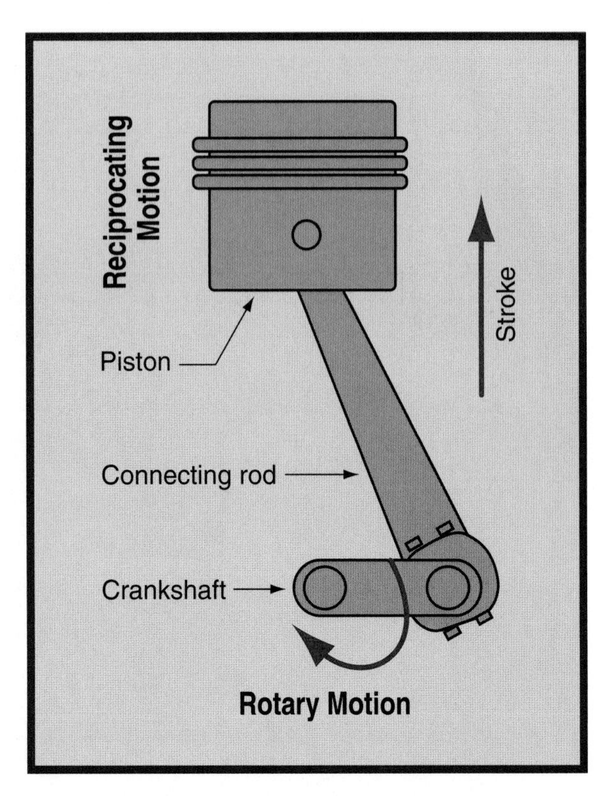

Figure 7-23. The reciprocating piston in an internal combustion engine turns a crankshaft.

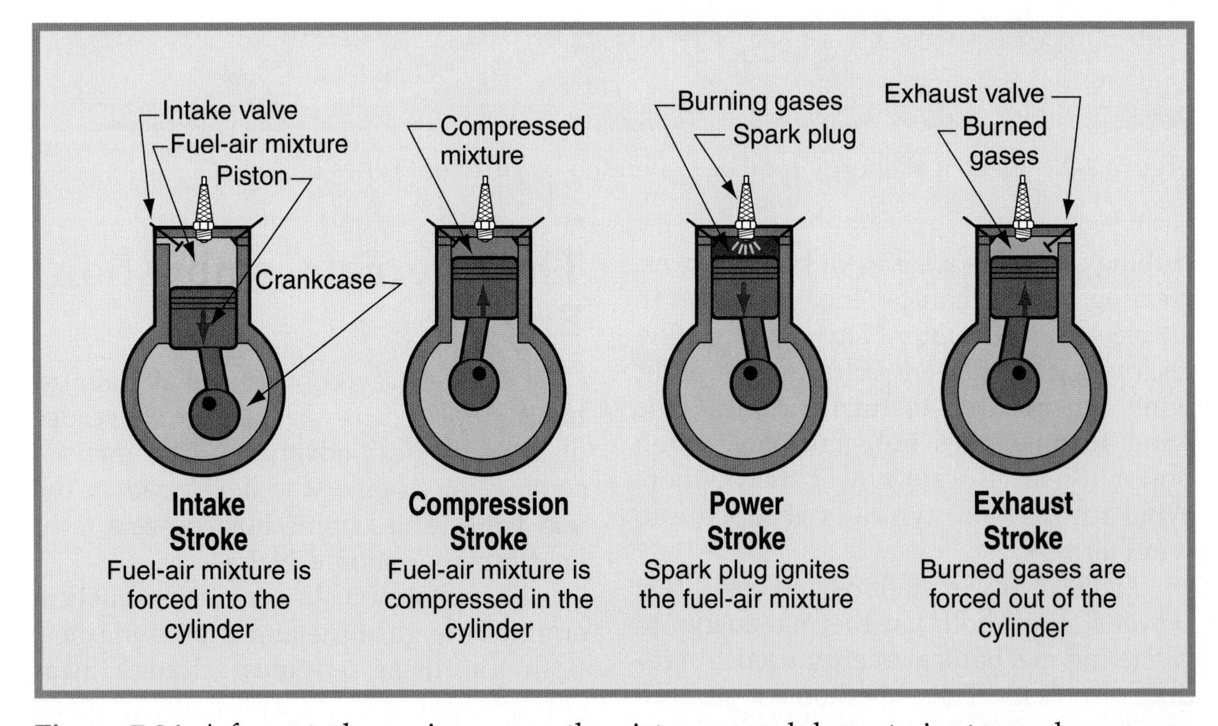

Figure 7-24. A four-stroke engine moves the piston up and down twice to produce a power stroke. The four actions are intake, compression, power, and exhaust.

- **Compression stroke.** The intake valve closes, and the piston moves upward. As the piston moves up, the fuel-air mixture is compressed in the small cavity at the top of the cylinder. The area forms a combustion chamber for the power stroke. The compression ratio of the engine tells you how much the fuel-air mixture has been compressed. The standard compression ratio is about 15:1. This ratio tells you that the cylinder volume at the beginning of the compression stroke is 15 times larger than at the end of the compression stroke. Therefore, the fuel-air mixture has been reduced to about 6.6% of its original volume.
- **Power stroke.** An electrical spark is produced between the two points of the spark plug. This action ignites the compressed fuel-air mixture. The burning gases produce temperatures as high as 4000°F (2200°C). This temperature expands the gas and generates pressure up to 1000 pounds per square inch (psi) or 7000 kilopascals (KPa). The pressure forces the piston downward in a powerful movement.
- **Exhaust stroke.** When the piston reaches the bottom of the power stroke, the exhaust valve opens. The piston moves upward to force exhaust gases and water vapor from the cylinder. These are the products of the combustion during the power stroke. At the end of this stroke, the cylinder is ready to repeat the four strokes.

Single-cylinder engines are common for low-horsepower applications such as lawn mowers, cement mixers, and portable conveyors. For more demanding applications, several cylinders are combined into one engine. The cycle of each cylinder is started at a different point in time so the engine has a series of closely spaced power strokes. Four-cylinder, six-cylinder, and eight-cylinder engines are common.

Information-Processing Machines

Processing and exchanging information are as old as human existence. ***Information processing*** is the gathering, storing, manipulating, and retrieving information. Early results of these actions can be seen in carvings found on cave walls and rocks. Today, the products of information processing are in books and photographs and on tape, film, and discs.

Modern information processing is built on a number of important technological advancements. Some of these are explored here. Others are introduced in Section 6 of this book.

The communication methods basic to all modern communication systems are printing, the computer, and telecommunications. Printing is based on movable type, developed in the 1400s. ***Computers*** are storage devices for information and are based on semiconductor circuits on microchips that process data. Telecommunications involves the communications achieved by telephones and radio technology. Each communication method is discussed in turn.

Printing

We have become accustomed to having printed materials at our fingertips. Books, magazines, newspapers, pamphlets, and brochures are everywhere. However, the abundance of printed materials is of recent history. Until the last 500 years, most people could not read and had never seen a book. Monks and scribes handwrote those

Standards for Technological Literacy

7 16 17

Extend

Differentiate among printing, telecommunication, and computer communication.

Standards for Technological Literacy

7 17

Extend

Discuss how the invention of movable type changed the world.

Demonstrate

Use a rubber stamp to demonstrate relief printing.

TechnoFact

Before Guttenberg's printing press of the 15th century, the Chinese used methods of printing. They carved into wood blocks or stone, covered those images with paper, and rubbed ink over them to produce a printed product.

books that did exist on parchment paper. As civilization grew, so did the demand for information. In the 1440s, this demand was addressed with a new printing process Johann Gutenberg, a German goldsmith, developed. This process used movable type. See **Figure 7-25.** Movable type is the blocks of individual characters that can be arranged for printing, then removed and rearranged.

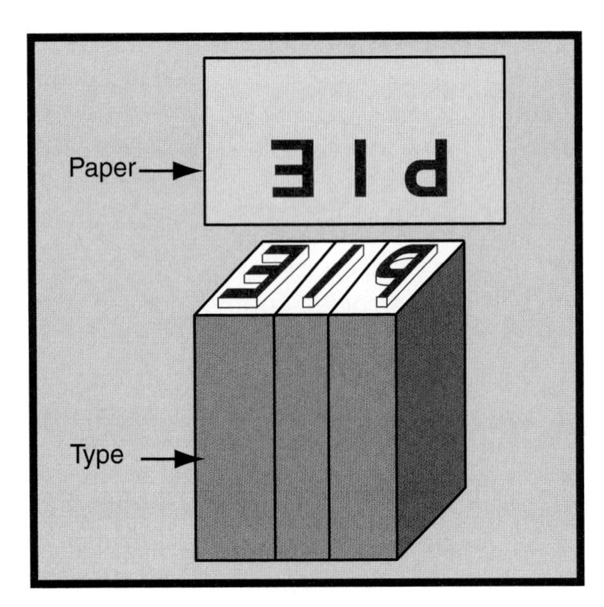

Figure 7-25. Gutenberg developed movable type, which revolutionized book production and started the printing industry.

Printing is the process of applying ink to a substrate, which is typically paper. A substrate is any material with a surface that can be printed on. Gutenberg's printing method (relief printing) is only one of six major printing processes that can be found in industry today. The six major processes are relief, lithographic, gravure, screen, electrostatic, and ink-jet. See **Figure 7-26.**

Relief printing uses raised image areas, which are inked and pressed against a substrate. The two types of relief printing are letterpress and flexography. Letterpress uses metal image carriers, while flexography uses rubber or plastic image carriers. Lithographic printing uses the concept of plates accepting ink and repelling water. The image area is inked on a plate, transferred to a cylinder, then set on a substrate. Image areas for gravure printing are etched into plates before being inked. Screen printing uses a stencil pressed up against a mesh screen. A substrate is placed behind the mesh screen and inked so the design in the

Process	How It Works
Relief	Raised image areas are inked and transferred to paper.
Lithographic	Flat plate accepts ink and repels water. Images are transferred from plates to cylinders, then to paper.
Gravure	Etched area on plate is inked and transferred to paper.
Screen	Uses stencil pressed up against mesh screen on top of substrate.
Electrostatic	Attaches toner to electrically charged image on paper.
Ink-jet	Computer-generated image is sprayed in droplets onto paper.

Figure 7-26. Six basic systems are used to operate an offset lithograph press.

stencil is applied to the substrate. Electrostatic printing uses a drum to electrically charge the image areas on a substrate before toner is applied. Image areas for ink-jet printing are computer-generated, then sprayed with fine droplets onto the substrate.

This type of printing is still being improved. Traditionally, film was used as an intermediate to create the printing plates. The computer-to-plate process eliminates the use for film and uses digital files and a laser. Another improvement is the elimination of the plate altogether in computer-to-press systems.

Computers

The computer is a major information-processing machine that has changed the way we handle information. Computers can store information and have programs that can be changed. These qualities provide great information-processing power.

All computers have five parts. These, as shown in **Figure 7-27**, are the input devices, processing unit, output devices, memory unit, and program. Input devices are used to enter data into the system and can include keyboards, mice, microphones, and Web cams. The processing unit is the *central processing unit (CPU)* or *microprocessor*, used to manipulate data. Output devices display and record results of the processing unit's actions and can include monitors, printers, and speakers. The memory unit has two types of memory. Read-only memory (ROM) contains the instructions that allow the computer to receive and manipulate data. This part of the memory cannot be changed. Random-access memory (RAM) temporarily stores data and feeds it to the CPU on command. RAM is constantly changing as the computer processes data. The program is the instructions the computer uses to process the data and to produce output.

Computers are at the heart of most information-processing systems. They are at every level of business life. Computers process financial information and prepare financial reports, maintain schedules and ticket records for airlines, operate point-of-purchase units at supermarket-checkout stands, guide spacecraft to distant planets, maintain the fuel-air mixture in automobiles, guide washing machines through their wash-rinse-spin cycles, help

Standards for Technological Literacy

17

Research

Have your students select a telecommunication system and investigate how it works.

Demonstrate

Show examples of the five units of a computer.

Extend

Discuss the types of operations that can be done on a computer.

TechnoFact

Although computers weren't in use until the 20th century, the idea was conceived in 1830 by Charles Babbage. He had spent most of his life searching for a way to perfect a mechanical calculating machine.

TechnoFact

Because computers are used for many activities in everyday life, they come in a variety of sizes including supercomputers, mainframes, minicomputers, and microcomputers. Each of these is used in a specialized capacity for business, industry, and home use.

©iStockphoto.com/sweetym

Figure 7-27. A computer has five parts: an input unit, a processing unit, a memory unit, an output unit, and a program.

Standards for Technological Literacy

17

TechnoFact

Although Alexander Graham Bell is credited with patenting the first working telephone in 1876, the German inventor Phillip Reis began working with the idea of a telephone in the early 1860s. His invention, however, did not produce consistent results as Bell's did.

prepare layouts for advertising, and control industrial machines. They provide Internet access and allow for worldwide communication.

Internet

The Internet allows people to gather information using the World Wide Web and exchange messages using e-mail (electronic mail). The Internet is a computer-based, global information system. The Internet connects people from all over the world, making this system an inexpensive and efficient communication tool. See Chapter 23 for further discussion on the Internet.

Common uses of the World Wide Web (WWW) are:

- Web browser
- E-mail
- Social networking
- Electronic commerce (e-commerce)

The WWW allows people to view information and images on the Internet using a Web browser. Most of the content is free, while some sites may charge a subscription fee. E-mail is a system that allows an individual to send a message to another computer or a number of different computers. Newsgroups and blogs allow for electronic conversations between two or more people. They are forms of social networking. Blogs allow everyone to see the messages, while newsgroups require people to subscribe in order to view messages. E-commerce involves selling products and services over the Internet.

Web-Enabled Mobile Devices

Two common types of telephones are traditional landline phones and cellular phones (cell phones). The telephone converts sound into electrical energy. The electrical energy is then transmitted from the sender to the receiver. At the receiver, electrical energy is converted back into sound waves. This is true of landline telephones and of cell phones. Cell phones use a cellular network to transmit and receive signals. A cellular network is a linked group of individual cells used for radio signals. The network allows cell phones to communicate with one another. Cell phones now have more applications than phone-to-phone communication. They include features such as e-mail and Internet services.

With Internet access and other applications available for cell phones, the cell phone has become a popular type of portable computer. In order to use the Internet from the cell phone, it must be equipped with a Web browser. Many cell phones come standard with a Web browser, and an option is typically offered in a cell phone plan to use the service. These browsers are typically connected through the cellular network.

Mobile browsers are the Web browsers used on cell phones. Mobile browsers are similar to traditional Web browsers used on computers, but they are specially designed for the smaller display. See **Figure 7-28.** Because cell phones are becoming more commonly used to browse the World Wide Web, some Web

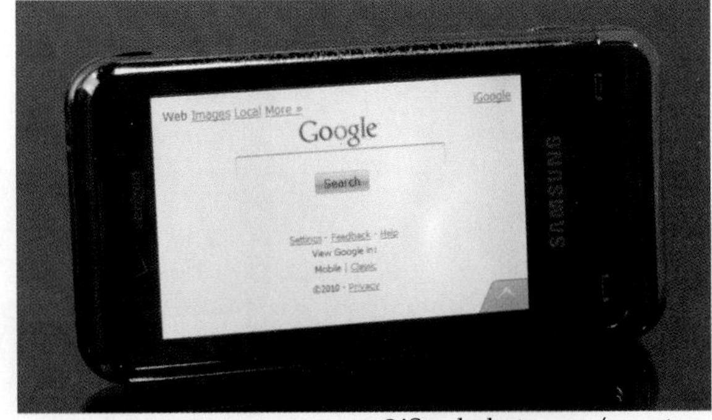

©iStockphoto.com/sweetym

Figure 7-28. Most cell phones come equipped with mobile browsers. When browsing Web sites on a cell phone, you may have the option of using the mobile version of the site.

sites now include a "mobile" viewing option. This option is often similar to the original Web site, but mobile sites display the information in a way that is easier to read on a cell phone, displays links in a more accessible way, and won't slow down the performance of the browser.

Cell phones are also capable of downloading and using programs called applications. An application is a type of program developed specifically for cell phone use. Some types of applications are gaming, educational, news, and organizational software. While some applications may be only condensed versions of software typically found on computers, others may be software used to enhance features already on the cell phone.

The telephone has allowed people to be connected. As already discussed, the Internet has allowed people to be connected worldwide. Cell phones with Web browsing capabilities offer both of these technologies in one package.

Radios

A radio communication system contains a transmitter to produce radio waves and a receiver to collect them. The transmitter changes sound into radio waves and imposes the sound information onto a carrier wave. This action is called *modulation*. Transmitters use either frequency modulation (FM) or amplitude modulation (AM) to place the information onto the carrier wave. FM changes the rate of cycles on the carrier wave. AM changes the strength of the carrier wave. The top of **Figure 7-29** shows how traditional radio transmitters and receivers work.

Standards for Technological Literacy

17

Research

Have your students investigate the advantages and disadvantages of AM and FM broadcasts.

Research

Have your students create a diagram illustrating the difference between frequency modulation and amplitude modulation signals.

TechnoFact

In the early 1900s, the first radio broadcast network was developed for the United Fruit Company. While this station and others like it used radio to help their businesses, the first commercial radio station wasn't established until 1920 by the Westinghouse Electric Corporation in Pittsburgh.

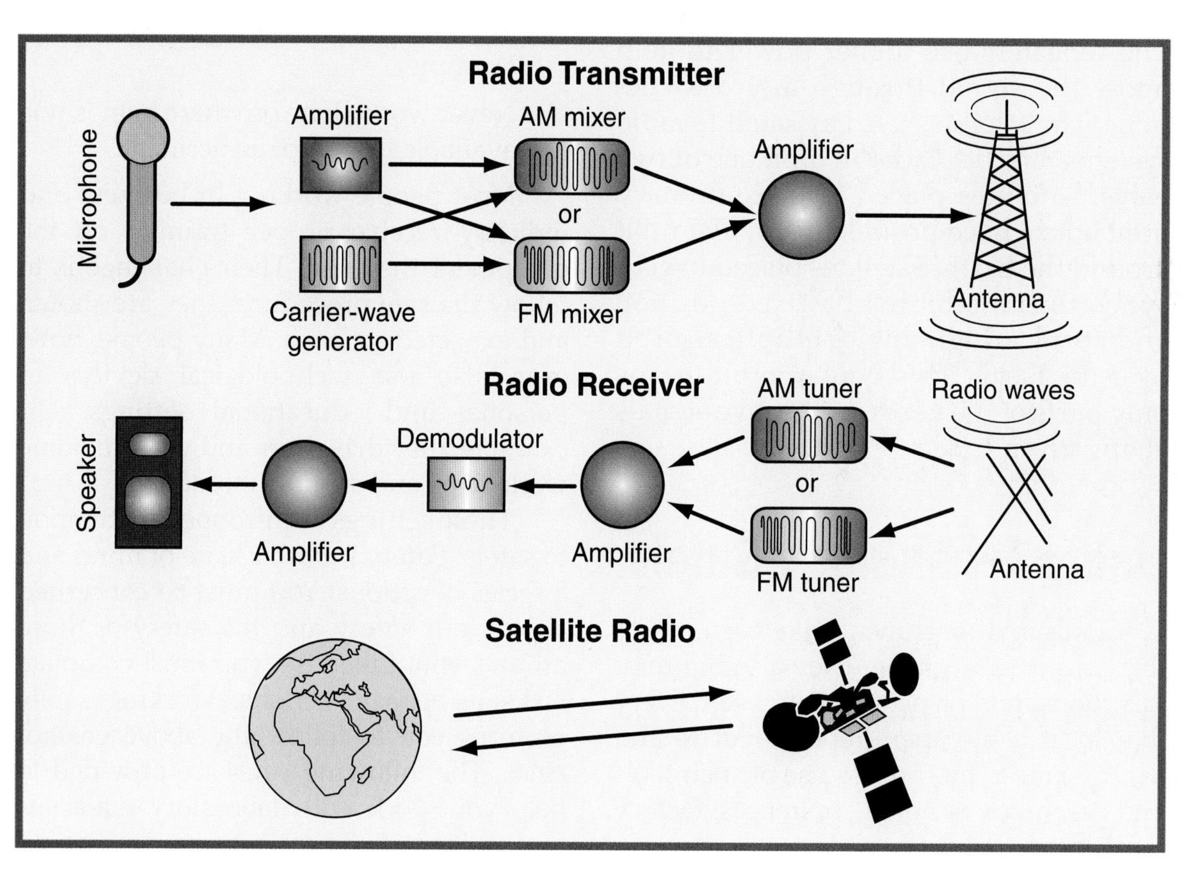

Figure 7-29. Satellite radio broadcasts signals from Earth stations to the satellites, which then bounce the signals back to receivers, such as this.

Standards for Technological Literacy

12 17

TechnoFact

Radio broadcasting's influence became greatest during World War II. For the first time, news from the battlefront was available to everyone, and leaders of the nations used this media to encourage their people.

Traditional radio systems are only capable of transmitting so many miles from their radio transmitters. Another type of radio is satellite radio. Rather than using local transmitters to broadcast AM or FM radio signals, satellite radio uses satellites in space to send digital information to the receivers. Satellite radio is a system that allows you to travel beyond the range of local stations and continue listening to the same radio station. The satellite radio providers have hundreds of channels for news, sports, and music, and these channels can be accessed anywhere by the appropriate receivers.

Satellite radio systems work by sending signals from stations or transmitters on Earth to satellites in orbit. The satellites then send the signals back to the ground. Refer to Figure 7-29. Ground repeaters are used to pick up the signals and use a higher level of power to distribute them to radio receivers in urban areas. The repeaters use higher power to help boost the signal through any obstructions. The satellites used by satellite radio systems may be placed in orbit one of two ways. Satellites placed in geostationary orbit hold a fixed position in circular orbit around the Earth. Satellites placed in geosynchronous orbit hold a fixed position and form a high-angle, elliptical orbit over the Earth. This type of orbit means only parts of the Earth can receive signals at any given time.

Using Technology Safely

Tools and machines make each of our lives more comfortable. They also injure or kill thousands of people each year. Every day, local newspapers report injuries and deaths caused by careless use of such tools and machines as home appliances, factory machinery, and automobiles.

If we want to understand technological devices and have them affect our lives positively, we must use them in a safe manner. These devices include, not just tools and machines that process materials, but also those that process energy and information. Operators should follow some basic guidelines when they select and use technological devices. These guidelines include the following:

- Select the correct tool for the job.
- Read about the proper operation of the device in owner's manuals and instruction books.
- Seek instruction and advice on the proper use of the device.
- Use the device only in the way described in operation manuals and for applications for which the device was designed.
- Never use a device when personal ability is impaired by medication, lack of rest, or distractions (loud music or people talking to you, for example).
- Never work alone or where help is not available in case of an accident.

Most people working in business and industry receive proper training on the equipment they use. Their challenge is to follow the safe procedures they are shown and expected to learn. Many people, however, also use technological devices in personal and educational settings. For example, they drive cars and work in home workshops and school laboratories.

These settings require special attention to safety. Safety is both a state of mind and a series of actions. You must be concerned about your safety and the safety of those around you. Likewise, you must complete tasks using safe actions. Working safely requires you to follow the above general rules. The following rules are provided to help you work with laboratory materials and tools safely. These rules cover working safely with people, materials, and tools and machines.

Safety with People

- Concentrate on your work. Watching other people or daydreaming can cause accidents.
- Dress properly. Avoid loose clothing and open shoes. Remove jewelry and watches.
- Control your hair. Secure any loose hair that might get caught in moving machine parts.
- Protect your eyes with goggles, safety glasses, or a face shield.
- Protect your hearing by using hearing protectors when working around loud machines or where high-pitched noises are present.
- Do *not* use compressed air to blow chips and dirt from machines or benches. Use a brush to sweep them away gently.
- Think before acting. Always think of what will happen before starting a task.
- Ask your teacher questions about any operation of which you are unsure.
- Seek first aid for any injuries.
- Follow specific safety practices your teacher demonstrates.
- Avoid horseplay. What you think of as harmless fun can cause injury to other people.
- Do *not* talk to anyone who is using a machine, except when necessary. You might distract her and cause an injury.
- Do *not* cause other people harm by carelessly leaving tools on benches or machines.

Safety with Materials

- Handle materials properly and with care.
    - Be careful when moving long pieces of material. Do not hit people and machines with the ends.
    - Use extreme caution when handling sheet metal. The thin, sharp edges can easily cut you and others.
    - Grip hot materials with pliers or tongs.
    - Wear gloves when handling hot or sharp materials.
    - Use extreme caution in handling hot liquids and molten metals.
- Check all materials for sharp burred edges and pointed ends. Remove these hazards when possible.
- If a material gives off odors or fumes, place it in a well-ventilated area, fume hood, or spray booth.
- Lift material properly. Use your legs to do the lifting, not your back. Do *not* overestimate your strength. If you need help, ask for it.
- Dispose of scrap material properly to avoid accidents.
- Clean up all spills quickly and correctly.
- Dispose of hazardous wastes and rags properly.

Safety with Tools and Machines

- Use only sharp tools and well-maintained machines.
- Return all tools and machine accessories to their proper places.
- Use the right tool or machine for the right job.
- Do *not* use any tool or machine without permission or proper instruction.

Casting and Molding Processes

- Do *not* try to perform a process that has not been demonstrated to you.
- Always wear safety glasses.

Standards for Technological Literacy

TechnoFact

The Occupational Safety and Health Administration (OSHA) was established by the Department of Labor to develop and enforce safety and health regulations for workers in businesses.

Standards for Technological Literacy

TechnoFact

The word techni was used by the Ancient Greeks to describe the art and skill used in making things.

- Wear protective clothing, gloves, and face shields when pouring molten metal.
- Do *not* pour molten material into a mold that is wet or contains any water.
- Carefully fasten two-part molds together.
- Perform casting and molding processes in a well-ventilated area.
- Do *not* leave hot castings or mold parts where they can burn other people.
- Constantly monitor material and machine temperatures during the casting and molding process.

Separating Processes

- Do *not* try to perform a process that has not been demonstrated to you.
- Always wear safety glasses.
- Keep hands away from all moving cutters and blades.
- Use push sticks to feed all material into woodcutting machines.
- Use all machine guards.
- Always stop machines to make measurements and adjustments.
- Do *not* leave a machine until the cutter has stopped.
- Whenever possible, clamp all work.
- Always unplug machines from the electrical outlet before changing blades or cutters.
- Remove all chuck keys or wrenches before starting the machine.
- Remove all scraps and tools from the machine before using it.
- Remove scraps with a push stick.

Forming Processes

- Do *not* try to perform a process that has not been demonstrated to you.
- Always wear safety glasses and gloves.
- Always hold hot materials with pliers or tongs.
- Place hot parts in a safe place to cool. Keep them away from people and from materials that will burn.
- Follow correct procedures when lighting torches and furnaces. Use spark lighters, *not* matches.
- *Never* place your hands or any foreign object between mated dies or rolls.

Finishing Processes

- Do *not* try to perform a process that has not been demonstrated to you.
- Always wear safety glasses and a respirator.
- Apply finishes in properly ventilated areas.
- *Never* apply finishing materials near an open flame.
- Always use the right solvent to thin finishes and clean finishing equipment.

Assembling Processes

- Do *not* try to perform a process that has not been demonstrated to you.
- Always wear safety glasses.
- Wear gloves, protective clothing, and goggles for all welding, brazing, and soldering operations.
- Always light a torch with a spark lighter. *Never* use a match.
- Handle all hot material with gloves and pliers.
- Perform welding, brazing, and soldering operations in well-ventilated areas.
- Use proper tools for all mechanical-fastening operations. Be sure screwdrivers, wrenches, and hammers are the proper size and are in good condition.

Fire Prevention and Safety

Many technological activities use flames or combustible materials. This requires the operators to understand the elements of fire protection, which is the study of the behavior and suppression of fire. Fire protection involves both fire prevention and fire extinguishing. Fire prevention minimizes the sources that can ignite a fire and educates people on the dangers and sources of fires. Within a total fire-protection system, there are the following:

- Fire-detection systems, including fire alarms or smoke-detection devices.
- Active fire-protection devices, including manual or automatic fire-detection and fire-suppression systems.
- Passive fire-protection activities, including using fire-resistant materials and separating and closing off fire-prone areas.

Firefighting is closely related to fire protection and all activities designed to extinguish destructive fires. These activities require one or more of the four components of combustion to be removed. The components are fuel, oxygen, heat, and a chemical chain reaction.

A common firefighting device available to the average citizen is the fire extinguisher. See **Figure 7-30.** This device is an active fire-protection device used to put out small fires. The common fire extinguisher is a handheld, cylindrical pressure vessel containing an agent that can be discharged to extinguish a fire. Most fire extinguishers are operated using a four-step process, with the operator standing a safe distance (about 4′–10′) from the fire. The acronym "PASS" describes this process:

P—Pull the safety pin.

A—Aim the nozzle at the base of the fire.

S—Squeeze the handle.

S—Sweep the nozzle from side to side.

Standards for Technological Literacy

12

©iStockphoto.com/molibra

Figure 7-30. Using a fire extinguisher can help put out small fires.

Answers to Test Your Knowledge Questions

1. Material processing, energy processing, and information processing
2. Evaluate individually.
3. Cutting motion is the action that causes material to be removed from the work. It causes the excess material to be cut away. Feed motion is the action that brings new material into the cutter. It allows the cutting action to be continuous.
4. True
5. scroll
6. Counterboring is producing two holes around the same center point. The larger diameter hole is drilled to a lesser depth than the smaller diameter hole.
7. The metal shaper moves the tool back and forth over the workpiece to produce the cutting motion. The metal planer reciprocates the workpiece under the tool to generate the cutting motion.
8. steam engine
9. An electric motor
10. offset lithographic printing
11. satellite radio
12. True
13. Evaluate individually.

Summary

Technological devices are designed to process materials, energy, and information. Tools and machines change the form of materials to make them more useful to people. They are also used to change the form of energy so we can do such things as heat buildings, power vehicles and machines, and light up dark areas. Information-processing tools and machines aid us in gathering and ordering data so we can better understand it.

Each one of these processing activities can make our lives better. Also, however, each can harm the environment and us. Processing activities must be carefully selected and properly and safely used.

Test Your Knowledge

Write your answers on a separate piece of paper. Please do not write in this book.

1. List the three major types of production tools.
2. Give an example of a multiple-point tool.
3. What is the difference between feed motion and cutting motion?
4. Lathes are primarily used to machine wood and metal, but plastics can be machined on lathes also. True or false?
5. The three basic types of saws are the circular saw, the band saw, and the ______ saw.
6. Define the term *counterboring*.
7. Describe the difference between shaping machines and planing machines.
8. The ______ was the first energy converter that did not simply change the form of mechanical energy.
9. ______ is an energy-converting device that uses the laws of magnetism and electromagnetism.
10. The most commonly used type of printing today is ______.
11. A(n) ______ allows you to travel beyond the range of local stations and continue listening to the same radio station.
12. Safety is both a state of mind and a series of actions. True or false?
13. Give one example of what *not* to wear when working with tools and machines.
14. List one rule with regard to working safely around other people.
15. Give one example of how to handle materials properly.
16. When working with tools and machines, wear safety glasses until you reach the assembling process. True or false?

STEM Applications

1. Select a product that has been manufactured using material-processing technology. List one to three operations used to make the product that used each of the following machine tools:

Machine Tools	Process or Operation
Turning machines	
Sawing machines	
Drilling machines	
Planing or shaping machines	
Grinding machines	

2. Select a major development in material processing, energy processing, or information processing. Write a two-page report on the development.
3. Select a major development in material processing, energy processing, or information processing. Build a model of the technological device.
4. Produce a simple product that uses material-processing technology. Write a summary of the procedure you used to make the product. List the tool or machine you used for each step of the procedure.
5. Communicate data or information using an information-processing device. Write a one-page report on the processes you used to complete the task.
6. Select a household device you use. Develop a set of safety rules for its use.

14. Student answers will vary but may include any of the following: avoid horseplay; do not talk to anyone who is using a machine, except when necessary; do not cause other people harm by carelessly leaving tools on benches or machines. Additional responses may be accepted at the instructor's discretion.
15. Student answers will vary but may include any of the following: be careful when moving long pieces of material; use extreme caution when handling sheet metal; grip hot materials with pliers or tongs; wear gloves when handling hot or sharp materials; use extreme caution in handling hot liquids and molten metals. Additional responses may be accepted at the instructor's discretion.
16. False

Chapter Outline
Measurement Systems: Past and Present
Qualities Measured
Types of Measurement
Measurement Tools
Measurement and Control

This chapter covers the benchmark topics for the following Standards for Technological Literacy:

2 3 7

Learning Objectives

After studying this chapter, you will be able to do the following:

- Recall the definition of *measurement*.
- Compare the U.S. customary and the metric measurement systems.
- Summarize the major physical qualities that can be measured.
- Recall the two main types of measurements.
- Recall common measuring tools to measure linear distances, diameters, and angles.
- Explain the relationship between measurement and quality control.

Key Terms

area
capacity
direct-reading measurement tool
distance
indirect-reading measurement tool
length
mass
measurement
metric system
micrometer
precision measurement
quality control
rule
square
standard measurement
temperature
U.S. customary system
volume
weight

Strategic Reading

As you read this chapter, think about the different types of measurement. Observe the objects around you, and determine the various ways they can be measured, such as weight or temperature.

When you travel from your home to school, how far is the trip? How long does the journey take? Do you travel by car or bus? If so, how much fuel does that vehicle's fuel tank hold?

This book has some physical qualities. How heavy is it? What size are the pages? How thick is it?

Think of how you compare with your classmates. How much do you weigh? How tall are you? How fast can you run the 100-yard dash?

These questions ask you to describe physical qualities. Physical qualities are characteristics of an object or event that can be described. If we say a tree is tall, but have nothing to compare it to, saying the tree is tall has no meaning. The tree can be a dwarf apple tree or a giant redwood. If it takes a long time to do something, what do we mean by a long time? Hours? Weeks? Years? Centuries?

To describe something to someone, we must have a common reference or standard. We use measurement to describe objectively the physical qualities of an item. ***Measurement*** is the practice of comparing the qualities of an object to a standard. See **Figure 8-1.**

Measurement Systems: Past and Present

All measurements compare the quality being described against a standard. In early times, people used the sizes of human body parts as standards for measurement. The biblical story of Noah records that he built an ark 300 cubits long, 50 cubits wide, and 30 cubits high. These are strange measurements to us. Early Egyptians, Romans, and Greeks used these terms, however. The standard was derived from the human

Standards for Technological Literacy

7

TechnoFact

Each country decides its own units of measurement because there is no universal standard system. For example, England took measurements from other nations and adjusted them in order to create its stable standard of measurement by the 1700s.

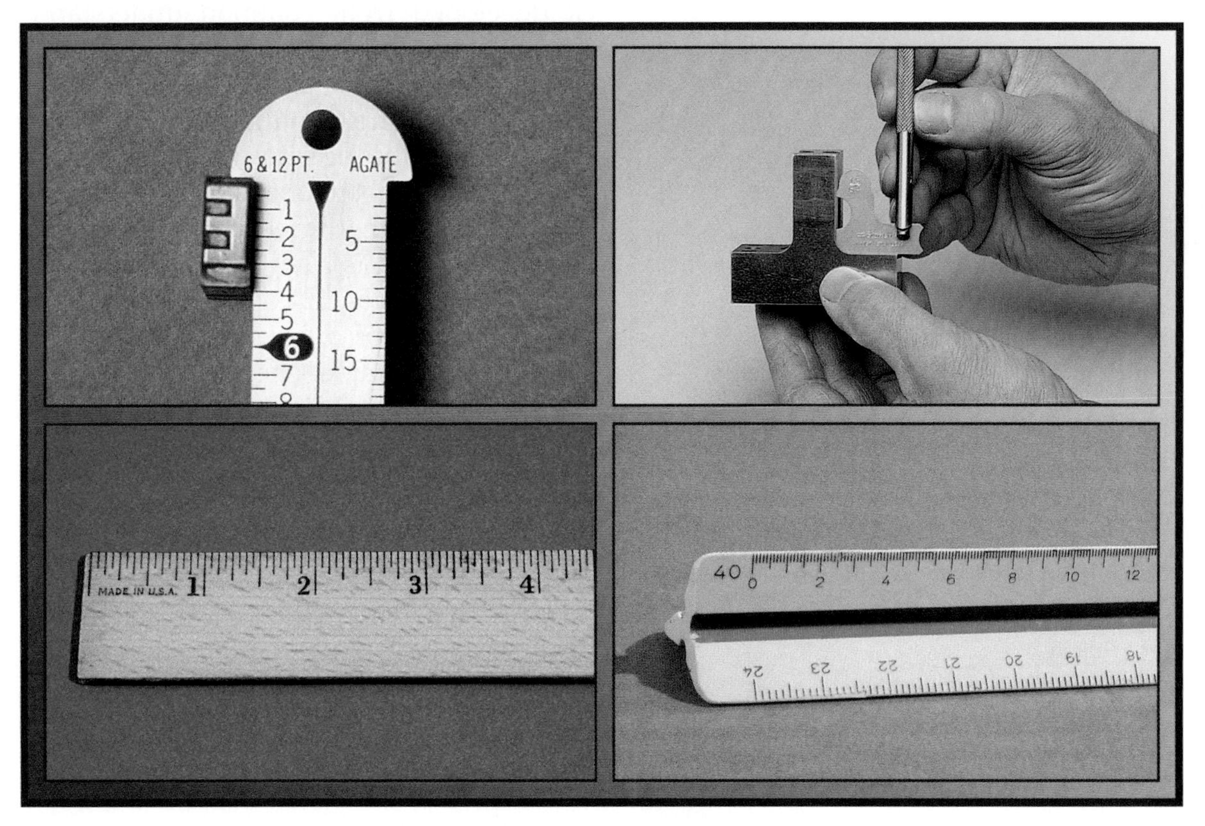

Figure 8-1. When we measure an object, we are comparing its qualities to a standard. Here, we see four different standards of measurement.

Standards for Technological Literacy

7

Demonstrate

Show how different tools are used to measure various qualities using the U.S. customary system.

Extend

Discuss the advantages of the SI Metric system over the U.S. Customary system.

Brainstorm

Ask your students why they think the United States has been slow to adopt the metric system of measurement.

TechnoFact

In an attempt to standardize its measurements, France developed the metric system of measurement or Système International, which established standard weights and measures.

arm. See Figure 8-2. A cubit was the distance from the tip of the middle finger to the elbow. A shorter measurement was the palm, which was the width of the four fingers. The width of a single finger was called a *digit*.

Human arms vary in size, however. As long as technology was simple, measurements did not have to be highly accurate. Now that technology is more complicated, measurements have been standardized. Each major physical quality is compared to a standard measurement that governments and international agreements have set.

Two measurement standards are in use today. One is the customary or English system, here referred to as the *U.S. customary* or *conventional system*. The more widely used system is called the *International System of Units*. This system is abbreviated *SI* (from the French name *Système international d'unités*). SI is more commonly known as the ***metric system***. See Figure 8-3. The United States is the only industrialized country that has not adopted the metric system for everyday use.

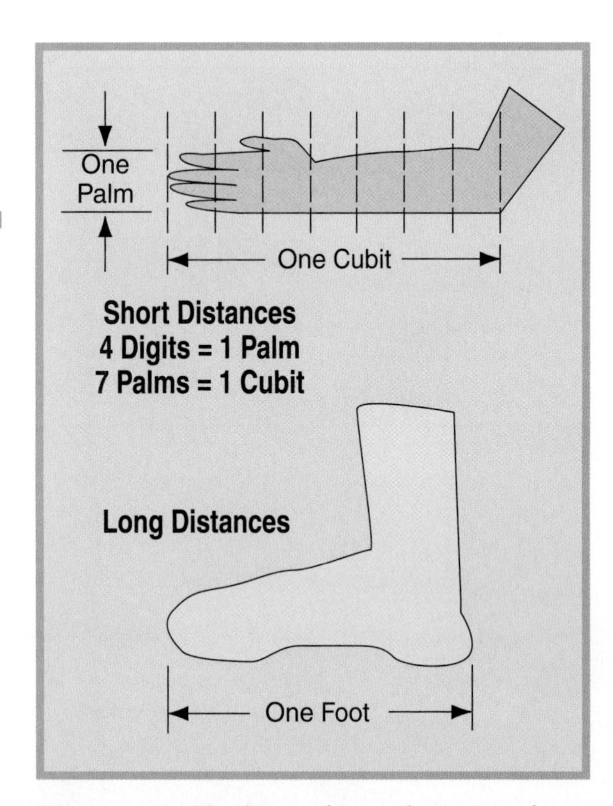

Figure 8-2. The bases for early units of measurement were parts of the human body.

The U.S. Customary System

The ***U.S. customary system*** of measurement is based on the system that developed in England from approximately the 1100s to the 1500s. As in earlier times, some of the system was based on the sizes of human body parts. For example, the term *inch*, a word meaning "thumb" in some languages, was used to represent the width of a human thumb. The foot was originally based on the length of a human foot. Other terms in the system were developed from Roman measurements. For example, the word *mile* originally meant the length of 1000 paces of a Roman legion.

Although it has some logic to it, in that it developed from common understandings of various kinds of measurement, this system also creates a great deal of confusion. For example, numerous terms are used to describe the same kind of measure. Twelve inches equal one foot, but thirty-six inches equal either three feet or one yard.

The system is also confusing because of inconsistencies in computing fractions and multiples of different base measurements. For example, one inch is 1/12 of a foot. One quart is 1/4 of a gallon.

The Metric System

In the 1790s, a group of French scientists assembled to create a measurement system that was more logical and exact. The SI metric system is based on decimals, or values of the number 10. There is a logical progression from smaller units to larger ones throughout the system because the base unit always decreases or increases by values of 10. Smaller units are decimal fractions (1/10, 1/100, 1/1000, and so on) of the base unit. Larger units are multiples of 10 (10, 100, 1000, and so on) times the base unit.

The metric system is also simpler because it uses a prefix to show us how the base unit is being changed. For example, the unit for ***distance*** is the meter. For large distances, we use the word *kilometer.* The prefix *kilo-* means "1000," so seven kilometers are equal to 7000 meters. For small distances, we use the word *millimeter.* The prefix *milli-* means "1/1000th." Twelve millimeters are

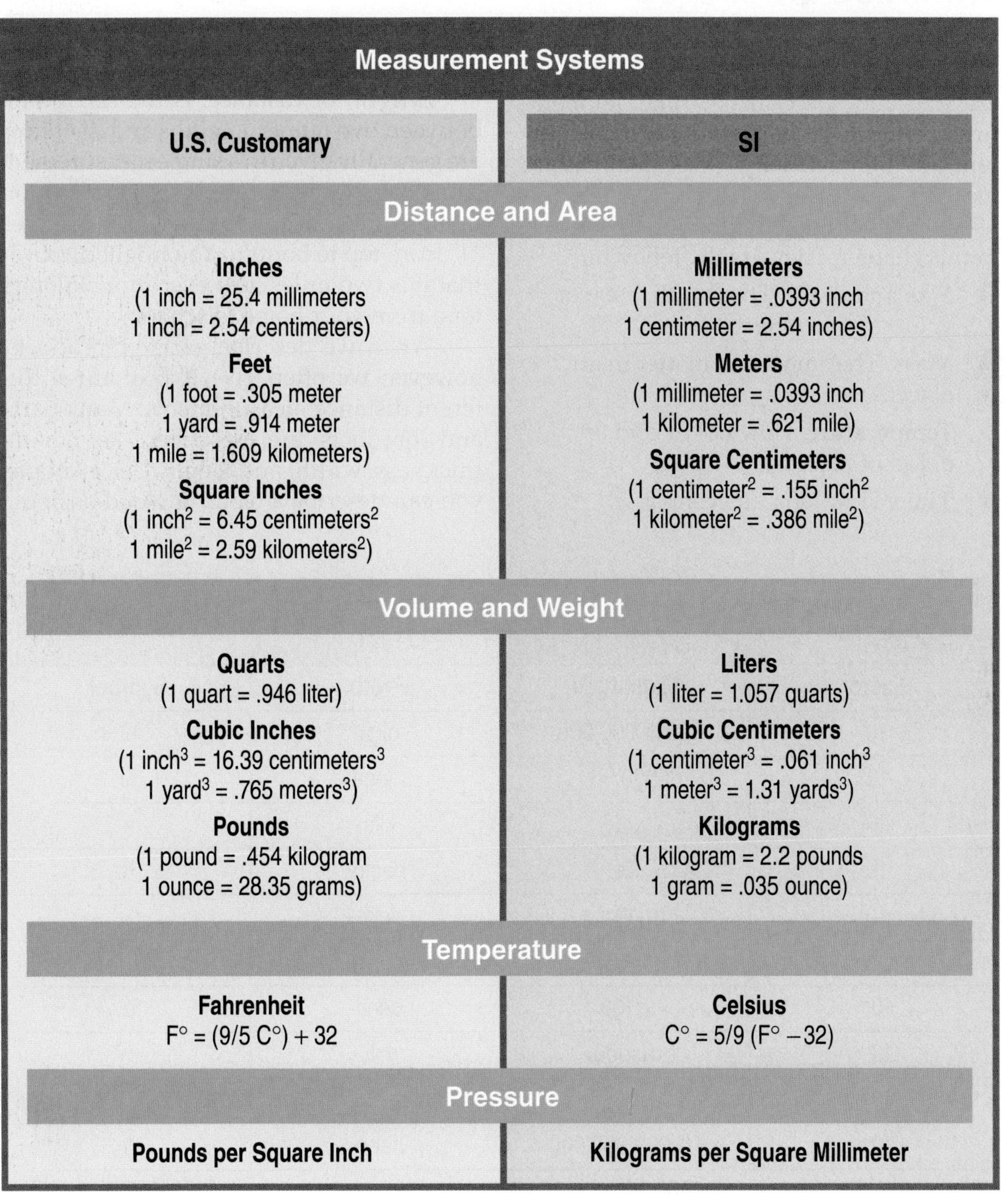

Measurement Systems

U.S. Customary	SI
Distance and Area	
Inches (1 inch = 25.4 millimeters 1 inch = 2.54 centimeters)	**Millimeters** (1 millimeter = .0393 inch 1 centimeter = 2.54 inches)
Feet (1 foot = .305 meter 1 yard = .914 meter 1 mile = 1.609 kilometers)	**Meters** (1 millimeter = .0393 inch 1 kilometer = .621 mile)
Square Inches (1 inch^2 = 6.45 centimeters^2 1 mile^2 = 2.59 kilometers^2)	**Square Centimeters** (1 centimeter^2 = .155 inch^2 1 kilometer^2 = .386 mile^2)
Volume and Weight	
Quarts (1 quart = .946 liter)	**Liters** (1 liter = 1.057 quarts)
Cubic Inches (1 inch^3 = 16.39 centimeters^3 1 yard^3 = .765 meters^3)	**Cubic Centimeters** (1 centimeter^3 = .061 inch^3 1 meter^3 = 1.31 yards^3)
Pounds (1 pound = .454 kilogram 1 ounce = 28.35 grams)	**Kilograms** (1 kilogram = 2.2 pounds 1 gram = .035 ounce)
Temperature	
Fahrenheit F° = (9/5 C°) + 32	**Celsius** C° = 5/9 (F° −32)
Pressure	
Pounds per Square Inch	**Kilograms per Square Millimeter**

Figure 8-3. The two measurement systems in use today are the U.S. customary system and the metric system. Some of the more common measurements are shown here.

Standards for Technological Literacy

7

Demonstrate

Show your students how to measure the size of a book using both measurement systems.

TechnoFact

The clocks in public squares were in wide use by the late 1300s. These new clocks met the need of people to know what time it was. However, clocks could not answer other questions, such as: How far is that from here? How heavy is that stone? How hot is the loaf of bread? The need for people to answer these and other similar questions gave rise to measurement systems.

12/1000 of a meter. The metric system lends itself to easy use in mathematical formulas. Look at **Figure 8-4** to see the most common metric prefixes. The metric system uses the same prefixes for all base units.

Standards for Technological Literacy

7

TechnoFact

The United States Geological Survey (USGS), founded by Thomas Jefferson in 1807, has mapped the country and determined that a ranch in Kansas is near the center of the 48 adjacent states.

Qualities Measured

We have developed standards for seven physical qualities. These qualities are size and space, mass, time, temperature, number of particles, electrical current, and light intensity. These qualities are measured as technology is developed and used. Four of these qualities are discussed in this chapter. They are the following:

- **Size and space.** The length, area, or volume of an object.
- **Mass.** The amount of matter in an object.
- **Temperature.** How hot or cold an object or place is.
- **Time.** How long an event lasts.

Size and Space

When we measure the size and space of an object, we might encounter three different, but related, measurements. These measurements are length, or distance; area; and volume. Each is discussed in turn below.

Length, or Distance

Length, or distance, is the separation between two points. Lengths and distances are generally given in a single measurement from Point A to Point B. For example, you might find that a page in a book measures 11″ from top to bottom. You might discover that it is two miles (just over three kilometers) from your home to school.

When we describe parts or products, however, we often give two or three different distance measurements. Many parts and products are described as having thickness, width, and length. For example, you can describe a block of wood as being

SI-Measurement Prefixes			
Factor	**Magnitude**	**Prefix**	**Symbol**
10^9	1,000,000,000	giga-	G
10^6	1,000,000	mega-	M
10^3	1000	kilo-	k
10^2	100	hecto-	h
10^1	10	deka-	da
10^{-1}	1/10	deci-	d
10^{-2}	1/100	centi-	c
10^{-3}	1/1000	milli-	m
10^{-6}	1/1,000,000	micro-	μ
10^{-9}	1/1,000,000,000	nano-	n
10^{-12}	1/1,000,000,000,000	pico-	p

Figure 8-4. The metric system uses the same prefixes for all base units.

3/4″ × 4″ × 12″. This means the block is three-fourths of an inch thick, four inches wide, and twelve inches long. Round parts are said to have thickness and diameter. A piece of pipe can be described as being 3/4″ in diameter and 24″ long. Finished products such as furniture, television sets, and cabinets are described by giving their height, width, and depth. See **Figure 8-5.** The common units for length, or distance, in the U.S. customary system are (in increasing size) inch, foot, yard, and mile. The common units for length, or distance, in the metric system are millimeter, centimeter, meter, and kilometer.

Area

The measurement of the separation between two points does not indicate the size, or ***area***, of an object, however. Determining the area of something involves measuring both the length and the width and then multiplying the two figures to see how much surface the object covers. See **Figure 8-6.** For example, you might find that the length of this page is 11″. Still, that

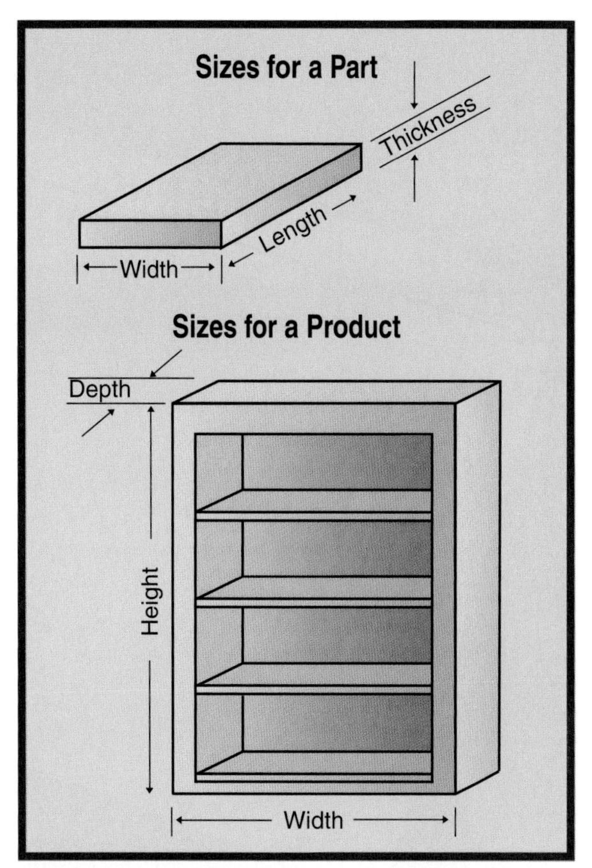

Figure 8-5. We use various distance measurements when describing parts and products.

Demonstrate

Show the techniques for calculating the area of squares, rectangles, and circles.

Career Corner

Architecture & Construction

Surveyors

Surveyors measure distances, directions, and angles between points and elevations along lines. They establish property boundaries and locate buildings on properties. Also, they write descriptions of land for deeds and leases. Surveyors use surveying instruments and electronic distance-measuring equipment to measure horizontal and vertical angles and linear distances. On some projects, surveyors use the global positioning system (GPS) to locate points on the land.

They generally work eight-hour days, five days a week. Seasonal demands might require them to work longer hours. Most licensed surveyors become trained by combining postsecondary school courses in surveying with extensive on-the-job training. Presently, technological advancements have caused some employers to require a four-year college degree.

Figure Discussion

Have your students use various formulas to calculate the volume or capacity of several objects, as shown in Figure 8-7.

TechnoFact

Different formulas are used to determine the areas of various shapes. For instance, the area of a circle is equal to pi multiplied by the radius squared (πr^2).

Figure 8-6. The area of an object is the measurement of how much surface the object covers. This measurement takes into account the length and width.

measurement does not tell you how much area the page covers because the page also has a width. To determine the size of the page, you must measure the width and then multiply that number by the length. Here, if the width of the page is 8″, the surface area of the page is 88 square inches. In the customary system, we express area measurements in such units as square inches, square feet, and square yards. The metric system uses such units as square centimeters and square meters.

Volume

Area does not indicate total size, however. A third consideration when measuring size and space is volume. ***Volume*** means the amount of space an object occupies or encloses. See **Figure 8-7.** Volume is

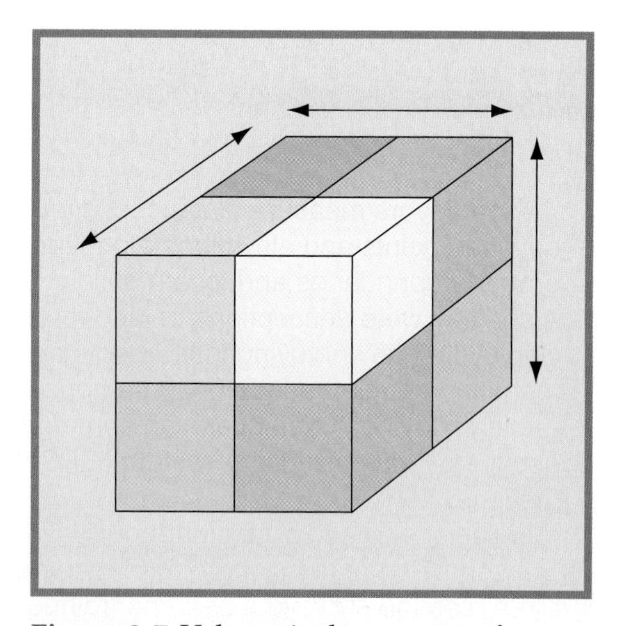

Figure 8-7. Volume is the amount of space an object occupies. This amount takes into account the length, width, and depth.

calculated by multiplying the length, width, and depth of an object. The measurement is expressed in cubic units.

For example, say we want to know how much space a room in a house will take up. We measure the length of the room (say, 10′), the width of the room (say, 8′), and the height (the measurement from the floor to the ceiling) of the room (say, 8′). The amount of space the room takes up in the house is then 640 cubic feet (10′ × 8′ × 8′).

Related to volume is ***capacity***, which is the amount of a substance an object can hold. Capacity is also measured in cubic units. We encounter the idea of capacity when we purchase such items as milk and soft drinks. An item described as a two-liter bottle of a soft drink, for example, indicates that the bottle can hold two liters of the liquid.

In the U.S. customary system, cubic units for volume are expressed in such terms as *cubic inch*, *cubic foot*, and *cubic yard*. In the metric system, volume is measured in such terms as *cubic millimeter*, *cubic centimeter*, and *cubic meter*. The cubic units for capacity are somewhat different. In the U.S. customary system, we find such terms as *pint*, *quart*, and *gallon*. In the metric system, we find such terms as *milliliter* and *liter*.

TechnoFact

Different formulas are used to determine the volumes of various solids. For example, to calculate the volume of a sphere, take four times pi times the radius of the sphere cubed divided by 3, which reads as $4/3\ \pi r^3$.

Mass

Mass describes the quantity of matter present in an object. Matter is anything that has mass and takes up space. Mass is not the same as weight, although they are related. ***Weight*** is the force of the Earth's pull on the mass. This force, therefore, can vary, depending on gravity. For example, by the time astronauts reach outer space, their weights have decreased dramatically because of the lack of gravity. The astronauts are, thus, able to float. Their masses, however, stay the same. Mass does not vary with temperature or location. See **Figure 8-8.**

Figure 8-8. The mass of this boulder remains the same, no matter what gravitational pull exists.

In most cases, measurement of *mass* and *weight* are close enough that we use the terms interchangeably. For example, when we purchase an apple, we are buying the matter of the apple, but we weigh the apple on a scale to determine its size. The scale weighs the apple, but it also gives us an idea of the apple's mass because everywhere on Earth the apple weighs the same. We buy products by the pound (fruit, vegetables, and sugar, for example), by the gram (prescriptions and precious metals, for example), or by the ton (hay, grain, sand, and gravel, for example). The U.S. customary system measures the weight of materials and uses such terms as *ounce* and *pound*. The metric system measures mass and uses such terms as *milligram* and *gram*.

Standards for Technological Literacy

Research

Have your students develop a conversion chart for the Fahrenheit and Celsius scales.

TechnoFact

The measurement of heat is called calorimetry. Using a calorimeter, it measures two units of heat: the calorie (the quantity of heat required to raise the temperature of one gram of water 1°C) and the Btu (the amount of heat required to raise the temperature of one pound of water 1°F).

Temperature

Temperature is the measurement of how hot or cold a material is. The terms *hot* and *cold* are relative, though, and therefore, must be compared to a standard. For example, the term *hot* has different meanings for human comfort and for melting metals. Temperatures that are hot enough to make you uncomfortable are considered cold for melting steel.

The range of temperature is broken into units. The U.S. customary system and the metric system use water as the basis to divide the temperature range into units called *degrees.* The range from the boiling point of water to the freezing point of water is divided into degrees. The customary Fahrenheit scale has 180° between the freezing point and boiling point. The freezing point of water at sea level is 32°F, and the boiling point of water is 212°F. The

STEM Connections: Mathematics

Measuring Area

Why do we need to use measurement systems in technology anyway? One answer is that measurement systems provide us with a common terminology when we wish to change or improve our environment. Suppose, for example, you want to paint your bedroom a different color. How much paint do you need to buy to cover all four walls?

To calculate the amount of paint needed, you need to determine the area to be covered. As noted in this chapter, in this case, area equals length times width, and the resulting amount is expressed in square feet. Say two walls of the bedroom are 9′ long and 8′ high and the other two walls are 11′ long and 8′ high. Also, take into account that you have one window measuring 36″ by 48″ and two doors measuring 36″ by 84″ each. Your calculations to figure the area to be covered, thus, look somewhat similar to this:

Two walls at 9′ long × 8′ high =
72 square feet × 2 walls = 144 sq ft
Two walls at 11′ long × 8′ high =
88 square feet × 2 walls = 176 sq ft
Subtract:
One window opening at 36″ × 48″
(3′ × 4′) = −12 sq ft
Two door openings at 36″ × 84″
(3′ × 7′) × 2 doors = −42 sq ft
Total = 266 sq ft

©iStockphoto.com/LUGO

Area measurements help painters estimate the cost and time for a painting job.

When you go to the store to purchase the paint, you find that one gallon of the paint you want covers 250 square feet. Therefore, you have to buy one gallon, plus another smaller can to cover the 266 square feet. How much paint do you need to purchase if the dimensions of your bedroom are as follows: two walls 10′ long and 8′ high, two walls 13′ long and 8′ high, two windows each measuring 32″ by 48″, and two door openings 30″ by 84″?

metric Celsius scale divides this range into 100°, with the freezing point at 0°C and the boiling point at 100°C.

Time

Time is how long an event lasts. The second is the basic unit of time in both the metric and the customary systems. Short-term measurements are given in *hours, minutes, seconds,* or *fractions of a second.* These terms measure everything from the time it takes to complete a race (minutes and seconds) to computer speeds (nanoseconds, or billionths of a second). Time can be measured in very long terms, such as a *millennium* (1000 years), a *century* (100 years), a *decade* (10 years), or a number of *years.* These terms are used to describe the age of civilizations, people, buildings, and vehicles.

Types of Measurement

The measurement of part and product size is important in technological design and production activities. Generally, this type of measurement can be divided into two levels of accuracy. See **Figure 8-9.**

- Standard measurement.
- Precision measurement.

Standard Measurement

Many production settings do not require close measurements. The length of a house, the width of a playing field, and the angle of the leg on a playground swing set need not be accurate to the utmost degree. If the product is within a fraction of an inch or a degree of angle, it works fine. This kind of measurement is called ***standard measurement.*** Standard measurement is often given to the foot, inch, or fraction of an inch in the customary system or to the nearest whole millimeter in the metric system.

The kind of production is only one factor affecting the type of measurement used. The material being measured is also important. For example, wood changes (expands or shrinks) in size with changes

Extend

Discuss where standard measuring is appropriate and where precision measurement should be used.

Brainstorm

Ask your students to list the ways they and their families use standard measurements in everyday life.

TechnoFact

The length of time called the second was invented and measured dividing the Earth's revolution around the Sun into equal parts. In the same way, the length of distance called the meter came from equally dividing the distance from the North Pole to the equator. Now, however, the second is defined using a clock that was invented in the 1950s called the cesium clock.

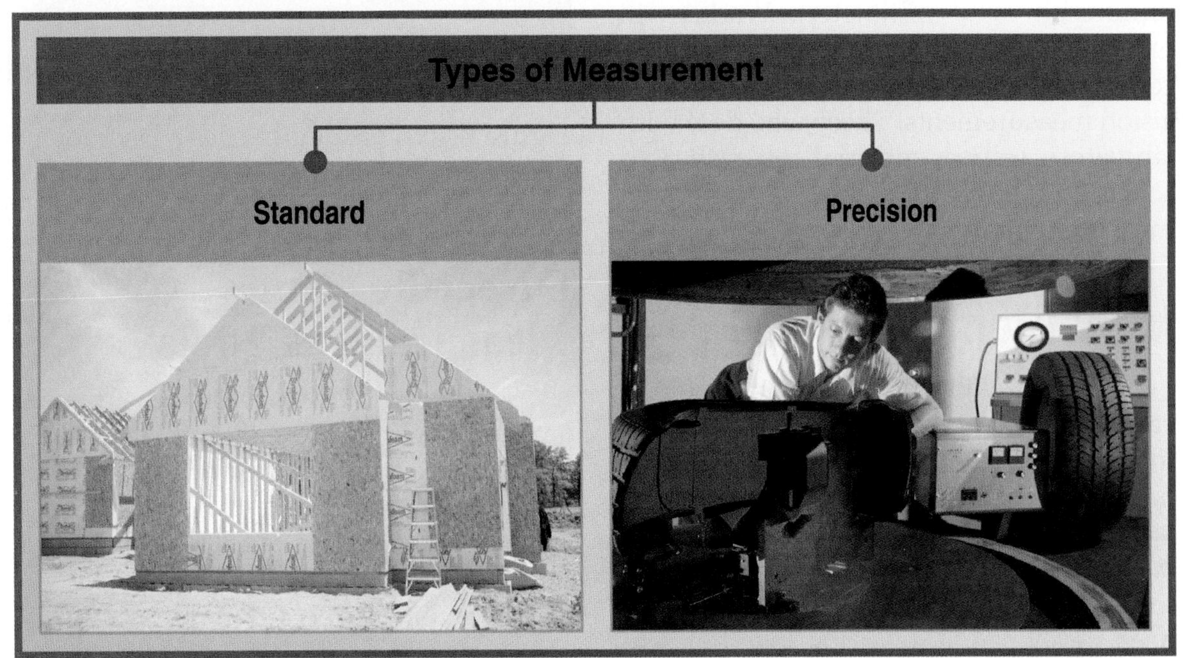

Goodyear Tire and Rubber Co.

Figure 8-9. Standard measurement is widely used in the construction industry. Precision measurement is found in many manufacturing applications.

Standards for Technological Literacy

7

Extend

Explain why precision measurement is used in making complex items, such as automobile engines, machine tools, and sewing machines.

TechnoFact

Although accuracy and precision may be mistaken for each other, measurements should reflect the qualities of both. Precision is the exactness of a measurement, while accuracy is ensuring that the measurement matches the quantity being measured.

in its moisture content and the atmospheric humidity. Measurements closer than 1/32 of an inch or 1 millimeter are not useful. Wood can change more than that amount in one day. Standard measurements are common in cabinet- and furniture-manufacturing plants, construction industries, and printing companies. The printing industry uses its own system of standard measurement based on the pica (1/6″) and the point (1/72″).

Precision Measurement

Standard measurement is not accurate enough for many production applications. Watch parts and engine pistons are useless if they vary by as much as 1/32″ (0.8 millimeter). These parts must be manufactured to a more precise size. For this type of production, ***precision measurement*** is required. Precision measurement measures features to 1/1000″ (one one-thousandth of an inch) to 1/10,000″ (one ten-thousandth of an inch) in the customary system. Metric precision measurement measures features to 0.01 mm (one one-hundredth of a millimeter). Workers in manufacturing activities involving metals, ceramics, plastics, and composites use precision measurements. Those who deal with laboratory testing, material research, and scientific investigation also use precision measurements.

Measurement Tools

People who work in technological areas have many types of measurement tools available. See **Figure 8-10.** The measurement tools an operator manipulates and reads are called ***direct-reading measurement tools***. In recent years, new measurement tools and machines have been developed. They bring together sensors and computers to automate measurement. These systems are called ***indirect-reading measurement tools***.

FMC Corp.

Figure 8-10. This device uses a laser for measurement.

Direct-Reading Measurement Tools

Three common uses of measurement are finding linear dimensions, diameters, and angles. Each of these three features can be measured using standard or precision devices. See **Figure 8-11.**

Linear Measurement Devices

The most common linear measurement device is the ***rule***. See **Figure 8-12.**

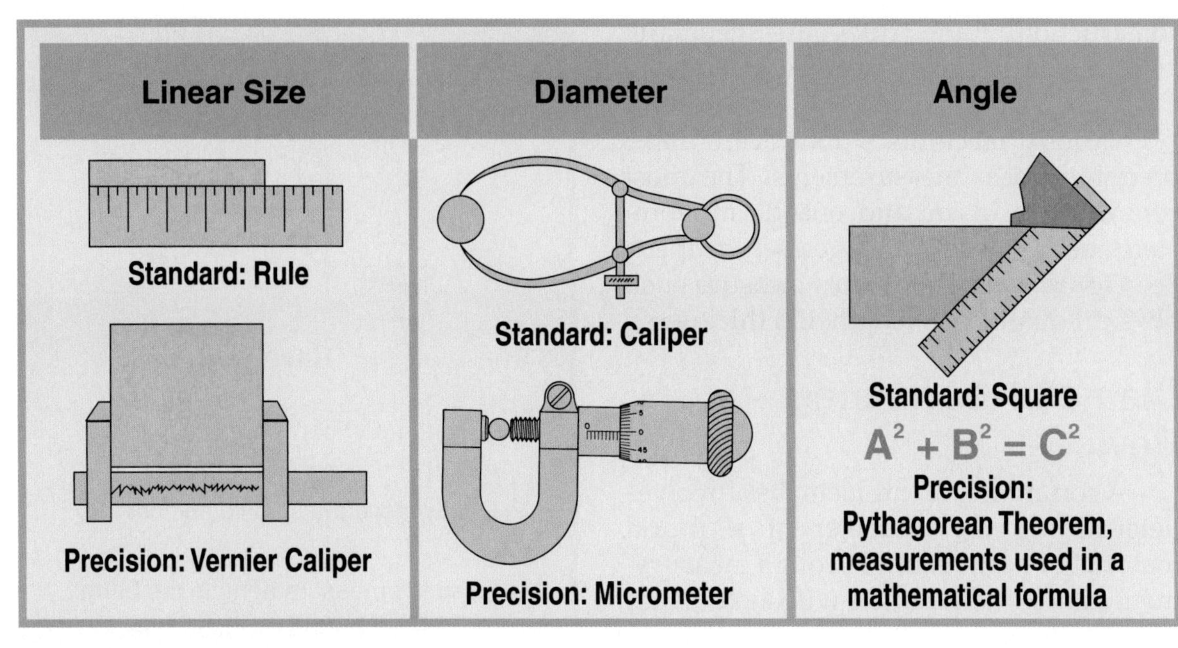

Figure 8-11. Both standard- and precision-measurement tools are used to determine linear sizes, diameters, and angles.

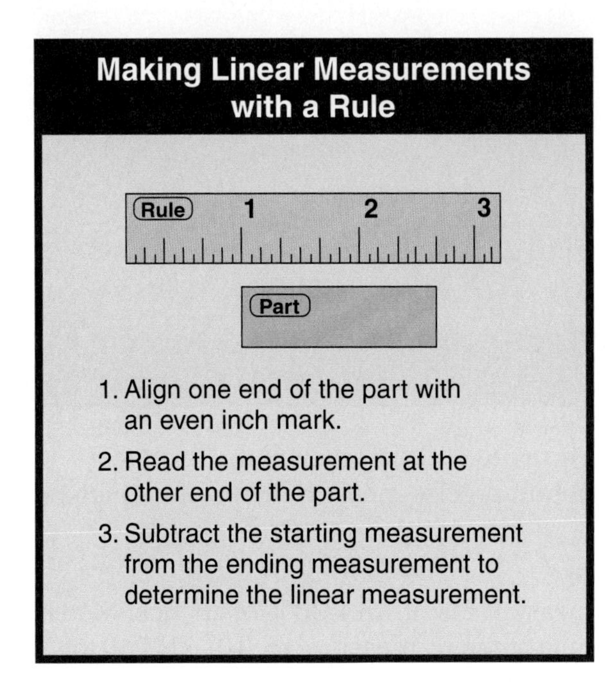

Figure 8-12. Most common linear measurements are made with a rule.

A rule is a rigid or flexible strip of metal, wood, or plastic with measuring marks on its face. Two types of rigid rules are used in linear measuring. These are the machinist's rule and the woodworker's or bench rule. A bench rule is generally divided into fractions of an inch. The most common divisions are sixteenths (1/16″). Metric bench rules are divided into whole millimeters. Machinist's rules are designed for finer measurements. Customary machinist's rules are divided into sixty-fourths (1/64″) or into tenths (1/10″) and hundredths (1/100″). Metric machinist's rules are divided into 0.5-mm increments. The part is measured with a rule by first aligning one end of the part with the zero mark. The linear measurement is then taken by reading the rule division at the other end of the part.

Flexible rules are often called *tape rules*. They are used in woodworking and carpentry applications. The catch at one end of the rule is hooked to the end of the board or structure. The tape is pulled out until it reaches the other end of the board or structure. A measurement is then taken.

Metric tape rules are divided into 1-mm increments. Commonly, the smallest division on a customary tape rule is 1/16″. Some tape rules highlight every 16″, which is the common spacing of studs in home

Standards for Technological Literacy

7

Demonstrate

Show your students the proper way to use common measuring tools.

Research

Have your students develop a conversion chart for the inch and centimeter scales.

TechnoFact

Precision in measuring is determined by the design of the measuring instrument. The instrument with smaller degrees on its scale is more precise. Accuracy is making the measurement match as closely to the quantity being measured as possible.

Standards for Technological Literacy

7

Demonstrate

Show your students the proper use of a micrometer to measure the thickness of a flat part and the diameter of a round part.

Demonstrate

Show your students the proper use of a square.

construction. Tape rules are generally available in lengths from 8′ to 100′ or from 2 meters to 30 meters.

Various machinist's tools can make precision linear measurements. The most common are inside and outside micrometers, depth gages, and vernier calipers. See **Figure 8-13.** These tools measure such physical qualities as length and thickness.

Diameter-Measuring Devices

A common measurement task involves determining the diameter of a round material or part. A simple, rough measurement can be determined with hole gages or circle templates. These devices have a series of holes into which the stock can be inserted. The smallest hole into which the material fits establishes the approximate diameter of the item.

More precise diameters can be established using a ***micrometer***. See **Figure 8-14.** The part to be measured is placed between the anvil (fixed part) and the spindle (movable rod). Turning the barrel moves the spindle forward. When the spindle and the anvil touch the part, a reading is taken on the barrel. Most customary micrometers measure to within 1/1000 of an inch. Metric micrometers measure to 0.01 (1/100) of a millimeter.

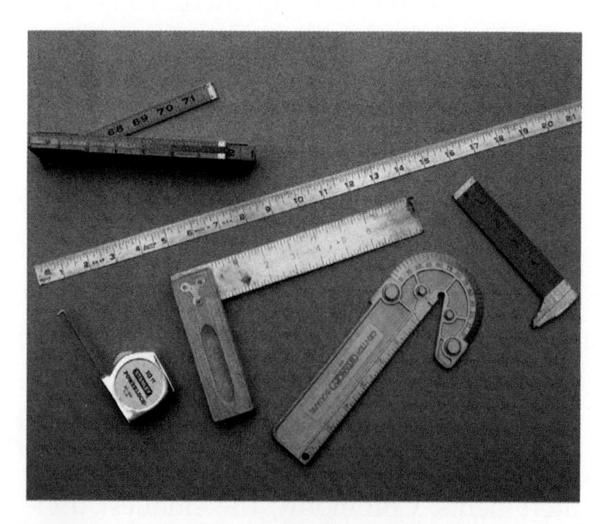

Figure 8-13. Common linear measurement tools include rigid rules, tape rules, micrometers, and calipers.

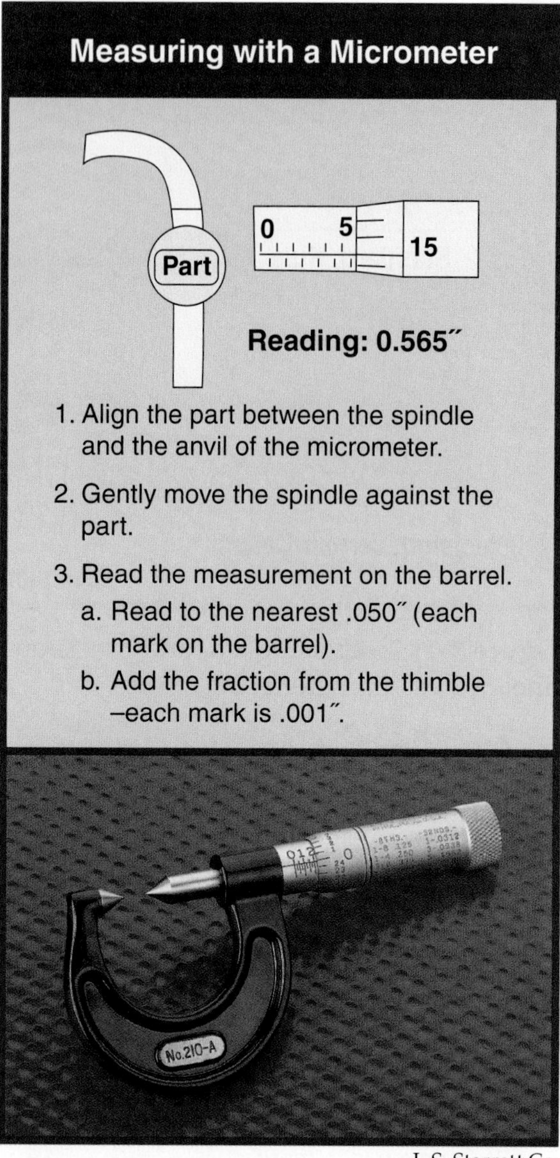

L.S. Starrett Co.

Figure 8-14. Micrometers are used to obtain precise measurements of diameters.

Angle-Measuring Devices

The angle between two adjacent surfaces or intersecting parts is important in many situations. The legs of a desk are generally square (at a 90° angle) with the top. The ends of picture-frame parts must be cut at a 45° angle to make a square frame.

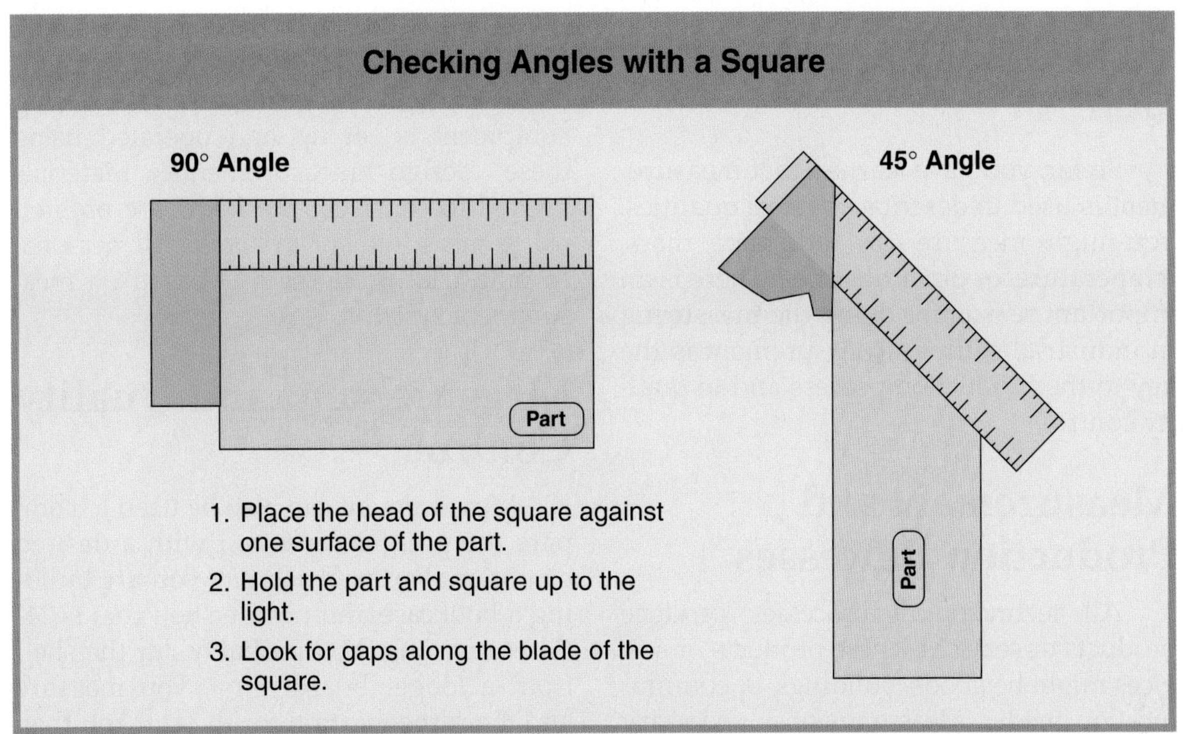

Figure 8-15. Angles that are 90° can be measured with a square. When an angle is not 90°, the blade of the square does not touch the surface of the work.

Standards for Technological Literacy

7

Demonstrate

Show your students the proper use of a protractor to establish angles.

TechnoFact

Pythagoras was a mathematician who lived before 500 B.C. His theorem of $a^2 + b^2 = c^2$ was easy to use and could be put to use to determine whether something was square.

Measurements at 90° angles are commonly done with ***squares***. See Figure 8-15. These tools have a blade that is at a right angle to the head. The head is placed against one surface of the material. The blade is allowed to rest on an adjacent surface. If the blade touches the surface over its entire length, the part is square. Parts that are not square allow light to pass under the blade.

Some squares have a shoulder on the head that allows the square to be used to measure 45° angles. This angle is important in producing mitered corners on furniture, boxes, and frames. The most common squares are the rafter or carpenter's square, the machinist's square, the try square, and the combination (90° and 45°) square.

Protractors and sliding T-bevels can be used to measure angles that a square cannot. A protractor allows for direct reading of angles. You place the protractor over the angle to be measured and then read the angle. The sliding T-bevel has an adjustable blade. You place the head on one surface and then clamp the blade along the angle of the second surface. With a protractor or by using a mathematical formula, you can then measure this angle.

Indirect-Reading Measurement Tools

In many modern measuring systems, humans no longer take measurements. Sensors gather the measurement data, which computers or other automatic devices process. The final measurement can be displayed on an output device, such as a digital readout, computer screen, or printout. These new systems include laser measuring devices, optical comparators, and direct-reading thermometers. See Figure 8-16. If you have weighed yourself on a digital bathroom scale, you have used an indirect-reading measuring device.

Standards for Technological Literacy

2 7

Extend

Discuss types of indirect measurement tools.

Brainstorm

Ask your students what types of measurements they would need to determine the quality of a bicycle seat.

TechnoFact

James Watt's definition of horsepower was how much work a horse could do in one minute. He determined that a horse could perform 33,000 foot-pounds of work in one minute. This allowed people to replace one horse with a one-horsepower engine.

TechnoFact

One of the earliest and most well-known clocks is the sundial. By studying the flow of air and water, however, a person named Ktesibios was able to regulate water flow by a feedback-controlled supply valve and create a new water clock.

Measurement and Control

So far, you have learned that measurement is used to describe physical qualities. You might measure size and shape, mass, temperature, or other qualities. There is an important reason for doing the measuring in industrial settings. Measurement is the key to the production process and to quality control.

Measurement and Production Processes

All technological processes produce products or services. These products or services might be goods, buildings, or communication media. Measurement is necessary in designing these artifacts. Their sizes, shapes, or other properties are communicated through measurements. Processing equipment is set up and operated using these design measurements. Materials needed to construct the items are ordered using measurement systems. All personal or industrial production is based on measurement systems.

Measurement and Quality Control

Measurement can also be used to compare the present condition with a desired condition. For example, say you are building a bookcase and need a shelf that is 24" (610 mm) long. You probably cut the shelf from a longer board. First, you measure and mark the location for the cut. You then

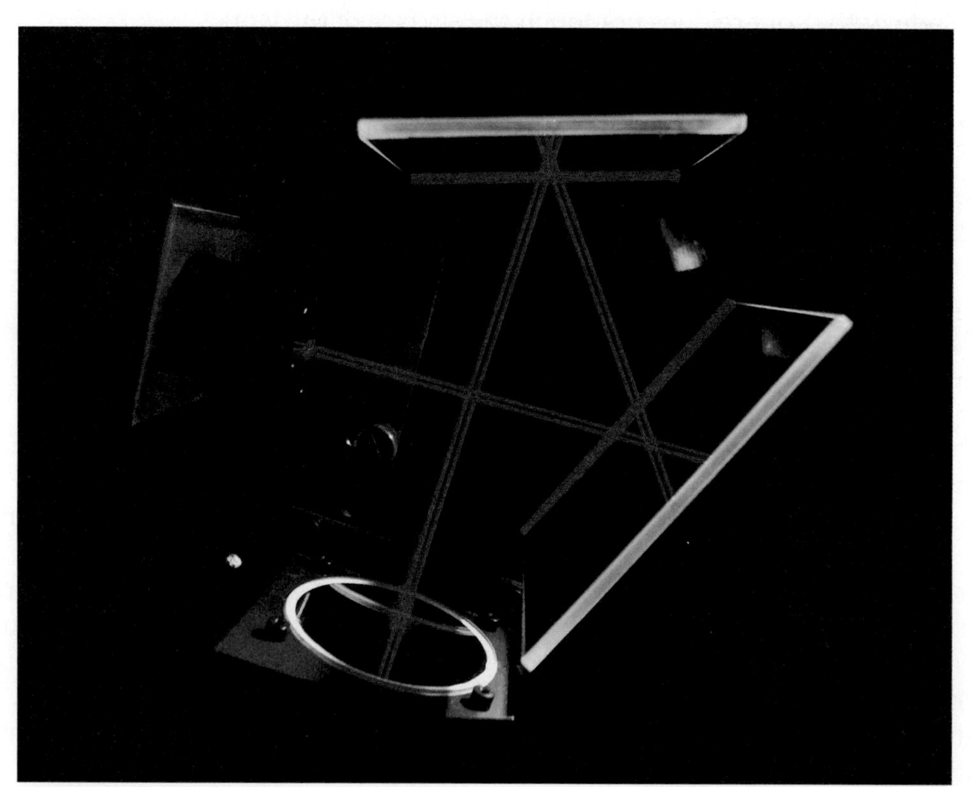

GM-Hughes

Figure 8-16. Machine operators use manual measurements for routine work. Indirect-measurement systems, such as this laser measurement system, are often built into continuous-processing and assembling operations.

saw along the line. If the board is too long, it will not fit into the case. A short board will fail to rest on the shelf supports. Measurement tells you if you have produced a 24″-long shelf.

A board of the correct length meets the intended purpose. This shelf meets your quality standards. The process of setting standards, measuring features, comparing them to the standards, and making corrective actions is called ***quality control.*** See **Figure 8-17.** This process is designed to ensure that products, structures, and services meet our needs. Quality control involves measuring and analyzing materials entering the system, work in process, and the outputs of the system. This control is an on-going process designed to ensure that resources are efficiently used and that customers receive functional products.

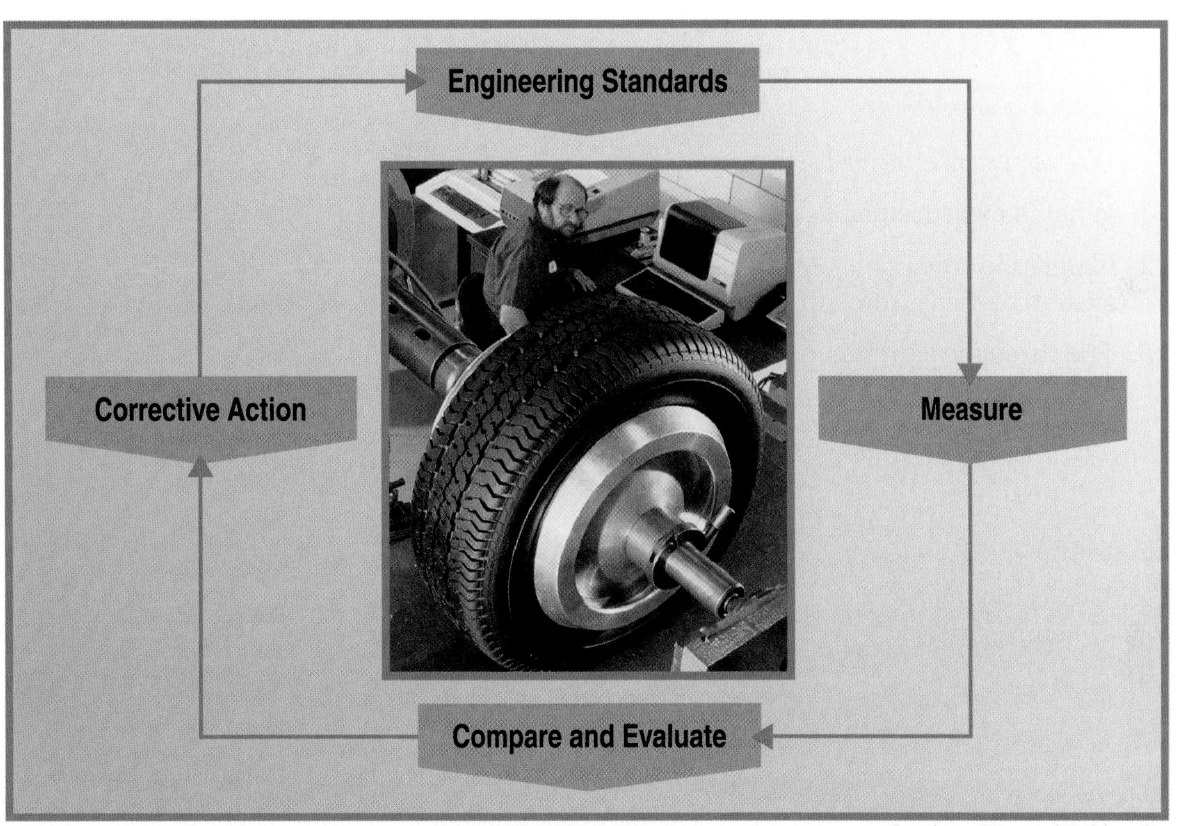

Goodyear Tire and Rubber Co.

Figure 8-17. The foundation of a quality control system is measurement and analysis.

Standards for Technological Literacy

2

Extend

Discuss how measurement is essential for quality control.

TechnoFact

While long distances of land may be measured in miles and kilometers, air and space are measured in units called nautical miles. The nautical mile is based on degrees of arcs taken from the Earth's circumference at the equator.

TechnoFact

While the storage disc in a computer hard drive is traveling by the magnetic head at 100 miles per hour, the heads ride as close as 50 nanometers from the surface of the disc.

TechnoFact

While temperatures may be measured using tenths, hundredths, and thousandths of a degree, the smallest increment of temperature a person can feel is one degree in either Fahrenheit or Celsius.

Answers to Test Your Knowledge Questions

1. Measurement is the practice of comparing the qualities of an object to a standard.
2. Student answers may include either of the following: There is a logical progression from smaller units to larger ones throughout the system because the base unit always decreases or increases by factors of ten. The metric system is also simpler because it uses a prefix to show us how the base unit is being changed.
3. volume
4. The common units for length or distance in the metric system are millimeter, centimeter, decimeter, meter, and kilometer. Other units may be accepted at the instructor's discretion.
5. Capacity is the amount of a substance an object can hold.
6. Mass describes the quantity of matter present in an object. Weight is the force of the earth's pull on the mass. Weight, therefore, can vary depending on gravity.
7. degree
8. second
9. Evaluate individually.

(Continued)

Summary

Measurement describes size and shape, mass, time, temperature, number of particles, electrical current, and light intensity and involves comparing a physical characteristic to an established standard. The common standards are the metric system and the U.S. customary system. These systems allow people to communicate designs, order materials, set up machines, fabricate products, and control quality.

Test Your Knowledge

Write your answers on a separate piece of paper. Please do not write in this book.

1. What is measurement?
2. Name one advantage of the metric system, as compared with the U.S. customary system of measurement.
3. The three measurements related to size and space are length, or distance; area; and ______.
4. Name one unit of measurement for length in the metric system.
5. Define the term *capacity*, as it relates to measurement.
6. What is the difference between mass and weight?
7. The unit of measurement for temperature in both the U.S. customary and the metric systems is the ______.
8. The basic unit of measurement for time in both the U.S. customary and the metric systems is the ______.
9. Name one industry in which standard measurements are commonly used.
10. Name one industry in which precision measurement is usually required.
11. The most common linear measurement device is the ______.
12. Precise diameters can be obtained by measuring with a micrometer. True or false?
13. You can measure 90° angles with tools called *squares*. True or false?
14. Give one example of an indirect-reading measurement tool.
15. What is the relationship between measurement and quality control?

10. Evaluate individually.
11. rule
12. True
13. True
14. Evaluate individually.
15. The process of setting standards, measuring features, comparing them to the standards, and making corrective actions is called quality control. Measurement is one means of determining if a resource fulfills its requirements.

STEM Applications

1. On a form similar to the one below, write the metric unit and the U.S. customary unit of measurement for each feature listed.

Measurement	Metric	U.S. Customary
Distance from Los Angeles		
Temperature on a hot day		
Weight of a loaf of bread		
Length of a pencil		
Capacity of a container		

2. Select an object, such as a desk or a bookcase. Describe its size using both the metric and U.S. customary measurement systems.
3. Check the squareness of a piece of furniture using a rafter, try, or combination square. Describe what you find and how any out-of-squareness can be corrected.

Section 3 Activities

Activity 3A

Design Problem

Background

All technology involves a machine or device to process materials, energy, or information. These machines are used to change the form of one or more of these resources into a new, more usable form.

Situation

You have been selected as the public relations director for a local citizens group. Your group is concerned about issues that local politicians and business leaders are not adequately addressing. This group wants to inform the public about these important issues.

Challenge

Choose an issue that you feel political and business leaders are not addressing. Some examples of this type of issue are the greenhouse effect, fossil fuel dependence, and nuclear power. Use the library for research, if necessary. Develop a one-page flyer to communicate your group's position on the issue. If your school has computers available, use one or more computer-software programs to help you produce the flyer.

Activity 3B

Fabrication Problem

Background

We live in a material world. All around us are products that have been developed using material-processing technology. Each of these products has been produced using a number of tools and processes. In this activity, you will change the form of materials to make a product that a number of people can use.

Challenge

Work with a partner to make a game that can be given to a local charity. See Figure 3B-1. This figure shows the layout for the game.

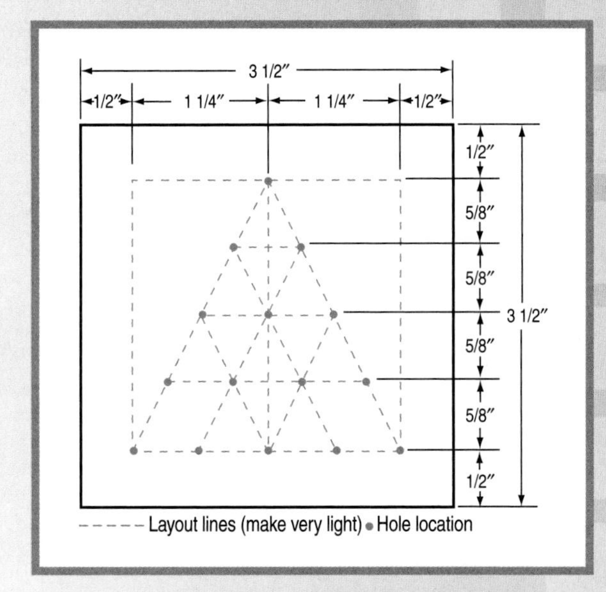

Figure 3B-1. The game-board layout.

Materials and Equipment

- One piece of 3/4″ × 3 1/2″ × 3 1/2″ clear pine, redwood, or western red cedar.
- 12 wood pegs or golf tees (1/8″ diameter × 1″).

Procedure

Carefully watch the demonstration your teacher gives, showing the proper use of the tools and machines needed to make the product. Follow the procedure given on the operation-process chart to construct the game. See Figure 3B-2.

Follow all safety rules your teacher discusses during the demonstration. Photocopy the directions for playing the game shown below. These will be packaged with the game.

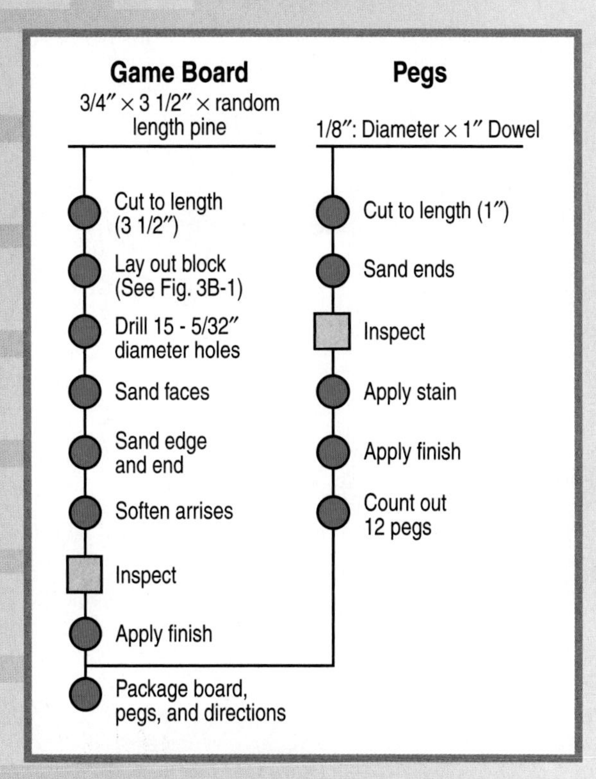

Figure 3B-2. The operation-process chart.

Analysis

Meet as a class. Analyze the material-processing activities you completed by answering the following questions:

1. Which material-processing tools did you use?
2. Which measuring tools did you use?
3. How could you have increased the speed of the manufacturing process?
4. What changes in the material-processing actions would you make to improve the quality of the product?

Mind Challenge

Directions

1. Put a peg in each hole, except one.
2. Select one peg. Jump an adjacent peg, ending in an empty hole.
3. Remove the jumped peg.
4. Continue jumping pegs until only one remains or no more jumps are possible.

Scoring

- One peg remains—Terrific—30 points.
- Two pegs remain—Good—20 points.
- Three pegs remain—Fair—10 points.
- Four pegs remain—Poor—0 points.

Section 4
Problem Solving and Design in Technology

Tomorrow's Technology Today

Nanotechnology

Many of today's technological products seem to be getting smaller and smaller—from machines to computer chips. It is estimated that, within the next 50 years, these products will continue to decrease in size so much that thousands of machines will be able to fit on the point of a pencil! In the coming decades, nanotechnology, or "the science of the small," will use these tiny machines for many amazing applications. One nanometer equals one-millionth of a millimeter. We are talking about very, very small machines.

All matter in the universe is made up of atoms. Cells are nature's tiny machines. We still have a lot to learn about constructing materials on such a small scale. With new developments in nanotechnology, however, we are getting closer. Many scientists believe we will be able to manipulate each individual atom of an object in the near future.

Nanotechnology combines engineering with chemistry. The goal of nanotechnology is to manipulate atoms and place them in a specific pattern. These patterns will produce a certain kind of structure. Depending on the atoms used and the patterns created, nanotechnology can be used in several different areas. The most promising uses for this technology are in medicine, construction, computers, and the environment.

This branch of technology is expected to greatly impact the medical industry. Nanotechnology devices will be programmed to attack cancer cells and viruses and rearrange their molecular structures to make them harmless. The tiny robots might also be able to slow the aging process. This might increase life expectancy. Since these machines will be able to perform work a thousand times more precisely than current methods, they can also be programmed to perform delicate surgeries.

The construction industry will benefit from nanotechnology as well. The tiny machines will be used to make much stronger fibers. There is hope that, one day, we will be able to replicate anything, from diamonds to food.

The parts of a computer continue to get smaller all the time. They will, however, soon reach their limit. Nanotechnology will be used to make much smaller storage devices with the capability to hold much more information.

Finally, nanotechnology can be used to better the environment. We are currently running out of many natural resources. With nanotechnology, we will be able to construct more of them. Manufacturing with nanotechnology produces much less pollution than traditional manufacturing. In addition, these devices can potentially be programmed to rebuild the ozone layer and clean up oil spills.

Nanotechnology has many exciting potential uses. This technology, however, also has its challenges. Similar to all new technology, it raises questions about its impacts on people, society, and the environment.

Discussion Starters

Some proposed nanotechnologies require self-replicating nanobots. Once the appropriate number of nanobots have been replicated, they need to begin performing the task for which they were designed. Ask your students how a microscopic nanobot might be designed (or programmed) to quit replicating after the required number of copies are made.

Group Activity

Divide the class into two groups. Have one group come up with a list of the possible benefits of nanotechnology. Have the second group develop a list of the dangers that nanotechnology may pose.

Writing Assignment

Have your students research the problem of supplying power to nanotechnology and write a short essay describing possible solutions.

Chapter Outline

This chapter covers the benchmark topics for the following Standards for Technological Literacy:

1 2 3 4 7 8 10 11

Learning Objectives

After studying this chapter, you will be able to do the following:

- Compare the problem-solving process and the design process.
- Explain the general problem-solving and design process.
- Explain the technological problem-solving and design process.
- Recall the definitions of *technological problem* and *technological opportunity.*
- Recall the major steps involved in solving technological problems and meeting technological opportunities.
- Recall the steps followed in identifying a technological problem or opportunity.
- Recall the types of criteria and constraints to be considered in the problem-solving and design process.
- Summarize the types of information gathered as a foundation for technological-development projects.
- Recall the general methods of gathering information for the problem-solving and design process.

Key Terms

critical thinking
descriptive method
design
ethical information
experimental method
historical information
historical method
human information
legal information
scientific information
technological information
technological opportunity
technological problem

Strategic Reading

As you read this chapter, make an outline of the details of the topics discussed, such as different types of and steps involved in the design and problem-solving processes.

All technology is created for a purpose. Technology has come from people who wanted to solve a problem or address an opportunity. For example, some people have developed technological artifacts that protect us from the physical environment. Others have developed artifacts that have helped us become more informed. Still others have created artifacts that allow us to travel with ease.

Most modern technological artifacts and systems are not developed or discovered by accident. Instead, as we noted in earlier chapters, people use problem-solving and design, production, and management processes to meet needs and wants. In this section, we focus on the activities that occur in the first stage, the problem-solving and design process. In this initial chapter of this section, we first learn about the problem-solving and design process in general and how that process applies to technology. We then focus on the initial step in the process. This step is identifying the technological problem or opportunity.

The Problem-Solving Process versus the Design Process

Many people use the terms *design* and *problem solving* interchangeably. This use, however, is not correct. Problem solving can be applied to all kinds of situations. For example, you might have trouble dealing with some of your friends. This is a social problem. You might have problems making your allowance stretch through the week. This is a financial problem. These and any number of other types of problems can be addressed through the problem-solving process. Problem solving can be described as a process people employ to think about and find solutions for a situation that has one or more reasonable answers. When problem-solving activities are applied to technical problems, the process is often called ***design***. This process can be defined as "planning and developing products or systems that meet human needs and wants."

Both problem solving and design use a special human ability called ***critical thinking***. This type of thinking uses mental abilities to identify, analyze, and evaluate a situation or problem. Using these processes, a person considers possible solutions to form a judgment about a course of action.

The Problem-Solving and Design Process in General

People usually realize a problem exists when they encounter a difficulty and are not sure how to resolve it. A problem can be as simple as opening a walnut when tools are unavailable. The problem can be as complex as moving 500 people from Los Angeles to Tokyo in two hours. See **Figure 9-1**. These problems have two things in common. The first thing is that a goal exists (cracking a walnut or moving people). The second thing is that no clear path is apparent from the present state (an intact walnut or people who need to be moved) to the goal.

Not all problems, however, are technological. You might have a problem getting along with a classmate or a family member. This is a social problem. You might have a problem identifying and describing the weather some types of clouds produce. This is a scientific problem. You might be trying to decide whether or not to keep some money you found on the street. This is an ethical problem.

Standards for Technological Literacy

1 2 8 10 11

Research

Have your students use the library or Internet to research problem solving methods.

Standards for Technological Literacy

1 2 8 10 11

Brainstorm

Ask your students why it is important to approach a problem in a systematic manner, using a set process.

Figure Discussion

Ask your students to clearly identify the problem in the photo on the left side of Figure 9-1.

NASA

Figure 9-1. Technology is born from a problem or an opportunity. This problem or opportunity can be as simple as how to clear snow from a road or as complex as how to land a person on the Moon.

Problem solving is a common human activity. We all use a fairly universal process as we approach problems. In short, we do the following:

1. Develop an understanding of the problem through observation and investigation.
2. Devise a plan for solving the problem.
3. Implement the plan.
4. Evaluate the plan.

Within this general framework, however, people in technological areas have created a specific approach to help solve problems and meet opportunities. This approach is called the *design process* or, as we call it here, the *problem-solving and design process*. The problem-solving and design process uses the term *design* in one of two ways. First, the term *design* describes the action, or process, used to create the appearance or operation of a technological device or system. Second, the term *design* can be used to describe the product of the design process. In this way, the word *design* is used to describe the plan, or drawing, showing the appearance or operation of a technological device or system.

The Problem-Solving and Design Process in Technology

The problem-solving and design process in technology is used in cases of attempting to satisfy people's technological needs and wants. People in technology use this problem-solving and design process, not only where they see a problem they might be able to solve, but also where they see opportunities for improvement. The difference between these two kinds of situations is described in more detail as follows.

Technological Problems

You face a ***technological problem*** if you need to develop tools, machines, or systems to help you do work. An example within the agricultural area highlights this point. Until the 1940s, most farmers raised their own hay and used it as feed for the animals on the farm. In recent years, however, the self-sufficient family farm has been disappearing. Family farms are being replaced with larger farms specializing in a specific crop. Some farmers operate large dairy farms that do not raise hay. Other farmers specialize in raising and selling hay to these dairy farmers and beef cattle feedlots. Hay now has to be shipped to other farms and to the feedlots. Having people gather and ship loose hay, however, is not very productive.

Added to this problem is the fact that farm labor is becoming more expensive. Many people do not want to do heavy manual labor. The solution to this problem was to develop farm implements allowing one person to do more work with the assistance of machines. See **Figure 9-2.** People learned how to collect hay into bales for easier transportation. This process includes the following steps:

1. The field of hay is cut with a mowing machine and allowed to dry on the ground.
2. A machine called a *side-delivery rake* is used to collect the hay into a narrow pile called a *windrow.*

Figure 9-2. Technology has changed farming from a labor-intensive activity to an equipment-intensive activity. A single farmer can now do the work of many people.

Standards for Technological Literacy

1 2 8 10 11

TechnoFact

The processes of design and problem solving are used to assess and meet human needs and wants. Gerry McGovern, Design Director of Land Rover, wrote: "Design is right at the core of every new product."

Career Corner

Science, Technology, Engineering & Mathematics

Engineers

Engineers apply science, mathematics, and technology knowledge to develop solutions to technical problems. They develop ways to extract and process natural resources. Engineers design products, machines to build products, production systems, and production facilities. They also design buildings, highways, and transit systems. Engineers develop ways to harness, convert, and use energy.

Some branches of engineering are aerospace, agricultural, biomedical, chemical, civil, computer hardware, electrical and electronics, environmental, industrial, mechanical, mining and geological, and nuclear. Most engineers work in office buildings, laboratories, and industrial plants or at construction sites, mines, and oil and gas production sites. A bachelor's degree in engineering is required for almost all entry-level engineering jobs. Most engineering education programs include study in an engineering specialty and courses in mathematics and science.

Standards for Technological Literacy

1 2 8 10 11

Extend

Discuss the differences among social (people), financial (money), and technological (thing) problems.

TechnoFact

Discovery and invention are equally responsible for the advancements in technology. To discover is to observe something already existing in nature. An invention is a creation of something that never existed before.

Quotes

"We are leaving a period when new ideas are exclusively associated with an individual, or functional departments, and entering a new age where mass involvement is not only desirable, but perhaps even essential."—PricewaterhouseCoopers, Innovation and Growth: A Global Perspective.

3. A tractor pulls the baler along the windrows. Here, it collects, compacts, and ties the hay into bales.
4. The bales are then pushed out of the machine onto the ground. Farmworkers pick them up from the ground and load the bales onto a truck.

More recent technology has further improved the machines used in harvesting hay. Harvesting hay was once labor-intensive. A *swather* has replaced mowing machines and side-delivery rakes. A swather cuts, conditions, and windrows hay in one operation. The hand loading and hauling of baled hay is being replaced with bale wagons that automatically pick up and stack bales into cubes. Hay is also gathered into large rolls. These cubes or rolls are loaded with forklifts onto trucks that haul the hay to customers. Thus, harvesting hay has become equipment intensive through the use of technological problem solving.

An examination of other technological areas reveals the same pattern. For example, in manufacturing operations, computer-aided design systems have increased the productivity of drafters. Robots perform many routine manufacturing tasks without human interference or monitoring. See **Figure 9-3.** In transportation, computer-controlled, driverless people movers speed people between terminals at airports. Hybrid-bus systems promise to increase the efficiency of urban transport. See **Figure 9-4.** Automated equipment on new commercial aircraft has reduced the number of flight deck officers from three to two.

Cincinnati Milacron

Figure 9-3. Machines, such as this robot, can perform repetitive tasks with high accuracy.

Daimler

Figure 9-4. Some of our newer transportation systems, such as this hybrid bus, use less energy and produce less emissions when moving people.

Technological Opportunities

Not all technology is developed to solve a problem, however. Many conditions exist that people do not view as a problem. Still, if a nonproblem condition can be improved

Academic Connections: History

The Origin of Radar

As indications of an impending world war grew in the late 1930s, various countries realized they had a major problem. Although scientists had discovered that radio echoes could detect ships and planes, the early devices made for this purpose were not reliable. Knowing that keeping track of an enemy's air and sea movements would be critical in a war, the various leaders kept urging scientists to improve the technology termed *radar* (radio detecting and ranging).

In the late 1930s and early 1940s, American and British researchers worked with each other to improve radar. Their problem was increasing the reliability and accuracy of radar. They knew their problem and developed and refined solutions to solve the problem. For example, the British developed an improved vacuum tube that had enough microwave energy to be used in radar systems. The Americans then took it over and developed units small enough for airplanes and boats. As the war continued, this improved radar indeed became a key element in the Allies' eventual victory. During the last half century, researchers have further developed the capabilities of radar. In what areas is it used today?

A radar dome.

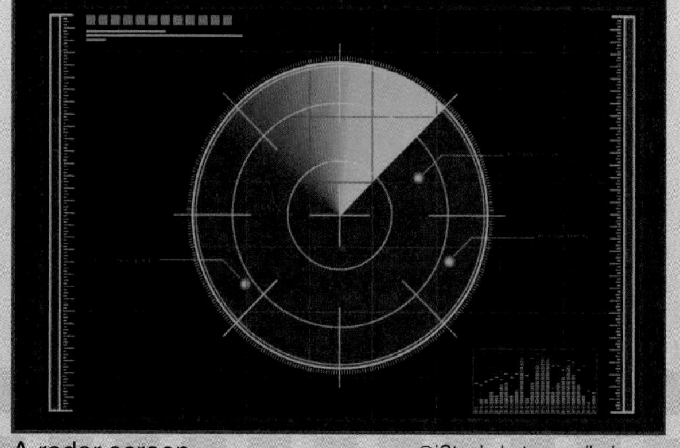

A radar screen.

with technology, then a ***technological opportunity*** has been discovered. People might want to use a new device, even though an old device was adequate for their needs. The automatic coffeemaker is an example of meeting a technological opportunity. There was no consumer cry for a new way to make coffee. The percolators and drip coffeemakers in use were adequate for the job. Nevertheless, the first Mr. Coffee® automatic-drip coffeemaker was an almost-instant success. Advertising helped many people develop a need for the new product. An opportunity was identified. A new product was born.

As we noted earlier, though, no matter whether the situation is a problem or an opportunity, those in technology use a specific problem-solving and design approach. That approach involves a series of steps following one from the other. They are described as follows.

Standards for Technological Literacy

1 2 8 10 11

Brainstorm

Have your students investigate the differences between a problem and an opportunity. Have them provide several examples of problems and several examples of opportunities.

TechnoFact

More than 2400 years ago, the Greek physician Hippocrates wrote: "Life is short, the art long, opportunity fleeting, experiment treacherous, judgment difficult." The "window of opportunity" may be applied to the brief period of time we have to address technological opportunities.

Standards for Technological Literacy

1 2 8 10 11

Extend

Differentiate between a problem and a solution (e.g. a chair is a solution to the problem of designing a device to hold a person in a seated position).

Brainstorm

Have your students list five problems that they face that could have technological solutions.

TechnoFact

While most inventions are developed with the intent of benefiting society, others, such as weapons, are harmful. Even those designed to help people sometimes produce undesirable outputs. While the automobile, for example, is an advantageous method of travel, it also produces damaging effects on the natural environment.

Steps in Solving Technological Problems and Meeting Opportunities

Solving technological problems and meeting technological opportunities require specific action. These challenges involve four major phases. See **Figure 9-5.** These four phases and the basic activities they include are the following:

1. **Identifying or recognizing a technological problem.** This phase of the technology development includes the following:
   - **Defining the problem.** This activity is describing the situation that needs a technological solution and establishing the criteria and constraints under which the device or system must operate.
   - **Gathering information.** This activity is obtaining background information needed to begin developing solutions to the problem or situation.
2. **Developing a solution.** This phase of the technology development involves the following:
   - **Developing possible solutions.** This activity is originating a number of different solutions that can solve the problem or meet the opportunity.
   - **Isolating, refining, and detailing the best solution.** This activity is picking out the most promising solution; integrating, modifying, and improving the solution; and creating detailed sketches of the best solution.
3. **Evaluating the solution.** This phase of the process involves the following:
   - **Modeling the best solution.** This activity is testing and evaluating the proposed solutions through graphic, mathematical, or physical modeling techniques.
   - **Selecting the best solution.** This activity is comparing the design solutions in terms of economic, market, technical, production, and environmental criteria to determine the best solution to the problem.

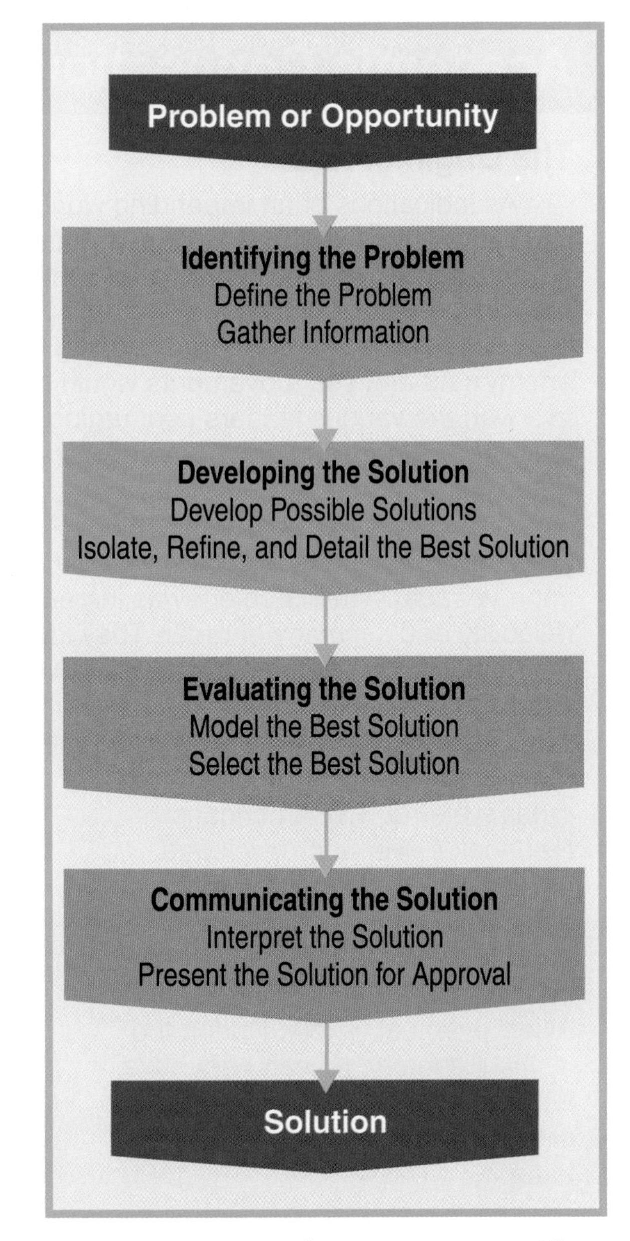

Figure 9-5. We need to carry out specific actions to solve technological problems and meet technological opportunities.

4. **Communicating the solution.** This phase of the process includes the following:
   - **Interpreting the solution.** This activity is communicating the final solution through such detailed documents and reports as engineering drawings, bills of materials, and specification sheets.
   - **Presenting the solution for approval.** This activity is presenting written and oral reports to obtain the appropriate approval (from management or government, for example) for implementing the solution.

The remainder of this chapter explores the first phase of the technology -development activity. This phase is identifying, or recognizing, a technological problem. The other phases are discussed in the following three chapters.

Identifying a Technological Problem or Opportunity

All technology starts when we want to create a new technological artifact or system to solve a problem or meet an opportunity. To achieve this goal, however, we must first identify what the problem or opportunity really is. Such a task is not always simple. For example, we know holes have developed in the ozone layer over the Antarctic and Artic regions. The holes are not the real problem, however. Instead, their presence indicates that a problem might exist. In this situation, we look further by gathering more information and discover that the holes are the consequence of our use of such items as polluting vehicles and aerosol paint and personal-care products. The problem, then, is to create such items as vehicles with cleaner fuels and products without the aerosol component.

Thus, if designers are to solve a technological problem or meet an opportunity, they must first define what the problem or opportunity really is and then gather information needed to begin developing solutions to the situation. In each of these steps, designers follow certain procedures. These procedures are described in more detail as follows.

Defining the Problem or Opportunity

The definition of a problem or an opportunity should complete two major tasks:

1. Describing the situation that needs a technological solution.
2. Establishing the criteria and constraints under which the artifact or system must operate.

Describing the Situation

Defining a problem or an opportunity involves describing it in a clear and concise statement. In this statement, we must take care that we do not confuse the problem or opportunity with the solution. For example, a bookend is a solution, not a problem. The problem is to hold books in an organized manner on a flat surface. Likewise, a chair is a solution. A chair can be described as a device with a seat, back, and four legs. With this description, a designer starting with the problem statement, "Design a chair," will never end up with a beanbag or a wicker basket suspended from a chain. Descriptions can restrict creativity or open doors to a variety of solutions.

Look back at our ozone layer problem. To close the hole is the solution. The problem is to develop technological devices emitting fewer pollutants into the atmosphere.

Standards for Technological Literacy

1 2 8 10 11

Extend

Explain the difference between an artifact and a system.

TechnoFact

It has been indicated recently that an essential employee in a business is a person who takes creative ideas and turns them into successful products and services.

Standards for Technological Literacy

1 2 4 8 10 11

Extend

Clearly differentiate between constraints and criteria.

Figure Discussion

Ask your students to provide additional examples of each of the five types of criteria and constraints shown in Figure 9-6.

TechnoFact

Although people may invent for different reasons, a successful invention is one that meets the needs and wants of society. The desire to make money from ideas or satisfy personal curiosity are other common reasons for inventing, but these alone may not make a useful invention and may not meet with as much success. The needs inventors strive to fill are economic, social, or military needs.

Establishing Criteria and Constraints

The problem or opportunity description leads directly to the next step, establishing a set of criteria and constraints. Designers must know how the effectiveness of the new technology will be evaluated. To do this, they establish criteria and constraints under which the product or system must operate. Criteria include a listing of the features of the artifact or system. Constraints deal with the limits on the design. See **Figure 9-6.** These criteria and constraints can be grouped into the following general categories:

- Technical or engineering criteria and constraints describe the operational and safety characteristics the device or system must meet. These criteria and constraints are based on how, where, and by whom the product will be used. An example of an engineering criterion for a new windshield wiper might be "Must effectively clear the windshield of water when the vehicle is traveling at highway speeds."
- Production criteria and constraints describe the resources available for producing the device or system. These criteria and constraints are based on the natural, human, and capital (machine) resources available for the production of the device or system. A production constraint for a product might be "Must be manufactured using existing equipment in the factory."
- Market criteria and constraints identify the function, appearance, and value of the device or system. These criteria and constraints are derived from studying what the user expects from the device or system. An example of a market constraint might be "Must be compatible with Early American decor."
- Financial criteria and constraints establish the cost-benefit ratio for the device or system. These criteria and constraints address the amount of money required to develop, produce, and use the technological device or system and the ultimate benefits from its use. An example of a financial constraint might be "Must be priced at 5% less than the major competitor's product."

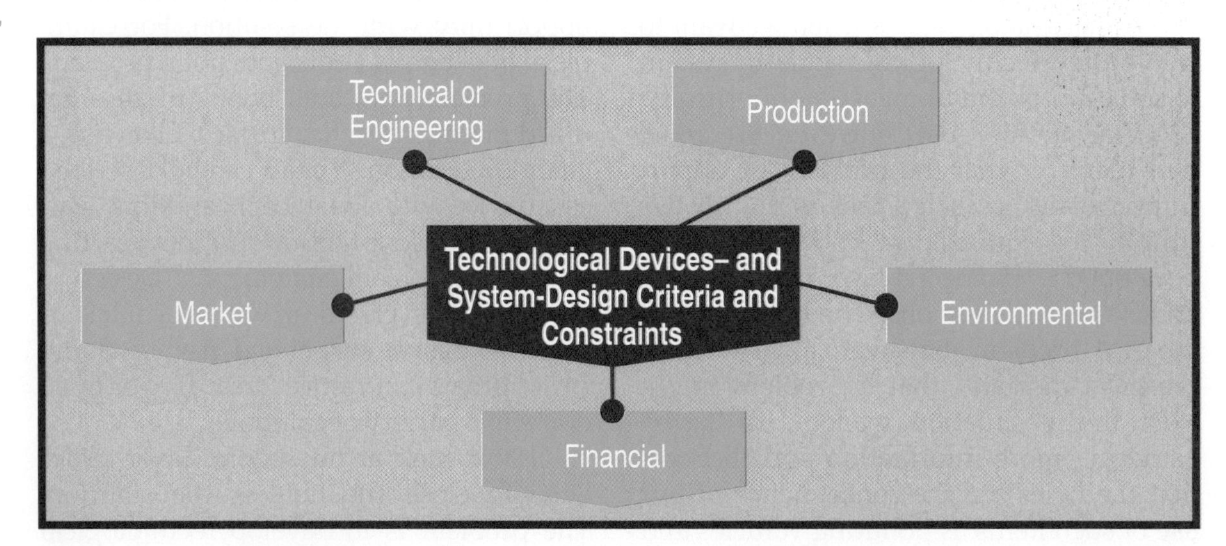

Figure 9-6. Technological devices and systems must meet both an individual's and society's expectations. These expectations can be communicated through five types of criteria and constraints.

- Environmental criteria and constraints indicate the intended relationship between the device or system and the natural and social environments. These criteria and constraints deal with the impacts of the technological device or system on people, societal institutions, and the environment. An environmental criterion might be "Must remove 95% of all sulfur dioxide emitted from the electric generating-plant smokestacks."

The problem or opportunity description and set of criteria and constraints allow the designers to start the next step in the identification process, gathering information.

Gathering Information

Designing technological devices and systems requires knowledge. This knowledge is derived from obtaining and studying information. A wide variety of information might be needed. See **Figure 9-7.** Typically, this information includes the following:

- ***Historical information*** about devices and systems developed to solve similar problems.
- ***Scientific information*** about natural laws and principles that must be considered in developing the solution.
- ***Technological information*** about materials and energy-processing techniques that can be used to develop, produce, and operate the device or system.
- ***Human information*** affecting the acceptance and use of the device or system. This information might include such factors as ergonomics, body size, consumer preferences, and appearance.
- ***Legal information*** about the laws and regulations controlling the installation and operation of the device or system.

Standards for Technological Literacy

1 2 4 8 10 11

Brainstorm

Ask your students what different types of insights they would get from using historical research methods, descriptive methods, and experimental methods.

Figure Discussion

Discuss why all six types of information, shown in Figure 9-7, are important as a foundation for design activities. Encourage class participation.

TechnoFact

Creativity is significant in all aspects of personal life and business, including technology. According to Douglas Ivester, Chief Operating Officer of Coca-Cola, "Everybody falls into the trap of looking at the latest gadget, or thinking that creativity has to be in the arts and sciences."

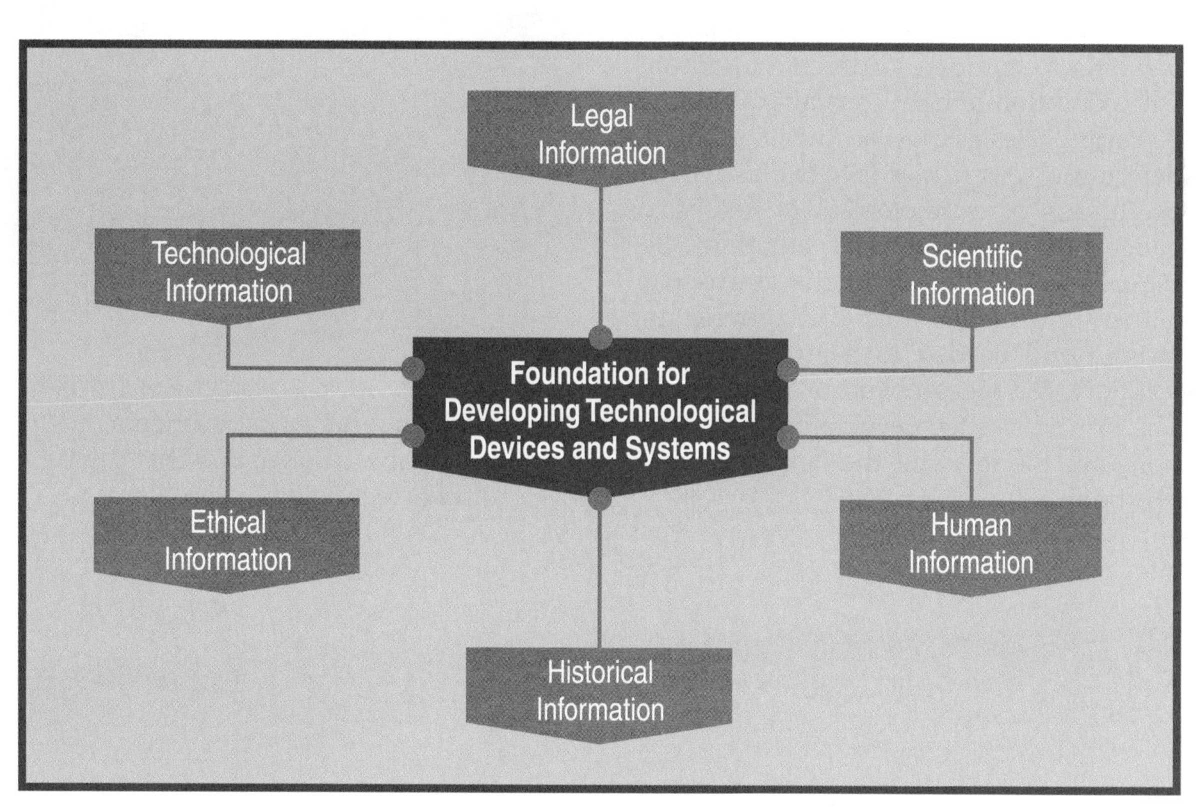

Figure 9-7. Many types of information are used as a foundation for technological-development activities.

Standards for Technological Literacy

1 2 3 4 7 8 10 11

Research

Have your students research how radar was a solution to a problem. Have them investigate the ways radar has been used to solve additional problems.

TechnoFact

It has been suggested by various governments that in today's global economy, companies and nations compete, make accomplishments, and thrive on the foundation of skills, knowledge, and creativity.

Quotes

"Design is about ideas, and through ideas we get change. The world only moves forward by changing...everyone can be part of that."—Wayne Hemingway, founder of Red or Dead.

Quotes

"Design is the engine of innovation, and the path to finding our desired future."—John Kao, CEO of The Idea Factory.

- ***Ethical information*** describing the values people have toward similar devices and systems.

Designers can obtain information through historical-, descriptive-, or experimental-research activities. They use ***historical methods*** to gather information from existing records. For example, they might consult books, magazines, and journals housed in libraries. See **Figure 9-8.** They can review sales records, customer complaints, and other company files. Designers can check judicial codes describing laws that might affect the development project.

They use ***descriptive methods*** to record observations of present conditions. For example, they might survey people to determine product preferences, opinions, or goals. They might observe and describe the operation of similar devices. Designers might measure and record physical qualities, such as size, weather conditions, and weight.

Finally, designers use ***experimental methods*** to compare different conditions. One condition is held constant. The other is varied. Designers use this method to determine scientific principles, assess the usefulness of technological processes, or gauge human reactions to situations. See **Figure 9-9.** All these methods are used to gather information that can provide the background needed to begin developing solutions for the problem or opportunity. The procedures for developing these solutions are the focus of the next chapter in this book.

©iStockphoto.com/dlewis33

Figure 9-8. Books, magazines, and journals provide a valuable source of information for solving technological problems.

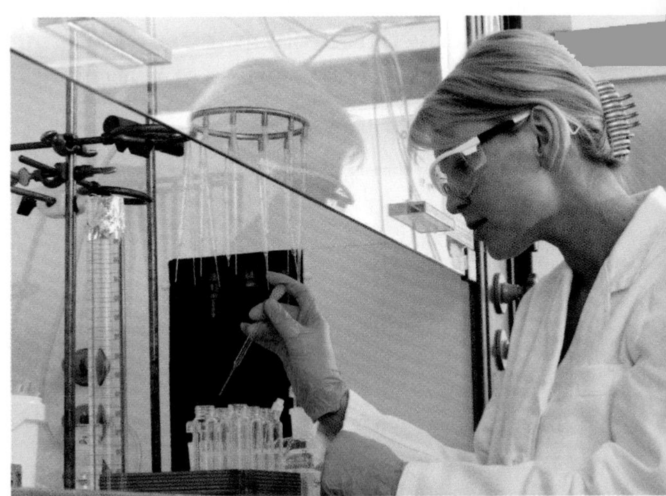

©iStockphoto.com/pidjoe

Figure 9-9. Laboratory experiments provide valuable information that can be used to solve technological problems.

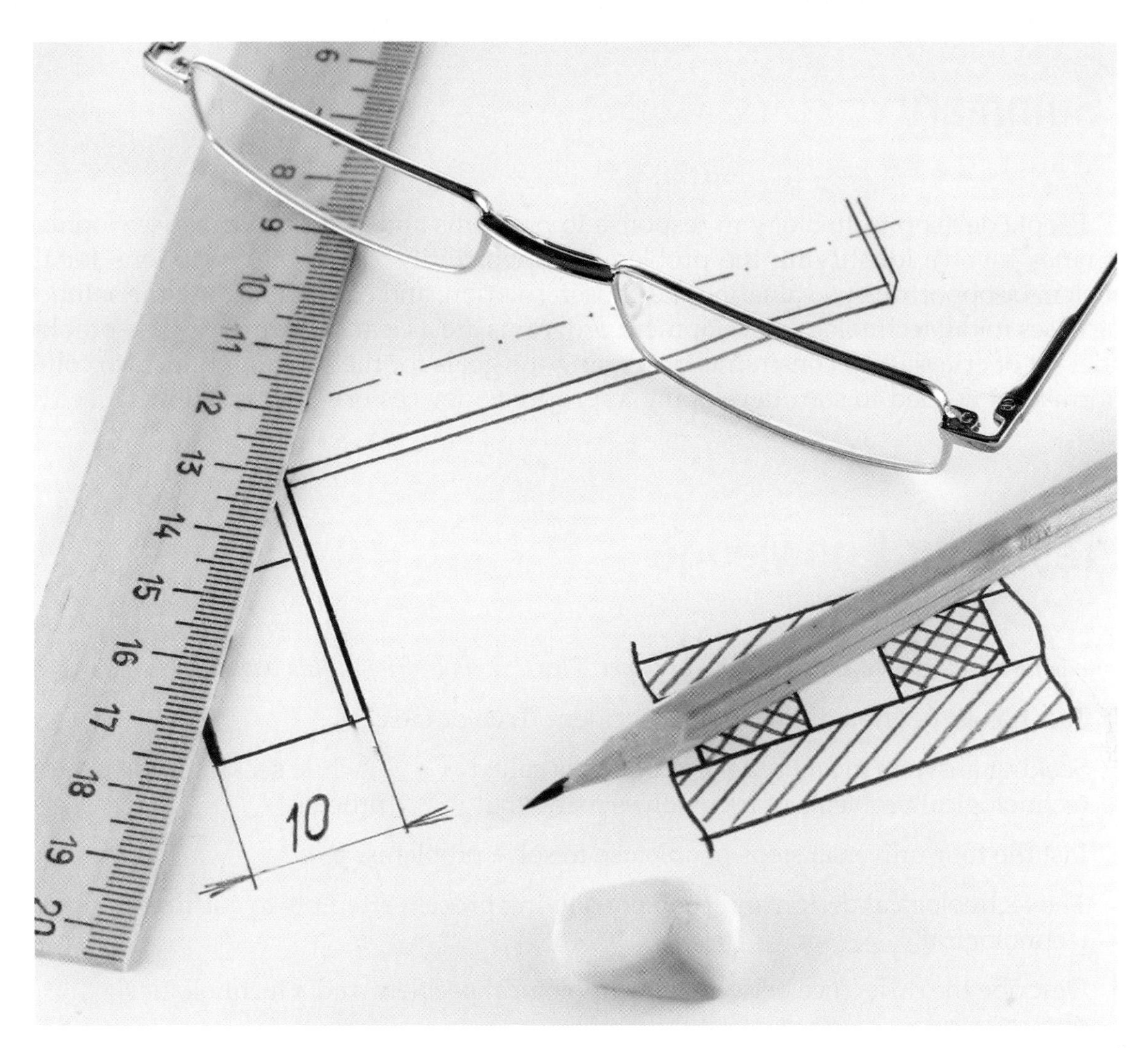

Solving technological problems involves four steps: Identifying problems, developing solutions, evaluating solutions, and communicating solutions. A sketch is one way of communicating a solution.

Summary

People develop technology in response to problems and opportunities. Development activities involve identifying the problem or opportunity, developing solutions for the problem or opportunity, evaluating the chosen solution, and communicating the solution. The bases for all technology-development programs are a clear description of the problem and a set of criteria and constraints describing the goals for the solution. Efforts to collect information needed to start developing appropriate devices or systems follow these two steps.

Test Your Knowledge

Write your answers on a separate piece of paper. Please do not write in this book.

1. Technology is often discovered by accident. True or false?
2. Seeking answers to any type of problem is called _____, while seeking answers to technological problems often involves using the _____ process.
3. List the four universal steps people use to solve problems.
4. The technological design and problem-solving process attempts to address technological _____ or _____.
5. Describe the difference between a technological problem and a technological opportunity.
6. List the four major steps used in technology to solve problems and meet opportunities.
7. What are the steps followed in identifying a technological problem or opportunity?
8. What is the first step in defining a problem or an opportunity?
9. The way a problem is described can restrict creativity. True or false?
10. Give one example of a production constraint in developing a technological artifact or system.
11. Give one example of a market criterion in developing a technological artifact or system.
12. List the six types of information that can help designers solve technological problems and meet technological opportunities.
13. List the three methods used to obtain information.

Answers to Test Your Knowledge Questions

1. False
2. problem solving, design
3. needs, wants
4. Develop an understanding of the problem through observation and investigation. Devise a plan for solving the problem. Implement the plan. Evaluate the plan.
5. You face a technological problem if you need to develop tools, machines, or systems to help you do work. If a "nonproblem" condition can be improved with technology, a technological opportunity exists.
6. Identifying or recognizing a technological problem. Developing a solution. Evaluating the solution. Communicating the solution.
7. Defining the problem or opportunity and gathering information.
8. Describe the situation that needs a technological solution.
9. True
10. Evaluate individually.
11. Evaluate individually.

12. Historical information, scientific information, technological information, human information, legal information, and ethical information.
13. Historical method, descriptive method, and experimental method.

STEM Applications

1. Select a problem or an opportunity technology can address. Use a form similar to the one shown below to help you do the following:
   A. Write a clear definition of the problem or opportunity.
   B. List the criteria and constraints you would consider in solving the problem or meeting the opportunity.

Problem or Opportunity:
Engineering criteria and constraints:
Production criteria and constraints:
Market criteria and constraints:
Financial criteria and constraints:
Environmental criteria and constraints:

2. Suppose you are given the following problem to solve: Design a device allowing you to generate enough electricity to power a television set while riding an exercise bicycle. Where would you go to gather information to start the design process?

Chapter Outline
System Design
Product Design
The Design Team
Steps for Developing Design Solutions

This chapter covers the benchmark topics for the following Standards for Technological Literacy:

Learning Objectives

After studying this chapter, you will be able to do the following:

- Recall the two major areas in which designers develop solutions to technological problems and opportunities.
- Explain what system design is.
- Recall the two major types of technological products and systems created with product design.
- Summarize some advantages of using a design team over an independent designer.
- Recall the three main steps followed in developing technological designs.
- Recall the ways in which ideas can be stimulated when developing design solutions.
- Compare the kinds of design sketches.
- Recall the three major types of information required for detailed sketches when building models.
- Recall the three types of pictorial sketches used in product design.

Key Terms

brainstorming	divergent thinking	refined sketch
classification	isometric sketch	rough sketch
convergent thinking	oblique sketch	synergism
detailed sketch	perspective sketch	what-if scenario

Strategic Reading

As you read this chapter, make a list of the key terms that are used in designing solutions. Can these terms be used in other contexts, or are they concepts specific to design?

Technological devices are designed to meet identified problems and opportunities. See **Figure 10-1.** These problems and opportunities can be either of the following:

- System-design problems and opportunities.
- Product-design problems and opportunities.

No matter what the area, though, developing design solutions involves three major steps. We will discuss those steps after we differentiate between system design and product design.

System Design

System design deals with the arrangement of components to produce a desired result. For example, automotive braking systems are a result of system-design efforts. Look at the drum brake system shown in **Figure 10-2.** This design brings

Standards for Technological Literacy

2 11

Reinforce

Review evaluating design solutions as part of the design process, which was presented in Chapter 5.

Extend

Use an example to differentiate between product design and system design.

TechnoFact

The system of division of labor was developed during the Industrial Revolution. It was at this time that the group of workers called designers emerged, being one of three groups of workers. Products were designed by the first group, built by the second, and sold by the third.

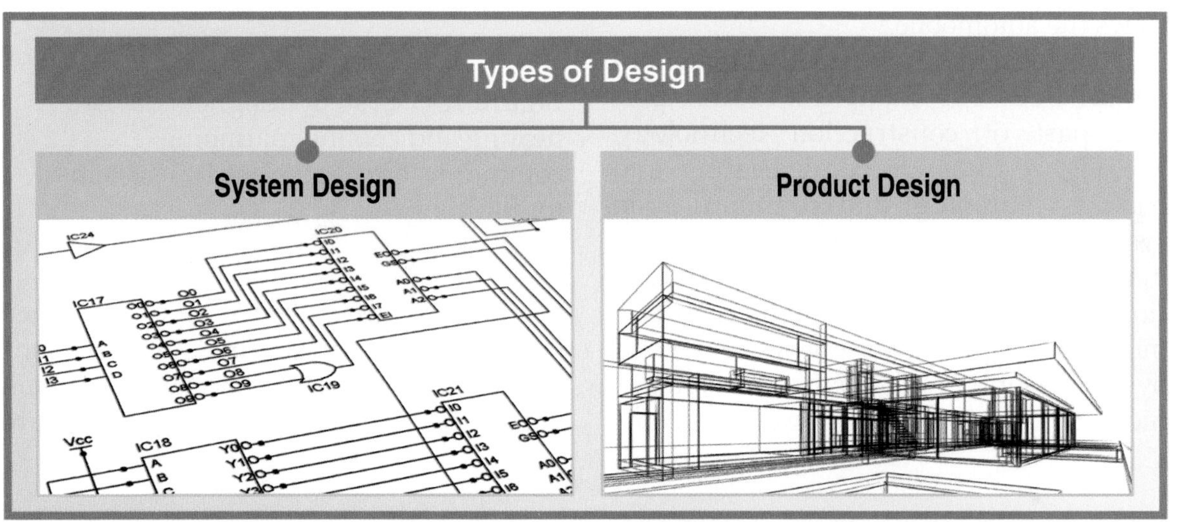

Figure 10-1. Designers create both systems and products.

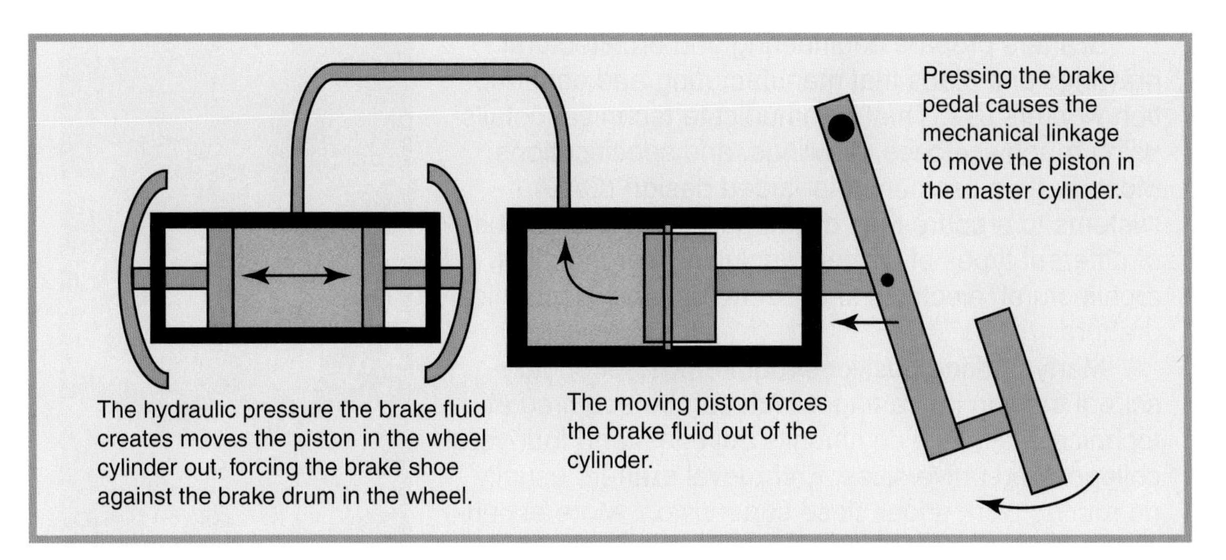

Figure 10-2. This brake system was the result of system-design efforts.

together mechanical and hydraulic components into a speed-reduction system. The brake-pedal unit is a mechanical linkage. When the pedal is depressed, a plunger in the master cylinder is moved. This motion causes the fluid to move in the hydraulic system connecting the master cylinder to the wheel cylinders. The fluid movement pushes the pistons outward in the wheel cylinders. These pistons are attached to the brake shoes. The piston movement causes the shoe to be forced against the brake drum. The mechanical action creates friction between the shoe and the drum, which slows the automobile.

This design can be used in all technological areas. For example, it is an important part of construction technology. See **Figure 10-3.** Electrical, heating and cooling, plumbing, and communication systems are designed for buildings. In manufacturing, the methods of production, warehousing, and material handling must be designed. Messages are carried over fiber-optic and microwave communication systems. Transportation systems combine manufactured vehicles and other components to move goods and passengers from place to place. Irrigation systems are used to water crops. Pipelines are part

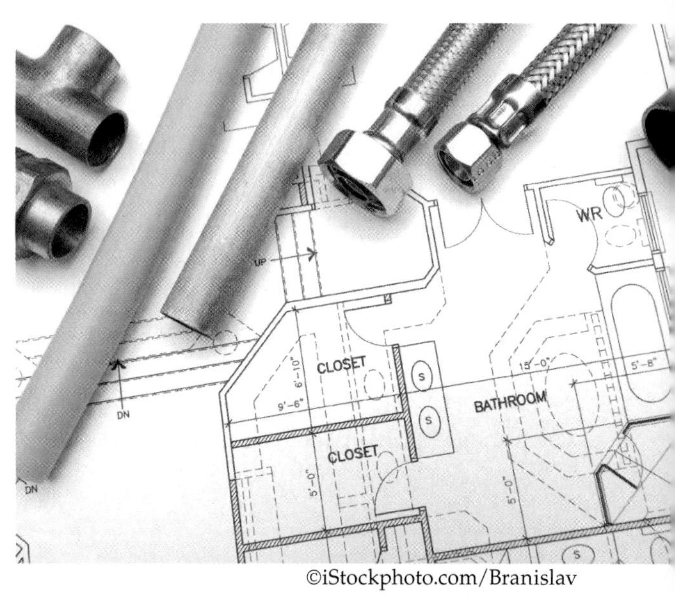

©iStockphoto.com/Branislav

Figure 10-3. System designers devise heating and cooling, plumbing, communication, and electrical systems for buildings.

Standards for Technological Literacy

2 8 9 11

Brainstorm

Have your students list the systems they have used today.

Research

Have your students investigate a simple system and diagram the way it works.

TechnoFact

A wider range of consumer products was offered by manufacturers during the Industrial Revolution. At this time, the primary concern of consumers was meeting their needs through these products, and the general interest in products' appearance decreased.

Career Corner

Drafters

Drafters prepare engineering and architectural drawings and plans that manufacturing and construction workers use. They communicate technical details using rough sketches, drawings, and specifications. Most drafters use computer-aided design (CAD) systems to prepare their drawings. There is a number of different types of drafters, including aeronautical, architectural, electrical and electronic, and mechanical drafters.

Many drafting positions require postsecondary-school training in drafting, which can be acquired at technical institutes, community colleges, and four-year colleges and universities. Entry-level drafters usually do routine work under close supervision. More experienced drafters do difficult work with less supervision.

of natural gas–distribution systems. Doctors and hospitals provide patient care in health-care systems.

Product Design

Product design deals with two areas: manufactured products (involving designers) and constructed structures (involving architects). See **Figure 10-4.** The goal of both activities is to develop a product or structure meeting the customer's needs. The product or structure must function well, operate safely and efficiently, be easily maintained and repaired, have a pleasant appearance, and deliver good value.

In addition, products and structures must be designed so they can be produced economically and efficiently. They must also be sold in a competitive environment. In short, the product or structure must be designed for the following:

- **Function.** The product or structure must be easy and efficient to operate and maintain.
- **Production.** The product or structure must be easy to manufacture or construct.
- **Marketing.** The product or structure must be appealing to the end user.

The Design Team

In early times, products were often designed by a single person or craftsperson. Paul Revere was a silversmith who designed and produced products he sold in his shop in Boston. As technology advanced, however, this single-person approach was replaced with design teams. This allowed various people to contribute their special talents to the project. For example, a design team for a large building might include architects, civil and mechanical engineers, interior designers, graphic designers, computer-aided drafting operators, and 3D illustrators. They would be involved with the many aspects of designing, planning, and constructing the building. No one person on the team could do

Standards for Technological Literacy

2 8 9 11

Extend

Select a product and point out its design for function, production, and marketing.

TechnoFact

By the 20th century, competition had swept the market. People could choose from various brands of very similar products. It was at this time manufacturers decided to use design to distinguish their products.

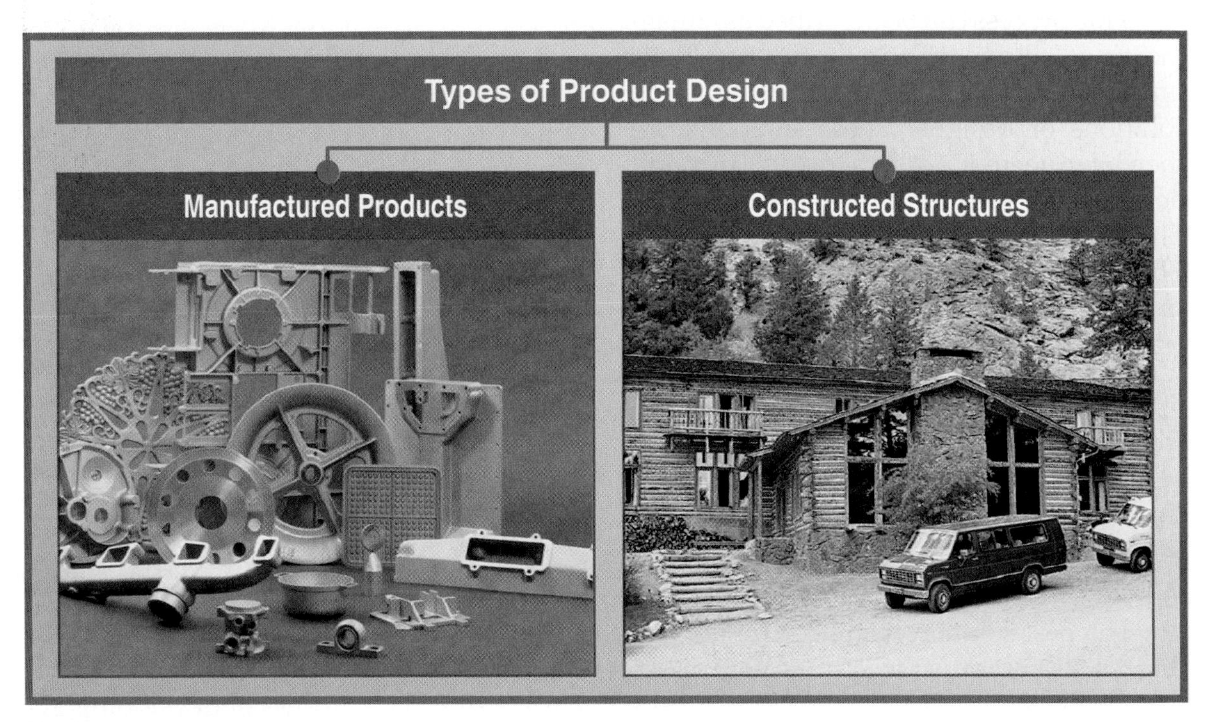

Figure 10-4. Manufactured parts and structures are designed using product-design procedures.

Standards for Technological Literacy

8 11

Brainstorm

Have your students list instances when they used divergent thinking and convergent thinking. Have them explain why they used the approaches they did.

TechnoFact

Industrial designers use the combination of their artistic ability and knowledge of costs, materials, engineering, marketing conditions, and manufacturing processes to plan and develop products and systems in order to meet the needs of consumers.

the job alone. The task would be too large and require too much knowledge. For the team, however, the job is manageable. See **Figure 10-5.**

Steps for Developing Design Solutions

System and product designs start with a clear definition of the situation or opportunity. We learned the procedures for developing the definition in the previous chapter. This problem definition leads to the next step in product design—developing design solutions. These solutions often evolve through three steps:

1. Developing preliminary solutions.
2. Isolating and refining the best solution.
3. Detailing the best solution.

As these three steps are completed, often, the problem is redefined or refined. The steps can then be repeated until a final design is developed. This circular system is shown in **Figure 10-6.**

This process can be described as "imagineering." First, the designers use their imagination to develop a number of unique solutions or designs. These solutions are then engineered back to reality through design-refinement and detailing activities. The first step starts with broad thinking. This kind of thinking is called ***divergent thinking***. Divergent thinking seeks to think of as many different (divergent) solutions as possible. The most promising solutions are then refined and reduced until one "best" answer is found. The refinement of ideas requires ***convergent thinking***. The goal is to narrow and focus (converge) the ideas until the most feasible solution is found.

The best solution might not be the one that works best or is the least expensive. As we noted earlier, criteria and constraints can compete with one another. Trade-offs often occur among appearance, function,

©iStockphoto.com/MarcusPhoto1

Figure 10-5. Members of an architectural design team look over a model of the building they are designing.

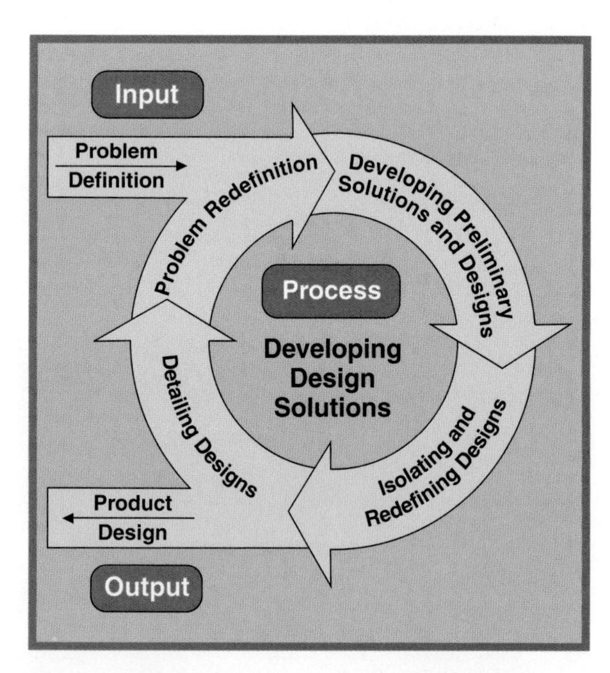

Figure 10-6. The process of developing design solutions has three major steps, plus a problem-redefinition phase.

and cost. Many of us cannot always afford the very best answer. Our budget and the length of time we expect to keep the product enter into our product choices. For example, you might find it unwise to purchase a $900 racing bicycle because you are only going to use the bicycle to ride occasionally to the park. If you regularly ride a bicycle to work or go on cycling vacations, you might be able to justify the expensive model. Likewise, the "snapshot" photographer probably does not need the most expensive digital camera. The professional photographer would choose such a camera, however. Product-design activities produce a wide range of products. This variety allows consumers to select one that meets their performance needs and financial resources.

Developing Preliminary Solutions

Designs start in the minds of designers, engineers, or architects. Ideas can be stimulated in various ways. Three popular techniques are brainstorming, classification, and what-if scenarios. After people have put forth their ideas, they then record the ideas in rough sketches.

Brainstorming

Brainstorming is a process requiring at least two people, although most people find that having three or more participants in the process is usually more productive. This process involves seeking creative solutions to an identified problem. Members of the group offer individual solutions that they think will work. See **Figure 10-7.** Proposed solutions often cause other members of the group to think of more ideas. The strategy uses a concept called ***synergism***. Synergism builds on the individual contributions of the participants to make a larger whole. The number of ideas the group generates is more than the number they could develop if each person worked alone.

Brainstorming activities work best when the group accepts some basic rules. These rules include the following:

- **Encourage wild, far-out ideas.** There are no bad or stupid ideas. Wild, but promising, ideas can always be engineered back to reality.
- **Record the ideas without reacting to them.** Many people stop offering some of their ideas if they are criticized. To avoid criticism, they provide only those ideas they think the group will like.
- **Seek quantity, not quality.** The chances of good ideas emerging are increased as the number of ideas increases.
- **Keep up a rapid pace.** A rapidly paced session keeps the mind alert and reduces the chance of judging the ideas.

©iStockphoto.com/track5

Figure 10-7. Members of a graphic design team brainstorm ways to improve a design for a national advertisement.

Standards for Technological Literacy

8 11

Brainstorm

Have groups of students use classification, brainstorming, or what-if scenarios to deal with a "desk clutter" problem.

Standards for Technological Literacy

3 8 11

TechnoFact

While consumers of the early 20th century had a choice between brands of products, product distinction had not yet been developed. Many manufactured goods were similar, and there was very little variety.

Quotes

"Design and innovation are about exploration and experimentation and accepting that your first solution may not be right. Creativity is about challenging your assumptions."
—Dr. Bettina von Stamm, London Business School.

STEM Connections: Mathematics

Integrated STEM Curriculum — Science, Technology, Engineering, Mathematics

Solid Geometry

Designers should be familiar with some basic geometric concepts in order to create effective pictorial sketches. They use the concepts of solid geometry, for example, when drawing such three-dimensional images as pyramids, cones, cylinders, and cubes. A pyramid, to use the first example, is a solid figure with a polygon as a base. A polygon, as you might recall from earlier geometry lessons, is a closed plane figure bounded by three or more straight lines. (The word *polygon* comes from the prefix *poly-*, meaning "many," and the suffix *-gon*, meaning "angle." Thus, a polygon is a many-angled figure.) The faces (surfaces) of the pyramid are triangles with a common vertex, or point where they intersect.

In a regular pyramid, the base is a regular polygon. The faces are congruent triangles. Again, as you might remember from earlier geometry, congruent means "equal in size and shape."

The other three-dimensional images share some characteristics, but they differ in others. For example, in what way or ways is a cone similar to a pyramid? In what way or ways is it different?

The Giza Pyramids in Egypt. ©iStockphoto.com/karimhesham

Classification

One person or a group of people can conduct classification. ***Classification*** involves dividing the problem into major segments. Each segment is then reduced into smaller parts. For example, buildings can be classified as business and commercial, homes, and industrial. Homes can be further classified as houses, apartments, and condominiums, for example. A house can be classified by its major features: foundations, floors, walls, ceilings, roof, doors, windows, and so on. Foundations can then be classified as poured concrete, concrete block, wood posts, timber, and so on. This process might result in a classification chart. A classification chart is often developed as a tree chart, with each level having a number of branches below it. See **Figure 10-8.** This chart ends up looking very similar to a family tree people use to trace their ancestors.

What-If Scenarios

What-if scenarios start with a wild proposal. The proposal's good and bad points are then investigated. The good points can be used to develop solutions. For example, peeling paint is a problem for housepainters. They must remove the old paint from a house before repainting. A wild solution suggests mixing an explosive material with the paint before it is applied. Whenever the building is ready to be repainted, the old paint can be blown off the building. Obviously, exploding house paint is ridiculous. The proposal, however, can lead to a solution. Paint sticks to a house through the adhesion between the paint and the siding. Maybe a material can be mixed with the paint that causes the paint to lose adhesion when a special chemical is applied. At repainting time, the chemical can be sprayed on to loosen the paint. The paint can then be easily removed from the siding.

Rough Sketching

Once designers have conceived of a number of ideas, they must record the ideas. The most common recording method is to develop ***rough sketches*** of the products, structures, or system components. See **Figure 10-9.** These sketches are as much a part of the thinking process as they are a communication medium. Designers are forced to think through concepts such as size, shape, balance, and appearance. The sketches then become a library of ideas for later design efforts.

©iStockphoto.com/nicolas_

Figure 10-9. Designers use rough sketches to record their ideas.

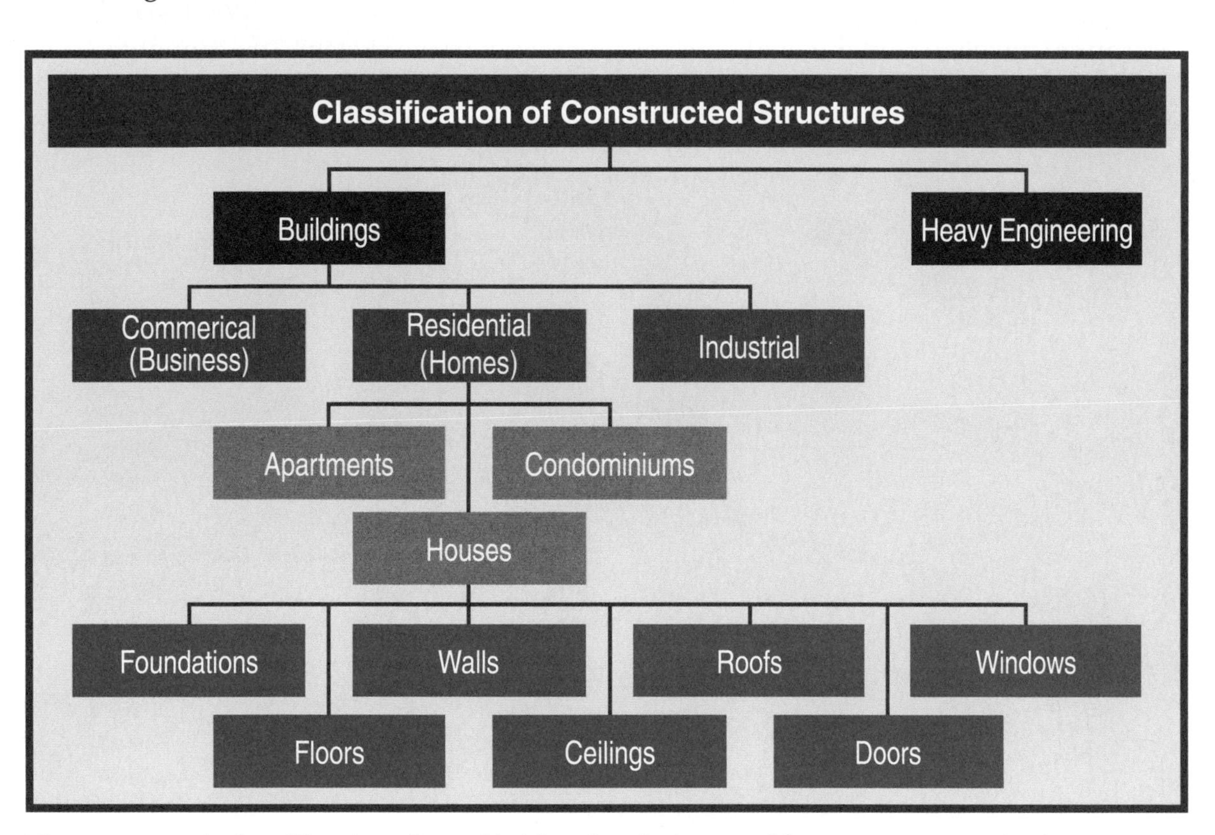

Figure 10-8. A classification chart divides the design problem or opportunity into segments to develop solutions more easily. This type of chart is called a *tree chart.*

Standards for Technological Literacy

8 11

Figure Discussion

Discuss Figure 10-8 and then have groups of students develop a classification of business/commercial buildings.

Figure Discussion

Discuss the quick techniques used in the rough sketches shown in Figure 10-9.

Quotes

"In a competitive world, high quality design is increasingly a differentiating factor. Good design is not just a matter of flair or an occasional flash of inspiration—it is an ongoing, systematic process."—George Cox, Institute of Directors.

Standards for Technological Literacy

8 11

Extend

Discuss why it is important to have many rough sketch ideas and a few refined sketches (divergent thinking followed by convergent thinking).

Quotes

"Whatever you do in your business in terms of innovation and technology, it must bring added value to your end customers. Without this, it just does not work."—Pierre-Yves Gerbeau, former Chief Executive of The Millennium Dome

The term *rough* is not used to describe the quality of the drawing. Rough sketches are not necessarily crude. They often represent good sketching techniques. The term *rough* describes the state of the design ideas. This term suggests that the designs are incomplete and unrefined.

Isolating and Refining Design Solutions

The rough sketches allow designers to capture a wide variety of solutions for the design problem or opportunity. The sketches are similar to books in a library. They contain a number of different thoughts, views, and ideas. These sketches can be selected, refined, grouped together, or broken apart.

Isolating and refining original designs in the "library of ideas" is the second step in developing a design solution. See **Figure 10-10.** Promising ideas are chosen and then studied and improved. This process might involve working with one or more good rough sketches. The size and shape of the product or structure might be changed and improved. Details might be added, and the shape might be reworked. In short, the design is becoming refined as problems are worked out and the proportions become more balanced.

Refined design ideas might also be developed by merging ideas from two or more rough sketches into a ***refined sketch.*** The overall shape might come from one sketch, and specific details might come from others. This approach is one of integration, which blends the different ideas into a unified whole. The new idea might not look at all similar to the original rough sketches.

Detailing Design Solutions

Rough and refined sketches do not tell the whole story. Look back at the sketches shown in **Figure 10-10.** What size is the product in the sketch? You cannot tell. The

©iStockphoto.com/nicolas_

Figure 10-10. Refined sketches are used to develop ideas captured with rough sketches.

sketches communicate shape and proportion. They do not communicate size. For this task, we need to add more details, and thus, we need a third type of sketch, called a ***detailed sketch***. This sketch communicates the information needed to build a model of the product or structure. Detailed sketches can also be used as a guide to prepare engineering drawings for manufactured products and architectural drawings for constructed structures. Engineering and architectural drawings are discussed in Chapter 12.

Detailed sketches are helpful when models of products or structures are made. Building models requires three major types of information. See **Figure 10-11:**

- **Size information.** This information explains the overall dimensions of the object or the sizes of features on an object. Size information might include the thickness, width, and length of a part, the diameter and depth of a hole, or the width and depth of a groove.
- **Location information.** This information gives the positions of features within the object. Location information might establish the location of the center of a hole, the edge of a groove, or the position of a taper.
- **Geometry information.** This information describes the geometric shapes or relationships of features on the object. Geometry information can communicate the relationship of intersecting surfaces (square or 45° angle, for example), the shapes of holes (rectangular or round, for example), or the shapes of other features.

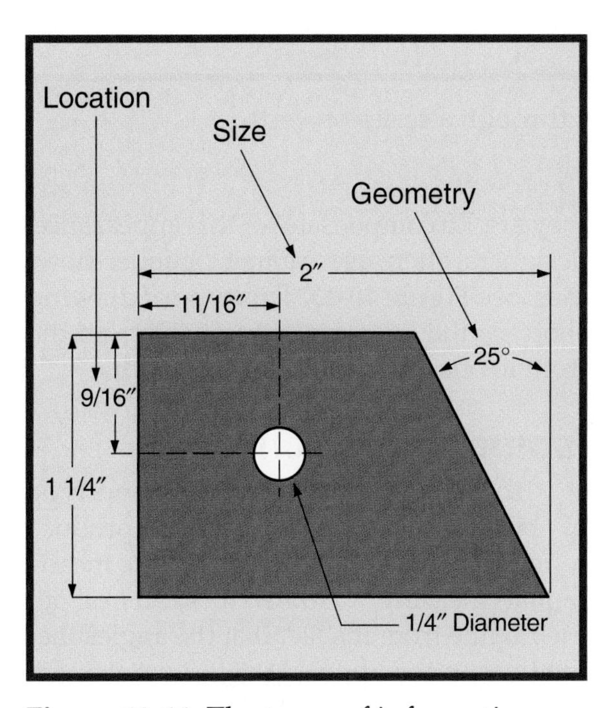

Figure 10-11. The types of information provided on detailed sketches are size, location, and geometry.

Designers often use pictorial-sketching techniques to capture and further refine product-design ideas. These techniques try to show the artifact very similarly to how the human eye will see it. Therefore, a single view is used to show how the front, sides, and top will appear. Designers produce three different kinds of pictorial sketches when refining ideas. These sketches are the following:

- Oblique sketches.
- Isometric sketches.
- Perspective sketches.

Oblique Sketches

Oblique sketches are the easiest pictorial sketches to produce. These sketches show the front view as if a person is looking directly at it. The sides and top extend back from the front view. They are shown with parallel lines that are generally drawn at 45° to the front view. The depth lines may be drawn at a different angle, such as 30° or 60°, depending on the intended result.

To produce an oblique sketch, the designer completes steps similar to those shown in **Figure 10-12.**

1. Lightly draw a rectangle that is the overall width and height of the object.

Standards for Technological Literacy

8 11

Quotes

"This is the new golden age of design. When industries are competing at equal price and functionality, design is the only difference that matters."—Mark Dziersk, President of the Industrial Design Society of America.

Standards for Technological Literacy

8 11

Figure Discussion:

Use Figure 10-12 as a basis to show your students how to make a simple oblique sketch.

Quotes

"It is probably a myth to imagine that it was ever enough for a company to start up, start making something (widgets perhaps), sell them and then carry on doing so as they had always done. If that world ever existed it has been gone for many years now. Businesses can only survive, prosper and grow in a voraciously competitive global economy if they continue to innovate."—Gordon Brown, Creativity Works.

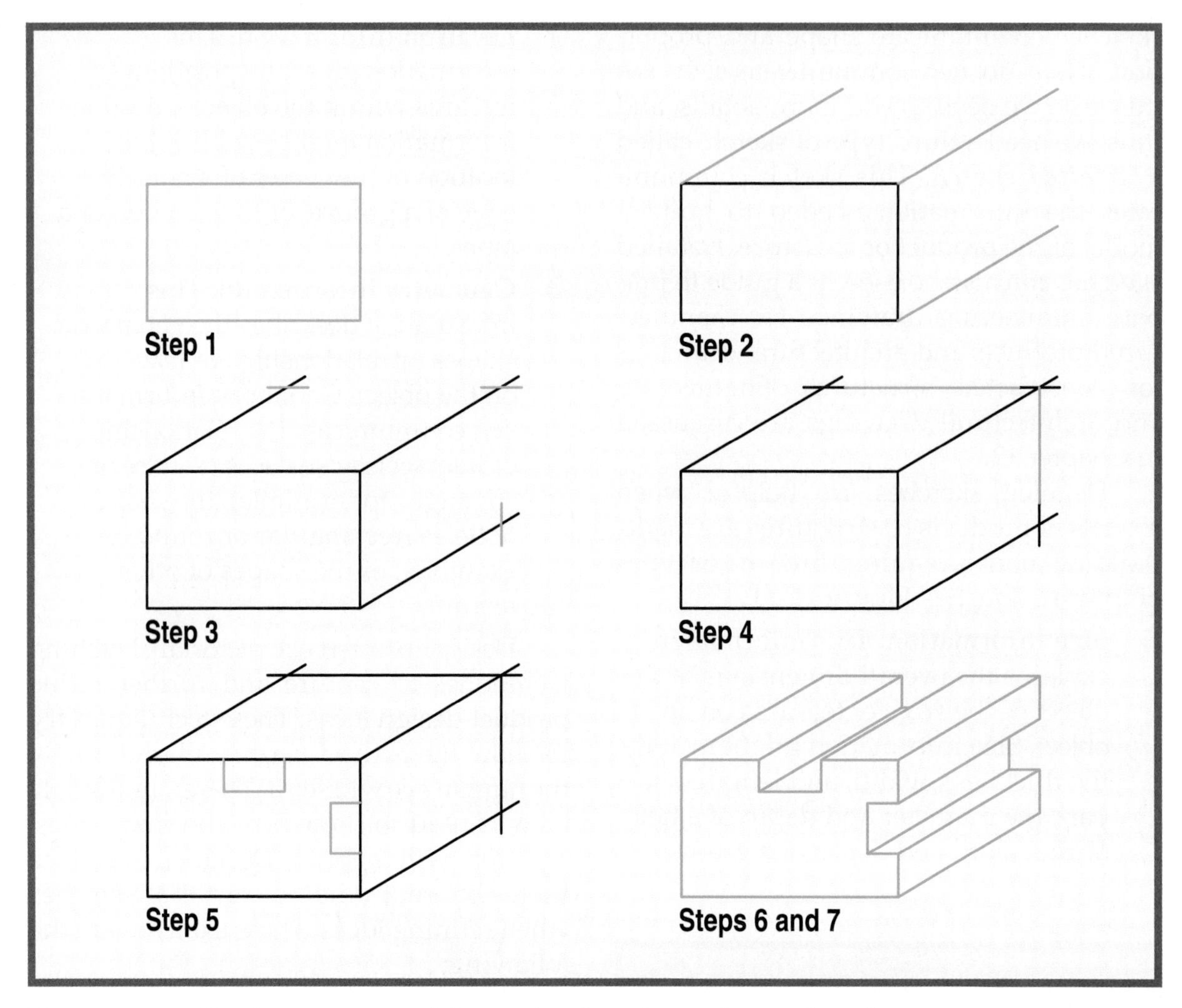

Figure 10-12. Designers create oblique sketches through a series of steps.

2. Lightly extend parallel lines from each corner of the box. Draw the depth lines at 45°, or at a different angle such as 30°.
3. Lightly mark the extension lines at a point equal to the depth of the object.
4. Lightly connect the depth lines to form a box.
5. Add any details, such as holes, notches, and grooves, onto the front view.
6. Extend the details the depth of the object.
7. Complete the sketch by darkening in the object and detail outlines.

The procedure listed above produces a cavalier oblique drawing. This type of drawing causes the sides and top to look deeper than they are. To compensate for this appearance, designers often use cabinet oblique drawings. See **Figure 10-13.** This type of drawing shortens the lines projecting back from the front to one-half their original lengths.

Isometric Sketches

Isometric sketches are the second type of pictorial drawings used to produce refined sketches. The word *isometric* means "equal measure." ***Isometric sketches*** get their name from the fact that the angles that the lines in the upper-right corner form are equal. Each angle is 120°. Designers use isometric sketching when the top, sides, and front are equally important. The object is shown as if it is viewed from one corner.

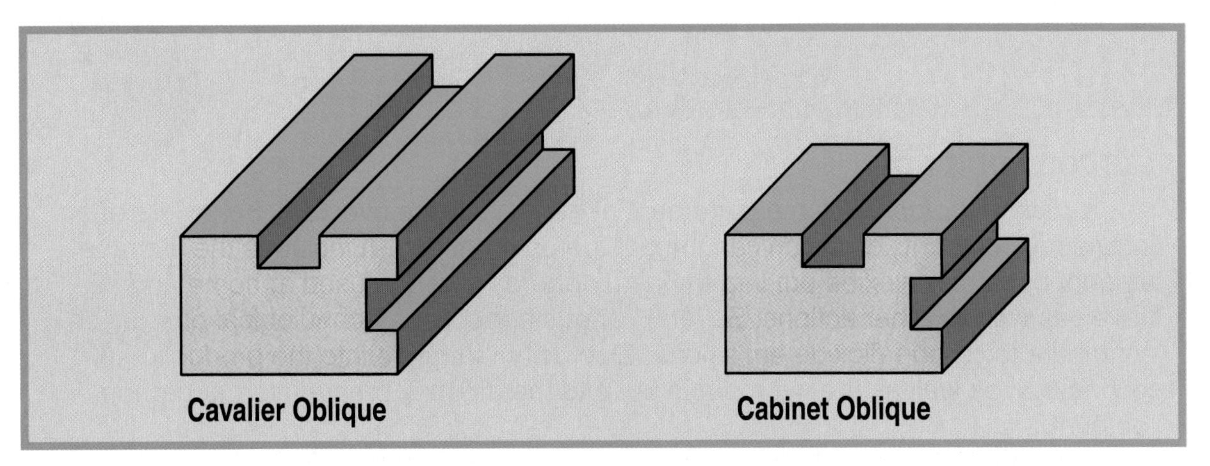

Figure 10-13. The two types of oblique drawings are cavalier oblique and cabinet oblique. Cabinet oblique drawings use one-half the depth of the object for a more natural appearance.

Designers follow four major steps when producing an isometric drawing. See **Figure 10-14.** These steps are the following:

1. Lightly draw the upper-right corner of an isometric box that will hold the object.
2. Complete the box by lightly drawing lines parallel to the three original lines.
3. Locate the major features such as notches, tapers, and holes.
4. Complete the drawing by darkening the features and darkening the object outline.

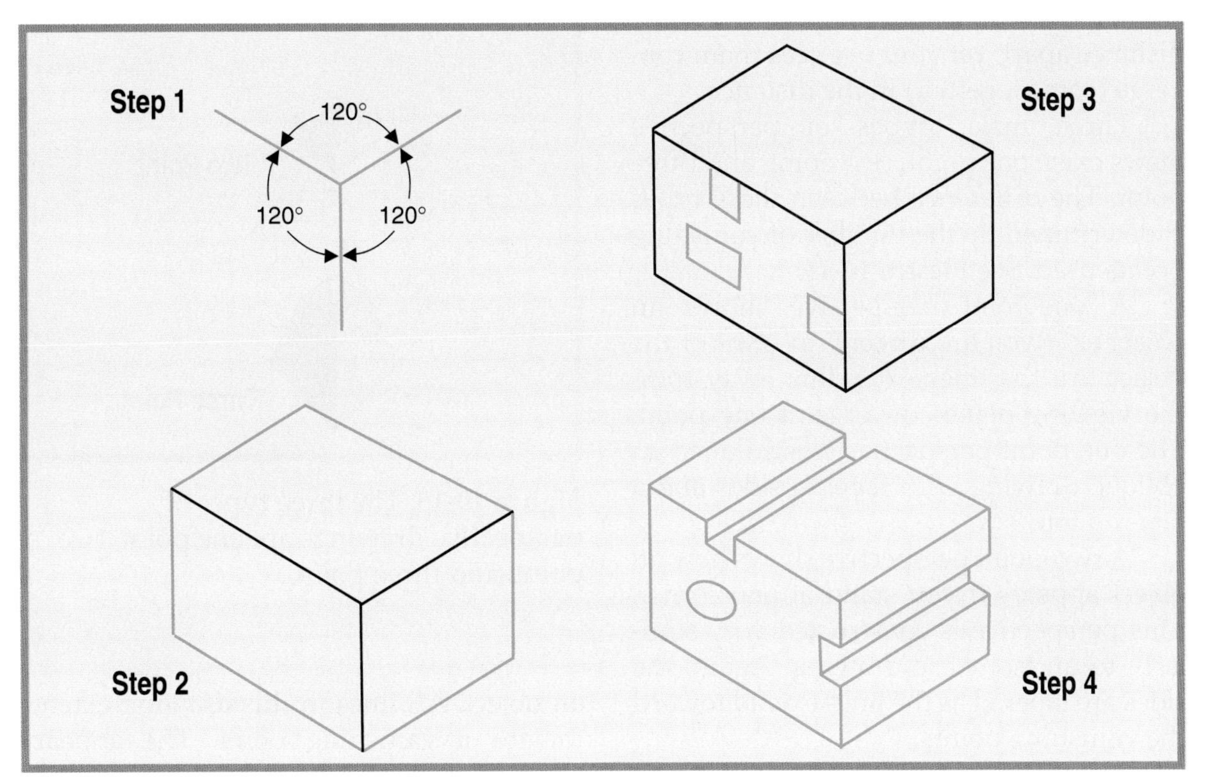

Figure 10-14. Designers follow four major steps when creating isometric sketches.

Standards for Technological Literacy

8 11

Figure Discussion

Explain why cabinet oblique techniques produce more appropriate kitchen cabinet sketches than cavalier oblique techniques.

Demonstrate

Show how to make a simple isometric sketch.

Quotes

"Today's consumers live in a world where rapid and constant change is now the 'norm', therefore, it is not a question of how much change, but rather ensuring that these changes are handled intelligently and articulately. Design must be used as a tool to facilitate accurate and engaging communication, as opposed to creating superficial camouflage."—Peter Knapp, Executive Creative Director, Landor Associates.

Standards for Technological Literacy

5 8 11

Figure Discussion

Use Figure 10-15 as a basis for discussing the types of perspectives and the uses for one-point, two-point, and three-point perspective sketches and drawings.

Demonstrate

Show your students how to make a simple two-point perspective sketch.

Think Green

Carbon Footprint

A *carbon footprint* is a measurement of how much the everyday behaviors of an individual, company, or nation can impact the environment. It includes the average amount of carbon dioxide put into the air by energy and gas used at home and in travel, as well as other actions. Several various aspects of technological production contribute to carbon dioxide emissions. Electricity that goes into the production of technology, as well as the byproducts put into the air, may create a larger carbon footprint.

Recently, companies have begun to determine their carbon footprints. Calculating either personal or business carbon footprints may be done with carbon calculators provided by various environmental organizations. By learning the details of their carbon footprints, people may be motivated to work toward reduction.

Perspective Sketches

Perspective sketches show the object as the human eye or a camera sees it. This realism is obtained by having parallel lines meet at a distant vanishing point. If you look down a railroad track, you see a similar effect. The rails remain the same distance apart, yet your eye sees them converge (come together) in the distance.

Three major types of perspective views exist: one point, two point, and three point. The difference between these types is determined by the number of vanishing points used. See **Figure 10-15.**

A one-point perspective shows an object as if you are directly in front of the object. All the lines extending away from the viewing plane converge at one point. The one-point perspective is similar to an oblique drawing with tapered sides and a tapered top.

A two-point perspective shows how an object appears if you stand at one corner. This perspective is constructed very similarly to an isometric drawing. Again, the sides are tapered as the lines extend toward the vanishing points.

A three-point perspective shows how the eye sees the length, width, and height of an object. All lines in this drawing extend toward a vanishing point. The appearance of a perspective drawing changes as the horizon changes. See **Figure 10-16.**

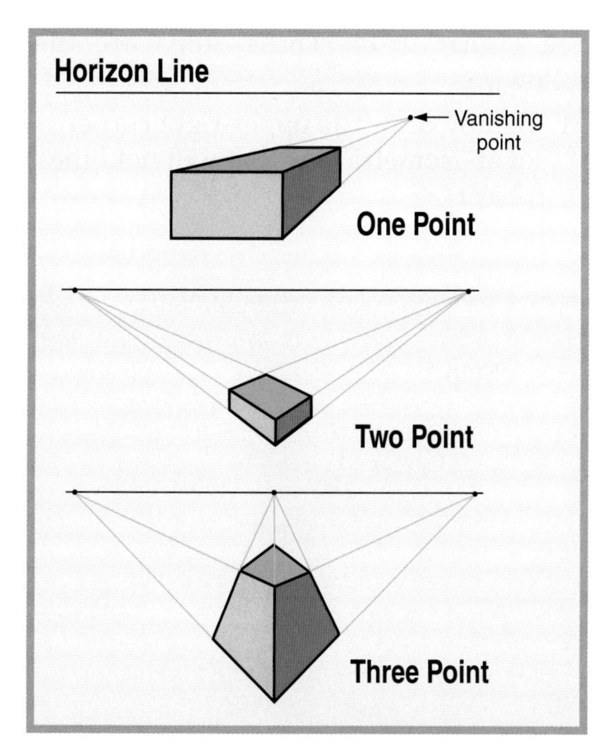

Figure 10-15. The three types of perspective drawings are one point, two point, and three point.

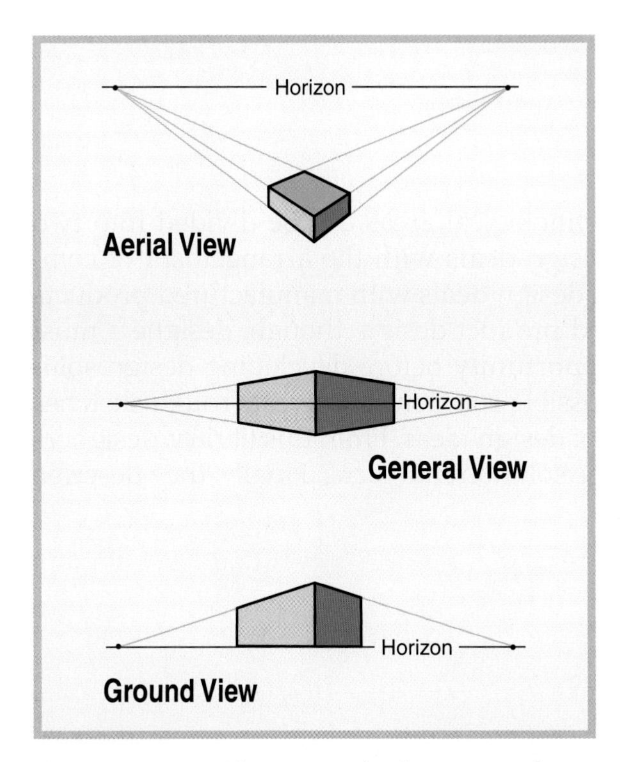

Figure 10-16. Changing the location of the horizon changes the appearance of a perspective drawing.

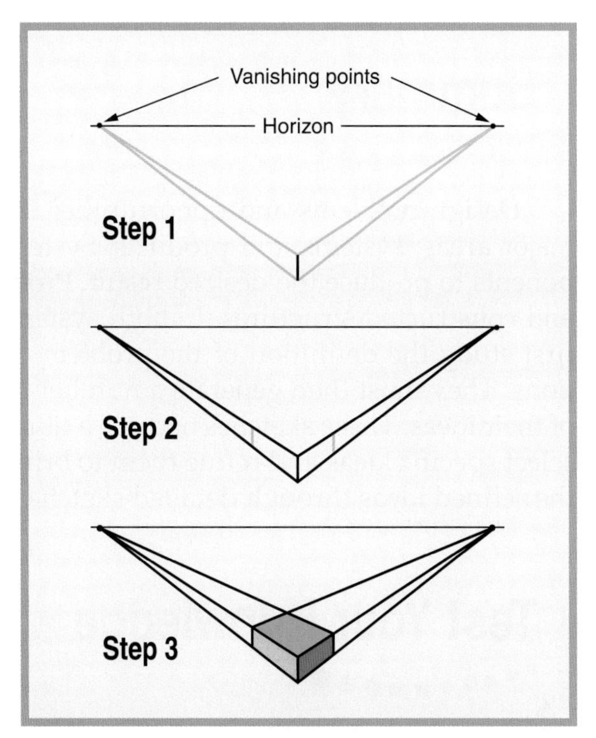

Figure 10-17. Designers follow three basic steps in developing perspective sketches.

Changing the position of the horizon line can cause the object to be seen as if the observer is looking down on the object (aerial view), directly at it (general view), or up at it (ground view). The designer must decide which of these views best suits the object and the audience who will see the sketch.

When developing the basic structure for one-, two-, or three-point perspective sketches, designers follow the same basic steps. See **Figure 10-17.** These steps are the following:

1. Establish the horizon line, vanishing point(s), and front of the object. Connect the front line(s) to the vanishing point(s).
2. Establish the depth of the objects along the lines extending to the vanishing point(s).
3. Connect the depth lines to the vanishing point(s). Darken in the object.

Designers then add details to complete the sketch. Perspective sketches are often shaded to add to their communication value. Developing the perspective, or "human eye," view is more difficult than developing the oblique or isometric views. Perspectives are, however, the most realistic of the three pictorial sketches.

Standards for Technological Literacy

8 11

Extend

Give examples of perspective sketches or drawings that are aerial view, general view, and ground view.

TechnoFact

Design is often considered more than just the appearance of a product; it is also the features of the product. Mark Foster, Business and Industry correspondent for the BBC said: "Design isn't just about the shape of a product, it's about how a company thinks."

TechnoFact

The development, production, and marketing process are all similar in that the design team needs to be a participant. The design of a product is something that is begun in the concept stage, and it should be constant throughout the process.

Summary

Design problems and opportunities in technological areas can be divided into two major areas: systems and products. System design deals with the arrangement of components to produce the desired result. Product design deals with manufactured products and constructed structures. In both system and product design, though, designers must first study the definition of the problem or opportunity before developing design solutions. They must then generate a number of possible solutions and create rough sketches of their ideas. These sketches become a library of design ideas. From this library, designers select specific ideas and refine them to bring the solution into focus. Finally, they describe the refined ideas through detailed sketches.

Test Your Knowledge

Write your answers on a separate piece of paper. Please do not write in this book.

1. What are the two major areas in which designers develop solutions to technological problems and opportunities?
2. Designing the arrangement of components to produce a desired result is called _____.
3. Product design deals with developing _____ products and _____ structures.
4. A group of people who work together to create a design is called a(n) _____.
5. List the three steps followed in developing design solutions.
6. What is the difference between divergent and convergent thinking?
7. The best solution is always the least expensive one. True or false?
8. Name one of the rules used for effective brainstorming.
9. Define the term *classification* as it is used in developing design solutions.
10. What is a *what-if scenario*?
11. Describe the differences among rough sketches, refined sketches, and detailed sketches.
12. List the three types of information required in detailed sketches when building models.
13. What are the primary differences among one-point, two-point, and three-point perspective sketches?
14. The appearance of a perspective drawing changes as the horizon changes. True or false?

Answers to Test Your Knowledge Questions

1. System design problems and opportunities. Product design problems and opportunities.
2. system design
3. manufactured, constructed
4. design team
5. Developing preliminary solutions. Isolating and refining the best solution. Detailing the best solution.
6. Divergent thinking is used to think of as many solutions as possible. Convergent thinking is used to narrow and focus the ideas until the most feasible solution is found.
7. False
8. Student answers may include any of the following: Encourage wild, far-out ideas. Record the ideas without reacting to them. Seek quantity, not quality. Keep up a rapid pace.
9. Classification involves dividing the problem into increasingly specific categories.
10. A what-if scenario is a wild hypothetical situation that can be used to help develop solutions.

Matching questions: For Questions 15 through 23, match each definition on the left with the correct type of pictorial sketch on the right. (Note: Answers can be used more than once.)

Definition

15. _____ Used when the top, sides, and front are equally important.
16. _____ Shows the object as the human eye sees it.
17. _____ Sides extend back at 45°.
18. _____ The angles that the lines in the upper-right corner form are equal.
19. _____ Shows the front view as if you are looking at it.
20. _____ Most difficult of three sketches to develop.
21. _____ Parallel lines meet at distant vantage point.
22. _____ Always has one corner made up of three 120° angles.
23. _____ The two types are cavalier and cabinet.

Pictorial Sketch

A. Oblique.

B. Isometric.

C. Perspective.

STEM Applications

1. Develop a set of rough sketches for the following design definition:
   Problem or opportunity: The director of the school cafeteria would like a holder containing a saltshaker, a pepper shaker, 20 rectangular (1″ × 1 1/2″) packages of sugar, and a bottle of ketchup. The holder should be easily removable from the table at the end of the lunch period.
2. Refine the best sketch you produce for the lunchroom-table organizer.
3. Develop a detailed sketch for your lunchroom-table organizer.
4. Select a device in the technology laboratory and develop a perspective sketch of it.

11. Rough sketches are sketches of designs that are incomplete and unrefined. Refined sketches are sketches of designs that have been improved through study, contemplation, or the merging of ideas. Detailed drawings communicate the information needed to build a model of the product or structure.
12. Size information, location information, and geometry information
13. The number of vanishing points used. A one-point perspective shows an object as if you were directly in front of it. A two-point perspective shows how an object would appear if you stood at one corner. A three-point perspective shows how the eye would see the length, width, and height of an object.
14. True
15. B. Isometric
16. C. Perspective
17. A. Oblique
18. B. Isometric
19. A. Oblique
20. C. Perspective
21. C. Perspective
22. B. Isometric
23. A. Oblique

Chapter Outline

This chapter covers the benchmark topics for the following Standards for Technological Literacy:

Learning Objectives

After studying this chapter, you will be able to do the following:

- Recall the definition of *modeling.*
- Explain how models are used in the design of technological artifacts and systems.
- Recall the three major types of traditional models.
- Summarize how computer models can be used with or in place of traditional models.
- Compare types of models.
- Summarize the types of analyses used to evaluate designs.
- Understand the role of redesigning.

Key Terms

analysis	graph	prototype
chart	graphic model	simulation
computer model	mathematical model	solid model
conceptual model	mock-up	surface model
diagram	model	wire-frame model
ergonomics	physical model	

Strategic Reading

Before you read this chapter, make a list of as many different types of models as you can think of. Check your list as you read the chapter, and identify models you were unfamiliar with when you made your list.

We use the products of technology every day. Each product started with a problem or an opportunity that could be defined. Designers were then challenged to solve this problem or meet the opportunity. They explored various solutions that started in their minds and were recorded on paper. The designers developed rough, refined, and detailed sketches containing design ideas, not product plans. At some point, the designers changed two-dimensional drawings into three-dimensional models that could be evaluated. See **Figure 11-1.** This conversion from drawings to models is the focus of this chapter. We discuss the three basic activities of this process:

- Modeling design solutions.
- Analyzing the design.
- Redesigning the product or structure.

Modeling Design Solutions

Everyone is familiar with models. Children play with dolls, model cars, and toy trains. They build towns out of wooden blocks, convert cardboard boxes into frontier forts, and have grand prix races with their tricycles. These children pretend they are dealing with real-life situations.

Likewise, museums use models to show how things worked in the past. See **Figure 11-2.** These models show a slice of historical life so we can better understand the present day.

Figure 11-2. This museum model shows how yarn was spun from wool in the Colonial period.

Ford

Figure 11-1. The design process changes product sketches into models that can be evaluated.

Standards for Technological Literacy

2 8 9

Reinforce

Review modeling and evaluating design solutions as part of the design process presented in Chapters 5 and 9.

Demonstrate

Show your students some typical everyday models (model airplanes, model cars, dolls, etc.).

Brainstorm

Ask your students why models are used in museums, schools, and libraries.

TechnoFact

Models can be used in many ways. Often, historic models tell us about past cultures. For example, models of terra-cotta houses that were buried in tombs in Japan tell researchers a great deal about prehistoric Japanese architecture.

TechnoFact

Models or simulations allow people to train to operate sophisticated equipment or machinery without endangering people or the equipment. One example is a flight simulator. This device simulates the experience of actual flight, including the use of cockpit instrumentation, without the risk. The simulator is tilted to simulate the roll, pitch, and yaw motions of an aircraft. The simulator software provides realistic video representations of various flight environments.

Standards for Technological Literacy

2 8 9

Extend

Discuss the relationship between sketches and models.

Research

Have your students investigate the way models are used to study aircraft and automobile aerodynamics.

TechnoFact

Many electronic games recreate certain real-world situations to challenge a person's reasoning or hand-eye coordination. These programs are often called simulations.

This activity of imitating reality is widely used in product and system design. People simulate expected conditions to test their design ideas in a process called *modeling*, or ***simulation***. Simply stated, ***modeling*** is the activity of simulating actual events, structures, or conditions. For example, architects might build a model of a building to show clients how it will look. They might use structural models to test a building's ability to withstand an earthquake and the forces of the wind. Economists might devise a model to predict how the economy will react to certain conditions. Weather forecasters use models to show the public the locations and movements of storms. They also use sophisticated ***computer models*** to predict the intensities and movements of hurricanes. See **Figure 11-3.**

A model allows us to reduce complex mechanisms and events into an easily understood form. Models allow us to focus on important parts of the total problem. This focus permits us to build understanding one part at a time. For example, an automobile is a very complex artifact. An automobile is almost impossible to study and understand as a whole. If you look at and understand the systems that make up the automobile one at a time, however, the whole becomes clear. For example, you can model and study its power train, cooling system, electrical system, lubricating system, or suspension system. See **Figure 11-4.**

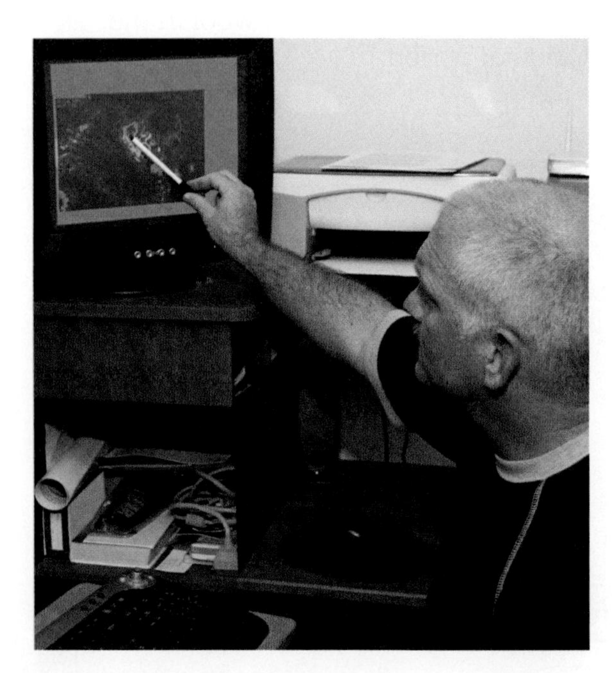

Figure 11-3. This forecaster is viewing a computer model of a storm.

Literally thousands of models are used each day. Each of them starts as a conception, an idea in the human mind. This idea is often communicated through verbal descriptions. These written or oral descriptions provide the foundations for the development of models.

Traditional models can be grouped into three types. See **Figure 11-5:**

- Graphic models.
- Mathematical models.
- Physical models.

The advent of the computer has provided new modeling techniques that can replace one or more of the traditional models. These new computer models make modeling much easier. In many cases, designers no longer need ***physical models***. They can develop and analyze the structure or product with computer modeling and simulation. This aspect of modeling is discussed in more detail after the three traditional types of models are described.

DaimlerChrysler

Figure 11-4. This graphic model shows an automotive torsion bar rear suspension system.

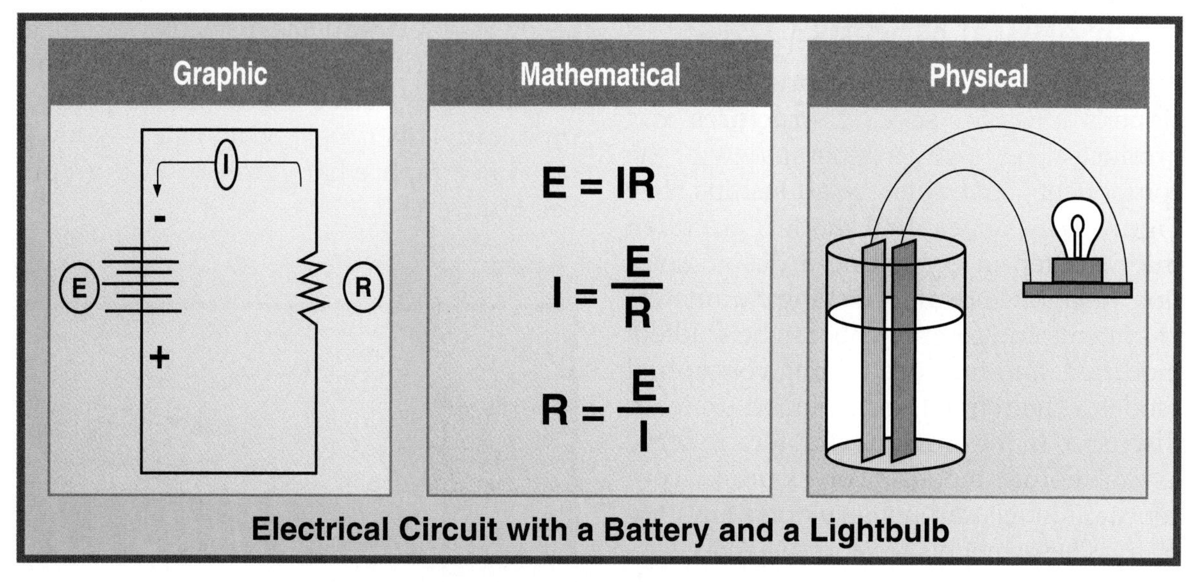

Figure 11-5. Models can be graphic, mathematical, or physical.

Standards for Technological Literacy

2 8 9

Example

Give additional examples of graphic, mathematical, and physical models to supplement those given in Figure 11-5.

Demonstrate

Show your students how to graph data, such as room temperature over time.

Graphic Models

Designers cannot make physical models early in the product- and structure-development process. They do not have enough information about the design to construct a physical model. The designers must explore ideas for components and systems, however. One way they do this is by creating ***graphic models***. Typical graphic models are conceptual models, graphs, charts, and diagrams. Each of these graphic models serves a specific purpose.

Career Corner

Market Researchers

Market researchers gather and analyze information about the potential sales of products and services. They use past sales data, information about competitors, and market trends to predict the market success of products. Market researchers develop methods for obtaining the data they need. They evaluate information they gather and make recommendations based on their findings.

People who hold bachelor's degrees in marketing and related fields might qualify as applicants for many entry-level positions. A master's degree is required for many private-sector market research positions. The applicants might have degrees in business administration, marketing, statistics, or communications.

Standards for Technological Literacy

2 8 9

Extend

Show your students examples of typical charts.

TechnoFact

According to Steve Brown, author of *Manufacturing the Future: Strategic Resonance for Enlightened Manufacture*, the dividing line between success and failure today is measured in months, not decades. The telephone book took almost 40 years to reach ten million customers in the United States. Yet, the Web browser reached ten million customers in 18 months.

Conceptual Models

Conceptual models capture the designer's ideas for specific structures and products. They show a general view of the components and their relationships. See **Figure 11-6.** Conceptual models are often the first step in evaluating a design solution. Relationships and working parameters of systems and components can be studied, modified, and improved using conceptual models. The refined and detailed sketches discussed in the previous chapter can serve as conceptual models. For example, conceptual models can be developed for a toy train. These models explore ways to connect the cars together, fabricate wheel-and-axle assemblies, and attach the car bodies to chassis assemblies.

Graphs

Graphs allow designers to organize and plot data. They display numerical information that can be used to design products and assess testing results. For example, a graph can be developed showing vehicle speed and braking distance for different types of brakes. The data can be charted on a line graph. This information helps designers select the type of braking system to be used in a specific vehicle. Likewise, plotting data on the colors of shirts purchased during a specific period can help designers select colors for next year's products. This type of data can be shown on a line graph, a bar graph, or a pie graph. See **Figure 11-7.**

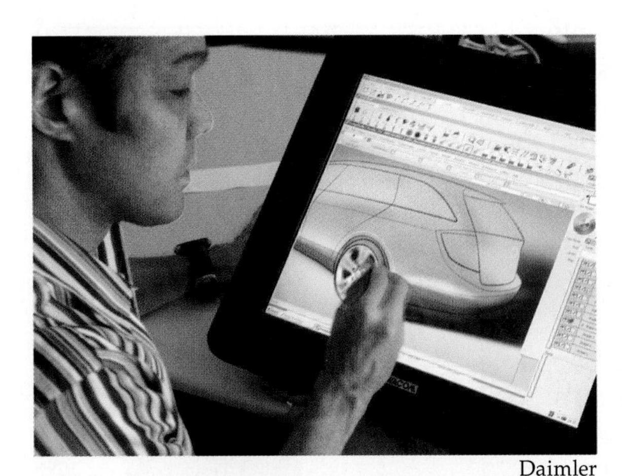

Daimler

Figure 11-6. This conceptual model shows a designer's ideas for an automotive system.

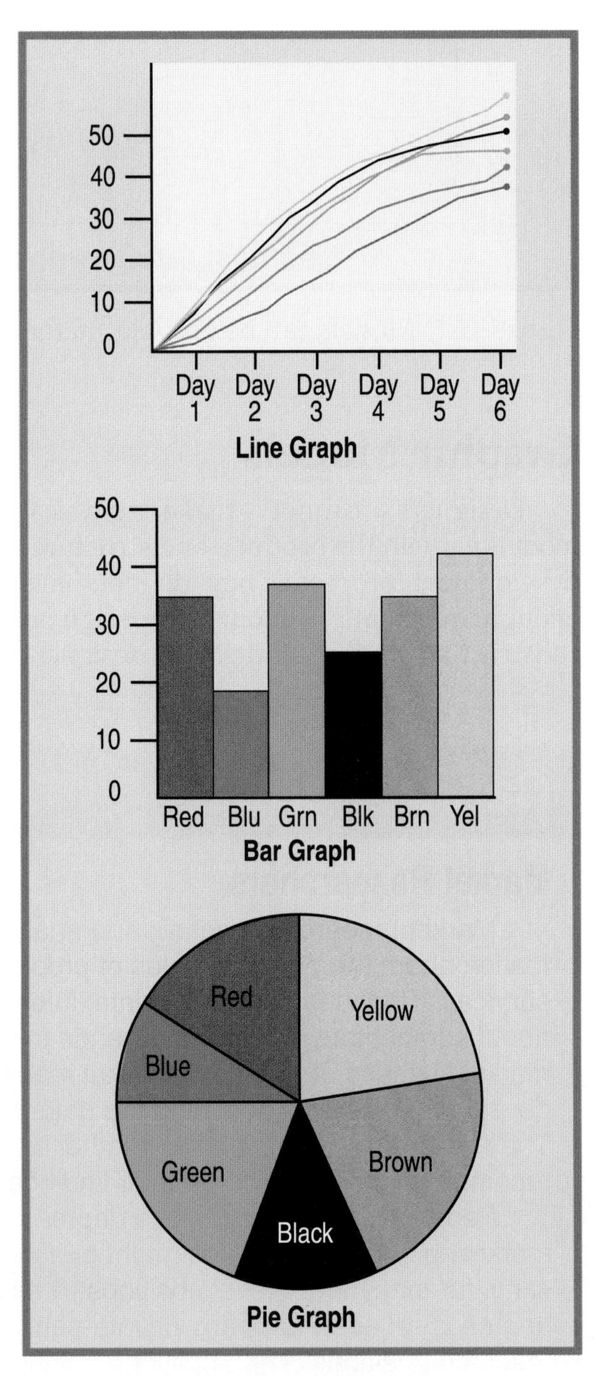

Figure 11-7. We use graphs to show the relationship among numerical data gathered about specific factors.

Charts

Charts show the relationship among people, actions, or operations. They are useful in selecting and sequencing tasks needed to complete a job. Various charts are used for specific tasks. Flow-process charts help computer programmers write logical programs. Flowcharts help the manufacturing engineer develop efficient manufacturing plants. Organization charts show the flow of authority and responsibility within a company, school, or business. See **Figure 11-8.**

Diagrams

Diagrams show the relationship among components in a system. A schematic diagram can be used to indicate the components in electrical, mechanical, or fluidic (hydraulic or pneumatic) systems. See **Figure 11-9.** Schematics do not show the specific locations of the parts. The relationship of the parts in the system is the important information communicated. Flow diagrams show how parts move through a manufacturing facility. Lines and arrows indicate the path the material takes as it moves from operation to operation.

Another type of diagram with which many people are familiar is a play diagram. For example, football coaches develop them to show how the players (components) should interact during an offensive play

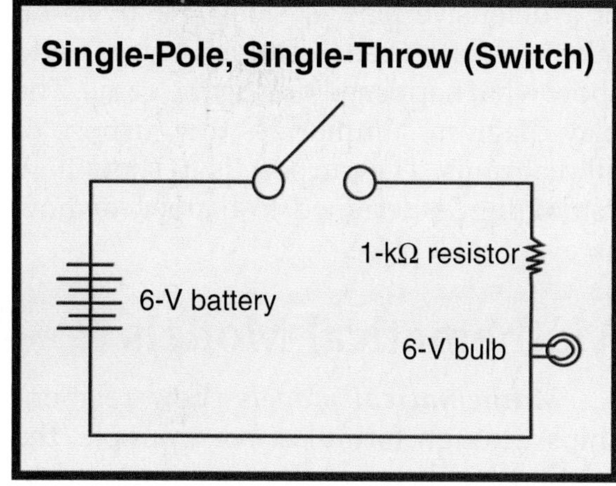

Figure 11-9. This schematic diagram shows the relationship among components in an electrical system.

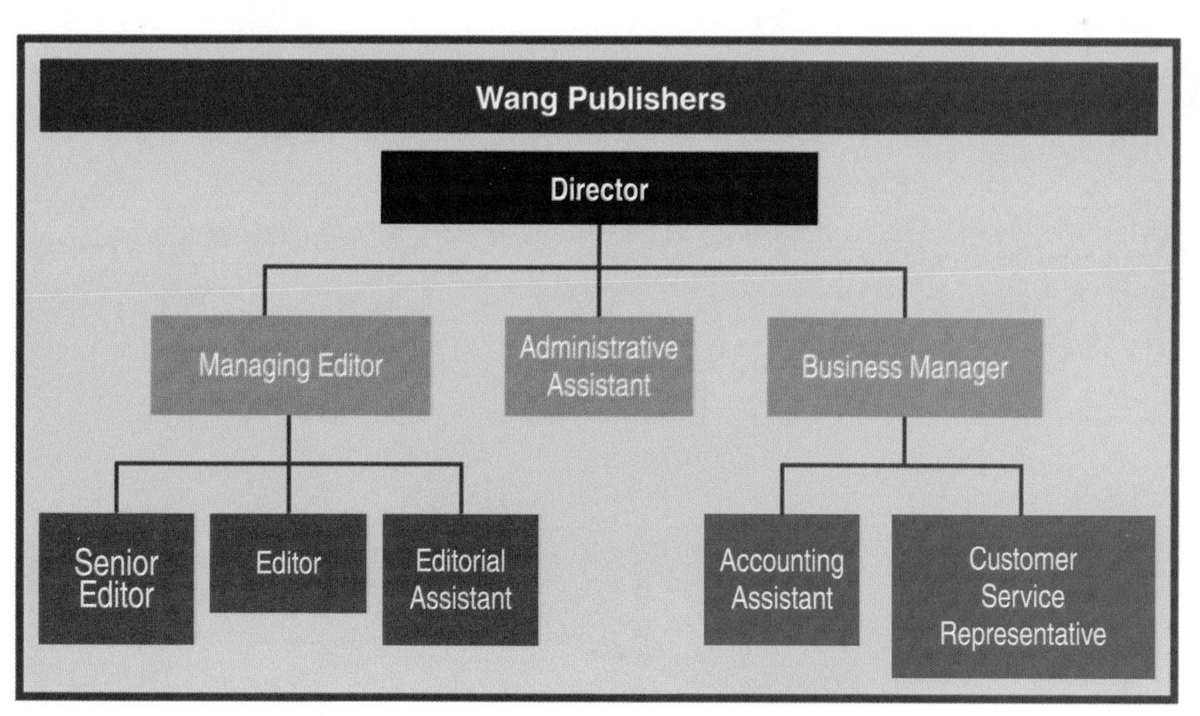

Figure 11-8. A chart can be used to show the structure of an organization.

Standards for Technological Literacy

2 8 9

Extend

Show examples of schematic diagrams from electrical or electronic devices, braking systems, etc.

Demonstrate

Show your students how to use Ohm's law to determine voltage, current, and resistance in an electrical circuit.

Demonstrate

Demonstrate the development of an organizational chart for the school.

TechnoFact

Program trading is one example of using models for economic ends. Program trading is a form of securities trading in which the differences between stock indexes and futures are tracked by computer models. Advances in computer and telecommunication technologies made program trading possible.

Standards for Technological Literacy

2 8 9

Brainstorm

Ask your students where they have seen physical models used.

TechnoFact

Theories can be tested with models. For example, Leonardo da Vinci had the idea of pulling himself into the air with a vertically mounted propeller. The machine he imagined is considered the forerunner of the helicopter. However, the idea was mostly dismissed until 1784, when a bowstring-driven model of the machine was developed in France. Helicopter models were built throughout 19th century Europe. Igor Sikorsky built the first real helicopter in 1939.

or a defensive play (system). Analysts on television often use this type of diagram to show what happened during the game. The play diagram summarizes the purpose of all diagrams. The purpose is to show how something is designed to happen or how an event took place.

Mathematical Models

Mathematical models show relationships through formulas. For example, the relationship between the force needed to move an object and the distance the object is lifted is shown in the formula for work. This formula is Work = Force × Distance. The formulas used to explain chemical reactions are also mathematical models. The following formula shows the chemical reactions that take place as an automotive lead-acid storage battery is charged and discharged:

$$PbO_2 + Pb + 2H_2SO_4 \underset{\text{charge}}{\overset{\text{discharge}}{\rightleftarrows}} 2PbSO_4 + 2H_2O$$

The examples given so far are for simple mathematical models. Individuals use more complex models with thousands of formulas to predict the results of complex relationships. For example, these formulas can be part of economic models that predict the economic growth for a period of time. Also, complex mathematical models track storms and spaceflights, predict ocean currents and land erosion, and help scientists conduct complex experiments.

Physical Models

Physical models are three-dimensional representations of reality. They are made of materials providing form and shape. The designer and other interested people can see and handle these models. There are two major types of physical models: mock-ups and prototypes.

Mock-Ups

A physical model showing people how a product or structure will look is called an *appearance model*, or ***mock-up***. This model is used to evaluate the styling, the balance, the color, or another aesthetic feature of a technological artifact. See **Figure 11-10.**

Figure 11-10. This mock-up shows the relationship among the different buildings.

Standards for Technological Literacy

2 3

Technology Explained

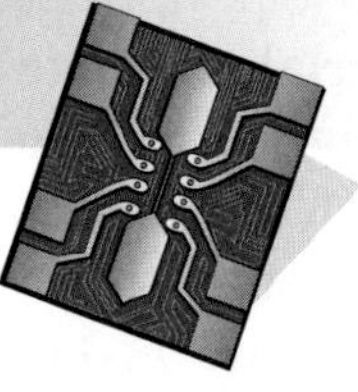

GPS: a system that relies on satellites to provide accurate time and location information to receivers on Earth.

GPS is an acronym for "global positioning system." The satellite system used for GPS was developed in the early 1970s by the United States Department of Defense. The satellite system is now operated by the United States Air Force. While GPS was originally developed for military use, it has now become a common civilian tool. Several people now have GPS receivers available to them, either as separate units or as a function on their cell phones.

The satellites send signals down to the receivers on Earth. See **Figure A.** There are typically 24 satellites in operation. These satellites are able to provide accurate time and location information. At least four of these satellites are used to determine three-dimensional positions. The satellites then send a one-way signal back to Earth with location and time information.

The satellites work by knowing their own location information. Their locations, relative to one another, determine the location of the GPS receiver. The satellites also must be able to determine the distance between themselves and the receiver to give exact location information. The satellites beam down their information using radio signals.

One common use of GPS receivers is navigation. See **Figure B.** People can use these receivers to instantly find directions from one location to another. Most receivers have the option of using an audio narration of the directions on the way to the new location. Some receivers are now able to inform the user of upcoming traffic congestion, accidents, and construction.

Location radar uses a narrow, flash-light-type beam. The beam is focused on an object so accurate elevation, distance, and speed data can be obtained. This type of radar has many applications. Most people are familiar with its use by police officers to enforce speed limits.

Applications for GPS are always changing and improving. GPS receivers are improving the efficiency with which traffic incidents are reported. Other uses for GPS receivers include collecting environmental and agricultural data and air and water navigation.

Figure A. A GPS satellite.

Figure B. A typical GPS receiver used for navigation.

Standards for Technological Literacy

2 8 9

Research

Have your students research the way models are used to test vehicle safety.

TechnoFact

Models can lead to new discoveries. In 1763, an instrument-maker at the University of Glasgow named James Watt was asked to repair a demonstration model of the Newcomen steam engine. While he worked with the model, he observed that, during each cycle, the engine wasted steam because a good portion of it condensed on the walls. He designed an engine with a separate condenser, which allowed the cylinder wall to remain relatively hot and the condenser to remain relatively cold. The new engine, known as the Watt steam engine, was more efficient than the Newcomen engine.

Mock-ups are generally constructed of materials that are easy to work with. Commonly, these materials include wood, clay, Styrofoam® product, paper, and paperboard (cardboard and poster board, for example). See **Figure 11-11.**

Prototypes

A ***prototype*** is a working model of a system, an assembly, or a product. Prototypes are built to test the operation, maintenance, or safety of the item. See **Figure 11-12.** They are generally built of the same material as the final product. In some cases, however, substitute materials are used. For example, some automobile manufacturers have found that a specific plastic reacts to external forces in the same manner as steel does. Plastic prototypes are used in place of steel ones because they are easier to fabricate.

Both types of physical models can be built full-size or to scale. Full-size models are needed to test the product's operation. For example, a full-size model is needed for people to evaluate the comfort of a new bus seat.

In other cases, full-size models are impractical. Building a full-size model of a new skyscraper is a waste of money. A scale model is used when the product or structure is too large to construct in full size just for a test. A scale model is proportional to actual size. This means the model's size is related to actual size by a ratio. A ratio of 4 to 1 (written as "4:1") means four units in actual size are equal to one unit on the model. A scale model of a new building, for example, is used to show a client how the structure will look, how it will fit on the site, and how it will be landscaped.

Computer Models

As we noted earlier, computers and computer models have affected the way we

Hasbro Inc.

Figure 11-11. These designers are developing mock-ups of new toys.

DaimlerChrysler

Figure 11-12. This prototype is being used to test the aerodynamics of an aircraft.

work with models. Computers can be used to develop three types of three-dimensional models: wire frame, surface, and solid. Two of these are shown in **Figure 11-13.**

Wire-Frame Models

The top view of **Figure 11-13** is a wire-frame model. A ***wire-frame model*** is developed by connecting all the edges of the object. The process produces a structure made up of straight and curved lines.

Surface Models

The bottom view of **Figure 11-13** is a surface model. This model can be thought of as a wire frame with a sheet of plastic drawn over it. ***Surface models*** show how the product will appear to an observer. They can be colored to test the effects of color on the product's appearance and acceptance.

Surface models are widely used in developing sheet metal products. The surface model is first developed. The computer then unfolds the model to produce a cutting pattern for the metal.

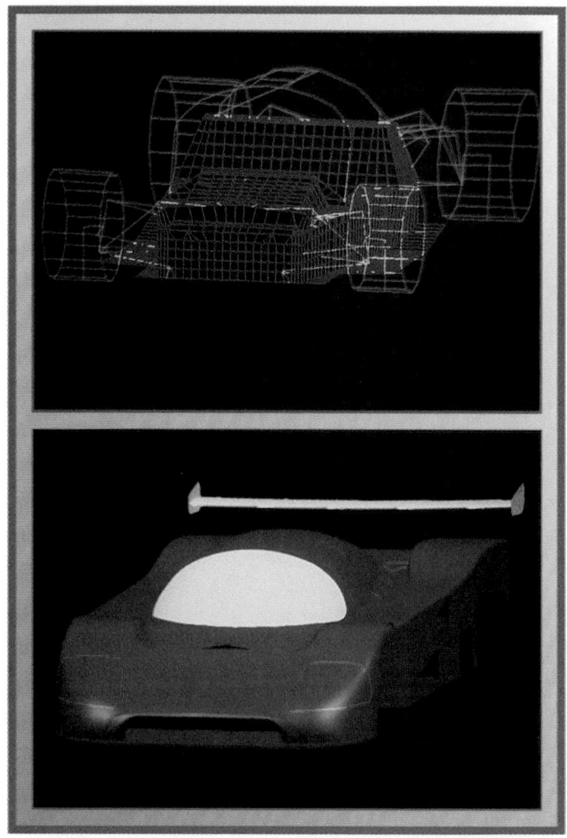

DaimlerChrysler

Figure 11-13. Two types of computer models are the wire-frame and surface models. They are used here in the design of a race car.

Solid Models

A ***solid model*** is the most complex of these models. Solid models look similar to surface models, except the computer "thinks" of them as solids. This allows the designer to direct the computer to cut away parts, insert bolts and valves, and rotate moving parts, for example. Solid models can also be used to establish fits for mating parts and to set up the procedures needed to assemble parts into products.

Other Computer Models

Other computer simulations allow designers to test strengths of materials and structures. See **Figure 11-14.** The designers create the structure on the computer and then apply stress. They might simulate

Standards for Technological Literacy

2 8 9

Demonstrate

Show your students examples of wire-frame, surface, and solid models.

Extend

Discuss how solid modeling can be used to test manufacturing processes.

TechnoFact

Today, meteorologists make extensive use of mathematical models to forecast the weather. L. F. Richardson was a pioneer of this technique. In 1922, he developed the basis for the mathematical prediction of the atmospheric circulation. C.G. Rossby contributed to the effort in 1938. However, the high-speed computer would have to be developed before their work would become practical. Computer forecast models became a reality in the mid-1950s and have been improving steadily ever since.

Quotes

"The whole process of conceiving an idea, designing it, looking at the practicalities of how to make it, what it's going to cost, who's going to buy it and how you get the commitment to buy is a continuum of thought that starts with a creative spark."—Lord Puttnam of England

Standards for Technological Literacy

2 8 9

Brainstorm

Ask your students to determine ways they could tell if a bicycle was functioning properly.

Quotes

"The most successful businesses appreciate the contribution design can make to their strategy, not just to the look and feel of their new product or the layout of their brochure. They know design can help confront previously unasked questions about what they do, how they do it and who they do it for."—Andrew Summers, Chief Executive of the Design Council.

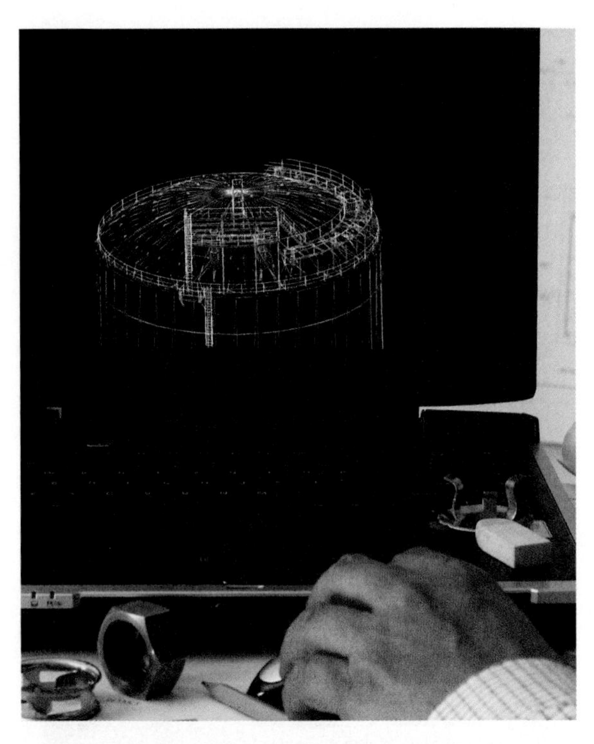

Figure 11-14. This computer program allows the designer to test a material's strength.

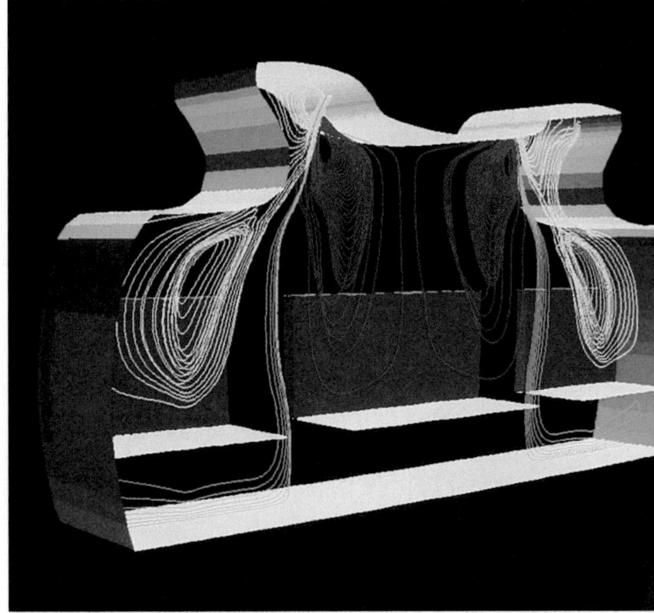

Boeing

Figure 11-15. This computer program shows a simulation of the air circulation in the interior of an aircraft. Note the seats (in purple) and the shape of the overhead bins.

forces created by weight, operating conditions, or outside conditions such as wind or earthquakes.

Finally, designers might use computer models to observe the product during normal operation. For example, computer programs simulate the flow of metals and plastics into molds. Other programs, such as the one shown in **Figure 11-15,** test air circulation in buildings and vehicles. This information is essential for the designers of heating and cooling systems.

Analyzing the Design

Modeling helps designers and decision makers enter an important design activity—design ***analysis***. This stage requires people to evaluate a design carefully in terms of such criteria as the design's purpose and likely acceptance in the marketplace. The evaluation helps designers and decision makers to choose the best design. See **Figure 11-16.** The five types of design analysis usually carried out are the following:

- Functional analysis.
- Specification analysis.
- Human-factors analysis.
- Market analysis.
- Economic analysis.

Functional Analysis

As we noted earlier, every product, structure, and technological system is designed to meet human needs or wants. Functional analysis evaluates the degree to which the product meets its goal. This analysis answers the basic question, "Will the artifact operate effectively under the conditions for which it was designed?" These conditions might relate to the outside environment (weather, terrain, or water, for example), operating conditions (stress, heat, or gases produced during use,

Ford

Figure 11-16. Product designs must be analyzed in five different areas.

Standards for Technological Literacy

8 9

Research

Have your students research ergonomics. Have them define ergonomics and determine ways that ergometric information is obtained and used in product design.

Quotes

"Design gives innovation real value and...design without innovation quickly becomes styling. You can have one or the other, but you should have both."—Clive Grinyer, Design Council.

for example), human use and abuse, and normal wear (material fatigue or part distortion, for example).

Specification Analysis

Every product must meet certain specifications. These specifications might be given in terms of size, weight, speed, accuracy, strength, or a number of other factors. The specifications for all new products and structures must be analyzed. They must be adequate for the function of the product. Excessive specifications, however, add to the cost of the product. Holding bicycle-handlebar diameters to 1/1000" (.025 mm) is foolish. That level of precision is not needed for the part to function. Tolerances (the amount a dimension can vary and still be acceptable) must be close for spark plug threads, however. The goal is to produce an economical, efficient, and durable product that will operate properly and safely.

Specifications must also relate to the material and manufacturing processes to be used. For example, holding wood parts to tolerances smaller than 1/64" (.397 mm) is impossible. Normal expansion and contraction due to changes in humidity can cause this much change. Likewise, specifying aluminum for the internal parts of a jet engine is unwise. The temperatures inside the engine will melt aluminum parts. Similarly, specifying green sand casting as a process to produce precision parts is a mistake. Green sand casting produces low-cost parts. These parts, however, cannot be held to close tolerances.

Human-Factors Analysis

To meet human needs, we design artifacts and structures for the people who will use them, travel in them, or live and work in them. Designing products and structures around the people who use them is the focus of human-factors analysis, more commonly known as ***ergonomics***. This science considers the size and movement of the human body; mental attitudes and abilities; and senses such as hearing, sight, taste, and touch.

Standards for Technological Literacy

8 9

Demonstrate

Show your students how to retrieve information from a book of ergonometric data.

TechnoFact

Because of the merging of the new economy and environmentalism, the future may reveal changing patterns of consumption. In turn, heavy emphasis could be placed on the ability to produce more goods with less energy.

Ergonomics also considers the type of surroundings that are the most pleasing and help people to become more productive. A good example of matching the environment to humans is an aircraft flight deck. See **Figure 11-17.** All the controls are within easy reach. Dials and indicator lights are within the pilot's field of vision. Windows are located so the pilots have a clear view of the sky ahead and above them.

Market Analysis

Most products of technology are sold to customers. These customers might be the general public, government agencies, or businesses. During design activities, the market for the product must be studied. The designs must then be analyzed in terms of that market. Market analysis includes finding customer expectations for the product's appearance, function, and cost. This analysis also includes studying present and anticipated competition. Market analysis is often conducted by taking surveys of potential customers and analyzing competing products available on the market.

Economic Analysis

As we said earlier in this book, private companies develop most technological artifacts and structures. They risk money to develop, produce, and market the items. In turn, they hope to make a profit as a reward for their risk taking. To increase their chances of success, individuals often conduct a financial analysis for the new products. The product is studied in terms of the costs of development, production, and marketing. These data are compared with expected sales income to determine the

Airbus Industries

Figure 11-17. This aircraft flight deck was designed with human movement and senses in mind.

financial wisdom for producing the product. Often, the product is judged on what is called its *return on investment (ROI)*. This figure indicates the percentage of return, based on the money invested in developing, producing, and selling the product. The higher the ROI is, the better the anticipated financial returns to the company will be.

Redesigning Products and Structures

The goal of all product- and structure-development activities is to design an artifact that will help people control or modify the environment. This goal requires the problem-solving and design process to be continual. Needs are identified and defined. Solutions are designed and modeled. The designs are analyzed. Flaws are identified. Redesign is then often required. Problems in the original design must be solved. New designs might be developed. The original designs might be altered. New models are built and evaluated until an acceptable product or structure emerges. See **Figure 11-18.**

This is not the end of the process. Products have a life expectancy. New technologies and changes in people's needs and attitudes make some products obsolete. Old products need redesigning. New product definitions appear. Product and structure design is a never-ending process.

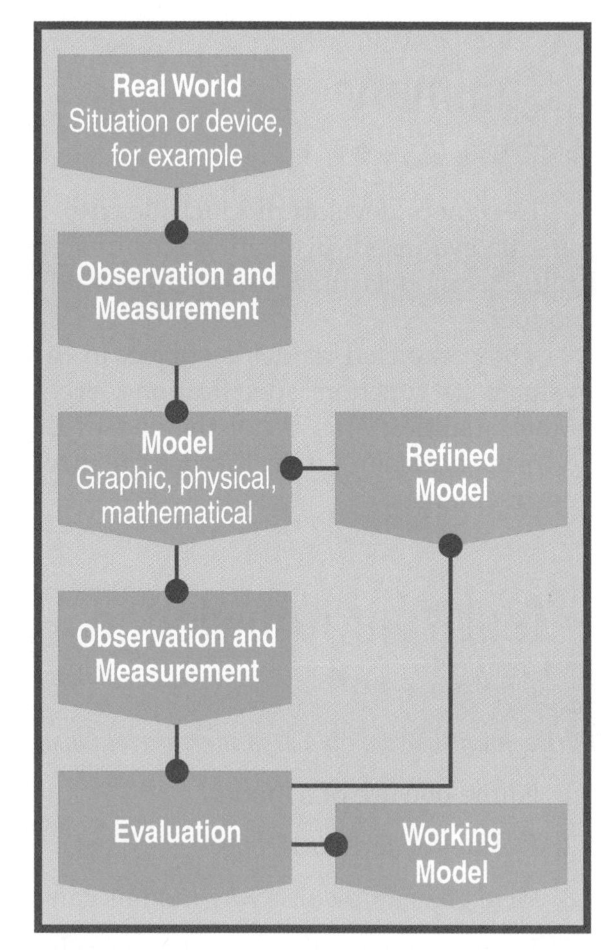

Figure 11-18. The role of a model in product design and redesign.

Standards for Technological Literacy

Figure Discussion

Discuss the iterative nature of the model shown in Figure 11-18.

Brainstorm

Ask your students why they think it is economically wise to design some products to fail after a set period of time or number of uses. Provide the students with examples such as paper cups and sandpaper.

Quotes

"The body and technology will become one. We aim to reduce technology in size so much that it can be worn on the body. In the future, a dress will be more than just a dress. It will monitor health, aid communication, provide instant access to information and more."—Alex Lightman, CEO of Charmed Technology.

"Good design brings high added value to practically every activity, whether it be mechanical design as an integral part of product development or aesthetic design illustrating the format and function of a product or service." —Richard Turner, Group Marketing Director, Rolls Royce.

Answers to Test Your Knowledge Questions

1. Evaluate individually.
2. Modeling is the activity of simulating actual events, structures, or conditions.
3. Test a building's ability to withstand earthquakes (or wind).
4. A. M
   B. P
   C. P
   D. M
   E. M
   F. P
5. Graphic models, mathematical models, and physical models
6. Evaluate individually.
7. Prototype
8. Wire-frame model, surface model, and solid model
9. Functional
10. Human factors analysis (ergonomics)
11. False
12. True

Summary

Designers develop product ideas to meet human needs and wants. These designs must be evaluated. Evaluation requires graphic, mathematical, or physical models to be built. The models allow designers to study and refine specific mechanisms and the total product.

The completed model is subject to careful analysis. Designers and decision makers evaluate its function, specifications, ergonomic qualities, market acceptance, and economic qualities. This is not the end of the design process, however. These individuals address shortcomings in the design through redesign activities. Finally, an acceptable design emerges.

Test Your Knowledge

Write your answers on a separate piece of paper. Please do not write in this book.

1. Name one use of a model.
2. Define the term *modeling*.
3. Give an example of a way models can be used in building design.
4. Below is a list of things that need to be tested or conveyed. Place an *M* next to the item, if you would use a mock-up, or a *P* next to the item, if you would use a prototype, as the model to do the task.
   A. Show the appearance of a new civic auditorium. ______
   B. Test the comfort of a new bus seat. ______
   C. Test the safety of a new can opener. ______
   D. Communicate a landscape plan. ______
   E. Determine consumers' color preference for a new lamp. ______
   F. Determine the ease of use for a new hammer design. ______
5. List the three types of models.
6. Name one difference between information shown on a graph versus that shown on a chart.
7. A working model of a system, an assembly, or a product is called a(n) ______.
8. List the three types of three-dimensional models a computer can display.
9. Evaluating whether or not a product or system will operate effectively under the conditions for which it was designed is the goal of ______ analysis.

10. The analysis of how a new artifact or structure will affect humans is commonly called ______.
11. In market analysis, a product is often judged by its ROI. True or false?
12. Product redesign is a common activity. True or false?

STEM Applications

1. Develop a graphic model and build a physical model of a microwave communication tower. Assume the toothpick represents a 15′-long piece of steel. Build the tower 10″ square at the bottom and 69″ high.
2. Select a simple game on the market. Analyze it in terms of the following:
   - The game's specifications. Do this by developing a set of manufacturing specifications.
   - The game's function and market acceptance. Do this by playing the game with classmates.

   Write a brief report summarizing your analysis.

TSA Modular Activity

Standards for Technological Literacy

12

This activity develops the skills used in TSA's Computer-Aided Design (CAD) 3D, Engineering event.

Computer-Aided Design (CAD), Engineering with Animation

Activity Overview

In this activity, you will create an animation illustrating the assembly sequence for a product.

Materials

- Paper.
- A pencil.
- A computer with CAD and animation software.

Background Information

- **Product selection.** Use brainstorming techniques to develop a list of possible products to model in your animation. Part modeling is easier if you have a sample of the product you can use for measurements or if you have actual part drawings. Some possible items to animate include the following:
  - A piece of self-assemble furniture, such as a computer desk.
  - A ballpoint pen or mechanical pencil.
  - A small construction project, such as a doghouse or shed.
  - Sports equipment, such as a swing set, tennis racket, or weight bench.
  - A mechanical device, such as a wheel-and-axle assembly.
- **Part modeling.** After selecting the object, create models for each part.
- **Animation.** Your animation is intended to illustrate the assembly sequence. You can begin the illustration with all parts shown. On the other hand, you can have the parts appear as they are assembled. To help the viewer anticipate the action, highlight a part before it is moved into place in the assembly. You can highlight the part in several ways, such as changing its color, outlining it, momentarily enlarging the part, or momentarily stretching the part.

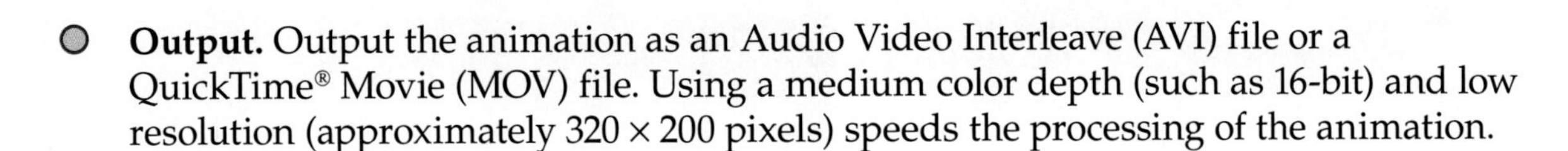

- **Output.** Output the animation as an Audio Video Interleave (AVI) file or a QuickTime® Movie (MOV) file. Using a medium color depth (such as 16-bit) and low resolution (approximately 320 × 200 pixels) speeds the processing of the animation.

Guidelines

- Your assembly must include at least five components. There must be at least five steps in the assembly procedure.
- Your animation should clearly show the assembly procedure.
- Your animation must be at least 10 seconds long.

Evaluation Criteria

Your project will be evaluated using the following criteria:

- Originality and creativity in design.
- Effectiveness of animation in illustrating the assembly sequence.
- Use of animation features.
- Technical animation skills.
- Smoothness and timing of animation.

Chapter Outline

This chapter covers the benchmark topics for the following Standards for Technological Literacy:

3 7 8 9 11 12

Learning Objectives

After studying this chapter, you will be able to do the following:

- Recall the three basic documents through which designers communicate product information.
- Recall the three types of engineering drawings.
- Summarize a CAD system.
- Recall the information found on a bill of materials.
- Recall the information found on a specification sheet.
- Recall the types of reports used to gain approval for designed products.

Key Terms

assembly drawing
bill of materials
centerline
computer-aided design (CAD) software
detail drawing
dimension line
engineering drawing
exploded view
extension line
geometry dimension
hidden line
location dimension
multiview method
object line
one-view drawing
orthographic assembly drawing
orthographic projection
pictorial assembly drawing
size dimension
specification sheet
standard view
systems drawing
technical data sheet
three-view drawing
tolerance
two-view drawing

Strategic Reading

As you read this chapter, make an outline of the main points. Be sure to include details of different types of drawings and specifications.

In the 1790s, Eli Whitney revolutionized the way products are made. He developed a system to mass-produce muskets for the army. The foundation of his new system was the concept of interchangeable parts. A part made for one gun fits all other guns of the same make and model.

Today, we take interchangeable parts for granted. If you break a part of a product, you expect to be able to buy a replacement. This ability requires a well-developed communication system between those who develop and engineer the product and those who make the parts and assemble the product. See **Figure 12-1.**

Communicating design solutions is a key element in technology. Those engaged in communicating design solutions are involved with two areas of communication. First, they create documents and reports specifying all the details of the product. Second, they prepare documents and reports designed to obtain approval for the solution from various decision makers. Each type of communication is discussed in turn as follows.

Product Documents and Reports

The workers who make the parts and assemble them into products must be well informed. They must have knowledge of manufacturing processes. These workers must be able to set up and operate machines, apply finishing materials, and perform assembly operations. Still, this is only part of the manufacturing knowledge they need. They must know the product. This means they must have knowledge about the materials to be used in the product's manufacture. The workers must also know the size and shape of each part. Finally, they must know how the product is assembled.

This knowledge of the product is delivered through three basic kinds of documents. See **Figure 12-2.** They are the following:

- Engineering drawings.
- Bills of materials.
- Specification sheets.

Standards for Technological Literacy

7 8 11

Research

Have your students trace the development of interchangeable parts.

Brainstorm

Ask your students why drawings are necessary for creating interchangeable parts and facilitating mass production.

Quotes

"The new economy has as its cornerstones knowledge, skills, innovation and enterprise. Its most valuable assets are knowledge and the creative skills of our people. The main source of value and competitive advantage in the modern economy is human and intellectual capital."—Stephen Byers, UK Secretary of State for Trade and Industry

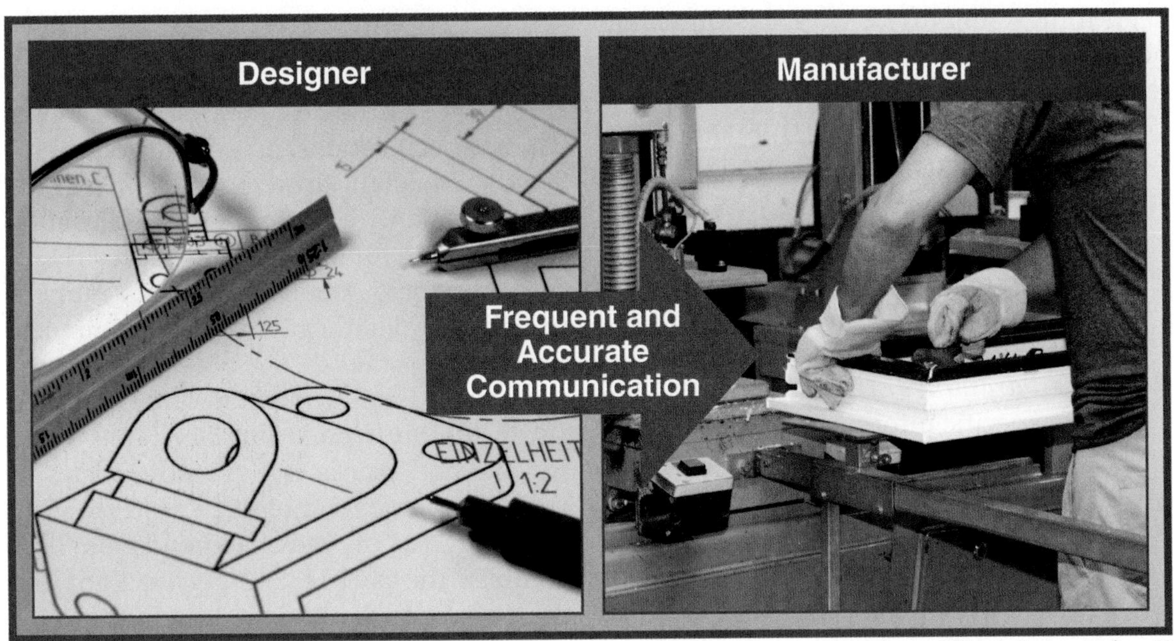

Figure 12-1. The design for the product must be communicated from the designers to the manufacturer.

Standards for Technological Literacy

8 11

Extend

Show examples of engineering drawings, bills of materials, and specification sheets.

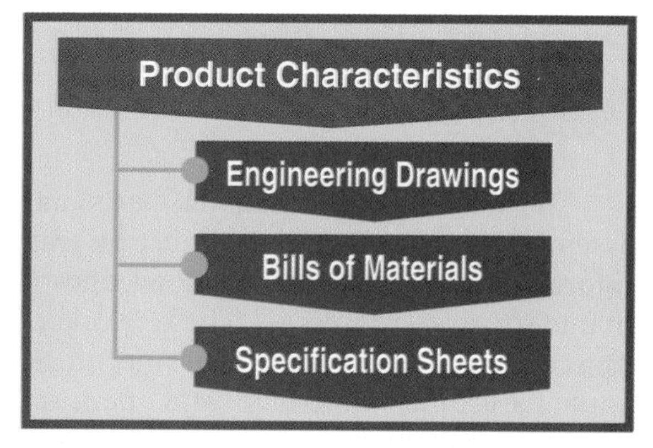

Figure 12-2. Designers use three types of documents to communicate product characteristics.

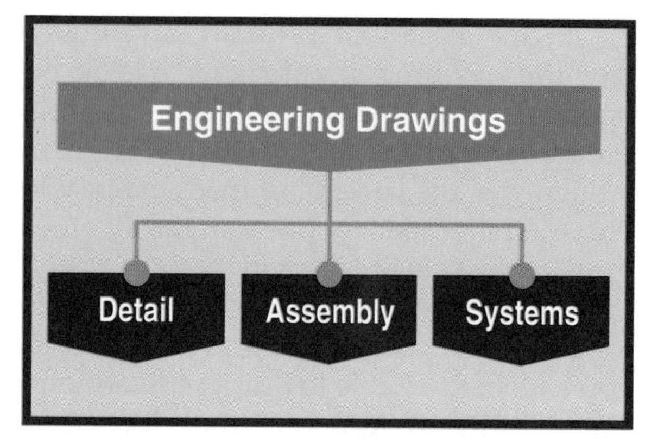

Figure 12-3. The three types of engineering drawings are detail, assembly, and systems.

Let us look at how each of these kinds of documents is developed and the information each one communicates.

Engineering Drawings

Engineering drawings communicate basic information needed to construct the product or structure. In manufacturing, they are called *engineering drawings* or *working drawings.* In construction, however, they are called *architectural drawings.* Here, to keep things simplified, we focus on engineering drawings. Keep in mind, however, that although architectural drawings are different from engineering drawings, the basic principles used to prepare both are similar.

Designers commonly use three types of engineering drawings to communicate product information. See **Figure 12-3.** These types are the following:

- ***Detail drawings.*** These drawings show specific information needed to produce a part.
- ***Assembly drawings.*** These drawings show how parts go together to make a subassembly or product.
- ***Systems drawings.*** These drawings show the relationship among electrical, hydraulic, or pneumatic components.

Before we go further here, however, we need to note the following. When we think about the term *drawing,* we might visualize someone with pencil in hand, illustrating the product on sheets of paper. Much drawing today, however, is accomplished with the use of computers. We explore the role of computers in this area after we discuss the three basic types of drawing documents.

Detail Drawings

Most products are made up of several parts. Each of these parts must be manufactured to meet the designer's specifications. These specifications are often communicated on detail drawings. Detail drawings commonly contain all the information needed to manufacture one part. Therefore, designers usually generate a number of different detail drawings for a complete product.

Most detail drawings are prepared using the ***multiview method.*** This drawing method places one or more views of the object in one drawing. See **Figure 12-4.** The number of views depends on how complex the part is. See **Figure 12-5.** The most common multiview drawings are the following:

- ***One-view drawings.*** These drawings are used to show the layout of flat,

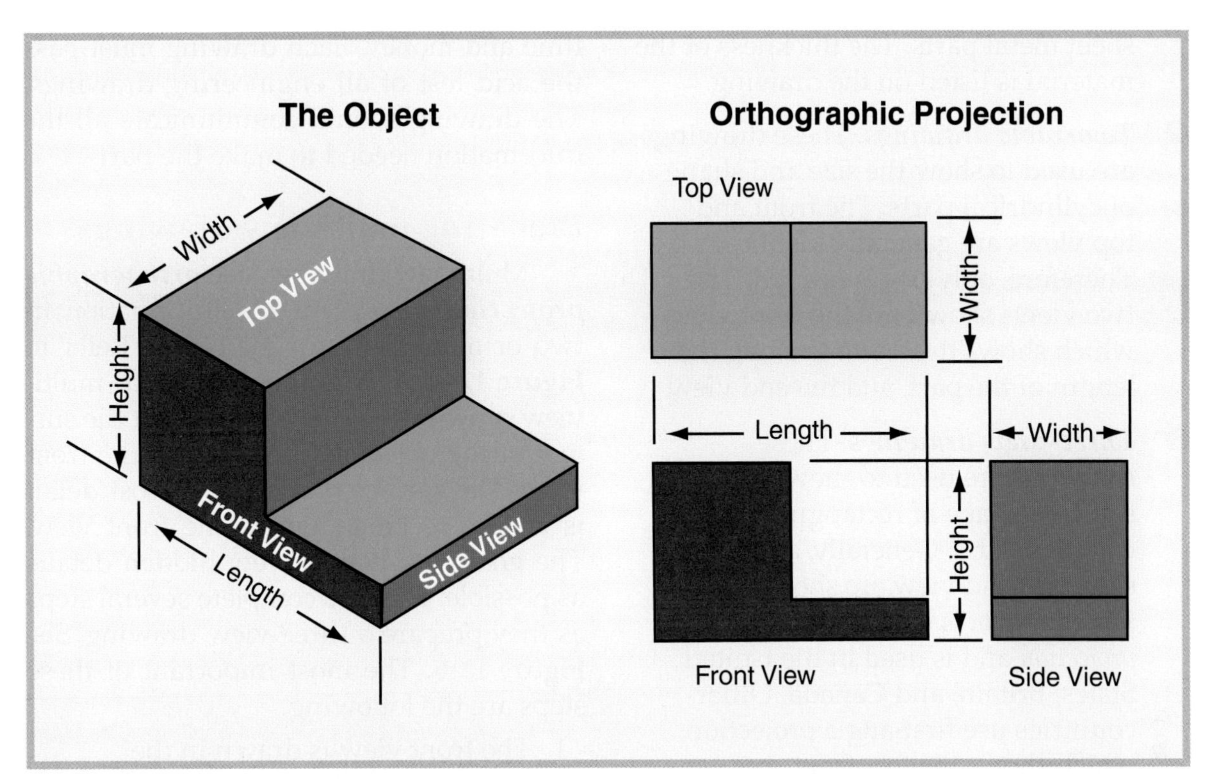

Figure 12-4. The orthographic projection shown on the right represents the object shown on the left.

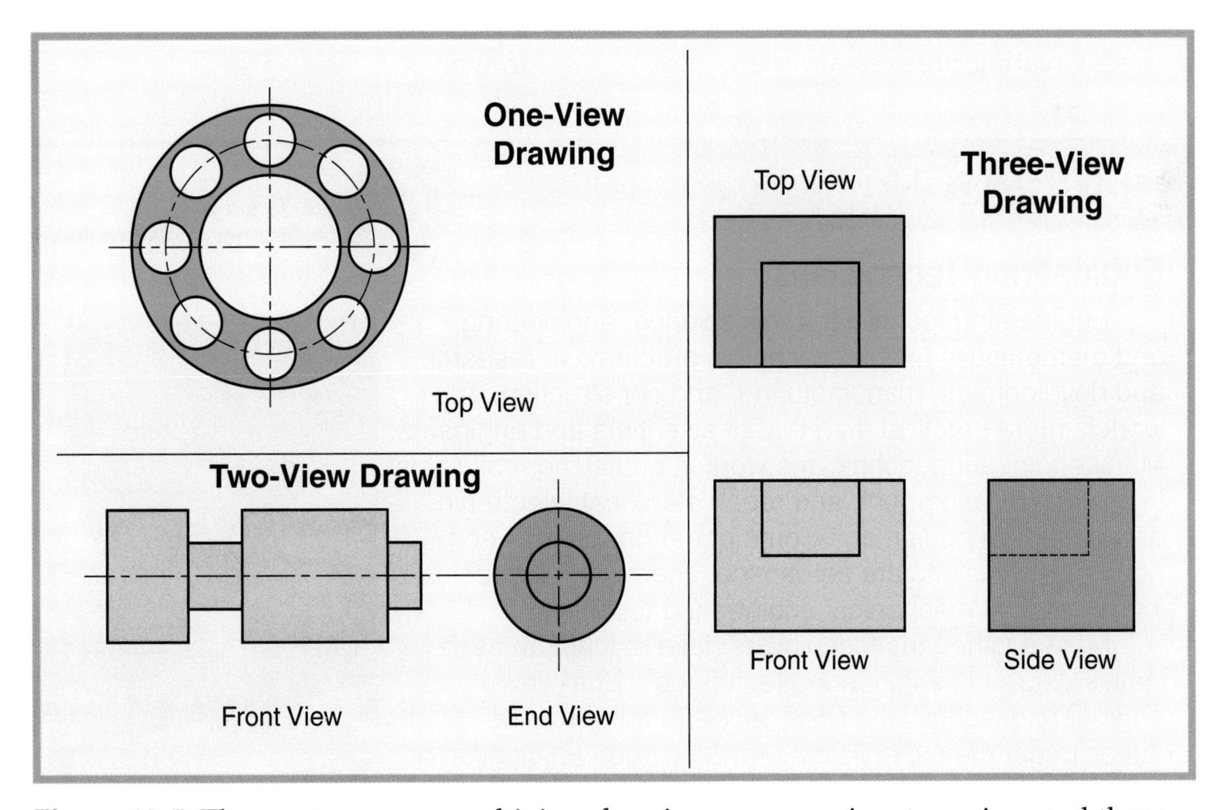

Figure 12-5. The most common multiview drawings are one-view, two-view, and three-view drawings.

Standards for Technological Literacy

8 11

TechnoFact

Engineering drawings and design sketches both communicate the appearance of an object. However, drawings are accurate and usually to scale. Drawings should provide clear information about sizes (dimensions) and angles. Drawings also frequently provide details about the surface finishes and production instructions. Sketches generally only convey generalized information about the appearance of the object.

Standards for Technological Literacy

8 11

Extend

Discuss why one-view, two-view, and three-view drawings are used.

TechnoFact

Engineering drawings generally contain a wealth of information. They describe the size and shape of parts and locate important features. They may also describe the surface finish for the part, the type of heat treatment used, and the specific machine settings required to produce the desired results.

sheet metal parts. The thickness of the material is listed on the drawing.

- ***Two-view drawings.*** These drawings are used to show the size and shape of cylindrical parts. The front and top views are generally identical. Therefore, only one is needed. The two views shown are the front view, which shows the features along the length of the part, and the end view.
- ***Three-view drawings.*** These drawings are used to show the size and shape of rectangular and complex parts. Generally, a top, right-side, and end view are shown. This arrangement is called *third-angle projection* and is used in the United States, Britain, and Canada. Other countries use first-angle projection, which shows the top, front, and left-side views. For very complex parts, designers might need to add more views, called *auxiliary views.*

In all cases, the least number of views is used. Creating unnecessary views costs time and money. Each drawing must pass the acid test of all engineering drawings. The drawing should communicate all the information needed to make the part.

Preparing multiview drawings

Multiview drawings use ***orthographic projection*** to present the information in two or more views of the object. Refer to **Figure 12-4.** Before beginning a multi-view drawing, drafters must select the surface of the object to be shown in the front view. The surface that has the most detail is chosen to be shown in the front view. The goal is to have as few hidden details as possible. Drafters complete several steps as they prepare a three-view drawing. See **Figure 12-6.** The most important of these steps are the following:

1. The front view is drawn in the lower-left quadrant of the paper. An accepted practice is to construct a box enclosing the object using light construction lines. Details on the object are then located and lightly drawn.

Career Corner

Science, Technology, Engineering & Mathematics

Engineering Technicians

Engineering technicians use science, engineering, and mathematics to solve technical problems in research and development, manufacturing, and construction. Their work is more practical than that of scientists and engineers. Many engineering technicians work in quality control; assist in product development; and repair electrical, electronic, or mechanical equipment. Some common types of engineering technicians are aerospace, chemical, electronics, industrial, and mechanical engineering technicians.

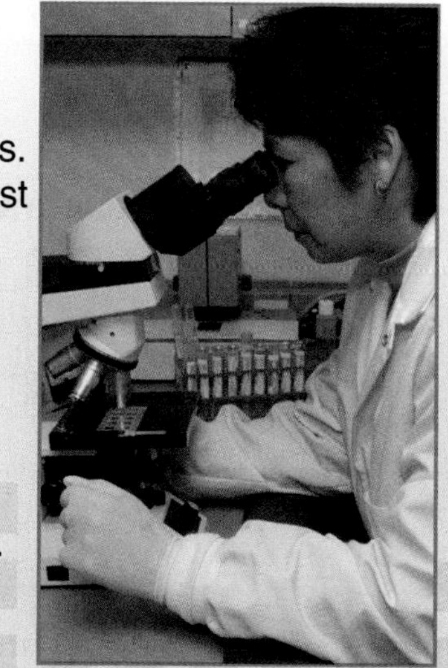

Most engineering technicians have at least an associate's degree in engineering technology. Entry-level engineering technicians begin by performing routine duties under the supervision of an engineer, a scientist, or another employee. With more experience, they are given more challenging assignments.

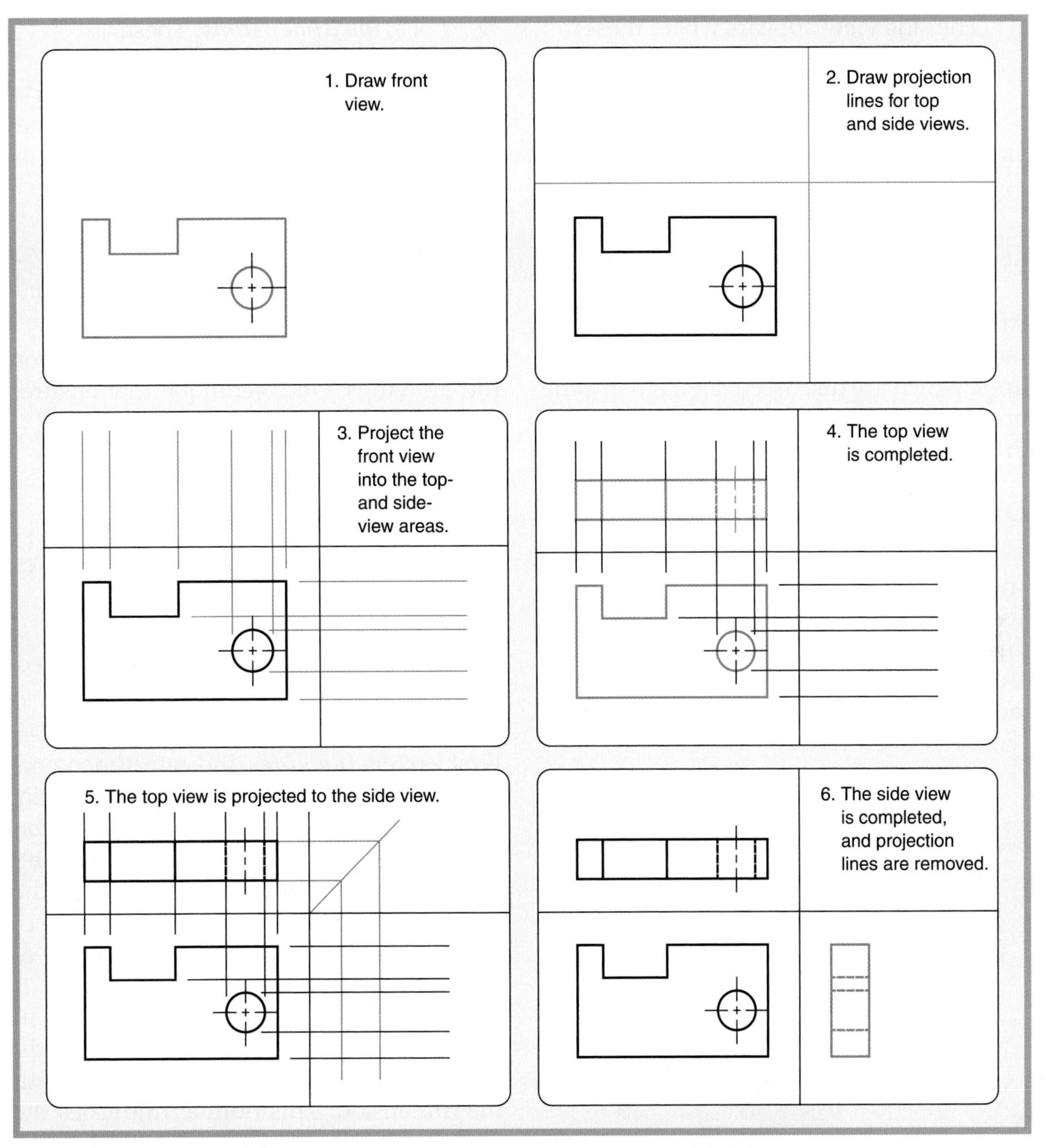

Figure 12-6. Creating a three-view drawing is a multistep process.

2. At the same distance above and to the right of the front view, lines are lightly drawn as the place to start the right-side and top views.
3. Projection lines are lightly drawn above and to the right of the front view. These lines are used to project the size of the object to the other views.
4. The outline and details of the top view are lightly drawn.
5. A 45° line is drawn in the upper-right quadrant. The outline and features of the top view are projected to the side view. This is done by projecting sizes and features to the 45° line and then down to cross the projection lines drawn earlier from the front view.

Standards for Technological Literacy

8 11

Demonstrate

Show your students how to make simple one-view, two-view, and three-view orthographic drawings.

TechnoFact

The textbooks that taught engineering drawing procedures remained unchanged for nearly a century. One such book written by Thomas French was first published in 1911 and was still in use long after the author's death in 1944. The development and proliferation of computer-aided drawing and design (CADD) software forced publishers to make drastic changes to their books.

Standards for Technological Literacy

8 11

Demonstrate

Show your students how to dimension overall sizes, curves and angles, and features such as holes, slots, and notches.

TechnoFact

Most designers, engineers, and drafters use computers to produce their engineering drawings. However, a few people still produce their drawings with manual drafting techniques and traditional tools, such as T-squares, triangles, architect's or engineer's scales, compasses, French curves, dividers, and templates.

The side view appears where these projection lines cross. Similar to those in step 2, these lines should be very light so they can be erased.

6. The side view is completed by constructing the overall shape from the front and top projections. Details are located. Dark object lines in all the views are made.

After drawing the three views of the object, the drafter adds dimension and extension lines, which are discussed later. Finally, the projection lines are erased to complete the drawing.

Dimensioning drawings

Detail drawings communicate the size and shape of an object. See **Figure 12-7.** To accomplish this task, designers include dimensions. The basics of dimensioning are presented in Chapter 10. In that chapter, you are introduced to three important types of dimensions:

- ***Size dimensions.*** These dimensions indicate the size of the object (length, width, and height) and the object's major features (diameter and depth of holes and width and depth of notches, for example).
- ***Location dimensions.*** These dimensions indicate the position of features on the object, such as center points for holes, edges of grooves, and starting points for arcs.
- ***Geometry dimensions.*** These dimensions indicate the shapes of features and the angles at which surfaces meet (round holes or square corners, for example).

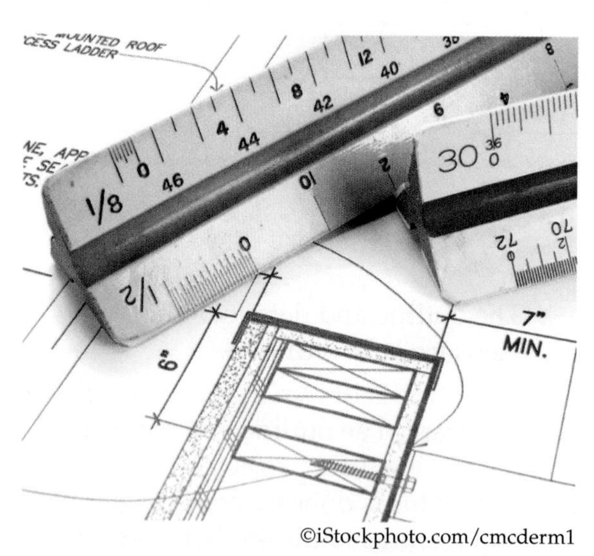

©iStockphoto.com/cmcderm1

Figure 12-7. Dimensions communicate the sizes on a drawing.

All these dimensions must be included on the drawings. One technique that ensures they are present suggests the following:

1. Dimension the size of the object first, followed by the sizes of all major features.
2. Dimension the locations of all features next.
3. Indicate any necessary geometric dimensions last. (Angles not indicated are assumed to be 90°.)

Dimensioning uses two kinds of lines. First, ***extension lines*** indicate the points from which the measurements are taken. Between the extension lines are ***dimension lines.*** Dimension lines have arrows pointing to the extension lines indicating the range of the dimension. The actual size of the dimension is shown near the center of the dimension line.

These dimensions can be given in fractions or decimals of an inch. In addition, a ***tolerance*** might be included with the dimension. This number indicates the amount of allowed deviation in the dimension. For example, a +/– (plus or minus) 1/64″ (0.4 mm) after the dimension indicates that the size can be 1/64″ larger or 1/64″ smaller than the dimension and still be acceptable.

The alphabet of lines for drawings

Users should be able to read the drawings easily. Therefore, a set of drafting standards has been developed so all drawings communicate well. These rules are very similar to the rules of grammar a writer uses. They allow each reader to interpret the prose (using grammar rules) or the drawing (using drafting standards) in a similar manner.

One set of essential drafting standards deals with lines and line weights. See **Figure 12-8.** The shape of the object is of primary importance. Therefore, the lines outlining the object and its major details must stand out. These solid lines are called ***object lines*** and are the darkest on all drawings.

Some details are hidden in one or more of the views. Their shapes and locations, however, are important to understanding the drawing. Therefore, they are shown, but with lighter, dotted lines called ***hidden lines***.

A third type of line locates holes in the part. These lines pass through the center of the hole and are, thus, called ***centerlines***.

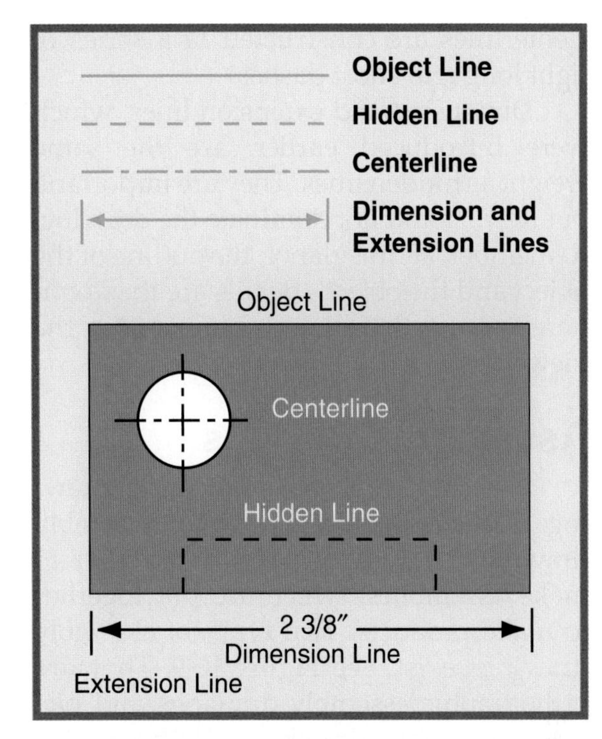

Figure 12-8. This alphabet of lines shows the types and weights of lines used in drawings.

Standards for Technological Literacy

3 8 9 11

TechnoFact

Most people associate the Industrial Revolution with the steam engine and the factory system of production that formed in England in the 1780s. However, the introduction of interchangeable parts and assembly-line techniques caused a second Industrial Revolution to spring up in the United States. The techniques that so drastically improved production also required accurate detail and assembly drawings for each product (and each part of the product).

Academic Connections: Communication

Principles of Design

This chapter describes the documents and reports designers use to gain approval for their solution to a technological problem or opportunity. Also important is the way in which these documents and reports are presented to the various decision makers. Designers have found that specific principles of design are effective in communicating a clear and pleasing message, thus leading to a successful presentation.

The basic principles of design a presenter needs to take into account are balance, unity, rhythm, contrast, and proportion. When the message is in proper balance, the images being presented are distributed evenly, and one element dominates. The message has unity when all elements are harmoniously balanced. When the message has the proper rhythm, the viewer's eyes are drawn to the area the presenter wants to emphasize. Proper contrast means variations in the elements exist (although not to the extent that the image becomes confusing). When an image is proportioned correctly, each element is proportional to all the others, thus creating a unified design.

These basic principles of design are used every day in the world of advertising. Look at some ads in current issues of newspapers or magazines. Can you find some examples of these design principles at work?

Standards for Technological Literacy

Figure Discussion

Discuss the difference between orthographic and pictorial assembly drawings as shown in Figure 12-9. Discuss the user of each (production worker vs. consumer).

Research

Have your students find and describe some assembly drawings that average consumers encounter with appliances, vehicles, etc.

Demonstrate

Show how to make a simple exploded-view assembly drawing for a product such as a wood pencil.

Extend

Discuss how to use exploded assembly drawings to order parts for and repair simple appliances and tools.

Centerlines are constructed of a series of light long and short dashes.

Dimension and extension lines, which were introduced earlier, are the same weight as hidden lines. They are important, but they should not dominate the drawing. Remember, at first glance, the outline of the object and the object's details are the dominant features that should "jump out" at the viewer.

Assembly Drawings

A second type of engineering drawing is the assembly drawing. Assembly drawings show how parts fit together to make assemblies, which are put together to make products. Two types of assembly drawings exist. See Figure 12-9. They are orthographic assembly drawings and pictorial assembly drawings. ***Orthographic assembly drawings*** use standard orthographic views to show parts in their assembled positions. ***Pictorial assembly drawings*** show the assembly using oblique, isometric, or perspective views similar to those discussed in Chapter 10.

Either type of drawing can be a standard view or an exploded view. See Figure 12-10. ***Standard views*** are constructed using the normal techniques for constructing orthographic or pictorial drawings. These views show the product in one piece, as it will be after it is assembled.

Exploded views show the parts making up a product as if it is taken apart. The parts are arranged in the proper relationships to each other on the drawing. This type of drawing is often found in owner's manuals and parts books. Exploded views are used to show the parts making up a product. Each part on the drawing generally has a code allowing the owners or repair person to order a replacement part. Most assembly drawings do not have dimensions. The exception is when the assembler must manually position the parts for assembly.

Systems Drawings

Systems drawings are used to show how parts in a system relate to each other and work together. They are used for electrical, hydraulic (fluid), and pneumatic (gas) systems. These drawings are often called *schematic drawings*. They do not attempt to show the actual positions of the parts in a product. Assembly drawings do this. Systems drawings are designed to show the connections for wires, pipes, and tubes.

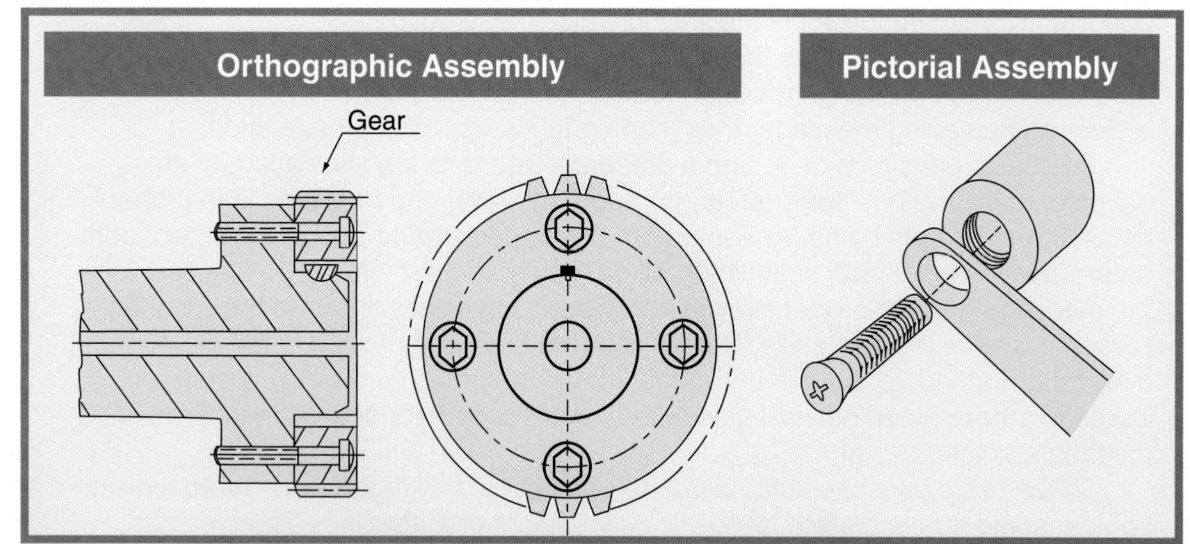

General Motors Corp.

Figure 12-9. Assembly drawings can be orthographic or pictorial.

TechnoFact

The word blueprint is commonly used to indicate a copy of a drawing. In fact, blueprinting has a far more specific meaning. It is a process for reproducing drawings that was invented in 1842. In the blueprinting process, a master drawing is created on translucent paper. The master drawing is then placed on top of chemically treated, light-sensitive paper. Bright light is shined down on the combination. The areas exposed to the light turn blue; the areas hidden from light by the lines of the master drawing remain white. The image is made permanent by washing the exposed copy in water.

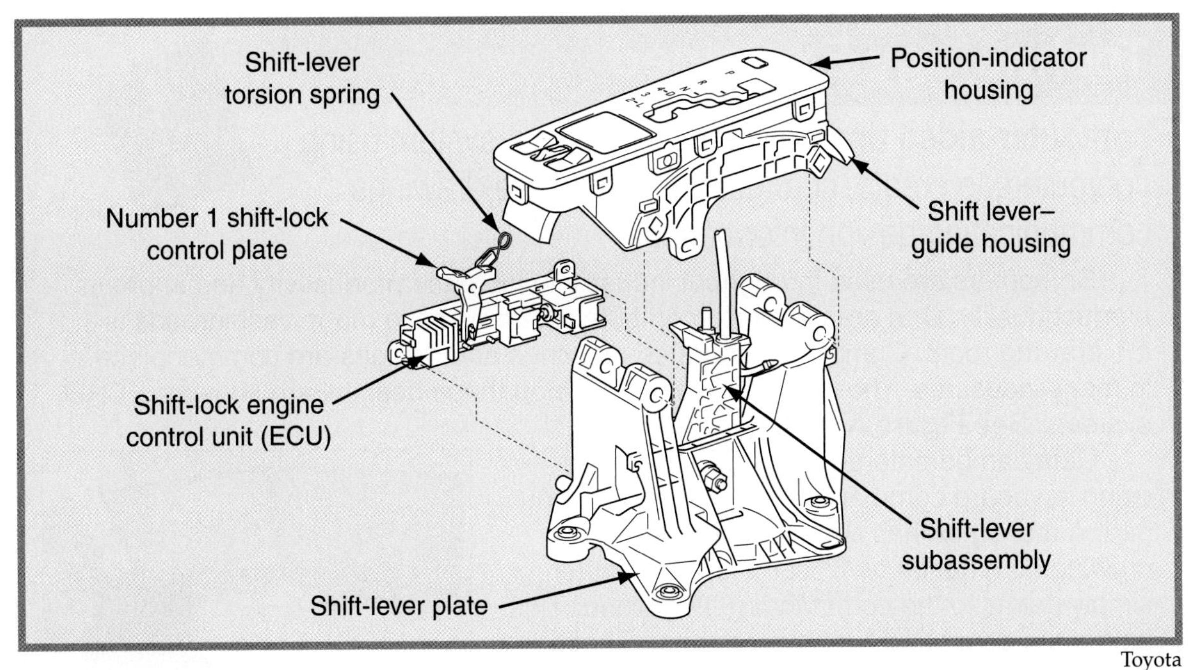

Toyota

Figure 12-10. An exploded-view assembly drawing shows the parts of a product as if the product is taken apart.

These drawings use symbols to represent the components. Standard symbols have been developed for electrical, pneumatic, and hydraulic parts. **Figure 12-11** includes some common symbols for electrical and electronic components.

Systems drawings are developed by first arranging the major components on the sheet. Connecting wires, pipes, and tubes are then indicated. Special drawing techniques are used to indicate when the lines connect or simply cross each other.

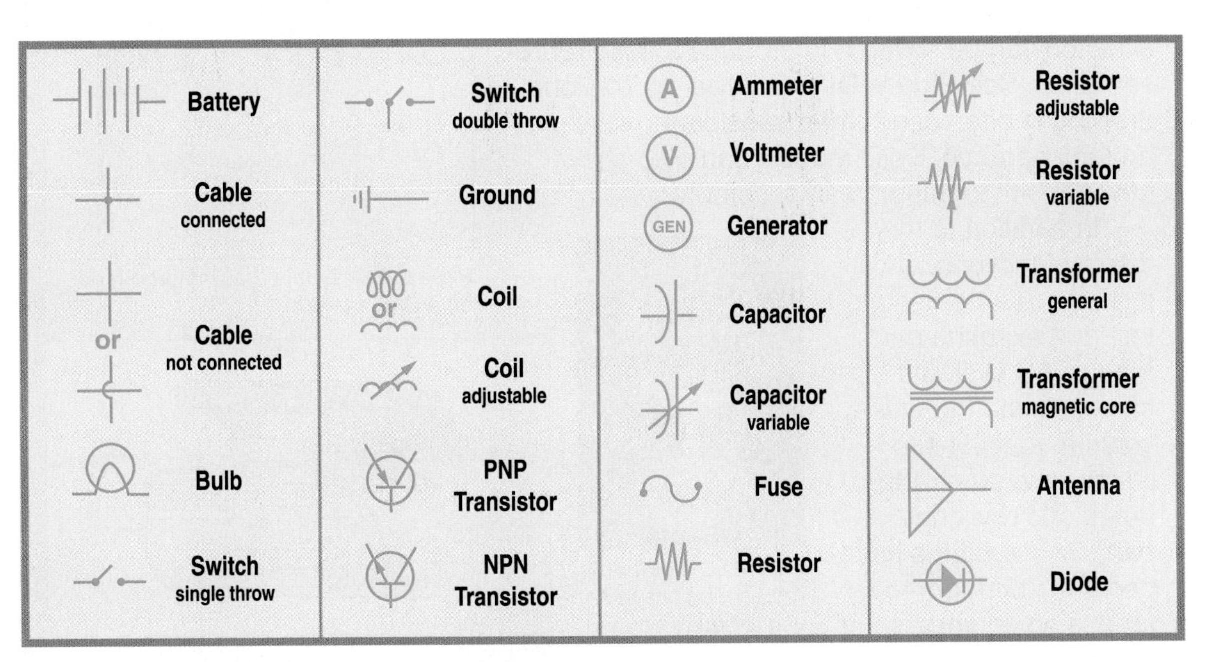

Figure 12-11. Some common symbols used on electrical and electronic systems drawings.

Standards for Technological Literacy

8 11

Figure Discussion

Show your students how to make a schematic drawing using the symbols in Figure 12-11.

Research

Have your students find and interpret a schematic drawing of a plumbing or electrical device used in a home.

TechnoFact

Large engineering drawings, schematic diagrams, and other large illustrations are frequently printed on a plotter. Most plotters have one or more pens that draw lines on wide paper as it is fed, off a roll, through the plotter. Most often, a computer controls the left and right movement of the pens and the feed-rate of the paper. Other plotter types include the inkjet plotter, which functions like a large inkjet printer, and the electrostatic plotter, which produces images in much the same way as a photocopier.

Standards for Technological Literacy

8 11

TechnoFact

Designers can create a three-dimensional model of an object with CAD software. Such a drawing resembles a photograph of the part or product. The designer can change the model's dimensions or rotate the object in space to create different views.

Technology Explained

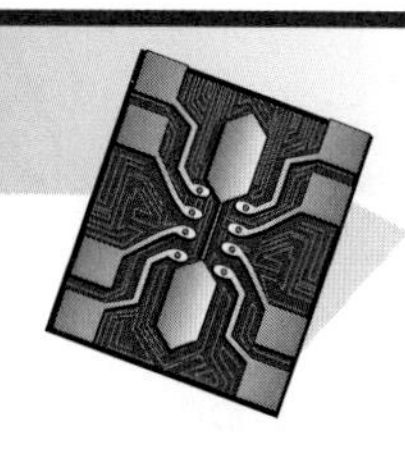

computer-aided design (CAD) system: a system using computers to create, change, test, and store drawings communicating design information.

Computers are used throughout industry to increase productivity and improve product quality. One area in which computer systems have made vast inroads is the drawing room. Computer-generated drawings and designs are commonplace in many industries. The systems used to develop these designs are known as *CAD systems*. See **Figure A.**

Data can be entered into a CAD system using keyboard commands, a mouse, or a menu pad. A menu pad has a number of common drawing commands on its surface. The drafter simply points to the commands with a wand. This enters the command into the computer. These controls allow drafters to produce standard two-dimensional detail and assembly drawings. Additional features and text can then be placed on any engineering drawing. See **Figure B.**

Figure A. A typical CAD station.

A growing use for CAD systems is in producing solid models (three-dimensional representations) of objects. This process uses a concept similar to weaving a rug. Strands are intertwined to produce a frame for the object. This produces a drawing called a *wire-frame representation*. See **Figure C.** The frame is the skeleton required to produce the object. Such a skeleton can be covered to produce a solid representation. See **Figure D.** This three-dimensional drawing is often used when considering styling and appearance. Solid models can also show how parts fit together in an assembly.

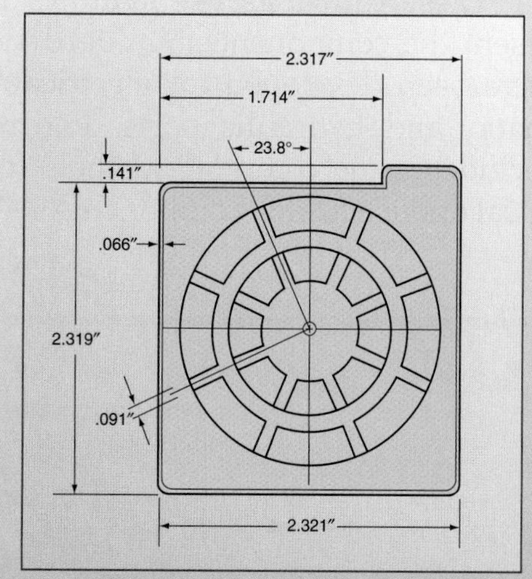

Figure B. A simple drawing produced on a CAD system.

In addition to their display functions, CAD systems can be used to test and examine parts. Many CAD systems can calculate the mass, volume, reactions to stress, and other attributes of a designed part. CAD systems have become important tools for design engineers.

Figure C. Wire-frame representations allow you to see through an object.

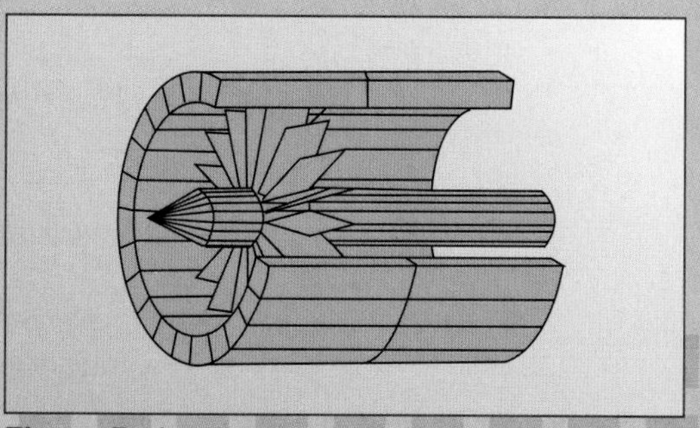

Figure D. A solid representation of a turbine.

Computer-Aided Design (CAD)

Computer systems are valuable technological tools that can be applied to a number of tasks. A common industrial application for computer systems is in preparing drawings and models. This application uses a computer, a plotter or printer, input devices (a keyboard, mouse, and graphics tablet, for example), and ***computer-aided design (CAD) software***. See Figure 12-12.

There is nothing magical about a CAD system. This system allows an operator to complete the steps of laying out and producing drawings following the methods a drafter uses. CAD systems, however, do the job more quickly and uniformly. Also, computer drawings are easier to correct, store, and communicate. Computer drafters can send their drawings across the country or around the world in seconds using the Internet.

Figure 12-12. Designers can use graphic tablets to enter data.

Bills of Materials

Not all information needed to produce a product can be contained on detail, assembly, and system drawings. Additional documents are needed to provide complete production information. One important document is a ***bill of materials***. The name of this document causes some confusion. We all pay bills for things we buy and use. A bill of materials does not contain cost information. Instead, it is a list of the materials needed to make one complete product. An example is shown in Figure 12-13.

Most bills of materials contain the following information for each part of the product:

- A part number that can be used on assembly drawings and for ordering repair parts.
- A descriptive name for the part.
- The number, or quantity (abbreviated *qty.*), of parts needed to manufacture one product.
- The size of the part, indicating the part's thickness, width, and length

Bookend

Bill of Materials

Part #	Part Name	Qty	Size			Material
			T	W	L	
BE-1	End	1	3/4	6	8	Oak
BE-2	Bottom	1	28ga	6	6	Steel
BE-3	Protection	1	1/32	6	6	Felt
	FH Screws	3	#6		1	Steel

Figure 12-13. A bill of materials for a simple bookend lists the materials needed to make one product.

Standards for Technological Literacy

8 11

Brainstorm

Ask your students to list the reasons they think CAD drawings are better than manually produced drawings.

Research

Have your students research the development of CAD systems.

Figure Discussion:

Refer to Figure 12-13 as you explain to your students how to develop a bill of materials for a simple product.

Brainstorm

Ask your students why a bill of materials is important.

TechnoFact

The terms computer-aided design (CAD) or computer-aided design and drafting (CADD) refers to using a computer to generate engineering drawings. A person using a CAD system enters software commands to produce lines and dimensions. The computer monitor displays the drawing as it is being developed. The completed drawing can be archived, transmitted across the Internet, or printed on a printer or plotter.

Standards for Technological Literacy

Research

Have your students find a specification (technical data) sheet for a finishing material or adhesive and list the types of information on it.

TechnoFact

Leonardo da Vinci sketched designs for inventions that were actually built much later in history. One device he envisioned in the 1500s was a bicycle. His design had a handlebar, spoked wheels, and a pedal-driven chain and sprocket. However, his sketch was not discovered until almost 100 years after the invention of the modern bicycle.

Quotes

"All innovation requires a committed champion. Someone to maintain the momentum when nervousness or uncertainty appear."—Danny Bone, New Product Innovation Manager, Black & Decker.

(for rectangular parts) or the part's diameter and length (for round parts). Sizes are given in the order shown: T × W × L or Dia. × L.

- The material out of which the part is to be made.

The items on a bill of materials are listed in a priority order. Manufactured parts are listed first. Parts purchased ready to use and fasteners are listed after the manufactured parts.

Specification Sheets

Not all materials can be shown on a drawing. Can you make a drawing of engine oil, an adhesive, or sandpaper? If you do make a drawing of any of these items, it will be of little value. These and thousands of other items are not chosen for their size and shape. Other properties are important in their selection.

For example, some important factors in adhesives are the working time (time between application and clamping), clamping time (time the work is held together for the glue to set), and shear strength. Window glass must be transparent. Insulating materials must stop heat from passing through them.

Specification sheets communicate the important properties a material must possess for a specific application. See **Figure 12-14.** These properties, which are introduced in Chapter 4, might include the following:

- **Physical properties.** These properties include moisture content, porosity, and surface condition.
- **Mechanical properties.** These properties include strength, hardness, and elasticity.
- **Chemical properties.** These properties include corrosion resistance.
- **Thermal properties.** These properties include resistance to thermal shock, thermal conductivity, and heat resistance.
- **Electrical properties.** These properties include resistance and conductivity.
- **Magnetic properties.** These properties include permeability.

GASOLINE
Unleaded

GM 6117-M
(For Factory Fill)

GM 6118-M
(For Immediate Use)

1. Scope.
These specifications cover two types of unleaded gasoline: one for factory-fill and one for immediate or normal use. The factory-fill gasoline is intended for use in vehicles which are stored for extended periods of time, or in tanks where the rate of turnover is low.

1.1 These specifications apply to samples takes directly from the tank car or tank wagon.

2 Chemical and Physical Properties

2.1 **LEAD** Lead shall not exceed 13.209 mg per liter (.05 gram per gallon).

2.2 **PHOSPHORUS** Phosphorus shall not exceed

2.3 **Sulfur** S...
(wei...

cupric acetate in benzene containing 0.20 milligram of copper, using the apparatus and procedure described in the standard method of test for oxidation stability of gasoline (induction period method).

Note 2: The oxidation stability test is designed to insure that gasoline purchased under these specifications is sufficiently stable in the presence of metal. If metal deactivator is added to the gasoline in order to meet the specification, extreme care is necessary to insure adequate blending of the metal deactivator with the gasoline.

In sampling gasoline for this test, it is desirable to collect the sample in a glass sample bottle since contact with metal surfaces may cause deterioration of stability. ...ereas for legal reasons the sample cannot

Figure 12-14. The characteristics of materials are described on specification sheets.

- **Acoustical properties.** These properties include sound absorption and sound conductivity.
- **Optical properties.** These properties include color, transparency, and optical reflectivity.

Technical Data Sheets

The specifications are included on two types of sheets. In the first type, manufacturers prepare ***technical data sheets*** to communicate the specifications for products they have on the market. These kinds of products are often called *standard materials and components* or *off-the-shelf materials and components*. They are generally kept in stock by the manufacturer and are often listed in a supplier's catalog.

For example, you can write to a manufacturer regarding your need for an adhesive. You would probably receive technical data sheets on several adhesives the manufacturer makes that might meet your needs. For each product, you would study the specifications and choose the one that meets your needs.

Specification Sheets for Suppliers

In the second type, large organizations might prepare their own specifications for materials and products they need. They send them to suppliers, who then compete to supply a specific item. One example is the military specification (Mil-Spec) system, in which the government lists its specifications. Large manufacturing companies also have specification systems.

Approval Documents and Reports

So far, we have discussed methods of communicating designs for manufacture. Before anything can be built, however, someone in charge must approve it. This "someone" might be company management, government agencies, or the customer.

The approval process will generally require two types of communication. First, designers prepare written reports. These reports might include need statements, proposed design solutions, cost estimates, marketing strategies, economic forecasts, and environmental-impact statements.

Oral reports to those who will give approval to the project support the written reports. In oral reports, designers present the highlights of the written report. They use graphs, illustrations, and other visual media to help communicate the design data. See **Figure 12-15.** This presentation is designed to make the final "sale" of the design idea. The result is official approval to proceed with the manufacturing or construction project. The actual practices used to produce the product or construct the structure are the focus of the next section of this book.

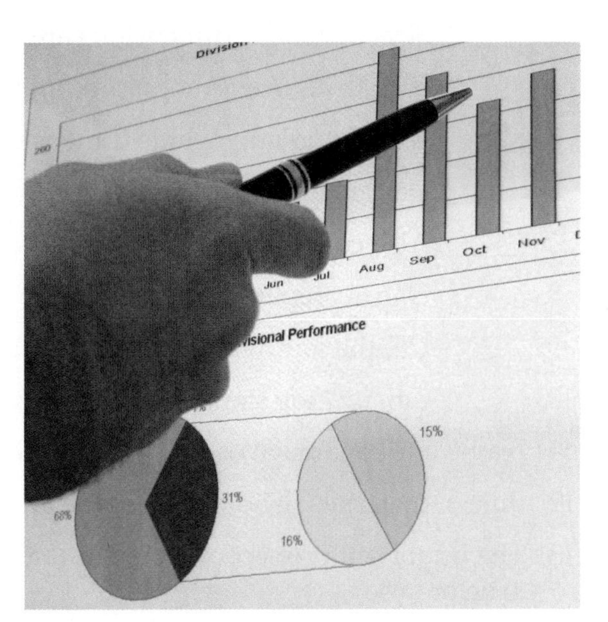

Figure 12-15. Graphic media are used to help communicate information about design solutions.

Standards for Technological Literacy

8 11

Extend

Discuss how a consumer could use a specification (technical data) sheet.

Extend

Discuss the differences between a report and an engineering drawing.

Brainstorm

Have your students discuss the reasons that drawings should be submitted for approval.

Quotes

"Patents and Intellectual Property Rights help people make money out of their inventiveness. They are also a dissemination mechanism. The whole point about giving people the right to exploit their invention is that, in return, they tell you how it works—thus allowing others to develop the idea further in a kind of ratcheting process." —Alison Brimelow, www.derwent.com.

"Businesses of all types will need to innovate to raise their productivity, create new products and high-value services, and stay ahead of the competition. Such a knowledge-driven economy could have significantly less environmental impact."—www.dti.gov.uk/sustainability.

Answers to Test Your Knowledge Questions

1. Engineering documents, bill of materials, and specification sheets
2. C. Systems
3. A. Detail
4. B. Assembly
5. B. Assembly (Note: C. Systems may also be accepted at the instructor's discretion.)
6. C. Systems
7. C. Systems
8. A. Detail
9. A. Detail
10. B. Assembly
11. Student answers may include any of the following: CAD systems do the job more quickly and uniformly. Also, computer drawings are easier to correct, store, and communicate. Computer drafters can send their drawings across the country or around the world in seconds using the Internet.
12. False
13. True
14. Technical data sheets are lists of specifications prepared by manufacturers for the products they have on the market.
15. Student answers may include: need statements, proposed design solutions, cost estimates, marketing strategies, economic forecasts, and environmental impact statements.

Summary

Communicating design solutions is a key element in creating successful technological systems. Designers must communicate vital information to manufacturing personnel and decision makers. They do this by preparing such documents as engineering drawings, bills of materials, and specification sheets. Engineering drawings communicate the details of each part, the way parts are assembled into products, and the arrangement of system components. Bills of materials list the parts needed to make one complete product. Specification sheets list the properties a material must possess for a specific application. Designers also prepare written and oral reports to gain approval for the product or structure.

Test Your Knowledge

Write your answers on a separate piece of paper. Please do not write in this book.

1. List the three basic kinds of documents through which designers communicate information about a product.

Matching questions: For Questions 2 through 10, match each definition on the left with the correct type of engineering drawing on the right. Answers can be used more than once.

2. ______ Use symbols to represent components drawings.
3. ______ Size and shape dimensions are included on the drawings.
4. ______ Two kinds are orthographic and pictorial drawings.
5. ______ Usually do not include dimensions.
6. ______ Designed to show connections for wires, pipes, and tubes.
7. ______ Also called schematic drawings.
8. ______ One is prepared for each part.
9. ______ Use the multiview method.
10. ______ Can include standard and exploded views.

A. Detail.
B. Assembly.
C. Systems.

11. Name one advantage of a CAD system.
12. Bills of materials list the costs involved in making a product. True or false?
13. The important properties a material must possess are listed on specification sheets. True or false?
14. What is a technical data sheet?
15. Give one example of a report designers need to prepare to gain approval for a project.

STEM Applications

1. Disassemble a simple product such as a ballpoint pen or a flashlight. Prepare the following:
   A. A detail drawing for one part.
   B. An assembly drawing for the product.
   C. A bill of materials for the product.
2. Select a simple game on the market. List the materials and parts that would need the following:
   A. A drawing to produce.
   B. Specification sheets prepared for them.

TSA Modular Activity

Standards for Technological Knowledge

12

This activity develops the skills used in TSA's Computer-Aided Design (CAD) 2D, Architecture event.

Computer-Aided Design (CAD), Architecture with Animation

Activity Overview

In this activity, you will create CAD drawings of an original residence design. The following drawings are required:

- A foundation plan with a section view of a typical footing and wall.
- A floor plan for each level.
- A front elevation.
- One full section illustrating the basic roof design.
- A detail sheet with at least three details.

Materials

- Paper.
- A pencil.
- A computer with CAD software.
- A printer or plotter.

Background Information

- **Basic layout.** Begin by considering the type of residence you plan to design. Will you design a single-level ranch style or multiple-level residence? What type of foundation system will the residence have? Develop rough sketches of room sizes and locations. Consider which rooms should be grouped together and the traffic flow through the residence. Generally, bedrooms are grouped in one section of the residence. The kitchen and dining room are normally adjacent to one another. If the residence has multiple levels, the upper-level bathroom is normally located directly above a lower-level bathroom to simplify the plumbing. Will the residence include other rooms and features, such as an attached garage, home office, exercise room, or home theater?

- **Floor plan.** Floor plans are normally drawn at a scale of 1/4″ = 1′0″. Use a scale that will allow your drawing to fit on the size of paper used for plotting. The floor plan should show the wall layout and dimensions, door and window locations, and room labels and sizes.
- **Foundation plan.** The foundation plan can be created from the floor plan. Offset the outside-wall line to locate the edges of the foundation wall and footing. Add square footings to support interior columns.
- **Front elevation.** Copy the wall locations from the floor plan to begin the elevation. Add doors, windows, and other elements. Select an exterior finish for the home.
- **Full section.** Determine the location of the section. Identify it on the floor plan. Develop the section, which shows cross sections of the walls, floor, ceiling, and roof.
- **Details.** Each of your details must be referenced on another drawing. Some common details are wall details (showing the interior and exterior components of the wall), kitchen cabinet details and elevations, and bathroom details and elevations.
- **Animation.** Your animation is intended to illustrate the walkthrough of your design.
- **Output.** Output the animation as an Audio Video Interleave (AVI) file or a QuickTime® Movie (MOV) file. Using a medium color depth (such as 16-bit) and low resolution (approximately 320 × 200 pixels) speeds the processing of the animation.

Standards for Technological Knowledge

12

Guidelines

- Create a separate CAD drawing file for each drawing.
- Plot each drawing.
- Your animation must be at least 10 seconds long.

Evaluation Criteria

Your project will be evaluated using the following criteria:

- Originality and creativity in design.
- Accuracy of drawings.
- Use of good drafting practice, including line-type, lettering, and symbol usage.
- Use of animation features.

Section 4 Activities

Activity 4A

Design Problem

Background

All technological devices are the results of design efforts by people. People define problems and opportunities, think up many solutions, and model the solutions selected. Finally, designers communicate the designs to production personnel.

Situation

Road Games Inc. is a company specializing in designing small compact games people can take with them on trips. You have been recently employed in the creative-concepts department of the company.

Challenge

Design a travel game using pegs and a 3/4″ × 3 1/2″ × 3 1/2″ wood board. Dice can be used in the game, but they are not required. Your boss expects you to produce rough sketches for five different ideas, a refined sketch for the best idea, a prototype, and a detailed drawing of the game board.

Note
Activity 4B is an example of a game fitting these criteria.

Activity 4B

Fabrication Problem

Background

Product designers often work to improve an existing product. They also develop new and improved products to meet the changing demands and requirements of customers.

Challenge

You are a product designer for the Acme Bookend Company. The company makes bookends for different markets. Each market gets unique graphics and special shapes for the bookends. Your boss has asked you to modify an existing product to meet these new criteria:

- A new shape.
- New graphics appealing to high school students.

Materials and Equipment

- Sketch paper.
- Drawing paper.
- Pencils.
- Felt-tip pens.
- A drafting ruler.
- A T square.
- Triangles.

Procedure

Redesigning a product requires a number of actions. You are to redesign the bookend to meet the new criteria by designing the new product shape, developing a new decoration for the product, and developing a set of drawings to communicate your new design. See **Figure 4B-1.**

Designing the shape

1. Photocopy the bookend layout sheet. See **Figure 4B-2.**
2. Sketch four new shapes for the bookend.
3. Select the best shape. Circle it with a colored marker.

Decorating the product

1. Make four layouts, using the shape you chose.
2. Sketch four new graphic designs. Make sure the design appeals to high school students.
3. Select the best design. Circle it with a colored marker.

Communicating the design

1. Obtain a piece of drawing paper.
2. Draw a border 1/2″ in from all edges of the paper.
3. Draw a title box. See **Figure 4B-3.**
4. Produce a dimensioned two-view orthographic drawing of your design for the new bookend.

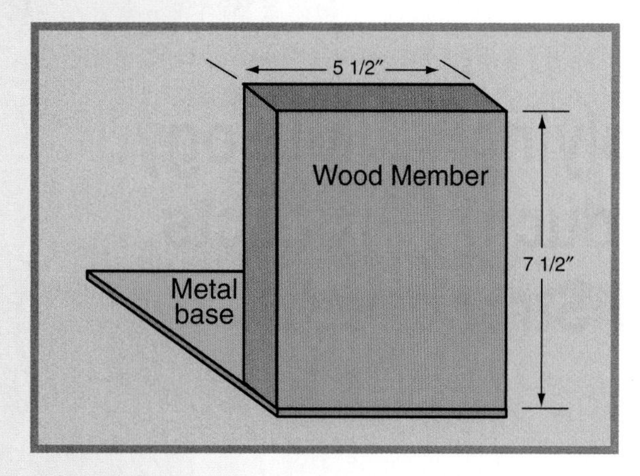

Figure 4B-1. A basic bookend.

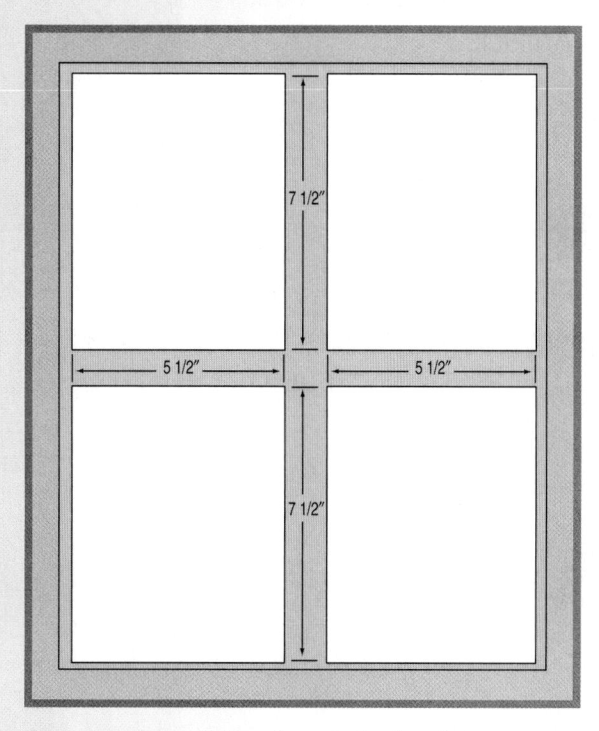

Figure 4B-2. A bookend-design layout sheet.

School:	Part Name:	Drawn By: Date:
Class:	Part Number:	Checked By: Date:

Figure 4B-3. A title block.

Section 5
Applying Technology: Producing Products and Structures

Tomorrow's Technology Today

Terraforming

What comes to mind when you think of an extraterrestrial? You may imagine the small, green beings you see on the covers of science fiction novels. Today, scientists are beginning to envision a different picture of life in space. Through the process of terraforming, they hope to someday sustain human life on another moon or planet.

Terraforming is the alteration of a planet or moon's surface to make it suitable for human life. In other words, it is an attempt to make it like Earth. Because the planets and moons are so diverse, terraforming would be a big challenge. It could take centuries, or even longer, to successfully complete a terraforming project and begin human habitation in a place other than Earth.

The process of terraforming is currently hypothetical. Many questions need to be answered before we can even consider such a project. Will a country, several countries together, or a private organization be responsible for the terraforming process? How much would it cost and who would provide funding? And most importantly, what planet or moon would be the best candidate for terraforming? Determining which is most Earth-like would ensure an easier and more successful conversion.

Atmospheric composition, distance from the Sun, and the presence of water are a few important factors to consider when evaluating a planet or moon's potential for terraforming. With a rotation rate and axial tilt similar to Earth, Mars has emerged as the leading contender. Water is frozen at Mars' polar caps while carbon, oxygen, and nitrogen are all present in its atmosphere. However, the atmosphere is very thin, oxygen levels low, and temperatures significantly cooler than on Earth. Scientists have suggested several methods to heat the planet, including placing large, sun-reflecting mirrors on its surface or building factories that would produce greenhouse gases. Greenhouse gases have proven to raise the temperature on Earth.

Terraforming an entire planet would be extremely costly and time-consuming. As a result, scientists are now exploring the option of paraterraforming. Paraterraforming would transform only a portion of the surface within an enclosed structure. The structure could then be expanded over time as financial capabilities and knowledge of the terraforming process increase.

Should the Earth ever suffer a major disaster or crippling overpopulation, terraforming would offer humans an opportunity to escape and thrive elsewhere in the solar system. Many questions still need to be answered, but scientists move one step closer every day to establishing human life on other moons and planets.

Discussion Starters

As a class, discuss why terraforming would be beneficial to mankind.

Group Activity

Divide the class into groups. Have each group discuss the various scientific and financial problems that would have to be taken into account before terraforming could become a reality. Have the groups brainstorm possible solutions to these problems.

Writing Assignment

Have your students research the work being done in the development of terraforming, specifically in the area of choosing the most likely candidate. Have them write a brief paper explaining whether they Mars is the best possible candidate and why.

Chapter Outline

Production Activities

Servicing and Repairing Products and Structures

This chapter covers the benchmark topics for the following Standards for Technological Literacy:

3 19 20

Learning Objectives

After studying this chapter, you will be able to do the following:

- Recall the two major types of production activities.
- Compare activities involved in resource processing, product manufacturing, and structure construction.
- Understand the meaning of servicing.
- Recall the definition of *repair.*

Key Terms

construction	industrial material	manufacture
consumer product	industrial product	repair

Strategic Reading

As you read this chapter, you will learn about how materials are processed for everyday use. Choose an everyday object and trace the process back to the origin of the material.

We developed some of our earliest technology for production systems. See **Figure 13-1.** These systems produce what is called *form utility.* They change the form of materials to make them more valuable. Look at **Figure 13-2.** Which is more useful and, therefore, more valuable to people: the lumber or the home built from the lumber? You probably answered the home. Wood in the form of a home is more useful to you than in the form of lumber. Therefore, the actions that change lumber into houses provide form utility.

Production activities have one of two major goals. These goals are to produce a product and to produce a structure. After the product or structure has been produced, the focus of activities turns to service and repair. Each is discussed as follows.

Figure 13-1. These people are practicing early production activities using technology of the Colonial period.

Weyerhaeuser Timber Co.

Figure 13-2. Lumber can be changed in form to become a home.

Production Activities

These activities produce either a product or a structure. The activities that make products are called *manufacturing activities*. The activities that produce structures such as buildings or roadways are called ***construction***. Construction can be described as producing a structure on the site where it will be used. These activities are different from ***manufacturing***, in that a manufactured product is produced in a factory and shipped to its point of use.

Literally thousands of different production activities fall under the categories of manufacturing and construction. These

Standards for Technological Literacy

19 20

Extend

Discuss the differences between manufacturing (form utility in a factory) and construction (form utility on a site).

Brainstorm

Ask your students why some houses are built in a factory (manufactured homes).

Extend

Trace the development of a loaf of bread from production (raising the grain, making flour, and baking loaves) to distribution (selling and delivering bread to stores) to the consumer (buying and eating bread).

TechnoFact

One important step in bringing goods and services to people is production. The other steps are distribution (the transfer of goods from the producer to the consumer) and consumption (the consumer using the goods and services to meet personal needs and wants).

activities, however, can be grouped into three major systems:

- Resource-processing systems (primary manufacturing).
- Product-manufacturing systems (secondary manufacturing).
- Structure-construction systems (construction).

Each of these systems plays a unique role in converting a material resource into a product or structure to meet human needs and wants.

Resource-Processing Systems

Few materials occur in nature in a usable state. Generally, materials must be converted into new forms before products and structures can be made. This conversion uses primary processing technology.

Typically, these processing systems involve two actions. First, the material must be located and obtained from the earth. This might involve growing and harvesting trees, crops, and domesticated animals. Locating the materials might involve searching for minerals or hydrocarbons (petroleum and coal). Drilling or mining might be required to extract these resources from the earth.

Once the natural resource has been obtained, it must be transported to a processing mill. There, the natural resource is changed into ***industrial materials***. These materials are the inputs to secondary manufacturing activities. For example, iron ore, limestone, and coke might be changed into steel. See **Figure 13-3.** The steel might then be processed into bars, rods, sheets, or pipes. This standard material becomes the raw inputs for systems that make products for industrial companies and retail consumers. The material might end up as part of a lathe, an automobile, a broadcast tower, an airline terminal, or one of thousands of other products and structures.

Specialized processing systems are used to manufacture food products, medicines, and chemicals. Other systems process petroleum into fuels and lubricants, natural gas into plastics, and coal into coke.

Standards for Technological Literacy

19 20

Example

Show your students some examples of industrial materials such as plywood, sheet aluminum, paper, etc.

TechnoFact

Natural resources are naturally occurring materials that help support life. This includes minerals, sunlight, water, soil, plants, and animals. Although natural resources are available for our use, they must be managed, protected, and used wisely.

TechnoFact

Materials are used to make products, structures, and systems. Manufacturers determine which material to use for a given product by evaluating the properties of a range of appropriate materials.

Career Corner

Architecture & Construction

Plumbers

Plumbers install, maintain, and repair pipe systems. They work on systems that move water into and throughout residential, commercial, and public buildings. Plumbers also work on systems that dispose of waste, provide gas to stoves and furnaces, and supply air conditioning.

They work from drawings that show the planned locations of pipes, plumbing fixtures, and appliances. Plumbers cut holes for pipes, assemble pipe systems, install fixtures and appliances, connect systems to outside lines, and check the operation of completed systems. Almost all plumbers are trained through an apprenticeship-training program. Local union-management committees administer these programs. These four- to five-year programs include on-the-job training, supplemented by classroom instruction.

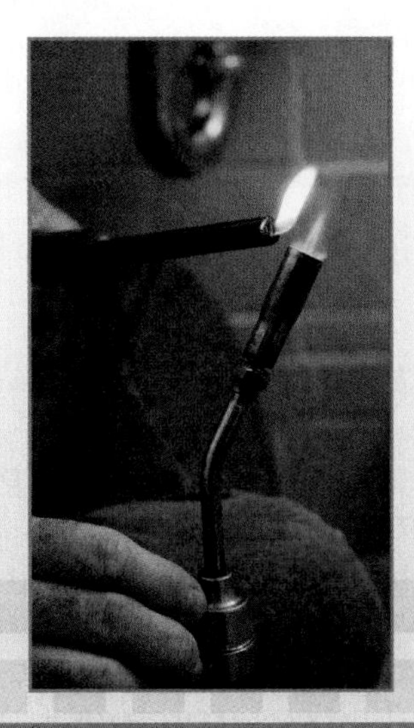

Inland Steel Co.

Figure 13-3. Primary processing activities change natural resources into industrial goods, such as the steel rods shown here.

Goodyear Tire and Rubber Co.

Figure 13-4. Process systems produce such products as these tires.

Standards for Technological Literacy

19 20

Brainstorm

Ask your students to list some examples of industrial products and consumer products.

See Figure 13-4. More detailed information on obtaining material resources and processing them into industrial materials is contained in Chapters 14 and 15 of this book.

Product-Manufacturing Systems

Most standard materials have limited use to the average person. Suppose someone gave you 10 sheets of plywood. You might think this is a good gift. What are you going to use them for, however? You will probably have to ask the question, "What should I build out of the plywood?" You might build a bookcase or a storage locker. This action uses secondary processing activities.

Secondary processing systems change industrial materials into products. See Figure 13-5. They cause the material to take on a desired size and shape. This might involve casting, applying force for forming, or machining using tools or other cutting devices. Secondary processing systems also assemble parts into products by welding, fastening, or gluing them together. Other secondary processing systems apply

Standards for Technological Literacy

19 20

TechnoFact

Before the Industrial Revolution, most manufacturing took place in craft shops in towns. The workers in these shops used simple tools to manufacture products such as cloth, hardware, leather goods, and silverware in limited quantities.

TechnoFact

Many early construction projects were inspired by structures found in nature. For example, early homes resembled caves. As early humans searched for food and water, they probably discovered the paths made by other animals. Such paths provided hunting opportunities, and undoubtedly led to water sources. As the humans began to use these paths, the earliest form of road system was created.

Hitachi, Bostitch

Figure 13-5. Secondary processing systems change industrial materials into products, such as this finishing sander and this power nailer.

coatings to the material, part, or product. These coatings protect the item from the environment and improve its appearance. Finally, secondary processing systems might change the properties of the material. This change might cause the material to be harder, stronger, or more resistant to fatigue.

The outputs of secondary processing systems are either industrial products or consumer products. See **Figure 13-6.** ***Industrial products*** are items companies use in conducting their businesses. For example, a computer terminal is an industrial product that can be used in the accounting activities of a company. Likewise, a furnace is an industrial product that becomes part of a building.

Consumer products are outputs developed for the end users in the product cycle. These end users are people such as homeowners, athletes, or students, just to name a few. Consumer products include such items as bath soap, television sets, lawn mowers, baseball bats, and furniture. The range of secondary manufacturing processes is introduced and discussed in Chapter 16.

Structure-Construction Systems

The three basic physical needs of all people are food, clothing, and shelter. Food comes to us from resource-processing systems. Clothing comes from manufacturing industries. Shelter is the output of construction activities.

Constructed works can be grouped into the categories of buildings and civil structures. Homes, factories, stores, and offices are typical buildings. Common civil structures are roads, railways, canals, dams, power-transmission lines, communication towers, and pipelines.

Each one of these structures is developed to meet a specific need. See **Figure 13-7.** The development involves such activities as preparing the site, building

Standards for Technological Literacy

19 20

Figure Discussion

Develop another model, like the one in Figure 13-6, for aluminum siding. Encourage your students to participate in its development.

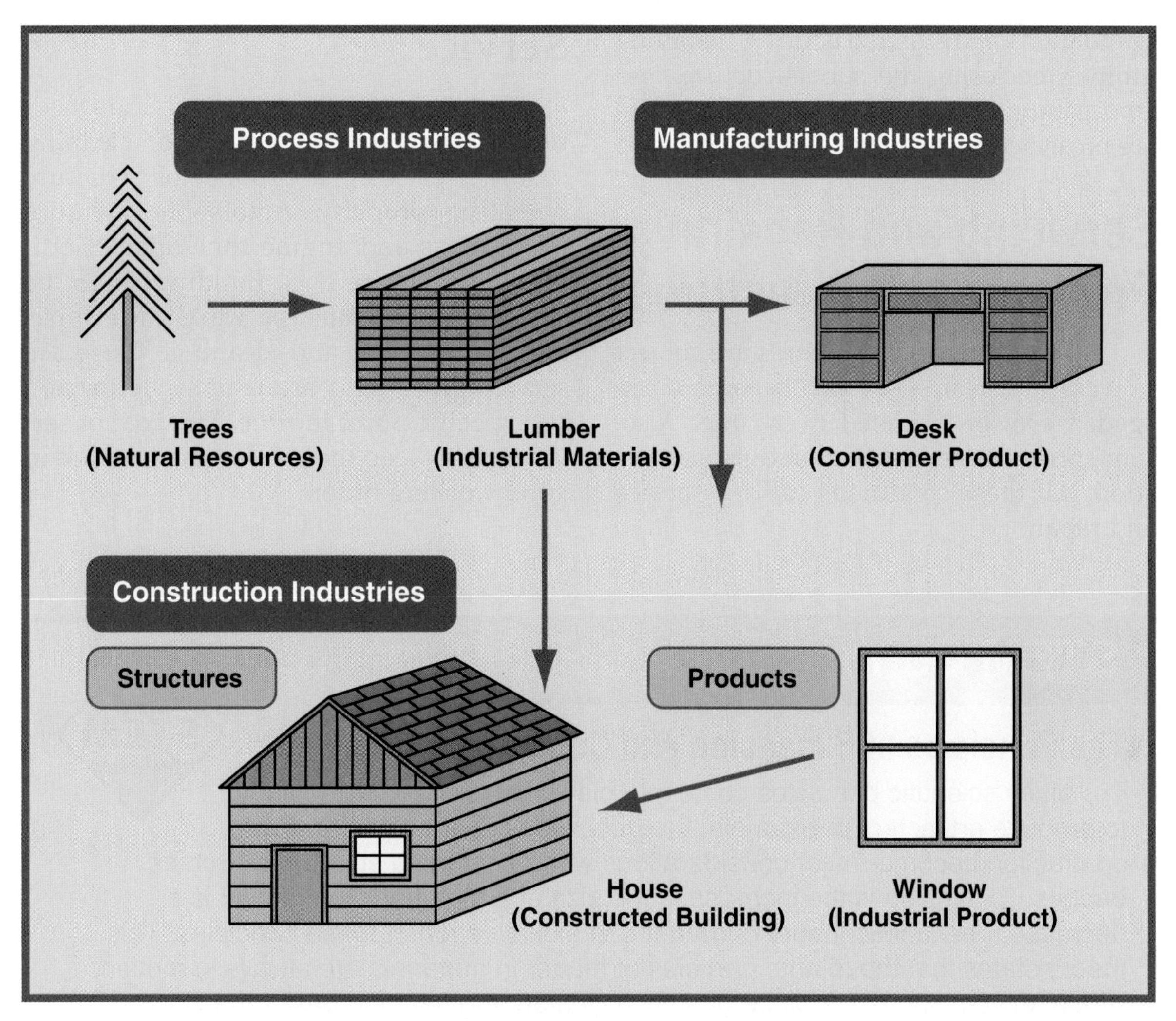

Figure 13-6. Production systems convert natural resources into industrial materials. These materials can be manufactured into industrial or consumer products. Materials and some manufactured goods are used in constructing structures.

Figure 13-7. Construction activities produce such structures as these grain-storage bins.

Standards for Technological Literacy

3 19 20

TechnoFact

Many modern household conveniences have far older origins than you may imagine. For example, the homes of many of ancient Rome's elite contained plumbing and central heating systems. The Romans were also the first builders to use glass windowpanes.

TechnoFact

Servicing offers an additional source of income for manufacturing companies. The Financial Times stated, "Manufacturers facing rising costs and shrinking margins must derive value not just from selling cars but from servicing, reselling and even recycling them."

foundations and superstructures, installing utilities, enclosing the superstructure, and landscaping the site. Construction activities are presented in detail in Chapter 17.

Servicing and Repairing Products and Structures

All products and structures are subject to wear and tear. They can become damaged, worn, or outdated over time. Also, some products need attention during operation. All these conditions call for service and repair.

Service

Servicing, or maintenance, is the scheduled adjustment, lubrication, or cleaning required to keep a product or structure operating properly. Automobiles require oil changes and engine tune-ups periodically. See **Figure 13-8.** Buildings must be cleaned. Floors must be waxed. Machines need adjustment and cleaning. These are servicing acts and are usually performed at a specific point in time. The goal of servicing is to keep the product or structure in good working order.

STEM Connections: Science

Integrated STEM Curriculum — Science, Technology, Engineering, Mathematics

The Principles of Expansion and Contraction

Many scientific principles come into play when we use technology to produce artifacts. For example, the principles of expansion and contraction become major considerations when designing structures such as bridges. Expansion is the increase in the size of a material. Contraction is a decrease. The kinetic theory of matter can explain each of these principles. This theory states that the minute particles of matter in materials are always in motion. When an item such as a piece of steel is heated, the heat increases the vibrations of the particles. The increased vibrations create more space between the particles, thus increasing the size of the item. The opposite happens when the piece of steel is cooled. The particles move closer together. The item contracts.

Designers add what are called *expansion joints* when determining how a long steel bridge is constructed because of the principles of expansion and contraction. The joints have interconnected seams that move closer together in the summer and move farther apart in the winter. Thus, the bridge will not buckle in the extreme heat of summer. Numerous other items would experience expansion and contraction problems, except that allowances have been made in their construction. Can you name two of these items?

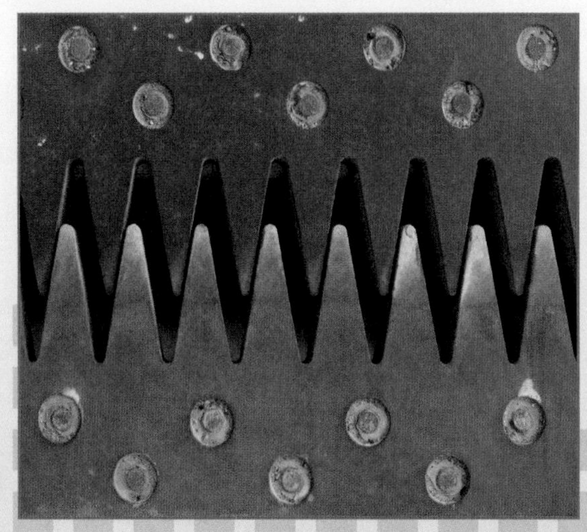

©iStockphoto.com/CheriJon

Figure 13-8. Servicing returns products to their original operating conditions.
A close-up shot of a bridge-deck expansion joint.

Repair

Repair, on the other hand, involves fixing a broken, damaged, or defective product or building and is designed to return a disabled product or structure to working condition. Product repair requires diagnosing (determining) the problem and fixing or replacing defective parts or materials. Repairing a building might include replacing damaged ceiling tiles, applying new wall coverings, or fixing a hole in a wall. Product repair can involve replacing worn-out or broken parts and adjusting mechanisms. The actions involved in servicing products and structures are presented in Chapter 18.

Standards for Technological Literacy

19 20

Figure Discussion

Discuss why do-it-yourself servicing and repair is easy because of readily available interchangeable parts like those shown in Figure 13-8.

TechnoFact

The term called value-added manufacturing describes one way to measure the effectiveness of manufacturing. It refers to the increased value of raw materials that have been converted into finished products. Such a measure reveals the effectiveness of manufacturing activities in financial terms.

Answers to Test Your Knowledge Questions

1. True
2. True
3. Evaluate individually.
4. False
5. True
6. True
7. construction
8. Evaluate individually.
9. Servicing is the scheduled adjustment, lubrication, or cleaning required to keep a product or structure operating properly.
10. Repair involves fixing a broken, damaged, or defective product or building.

Summary

Production activities have one of two major goals. These goals are to produce a product and to produce a structure. Although numerous kinds of production activities exist, they can be classified into three major groups or systems. These systems are resource processing, product manufacturing, and structure construction. After the produce or structure has been produced, we turn our attention to service and repair.

Test Your Knowledge

Write your answers on a separate piece of paper. Please do not write in this book.

1. Manufacturing activities are designed to produce products. True or false?
2. Producing a roadway is an example of a construction activity. True or false?
3. Give one example of a resource-processing system.
4. Iron ore is an industrial material. True or false?
5. Secondary processing systems change industrial materials into products. True or false?
6. An item a company uses can be considered an industrial product. True or false?
7. Buildings are outputs of ______ activities.
8. Name a common type of civil structure.
9. Define the term *servicing,* as used in this chapter.
10. Define the term *repair,* as used in this chapter.

STEM Applications

1. Select two products or structures you see around you. Complete a chart similar to the following one for each product.

Product		
Primary natural resource or resources		
Industrial material or materials used		
Product of manufacturing or construction		
Service or maintenance required		

Chapter 14 The Types of Material Resources and How They Are Obtained

Chapter Outline

Types of Natural Material Resources
Locating and Obtaining Natural Resources

This chapter covers the benchmark topics for the following Standards for Technological Literacy:

3 5 19

Learning Objectives

After studying this chapter, you will be able to do the following:

- Recall the types of natural resources used as inputs to production systems.
- Recall the types of genetic materials used in production systems.
- Recall the types of fossil fuel materials used in production systems.
- Recall the types of minerals used in production systems.
- Explain how genetic materials are obtained for use in production systems.
- Explain how fossil fuel materials are located and obtained for use in production systems.
- Explain how minerals are located and obtained for use in production systems.

Key Terms

blowout	germinate	proven reserve
buck	log	seed-tree cutting
ceramic mineral	maturity	seismographic study
clear-cutting	mineral	selective cutting
coal	mud	shaft mining
drift mining	natural gas	slope mining
evaporate	nonmetallic mineral	surface mining
fell	open-pit mining	timber cruising
fluid mining	ore	underground mining
fossil fuel	petroleum	yard
gem	potential field	

Strategic Reading

As you read this chapter, take note of the main points about natural resources. What types of natural resources are available? How are they found?

You have learned that technology involves people designing and using tools and artifacts. These actions extend human abilities to control or modify the environment. When no object or technical means exists, no technology is present. All technological objects (human-made objects, called *artifacts*) are made of materials. The materials each object is made from can be traced back to one or more natural resources. For example, plastic materials are made from plant fiber (cellulose) or natural gas. Glass is made from silica sand and soda ash. Cotton is grown and harvested from plants. Plywood comes from trees in the forest. Steel is made from iron ore, limestone, and coal.

Materials form the foundation for all production activities. Without material resources, production is not possible. In this chapter, we explore the types of natural resources we have available for production and how we obtain them.

Types of Natural Material Resources

Production technology uses materials and energy as inputs and makes products and structures as outputs. Three types of natural resources can become the inputs to production systems. See **Figure 14-1.** These materials are the following:

- Genetic materials.
- Fossil fuel materials.
- Minerals.

Genetic Materials

Many resources come from living things. These resources, as you learned in Chapter 4, are called *organic materials.* Some organic materials are from organisms that have been dead for hundreds of years. This type of organic material includes fossil fuels,

Standards for Technological Literacy

19

Example

Give your students examples of technological vs. non-technological activities (i.e. walking vs. riding a bicycle, talking vs. e-mailing, looking vs. x-raying a bone, etc.).

TechnoFact

Natural resources are commonly categorized by their ability to be replenished. Inexhaustible resources have an unlimited supply. Renewable resources have a limited immediate supply, but can be replenished over time. Nonrenewable resources have a limited supply and cannot be replaced, or can only be replaced over extremely long periods of time. Recyclable resources can be used more than once.

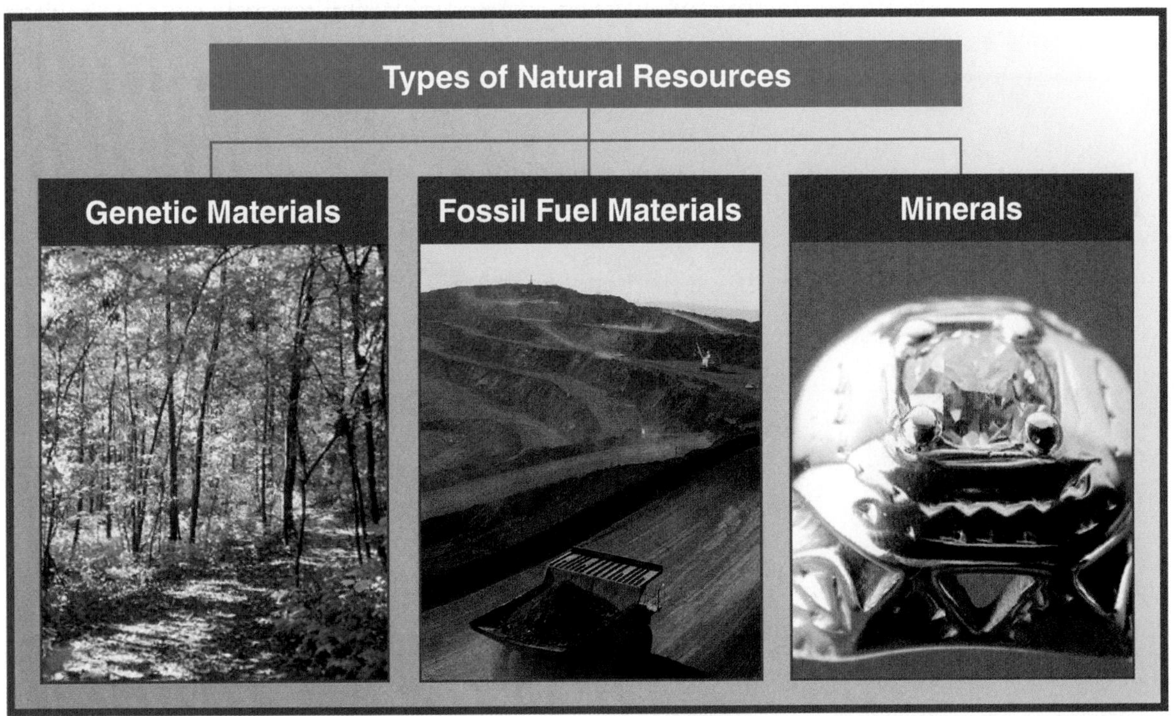

U.S. Department of the Interior

Figure 14-1. Genetic, fossil fuel, and mineral materials are used as inputs to production systems. Here, we see them in the form of trees, a coal mine, and a diamond, respectively.

Standards for Technological Literacy

19

Research

Have your students trace the typical lifecycle of a commercially valuable tree.

Research

Have your students investigate the advantages and disadvantages of using fossil fuels.

TechnoFact

Genetic, or biological, material is a classification of natural resources that includes all substances that are developed by plants (e.g. wood, cotton, flax, jute, etc.) or animals (e.g. leather, fur, wool, silk, etc.).

TechnoFact

Fossil fuels provide 90% of the energy people use today.

which are discussed later in this chapter. Other organic materials are obtained during the normal life cycles of plants or animals. These materials can be called *genetic materials.*

We obtain genetic materials through farming, fishing, and forestry. Each of these activities works directly with nature, as plants and animals move through the stages of life. See **Figure 14-2.**

Typical genetic materials used in production systems are grains (wheat, oats, barley, and corn, for example), vegetable fibers (wood, flax, and cotton, for example), and animals (meat, hides, and wool, for example).

The origin of all genetic materials is in birth or ***germination***. The appearance of animal life is called *birth*. Plant life generally starts with the germination of seeds or spores.

Young plants and animals grow rapidly early in their life cycles. Their growth generally slows down as they reach older age. This period is called ***maturity***. The organism is still healthy at maturity, but it stays about the same size. The length of this maturity stage can be a matter of days, as with mushrooms, or centuries, as is the case with redwood trees. All organic life ends at some point, however. The plant or animal dies from old age or disease.

Fossil Fuel Materials

Fossil fuels are mixtures of carbon and hydrogen. They are called *hydrocarbons*, and they include a vast number of products in use today, from fuels to medicines. Most hydrocarbon products, however, come from only three fossil fuel resources. These are the following:

- Petroleum.
- Natural gas.
- Coal.

Petroleum

Petroleum is an oily, flammable mixture of hydrocarbons that has no specific composition. Instead, it is a mixture of a

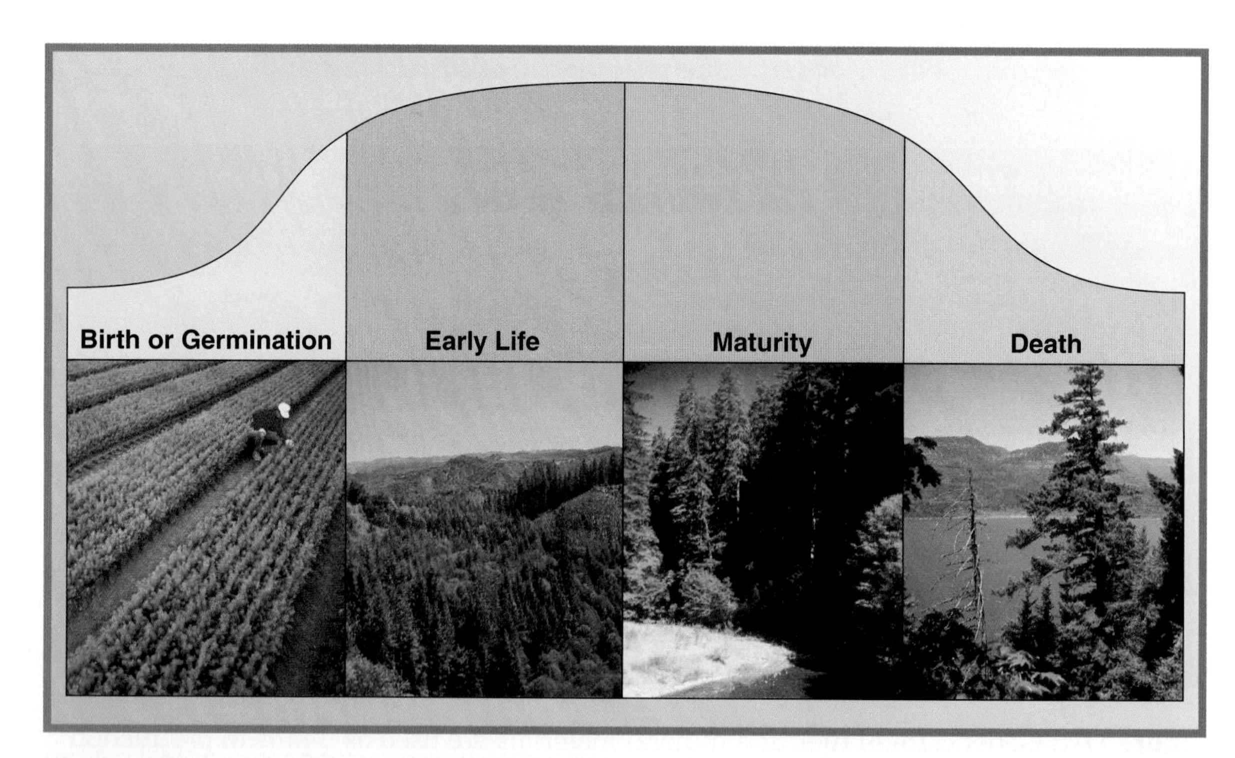

Figure 14-2. All genetic materials have a life cycle, starting with birth, or germination, and ending with death.

number of different solid and liquid hydrocarbons. The composition of petroleum varies with where the petroleum is found on earth. Petroleum is the principal source of fuels, such as gasoline, diesel fuel, and heating oil. See **Figure 14-3.** Lighter hydrocarbon products that come from petroleum are gases. They include methane, ethane, butane, and propane. These gases are widely used in producing plastic resins such as polypropylene and polyethylene.

Natural Gas

Natural gas is a combustible gas occurring in porous rock. This gas is composed of light hydrocarbons. Typically, natural gas is about 85% methane, and the rest is made up of propane and butane. Natural gas is used as a fuel for homes and industry. This gas is also used to make plastics, chemicals, and fertilizers.

Compared with other fossil fuels, natural gas burns cleanly. This gas requires complex pipeline networks to distribute it to potential users. In some cases, it is compressed and used as a fuel for vehicles. In this form, it is called *compressed natural gas (CNG).*

©iStockphoto.com/pic4you

Figure 14-3. Petroleum is a vital natural resource that provides the fuels to power vehicles and other machines.

Coal

Coal is a combustible solid composed mostly of carbon. This solid started as plant matter thousands of years ago. In moist areas, the plant matter did not decay easily and layered up to make peat. Peat is brownish-black plant matter that looks similar to decayed wood. Dry peat burns, but it gives off a great deal of smoke. When sediment buried peat thousands of years ago, pressure and heat changed the peat into coal.

The principal types of coal are the following:

- **Lignite.** This soft and porous material is made from peat that has been pressed by natural action. Lignite gives off more heat than peat and is used in electrical generating plants and for industrial heating.
- **Bituminous coal.** This coal is the most commonly found coal, which is harder than lignite. Bituminous coal is sometimes called *soft coal* because it can be easily broken into various sizes. See **Figure 14-4.** This coal is widely used for power generation and heating. Bituminous coal can also be used for coal gasification and chemical processes.
- **Anthracite coal.** This coal is the hardest coal and is a shiny black material that burns without smoke. Anthracite has the highest carbon content of all the types of coal. Anthracite coal is used for heating and to produce coke for steelmaking.

Coal does not burn cleanly, and its sulfur content is a source of chemicals that make acid rain. Also, its bulk makes it costly to ship.

Standards for Technological Literacy

19

Research

Have your students research the locations of coal deposits in North America and around the world.

TechnoFact

Coal-burning steam engines were partly responsible for driving the Industrial Revolution. Coal provided most of the power for factories from the early 1800s to the early 1900s.

Standards for Technological Literacy

19

Extend

Give your students examples of products made from ores, ceramic minerals, nonmetallic minerals, and gems.

TechnoFact

Minerals have long been regarded as valuable resources. Early Egyptian paintings, more than 5000 years old, illustrate the use of minerals in weapons and jewelry. Early Greek and Roman texts describe metals, ores, and stones.

Jeffboat Shipyard

Figure 14-4. Bituminous coal is sometimes called *soft coal.*

Minerals

Minerals are any substance with a specific chemical composition that occurs naturally. They are different from fossil fuel resources, which are all chemical mixtures. Typical minerals are iron ore, bauxite (aluminum ore), and sulfur.

We can classify minerals in various ways. One way is to group them by their chemical compositions. This grouping includes native elements (elements that occur naturally in a pure form), oxides, sulfides, nitrates, carbonates, borates, and phosphates.

There is a more useful way to group minerals available, however. This method groups the minerals that have economic value into families with similar features. These groups include the following:

- ***Ores.*** These minerals have a metal chemically combined with other elements. Ores can be processed to separate the metal from the other elements.
- ***Nonmetallic minerals.*** These substances do not have metallic qualities, such as sulfur.

Career Corner

Forest and Conservation Workers

Forest and conservation workers perform tasks associated with replanting and conserving timberlands. Also, they work to maintain forest facilities, such as roads and campsites. Forest workers plant seedlings, remove diseased or undesirable trees, spray trees to kill insects, and apply herbicides on undesirable brush and trees. They might manually clear brush and debris from hiking trails and camping sites and help maintain recreational facilities and campgrounds.

Forest and conservation work is concentrated in the West and Southeast of the United States. Most forest and conservation jobs require little formal education. Many secondary schools and community colleges in timber states offer courses in general forestry, wildlife, conservation, and forest harvesting.

- *Ceramic minerals.* These fine-grained minerals are formable when wet and become hard when dried or fired.
- *Gems.* These stones are cut, polished, and prized for their beauty and hardness.

Locating and Obtaining Natural Resources

All natural materials are found on the earth. They are not all visible as we travel across the land and water or through the air, however. In fact, we need to perform two major actions to obtain natural resources. First, we must locate them, and then we must extract or gather them.

Locating and Obtaining Genetic Materials

Most genetic materials are easy to find. Trees and farm crops are on easily used plots of land. Domesticated animals and the fish raised on fish farms are contained in specific locations. Only those who fish commercially must seek genetic resources that are sometimes hard to find.

The major challenge for people dealing with genetic resources is to harvest the plant or animal at the proper stage of growth. This stage varies with the growth cycle and growing habits of the organism. Many trees are harvested during their mature phase. Some young trees, however, do not grow in the shade of older trees. Thus, sometimes all the trees in a single plot must be harvested at one time. New trees can then be planted, which will grow in the cleared area.

Likewise, most farm crops are planted at one time in fields. The plants can be fertilized and irrigated to stimulate their growth. See Figure 14-5. Most of the crop matures at the same time. Therefore, the entire crop can be harvested at one time.

©iStockphoto.com/Mur-Al

Figure 14-5. Intensive farm management, such as this irrigation process, can increase crop yields.

Let us look at harvesting a genetic material—trees. The forest-management process requires each stand of trees to be evaluated and designated for a specific use. Wilderness areas are set aside. Roads and logging are not permitted in a wilderness area. National parks, and many state and provincial parks, protect scenic beauty. Roads are allowed in the parks, but logging is not. National, state, and provincial forests, generally are multiuse lands. Lakes, hiking trails, and camping areas are set aside for recreational use. ***Logging*** is permitted in selected areas to harvest mature trees. Private forests are generally designed and managed to maximize the amount of harvested material.

Standards for Technological Literacy

19

Research

Have your students research the varying opinions about the role of national forests.

Brainstorm

Have your students discuss whether all the aims of a multiuse forest can be met.

Standards for Technological Literacy

19

TechnoFact

A multiuse forest is actively managed to provide a sustainable source of timber, recreation for humans, food and shelter for wildlife, grazing for livestock, and water for communities.

The logging process requires both planning and action. Planning for the removal of trees involves several steps. First, the forest is studied to determine whether or not it is ready for harvesting. This study involves a process called ***timber cruising***. Teams of two or three foresters measure the diameters and heights of the trees. Their task is to find stands of trees that can be economically harvested. These teams also prepare topographical maps showing the locations and elevations of the features on the potential logging site.

Forest engineers must then plan the proper way to harvest the trees. They plan logging roads and loading sites. These engineers also select the type of logging to match the terrain and the type of forest. Three logging methods exist:

- ***Clear-cutting.*** All trees, regardless of species or size, are removed from a plot of land that is generally less than 1000 acres. See **Figure 14-6.** This process allows for replanting the area with trees that cannot grow in competition with mature trees. Also, the number of tree species can be controlled.
- ***Seed-tree cutting.*** All trees, regardless of species, are removed from a large area, except three or four per acre. These trees are used to reseed the area. Again, the type of seed trees left controls the number of reseeded species.
- ***Selective cutting.*** Mature trees of a desired species are selected and cut from a plot of land. This technique is used in many pine forests, where tree density is limited.

These steps are the prelude for the main activity called *logging*. See **Figure 14-7.** Workers move equipment into the forest to remove the trees. Loggers called ***fellers*** use a chain saw to cut down (fell) the appropriate trees. A machine that shears the trees off at ground level can harvest smaller trees. Loggers are careful to drop the trees into a clear area so they are not damaged and so they do not cause damage to other trees.

Figure 14-6. This clear-cutting increases the yield of Douglas firs.

Workers called ***buckers*** remove the limbs and tops. These parts of the tree are called *slash*. The slash is piled for later burning or chipping so nutrients are returned to the soil. The trunk is then cut into lengths called *logs*.

Workers gather the logs in a central location called the *landing*. This process of gathering is called ***yarding*** and can be accomplished in several ways. First, workers use chokers (cables) to bind the logs into bundles. They can then use cables to drag logs using high-lead and skyline yarding. See **Figure 14-8.** These systems use a metal spar (pole), cables, and an engine. High-lead yarding drags the logs to the landing. Skyline yarding lifts and carries the logs over rough or broken terrain.

Boise Cascade

Figure 14-7. Logging starts with selecting and cutting trees. The felled trees are collected using yarding. The trees are then hauled to a mill for storage and further processing.

Standards for Technological Literacy

19

Research

Have your students research the timber output per acre from private forests and national forests and explain why the numbers are so different.

TechnoFact

Logging, particularly on tree plantations, has been greatly simplified by the development of modern machines. Hydraulic tree shears are capable of slicing through a tree like a pair of giant scissors. A single machine called a tree harvester is capable of felling a tree, stripping off its branches, and cutting it into logs.

On gentle terrain, ground yarding is used. This system uses tractors and an implement called an *arch*. These items drag the logs, bound together with a choker, to the landing. Very steep terrain might require helicopter yarding. Refer to **Figure 14-7.**

Once logs arrive at the landing, workers load them on trucks and move them to the processing plant. This plant might be a lumber, plywood, particleboard, hardboard, or paper mill. The logs are often stored at the mill in ponds or stacked up and sprayed with water to prevent cracking and insect damage.

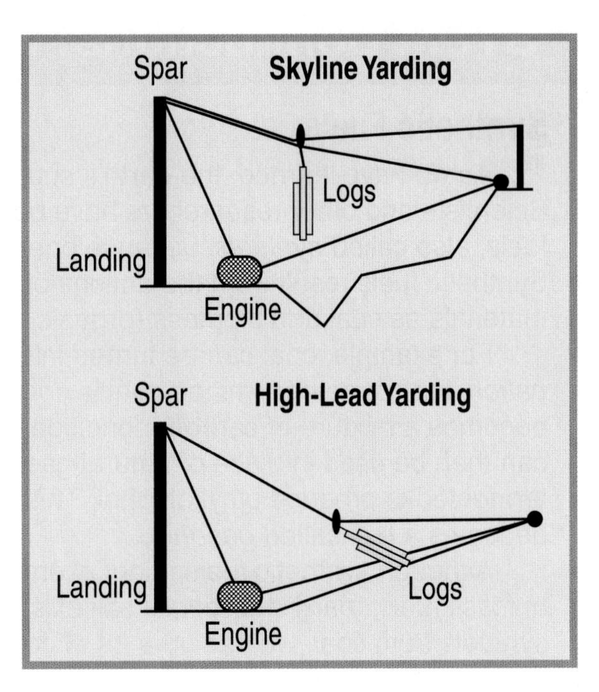

Figure 14-8. Two ways used to move logs to the landing are high-lead and skyline yarding.

Locating and Obtaining Fossil Fuel Materials

Most fossil fuel resources are buried under the surface of the earth. Fossil fuels

Standards for Technological Literacy

3 19

Extend

Discuss the role of fossil fuels in modern life.

TechnoFact

Although petroleum has been in use for thousands of years, it didn't become an important fuel until the 1800s. The development of the kerosene lamp and the automobile increased demand for gasoline and kerosene, both derived from petroleum. These products are still in high demand today.

can be pools of petroleum, pockets of natural gas, or veins of coal. Locating and extracting these resources are major challenges. The techniques used for natural gas and petroleum are much different from those used for coal. Therefore, let us look at them separately.

Locating and Obtaining Petroleum and Natural Gas

People do not look for oil (petroleum) and gas directly. They look for rock formations that might contain deposits of oil and gas. Finding these rock formations is the job of geologists and geophysicists. See **Figure 14-9.**

Petroleum comes from decayed plant and animal matter that layers of sediment from rivers covered. The layers built up and created great pressure and heat. Over millions of years, the pressure and heat turned the organic matter into oil. Oil and gas are generally found in porous rock under a

BP

Figure 14-9. This geologist is seeking deposits of petroleum.

STEM Connections: Science

Integrated STEM Curriculum — Science, Technology, Engineering, Mathematics

Synthetic Fuels

As we have learned, the earth's supply of fossil fuels is limited. Scientists and other researchers have been working on creating synthetic fuels, also called *synfuels*, because one day, fossil fuels will be exhausted. Synthetic fuels result from the actions of various chemical processes on such materials as coal and biomass (organic matter).

For example, coal can be turned into gas through a process called *gasification.* In this method, mined coal is combined with steam and oxygen and becomes a mixture of carbon monoxide, hydrogen, and methane. This mixture can then be used in place of natural gas. In another process, organic matter is fermented to produce ethyl alcohol. This product is then mixed with gasoline and becomes a fuel called *gasohol.*

Although synthetic fuels might eventually be the answer to our limited supply of fossil fuels, many drawbacks still exist. For example, in the process of creating synfuels from coal, we use up a lot of coal, thus reducing our supply. Producing synfuels is also extremely costly at this time. Scientists continue to work on these and other problems associated with synfuels. Can you name a synfuel currently being investigated and a problem being addressed in the research?

layer of impervious (dense) rock. The oil and gas collect under the dense rock. These deposits can be under oceans, mountains, deserts, or swamps. They can be near the surface, as they are in the Middle East, or several miles beneath the land or sea.

We have developed various ways to explore for petroleum and natural gas. One of the most accurate ways is ***seismographic study***. See **Figure 14-10.** This technique uses shock waves, such as those in an earthquake. A small explosive charge is detonated in a shallow hole. The shock waves from the explosion travel into the earth. When the waves hit a rock layer, they reflect back to the surface. Seismographic equipment uses two listening posts to measure the shock waves bouncing off various rock layers. Measuring the time it takes the waves to go down to the layers and reflect back allows the geologist to construct a map of the rock formations.

Other methods use geological mapping, fossil study, and core samples from drilling to search for the deposits. See **Figure 14-11.** Geological mapping measures the strength of magnetic forces.

The geological study helps people select a good site for exploration. If an area has never produced oil or gas, it is called a ***potential field***. Producing fields are called ***proven reserves***.

Standards for Technological Literacy

19

Research

Have your students research the instruments, other than a seismograph, that are used to explore for oil.

TechnoFact

A variety of technological devices are used in the search for petroleum deposits in rock formations. Commonly used devices include the seismograph, the gravimeter, and the magnetometer.

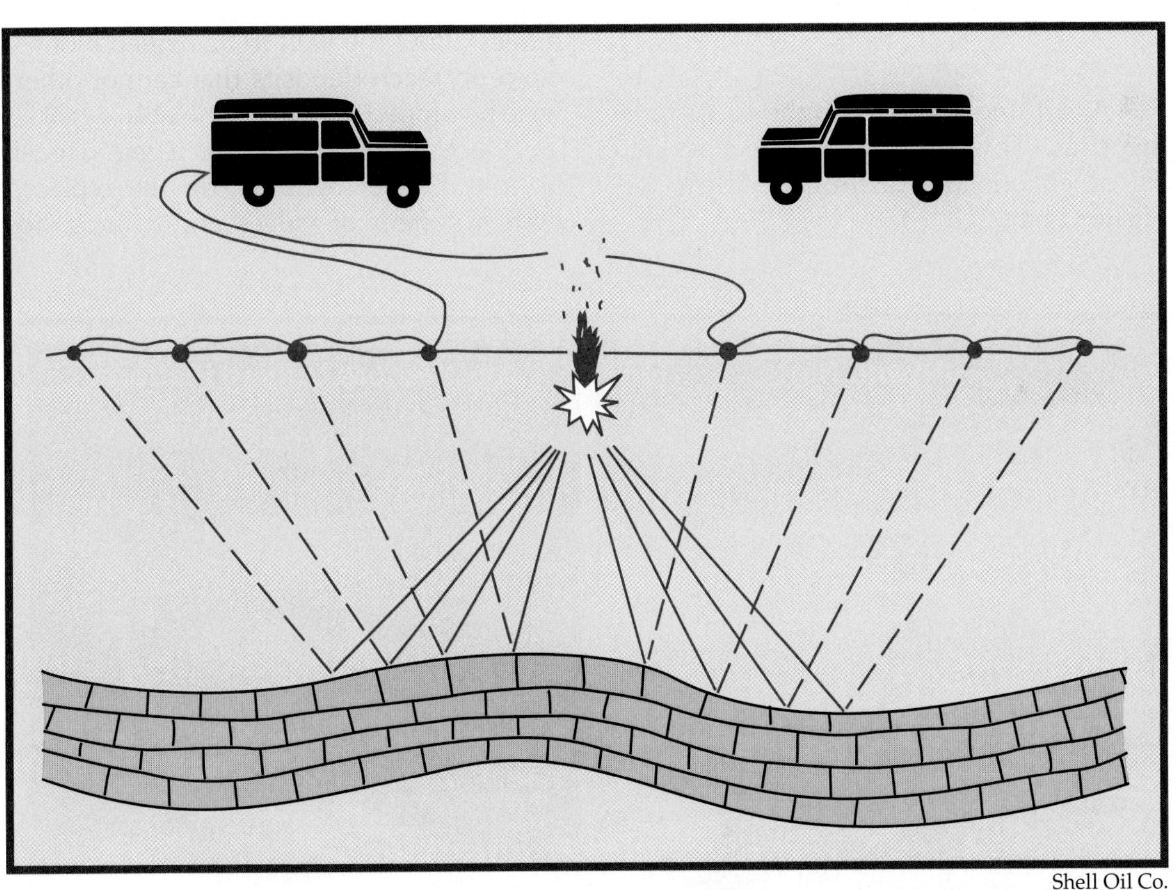

Shell Oil Co.

Figure 14-10. Seismographic surveys are used to determine the shape of underground rock formations that might hold oil and gas. An explosive charge is detonated, creating a shock wave that travels down into the earth. The boundaries between different types of rock reflect the shock waves back to the detectors on the surface. Geologists measure the time it takes for the waves to reach the detectors. The timing information gives the geologists an idea of the shape of a formation.

Standards for Technological Literacy

19

Research

Have your students research and identify the major petroleum- and natural gas-producing areas in the world. Have them determine the percentage of worldwide production that each area represents.

TechnoFact

Some geologists specialize in the search for mineral and fuel resources. Geologists may seek fuel sources, such as oil, coal, gas, and uranium deposits. They may also search for ores, including those containing copper, iron, gold, lead, silver, and tin. Other geologists locate clay, building stone, and underground water sources.

Figure 14-11. This scientist is studying fossils to search for petroleum.

A drilling rig is brought to promising sites. This rig can be either a land rig or an offshore drilling platform. See **Figure 14-12.**

Drilling involves rotating a drill bit on the end of a drill pipe. Lengths of pipe are added to the drill pipe as the hole gets deeper. See **Figure 14-13.**

During drilling, a mixture of water, clay, and chemicals is pumped down the drill pipe. This mixture is called ***mud***. The mud flows out through holes in the drill bit to cool and lubricate. This mixture also picks up the ground-up rock and carries it to the surface. Finally, it seals off porous rock and maintains pressure on the rock. Pressure is maintained to prevent a blowout. A ***blowout*** occurs when oil surges out of the well. Blowouts are very dangerous and waste large quantities of oil and gas.

Early oil wells were drilled straight down or at a specific angle. Modern techniques allow the well to be drilled along a curve to reach deposits that cannot otherwise be tapped. See **Figure 14-14.**

Once an oil deposit or a gas deposit is found, the drilling rig is replaced with a system of valves and pumps. See

BP, Gulf Oil Co.

Figure 14-12. Oil and gas rigs allow people to drill wells on land and under lakes and seas.

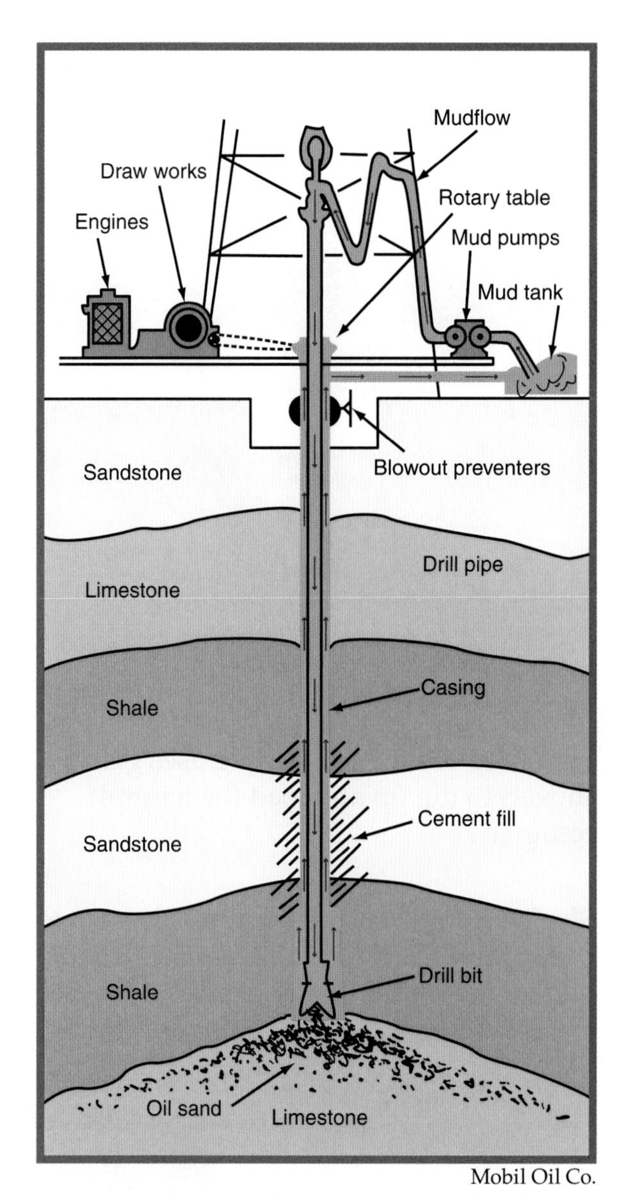

Mobil Oil Co.

Figure 14-13. Drilling involves rotating a drill bit on the end of a drill pipe.

BP

Figure 14-14. Modern techniques allow wells to be drilled along curved lines to tap difficult-to-reach deposits of oil and gas.

Standards for Technological Literacy

19

Figure Discussion

Discuss the importance of angular drilling for off-shore drilling.

TechnoFact

Although most people use petroleum-based products on a daily basis, very few have ever seen petroleum. Petroleum is a liquid that is found in a variety of colors and viscosities, ranging from relatively thin and clear to thick and black.

Figure 14-15. The recovered resource flows through pipes into storage tanks. From the well, the petroleum is transported to refineries. Natural gas is compressed and sent to petrochemical plants. Pipeline companies sell natural gas to home-heating and electric-power customers.

Locating and Obtaining Coal

Coal is the most abundant fossil fuel and is found on every continent. Most of the

AMAX

Figure 14-15. A valve system, called a *Christmas tree,* is set on top of an oil well.

known reserves are in the Northern Hemisphere, however. These reserves are generally recovered through mining operations. There are three types of mining: surface or open pit, underground, and fluid. The first two are used to mine coal. The third is used to obtain minerals and is discussed in the next section.

Surface mining, or ***open-pit mining***, is used when the coal vein is not very deep underground. This type of mining generally involves four steps:

1. The surface layers of soil and rock are stripped from above the coal. This material is called *overburden*. The overburden is saved for later use in reclaiming the land.
2. The coal is dug up with giant shovels. See **Figure 14-16.**
3. The coal is loaded on trucks or railcars to be transported to a processing plant.
4. The site is reclaimed by replacing the topsoil and replanting the area.

Underground mining requires shafts in the earth to reach the coal deposits. The three types of underground mining are shaft, slope, and drift. See **Figure 14-17.**

American Electric Power

Figure 14-16. Surface mining uses giant shovels to dig up and load the natural resource.

Shaft mining requires a vertical shaft to reach the coal deposit. Workers then dig horizontal shafts to remove the coal. See **Figure 14-18.**

Standards for Technological Literacy

5 19

Brainstorm

Ask your students to list the environmental challenges presented by surface, or open-pit, mining.

TechnoFact

Since petroleum is a nonrenewable resource, new fuels must ultimately replace petroleum-based fuels. Concern about the instability of the petroleum market and dwindling supplies has recently sparked a growing interest in synthetic fuels (synfuel). Synfuels are fuels that can be substituted for petroleum and natural gas. They are generally produced from coal, oil shale, bituminous sands, or biomass.

Think Green

Green Materials

We have already discussed the ideas of recycling and sustainability. Now, let's talk about materials that support those ideas: *green materials*, or materials that are considered environmentally friendly. Green materials can be defined in many different ways. The most common criteria say that for a material to be considered green, it must not tax exhaustible natural resources, it must not cause carbon dioxide emissions when processed, and it must either be recyclable or be biodegradable. Manufacturers have been making strides to not only make their products recyclable, but also to make the products out of recycled materials while emitting less carbon dioxide in processing.

For example, in recent years, manufacturers have begun using plant fiber rather than petroleum to create plastic, or bio-plastic. These materials are made of renewable resources, are cleaner to process, are recyclable, and are biodegradable. Bioplastic is just one innovation of the search to find alternative green materials.

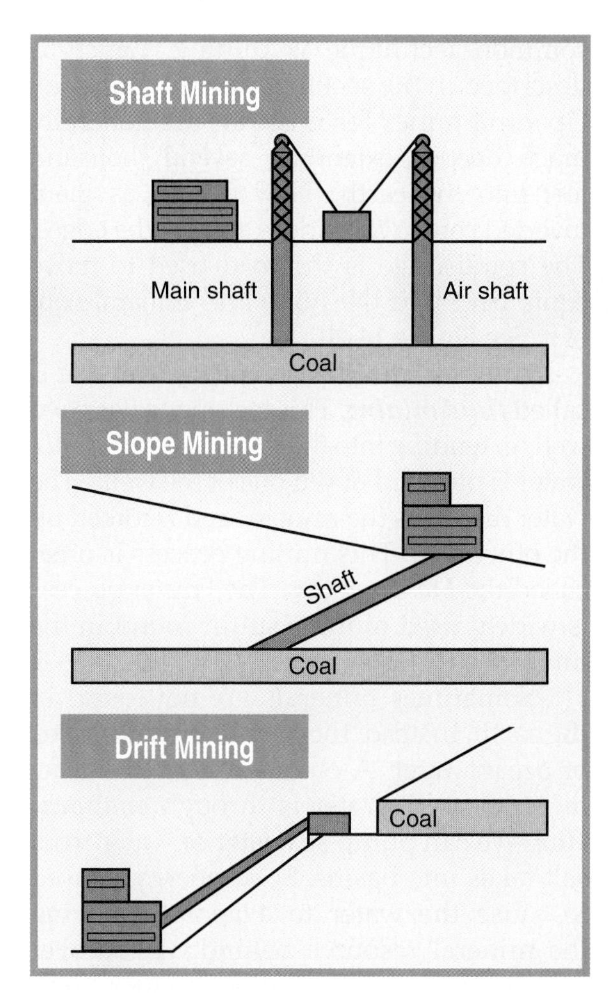

Figure 14-17. The three types of underground mining are shaft, slope, and drift.

FMC Corp.

Figure 14-18. This long-wall mining machine is recovering coal from an underground mine.

Miners use ***slope mining*** when the vein is not too deep under the ground. They dig a sloping shaft to reach the coal. The miners then dig a horizontal shaft to follow and remove the coal vein.

They use ***drift mining*** when the coal vein extends to the surface of the earth. The miners then dig a horizontal shaft directly into the vein. Those miners involved with underground mining use elevators to remove the mined coal. The same elevators move people and equipment in and out of the mine. Those working with slope and drift mines often use coal cars or conveyors to remove the coal. See **Figure 14-19.**

AMAX

Figure 14-19. Coal is removed from slope and drift mines on tractors or conveyors.

Standards for Technological Literacy

19

Figure Discussion

Discuss where each type of mine shown in Figure 14-17 is used.

TechnoFact

For most of our history, humans have been mining. Some early inhabitants of the European continent dug pits and tunnels to find flint for their tools and weapons. Humans started mining copper and tin about 5500 years ago. Mining was an integral part of the Roman economy. The Romans seized the mines of every country they conquered. Europeans began actively mining for coal in the 1400s.

Standards for Technological Literacy

19

Research

Have your students research the ways in which salt is removed from seawater.

Extend

Explain why solar salt is very pure.

TechnoFact

The salt in seawater, roughly 3% by weight, can be extracted through solar evaporation. The water is moved through a series of evaporating ponds. The concentration of salt increases as the water evaporates. When the salt concentration reaches approximately 26% by weight, the salt begins crystallizing and falling to the bottom of the pond where it can be harvested. Such evaporation processes require hot, dry climates and can be found in the United States near the San Francisco Bay and the Great Salt Lake.

TechnoFact

The word petroleum comes from two Latin words: petra, which means rock, and oleum, which means oil. The word owes its origins to the fact that petroleum was first discovered seeping from cracks in rocks.

Locating and Obtaining Minerals

Over the years, valuable minerals have been found by one of two methods. The historical way was the prospector, who looked for interesting rock formations or mineral deposits in streams. Prospectors staked claims on promising locations and then either worked the claims or sold them to mining companies. Today, modern prospectors are usually trained geologists who study earth formations to discover deposits of a specific mineral. The geologists first identify an area that might contain the mineral. They select this area using their knowledge of how ores are formed and where they occur. The selected area is then reduced in size through geological investigations that can include testing of the surface and subsurface geology (the earth's physical structure). This investigation might include studying variations in gravity, magnetism, and other variables. The end result is locating a deposit of ore that can be extracted.

We can extract minerals from the earth in a variety of ways. Probably the most common technique is mining, which is described in the section on recovering coal. Open-pit mines for minerals are generally much deeper, extending several thousand feet into the earth. They appear as giant inverted cones with ridges around the edges. The spiral ridge is the road used to move equipment into the mine and minerals out of it. See **Figure 14-20.**

Another mining method we can use is called ***fluid mining.*** This technique uses two wells extending into the mineral deposit. Hot water is pumped down one of the wells. The water dissolves the mineral and is forced up the other well. This mining process is often called the *Frasch process.* The Frasch process is widely used to mine sulfur found in the limestone rocks covering salt domes.

Sometimes minerals are not found in the earth. Instead, they are in ground-, lake, or ocean water. A common way to extract minerals from water is through ***evaporation.*** We can pump seawater or water from salt lakes into basins. Solar energy is used to cause the water to evaporate, leaving the mineral resource behind. We recover a number of minerals in this manner from the Great Salt Lake in Utah.

Brush-Wellman

Figure 14-20. Open-pit mineral mines can extend several thousand feet into the earth.

This lumber comes from a natural resource. Before it is processed into lumber, trees (genetic materials) must be harvested.

Answers to Test Your Knowledge Questions

1. minerals
2. True
3. True
4. Petroleum, natural gas, and coal
5. Minerals are naturally occurring inorganic materials with homogenous composition, usually a crystalline structure, and distinct physical properties. Minerals differ from fossil fuels in that they have inorganic origins and are composed of a single element or compound.
6. Evaluate individually.
7. The major challenge for people dealing with genetic resources is to harvest the plant or animal at the proper stage of growth.
8. yarding
9. Locating and extracting the material

Summary

All production systems use materials. Common production materials come from genetic, fossil fuel, and mineral resources. To use these resources, we must first locate them and then know how to extract them. Extraction can involve harvesting trees, farm crops, and animal life and also includes drilling oil and gas wells. Finally, we can extract material resources from the earth using mining techniques.

Test Your Knowledge

Write your answers on a separate piece of paper. Please do not write in this book.

1. The three types of natural resources that are inputs to production systems are genetic materials, fossil fuel materials, and ______.
2. We obtain genetic materials through farming, fishing, and forestry. True or false?
3. Fossil fuels are mixtures of carbon and hydrogen. True or false?
4. List the three main fossil fuel resources.
5. What is the main difference between fossil fuel resources and minerals?
6. Name a typical mineral.
7. What is the major challenge involved in obtaining genetic materials?
8. The process of gathering logs is called ______.
9. What are the major challenges involved in obtaining fossil fuel materials?
10. Describe the method of seismographic study.
11. What is the difference between a potential field and a proven reserve?
12. Coal is the most abundant fossil fuel. True or false?
13. Name one type of mining used to obtain coal.
14. Name one way we can obtain minerals from water.
15. What is fluid mining?

STEM Applications

1. Select a resource from the categories of genetic materials, fossil fuels, and minerals. Research how it is located (exploration) and recovered from the earth (production). Use a chart similar to the one below to record your findings.

Resource:		
Exploration processes:		
Extraction processes:		

2. Build a model to show a selected mining or drilling technique.
3. Read an article and write a report on a controversial resource-recovery issue. The report can deal with a number of issues, including protecting the environment, wildlife, or the supply of the resource.

10. This technique uses shock waves such as those in an earthquake. A small explosive charge is detonated in a shallow hole. The shock waves from the explosion travel into the earth. When the waves hit a rock layer they reflect back to the surface. Seismographic equipment uses two listening posts to measure the shock waves that bounce off various rock layers. Measuring the time it takes the waves to go down to the layers and reflect back allows the geologist to construct a map of the rock formations.
11. If an area has never produced oil or gas it is called a potential field. Producing fields are called proven reserves.
12. True
13. Surface or open-pit mining and underground mining
14. Evaporation
15. Fluid mining uses two wells that extend into the mineral deposit. Hot water is pumped down one of the wells. The water dissolves the mineral and is forced up the other well.

Chapter Outline

Mechanical Processes

Thermal Processes

Chemical and Electrochemical Processes

This chapter covers the benchmark topics for the following Standards for Technological Literacy:

3 5 16 19

Learning Objectives

After studying this chapter, you will be able to do the following:

- Recall the types of primary processes.
- Give examples of materials resulting from mechanical processing.
- Explain how mechanical processing techniques are used to process a resource.
- Give examples of materials resulting from thermal processing.
- Explain how thermal processing techniques are used to process a resource.
- Give examples of materials resulting from chemical and electrochemical processing.
- Explain how electrical or electrochemical processing techniques are used to process a resource.

Key Terms

alumina
bauxite
billet
blast furnace
bloom
cant
chemical process
conversion
core
cracking
cross band
edger saw
electrochemical process
face
fiberglass
fiber-optic cable
float glass
fractional distillation
fractionating tower
galvanized steel
glass
hardwood lumber
head rig
lehr
lumber
lumber-core plywood
mechanical process
particleboard-core plywood
pig iron
polymerization
refine
resaw
separation
skelp
slab
smelt
softwood lumber
steel
tapping
thermal process
tin plate
treat
trim saw
veneer
veneer-core plywood

Strategic Reading

As you read this chapter, think about the different processes used to produce products. What types of materials are processed mechanically? Are they different from those processed thermally?

Natural resources are used to produce products and structures. The production process involves several steps. One of these steps comes between obtaining the resource and manufacturing the product or constructing the structure. This step can be called *primary processing*. See **Figure 15-1.** The goal of primary processing is to convert material resources into industrial materials. Industrial materials are often called *standard stock*. For example, primary processing converts wheat into flour; aluminum ore into aluminum sheets, bars, and rods; logs into lumber, plywood, particleboard, hardboard, and paper; natural gas into plastic pellets, film, and sheets; and silica sand and soda ash into glass. These industrial materials are used as inputs to further manufacturing or construction activities.

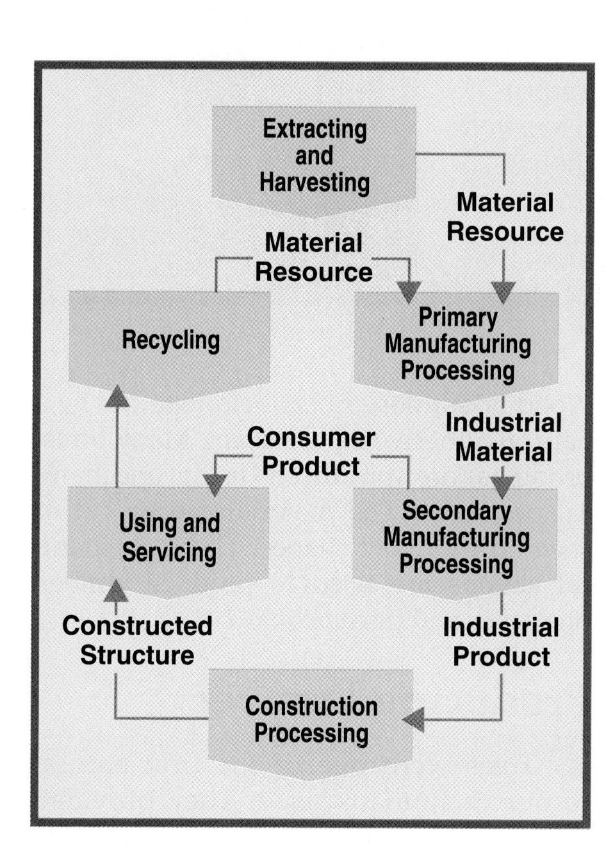

Figure 15-1. This model shows the production process. Note how primary processing fits into the cycle.

Primary processing uses many production actions. These actions can be grouped in a number of ways. One useful way to group processes is by the type of energy used. See **Figure 15-2.** This grouping includes the following:

- Mechanical processes.
- Thermal (heat) processes.
- Chemical and electrochemical processes.

This grouping provides a general way to understand most primary processing activities. Many materials are produced using more than one type of process. For example, steel is made from iron ore, coke, and limestone using a thermal process. Some steel that is produced is formed into bars, rods, and sheets using mechanical processes. Some sheets of steel are coated with zinc to produce galvanized steel. This process uses an ***electrochemical process***.

In this chapter, you are introduced to the manufacture of several materials you come into contact with daily. These materials include lumber, plywood, steel, glass, petroleum products, and aluminum. The primary processes used to produce these materials are viewed in terms of the first process used to change them from a raw material to an industrial material. The other processes used after the initial action are briefly covered during each specific discussion.

Standards for Technological Literacy

19

Reinforce

Review the role of primary processing in production.

Extend

Discuss the meaning of the words mechanical, thermal, and chemical.

Figure Discussion

In Figure 15-1, describe to the students each stage of the cycle of a natural resource, such as trees.

TechnoFact

During the 1700s, most manufacturing took place in the home, using simple tools and hand labor. Families worked together to produce goods to sell in the village market.

TechnoFact

Until the early 1900s, Western Europe was the major seat of manufacturing. During World War I (1914–1918) the United States became the leading manufacturer. The United States remains the world's leading manufacturer, but its dominance in manufacturing is challenged by several Asian nations.

Standards for Technological Literacy

19

Demonstrate

Show the students how mechanical forces are used in sawing, pounding, etc.

TechnoFact

As you might expect, sawing was one of the common mechanical processes found in sawmills. Early mills produced boards over open pits in the ground. One person stood above the log while the other person stood in the pit beneath it. These men operated the ends of a two-handled saw. Crude, water-powered sawmills replaced this tedious, hand operation in the early 1400s. By the early 1800s, steam-powered mills replaced the water-powered ones.

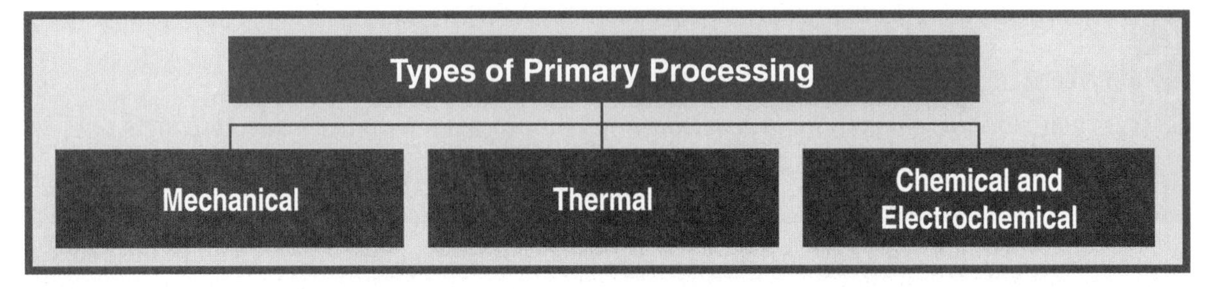

Figure 15-2. The three types of primary processes are mechanical, thermal, and chemical and electrochemical.

Career Corner

Manufacturing

Millwrights

Millwrights install and repair machinery used in many industries. They unload, inspect, and move new equipment into position. Millwrights assemble new machines by leveling the bases, fitting bearings, aligning gears and wheels, attaching motors, and connecting drive belts.

Also, millwrights work with maintenance workers to repair and maintain machines and processing equipment. Millwrights generally are trained through four-year apprenticeship programs, which combine on-the-job training with classroom instruction. Most employers want trainees with high school diplomas and some vocational training or experience.

Mechanical Processes

Mechanical processes use mechanical forces to change the forms of natural resources. They might use compression (pressure) to crush the material to reduce its size or change its texture. Other processes use shearing forces to cut and fracture the material. Still other processes run the material over screens to sort it by size.

A number of natural resources are first processed using mechanical means. The most common of these produces forest products from trees. As you learned in Chapter 4, wood is a natural composite. Wood is cellulose fibers held together by a natural adhesive called *lignin*. Many forest products are produced using mechanical processes. The material can be cut or sheared into new shapes. These mechanical actions are used to produce lumber, plywood, and particleboard.

Producing Lumber

Trees were one of the first natural resources humans used. They provided the raw material for shelters and crude tools. Humans also used wood as the primary fuel for cooking and heating. In fact, even today, wood is a major energy source in developing countries around the world. In ancient Egypt, wood had become a basic material for carpentry and boat-building. This use of wood, a natural resource, continues in today's modern civilization.

One form of widely used wood is ***lumber.*** A piece of lumber is a flat strip, or slab, of wood. Lumber is available in two types:

- ***Softwood lumber*** is produced from needle-bearing trees, such as pine, cedar, and fir. This lumber is used for construction purposes, for shipping containers and crates, and for railroad ties. Softwood lumber is produced in specific sizes called *nominal sizes.* Typical sizes range from 1 × 4 (3/4″ × 3 1/2″ when finished, or smoothed) and the common 2 × 4 (1 1/2″ × 3 1/2″ when finished) to as large as 4 × 12 (3 1/2″ × 11 1/2″ when finished). This material is available in standard lengths in 1′ increments. Generally, it is available in lengths from 6′ to 16′. You might want to look at a building-supply catalog to see the many available standard sizes of lumber.
- ***Hardwood lumber*** is produced from deciduous, or leaf-bearing, trees that lose their leaves at the end of each growing season. Hardwoods are widely used for cabinetmaking and furniture making, for making shipping pallets, and for manufacturing household decorations and utensils. Hardwood lumber is produced in standard thicknesses. These range from 5/8″-thick to 1″-thick rough boards, known as *4/4* (pronounced "four-quarter"), to as large as 4″-thick rough boards, or 16/4. The boards are available in random widths and lengths. Hardwoods are usually not cut to specific widths and lengths, as softwood lumber is.

The largest amount of lumber is produced from softwood trees. Therefore, let us look at how lumber is manufactured from logs.

In the previous chapter, you learned how trees are harvested and shipped to the mill. See **Figure 15-3.** At the mill, logs are stored in ponds to prevent checking (cracking) and to protect them from insect damage. These logs are the material input for lumber manufacturing.

Standards for Technological Literacy

19

Demonstrate

Show your students that the classification of hardwood and softwood has nothing to do with material hardness by showing balsa as a hardwood and fir as a softwood.

Research

Ask your students to search the Internet to identify typical commercial hardwoods and softwoods and their uses.

TechnoFact

The United States, Canada, and Russia are leading lumber-producing countries. In the United States, lumbering started in the New England area, with Maine being an early leader. The industry spread first to the upper Midwest and on to the West Coast states, with Oregon, Washington, and California becoming leading producers. In recent years, reduced cutting in the western national forests has caused the South to take over a large share of lumber production.

Boise Cascade

Figure 15-3. Logs are shipped to the mill on trucks or floated down rivers. Lumber production starts when the logs arrive at the mill.

Standards for Technological Literacy

19

Extend

Show the class a videotape, film, slide, or computer presentation on lumber production.

Research

Have the students use the Internet to discover how lumber is manufactured. You may suggest the Web sites of the following companies: Pacific Lumber Company, Griffith Lumber Company, and L.L. Johnson Lumber Manufacturing Company.

TechnoFact

All products made from trees can be described as forest products. Early humans used forest products for shelter, food, clothing, and fuel. They ate fruit, nuts, and berries; they built shelters from branches and boughs, and made clothing from bark and other plant materials. Today, forests provide the raw materials for a wide variety of products, including building materials, maple syrup, paper, and photographic film.

We follow some basic steps when we change the logs from a natural resource into the industrial material of lumber. See **Figure 15-4.** These steps include the following:

1. The log is removed from the pond and cut to a standard length. This length is established to give the mill a uniform input and maximum yield from the log.
2. The log is debarked. In this process, the operator might use mechanical trimmers or high-pressure water jets to remove the bark. The bark is a by-product that can be used as fuel for the mill or sold as landscaping mulch.
3. The log is cut into boards and cants at the head rig. A ***head rig*** is a very large band saw that cuts narrow slabs from the log. When the square center section (called a ***cant***) remains, a decision is made. The cant might remain at the head rig to be cut into thick boards, or it might move to the next step.
4. The cant is cut into thin boards at a ***resaw.*** This machine is a group of circular or scroll saw–type blades evenly spaced to cut many boards at once. Small logs often move directly from the debarker to the gang saw, bypassing the head rig.
5. The boards are cut to standard widths at an ***edger saw.*** This machine has a number of blades on a shaft. The blades can be adjusted at various locations to produce standard widths from 2″ to more than 12″.
6. The edged boards are cut to standard lengths at a ***trim saw.*** This machine has a series of blades spaced 2′ apart. The operator can actuate any or all the blades. This allows for cutting out defects and producing standard lengths of lumber. The boards can be 6′, 8′, 10′, 12′, 14′, or 16′ long. All the blades cut low-quality boards.

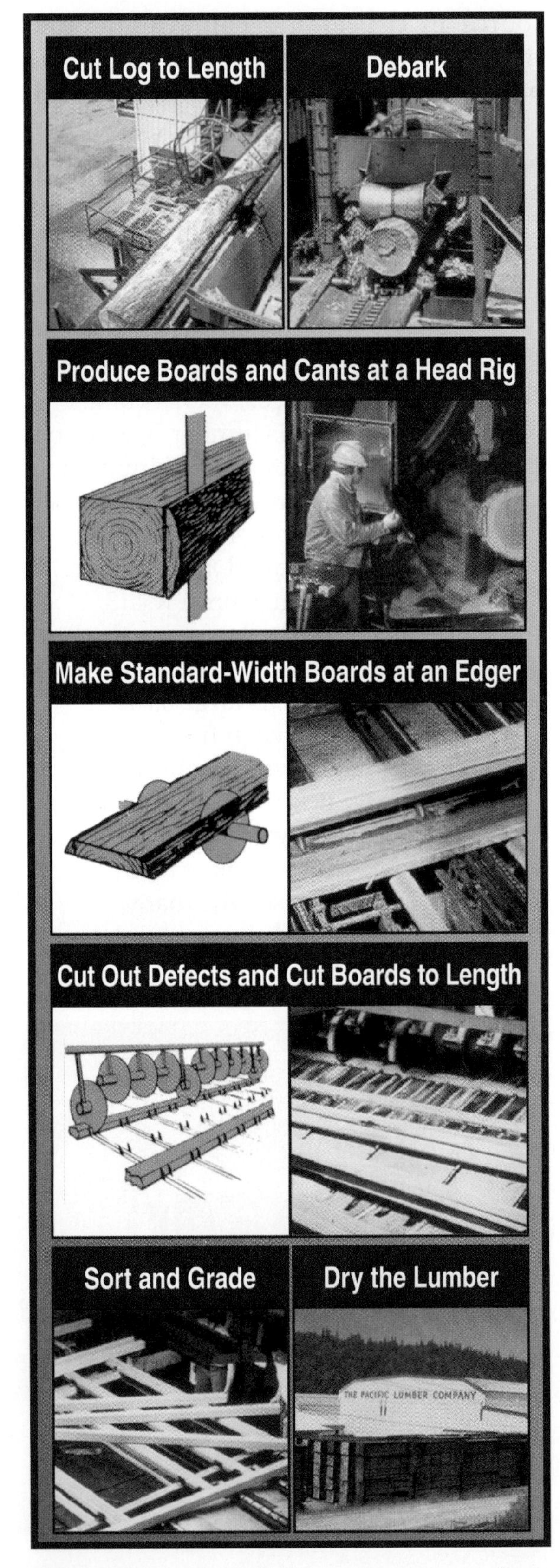

Figure 15-4. We follow certain steps when we change logs into lumber.

This cuts the board into 2′-long scrap pieces. The scrap from the edgers and trim saws is used as fuel for the sawmill or becomes the raw material for board products and paper.

The processed lumber then moves onto the green chain, where the boards move down and are inspected and sorted by quality. The lumber is air- or kiln (oven) dried to make it a more stable product.

The dried lumber is shipped as an industrial material. Some lumber receives special processing. Short boards can be processed into longer boards using finger joints cut in the ends of the boards. See **Figure 15-5.** The boards are then glued end to end to form a continuous ribbon of lumber. The ribbon is cut into standard lengths as it leaves a glue-curing machine.

Producing Plywood

Another common forest product is plywood. Plywood is a composite material made up of several layers of wood. See **Figure 15-6.** Plywood is more stable than solid lumber because cross-grained layers in the plywood reduce warping and expansion.

Figure 15-5. These operations are changing short lengths of redwood lumber into long boards. Later, the boards are made into molding for homes.

The outside layers are called ***faces.*** Between the faces are layers called ***cross bands.*** The grain of the cross bands is at a right angle (90°) to the face grain. The layer in the center is called the ***core.*** The core's grain is parallel with the face grain. Plywood with only three layers does not have cross bands. Three-layer plywood has a core with its grain running at a right angle to the face layers.

Three types of cores are used for plywood. See **Figure 15-7.** The most common

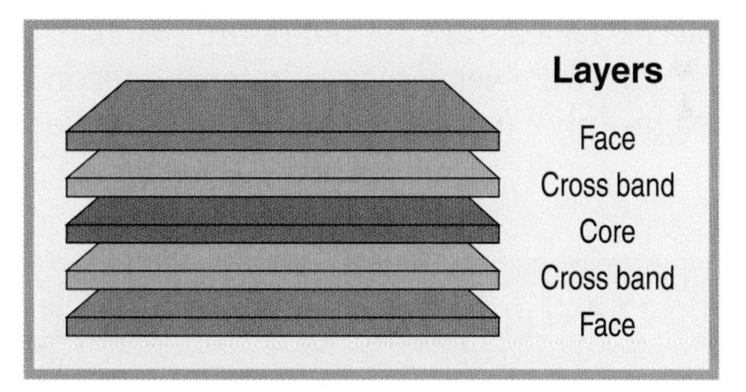

Figure 15-6. Plywood is made of layers of wood glued together under pressure. The number and the thickness of the layers are varied to make different types of plywood.

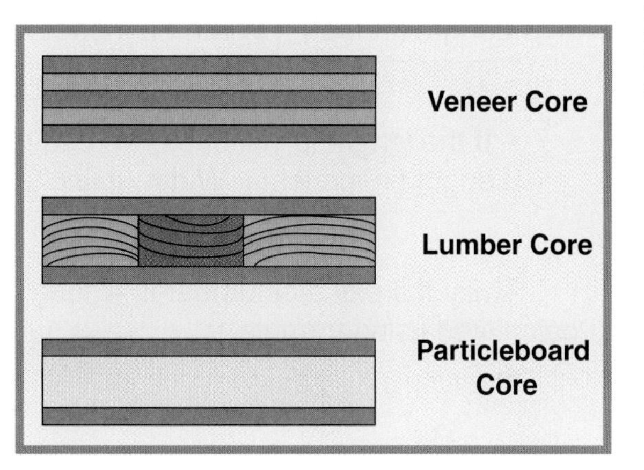

Figure 15-7. Plywood is available with three types of cores: veneer, lumber, and particleboard.

Standards for Technological Literacy

Extend

Discuss why lumber is dried.

Research

Have the students look up how lumber is dried and the effects of drying on the product.

Research

Ask the students to examine the difference between interior and exterior plywood and write a short report on this subject.

Extend

Discuss the system used to grade plywood and the differences between A, B, C, and D grade faces.

TechnoFact

Plywood is created by gluing together an odd number of sheets of veneer wood under pressure. The first known instances of plywood manufacturing took place in ancient Egypt and China. Plywood was first manufactured in the United States shortly after 1850.

Standards for Technological Literacy

3 19

Demonstrate

Make a sheet of plywood from small sheets of veneer during class or lab.

Research

Have the students use the Internet to research how plywood is manufactured. Suggest the Web sites of the following companies and associations: the Canadian Plywood Association, the Canadian Hardwood Plywood and Veneer Association, and the Columbia Forest Products Company.

TechnoFact

Plywood can be categorized by its use (interior or exterior), the wood it is made from (softwood or hardwood plywood), or by the type of cores used (veneer, lumber, or particleboard cores).

is ***veneer-core plywood***. A ***veneer*** is a thin sheet of wood sliced, sawed, or peeled from a log. Plywood used for cabinetwork and furniture usually has a lumber or a particleboard core. ***Lumber-core plywood*** has a core made from pieces of solid lumber that have been glued up to form a sheet. ***Particleboard-core plywood*** has a core made of particleboard. Particleboard is made up of wood chips that are glued together under heat and pressure. Plywood is typically available in 4′ × 8′ sheets. Thicknesses from 1/8″ to 3/4″ are available.

Veneer-core plywood is produced in two stages. The first stage makes the veneer. The veneer is sliced or peeled from the log and moves through a dryer. The dried veneer is sheared into workable-size pieces. Defects are cut out or are patched.

Now the veneer is ready for the second stage—plywood production. Glue is applied between the layers. The layers of veneer are stacked up. The sheet is then placed in a heated press. The press is closed, and the glue cures under heat and pressure.

After pressing, the sheet is removed, trimmed to size, and sanded. The completed sheets are inspected and loaded for shipment. Look at **Figure 15-8.** This series of photos shows the steps used in manufacturing common plywood.

Thermal Processes

Many industrial materials are produced by processes that use heat to melt and reform a natural resource. These types of processes are called ***thermal processes***. The thermal process is widely used to extract metals from their ores. This process is often called ***smelting***. Other thermal processes are used to make glass and

STEM Connections: Mathematics

Integrated STEM Curriculum — Science, Technology, Engineering, Mathematics

Calculating Board Footage

As mentioned in this chapter, hardwood lumber is usually not cut to specific widths and lengths. Instead, this type of lumber is sold by volume. The unit of measurement for the volume of lumber is the board foot.

We generally use one of the two following formulas to calculate board footage:

- If the length is in feet, we use the following formula:

$$\frac{\text{Length (in feet)} \times \text{Width (in inches)} \times \text{Thickness (in inches)}}{12}$$

- If the length is in inches, we use the following formula:

$$\frac{\text{Length (in inches)} \times \text{Width (in inches)} \times \text{Thickness (in inches)}}{144}$$

Thus, if a piece of lumber is 4′ long, 6″ wide, and 1″ thick, the board footage is calculated using formula 1:

$$\frac{4' \times 6'' \times 1''}{12} = \frac{24}{12} = 2$$

The piece of lumber is, thus, 2 board feet. What is the board footage for a 96″-long, 12″-wide, and 2″-thick piece of lumber?

cement. For this discussion, we look at three common thermal processing activities: steelmaking, glassmaking, and petroleum refining. These use a combination of thermal and chemical processes. The thermal energy melts the materials. During the melting process, chemical reactions take place to produce a new material. The new material is then shaped into standard stock. The shaping process uses mechanical techniques. The material is cast, drawn, rolled, or squeezed into new sizes and shapes.

Standards for Technological Literacy

19

TechnoFact

Early colonists from England brought with them the knowledge of iron making. The first North American iron works was set up at the James River settlement; it was destroyed in 1622. In 1646, the Massachusetts Bay Colony established the first integrated iron works in what is now Saugus, Massachusetts. A modern reconstruction of this site is part of the National Park System.

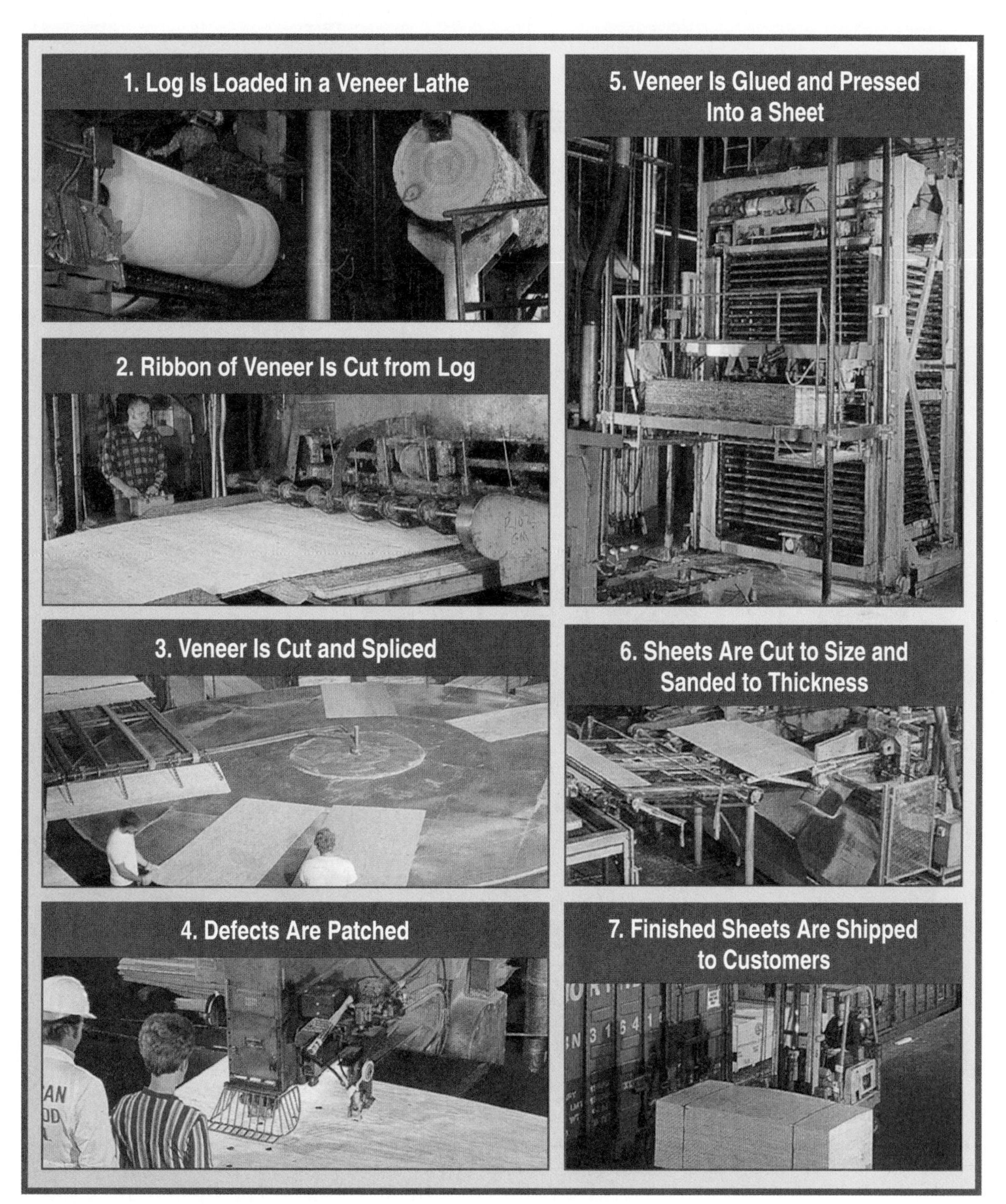

Figure 15-8. We follow certain steps in plywood manufacture.

Standards for Technological Literacy

19

Extend

Discuss why steel may be considered the refined product of iron.

Reinforce

Review material properties as presented in Chapter 4 and relate them to steel.

Figure Discussion

Describe the three steps in steel making, using Figure 15-9 as a guide.

TechnoFact

Many iron alloys are corrosion resistant. Stainless steels are alloys of iron containing metals such as chromium and nickel. They do not corrode because the added metals help form a hard oxide coating that blocks corrosion. Plating is another method of protecting steel from corrosion. Chromium, nickel, tin, and zinc are among the metals used for plating steel. Zinc plating is known as galvanizing.

Producing Steel

Steel is an alloy, a mixture of iron and carbon. Adding other elements in small amounts gives the steel specific qualities. These elements include manganese, silicon, nickel, chromium, tungsten, and molybdenum. Steel with nickel and chromium is called *stainless steel.* Adding molybdenum increases the hardness of the steel. Molybdenum steels are widely used for tools. Adding tungsten makes the steel more

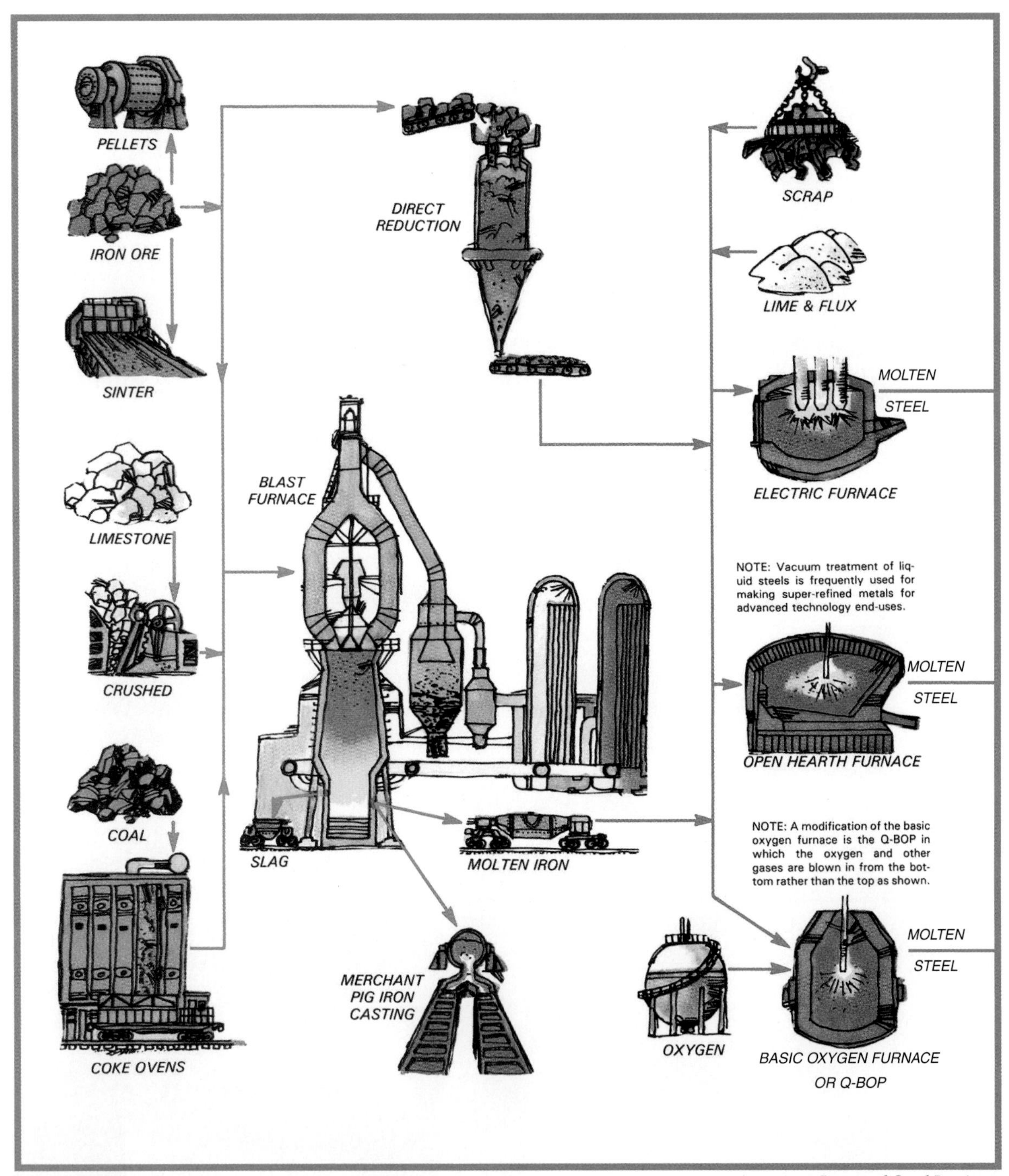

American Iron and Steel Institute

Figure 15-9. The steps it takes to turn iron ore into steel products. The three main stages are smelting, steelmaking, and steel finishing.

heat resistant. See **Figure 15-9.** The making of steel requires three steps:

1. Iron smelting.
2. Steelmaking.
3. Steel finishing.

Iron Smelting

Iron smelting produces ***pig iron,*** the basic input for steelmaking. Pig iron results from thermal and chemical actions that take place in a ***blast furnace.*** The blast

Standards for Technological Literacy

19

TechnoFact

In 1713, Abraham Darby became the first person to successfully use coke to smelt iron. Before that, charcoal smelting was the primary technique used in iron production. Coke-based smelting grew in popularity slowly until a method of making wrought iron using the coke smelting technique was developed.

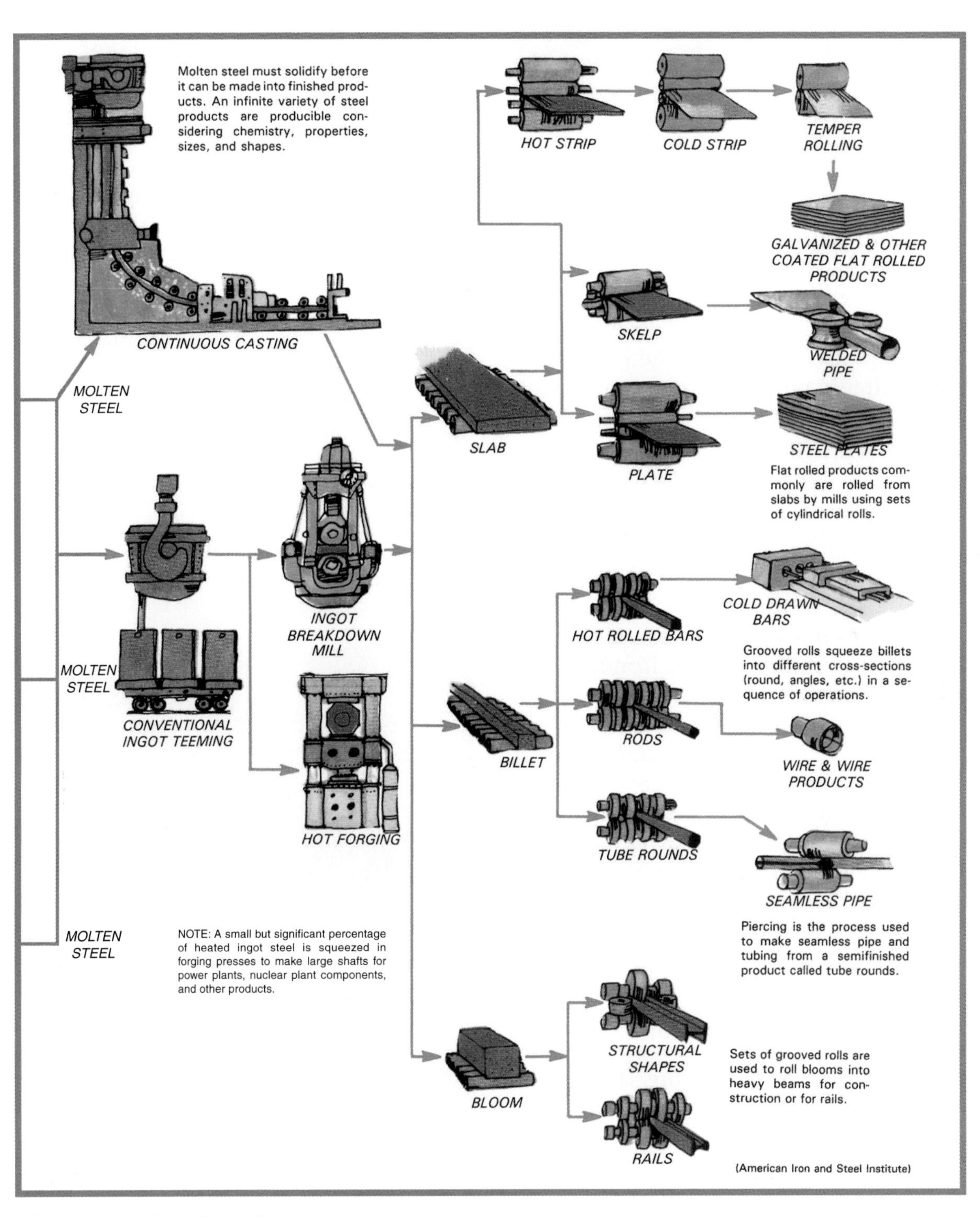

Figure 15-9. Continued.

Standards for Technological Literacy

19

Brainstorm

Ask the students where they have seen galvanized steel and tin plate used.

TechnoFact

The rusting of iron is the most recognizable and important form of corrosion. Basically, rust results from the combination of iron with oxygen (from water) in a process known as oxidation. Rust (ferric oxide) is a reddish-brown crust that is brittle and flaky. Rust flakes off easily, exposing the base metal to further oxidation in a continual and vicious cycle.

furnace uses a continuous process. At no time is the blast furnace empty during the smelting process. Every so often, raw materials are added to the top, and molten pig iron is removed from the bottom.

The operation of a blast furnace is simple. Alternating layers of iron ore, coke (coal with the impurities burnt out), and limestone are added to the furnace. The blast furnace is charged with four parts iron ore, two parts coke, and one part limestone.

Very hot air is blown into the bottom of the furnace. The coke burns, causing the iron ore to melt. The oxygen in the iron ore combines with the carbon to make carbon monoxide gas. During the melting, limestone joins with impurities to form slag. The slag floats on the molten iron and can be drawn off. This leaves molten iron with carbon dissolved in it. This material is called *pig iron*, which is iron with 3%–4.5% carbon. Pig iron also has 1%–2% other elements, including manganese, silicon, sulfur, and phosphorus.

Steelmaking

The steelmaking process starts with the pig iron produced in the iron-smelting step. Steelmaking actually removes some carbon from the iron. Heat and oxygen are used to take some of the carbon out of molten pig iron.

The most common steelmaking process uses the basic oxygen furnace. Making steel in a basic oxygen furnace involves three steps. The first is charging. The furnace tilts to one side to receive pig iron, scrap steel, and flux, a material that combines with impurities to form slag. This charge provides the basic ingredients for steel.

The second step is ***refining***. The furnace moves into an upright position, and the charge is melted. A water-cooled oxygen lance is then placed above the molten material. Pure oxygen is forced out of the lance into the iron at supersonic speeds. The oxygen causes the part of the carbon to burn away, producing steel and slag.

The final step is ***tapping***. See **Figure 15-10.** The floating slag is skimmed off the melt. The entire furnace tips to one side, and the steel is poured out. The steel is now ready to enter the steel-finishing cycle.

Steel Finishing

Steel finishing changes molten steel into sheets, plates, rods, beams, and bars. The first step involves pouring the steel into ingots or into the head end of a continuous caster. A continuous caster solidifies the molten steel into shapes called *slabs*, *billets*, and *blooms*. ***Slabs*** are wide, flat pieces of steel. See **Figure 15-11.** Sheets, plates, and skelps are produced from them. Sheets are wide, thin strips of steel, whereas plates are thicker. ***Skelps*** are strips of steel used to form pipe.

Billets are square, long pieces of steel. Bars and rods are produced from them. ***Blooms*** are short, rectangular pieces used to produce structural shapes and rails.

American Iron and Steel Institute

Figure 15-10. A steel furnace is being tapped. The furnace is tilted, and the steel pours out.

Inland Steel Company

Figure 15-11. These hot slabs are waiting to be rolled into sheets or plates.

Steel is prone to rust if it is left exposed to the atmosphere. Therefore, many steel shapes are finished. They might receive a zinc or tin coating. See **Figure 15-12.** Zinc-coated steel is called ***galvanized steel***. This steel is used for such automobile parts as fenders and for such containers as buckets and trash cans. Tin-coated steel is called ***tin plate***. Tin plate is widely used to make food cans.

Producing Glass

Glass is another material produced using thermal processes. This material is made by solidifying molten silica in an amorphous state. An amorphous material has no internal structure similar to the regular, uniform lattice structures of metals. Glass is made from sand (silica), soda ash (sodium carbonate), and lime (from limestone). These ingredients are weighed and mixed to form a batch. For sheet glass, the mixture contains about 70% silica, 13% lime, and 12% soda.

The batch is moved into a melting furnace. A typical furnace for flat glass can be 30′ (9 m) wide, 165′ (50 m) long, and 4′ (1.2 m) deep. This furnace holds about 1200 to 1500 tons (1100 to 1400 metric tons) of glass at one time. The melting end of the furnace reaches temperatures of 2880°F (1580°C). This heat causes the material to melt and flow together.

From the furnace, the glass might go to a secondary manufacturing process. Products such as jars, bottles, dishes, glasses, and

American Iron and Steel Institute

Figure 15-12. This massive machine applies a tin coating to steel to make tin plate.

Standards for Technological Literacy

19

Research

Have each of the students select and research one of the following kinds of glass products: flat glass, glass containers, optical glass, or fiberglass.

TechnoFact

It is unclear when glass making first appeared. Many experts believe that glass was first created about 5000 years ago, in the form of a glaze (glass coating) applied to pottery. About 3500 years ago, the Egyptians created the first glass containers. The invention of the blowpipe, over 2000 years ago, revolutionized the slow and expensive process of glass making, allowing craftsmen to produce glass objects more quickly and cheaply.

Standards for Technological Literacy

19

Research

Have the class investigate and identify the major products of a petroleum refinery.

TechnoFact

In 1859, Edwin Drake used steam power to drill the first successful oil well in the United States. The creation of this first well, near Titusville, Pennsylvania, is often considered the birth of the modern petroleum industry.

cookware are manufactured using casting and forming techniques. See **Figure 15-13.**

Primary manufacturing lines might change the glass into sheets for windows and similar products. See **Figure 15-14.** Most of this glass is called ***float glass*** because it is formed by floating the molten glass on a bed of molten tin. The glass flows out of the furnace onto the tin. This material forms a ribbon as it cools and moves toward the end of the float tank. A typical float tank is 150′ (46 m) long and can form glass 160″ (4 m) wide.

The formed glass moves from the float tank to an annealing oven, called a *lehr.* The temperatures in the lehr start at 1200°F (650°C) and fall to 400°F (200°C). This gradual temperature drop relieves internal stresses in the glass. From the lehr, the ribbon of glass is cut and packed for shipment.

Owens-Brockway

Figure 15-13. Large quantities of glass are processed into food containers such as these jars.

Other glass products are fiberglass and fiber-optic cables. These strands of glass are formed, cooled, and annealed. ***Fiberglass*** is used as the matrix for composite materials and for insulation. ***Fiber-optic cables*** are used to transmit voice, television, and computer data at high speeds.

Refining Petroleum

Petroleum is another resource processed using thermal actions. As you learned earlier, petroleum is not a uniform material. Instead, it is a mixture of a large number of different hydrocarbons. The process used to isolate these hydrocarbons is called *refining.*

Most petroleum refineries use three processes: separation, conversion, and treating. ***Separation*** does exactly what its name says. This process breaks petroleum into major hydrocarbon groups. ***Fractional distillation*** is the process used. The petroleum is pumped through a series of tubes in a furnace. There, it is heated to about 725°F (385°C). The petroleum becomes a series of hot liquids and vapors. They pass into the bottom of a ***fractionating tower.*** See **Figure 15-15.**

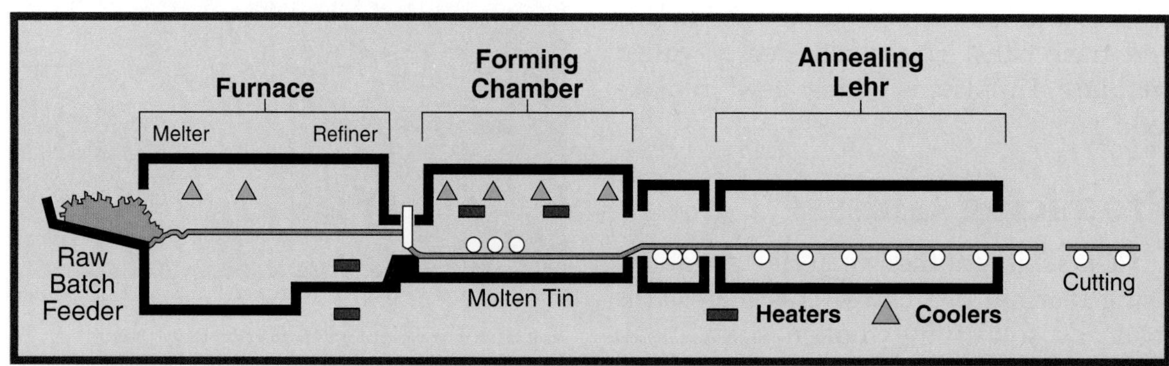

PPG

Figure 15-14. A float glass production line.

American Petroleum Institute

Figure 15-15. These fractionating towers are in Saudi Arabia.

Standards for Technological Literacy

19

Brainstorm

Ask the class how we could reduce our use of petroleum products.

This tower can be a tall as 100′ (30 m). Within the tower are a series of pans or trays that can hold several inches of liquid. Each pan and its fluid are maintained at a specific temperature. The higher the level in the tower is, the lower the temperature is.

The hot vapors coming from the furnace rise in the tower. As they rise, they are forced to bubble through the liquid in the pans. This action cools the vapors to the temperature of that pan. Hydrocarbons with a boiling point at or below the pan's temperature condense and stay in the pan. Other vapors continue to rise. The condensed liquids are continuously drained from the trays.

Typically, the lighter fractions (products) are taken off the top. These are gasoline and gases such as propane and butane. Other fractions in descending order of temperature are jet fuel, kerosene, diesel, fuel oil, and asphalt. **Figure 15-16** shows a diagram of two distillation towers.

The process of ***conversion*** changes hydrocarbon molecules into different sizes, both smaller and larger. For example, heavier hydrocarbons can be broken into smaller ones. This is called ***cracking***. Thermal cracking heats heavier oils in a pressurized chamber. See **Figure 15-17.** The heat and pressure cause the hydrocarbon molecules to break into smaller ones. Catalytic cracking does the same by using a chemical called a *catalyst,* which helps a reaction to take place. The catalyst is not used up during the reaction. Both catalytic and thermal cracking are used to increase the amount and the quality of products produced from a barrel of petroleum.

A second type of conversion process is ***polymerization***. This process is the opposite of cracking. Polymerization causes small hydrocarbon molecules to join together. Refinery gases are subjected to high pressures and temperatures in the presence of a catalyst. They unite (polymerize) to form hydrocarbon liquids. This increases the yield of petroleum products.

Standards for Technological Literacy

19

Figure Discussion

Use Figure 15-16 to explain the fractional distillation process.

TechnoFact

Although the increasing use of petroleum products has raised the standard of living for the people of developed countries, they have paid a price in the form of pollution, rising energy costs, and periodic energy shortages.

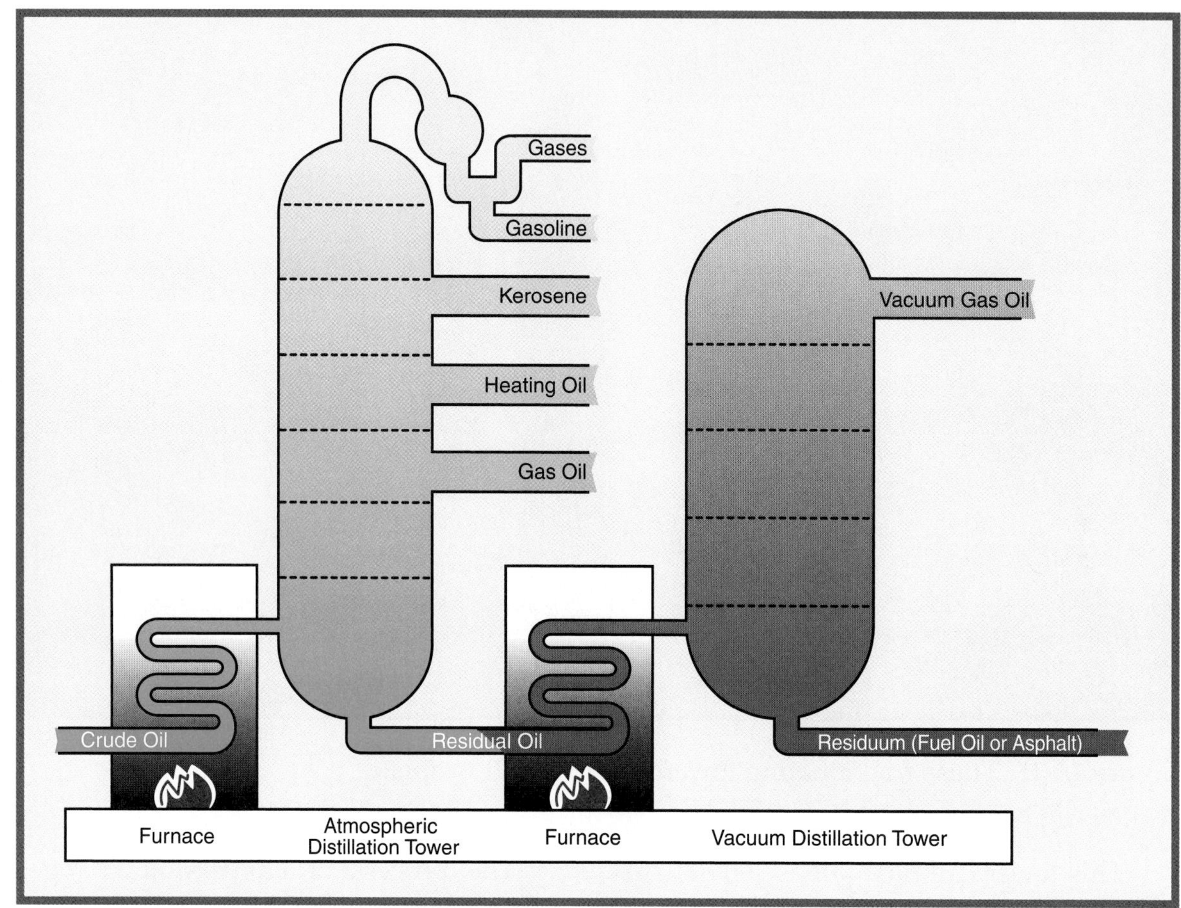

Chevron-USA

Figure 15-16. A two-stage distillation unit.

American Petroleum Institute

Figure 15-17. This cracking facility in Venezuela processes more than 50,000 barrels of heavy oil a day.

Standards for Technological Literacy

16 19

Technology Explained

nuclear energy: the energy holding atoms together, which is released by using controlled reactions.

Humans have always marveled at the Sun. The Sun is a giant ball of hydrogen gas. The Sun's intense heat warms our planet and creates Earth's weather patterns. Reactions that turn millions of tons of hydrogen into helium every second create this heat. This process is called *nuclear fusion*.

Fusion combines the nuclei (the centers of atoms) of two atoms with low atomic numbers. Nuclear fusion produces heavier nuclei and releases neutrons and great amounts of energy. The most important fusion process unites atoms of hydrogen. See **Figure A.** This element is the lightest of all. Most hydrogen atoms contain one proton in the nucleus and one orbiting electron. Other forms of hydrogen, however, do exist. These forms are called *isotopes*. The two isotopes used in fusion are deuterium and tritium. Every atom of deuterium has one neutron in its nucleus, while each tritium atom has two neutrons. Fusion of these two isotopes creates one helium nucleus. Helium, in its normal condition, has two protons and two neutrons. One neutron is left over. Also, over 17 million electron volts of energy are released.

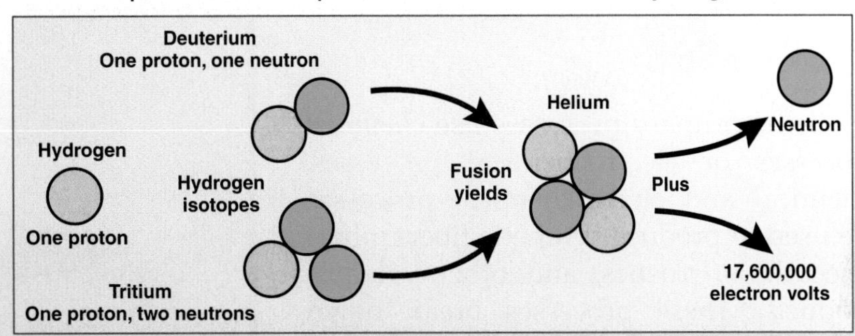

Figure A. Nuclear fusion forces the nuclei of atoms together. The extra neutrons can be a problem to contain.

Creating a controlled fusion reaction is very difficult. The atoms are very small, and the nuclei carry positive charges. Since like charges repel, the nuclei resist coming close together. Therefore, they must be accelerated to high speeds to overcome the repelling forces. Also, the extra neutrons cause a serious problem. These particles can break free and damage the vessel in which the reaction is taking place.

Fusion is not the only way to use the energy of the atom. A second method is called *fission*. See **Figure B.** This process breaks down radioactive materials into smaller molecules. Fission releases some of the energy holding the atom together.

The most common material used in fission is uranium. Uranium has three main isotopes: U-238, U-235, and U-234. U-235 is the isotope used in nuclear reactors. When a U-235 molecule is bombarded with neutrons, the nucleus breaks apart. The nucleus forms two new molecules. The extreme heat released is used to create steam, which turns turbines in electric generating plants. Fission reactors also provide power on submarines and ships.

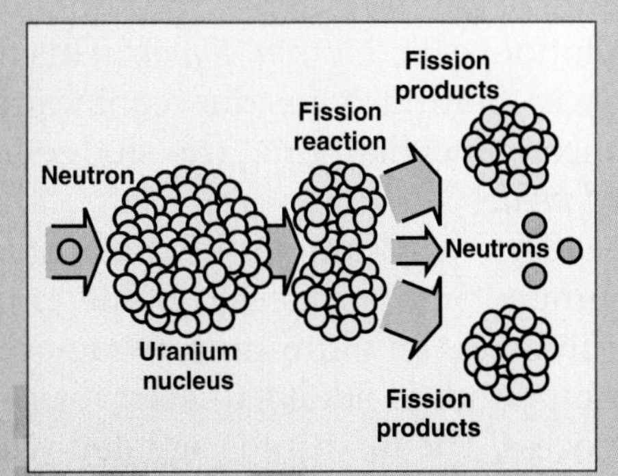

Figure B. Nuclear fission breaks apart atoms of radioactive materials. The waste products from fission reactors are, however, very harmful to living things.

Standards for Technological Literacy

19

Brainstorm

Ask the students where they have seen aluminum used.

Extend

Discuss why recycling aluminum saves energy.

Figure Discussion

Use Figure 15-18 to review and summarize the production of aluminum.

TechnoFact

One of the first uses for aluminum was as a mordant for dyes. Ancient people used naturally occurring aluminum compounds to fix dyes to cloth so that the colors would not wash out.

Treating is the third petroleum-refining process. This process adds or removes chemicals to change the properties of petroleum products. Sulfur can be removed so kerosene burns cleaner and smells better. Additives improve the lubrication properties of oils. Other additives help fuels burn quickly and cleanly.

Chemical and Electrochemical Processes

Some primary processes use ***chemical processes*** or electrochemical processes. Chemical and electrochemical processes are used to produce synthetic fibers, pharmaceuticals, plastics, and other valuable products. These processes break down or build up materials by changing their chemical compositions. The most common chemical or electrochemical process in use is the one to make aluminum, which is the focus of this section.

The process of making aluminum is carried out in two stages. See **Figure 15-18.** The first stage chemically changes aluminum ore, known as bauxite, into aluminum oxide, or alumina. The ***bauxite*** is crushed and mixed with a caustic soda, sodium hydroxide. The soda dissolves the aluminum oxide, forming a sodium-aluminate solution called *green liquor.* This process leaves behind a residue containing iron, silicon, and titanium. The sludge is called *red mud.*

The sodium-aluminate liquid is pumped into a digester. Some aluminum trihydrate, an aluminum oxide-and-water compound, is added to start, or seed, the process. The mixture is agitated with compressed air and cooled. During the process, a chemical action takes place. Aluminum trihydrate and caustic soda are formed. The aluminum compound settles out of the solution. This compound is removed and dried in a kiln. The drying drives the water out of the aluminum trihydrate, leaving pure aluminum oxide. The aluminum

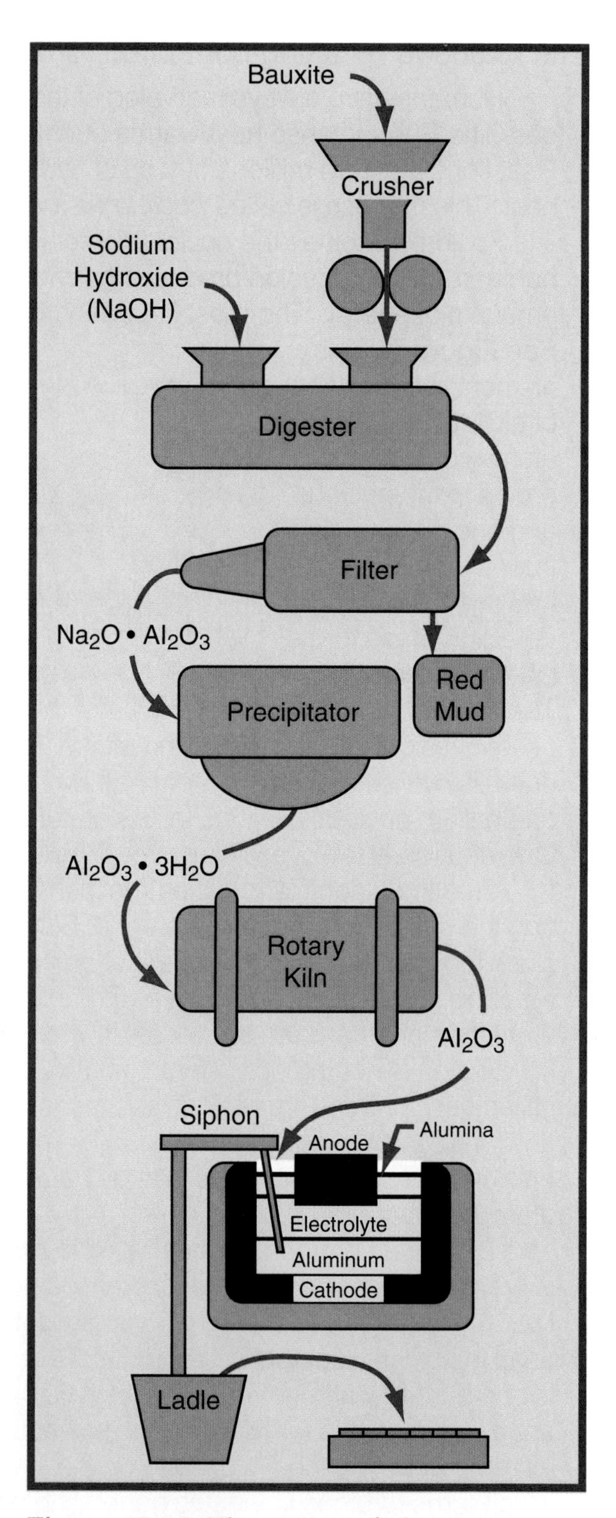

Figure 15-18. The stages of aluminum production from bauxite to pure aluminum ingots.

Think Green

Volatile Organic Compounds

Volatile Organic Compounds (VOCs) are toxic substances that evaporate into the atmosphere. The evaporation of VOCs contributes to the development of such environmental hazards as smog, but VOCs can also reduce the quality of air indoors. Not only are VOCs harmful to the environment, but there are also health concerns associated with them. Continued exposure to these toxins can produce results as mild as headaches and nausea or as severe as organ damage and cancer. Products that contain VOCs include substances like paint and cleaning supplies, but they are also found in fabrics or carpets.

You can reduce your exposure to VOCs being released into your home. Many VOC-containing products are offered in more organic alternatives. These products use plant-based materials rather than chemicals like benzene. Also, if you are buying new furnishings, you can look for pieces with natural finishes, or fabric that is made from organic cotton.

oxide is called ***alumina*** and is the input to the second phase of making aluminum.

In the second stage, alumina from stage one is converted into pure aluminum. This process takes place in large electrolytic cells called *pots*. The cell has a carbon lining, which is the cathode. The anode is also made of carbon.

The pot is partially filled with an electrolyte, a material that conducts electricity. This material is a mixture of molten aluminum fluoride (a salt) and molten cryolite. The electrolyte is kept at 1650°F (900°C). The alumina is dissolved in the electrolyte. Large quantities of electrical current are passed between the anode and cathode. The electrical energy causes the alumina to break into pure aluminum and oxygen. The aluminum settles to the bottom of the pot, where it is gathered, purified, and cast into ingots. The ingots are then formed into sheets, bars, rods, or other shapes. These shapes become industrial materials used for numerous secondary manufacturing processes.

Standards for Technological Literacy

5 19

TechnoFact

In 1825, Hans Christian Oersted became the first person to produce metallic aluminum. However, a relatively inexpensive method of producing aluminum was not developed until 1886.

TechnoFact

In 1886, two scientists, Charles Martin Hall and Paul L. T. Heroult, working completely independently and unaware of the other's work, both developed the same method of producing aluminum. First, they dissolved alumina in cryolite and then they separated aluminum from the mixture with an electrolytic reduction process. Three years later, Karl Bayer invented an inexpensive process for obtaining alumina from bauxite.

Summary

Primary manufacturing processes change natural resources into industrial materials. They accomplish this through mechanical, thermal, and chemical and electrochemical means. Mechanical processes change the forms of natural resources through the use of mechanical forces. Thermal processes use heat, and chemical and electrochemical processes change materials by changing the chemical compositions. The resulting materials are used to manufacture products and construct structures.

Test Your Knowledge

Write your answers on a separate piece of paper. Please do not write in this book.

1. List the three types of primary processes.
2. What kind of primary processing produces lumber?
3. Most lumber comes from softwood trees. True or false?
4. List the three types of cores used in plywood manufacture.
5. What kind of primary processing changes iron ore into steel products?
6. Steel is an alloy. True or false?
7. Pig iron is produced in a(n) ______ furnace.
8. Zinc-coated steel is called ______.
9. Glass is produced by electrochemical means. True or false?
10. What is float glass?
11. List the three processes most petroleum refineries use.
12. What is fractional distillation?
13. What kind of primary processing produces aluminum?
14. Bauxite is the raw material for producing aluminum. True or false?
15. What is alumina?

Answers to Test Your Knowledge Questions

1. Mechanical processes, thermal (heat) processes, and chemical and electrochemical processes
2. Mechanical
3. True
4. Veneer, lumber, and particleboard
5. Thermal and chemical processes
6. True
7. blast
8. galvanized
9. False
10. Float glass is formed by floating the molten glass on a bed of molten tin.
11. Separation, conversion, and treating
12. Fractional distillation breaks petroleum into major hydrocarbon groups.
13. Chemical and electrochemical processes
14. True
15. Pure aluminum oxide, the input to the second phase of making aluminum

STEM Applications

1. Build a scale model of a plant to produce an industrial material from a natural resource.
2. Select a material not discussed in this chapter. Examples are paper, hardboard, copper, gold, nylon, and polyethylene. Research the processes used to produce the material you choose. Report your results in a chart similar to the one shown below.

Material:
Primary natural resource used to make the material:
Primary production process (steps):
Uses for the material:

3. Develop a list of materials you see used in the room in which you are. List the type of process used to produce each one.

Chapter Outline

Types of Manufacturing Processes

Automating and Controlling Processes

This chapter covers the benchmark topics for the following Standards for Technological Literacy:

5 19

Learning Objectives

After studying this chapter, you will be able to do the following:

- Recall the basic steps involved with casting and molding.
- Explain methods used in casting and molding processes.
- Recall safety rules involved with casting processes.
- Understand the basic concepts of forming processes.
- Recall the types of separating processes.
- Explain methods used in separating processes.
- Recall safety rules involved with using machine tools.
- Recall the three types of conditioning processes.
- Explain methods used in conditioning processes.
- Explain methods used in assembling processes.
- Explain methods used in finishing processes.
- Recall safety rules involved in finishing processes.
- Summarize the characteristics and major parts of robots.

Key Terms

adhesive bonding
anneal
artificial intelligence (AI)
assembling process
bonding
bonding agent
casting and molding processes
chemical action
chemical conditioning
chemical machining
chip removal
cold bonding
compound
computer-controlled machining
conditioning process
converted surface finish
die
dip
drawing machine
dry
elastic range
electrical discharge machining (EDM)
enamel
expendable mold
extrusion
fasten
finishing process
firing
flame cutting
flow bonding
flow coating
forming process
fracture point
fusion bonding

hammer
hardening
heat-treat
joint
lacquer
laser machining
mated die
mechanical conditioning
mechanical fastening
milling machine
nontraditional machining process
open die
paint
permanent mold
pickle
plastic range
plating
press
press fit
pressure bonding
programmable logic controller (PLC)
robot
rolling machine
secondary manufacturing process
separating process
shaped die
shear
spray
temper
thermal conditioning
varnish
yield point

Standards for Technological Literacy

19

Reinforce

Be certain to differentiate between primary and secondary manufacturing for the class (Chapter 13).

Strategic Reading

As you read this chapter, make an outline of the different types of manufacturing processes. Be sure to include specific details, such as the steps involved in each process.

In Chapter 15, you learned that primary processing produces industrial materials. These materials have little worth to the average person. What can you do with a sheet of steel, a 2 × 4 stud, a pound of polypropylene pellets, or an ingot of pure aluminum? These materials must be changed into products before they are useful to you. We use the term ***secondary manufacturing processes*** to describe the actions used to change industrial materials into products. See **Figure 16-1.**

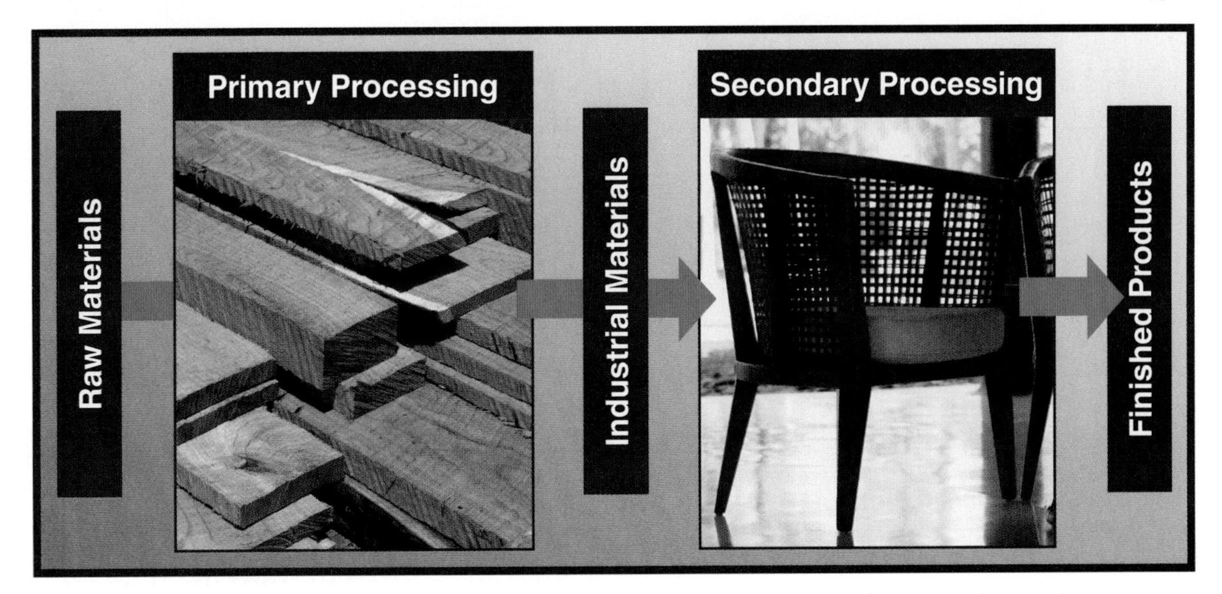

Figure 16-1. Secondary manufacturing processes turn industrial materials, such as lumber, into useful products, such as chairs.

Standards for Technological Literacy

19

Figure Discussion

Trace some common products through the processes shown in Figure 16-2.

TechnoFact

The manufacturing industry creates a virtually unimaginable range of products, including automobiles, books, lava lamps, soap, and dog chewy toys. The word manufacture is a combination of the Latin words manus (hand) and facere (to make).

Types of Manufacturing Processes

Thousands of manufacturing processes exist. They are used to change the size and shape of materials, to fasten materials together, to give materials desired properties, and to coat the surfaces of products. It is very difficult to study and comprehend all the individual ways to process materials. See **Figure 16-2.** Instead, we can understand secondary manufacturing processes more easily when we classify them into six groups:

- Casting and molding.
- Forming.
- Separating.
- Conditioning.
- Assembling.
- Finishing.

Each process has actions or concepts common to all the other processes in its group. Within these groups are specific processes or techniques. Each one differs in some way from the other processes in the group. In other words, these groups are similar to the members of a family. They do many things alike and look alike, but each member is unique.

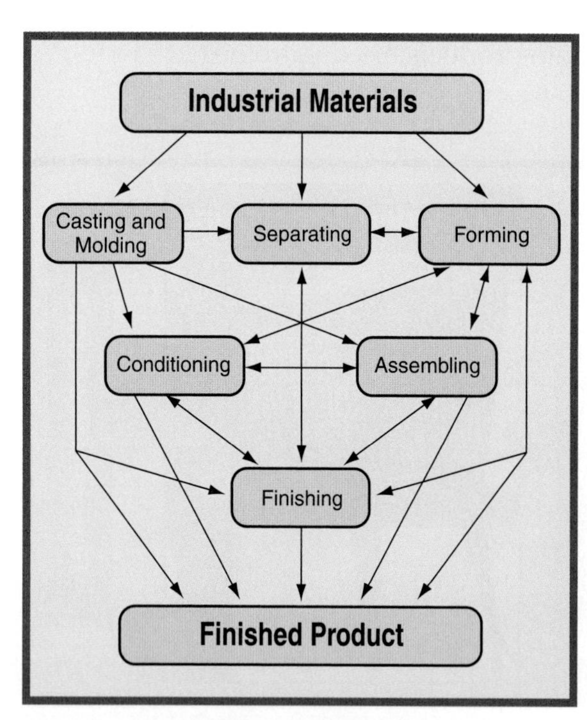

Figure 16-2. The six groups of manufacturing processes are all related.

Casting and Molding Processes

Three groups of processes give size and shape to pieces of material. See **Figure 16-3.** The first of these is casting and molding. ***Casting and molding processes*** give materials shape by introducing a liquid material into a mold. The mold has a cavity of the desired size and shape. The liquid material is poured or forced into the mold, where it is allowed to solidify before being removed.

In this discussion, we refer to molten and fluid materials as *liquids.* Molten refers to materials heated to a fluid state. These materials are normally solid at room temperature. Fluid materials, such as water and casting plastics, are liquid at room temperature.

All casting and molding processes involve five basic steps. See **Figure 16-4.** These steps are the following:

1. Producing a mold of the proper size and shape.

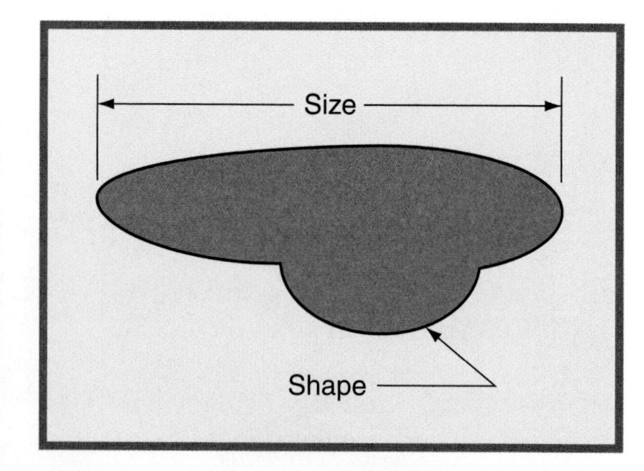

Figure 16-3. Size is the dimensions of an object, and shape is the object's form and outline.

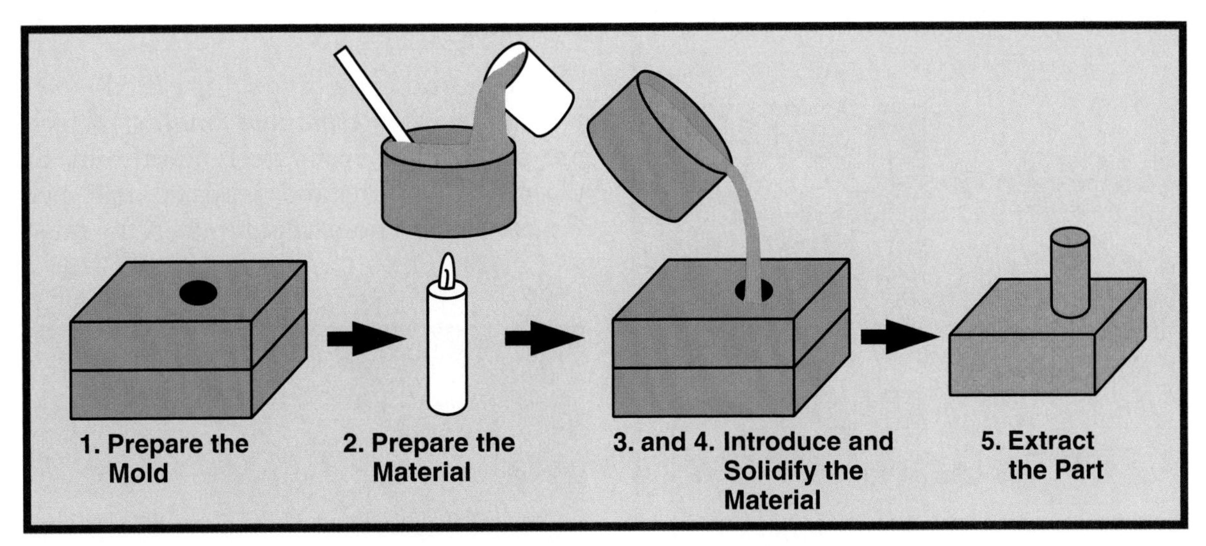

Figure 16-4. All casting and molding processes produce parts and products using the five basic steps shown here.

2. Preparing the material.
3. Introducing the material into the mold.
4. Solidifying the material.
5. Extracting the product from the mold.

Producing a Mold

These processes require a mold to hold the liquid material until the material becomes solid. Inside the mold is a cavity of the proper size and shape. Channels, called *gates,* guide the liquid into the cavity. Two major types of molds are used in casting and molding processes. These are expendable molds and permanent molds.

Expendable molds

Many cast products are made in molds that are used once. The mold is destroyed to remove the cast item. These molds are called ***expendable molds.*** Molds can have two or more parts. They are generally made in two steps. First, a pattern of the same shape as, but slightly larger than, the finished product is made. The extra size allows for shrinkage. Most materials shrink when they change from a liquid to a solid.

The pattern is the foundation for making an expendable mold. This foundation is surrounded with an inexpensive substance, such as sand or plaster. Sand is tamped into place, and plaster is allowed to dry around the pattern. When the mold is completed, the mold parts are separated, and the pattern is removed. This leaves a cavity of the correct shape to be filled by the liquid material. The mold parts are put back together to make a ready-to-use mold. See **Figure 16-5.**

Safety with Casting Processes

People who use casting processes need to observe some basic safety rules:

- Wear protective clothing and goggles when pouring a casting.
- Be sure everything to be used in the casting process is free of moisture.
- Place hot castings where they will not start a fire or burn someone.
- Keep the casting area orderly.
- Do not talk to anyone while pouring.
- Do not stand over the mold when pouring molten metal.

The key to safety in the technology-education laboratory is a safe attitude.

Standards for Technological Literacy

19

Demonstrate

During class, make a candle using a sand mold. Correlate the actions with Figure 16-4.

Example

Show the class some common molds for making shaped ice cubes, candles, candy, and other simple products.

TechnoFact

Casting is one of the oldest manufacturing processes. The Egyptians had mastered the art of bronze casting more than 3500 years ago. The modern manufacturing industry uses casting to create a variety of products, including tools, toys, and machine parts. More artistic types use the same processes to create jewelry, statues, and other decorative items.

Standards for Technological Literacy

19

Research

Ask the students to investigate how the Liberty Bell was made.

Demonstrate

Using a permanent mold, make a candle or a piece of chocolate candy in class. Correlate the actions with Figure 16-4.

TechnoFact

The Liberty Bell is one familiar cast product. The art of casting bells was developed in the 13th century. Most bells have a somewhat flat top and curve down to a thick lip. An English bell foundry originally made the Liberty Bell to celebrate Pennsylvania's 50th anniversary, long before the American Revolution.

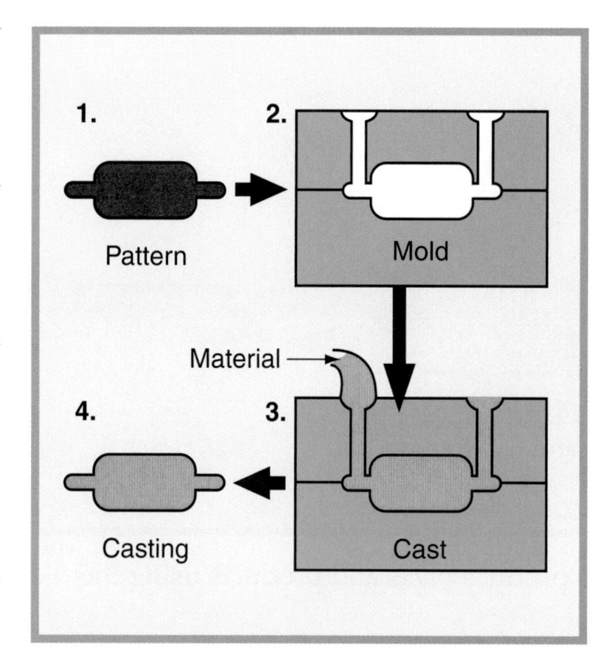

Figure 16-5. Four stages are followed in making an expendable-mold casting: 1. Making the pattern. 2. Surrounding the pattern with sand or plaster. 3. Introducing the liquid material and allowing it to solidify. 4. Breaking up the mold and removing the casting.

Permanent molds

A second type of mold is called a *permanent mold*. ***Permanent molds***, which are often made from steel, aluminum, or plaster, can withstand repeated use. See **Figure 16-6.** These mold materials must

©iStockphoto.com/ideabug

Figure 16-6. These permanent molds are used to produce cast plastic parts.

Career Corner

Automobile Assembly Workers

Automobile assembly workers put together parts to form subassemblies. Also, they assemble parts and subassemblies to build automobiles, trucks, and buses. They might mount tires, adjust brakes, and add fluids as the vehicles are assembled. These workers complete a number of tasks, including welding metal, cutting and molding plastic and glass, and sewing fabric. Many assemblers work with robots, computers, and other forms of automation as they complete vehicle assembly.

They often work more than 40 hours per week. Overtime is common during periods of peak demand. Some production work involves uncomfortable conditions, with heat, fumes, noise, and repetition. Newer plants are more automated and have safe working conditions.

withstand temperatures above the melting point of the material being cast. Most metal molds are produced by machining out the cavity. Plaster molds are often produced by casting the mold material around a pattern.

Processes that use permanent molds are die casting, injection molding, and slip casting. Die casting is used to produce aluminum and zinc parts. Injection molding can produce a wide range of plastic parts in many different resins. Slip casting is used to produce clay products.

Preparing the Material

All casting and molding processes require a material in a liquid state so the material can flow freely into the mold. Three types of liquids are used in casting. These liquids are solutions, suspensions, and molten materials.

Solutions

A solution is a uniform mixture of two substances. Casting plastics and frozen treats are examples of solutions. They are mixtures that give desired properties when solid. The act of mixing the parts of a fluid for casting is called ***compounding***.

Suspensions

Suspensions are mixtures in which the particles settle out. To keep the components mixed, you must shake the suspension. When you stop shaking the mixture, the particles begin to settle out. Slip, which is used in casting ceramics, is a suspension of clay and water.

Molten materials

Some materials, such as metals, must be heated to a molten state in order to be cast. At room temperature, steel is a solid and cannot be poured into a mold. Injection -molded plastics are heated to a liquid state so they can be forced into a mold.

Introducing the Material

A liquid material can be introduced into the mold. Most expendable molding processes use gravity to fill the mold. The material is poured into the mold and allowed to fill the cavity.

Many permanent molding processes use force to introduce the material into the mold. An example of this technique is shown in **Figure 16-7.** This figure depicts the action of an injection-molding machine used with plastic resins. The plastic is heated to a liquid state. A ram then forces the resin into the mold cavity. After the resin cools and solidifies, the mold opens, and the finished part is ejected. The mold then closes, and the process repeats. Die casting with metals uses the same procedure in a machine similar to an injection molder.

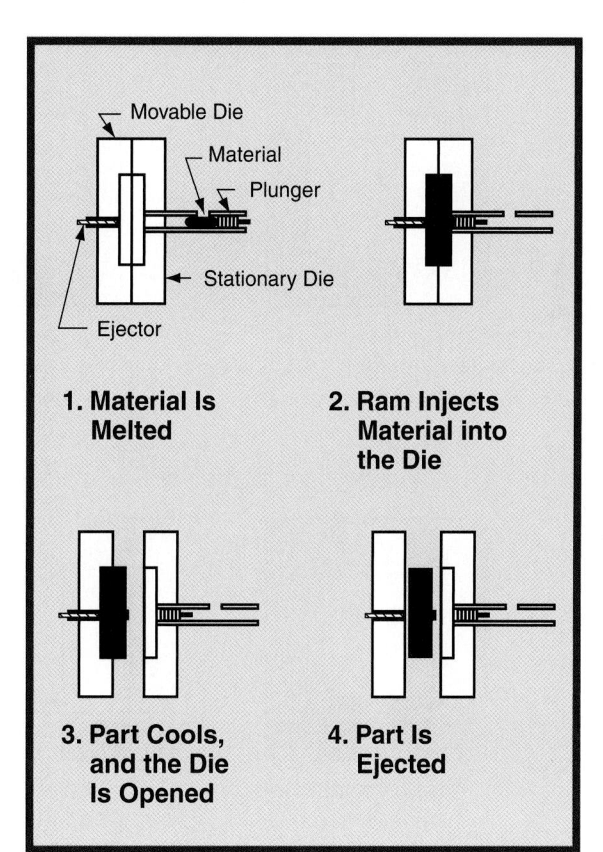

Figure 16-7. The action of an injection-molding machine.

Standards for Technological Literacy

19

Extend

Identify some typical products familiar to students that are die cast and injection molded.

TechnoFact

Many materials, including metals, plastics, and ceramics, can be cast. The glass blanks that formed the optics of powerful telescopes were some of the largest castings ever made. For example, the Mount Palomar Observatory in California required a 16.6-foot diameter mirror. In 1934, the Corning Glass Company created the mirror by melting a mass of glass for six days at 2700°F. The glass was then poured into a mold where it slowly cooled for eight more months. In the final stage, the mirror was ground by hand.

Standards for Technological Literacy

19

Brainstorm

Ask the class what products in their home were made by slip casting.

Research

Have the students use the Internet to find out why slip casting is used to produce ceramic products.

TechnoFact

Slip casting is one common method used to produce a large number of identical pieces of pottery. Aside from the potter's wheel, slip casting is the most well-known and widely used technique in the ceramics field. If you buy a mug in a tableware shop, chances are very good that it was produced with this technique.

Solidifying the Material

Once the liquid material is introduced into the mold cavity, it must become a solid. We can make materials solid in three ways:

- Cooling.
- Drying.
- Chemical action.

In cooling, the heat energy that caused the solid material to melt leaves. This makes the material return to its solid state. The mold can be cooled by letting the heat radiate into the air. Water can be pumped through passages in the mold to carry the heat away.

Water-based products are also solidified through cooling. Many food products, such as ice pops and ice-cream bars, are given their shape in a mold by cooling. This action we call *freezing*.

Suspended materials are solidified by allowing the solvent to be absorbed into the mold. This action is caused by ***drying***. An example of this technique is slip casting. See **Figure 16-8.** Slip, a clay-and-water mixture, is poured into the mold. The plaster mold absorbs water from the slip. The action causes a layer of solid clay to build up on the mold walls. The longer the slip stays in the mold, the more moisture it loses, and the thicker the walls become. When the proper wall thickness is reached, the remaining slip is poured out. The product is allowed to dry partially. The mold is then opened, and the product is removed. The product is placed in a humidity-controlled cabinet for further drying.

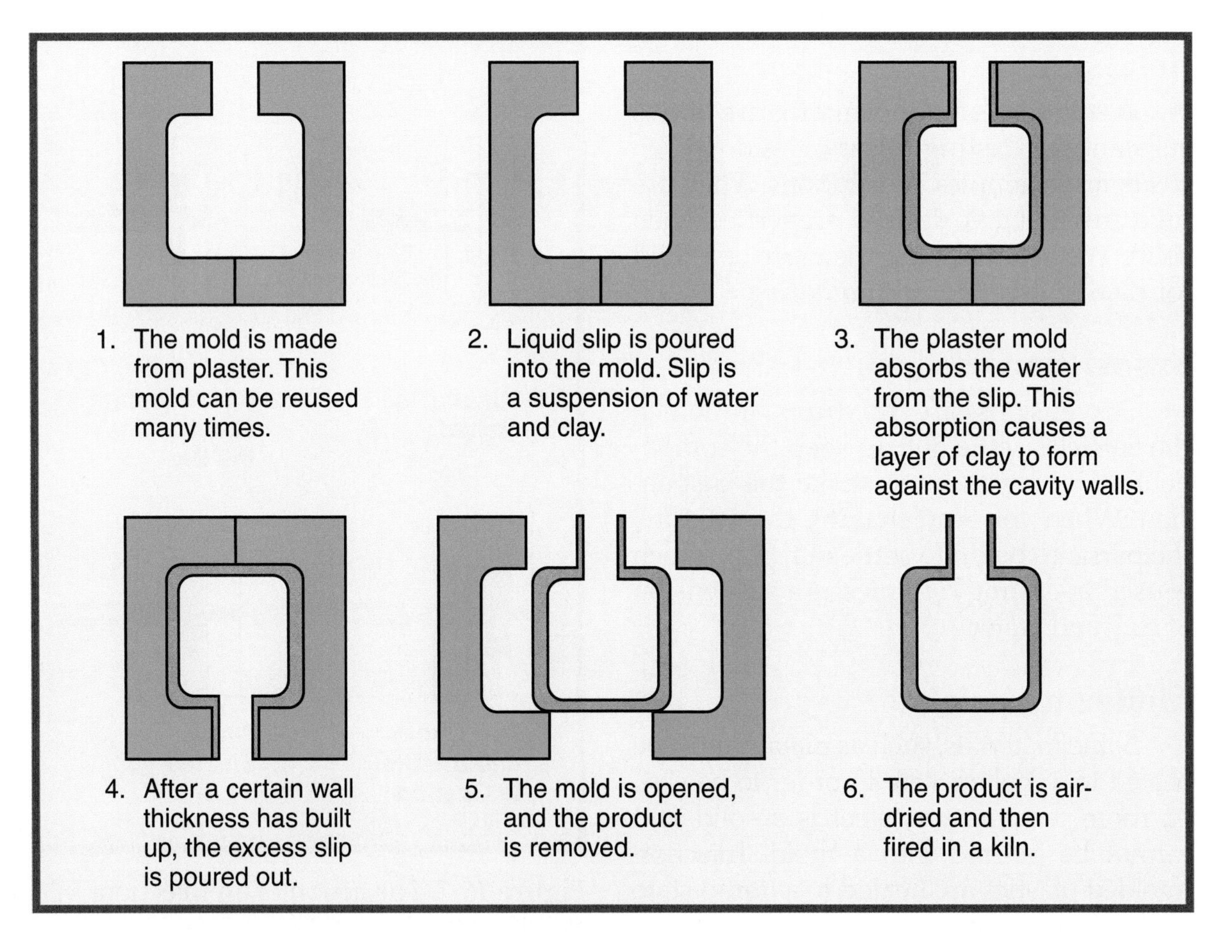

Figure 16-8. The steps in making a slip-cast product.

The third method of making a material solid is by ***chemical action***. For example, casting plastics have hardeners added. These chemicals cause the material to become solid, or set up. The short molecules in the plastic resin become longer, more complex molecules. This action, called *polymerization*, changes the liquid plastic into a solid.

Extracting the Product

The solidified product must be removed from the mold. See **Figure 16-9.** Expendable molds are destroyed to remove the part. Permanent molds are designed for easy opening. Often, the molding machine automatically introduces the material, causes it to solidify, and then opens and ejects the finished part.

Forming Processes

Forming is the second family of processes that give materials size and shape. All ***forming processes*** apply force through a forming device to cause the material to change shape. This force must be in a specific range, above the material's yield point and below the material's fracture point.

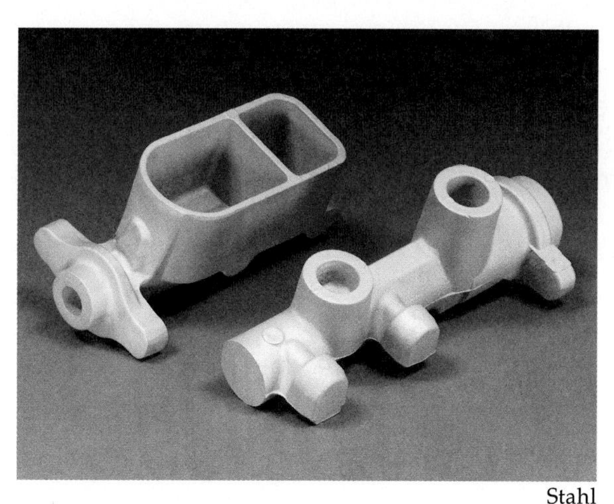

Stahl

Figure 16-9. These completed castings have been removed from the mold. They can be machined and assembled with other parts to make useful products.

All materials react to outside forces. See **Figure 16-10.** Small forces cause a material to flex (bend). When the force is removed, the material returns to its original shape. If the force increases, a point is reached at which the material does not return to its original shape. This point is called the ***yield point***. The range between rest and the yield point is called the material's ***elastic range***.

Above the yield point, the material is permanently deformed. The greater the force is, the more the material will be stretched, compressed, or bent. This range is called the material's ***plastic range***. At a certain point, the material cannot withstand any more force, and it breaks. This point is called the ***fracture point***. All forming processes operate in the plastic range of the material. These processes have three things in common:

- The presence of a forming device.
- The application of force.
- The consideration of material temperature.

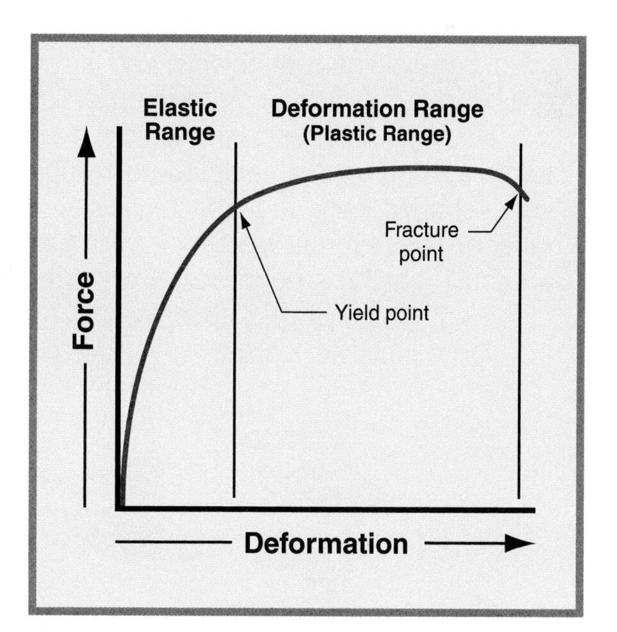

Figure 16-10. This stress-strain chart shows the elastic and plastic ranges for a material. Note the yield and fracture points.

Standards for Technological Literacy

19

Reinforce

Discuss forming in terms of the material properties presented in Chapter 4.

TechnoFact

Many artists have produced works of art with casting techniques. For example, Frederick Remington used wax casting technologies to create sculptures depicting life in the Old West.

Standards for Technological Literacy

19

Technology Explained

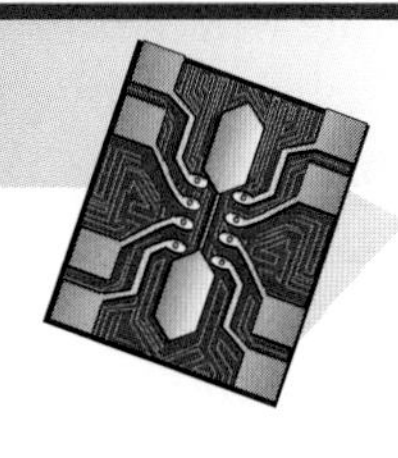

robot: a programmable part-handling or work-performing device, often used to replace human labor in industrial settings.

Robots are industrial devices used to increase the productivity of operations. They also remove humans from dangerous and undesirable working conditions. The Czech author Karel Capek first coined the word *robot* in the play *Rossum's Universal Robots*. He derived the word *robot* from the Czech word *robota*, which means "work."

Some of the first robots were designed in the 1940s to handle radioactive materials. The first industrial robot was developed in 1962. The robot's functions were limited to picking up an object and setting it down in a new location. This simple type of robot is called a *pick-and-place robot*. Today, there are many different types of robots. See **Figure A.** They are used in a wide variety of applications, including parts handling, welding, and painting.

Figure A. A common industrial robot.

A robot contains three important units: a mechanical unit for performing a task (the manipulator), a power unit to move the robot arm (the power supply), and a control unit to direct the robotic movement (the controller). Electric and hydraulic power supplies are the most common units used to raise, lower, and pivot robot arms into various positions. Hydraulic units are usually used to handle heavier objects, and they are generally considered faster than electric units. Electric units, though, take up less floor space and run more quietly.

Most robots operate in multiple planes. See **Figure B.** The simplest robots rotate around a single axis. More complex units can produce motion in two directions—horizontally and vertically. Even more complex units add an in-and-out motion. These three basic motions, or degrees of freedom, can be combined with an additional three degrees of freedom (six total) in the end effector (the device at the end of the arm). The six degrees of freedom can place the robot's effector anywhere in its work area.

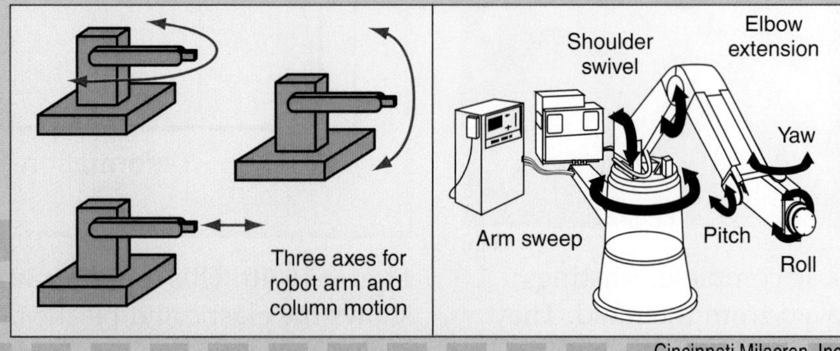

Figure B. Typical motions of an industrial robot.

Forming Devices

Forming processes are used to produce specific shapes in a material. They must have a way to ensure that the shape is correct and consistent. This goal is achieved by using one of two devices:

- Dies.
- Rolls.

Dies

Dies are forming tools made of hardened steel. They can be used to form any material softer than they are. The three types of dies are open dies, mated dies, and shaped dies. See **Figure 16-11.** ***Open dies*** are the simplest of all dies. They consist of two flat die halves. The material is placed between the halves, and the dies are closed. This action presses the material into a new shape. ***Mated dies*** have the desired shape machined into one or both halves of the die set. The material is placed between the die halves. The die set is closed, and the material is forced to take on the shape of the die cavities. One-piece ***shaped dies*** are widely used to form plastic objects. A common process using this type of mold is vacuum forming. A sheet of heated plastic material is placed over the mold. The air between the mold and the sheet is sucked out. This allows atmospheric pressure to force the material tightly over the mold. When it cools, the product is removed. See **Figure 16-12.**

Brush-Wellman

Figure 16-12. This ice-cube tray was molded using the mold shown at the top.

Rolls

Some processes use rolls to form the material. Two types of rolls exist: smooth and shaped. Smooth rolls are used to make curved shapes from sheets and shapes of metal. Shaped rolls have shapes machined into their surfaces. When material passes between shaped rolls, the material is squeezed into the shape of the roll. Shaped rolls are used to make pipe, tubing, and corrugated metal.

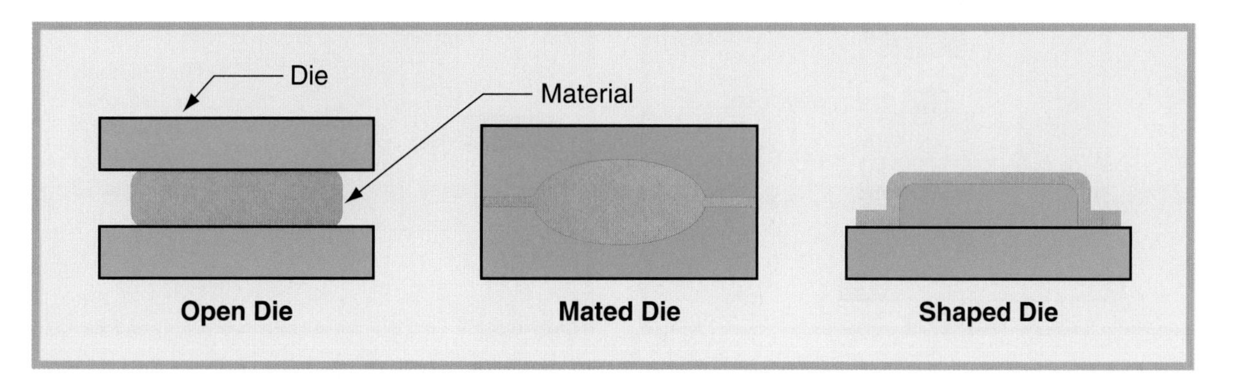

Figure 16-11. The three types of dies are open, mated, and shaped.

Standards for Technological Literacy

19

Research

Ask the students to find out how glass or plastic bottles are formed.

Demonstrate

In class, use a cookie press to demonstrate die forming for the students.

Standards for Technological Literacy

19

Figure Discussion

Use Figure 16-13 to describe the action of both a forging press and a forging hammer.

TechnoFact

The village blacksmith was an artisan. He not only had to master metalworking, but also design and engineering. The local blacksmith created a wide variety of products, including horseshoes, tools, hardware, cooking utensils, and cutlery. The trade got its name by combining two words, smith (to smite or hit) and black for the black iron he worked with.

TechnoFact

Drop forging is a technique used to produce small- and medium-size objects, such as gear blanks, machine parts, tools, and hardware.

The Application of Force

The force needed to complete a forming action can be delivered in a number of ways. The four most common are the following:

- Presses.
- Hammers.
- Rolling machines.
- Drawing machines.

Presses and hammers are very much alike. See **Figure 16-13.** They have a bed onto which a stationary die can be attached. The other die part is attached to a ram. A power unit lifts the ram, and the material to be formed is placed between the die halves. ***Presses*** slowly close the die halves by lowering the ram to produce a squeezing action. ***Hammers*** drop or drive the ram down with a quick action. This motion causes a sharp impact, which creates the forming force.

Rolling machines use two rolls rotating in opposing directions to form the material. The rolls draw the material between them and squeeze or bend it into a new shape. Forming rolls can be smooth, or they can have a pattern machined into them. Smooth rolls are used to produce metal sheets and foils and to shape sheets and strips into curved products, such as tanks and pipes. Shaped rolls produce patterns or bend materials. Typical products produced with shaped rolls are corrugated roofing, bent metal siding, and metal trim.

Drawing machines pull or push materials through die openings. They make wire from rods and extrude other shapes. Wire drawing pulls the rods through a series of dies that have progressively smaller openings. At each die, the material is reduced in diameter and increased in length.

Extrusion works very similarly to squeezing toothpaste from a tube. A shaped die is placed over the opening of the machine. The material is held in a closed cavity behind the die. A ram opposite the die forces the material out of the cavity. See **Figure 16-14.** As the material passes through the die, it takes on the shape of the opening. Extrusion is used to produce complex shapes from plastic, ceramic, and metallic materials. Air pressure, electrical fields, and explosive charges are additional ways forming forces can be generated.

Material Temperature

Something to consider about forming processes is the temperature of the material while it is being formed. Metals can be

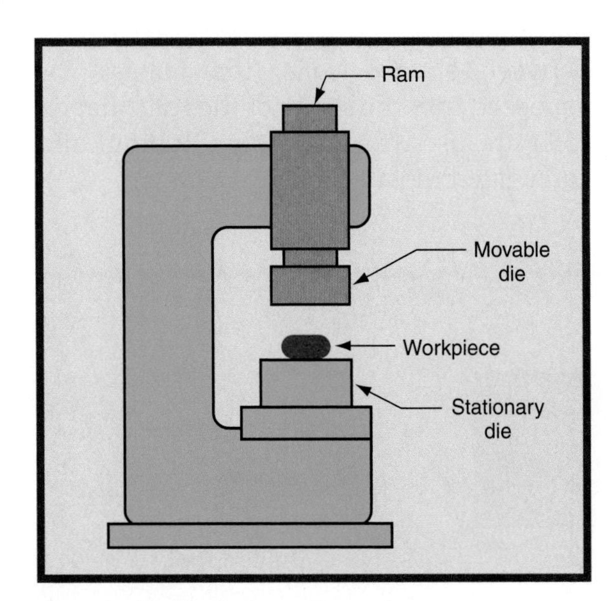

Figure 16-13. The parts of a forging hammer.

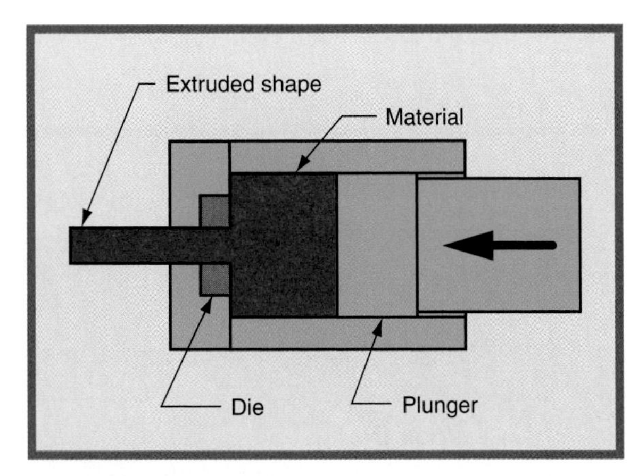

Figure 16-14. How extrusion works. The material takes on the shape of the die.

formed hot or cold. Plastics and glass are formed hot, and ceramics are formed cold.

All materials have a temperature at which their properties change. The internal structure changes in all materials as they are heated. We must remember that the terms *hot* and *cold* are relative. A "cold" slab of steel at a steel mill might be hot enough to burn you.

When a metal is cold formed, internal stresses are built up. These stresses can cause the material to become brittle. This is called *work hardening*. Work hardening is relieved by heat-treating the metal. ***Heat-treating*** involves heating the metal and allowing it to cool slowly.

When a metal is hot formed, it is important that the metal be heated above its recrystallization point. Hot forming takes place above this point. Cold forming takes place below this point. Hot forming prevents work hardening of the metal. As the material cools, it forms a normal structure. Therefore, the minimum temperature for hot forming is different for each material. Also, material shaped by hot forming is stress free, whereas cold forming builds internal stress in the material.

Separating Processes

Separating processes remove excess material to make an object of the correct size and shape. Casting and forming processes change the shape and size of materials without any removal. Separating processes remove material by either machining or shearing. See **Figure 16-15.** Machining is based on the motion of a tool against a workpiece to remove material. ***Shearing*** uses opposing edges of blades, knives, or dies to fracture the unwanted material away from the work.

Machining

Machining removes excess material in small pieces. This process cuts material away using three methods. These methods are chip removal, flame cutting, and nontraditional machining.

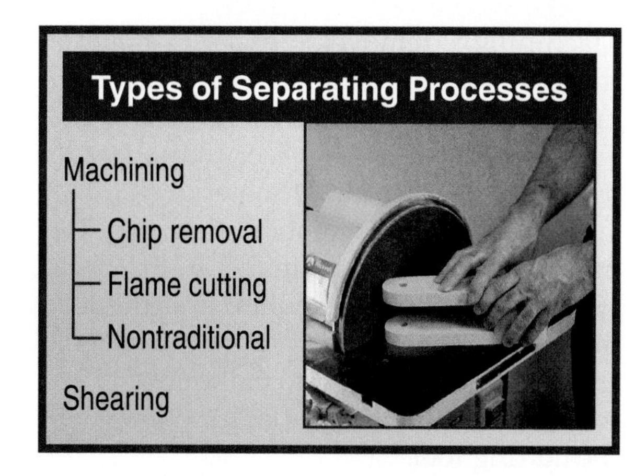

Figure 16-15. Separating processes include machining and shearing. The photo shows a sanding machine, which uses chip removal to take off unwanted material.

Chip removal

Chip removal is the most common separating process. All chip-removal processes have three things in common:

- A tool or other cutting element is always present.
- Motion occurs between the tool and work.
- The work and the tool are given support. See **Figure 16-16.**

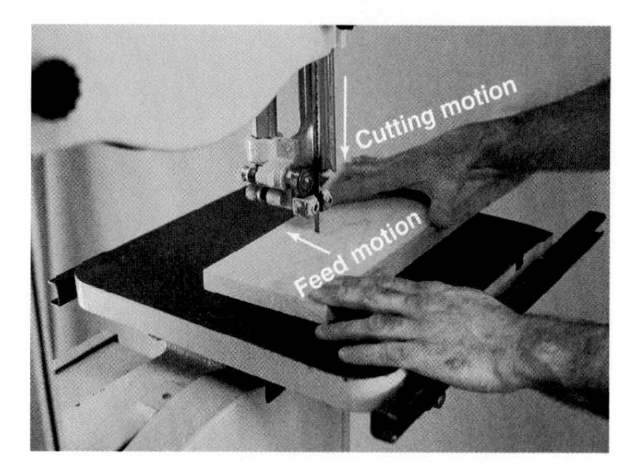

Figure 16-16. The cutting and feed motions of a band saw.

Standards for Technological Literacy

19

Demonstrate

Use a simple machine or hand tool to demonstrate feed and cutting motion.

TechnoFact

Grooves were added to steel rolling mills in the 1720s. This allowed manufacturers to produce iron shapes as well as sheets.

Standards for Technological Literacy

19

Extend

Contrast machining with shearing.

Figure Discussion

Use Figure 16-17 to discuss the cutting element, cutting motion, and feed motion of common machines.

TechnoFact

Forging presses squeeze the workpiece slowly so force is evenly distributed. Such action is necessary for forging large objects. Press forges are available in many sizes, with the largest having more than 120,000-ton capacity. Small presses produce their force either mechanically or hydraulically. Large presses are hydraulically powered.

Machining removes excess material through cutting actions. The most common cutting device is a chip-removing tool. Tools are based on the fact that a harder material will cut a softer material.

With regard to motion, two types of motion are needed to make a cut. These types are the cutting motion and the feed motion. Cutting motion moves a cutting tool through a material to make chips. Feed motion brings new material in contact with the cutting tool.

Safety with Machine Tools

People who work with machine tools use various machines. Some basic rules should be observed during the operation of these machines:

- Wear eye protection.
- Avoid loose-fitting clothing and keep long sleeves rolled up.
- Wear hearing protection when loud or high-pitched noises are present.
- Keep the laboratory clean. Wipe up any spills immediately.
- Wait until the machine has stopped to make adjustments, remove the workpiece, or clean up any scrap. Use a brush to clean chips off a machine.
- Do *not* talk to anyone while running a machine.
- Most of all, if you do not understand how to run a machine, ask first!

The key to safety in the technology-education laboratory is a safe attitude.

Common separating machines can be grouped by the cutting and feed motions they use. See **Figure 16-17.** These groupings include the following:

- ***Milling machines*** use a rotating cutter for the cutting motion. The

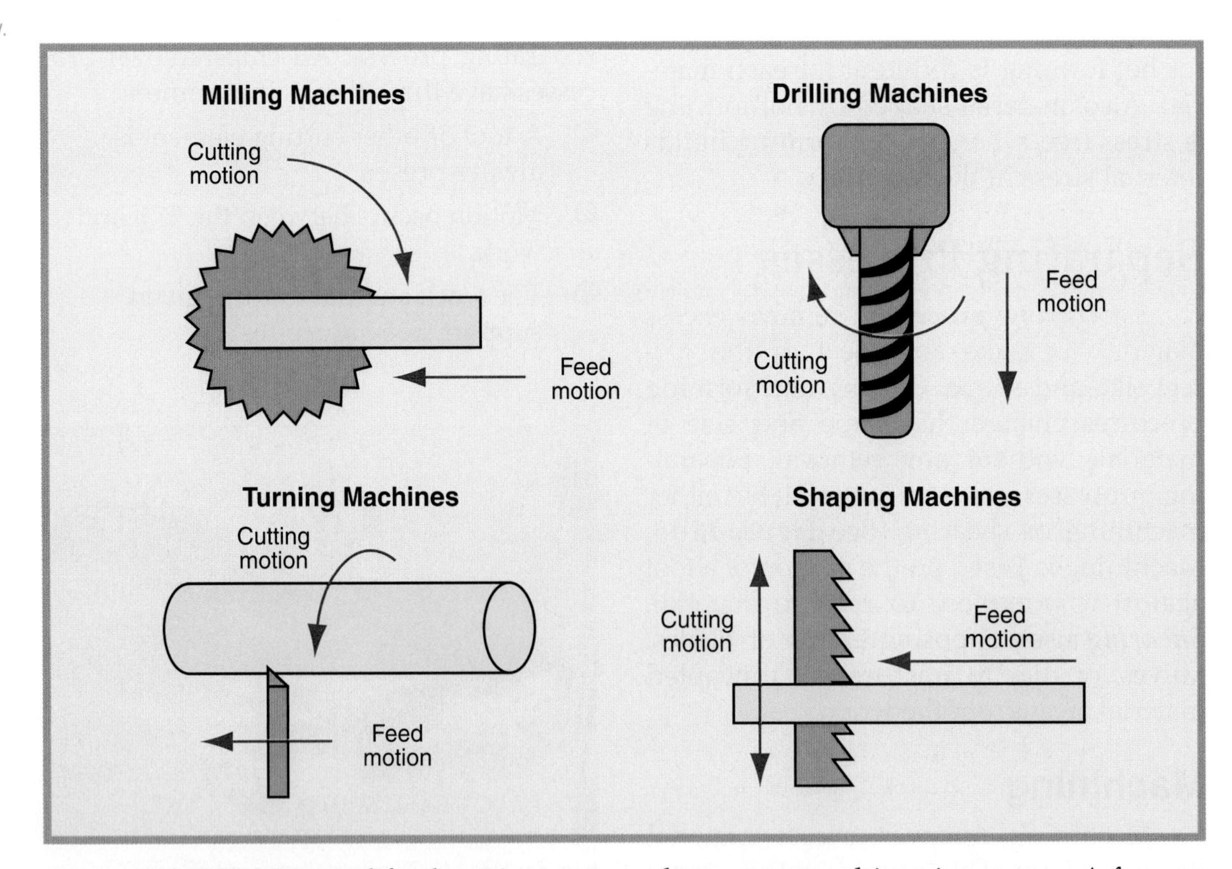

Figure 16-17. Cutting and feed motions are used to group machines into types. A few examples of these types are shown here.

feed motion is linear. The material is pushed into the cutter in a straight line. Machines that use the milling principle are horizontal and vertical milling machines; table saws; wood shapers, planers, and jointers; routers; disc sanders; and pedestal and bench grinders. The radial saw and the cutoff saw are variations of this group. Rotating multiple-point tools, called *blades*, are rotated. The blade, however, moves to produce the feed and cutting motions at the same time.

- Drilling machines rotate a cutter for the cutting motion. The cutter is fed in a linear manner for the feed motion. Machines that use this principle are the drill press and the portable electric drill.
- Turning machines rotate the work against a single-point tool to produce the cutting motion. The tool is fed along or into the work to produce the feed motion. Woodworking and metalworking lathes are turning machines.
- Shaping machines use a single-point tool that moves back and forth across the work to produce the cutting motion. The cut is usually made on the forward stroke. The tool is lifted slightly for the backstroke. The work is fed linearly under the tool. Metalworking shapers are the primary machines using this cutting action. Small saws, such as hacksaws and scroll saws, use a reciprocating motion also.
- Planing machines move the work under the tool to make the cutting motion. Moving the tool across the work in small steps creates the feed motion. The metal planer is the machine that uses this technique. Both metal shapers and planers are used to make large machinery.

Flame cutting

Burning gases can be used to remove unwanted material from a workpiece. See **Figure 16-18.** A fuel gas and oxygen are mixed in a torch, ignited, and allowed to heat the surface of the metal. When the metal is hot enough, a stream of pure oxygen is forced out of the torch. The oxygen causes the metal to burn, which separates the workpiece from the scrap. Cutting material to size and shape using burning gases is called ***flame cutting***.

Acetylene gas and natural gas are used as fuel gases for cutting metals. The fuel gas and oxygen are fed through separate regulators to control the pressure. Each gas flows through a hose to the torch. The torch controls the volume and mixture of the gases. The mixture then moves out of the torch. The blast of pure oxygen is controlled separately from the fuel gas and oxygen mixture.

Bystronic, Inc.

Figure 16-18. This computer-controlled flame-cutting machine is cutting out parts from a steel plate.

Standards for Technological Literacy

19

Research

Assign small groups of students to select a machine and prepare a diagram that shows its tool, feed motion, and cutting motion. Be sure to remind them to label the machine's major parts.

TechnoFact

Early machines were crude and the parts they produced lacked precision, which led to product failures and accidents. Machine tools were developed to combat this problem. The term *machine tools* refers to tools that are designed to produce other machines.

TechnoFact

The basic machine tools operations were developed to mimic the function of woodworking hand tools. In 1775, John Wilkinson invented the first machine tool, a boring machine for accurately drilling holes in metal.

Standards for Technological Literacy

19

Figure Discussion

Use Figure 16-19 to explain the operation of an electrical discharge machine.

TechnoFact

The development of new machine tools helped drive the Industrial Revolution. Some of the machine tools that were developed included boring machines (1775), gear cutters (1820), grinders (1834), metal shapers (1836), and screw-cutting machines (1880).

Nontraditional machining

Chip removing and flame cutting have been used for a long period of time. In the 1940s, people started developing a series of new separating processes. These processes use electrical, sound, chemical, and light energy to size and shape materials. They are often called ***nontraditional machining processes***. This term is not very accurate today because many of these processes are widely used. They are rapidly becoming traditional ways to machine materials. Three common nontraditional processes are electrical discharge machining (EDM), laser machining, and chemical machining.

Electrical discharge machining (EDM) uses electric sparks to make a cavity in a piece of metal. This process requires an electrode, a power source, a tank, and a coolant. See Figure 16-19. The workpiece is connected to one side of the power supply and placed in the tank. The electrode, which is made in the shape of the desired cavity, is connected to the other side of the power supply. The tank is filled with coolant that is a dielectric material. A dielectric resists the flow of electricity. The electrode is lowered until a spark jumps between the electrode and the work. When the spark jumps, the dielectric quality of the coolant has been overcome. The spark dislodges small particles of material that the coolant carries away. A cavity the same shape as the electrode is created. The electrode is lowered as the cavity is produced until the proper depth is achieved.

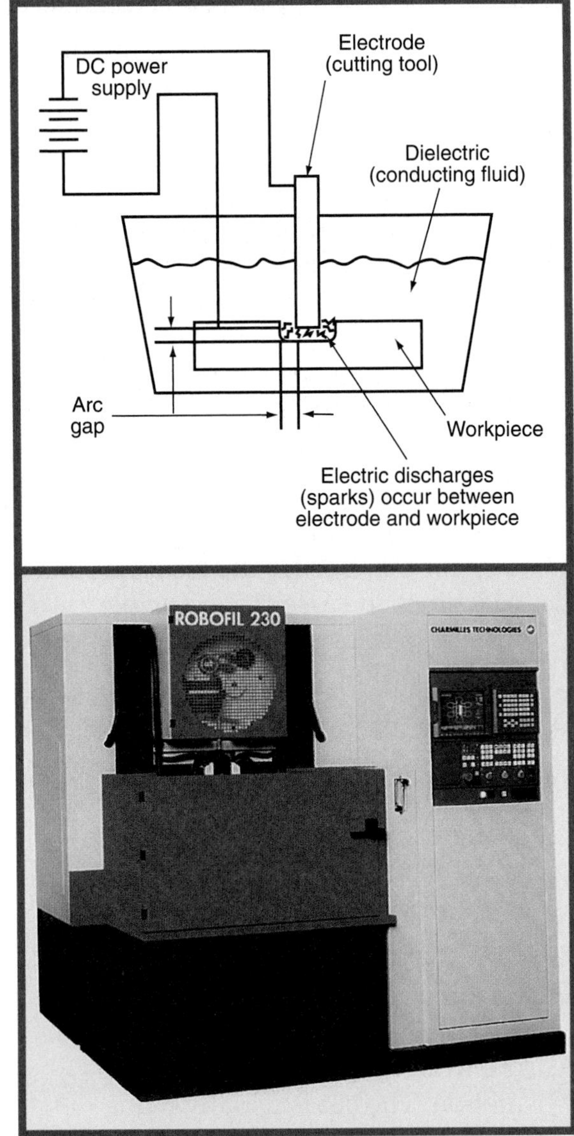

Caterpillar Tractor Co., Charmilles Technologies

Figure 16-19. The diagram shows how EDM works, and the photo shows an EDM machine.

Laser machining uses the intense light a laser generates to cut material. See Figure 16-20. A laser converts electrical energy into monochromatic (single-color) light. The light is focused onto the work, where it melts the material. The beam of light is guided along a path to produce the desired shape.

Chemical machining uses chemical reactions to remove material from a workpiece. The workpiece is coated with a resist, which does not react to the chemicals used as the cutting agent. The resist is removed from the areas to be machined. The workpiece is then placed in a tank and covered with the cutting chemical. Areas coated with the resist are untouched. Where the workpiece is exposed to the chemical, material is removed.

Two major chemical machining processes are chemical blanking and chemical milling. The process is the same for both.

Figure 16-20. A close-up view of a laser-cutting operation.

The only variable is in how long the workpiece is etched. In chemical blanking, the etching continues until only the part the resist covers remains. Chemical blanking is used for very thin, delicate parts or for short runs (small numbers of parts). Chemical milling uses the same process, but the part is not totally etched away. This process is used to remove excess metal in aircraft parts to save weight.

Shearing

The second group of separating processes is shearing. Shearing cuts material to create the desired size and shape. These processes can be used to cut material to length, produce an external shape, or generate an internal feature. See **Figure 16-21.** Cutting to length is generally accomplished with opposing blades. The upper blade moves downward to deform and fracture the material where it contacts the lower blade.

Internal and external shapes are often made with a punch and a die. The die has a cavity of the desired shape. The punch fits into this cavity. The material is placed on the die. When the punch moves downward, the material is sheared into the shape of the die cavity. Punches and dies are used to produce holes, slots, and notches.

Conditioning Processes

Casting, molding, and separating operations change the external features of a workpiece. The material is given a new size and shape. In some cases, this is not

Standards for Technological Literacy

19

Demonstrate

In class, use tin snips or scissors to show how the shearing action works.

Figure Discussion

Use Figure 16-21 to explain how opposed edges can shear materials.

TechnoFact

In 1873, an American named C. M. Spencer invented the first fully automatic lathe.

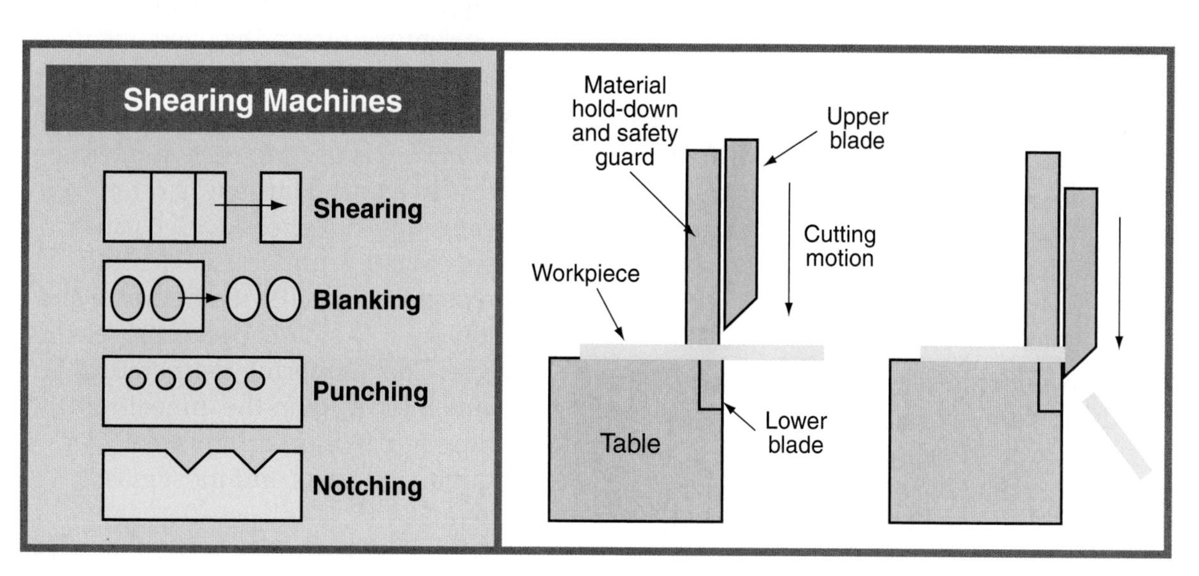

Figure 16-21. Common shearing operations and how shearing works.

Standards for Technological Literacy

19

Brainstorm

Ask the students what products (or parts of products) that they have used have been thermally conditioned.

TechnoFact

Today there are approximately 500 types of machine tools in use. Some machine tools are designed to complete a single task. Others, called machining centers, may contain more than 100 tools that are automatically switched depending on the operations the machine must perform.

enough. The internal structure of the material needs to be changed. The material might need to be harder, softer, stronger, or more easily worked. To change internal properties, ***conditioning processes*** are used. The three types of conditioning processes are mechanical, chemical, and thermal conditioning.

Mechanical Conditioning

Mechanical conditioning uses mechanical forces to change the internal structure of the material. Most metals become harder as they are squeezed, stretched, pounded, or bent. This action, as we learned earlier in this chapter, is called *work hardening*. In most cases, work hardening is not desired, but it is unavoidable. If you have ever bent a wire back and forth to cause it to break, you have used work hardening. The repeated bending makes the wire harder and more brittle.

Bethlehem Steel Co.

Figure 16-22. These hot parts are leaving a heat-treating furnace. They will be quenched to harden them.

Chemical Conditioning

Chemical conditioning uses chemical actions to change the properties of a material. The lenses of many safety glasses are chemically treated to make them more shatterproof. Likewise, the resins used in fiberglass layups undergo chemical action as they cure and become hard.

Thermal Conditioning

The most common conditioning processes use heat. These are called ***thermal conditioning*** and include heat-treating, firing, and drying. *Heat-treating* is a term used to describe the thermal conditioning processes used on metals. These processes include the following:

- ***Hardening*** is used to increase the hardness of a material. See **Figure 16-22.** Hardening steel requires the part to be heated to a specific temperature and allowed to soak. This ensures that the entire part is at a uniform temperature. The part is then rapidly cooled in a tank of oil or water (quenched).
- ***Annealing*** is used to soften and remove internal stress in a part. The part is heated to a specific temperature, allowed to soak, and then removed from the oven and allowed to cool slowly to room temperature.
- ***Tempering*** is used to relieve internal stress in a part. Hardening often creates internal stress, which causes a part to crack under use. Heating the part to a specific temperature and allowing it to slowly cool removes this stress. The tempering temperature is much lower than the annealing temperature. Tempering is used for many metal parts and most glass products.

Firing is a thermal conditioning process used for ceramic products. Most ceramics are made from clay materials that are plastic when wet. After drying, the clay can be heated to a high temperature. The water is driven out of the clay particles, and the grains bond together to make a solid structure.

Likewise, certain coatings are fired. Porcelain enamels are fired to give a glass-like finish. The ***enamel*** is applied to the part by spraying or dipping. The product is heated, and the coating fuses with the part.

Drying is a common thermal conditioning process. This process removes excess moisture from materials. Ceramic materials and wood products must be dried before they are useful. Drying can happen naturally or be helped by adding heat. For example, lumber is air-dried or kiln dried. Air-dried lumber is stacked outdoors to dry after cutting, while kiln-dried lumber is carefully heated in special ovens called *dry kilns*.

Assembling Processes

Look around you. How many products with one part do you see? You might notice such items as paper clips or straight pins. Most of the things you see, however, are made from more than one part. These products are assembled from two or more parts. See **Figure 16-23.** The word *assemble* means "to bring together." Through ***assembling processes***, for example, a simple product, such as a lead pencil, is created from five parts. The barrel is two pieces of wood glued around the graphite "lead." The eraser is held onto the barrel with a metal band. In fact, a lead pencil uses the two methods by which products are assembled: bonding and mechanical fastening.

Bonding

Bonding holds plastic, metal, ceramic, and composite parts to each other. This method uses cohesive or adhesive forces to hold parts together. Cohesive forces hold the molecules of one material together. Adhesive forces occur between different kinds of molecules. They cause some materials to be sticky. The following affect the way in which parts are bonded together:

- The bonding technique.
- The bonding material.
- The type of joint.

Bonding techniques

Five bonding techniques are used for assembly. The first is ***fusion bonding***. This technique uses heat or solvents to melt the edges of the ***joint***. The surfaces are allowed

Figure 16-23. Products are assembled using both mechanical and bonding techniques. These washing machines have mechanisms bolted into place and sound insulation bonded inside.

Standards for Technological Literacy

19

Research

Assign small groups of students to find out how one welding process works and diagram its operation.

Demonstrate

During class, show the difference between adhesion and cohesion by gluing paper together (adhesion) and putting two pieces of wet glass together (cohesion).

TechnoFact

Laser cutting is one of the newest and most advanced methods of machining. This process is capable of making extremely precise cuts and holes in material. The level of precision is so high because a very thin laser beam is focused on the workpiece, causing controlled melting in an extremely localized area.

Standards for Technological Literacy

19

TechnoFact

Vulcanization, a process developed by Charles Goodyear in 1839, combines chemical and thermal conditioning of rubber. The vulcanization process increases the rubber's strength, elasticity, and resistance to solvents. Vulcanization greatly increased the usefulness of rubber.

to flow together to create a bond. In some cases, more material is added to the joint to increase the strength. Oxyacetylene, arc, inert gas, and plastic welding are examples of fusion-bonding methods.

A second method is flow bonding. ***Flow bonding*** uses a metal alloy as a bonding agent. The base metal is cleaned and then heated, but it is not melted. The alloy is applied where the two parts meet. This metal alloy melts, flows between the parts, cools, and creates a bond. Soldering and brazing are examples of flow bonding.

Pressure bonding applies heat and pressure to the bond area. This method is used on plastics and metals. Resistance (spot) welding is an example of pressure bonding. See Figure 16-24. Spot welding has four stages in its cycle:

1. The parts are held together between two electrodes.
2. Electrical current is passed between the electrodes. The current melts the metal between the electrodes.
3. The current is stopped, and the molten metal solidifies.
4. The electrodes are released, and the welded parts are removed.

Plastics can be pressure bonded using a similar process. No electric current passes between the parts. Heated jaws hold the parts and bond them together. The jaws release, and the parts are removed. Home sealers for plastic storage bags use pressure bonding.

Cold bonding uses extreme pressure to squeeze the two parts to create a bond. This process is not used often. Cold bonding is used only for small parts made of soft metals, such as copper and aluminum.

Adhesive bonding uses substances with high adhesive forces to hold parts together. These techniques are gaining wide use beyond their original use in woodworking. The advent of synthetic adhesives allows a wide range of plastics, metals, and ceramics to be joined together.

Ford

Figure 16-24. This automobile assembly is being put together using resistance welding. Resistance welding is a type of pressure bonding.

Bonding materials

All bonding techniques use a bonding material. This material is referred to as a ***bonding agent***. A bonding agent can be one of three kinds:

- **The same material.** The same material as the base material holds parts together. For example, two metal parts might be welded together, or a steel welding rod might be used to strengthen the bond between two steel parts. See Figure 16-25.
- **Similar material.** A similar type of material holds parts in position. For example, a metal (solder) might be used to bond metallic parts (copper wires).
- **Different material.** A different type of material holds parts together. For example, white glue (a plastic) might hold wood together, or an epoxy (a plastic) might hold ceramic parts together.

Figure 16-25. This welding operation uses filler rods made from the same metal as the parts being welded.

Types of joints

Joints are where parts meet. They can occur at the ends, sides, and faces of the parts being assembled. Joints are used to add length, width, or thickness and to make corners. The type of joint chosen affects the type of bonding process used. A simple joint can be modified to increase its strength or improve its appearance. See Figure 16-26.

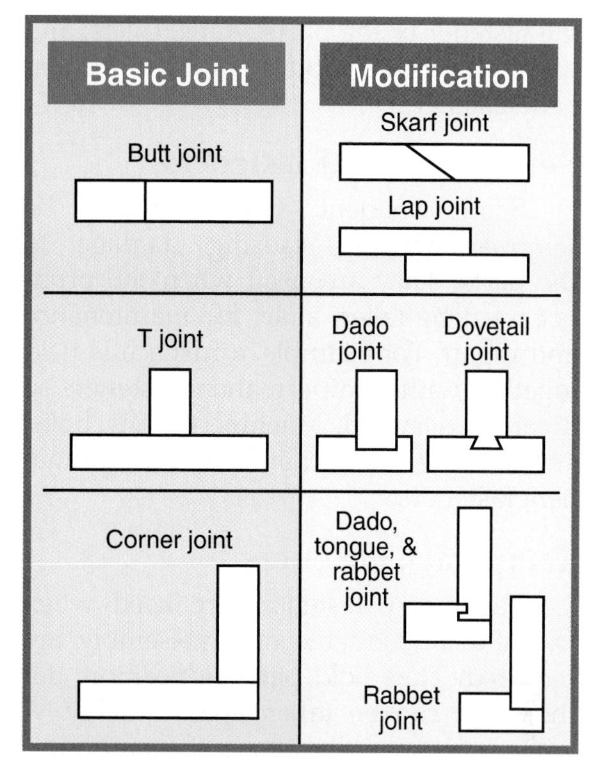

Figure 16-26. Simple joints can be changed for higher strength and better appearance.

Mechanical Fastening

Mechanical fastening uses mechanical forces, such as friction, to hold parts together. For example, a part might be pressed or driven into a hole slightly smaller than it is. The friction between the parts causes the parts to remain together. This type of fit is called a ***press fit***. Press fits can be used to hold bearings in place on a shaft.

In other cases, the parts might be bent and interlocked to hold the parts together. This type of joint is called a *seam*. Many sheet metal parts are held together using seams.

The most widely used method to hold parts together is mechanical ***fasteners***. See Figure 16-27. Examples of fasteners are staples, rivets, screws, nails, pins, bolts, and nuts. Mechanical fasteners can be described as permanent, semipermanent, or temporary.

Figure 16-27. The lug nuts on this wheel are mechanical fasteners.

Permanent fasteners

Permanent fasteners are not intended to be removed. Their removal damages

Standards for Technological Literacy

19

Example

Show examples of parts that have been bonded with the same material (welded), similar material (soldered), and a different material (glued).

Figure Discussion

Use Figure 16-26 to find examples of joints in the classroom/ laboratory.

TechnoFact

Early potters fired their work in an open fire to harden it. Today, ceramic ware is kiln-fired at temperatures between 1,100°F to 2,800°F. This is usually done in a two-step process. The first firing, called biscuit firing, hardens the clay. The second firing fuses the glaze to the ceramics, forming a glasslike shell on the piece.

Standards for Technological Literacy

5 19

Brainstorm

Have each student select three products and explain what finish was used on it and why it was used.

TechnoFact

The assembly of products became much easier following the advent of interchangeable parts. Eli Whitney, an American, is widely (but mistakenly) credited with the invention of interchangeable parts. The use of interchangeable parts in mass-production became known as "the American system."

TechnoFact

Sewing is one of the oldest and most common mechanical fastening systems. Early humans created clothing by sewing hides together with sinews and bone needles. Elias Howe's 1846 invention of the lockstitch sewing machine marked a new era in the garment industry. This was quickly followed by Isaac M. Singer's 1851 invention of the first practical domestic sewing machine. Singer's machine was a great time-saver and relatively affordable. For these reasons, it became the first major consumer appliance.

the fastener or the parts. Nails, rivets, and wood screws are good examples of permanent fasteners.

Semipermanent fasteners

Semipermanent fasteners can be removed without causing damage to the parts. They are used when the product must be taken apart for maintenance and repair. For example, a furnace is held together with semipermanent fasteners so it can be serviced. Machine screws, bolts, and nuts are good examples of semipermanent fasteners.

Temporary fasteners

Temporary fasteners are used when frequent adjustments or disassembly are required. They hold parts in position, but the parts can be taken apart quickly. A wing nut is a good example of a temporary fastener.

Finishing Processes

Finishing processes are the last of the secondary processes most products go through. These techniques protect products and enhance their appearances. Finishing processes can be broken into two types. One group changes the surface of the product. The other group applies a coating.

Most metals begin to corrode if they are not protected in some way. Metals are easy to protect by changing the surface chemically. For example, anodizing converts the surface of aluminum products to aluminum oxide. This type of finish is called a ***converted surface finish***.

The other type of finishing applies a coating to the product. A film of finishing material is applied to the product or base material. These coatings protect the surface and can add color. Finishing processes involve cleaning the surface, selecting the finish, and applying the finish.

Safety with Finishing Materials and Equipment

People who work with finishing processes observe basic safety rules:

- Wear eye and face protection.
- Apply finishes in a well-ventilated area. Wear a respirator and protective clothing if solvents are toxic.

Think Green

Reduction

Reduction refers to finding ways to reduce your impact on the environment. You have probably heard the phrase "Reduce, reuse, recycle." Reusing and recycling are two ways to reduce your environmental impact. The first step you can take to reduce your impact is to learn about the products and processes that have harmful effects on the environment. You may be surprised at how small changes in your everyday life may help change your impact. Being more mindful of green alternatives to products or processes is also a good way to work on reduction.

Companies and individuals alike are taking these steps toward going green. People have begun to calculate waste production and carbon dioxide emissions. Also, the amount of resources like water and energy must be taken into consideration. The strides being made to reduce waste has been studied by various organizations and has been found to have a great impact on the environment so far.

- Never apply a finish near an open flame.
- Keep the finishing area orderly, and clean up any spills immediately.
- Store all rags and chemicals in approved containers.
- Use a spray booth (an enclosed area with fans to remove fumes) to remove toxic fumes.
- Dispose of any waste solvents and finishes properly. Wash your hands to remove any finishes you have handled.
- Most of all, if you do not understand how to apply a finish, ask first!

The key to safety in the technology-education laboratory is a safe attitude.

Cleaning the Surface

Finishing materials are applied to the surface of a product. This surface must be free of dirt, oil, and other foreign matter. Mechanical means or chemical cleaning can remove these unwanted materials. Chemical cleaning is often called ***pickling***. Pickling involves dipping the material into a solvent that removes the unwanted materials. The clean part is then rinsed to remove the solvent. Mechanical cleaning includes abrasive cleaning (sandblasting or sanding), buffing, and wire brushing.

Selecting the Finish

Hundreds of finishing materials exist. We can classify them as either inorganic or organic materials, however, to understand them better. Inorganic materials include metal and ceramic coatings. For example, steel is coated with zinc to produce galvanized steel or with tin to make tin plate. See **Figure 16-28.** Clay products, such as floor and wall tile, are coated with a glaze. Electric-range tops are coated with a ceramic material called *porcelain enamel.* All of us are familiar with the organic finishes. These materials include the following:

American Iron and Steel Institute

Figure 16-28. This roll of steel has been coated with tin to make tin plate. Tin plate is used to make cans for food.

- ***Paint.*** This material is any coating that dries through polymerization (hardening).
- ***Varnish.*** This material is a clear finish made from a mixture of oil, resin, solvent, and a drying agent.
- ***Enamel.*** This material is a varnish that has color pigment added.
- ***Lacquer.*** This material is a solvent-based, synthetic coating that dries through solvent evaporation.

Applying the Finish

Finishing materials can be applied in many ways. Metallic coatings are often applied through dipping or plating.

Standards for Technological Literacy

19

Research

Have each student select a finishing material and find out how it is made and used.

TechnoFact

In 1836, Edmund Davy became the first person to produce acetylene, a common welding gas. Acetylene is produced by a chemical reaction between calcium carbide and water. An oxygen and acetylene flame burns at about 6000° F (3200° C), hot enough to weld and cut metals.

Standards for Technological Literacy

19

Extend

Discuss the advantages and disadvantages of the common ways of applying a finish coating.

TechnoFact

Humans have been using paint for at least 30,000 years. Prehistoric people mixed vegetable and earth pigments with water or animal fat to create crude paints. They used these paints to decorate the walls of their caves and their bodies.

Dipping involves running the stock through a vat of molten metal. This technique is widely used to produce galvanized steel.

Plating is an electrolytic process. The parts are hung on racks and lowered into a cell full of electrolyte. See **Figure 16-29.** The plating metal is the anode, and the part is the cathode. Electrical current moving through the cell causes the metal to move from the anode into the electrolyte. From there, the metal moves across to the cathode and is deposited as a uniform coating on the product.

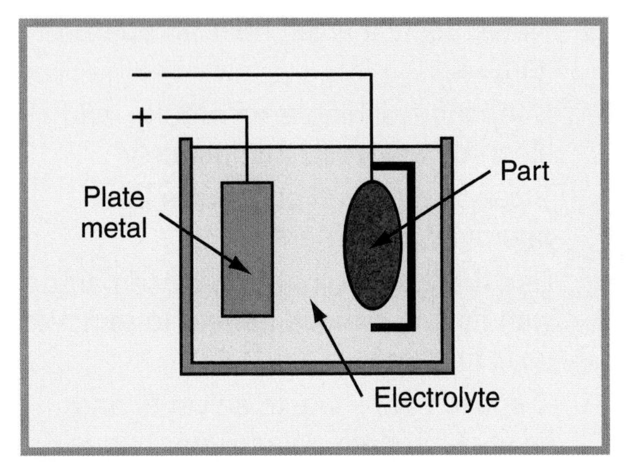

Figure 16-29. An electrolytic cell. This cell is used to apply metal plating to parts.

Organic materials can be applied through brushing, rolling, spraying, flow coating, and dip coating. Most of us are familiar with brushing and rolling. The material is gathered in the brush or on the roller and is then applied by wiping the brush or rolling the roller across the surface of the material.

Spraying uses air to carry fine particles of finishing materials to the surface of the product. See **Figure 16-30.** Some of these processes are manual, in which a worker uses a spray gun to apply the finishing material. Many products, however, are coated by automatic spraying systems.

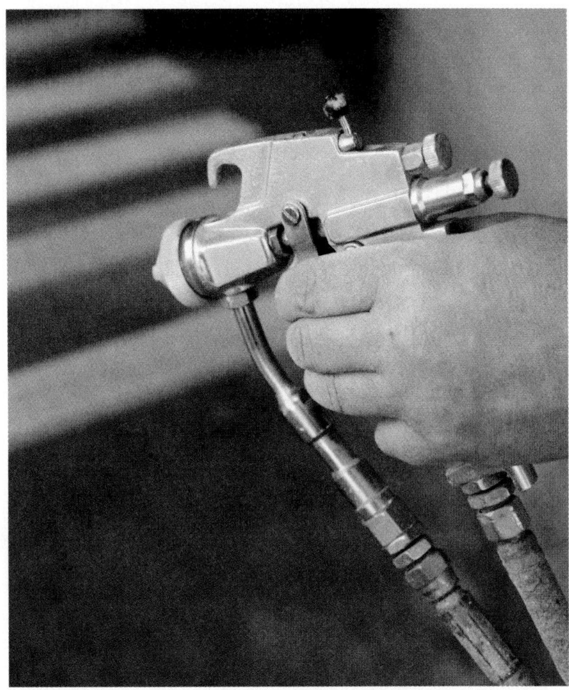

©iStockphoto.com/OwenPrice

Figure 16-30. Paint being applied to a metal part using a hand-operated spray gun.

Flow coating passes the product under a flowing stream of finishing material. The material flows over the surface. The excess runs into a catch basin to be used again.

Dip coating is the same process as that used for galvanizing. In both cases, excess coating is allowed to drain off the product for reuse. After parts have a finish applied, they are allowed to air-dry, or they can be run through a heated drying oven.

Automating and Controlling Processes

Historically, skilled workers completed the primary and secondary manufacturing processes using simple tools. During the Industrial Revolution, much of this handwork was moved to machines in factories. Humans who had limited skills operated most of these machines. The operators loaded materials into the machines, adjusted tool feeds and speeds, monitored the machines' operation, and unloaded the finished work. In recent years, a new breed of machines has replaced many humans on the factory floor. These machines make up systems called *automation*. This term refers to the automatic systems used to operate or control equipment or a process. Automation has many parts, which include robots, programmable logic controllers (PLCs), computer-controlled machines, and artificial intelligence (AI) systems.

Robots

Robots are mechanical devices that can perform tasks automatically or with varying degrees of direct human control. Industrial robots are a very common type of robot. See **Figure 16-31.** Industrial robots can be described as programmable, multipurpose manipulators. They can operate using computer programs, and they can manipulate (hold and rotate) parts or devices. Other robots can mow your lawn, vacuum your house, gather samples on the Moon, clean up toxic waste, and search the ocean's floor.

Some common applications of industrial robots are welding, painting, assembling, moving materials and parts (pick and place actions), and inspecting products. A typical industrial robot has five main parts: an arm, a controller, a drive, an end effector, and a sensor. See **Figure 16-32.**

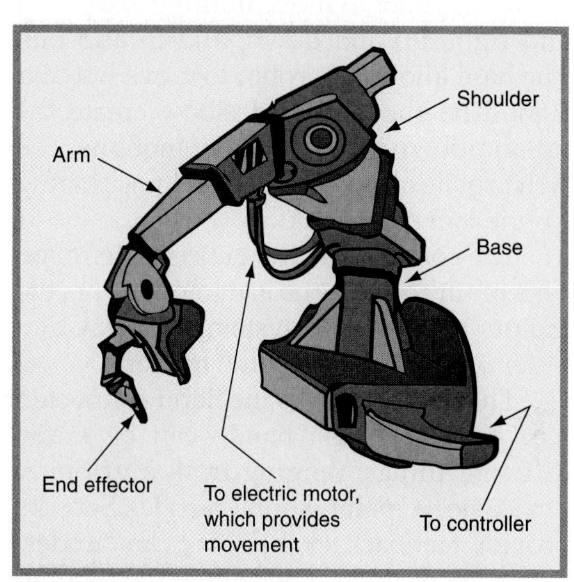

Figure 16-32. Parts of a common robot.

Standards for Technological Literacy

Research

Have each student explore the Internet to find ways robots are used in nonindustrial applications, such as in the home or for hobbies.

TechnoFact

Coating metals to protect and beautify them is not new. One common process is called silver-plating, which was developed in 1742 in Sheffield, England. Sheffield silver-plating involved pressing thin strips of silver onto copper using rolling machines. In 1880, the process of electroplating was developed and roll-formed silver-plating soon became obsolete.

Figure 16-31. One type of industrial robot.

Standards for Technological Literacy

19

TechnoFact

Over 5000 years ago, Egyptians decorated their tombs with materials that were very similar to modern paints. These paints contained three key components found in modern paints: pigments, resins, and drying oils.

The drive system moves and positions the arm of the robot. The drive is composed of hydraulic, pneumatic, or electric devices that move and position the arm and end effector to a desired position. Most arms resemble a human body and arm, in that they have a body, a shoulder, an elbow, a forearm, and a wrist. These parts of the arm allow the robot to move in three axes—left and right, up and down, and in and out. The base allows the robot to move left and right. The shoulder and elbow create the up-and-down and in-and-out motions. The wrist rotates to create the final positioning for the tool or material.

The controller is the brain of the robot. This brain controls the arm movements the motors or hydraulic systems create. Computer commands direct the motions.

The end effector is the hand connected to the wrist. This "hand" can be many different things, ranging from a gripping device to a paint spray nozzle. Sensors provide feedback so the robot can "understand" its position. They allow the robot to know where its end effector is and determine the moves needed to move the effector into a new position.

Programmable Logic Controllers (PLCs)

Programmable logic controllers (PLCs), or programmable controllers, are devices that use microprocessors to control machines or processes. A technician on the factory floor generally creates the program for the controller, rather than a computer programmer located elsewhere. PLCs have three basic functions, which are input, control, and output. The PLC works by analyzing its inputs and, depending on their states, turning on or off its outputs. For example, the controller might receive input such as temperature, liquid levels, or shaft speed. Based on this input and the logic written into the controller, it activates appropriate outputs. A flame might be reduced to lower the temperature, a valve might be opened to fill a tank, or the speed of a motor might be changed to improve a machine's operation.

Computer-Controlled Machines

Computer-controlled machining adds automatic control to any basic machine. See **Figure 16-33.** A computer program controlling the motions of the machine generally provides this automatic control. The control is called *CNC*. The basic function of a CNC machine is to produce automatic, precise

Figure 16-33. A CNC machining center can be programmed to do many different tasks.

motion of the machine's tools or the material being processed. This motion is in two or more directions, called *axes*. The tool or workpiece can be precisely and automatically positioned along the length of travel on any axis.

For example, a person turning a handwheel attached to a lead screw positions the table holding a workpiece on a milling machine. In CNC milling machines, these motions are still created by turning a lead screw. A motor responding to computer-programmed commands, however, rotates the lead screw.

Artificial Intelligence (AI) Systems

Artificial intelligence (AI) is a field that is starting to be more widely used in manufacturing. This field can be described as intelligence a manufactured device or system exhibits. Most AI systems involve a computer to act as the brain for the system. These systems often use a type of AI called *machine vision*. The purpose of a computer vision system is to program a computer to "see and understand" a scene, using digital cameras or other sensors. The system then acts on this understanding to inspect a part, locate a cutter, or pick up a randomly positioned part. Other uses for AI include handwriting recognition for handheld communication devices, optical character recognition (OCR) programs for scanners, speech-recognition systems for automated customer call centers, and face recognition for security systems. See **Figure 16-34.**

Figure 16-34. Many types of modern devices use AI.

Standards for Technological Literacy

19

TechnoFact

British colonies in the New World were the first sites of industrialization outside Europe. Among the industries that sprung up in the colonies, shipbuilding was the most successful. By 1776, America built about a third of Britain's ships.

Summary

Secondary manufacturing processes convert industrial materials into finished products. They add worth to materials by making them useful to consumers. The secondary processes can be grouped as casting and molding, forming, separating, conditioning, assembling, and finishing. The first three change the size and shape of the materials. Conditioning changes the properties of materials by altering their internal structures. Assembling processes attach parts together to make assemblies and products. Finishing produces coatings that protect and improve a product's appearance.

Test Your Knowledge

Write your answers on a separate piece of paper. Please do not write in this book.

1. List the five basic steps involved in casting and molding.
2. What is an expendable mold?
3. List two basic safety rules to observe when using casting processes.
4. Name the three ways we can make materials solid.
5. What is a material's plastic range?
6. Name the three things all forming processes have in common.
7. What are mated dies?
8. Name one product shaped rolls produce.
9. Plastics and glass are formed hot, but ceramics are formed cold. True or false?
10. The two kinds of separating processes are machining and ______.
11. What are the three things all chip-removal processes have in common?
12. The two types of motion needed to make a cut are cutting motion and ______ motion.
13. List two basic safety rules to observe when using machine tools.
14. Burning gases to remove unwanted material from a workpiece is a nontraditional form of machining. True or false?
15. What are the three types of conditioning processes?
16. Name two kinds of bonding techniques.
17. Give one example of a mechanical fastener.
18. List two basic safety rules to observe when working with finishing processes.

Answers to Test Your Knowledge Questions

1. Producing a mold of the proper size and shape. Preparing the material. Introducing the material into the mold. Solidifying the material. Extracting the product from the mold.
2. A mold that is destroyed to remove the cast item
3. Student answers may include any two of the following: Wear protective clothing and goggles when pouring a casting. Be sure everything to be used in the casting process is free of moisture. Place hot castings where they will not start a fire or burn someone. Keep the casting area orderly. Do not talk to anyone while pouring. Do not stand over the mold when pouring molten metal.
4. Cooling, drying, and chemical action
5. It is the pressure range between the yield point and the fracture point, which is the pressure range at which all forming processes operate.
6. The presence of a forming device, the application of force, and the consideration of material temperature
7. Mated dies are dies that have the desired shape machined into one or both halves of the die set. The material is placed between the die halves. The die set is closed, and the material is forced to take on the shape of the die cavities.
8. Evaluate individually. Common student answers include: pipe, tubing, and corrugated metal.
9. True
10. shearing
11. A tool or other cutting element is always present. Motion occurs between the tool and work. The work and the tool are given support.

19. Chemical cleaning of product surfaces is often called *pickling*. True or false?
20. Name one type of organic finish.
21. List the five main parts of a robot.

STEM Applications

1. Build a simple product, such as a kite, using secondary manufacturing processes. List each step and the type of process used.
2. Select a simple product made from more than one part that you see in the room.
   A. List the parts it is made of.
   B. Select one part. List the steps you think were used to manufacture it.
   C. Complete a form similar to the following for one of the steps.

Product:
Part name:
Production process:
Step needed to complete the process:

3. Use a robot with a teach pendant to move a peanut over a set part. Document the initial and final locations of the peanut and the moves used to move the object.

12. feed
13. Student answers may include any of the following: Wear eye protection. Avoid loose-fitting clothing and keep long sleeves rolled up. Wear hearing protection when loud or high-pitched noises are present. Keep the laboratory clean. Wipe up any spills immediately. Wait until the machine has stopped to make adjustments, remove the workpiece, or clean up any scrap. Use a brush to clean chips off of a machine. Do not talk to anyone while running a machine. Most of all, if you do not understand how to run the machine, ask first.
14. False
15. Mechanical, chemical, and thermal conditioning
16. Student answers may include any two of the following: fusion bonding, flow bonding, pressure bonding, cold bonding, and adhesive bonding.
17. Student answers may include any of the following: staples, rivets, screws, nails, pins, bolts, and nuts. Other answers may be accepted at the instructor's discretion.
18. Student answers may include any two of the following: Wear eye and face protection. Apply finishes in a well-ventilated area; wear a respirator and protective clothing if solvents are toxic. Never apply a finish near open flame. Keep the finishing area orderly, and clean up any spills immediately. Store all rags and chemicals in approved containers. Use a spray booth (an enclosed area with fans to remove fumes) to remove toxic fumes. Dispose of any waste solvents and finishes properly; wash your hands to remove any finishes you have handled. Most of all, if you do not understand how to apply a finish, ask first.
19. True
20. Student answers may include any of the following: paint, varnish, enamel, and lacquer.
21. Arm, controller, drive, end effector, and sensor

Chapter Outline
Buildings
Heavy Engineering Structures

This chapter covers the benchmark topics for the following Standards for Technological Literacy:

3 5 20

Learning Objectives

After studying this chapter, you will be able to do the following:

- Recall the two types of construction.
- Summarize the characteristics of the major types of constructed buildings.
- Give examples of types of buildings other than residential, commercial, and industrial.
- Recall the steps involved in constructing a structure.
- Recall the types of foundations.
- Summarize the characteristics of the frameworks used in buildings.
- Explain how buildings are enclosed.
- Summarize the types of utility systems used in buildings.
- Summarize the characteristics of materials used in finishing buildings.
- Summarize the characteristics of heavy engineering structures.

Key Terms

arch bridge
beam bridge
buttress dam
cantilever bridge
ceiling
ceiling joist
commercial building
drywall
fascia board
floor joist
forced-air heating system
gravity dam
header
heat pump
hot water heating
industrial building
landscape
manufactured home
pile foundation
potable water
rafter
reinforced concrete
residential building
sheathing
sill
slab foundation
soffit
sole plate
spread foundation
stud
subfloor
suspension bridge
top plate
truss
truss bridge
wastewater

Strategic Reading

As you read the following chapter, make a list of the different types of buildings mentioned. Think of an example in your city of each of these buildings.

Standards for Technological Literacy

20

Research

Ask the students to find out what construction is and how projects are divided into categories.

TechnoFact

When architects design buildings, they use a combination of art and technology. The design of the structure and the materials used to build it must be functional and pleasing.

Human beings have three basic needs: food, clothing, and shelter. Each of these needs can be satisfied using technology. Agriculture and related biotechnologies help us grow, harvest, and process food. Manufacturing helps us to produce natural and synthetic fibers. These fibers become the inputs to clothing and fabric manufacture. Materials and manufactured goods can be fabricated into dwellings and buildings using construction technology. Construction uses technological actions to erect a structure on the site where the structure will be used.

Construction technology builds two types of structures. These are buildings and heavy engineering structures. See Figure 17-1. Buildings are enclosures to protect people, materials, and equipment from the elements. They also provide security for people and their belongings. Heavy engineering structures help our economy function effectively.

Buildings

Buildings are grouped into three types: residential, commercial, and industrial. See Figure 17-2. These groupings are based on how the buildings are used. Other types of buildings, however, exist. These special buildings follow the same construction steps as the other buildings. This section describes each of the main types, provides an overview of the other types, and discusses the general steps involved in constructing buildings.

Types of Buildings

The types of buildings differ because people have different uses for structures.

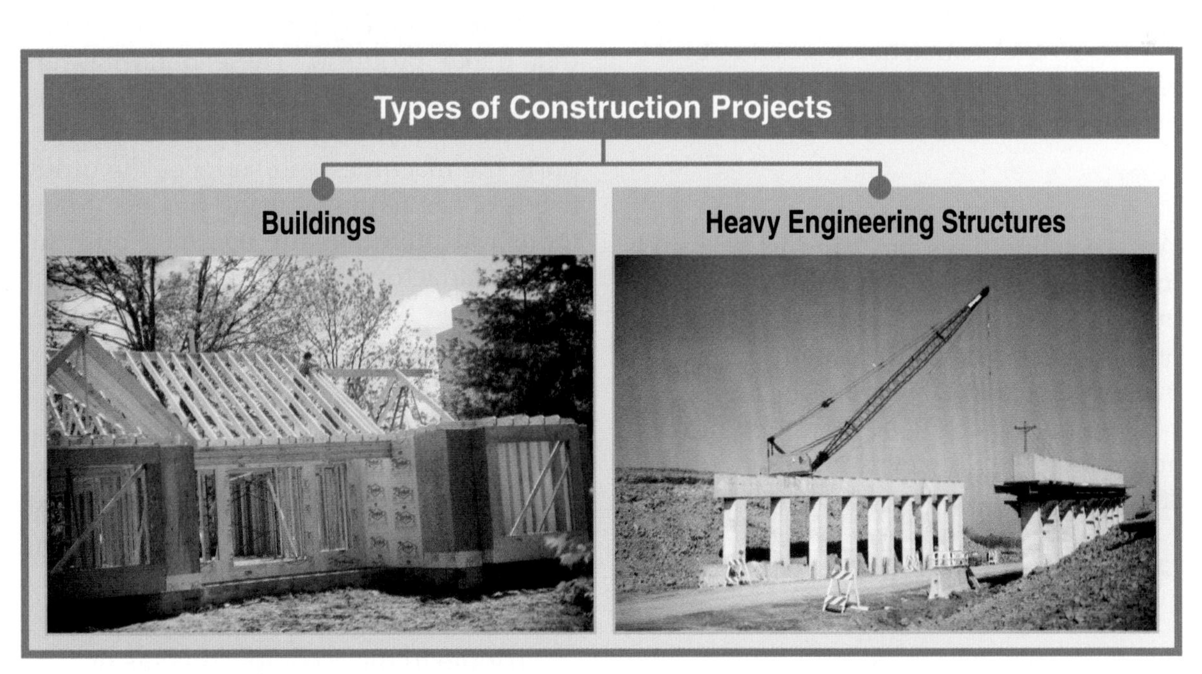

Figure 17-1. Construction erects buildings and heavy engineering structures.

Standards for Technological Literacy

20

Brainstorm

Ask the students to list and discuss the advantages and disadvantages of single dwellings, apartments, and condominiums.

Extend

Discuss the types of monuments in your city or state.

TechnoFact

As veterans returned from World War II, housing demand grew as the veterans sought affordable places to live. William Levitt saw the increased housing demand as a golden opportunity. In 1951, he created the first major planned community, Levittown in Long Island. For $7,000, a person could buy one of Levittown's 17,500 two-bedroom homes.

TechnoFact

One of the most familiar of the world's monuments is the Washington Monument. This granite structure weighs 81 thousand tons and was finished in 1885. At 555 feet tall, it is the tallest masonry structure ever built.

Marvin Windows and Doors, Inland Steel Company

Figure 17-2. Construction is used to build residential, commercial, and industrial buildings.

We live in buildings, buy products in buildings, work in buildings, and worship in buildings, to name a few activities we do in buildings. To help understand the types of buildings people design, build, and use, we will explore three major types of buildings and then briefly review some more unique structures.

Residential Buildings

Residential buildings are buildings in which people live. These buildings can be single-family or multiple-unit dwellings. The multiple-unit dwellings include apartments, town houses, and condominiums.

A residential building can be either owner occupied or rented from the owner. The owner of a dwelling is responsible for the dwelling's upkeep. In some types of dwellings, such as condominiums, the costs of upkeep are shared between the owners. Each owner belongs to and pays fees to an association. This group elects officers who manage the maintenance of common areas such as entryways, garages, parking areas, and lawns. The association is also responsible for exterior repairs and insurance on the building. The individual owners maintain their own living quarters and insure their personal belongings against fire and theft.

Commercial Buildings

Commercial buildings are used for business purposes. These buildings can be publicly or privately owned. Commercial buildings range in size from small to very large. Retail stores, offices, and warehouses are commercial buildings.

Industrial Buildings

Industrial buildings house machines that make products. These buildings are used to protect machinery, materials, and workers from the weather. The building supports the machines and supplies the utility needs of the manufacturing process. Many industrial buildings are specially built for one manufacturing process.

Other Types of Buildings

You see commercial, industrial, and residential buildings all around you. If you look around your town or city, however, you probably see other types of buildings. See Figure 17-3. These might include the following:

- **Monuments.** These structures pay tribute to the accomplishments or sacrifices of people or groups.

Figure 17-3. Buildings other than the three main types can be found in many forms.

- **Cultural buildings.** These buildings house theaters, galleries, libraries, performance halls, and museums. They host musical, dramatic, and dance performances; literary activities; and art exhibits.
- **Government buildings.** These buildings house government functions. Examples include city halls, post offices, police stations, firehouses, state capitols, courthouses, and government office buildings.
- **Transportation terminals.** These buildings are used to aid in the loading and unloading of passengers and cargo from transportation vehicles. Examples are airports, train and bus stations, freight terminals, and seaports.
- **Sports arenas and exhibition centers.** These facilities are used for sporting events, concerts, trade shows, and conventions.
- **Agricultural buildings.** These structures include barns and storage buildings used to house livestock, shelter machinery, and protect farm products (grain and hay, for example).

As noted earlier, these special buildings are built using the same construction steps used for a single-family home.

One special type of building is the ***manufactured home***. As you remember, manufacturing produces products in a factory. The completed product is transported to its place of use. This is exactly how manufactured homes are produced. See **Figure 17-4.** Most of the structure is built in a factory. This type of home is usually built in two halves. The floors, walls, and roof are erected. The plumbing and electrical systems are then installed. The structure's interior and exterior are enclosed and finished. This step includes installing flooring, painting walls, setting cabinets and plumbing fixtures, and installing appliances and electrical fixtures.

Standards for Technological Literacy

20

Figure Discussion

Use Figure 17-3 to point out other types of buildings in your city.

Extend

Have your students use the Internet to explore the types, interiors, and exteriors of manufactured homes. Have them discuss what they found.

TechnoFact

Many Americans live in what are called suburbs. Suburbs are not new phenomena. They have existed in one form or another throughout history. In ancient times, many cities were surrounded by walls. The villages outside of the city walls were smaller and often considered lower in the social hierarchy. Now, suburbs are often considered the opposite of that: upscale places for upper- and middle-income families. This "suburb" may have appeared over twenty-six hundred years ago in Babylon. The word *suburb* comes from the Latin *suburbani*, which Roman Emperor Cicero used in the first century B.C. to describe luxurious estates outside Rome.

Standards for Technological Literacy

20

Extend

Invite a real estate broker or bank officer to discuss finding and financing a homesite.

TechnoFact

The money spent on construction comes from many sources. Home construction is frequently financed through banks, lending institutions, or personal savings. Governments may issue bonds to finance construction projects.

Figure 17-4. The steps in building a manufactured home.

The two halves of the structure are transported to the site, where a foundation is already in place. Each half is lifted from its transporter and placed on the foundation. The two halves are finally bolted together. The final trim that connects the halves is installed. The utilities are hooked up, and the home is ready for the homeowner. Similar techniques are used to produce temporary classrooms, construction offices, and modular units that can be assembled into motels or nursing homes.

Constructing Buildings

Constructed structures start with architectural and engineering plans that are a result of the designing process discussed earlier. The owner's needs and budget constrain these plans. Also, three other factors constrain the plans. These are the following:

- **Zoning laws.** These laws are government regulations restricting how a piece of land can be used.
- **Building codes.** These codes are regulations controlling the design and construction of a structure to provide for human safety and welfare.
- **Best (professional) practices.** These practices are the accepted methods or processes the profession considers to be the most appropriate ways to complete an activity or build a structure.

Within the constraints the design provides, most construction projects follow the same basic steps. See Figure 17-5. These steps include the following:

1. Preparing the site.
2. Setting foundations.
3. Building the framework.
4. Enclosing the structure.
5. Installing utilities.
6. Finishing the exterior and interior.
7. Completing the site.

Each type of structure needs to have specific actions taken during each step.

Gehl Co.

Figure 17-5. Most construction projects follow the same basic steps.

Standards for Technological Literacy

20

Brainstorm

Ask the students what wants and needs a home fulfills.

This helps complete the structure on time. We will look at the steps used to construct a small single-family home. Later in the chapter, other construction activities are discussed.

A common type of building is a single-family home. This building is designed to meet a number of needs of the owners. These needs include protection from the weather, security, and personal comfort. See **Figure 17-6.** To meet these needs, a home must be properly designed and constructed. The construction process starts with locating, buying, and preparing a site.

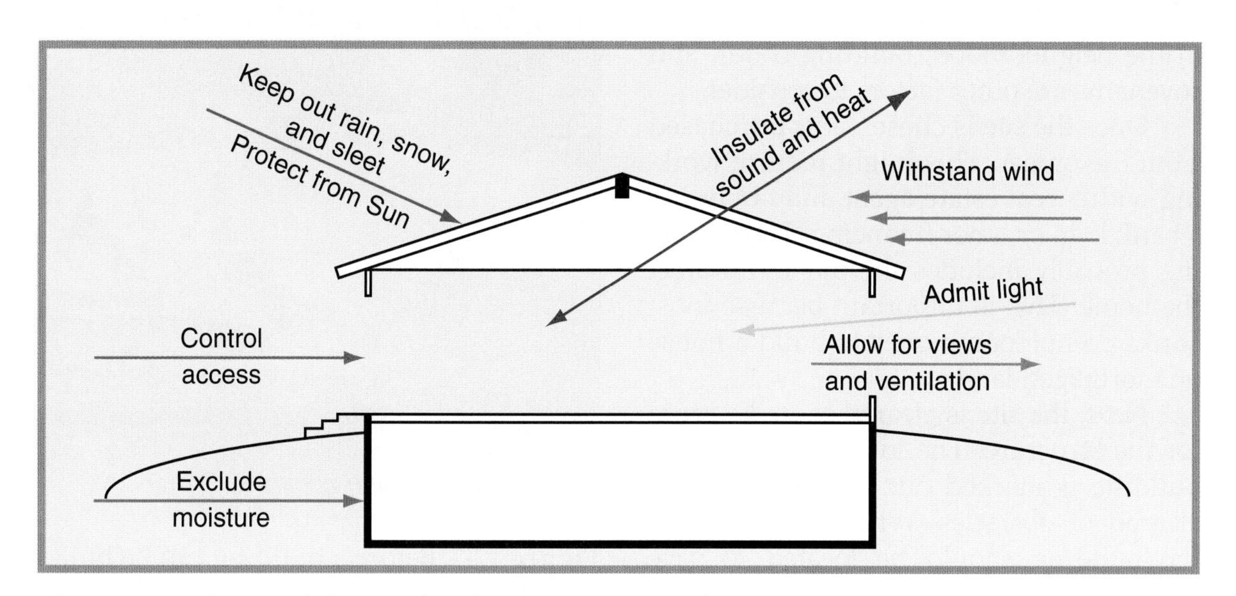

Figure 17-6. Some of the needs a home must meet for its owner.

Standards for Technological Literacy

20

TechnoFact

Concrete is a common material used in foundations. It is a structural material made by mixing gravel, sand, cement, and water. The Romans used a material very similar to modern concrete about twenty-two hundred years ago for constructing roads, buildings, and other public works, such as their aqueducts.

Career Corner

Carpenters

Carpenters help construct buildings, highways, bridges, factories, and other structures. Some carpenters do all types of work, while others specialize in doing a specific job, such as setting forms for concrete, framing walls and partitions, laying hardwood floors, or installing interior and exterior trim. Carpenters must know local building codes and be able to work from blueprints or instructions from supervisors. Those who remodel homes must be able to do all aspects of a job and, therefore, require a good basic overall training.

Carpentry work is somewhat strenuous and requires standing, climbing, bending, and kneeling. Carpenters must be competent in using tools and power equipment. Carpentry is considered a skilled trade. Most carpenters learn their trade through on-the-job training, vocational education, or apprenticeships.

Preparing the Site

The location for a home needs to be carefully selected. This location should meet the needs of the people who will live there. For example, a family with children might think about the schools serving the area. The parents consider the distance to work and shopping, recreation, and cultural facilities. The condition of other homes in the neighborhood, building codes, and covenants are other factors to consider.

Once the site is chosen, it is purchased from the owner. This might require working with a real estate agent and obtaining a bank loan or other financing. The financing probably includes the money to erect the home. This is important because most banks do not loan money to build a house on mortgaged land.

Next, the site is cleared to make room for the structure. The location of the new building is marked out. The area is then cleared of obstacles. When it is possible, the building should be located to save existing trees and other plant life. The site might require grading to level the site. See Figure 17-7. Grading prepares areas for sidewalks and landscaping and helps water to drain from the site. These preparations are needed for the next step, setting foundations.

Figure 17-7. Before a building can be built, the site must be cleared and graded.

Setting Foundations

The foundation is the most important part of any building project. This part serves as the feet of the building. Try to stand on just your heels. You will be unstable and wobble. Likewise, a building without a proper foundation settles unevenly into the ground. Such a building leans, becomes unstable, and might collapse. The Leaning Tower of Pisa in Italy is an example of a building that has a poor foundation. Over time, the tower has settled and is leaning several feet to one side.

A complete foundation has two parts: the footing and the foundation wall. See **Figure 17-8.** The footing spreads the load over the bearing surface. The bearing surface is the ground on which the foundation and building will rest. This can be rock, sand, gravel, or a marsh. Each type of soil offers unique challenges for the construction project.

The type of foundation to use is selected to match the soil of the site. See **Figure 17-9.** Three types are the following:

- ***Spread foundations.*** These types of foundations are used on rock and in hard soils, such as clay. The foundation walls sit on a low, flat pad called a *footing.* On wide buildings, posts support the upper floor between the foundation walls. These posts also rest on pads of concrete called *footings.*
- ***Slab foundations.*** These types of foundations are used for buildings built on soft soils. They are sometimes called *floating slabs.* The foundation becomes the floor of the building. Such foundations allow the weight of the building to be spread over a wide area. This type of foundation is used in earthquake areas because it can withstand vibration.

Standards for Technological Literacy

20

Demonstrate

In class or lab, use various widths of boards and a tray of sand to show the role of a footer.

TechnoFact

The Leaning Tower of Pisa is a marvel to see, but it also shows what a poor understanding of foundations can lead to. Construction was started in 1173 and completed about two hundred years later. The tower is over 180′ tall and sits on soft soil. Because of its poor foundation, the tower started to tilt, even as it was being built. It has continued to sink and tilt over the last eight hundred years.

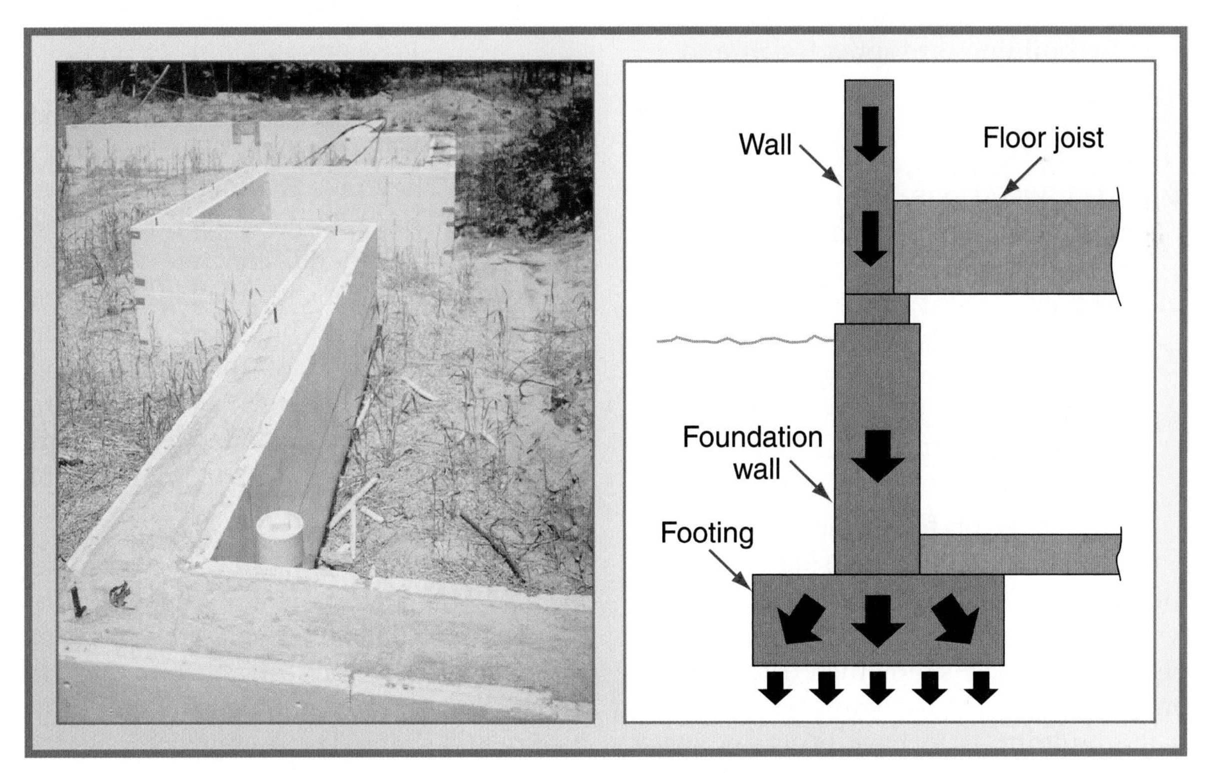

Figure 17-8. The foundation wall and footing spread the building's weight onto the bearing surface. The concrete foundation shown in the photo has been insulated to reduce heat loss.

Standards for Technological Literacy

20

Extend

Compare the foundation of a building with the roadbed for a highway.

TechnoFact

Because of the loads that they are subjected to, major civil structures rest on foundations of solid rock. For example, a bridge must be able to support its own weight and the weight of vehicles crossing it.

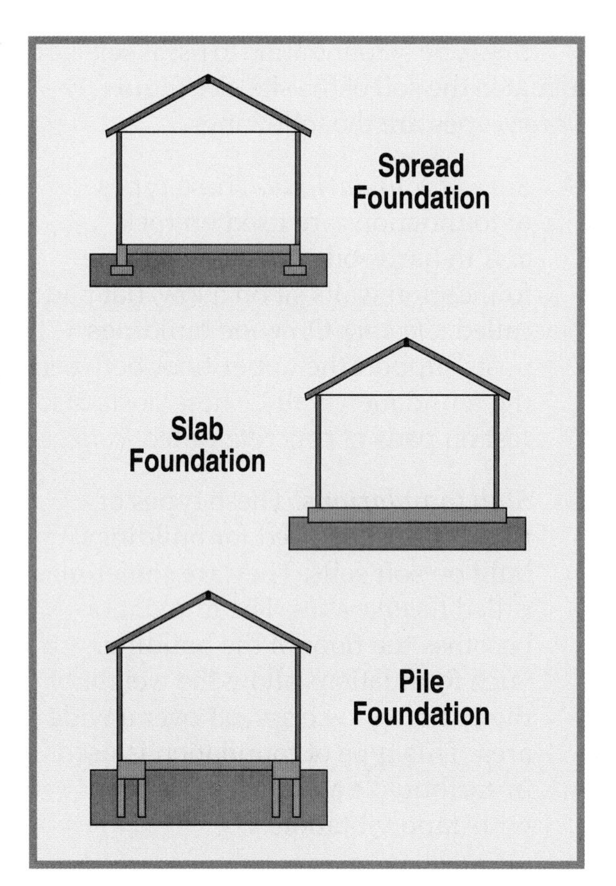

Figure 17-9. Three types of foundations used for buildings are spread, slab, and pile.

Inland Steel Co.

Figure 17-10. This worker is surveying a site for a new building. The survey locates where the foundation will be placed.

- ***Pile foundations.*** These types of foundations are used on wet, marshy, or sandy soils. Piles are driven into the ground until they encounter solid soil or rock. They are large poles made of steel, wood, or concrete. Piles are widely used for high-rise buildings, marine docks, and homes in areas that flood easily.

Each type of foundation is built in a unique way. Let us consider a spread foundation. The site is surveyed to locate where the foundation will be placed. See **Figure 17-10.** The site is then excavated in preparation for the footings and the walls. If the building is to have no crawl space under the first floor, excavation does not go as deep. Buildings with basements require deeper excavations. Footing forms are set up next. Forms are a lumber frame to hold the wet concrete until the concrete cures (hardens). They give the footings or slabs height and shape. Concrete is poured and leveled off. When the concrete is cured, the forms are removed. Walls of poured concrete or concrete block are built atop footings. Slabs are ready for aboveground superstructures. **Figure 17-11** shows an excavation for a pool. Wooden foundations use no concrete for either footings or walls.

Building the Framework

The foundation becomes the base for the next part of the building, the framework. Erecting the framework gives the building its size and shape. The framework includes the floors, interior and exterior walls, ceilings, and roof. Also, the locations of doors and windows are set up at this time.

Figure 17-11. This worker is excavating a hole for a pool.

The framework can be built out of three different materials. See Figure 17-12. Small and low-cost buildings have frameworks made from lumber. Most industrial and commercial buildings have either steel or ***reinforced concrete*** frameworks. Building the framework involves three steps.

Floors

First, the floor is constructed. See Figure 17-13. Homes with slab foundations use the surface of the slab as the floor. Those with basements or crawl spaces use lumber floors.

Lumber floors start with a wood ***sill*** bolted to the foundation. ***Floor joists*** are then placed on the sill. They extend across the structure. Floor joists carry the weight of the floor. The span (distance between outside walls) and the load on the floor determine the size and spacing of the joists. On top of the joists, a ***subfloor*** is installed, usually made from plywood or particleboard. After the building is enclosed, flooring material is installed on top of the subfloor.

Standards for Technological Literacy

20

Extend

Explain what is meant by the term reinforced concrete.

TechnoFact

Before the mid-1800s, houses were built with heavy interlocking beams. A construction technique was developed in Chicago in the 1830s in which a house was framed with 2″ x 4″ or 2″ x 6″ studs, crisscrossed with pieces of wood. This technique was called balloon framing and remained popular for more than a hundred years because it allowed people to build lighter, stronger buildings.

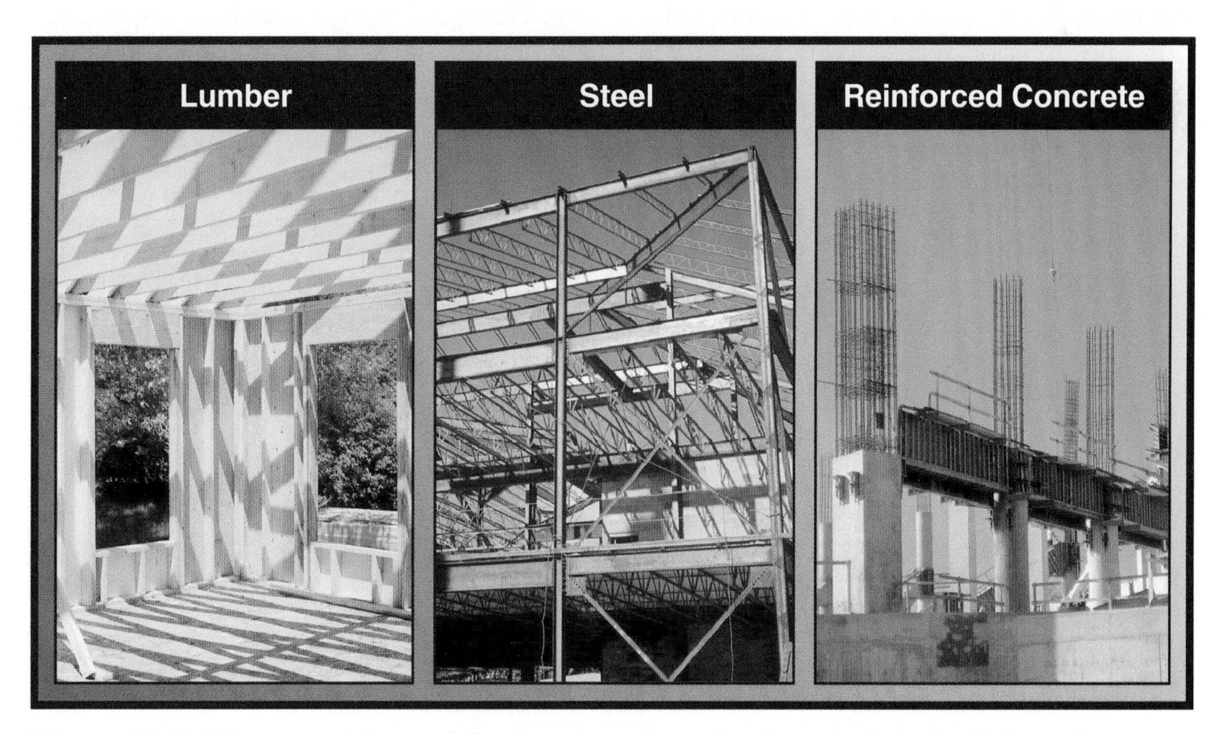

Figure 17-12. The materials used for framework are lumber, steel, and reinforced concrete.

Standards for Technological Literacy

20

Research

Ask the class to look up why many houses have 2 × 6 walls instead of the traditional 2 × 4 walls.

TechnoFact

Some houses serve both public and private purposes. The White House is one example. Although it houses the offices for the President of the United States and staff members, it also provides living quarters for the president's family. The White House is a 132-room mansion that sits in the center of an 18-acre plot. Since the 1800s, the nation's executive mansion has been known as the White House. However, it did not receive that official title until 1901.

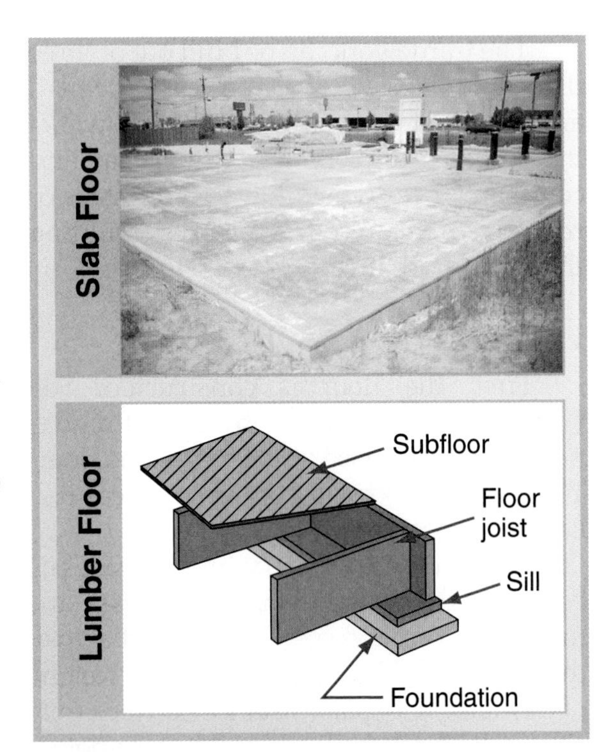

Figure 17-13. The floors in single-family homes are either concrete slabs or lumber.

Walls

The wall frames are placed on top of the floor. These frames support both exterior and interior walls. Wall framing is often made of 2 × 4 or 2 × 6 construction-grade lumber. See **Figure 17-14.** A framed wall has a strip at the bottom called the ***sole plate***. Nailed to the sole plate are uprights called ***studs***. The length of the studs is set by how high the ***ceilings*** will be. At the top of the wall, the studs are nailed to double ribbons of 2 × 4s called a ***top plate*** or wall plate. Door and window openings require headers above them. ***Headers*** carry the weight from the roof and ceiling across the door and window openings. Shorter studs called *trimmer studs* hold up the headers.

Ceilings and roofs

The walls support the ceiling and roof. See **Figure 17-15.** The ceiling is the inside surface at the top of a room. The roof is the top of the structure that protects the house from the weather.

Ceiling joists support the ceiling. These joists rest on the outside walls and some interior walls. Interior walls that help support the weight of the ceiling and roof are called *load-bearing walls* or *bearing walls.*

The roof forms the top of the building. There are many types of roofs, including gable, flat, hip, gambrel, and shed.

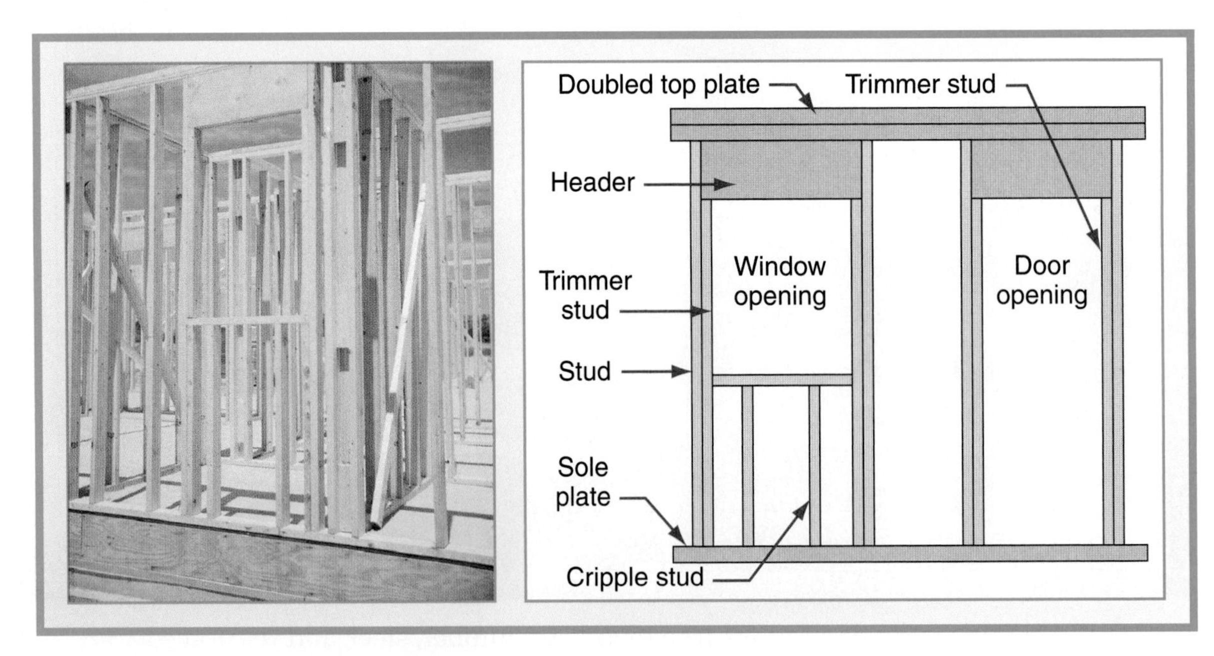

Figure 17-14. Many of the parts of a wood-framed wall.

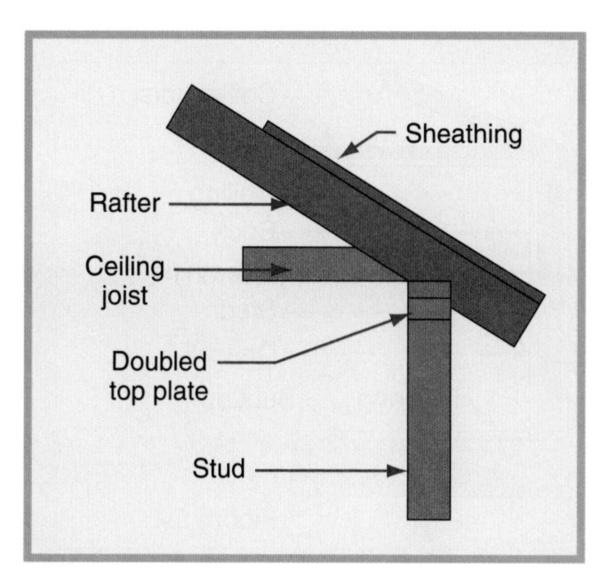

Figure 17-15. How the roof meets the wall frame.

See Figure 17-16. The type of roof is chosen for its appearance and how it withstands the weather. For example, flat roofs are poor choices in areas with heavy snow. This type of roof cannot easily support the weight of deep snow. Likewise, a hip roof looks out of place on a Spanish-type home. This kind of roof does not give the "Spanish-style" look.

Roof construction involves two steps. First, the roof frame is built with rafters. ***Rafters*** are angled boards resting on the top plate of the exterior walls. Often, a special structure called a *truss* is used. A ***truss*** is a triangle-shaped structure including both the rafter and ceiling joist in one unit. Trusses are manufactured in a factory and then shipped to the building site. The rafters or trusses are covered with plywood or particleboard sheathing. This step completes the erection of the frame.

Enclosing the Structure

After the framework is complete, the structure needs to be enclosed. The roof and wall surfaces need to be covered. This process involves enclosing the walls and installing the roof. With regard to enclosing the walls, all homes have both interior and exterior wall coverings. These coverings improve the looks of the building and keep out the elements (rain, snow, wind, and sunlight).

Enclosing the exterior walls involves ***sheathing*** (covering) all the exterior surfaces. See Figure 17-17. Plywood, fiberboard, or rigid foam sheets are used to sheath the

Standards for Technological Literacy

20

Figure Discussion

Discuss where students might have seen each type of roof shown in Figure 17-16.

TechnoFact

Climate plays an important role in determining the type of roof that a building will have. For example, the heat and lack of rain in ancient Egypt and Syria caused those cultures to develop flat roofs. The wetter climate and subsequent need to shed water caused Central Europeans to develop steep, sloping roofs. Houses built in areas that receive heavy snowfall also need steep roofs, but their roofs require additional reinforcement.

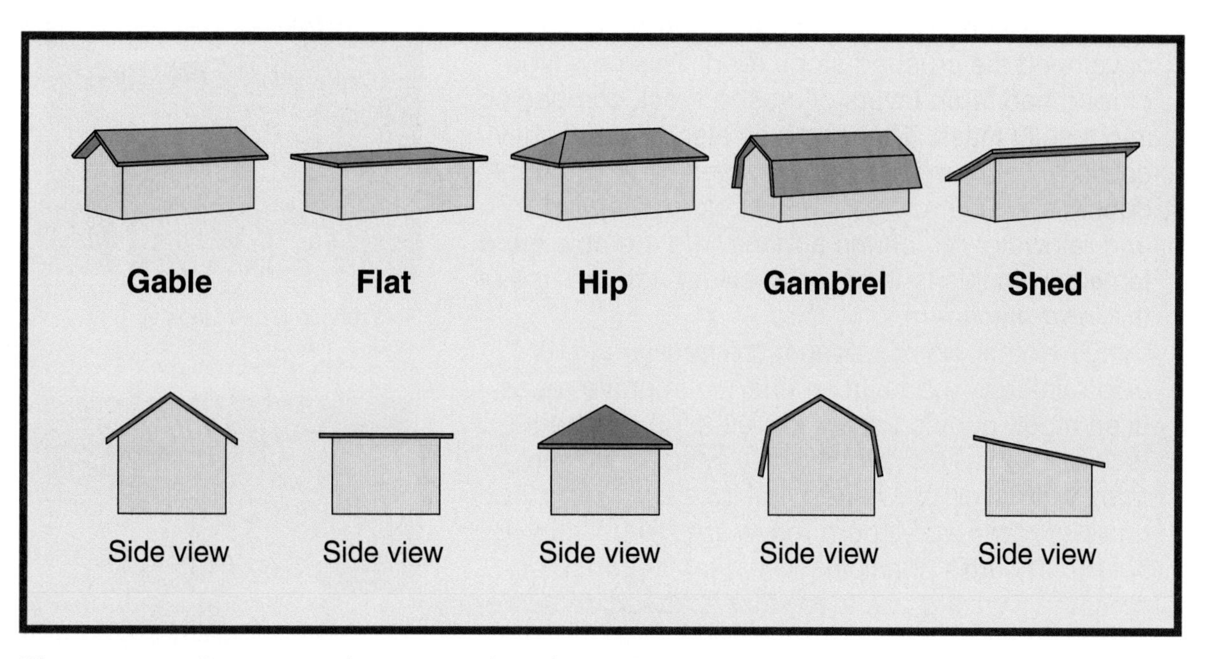

Figure 17-16. Some popular types of roofs used on homes.

Standards for Technological Literacy

3 20

Extend

Discuss the role of building paper under siding and shingles.

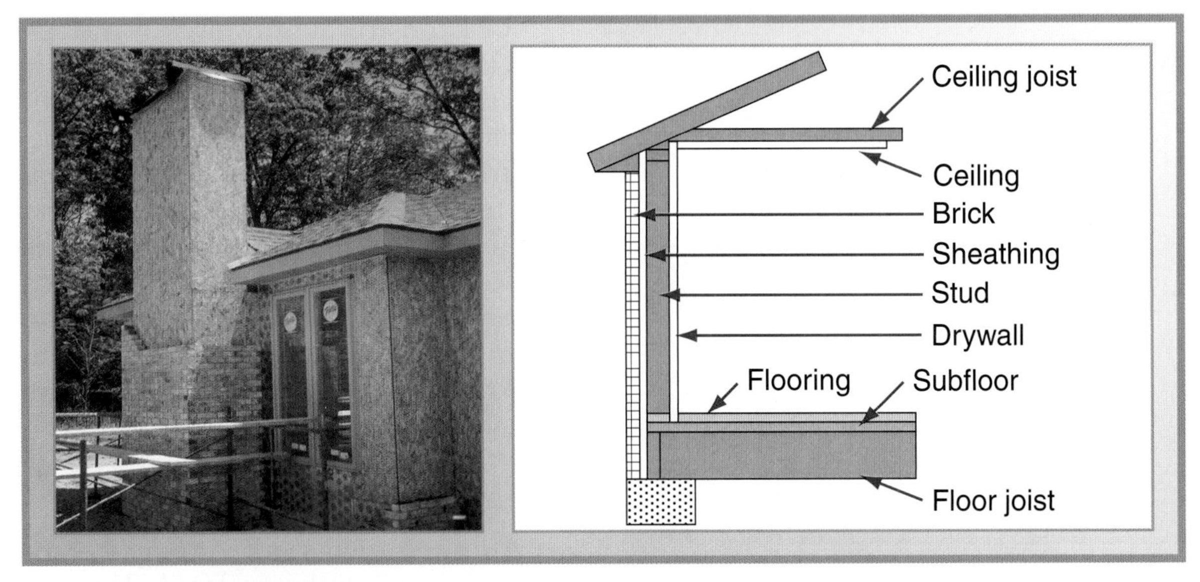

Figure 17-17. A cross-sectional diagram of a finished wall. The photo shows brick being applied as a siding material.

Academic Connections: Communication

Word Origins

Probably most people would be happy to have their names become part of everyday language as a result of their inventions. John McAdam, a Scottish engineer who experimented with road construction, might be doubly pleased. His work with roads has resulted in two words in common use today.

As described in this chapter, John McAdam developed the crushed-stone road. This new type of road had three layers of crushed rock compacted into a solid mass. The road was also made slightly convex. McAdam's design improved roads tremendously because now the traffic load was spread, and rainwater ran off the surface. We are now more familiar with this type of roadway through the use of the term *macadam.*

©iStockphoto.com/resonants

An asphalt road, a blacktop road, or a tarmac road.

The other word is even more familiar and is also related to roads. In an effort to improve roads even more, people used tar to bind the crushed rock together. This process was given the name *tarmacadam* or, as we now call it when we use it on runways, *tarmac.* Modern roads are still built using John McAdam's principles. Can you find another common word we use today that is based on someone's name and invention?

©iStockphoto.com/PeskyMonkey

Fresh asphalt (tarmac).

walls. Most foam sheets have a reflective backing to improve the insulation value of the sheet. Most homes constructed today have a layer of plastic over the sheathing to prevent air from leaking in.

Normally, the roof is put in place before the utilities are installed. See Figure 17-18. The actual roof surface has two parts. Plywood or wafer-board sheathing is applied over the rafters. Builder's felt is often applied over the roof sheathing. Wood or fiberglass shingles, clay tiles, or metal roofings are then installed over the sheathing and felt. Flat and shed roofs often use a built-up roof. A built-up roof starts with laying down sheets of insulation. Roofing felt is laid down, followed by a coat of tar, which is covered with gravel.

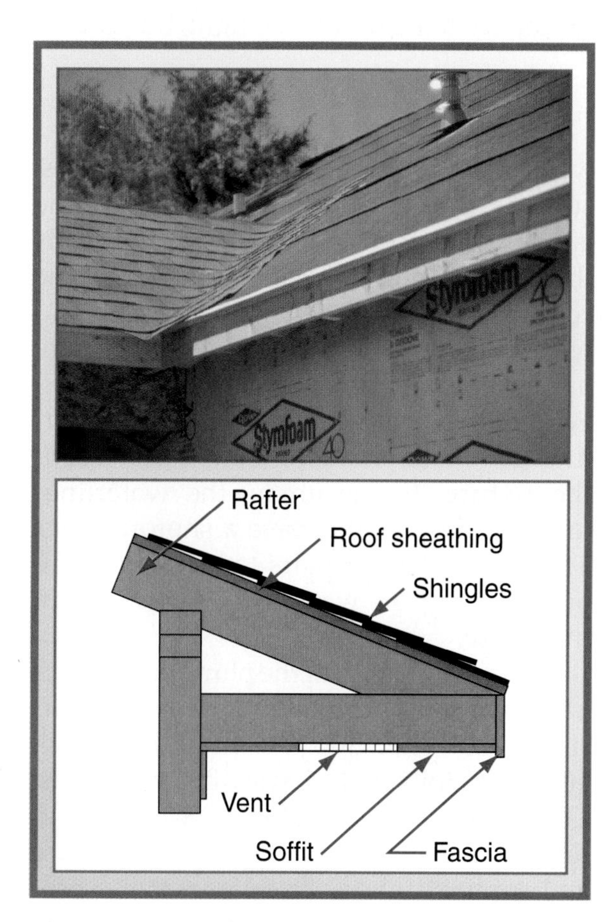

Figure 17-18. The parts of a finished roof. The photo shows asphalt shingles being installed on a new roof.

On many structures, the overhang of the roof is also finished. A ***fascia board*** is used to finish the ends of the rafters and the overhang. The ***soffit*** is installed to enclose the underside of the overhang. Soffits can be made of aluminum, vinyl, or plywood. They must have ventilation holes or vents to prevent moisture and heat buildup in the attic.

Once the sheathing and roof are installed, the openings for doors and windows are cut out. The doors and windows are then set in place. Now the house is secure and weather tight.

Installing Utilities

Normally, the utilities are installed after the building has been enclosed. This prevents theft and damage from the weather. Some parts of the utilities are installed earlier, such as large plumbing lines. The utility system includes four major systems:

- Electrical.
- Plumbing.
- Climate control.
- Communications.

Electrical systems

The electrical system delivers electrical power to the different rooms of the home. The power is brought into the house through wires to a meter and distribution panel. This panel splits the power into 110-volt and 220-volt circuits. Each circuit has a circuit breaker to protect against current overloads.

Appliances such as clothes dryers, electric ranges, water heaters, and air conditioners require 220-volt power. Circuits for smaller appliances, lighting, and wall outlets use 110 volts. Outlets might have power fed to them at all times. Switches can also control outlets. Figure 17-19 shows a 110-volt circuit with wall (duplex) outlets and a ceiling light. The outlets always have power. The circuit to the light has a switch, however.

Standards for Technological Literacy

20

Demonstrate

Show the class how to wire a single-throw switch to a light.

TechnoFact

The electric lamp was one of the most influential inventions ever created. During the middle part of the 19th century, many inventors seemed to comprehend the potential of electric lighting and were working toward developing a practical electric lightbulb. Unfortunately, developing an electric lamp was only half of the problem. Electric power also had to be delivered to individual homes to allow people to use their new lights. Thomas A. Edison developed solutions to both problems and won recognition as the inventor of the electric light.

Standards for Technological Literacy

20

TechnoFact

The first plumbing system designed to remove human waste was developed about 4500 years ago in the areas now known as Pakistan and western India. Five hundred years later, a method for delivering drinking water was built into a palace on the island of Crete. The ancient Romans contributed faucets and sewer systems to the history of plumbing.

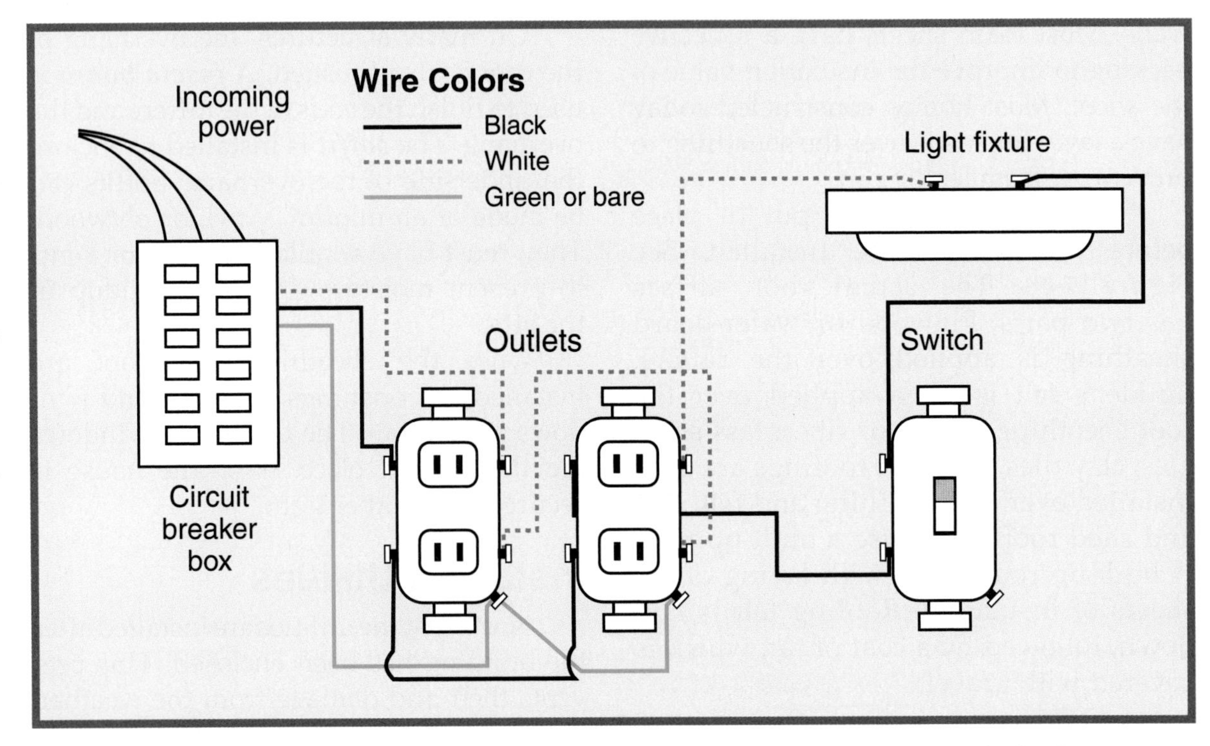

Figure 17-19. A 110-volt electric circuit. Note how the switch controls the light, but not the outlets.

Most 110-volt circuits are limited to 15 or 20 amps. Therefore, a number of different circuits are required to supply various parts of the home. A kitchen might have one or two circuits because of how many appliances are used there. One circuit might feed two bedrooms because there are few appliances in these rooms.

Plumbing systems

The plumbing system has two separate parts. One part supplies potable water. See **Figure 17-20.** ***Potable water*** is safe for drinking. The other part of the system carries away ***wastewater.*** Plumbing fixtures and systems are designed to prevent mixing of potable water and wastewater and to stop sewer gas from leaking into the dwelling.

The potable-water system starts with a city water supply or a well for the house. The water enters the house through a shutoff valve and might pass through a water conditioner to remove impurities, such as iron and calcium. The waterline is split into two branches. One line feeds the water heater. The other line feeds the cold-water system. Separate hot and cold waterlines feed fixtures in the kitchen, bathrooms, and utility room. Toilets, however, receive only cold water. Most waterlines have shutoff valves before they reach the fixture. For example, the waterlines under a sink should have a shutoff valve. The valve allows repairs to be made without stopping the water flow to the rest of the house.

The second part of the plumbing system is the wastewater system. This system carries used water away from sinks, showers, tubs, toilets, and washing machines. The wastewater is routed to a city sewer line or to a septic system. At each fixture and appliance, a device called a *trap* is provided. A trap is a *U*-shaped piece of pipe that remains full of water. The water in the line stops gases from the sewer system

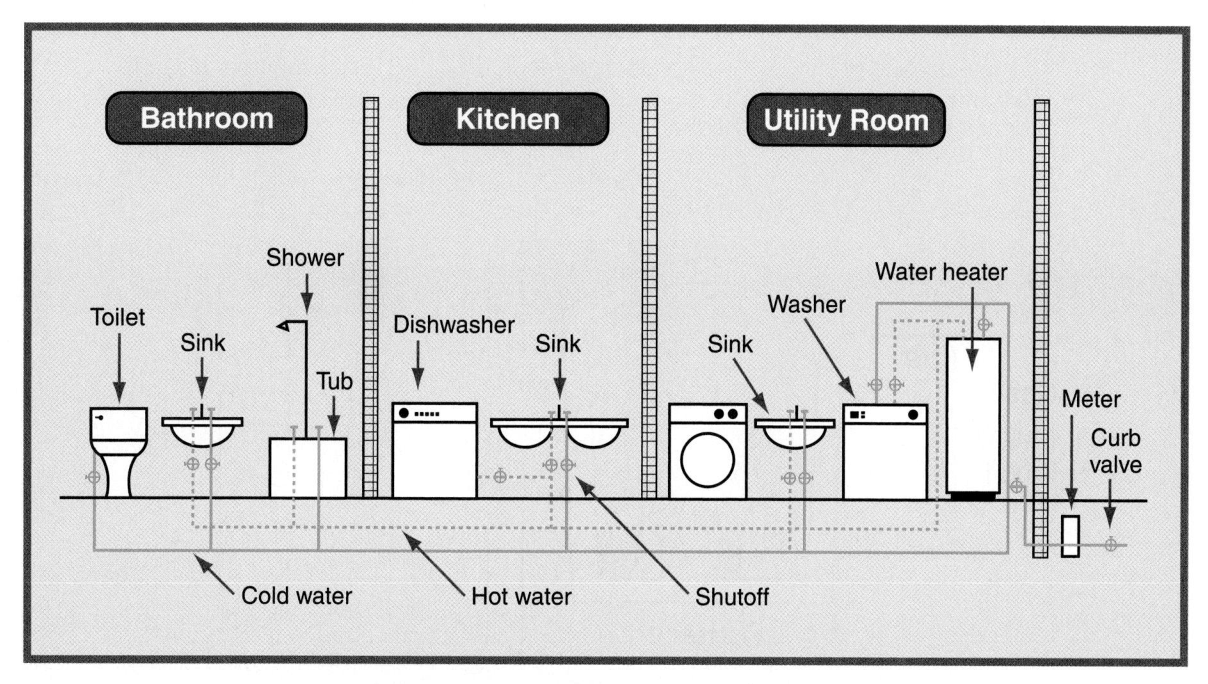

Figure 17-20. The potable-water system for a home. Each fixture has a shutoff valve.

from leaking into the home. Wastewater systems have a network of vents to prevent the water from being drawn out of the traps. The vents also allow sewer gases to escape above the roof without causing any harm.

Homes that use natural gas have a third type of plumbing. Gas lines carry natural gas to furnaces, stoves, water heaters, and other appliances. Shutoff valves are installed at the entrance and at the major appliances.

Climate-control systems

In many homes, the climate-control system is used to heat the building in winter and cool it in summer. This can be done with a single unit or with separate heating and cooling units. Heating systems can directly or indirectly heat the home. In a direct heating system, the fuel is used in the room to be heated. Direct heating might use a stove or a fireplace that burns wood or coal.

Other direct heating methods use electrical power. These systems use resistance heaters installed in the walls or along the baseboards. Also, ceiling radiant wires or panels might supply the heat.

Indirect systems heat a conduction medium, such as air or water. This medium carries the heat from a furnace to the rooms. The heat is then given off to the air in the room. See **Figure 17-21.** The energy sources for these systems are electricity, coal, oil, wood, natural gas, or propane.

Furnaces that heat air as a conduction medium are called ***forced-air heating systems.*** Forced-air furnaces draw air from the room. This air is heated as it moves through the furnace. A fan delivers the heated air through ducts to various rooms.

Hot water heating uses water to carry the heat. The water is heated in the furnace, pumped to various rooms, and passes through room-heating units that have metal fins surrounding the water pipe. The fins dissipate the heat into the room.

Some homes are heated with active or passive solar-energy systems. Passive solar systems use no mechanical means to collect and store heat from the Sun. Active solar systems use pumps or fans to move a

Standards for Technological Literacy

20

Figure Discussion

Discuss Figure 17-20 and show the class a faucet. Be sure to explain how it works in the system.

TechnoFact

A room's comfort level is determined by a combination of temperature and humidity of the room. A temperature between 68° F and 75° F and a relative humidity between 30% and 60% produce a comfortable environment for most people. If the temperature remains constant but the humidity increases, the room will feel warmer than a room with the same temperature but a lower humidity.

Standards for Technological Literacy

20

Extend

Discuss how convection, conduction, and radiation heat a room.

Research

Have your students use the Internet to research and diagram how a "smart house" would deal with voice, Internet, and television communication.

TechnoFact

There are two general approaches to heating a building. The first method distributes heat from a single source through the entire building. The second method uses multiple localized heat sources, such as fireplaces or electric baseboard heaters, to distribute heat throughout a building.

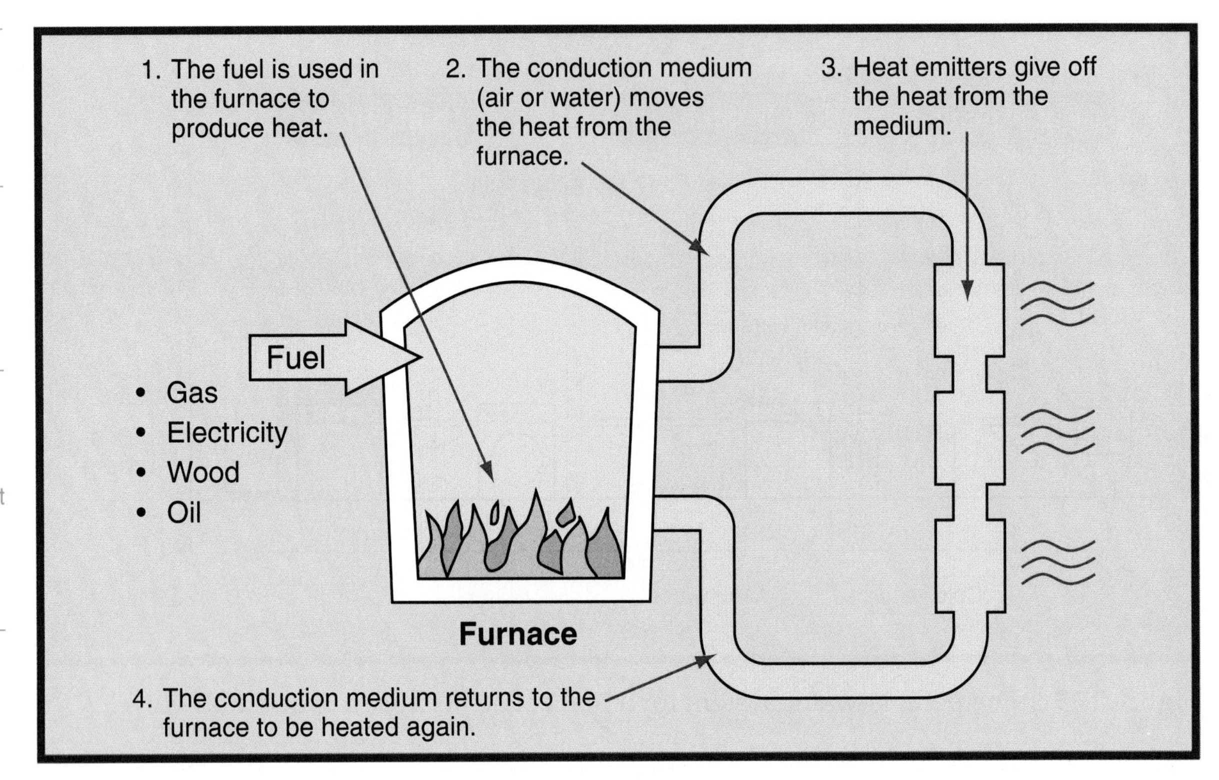

Figure 17-21. Indirect heating systems heat a conduction medium. The resulting heat is then put into the air in the room.

liquid or air to collect solar heat. After the heat is collected, the liquid or air is moved to a storage device. Solar heating systems are very effective in areas that have ample sunshine.

Many buildings have cooling systems to cool the air during the warm parts of the year. Cooling systems use compressors, evaporators, and condensers, very similar to a refrigerator. The system has a fan that draws the air from the room and passes it over a cold evaporator. This is similar to a forced-air furnace. Instead of the air being heated, it is cooled before it is returned to the room.

Another system used in climate control is a unit called a ***heat pump***. A heat pump works as a cooling and heating system. This pump can be operated in two directions. Operating in one direction, the heat pump acts similar to an air conditioner. The heat pump takes warmth from inside the house and discharges it outside. In the winter, the heat pump works in reverse. The pump takes warmth from the outside and brings it into the house. Heat pumps can use air, water, or the ground as a heat source. Those that use air work best in areas that do not get very cold, such as the southern and central United States. Groundwater heat pumps use well water as the heat source. The water is pumped from the well to the heat pump, has heat removed or receives excess heat, and then returns to the well. Ground-coil systems are buried in the soil to take on or give off heat. Groundwater and ground-coil systems can be used in colder climates. Otherwise, heat pumps need a small furnace or other auxiliary heat source as a backup.

Communication systems

Most homes have communication systems, such as telephone, radio, and television, which require special wiring. Telephone wiring and television cables are normally installed during the construction of the building. Installing them after a building is finished takes considerable work to feed the wires through attics, under floors, and inside walls. Many houses are now being built with sound systems that allow music to be played throughout the home.

Finishing the Exterior and Interior

The final exterior finishing step is installing siding and trim. Siding is the finish covering used on a wood building. Wood shingles and boards, plywood, hardboard, brick, stone, aluminum, vinyl, steel, and stucco can be used as siding. Look back at **Figure 17-17.** You see bricks being installed over plywood sheathing. Trim is the strips of wood covering the joints between window and door frames and the siding.

The interior walls are the next walls to be finished. Insulation is placed between the studs and around the windows and doors of all exterior walls and reduces heat loss on cold days and heat gain on hot days. The most common type of insulation is fiberglass blankets or batts. A vapor barrier of polyethylene film is attached to the studs over the insulation. This barrier prevents moisture from building up in the insulation.

Once insulation and utilities are in place, the interior wall surfaces can be covered. The most widely used interior wall covering is gypsum wallboard, commonly known as ***drywall***. Drywall is a sheet material made of gypsum bonded between layers of paper. The sheets of drywall are nailed or screwed onto the studs and ceiling joists. The fastener heads and drywall seams are then covered with a coating called *joint compound*. The compound is applied in several thin coats. This is done to make smooth surfaces and joints between the sheets of drywall.

The inside and outside of the house are now ready for the finishing touches. Interior wood trim is installed around the doors and windows. Kitchen, bathroom, and utility cabinets are set in place. Floor coverings, such as ceramic tile, wood flooring, carpet, or linoleum, are installed over the subflooring. Baseboards are installed around the perimeters of all the rooms. The exterior siding and wood trim are painted. Interior trim is painted or stained. The walls are painted or covered with wallpaper or wood paneling. Lighting fixtures, switch and outlet covers, towel racks, and other accessories are installed. The floors and windows are cleaned. Now the home is finished and ready to be occupied.

Completing the Site

Completing the building is the major part of the project. Other work remains to be done, however. The site must be finished. Earth is moved to fill in areas around the foundation. Sidewalks and driveways are installed.

The yard area needs to be landscaped. ***Landscaping*** is trees, shrubs, and grass that are planted to help prevent erosion and improve the appearance of the site. These trees, shrubs, and grass can divide the lot into areas for recreation and gardening. Landscaping can be used to screen areas for privacy, direct foot traffic, and shield the home from wind, sunlight, and storms.

Look at **Figure 17-22.** The top view shows dirt being moved onto the site for landscaping activities. The bottom view shows a finished landscaped area. Notice how the trees and lawn improve the appearance. Also, a grassy mound is used to guide people onto the sidewalk.

Heavy Engineering Structures

Construction activities do not always produce buildings. They can be used to

Standards for Technological Literacy

3 20

Research

Have your students find out what gypsum plaster is and how it is used in drywall.

Brainstorm

Ask your students what the value of landscaping is.

TechnoFact

Insulation blocks heat from passing through the walls, floors, and ceilings of a building. In the summertime this keeps the heat outside the house; in the wintertime it keeps the heat inside the house. A wide variety of materials can be used in insulation, including cellulose, rock wool, gypsum, and fiberglass. The type of insulation used depends on the climate and the part of the house it is being installed in.

TechnoFact

Many ancient cultures practiced the art of landscape gardening. The ancient Egyptians and Mesopotamians created grand gardens. The Persians decorated their landscaped areas with water effects. The Japanese practiced the art of bonsai to create elaborate dish gardens. Landscape gardening has continued to evolve in technique and popularity since these early examples.

Standards for Technological Literacy

20

Research

Have the students investigate and diagram the cross-section of a modern highway, labeling all parts.

TechnoFact

People began building roads shortly after the invention of the wheel, about 5000 years ago. The first roads were likely constructed in the eastern Mediterranean area. Road building grew out of a need to facilitate trade between different settlements. The extensive trade route known as the Silk Road consisted of more than 5000 miles of road and connected China and Europe.

Figure 17-22. This site was finished by grading the lot and planting landscaping.

produce structures that are sometimes called *civil structures*, or *heavy engineering structures*. These structures include highways, rail lines, canals, pipelines, power-transmission and communication towers, hydroelectric and flood-control dams, and airports. They provide the paths for the movement of water, people, goods, information, and electric power. These projects can be grouped in various ways. For this discussion, we group them into transportation, communication, and production structures.

Transportation Structures

Transportation systems depend on constructed structures such as railroad lines, highways and streets, waterways, and airport runways. Other constructed works help vehicles cross uneven terrain and rivers. These structures include bridges and tunnels. Pipelines are land-transportation structures used to move liquids or gases over long distances. Let us look at some examples of these constructed works. We discuss roadways and bridges.

Roadways

Roads are almost as old as civilization. People first used trails and paths to travel. Later, they developed more extensive road systems. The Romans built the first engineered roads more than 2000 years ago. Their influence remained until the 1700s, when modern road building started. Today's roads have their roots in the work of the Scottish engineer John McAdam. He developed a crushed-stone road built of three layers of crushed rock, laid in a 10″ (25 cm)–thick ribbon. Later, this roadbed was covered with an asphalt-gravel mix that is very common today. A more recent development is the concrete roadway.

Building a road starts with selecting and surveying the route. Next, the route is cleared of obstacles such as trees, rocks, and brush. The roadway is graded so it will drain. Drainage is important to prevent road damage from freezing and thawing. Also, a dry roadbed withstands heavy traffic better than a wet, marshy one. Another reason for grading is to keep the road's slope gentle. Elevation changes are described using the term *grade*. Grades are expressed in percentages. A road with a 5% grade gains or loses 5′ of height for every 100′ of distance. Most grades are kept below 7%.

Once the roadbed is established, the layers of the road are built. See **Figure 17-23.** The graded dirt is compacted, and a layer of coarse gravel is laid. This is followed with finer gravel that is leveled and compacted. Next, the concrete or asphalt top layer is applied. Concrete roads are laid in

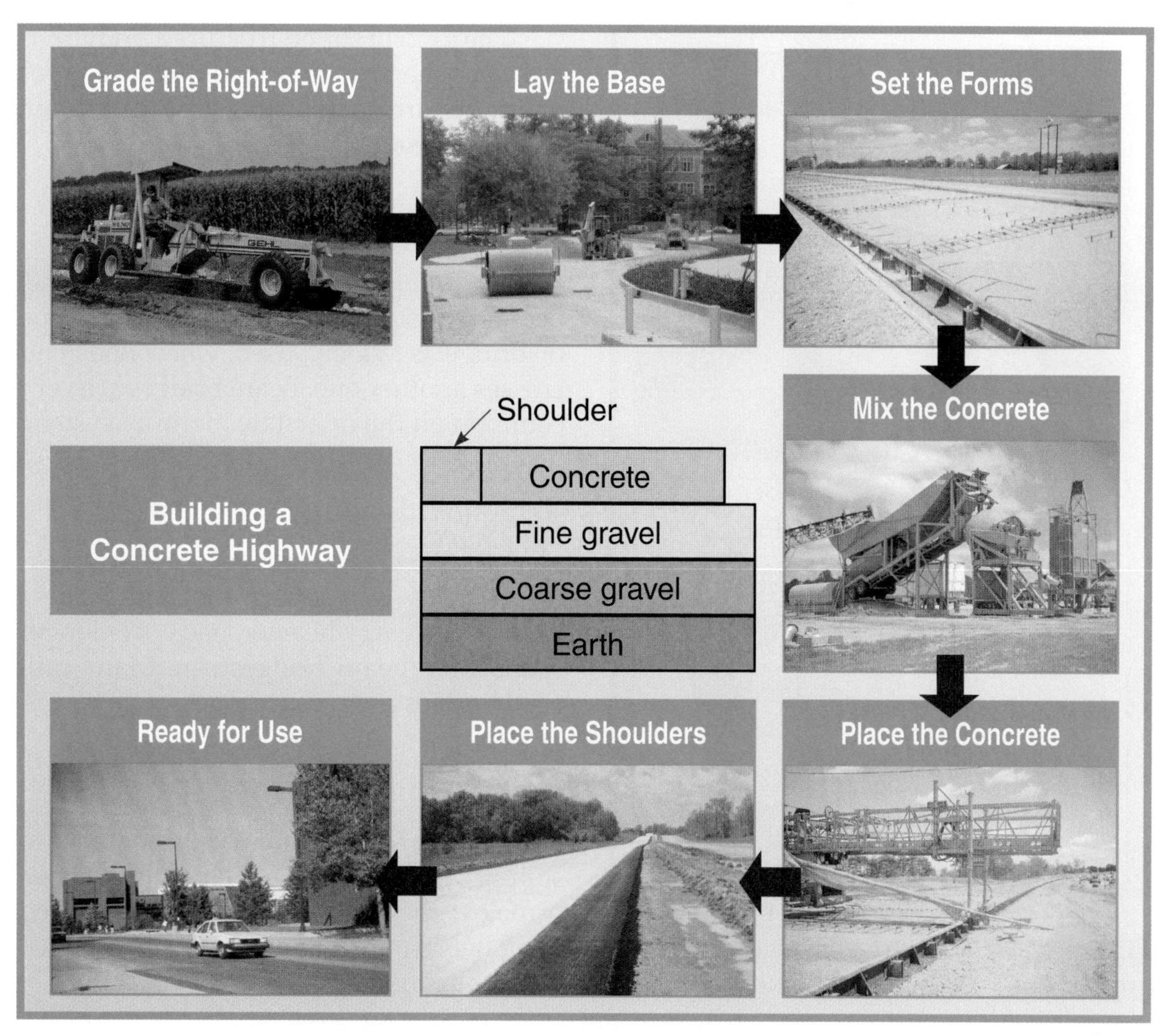

Figure 17-23. Road building is a step-by-step process.

Standards for Technological Literacy

20

Extend

Discuss why the government builds streets, roads, and highways.

one layer. Asphalt is generally applied in two layers: a coarse undercoat and a finer topcoat. Finally, the shoulders, or edges, of the road are prepared. The shoulders can be gravel or asphalt.

Bridges

Another constructed structure vital for transportation is the bridge. Bridges provide a path for vehicles to move over obstacles. These obstacles include marshy areas, ravines, other roads, and bodies of water. Bridges can carry a number of transportation systems. These systems include highways, railroads, canals, pipelines, and footpaths.

Generally, there are fixed and movable bridges. A fixed bridge does not move. Once the bridge is set in place, it stays there. Movable bridges can change their positions to accommodate traffic below them. This type of bridge is used to span ship channels and rivers. The bridge is drawn up or swung out of the way so ships can pass.

Bridges have two parts. See **Figure 17-24.** The substructure spreads the load of the bridge to the soil. The abutments and the piers are parts of the substructure. The superstructure carries the loads of the deck to the substructure. The deck is the part used for the movement of vehicles and people across the bridge.

Standards for Technological Literacy

5 20

TechnoFact

The Romans were outstanding road builders. They built approximately 50,000 miles of road to connect Rome to the remote parts of the empire. The Roman roads were well constructed and paved with rock or concrete. The roads ran in nearly straight lines and passed over hills instead of curving around them.

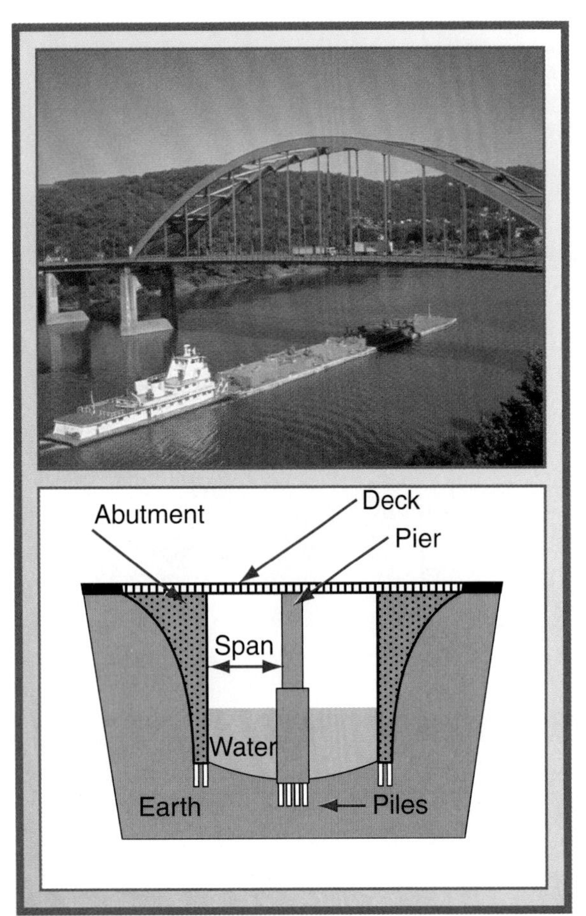

American Electric Power Co.

Figure 17-24. The parts of a bridge. An arch bridge is shown in the photo.

The kind of superstructure a bridge has indicates the type of bridge it is. The types of bridges are beam, truss, arch, cantilever, and suspension. See **Figure 17-25.**

Beam bridges

Beam bridges use concrete or steel beams to support the deck. This type of bridge is widely used when one road crosses another one. Beam bridges are very common on the interstate highway system.

Truss bridges

Truss bridges use small parts arranged in triangles to support the deck. These bridges can carry heavier loads over longer spans than beam bridges can. Many railroad bridges are truss bridges.

Arch bridges

Arch bridges use curved members to support the deck. The arch can be above or below the deck. Arch bridges are used for longer spans. One of the longest arch bridges spans more than 1650′ (502 m).

Think Green

Green Architecture

You may not think about this, but buildings consume a lot of resources. Buildings consume a great deal of electricity, water, and raw materials. They also output a large amount of waste and carbon dioxide. Through the practice of green architecture, the resources and waste are greatly reduced. Green architecture is a means of being more environmentally responsible with buildings from design to construction and landscaping. Buildings are designed to be more efficient with their use of resources, starting with the building materials.

Green architecture may use recycled materials in construction. This type of architecture also uses more energy-efficient resources and renewable energy sources. It is responsible in planning for landscaping, in order to consume less water. And while green architecture benefits the environment in general, even the air and water quality within the building may improve.

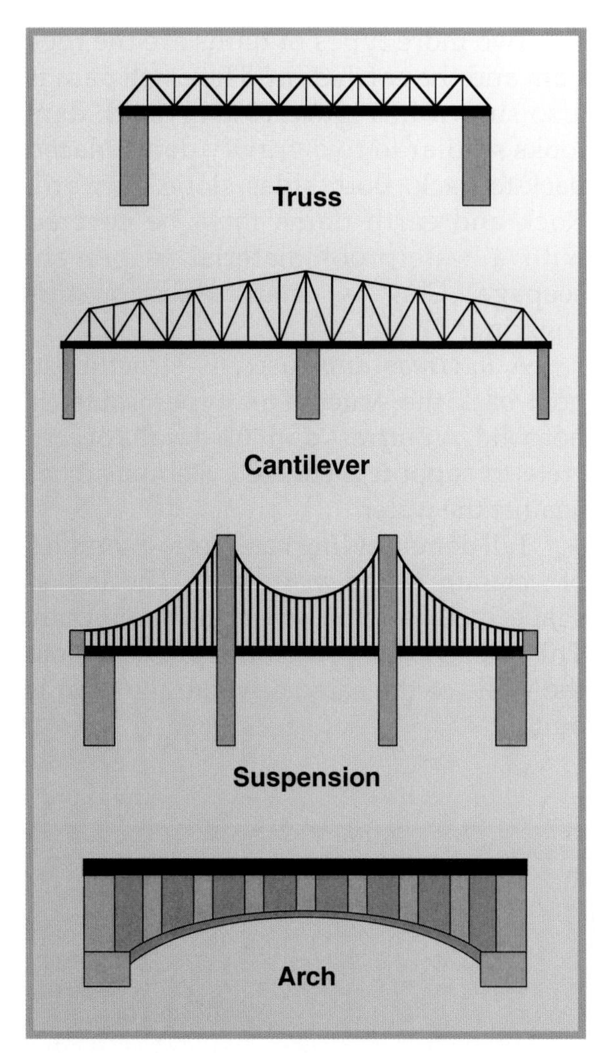

Figure 17-25. Four of the five types of bridges are the truss, the cantilever, the suspension, and the arch. The fifth is the beam bridge.

Cantilever bridges

Cantilever bridges use trusses extending out in both directions from the support beams, similar to arms. The ends of the arms can intersect with the road leading up to the bridge or hook up to another truss unit to form a longer span. The arms transmit the load to the center.

Suspension bridges

Suspension bridges use cables to carry the loads. A large cable is suspended from towers. From the large cable, smaller cables drop down to support the deck. Suspension bridges can span distances as great as 4000′ (1220 m) and longer.

Communication Structures

Most telecommunication technology relies on constructed towers to support antennas. These towers are usually placed on a concrete foundation. A steel tower is built on top of the foundation. Once the tower is complete, the signal wiring can be installed. Similar techniques are used to construct towers for power-transmission lines. See **Figure 17-26.**

American Electric Power Co.

Figure 17-26. This helicopter is helping to construct a tower for an electricity-transmission line.

Standards for Technological Literacy

20

Extend

Discuss or show a video on how a suspension bridge, like the Golden Gate Bridge, was built.

Research

Have your students use the Internet to find examples of each type of bridge shown in Figure 17-25.

TechnoFact

In the early 1800s, John McAdam revolutionized road building with his development of a new road surface. A modern variation of McAdam's road surface is known as tarmacadam, or simply tarmac. He also emphasized the importance of proper drainage to maintain a solid foundation for the roads.

Standards for Technological Literacy

20

Extend

View and discuss a video on building a major dam like the Hoover Dam or the Grand Coulee Dam.

TechnoFact

Ancient humans constructed many types of tunnels. These types included catacombs and crypts, in which to bury the dead, and many transportation tunnels. Over twenty-five hundred years ago, the Greeks constructed a 3500′ long, 6′ square tunnel to carry water. Over two thousand years ago, the Romans built a one mile long, 30′ wide highway tunnel connecting two towns.

TechnoFact

The movement of the affluent from urban centers to the suburbs has been going on for more than a hundred years. In the 19th century, many of London's wealthy families owned weekend villas outside the city. One by one, they made their suburban estates their primary residences. The newly developing middle class soon followed.

Production Structures

Some structures used for production activities are not buildings. For example, petroleum refineries are mixes of machinery and pipelines. Irrigation systems are constructed to bring water to farms in dry areas. Evaporation basins are built to recover salt and other minerals from seawater.

Another important production structure is the dam. Dams are used for controlling floods, supplying water, making recreational lakes, or generating electricity. Several types of dams exist. One type is called a ***gravity dam***. This dam's lakeside is vertical, whereas the other side slopes outward. The sheer weight of the concrete the dam is made from holds the water back. The dam on the left in **Figure 17-27** is a gravity dam.

Two more types of dams are the rock dam and the earth dam. The earth dam is also shown in **Figure 17-27.** A rock dam looks similar to two gravity dams placed back-to-back. Both sides slope outward. Rock and earth dams must be covered with a waterproof material to prevent seepage. Clay is often used for this covering.

A ***buttress dam*** uses its structure to hold back the water. This type of dam is not solid. A buttress dam uses walls of concrete to support a concrete slab or arches against the water.

Tall dams holding back large quantities of water are called *arched dams*. The arched shape increases the strength of the dam. This shape also spreads the pressure onto the walls of the canyon where the dam is built.

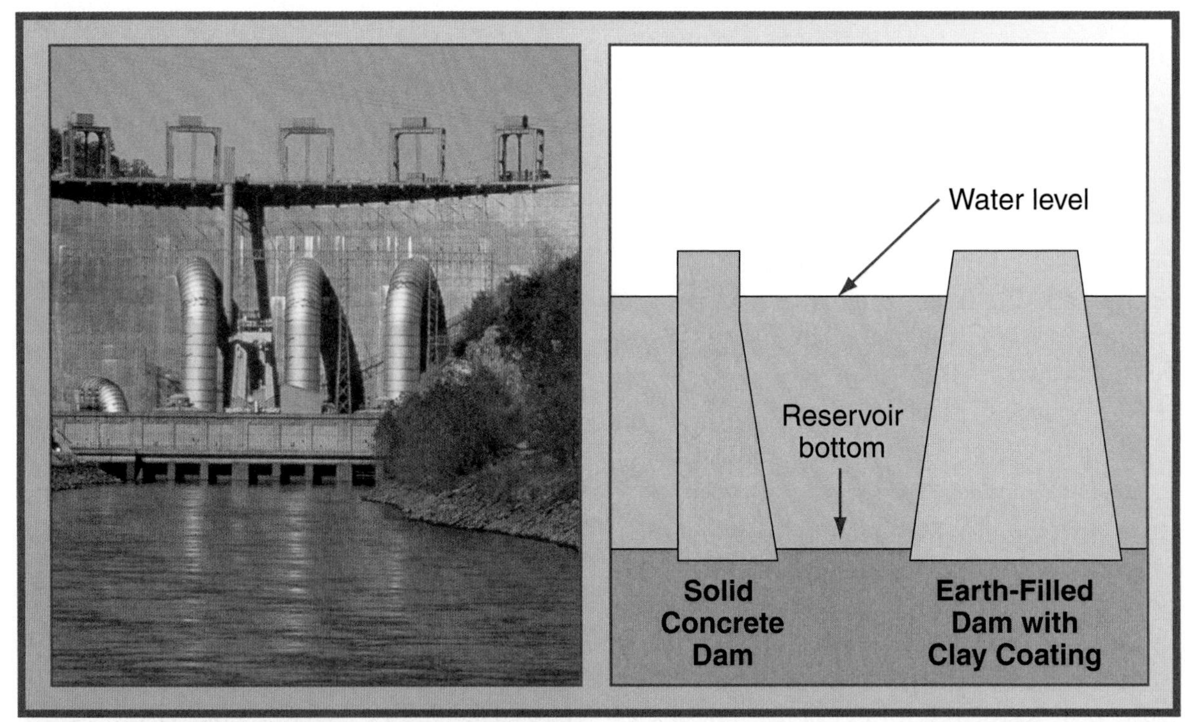

American Electric Power Co.

Figure 17-27. The drawing shows two common types of dams. The photo shows a gravity dam.

Summary

Construction is a vital production activity. This activity provides us with homes, offices, factories, highways, railroads, pipelines, bridges, dams, and other structures. Construction can be divided into two kinds of projects: those that produce buildings and those that produce heavy engineering structures.

Most buildings are constructed using the same steps. The site is cleared and prepared for construction. The foundation for the structure is set. The framework or superstructure is then erected. Utilities are installed, and the structure is enclosed. The building is finished, and the site is landscaped.

The construction of heavy engineering structures varies according to type. We can group these kinds of projects into three categories. These categories are transportation structures, communication structures, and production structures.

Test Your Knowledge

Write your answers on a separate piece of paper. Please do not write in this book.

1. List the two kinds of constructed works.
2. A condominium is a residential structure that the people living in it generally own. True or false?
3. Give two examples of a government building.
4. A home built in a factory is called a(n) _____.
5. You should use spread foundations on wet or sandy soils. True or false?
6. The two types of floors in single-family homes are concrete slab or _____.
7. What is a fascia board used for?
8. The two types of water systems that are part of a plumbing system are called _____ and _____.
9. The most common type of insulation is _____.

Answers to Test Your Knowledge Questions

1. Buildings and heavy engineering structures
2. True
3. city halls, post offices, police stations, firehouses, state capitols, courthouses, and government office buildings
4. manufactured home
5. False
6. lumber
7. A fascia board is used to finish the ends of the rafters and the overhang.
8. potable water, wastewater
9. fiberglass
10. D. Enclosing the structure
11. F. Finishing interior and exterior
12. A. Preparing the site
13. B. Setting foundations
14. G. Completing the site
15. E. Installing utilities
16. C. Building the framework
17. C. Building the framework
18. B. Setting foundations
19. A. Preparing the site
20. C. Building the framework
21. F. Finishing interior and exterior
22. G. Completing the site
23. Elevation changes expressed in percentages
24. False
25. True
26. arched

Matching questions: For Questions 10 through 22, match each description on the left with the correct construction step on the right. (Note: Answers can be used more than once.)

10. ______ Sheathing the walls.
11. ______ Putting up drywall.
12. ______ Grading.
13. ______ Putting in footings.
14. ______ Landscaping.
15. ______ Putting in a heat pump.
16. ______ Installing a subfloor.
17. ______ Installing the roof.
18. ______ Driving in piles.
19. ______ Marking the building site.
20. ______ Placing floor joists.
21. ______ Adding baseboards.
22. ______ Installing a sidewalk.

A. Preparing the site.
B. Setting foundations.
C. Building the framework.
D. Enclosing the structure.
E. Installing utilities.
F. Finishing the exterior and interior.
G. Completing the site.

23. What does the term *grade* mean, as used in this chapter?
24. Most railway bridges are beam bridges. True or false?
25. With cantilever bridges, the ends of the arms do not carry any of the load. True or false?
26. Tall dams holding back large quantities of water are called ______.

STEM Applications

1. Use a chart similar to the one below to list and describe a few of the constructed structures you see as you travel from your home to school.

Structure	Type of Construction	Description–Use

2. Select one structure you saw in completing the previous assignment. Make a drawing or model of the structure. Label the major parts.
3. Develop a preventive maintenance schedule for your home. List the items or portions of the structure that need maintenance and the type of maintenance each needs.

This activity develops the skills used in TSA's Structural Engineering event.

Structural Engineering

Activity Overview

In this activity, you will create a balsa-wood bridge and determine its failure weight (the load at which the bridge breaks).

Materials

- Grid paper.
- 20′ of 1/8″ × 1/8″ balsa wood.
- A 3″ × 5″ note card.
- Glue.

Background Information

- **General.** There are several types of bridges: beam, truss, cantilever, suspension, and cable stayed. The length of the span and available materials generally determine the type of bridge used in a particular situation. For this activity, a truss design is considered the most efficient.
- **Truss bridges.** The truss bridge design is based on the assumption that the structural members carry loads along their axes in compression or tension. The members along the bottom of the bridge carry a tensile load. Those along the top of the truss carry a compressive load. The members connecting the top and bottom chords (members) can be in tension or compression.
- **Gussets.** Gussets are plates connected to members at joints to add strength. They are normally used in steel construction. The structural steel members are welded or bolted to the gusset. When designing your bridge, include a gusset at each joint, if possible.
- **Wood properties.** Due to its molecular structure, wood can normally carry a larger load in tension than it can in compression. Also, a shorter member can carry a greater compressive load than a longer member.

Guidelines

- You must create a scale sketch of the bridge before building.
- Two pieces of balsa wood can be glued together along lengthwise surfaces. No more than two pieces of balsa can be glued together. You cannot use an excessive amount of glue.
- Gussets cut from the 3″ × 5″ card can be no larger than the diameter of a U.S.-quarter coin. A gusset cannot touch another gusset. This plate cannot be sandwiched between two pieces of balsa wood.
- The bridge design must take into account the loading device. Your teacher will provide specific guidelines for bridge length, bridge width, and the required details for attachment of a loading device.
- Your bridge will be weighed before being loaded.

Evaluation Criteria

Your project will be evaluated using the following criteria:

- Accuracy of sketch, compared to completed bridge.
- Conformance to guidelines.
- Efficiency (failure weight ÷ bridge weight).

Chapter Outline
Selecting Technological Products
Installing Technological Products
Maintaining Technological Products
Repairing Technological Products
Altering Technological Products
Disposing of Technological Products

This chapter covers the benchmark topics for the following Standards for Technological Literacy:

3 5 15 19 20

Learning Objectives

After studying this chapter, you will be able to do the following:

- Summarize the factors to consider when selecting a product or structure.
- Recall the five steps used in installing a product.
- Summarize the factors involved in maintenance.
- Recall the steps used in repairing.
- Summarize the activities involved in altering products and structures.
- Summarize the importance of recycling.
- Summarize ways to dispose of materials.

Key Terms

adjustment	maintenance	replacement
alter	preventive maintenance	testing
appearance	recycle	value
function		

Strategic Reading

Before you read this chapter, look at the titles of the main headings. Try to guess the main points in each section. Write down your guesses so you can check as you read.

People use manufactured products and constructed structures every day. They choose the items meeting their needs. People make their choices to live better. Sometimes, however, people misuse some of their belongings. People can also choose the wrong item, use it improperly, or maintain it poorly.

If technology is to help us live better, we must choose the correct product or structure for our needs or for the task at hand. This involves six steps. These steps are selecting, installing, maintaining, repairing, altering, and disposing of technological products. See **Figure 18-1.**

Selecting Technological Products

The next time you go to a store, stop and look around you. See **Figure 18-2.** You will see thousands of products on display. Each of these products is developed for a specific use. Some products help fulfill basic needs or perform essential tasks. For example, stoves cook our food. Automobiles and buses transport us to work and school. Other products are designed to make life more pleasant. For example, games, DVDs, and recorded music help to entertain us. Colorful pictures and works of art brighten our homes and workplaces.

Choosing the best product to fill a need or want is quite a challenge. Needs come first because they are the things necessary for living. Wants are requirements that are not necessary for living or to complete a task, but they are nice to have. It is not always easy to determine what is a need and what is merely a want. Once you have separated needs and wants, you must decide how much money can be spent on

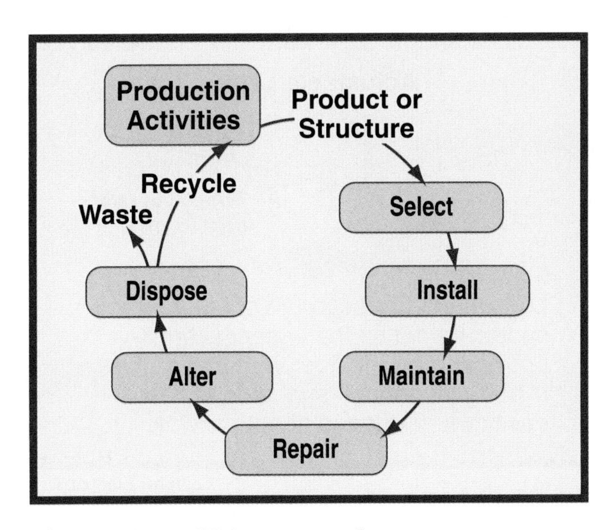

Figure 18-1. Using a product or structure in the proper way involves six steps.

©iStockphoto.com/fstop123

Figure 18-2. Thousands of products are made to meet our needs and wants.

meeting the needs. You probably cannot afford to pay for all your wants at one time. Therefore, you have to rank them by their importance to you. Such essentials as food and clothing will rank high on your list. Decorative items might rank lower on your list.

Knowing what you need leads to the next step. This step is selecting the best of the many products fitting your needs. Some people buy products on impulse. A better way is to analyze the products. This allows you to think about the products you want to buy. You should consider three important factors:

- ***Function.*** How well does the product meet your needs? How well does it work? Is the product durable and easy to maintain? How well does it do the job you have in mind?

Standards for Technological Literacy

19

Brainstorm

Ask the students what outside influences cause them to want to buy a product.

Extend

Explain the difference between a need and a want.

Figure Discussion

Trace a product, such as a kitchen appliance, lighting fixture, or lawn mower, through the cycle shown in Figure 18-1.

TechnoFact

Advertising is developed to encourage people to select one product over another. Unfortunately, the arguments they make can be deceptive. In a recent study of more than 300 weight-loss advertisements, many were found to make claims that they likely could not deliver on. Many of these ads supported their claims with misleading testimonials and endorsement rather than verifiable scientific evidence.

TechnoFact

The Web site of the United Way of Central New Mexico put human needs and wants in perspective by writing, "What do you really need to survive? Some of you may be surprised to find out that a car, video game, VCR, and TV do not even make the list. Only the most basic needs are necessary for life. You may be able to do without one or the other for some time, but not for long. The basic needs for life are air, water, food, shelter, and clothing."

Standards for Technological Literacy

19

Extend

Review the installation directions for an appliance, plumbing fixture, or lighting fixture.

TechnoFact

There are several considerations that should be thought out before appliances are installed. For example, the appliance should be easy to remove for service or repair. Flexible hookups make appliances easier to move, thereby lowering repair costs.

- *Value.* Does the performance of the product match the product's price? Do other products meet your needs as well, but cost less? Is this product worth the selling price?
- *Appearance.* Are the design and color of the product pleasing to you? Is the product something you would be proud to own?

Installing Technological Products

Some products are ready to use when you buy them. Other products must be installed. For example, items such as cookware, clothing, and tools are sold ready to use. No setup is required. Other products require installation, such as the dishwasher shown in **Figure 18-3.** The purchaser or a service person must complete several tasks to make a product ready to use.

First, products must be unpacked. Products are shipped from the factory in protective crates and boxes. Boxes, crates, and rigid-foam components protect the product from damage during shipment. The boxes and packing materials must be removed and properly discarded or recycled.

Second, some products require utilities in order to work. The products are attached to electrical lines, waterlines, or wastewater lines. Natural gas or compressed air might also be needed. Third, the product might need to be positioned and secured to the floor or inside a cabinet. The product might also need to be leveled to operate properly.

Fourth, the product might require ***adjustment***. Electrical meters might need to be set on zero. Clearances between doors might need to be adjusted. The space between moving parts should always be checked.

Finally, most products must be tested. For example, a set of digital images might be shot to test a camera's operation and adjustments. A water conditioner might be cycled to see that its controls are operating.

Directions for the installation and use of many products are in the owner's manuals. See **Figure 18-4.** This information

White Consolidated Industries

Figure 18-3. This dishwasher must be installed before it can be used.

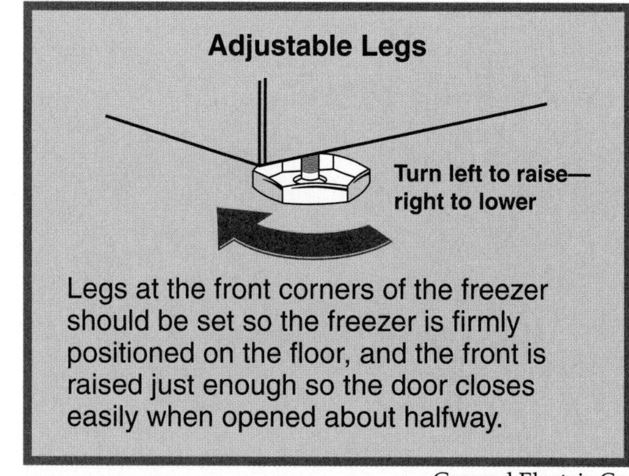

General Electric Co.

Figure 18-4. An example of installation instructions for a product.

might be on one sheet of paper or in a large book. The complexity of the product determines the size of the owner's manual.

Maintaining Technological Products

We want products to work properly when we need them. This often requires a maintenance program. The goal of ***maintenance*** is to keep products in good working order. See **Figure 18-5.** Clothing and dishes are washed to maintain their usefulness and extend their lives. Filters in furnaces and air-conditioning units are changed so the units function properly. The oil in automobile engines is changed to maintain the oil's ability to lubricate moving parts. Locomotives and buses are serviced to maintain their performance.

United Parcel Service

Figure 18-5. This mechanic is performing routine maintenance on an aircraft engine. The maintenance keeps the engine in good working order.

Standards for Technological Literacy

19

Research

Have your students refer to an owner's manual for an automobile to determine the routine maintenance requirements.

TechnoFact

Upkeep is a term that is often used in conjunction with or in place of the term *maintenance*. Upkeep refers to all the costs and actions required to keep products and systems operating properly.

Career Corner

Building Inspectors

Building inspectors examine the construction work done on various types of buildings. Their job is to make sure the structures meet building codes and zoning regulations. Almost half of all inspectors work for local governments. Other inspectors conduct home inspections as part of the home-purchasing process. Building inspectors generally work alone and typically work regular hours. They spend time inspecting construction work sites and in field offices, reviewing blueprints, answering letters, writing reports, and receiving telephone calls.

Inspectors should have technical knowledge, experience, and education. Many inspectors have formal training and experience in the aspect of construction they inspect. Generally, inspectors must have at least a high school diploma and might be required to have studied engineering or architecture or taken courses in building inspection. Most states and cities require some type of certification for employment. This requires proof of appropriate construction experience and education and passing an examination on code requirements, construction techniques, and materials.

Standards for Technological Literacy

19 20

Extend

Differentiate between maintenance and repair.

Brainstorm

Ask the students to list all the information someone would need to repair a defective product such as a lamp or bicycle.

TechnoFact

In the future, cars may be able to self-diagnose impending failures before they happen. Such cars would be linked through wireless communications to service agencies that would then alert roadside assistance and service facilities as needed.

Most maintenance is done on a schedule. See **Figure 18-6.** The schedule is designed to keep the product working properly. Therefore, maintenance is sometimes called ***preventive maintenance.*** Maintenance is designed to prevent breakdowns. Many products come with a maintenance manual. This document lists the following:

- The types of maintenance needed.
- Methods for performing maintenance.
- A time schedule for each maintenance task.

Buildings and other constructed structures need maintenance, just as manufactured products do. Buildings must be cleaned and painted. See **Figure 18-7.** Windows are cleaned. Roofs are sealed to prevent leaks. Bridges and communication towers are painted to prevent rusting. Railway tracks are leveled, and switches are lubricated. Streets and driveways receive periodic coatings to seal out water.

©iStockphoto.com/JaniceRichard

Figure 18-6. Many products need maintenance done periodically. This technician is sliding a new air filter into a gas furnace.

Repairing Technological Products

No product or structure works all the time or lasts forever. Some products are used until they stop functioning and are then discarded. For example, few of us try to salvage bent paper clips or bolts with stripped threads.

Many products, however, are too costly to discard the first time they stop working. Throwing away a bicycle or a car every time a tire goes flat would be very expensive. It costs less money to repair the product than to buy a new one. Repair is the process of putting a product back into good working order. See **Figure 18-8.** This requires three steps:

1. **Diagnosis.** The cause of the problem is determined.
2. ***Replacement* or adjustment.** Worn or broken parts are replaced. Misaligned parts are adjusted.

Figure 18-7. Painting is an example of home maintenance.

3. *Testing.* The repaired product must be tested to ensure that it works properly.

The information needed to repair a product is contained in the product's service manual. This manual provides a parts list so repair parts can be ordered. See **Figure 18-9.** The service manual gives directions for completing common repairs. See **Figure 18-10.**

Standards for Technological Literacy

19

Demonstrate

Use a nonfunctioning flashlight or table lamp to show the class how to diagnose, repair, and test a product.

Brainstorm

Ask the students how diagnosing an illness is like diagnosing a product malfunction.

TechnoFact

Product safety requires both a safe product and appropriate operator actions. One government agency that deals with product safety is the National Highway Traffic Safety Administration. This agency sets and enforces safety performance standards for motor vehicles and related equipment in order to reduce deaths, injuries, and economic losses resulting from motor vehicle crashes.

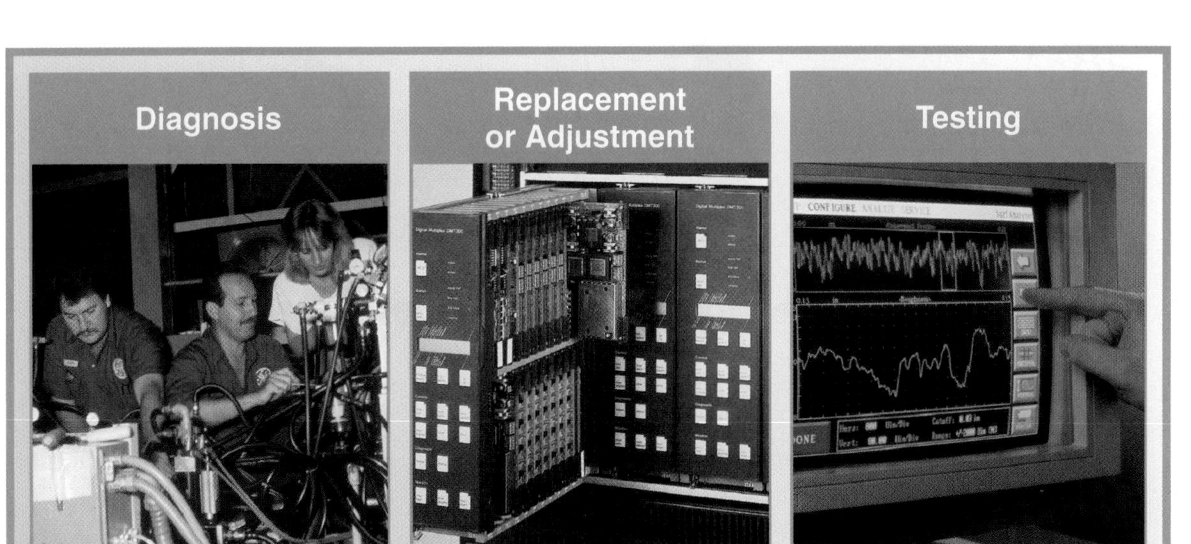

General Electric, Hewlett-Packard, Federal Products Corp.

Figure 18-8. Three steps are followed when repairing a product.

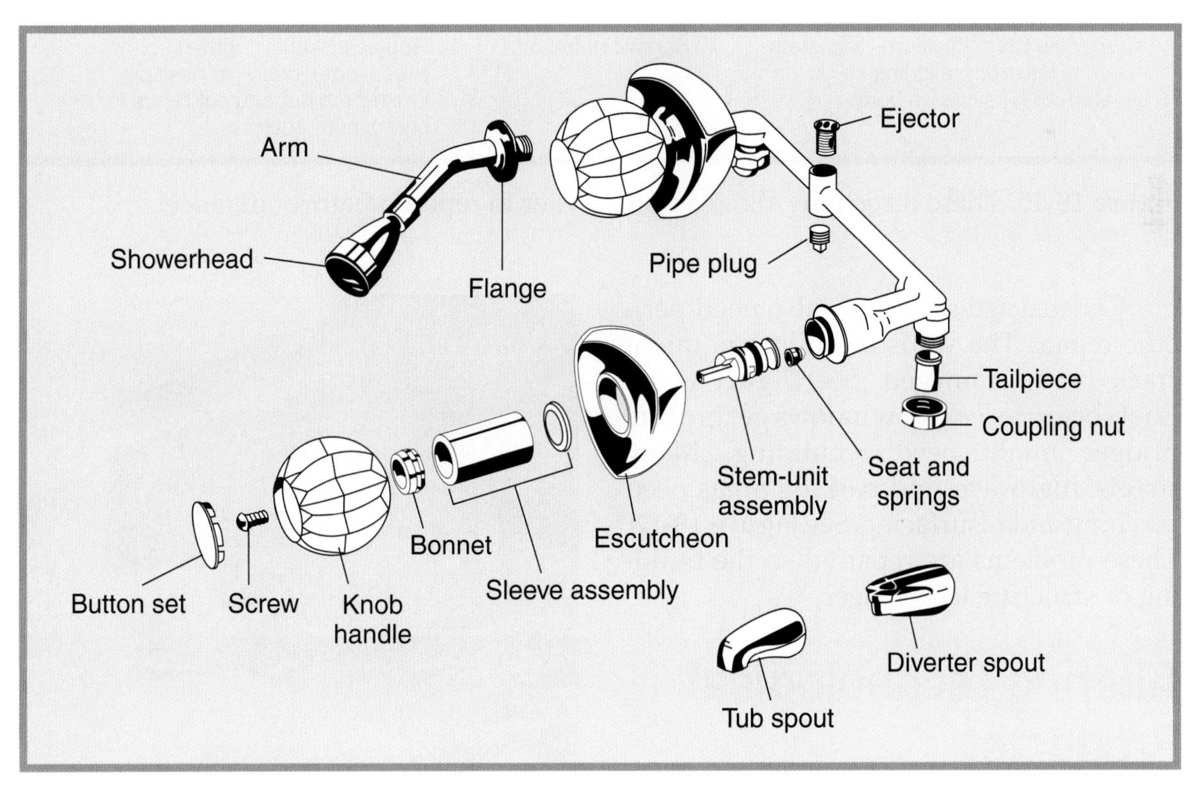

Figure 18-9. This exploded pictorial view provides a parts list. The list helps a service person order repair parts.

Standards for Technological Literacy

19 20

Figure Discussion

Discuss why pictorial drawings, like the one shown in Figure 18-10, are used in maintenance and repair manuals.

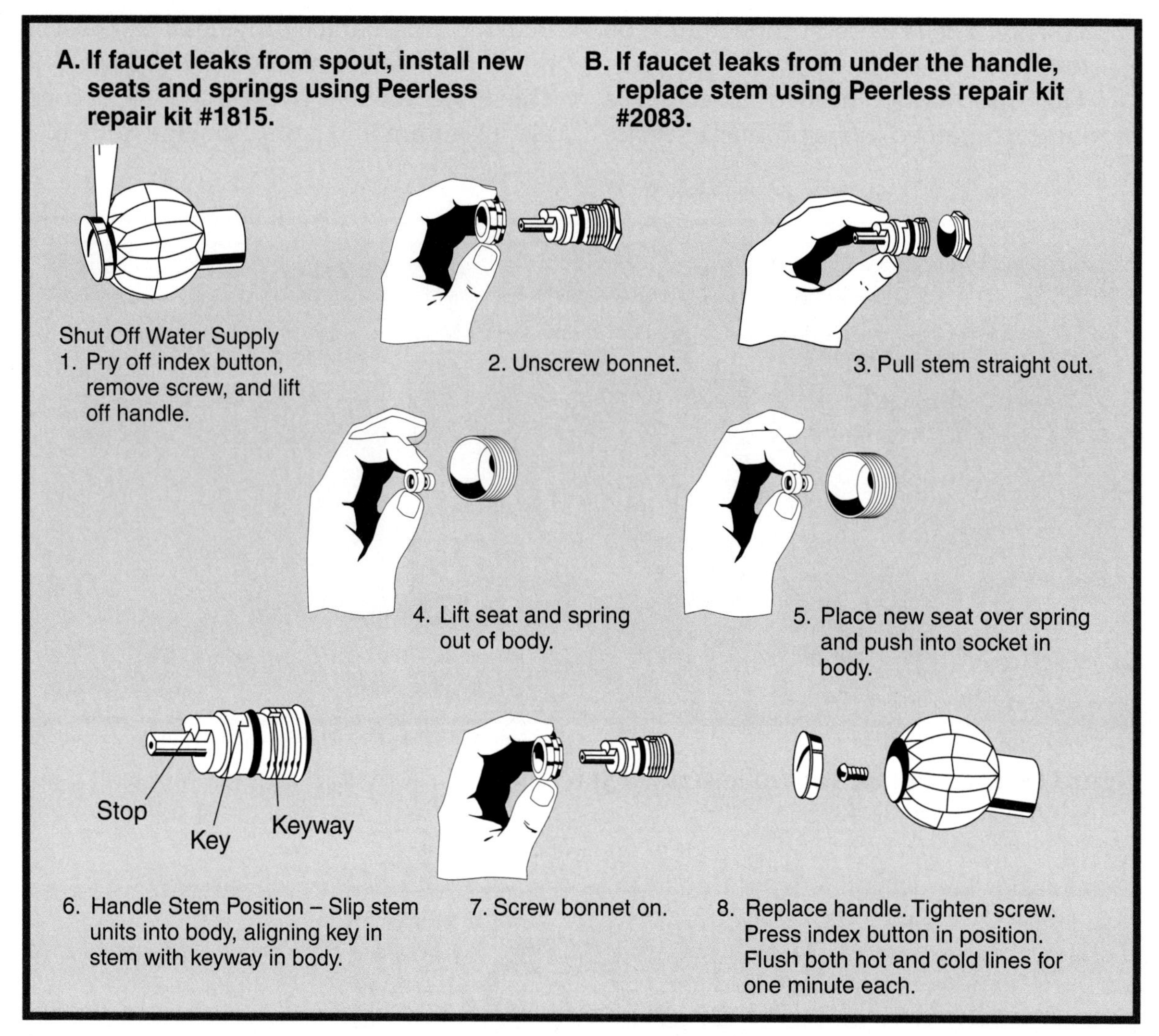

Figure 18-10. These directions allow a homeowner to repair a bathroom faucet.

Constructed structures also need periodic repair. The walls of buildings might crack or be damaged. See **Figure 18-11.** Roofs begin to leak, or windows get broken. Bridges might need rebuilding. Many streets, highways, and parking areas need patching and resurfacing. See **Figure 18-12.** These problems are repaired so the building or structure lasts longer.

Figure 18-11. Workers sometimes need to repair a faulty wall of a building.

Altering Technological Products

Some products become obsolete as time passes. ***Altering*** the products can extend

STEM Connections: Science

Materials Science

Knowledge derived from the scientific discipline called *materials science* has greatly aided us in our efforts to dispose of products properly and efficiently. *Materials science* can be defined as "the scientific study of the structures and properties of manufacturing and construction materials." Materials scientists analyze the molecular structures inherent in various materials to determine the materials' specific compositions. Once a material's composition is known, we can determine its best uses and reuses.

Scientists have placed these materials into four specific categories for study: metals, plastics, ceramics, and composites. Each exhibits a specific set of properties. For example, most glass (a ceramic) is composed primarily of silica, sodium oxide, and calcium oxide. These properties make glass easy to melt and reshape. Thus, glass is a good material to recycle because it can be easily heated and shaped into new forms.

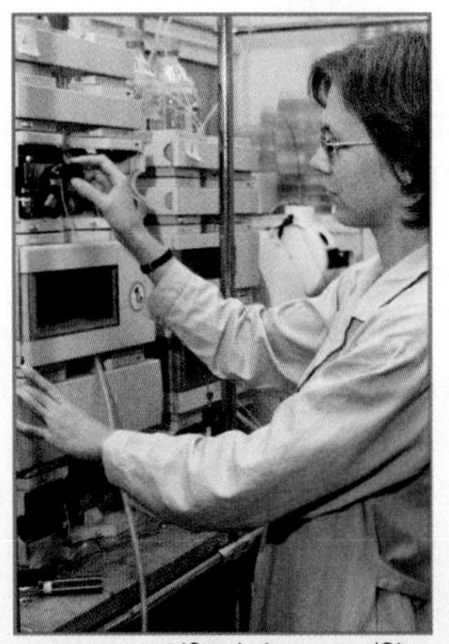

This scientist is conducting experiments on industrial materials.

On the other hand, plastics are created from fractions of natural gas or crude oil changed with chemicals into solid form. These chemicals include benzene, cadmium compounds, and carbon tetrachloride. Plastics do not degrade easily and never completely decompose. Furthermore, because of their chemical properties, plastics, when burned, release many pollutants into the air. Thus, recycling efforts are focused on reusing the material as is and in reducing its volume through such methods as pulverizing and dissolving the material with a solvent. In what ways are the other two types of materials (metals and composites) disposed of, and why are these methods used?

Standards for Technological Literacy

3 19 20

TechnoFact

The Federal Trade Commission's Bureau of Consumer Protection was established to protect consumers against deceptive, unfair, or fraudulent business practices. They accomplish this goal by providing the public the information they need to make informed decisions. They are also responsible for enforcing a variety of federal consumer protection laws.

Figure 18-12. The parking lot shown here is being repaired by applying a cold-mix asphalt and compacting the asphalt.

Standards for Technological Literacy

5 19 20

Brainstorm

Ask the students how they would remodel the school to make it look nicer and serve all the students better.

Extend

Discuss how remodeling extends the life of products and systems.

TechnoFact

According to the National Association of the Remodeling Industry, changing a room's color is one of the most cost-effective ways of revitalizing it.

their useful lives. For example, a person might gain or lose weight, so his clothes might not fit properly. A tailor can alter the clothes to fit the person better. Buildings are altered also. An outdated building can be changed. The needs of the owners might change. The rooms might be too small, or the windows might be too large. Contractors can alter (remodel) the building to meet current needs. See **Figure 18-13.** Remodeling can involve restoring or changing the appearance of a building.

Some altering is done to change the performance of a product or structure. Memory chips might be added to improve a computer's speed. A better set of tires and wheels can be installed on a car to improve the car's handling. New lighting fixtures and better heating and cooling systems can lower a building's energy use. Scrubbers are added to electrical generating plants to reduce sulfur emissions. These actions help products and structures to last longer and work better.

Figure 18-13. The exterior of this historic building has been preserved. The building's interior has been altered to provide modern offices for a law firm.

Disposing of Technological Products

All products and structures eventually reach the end of their useful lives. They become so worn that they no longer can be repaired or altered. At this point, the owner has the responsibility to dispose of them properly. See **Figure 18-14.**

The first choice for disposal should be ***recycling***. This means that the materials in the product or structure are reclaimed. The reclaimed material is used to make new products.

Recycling can reduce the strain on both resources and landfill disposal sites. The average home produces an enormous amount of waste. Typically, household garbage contains the following materials:

- Paper and paperboard—42%.
- Food and yard wastes—24%.
- Glass—9%.

Figure 18-14. This woman has chosen to recycle her paper and plastic.

- Metals—9%.
- Plastics—7%.
- Other—9%.

Newsprint is the largest single component in landfills and accounts for nearly 15% of the volume. The paper can be deinked (have the ink removed) and used to make cardboard and new newsprint. Magazines and coated book paper are more difficult to recycle because of the coatings used to give the paper a glossy finish.

Some food wastes and most yard wastes can be placed in a compost bin. The waste decomposes and becomes excellent garden fertilizer. Glass is easily recycled into new glass. This material readily melts and can be reformed into containers and sheet glass. Metal is also easily recycled. Nearly half of all aluminum is now coming from recycled beverage cans.

Plastics provide a unique recycling challenge because there is not just one plastic. Hundreds of types are in use today. Household packaging is divided into seven categories for recycling. A code number inside a triangular symbol identifies each of these. See **Figure 18-15.** Many of these plastic materials can be converted into completely different products, such as the deck materials shown in **Figure 18-16.**

Not all materials are easily recycled. For example, some materials in household waste are very dangerous. These materials are called *household hazardous waste*. Paints, solvents, engine oil, batteries, and other hazardous chemicals are examples of this type of waste. They should be disposed of through a hazardous-waste center.

In addition to consumers doing their part, industry has joined the recycling campaign. Steelmakers use recycled iron and steel as base material in the steelmaking process. Lumber mills use scraps from their processes to make paper, particleboard, and hardboard. They also burn scrap lumber to provide power for the mills.

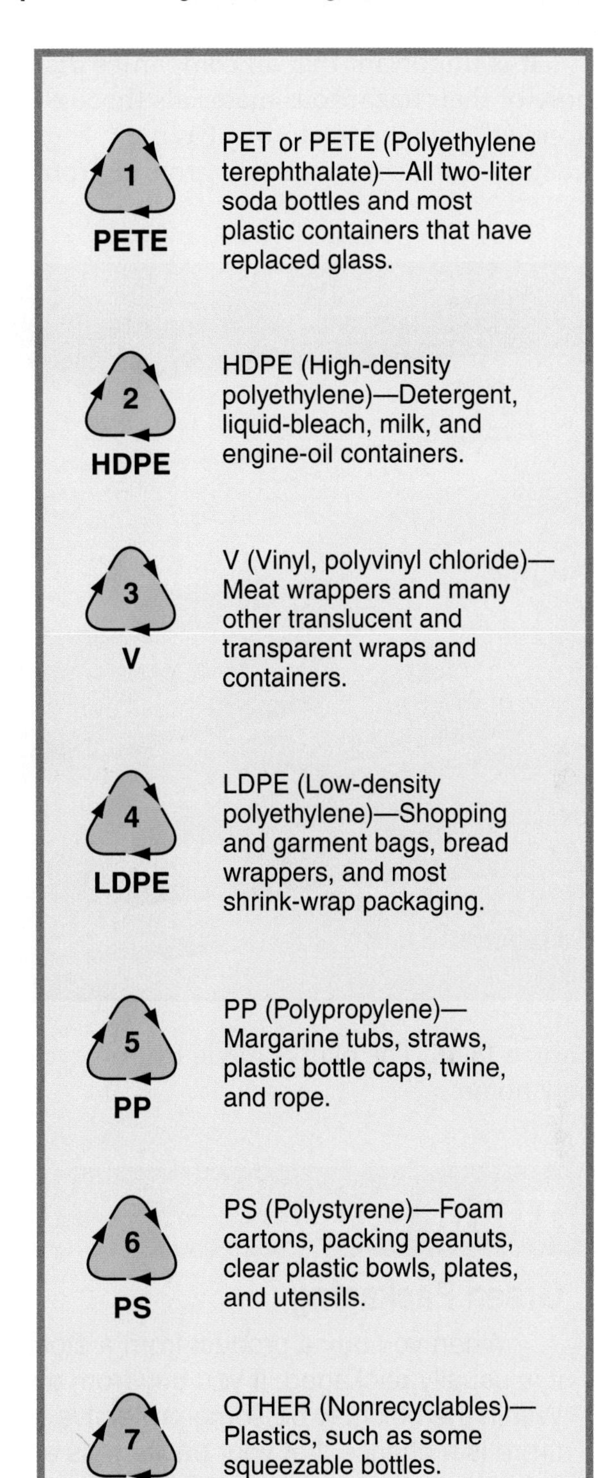

Figure 18-15. These symbols are used on plastic containers and materials to aid in sorting for recycling.

Standards for Technological Literacy

5

Research

Ask the students to find out how much material is recycled in their city, state, or the nation.

Figure Discussion

Discuss where the materials listed in Figure 18-15 can be recycled.

TechnoFact

Developed nations dispose of a tremendous number of plastic bottles every day, and disposing of them has become quite a problem. The bottles are not biodegradable, so they must be recycled or buried in a landfill. Recycling is growing in popularity, and new techniques are being developed to reduce the cost and increase the usefulness of recycled plastics.

TechnoFact

A concept called zero waste goes a step beyond recycling. In a zero-waste model, the flow of raw materials and waste through a society are managed as a system. The reusability of products is one of the key considerations during the design phase. This approach ensures that all products that are created, used, and discarded can be reintroduced into the front end of the system, eliminating waste.

Standards for Technological Literacy

5

Extend

Ask your students why not all products and materials can be recycled.

Brainstorm

Ask the students which items they use should be reworked, repaired, or recycled.

TechnoFact

Recycling, organic composting, and waste prevention are all forms of waste management. The responsibility for recycling used products falls on the consumer. Similarly, communities bear the responsibility for composting community-generated organic materials, such as grass trimmings and leaves, and safely reintroducing them into the environment.

Quotes

"There is growing recognition that influencing consumer behavior is key for achieving the goals of sustainable development. Twenty-five percent of vacuum cleaners, 60 percent of stereos, and 90 percent of computers still function when people dispose of them."—Evan Hinte, Eternally Yours: Visions on Product Endurance.

It is important that all companies dispose of their hazardous materials through licensed subcontractors. Paving contractors are using material ground from highway surfaces as aggregate for new blacktop. Old concrete becomes excellent flood-control fill and is also used as aggregate.

©iStockphoto.com/jjshaw14

Figure 18-16. The planks made from recycled plastic will be used to cover the deck on a new home.

Think Green

Green Packaging

When you buy a product from a store, or buy something and have it delivered, it is usually packaged. If you buy from a store, there might be a box, tape, or plastic. Within the box may be some protective insulation. If something is delivered to you, there is a chance that your package is a box within the shipping box, and the space in between the two boxes is probably a protective material, such as foam peanuts or an air-filled plastic cushioning material.

The companies associated with these packaging materials have been taking steps for years now to make their packaging more environmentally friendly. For example, foam peanuts are now often made of a biodegradable material rather than polystyrene. The air-filled plastic cushioning materials are now being made of recycled plastic or a biodegradable plastic material. Even shipping boxes may be made of recycled materials.

Summary

Using products makes our lives better. We must use and dispose of products wisely, however, and should select products meeting our needs and our ability to pay for them. Instruction manuals should be carefully read, and the instructions should be followed. All products should be used only for the purposes for which they were designed. Each product and structure should receive periodic maintenance and necessary repairs. Finally, each product should be disposed of properly after it has served its purpose. Whenever possible, products should be recycled. Recycling helps to reduce the strain on natural resources and disposal sites.

Test Your Knowledge

Write your answers on a separate piece of paper. Please do not write in this book.

1. What are the three major factors to consider when analyzing whether or not to purchase a product?
2. List the five steps followed in installing a product.
3. Changing the oil in an engine is part of a(n) ______ maintenance program.
4. Name the three major items discussed in a maintenance manual.
5. List the three steps involved in repairing a product.
6. Service manuals contain parts lists. True or false?
7. Give one example of altering that can be done to change the performance of a product or structure.
8. Why is recycling important?
9. Give one example of a specific material and a manner in which it can be disposed.
10. All products can be recycled. True or false?

STEM Applications

1. Select one day in your life. List all the items you throw away. Determine their types and whether or not they can be recycled.
2. Select a complex product you use often. List the following:
   A. The preventive maintenance it requires.
   B. The repairs it needs now or might need in the future.

Answers to Test Your Knowledge Questions

1. Function, value, and appearance
2. Products are unpacked. Products are attached to utilities. Products are positioned and secured. Products are adjusted. Products are tested.
3. preventive
4. The types of maintenance needed. Methods for performing maintenance. A time schedule for each maintenance task.
5. Diagnosis, replacement or adjustment, and testing
6. True
7. Evaluate individually.
8. Recycling can reduce the strain on both resources and landfill disposal sites.
9. Evaluate individually.
10. False

Section 5 Activities

Activity 5A

Design Problem—Manufacturing Technology

Background

A major task in developing a manufacturing system is to select and properly sequence the production operations.

Situation

Balsum Manufacturing Company has purchased the rights to a game. See **Figure 5A-1.** Golf tees or 1″ long, 1/4″-diameter dowels are used as pegs. You need five light-colored and five dark-colored pegs. This game is a modification of tic-tac-toe developed in central Africa. The rules are as follows:

Each player uses four pegs. The object of the game is to get three pegs in a horizontal, vertical, or diagonal row. Play begins with each player alternately placing a peg in a hole. If no one has completed a row when all eight pegs are in place, the second phase starts. Each player alternately moves a peg into an empty hole adjacent to one of his pegs until one player makes a row and wins.

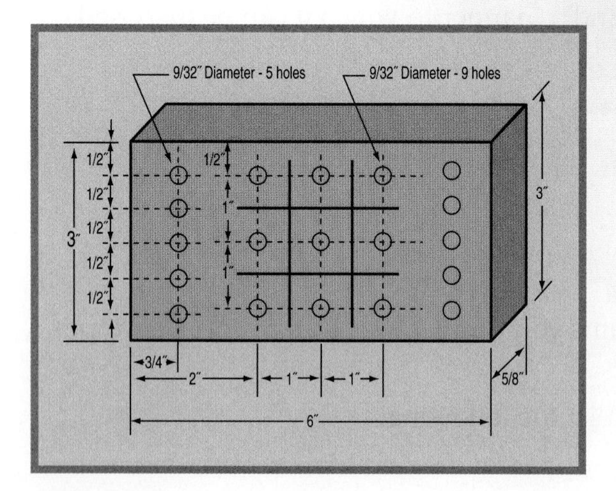

Figure 5A-1. The product drawing.

Challenge

List the operations you would use to produce the product. Set up a simple production line to make one product for each member of the class.

Safety

Use the safe and appropriate procedures your teacher demonstrates.

Activity 5B

Fabrication Problem—Manufacturing Technology

Background

Manufacturing adds worth to materials by changing their form. Many different techniques can do this.

Challenge

Study the drawing and operation sheet for the recipe-card holder. See **Figure 5B-1.** The recipe cards hang from the dowel on split key rings. Build a holder individually or use a continuous production line.

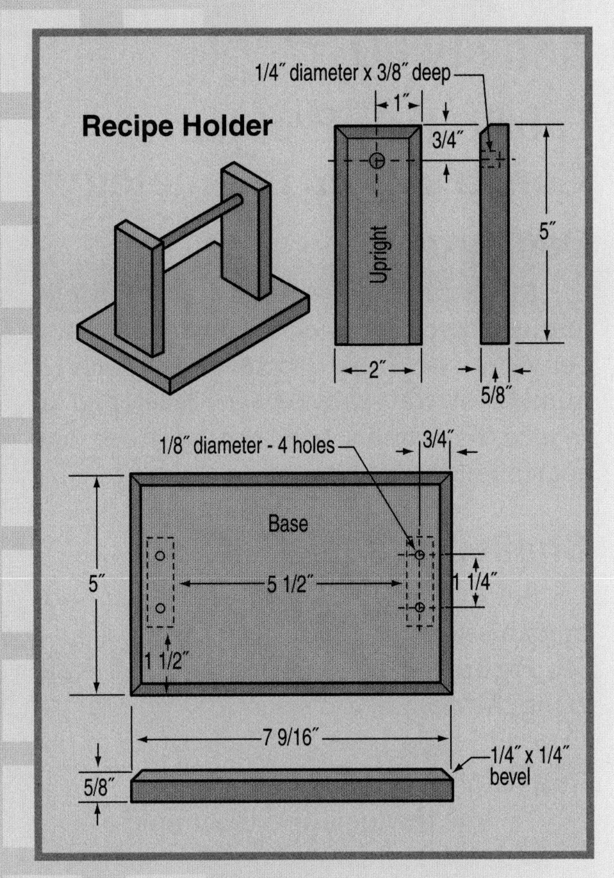

Figure 5B-1. The product drawing.

Safety

Be sure to use the safe and proper practices your teacher demonstrates.

Materials and Equipment

- 5/8″-thick × 5″-wide, random-length pine (the base).
- 5/8″-thick × 2″-wide, random-length pine (the uprights).
- A 1/4″-diameter birch dowel.
- 3″ × 5″ note cards.
- Split key rings.

Procedure

Use the procedure shown on the operation-process chart. See **Figure 5B-2.** The parts are finished before assembly. This makes it easier to produce a high-quality finish.

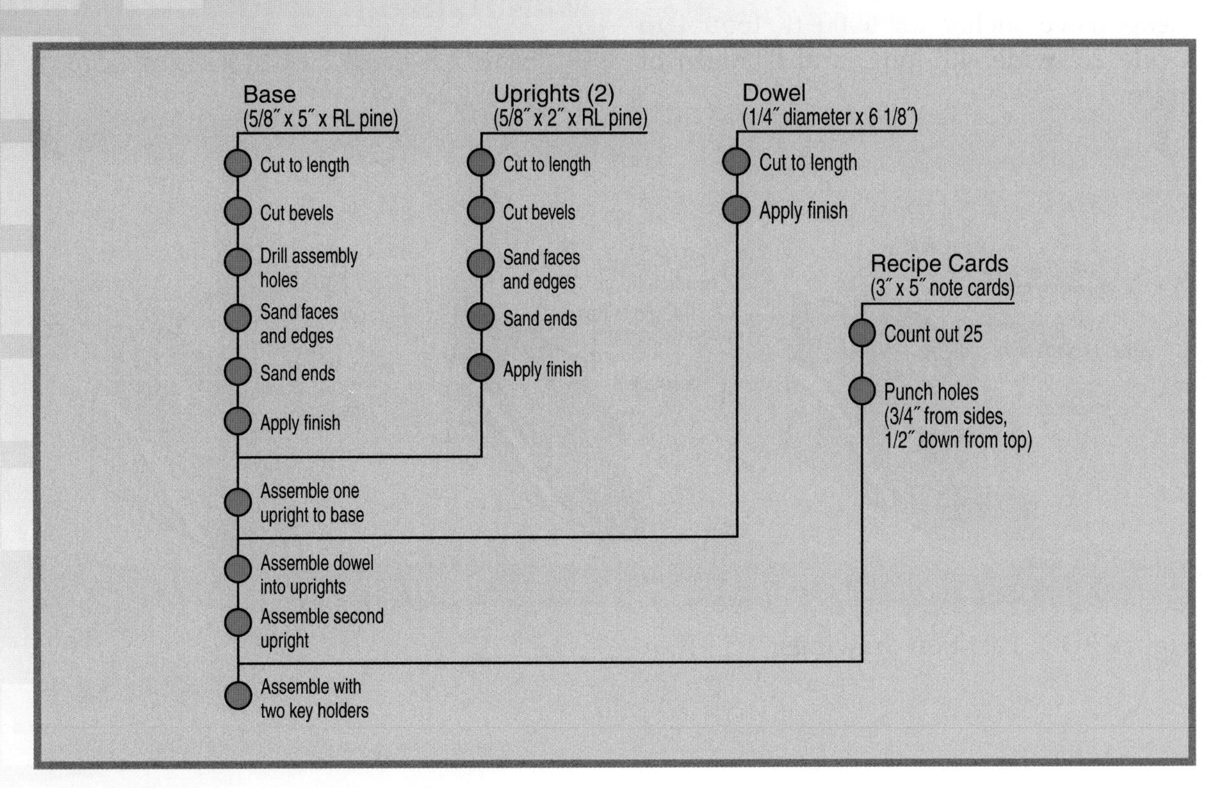

Figure 5B-2. The operation-process chart for the recipe holder

Activity 5C

Design Problem—Construction Technology

Background

During harvest, farmers haul their grain to local and regional elevators. Often, more grain is brought in than can be processed. This extra grain is stored on the ground next to the elevator until it can be processed.

Situation

The managers of Southwest Grain Elevators expect to receive 20% more grain than they will be able to process during the month of August. In previous years, they stored the excess grain on the ground. Storing the grain in this manner results in a 10% reduction in quality.

Challenge

Design and construct a model of an inexpensive shelter for 8000 ft^3 (equal to a pile 25′ wide, 40′ long, and 8′ high) of grain.

Activity 5D

Fabrication Problem—Construction Technology

Background

More people are living in multiple-family dwellings. Sound transmission between dwellings is a major concern. A number of ways have been developed to reduce the sound transmitted from one apartment to another.

Challenge

Select one of the high sound-transmission class (STC) partition designs. See **Figure 5D-1.** Construct a model wall, using the design you selected.

Safety

Use the tools and techniques your teacher demonstrates.

Materials and Equipment

- Scale 2 × 4 and 2 × 6 stock (scale can be 1/2 or 1/4).

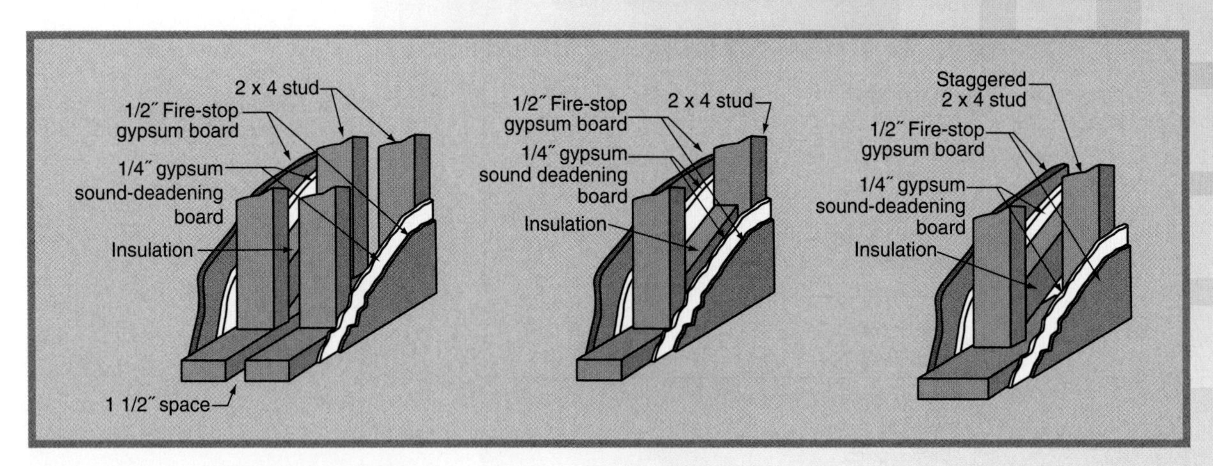

Figure 5D-1. Partition drawings.

- Scale 1/4″ and 1/2″ gypsum board.
- Scale 2 1/2″ fiberglass insulation.
- Pencils.
- Rulers.
- A backsaw.
- Brads.
- A hammer.

Procedure

1. Select the type of wall you will construct.
2. Build the model according to the design.
3. Test the sound transmission by placing a speaker on one side of the wall and a tape recorder and microphone on the other. Create a sound with the speaker. Observe the level of sound on the microphone side by using the level meter on the recorder.

Section 6
Applying Technology: Communicating Information and Ideas

Tomorrow's Technology Today

Smart Materials

What makes a smart material "smart"? It is not good grades or high test scores; smart materials are extraordinary because they have the ability to dramatically change their original state when external forces are applied. Smart materials react differently to these forces, or stimuli. No two smart materials are the same.

You might be wondering how smart materials are different from "normal" materials. A common material, such as wood, can be exposed to external stimuli such as fire. As wood burns, its properties change, and it is eventually reduced to ash. Extinguish the fire, or remove the stimulus, and the ash remains—the wood will never again return to its original state. Compare this to a smart material called *magneto-rheostatic*. Under normal circumstances, magneto-rheostatic is a liquid. However, if placed in a magnetic field, the liquid transforms into a solid—a change that occurs in less than a second. Remove the magneto-rheostatic from the magnetic field, and it is a liquid once more.

While smart materials may seem like the wave of the future, they actually have many common applications. Piezoelectric smart materials are used with the airbag sensor in your car. The force of an impact creates an electric pulse that deploys the airbag. The wire used in dental braces is a shape memory alloy whose form changes with temperature. The wire change follows a predetermined pattern designed for each patient's teeth. Over time, the shape memory alloy slowly corrects imperfect teeth. When designing new products, smart materials can be more easily manipulated to serve a specific purpose than normal materials.

For example, shape memory alloys change their shape when exposed to extreme temperature change. Some shape memory alloys also possess pseudo-elasticity, a quality that enables shape change through the application of force or pressure. Pseudo-elastic shape memory alloys have been used in eyeglass design for many years. Special frames are marketed as "bendable"—good for accident-prone people whose normal frames break easily. When these special frames are bent, or force is applied to them, they adapt to the pressure. Once the force is removed, they return to their original shape. What seems like a magic pair of glasses is really the application of a smart material.

As our understanding of smart materials grows, scientists will discover even more impressive applications. It is hard to imagine what revolutionary new products they will discover in the future!

Discussion Starters

Explain to your students how smart materials work. Discuss the differences between smart materials and conventional materials.

Group Activity

Have the class imagine the ways society might change when smart materials are more commonly used. Ask the class if they can think of any drawbacks to the implementation of this technology.

Writing Assignment

Smart materials have been implemented in small ways in recent years. Have your students write a brief paper giving examples of smart materials currently in use and explaining how using smart materials may have improved those products.

Chapter Outline

This chapter covers the benchmark topics for the following Standards for Technological Literacy:

Learning Objectives

After studying this chapter, you will be able to do the following:

- Recall the definition of *communication technology.*
- Summarize the characteristics of communicated items.
- Recall the three goals of communication.
- Recall features of the communication model.
- Recall the types of communication.
- Recall the five types of communication-technology systems.

Key Terms

channel
communication technology
edutainment
emotion
entertain
human-to-human communication
human-to-machine communication
idea
inform
infotainment
interference
Internet
machine-to-human communication
machine-to-machine communication
noise
persuade
photographic communication
printed graphic communication
receiver
technical graphic communication

Strategic Reading

Before you read this chapter, list all the types of communication you are familiar with, and give an example of each. Add to your list, if necessary, as you read the chapter.

Earlier in the text, we looked at data, information, and knowledge. You discovered that data are unorganized facts, information is organized data, and knowledge is information applied to a task. How do you obtain information? From where does it come? These are important questions.

Think back over the day. Did you read a newspaper or listen to a radio newscast? Did you see a traffic signal, or sign? Did you look at a speedometer? Did you listen to music or watch television? If you did any of these, you were using communication technology. ***Communication technology*** is a system that uses technical means to transmit information or data from one place to another or from one person to another.

Communication has always been an essential part of human life. Humans first communicated with gestures and grunts. Later, they developed language. Language increased their ability to communicate. These forms of communication, however, do not involve technology. There is no technical means between the sender of the message and the receiver.

Possibly the first use of communication technology was cave paintings. The "artists" used sticks, grass, or their fingers to apply paint to cave walls. See **Figure 19-1.**

Figure 19-1. A reproduction of a painting found in an ancient kiva (a Pueblo Indian structure) in the southwest part of the United States.

The result was a message that could be stored. At some later date, a person could retrieve the message.

Over time, communication technology has expanded in range and complexity. To understand communication technology better, in this chapter, we discuss five aspects of communication. These aspects are the major kinds of communicated items, the goals of communication, the communication model, types of communication, and communication systems.

Standards for Technological Literacy

17

Research

Have the students investigate how people have communicated over the different centuries.

Extend

Have the students search the Internet for pictures of ancient cave and tomb paintings. If the students have access to printers, have them bring in prints of the paintings.

TechnoFact

All people use some form of communication system on a daily basis. Private communication systems allow one individual to communicate with another individual, as in the case of the telephone. Mass communication systems allow a person to communicate with a large number of people simultaneously, as in the case of broadcast radio or television.

Career Corner

Marketing, Sales & Service

Advertising

Careers in advertising include advertising specialists, copywriters, media planners, media buyers, creative directors, and public-opinion researchers. Jobs in advertising require good oral- and written-communication skills. Many people in the advertising field have college degrees in business, journalism, or mass communication. In college courses in advertising, people learn about the research used in developing advertising programs, how advertising campaigns are produced, about the role of advertising in marketing, and related computer skills.

Standards for Technological Literacy

17

Example

Show the class a printed advertisement and discuss its ability to inform, persuade, and entertain readers.

TechnoFact

The development of writing was a tremendous step forward for humankind. This system allowed people to encode and save messages for later retrieval. The meanings of verbal messages can be distorted as the messages pass through the population. However, once a message is written down, it remains the same regardless of the number of people that read it and pass it on.

Communicated Items

Communication is used for a number of purposes. Most often, it is used to convey ideas, exchange information, and express emotions. See **Figure 19-2.**

Ideas

An ***idea*** is a mental image of what a person thinks something should be. You probably have ideas about what kind of music is good, how people should behave, or what activity is fun. Also, you probably have opinions on how to perform various tasks, such as washing an automobile, riding a bicycle, and mowing a lawn. People also have ideas on such issues as how to protect the environment and whether or not to allow capital punishment. Communication media can be used to share these ideas.

Information

Information is vital to taking an active part in society. This data provides a concrete foundation for decision making and action. Information can be as simple as the serving time for lunch. This data can be as complex as the Moon's effect on the tides.

Emotions

Finally, you might want to communicate ***emotions***. Ideas and information are important. Feelings, however, are just as vital to many people. Communication media can convey these feelings. For example, a photograph can communicate the excitement of a sporting event. People can communicate affection for each other through greeting cards. Communication media can make us laugh or cry, be excited or calm, or feel good or bad.

Goals of Communication

Each communication message is designed to impact someone. The communication can meet one or more of three basic goals:

- To ***inform***, by providing information about people, events, or relationships. We read books,

Figure 19-2. Communication technology can be used to convey ideas, information, or emotions.

magazines, and newspapers to obtain information. Radio news programs, television news programs, and documentaries are designed to provide information.

- To ***persuade*** people to act in a certain way. Examples include political-campaign advertisements and the "Give a Hoot, Don't Pollute" campaign. Print and electronic advertisements, billboards, and signs are typical persuasive communication media.
- To ***entertain*** people as they participate in or observe events and performances. Television programs, movies, and novels are common entertainment-type communication.

These three goals can be merged. Two new words in our language arise from this merging of goals. The first is *infotainment*. ***Infotainment*** means "providing information in an entertaining way." You might learn as you watch a quiz show on television or use a computer simulation. Both of these are enjoyable ways of gaining new and useful information.

The second term, ***edutainment***, takes communication one step beyond infotainment. Edutainment is more than allowing the information to be available in an entertaining way. This term creates a situation in which people want to gain the information. The television program *Sesame Street* is a good example of edutainment.

Standards for Technological Literacy

17

TechnoFact

Around 2000 B.C., Egyptians developed papyrus scrolls, a major advancement in writing technology. This writing material, made from the fibers of the papyrus plant, replaced the bulky clay tablets of the time.

The Communication Model

Communication can be thought of as a simple process. See **Figure 19-3.** The action starts with the encoding of a message.

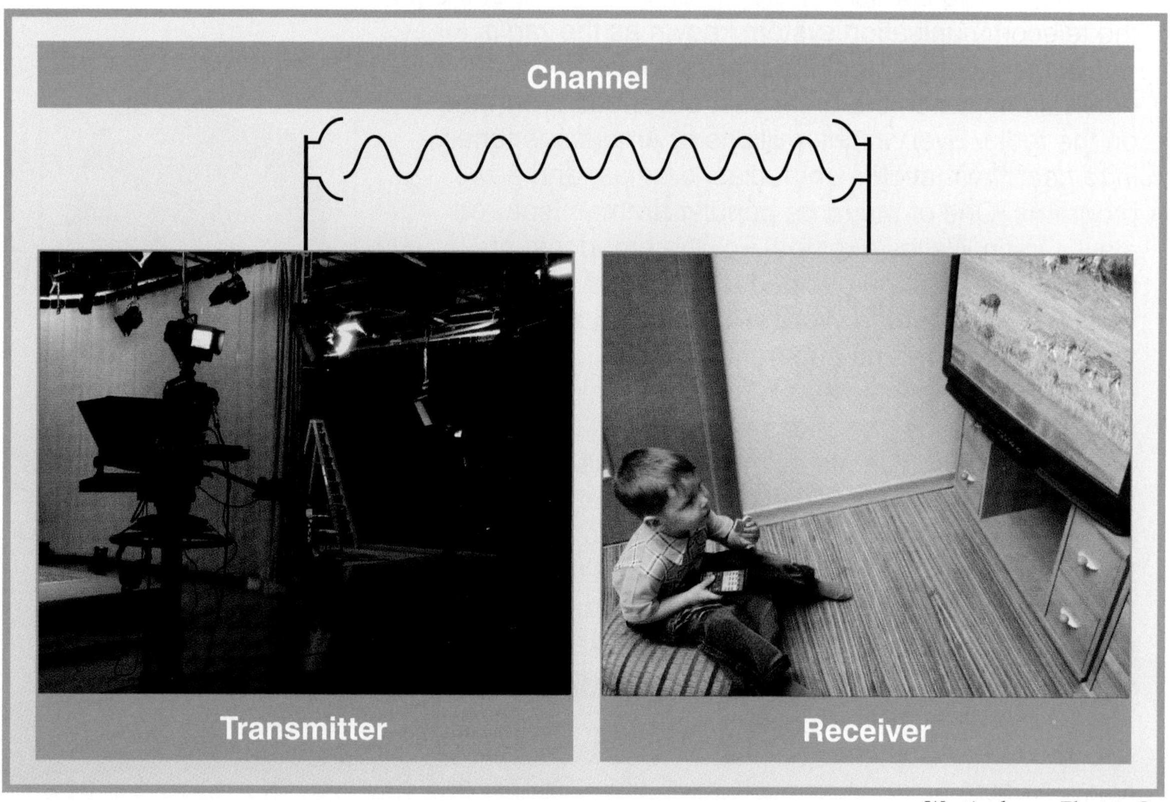

Westinghouse Electric Co.

Figure 19-3. All communication systems have transmitters, channels, and receivers. This shows the television communication system.

Standards for Technological Literacy

3 17

TechnoFact

In 59 B.C., Julius Caesar ordered the regular publication of what may be considered the first newspaper, the Acta Diurna. The Acta Diurna was written on a large white board in a public area and contained information about current events, senate proceedings, births, deaths, marriages, financial news, and special interest articles. Some entrepreneurs had slaves copy the news onto paper sheets that were then distributed to paying subscribers.

Encoding means that the message is changed into a form that can be transmitted. The encoding can be a series of bumps on a CD or DVD, an arrangement of letters on a printing plate, or exposed chemicals on photographic film.

The message is then transmitted to the receiver. Transmission involves a communication channel or carrier. The ***channel*** might be electromagnetic waves broadcast through the air, electrical signals carried by a wire, pulses of light on fiber-optic cable, or printed text on paper.

The other end of the communication channel is the ***receiver***. The receiver gathers and decodes the message. Examples of receivers are radios and television sets. Radios change electrical impulses into sound. Television sets change electrical impulses into a series of images. The human mind decodes the written and graphic messages contained in photographs and printed media.

Communication models can be more complex than the one shown in **Figure 19-3.** These models show interference in the

Academic Connections: Communication

The Power of Radio

The author H. G. Wells wrote one of his most popular science fiction books, *The War of the Worlds*, in 1898. The book is about an invasion from Mars. Little did Wells know the uproar the book would cause 40 years later, as a result of the power of the telecommunication system known as the *radio*.

In the 1930s, one of the more popular forms of entertainment was the broadcast of various programs on the radio. Every night, millions of Americans tuned in to hear their favorite comedies, dramas, and news programs. One of the more popular dramas featured plays Orson Welles directed. For the broadcast of October 30, 1938, he decided to feature an adaptation of *The War of the Worlds*. In adapting the book for the radio, however, he apparently wanted to add more drama. He made the play sound more similar to a news broadcast. Welles had fake news bulletins interrupt the play's music. Various actors read these bulletins in dramatic tones. The "bulletins" warned that Martians had invaded New Jersey and were intending to destroy the United States. Even though Welles had noted before the program that this was just a play, many had not heard the disclaimer. The program sounded so realistic that some people thought an actual invasion had occurred. They panicked. Some jumped into their cars to run. Others hid in their basements. Still others put wet towels on their heads to avoid the Martians' poison gas.

An old radio.

The disturbance caused a scandal. Some called for more government regulation. The commotion eventually died down. To this day, however, the incident shows the tremendous power involved in using technology to communicate. Can you give another example of this power?

communication channel. ***Interference*** is anything impairing the accurate communication of a message. Static on a radio is interference. Noise in a movie theater is interference. Smudged type is interference in printed messages.

It is important to be aware of the difference between information and noise. ***Noise*** is unwanted sounds or signals that become mixed in with the desired information. When you listen to the radio, both information and noise are present. In this example, they are both in the form of sound. The information is the sound you want to hear. The noise is the sound you do not want to hear. Noise, then, is a type of interference. Also, noise can involve personal taste. Some people call the music you listen to *noise*. You might feel the same way about the music they prefer. Both of you are correct because unwanted sound is noise.

Types of Communication

One way to look at communication is in terms of the sender and receiver. See **Figure 19-4.** We are all familiar with people communicating with people. This type of

Standards for Technological Literacy

17

Figure Discussion

What are examples of each of the four types of communication in Figure 19-4 that you use?

TechnoFact

In ancient times, communication was limited by the population's ability to read and write and the cost of writing supplies, such as papyrus scrolls and parchment (treated animal skin).

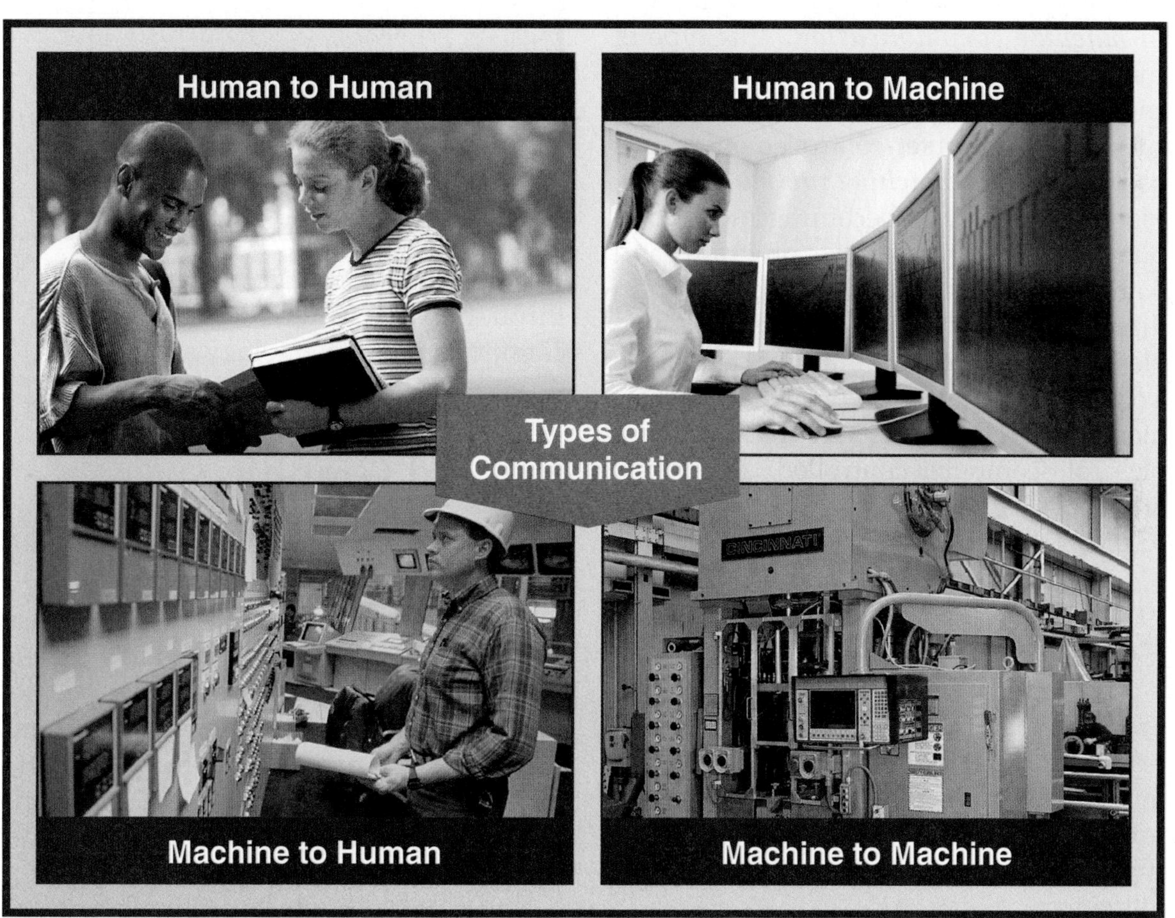

United Parcel Service; Siemens; Inland Steel Co.; Cincinnati, Inc.

Figure 19-4. The type of communication is based on the sender and the receiver. Human-to-human communication includes relaying information to others through printed media. Human-to-machine communication includes relaying information to a computer. Machine-to-human communication includes checking a readout from a plant system. Machine-to-machine communication includes controlling a manufacturing system through a computer.

Extend

Discuss the evolution of four types of communication in the following order: human-to-human, human-to-machine, machine-to-human, and machine-to-machine.

Research

Assign one of the five basic types of communications shown to Figure 19-5 to each student and have him or her find an example or picture that represents it.

TechnoFact

Mass communication began with the invention of the printing press in 15th-century Europe. Because book and other manuscripts could be mass-produced on a printing press, information became available to a greater segment of the population.

communicating is very common and works through our electronic media and printed products. This type of communication is called ***human-to-human communication***. Human-to-human communication is used to inform, persuade, and entertain other people.

Other types of communication also exist. Have you ever reacted to the bell indicating the end of a class period? If so, you have participated in ***machine-to-human communication***. This type of communication system is widely used to display machine operating conditions.

Have you keyed material into a computer or set the temperature on a thermostat? If so, you have engaged in ***human-to-machine communication***. This type of communication system starts, changes, or ends a machine's operations.

Finally, computer-controlled operations use ***machine-to-machine communication***. Modern industry is becoming more computer based. Humans enter programs and data into the computer. The computer then directs and controls an apparatus. Examples of this type of communication include a computer controlling a printer and a thermostat controlling a furnace. Typical examples of computer-controlled operations are CAD, computer-aided manufacturing (CAM), CIM, and robotics.

Communication Systems

We are bombarded with information every day. The information comes in many printed and electronic forms. Still, the communication technology used to deliver the information can be divided into five main types. See **Figure 19-5.** These types are the following:

- Printed graphic communication.
- Photographic communication.
- Telecommunication.
- Technical graphic communication.
- Computer and Internet communication.

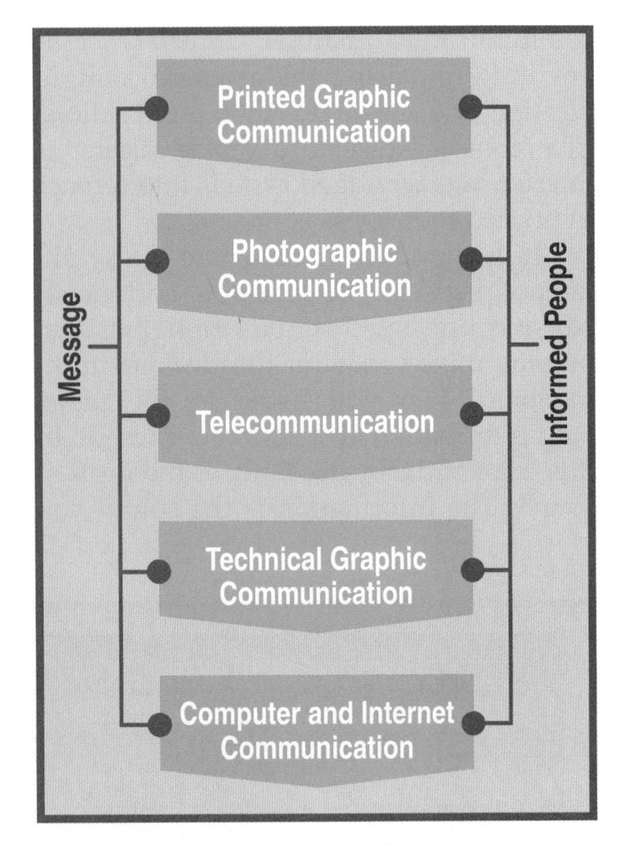

Figure 19-5. We receive information through five basic types of communication-technology systems.

Printed Graphic Communication

Much of communication technology was developed to satisfy the need for mass communication. People wanted to tell their messages to large numbers of other people. The first mass-communication system developed was printing, or ***printed graphic communication***. See **Figure 19-6.** In Chapter 7, you were introduced to movable type, the foundation of printing.

The term *printing* originally meant "putting an image on paper with inked type." Printing now includes all the processes used to reproduce a two-dimensional image on a material. The material can be paper, metal, plastic, cloth, or wood. These processes are discussed further in Chapter 20.

Standards for Technological Literacy

17

TechnoFact

The development of the first telegraph system (1844) allowed the nearly instantaneous transmission of information. The development of the telegraph was quickly followed by the invention of the telephone in 1876. Marconi's 1895 invention of the radio introduced the era of wireless communications. The first commercial radio station began operation in 1920.

Technology Explained

digital video disc (DVD): an high-capacity storage format used for data storage for video.

Storage media are always being improved. Commercially available digital storage media began with compact discs (CDs). CD systems use a laser to read optically encoded information, changing the optical signals into digital signals before re-creating the information.

CDs were first used for audio recordings before being used to store data. In much the same way, digital video discs (DVDs) were first used for video before becoming a common storage medium. DVDs have a higher storage capacity than CDs. Part of the reason DVDs can have more than 25 times the amount of storage is the way information is recorded onto the disc. The pits imprinted on the disc of a CD are longer and not packed closely together. Because of the way DVD's are read, they can have smaller and greater amounts of pits packed more closely together. See **Figure A.**

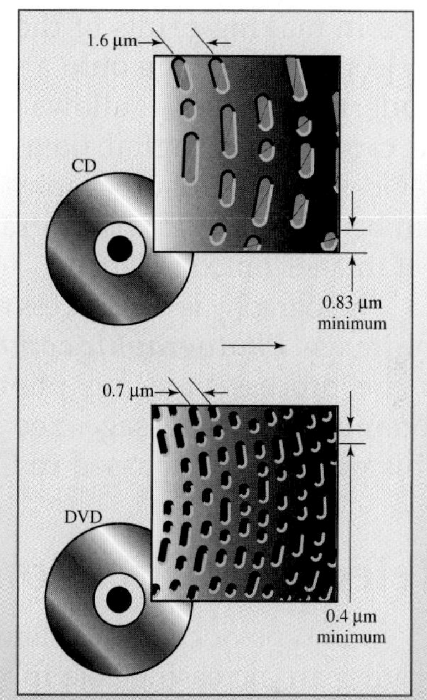

Figure A. The shorter wavelength of a DVD laser allows for more compressed data.

DVD players are similar to CD players. The physical size and shape of the discs is typically the same, and both are optically read using lasers. The laser used to read DVDs is different from the laser used on CD players. A wavelength that is near infrared is used on CDs, while a shorter wavelength is used on DVD players. However, several DVD players can read both DVDs and CDs. See **Figure B.**

DVDs are now also used for data storage. A recordable DVD may hold up to 17 GB of information, whereas a CD can typically store up to 700 MB. The higher-capacity DVDs are dual-layer discs and have twice the storage capacity as typical DVDs. However, most DVDs are single-layer discs and can store only 4.7 GB of data.

Figure B. A typical DVD player can play DVDs and CDs.

Another optical storage medium that functions similarly to the DVD is the Blu-ray disc. The Blu-ray disc is commonly used for video. It is typically the same physical size and shape of a DVD, but its much higher storage capacity allows for high-definition video to be recorded onto it. Blu-ray discs are read the same way CDs and DVDs are read, using a laser of a shorter wavelength to read and write the data. As with the DVD, the shorter laser allows for data to be more compressed, allowing for its higher storage capacity. Dual-layer Blu-ray discs may hold up to 50 GB of information. Another difference between DVDs and Blu-ray discs is that Blu-ray players with an Internet connection can access more features of a video.

Standards for Technological Literacy

17

Figure Discussion

What emotions or feelings does Figure 19-7 communicate?

TechnoFact

Mass communication was given a boost by the invention of photography in the 1830s and motion pictures in the 1890s. These technologies allowed people to capture images and share them with people around the world. Cornerstones in the development of modern mass communication include the introduction of the television (1930s) and the formation of the Internet (1960s).

Photographic Communication

Some communication technologies use light to convey their messages. The most common of these is photography. Photography captures light as an image on a recording medium we call *film*. When the film is developed, the image becomes permanent. The film can be used in making prints of the image or in projecting the image onto a screen. More modern technology allows the light to be captured as digital data on memory devices (digital photography). This data can be used to print photographs or project images onto screens.

Photography is the process of producing the image. ***Photographic communication*** is the process of using photographs to communicate a message. See **Figure 19-7.** This process is discussed in Chapter 21.

Figure 19-6. Books are an example of printed graphic communication media.

Figure 19-7. This photograph communicates specific emotions.

Telecommunication

One type of communication system is playing an increasing role in modern life. This system is telecommunication, communication at a distance. See **Figure 19-8.** Telecommunication includes a number of specific types of communication. Probably the most widely used are radio, television, and the telephone. Telecommunication is the subject of Chapter 22.

Technical Graphic Communication

People and companies often want to communicate specific information about a product or the product's parts. This information might convey the size and shape of a part, suggest how the parts are assembled to make a product or structure, or indicate how to install and operate a product. Also, it might tell how to adjust and maintain the product. This type of information is

Figure 19-8. These customer service representatives are using telecommunication to help customers.

Standards for Technological Literacy

5 17

TechnoFact

A drawing shows the viewer how an object should look and communicates shape and sizes (dimensions). Engineering drawings provide the information needed to produce an object, structure, or system.

Think Green

Forest Stewardship Council

Paper products are used for a variety of reasons, and consequently, a great deal of it is used. Today, most paper is recycled. While recycling is important, another consideration for using paper should be where the paper comes from. The Forest Stewardship Council (FSC) is an organization that has created standards meant to lessen the environmental impact of practicing forestry. The FSC certifies specific forests to be used to create wood products and paper.

For companies wishing to use certified materials, the FSC issues a chain-of-custody certificate to verify the materials came from an FSC-certified location. This chain-of custody certificate allows companies to be more responsible with materials by giving them the ability to track the origin of the materials. The safe and responsible handling of the materials may also be checked with the chain-of-custody certification. As a result of using certified materials, companies may be printed to say they use FSC-certified materials.

Standards for Technological Literacy

17

Extend

Discuss the difference between artistic drawing and engineering drawing.

TechnoFact

An artistic drawing can be created with line or tone. The images an artist produces may help them develop other works, such as paintings or sculptures. Drawings may also be created as works of art in their own right.

often communicated through engineering drawings or technical illustrations. See **Figure 19-9.** The methods that prepare and reproduce this media are called *technical graphic communication systems*. ***Technical graphic communication*** is discussed at length in Chapter 12.

Ohio Art Co.

Figure 19-9. This designer is preparing technical illustrations for a new product.

Computer and Internet Communication

A new system of communication is rapidly changing how people gather and use information. This system is based on a key invention, the personal computer. The new type of communication uses a network of computers to share information. This network is called the ***Internet***. The Internet allows people to gather information using the World Wide Web and exchange messages using electronic mail. Using the Internet for communication is discussed in Chapter 23.

With the development of new technology, people have found new ways to communicate with one another.

Answers to Test Your Knowledge Questions

1. Communication technology is a system that uses technical means to transmit information or data from one place to another or from one person to another.
2. Ideas, information, and emotions
3. Inform, persuade, and entertain
4. True
5. noise
6. Human-to-human, machine-to-human, human-to-machine, and machine-to-machine
7. A. Printed graphic
8. C. Telecommunication
9. B. Photographic
10. E. Internet
11. D. Technical graphic
12. B. Photographic
13. D. Technical graphic
14. E. Internet
15. C. Telecommunication

Summary

We are now living in a new period of history called the *information age.* This new age is characterized by the wide-scale availability of communication media. People receive and send a constant barrage of data and information every day. Each communication message is designed to inform, persuade, or entertain or to achieve some combination of these goals.

Transfer of communication is done in terms of human-to-human, human-to-machine, machine-to-human, and machine-to-machine communication. Each of these systems has a transmitter that encodes and sends the message. The message then travels over a carrier to a receiver. The receiver collects and decodes the message. The most common forms of communication systems are printed graphic communication, photographic communication, telecommunication, technical graphic communication, and computer and Internet communication. Our lives are affected daily through these systems.

Test Your Knowledge

Write your answers on a separate piece of paper. Please do not write in this book.

1. What is communication technology?
2. What are the three major items we try to communicate?
3. List the three goals of communication.
4. All communication systems have transmitters, channels, and receivers. True or false?
5. An unwanted sound or signal is called ______.
6. List the four types of communication.

Matching questions: For Questions 7 through 15, match each example on the left with the correct communication system on the right. (Note: Answers can be used more than once.)

7. ______ Newspaper.
8. ______ Radio.
9. ______ Motion picture.
10. ______ Electronic mail.
11. ______ Blueprint of a house.
12. ______ Vacation photo.
13. ______ Parts drawing in an owner's manual.
14. ______ Web page.
15. ______ Television.

A. Printed graphic communication.
B. Photographic communication.
C. Telecommunication.
D. Technical graphic communication.
E. Internet communication.

STEM Applications

1. Select a piece of information or an emotion you want to communicate. List four ways you can communicate it and the communication system you would use (printed graphic communication, photographic communication, telecommunication, technical graphic communication, or Internet communication).
2. Design a communication message persuading a person to your point of view on an issue.

Chapter 20 Printed Graphic Communication

Chapter Outline

Printing Methods

Steps in Producing Printed Graphic Messages

Computer-Based Publishing

This chapter covers the benchmark topics for the following Standards for Technological Literacy:

Learning Objectives

After studying this chapter, you will be able to do the following:

- Summarize the characteristics of printed graphic communication.
- Give examples of products that printed graphic communication processes produce.
- Recall the six major types of printing processes.
- Recall the steps involved in designing a printed graphic message.
- Recall the principles of design.
- Explain how printed messages are prepared for production.
- Explain how printed graphic messages are produced.
- Recall activities involved with finishing printed graphic communication products.
- Summarize the characteristics of computer-based publishing.

Key Terms

audience assessment
balance
comprehensive layout
contrast
copy
desktop publishing
electronic publishing
electrostatic printing
flexography
gravure
harmony
illustration
illustration preparation
image carrier
ink-jet printing
layout
letterpress
lithographic printing
offset lithography
pasteup
printing
proportion
relief printing
rhythm
screen printing
silk screening
substrate
thumbnail sketch
typesetting
variety

Strategic Reading

Before you read this chapter, make a list of the different types of printed communications you are familiar with. As you read the chapter, revise your list if necessary and list the steps involved in creating each type.

Communication means many things to many people. We talk about communication messages, communication media, communication networks, and communication systems. Central to all these is the word *communication*. Simply stated, communication is the passing of data and information from one location to another or from one person to another. When the spoken word is used to communicate, we call it *language*, or *verbal communication*. When a technical means is used to convey information, it is called *communication technology*.

As you learned in the last chapter, a number of communication systems use technology. The oldest of these systems is printed graphic communication, or ***printing***. Most printed communications use an alphabet to convey the message. An alphabet is a series of symbols developed to represent sounds. The first alphabet was developed in Syria around 1200 BC. Today, various alphabets exist, including the 26-letter one used for the English language. These letters can be arranged to convey a message through writing.

The process of printing takes written words and places their images on a material called the ***substrate***. Originally, almost all printing was done on paper. Today, printing is done on a variety of substrates, including paper, glass, plastic, cloth, ceramics, metal, and wood. The result is a broad range of printed products, including newspapers, magazines, books, brochures, pamphlets, labels, stickers, clothing designs, and signs. Each of these is carefully designed and produced by a specific printing process. In this chapter, we look at printing methods, the steps in producing printed graphic messages, and computer-based publishing.

Printing Methods

Human beings first printed by carving an image into a block of some material. Next, they covered the block with ink. They then pressed the inked block on paper to produce the "printed word." This technique was the forerunner for all printing processes. See **Figure 20-1.** Today, six

Figure 20-1. This museum exhibit shows early block printing.

Standards for Technological Literacy

9 17

Research

Assign the students to find out when and where printing started.

Brainstorm

Ask your class why printing is important to modern society.

TechnoFact

In addition to being an essential communication medium, printing is also an important cornerstone in all democracies. It is an irreplaceable means of educating the populace about important topics, allowing them to make informed decisions.

Standards for Technological Literacy

9 17

Demonstrate

In class or lab, use a rubber stamp to demonstrate relief printing.

TechnoFact

Books have evolved over thousands of years. The earliest forms were printed on clay tablets. This medium was impractical for texts because the printing process was slow and the texts were hard to read. In addition, the texts created on this medium were brittle and difficult to store. Despite this, several texts of this type have survived. Perhaps the oldest surviving example of this type of text is a poem printed on twelve tablets for an Assyrian king about 2650 years ago.

major printing processes exist. These processes are the following:

- Relief printing.
- Lithographic printing.
- Gravure printing.
- Screen printing.
- Electrostatic printing.
- Ink-jet printing.

Relief Printing

Relief printing is the oldest of all printing processes. This printing uses an image on a raised surface. See **Figure 20-2.** Ink is applied to this raised surface, and it is pressed against the substrate. The pressure forces the ink to adhere to the substrate, producing the printed message.

This printing requires the image to be reversed, or "wrong reading," on the printing block, or ***image carrier.*** Once the inked image has been pressed against the substrate, the message is forward, or "right reading." Look at a rubber stamp, and you see an example of a reversed, or wrong-reading, relief image. The two main types of relief printing are letterpress and flexography.

Letterpress

Letterpress uses metal plates or metal type as the image carrier. Until the last 40 years or so, letterpress was the most common printing process. Letterpress now accounts for a very small percentage of all printing.

Flexography

Flexography is an adaptation of letterpress. This type of printing uses a rubber or plastic image carrier. Flexography is widely used for printing packaging materials.

Lithographic Printing

Lithographic printing, or offset lithography, is the most widely used method of printing today. This process prints from a flat surface. See **Figure 20-3.** ***Offset lithography*** is based on the principle that oil and water do not mix. Alois Senefelder discovered this printing method. Senefelder used a limestone slab as his image carrier. He first wrote a reverse, wrong-reading, image on the stone with a grease pencil. Next, he wet the stone, which dampened the stone surface, except for the

Career Corner

Arts, A/V Technology & Communications

News Reporters

News reporters gather information, prepare stories, and make broadcasts that inform an audience about recent events. Reporters investigate leads, read documents, observe events, and interview people. They take notes, organize the material, determine the story's focus, write the story, and edit accompanying video material.

Most employers prefer individuals with a bachelor's degree in journalism or mass communications. Many larger television stations prefer candidates with a degree in a subject-matter specialty, such as economics, political science, or business. Most large broadcasters hire only experienced reporters.

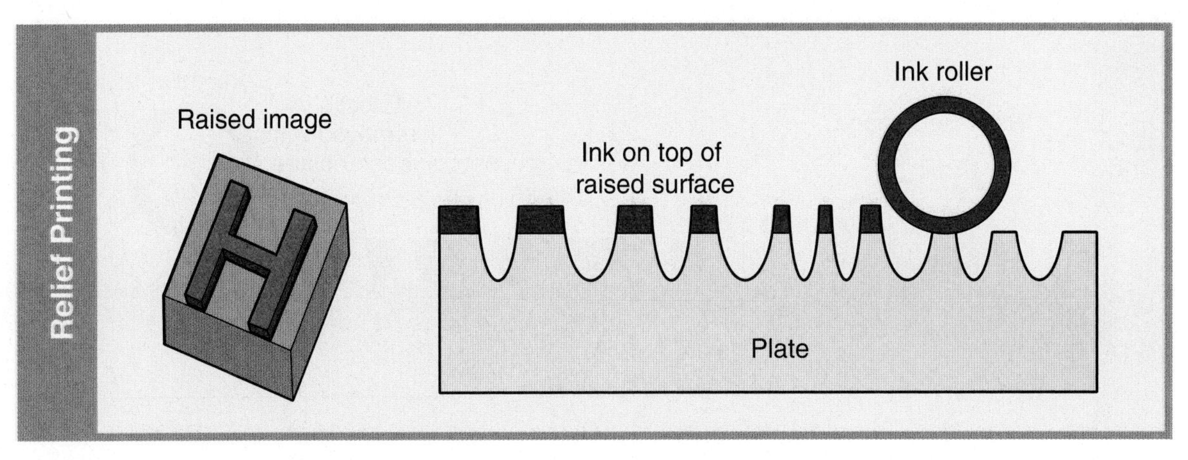

Figure 20-2. Relief printing. Note the raised printing surface.

Figure 20-3. Lithographic printing. Note the flat printing surface of this method.

Standards for Technological Literacy

9 17

Figure Discussion

Compare and contrast the printing processes shown in Figure 20-2 and Figure 20-3.

TechnoFact

The history of printing technology involves two major technological breakthroughs, the development of relief printing and the invention of paper. The earliest form of relief printing was born when people began pressing carved designs into clay. The Chinese invented the other major breakthrough, paper, more than 2000 years ago.

areas with grease marks. He then rolled ink over the stone surface. The ink adhered to the grease-marked areas, leaving an inked image, which could be transferred to paper. Finally, he placed paper over the stone and applied pressure, causing a direct transfer of the image from the carrier to the substrate. Today, some artists still use Senefelder's basic process to produce prints of their work.

The basic lithographic process has been refined. A right-reading image is produced on an image carrier called an *offset plate.* The inked image on the plate is first transferred to a special rubber roller called an *offset blanket*. This reverses the image, but the image becomes "right reading" again when it is transferred to paper as it is fed through the press. The process is called *offset lithography* because the original image is first offset to a blanket. Modern offset presses can produce more than 500,000 copies from a single offset printing plate.

Gravure Printing

Finely detailed items, such as paper money and postage stamps, are usually printed using the gravure, or intaglio, process. See **Figure 20-4.** This process is the opposite of relief printing. In gravure printing, the message is chemically etched or scribed into the surface of the image carrier. The carrier is coated with ink, and then the surface ink is scraped off with a doctor blade. This leaves ink only in the recessed areas of the carrier. When paper is pressed

Standards for Technological Literacy

9 17

Demonstrate

Show the students in class or lab how a silkscreen image can be printed on paper or fabric.

TechnoFact

Screen-printing techniques can be used to print messages on almost any substrate, regardless of its shape and size. For this reason, it is the most common method on printing designs and text on variety of products, including bottles, CDs, T-shirts, brass plaques, and computer keyboards.

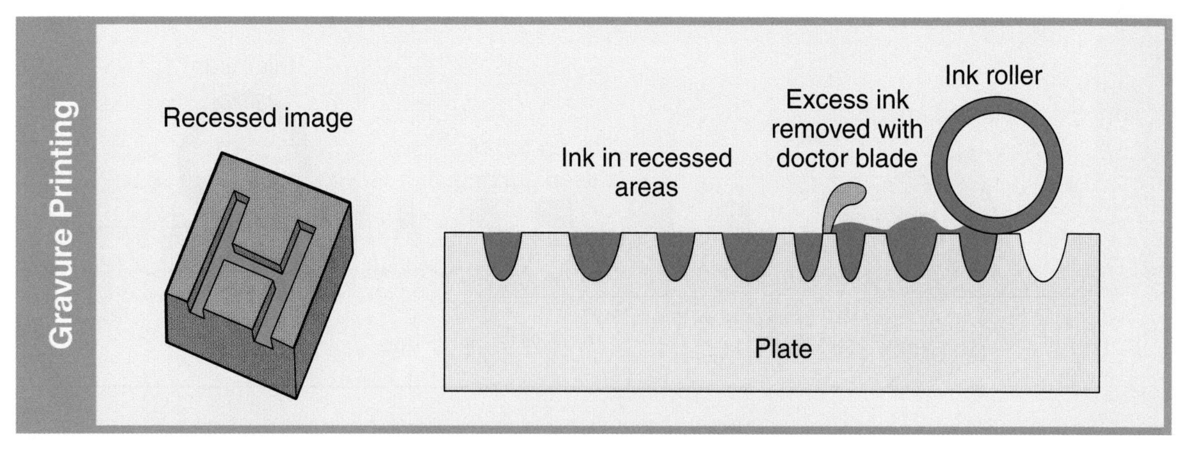

Figure 20-4. The gravure printing process. The inked image is inset into the image carrier.

very tightly against the carrier, the paper picks up the ink in the cavities.

Gravure is an expensive process. This process becomes economical for very long production runs, however, because one gravure plate can withstand several million impressions. In addition to money and stamps, some magazines are printed with this process.

Screen Printing

Screen printing is a very old printing process, dating back more than 1000 years. This printing uses a stencil with openings that are the shape of the message. The stencil is mounted on a synthetic fabric screen. Paper is then placed beneath the screen, and ink is applied to the screen's upper surface. Finally, a squeegee is pulled across the stencil, forcing ink through the openings in the stencil to produce the printed product. See **Figure 20-5.**

This printing is used to print on fabrics, T-shirts, drinking glasses, printed circuit (PC) boards, and many other products. Screen printing is also used to print small quantities of very large products, such as posters or billboards. This printing is often called ***silk screening***. Originally, the stencils were mounted on silk fabric. Currently, however, stronger and more durable fabrics have almost completely replaced silk as the screen fabric used in the process.

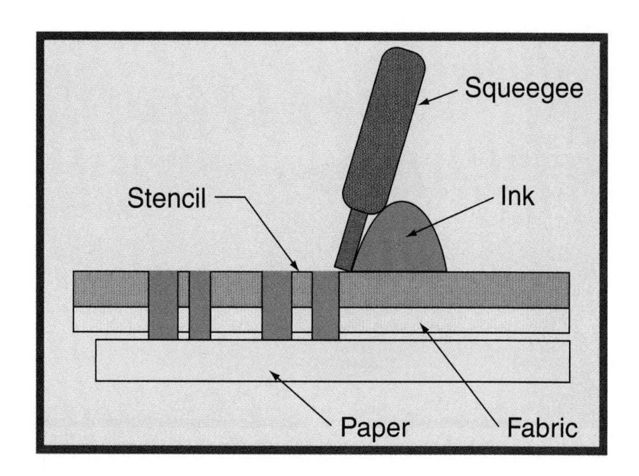

Figure 20-5. Screen printing, or silk screening.

Electrostatic Printing

A commonly used process today is ***electrostatic printing***. Electrostatic printing uses a machine with a special drum. The image to be copied is placed on this machine. A strong light is then reflected off the image and onto the drum. The reflected light creates an electrically charged likeness of the image on the drum's surface. This charge is transferred to a sheet of paper, which is then passed over fine particles called *toner*. The toner is attracted to the charged image on the paper. Finally, the paper passes through a heating unit, which fuses the toner onto the paper.

The early office copiers were fairly slow. Today, high-speed electrostatic copiers can produce several thousand copies an hour. Some of the newest electrostatic copiers can be directly interfaced with (joined to) microcomputers. The computer sends a chosen image directly to the copier. The paper original is eliminated, and even the most complex page ***layouts*** are easily and quickly reproduced.

Ink-Jet Printing

The newest printing process is ***ink-jet printing***. In this process, a computer generates the printed message. The computer then directs a special printer, which sprays very fine drops of ink onto the paper. A wide range of materials can be printed by this process because the printing head never touches the substrate. Ink-jet printing is used for coding packages and producing mailing labels. The process can use multiple printing heads, allowing the printer to produce several thousand characters per second.

The Steps for Producing Printed Graphic Messages

Printed messages and products are the results of a series of planning and production activities. These activities can be grouped into three main steps. See **Figure 20-6:**

1. Designing the message.
2. Preparing to produce the message.
3. Producing the message.

Designing the Message

Communication is successful only when the intended audience receives the given data or information. We have various ways to communicate an idea or a concept. Think of the concept *church*. See **Figure 20-7.** We can say the word and cause people to form a mental image of a church. This process, however, requires us to be near the audience. A larger audience over a greater distance can be reached if we use printed graphic communication.

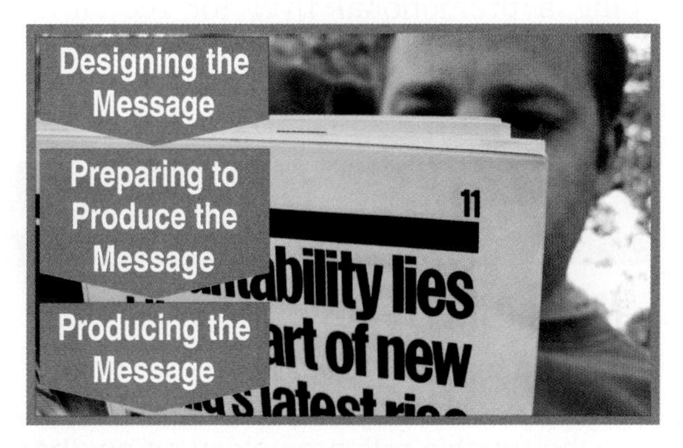

Figure 20-6. The three main steps in creating printed communication are designing the message, preparing the message, and producing the message.

Figure 20-7. We can communicate the idea "church" in several ways.

Standards for Technological Literacy

9 17

Demonstrate

Show the students how to produce copies using electrostatic and ink-jet printers and an electrostatic (photo) copier. Compare the results in class.

TechnoFact

Printing technology as we know it today is less than 600 years old. Prior to the invention of the movable type printing press, all books were either handwritten or printed in small quantities from hand-carved wooden blocks. The invention of the movable type printing press allowed books to be mass-produced quickly and cheaply.

Standards for Technological Literacy

3 9 17

Brainstorm

Ask the students why information gained from audience assessment is important to the message designer.

TechnoFact

Recorded history began with the invention of writing. Widespread information sharing was made possible with the rise of book, magazine, and newspaper publication.

Printed graphic communication can include additional words, which we call *text*, or ***copy***. We can also illustrate the message. ***Illustrations*** are pictures and symbols that add interest and clarity to the printed communication.

The design of each communication follows a set procedure. This process starts with an ***audience assessment***. For an audience assessment, the designer must determine, among other factors, who the audience is, what the audience likes and dislikes, and what the audience's interests are. See **Figure 20-8.** In short, graphic designers must get to know their audience. This background information helps the designer decide what information to tell and how to tell it. The designer generally has a great deal of information available. The designer's challenge is to select the useful information to include in the message. For example, suppose you are developing a promotional flyer for recycling. You first have to decide who your target

Figure 20-8. An audience assessment allows this designer to tailor her message to the audience.

STEM Connections: Mathematics

Integrated STEM Curriculum — Science, Technology, Engineering, Mathematics

Measuring Type

We learned earlier about two different measurement systems, the U.S. customary system and the SI metric system. Some of those who work in the graphic arts area also use another system, often referred to as the *American point system*. This system is used to measure such items as the size of type, the lengths of typeset lines, and the space between lines of type.

The two main units of measurement in this system are the point and the pica. A point measures approximately 1/72″. A pica is approximately 1/6″. One pica equals 12 points, and six picas equal approximately 1″.

Generally, points are used to measure the vertical height of type sizes, and picas are used to measure the lengths of lines. The vertical size of a line of type is the measurement from the tops of the letters (ascenders) to the bottoms (descenders). For example, for the word *deep*, one measures from the top of the *d* to the bottom of the *p*. Line length is measured from the beginning of the first word of the line to the end of the last word of the line.

Those who work with type use a typographer's rule to measure points and picas. After locating one of these rules, use it to measure a line of type in this paragraph. How many picas long is the line? What size type is being used?

audience is. Is it private citizens, business, industry, or government? Each of these groups has its own set of interests and behaviors. Your message has to appeal to your target audience's individual interests in order to help change the personal behavior of the audience members. It is difficult to write one message for all people. Most communication messages are targeted to a specific audience.

After getting to know your audience, you must next get them to receive the message. This prompts the "how" part of the design. To promote your recycling program, should you use a humorous appeal or an environmental-concerns appeal? A message needs an attention-getting aspect, or a hook, to have a strong chance of making an impact.

Designing the message involves two main steps, design and layout. In the design stage, the message is developed. In the layout stage, the message is put together.

Design

Graphic design deals with the appearance, or look, of the page. The design should attract readers and hold their attention because a message that is never read is useless. To develop a message, you must consider a number of design principles. See **Figure 20-9.** These principles include proportion, balance, contrast, rhythm, variety, and harmony.

Proportion

The principle of ***proportion*** deals with the relative sizes of the parts of the design. Proportion is concerned with the height-width relationships of the parts within the design. Good proportion has an eye-pleasing relationship between large and small elements within the message.

Balance

People seem to enjoy a visual ***balance***, so strong media messages are designed with this principle in mind. Balance is accomplished by having the information on both sides of a centerline appear equal in visual weight. You can achieve balance in two ways. See **Figure 20-10.** The first way is formal balance. With formal balance, both sides of the centerline are close to mirror images of each other. The second way is informal balance, in which there are equal amounts of copy and illustrations on each side of the layout, yet both sides of the

Figure 20-9. Publishers of printed graphic communications use design principles in their publications.

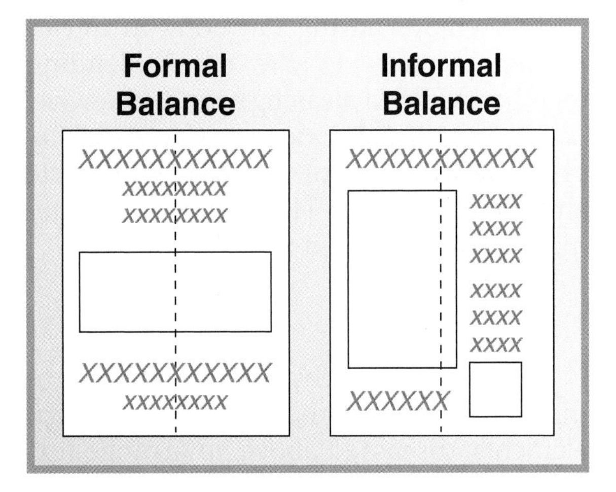

Figure 20-10. Representations of formal and informal balance. Notice how there is a similar visual weight on both sides of the centerline.

Standards for Technological Literacy

9 17

Figure Discussion

Use Figure 20-10 and pages from magazines to show formal and informal balance.

Brainstorm

Ask the students whether they like formal or informal balance in layouts. Why?

TechnoFact

To combat the counterfeiting, a new division was created within the Department of the Treasury. This division, the Bureau of Engraving and Printing, was given the job of designing, engraving, and printing valuable official documents, including stamps, currency notes, bonds, certificates, and security documents.

Standards for Technological Literacy

9 17

Demonstrate

Show how to make refined, rough, and comprehensive layouts.

TechnoFact

Type is available in three different forms: metal (hot) type, photographic type, and digital type. Metal type is composed of metal slugs with raised letters. Photographic type is the image of letters on film. Digital type is computer-stored instructions for forming each character in a font. Today, digital type is the most common form used.

centerline do not mirror each other. This layout still produces a feel of equal weight, or balance, when viewed.

Contrast

Contrast is used to emphasize portions of the message. You can achieve contrast by changing the color of the important elements. Contrast can also be achieved by printing the important elements in bold, italic, or enlarged type.

Rhythm

The principle of ***rhythm*** deals with the flow of the communication. Rhythm in an advertisement might repeat certain elements in the design. This principle produces a sense of motion and guides the eye to an important feature.

Variety

People receive thousands of messages each day. ***Variety*** is the technique that makes the message unique and interesting. For example, in a full-color publication, a black-and-white message stands out. Unique layouts also catch the reader's eye.

Harmony

We enjoy hearing harmony in music. ***Harmony*** refers to the notes blending together to form a pleasing sound. Likewise, harmony in graphic design is achieved by blending the parts of the design to create a pleasing message. The parts of a harmonized message fit and flow together.

Layout

Layout is the physical act of designing the message. This act uses the design elements, discussed above, to arrange text and pictures on a page. The positioning of the copy and illustrations forms the overall message. The layout selected ultimately determines if the printed document effectively communicates the message.

Sketches

This task of laying out a graphic communication product starts with sketching. First, the designer prepares a series of ***thumbnail sketches***. See **Figure 20-11.** These allow the graphic designer to experiment with various arrangements of copy and illustrations. The designer then uses these thumbnails in preparing rough sketches, which integrate and refine the ideas generated in the thumbnails. Rough sketches are drawn in the final proportion.

Ohio Art Co.

Figure 20-11. Typical thumbnail sketches used in designing a graphic communication.

Comprehensive layouts

Finally, the designer makes a ***comprehensive layout*** of the best rough sketch. See **Figure 20-12.** The comprehensive, or "comp," is the layout for the final design. The comp deals with several elements. One is typography. Typography is the typeface, size, and style of type that will be used in the layout. Also, the designer must size illustrations and crop them to fit into the layout. Cropping is selecting and separating the important information in a photograph. Finally, the designer must specify the colors for the various parts.

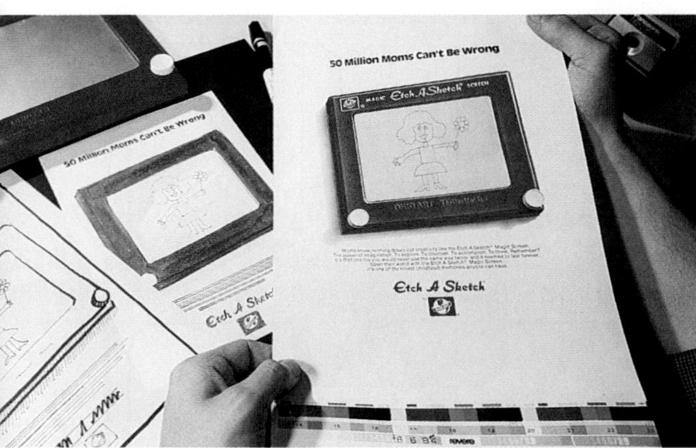

Ohio Art Co.

Figure 20-12. Comprehensive layouts for a communication product.

The large brown dog ran fast.

The large brown dog ran fast.

The large brown dog ran fast.

The large brown dog ran fast.

The large brown dog ran fast.

THE LARGE BROWN DOG RAN FAST.

Figure 20-13. The same words can be shown in many different typefaces. Designers choose a typeface on the basis of the message they are trying to convey.

Preparing to Produce the Message

After the design has been developed, it must be made ready for printing. This involves three basic steps. These steps are composition, page layout, and image-carrier preparation.

Composition

Composition starts the process of moving a graphic designer's proposed layout into a printed message. This step consists of the activities that change written words and illustrations into forms that can be used in various printing processes. In most cases, this involves text preparation (typesetting) and illustration preparation.

Typesetting

Typesetting produces the words of the message. This includes generating, justifying, and producing the type. A typeface is specified in the design and selected from hundreds of different available styles. See **Figure 20-13.** Today, the written words are usually developed with word processing and graphic design software. Whether to use letters with serifs or letters without serifs (sans serif letters) is one option in choosing a typeface. Serifs are the small flares at the ends of printed letters. **Figure 20-14** shows words printed with and without serifs. Additional groupings include traditional, contemporary, decorative, and script.

Justification (alignment of the copy) on the left, on the right, in the center, or on both edges (also called *full justification*) of the paper is another part of typesetting. Refer to **Figure 20-14.** Letters and informal messages are generally left justified, leaving an uneven right margin, or a ragged right margin. Books, magazines, and newspapers justify both edges of the copy. Artistic messages are sometimes right justified or center justified to attract attention.

Finally, the type font must be selected. Fonts include plain, bold, italic, condensed, and expanded, among other styles. Refer to **Figure 20-14.**

Illustration preparation

Illustration preparation is often required for communication media. This preparation might include sizing and converting line art or photographs. These pieces might need to be enlarged or reduced to fit the space in the layout. Other than sizing, line art often needs no preparation to be placed in a layout. Photographs, however,

Standards for Technological Literacy

9 17

Research

Assign the students to find examples of serif and sans serif type in magazine and newspaper advertisements.

Demonstrate

Show the class examples of type and justification methods.

TechnoFact

Type is grouped into four general classes: serif, sans serif, script, and decorative. Serif typefaces have small finishing strokes called serifs. Serif typefaces are frequently used in books, magazines, and newspapers. San serif typefaces, which have no serifs, are frequently used for advertisements, headings, and texts. Script typefaces resemble handwriting and are widely used in advertising. Decorative typefaces vary widely, but are generally used for their appearance rather than their legibility.

Standards for Technological Literacy

9 17

Research

Assign each student to find out how color photographs are prepared for printing.

TechnoFact

Around 1040 A.D., the Chinese began printing with carved woodblocks. In the 14th century, Korea developed movable copper type. Johann Gutenberg independently developed a similar system in Europe in the mid-15th century. The Mazarin Bible, also known as the Gutenberg Bible, was published around 1455. It is the oldest existing Western book printed with movable type.

Typestyle

Serif

Johannes Gutenberg began the printing revolution more than 500 years ago. Gutenberg's method of printing is called letterpress.

San Serif

Johannes Gutenberg began the printing revolution more than 500 years ago. Gutenberg's method of printing is called letterpress.

Left Justification

Johannes Gutenberg began the printing revolution more than 500 years ago. Gutenberg's method of printing is called letterpress.

Full Justification

Johannes Gutenberg began the printing revolution more than 500 years ago. Gutenberg's method of printing is called letterpress.

Plain

Johannes Gutenberg began the printing revolution more than 500 years ago. Gutenberg's method of printing is called letterpress.

Bold

Johannes Gutenberg began the printing revolution more than 500 years ago. Gutenberg's method of printing is called letterpress.

Figure 20-14. Notice the contrasts in these three typesetting options.

must be changed into different forms. They cannot be directly used in printing processes. All photographs are first changed into halftones.

Most photographs used today are digital, which means they are already halftones. No conversion is necessary for digital photos. Photos may be raster images or vector images. Raster images are made of pixels, with information assigned to each pixel. These images are difficult to resize because they may produce jagged edges. Vector images are generated mathematically, with specific points in the image defining its shape. All photographs must be separated into process colors.

Color photographs are separated into four process colors. These process colors are yellow, magenta (red), cyan (blue), and black. See **Figure 20-15.** During the separation, the four are run through a halftone screen. Later, each color is printed independently to re-create the color photograph.

Figure 20-15. A color photograph, such as this one of a flower, is separated into four single-color process colors.

Line drawings for multicolor screen prints are produced in several layers. One layer is necessary for each different printed color. See **Figure 20-16.**

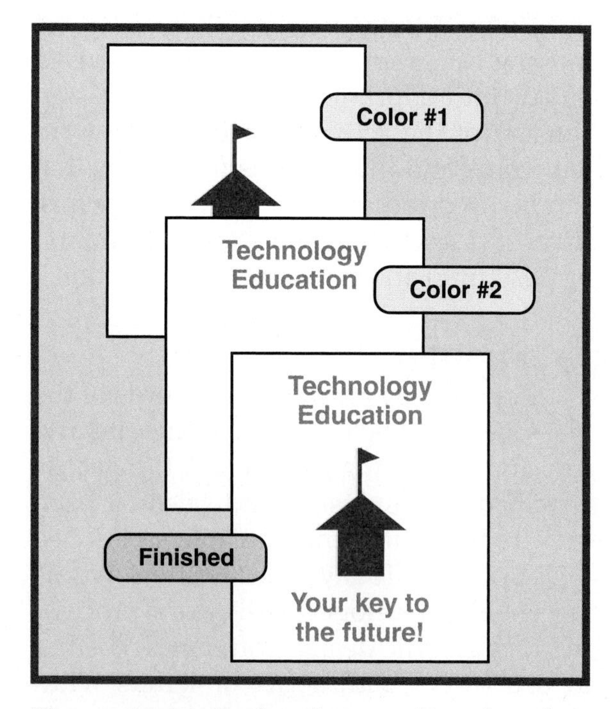

Figure 20-16. Each color on a line drawing is printed separately.

Page Layout

The second step in preparing the message is layout. Before the advent of computer-generated publications, this step involved preparing a pasteup. In this process, all the type and illustrations were assembled onto a sheet that looked exactly as the finished message would look. This assembled layout was called a ***pasteup***.

Computer-driven page-composition systems generate the page layouts electronically (digitally). The computer operator edits the text, manipulates the copy, and locates the spaces for illustrations right on the computer screen. The layout person places (brings in) three different types of computer-generated materials. See **Figure 20-17**. These items are text, line art, and photos.

The text is generally the product of a word processing program, such as Microsoft® Word software. Line art might be from a graphics program, such as the Adobe® Illustrator® program, or from a computer-aided drafting program. Photos can be digitized using a scanner or come directly from a digital camera. These images are often first edited using programs such as the Adobe Photoshop® editor.

The layout is sent electronically to the printer, which then produces a finished page layout. With these computer systems, the need for manual locating and pasting down of type and illustrations is eliminated. Also, the layouts are stored in the computer's memory, so changes are quick and easy. The use of these systems, often called *electronic publishing*, is further discussed later in this chapter.

Image-Carrier Preparation

Most printing processes use an image carrier. The image carrier is the feature making each process unique. Let us look at how relief, offset-lithographic, gravure, and screen image carriers are prepared.

Standards for Technological Literacy

9 17

Demonstrate

Show the class how an offset plate is prepared for printing.

Figure 20-17. A graphic designer can bring in three types of computer-generated materials: text, line art, and photos.

Standards for Technological Literacy

9 17

TechnoFact

The Kentucky Legislature established the American Printing House for the Blind in 1858. It is a nonprofit corporation that seeks to promote the independence of the visually impaired by making printed materials available and accessible to them. The American Printing House for the Blind publishes Braille, large type, and recorded versions of books and magazines.

TechnoFact

The first type used in European printing was known as black letter or Gothic typeface, which imitated the ornate lettering of religious scribes. This type was replaced by Roman (upright) typefaces developed by Nicholas Jenson in 1470. Aldus Manutius created the first italic type in 1501. In the mid-1500s, the first script styles were produced. Around 1816, William Caslon IV created the first sans serif type.

Technology Explained

fax machine: a device that sends copies of documents over telephone lines or radio waves, using digital signals.

A common communication system that serves both individuals and businesses is the facsimile machine, more commonly known as the *fax machine*. A fax machine is similar to a copier, except it has two parts. The two parts, the scanner and the printing unit, work independently. The scanner on one fax machine sends the image of a page to the printing unit of another fax machine. The two fax machines can be many miles apart.

A fax machine changes the black-and-white images on a page into digital information. Digital information has two states, represented by *1*s and *0*s. The two states can be on and off or a higher voltage and a lower voltage. A fax machine uses an optical scanner that looks at the image on a page. See **Figure A.** The scanner breaks the page into small lines. Each line is broken into small dots called *pixels*. The scanner examines each pixel to see if it is black or white. The black and white pixels generate digital signals. A *1* might represent a white pixel, and a *0* might represent a black pixel (or vice versa).

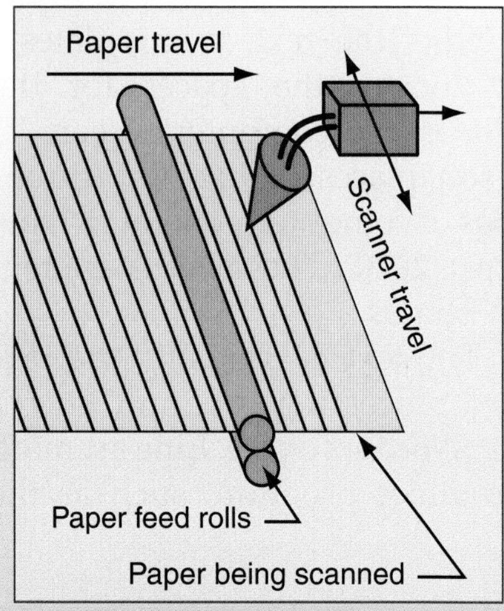

Figure A. How a fax scanner works. Some fax machines use moving scan heads. Others move the paper over a row of sensors.

A modem is used to convert the digital signals into audible tones that can be sent over telephone lines or radio waves. Newspapers and police departments have sent wirephotos over the phone lines for years. Amateur radio hobbyists use fax machines to send documents over the airwaves. Weather maps are also sent via radio. Instead of using telephone lines, the tones from the modem are fed into a transmitter and broadcast. Another person with a receiver feeds the tones back into a fax machine for printing. Today, anyone with a fax machine can send letters, pictures, and other printed information across town or around the world.

The receiving fax machine has a modem converting the tones back into digital signals. A printing system changes the digital signals back into black and white pixels. The pixels are grouped back into lines, and the lines are printed onto a page. The printing system in a fax machine can be a thermal printer, or it might use a photosensitive drum and toner, similar to an office copier. See **Figure B.**

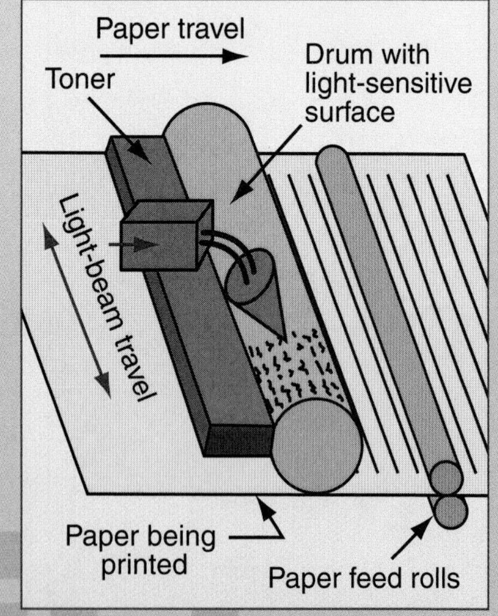

Figure B. The printing unit of a fax turns the digital signals from the modem into black and white pixels on paper. This diagram shows a photosensitive-drum printing unit.

Preparing relief image carriers

A relief image carrier has a raised printing surface. This surface can be produced by setting type, as described in Chapter 7. The letterpress (metal-type) printing process, however, is becoming less common each year.

As you learned earlier in the chapter, flexography is an emerging relief printing process. Flexography uses a flexible plastic or rubber printing plate. This image carrier is produced in one of two basic ways. These ways are casting and etching. Casting uses a mold formed and cured around metal type. The mold is then removed from the type and filled with a plastic material. The plastic cures (hardens) into a sheet, which is removed from the mold. This becomes the printing plate.

Etching is a photochemical process. A negative of the desired image is placed over a light-sensitive material. The material and negative are held together tightly and placed under a strong light. The light-sensitive material is exposed through the open areas on the negative. A chemical developer hardens the exposed material. A detergent is then applied to wash away the unexposed material. A raised image of the message is left.

Preparing offset-lithographic image carriers

Offset lithography may use a flat printing plate, produced through a photochemical process. See **Figure 20-18.** There are several other ways to create offset lithographic plates. Photo direct plates are imaged from the copy with no film intermediate. The plate is ready to use after exposure and processing. Thermal plates use heat to expose the image. Laser plates use lasers to digitally engrave the plate. Photopolymer plates are exposed with ultraviolet light. Computer-to-plate systems use digital files to image heat- or light-sensitive plates directly from a computer. Waterless plates may also be used for environmental reasons. They omit the use of solvents and are recyclable.

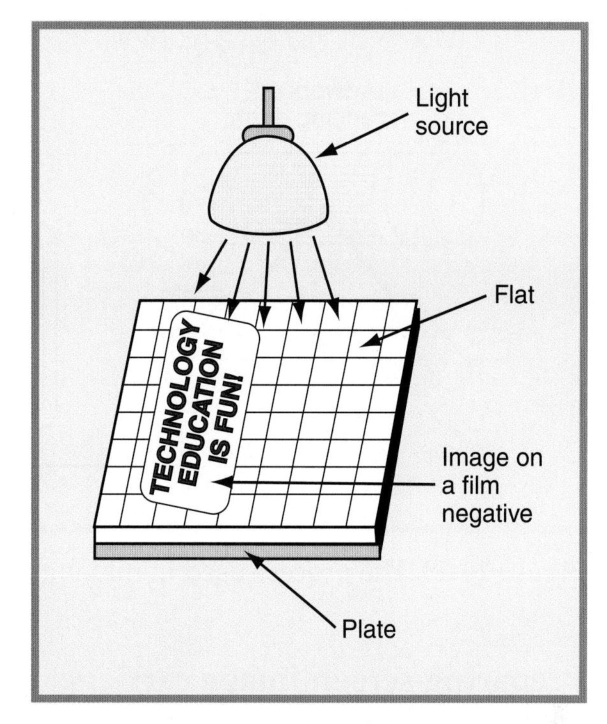

Figure 20-18. Offset plates are produced by exposing the plate through a photographic negative and then developing the resulting image.

Preparing gravure image carriers

The gravure image carrier, or printing cylinder, has the image engraved into it. This is generally done with an electromechanical scanner. See **Figure 20-19.** This computer-controlled machine reflects light off the copy. A sensor then picks up the light intensity. Black areas reflect little or no light, whereas white areas reflect a high percentage of light. The gray areas reflect varying amounts of light, in proportion to their shade. The amount of light reflected is communicated to the computer through the sensor. The computer processes the information and then uses the information to drive a stylus touching the gravure cylinder. The stylus carves the desired pattern into the cylinder, producing the image carrier.

Standards for Technological Literacy

9 17

Demonstrate

During class or lab, show the students how to prepare a screen image carrier.

Standards for Technological Literacy

9 17

TechnoFact

These days, most printing is done on paper. The word paper is derived from papyrus, a reed the ancient Egyptians used to make their scrolls. However, a papyrus scroll was not paper as we know it today. The first true paper, produced from hemp or mulberry bark, was made by the Chinese more than 2000 years ago.

TechnoFact

Paper consists of matted cellulose fibers, from plants, rags, and other fibrous materials, in sheet form. Today, wood fiber and recycled paper are used to make most paper. Some specialty papers (such as resume paper) are made entirely or partly from rags.

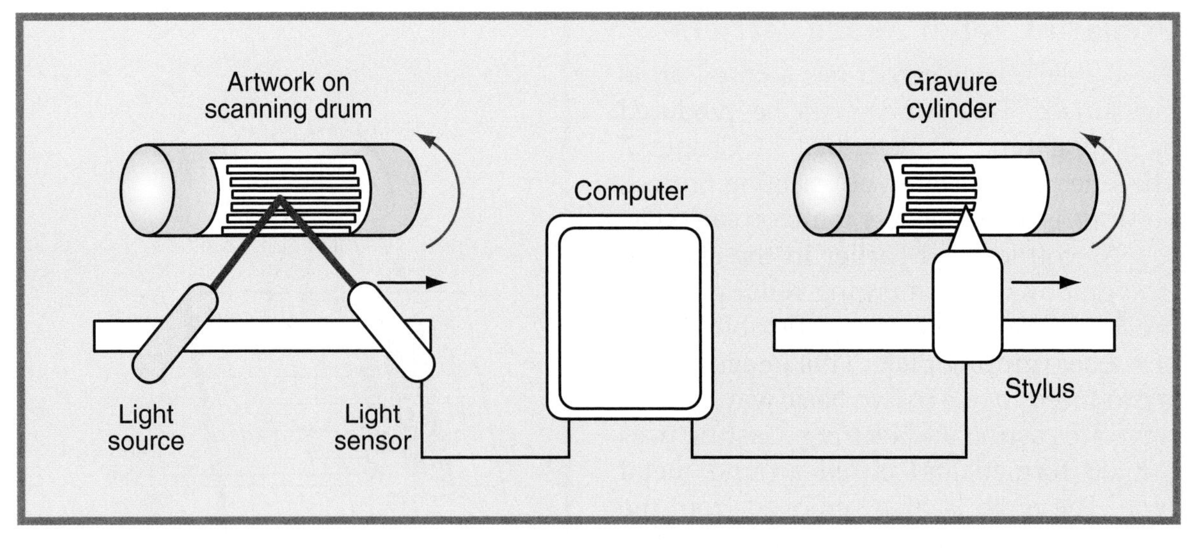

Figure 20-19. The process used in making a gravure plate.

Preparing screen image carriers

Most screen carriers are produced using a photochemical process. First, a positive transparency of the image is produced. This transparency is placed on top of photosensitive film with a clear plastic backing. Light is passed through the positive, exposing the film. When the film is developed, the portion that was exposed to light hardens. The remaining emulsion is then rinsed away. The wet film is placed under the screen fabric. The fabric is pressed into the film, and it is allowed to dry. The clear plastic film backing is then removed, leaving a negative of the image on the screen.

Producing the Message

The third activity in printed graphic communication is producing the message. Production involves printing the message (message transfer) and finishing the product (product conversion). Two basic forms of substrate are used in the production of printed graphic messages. See **Figure 20-20.** The first form is sheet, which includes all the flat, rectangular pieces of substrate. They can be sheets of paper, glass, plastic, or metal, for example. The other form is web, which uses a roll of substrate that is continuously fed into the printing process. Both of these forms require specific printing press equipment. A typical press has a feeder unit, which moves the substrate into the press; a registration unit, which positions the material for correct placement (register) of the message; the printing unit, which transfers the message from the carrier onto the substrate; and the delivery unit, which stacks the printed product.

Web presses have two added features, drying units and sheeting units. Drying units set the ink so it does not smear, and then the sheeting, or cutting, units cut the continuous ribbon of material into sheets. Some printing equipment also includes automatic folding machines, which fold and assemble the sheets into products such as newspapers or magazines.

Message Transfer

The action of placing the message on the substrate is called *message transfer.* The transfer process varies with each printing process. We will look at the message-transfer process for four printing processes: relief printing, offset lithography, gravure printing, and screen printing.

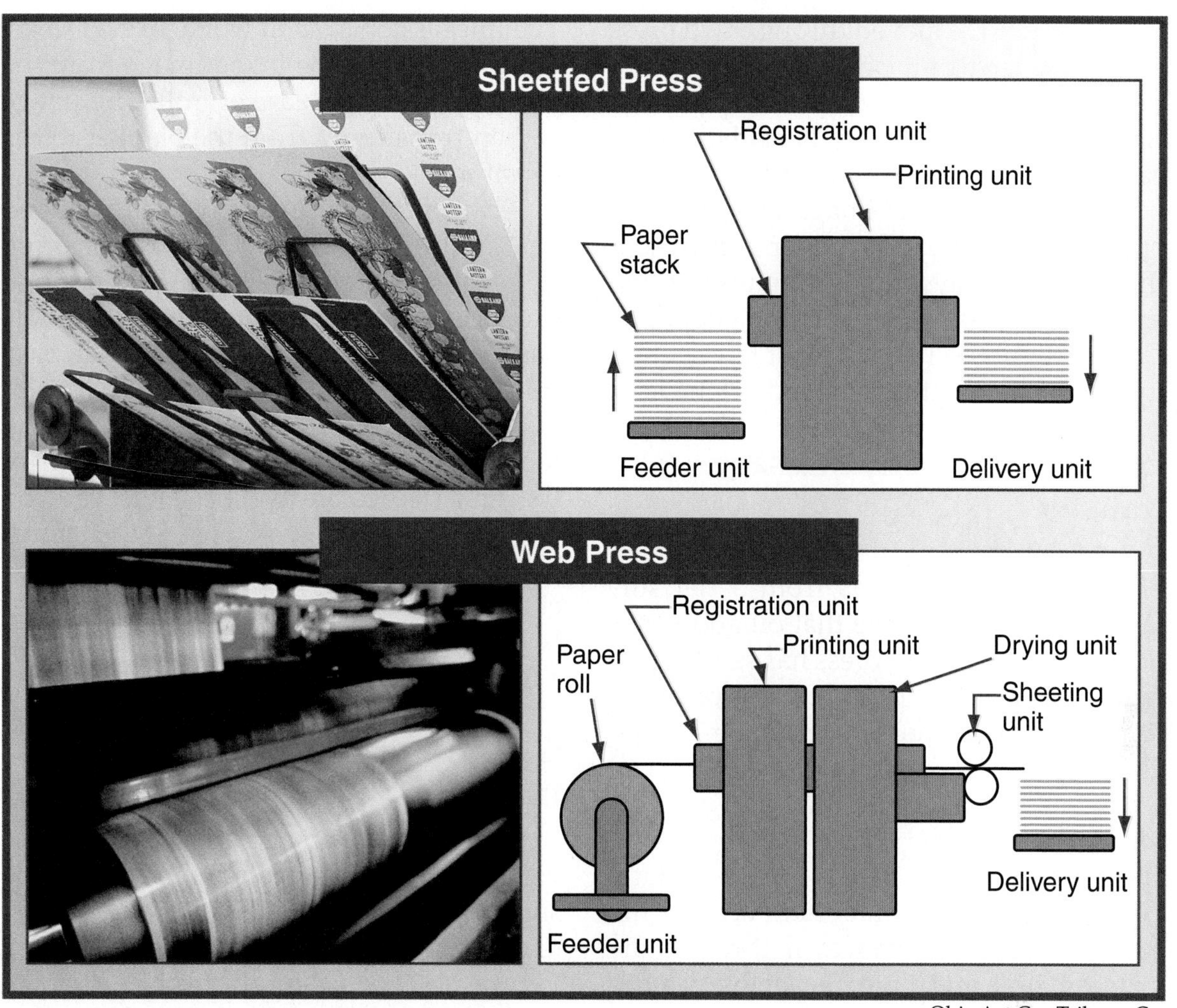

Ohio Art Co., Tribune Co.

Figure 20-20. Two printing press systems. The top system is printing on a sheet substrate. The bottom system is printing on a web (roll) substrate.

Standards for Technological Literacy

9 17

Figure Discussion

Discuss the advantages and disadvantages of sheet-fed and web-fed presses shown in Figure 20-20.

Figure Discussion

Discuss flexography printing as shown in Figure 20-21.

Relief printing message transfer

Most relief printing is done on flexographic presses. This press transfers very thin ink from a reservoir to an application, or anilox, roll. See **Figure 20-21.** This roll distributes a thin coating of ink on the raised surface of the rotating plate. As the plate revolves, it contacts the web of paper passing over the impression roll. The pressure between the plate and impression roll transfers the image to the substrate.

A special flexography press, called the *Cameron press*, is sometimes used to print books. This press prints with a flexography "belt," which contains all the pages of the book. The paper and the flexography

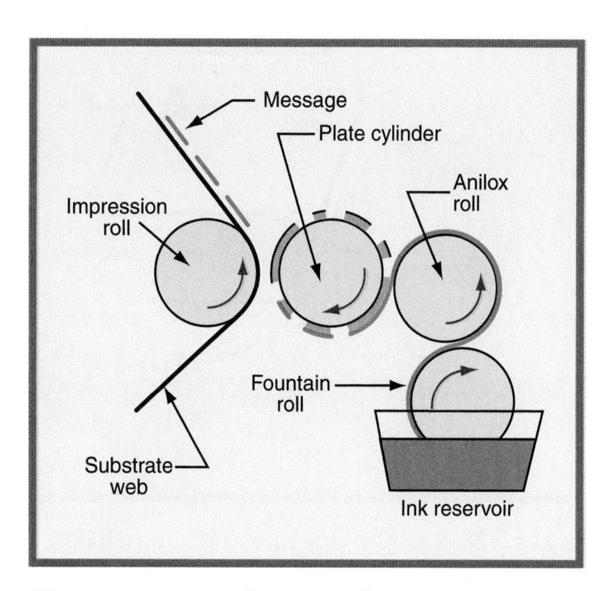

Figure 20-21. A flexography press.

Standards for Technological Literacy

9 17

Figure Discussion

Show how lithography changes "right-reading" images on the plate to "wrong-reading" images on the blanket and back to "right-reading" images on the substrate.

Figure Discussion

Explain the gravure process that is shown in Figure 20-23.

TechnoFact

Carved wooden blocks were the first means of mechanically reproducing illustrations. The image was drawn on a block and the background area was carved away. The resulting block was smeared with ink and pressed against the paper. The high areas of the block created an image on the paper. The process could be repeated as many times as needed, producing identical results.

TechnoFact

Harvard University was home to America's first printing press, which began operation in 1638. The first items published were an 8-page almanac titled *The Freeman's Oath* and the now famous *Bay Psalm Book*. In 1663, the same press was used to print 1500 copies of a Bible in the native language of the Algonquian tribe.

image carrier are moved through two press units. The first unit prints on one side of the paper, while the second unit prints on the reverse side. A dryer is at the end of each printing unit to set the ink.

Each cycle of the image carrier prints one complete book. The paper is then sent through assembly and binding units at the end of the press to complete the book. The Cameron press is used to print lower-quality publications, such as paperback novels.

Offset-lithography message transfer

As you learned earlier in the chapter, offset lithography prints from a flat surface, using the principle that oil and water do not mix. The offset press has several systems in its printing unit. See **Figure 20-22.** A plate cylinder holds the image carrier. The dampening system applies a special water-based solution, called the *fountain solution*, wetting the plate. Next, the inking system applies ink to the plate's surface. Finally, the impression system accepts the substrate and then moves it through the printing unit and onto the delivery unit. The impression system works in two steps. First, it transfers the inked image from the plate to an offset (blanket) roll. The image is then transferred from the blanket to the substrate.

Two or four offset units can be arranged in a single press. This allows color printing to be done efficiently. A four-color press can print full-color materials in a single pass. A two-color press can print the base color (usually black) and a highlight color in one pass. This press can also be used for four-color work, but the press must print the first two colors (yellow and cyan) on one pass. The two-color press then must be cleaned, and the ink colors must be changed. The second pass prints the other two colors (magenta and black) needed for a four-color run.

Gravure-printing message transfer

The gravure image area is a series of very small ink reservoirs engraved in the surface of the carrier. This carrier is a metal drum partly suspended in a tray of ink. See **Figure 20-23.** As the roll revolves, it picks

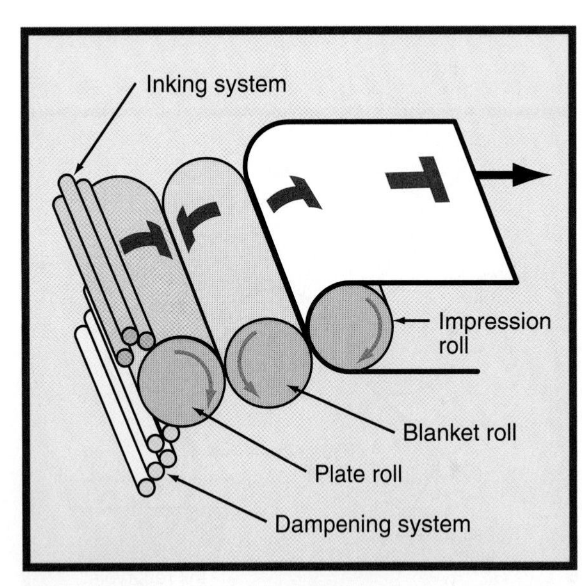

Figure 20-22. The operation of an offset-lithography press. Notice how the image is first offset, before it is printed.

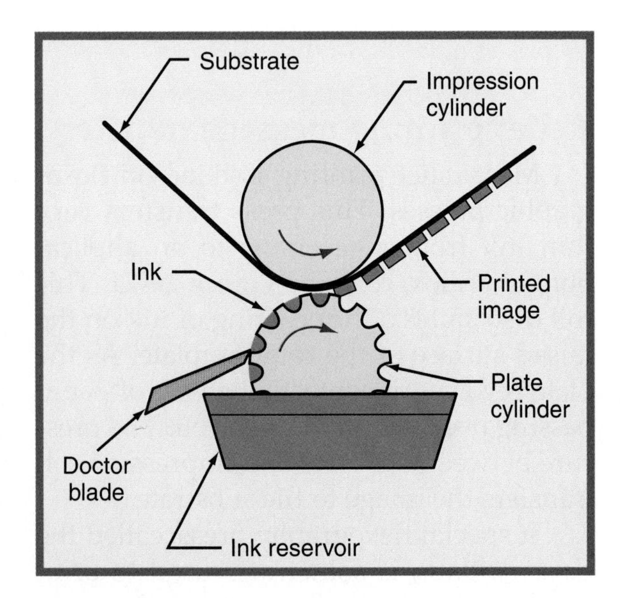

Figure 20-23. How a gravure press operates.

up a layer of ink. A doctor blade wipes off the excess, leaving ink only in the cavities in the roll. The roll then contacts the substrate under high pressure. This action transfers the ink onto the substrate.

Often, several gravure units are assembled into a single press. Each unit is used to print a different color of a multiple-color run. The image on a gravure roll can be a continuous, repeated pattern. Therefore, this process is often used to print wallpaper, hardboard wall paneling, vinyl flooring, and vinyl shelf covering.

Screen-printing message transfer

Many people have done screen printing in art classes or as a hobby. The process, as described earlier in this chapter, is quite simple. Ink is pressed through a stencil onto the substrate. Production-screen printing presses use one or more stations. The substrate is automatically fed into position under the screen. A screen is lowered onto the substrate, and a squeegee automatically moves across the image. The screen is then lifted, and the substrate is removed. In other cases, if more than one color is being used, the substrate is moved to another printing station. Each station lays down a different color, until the final image is produced on the product.

Product Conversion

Most printed graphic communication products are not finished when they leave the printing press. They require additional work. This work might include the following:

- **Cutting.** This work is trimming the substrate into rectangular sheets or cutting assembled products to their final sizes.
- **Die cutting.** This work is shearing irregular shapes or openings in the material. See **Figure 20-24.** This process is often done with cutting dies on a relief press.
- **Folding.** This work is creasing and folding the product to form pages. Often, each printed sheet contains a number of pages. See **Figure 20-25.** Properly folded sheets are called *signatures.*
- **Drilling or punching.** This work is creating holes in the substrate for insertion in binders.

Figure 20-24. The irregular shape of this package is die cut.

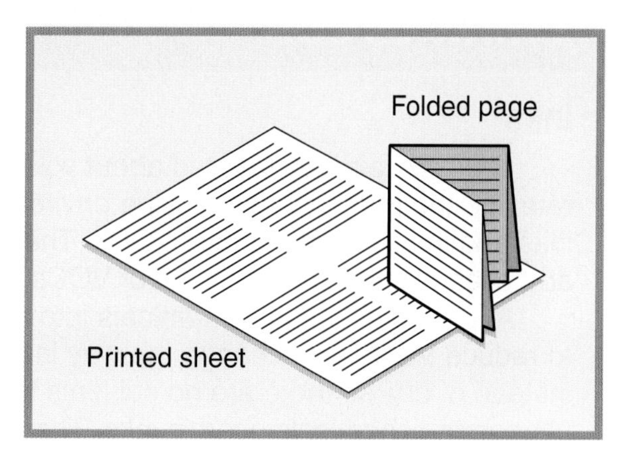

Figure 20-25. A single sheet containing multiple pages is folded to make a signature.

Standards for Technological Literacy

9 17

Demonstrate

During class or lab, use several unfolded paperboard packages to show finishing operations, such as creasing, die cutting, etc.

TechnoFact

Gravure printing, also called intaglio or recess printing, is used for high-volume, high-speed printing. Presses using this process can operate at speeds up to 3000 feet per minute (50 feet per second). The gravure process is used to print many products, including money, postage stamps, stock certificates, giftwrap, and catalogs.

Standards for Technological Literacy

9 17

Demonstrate

Show how to merge text and clip art using desktop publishing software.

TechnoFact

The first role of computers in publishing was to create strips of typeset text. The strips were then pasted to a board and photographed. Eventually, computers allowed people to directly edit text and create layouts. Today, publishers can even use computers to print the layouts to film or directly to printing plates.

- **Assembling.** This work is gathering and placing sheets or signatures into the final order. See **Figure 20-26.** This can be done manually or with automated machines.
- **Binding.** This work is securing sheets and signatures into one unit. Binding can be done by gluing, stapling, or sewing the units to form a book or magazine.
- **Stamping.** This work is transferring an image to a book or an album cover with a hot-stamping technique. Stamping is often done using heated type and colored foil. The heat melts the foil and transfers the type image onto the product.
- **Embossing.** This work is pressing a raised pattern into the substrate. Paper or other material is pressed between mated dies to produce a pattern.

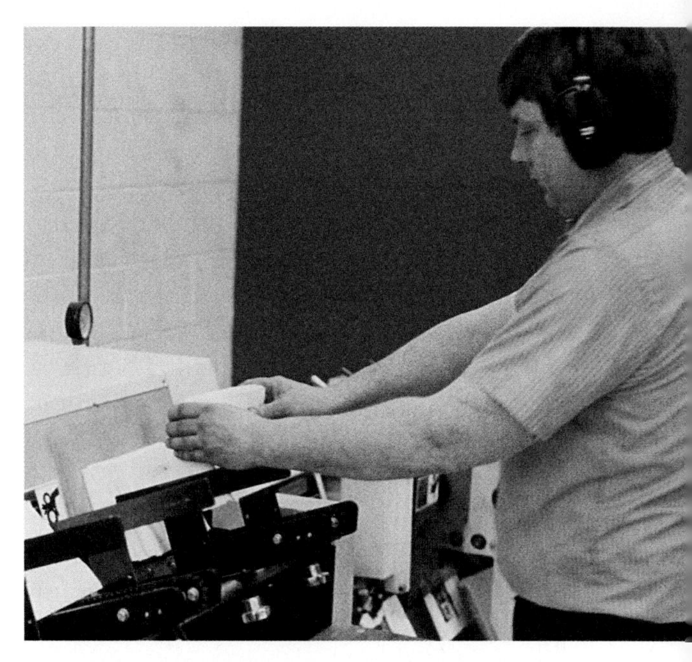

Figure 20-26. This worker is using an automatic machine to assemble signatures into a book. Note his ear protection to prevent hearing loss.

Computer-Based Publishing

Computers have changed the way people do work in many areas. In graphic communications, the desktop computer has brought in new ways to prepare and print documents and other media. The computer has given rise to new categories of publishing. They are most commonly referred to as desktop publishing and electronic publishing.

Think Green

Inks

You have already learned about volatile organic compounds (VOCs) that evaporate into the atmosphere, causing environmental and health problems. Conventional inks and ink solvents contain VOCs. The printing industry is taking steps to reduce and even eliminate the amount of VOCs it produces.

As ink dries, it emits chemicals into the atmosphere. One of the first steps taken to reduce VOCs is to change the way inks dry. For instance, UV-curable inks cure instead of dry, so there are no solvents that evaporate into the air. Another step is to change the chemicals used in inks. There are now vegetable oil–based inks, or soy inks, used as an alternative to conventional inks. These newer inks dry more slowly, but since they do not use petrochemical solvents, they are used in green printing facilities.

The term ***desktop publishing*** has often been used to describe a fairly simple computer system that produces type and line-illustration layouts for printed messages. The type and line illustrations are generated using two separate software packages.

More complex systems can function as typesetting and layout systems. These systems are referred to as ***electronic publishing***. They produce and combine text and illustrations into one layout. Electronic publishing produces a higher-quality product than desktop publishing produces.

The functions of desktop publishing and electronic publishing are essentially the same. They are both types of computer-based publishing, but they may be meant for different sizes of production. A common computer-based publishing system includes the following. See **Figure 20-27:**

- A computer keypad to input text copy and to provide commands to the system.
- A processing unit (microcomputer) to store and process data.
- A monitor to display the text and illustrations being manipulated.
- Page-layout software to direct and control computer and printer actions.
- A mouse to send commands to the computer easily.
- An ink-jet printer to run proofs and sample layouts.
- A laser printer to produce high-quality single-page layouts.
- A scanner to digitize illustrations and input them into the computer.

Standards for Technological Literacy

9 17

Extend

Discuss the advantages of electronic publishing over other types of publishing.

Brainstorm

Ask the students what they like and dislike about electronic media.

TechnoFact

Electronic publication can incorporate features unavailable in printed products, including sound and animation. Electronic publications are easy to revise and distribute, cost less to publish than printed products, and can be stored and retrieved quickly. Also, because CDs hold up to 250,000 pages of text and are cheap to produce, publishers may include more material than could be found in a comparable printed product.

Figure 20-27. The operator is using a computer system to make a page layout for a publication.

Standards for Technological Literacy

TechnoFact

Government Printing Office (GPO) was established in 1860. Today, it is one of the largest printing plants in the world. The office prints and binds publications for the United States government and distributes them online, through mail order, and through a series of libraries. Each year, the GPO prints or contracts the printing of approximately two billion copies of various publications.

The system might also use several additional software programs. Many systems include illustration software to create line illustrations. Also, computer clip-art files might be used as a source of additional line illustrations.

The layout software used in computer-based publishing allows the operator to merge text and illustrations very accurately. See **Figure 20-28.** The abilities of the layout software may differ between one type and another. These abilities might include cropping or size adjustments to photographs, color enhancement, image-grade enhancement, wrapping text around images, and complete layout control. Often, the output of the system can be directly converted to film or printing plates.

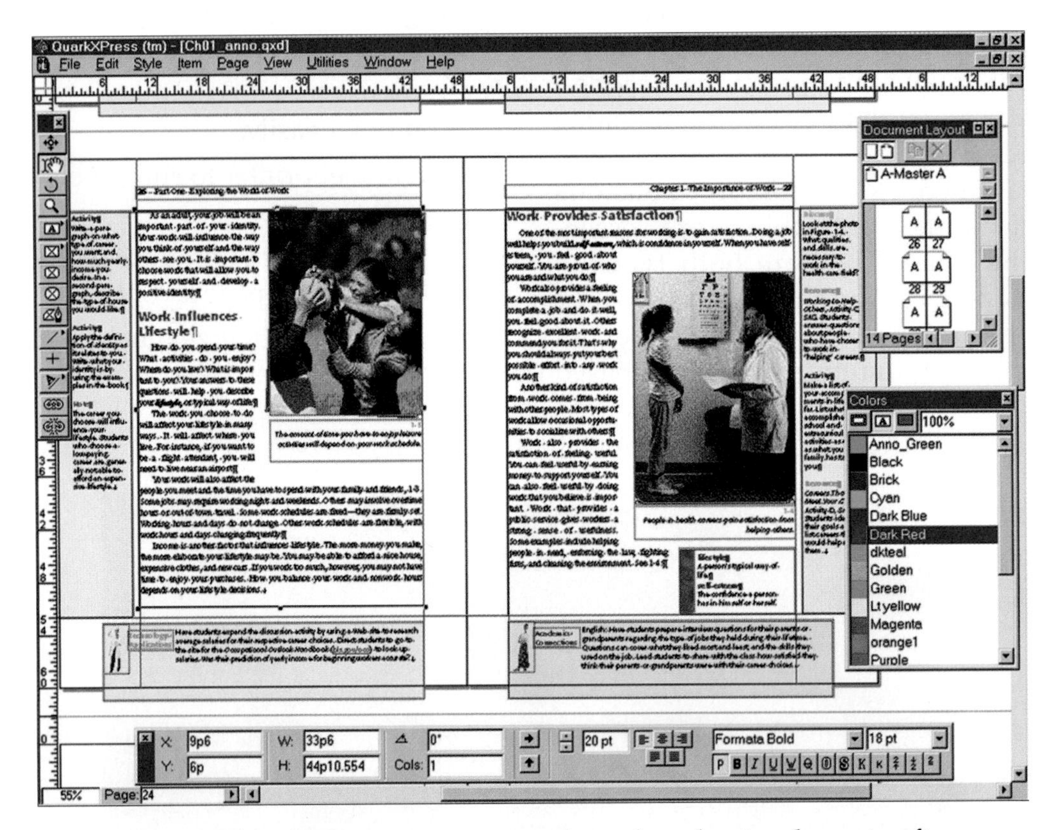

Figure 20-28. This design was set up and produced using layout software.

Summary

Printed graphic communication products constantly serve and impact each of us. These products can be books, magazines, flyers, pamphlets, signs, and posters. Each printed product is produced using one of six common printing methods. These methods are relief printing, offset lithography, gravure printing, screen printing, electrostatic printing, and ink-jet printing. Each product follows a three-step production process. The product is designed. The design is prepared for printing. The product is printed and finished. Computers have led to new graphic communication systems. These systems are desktop publishing, electronic publishing, and pagination systems.

Test Your Knowledge

Write your answers on a separate piece of paper. Please do not write in this book.

1. Define *printed graphic communication.*
2. Name three products that printed graphic communication processes produce.

Matching questions: For Questions 3 through 12, match each description on the left with the correct printing method on the right. (Note: Answers can be used more than once.)

3. _____ Uses a stencil.
4. _____ An office copier is one type.
5. _____ Requires the image to be reversed on the printing block.
6. _____ A computer generates the printed message.
7. _____ Is based on the principle that oil and water do not mix.
8. _____ Is also known as the *intaglio process.*
9. _____ Is most widely used today.
10. _____ Is used to print on T-shirts.
11. _____ Is used for coding packages.
12. _____ Is used to print paper money.

A. Relief.
B. Lithographic.
C. Gravure.
D. Screen.
E. Electrostatic.
F. Ink-jet.

13. What are the three main steps in producing printed products?
14. Name two principles of design.
15. The physical act of designing a message is called _____.
16. Other than sizing, line art usually needs no other preparation for printing. True or false?

Answers to Test Your Knowledge Questions

1. Printed graphic communication is the process of communicating information through symbols printed on a medium.
2. Student answers may include any three of the following: newspapers, magazines, books, brochures, pamphlets, labels, stickers, clothing designs, and signs. Additional answers may be accepted at the instructor's discretion.
3. D. Screen
4. E. Electrostatic
5. A. Relief
6. F. Ink-jet
7. B. Lithographic
8. C. Gravure
9. B. Lithographic
10. D. Screen
11. F. Ink-jet
12. C. Gravure
13. Designing the message. Preparing to produce the message. Producing the message.
14. Student answers may include any two of the following: proportion, balance, contrast, rhythm, variety, and harmony.
15. layout
16. True
17. Evaluate individually.
18. Student answers may include any three of the following: cutting, die cutting, folding, drilling or punching, assembling, binding, stamping, and embossing.
19. The term desktop publishing is most often used to describe a fairly simple computer system that produces type and line illustration layouts for printed messages. The type and line illustrations are generated using two separate software packages. Electronic publishing employs more complex systems that can function as typesetting and layout systems. These systems produce and combine text and illustrations into one layout. Electronic publishing produces a higher quality product than desktop publishing produces.

Standards for Technological Literacy

11 12

17. Briefly describe how an offset-lithography message is produced.
18. List three activities involved with finishing a printed graphic communication product.
19. What is the difference between desktop publishing and electronic publishing?

STEM Applications

1. Choose or invent a school event to promote. Complete a chart of the event, similar to the one below. Prepare rough and refined sketches for a poster to promote the school event.

What is the name of your event?
Who is your audience?
What is the theme of your event?
What is your design approach (humorous, for example)?
What type of promotional material (flyer or poster, for example) will you use?

2. Design a logo, or symbol, for an organization or a cause. Make a block-cut relief image carrier to print your image.

This activity develops the skills used in TSA's Promotional Graphics event.

Standards for Technological Literacy

11 12

Promotional Graphics

Activity Overview

In this activity, you will create a color graphic design appropriate for posters and T-shirts. The theme of the design can be either your school or your community.

Materials

- Paper.
- A pencil.
- A computer with graphic design software and clip art.
- A color printer.

Background Information

- **General.** Consider the principles of design as you develop your project. Locate and size elements to achieve balance and proportion. Use contrast to emphasize key elements. Select elements to provide rhythm.
- **Elements.** The design can include graphic elements, such as photographic images, clip art, or original design elements. When selecting illustrations, consider the content of the illustration (is it appropriate for the cover?) and the illustration's final printed size (will the illustration be printed so small that details are lost?). Try to incorporate elements that are generally connected with and related to your theme.

Guidelines

- Your design cannot exceed 8″ × 10″ (20 cm × 25 cm).
- Identify good design examples from other similar graphics.
- Develop sketches and a rough layout using paper and a pencil.
- After you have developed the rough layout, create the design using the graphic design software.
- Print the final design on a color printer.

Evaluation Criteria

Your project will be evaluated using the following criteria:

- Design elements.
- Attractiveness and impact.
- Relation of design to theme.

TSA Modular Activity

Standards for Technological Literacy

11 12

This activity develops the skills used in TSA's Desktop Publishing event.

Desktop Publishing

Activity Overview

In this activity, you will create one of the following items:

- A trifold pamphlet designed as a recruitment tool for potential new students.
- A three-column newsletter providing information about a current event or issue in your community.
- A poster advertising an upcoming school function or activity.

Materials

- Paper.
- A pencil.
- A computer with desktop publishing software and clip art.
- A color printer.

Background Information

- **General.** Consider the principles of design as you develop your project. Locate and size elements to achieve balance and proportion. Use contrast to emphasize key elements. Select elements to provide rhythm.
- **Elements.** Do not use too many variations of font type and size. When selecting illustrations, consider the content of the illustration (is it appropriate for the piece?) and the illustration's final printed size (will the illustration be printed so small that details are lost?). Finally, consider the use of white space.
- **Type.** Too many fonts and type sizes can make a piece unattractive. Vary the size and font based on the function of the text. Titles are meant to draw attention. Body type should blend in to the overall design and use a typeface, a size, spacing, and a justification comfortable for reading. If using a justified layout, it might be necessary to break words (use a hyphen to divide words onto two separate lines) to avoid excessive spacing between words.
- **Trifold pamphlet.** Begin with the content of the trifold. What information should you provide in it? What would a potential student like to know about your school? The pamphlet cover should be designed to be visually appealing. Inside the pamphlet, use several sections of body type, each introduced with a heading. Add illustrations where appropriate.

- **Three-column newsletter.** The title of the newsletter should run the entire width of the page. You might want to use headings running across all three columns or, perhaps, two columns. Illustrations should be sized to fit in either one or two columns. They should be selected or created to both communicate information and provide visual interest. When placing illustrations, do not disrupt the flow of text for the reader.
- **Poster.** Include all the critical information on the poster: what, where, and when. The "what" is the attention grabber. The "where" and "when" are also critical elements. Include some advertising copy to persuade the reader to attend the event. Why should the reader attend? Will the reader be entertained, educated, or involved? What will the reader miss by not attending? Pay close attention to rhythm when designing the poster. Your selection of elements determines how the reader's eye moves when viewing.

Guidelines

- Identify good design examples from samples of trifold pamphlets, newsletters, and posters before beginning your design.
- Develop sketches and a rough layout using paper and a pencil.
- After you have developed the rough layout, create the design using the desktop publishing software.
- Your project must be designed to be printed on 8 1/2″ × 11″ paper.
- Your project must include a sample of text wrapping (text flowing around an illustration).
- Your project must use at least one clip-art illustration and one original illustration or graphic created using a graphics program.
- Print your project in color.

Evaluation Criteria

Your project will be evaluated using the following criteria:

- Use of design principles.
- Page layout.
- Selection and placement of clip art.
- Font selection and usage.

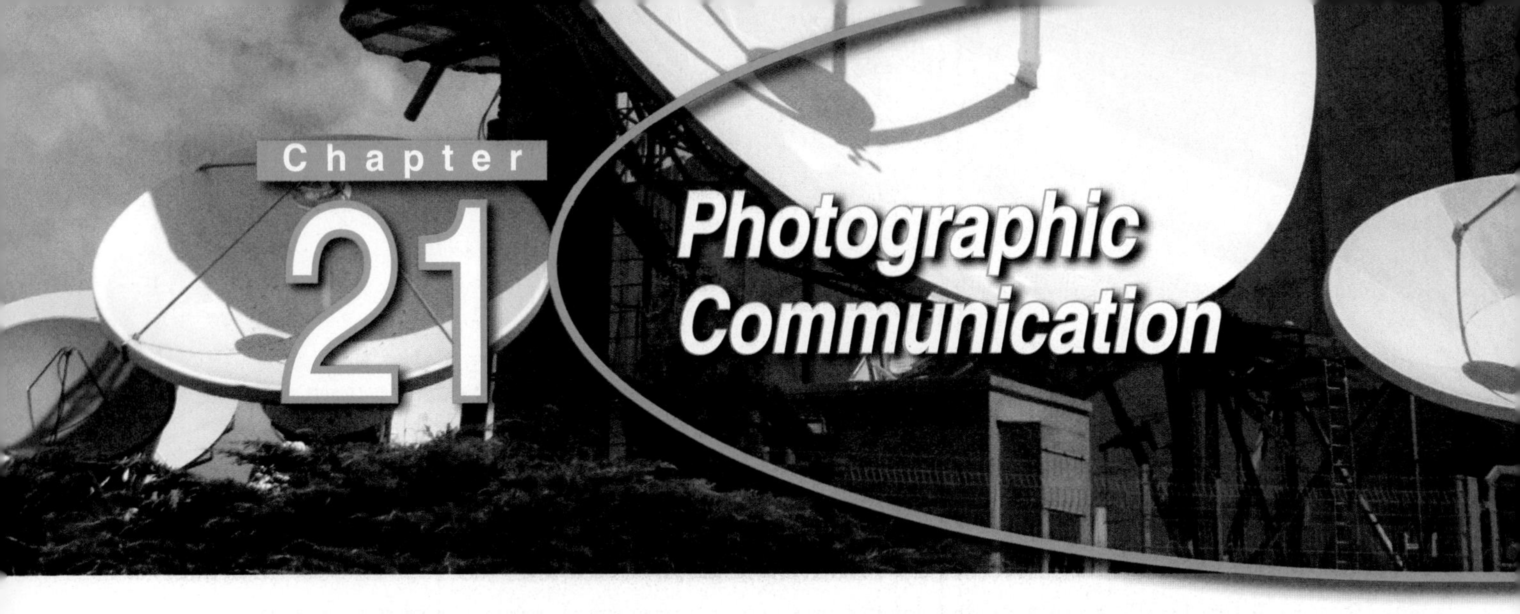

Chapter 21 Photographic Communication

Chapter Outline
- Light and Photography
- Fundamentals of Photographic Communication
- Types of Photographic Communication

This chapter covers the benchmark topics for the following Standards for Technological Literacy:

3 5 7 11 17

Learning Objectives

After studying this chapter, you will be able to do the following:

- Recall the definition of *photographic communication.*
- Compare photography and photographic communication.
- Recall the three characteristics of light waves.
- Recall the three steps in producing a photographic communication message.
- Explain how photographic communication messages are designed.
- Compare digital and traditional film photography.
- Summarize the principles of a camera.
- Compare types of film.
- Explain how photographic messages are reproduced.
- Summarize the characteristics of at least two types of photographic communication.

Key Terms

absorb	fixing	positive transparency
amplitude	f-stop number	projection printing
body	lens	reflect
composed	motion picture	scope
contact print	negative film	shutter
developing	panchromatic film	slide
diaphragm	photograph	stop
direction	photography	stop bath
film speed	point of interest	wavelength

Strategic Reading

Before reading this chapter, think of several examples of photographic communication. As you read, see if there are any types of photographic communication you weren't familiar with.

"A picture is worth a thousand words" is an old saying. See **Figure 21-1.** This statement suggests that it is often more effective to convey your message visually than to describe it. Pictures are an efficient method of communication. In some cases, a picture alone is used to communicate an idea or a feeling. In other cases, the picture is used to supplement the written word.

The word *picture* once was used to refer only to paintings. Now, just about any two-dimensional visual representation is called a *picture.* A ***photograph*** is a common type of picture. The act of producing a photograph is called ***photography***. Using these photographs to convey an idea or information is called *photographic communication.*

To distinguish photography from photographic communication, consider the following. Your family might take snapshots at family events and during vacation travels. The main goal of the snapshots is to capture a moment in time. Later, you can look at these pictures and remember those moments. This type of photograph is designed to provide a historical record, rather than to communicate information.

On the other hand, if you have flown on a commercial jet, you have seen a safety information card. This card contains a series of photographs showing how to fasten the seat belt and exit the plane during an emergency. These photographs are designed to communicate a specific procedure. They are used as photographic communication.

Light and Photography

Photography is based on the understanding and appropriate use of light. Light is a wave of energy. We define *light* as "electromagnetic radiation with wavelengths falling in a range including infrared, visible, UV, and X ray." Light waves have three central characteristics. See **Figure 21-2:**

- ***Amplitude*** is the height of the wave. This height indicates the wave's strength, or intensity. Amplitude is measured from the center of one wave to the peak of that wave.

Figure 21-1. A replica of an early Egyptian hieroglyph.

Standards for Technological Literacy

17

Brainstorm

Have your students list the different ways they have seen photographs used. Ask them why they think photographs were chosen for each use.

Extend

Discuss why portrait painting was important in early times and how this technique has been replaced by photography.

TechnoFact

The modern camera evolved from two previous inventions, the camera obscura and the photogram. A camera obscura, invented about 2000 years ago, is a box or room with a small hole in one wall. Light shines through the hole, creating an upside-down image of the outside scene on the opposite wall. Thomas Wedgwood invented the photogram about 160 years ago. He was able to capture an image on paper coated with silver nitrate. Unfortunately, he had no way to fix the images, so they lasted only seconds.

TechnoFact

2300 years ago, the ancient Greek philosopher Aristotle observed that light passing through a small hole in one wall of a dark room would create an inverted image of the outside scene on the opposite wall. More than 1000 years ago, the Arab scholar Alhazen applied this principle to watch solar eclipses. In the 1400s, Leonardo da Vinci wrote a detailed description of the principle.

Standards for Technological Literacy

17

Demonstrate

Show your students the difference between transparent, translucent, and opaque materials.

Example

Give examples of the use of materials that transmit, reflect, and absorb light.

TechnoFact

The lens of a modern photographic camera focuses the light passing through a lens onto a photosensitive film. In the 1500s, a filmless version of a camera was used to project scenes onto a screen, where they could be traced. The device's name, *camera obscura*, literally means dark chamber. It was a lightproof box with a pinhole (lens) on one side and a translucent screen on the opposite side. Light entered the box through the pinhole and projected an inverted image on the screen.

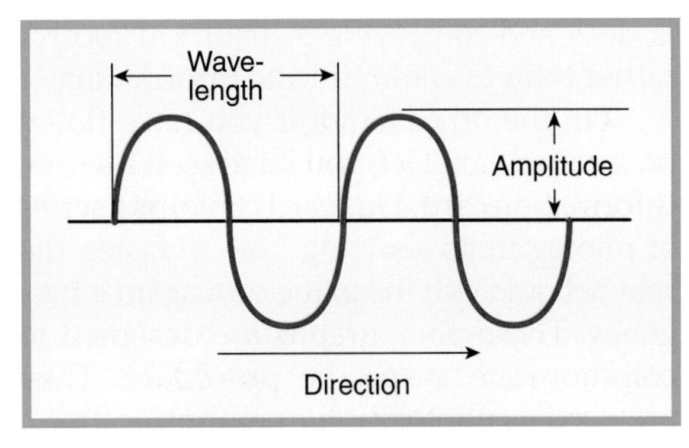

Figure 21-2. The wavelength, amplitude, and direction are three descriptive characteristics of a light wave.

- ***Wavelength*** is the distance from the beginning to the end of one wave cycle. The wavelength of the light determines whether or not the wave falls into the visible range, and if it does, what color it is. Color is more properly referred to as *hue*. Each hue, or color, has a set wavelength. For example, red waves are 650 nanometers (nm), or 650 billionths of a meter, in length. Closely associated to wavelength is frequency. Frequency is the number of waves passing a point in one second.
- ***Direction*** refers to the path the light wave travels. Light waves travel in straight lines in all directions from their source. The direction might change when the wave encounters another material or force. See **Figure 21-3.**
  - If the wave passes through the material, the wave is said to be transmitted. Smooth, clear glass allows waves to pass directly through it without bending. This is called *specular transmission*. Etched glass allows the light to pass. This glass, however, randomly bends the light's path. This is called *diffuse transmission* because the light is diffused, or spread out.

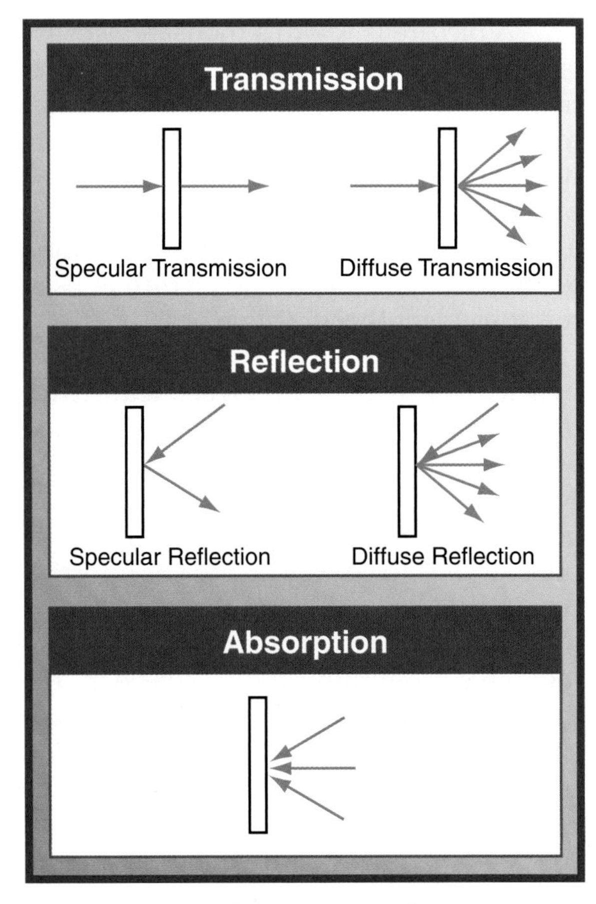

Figure 21-3. Light waves can be transmitted, reflected, or absorbed when they strike an object.

- Some materials ***reflect*** the light that strikes them. If the angle of reflection is equal to the angle at which the wave strikes the object, it is called *specular reflection*. Good mirrors exhibit specular reflection. Other substances reflect and diffuse the light. Rough surfaces exhibit this property, called *diffuse reflection*.
- Finally, many materials ***absorb*** light waves. For example, black objects absorb all light that strikes them. Colored objects absorb only certain wavelengths. The other wavelengths are reflected. Your eyes receive them. This property allows your eyes to see the object's color.

Light is but one of a number of different types of electromagnetic waves. Electromagnetic waves include gamma rays, X rays, UV light, visible light, infrared rays, and radio waves. Each of these has a specific frequency range and can be divided into even more specific groups. See **Figure 21-4.**

When all the wavelengths of visible light are present, we have white light. Sunlight is an example of white light. This light, however, can be divided into a group of wavelengths, each a different color. Have you ever seen a rainbow? If so, you have seen white light divided into its six basic color components: violet, blue, green, yellow, orange, and red. The raindrops in the air cause this separation. As light passes through them, they reflect and separate the various wavelengths. The shorter violet rays appear on the bottom of the rainbow. The longer red waves are on top.

Standards for Technological Literacy

17

TechnoFact

In 1826, Joseph Nicephore Niepce produced the world's first photograph. To create the image, he took a metal plate and coated it with light-sensitive chemicals. He placed the plate inside a camera and exposed it for about eight hours.

TechnoFact

Building on the Niepce's work, Louis Daguerre created his photographs on sheets of silver-coated copper. He developed the images by exposing the plate to mercury vapors and used table salt to permanently fix the image. These types of photographs, called daguerreotypes, required only a 30-second exposure time and were sharper and more detailed than previous types.

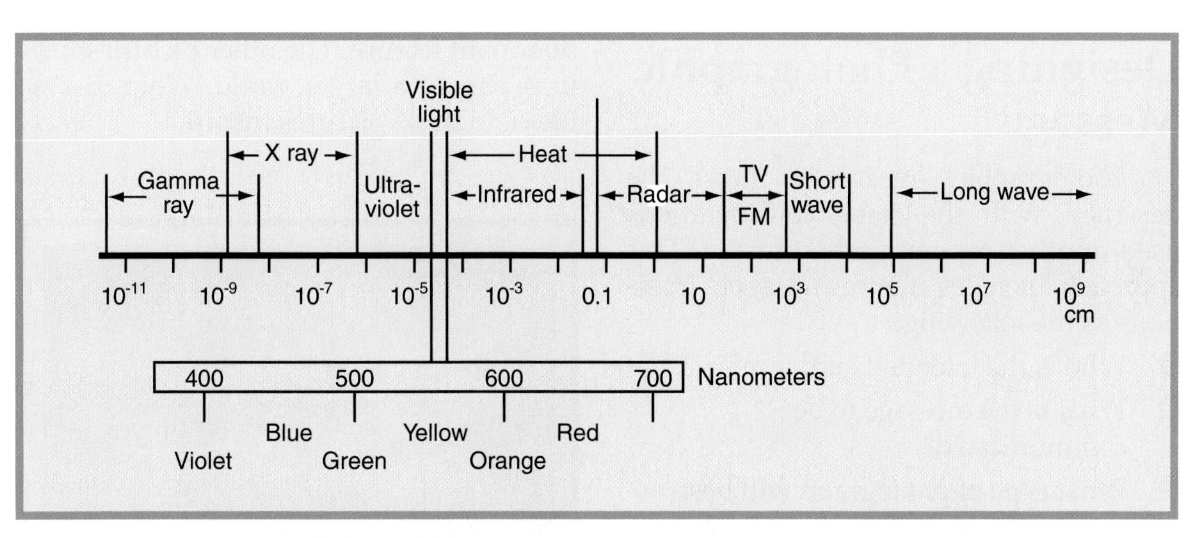

Figure 21-4. Electromagnetic waves grouped by their wavelengths. Visible light is only a small set of electromagnetic radiation.

Career Corner

Arts, A/V Technology & Communications

Commercial and Industrial Photographers

Photographers produce and preserve images that capture a scene, tell a story, or record an event. To create commercial-quality photographs, photographers must be creative and technically competent. A photographer must be able to choose appropriate subjects and present them in interesting ways.

Commercial and industrial photographers take pictures of buildings, machinery, products, and workers. Generally, they must be able to take and process standard film images and produce, manipulate, and transmit digital images. Most positions in industrial and scientific photography require a college degree in journalism or photography.

Standards for Technological Literacy

17

Example

Show your students examples of photographs with different points of interests. Magazine images could provide a wide variety of examples.

Brainstorm

Ask your students why they think having a point of interest is important in composing a photograph.

TechnoFact

In 1839, William H. Fox Talbot developed a process for making photographic negatives. Countless positive images could be printed from the negatives. One of Talbot's friends called the invention "photography."

Fundamentals of Photographic Communication

Photographic communication media are produced using a common set of steps. These steps are the following:

1. Designing the message.
2. Recording the image.
3. Reproducing the message.

Designing a Photographic Message

Photographic messages must be designed with the same considerations used in other communication media. This approach includes answering such questions as the following:

- Who is the intended audience?
- What is the message to be communicated?
- What type of photograph will best communicate the information or idea?

A photograph shows a record of light. The picture captures the light reflected from a series of objects the instant before the film is exposed. Some people see this as a window to reality. Others see it as a flat representation of a scene. The goal in designing a photographic communication message is to make the message a realistic window, rather than the flat representation.

To be photographic communication, the photograph must describe something. This can be a process (a way of doing something), an idea (a mental image), or an event. Several techniques are used in helping photographs communicate. First, photographs are ***composed***, or designed, with a point of interest. The ***point of interest*** is the place where your eye is drawn. This is where the central message is communicated. Look at the two photographs in **Figure 21-5.** Where is your eye drawn in each of them? In one photograph, the sky is the central focus. The other uses the sky simply as a background to the central focus, the horseback riders. What message does each point of interest communicate to you?

Another technique uses distance and ***scope*** (panorama) to add interest. Look at the photographs in **Figure 21-6.** They both show the same mountain from approximately the same vantage spot. The mountain is the point of interest for both pictures. In one picture, however, the mountain is the dominant feature. The other picture shows it as part of a larger world. What diverse ideas do these pictures inspire?

Figure 21-5. In these photographs, we see two different points of interest. In the first photograph, the eye is drawn to the sky. In the second photograph, the riders are the central point of interest.

Figure 21-6. In these photographs, the mountain is the main point of interest. The change in distance and scope, however, creates two different images.

A third technique uses color to help communicate. Look at the photographs in Figure 21-7. Both are pictures of food. Which one communicates the idea more clearly to you? Why?

A photographic message must meet all the design principles discussed in Chapter 20. The message's parts must be in balance and in harmony with one another. In addition, the message's size and proportion should be pleasing. Once the shot is composed, most photographic messages are produced using either film or digital techniques. This involves using a camera, capturing the image, and reproducing the image.

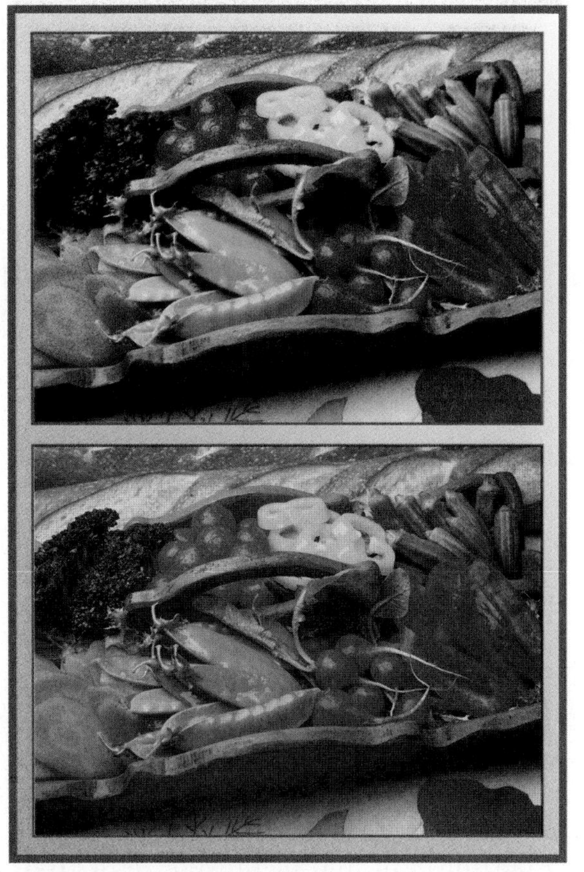

Figure 21-7. Notice how color adds interest to this photograph.

Recording a Photographic Image

Digital and film photography differ primarily in the way the image is captured. Film photography uses chemistry for this task. Digital photography uses electronics.

Cameras

A camera is the common way to capture an image for photographic communication. Both film and digital cameras have many common parts or systems. The main difference between these two types of cameras is the media used to retain the image. Film cameras use photographic film. Digital cameras use memory devices. Almost all cameras have a body, diaphragm, shutter, lens, and viewing system. See Figure 21-8.

Standards for Technological Literacy

17

Brainstorm

Ask your students what is meant by a photographic message and how is this different from a photograph.

Standards for Technological Literacy

17

Research

Have your students pick several events, activities, and settings and research what film speed should be used to photograph them.

Extend

Explain why slower speed film should be used to obtain the most detail in a photograph.

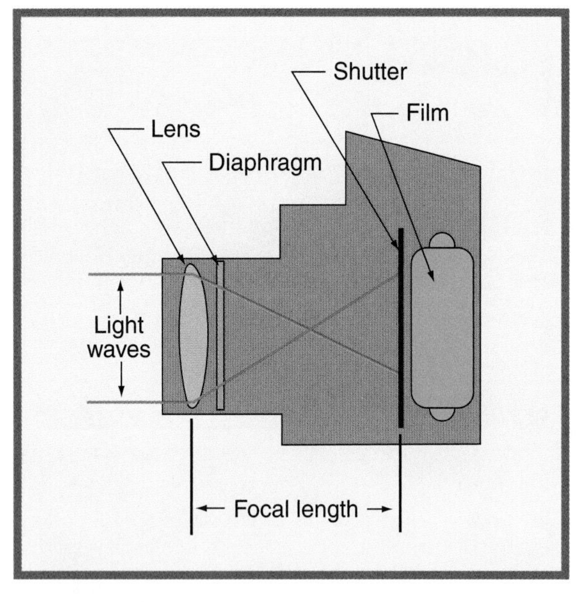

Figure 21-8. The major parts of a camera.

©iStockphoto.com/Inok

Figure 21-9. A typical camera diaphragm.

The ***body*** provides the overall structure for all other parts of the camera. This part is the basic framework of the unit. The ***shutter*** is a device that opens and closes to permit light into or prevent light from entering the camera. The length of time the shutter is open controls the length of time light can strike the film. Many cameras have adjustable shutters with speeds that can range from 1 second or more, down to 1/1000 of a second.

The ***diaphragm***, or aperture control, regulates the amount of light that can enter the camera at any given moment. Generally, it is made of a series of very thin metal leaves that can be adjusted to change the size of the hole (aperture) behind the camera lens. See **Figure 21-9.** An ***f-stop number*** identifies the size of the opening. The f-stop numbers are the reverse of the apertures. As the f-stop number increases, the size of the aperture decreases. The amount of change is specific. The amount of light doubles or halves from stop to stop. Thus, going from f/11 to f/16 decreases the light by one-half. Moving from f/11 to f/8 doubles the amount of light.

The ***lens*** focuses the light on the film or an image sensor. The focal length of the lens determines the size of the image. A typical 35-mm camera has a lens with a 50-mm focal length. This means the front of the lens is 50 mm away from the film. Different lenses are used for varying effects. A 28-mm, wide-angle lens produces a wide view for close objects. A 300-mm telephoto lens has a narrow angle of view and enlarges distant objects.

A camera captures only the light reflected from the scene in front of the lens. Therefore, it is essential that the camera is pointed precisely at the desired scene. Also, it is important that the object is in focus. Both of these conditions can be achieved using the viewing system.

Film cameras use one of two systems to select and adjust the view. These systems are the viewfinder system and the single-lens reflex (SLR) system. In the viewfinder system, the operator looks through a viewing port on the back or top of the camera. See **Figure 21-10.** The scene is viewed through a lens on the camera's front that is separate from the camera lens. The view the operator sees is approximately what the camera views. The view is not exactly what will be captured on film, however, because the viewfinder lens and camera lens are separate.

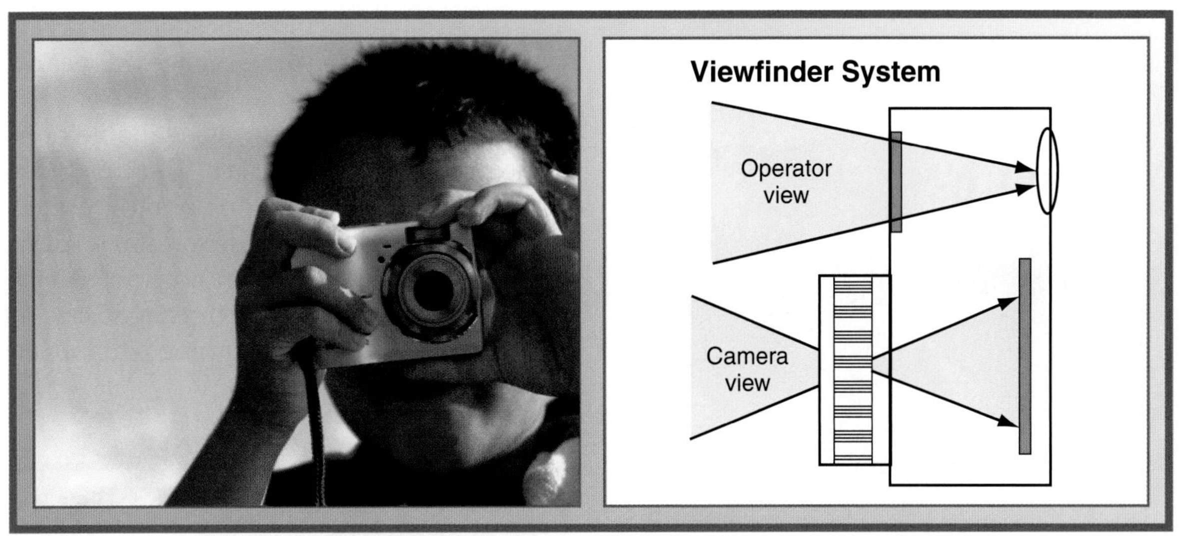

©iStockphoto.com/dejanaq

Figure 21-10. A viewfinder type of viewing system.

Academic Connections: History

The Beginning of Photojournalism

The creation of cameras and experiments with photography date back to the 1500s. Not until the mid-1800s, however, did the use of cameras begin to play a major role in how society viewed events. Perhaps the best example of the new power of photographic communication can be seen in the work of Matthew Brady, America's first photojournalist.

When the Civil War broke out in 1861, Brady was well-known as a portrait artist. He asked for and received permission to create a photographic record of the war. Brady hired a crew and placed people at various sites where soldiers were camped or battles were expected to break out. The resulting photographs were the first visual images ordinary American citizens had ever seen of war as it was occurring.

The images put a human face on the war. People saw pictures of soldiers at their daily chores, as well as pictures of dead combatants lying on the field after a battle. The photographs were dramatic and changed Americans' perceptions. Previous artwork had glamorized and romanticized war. The new images awakened Americans to war's other side, the side that includes both acute boredom and terrible suffering. Can you think of another example when new technology affected people's perceptions of reality?

More accurate viewing is achieved using an SLR system. See **Figure 21-11.** This system uses a series of prisms and mirrors that allow the operator to look directly through the camera lens. The view seen through the SLR is exactly the view captured on film. Digital cameras can use either a range finder system or an SLR viewing system. They also have a small video display showing the view before and after the picture has been taken.

Standards for Technological Literacy

3 7 17

Discussion

Use a camera to show your students the parts.

TechnoFact

In 1851, Frederick Archer reduced required exposure time and improved print quality with a new photographic process. In his process, a glass plate was coated with collodion, a wet, sticky emulsion, and silver salts. The collodion had to remain moist at all times; the photographer had to prepare a plate, expose it, and process the image immediately. For this reason, photographers using this process either worked in a studio or traveled in wagons with both a darkroom and a developing laboratory.

Standards for Technological Literacy

17

Discussion

Discuss the differences between panchromatic and color film.

Extend

Describe why depth of field is an important consideration in making a photograph. Show your students photographs that use depth of field effectively.

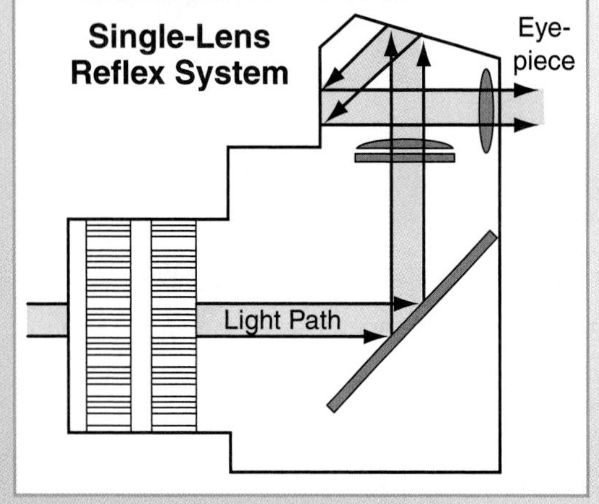

©iStockphoto.com/Kwanisik

Figure 21-11. An SLR viewing system.

Capturing an Image on Film

Film photography first captures the image or message on film. The film is then used to reproduce the message. Recording the message requires two procedures. These procedures are selecting the medium and exposing the film.

Selecting a film

The film used in producing photographs can be grouped according to type, format, size, and speed. The two major types of film are black-and-white film and color film. ***Panchromatic film*** is the most common form of black-and-white film. This film reacts to all colors of visible light and records those colors as shades of gray. Color film captures the image in nearly true colors.

Black-and-white film is available only in negative form. ***Negative film*** produces a reverse image of the photographed scene. This means the negative is dark where the subject is light and light where the subject is dark. Later, the negative can be used to make photographic prints. See **Figure 21-12.**

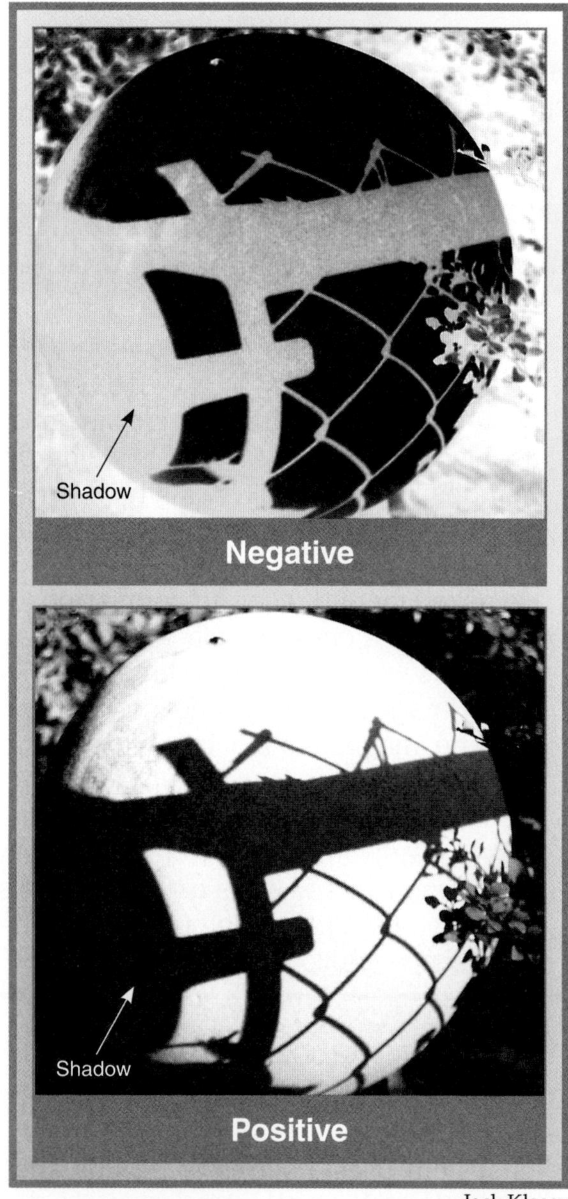

Jack Klasey

Figure 21-12. The negative and positive are of the same photograph.

Color film is available in a positive (transparency) or a negative form. ***Positive transparencies***, or *slides* (as small transparencies are called), produce the actual scene that can be viewed directly through a slide or movie projector. Color negatives are used to make color photographic prints.

Film comes in rolls and sheets. Roll film is loaded in a lighttight canister and is drawn from its container as it is exposed in a camera. Later, it is drawn back into the container and removed from the camera for ***developing***. Both sheet and roll film come in a variety of sizes. For example, you can buy 4″ × 5″ sheet film or 35-mm (width) roll film, to name just two options.

Both panchromatic and color films are available in a number of ***film speeds***. An American Standards Association (ASA) or International Organization for Standardization (ISO) number gives these speeds. Film speeds indicate the film's sensitivity to light. Fast film is very light sensitive. A slow-speed film takes a considerable amount of light to generate an image. Fast film is used to photograph objects and events in poorly lit situations or in photographing fast-moving objects. Slow film is used in well-lighted areas and outdoors. Films with ASA ratings of 25 to 100 are considered slow films. ASA ratings of 200 and above indicate fast films.

Exposing film

Exposing film requires the photographer to set the f-stop (aperture opening) and the speed and then focus the lens. Some of the lower-priced "point-and-shoot" cameras do these tasks automatically. To expose the film, the shutter-release button is depressed to cause the shutter to activate (open and close). While the shutter is open, light passing through the lens and diaphragm strikes the films. This causes a chemical reaction to take place on the film coating. Later, the film can be chemically developed to produce a copy of the original scene (image).

Capturing a Digital Image

Most digital cameras operate very similarly to simple automatic, point-and-shoot film cameras. The image to be captured is shown on a small screen on the back of the camera. See **Figure 21-13.** When the photographer is satisfied with the image shown, a button is pressed to activate the shutter. The shot is captured. Newer, more expensive digital cameras are the SLR cameras described earlier. They allow the operator to view the image to be captured through the camera lens.

Film cameras use light to expose film (make a photochemical record of an image). Digital cameras record the image on a digital-sensor array. This array is a specialized computer chip. The digital sensor has thousands of tiny square points that can sense light. These points are called *picture elements (pixels)*. A *pixel* is defined as "the information stored for a single grid point in the image." Essentially, a pixel is a square dot of light that has a given color and brightness value.

Figure 21-13. Most digital cameras have a viewing screen to see the image before and after it has been captured.

Standards for Technological Literacy

17

Extend

Discuss why some photographs are better in black and white and why some are better in color.

Demonstrate

Show your students how to load film and expose an image using a camera.

TechnoFact

In 1871, Richard Maddox's dry-plate process used a gelatin emulsion that dried on the plate but did not harm the silver salts. Photographers who used these new dry plates did not have to process a picture immediately.

Standards for Technological Literacy

17

TechnoFact

In 1888, George Eastman designed a new camera that was lightweight, easy to operate, and inexpensive. It was the first mass produced camera design intended for the amateur photographer. The new camera was called the Kodak.

Technology Explained

digital theater: a movie theater receiving and displaying digitally formatted movies.

Movie film is a series of pictures on a long strip of plastic. As the film is fed through a movie projector, the series of still pictures creates the illusion of motion. Providing each theater with a copy of the film, however, is expensive. Each copy can cost more than a thousand dollars to produce. If a movie is showing in several thousand theaters on any day, this cost is substantial.

A new system that can replace film movies is called *digital theater.* This process uses technology similar to home DVD players. A special device—a telecine scanner—converts the films to digital video for theaters. At this stage, the operators can correct the film for color. They can add sound and captions to the film, using computerized equipment. These operators can also produce alternate sound tracks for foreign languages.

At the local theater, the digital film is received and loaded on the local server. See **Figure A.** The server can play several movies into different auditoriums at the same time. A single server can service about four different auditoriums at once. Therefore, a 12-screen theater needs three servers.

The theater server directs the video part of the film to a special digital projector. The audio portion is sent to a cinema audio processor. Advanced systems automatically dim the lights, open the curtain, and start the film without human operators.

As a result of changes in digital technology, 3-D films are now more advanced and prevalent in theaters. Digital theaters can now accept different types of 3-D films while supplying the special glasses or headgear to accompany the film.

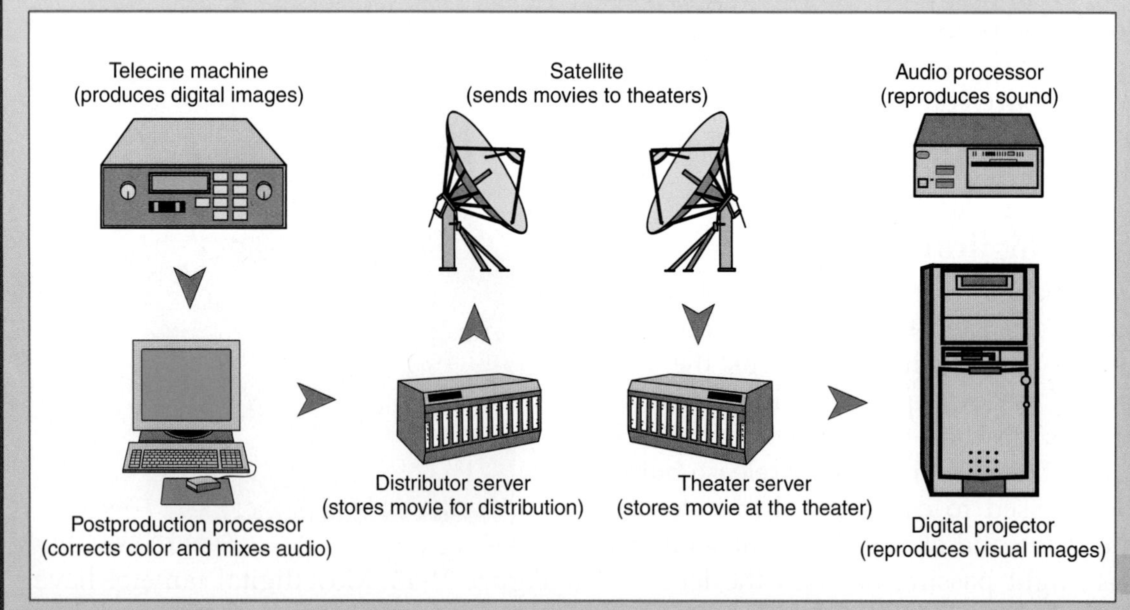

Figure A. How a digital movie is delivered to a theater.

Digital images are made up of thousands of pixels, just as a mosaic is made up of many colored pieces of glass. See **Figure 21-14.** The complete image is developed by an arrangement of pixels in rows and columns. The varying colors and brightness of the many pixels present the image. Digital cameras are available with differing numbers of pixels. The following are some common camera resolutions:

- 1216 × 912 pixels, or 1 megapixel (1 million pixels). This resolution provides good family photos.
- 1600 × 1200 pixels, or 2 megapixels. This resolution produces 4″ × 5″ photo lab–quality prints.
- 2240 × 1680 pixels, or 4 megapixels. This resolution provides the ability to print larger high-quality prints.
- 4064 × 2704 pixels, or 11 megapixels. This resolution produces large, high-quality prints.

The camera's circuitry processes the information the digital sensor has gathered and stores the information on a memory device.

Reproducing a Photographic Image

Communication technology is often used to produce messages for a large, diverse audience. Therefore, the message must be produced in quantity. Reproducing photographic-communication media requires developing film and making prints, printing images from digital files, or arranging digital images for viewing on electronic devices.

Reproducing an Image on Film

Developing film involves chemically treating the light-sensitive materials in the film emulsion to bring out the image. Black-and-white film is developed with a fairly simple process involving a series of chemical reactions. See **Figure 21-15.** These reactions are the following:

1. Developing.
2. Stopping.
3. Fixing.

Standards for Technological Literacy

17

Demonstrate

Show your students how to develop film.

Extend

Explain to your students why a red light can be used in a black-and-white darkroom but not in a color darkroom.

TechnoFact

Thanks to George Eastman's Kodak system, photographers no longer had to process their own pictures. The Kodak camera would capture 100 pictures on its roll of gelatin-coated film. The owner sent the entire camera, exposed film and all, to Kodak processing plants. There, they developed the film, made prints, and reloaded the camera. The reloaded camera and the prints were shipped back to the owner.

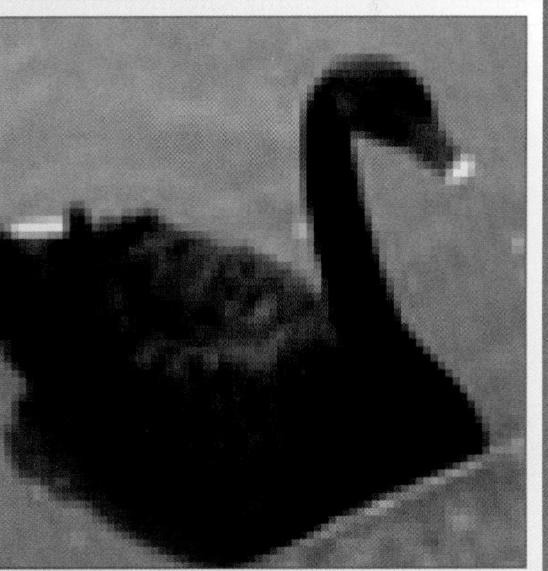

Figure 21-14. The picture on the left is made up of colored square pixels, as seen on the right.

Standards for Technological Literacy

17

Demonstrate

Show how to make a contact print from a negative.

Brainstorm

Ask your students why they think cleanliness is an essential element in photography.

TechnoFact

In 1924, the German-made Leica 35mm camera was introduced. It was a small camera that produced sharp, detailed photographs. For the first time, photographers could take pictures without their subject noticing, a practice that became known as candid photography.

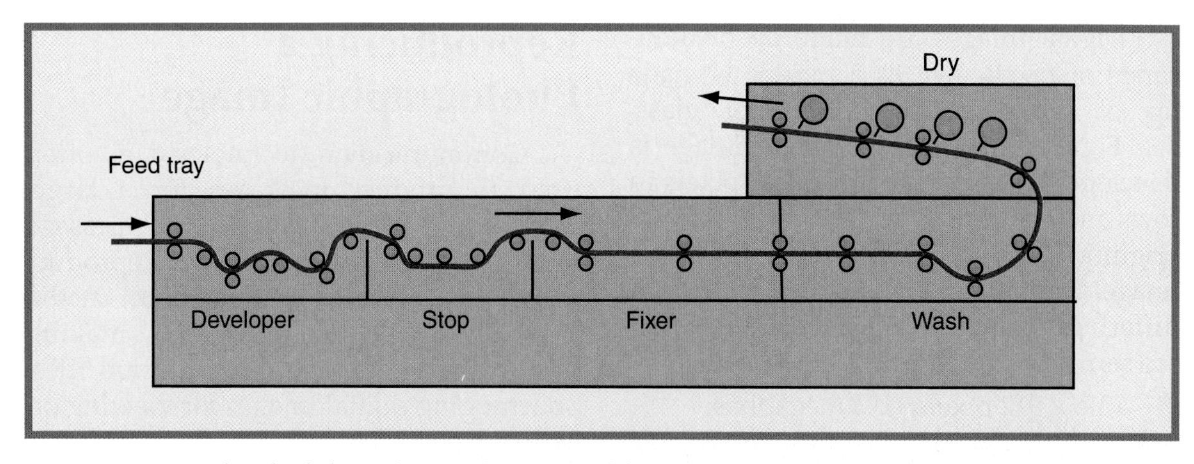

Figure 21-15. The dark line shows the path of film in an automatic film-developing machine.

The first step, developing, uses chemicals (developers) to alter the light-sensitive crystals in the emulsion. The crystals are changed from silver halide to black metallic silver. Silver halide cannot be seen. Black metallic silver can be seen. The next step, ***stopping***, stops the chemical action of the developer. Stopping uses an acid solution, called a ***stop bath***, to neutralize the developer. ***Fixing*** then removes the unexposed silver-halide crystals remaining in the film. This makes the image permanent. After these three steps are complete, the film is washed to remove any chemical residue. The film is then dried so it can be stored or used to make prints.

Safety with Darkroom Chemicals

Use care when mixing darkroom chemicals. They can cause skin and eye irritation. Also, use tongs and an apron when transferring prints from one solution to another. Wash your hands if they come in contact with the chemicals.

Most types of black-and-white film produce negative (reverse) images. A negative is used to make photographic prints by either contact printing or projection printing. In ***contact printing***, the negative is placed directly on top of a piece of light-sensitive, photographic paper. The film and paper are held tightly under a sheet of glass while they are exposed to a bright light. The paper is then removed and developed, with the photographer following the same steps used for developing film. These steps are developing, stopping, fixing, and washing. A different chemical, however, is used for developing prints than is used for film.

Projection printing is done to produce enlarged prints (prints larger than the negative). This is done by projecting (shining) light through the negative onto a piece of photographic paper. See **Figure 21-16.** Projection printing requires an enlarger to adjust and control the print size and focus.

Reproducing a Digital Image

The digital images stored on memory devices are generally downloaded directly into a computer system. There, the information can be manipulated (color balanced, cropped, and sized, for example) using various computer-software programs. The images can then be printed on an ink-jet printer or professionally printed through a camera store or other retail business. See **Figure 21-17.** They can also be viewed on a number of electronic devices. These devices include computer monitors and TV sets.

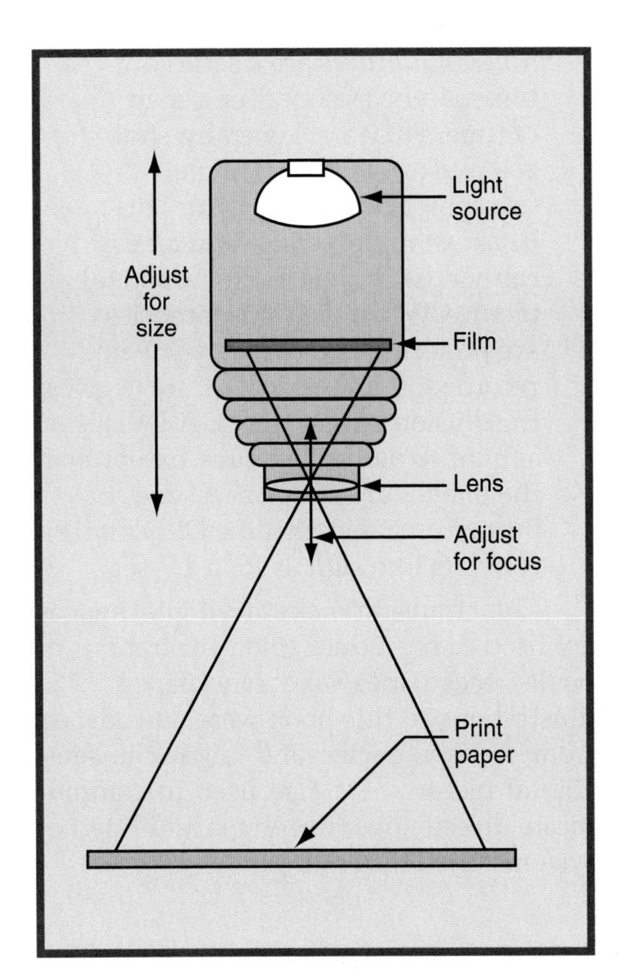

Figure 21-16. In projection printing, the operator shines light through a negative onto a piece of photographic paper.

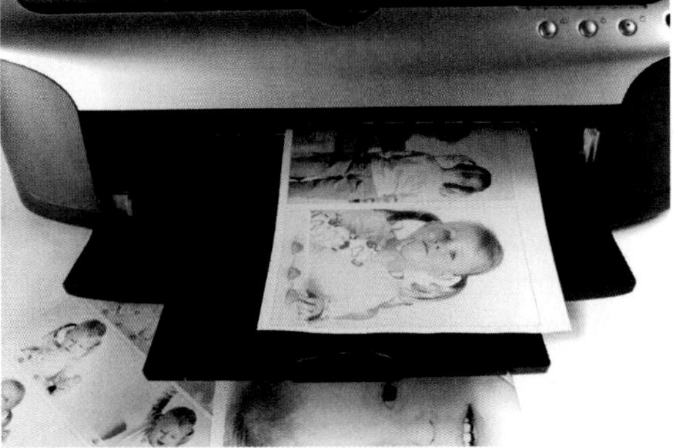

©iStockphoto.com/gmnicholas

Figure 21-17. Digital images are often printed using computer systems and ink-jet printers.

Types of Photographic Communication

Prints are one common use for photography. They can be used to communicate directly or integrated into a printed graphic message. This was discussed in Chapter 20.

Photography is also used to produce transparencies. As you learned earlier, these are positives of the recorded scene.

Standards for Technological Literacy

5 17

Research

Have your students research how a digital camera works.

Extend

Discuss the advantages and disadvantages of digital photography as compared to traditional photography.

TechnoFact

The electric flashbulb was introduced in 1929, followed by the electronic flash in 1931. These inventions allowed photographers to take pictures in places or under conditions that had previously been too dark.

TechnoFact

Modern filmmaking is built of the efforts of a number of pioneers, including Georges Méliès and Edwin S. Porter. They found that they could warp the motion in the movie by shooting film at a slower speed than the speed at which it was played back. This produced the unnaturally jerky and fast motion seen in early comedies.

Think Green

Wet Chemistry

Reproducing a photographic image on film is different from reproducing digital images. The changes in technology leading to the widespread use of digital photography has begun to reduce much of the wet chemistry used throughout the process of film development and image reproduction. Although the change to digital photography may have more to do with ease of use and to save time, it also removes the use of chemicals during the steps of processing film.

The main concern of using wet chemistry in film development has to do with its disposal. Developers and fixers contain toxins that could contaminate the ground and water supply. Environmental regulations are now in place for disposal of chemicals. Film developers are prohibited from pouring certain chemicals down the drain or on the ground. Digital photography has less of an impact on the environment.

Standards for Technological Literacy

17

Research

Have your students investigate how a motion picture camera and projector work.

TechnoFact

The motion picture camera captures 18 or 24 still images a second. When projected at the same rate, the still images produce the illusion of fluid, natural movement.

TechnoFact

In 1947, Edwin Land developed the first practical process for producing instant photographs. His Polaroid camera could capture a scene and develop a black-and-white photo in less than a minute.

They show the image as the eye sees it. Two common uses of transparencies in communication are the following:

- ***Slides*** are single transparencies designed to be viewed independently. They are cut from the film and mounted in frames. Slides can be arranged into sets called *slide series.* Digital images can also be arranged in a "slide series" using computer software and projected using a laptop computer and digital projector. See **Figure 21-18.**
- ***Motion pictures*** are a series of transparencies shot over a span of time. They are, generally, shot at a rate of 24 frames (single transparencies) per second. This rate is fast enough that the human eye cannot distinguish each individual picture. When the transparencies are projected at the rate they were recorded, the series of still shots gives the illusion of movement. A DVD is similar to motion pictures, except that the images are stored as a series of bumps on a disk that the DVD player converts into signals for a TV set.

Color transparencies and digital images are used to reproduce color illustrations in books, magazines, and newspapers. The illustrations in this book were reproduced from transparencies and digital images. Digital pictures are also used to communicate directly over telephone and Internet systems. See **Figure 21-19.**

©iStockphoto.com/Michal_edo, ©iStockphoto.com/vm

Figure 21-18. Laptop computers and digital projectors similar to those shown in the top photo can be used to project transparency-like images for meetings.

©iStockphoto.com/mbbirdy

Figure 21-19. A newer use for digital images is communication using cell phones and other similar technologies.

Using digital cameras, photographers can take photos and review them almost immediately afterwards.

Summary

Communication through photographic images is another method human beings have to express ideas, information, and feelings. To communicate effectively, however, we must know how to design, record, and reproduce the image or images correctly. Design involves knowing the audience and what message needs to be communicated. Recording involves choosing the proper film and exposing that film, based on the type of camera used. Reproducing involves developing film and making prints or transparencies.

Test Your Knowledge

Write your answers on a separate piece of paper. Please do not write in this book.

1. What is photographic communication?
2. List the three characteristics of light waves.
3. List the three steps involved in producing a photographic communication message.
4. Give one example of a question that must be addressed when designing a photographic message.
5. Name one design technique used to add interest to a photographic message.
6. The spot to which your eye is drawn in a photograph is called the _____.
7. What is one of the main differences between digital and film photography?

Matching questions: For Questions 8 through 12, match each description on the left with the correct part of a camera on the right. (Note: Answers can be used more than once.)

8. ______ Opens and closes to control light.
9. ______ F-stops measure its setting.
10. ______ Its speed ranges from 1 second to 1/1000 of a second.
11. ______ Its purpose is to focus light on the film.
12. ______ Is also called an *aperture control.*

A. Shutter.
B. Diaphragm.
C. Lens.

13. What is the difference between negative film and positive transparencies?
14. List the three steps in developing film.
15. Name the two basic techniques used to produce prints.
16. Name three common uses of transparencies in photographic communication.

Answers to Test Your Knowledge Questions

1. Photographic communication is the process of using photographs to convey an idea or information.
2. Amplitude, wavelength, and direction
3. Designing the message. Recording the image. Reproducing the message.
4. Student answers may include any of the following: Who is the intended audience? What is the message to be communicated? What type of photograph will best communicate the information or idea?
5. Student answers may include any of the following: using point of interest, using distance and scope, and using color.
6. Point of interest
7. Evaluate individually.
8. D. Light meter
9. A. Shutter (Note: B. Diaphragm may also be accepted at instructor's discretion.)
10. C. Lens
11. A. Shutter
12. C. Lens
13. Negative film produces a reverse image of the scene that is photographed. Positive transparencies produce the actual view, such that it can be viewed directly through a slide or movie projector.
14. Developing, stopping, and fixing
15. Contact printing and projection printing
16. Slides, filmstrips, and motion pictures

STEM Applications

1. Design and shoot a six-photograph series. Have the series describe how to do something or communicate an idea about pollution, energy conservation, or another important issue. Develop a layout, similar to the one below, to describe the theme and each shot you will use. Sketch the location of the major elements as you compose each photograph.

Theme or topic:		
Slide 1 (Description)	Slide 2 (Description)	Slide 3 (Description)
Sketch of the shot	Sketch of the shot	Sketch of the shot
Slide 4 (Description)	Slide 5 (Description)	Slide 6 (Description)
Sketch of the shot	Sketch of the shot	Sketch of the shot

2. Shoot and develop photographs that show the following:

   A. Happiness.

   B. Sadness.

   C. Quietness.

   D. Harmony with nature.

TSA Modular Activity

Standards for Technological Literacy

11

This activity develops the skills used in TSA's Photographic Technology event.

Imaging Technology

Activity Overview

In this activity, you will create a photographic display. The theme of the display can be provided by your teacher or chosen from one of the following:

- Your school.
- Your local community.
- Nature.
- Technology.

Materials

- A digital camera.
- A computer with photo-editing software and color printer.

Background Information

- **General.** Check your camera's settings before you take pictures. Most often, using the camera's automatic setting allows you to "point and shoot." This means you can concentrate on your subject, instead of worrying about camera controls. To avoid blurry photos, hold the camera firmly with both hands. Press the shutter button with a slow, steady motion. Do not jab the button with your finger. Jabbing causes the camera to shake.
- **Portraits.** Get close to your subject. Fill most of the viewfinder with your subject, instead of the background. Press the shutter button halfway to bring the subject into focus. Watch for a good expression. Take the photo. Try both people and pet portraits, with and without using the camera's flash. If your subject is against a bright background (such as a window), you probably need to use a flash, unless you want to do a silhouette.
- **Landscapes.** If the camera has a zoom lens, use the wide-angle setting to show a broad area. Try to get an interesting object (for example, a rock, bush, or fence) in the foreground to add depth to your picture. The wide-angle lens makes it easier to keep both near and distant subjects in focus.
- **Still life.** Try for a pleasing arrangement of objects that fit the theme. Do not use your flash, since the flash causes harsh shadows. Instead, use room lighting, possibly with a table or desk lamp placed just outside the picture to provide additional light. You can move the lamp around to place shadows where you want them.

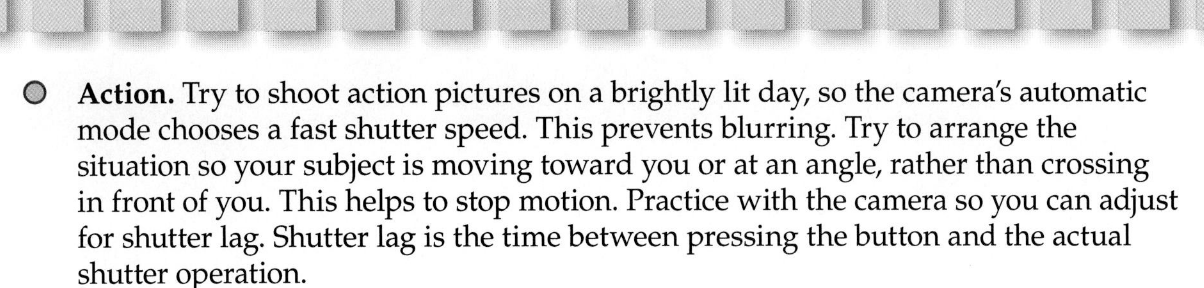

Standards for Technological Literacy

11

- **Action.** Try to shoot action pictures on a brightly lit day, so the camera's automatic mode chooses a fast shutter speed. This prevents blurring. Try to arrange the situation so your subject is moving toward you or at an angle, rather than crossing in front of you. This helps to stop motion. Practice with the camera so you can adjust for shutter lag. Shutter lag is the time between pressing the button and the actual shutter operation.
- **Photo editing and enhancement.** Use your computer's photo-editing software to adjust the brightness of (darken or lighten, as necessary) your picture. If the picture has a color cast, such as a very blue or very yellow appearance, your software should have the ability to remove the cast. Do not be afraid to crop away part of the picture. Many pictures can be greatly improved by eliminating extra background and focusing attention on the subject.

Guidelines

- Your display must include 15 prints. At least 5 prints must be color prints, and at least 5 prints must be black-and-white prints.
- None of the prints can be smaller than 3″ × 5″ or larger than 11″ × 14″.
- Your display should include a variety of prints, such as still life, action, product, portrait, landscape, and special effects.
- Prints can be single matted. Matting and display board must be a single color—white, black, or gray.

Evaluation Criteria

Your project will be evaluated using the following criteria:

- Composition.
- Lighting.
- Relation of images to theme.
- Processing and finishing.
- Creativity.
- Display quality.

Chapter Outline

This chapter covers the benchmark topics for the following Standards for Technological Literacy:

3 5 6 7 11 17

Learning Objectives

After studying this chapter, you will be able to do the following:

- Recall the definition of *telecommunication.*
- Summarize the major areas of physics that contribute to an understanding of telecommunications.
- Recall the major types of telecommunication systems.
- Compare FM and AM.
- Compare the two types of television stations.
- Recall the major activities involved in designing a broadcast message.
- Recall the activities involved in producing a broadcast message.
- Explain how broadcast messages are delivered.
- Summarize the characteristics of a remote-link system.
- Recall the three types of mobile communication systems.

Key Terms

alternating current (AC)
amplitude modulation (AM)
analog signal
audio message
broadcast frequency
broadcast system
carrier frequency
casting
conductor
digital signal
direct current (DC)
duplex system
frequency
frequency division multiplexing (FDM)
frequency modulation (FM)
hardwired system
hertz
induction
kilohertz (kHz)
megahertz (MHz)
multiplex
multiplex system
radio wave
script
simplex system
time division multiplexing (TDM)
transducer
ultrahigh frequency (UHF) station
very high frequency (VHF) station
video recording

Standards for Technological Literacy

3 17

Reinforce

Present telecommunication as part of communication technology (Chapter 19).

Brainstorm

Ask your students how telecommunication has impacted their lives.

Strategic Reading

Before reading this chapter, look at all the main headings and see if you can make a list predicting what the sections will be about. As you read, check your list to see if you were right.

Humans have always communicated their ideas and feelings. Early communication included people speaking to one another. Later, humans developed writing to record and transmit information. Writing allows us to express and store information, opinions, and concepts. These types of communication, however, do not meet all human needs. People wanted to hear the human voice beyond the limits of face-to-face communication. They wanted to communicate their thoughts and knowledge over great distances. Out of this desire came telecommunication. Telecommunication means communication over distance. See **Figure 22-1.** This communication implies that a message exists and that hardware (technology) is available to deliver the message.

Figure 22-1. Telecommunication means communicating over distance. This satellite-dish receiver is a telecommunication system that aids in capturing messages sent from great distances.

In this chapter, we will study various aspects of telecommunication today. Included are discussions of the physics behind telecommunication and the different types of major telecommunication systems that exist. We will also examine how we communicate with major telecommunication systems and look at other telecommunication technologies.

The Physics of Telecommunication

There are scientific principles behind technological devices and systems. One important branch of science providing a foundation for communication is physics. This area of science describes and explains the nature of matter, energy, motion, and force. The principles of electricity and electromagnetic waves are a part of physics and help people understand technology. These principles can help you understand the technology you use daily.

Electrical Principles

You have probably learned about the atom in your science classes. All matter is made up of atoms. Each atom has a nucleus, or center. This center is made up of positively charged particles, called *protons,* and neutral particles, called *neutrons.* A group of negatively charged particles called *electrons* orbit the nucleus. The electrons are held in orbit by their attraction to the positively charged protons. See **Figure 22-2.**

Standards for Technological Literacy

3 17

TechnoFact

The inventions of the telegraph, telephone, radio (wireless), and television have allowed instant communications over long distances (telecommunication).

TechnoFact

In 1864, James Clerk Maxwell predicted that electromagnetic waves existed and identified visible light as a band of frequencies in the electromagnetic spectrum. He also theorized that other, invisible, frequencies of electromagnetic waves existed. In the late 1880s, Heinrich R. Hertz verified Maxwell's hypotheses through a series of experiments. Hertz's work was the foundation for the later inventions of the radio and television.

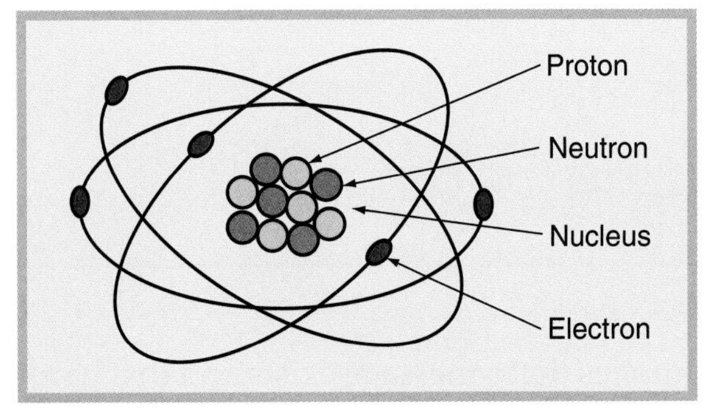

Figure 22-2. Atoms are made up of protons, neutrons, and electrons. Protons and neutrons form the nucleus. Electrons orbit the nucleus.

In certain situations, electrons travel from one atom to another. This movement is called *electricity* and often takes place in a metal called a ***conductor***. When electrons are flowing in one direction along the conductor, it is called ***direct current (DC)***. If the electrons flow in both directions along the conductor, reversing at regular intervals, it is called ***alternating current (AC)***.

Movement of electrons in a conductor creates magnetic lines of force. See **Figure 22-3.** This force is known as an *electromagnetic force*. As these lines of force increase and decrease in strength, they can cause electrons to flow in an adjacent wire. This process is called ***induction***. The principle of induction is commonly used to change sounds into electrical signals or to change electrical signals into sound. This process is used in microphones and speakers. These technological devices are examples of ***transducers***. Transducers change energy of one form into energy of another form.

Electromagnetic Waves

Earlier in this book, you were introduced to basic electromagnetic wave theory. Two important characteristics of waves are frequency and amplitude. See **Figure 22-4.**

Frequency

Frequency is the number of cycles (complete wavelengths) passing some point in one second. The number of cycles per second is measured in ***hertz***. The basic units of measurement in telecommunication are kilohertz (kHz) and megahertz (MHz). One ***kilohertz (kHz)*** is 1000 cycles per second. One ***megahertz (MHz)*** is 1 million cycles per second.

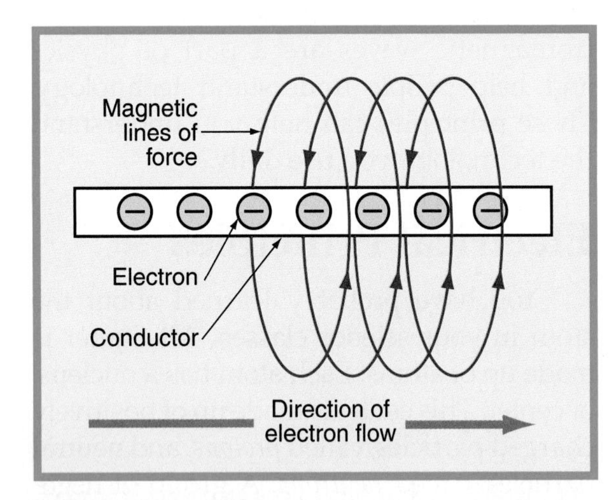

Figure 22-3. Electron movement through a conductor is called *electricity*. This movement creates magnetic lines of force around the conductor.

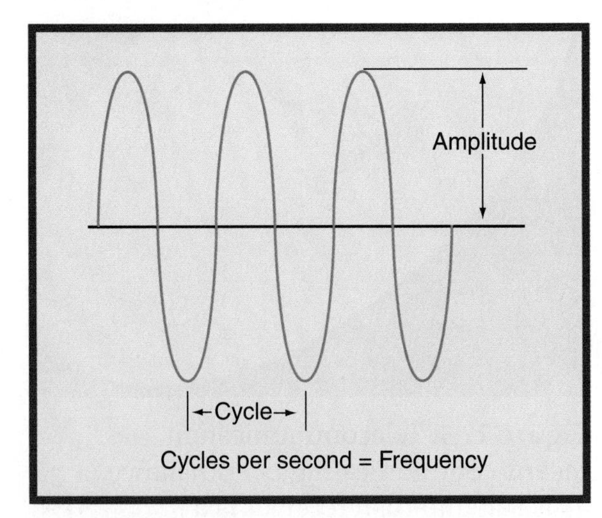

Figure 22-4. The information in radio waves is coded through varying the frequency and amplitude of the wave.

Standards for Technological Literacy

3 17

Extend

Discuss how sections of the electromagnetic spectrum are assigned to different types of communications.

TechnoFact

Alexander Graham Bell, who filed for his patent in 1876 and 1877, is credited as the inventor of the telephone. However, many different inventors have improved and modified the phone over the years, helping it evolve into the phone we use today.

Career Corner

Arts, A/V Technology & Communications

Radio and Television Broadcasting

Broadcasting is experiencing rapid change, brought about by an expanded use of computers and interactive media. Although the on-air positions are most recognized, there are more off-camera and off-microphone jobs than on-air positions. Many of these technical, sales, and administrative positions pay as well as or better than the ones on-air performers hold. Nonbroadcast careers in electronic media, radio, television, and broadcasting include broadcasting station manager, film or tape librarian, community-relations director, unit manager, film editor, news director, news writer, transmitter engineer, technical director, advertising-sales coordinator, producer, account executive, and floor manager.

Within the electromagnetic spectrum is a series of frequencies we call ***radio waves***. These frequencies extend from around 30 hertz to 300 gigahertz (300 billion cycles per second). Below this range of frequencies is a series we call *extremely low frequency (ELF)*. The series extends from about 10 Hz to 13.6 Hz. ELF is used for underwater communication.

Radio waves are part of a series of frequencies known as ***broadcast frequencies***. These frequencies are used for a wide range of communication systems. These systems include police- and fire department radio, broadcast radio, cellular telephone, and television communication.

The Federal Communication Commission (FCC) assigns each type of communication to a range of frequencies. For example, 160.215 to 161.565 kHz is assigned for railroad communication. For 6-meter amateur radio, 50.0 to 54.0 MHz is assigned.

Amplitude

Amplitude measures the strength of the wave. The higher the amplitude is, the stronger the signal is. Telecommunication uses changes in the amplitude of the waves, the frequency of the waves, or both to carry a communication message.

Types of Signals

Signals are fundamental to communication technology. They are used to carry information from a source to a receiver. All signals have two basic properties:

- They contain energy and power.
- They have a pattern carrying information.

The basic types of signals used to communicate information are analog and digital signals.

Analog signals

Thomas Edison developed the first workable device that could record sound in 1877. His machine used a diaphragm attached to a needle to record sound on a foil cylinder. When a person spoke into the diaphragm, it vibrated, causing the needle to scribe a groove into the foil. In the playback mode, the opposite happened. The needle traveled along the groove of the rotating cylinder. The movement of the needle vibrated the

Standards for Technological Literacy

3 17

Research

Have each student listen to an FM broadcast and an AM broadcast and write a short report comparing the sound qualities of the two broadcasts.

TechnoFact

The telegraph is an example of a hard-wired communication system. Samuel F. B. Morse developed the first operational telegraph system. His system used a single overhead wire and the earth as the other conductor. Only one message could be sent at a time on the Morse system. In 1872, J. B. Stearns devised a method for sending two messages simultaneously over the same wire (duplex telegraphy).

diaphragm and reproduced the original sound. Edison's invention was improved by a number of people over the years and became the common way to record speech and music. This type of recording equipment stores the sound using analog signals.

An ***analog signal*** is a continuous electronic signal carrying information in the form of variable physical values. Information is added to the base signal by amplifying the signal's strength (AM) or varying the signal's frequency (FM). See **Figure 22-5.** AM and FM are discussed later in this chapter.

Until recently, nearly all telephone, television, and radio signals used analog signals because analog signals are fairly easy to create and transmit. Analog signals have, however, one major disadvantage. Every detail in the signal pattern is important. Outside forces can alter the pattern. These alternations create noise and distortion that make the output different from the input. This major problem has led to the development of a newer type of signal—*digital signals.*

Digital signals

Digital technology generates, stores, processes, and transmits data using positive and nonpositive electrical states. The positive state (on) is expressed by a *1.* The nonpositive state (off) is expressed by a *0.* All digital data is, therefore, expressed as a string of *1*s and *0*s. See **Figure 22-6.**

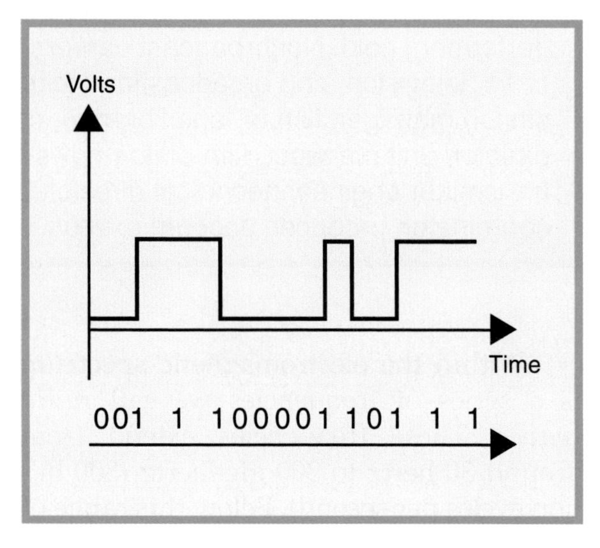

Figure 22-6. A digital signal is a series of *1*s (on) and *0*s (off).

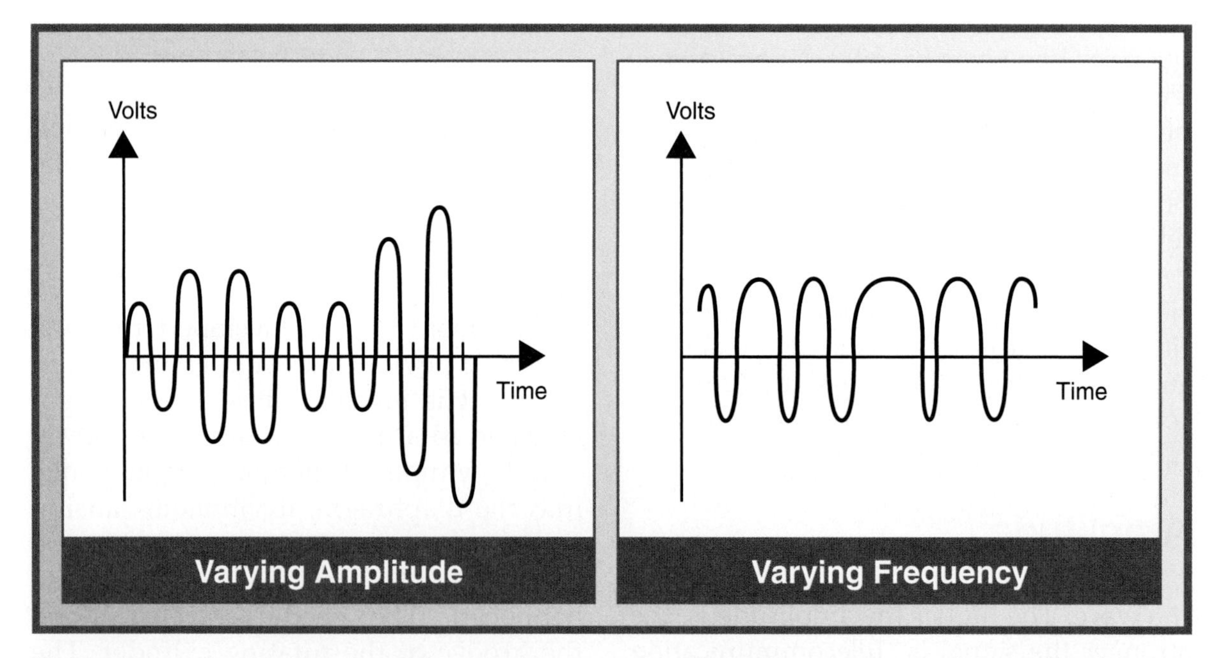

Figure 22-5. An analog signal is a continuous signal varying in strength (amplitude) or in frequency.

This type of signal has several advantages over analog signals. Digital signals can be transmitted faster, therefore allowing more information to be moved in a given period of time. Also, they are more accurate and less prone to outside interference (noise and distortion). This makes digital music recordings (such as those found on compact discs) clearer than older analog long-playing (LP) recordings and cassettes. Digital recordings produce sound that is very close to the reproduced signal. They also sound the same no matter how many times they are played. Likewise, digital TV programs have clearer pictures and better sound than analog programs.

Types of Telecommunication Systems

There are many types of telecommunication systems. These systems can be divided into two major types. The types are hardwired systems and broadcast systems. See **Figure 22-7.**

Hardwired Systems

Communication systems have three major parts. These parts are a sender, a communication channel, and a receiver. A typical ***hardwired system*** is the telephone system. The telephone system is described briefly in Chapter 7. See **Figure 22-8.** In this system, the microphone in the mouthpiece changes sound waves into electrical impulses. The frequency and duration of the electrical impulses are the coded message.

These electrical codes are usually conducted over a permanent waveguide connecting the sender and the receiver. This guide might be a copper wire or fiber-optic (glass fiber) conductor. In some cases, microwave radio signals take the place of a waveguide for a portion of the circuit. Microwaves are often used to send a message between major cities. In these cities, the signals are transferred back onto wire or cable.

A special system called ***multiplexing*** is frequently used to increase the capacity of the waveguide. This system allows several unrelated messages to travel down

Standards for Technological Literacy

3 7 17

Brainstorm

Have your students list examples of hardwired and broadcast communications.

TechnoFact

The radio is based on Maxwell's and Hertz's work in electromagnetic waves. In 1895, Guglielmo Marconi applied these theories in designing a wireless communication system he called the wireless telegraph, which is considered the first successful radio. He continued to improve the system. In 1901, a radio of his design successfully transmitted the letter S across the Atlantic Ocean.

Research

Have your students diagram a telegraph circuit as used in the early 20th century.

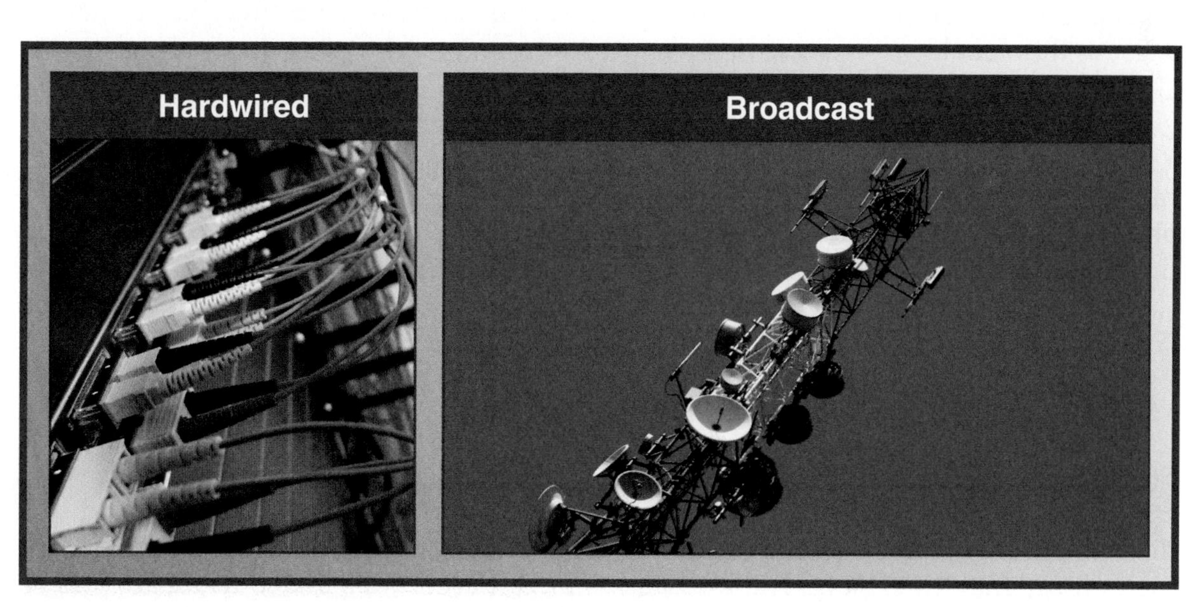

Figure 22-7. Telecommunication systems can be grouped as hardwired or broadcast systems. Shown on the left is a fiber-optic (hardwired) unit. On the right is a communications tower (broadcast).

Standards for Technological Literacy

17

TechnoFact

In the 1910s, 1920s, and 1930s, Edwin Armstrong worked tirelessly to improve and expand the applications of the radio. In 1912, he invented the first radio amplifier. During World War I, he developed the superheterodyne circuit, the basic receiving circuit used in most modern radios and television sets. Perhaps his greatest accomplishment was the development of a wide-band frequency modulation (FM) system in 1933.

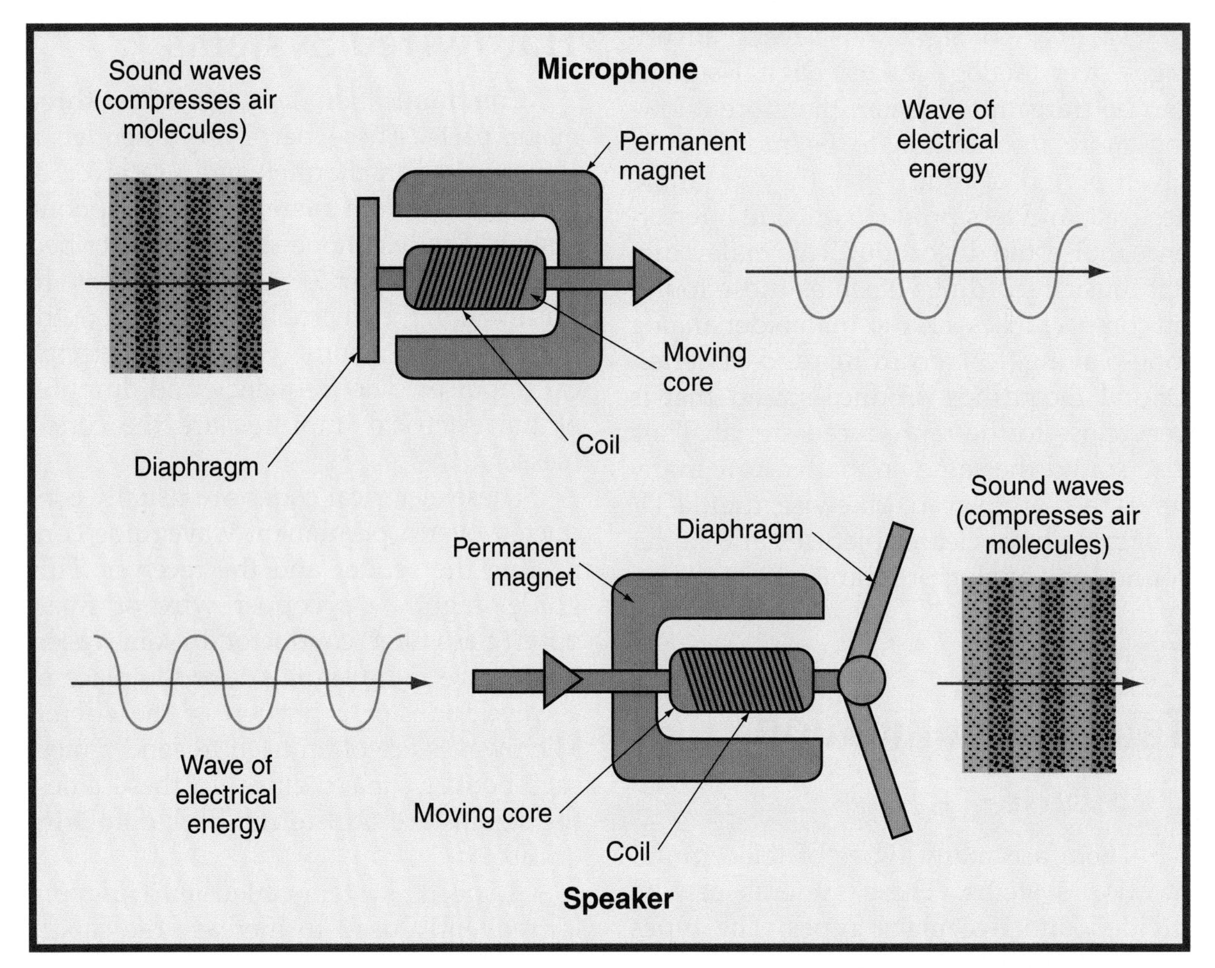

Figure 22-8. Telephone systems use coded electrical messages to communicate the spoken word. Shown here are sound waves being transferred into electrical waves by a microphone. The electrical wave is then channeled to a receiver. In the receiver, a speaker decodes the signal.

a single conductor at the same time. This can be done through time division multiplexing (TDM) or frequency division multiplexing (FDM).

Time Division Multiplexing (TDM)

Time division multiplexing (TDM) divides time into very brief segments. Several messages are also divided into small discrete bits. Electronic equipment assigns each bit separate time slots. The message bits are transmitted. The receiver collects all the bits, sorts the bits, and then assembles each message back into its original form.

Frequency Division Multiplexing (FDM)

Frequency division multiplexing (FDM) uses a separate frequency to transmit each message. Several messages are transmitted at the same time over the same waveguide. They are blended together and channeled down the carrier. At the receiving end, the various frequencies are separated. Each message is delivered to a separate receiver. The speaker in the earpiece decodes telephone messages. There, the electrical impulses are changed back into an audible sound.

Broadcast Systems

Broadcast systems send radio waves carrying the signal through the air from the sender to the receiver. The transmitter (sender) changes sound into a signal containing the message. This signal radiates into the atmosphere from an antenna. Another antenna attached to a receiver gathers the signal. The receiver separates the desired signal from other signals and changes it back into audible sound. See **Figure 22-9.**

Generally speaking, radio signals radiate in all directions from an antenna. Telephone-microwave communication systems use directional antennas to focus the signal to receiving antennas. Two common broadcast systems are radio and television.

Radio Broadcast Systems

Radio communication was the first widespread broadcast medium. Originally, it was called *wireless* because no hardwired connection exists between the sender and the receiver (radio set). See **Figure 22-10.**

©iStockphoto.com/Adrio

Figure 22-10. An example of an early wireless (radio).

All radio broadcast systems use a ***carrier frequency*** radiating from the transmitter. This is the carrier frequency you tune your radio to, to receive a station. The code for the audible sound is imposed onto this frequency. See **Figure 22-11.** The earliest radios used ***amplitude modulation (AM)*** to code the carrier frequency. These systems merged the message onto the carrier wave by changing the strength of the

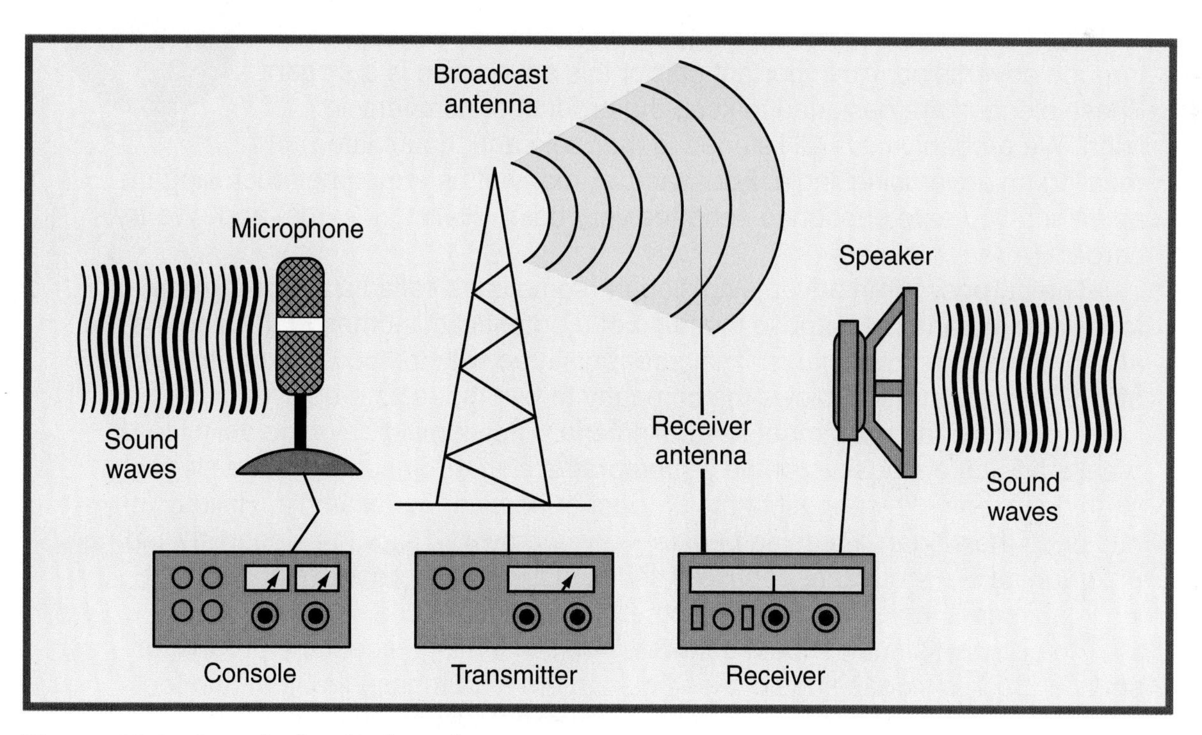

Figure 22-9. A typical radio broadcast system.

Standards for Technological Literacy

3 7 17

Research

Have your students determine the call letters and frequencies of the radio stations in your area.

Figure Discussion

Discuss the parts of a radio broadcast system shown in Figure 22-9 in terms of encoding, transmitting, receiving, and decoding messages.

TechnoFact

In 1904, Sir John Fleming developed the thermionic radio valve, a vacuum tube that could detect radio signals. Two years later, Lee de Forest added a grid to the vacuum tube. A weak signal applied to the grid could control much greater currents through other elements of the tube. This new tube, called a triode, could both detect and amplify radio waves.

Standards for Technological Literacy

3 6 7 17

TechnoFact

In 1919, General Electric Company formed the Radio Corporation of America, which acquired patents from U.S. manufacturers of radio equipment. In 1920, Westinghouse launched the first radio broadcast station, KDKA, in Pittsburgh.

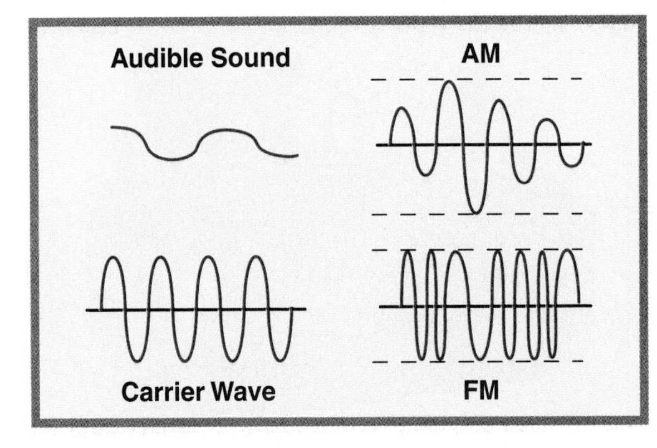

Figure 22-11. Radio waves are changed to carry the message through AM or FM. Notice how the amplitudes of the sound wave and the carrier wave have been blended in AM. The amplitude of the combined waveform oscillates in a pattern similar to the sound wave. Notice, in the frequency modulated waveform, how the frequency varies in a pattern with the initial sound wave.

carrier signal. This type of broadcast radio is assigned the frequencies between 540 and 1600 kHz.

Later, radio broadcast systems using ***frequency modulation (FM)*** were developed. These systems encode the message on the carrier wave by changing the wave's frequency. The 200 separate FM radio-broadcast frequencies range from 88.1 to 107.9 MHz. Look back to Chapter 7 for an explanation of the operation of radio communication systems. Included in that discussion is an explanation of the operation of radio transmitters and receivers.

Television Broadcast Systems

Television broadcast systems are really two systems in one. Each channel is assigned a bandwidth. See **Figure 22-12.** A

Academic Connections: Communication

Advertising

Many commercial telecommunication broadcasts are supported through advertising. An important part of this advertising is a slogan. These slogans are designed to keep the product or company in mind. We probably have all listened to the radio and, hours later, recalled an advertising jingle. For example, many of us remember such slogans as "When you care enough to send the very best" (Hallmark cards) and "We try harder" (Avis rental cars).

The purpose of an advertising slogan (sometimes called a *tag* or *tag line*) is to leave the brand message in the mind of a potential customer. Many slogans are also protected as trademarks. They are registered with the government trademark office. This registration allows the company to use the registered symbol (®).

A good slogan must meet several criteria. First, it must have the ability to be recalled. Such a slogan is called a *memorable slogan.* Second, a good slogan should present a key benefit of the product or service. For example, Holiday Inn has used the slogan "Pleasing people the world over." Third, the slogan should differentiate the product or service. The Hallmark slogan above suggests that Hallmark cards are better than any others.

An effective slogan should generate positive feelings about the product or service. Such a slogan should cause customers to purchase the item and feel good about owning it. Can you think of three examples of effective slogans?

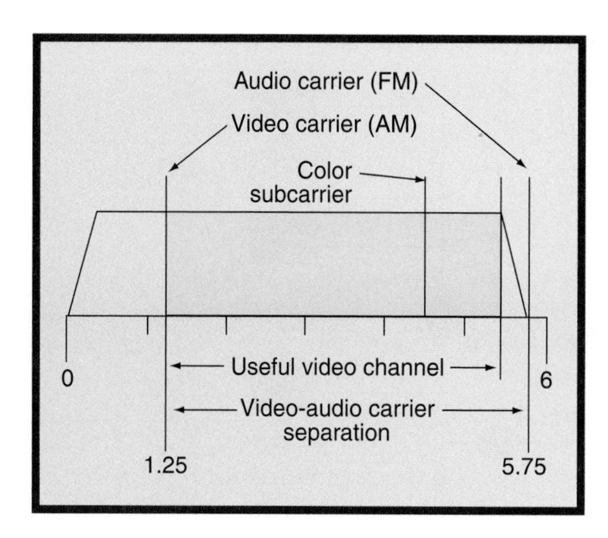

Figure 22-12. This diagram of a traditional (analog) television broadcast channel shows the audio and visual broadcast frequencies.

radio-like system uses one portion of the band to send and receive the audio (sound) portion of the message. This portion uses FM to impose the ***audio message*** onto the carrier wave.

The larger portion of the band is assigned to a second system. This system communicates the video (visual) part of the program and uses AM to send the picture portion of the message. At the ends of each channel are unused frequencies. These buffer zones keep the signals from one channel from disturbing adjoining channels.

Television systems use a microphone to capture the sound. A camera generates the picture. The television receiver reproduces the program using a speaker for the sound and, usually, a cathode-ray (TV) tube to display the picture. The operation of these devices is discussed later in this chapter.

Traditional television systems

Two basic types of television stations exist. ***Very high frequency (VHF) stations*** broadcast on channels 2–13. Each channel is assigned a specific 6-MHz range of frequencies. Channels 2–6 broadcast on frequencies between 54.0 and 88.0 MHz. Channels 7–13 use the range between 174 and 216 MHz. The frequencies between these two broadcast ranges are used for FM radio, aircraft-navigation and aircraft-communication operations, weather satellites, and amateur (ham) radio.

The ***ultrahigh frequency (UHF) stations*** broadcast on channels 14–83. They use the frequency range from 470 to 890 MHz. VHF channels are generally assigned to the major television-network outlets and large local stations because they can broadcast a greater distance than UHF channels can. Public broadcasting and smaller independent stations are often found on the UHF channels.

Cable television systems

For many years, there were three major television networks in the United States. These networks had broadcast stations in major cities that produced signals that could be received only if a home's antenna was in a line of sight from the transmitting antenna. People living in remote and mountainous areas could not receive programs that were becoming a part of everyday American life. In 1948, people living in the valleys of the Pennsylvania mountains could not receive a television signal. They started putting antennas on hilltops and running cables to their houses so they could receive their favorite television programs. This was called a *community antenna television (CATV) system* and was the start of cable television as we know it today. See **Figure 22-13.**

Cable television is a system that transmits signals to televisions through fixed optical fibers or coaxial cables, as opposed to the through-the-air method used in traditional television broadcasting. Today, cable systems deliver hundreds of channels of television, along with high-speed Internet access.

Standards for Technological Literacy

17

Brainstorm

Ask your students why Americans have more television stations to watch than people in any other country.

TechnoFact

Narinder Kapany coined the term fiber optics in 1956 to describe flexible glass fibers with a reflective coating.

Standards for Technological Literacy

17

Brainstorm

Ask your students to list ways the electronic communication has improved their lives.

TechnoFact

The television camera dates to the work of V. K. Zworykin. In 1923, he developed the iconoscope, the first successful television camera tube. It used many of the same principles as modern television cameras.

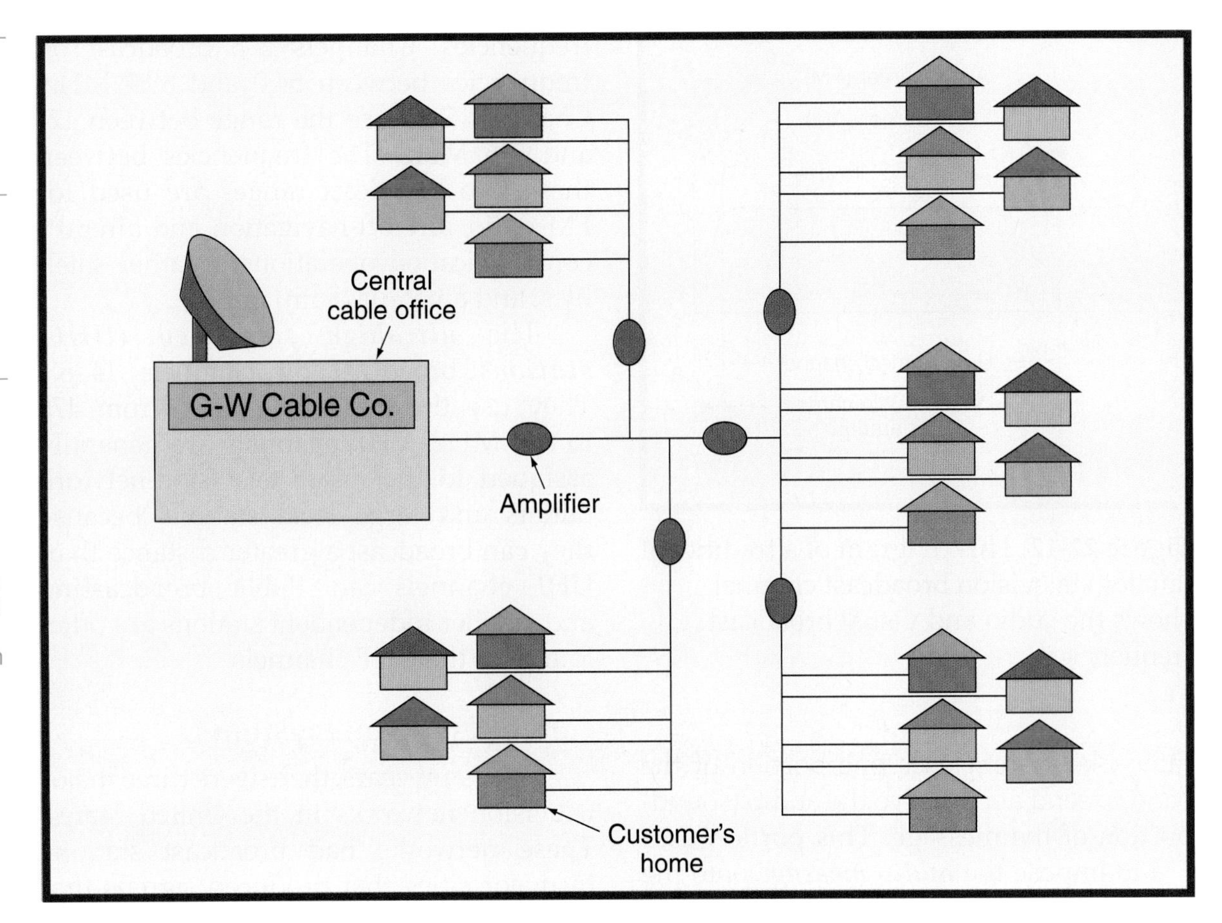

Figure 22-13. This diagram shows a typical cable television system.

Satellite television systems

There are parts of the country that cannot get line of sight reception from television stations. They also do not have cable television service. This challenge gave rise to the third type of television broadcast (delivery) system. The third type of system is satellite television. The common satellite TV systems use an all-digital signal to deliver audio and video. The system's major actions, very similar to satellite radio's, are the following:

1. Programming to be delivered to customers is developed or selected.
2. Broadcast centers receive various programs and beam them to satellites in geosynchronous orbit.
3. Satellites receive signals from the broadcast station and rebroadcast them to the customers.
4. Dishes receive the signals from the satellite and relay them to the customer's receiver.
5. Receivers process the signals and relay them to television sets for display.

Communicating with Telecommunication Systems

Earlier, you learned that telecommunication includes a message and a delivery system. Communicating with telecommunication systems requires three distinct actions. See **Figure 22-14:**

1. Designing the message.
2. Producing the message.
3. Delivering the message.

Figure 22-14. Telecommunication messages move through three stages. They are first designed, then produced, and finally broadcast.

Designing the Message

Many communication messages are not carefully designed. When you talk on the telephone, you often have to adjust to circumstances. You have a basic message, or reason for the call. Yet, spontaneous interaction generally occurs between the caller and the receiver. Callers adjust their messages as they receive feedback from the receivers. The format and sequence of the messages are not fixed.

This scenario is not typical for broadcast telecommunications. Most broadcast messages are the result of focused design activities. These activities are the following:

1. Identifying the audience.
2. Selecting the approach.
3. Developing the message.

To understand these steps, let us explore the development of a television program. A radio program follows a similar procedure, except for the use of visual impressions.

Standards for Technological Literacy

17

Extend

Discuss why different kinds of programs run at different times of the day and on different channels.

TechnoFact

Two different color television systems were developed in the 1950s. The Columbia Broadcasting System (CBS) developed one system, and the Radio Corporation of America (RCA) developed the other. In 1953, the FCC revised the standards for color television broadcasts to eliminate the problem of incompatible systems. The RCA system met these standards and the CBS system did not.

Standards for Technological Literacy

17

Extend

Discuss the history of fiber-optic communication, including Claude Chappe's optical telegraph and Alexander Graham Bell's optical telephone system.

Research

Have each student prepare a chart comparing fiber-optic and copper-wire systems, in terms of the cost, the weight of the conductor, the speed of the signal transfer, and the distance signals can be moved.

TechnoFact

Audio (sound) recording converts acoustic energy (audible sound) into a form that can be stored and reproduced at any time. In 1855, Leon Scott developed a very early recording device called a phonautograph.

Technology Explained

fiber optics: channeling messages, in the form of light, through glass fibers.

Information has been transmitted with light for many years. One of the first methods used was smoke signals. More recently, flags and flashing lights have been used to convey information. All these techniques, though, are limited to the line of sight between the sender and receiver. Practical use of long-distance light communication came with the invention of optical cables.

Optical cables are channels guiding light waves, through internal reflection, over some distance. Internal reflection means that, when the light waves strike the outer edge of the fiber, they are reflected back toward the center. Optical communication of this type is called *guided optical transmission*. The development of glass fibers for guided optical transmission began in the 1960s.

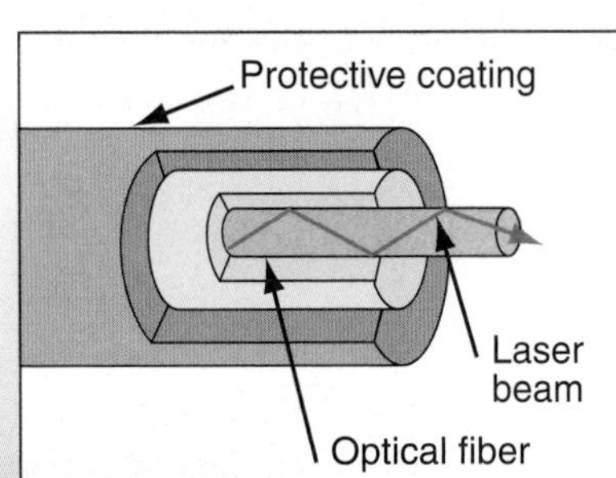

Figure A. A diagram of a fiber-optic cable.

A typical fiber-optic cable has three layers. See **Figure A.** The outside layer is a protective plastic coating. The middle layer is called *cladding*. This layer reflects the light waves back into the glass fiber. The inner layer is a strand of glass called the *core*. The individual core strands are as thin as human hairs. Typically, they are about 0.0005″ in diameter.

Several hundred of these strands are bundled into larger cables. Each of these fiber-optic waveguides can carry numerous messages. Billions of bits of information per second can move down an optical fiber. See **Figure B.** This is the same amount of information contained in thousands of independent telephone conversations. As you can see, optical fibers are capable of carrying encyclopedias of information per second.

Fiber-optic cables are rapidly replacing copper circuits in telephone systems. They are less expensive to install, smaller in diameter, use less power, and are much more resistant to interference. Phones linked with fiber-optic cable remain unaffected by the flashes of lightning. Fiber-optic systems are especially suited to carrying the digital signals widely used in cable television systems and computer networks.

Fiber optics is finding uses in areas other than communication. They are used extensively in imaging applications such as bronchoscopes and endoscopes in medicine and welding, jet engine, and plumbing-pipe inspections in industry.

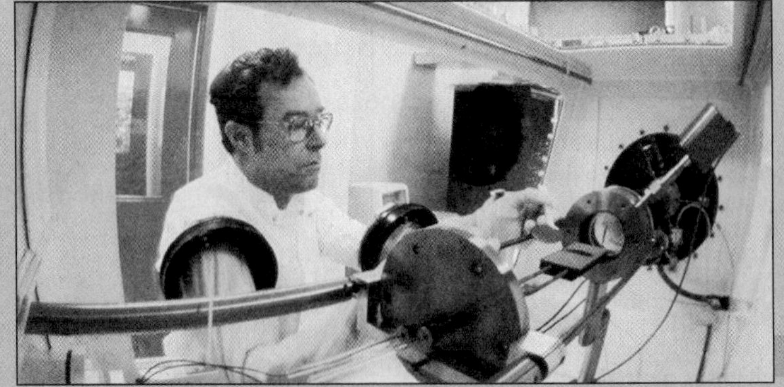

Northern Telephone Co.

Figure B. A worker testing an optical fiber.

Identifying the Audience

The first step in designing a television program or an advertisement is to determine who the audience is and what they like. Different people find different things important. Many teenagers want to be part of "the group." Therefore, advertisements and programming directed toward this age group emphasize this factor. Advertisements often try to convince teenagers that wearing the right clothes or driving the right car increases popularity.

The aims of the communication are chosen through research. Actions that study audience needs and wants are called *audience assessment*. Audience assessments are discussed in Chapter 20.

Selecting the Approach

Once the audience has been identified and described, designers select the approach most effectively communicating their message. In advertising, this might involve a catchphrase such as "All news, all the time," or "Quality people, quality products." Included in the approach is the tone of the message. The message can be expressed in a serious or humorous way. A regal (dignified) technique is another option. In another case, a contemporary (present-day), historical, or futuristic theme can be used. Whatever the style, the tone must appeal to the audience.

The tone must fit the situation. A program covering the effects of air pollution will probably not use a humorous approach. On the other hand, a program on circus life might have humor as an important component.

Developing the Message

The number of people who receive and act on a broadcast message determines the ultimate success of the message. Therefore, the message must be carefully crafted. This most often requires a ***script*** identifying the characters, developing a situation, and communicating a story. Also, a script provides the dialog for the characters and describes their movements on the stage.

The previous knowledge level of the audience must be considered. Those involved in local programming assume the audience has an understanding of local politics and geographical features (such as towns and rivers). People who deal with national programming might need to explain these details. Four common types of scripts are used in television:

- Full scripts
- Partial-format scripts
- Show-format scripts
- Fact scripts

Full, or detailed, scripts are complete scripts containing the following:

- Every word to be spoken.
- Sound effects and music information.
- All major visual effects.
- Production notes (timing and camera angles, for example).

See **Figure 22-15.** Generally, this script is written in a two-column format. The left column contains video and production information. The right column contains audio information. This information

Video	Audio
Show title	Music theme up, and then fade.
Host seen– head and shoulders	Art has long been an important communication medium. In recent years, it has grown in popularity as the availability of reasonably priced, quality paintings has grown.
Background light as the camera dollies past host to pan art on back wall	In our city, this trend is easily seen with a visit to the Artists' Guild show at the convention center.
Zoom in on watercolor #1	Artists using many media are exhibiting the results of their talent.

Figure 22-15. An example of a full-script format for a local public-interest program.

Standards for Technological Literacy

17

Brainstorm

Ask your students why scripts are necessary.

TechnoFact

Thomas Edison was the first person to record a sound and play it back. He patented the machine he used to do this, the phonograph, in 1877. The quality of sound recordings improved drastically in 1925, when microphones and electronic amplifiers were added to the recording process.

Standards for Technological Literacy

17

Demonstrate

Show your students how to develop a storyboard.

TechnoFact

In 1947, Peter Goldmark developed the LP (long-playing) record. Much more music could be recorded on an LP than on the 78-rpm discs of the day. In 1970, James Russell developed the first compact disc system.

includes the dialog (words to be spoken) and sound effects (background sounds).

Full scripts for commercials often develop from a storyboard. See **Figure 22-16.** A storyboard is a series of sketches or pictures showing all the scenes in the commercial. The commercial's dialog and production notes are under each sketch.

Partial-format scripts contain complete scripts for the introduction and conclusion. The remaining parts of the show are simply outlined. This type of script is used for talk shows and sports events, in which actions or responses give direction to the show. The script simply provides a skeleton around which the show develops. Often, it contains time cues and other production notes.

Show-format scripts develop a list of the various film or show segments. Production notes and timing cues are included. This type of script is used for programs following a specific format for each show. Programs using this type of script include morning and nighttime news programs. See **Figure 22-17.**

Fact or rundown scripts develop a list of facts or characteristics. This type of script is used when the performer ad-libs a part in an advertisement or a sportscaster uses statistics for broadcasting color commentary during a sportscast.

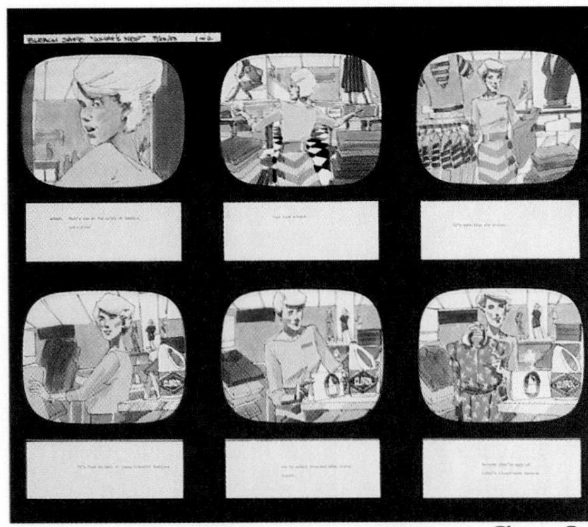

Clorox Co.

Figure 22-16. An example of a storyboard for a commercial.

Figure 22-17. News programs often use show-format scripts.

These scripts are designed to do three things for each program. They establish the format, contain production direction, and communicate the content.

Producing the Message

Designing a message is important. A well-designed message is of little value, however, unless a person can benefit from it. The next stage in helping a message reach the audience involves producing the message. This stage moves the message from design to reality. The major steps in this production activity include casting, rehearsing, performing, and recording.

Casting

Developing a television program is a controlled activity. Many different people are involved. They include creative personnel, production personnel, and performers. These people are all employed to deliver a service. The behind-the-camera personnel are hired similarly to most other workers.

They seek a job. The production company interviews them. On-camera performers are employed through a process called ***casting***. They are selected for their appearances and talents. Performers have to look the parts of the characters they are playing and be able to deliver believable performances.

Rehearsing

The production team and performers must rehearse. This involves the actors and musicians practicing their performances. Rehearsing also involves the camera and sound technicians running through their work, moving the camera and sound booms (microphones on long shafts) through their various stations.

Performing

Once the performers have rehearsed their production, it is time for them to perform. In general, a performance involves one group of people (the performers) presenting an event for another group of people (the audience). This performance can be acting, music, or dance, or any combination of these three arts. Performances can be musical concerts or recitals, theatrical plays, operas, ballets, circus acts, poetry or prose readings, magic acts, or storytelling, to name a few types.

The performance might be live. The audience might be present to enjoy it. Many times, however, the performance is recorded so it can be viewed at a later date or in some distant location.

Recording

The goal of a television project is to capture sound and sight for later broadcast. This process uses two important transducers. A microphone changes audible sound into electrical signals. Likewise, the television camera changes light into electrical signals. Together, these coded messages are captured either in digital format or on videotape. Let us look at how these technological devices work.

Audio recording

Recording sound is essentially the same for radio, audio recordings, and television and usually involves a microphone, a mixer, and a recording device. The microphone changes sound waves into electrical waves of a similar pattern. Each increase in the sound wave corresponds to an increase in the electrical wave. These waves form a continuous, flowing pattern we call *analog*. The waveform moves upward and downward as the amplitude changes.

Video recording

Video recording uses a camera to capture the light reflecting from a scene. The lens in the camera focuses the light on the image plate in a vidicon tube. Color television cameras use a three-tube system. See **Figure 22-18.** A prism breaks the beam of light into red, green, and blue segments. Each segment is focused on a different tube. This action produces separate red, green, and blue signals. The signals are then transmitted to the television receiver. This receiver uses three guns to project the three colors onto the tube screen, producing a picture.

Storing signals

Audio and video signals are often recorded on tape or disk. Magnetic tape recording uses a specially coated plastic tape. The coating is a metal-oxide material made up of needlelike crystals. The signal from a receptor (a microphone or video recorder, for example) is fed into an amplifier and then to a recording head. The electrical impulses from the signal generate magnetic forces in the recording head. These magnetic forces realign the metallic crystals in the tape. This new arrangement forms a code corresponding with the sound or picture the receptor captured.

Standards for Technological Literacy

17

Figure Discussion

Discuss the advantages of digital recordings over analog recordings.

Brainstorm

Ask your students to list ways to store audio and video recordings (media and processes).

TechnoFact

Global communications rely on satellites. The first communication satellite was Echo 1, which was launched in 1960. It was simply an orbiting balloon that bounced radio signals back to Earth. Later, satellites were equipped with electronics that allowed them to receive, amplify, and retransmit signals.

Standards for Technological Literacy

17

Figure Discussion

Use Figure 22-18 to explain how a color TV camera works.

Figure Discussion

Use Figure 22-19 to discuss the "footprint" of a satellite.

TechnoFact

In 1962, the successful launch of Telestar I, the first active communications satellite, made live television broadcasts between the United States, Europe, Japan, and South America possible for the first time.

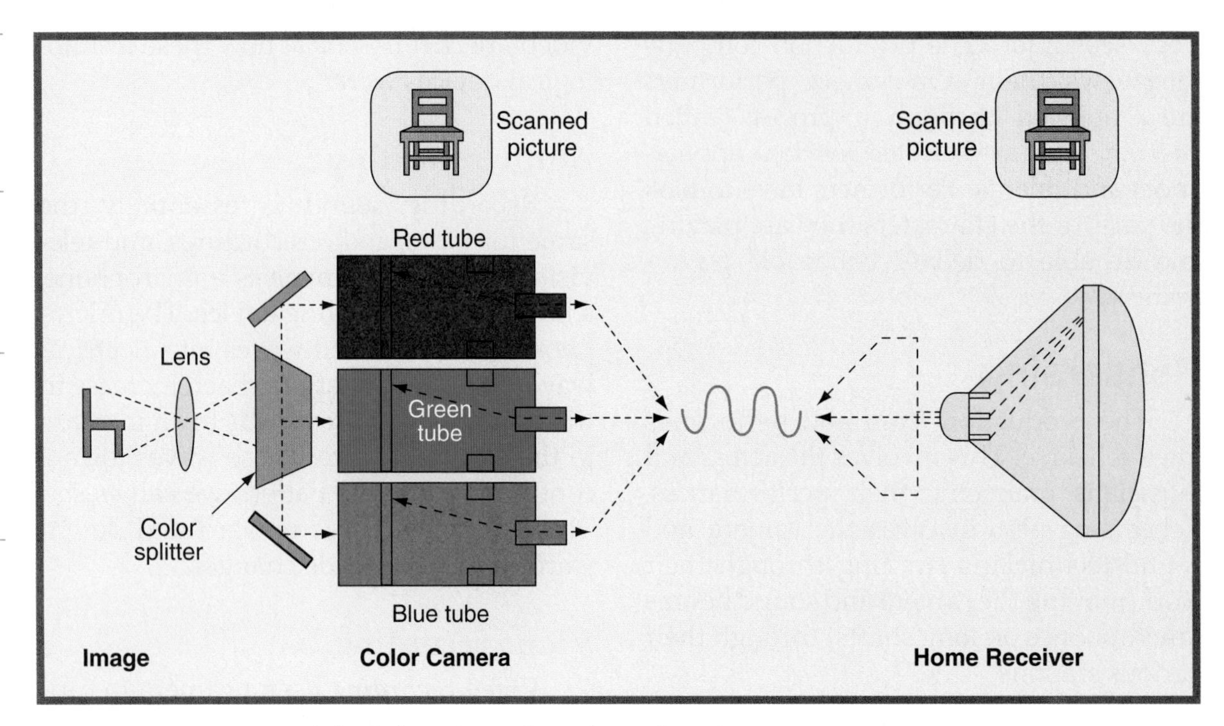

Figure 22-18. A simplified drawing of a color television-camera system.

Later, this code can be used to reproduce the information. Another recording system is the DVD. The DVD uses bumps and pits on a disk that record the digital code for both pictures and sound. Later, a laser can read these codes.

Delivering the Message

All broadcast media (radio and television) use a carrier wave with a code imposed on the wave. This coded wave is sent through the air or over a cable from a transmitter, or broadcast site, to receivers at distant locations. Radio waves travel in straight lines. Two types of broadcast systems have been developed that allow these waves to travel great distances.

Direct-Wave Systems

The first system uses direct waves. See **Figure 22-19.** Television and FM stations use direct waves. Direct-wave transmitting and receiving antennas must have an open line of sight. These antennas must be placed within approximately 60 miles

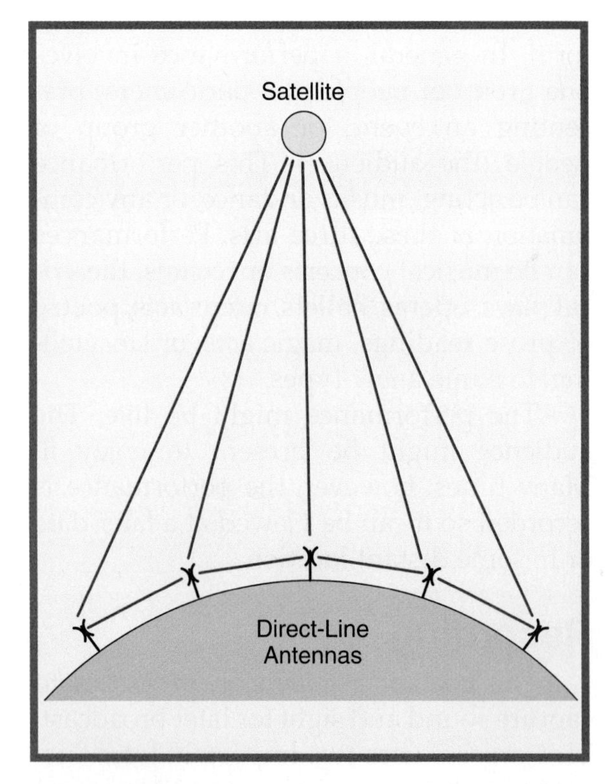

Figure 22-19. The curvature of the Earth requires direct-line antennas to be placed within 60 miles of one another. Satellites present another way to allow direct waves to cross long distances.

of each other because of the curvature of the Earth. The Earth blocks the signal from reaching an antenna beyond this distance. This is why you have trouble receiving distant television and FM stations.

One broadcast station and antenna can be used to cover a specific area. This type of system is used for some television and radio broadcasts. In other cases, a series of receivers and transmitters (relays) are used to transfer the waves between broadcast towers. This type of system is used for microwave telephone communication. See **Figure 22-20.**

Another way to overcome the problem the curvature of the Earth causes is to use satellite communication systems. See **Figure 22-21.** The signal is transmitted to a satellite orbiting in space. One part of the satellite is a receiver. Another part is a transmitter. The transmitter portion takes

NASA

Figure 22-21. Communication satellites can send information in straight-line waves over a large part of the Earth. This photo shows a communication satellite being placed in orbit from the space shuttle.

Figure 22-20. A microwave tower. This tower transmits telephone messages in straight lines.

the message from the receiver and transmits it back to Earth. This system allows a very large area of the Earth to receive a message from one transmitter site.

Reflected-Wave Systems

The second broadcast system uses reflected waves. In this system, the signal is bounced off the ionosphere (upper part of the atmosphere). A distant antenna then receives the reflected wave. AM radio signals are transmitted through reflection. When atmospheric conditions are just right, you can receive radio stations from thousands of miles away.

Parts of Broadcast Systems

Broadcast systems, as you have learned, also have several major parts. The first is a transmitter. The transmitter imposes the message on the carrier. A broadcast antenna radiates the resulting signal into the atmosphere. A receiving antenna captures this signal. The signal is fed into a receiver. In the receiver, the coded message is separated from

Standards for Technological Literacy

17

Research

Have your students research the way commercial broadcast satellites work.

TechnoFact

Geostationary orbits are frequently used by communication satellites. To achieve a geostationary orbit, a satellite is launched into an orbit that is 22,300 miles above Earth. At such a distance, the satellite will make one revolution around Earth every 24 hours, the exact time that the Earth takes to rotate once on its axis. This causes the satellite to remain over a fixed point on the Earth at all times.

the carrier wave. A transducer then converts the electrical impulses of the message. Speakers convert audio messages to the sound you hear. Video messages are usually changed into pictures with cathode-ray tubes.

Standards for Technological Literacy

5 17

Extend

Describe the types of commercial and private broadcast systems.

TechnoFact

Tokyo became home to the first cellular telephone system in 1979. In 1983, Chicago became the first city in the United States to have commercial cellular phone service.

Remote-Link Systems

A fairly new broadcast system is the remote-link system. See **Figure 22-22.** This system allows the signal to be produced at a remote location, such as a distant sporting event, the site of a natural disaster, or the place of some other newsworthy incident. The on-site camera and transmitter generate the signal. The signal is transmitted from an antenna on a truck to a relay antenna. The relay antenna might be on a building or a satellite. Finally, the relay antenna relays the signal to the receiving antenna.

Figure 22-22. Remote relay communication allows us to see live television pictures from almost any spot on the globe.

Broadcasting with Cables

As we discussed earlier, cable television systems also overcome the curvature-of-the-Earth broadcast problem. These systems usually contain several antennas receiving signals (programs) from satellites. Refer to **Figure 22-13.** These signals are processed, assigned frequencies (channel numbers), and sent through long cables to a large number of television sets. As the signals travel out from the central cable station, they become weaker. Amplifiers,

Think Green

Efficient Energy Use

Energy efficiency is a kind of reduction. An energy-efficient appliance is something that still works in the same way as other appliances while using less energy. There are several ways of achieving a more efficient use of energy. Examples of products with energy-efficient alternatives include insulation, lamps, automobiles, and household appliances. There are various reasons energy efficiency is important. Two of the most important reasons are that it helps reduce our use of depleting fossil fuels, and it emits less carbon dioxide into the atmosphere.

A well-insulated home will ease help regulate the temperature inside. Therefore, the use of a furnace or air conditioner is reduced. Automobile bodies are being redesigned to minimize the drag, helping reduce fuel usage. Incandescent lamps are being replace with compact fluorescent lamps (CFLs), which use less energy but give off more light. You will learn more details of these energy-efficient products in upcoming chapters.

placed at regular intervals along the cable, boost the strength of the signal and make the signal acceptable for viewing.

Other Communication Technologies

Broadcast radio and television are not the only telecommunication technologies available. Mobile radio and cellular communication systems have become a vital part of personal and business communications. So, too, have satellite communication systems and sound recordings.

Mobile Communication Systems

Three basic systems are used in mobile communication. These are shown in Figure 22-23. They are the following:

- *Simplex systems* use the same frequency, or channel, for both the base and mobile transmissions. The systems allow only one-way transmission at any given time. The mobile unit cannot break into a base transmission. This unit must wait until the base transmission is completed.
- *Duplex systems* use two frequencies, or channels. Generally, the base unit transmits on one frequency. The mobile unit replies on another. The systems allow one mobile unit to transmit information to the base station, while another unit is receiving a signal from the home base. Most of these systems, however, do not allow the mobile units to talk to one another. Their messages must be relayed through the base station.
- *Multiplex systems* use multiple frequencies to accommodate different types of units. The base might broadcast on one frequency. Vehicle units might use a second frequency. Handheld units might use a third frequency. These systems ensure that the low-power handheld units can get through to the base station.

Sound-Recording Systems

Still another communication system affecting almost all of us is sound recording. The fundamentals of this system are explained in the earlier parts of this chapter dealing with audio recording. Compact disc (CD) recordings are generally the system of choice today.

Standards for Technological Literacy

17

Extend

Explain how a cell phone works.

Brainstorm

Cellular telephony systems make personal communications widely available and inexpensive. Ask your students whether they think this is good or bad and why.

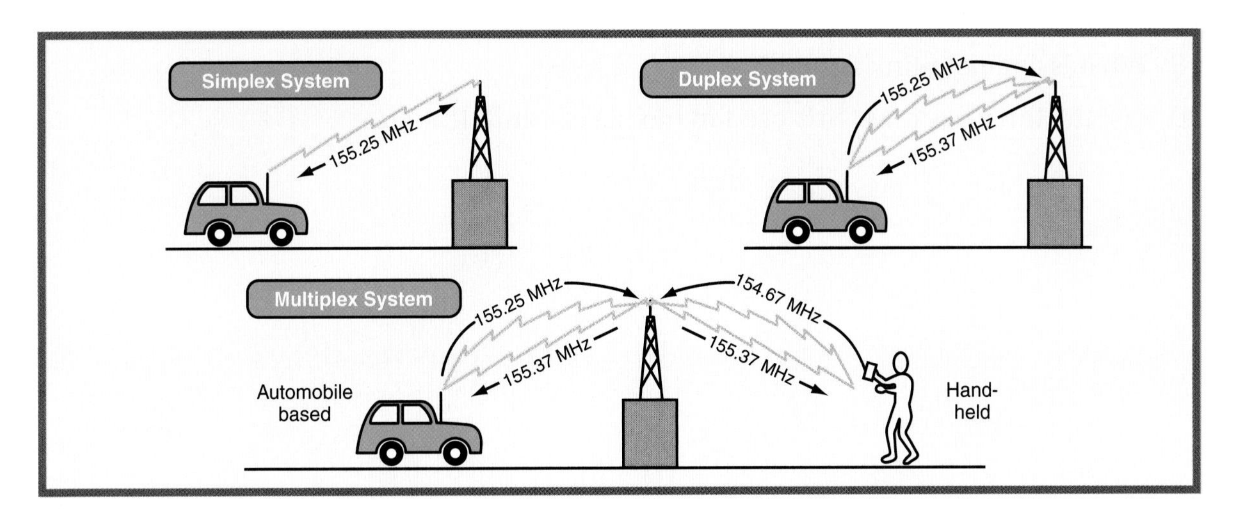

Figure 22-23. The three basic mobile communication systems vary according to the number of frequencies used.

Answers to Test Your Knowledge Questions

1. Tellecommunication means communicating over a distance.
2. direct current
3. Hard-wired and broadcast
4. Evaluate individually.
5. UHF
6. Identifying the audience, selecting the approach, and developing the message
7. Casting, rehearsing, performing, and recording
8. reflected waves
9. A broadcast system in which the signal is generated at a remote location, transmitted to a relay antenna, and then relayed to the receiving antenna
10. Simplex system, duplex systems, and multiplex systems

Summary

Telecommunication is an important part of modern life. The two major types of telecommunication are hardwired systems, such as the telephone, and broadcast systems. Broadcast systems are primarily radio and television broadcasting. Telecommunication usually involves three steps. First, we design the message through audience assessment, tone selection, and script writing. Second, we produce the message through casting, rehearsing, performing, and recording. Third, we broadcast the message using direct or reflected waves. Some newer telecommunication technologies are mobile radios, cellular telephones, and satellite communication systems.

Test Your Knowledge

Write your answers on a separate piece of paper. Please do not write in this book.

1. What does telecommunication mean?
2. The term used to describe electrons flowing in one direction along a conductor is ______.
3. List the two major types of telecommunication systems.
4. What is one difference between AM and FM radio stations?
5. Television stations that broadcast on channels 14–83 are called ______.
6. List the major activities involved in designing a broadcast message.
7. What are the four steps in producing a message?
8. The two types of wave communication used in broadcasting are direct waves and ______.
9. What is a remote-link system?
10. List the three basic systems used in mobile communication.

STEM Applications

1. Select a product or an event. Conduct an audience assessment for it. Record your data on a chart similar to the one below.

Product or event:
Intended audience:
Size of the audience:
What has high value to the audience?
What approaches are most likely to attract the audience's attention?

2. For the product or event selected above, do the following:
   A. Select a theme or an approach.
   B. Develop a full script or a storyboard for an advertisement to promote it.
3. Produce an advertisement using the script developed above by doing the following:
   A. Selecting production personnel and performers.
   B. Rehearsing the production.
   C. Recording the advertisement.

TSA Modular Activity

Standards for Technological Literacy

11

This activity develops the skills used in TSA's Digital Video Production event.

Film

Activity Overview

In this activity, you will develop a storyboard, script, and finished film or video for one of the following areas:

- The arts.
- Social studies.
- Science.
- Technology.

Possible subjects include social-study documentaries, nature films, advertisements, comedies, and dramas.

Materials

- Paper.
- A pencil.
- 3″ × 5″ note cards.
- A three-ring binder.
- A video camera.
- A computer with video-production software.

Background Information

- **Planning the video.** Plan your film as a story, making sure it has a beginning, a middle, and an ending. List the important points you want to cover. Use those points to then create a simple outline. Think visually. Remember that your story will be told mostly with pictures, instead of words.
- **Storyboard.** To help yourself and the others involved in the project visualize the video, make a storyboard. A storyboard is a series of simple pictures, similar to a comic book. These pictures show each change in what the viewer will see. Make each individual storyboard sketch on a separate note card. This allows you to try different combinations to get the most effective sequence for the video. When you have a final sequence, number your cards.
- **Scripting.** Even though the pictures (video) tell most of the story, the spoken words (audio) tie the pictures together. Your script will follow the sequence of the storyboard, with a brief description of each numbered shot and the actual words to be recorded to accompany that shot. Sometimes, audio is not fully scripted.

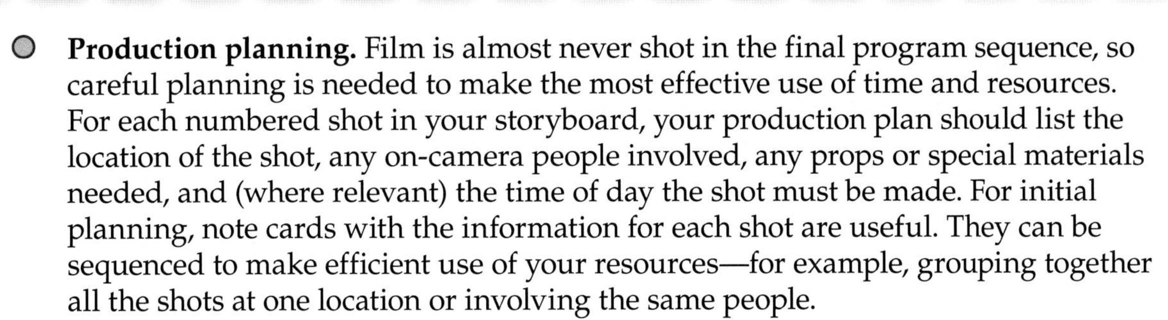

Standards for Technological Literacy

11

- **Production planning.** Film is almost never shot in the final program sequence, so careful planning is needed to make the most effective use of time and resources. For each numbered shot in your storyboard, your production plan should list the location of the shot, any on-camera people involved, any props or special materials needed, and (where relevant) the time of day the shot must be made. For initial planning, note cards with the information for each shot are useful. They can be sequenced to make efficient use of your resources—for example, grouping together all the shots at one location or involving the same people.
- **Shooting the film.** To provide visual interest in the finished video, be sure to vary your types of shots. Do not shoot all close-ups or all wide shots. Suit the shot type to the subject and to the flow of the video as shown on your storyboard. To help identify each shot, make a large card with the scene number written in bold marker. Shoot a few seconds of the card at the beginning of the shot. To provide editing flexibility, shoot a variety of cutaway shots. These are used as transitions between other shots. For example, if you are showing a school-assembly program, include a few close-ups of members of the audience listening to the speaker. When shooting interviews, you normally concentrate on the person being interviewed. When the interview is finished, shoot some close-up shots of your reporter asking questions and some in which she is just looking interested and nodding.
- **Camera and microphones.** If you are shooting in a location where you depend on battery power for the camera, be sure to have one or more fully charged backup batteries available. When lighting conditions change (such as moving from indoor to outdoor settings), be sure to check the camera's white balance and adjust it if necessary, to avoid adding a color cast to your video. When shooting general scenes, such as a cafeteria or sporting event, record the "live" audio. This audio can later be used in postproduction. For recorded narration or interviews, avoid using the camera's built-in microphone, if possible. Much better audio quality results from using a microphone (wired or wireless) placed close to the subject being recorded.
- **Postproduction.** When editing your scenes, use various visual effects, such as dissolves, fades, and zooms. Music, sound effects, and additional voice-over narration can be added during postproduction. Use cutaway shots to avoid disturbing jump cuts in the video. These occur, for example, when an interview must be edited to eliminate unwanted material. If the camera is focused on the speaker, changes in expression or head position between the adjoining shots make the cut obvious. By inserting a second or so of video showing the interviewer looking interested, the change will not be noticed.

Standards for Technological Literacy

11

Guidelines

- The final video must be three to five minutes in length.

The following items must be included in your final printed report:

- A cover page.
- A table of contents.
- A storyboard.
- A script.
- A description of editing techniques.

Evaluation Criteria

Your project will be evaluated using the following criteria:

- Creativity and originality.
- Correlation of storyboard and script to finished video.
- Technical quality and skill.

The technological devices developed for telecommunication have changed the way information is transmitted and received. It has also brought about telecommuters, or people who work from home using telecommunication technology.

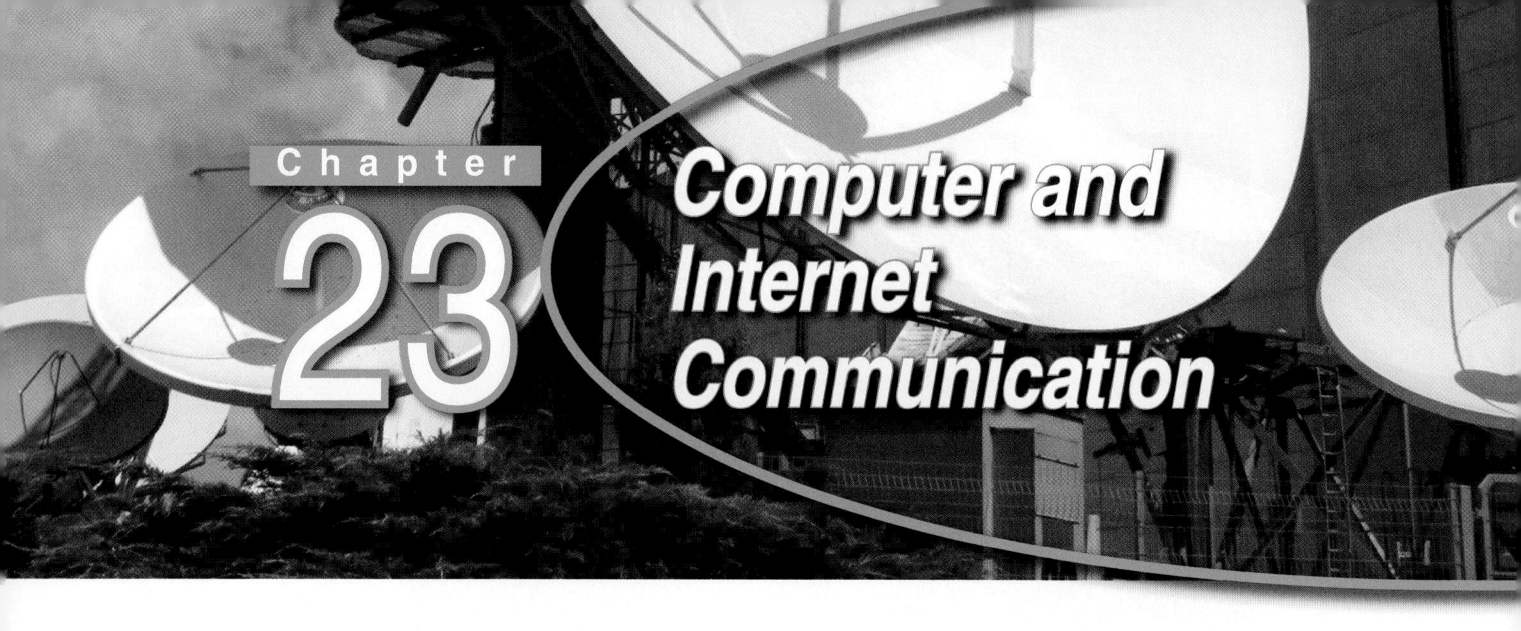

Chapter Outline
Computer Systems
Networks
The Internet

This chapter covers the benchmark topics for the following Standards for Technological Literacy:

3 5 7 11 12 17

Learning Objectives

After studying this chapter, you will be able to do the following:

- Recall the parts of a computer system.
- Recall the types of computer networks.
- Recall common uses of the Internet.
- Explain how computers are connected using the Internet.
- Recall the definitions of terms used in accessing the Internet.
- Recall the parts of domain names.
- Explain how Web browsers and Web servers work.
- Summarize what hypertext markup language (HTML) tags are.
- Summarize what a hyperlink is.
- Explain how search engines work.
- Explain how electronic mail (e-mail) works.

Key Terms

backbone
browser
central processing unit (CPU)
dedicated access
dial-up access
domain name
electronic commerce (e-commerce)
external storage device
hyperlink
hypertext markup language (HTML) tag
input device
Internet access
Internet Protocol (IP) address
Internet service provider (ISP)
link
local area network (LAN)
memory
network
network access point (NAP)
output device
page
point of presence (POP)
random-access memory (RAM)
read-only memory (ROM)
router
search engine
server
uniform resource locator (URL)
wide area network (WAN)
World Wide Web (WWW)

Strategic Reading

As you read this chapter, think about all the ways you use technology to communicate. Think about how different your relationships would be if such technology didn't exist.

Standards for Technological Literacy

3 17

Research

Have your students investigate the meaning of the terms *Information Age* and *Computer Age*.

Brainstorm

Ask your students to list the advantages of living in the Information Age.

TechnoFact

The development of the modern computer began about 5000 years ago with the invention of the abacus. Widely held as the earliest counting device, the abacus was probably invented by the ancient Babylonians. In its most basic form, the abacus represented units, tens, hundreds, and thousands with simple pebbles. The model developed by the Romans used beads, which were moved in metal trays.

The world entered a new age with the advent of the widespread use of computers. This age has been called by different names. These names include the *computer age* and the *information age*. Unlike other ages, a single technological advancement launched this age. This advancement was the invention and development of the personal computer.

A computer is a machine performing a number of tasks controlled by a set of instructions. Computers are used in all areas of personal and business life. See **Figure 23-1.** For example, computers are used to scan the prices of items at the grocery-checkout area, maintain financial records, and operate banking machines. They select channels on the television, create images from DVD players, and operate game players. Computers maintain temperatures in homes, control microwave ovens, and sequence the cycles in washing machines. They control automobile systems, operate home-security systems, and manage sports scoreboards.

In this chapter, we explore the use of computers as a primary communication tool. First, we look at computer systems in general. We then examine the expansion of communication resulting from the networking of computer systems, with particular attention paid to the types of communication allowed by way of the Internet.

Computer Systems

You learned in Chapter 7 that the computer includes physical equipment (hardware) and operating instructions (software). Hardware includes the computer itself and

Figure 23-1. Computers are used in all areas of life.

Standards for Technological Literacy

17

Demonstrate

Show your students the parts of a computer.

Research

Have your students draw a simple diagram of a computer system.

TechnoFact

In 1624, Wilhelm Schickard developed the calculator clock, a machine that could add, subtract, multiply, and divide numbers.

Figure 23-2. Computer systems have input devices, a CPU, memory, and output devices.

devices attached to it. See **Figure 23-2.** The four main units of hardware are the CPU, the memory, the input devices, and the output devices.

The ***central processing unit (CPU)*** is the heart of a computer. This unit is the working part of the computer that carries out instructions. The CPU is a microprocessor chip. This unit is a piece of silicon containing millions of electrical components.

The other major part of a computer itself is called ***memory***. This part is where the computer stores its data and operating instructions. The computer has two types of memory:

- ***Read-only memory (ROM).*** The computer can read this type of memory. ROM cannot, however, be changed.
- ***Random-access memory (RAM).*** Both the computer and the user can read or change this memory.

In addition to the internal memory, data can be stored outside the basic computer circuitry on devices called ***external storage devices***. These devices include hard drives, floppy discs, compact discs read-only memory (CD-ROMs), and DVDs.

Attached to the computer are input and output devices. ***Input devices*** allow the operator to enter data into the computer's operating system. Typical input devices are the keyboard and mouse. Other input devices are the following:

- A joystick used in video and computer games.
- A scanner that converts images, such as drawings and photographs.
- A microphone that can be used to gather sound information.
- A touch panel that senses where a person places his finger. See **Figure 23-3.**
- A pen with a light that can be used to draw images or select objects from a video display.

The data that computer operations generate are called *output*. Output can be viewed on a number of ***output devices***. The most common is the video-display

Deere and Co.

Figure 23-3. This farmer is using a touch screen to enter data.

monitor, or the flat-screen liquid crystal display. Other output devices include the following:

- Dot matrix, ink-jet, and laser printers. See **Figure 23-4.**
- Speakers.
- Overhead projectors.

Deere and Co.

Figure 23-4. An ink-jet printer produced these pages.

The other major part of a computer system, as we noted earlier, is called *software.* This term is used to describe the instructions directing the computer to perform specific tasks. For example, software programs allow people to use computers to write text, prepare drawings, maintain financial records, and perform hundreds of other functions. See **Figure 23-5.**

Networks

Computers can be used alone to do specific tasks. Also, they can be connected into ***networks***. Through a network connection, a computer can communicate with other computers. Networking allows people to exchange data rapidly and to share output

Figure 23-5. Software enables the computer to produce this graphic layout.

devices and hard disk storage. Networks are essential for computer communication techniques. Two basic types of networks exist.

Local Area Networks (LANs)

The first type of network is called a ***local area network (LAN)***. This system is generally used in a single building or site. A LAN connects several personal computers, or workstations, to a special computer called the ***server.*** See **Figure 23-6.** The server is used to store programs and data for the network.

Wide Area Networks (WANs)

The second type of network is called a ***wide area network (WAN)***. These

Standards for Technological Literacy

17

Extend

Discuss the role of software in computer systems.

Example

Show your students some examples of software.

TechnoFact

In 1642, Blaise Pascal invented the first true calculating machine, called a Pascaline. This machine added and subtracted numbers through the mechanical action of a series of gears. Numbered wheels that were linked to the gears allowed the user to enter the numbers and also display the results. The first wheel represented the numbers 0 to 9, the second wheel represented 10s, and so on. It operated by moving an adjacent wheel one notch every time a wheel turned ten notches.

Standards for Technological Literacy

17

Research

Have your students research the development of the Internet.

TechnoFact

Gottfried Wilhelm Leibniz developed the binary number system in the 1670s. In 1800s George Boole invented a new type of mathematics for performing complex mathematical operations with only 0s and 1s, the two digits in the binary number system. These developments were essential for the development of the digital computer.

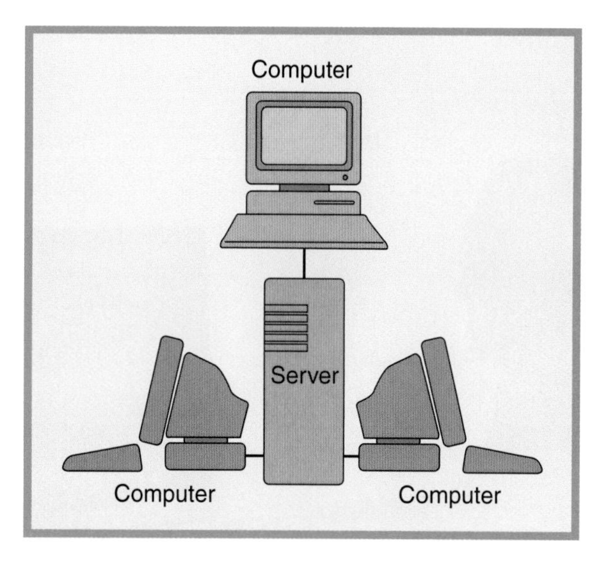

Figure 23-6. A LAN connects personal computers to a server.

networks cover large geographical areas. They are used to connect computers in distant cities and countries. The largest WAN is the Internet. We will now examine the Internet in more depth.

The Internet

The term *Internet* means "interconnected networks." The Internet is a computer-based, global information system. No one really owns it because much of its creation and initial workings emerged from federally funded research. The Internet is a collection of large and small independent networks called *intranets*. Each of these intranets can link hundreds of computers in a company or another institution. The Internet connects all these intranets so they can share selected information. This system has become a system providing inexpensive and efficient communication for people all over the world.

People use the Internet for various activities. See **Figure 23-7.** Four common uses are the following:

- **The *World Wide Web (WWW)*.** The WWW is a way to find and display multimedia documents.

Career Corner

Computer Programmers

Computer programmers write, test, and maintain the programs computers follow to perform their functions. These programs are written to specifications that computer-software engineers and systems analysts establish. A program is designed. The programmer then converts the design into a logical series of instructions the computer can follow. These instructions are coded in a conventional programming language. Programmers update, repair, modify, and expand existing programs. They test a program and make changes until it produces the correct results.

Programmers often are grouped into two broad types. These types are applications programmers and systems programmers. Applications programmers write programs to handle a specific job. Systems programmers write programs to maintain and control computer systems' software. A bachelor's degree is commonly required for programmers. Some positions, however, require only two-year degrees or certificates.

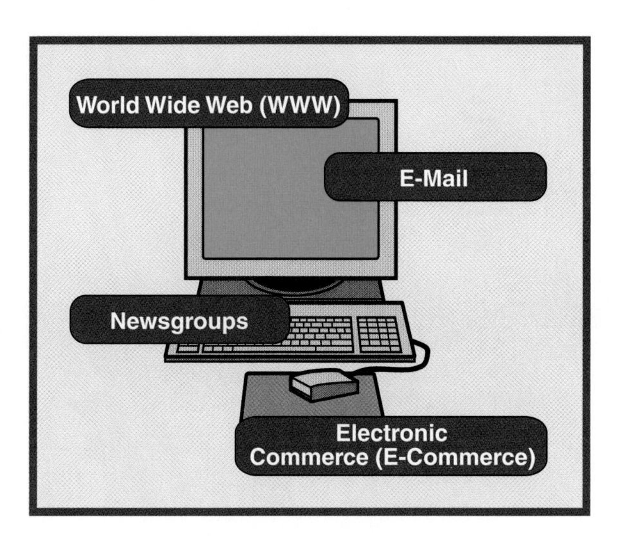

Figure 23-7. Through the Internet, an individual can access such features as the WWW, e-mail, newsgroups, and e-commerce.

- **E-mail.** This use is a way to send and receive written messages.
- **Newsgroups and chat rooms.** These uses are ways to carry on interactive discussions using written text.
- ***Electronic commerce (e-commerce).*** E-commerce is a way to buy and sell goods.

Every computer connected to the Internet is part of a network. Companies have their own LANs. A home computer is connected into a network formed by the ***Internet service provider (ISP).*** The Internet can be described as a network of networks. This complex network is a collection of backbones, access points, and routers.

Large communications companies have built their own communication lines called ***backbones.*** These backbones are typically fiber-optic lines connecting regions in the companies' information technology (IT) systems. The companies have also established connection points in each region called a ***point of presence (POP).*** These connection points let local users access the company's network. The POP is often a local or toll-free phone number or a dedicated line. The various communication companies connect their systems at ***network access points (NAPs).*** NAPs allow a customer of one company to connect with a customer of another company. Many large ISPs interconnect at NAPs in various cities. Therefore, the Internet is a collection of large corporate networks that agree to intercommunicate at the NAPs.

The Internet also contains many ***routers.*** These devices are specialized computers that determine how to send information from one computer to another. To use the Internet, an individual must first have access and a domain name. These features are discussed in more detail as follows.

Standards for Technological Literacy

17

Extend

Discuss the advantages and disadvantages of newsgroups and chat rooms.

Brainstorm

Direct your students to discuss the reasons they should be careful providing personal information over the Internet and to people in chat rooms.

TechnoFact

In 1801, Joseph Marie Jacquard, a French weaver, developed a control system for his loom that read punch cards to create different weave patterns. In later systems, holes were punched in cards to represent 0s or 1s (unpunched areas represented the other digit). This process allowed data to be encoded in binary form on the cards.

Internet Access

Internet access is the term referring to the way a computer is connected to the Internet. This access is provided in two basic ways. The first is ***dial-up access.*** In this system, computers are connected to the Internet through a modem. A modem is a device that can convert data into signals a telephone system can recognize. This device also converts these signals back into data. Special software accesses the Internet. This software places a telephone call to a company that provides Internet service. This company is the ISP. See Figure 23-8. The ISP modem answers calls from the user's modem. This modem receives and transmits the signals through the telephone lines.

The second system is called ***dedicated access.*** In this system, the subscriber's computer is directly connected to the Internet at all times. Many companies have their own dedicated Internet connections. Digital subscriber lines (DSLs) or cable modems often connect smaller business and home computers to the Internet. A DSL connection uses telephone lines for the connection. Cable modems use cable television lines to connect to the Internet. Both DSL and cable modems send data over the wires using unique frequencies or channels. This

Standards for Technological Literacy

17

Research

Have your students determine the URL for your school's homepage.

Extend

Explain how a domain name is obtained.

TechnoFact

In the 1820s, Charles Babbage began to develop a mechanical computer, a concept was inspired by Jacquard's punch card loom. For almost 50 years he modified and improved his designs, but he lacked the money to make a working model. Babbage's analytical engine designs share many features with the modern computer. Both have means of input and output, memory, data storage, and a means of moving data between memory and storage.

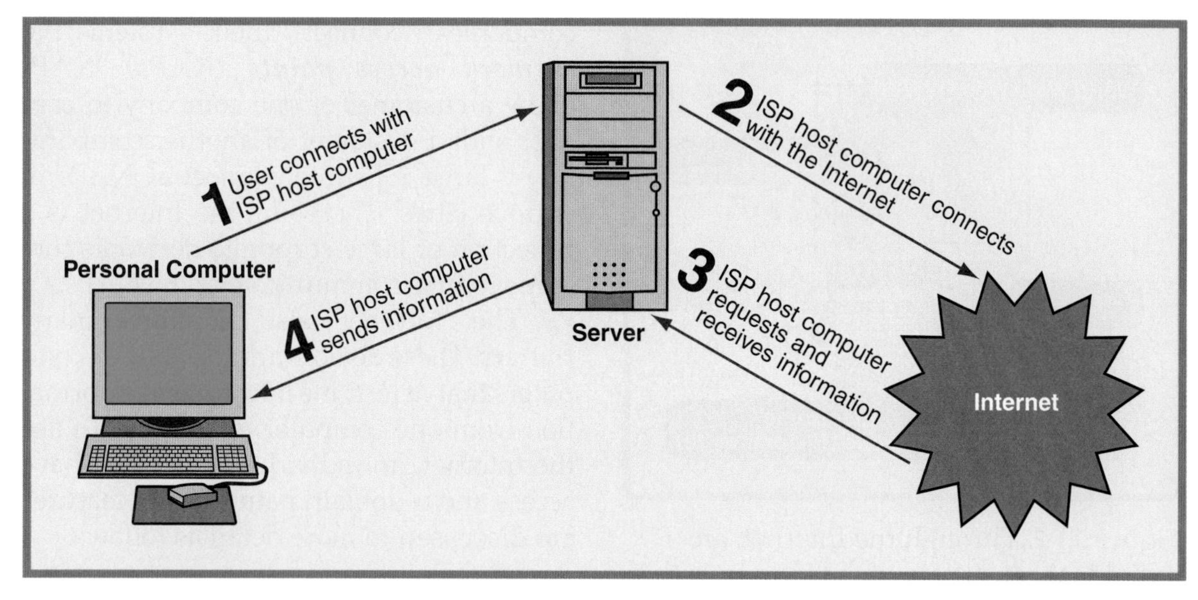

Figure 23-8. Through an ISP, an individual can use a personal computer to access information from Internet sources.

keeps the Internet communication from interfering with other signals on the wires or cables.

Every computer connected to the Internet has its own identifying number. This number is called an ***Internet Protocol (IP) address***. IP is the computer language used to communicate over the Internet. A typical IP address is a series of numbers, such as 12.225.103.96. The numbers are used to create groups of IP addresses that can be assigned to specific businesses, government agencies, or other entities. These entities include individuals. This number contains two sections. The first is the net section. This section identifies the network to which the computer is connected. The second is the host section. This section identifies the computer on the network.

Internet Domains

Every time a person uses the Internet, she uses a ***domain name***. See **Figure 23-9.** Each site on the WWW has a ***uniform resource locator (URL)***. This code is a method of naming documents or sites on the WWW. A URL is a series of characters identifying the type and name of a document. This locator also includes the domain name of the computer on which the ***page*** is located. For example, the URL of the ITEEA home page is http://www.iteaconnect.org. If you use e-mail, you have an address. For example, the e-mail address for the ITEEA is itea@iteaconnect.org.

The domain name server (DNS) translates domain names. This server changes the name humans can read into machine language a computer can read. A domain name might have three or four segments. The first series of letters (working backward) is called a *top-level domain*. In four-segment codes, such as http://www.technology.org.uk, the group of letters on the far right identifies the country. (For example, *uk* means "the United Kingdom," and *au* means "Australia.") The next set of letters from the right identifies the type of site. There were seven original types:

- **com.** This type signifies commercial organizations.
- **edu.** This type signifies four-year colleges and universities.

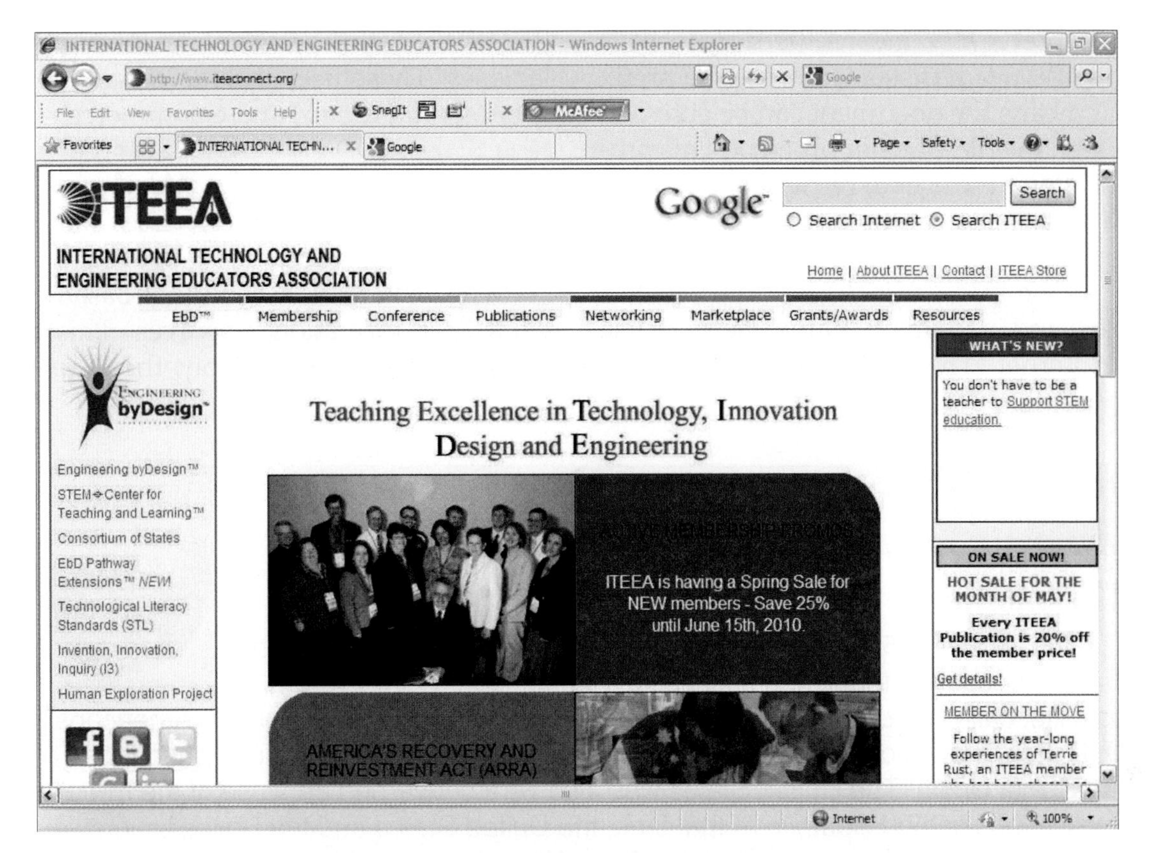

Figure 23-9. Many individuals and organizations have Internet domain names and e-mail addresses.

Standards for Technological Literacy

17

Demonstrate

Show your students how a Web site can be accessed using an address.

TechnoFact

In 1888, Herman Hollerith designed a punch card system for computing census data for the United States. His system required the data to be punched into cards, which were then fed into a tabulating machine. Pins inside the machine would be blocked by the card, but would pass through the holes in the card to complete a circuit. Each time a pin completed the circuit, a wheel moved one notch to record an entry.

- **gov.** This type signifies government institutions.
- **int.** This type signifies international organizations.
- **mil.** This type signifies military groups.
- **net.** This type signifies organizations directly involved in Internet operations.
- **org.** This type signifies organizations other than those above.

The group called the *Internet Corporation for Assigned Names and Numbers* is in charge of assigning domain names. The Internet Corporation for Assigned Names and Numbers has added more codes because of the need to increase the range of top-level domains available. These codes include aero (for the air-transport industry), biz (for businesses), and info (for unrestricted use). Most likely, even more codes will be developed to allow for new domain names.

In three-segment names, such as http://www.dtonline.org, the country code is not used. Instead, the name ends with the three-letter type code. Following from right to left, after the top-level domain code is a second-level domain code. This code identifies the specific organization, agency, or business. In the ITEEA Web address, the "iteaconnect" is the organization identifier.

The left segment of the address contains the host name. In the ITEEA example, the "www" is the host. The host specifies the computer at the site that will receive the message.

Finally, the prefix to the domain name is important. The part of the URL before the colon indicates the format used to retrieve the document. Hypertext transfer protocol,

Standards for Technological Literacy

3 7 17

Demonstrate

Show your students how to use a browser to locate information on the Web.

or the prefix *http*, means the document is on the WWW. If instead, you see the prefix *ftp*, it means the document can be accessed through the file-transfer protocol. This protocol allows the user to retrieve and modify files on another computer connected to the Internet.

As we said earlier, once a person has access and a domain name, he can use the Internet. The first use we look at is the WWW. We then explore other popular uses. These uses include e-mail, newsgroups, chat rooms, and e-commerce.

The World Wide Web (WWW)

The WWW is a computer-based network of information resources. This network is often called, simply, the *Web*. The

Academic Connections: History

The Internet

In October 1957, the former Soviet Union stunned the world with its announcement that it had successfully launched the first artificial satellite, called *Sputnik*, into space. No one was more surprised than the Americans. Until then, they had assumed their scientific achievements greatly surpassed those of the Soviet Union. The United States was also very concerned because of the military implications behind *Sputnik*. At that time, the United States and the Soviet Union were engaged in what has been called the *Cold War*. Whoever achieved technological superiority might be able to do great damage to the other country's military capabilities. One immediate response by the United States, therefore, was to put more funds into the research and development efforts of the Department of Defense.

One result was the department's creation of the Advanced Research Projects Agency (ARPA). Among other goals, the agency was interested in helping to create technology that would enable computers to communicate with one another and to do so in such a way that some computers could still be in contact, even if others had been shut down because of a military attack. This agency funded the research efforts of scientists at some major corporations and at four universities. These scientists were also working on the idea of computer networking.

After years of testing and retesting various methods, in 1972, the scientists unveiled the ARPA Network (ARPANET), a computer-networking system based on the idea of breaking data down into labeled packets that can be forwarded from computer to computer. This effort led to the development of the network control protocol (NCP) to transfer data. The NCP allows communication between hosts on the same network. Development continued, eventually resulting in the Transmission Control Protocol/Internet Protocol (TCP/IP) technology. Now, computer networks can interconnect and communicate with one another.

Thus, what started as a military concern became a new way of communication for millions of people. ARPANET became the Internet. The Internet, today, continues to expand.

In 1972, ARPA became the Defense Advanced Research Projects Agency (DARPA). DARPA is still in operation. Can you identify some projects in which it is involved today?

WWW was developed in 1993 to allow people to view information and images on the Internet. The Web provides companies, universities, government agencies, and other organizations and individuals a way to present information. Most information on the Web is free. Some sites charge a subscription fee, however, for user access.

The Web has rapidly become a source of information for millions of people. This network allows individuals to read text, view digital images (photos and drawings, for example), listen to sounds (music and speeches, for example), and access multimedia presentations (images with integrated sound). The Web has numerous features. Here, we look at browsers, servers, pages, links, and search engines.

Web Browsers

Individual computers are connected to the Web through one of the Internet connections already discussed. Sites on the WWW are accessed using a Web ***browser***. See **Figure 23-10.** A browser is a software program such as Netscape® software or the Internet Explorer® Internet browser. The program acts as the interface between the user and the WWW. Browsers know how to find a Web server on the Internet. They can request a page and deliver it to a personal computer. Finally, they can format the page so it is correctly displayed on the computer monitor.

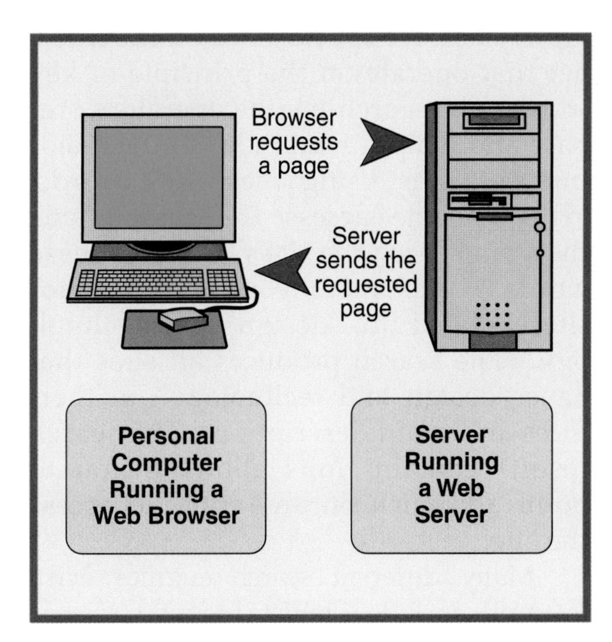

Figure 23-10. People access the WWW through a Web browser.

Most browsers are graphic media. See **Figure 23-11.** They allow the user to view images on their computer. "Point-and-click" or button functions allow the person to access resources on the Internet.

Web Servers

As noted earlier, servers are special computers used to store programs and data for the network. The organizations that develop and share information operate Web servers. The servers hold Web documents and related media. They contain computer software that can respond to a browser's request for a page. These servers can deliver the selected page to the Web browser through the Internet.

Each document on the server has an address. This address is also a URL. If the site has 100 Web pages, it has 100 URLs. For example, the URL of ITEEA's home page is http://www.iteaconnect.org/index.html. The URL for the page that appears when the "About ITEEA" link is clicked is http://www.iteaconnect.org/About ITEEA/about.htm.

Web Pages

A Web page is a text file someone creates to share information or ideas. This page contains the text of the message and a set of ***hypertext markup language (HTML) tags***. These tags or codes tell the receiving computer how a page should look. Look at **Figure 23-11** again to see a format that HTML tags created. The tags allow the developer to specify fonts and colors, create headlines, format text, and present graphics on a page.

Standards for Technological Literacy

17

Research

Have your students use a browser to find information about the development of the World Wide Web.

TechnoFact

In 1896, Herman Hollerith formed the Tabulating Machine Company to improve, manufacture, and market electric tabulating machines. In 1911, he sold his interest in the company. The company merged with two other companies and became the Computing-Tabulating-Recording Company. In 1924, the company changed names again, becoming the International Business Machines Corporation (IBM).

Standards for Technological Literacy

Demonstrate

Show your students how to use links on a Web page to locate additional information on a topic.

Research

Have your students select a technological topic and access a Web page that has links. Have them open and evaluate at least five of these links.

TechnoFact

In 1930, Vannevar Bush built the differential analyzer, the first general-purpose analog computer. In 1939, John Atanasoff and Clifford Berry built the first electronic digital computer (ABC). During World War II, Konrad Zuse of Germany built the world's first program-controlled digital computer (Z3). In 1944, Howard Aiken created the Mark 1 (also known as the ASCC), a digital computer controlled by electromechanical relays.

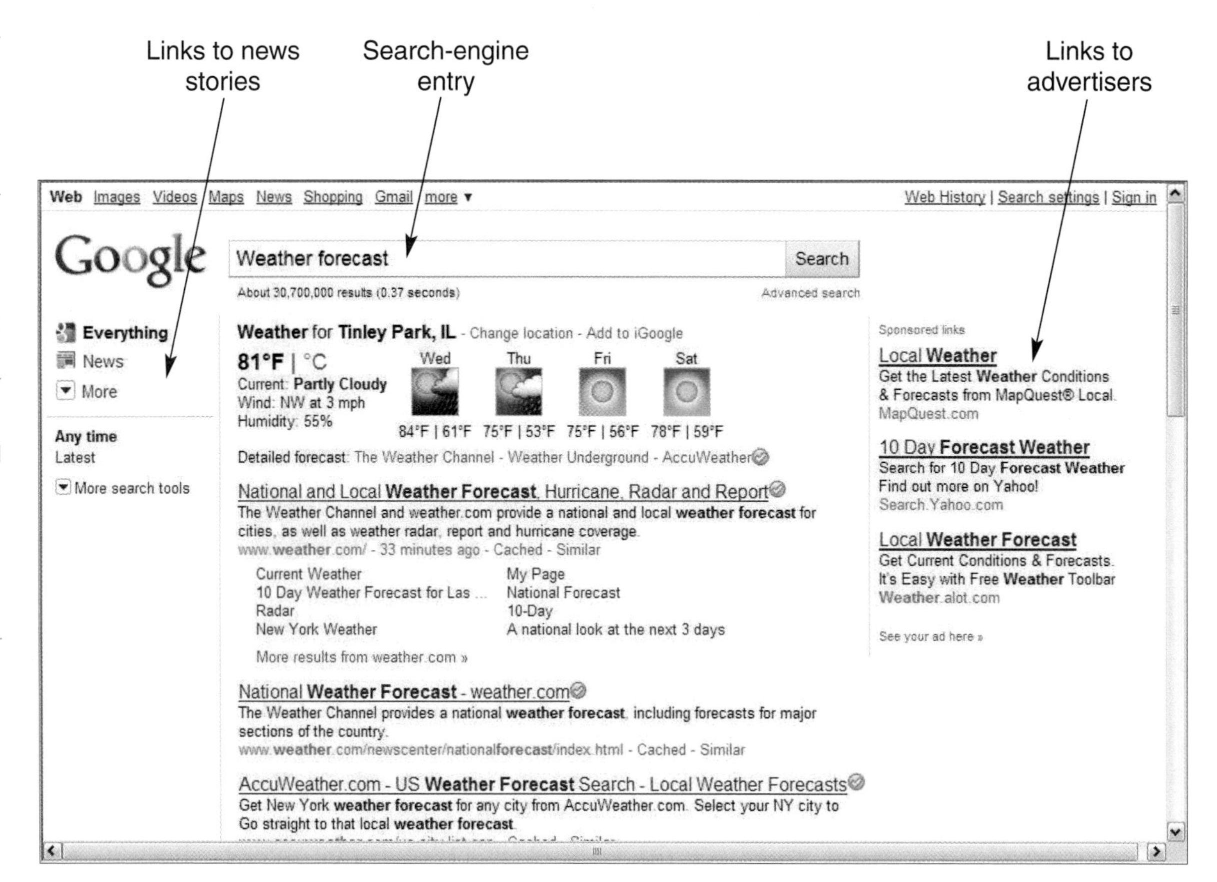

Figure 23-11. The browser page for a search engine.

Links

The URLs and HTML tags allow Web sites to connect a Web page with other pages and Web sites. These connections are called ***hyperlinks***. They are generally referred to as, simply, ***links***. The connections can be underlined phrases, buttons, or other means that can be selected. See **Figure 23-12.** They allow the operator to point and click on the link and be connected to the selected site or page. Hyperlinks allow users to move between Web pages in no particular order.

Search Engines

Often, people want to find information about a specific subject. They, however, do not know where this information is on the Web. ***Search engines*** allow individuals to search the Web by topic. They are special sites on the Internet that operate on the principle of key words. The search-engine operators prepare and maintain an index of the major Internet sites using these key words. When people access the search site, they can search the site's index for these words. For example, you can search for sites related to "design and technology." The search produces all sites that have "design and technology" in their titles and main descriptions. The search results contain links allowing you to point and click on any entry to access the site.

Many different search engines exist. Examples include AltaVista® services, Excite® services, Google® services, and Yahoo!® services. One single search engine cannot cover

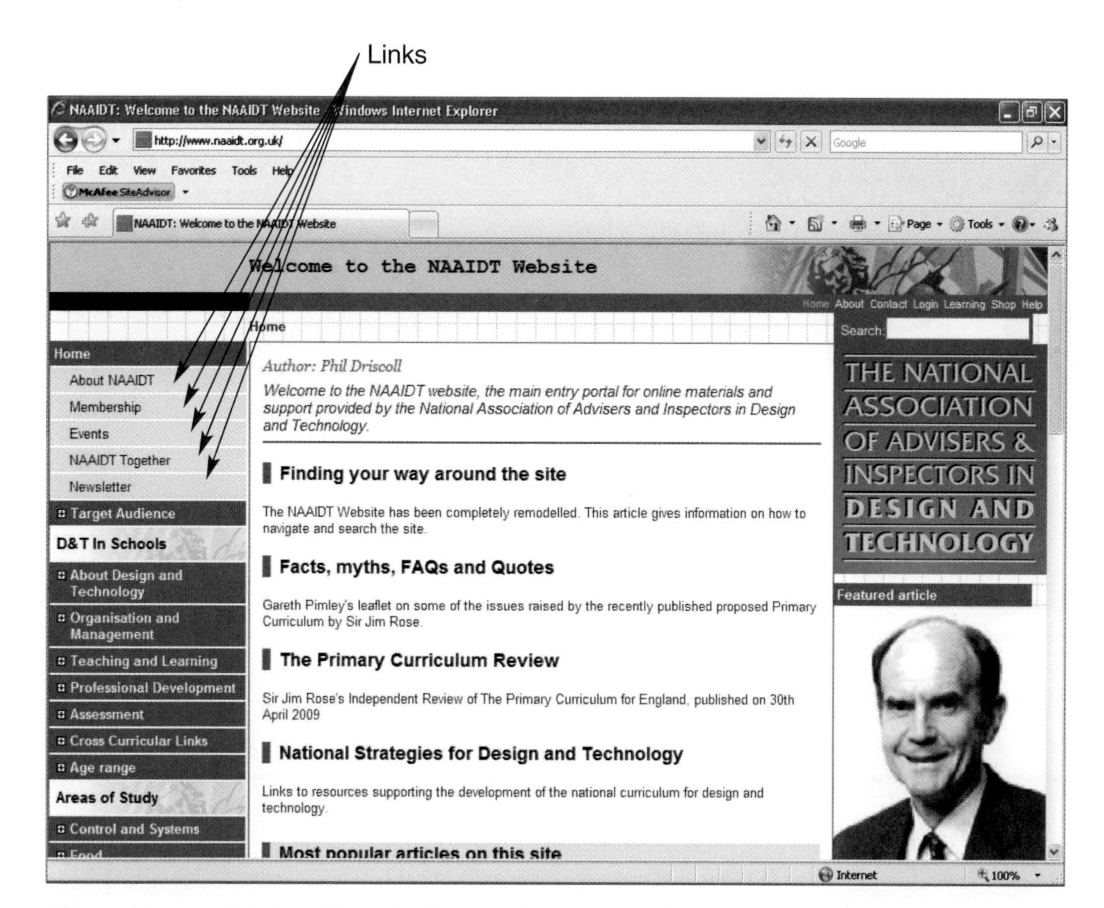

Figure 23-12. Links allow individuals to move (navigate) from one Web page to another.

every available Web resource. The search engines' operators select what they think are the most appropriate sites for each key word in their indexes. Therefore, the results from using the same key word vary from one search engine to the next.

To use a search engine, the computer operator must first open the Internet software that accesses the ISP's service. She should then do the following:

1. Determine the key words that will be used to find appropriate information. See **Figure 23-13.**
2. Review the names and descriptions of the sites the search engine displays.
3. Select and click on sites that seem promising.

E-Mail

Every day, millions of personal messages travel over the Internet. These messages can move over the Internet because of a major computer-based communication tool called *e-mail*. The system allows an individual to send a message to another computer or a number of different computers. See **Figure 23-14.** E-mail operates on a set of Internet rules called the *simple mail transfer protocol (SMTP)*.

This protocol establishes a client-and-server relationship. A program such as Outlook® Express messaging and collaboration client on an individual's computer allows the person (client) to interact with a server computer. This program lets the client compose, send, receive, read, and reply to messages.

Standards for Technological Literacy

17

Extend

Describe and discuss the major search engines.

Research

Assign the class a topic and have each student use at least three search engines to research it. Have the students compare the top ten sites each engine displays.

TechnoFact

In 1945, J. Presper Eckert, Jr. and John Mauchly built one of the earliest general-purpose electronic digital computers. This machine was known as the electronic numerical integrator and computer, or ENIAC. The ENIAC was approximately 1000 times faster than the Mark 1 and could complete about 5000 addition tasks and 1000 multiplication tasks per second.

Standards for Technological Literacy

17

Brainstorm

Have your students list the advantages and disadvantages of e-mail compared to traditional mail.

Demonstrate

Show your students how to send a message using e-mail.

Ask

Figure 23-13. A typical search-engine page.

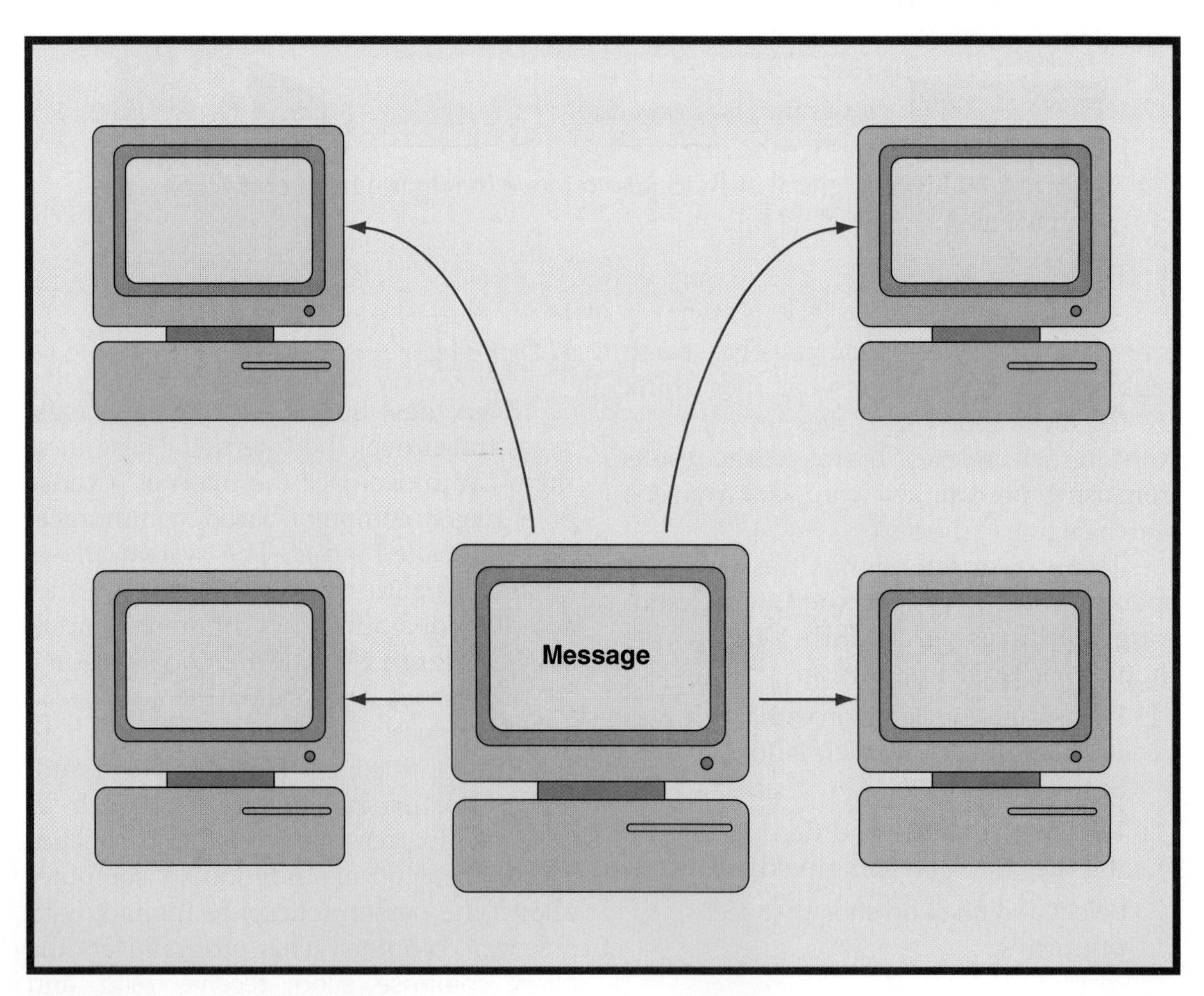

Figure 23-14. E-mail systems can deliver messages between two or more computers.

Technology Explained

Virtual Reality: a computer interface allowing a user to interact with three-dimensional, computer-generated images.

Computer technology has taken one giant step after another, over the past 30 years. As the computer has become more complex, it has branched off into many different areas. Virtual reality is one of the more spectacular areas of computer technology. See **Figure A.**

Virtual reality was born in the 1960s, though in a very crude form. In 1981, the first practical application of a virtual reality system was exhibited. The Visually Coupled Airborne Systems Simulator trained pilots to fly complex, high-speed aircraft. The term *virtual reality* did not, however, appear until the mid-1980s. Jaron Lanier, the founder of VPL Research, coined the term.

Caterpillar Inc.

Figure A. Employees manipulate a virtual front-end loader.

This interface allows a user to interact with an environment that a computer creates. Therefore, complex equipment is needed, in addition to the computer. Computer software, a headset, and some tool relaying the user's movements are required. This tool is often a glove transmitting its relative position and many finger movements to the computer.

Uses for virtual reality are many, with new uses being developed everywhere. The military has used virtual reality to train fighter pilots for some time. Companies use virtual reality to train new employees. Architects allow clients to tour new homes during the design stage. See **Figure B.** Clients can make changes based on what they see before construction begins. Of course, virtual reality has recreational uses as well. Complex systems can be found in expensive arcades. Simpler systems are available for home video games.

Figure B. Virtual objects can be moved about in this virtual office space.

There are impressive prospects for the future of this technology. Soon, surgeons might have virtual reality training facilities. They could train on virtual patients with virtual scalpels. Scientists, as well, can do research with virtual reality. Areas such as molecular modeling and engineering might see significant benefits.

Standards for Technological Literacy

17

TechnoFact

At the time it was built, the ENIAC computer was the largest single electronic device in the world. It contained almost 18,000 vacuum tubes, covered 1500 square feet of floor space, weighed 30 tons, and required more than 150 kilowatts of power to operate.

Standards for Technological Literacy

17

Extend

Describe an e-mail post office and the role of the post office administrator.

TechnoFact

In 1975, the first personal computer, called Altair, was sold to electronics hobbyists in kit form. Steven Jobs and Stephen Wozniak offered the first preassembled personal computer, the Apple II, through their Apple Computer Company in 1977. The relatively inexpensive Apple II brought the power of computing to people that didn't have the resources to lease a mainframe or the expertise to build an Altair.

E-Mail Systems

The e-mail systems most people use have two servers. The first one handles outgoing mail. This server is called the *SMTP server* because it operates on the SMTP. The second server handles incoming mail and is called the *POP3 server.* This server operates on the post office protocol. The two protocols are the rules that allow the computers handling incoming and outgoing mail to talk to one another. They allow the computers to become a communication system.

E-Mail Messages

E-mail messages are simple text messages. They are letters in electronic form. Special computer software prepares and processes them. This software has several features. The most important features are the message-generation window, the mailbox, and the address book.

The message-generation window allows the user to compose, edit, and spell check messages. The mailbox is a feature containing the sent and received messages. See **Figure 23-15.** An address book contains

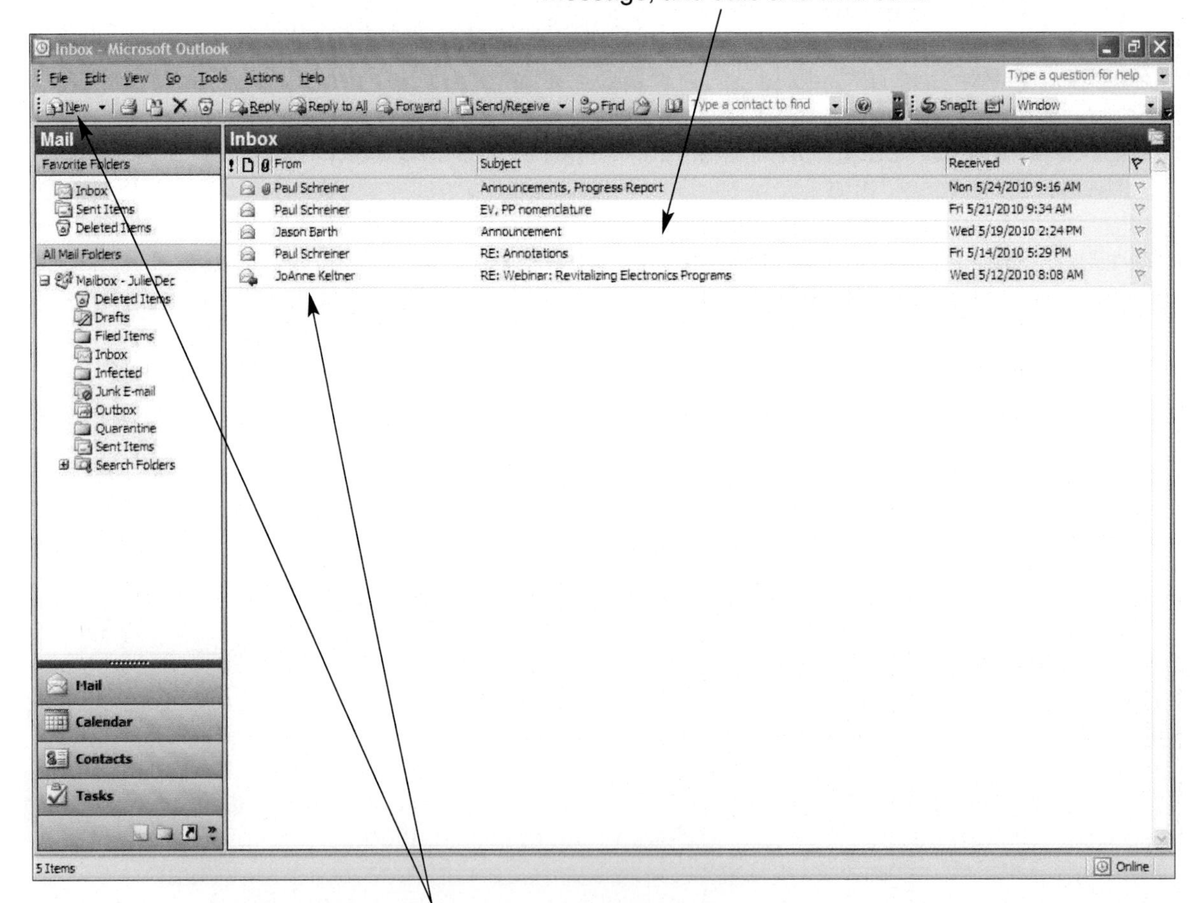

Figure 23-15. A typical e-mail software page.

the e-mail addresses of people whom the client communicates with regularly.

In addition, most of this software allows the user to make attachments to the message. That is, the person can attach photos, documents, sound files, and other electronic media to the e-mail message. To send a message, the user must complete several tasks. These tasks include the following:

1. Open the e-mail software.
2. Select "New Message" on a menu or dialog box.
3. Type an e-mail address or select one from the address book.
4. Type a title for the message.
5. Prepare, edit, and spell check the message.
6. Select any appropriate attachments.
7. Send the message.

To receive a message, the user can point and click on the message in the in box. The message opens in the message window. Attachments can be opened in the same manner or by way of other computer software, such as graphics or word processing programs.

E-Mail Mailing Lists

Sometimes, a single message needs to be sent to a number of people. In this case, the sender uses a mailing list. The sender can develop mailing lists. In other cases, people can join a list so they receive regular news or product updates.

A mailing list works fairly simply. First, a message is sent to a single e-mail address. This address is called the *mailing-list address.* The same message is then re-sent or reflected to the other people on the list.

Forms of Public Internet Discussion

E-mail is the most popular means of communicating over the Internet. This means of communicating allows people to send messages directly to an individual or a group. E-mail works when the recipient of the message is identified and an immediate answer is not needed. Several other communication approaches are available on the Internet, however.

Standards for Technological Literacy

5 17

Demonstrate

Show your students how to develop an address book for e-mail and how to send a message to one person on the list and to all the people on it.

TechnoFact

Bill Gates and Paul Allen founded Microsoft Corporation in 1975 to develop programs for the Altair personal computer. In 1981, IBM introduced the PC, which was constructed from off-the-shelf parts, and therefore could be easily copied by competitors. Microsoft was hired to develop the operating system and applications for the PC, and, by default, all of its "clones."

Think Green

Electronic Media Waste

The switch to digital technology has helped reduce production and paper waste. This change to a more technological society may mean you use a great deal of electricity. However, there is also another pressing environmental impact as a result of the continuing use and constant growth of technology. Many technological products are being upgraded on at least an annual basis. If you invest in an upgrade, what do you do with your older technology? What happens when you throw out CDs or DVDs?

Any types of electronic technology that are thrown out are called *e-waste.* This includes computers and related technology; audio technology; and video technology. E-waste is one of the fastest growing causes of toxic waste in the United States. These toxins may cause harm to the environment, as well as to personal health. To reduce e-waste, several electronics manufacturers have initiated recycling programs for their products. Local communities may also collect and recycle your e-waste.

Standards for Technological Literacy

17

Extend

Discuss the differences between newsgroups and chat rooms.

Extend

Differentiate between e-commerce sites and auction sites.

TechnoFact

In 1993, Marc Andreessen developed a graphical browser for the Web called Mosaic. Although Mosaic was originally run from the UNIX platform, Windows and Macintosh versions were made available in 1994. This new browser made the World Wide Web accessible to more people. As more people used the Web, the content it contained grew and evolved. As the Web became more useful, it attracted more users. It is this continuous growth cycle that has made the World Wide Web the vital medium it is today.

They include the following:

- **Chat.** Electronic conversation between two or more people. "Chat rooms" allow people to key in messages that everyone in the "chat room" can see. Some of the more popular types of chat rooms are blogs or Twitter™.
- **Newsgroups and forums.** Ongoing public discussions about particular topics. All people who subscribe to a group have access to all messages and can reply to any or all of them.
- **Instant messaging.** Allows people to develop a list of individuals with whom they wish to interact. They can send messages to and receive messages from any of the people on the list who are on-line at the time.
- **Social networking.** Allows people to share information with one another for different purposes. Examples include LinkedIn®, Facebook, MySpace™, YouTube, and Wikipedia.

Internet Protocol (IP) Telephony

A rapidly developing alternative to traditional telephone communication systems is IP telephony. This new system uses traditional Internet connections for voice, facsimile (fax), and voice-messaging applications. IP-telephony systems have three major steps. First, the voice message is converted from an analog voice signal into a digital format. Second, the digital signal is compressed and translated into IP packets. These packets are transmitted over the Internet. Finally, the first two steps are reversed at the receiving end of the system. The digital signal is decompressed and translated into analog (voice) sounds. The IP-telephony system allows people to make and receive calls as they do over a traditional landline (Public Switched Telephone Network, or PSTN). There are three common ways to make residential connections for Voice over Internet Protocol (VoIP) communications.

Analog Telephone Adapter (ATA) Systems

Analog telephone adapter (ATA) systems use standard telephones that plug into ATAs. The company providing the VoIP service generally supplies this adapter. The ATA converts the analog phone signal into a digital signal. The digital signal is routed through a broadband connection the local telephone company provides.

Computer-to-Computer Systems

Computer-to-computer systems require a computer connected to the Internet at both ends of the communication channel. Both computers must have hardware allowing the operators to speak and listen (a headset or speakers and a microphone). Also, both computers need voice-communication software loaded and operating.

Internet Protocol (IP)-Phone Systems

IP-phone systems use IP phones that look similar to normal telephones. The IP phone is connected to the Internet, however, using an Ethernet cable or a wireless connection. The Ethernet cable attaches to a router or a gateway. This device performs the necessary tasks to transmit a VoIP message.

Electronic Commerce (E-Commerce)

People have participated in commerce for hundreds of years. Simply put, commerce is the exchange of goods and services for money. In recent years, a new type of commerce has developed. This type is called *e-commerce.* E-commerce involves selling products and services over the Internet. This system connects the two major players in commerce. These players are the buyers who want to purchase a good or service and the sellers who have goods and services to sell.

An e-commerce site on the Internet must provide the essential elements of commerce. The first element is merchandising. The product must be displayed so the potential customer can view it. Pictures and text displays describe the features of the product. Often, a series of links allows the customer to narrow the choices until a single product is selected. See **Figure 23-16.**

Second, the customers must be able to select the items they want to buy. The site generally provides an order form so the customers can purchase the products. Third, the site must have a way to close the sale. This site must have a way to collect money and issue a receipt. Generally, the site provides a way for the customer to use a credit card to pay for the product. An electronic receipt is usually sent to the customer confirming the order.

The e-commerce company must then fulfill the order. The merchandise must be sent to the customer. The company might send packages through the U.S. Postal Service or by some private package service. This e-commerce company might deliver printed matter and software over the Internet. A way to return unwanted, damaged, or defective merchandise must also be in place. Finally, the company should provide technical support for complex products.

Similar to all businesses, e-commerce companies must attract customers. They must advertise their sites and their products. This can be done by traditional ways, such as print, radio, and television advertising.

Another avenue is called *banner advertising.* These advertisements are small rectangular messages appearing on Web pages. A banner ad is a special hypertext link instructing a Web server to display a particular Web page containing graphics and, sometimes, animation. The banner ad encourages the Internet user to click on the banner ad and go to the advertiser's Web site. The ad also builds name identification for a product, company, or Web site.

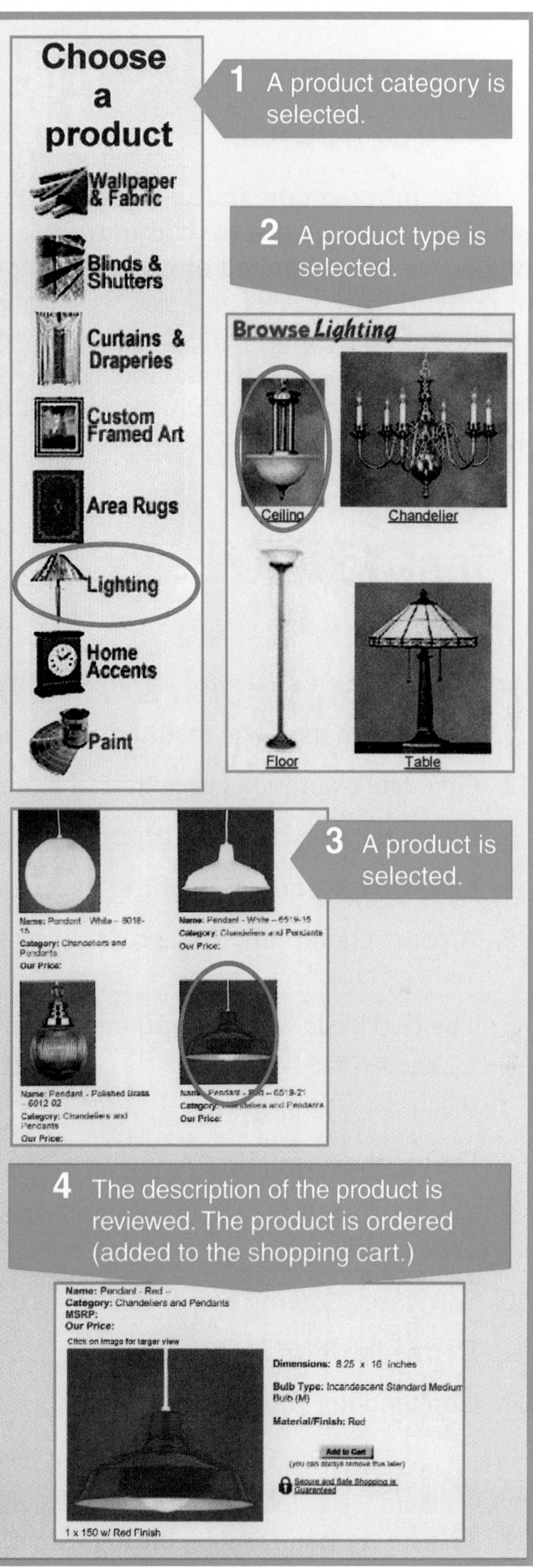

American Blind and Wallpaper Factory

Figure 23-16. An e-commerce site uses a series of links to guide the customer to a product to buy.

Standards for Technological Literacy

17

Demonstrate

Show your students how an e-commerce site works.

Discussion

Discuss how an e-commerce site allows the customer to view and select merchandise.

TechnoFact

In 1959, two inventors, Robert Noyce and Jack Kilby, working independently, simultaneously invented the integrated circuit. In 1968, Ted Hoff led a team of engineers in the development of the first microprocessor. This microprocessor, called the 4004, had the same processing power as the 30-ton ENIAC computer.

TechnoFact

In 1991, Tim Berners-Lee developed a system for linking documents to other relevant documents on the Internet. He called his system the World Wide Web. Prior to the introduction of Marc Andreessen's Mosaic software in 1993, images and text had to be viewed as separate documents. Andreessen gave his browser the ability to display pages that integrated text and graphics.

Answers to Test Your Knowledge Questions

1. personal computer
2. Student answers may include any of the following: hard drives, floppy discs, CD-ROMs (compact disc—read-only memory), and DVDs. Additional answers may be accepted at the instructor's discretion.
3. Local area networks and wide area networks
4. Student answers may include any two of the following: the World Wide Web (WWW), e-mail, newsgroups or chat rooms, and e-commerce.
5. True
6. dial-up
7. An Internet protocol address is a number that identifies an individual computer connected to the Internet.
8. A uniform resource locator (URL) is a code that identifies a document or site on the World Wide Web.
9. True
10. Web browser
11. Hypertext markup language is a set of tags or codes that tell the receiving computer how a Web page should look.
12. hyperlinks (or links)
13. A search engine is a special site that allows users to search for Web content by topic.
14. POP3
15. IP telephony

Summary

The introduction and use of the personal computer dramatically changed our lives, especially in the ways we communicate. By connecting computers together through networks, we have changed our methods of communication even more. The two basic types of networks are LANs and WANs. The largest WAN is the Internet, a collection of both small and large independent networks. We can use the Internet by accessing it correctly. Once on the Internet, we can engage in such activities as obtaining information from the WWW, sending and receiving e-mail, joining newsgroups, and buying and selling commercial products.

Test Your Knowledge

Write your answers on a separate piece of paper. Please do not write in this book.

1. The information age resulted from the invention of the ______.
2. Give one example of an external storage device for a computer system.
3. Name the two common types of computer networks.
4. List two common uses of the Internet.
5. A connection point that a company establishes in a particular region is called a *POP.* True or false?
6. The two basic ways computers are connected to the Internet are dedicated access and ______ access.
7. What is an IP address?
8. Define the term *URL.*
9. The prefix *http* in a domain name means the document is on the WWW. True or false?
10. Software allowing a user to connect with the WWW is called a(n) ______.
11. Define the term *HTML tag.*
12. Connections allowing users to move from one Web page to another are called ______.
13. What is a search engine?
14. The server handling incoming e-mail is called the ______.
15. The system allowing computers to carry voice messages is called ______.

STEM Applications

1. Use a search engine, such as AltaVista® services, Excite® services, or Yahoo!® services, to locate Internet sites explaining how things work. Select one device and print out the information the site supplies. Use the same search of key words on a second search engine and compare the results you get.
2. Design a layout for an e-commerce page to sell school supplies. Sketch what each page will contain and the links you will use so people can navigate your site.

TSA Modular Activity

Standards for Technological Literacy

11 12

This activity develops the skills used in TSA's Webmaster event.

Cyberspace Pursuit

Activity Overview

In this activity, you will create a Web site composed of four components:

- An overview of your school's technology-education program.
- General information about your school.
- Historical information about your school.
- A page of links to related and interesting Web sites.

Materials

- Paper.
- A pencil.
- A computer with Internet access and Web page–development software.

Background Information

- **Design.** Careful planning is critical when developing a Web site. Create a sketch for each Web page. List the elements to be included on each page. Include lines showing the links between Web pages. Consider how a user will navigate within the Web site to be sure you have the necessary links.
- **Navigation and functionality.** The visual appearance of your Web page is important. This appearance is not, however, as important as easy navigation and functionality. Design your Web page so links can be easily located. Select easy-to-read type for links.
- **Type.** Too many fonts and type sizes can make your Web page unattractive. Vary the size and font based on the function of the text. Titles are meant to draw attention. Body type should blend into the overall design and use a typeface, size, spacing, and justification comfortable for reading. If you use an unusual type font, people viewing your Web page might not have that font on their computers. In this situation, another font is substituted. This causes your Web page to have an unintended (and most likely, less attractive) appearance. Use common type fonts such as Arial, Times New Roman, Tahoma, and Courier. If you want to use an unusual font, create the text as an image file. Insert the file into the Web page.
- **Images.** Images can add to the visual appeal of your Web pages. Using too many images, however, can clutter a page and cause it to load slowly. Use a compressed-image file format (such as .jpg) and the lowest acceptable resolution for faster loading.

Guidelines

- Review at least five Web sites, evaluating the designs of the sites, in terms of attractiveness and usability.
- Your Web site must have a home page containing separate links to each of the four components.
- There is no minimum or maximum number of pages for the individual components.
- The home page and each of the four components must contain both text and graphics.
- All pages must include a link to the home page.
- Use a pencil and paper to prepare a rough sketch for each page and an organizational chart showing how pages are linked.
- After you have developed the rough sketches, create the Web pages.
- Test the completed design to make sure all links work properly.

Standards for Technological Literacy

11 12

Evaluation Criteria

Your Web site will be evaluated using the following criteria:

- Web-page design.
- Originality.
- Content of Web pages.
- Functionality.

Section 6 Activities

Activity 6A

Design Problem

Background

Communication technology is used to deliver information, project ideas, and generate feelings. Often, these goals are incorporated in advertising. Communication can promote a product or an idea by delivering information and persuading people to act in a certain way.

Situation

You are employed as an advertising designer for Breckinridge and Rice Agency. One of your clients wants to promote technological literacy as a public-service effort.

Challenge

Design a full-page (6 3/4″ × 9 1/2″) magazine advertisement promoting technological literacy. The ad should encourage students to select at least one class in technology education to help them understand technology as it impacts their lives. In this process, develop a theme; prepare a layout; and specify the type size and style, ink colors, and type of and location of the photograph for your design.

Optional

1. Print the advertisement as a flyer that can be sent home with students in your school.
2. Convert the layout to a poster. Print copies. Post them on bulletin boards in your school.

Activity 6B

Production Problem

Background

Photographs are used either to capture an event for historical purposes or to communicate information and feelings. The first use allows people to record family experiences, chronicle trips, or capture specific happenings. The second use is communication. This use attempts to give directions, develop attitudes, or communicate feelings.

Challenge

You are a communication designer for a publisher of children's instructional books. The publisher is developing a series of simple how-to booklets for children four to eight years old. Select a task a child of this age must master. Design and produce a six-picture set showing the child how to systematically complete the task.

Materials and Equipment

- Layout sheets. (Photocopy the form included with this activity.)
- Pencils.
- Felt-tip pens.
- A digital camera.
- A computer system with photo software, such as Photoshop Elements.
- An ink-jet printer.

Procedure

Designing the product

1. Select a task.
2. List the steps needed to complete the task.
3. Group steps that can be shown in a single photograph (shot).
4. Develop a layout for each shot, using the layout sheets. See Figure 6B-1.

Producing the product

1. Gather the items (props) needed for each shot.
2. Set up shot #1.
3. Shoot at least two photographs of the shot.
4. Repeat steps 2 and 3 until all six shots are taken.
5. Download the images you shot onto a computer.
6. Select the best negative for each shot.
7. Use the photo software to process each image by doing the following:
   A. Correcting the color and brightness.
   B. Cropping the image to size.
8. Make a print of each of the six selected negatives.
9. Mount the prints for display.

Optional

1. For black-and-white prints, make a storybook by producing a printed narrative for each picture. Print the story for each photograph on a separate page, and then mount the picture above the narrative. Design and produce a cover for your book.
2. For slides, prepare a script and record a narrative for the series. Produce a title slide and an end slide for the series.

Series Title: ______________________

Shot #: ______________________

Description: ______________________

Figure 6B-1. The shot layout sheet.

Section 7
Applying Technology: Transporting People and Cargo

Tomorrow's Technology Today

Invisibility Cloaks

From the imaginative minds of authors and screen writers, stories are created that possess elements of fantasy and magic, oftentimes exceeding the scope of possibility. In these narratives, rebellious teenagers slip on an invisibility cloak and sneak past unsuspecting adults on their way to a grand adventure—a scenario worthy of a year's allowance. Although an invisibility cloak might seem like a magical device, it may soon become a reality if today's scientists continue their successful research and development.

Scientists have explored two options for achieving invisibility. The first, optical camouflage, creates the illusion of invisibility by blending the intended object into its background. Optical camouflage employs technology similar to a weather forecaster's blue screen. A camera records the background image which will appear behind the newscaster. The image will simultaneously be projected onto the person who is also wearing a special silver coat that enhances the quality of the projection. Optical camouflage makes a person or object "invisible" from only one vantage point. In reality, it masks an object rather than making it invisible.

Another method currently being developed is the creation of a covering, or cloak, that would divert light rays around the desired object. When we "see" an object, we actually see light that has struck the object and reflected off it into our eyes. If light did not strike the object, but instead flowed around it, we could not "see" the object. In order to construct such a material, scientists have turned to the technology of metamaterials. Metamaterials are materials created by scientists to exhibit characteristics that cannot be found in nature.

In order to achieve invisibility, the structure of the metamaterial must be smaller than the wavelength of light—blocking light ray access to the material covered by the cloak. Researchers have created a successful prototype of this type of invisibility cloak, but it is extremely small—measuring only ten micrometers wide. That is equivalent to .00001 meters!

The small size is not the only reason this invisibility cloak has restricted applications. As it stands, the cloak can only cover still objects, so it could not actually be worn like a cloak. However, it could make a building or motionless army tank invisible. Additionally, following current design and construction, an invisibility cloak large enough to cover your bike would be incredibly heavy.

While a true invisibility cloak has yet to be created, the technology is in the works. Because light rays and sound waves work in a similar way, scientists are also attempting to create a sound barrier that would serve much like an invisibility cloak for the ears. With invisibility cloaks in hand, a future generation of high school students will keep teachers on their toes.

Discussion Starters

Devices that can be used to cloak people and objects could have several potential applications. Have the class suggest examples of those applications.

Group Activity

Ask the class if they can think of any drawbacks to the implementation of technology that achieves invisibility.

Writing Assignment

Several books and films have told of devices that can be used to make people and objects invisible. Have your students write a brief paper discussing how the technology behind actual invisibility cloaks differs from fiction.

Chapter Outline
Transportation: A Definition
Transportation as a System
Types of Transportation Systems
Transportation-System Components

This chapter covers the benchmark topics for the following Standards for Technological Literacy:

3 18

Learning Objectives

After studying this chapter, you will be able to do the following:

- Recall the definition of *transportation.*
- Summarize why transportation can be seen as a system.
- Recall the types of transportation systems.
- Recall the components of a transportation system.
- Recall what place utility means.
- Summarize the characteristics of transportation pathways.
- Recall the types of transportation vehicles.
- Recall the types of transportation support structures.

Key Terms

air-transportation system
intermodal shipping
land-transportation system
place utility
space-transportation system
transportation
vehicle
water-transportation system

Strategic Reading

Before you read this chapter, make a list of all the different types of transportation you're familiar with. As you read the chapter, check your list and add to it if necessary.

The development of transportation and civilization are closely related. See **Figure 24-1.** Without transportation, humans are restricted to a very small area. In prehistoric times, for example, travelers had to make their journeys out from home and back in one day. This distance was less than 25 miles (40 km). The domestication of animals and their use as beasts of burden and for riding enlarged this travel area. People could then travel 40 to 50 miles (64 to 80 km) in a day. More travel led people to design and build roads. As early as 30,000 BC, established transportation routes existed. Only the very brave ventured far from home, however. In fact, human beings tended to stay within a restricted area of travel until recent times.

The development of sailing ships, followed much later by the railroad, enlarged the area of travel. Still, worldwide travel was available only to wealthy or adventurous people. Travel, as we know it today, is a recent development. The advent of jet-powered aircraft after World War II made the far reaches of the globe available to many people. In one day, we can eat breakfast in New York, eat lunch in Los Angeles, and be back in New York for a late dinner.

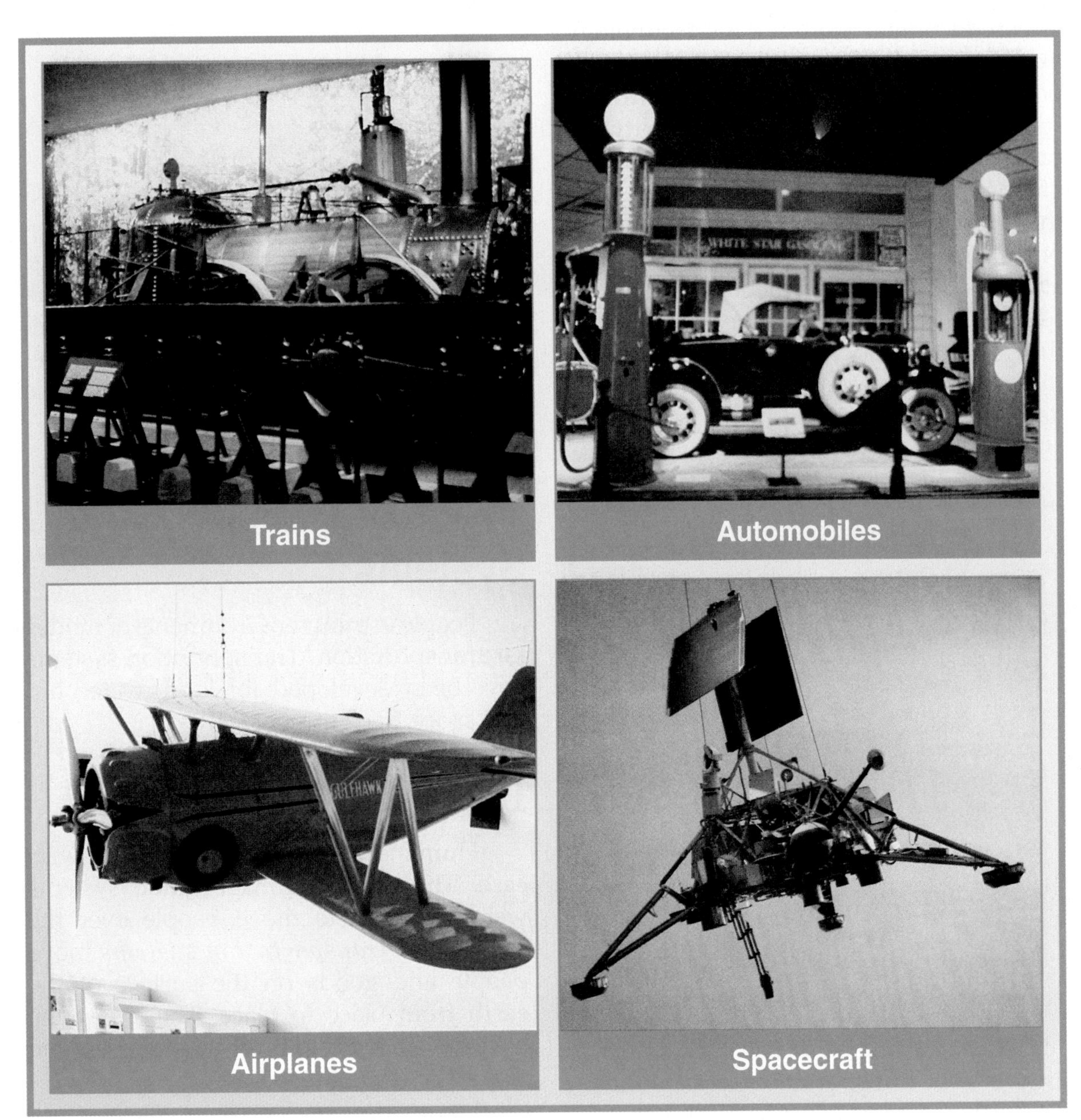

Figure 24-1. Transportation systems have evolved to meet our changing needs. All the transportation vehicles shown above were new transportation vehicles at one time.

Standards for Technological Literacy

Research
Have the students search (encyclopedia, library, Internet) and identify the significant milestones in the development of transportation technology.

Extend
Discuss the importance of transportation to the development of civilizations.

Reinforce
Describe transportation as a technological process (Chapter 5).

Research
Ask the students what transportation contributes to their state's economy.

TechnoFact
The donkey and the ox were domesticated (tamed) to meet people's needs for food products and heavy labor. Donkeys and oxen were first used as pack animals about 7000 years ago.

TechnoFact
A single person in a single place did not invent the wheel. Most likely, early innovators placed rollers underneath heavy objects, allowing them to move the objects more easily. Over time, people probably reduced the mass of the rollers in order to minimize weight and rolling resistance. The axle was likely developed as a means of fixing narrow rollers to the load, completing the evolution of the basic wheel and axle.

Standards for Technological Literacy

18

Extend

Discuss transportation as a system.

TechnoFact

Wagons evolved naturally from the wheel and axle. The first stage in the development of the wagon was the one-wheel cart. A cart is simply a frame fixed to a single wheel and axle assembly. The two-wheel cart was the next stage in wagon development. In the final stage of development, axle and wheel assemblies were added to each end of a box frame to create a wagon.

Transportation: A Definition

Transportation is one of the basic areas of technological activity. We use transportation systems today. In the future, we will continue to use them. Simply stated, the word ***transportation*** can be defined as "all acts that relocate humans or their possessions." Transportation technology provides for this movement, using technical means to extend human ability. This technology extends our ability beyond our own muscle power and our ability to walk and run.

The area of technological activity we call *transportation* is also an interaction among physical elements, people, and the environment. See **Figure 24-2.** Transportation provides mobility for people and goods. This interaction uses such resources as materials, energy, money, and time. Transportation has a level of risk to the cargo and passengers that can result in damage, injury, or death. This interaction has a societal impact, in that it affects such areas as employment and pollution.

Transportation is so important that it has become a part of human culture. Try to imagine life without well-developed transportation systems. We think of transportation in the same light as food, clothing, and shelter. Transportation has become a basic need. For example, transportation takes us to work, opens up areas for recreation, allows for easy shopping, and helps keep families in touch.

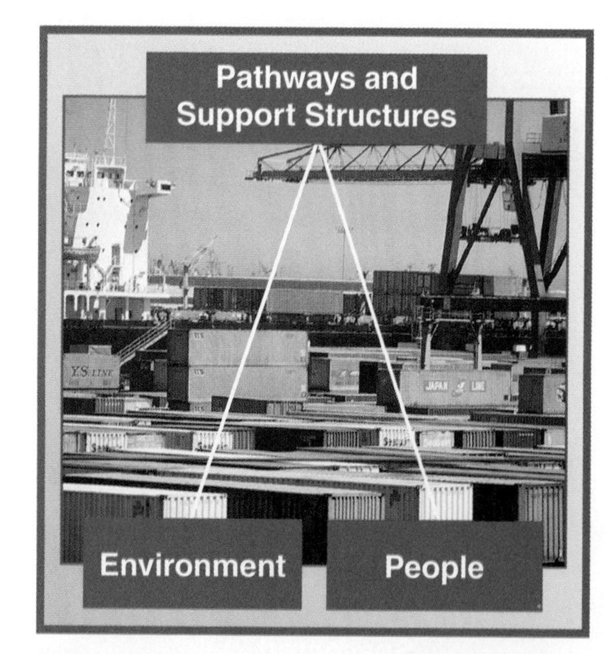

Figure 24-2. Transportation systems are interactive systems of physical elements (vehicles and pathways, for example), people, and the environment.

Transportation as a System

Similar to all other technologies, transportation can be viewed as a system. Transportation is a series of interrelated parts. The parts work together to meet a goal. Transportation uses people, artifacts, vehicles, pathways, energy, information, materials, finances, and time to relocate people and goods.

Types of Transportation Systems

People use four environments or modes for transportation. Transportation systems have been developed for land, water, air, and space. See **Figure 24-3.**

Land

Humans can move over land with ease. The earliest transportation systems were designed to move people over the land. ***Land-transportation systems*** move people and goods on the surface of the earth from place to place. These systems have developed into three major types. The major types are the following:

- **Highway systems.** These systems include automobiles, buses, and trucks.

Standards for Technological Literacy

18

Example

Give local examples of land, water, and air transportation.

MITS, BEA Systems, NASA

Figure 24-3. Transportation systems operate on land, in water, and through air and space.

- **Rail systems.** These systems include freight, passenger, and mass-transit systems.
- **Continuous-flow systems.** These systems include pipelines, conveyors, and cables.

Water

The next system developed was water transportation. From the humble hollow-log canoe, water transportation has grown to be an important mode for moving people and cargo. ***Water-transportation systems*** use water to support the vehicle. Water transportation includes inland waterways (rivers and lakes) and oceangoing systems.

Air

Air transportation became practical in the twentieth century. Orville and Wilbur Wright made the first successful flight in a power-driven airplane

Standards for Technological Literacy

18

Research

Have the students search the NASA web page to find out what the future for space transportation may be.

TechnoFact

Transportation has been on Earth as long as people have. In the Americas, ancient South and Central American people built extensive roads and bridges for transporting goods. The plains tribes of North America transported their belongings in pole and hide, horse-drawn sledges.

in 1903. This flight took place in Kitty Hawk, North Carolina. During the succeeding years, air travel became a large industry. ***Air-transportation systems*** use airplanes and helicopters to lift passengers and cargo into the air so they can be moved from place to place. Today, air transportation includes commercial aviation (passengers and freight) and general aviation (private and corporate aircraft).

Space

Space-transportation systems are the fourth transportation mode we use. Space transportation can best be described as emerging. This transportation started in 1957 with the launch of *Sputnik*, a Soviet satellite. Since 1957, space exploration has expanded our knowledge of the universe. Humans have traveled to the Moon. The space shuttle now lifts satellites and scientific experiments into space.

On the horizon is personal space travel. Hypersonic aircraft will merge air-travel and space-travel technologies. People will then be able to travel anywhere on the globe in a matter of hours.

Transportation-System Components

Within each of the transportation systems are three major components. See **Figure 24-4.** These are the following:

- Pathways.
- Vehicles.
- Support structures.

Pathways

Transportation systems are designed to move people and cargo from one place to another. This movement provides ***place utility***. People value being able to move things from one place to another. The things that are moved can be food, products, or people. We are willing to pay someone else to transport us or our goods from place to place.

To deliver place utility, a transportation system links distant locations with a network of pathways. See **Figure 24-5.** Pathways on land are readily visible. Land pathways are the streets, roads, highways, rail lines, and pipelines people

Career Corner

Automotive Mechanics

Automotive mechanics use skill and knowledge to inspect, maintain, and repair automobiles and light trucks. They review the description of a problem, use a diagnostic approach to locate the problem, repair the vehicle, and test the repair. Automotive mechanics must be able to use a variety of testing equipment. This equipment includes onboard and stationary diagnostic equipment and computers. Most automotive-mechanic jobs require formal training in high school or in a postsecondary vocational school. Some mechanics learn the trade by assisting and learning from experienced workers.

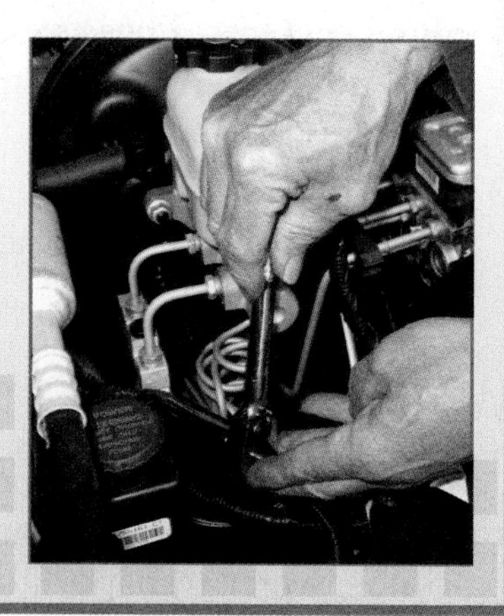

Daimler

Figure 24-4. Transportation systems use pathways, vehicles, and support structures. This photo shows a guided-bus system. The pathway is the concrete guideway. The vehicle is the bus. Support structures are the stations where passengers board the bus.

have built. These pathways form a network connecting most areas of the United States. Land pathways are the result of construction activities used to fulfill the needs of transportation.

Many water pathways are less visible. They include all navigable bodies of water. Navigable means a body of water is deep enough for ships to use. Water pathways can be grouped into two categories. These categories are inland waterways and oceans. Inland waterways are rivers, lakes, and bays. They also include human-made canals. These canals allow water transportation across areas lacking navigable rivers and lakes. Often, the canals use aqueducts

©iStockphoto.com/buzbuzzer, ©iStockphoto.com/photosfromafrica, ©iStockphoto.com/Elerium

Figure 24-5. Pathways include roads, rail lines, rivers, canals, oceans, and air and space routes.

Standards for Technological Literacy

Extend

Discuss the reason airliners follow fixed flight paths.

Research

Have the students find a route map for an airline or railroad.

TechnoFact

Water transportation had its beginnings in prehistoric times. Early humans used natural materials such as reeds and logs to build rafts. Later water transportation vehicles included dugouts and bark canoes, which could be propelled by paddles, poles, or even hands. These early types of watercraft lacked strength and were mainly used in inland waterways.

Standards for Technological Literacy

18

TechnoFact

We can tell from drawings that early Egyptians constructed rafts by tying bundles of reeds together. Such rafts carried goods and people up and down the Nile River about 9000 years ago. By 5000 years ago, the Egyptians had developed sailing ships that were sturdy enough for sea voyages.

TechnoFact

Water transportation is ancient technology. More than 5000 years ago, the ancient Egyptians were traveling on the Mediterranean in ships with square sails.

to cross streams. Canals use locks to raise or lower watercraft as the terrain changes. See **Figure 24-6.**

The oceans and seas provide a vast water pathway. An ocean, or a sea, touches the major landmasses on Earth. Therefore, an ocean transportation network can easily serve the continents.

Some pathways are hard to see. Air and space routes are totally invisible. They are defined by humans and appear only on maps. Airplanes and spacecraft can travel in almost any direction. Only human decisions establish the correct pathway. The pathway is chosen to ensure efficient and safe travel. Airplanes use pathways to take us easily over land and water barriers. They allow us to travel thousands of miles in a few hours. Covering these distances before airplanes or railroads took months of travel by wagon.

Vehicles

All transportation systems, except for continuous-flow systems (such as pipelines and conveyors), use vehicles. ***Vehicles*** are technological artifacts designed to carry people and cargo on a pathway. They are designed to contain and protect the cargo as it is moved from place to place. The demands on the vehicle change with the type of cargo. See **Figure 24-7.**

People need to be protected from the environment through which the vehicle travels. They desire to travel in comfort. Passenger vehicles should provide such features as proper seating and good lighting. Vehicles that carry people long distances should provide food and rest rooms. Also, passenger vehicles should include safety features because of the possibility of accidents. Impact-absorbing construction, seat

Figure 24-6. Canals are constructed waterways. They include waterways, locks to raise or lower ships, and aqueducts to cross streams and gullies.

Alaska Airlines, United Airlines

Figure 24-7. Vehicles such as these aircraft are designed differently for cargo and people.

Standards for Technological Literacy

3 18

Figure Discussion

Compare the cargo compartment with the passenger compartment shown in Figure 24-7.

TechnoFact

Road systems have a long history. The Chinese connected their major cities by road about 3000 years ago. Earlier road systems were developed in the Middle East about 5000 years ago, but those roads were basically dirt tracks.

STEM Connections: Science

Newton's First Law of Motion

Many scientific principles come into play when we use the various kinds of transportation systems now available. Among those principles are Newton's laws of motion. For example, we can use Newton's first law of motion to explain why we need seat belts in our automobiles. Newton's first law of motion states that a moving object continues moving at the same speed, in the same direction, unless some force acts on it. This tendency of an object to resist change is also called *inertia*.

Thus, when we are traveling in an automobile at, say, 50 miles per hour, we continue moving at this speed, even when we forcefully apply the brakes of the car. We have applied force (the brakes) to stop the car. Such force works only on the car, however, not on us. Therefore, we need the seat belt because it provides the force needed to stop us from continuing to move and perhaps hitting the windshield. Newton's first law of motion also involves the concept of friction. Using another method of transportation as an example, can you think of a way friction affects this object when it is moving?

©iStockphoto.com/JohnnyLye

Inertia keeps this gyroscope spinning.

Standards for Technological Literacy

18

Research

Have the students find out how the comfort of passenger areas in vehicles is tested.

Extend

Discuss container shipping as an example of intermodal shipping.

Research

Have the students find out how intermodal shipping got started.

belts, fire protection, and other safety measures should be designed and built into the vehicle.

Cargo vehicles are designed to protect the cargo from damage by motion and the outside environment. Cargo should be cushioned so it does not become broken, dented, or scratched. Fumes, gases, or liquids that can affect the cargo must be kept at a distance. Finally, the cargo should be protected from theft and vandalism.

Support Structures

All transportation systems use structures. Many transportation pathways use human-built structures. Rail lines, canals, and pipelines are examples of constructed works.

Transportation systems need structures other than just pathways. They also need structures called *terminals*. Terminals are where transportation activities begin and end. They are used to gather, load, and then unload passengers and goods. See **Figure 24-8.**

Figure 24-8. Transportation terminals are at the start and the end of transportation pathways. This photo shows an air terminal in Chicago.

Terminals provide passenger comfort and cargo protection before the people and cargo are loaded into transportation vehicles. They also provide connections for various transportation systems. Some allow truck shipping to connect with air, rail, or ocean shipping. Other terminals connect automobile, rail, and bus systems with air-transportation systems. These types of terminals allow for ***intermodal shipping***. See **Figure 24-9.** Intermodal shipping means people or cargo travel on two or more modes of transport before they reach their destination. An example of intermodal shipping is semitruck trailers hauled on railcars. Other structures are used to control transportation systems. These structures are communication towers, radar antennae, traffic signals, and signs. These structures help vehicle operators to stay on course and observe rules and regulations.

©iStockphoto.com/GreenStock

Figure 24-9. These oceangoing shipping containers are now traveling on rail. This is an example of intermodal shipping.

Different transportation systems use different support structures. This train station is used by passengers arriving and departing by train.

Answers to Test Your Knowledge Questions

1. Transportation can be defined as all acts that relocate humans or their possessions.
2. False
3. parts
4. Land, water, air, and space
5. Value provided by the ability to move people or cargo from one place to another
6. True
7. oceans
8. Evaluate individually.
9. Places where transportation activities begin and end
10. intermodal shipping
11. True

Summary

Transportation is essential to modern-day life. We all use it and expect it to make our lives better. Each of us can choose from a variety of transportation systems to get to a place. These systems include land, water, and air transport. In the future, we might also be using space as a transportation link. All transportation systems have pathways and support structures. Most of them use vehicles to carry people and cargo from one point to another.

Test Your Knowledge

Write your answers on a separate piece of paper. Please do not write in this book.

1. Define the term *transportation*, as used in this chapter.
2. Walking to school is an example of transportation technology. True or false?
3. Transportation is a system because it is made up of a series of interrelated ______.
4. List the four types of transportation systems.
5. Define the term *place utility*.
6. Water pathways can be grouped into the two categories of inland waterways and ______.
7. Pathways can be visible or invisible. True or false?
8. Name one type of transportation vehicle.
9. What is a terminal?
10. Using more than one mode of transportation to ship cargo is called ______.
11. A traffic signal is an example of a transportation support structure. True or false?

STEM Applications

1. Select a transportation system you use. Describe it in terms of the following:
   A. The pathway.
   B. The vehicle.
   C. The structures used.

Enter your data on a chart similar to the one below.

Type of Transportation System:		
	Name	Description
Pathway:		
Vehicle:		
Structure:		

2. Obtain a number of bus-, rail-, ship-, or airline-route maps and road maps. Choose a place you would like to visit. Plan a trip that uses intermodal transportation. Be sure your route uses at least two different transportation modes.

This chapter covers the benchmark topics for the following Standards for Technological Literacy:

3 5 7 18

Learning Objectives

After studying this chapter, you will be able to do the following:

- Recall the five systems present in a transportation vehicle.
- Recall the two major components of a structural system.
- Recall common propulsion systems used in transportation vehicles.
- Summarize the characteristics of common suspension systems used in transportation vehicles.
- Recall common guidance and control systems used in transportation vehicles.
- Recall the major systems of a land-transportation vehicle and their functions.
- Recall the major systems of a water-transportation vehicle and their functions.
- Recall the major systems of an air-transportation vehicle and their functions.
- Recall the major systems of a space-transportation vehicle and their functions.

Key Terms

airfoil
apogee
automatic transmission
aviation electronics (avionics)
ballast
blimp
bow thruster
buoyancy
commercial aviation
commercial ship
containership
control system
degree of freedom
direction control
dirigible
dry-cargo ship
Earth-orbit travel
exosphere
fixed-wing aircraft
fuselage
general aviation
geosynchronous orbit
guidance system
hovercraft
hydrofoil
inland-waterway transportation
internal combustion engine
jet engine
lift
lighter-than-air vehicle
liquid-fuel rocket
locomotive
manned spaceflight
maritime shipping
mechanical transmission
merchant ship
mesosphere
military ship
outboard motor
ozone layer
passenger ship
perigee
pleasure craft
power-generation system
power-transmission system

propulsion system
rotary-wing aircraft
rudder
solid-fuel rocket
speed control
stratosphere
structural system
submersible
suspension system
tail assembly
tanker
thermosphere
tractor
trailer
troposphere
turbofan engine
turbojet engine
turboprop engine
unmanned spaceflight
wing

Strategic Reading

As you read this chapter, note the different vehicular systems shared by vehicles of all four environments. List their similarities and differences.

As we noted in the last chapter, *transportation* can be defined as "all acts that relocate humans or their possessions." Most transportation activities use technology to make them more efficient. This technology often takes the form of a vehicle. See **Figure 25-1.** A simple definition of a *vehicle* is "a powered carrier that supports, protects, and moves cargo or people within a transportation system." Vehicles are used in all four environments of travel. They are used on land and in water, air, and space. Each of these areas is discussed in turn, after we examine vehicular systems in general.

©iStockphoto.com/LUGO

Figure 25-1. Most transportation systems require a vehicle.

Vehicular Systems

All vehicles share some common systems. These systems are the structural, propulsion, suspension, guidance, and control systems. See **Figure 25-2.**

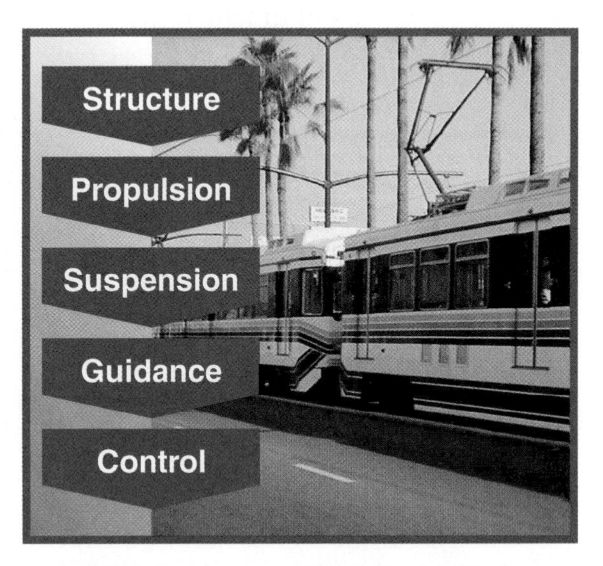

Figure 25-2. Transportation vehicles have five basic systems.

Standards for Technological Literacy

Research

Have each student select a vehicle and determine who invented it and when it came into use.

TechnoFact

Transportation systems have made it possible for individuals to explore distant parts of the world. The time period from the late 1400s through the 1500s could be called the *age of exploration and expansion*. Many European explorers made remarkable sailing voyages, such as Sir Francis Drake, James Cook, Ferdinand Magellan, and Christopher Columbus. They located and claimed land in North and South America, the South Pacific, Australia, and New Zealand during these journeys.

TechnoFact

Overseas commerce increased as the globe was explored and mapped. Shipbuilders designed and constructed increasingly bigger cargo ships to aid the rising business during the 1600s and 1700s. These larger vessels had more sails and went faster than their predecessors. By the mid-1800s, quick commercial craft, such as the *Cutty Sark*, had as many as thirty-five sails and moved at speeds up to 20 knots.

Standards for Technological Literacy

18

Reinforce

Discuss vehicles and their systems as part of a transportation system (Chapter 24).

Brainstorm

Ask your students what the propulsion system is for a bicycle, sailboat, automobile, and space shuttle.

The Structural System

All vehicles are designed to meet a common goal—to contain and move people and goods. People want to arrive safely and in comfort. Cargo must be protected from the weather, damage, and theft. The ***structural system*** of the vehicle helps to do these things. This system is composed of the physical frame and covering. The structural system provides spaces for people, cargo, power and control systems, and other devices.

The Propulsion System

Transportation vehicles are designed to move along a pathway. This pathway might be a highway, a rail line, a river, an ocean, or an air route. A vehicle must have a force to propel it from its starting point to its destination. The ***propulsion system*** produces this force. This system uses energy to produce power for motion.

Propulsion systems range from the simple pedal, chain, and wheel system of a bicycle to complex heat engines, such as gasoline, diesel, and rocket engines. See **Figure 25-3.** Several factors determine the type of engine to be used in a vehicle. These factors include the following:

- The environment in which the vehicle travels.
- Fuel availability and cost.
- The forces that must be overcome, such as vehicle and cargo weight, rolling friction, and water or air resistance.

The engine must match the job. For example, using a jet engine to propel an automobile is overkill. The capabilities of a jet engine do not match the job of moving a car. Likewise, using a large diesel engine to power an airplane is not the best choice. The diesel engine is too heavy. Therefore, we use many different sizes and many different types of engines in transportation vehicles.

The Suspension System

All vehicles and cargo have weight. This weight must be supported as the vehicle moves along the pathway. A ***suspension system*** produces proper support. Suspension systems include the following:

- Wheels, axles, and springs on land vehicles.
- The wings on an airplane.
- The hull of a ship.

The Guidance System

Handling any vehicle requires information. Operators must know their locations,

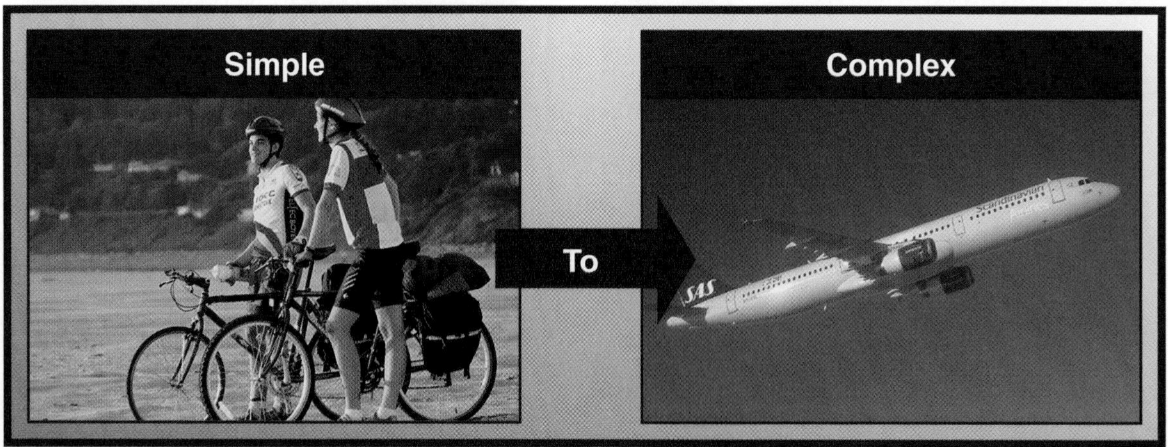

Airbus

Figure 25-3. Propulsion systems range from simple to complex.

speeds, and directions of travel. Information about traffic conditions and rules is also required. The ***guidance system*** provides this information.

Guidance information can be as simple as a speedometer reading or a traffic light. Guidance systems can also be quite complex. Instrument landing systems (ILSs) and land-based satellite-tracking stations are examples of complex guidance systems.

The Control System

A vehicle moves from its origin to its destination to relocate people and cargo. As a vehicle moves, it requires control. See Figure 25-4. The ***control system*** consists of two types: speed control and ***direction control***. The vehicle can be made to go faster through acceleration or slowed down by braking or coasting. How a vehicle changes its direction depends on its environment of travel. Land vehicles turn their wheels or follow a track. Ships move their ***rudders***. Airplanes adjust ailerons, rudders, and flaps. Spacecraft use rocket thrusters.

The system of controls used depends on the ***degrees of freedom*** a vehicles has. For example, rail vehicles have one degree of freedom. They move forward or backward by using ***speed control***. Gravity and the rail eliminate the possibility of up-down and left-right movement.

Automobiles and ships have more freedom. They can move forward and backward and can change their directions by turning left or right. These vehicles have two degrees of freedom.

Aircraft, spacecraft, and submarines have the most freedom. This freedom is three degrees. These vehicles can move forward and backward, move left and right, and change their altitudes by moving up and down.

Land-Transportation Vehicles

Land transportation includes all movement of people and goods on or under the surface of the Earth. This transportation includes highway-, rail-, material-, and

Standards for Technological Literacy

18

Extend

Discuss guidance and control as it applies to operating a vehicle.

TechnoFact

During the 1700s, England and France opened the first well-built paved roads since the first ones were established in ancient Rome. The first major U.S. highway was built in the mid-1800s, and it was a gravel road called the National Road. It extended from Vandalia, Illinois to Cumberland, Maryland.

©iStockphoto.com/grahamheywood

Figure 25-4. Transportation vehicles require control in three degrees of freedom. The degrees are forward and backward, left and right, and up and down. The modern passenger train in the photo has one degree of freedom.

Freightliner Corp.

Figure 25-5. All vehicles have operator, power, and cargo or passenger units.

Standards for Technological Literacy

18

Brainstorm

Ask your students what types of land transportation systems are used in your community.

TechnoFact

American pioneers who wanted to travel west of the Mississippi River had to travel through the wilderness. There were no roads leading to the Pacific Ocean, so these pioneers simply traveled on dirt paths. Because so many people were heading west, these paths became common travel routes, such as the Santa Fe and Oregon Trails.

on-site transportation systems. Highway systems use automobiles, trucks, and buses to move people and cargo. Rail systems move people and cargo from place to place. Material-transportation systems include pipelines and conveyors. On-site transportation systems include systems commonly found in such places as factories, stores, and hospitals. These systems might be forklifts, tractors and carts, chutes, elevators, and conveyors.

The Structure of Land-Transportation Vehicles

Each land-transportation system requires special vehicles. The structure a vehicle has is based on the vehicle's use. Passenger vehicles are different from cargo-carrying vehicles. See **Figure 25-5.** All the vehicles, however, have three basic structural units:

- A passenger or operator unit.
- A cargo unit.
- A power unit.

The sizes and locations of these units or compartments vary with the type of vehicle. Freight and some passenger rail systems place the power and operator units in one vehicle called a ***locomotive***. The locomotive pulls the cargo and passenger units. Mass-transit systems have power units (electric motors) and passenger units in each car. The operator unit is located in the front car. See **Figure 25-6.**

Long Beach Light Rail, Norfolk Southern

Figure 25-6. The light-rail vehicles integrate operator and passenger areas. The freight rail system uses a locomotive that combines operator and power units. The locomotive pulls railcars, the cargo units.

Standard automobiles and delivery trucks have all three units in one vehicle. Most long-distance trucks, however, place the power and operator units in the ***tractor***. The cargo unit is the ***trailer*** attached to the tractor.

The passenger and operator units must be designed for comfort and ease of operation. This requires the use of ergonomic principles. Ergonomics is the study of how people interact with the things they use. Ergonomic vehicle design requires the seats to adjust for people of different sizes. The instruments providing guidance data must be easily seen. See **Figure 25-7.** The operating controls, such as for steering and braking, must be easy to reach and operate.

The vehicle must also have an appropriate structural design. This carrier must provide for safety and for operator, passenger, and cargo protection. Most vehicles have a reinforced frame with a skin. The skin protects the vehicle interior from the

Boeing

Figure 25-7. This aircraft flight deck was designed using ergonomic principles for easy operation. Ergonomic principles are also used in land, water, and space vehicles.

outside environment. The frame supports the skin and carries the weight of passengers and cargo. This reinforced frame also provides crash protection by absorbing the impact when vehicles are involved in accidents. See **Figure 25-8.**

Arvin Industries

Figure 25-8. This vehicle is being crash tested to evaluate its structural design.

The Propulsion of Land-Transportation Vehicles

Most land-transportation vehicles move along their pathways on rolling wheels. Two systems produce the rotation. These systems are power generation and power transmission.

The Power-Generation System

The ***power-generation system*** uses an engine as an energy converter. The engine produces the power needed to propel the vehicle. The most common engine in land vehicles is the ***internal combustion engine.*** This name means fuel is burned inside the engine to convert energy from one form to another. The chemical energy in the fuel is first changed to heat energy. The piston and crankshaft then convert the heat energy into mechanical (rotating) energy to move the vehicle.

Land vehicles normally use either a four stroke–cycle gasoline engine (described in Chapter 7) or a diesel engine. Some rail vehicles use electric motors for propulsion. Electrically powered vehicles for highway use are on the horizon.

As noted in Chapter 7, most internal combustion gasoline engines use four strokes to make a complete cycle. The first stroke is called the *intake stroke.* The intake valve opens, the piston moves down, and a fuel-air mixture is drawn into the cylinder. When the piston is at the bottom of the intake stroke, the valve closes. The piston moves upward, beginning the compression stroke that compresses the fuel-air mixture. When the piston is at the top of its travel, a spark plug ignites the fuel-air mixture. The burning fuel expands rapidly and drives the piston down. This is called the *power stroke.* At the end of the power stroke, the exhaust valve opens. The piston moves upward to force burnt gases from the cylinder. The last stroke is called the *exhaust*

Standards for Technological Literacy

18

Extend

Discuss the types of engines that have been used over the years to power land transportation vehicles.

Research

Have students identify the limits to trailer size and weight for over-the-road trucks and determine why these limits were established.

Standards for Technological Literacy

18

TechnoFact

Most vehicles have structures to protect their passengers. One of the first examples of this is the horse-drawn carriage, which appeared as early as the Bronze Age. The Roman chariot was another early passenger carriage. Other forms of transportation were invented for transporting goods. Adjustments had to be made to all of these early vehicles once roads became commonplace.

stroke. The exhaust valve closes, and the engine is ready to repeat the four strokes—intake, compression, power, and exhaust. You might want to look back to **Figure 7-22** to see a drawing of these four strokes.

Diesel engines power many large vehicles, such as buses, heavy trucks, and locomotives. The most common diesel engine uses a four-stroke cycle, similar to a gasoline engine. See **Figure 25-9.** During the intake stroke, however, only air is drawn in. As the air is compressed, it becomes very hot. When the piston is at the top of the compression stroke, fuel is injected directly into the cylinder. The fuel touches the hot air and burns rapidly. This causes

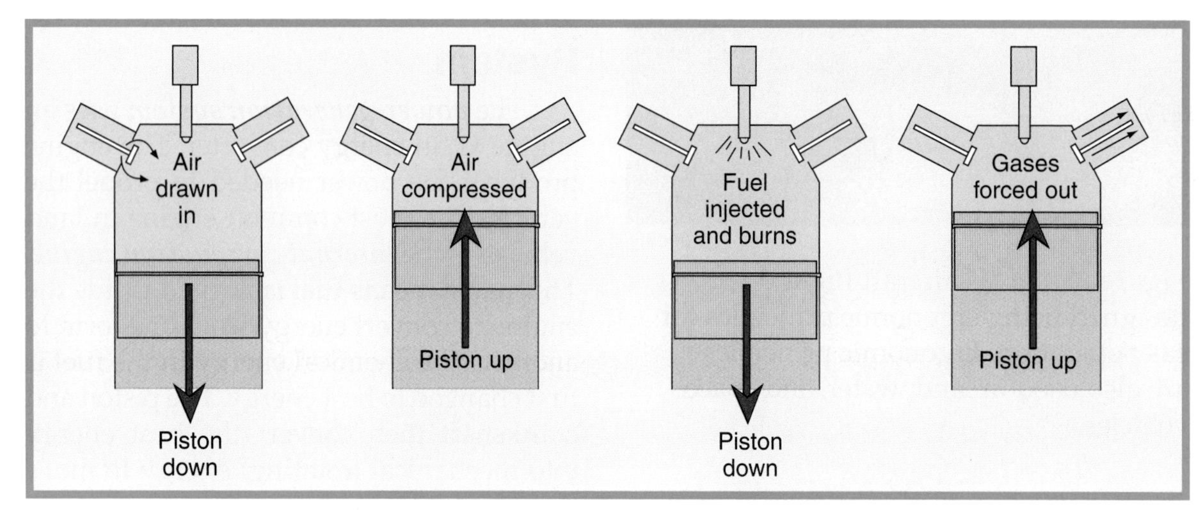

Figure 25-9. How a diesel engine works.

Career Corner

Transportation, Distribution & Logistics

Railroad Conductors

Railroad conductors coordinate the activities of train crews. Freight-train conductors review schedules, switching orders, waybills, and shipping records to obtain cargo-loading and -unloading information. Passenger-train conductors ensure passenger safety and comfort as they go about collecting tickets, making announcements, and coordinating passenger-service activities.

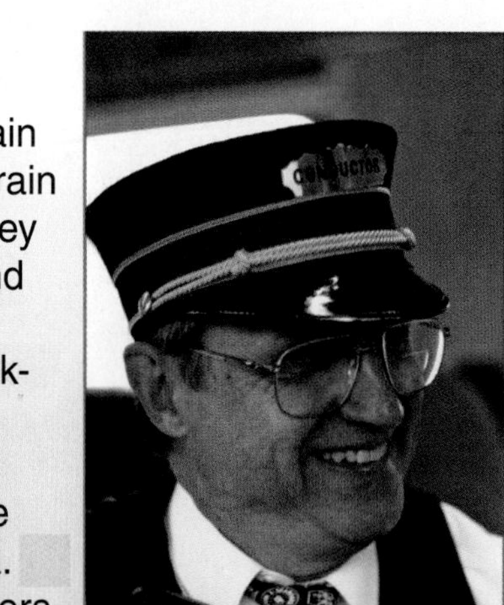

Conductors might be expected to work nights, weekends, and holidays because trains operate around the clock, seven days a week. The more senior conductors receive the more desirable shifts. Railroads require applicants to have a minimum of a high school diploma. Almost all railroad workers begin working as yard laborers and, later, might have the opportunity to train to become conductors.

the power stroke, which is followed by the exhaust stroke. These last two strokes are very similar to those of a gasoline engine.

Newer types of propulsion systems are being developed because of the rising cost of petroleum-based fuels and a growing environmental awareness. One recent development uses a power-generation system called a *hybrid electric system*. See **Figure 25-10.** This system is at the heart of hybrid electric vehicles (HEVs), which are discussed in the Technology Explained feature on page 107. In that feature, you learned that the gasoline-electric hybrid propulsion systems contain three major parts:

- A gasoline engine very similar to the one in standard cars, except it is smaller.
- A high-voltage battery that can be used to provide energy to move the vehicle and store energy for later use.
- A sophisticated electric motor that can be used as a power source to move the vehicle and a generator to charge the vehicle's battery. When the motor is acting as a motor, it draws energy from the battery to power the vehicle. When the vehicle slows down, the motor acts as a generator and returns energy to the battery and also provides a braking force (regenerative braking).

Many HEVs reduce fuel use and emissions by shutting down the engine at idle and restarting it when needed.

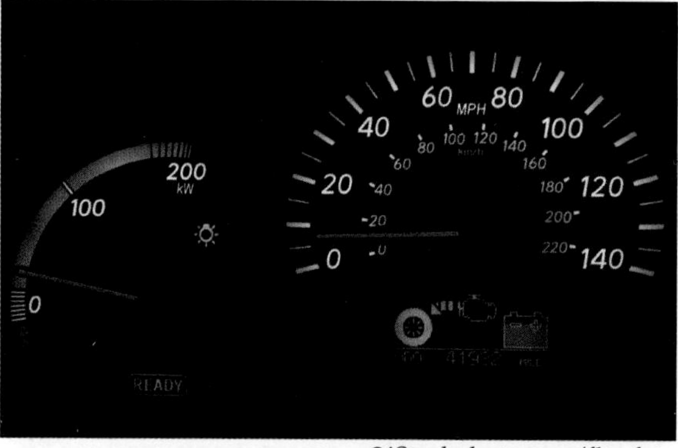

©iStockphoto.com/fluxfoto

Figure 25-10. A dashboard readout on a hybrid car showing kilowatts, a speedometer, engine and battery statuses, and an odometer.

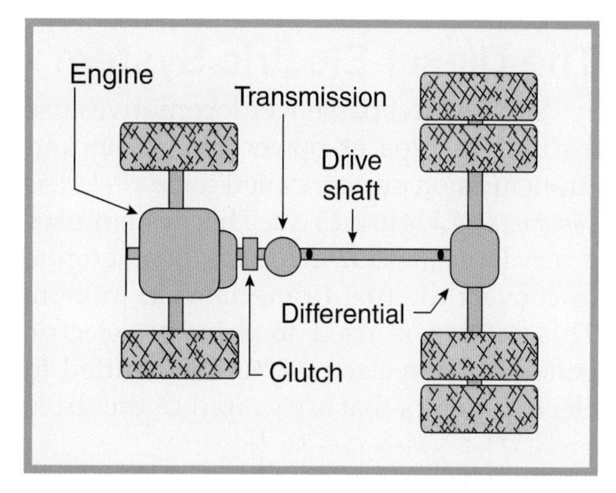

Figure 25-11. The transmission joins the engine to the rest of the drivetrain.

The Power-Transmission System

A transmission is often an important component in a ***power-transmission system***. This component is located between the engine and the drive wheels. See **Figure 25-11.** Most transmissions in automobiles and trucks are either mechanical or fluid devices. Often, a transmission has several input-to-output ratios, commonly called *speeds*. A five-speed transmission has five ratios. The different ratios allow the power of the engine to be used efficiently. The transmission provides high torque (rotating force) at low speeds, as the vehicle starts to move. Later, as gravity, rolling resistance, and air resistance are overcome, low-torque, high-speed outputs are used.

A ***mechanical transmission*** (also called a *manual transmission*) has a clutch between the engine and the transmission. This allows the operator to disconnect the engine so the transmission ratio can be changed. The action is called *shifting gears*.

Standards for Technological Literacy

18

Extend

Discuss how gears can speed up or slow down the speed at which a shaft turns.

Brainstorm

Ask the students why people are developing alternate-fuel engines and hybrid cars.

TechnoFact

The prototype of the first automobile was built in 1855 by Karl Benz. His vehicle was powered with an Otto four-stroke cycle internal combustion engine. In the same year, Gottlieb Daimler patented a design for a gasoline engine. Several years later, in 1892, Rudolf Diesel patented the engine carrying his name.

Standards for Technological Literacy

18

Brainstorm

Ask your students to name the advantages of an automatic transmission.

Figure Discussion

Explain how a diesel-electric locomotive works.

TechnoFact

The first U.S. railroad was built in 1826 in Massachusetts. It was not like the railroads we know today, however; this railroad used horse-drawn wagons that followed iron rails. The Baltimore & Ohio Railroad (B&O) was the first "long-haul" railroad in the United States. It began operating in 1828 with horse-drawn cars, but it began testing steam-powered locomotives soon afterward. The successful run of the *Tom Thumb* locomotive on the B&O Railroad in 1830 paved the way for steam-powered locomotives everywhere.

Automatic transmissions use valves to change hydraulic pressure so the transmission shifts its input-to-output ratios.

When a vehicle makes a turn, the wheels on the outside must be able to turn faster than those on the inside. The outside wheels turn faster because they have farther to travel. A device called a *differential* allows the wheels to turn at different speeds. The differential is a set of gears that independently drive each axle. The axles are connected to the vehicle's drive wheels. In rear wheel–drive automobiles and trucks, the differential is separate from the transmission. Front wheel–drive cars combine the transmission and differential into a single unit called a *transaxle.*

The Diesel-Electric System

Freight and passenger locomotives use a different type of power-generation and -transmission system called a *diesel-electric system.* See **Figure 25-12.** This system uses a very large (8- to 12-cylinder) diesel engine to convert the fuel to mechanical motion. This motion is used to drive an electric generator. The electricity is transmitted to electric motors that are geared to the drive wheels of the locomotive. This motor-generator unit is relatively light. Therefore, extra weight, called ***ballast,*** is added to the locomotive to give it better traction. Making the frame of the locomotive out of heavier steel plates than needed creates this extra weight.

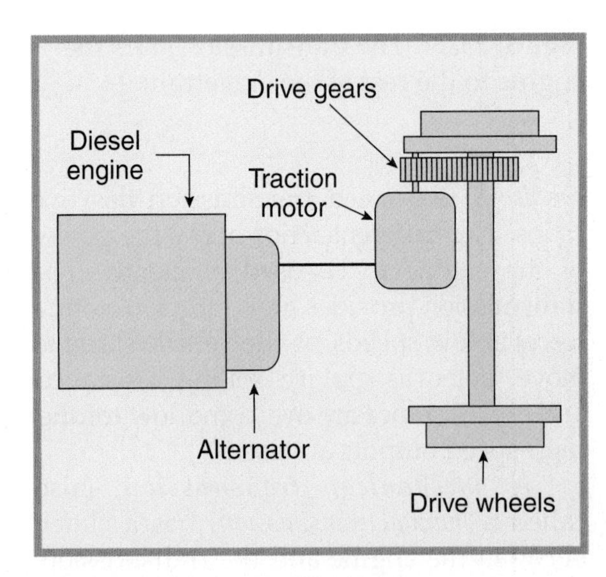

Figure 25-12. A diesel-electric traction system used to power railroad locomotives.

The Electric System

Some rail vehicles are powered by only an electrical system. The vehicle has a moving connection that remains in contact with the electric conductor. See **Figure 25-13.** Overhead wires are used, or a third rail is used, as the conductor. The electricity is conducted to motors that turn the vehicle's wheels.

The Suspension of Land-Transportation Vehicles

Suspension systems keep the vehicle in contact with the road or rail. They also separate the passenger compartment from the drive system to increase passenger and

©iStockphoto.com/Sparky2000

Figure 25-13. Electric-powered light-rail cars in San Francisco.

operator comfort. The suspension system of a vehicle has three major parts:

- Wheels.
- Axles.
- Springs and shocks.

The wheels provide traction and roll along the road or rail. They also spread the vehicle's weight onto the road or rail. A wheel can be used to absorb shock from bumps in a road or rail. Often, rubber tires are attached to the wheels to absorb shock and increase traction. Most rail systems use steel wheels and depend on friction between the wheels and rails for traction.

Axles carry the load of the vehicle to the wheels. They support a set of springs that absorb movements of the axles and wheels. Springs, however, give a very bouncy ride, which the use of shock absorbers softens. The shock absorbers are attached between the axle and the frame to dampen the spring action. See **Figure 25-14.** The result is a controlled, smooth ride.

A variation of the suspension system uses torsion bars. Instead of a spring, a steel bar absorbs the movement of the wheels and axles by twisting. The torsion bar untwists to release the absorbed energy. Shock absorbers are used to dampen the motion of the torsion bar.

The Guidance and Control of Land-Transportation Vehicles

An operator manually controls the speed and direction of most land-transportation vehicles. Throttles and accelerator pedals are used to control engine speed. Transmissions are shifted to increase the power being delivered to the drive wheels. Brakes can be applied to slow the vehicle. The wheels can be turned to take the vehicle in a new direction.

These actions require the operator to make decisions. These decisions are made on the basis of visual information and judgment. The operator receives information from signs and signals. See **Figure 25-15.** The operator also observes traffic conditions. Gages and instruments provide information on vehicle speed and operating conditions. The operator's judgment also becomes important because all this information must be considered and acted on. Inexperienced operators and those influenced by drugs and alcohol use might make poor judgments. Their actions can cause accidents and human injury or death.

Standards for Technological Literacy

18

Research

Ask your students how vehicle-sensing traffic lights work.

Extend

Discuss how a braking system works.

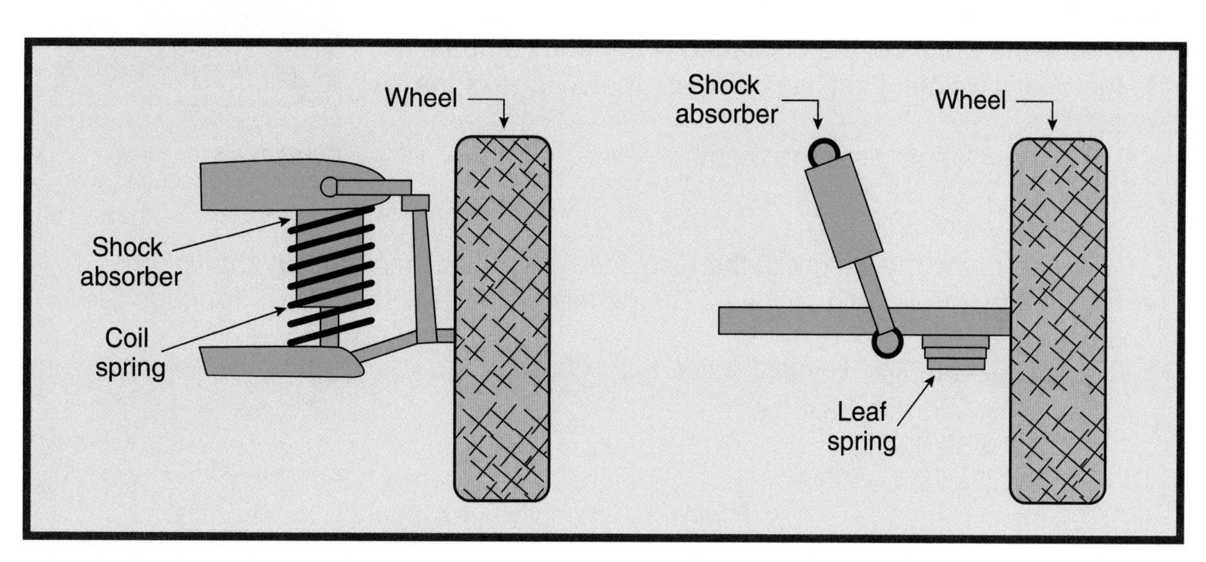

Figure 25-14. Two types of suspension systems used on automobiles.

Standards for Technological Literacy

3 18

TechnoFact

Many countries have developed high-speed rail systems that go as fast as 185 miles per hour. In 1964, Japan established the first such system with the *Shinkansen* or "bullet train." The French brought the high-speed train to Europe in 1981. Following France were the United Kingdom in 1984, Italy in1988, Germany in 1991, and Spain in 1992. In 2000, the United States introduced its fast train, *Acela Express*, which runs between Washington, D.C. and Boston.

Figure 25-15. Signs and signals, such as highway signs and rail signals, give operators guidance.

STEM Connections: Mathematics

Calculating Buoyant Force

As we discussed in this chapter, the buoyant force, which causes objects to float, is equal to the weight of displaced water. This force can be calculated by multiplying the volume of displaced water by the density of water. If this force is greater than the weight of the boat, the boat will float.

For example, a fully loaded flat-bottom boat weighs 50,000 lbs. The boat is 20′ long, 8′ wide, and 6′ deep. To determine whether or not the boat will float, we need to calculate the weight of the displaced water and compare it to the boat's weight. First, we calculate the volume of the boat:

$$\begin{aligned} \text{volume} &= \text{length} \times \text{width} \times \text{depth} \\ &= 20' \times 8' \times 6' \\ &= 960 \text{ ft}^3 \end{aligned}$$

©iStockphoto.com/Swan-City-Church

Buoyancy is the force allowing this canoe to float.

To determine the weight of the displaced water, we then multiply the volume of the boat by the density of water. The density of water is 62.4 lbs/ft^3 (pounds per cubic foot):

$$\begin{aligned} \text{weight} &= \text{volume} \times \text{density} \\ &= 960 \text{ ft}^3 \times 62.4 \text{ lbs/ft}^3 \\ &= 59{,}904 \text{ lbs.} \end{aligned}$$

The buoyant force (59,904 lbs.) is greater than the weight of the boat (50,000 lbs.). Therefore, the boat will float. If a fully loaded flat-bottom boat is 25′ long, 10′ wide, and 6′ deep and weighs 75,000 lbs., will it float? Why or why not?

Water-Transportation Vehicles

Water covers more than 70% of the Earth's surface. Thus, water transportation is an important form of travel. Water transportation includes all vehicles that carry passengers or cargo over water or underwater. This transportation on rivers, on lakes, and along coastal waterways is called ***inland-waterway transportation***. Water transportation on the oceans and large inland lakes, such as the Great Lakes in the United States, is called ***maritime shipping***.

There are three kinds of water-transportation vehicles (called *vessels*). See **Figure 25-16.** Vessels that private citizens own for recreation are called ***pleasure craft***. They allow people to participate in activities such as waterskiing, fishing, sailing, and cruising. Large ships used for transporting people and cargo for a profit are called ***commercial ships***. Some ships provide for the defense of a country. These vessels are called ***military ships*** (or *naval ships*) and are owned by the government of a country.

The Structure of Water-Transportation Vehicles

A ship's intended use influences how the ship is designed. All ships have two basic parts. See **Figure 25-17.** The hull forms the shell that allows the ship to float and contain a load. The superstructure is the part of the ship above the deck. This part contains the bridge and crew or passenger accommodations.

©iStockphoto.com/CaraMaria

Figure 25-17. The two major parts of a ship are the hull and the superstructure.

Standards for Technological Literacy

18

Figure Discussion

Differentiate between pleasure, commercial, and military vessels as shown in Figure 25-16.

TechnoFact

People have used oceangoing ships since ancient times. Oars or sails, and sometimes both, drove these ships. Over time, people developed specialized vessels with distinctive hull shapes to serve various purposes. Heavy boats with curved undersides were used to carry grain, while craft with slim keels were used in combat.

©iStockphoto.com/Terraxplorer, ©iStockphoto.com/Toprawman, ©iStockphoto.com/DomD

Figure 25-16. The three types of boats and ships are pleasure, commercial, and military.

Standards for Technological Literacy

18

Extend

Discuss the types and design of commercial ships.

Similar to land vehicles, ships can be designed to carry either passengers or cargo. See **Figure 25-18.** Vessels that carry people are called ***passenger ships*** or *passenger liners.* Cargo-carrying ships are called ***merchant ships***.

Passenger Ships

Passenger ships include oceangoing liners, cruise ships, and for short distances, ferries. Ocean liners and cruise ships are miniature cities. They must supply all the needs of the passengers. These ships must, thus, include such areas as sleeping quarters, kitchens, dining space, recreational areas, and retail shops, such as haircutting and clothes-cleaning establishments.

Merchant Ships

Merchant ships are usually designed for one type of cargo. The types of cargo can include dry cargo, liquids, gases, and cargo containers. See **Figure 25-19.** ***Dry-cargo ships*** are used to haul both crated and bulk cargo. The term *bulk cargo* describes the loose commodities, such as grain, iron ore, and coal, loaded into the

Passenger Ship

Merchant Ship

Figure 25-18. Commercial ships can be either passenger or merchant ships.

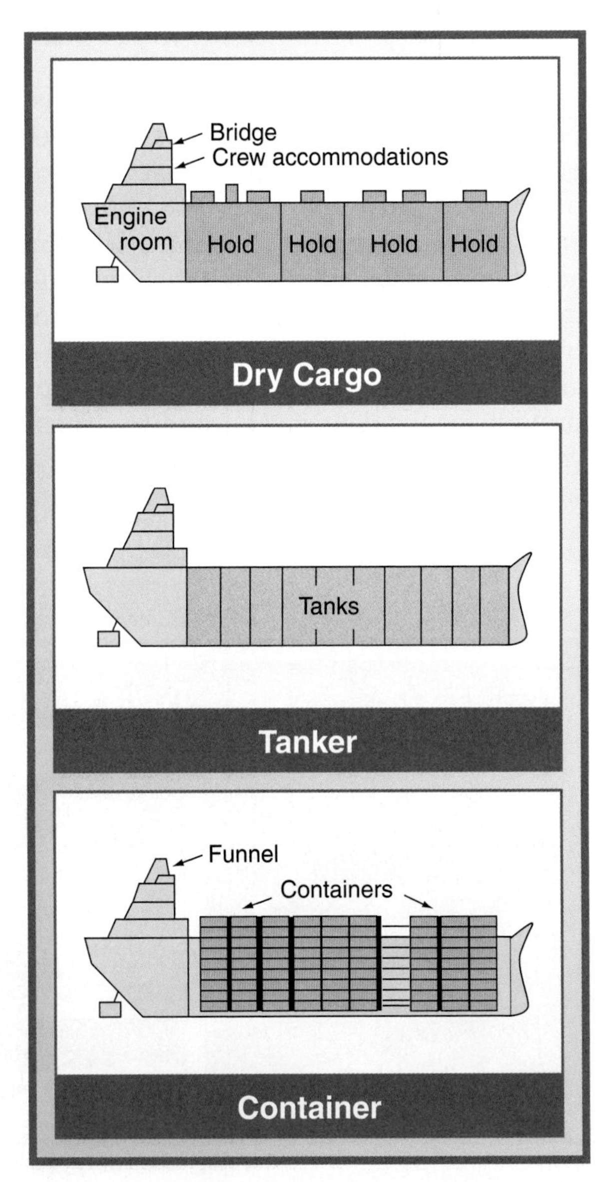

Figure 25-19. The common types of merchant marine vessels.

holds (compartments) of ships. Other dry cargo is contained in crates loaded into various holds of a ship. Land vehicles, such as cars and trucks, are another bulk cargo often transported on special vessels called *roll-on, roll-off (RORO) ships*. The vehicles are driven onto the ship and off the ship through large hatches (doors) in the hull.

Tankers are used to move liquids, such as petroleum, liquefied natural gas, and other chemicals, in a series of large tanks. The liquid cargo can be pumped into and out of the tanks. Tankers have also been built to move gases.

Containerships are a newer and faster way to ship large quantities of goods. See **Figure 25-20.** The shipments are loaded into large steel containers resembling semi-truck trailers without wheels. The containers are sealed, loaded into the hold, and then stacked on the deck of the ship. The loading process that once took days is now done in hours.

Large ships sail the oceans, whereas smaller ships are used on such lakes as the Great Lakes in North America. Most shipping on inland waterways, however, is done with tugboats and barges. See **Figure 25-21.** These vessels are "water trains." The tugboat acts similar to the locomotive. This vessel contains the power-generation and -transmission systems and the operation controls. The barges are the waterway "railcars." Various types of cargo have specifically designed containers. The tug pushes or pulls a group of barges lashed together.

Figure 25-21. Tugboats pull or push barges to carry large loads on inland waterways.

Standards for Technological Literacy

18

TechnoFact

In the Middle Ages, Viking ships used both oars and sails to travel over the water. One of these large ships brought Leif Eriksson to America. Researchers unearthed a Viking ship in 1880, and it was 80′ long, 16.5′ wide, and nearly 7′ deep.

Figure 25-20. Containers are often shipped from inland points and then unloaded at the dock. Later, the containers are loaded onto containerships.

Standards for Technological Literacy

18

TechnoFact

In the 1700s and 1800s, numerous kinds of wooden sailing vessels were developed. Some of these were 300′ long. Commercial fleets developed the brig, clipper, and schooner, while navies developed men-of-war, including the frigate and corvette. Each style of boat had its own design, with varying numbers and locations of poles and sails.

Technology Explained

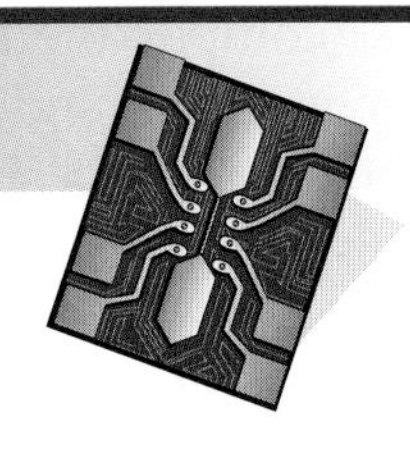

magnetic levitation (maglev) train: a transportation vehicle using magnetism to suspend and propel itself along a guideway.

The rough ride and noise traditional rail systems produce have been a drawback to their use for passenger travel. The steel wheels and rails used in most rail systems cause these drawbacks. Improvements in rails, wheels, and maintenance have made trains quieter. Some trains, such as monorails, use rubber tires or inserts in steel wheels. Some noise is still produced, however, because of the contact between the wheels and rails. In the 1960s, a new land vehicle appeared that did not touch the rails. This type of vehicle is called a *maglev train.* Maglevs use magnetic forces to support and move along special pathways called *guideways.*

Trapsrapid International

Figure A. The Transrapid® 06 car. This car uses magnetic attraction (magnets pulling toward each other) to suspend the vehicle.

There are two types of maglevs: attraction and repulsion. An attraction maglev uses magnets pulling toward each other to support the train. See **Figure A.** A repulsion maglev uses magnets pushing away from each other to support the train. See **Figure B.** The guideways differ between the two types of maglevs. Attraction maglevs wrap around the guideways. Repulsion maglevs sit in troughlike guideways.

The maglev train includes a passenger compartment providing a quiet and comfortable environment for the passengers. Beneath the passenger compartment is the suspension system. Attraction maglevs use electromagnets to support the train. The electromagnets pull toward a rail in the guideway, lifting the train. The amount of electricity is controlled so the electromagnets and rail never touch. Repulsion maglevs use superconducting magnets to induce magnetic fields in coils in the guideway. The maglev floats on these magnetic fields.

Japan Railways Group

Figure B. Magnet repulsion can also be used to suspend a maglev. The MLU-002 maglev uses magnets pushing away from each other for suspension.

A device called a *linear induction motor (LIM)* propels maglevs. A LIM works similarly to a standard electric motor, using the principle that like poles of a magnet repel each other, while unlike poles attract. A standard electric motor produces rotary motion. A LIM is laid out flat in a straight line. Thus, a LIM produces linear, or straight-line, motion.

A LIM uses alternating electric current to create magnetic fields. The direction of the electric current changes many times a second. When this happens, the lines of magnetic force collapse and change direction. This causes the rail to be attracted to the magnet, and the vehicle moves along the rail. The current then switches direction, causing the magnet to repel the rail. This cycle repeats over and over and pushes the maglev along the guideway.

The Propulsion of Water-Transportation Vehicles

A propeller (prop) driven by a steam turbine or diesel engine propels most commercial ships through the water. Often, the engine is located inside the ship or boat. A motor (an internal combustion gasoline engine) attached to the stern (the back) of the boat, however, might power small boats. This type of power source is called an ***outboard motor.***

Nuclear reactors power some military ships and submarines. Heat from the nuclear reactor is used to turn water into steam. This steam is used to turn a turbine. The turbine turns the prop using a shaft.

The prop is a device with a group of blades radiating out from the center. See **Figure 25-22.** Props can have from two to six blades. The blades attach to a shaft at the center. Props range in size from 2″ (51 mm) to more than 30′ (9 m) in diameter. Each blade of the prop is shaped to "bite" into the water, very similar to how a window fan "bites" into the air. The rotation of the prop forces the water past the prop. As we learned earlier, one law of physics tells us that, for every action, there is an equal and opposite reaction (Newton's third law of motion). Therefore, the action of forcing the water through the spinning prop causes an opposite reaction that pushes the boat forward.

©iStockphoto.com/Folscheid

Figure 25-22. A marine prop on a large ship.

Engines do not power all boats. Some boats use sails to capture wind, which is one of the earliest sources of propulsion. A sail catches the wind and pushes the boat through the water. See **Figure 25-23.**

The Suspension of Water-Transportation Vehicles

For water-transportation vehicles, the hull is the primary component of the suspension system. The hull must be carefully designed to ensure that the boat remains stable and afloat. If the hull is not properly designed, a boat can roll onto its side and sink.

The shape of the hull has a great effect on the stability of the boat. Boats with

©iStockphoto.com/DanCardiff

Figure 25-23. Sails are widely used on pleasure craft. Wind power was once the leading way to propel ships.

Standards for Technological Literacy

18

Extend

Differentiate between an inboard and an outboard engine.

Research

Have your students find out what the pollution problems associated with outboard engines are.

Demonstrate

During class or lab, use a tub of water and a wood block to demonstrate buoyancy.

TechnoFact

Fulton's steamship, the *Clermont*, constructed in 1807, started the shift from sails to steam power. Steel craft started to take the place of wooden boats later in the nineteenth century. Also, the steam turbine replaced the steam engine in this period. The diesel engine came into nautical use in the early twentieth century. Nuclear power was introduced as a power source for ships during the 1950s.

Standards for Technological Literacy

18

Research

Have the students determine what is meant by displacement. Have them research the range of displacement of common merchant ships.

Figure Discussion

Discuss hull design in terms of displacement (buoyancy) and speed.

Extend

Describe multihull vessels (catamaran, etc.)

Research

Have the students trace the evolution of hull designs.

Extend

Discuss the uses for hovercraft.

round-bottom hulls are relatively unstable. V-bottom hulls, which are shaped similar to the letter *V*, provide greater stability than round-bottom hulls. Flat-bottom hulls are the most stable. See **Figure 25-24.**

The weight distribution in the boat also affects stability. If the boat's load is evenly spread over the hull, the boat should be very stable. If the boat is unevenly loaded, it might lean in the water, and the propulsion system and control system will not work efficiently. In addition, an unevenly loaded boat is more likely to sink.

When an object is immersed in a fluid, the fluid tends to push the object up. If you lift your arms away from your body while swimming, you will notice that the water tends to push your arms toward the surface. This tendency is called *buoyancy.* ***Buoyancy*** is the upward force exerted on an object immersed in a fluid.

You can think of the buoyant force as a force trying to hold an object up in the water. In order for the object to float, the buoyant force must be equal to the weight of the object. If the weight of the object is greater than the buoyant force, the object sinks.

When an object is placed in a fluid, the fluid must move to make room for the object. Displaced fluid is the fluid that had previously occupied the space that the immersed object now occupies. The volume of displaced fluid is identical to the volume of the submerged object. The weight of the displaced fluid can be calculated by multiplying the volume by the density of the fluid. The buoyant force is equal to the weight of the displaced fluid.

To determine whether or not an object will float, you simply compare the weight of the object to the weight of the displaced water (buoyant force). If the weight of the object is greater, the object sinks. If the buoyant force is greater, the object floats.

Three special types of boats use unique suspension principles. The first is the hovercraft. See **Figure 25-25.** A ***hovercraft*** is suspended on a cushion of air. Large fans force the air into a cavity under the boat. As the air escapes this pocket, the boat is lifted above the water. Hovercrafts are used over shallow water, swamps, and marshy land and where speed is important.

The second type of special boat is the hydrofoil. A ***hydrofoil*** has a normal hull

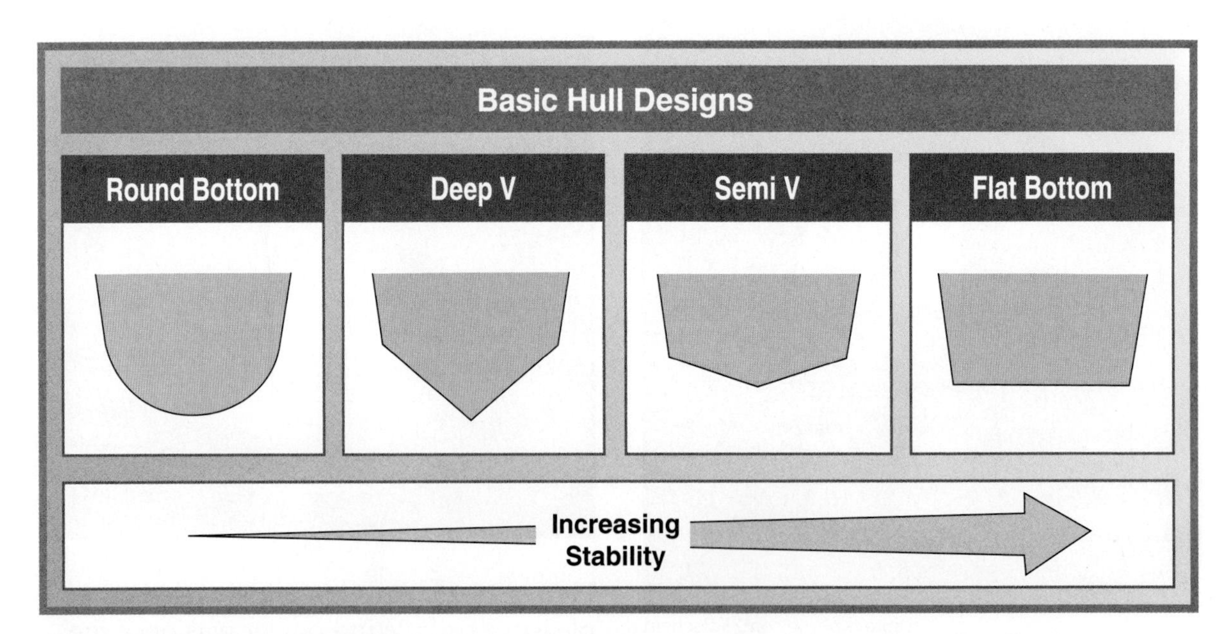

Figure 25-24. Basic hull designs range from the round-bottom kind to the flat-bottom kind. The flat-bottom hull is the most stable.

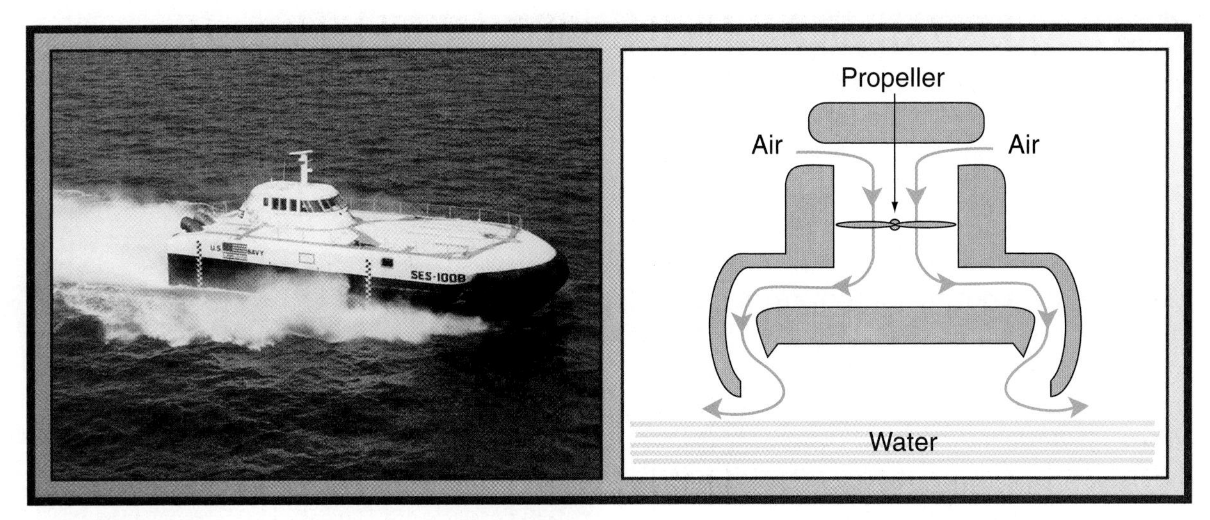

Figure 25-25. A hovercraft is suspended above the water on a cushion of air.

and set of underwater wings. These ***wings*** are called *hydrofoils.* A ***jet engine*** provides power for the boat. As the boat's speed increases, the water passing over the hydrofoils produces ***lift***. The lift causes the hull to rise out of the water. The reduced friction between the water and the hull allows the boat to travel faster, while using less fuel.

The third type of special boat is a ***submersible*** or submarine. This vessel can travel on the surface of water and underwater. Allowing water to enter or forcing it out of special tanks adjusts the boat's buoyancy. As the tanks fill with water, the vessel becomes heavier, or less buoyant. Therefore, the submarine sinks into the water. Compressed air can be used to force the water out of the tanks and increase the buoyancy. The submarine then rises to the surface.

The Guidance of Water-Transportation Vehicles

The operator (called a *skipper,* or *captain*) of a ship is responsible for its course (path) and safety. Skippers obtain guidance from a number of sources. A compass and the charts of rivers, harbors, and oceans are the bases for navigating ships of any kind. In or near a harbor, lighthouses and lighted buoys identify safe channels for navigation or mark hazards. See **Figure 25-26.** Flags and radio broadcasts communicate weather conditions. Special electronic systems help to pinpoint the ship's location and indicate the depth of the waterway.

©iStockphoto.com/Fitzer

Figure 25-26. Lighted buoys and lighthouses mark safe channels.

Standards for Technological Literacy

18

Research

Have the students describe and sketch the system a submarine uses to submerge and surface.

TechnoFact

Navigation is using guidance systems to locate and manage the routes of boats and airplanes. The earliest mariners found their way using markers along the shore and by the Sun and stars. The Phoenicians were perhaps the most highly developed of the early navigators. They constructed large vessels and journeyed out of view of land both day and night. The Polynesians navigated from island to island by means of "guiding stars" and information passed from generation to generation.

TechnoFact

A major improvement in navigation occurred along with the development of the compass. Coupled with navigation charts developed in the 1400s, the compass allowed people to investigate the distant reaches of the world. Later on, lighthouses assisted coastal navigation. The first lighthouses used tallow candles, coal fires, and oil lanterns to generate light. Gas and acetylene flames later took the place of these light sources. Electrical energy was first used for navigation in 1858 at the South Foreland Lighthouse in England.

Standards for Technological Literacy

18

Figure Discussion

Use Figure 25-27 to discuss steering a ship.

TechnoFact

A change in navigation took place in the mid-twentieth century with the extensive use of radio, radar, loran radio navigation (long-range navigation), and radio direction-finding technologies. Later developments, such as the satellite-based global positioning systems, made it possible for navigators to establish their locations and pathways to within a few feet.

The Control of Water-Transportation Vehicles

This information is used as the operator makes judgments about the best route. The path a ship follows is controlled in various ways. See **Figure 25-27.** Ships that have their own power sources are generally guided with rudders. See **Figure 25-28.** A rudder is a large flat plate at the stern of the ship. When it is turned away from the ship's course, it deflects the water passing under the hull. This deflection forces the ship into a turn. Unlike most land vehicles, in which the front of the vehicle contains the guidance and control systems, the back of the ship changes the ship's path and causes the vehicle to turn.

Figure 25-28. Note the red rudder under the red and white hull. Also, see the depth (water level) markings of the hull.

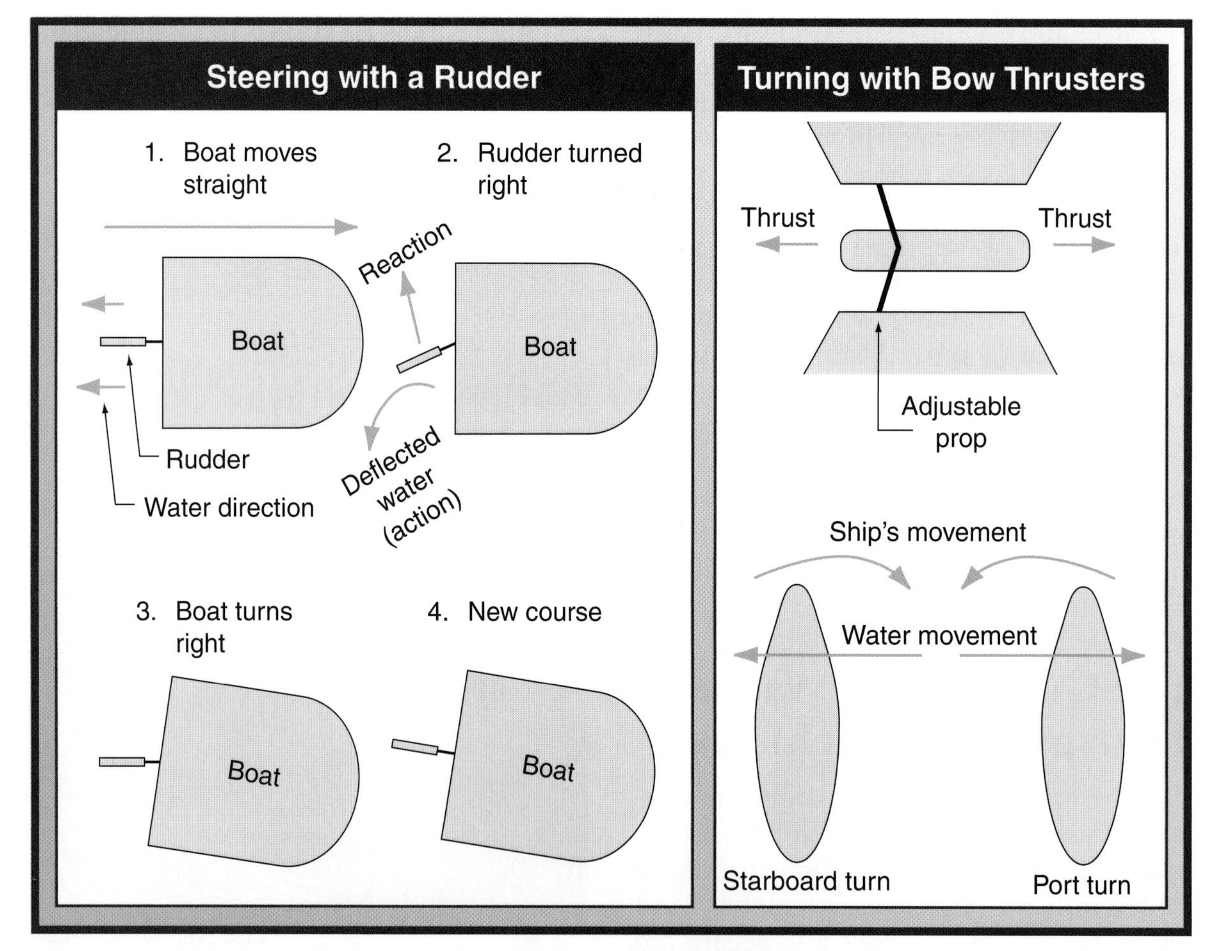

Figure 25-27. Boats and ships use rudders to change their directions. Some large ships have bow thrusters to make small changes while docking.

Some large ships have two props or "twin screws." Increasing the speed of one prop creates additional force on that side of the ship. The ship then turns toward the side with the slower prop.

Another way large ships can be turned, especially in docking, is with bow thrusters. A ***bow thruster*** is a prop mounted at a right angle to the keel. The blades of the prop can be adjusted to provide thrust in either direction. The ship turns opposite the thrust.

A combination of a rudder and sails controls sailboats. The rudder uses the force of the water to turn the boat. Sails use forces that the wind produces to aid the rudder in guiding the boat.

Air-Transportation Vehicles

The newest transportation vehicle in widespread use is the airplane. In the past 60 years, the airplane has changed from a vehicle for the wealthy to an everyday transportation device. Airplanes can be divided into two groups: general and commercial. ***General aviation*** is travel for pleasure or business on an aircraft a person or a business owns. The airplane is not available for the use of the general public. ***Commercial aviation*** includes the airplanes businesses use that make money by transporting people and cargo. These companies include scheduled airlines, commuter airlines, airfreight carriers, and overnight package companies.

The Structure of Air-Transportation Vehicles

Three major types of air-transportation vehicles exist. See **Figure 25-29.** They are lighter-than-air, fixed-wing, and rotary-wing aircraft.

The Structure of Lighter-than-Air Vehicles

Lighter-than-air vehicles use either a light gas (such as helium) or hot air to produce lift. These were the first air vehicles to be used by people. The earliest was a hot air balloon developed in France in 1783. This balloon, similar to present-day hot air balloons, was basically a fabric bag with an opening in the bottom. The

Standards for Technological Literacy

7 18

Extend

Review air transportation as a technological system.

Figure Discussion

Discuss and give examples of the three types of aircraft shown in Figure 25-29.

TechnoFact

Emmanual Swedenborg developed the first plans for an air-cushion vehicle in 1716, but it was never constructed. Sir John Thornycroft worked on air-cushion effects and filed patents involving air-lubricated hulls in the mid-1800s. A British air-cushion vehicle was the first one to be put to functional use, in 1962. It serviced a 19-mile ferry run.

Goodyear, Grumman Corp., Bell Helicopter

Figure 25-29. The three types of aircraft are the following: lighter-than-air, fixed wing, and rotary wing.

Standards for Technological Literacy

18

Figure Discussion

Discuss the three types of jet engines shown in Figure 25-30.

TechnoFact

Humans first took flight in hot air balloons. Jacques and Joseph Montgolfier launched a smoke-filled balloon in 1783. It was 35′ in diameter and made of fabric lined with paper. A duck, a rooster, and a sheep were carried on an eight-minute balloon trip later that same year. Soon afterward, Jean Francois Pilatre de Rozier became the first human to make a hot air balloon flight in a bound Montgolfier balloon. One month later, he and the Marquis Francois Laurant d'Arlandes made the first free flight in a Montgolfier balloon, becoming the first people to fly.

opening allows warm air from a burner to enter the balloon and displace heavier, cold air. Warm air rises to the top of the bag because it is less dense than cold air. When enough warm air has built up in the balloon, the balloon rises off the ground.

Blimps and dirigibles are more advanced lighter-than-air vehicles. A ***blimp*** is a nonrigid aircraft, meaning it has no frame. The envelope (the bag filled with a light gas, usually helium) determines the blimp's shape. Slung beneath the envelope is an operator-and-passenger compartment. One engine or more is attached to the compartment to give the blimp forward motion.

Dirigibles are rigid airships with a metal frame covered with a skin of fabric. They use hydrogen gas to give maximum lift. Dirigibles were used for transatlantic passenger service in the 1930s. Hydrogen is highly flammable, however, and several disastrous accidents occurred because of its use. As a result, dirigibles ceased to be used.

Figure 25-30. These historic aircraft, the Wright *Flyer* and the *Spirit of St. Louis,* are on display at the Smithsonian Institution.

The Structure of Fixed-Wing Aircraft

Today, most passenger and cargo aircraft are ***fixed-wing aircraft***. They all use similar structures. The flight-crew, passenger, and cargo units are contained in a body called the ***fuselage***. One or more wings attached to the fuselage provide the lift necessary to fly. A ***tail assembly*** provides steering capability for the aircraft.

As we noted in the last chapter, the Wright brothers developed the first successful fixed-wing aircraft. Their *Flyer* was first flown in 1903. Later, Charles Lindberg made his historic transatlantic flight in a fixed-wing aircraft called the *Spirit of St. Louis.* See **Figure 25-30.**

The Structure of Rotary-Wing Aircraft

The helicopter is the most common ***rotary-wing aircraft*** in use today. The body of the craft contains the operating and cargo unit and encloses the engine. Above the engine is a set of blades. The blades are adjustable to provide both lift and forward and backward motion. The tail has a second, smaller rotor that keeps the body from spinning in response to the motion of the main rotor.

Helicopters are widely used by the military. In civilian life, they are used for emergency transportation, law enforcement, and specialized applications. These special jobs include communication-tower erection, high rise–building construction, flying to remote areas, and aerial logging.

The Propulsion of Air-Transportation Vehicles

Aircraft use two major types of propulsion systems: props and jet engines. Smaller aircraft use props attached to internal combustion engines. The engine operates on the same principle as the automotive internal combustion engine. Some large aircraft use a variation of the jet engine to turn the prop.

Most business and commercial aircraft are powered by one of three types of jet engines. These types are turbojet, turbofan, and turboprop. See Figure 25-31.

Turbojet Engines

The first type to be used was the ***turbojet engine***. This engine was developed during World War II. The turbojet engine operates through the following steps:

1. Air is drawn into the front of the engine.
2. Air is compressed at the front section of the engine.
3. The compressed air is fed into a combustion chamber.
4. The fuel is mixed with the compressed air.
5. The fuel-air mixture is allowed to ignite and burn rapidly.
6. As a result, the rapidly expanding hot gases exit the rear of the engine.

The exiting hot gases serve two functions. The gases turn a turbine operating the engine-compression section. They also produce the thrust to move the aircraft. The turbojet engine operates at high speeds and is used in military aircraft. Early commercial jet airliners also used these engines.

Turbofan Engines

The ***turbofan engine*** is the engine of choice for most commercial aircraft in use today. See Figure 25-32. The turbofan operates at lower speeds than a turbojet

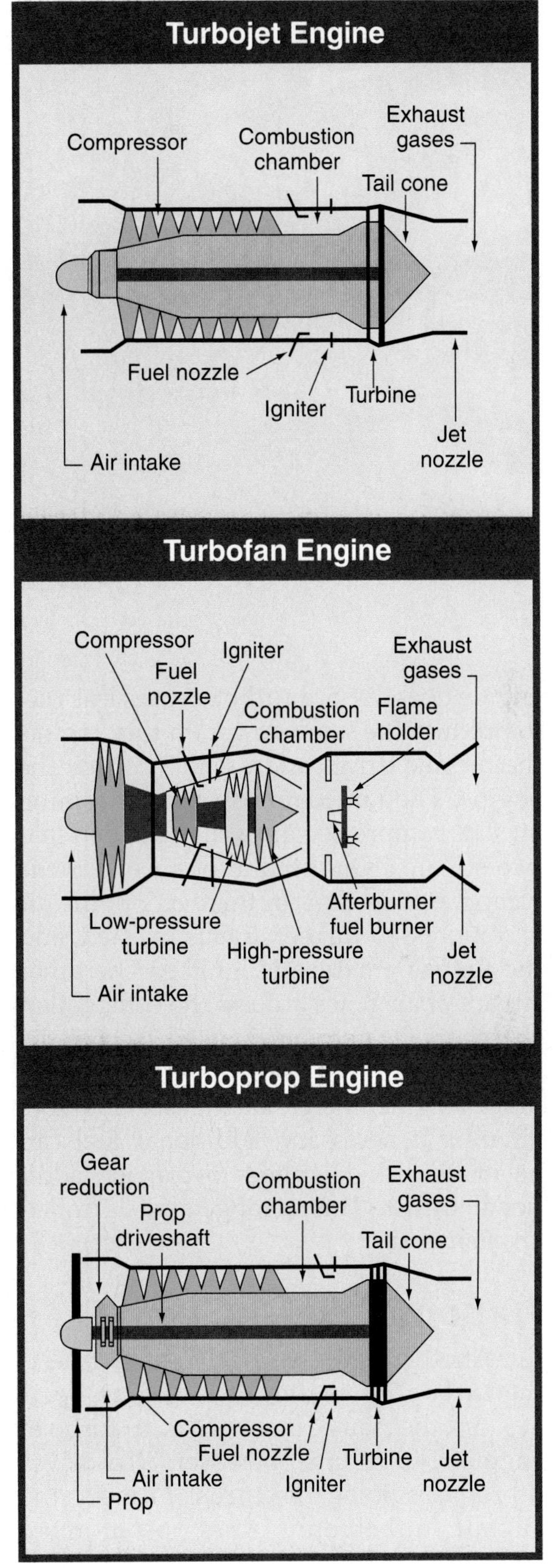

Figure 25-31. Three types of jet engines are in use today.

Standards for Technological Literacy

18

Extend

Identify the types of propulsion systems used in aircraft.

TechnoFact

William Henson patented plans in 1843 for an aircraft with an engine, propellers, and a fixed wing. He gave up on the design, however, after building one ineffective model. Samuel Langley made a steam-powered prototypical plane during the 1890s that flew more than 1/2 mile, but his later, full-sized models crashed. Orville and Wilbur Wright developed their first airplane, the Flyer, in 1903. Orville Wright became the first individual to successfully fly a heavier-than-air machine driven by an engine, on December 17th of that year.

Standards for Technological Literacy

18

Demonstrate

During class or lab, use a balloon to demonstrate how a reaction engine works.

TechnoFact

Louis Bleriot flew across the English Channel from France to England in 1909, making the first international airplane journey. Calbraith P. Rodgers completed the first airplane voyage across the United States in 1911, flying from Sheepshead Bay, New York to Long Beach, California. Charles A. Lindbergh made the first unaccompanied, nonstop flight across the Atlantic Ocean in 1927.

©iStockphoto.com/the_guitar_mann

Figure 25-32. A close-up of a jet engine.

engine does. Also, a turbofan uses less fuel to produce the same power. In this engine, the turbine drives a fan at the front of the engine. The fan compresses the incoming air. The compressed air is then divided into two streams. One stream of air enters the compressor section. In this section, the air is compressed further, fuel is injected, and the fuel-air mixture is ignited. The other stream of air flows around the combustion chamber. This stream is used to cool the engine and reduce noise. In the rear section of the engine, the exhaust gases and cool air mix. If necessary, additional fuel can be injected and ignited to provide additional thrust. This arrangement is called an *afterburner.*

Turboprop Engines

Another variation of the jet engine is the ***turboprop engine.*** This engine operates in the same manner as a turbojet engine. The turbine, however, also drives a prop providing the thrust to move the aircraft. Turboprop engines operate more efficiently at low speeds than turbojet or turbofan engines do. Therefore, they are widely used on commuter aircraft.

Prop-Fan Engines

A new type of jet engine has been developed also. This engine is called a *prop fan.* The prop fan differs from a turboprop in that two props are driven. The props rotate in opposite directions. Prop-fan engines promise to be fuel efficient while operating at high speeds.

The Suspension of Air-Transportation Vehicles

Air-transportation vehicles are suspended in the atmosphere. Such a state requires knowledge of the principles of physics. All air-transportation vehicles depend on the fact that an area with less dense air allows heavier air to force the vehicle up. This force must be greater than the force of gravity drawing the craft toward the Earth. As with structure, suspension varies according to the type of vehicle.

The Suspension of Lighter-than-Air Vehicles

Lighter-than-air vehicles use air-weight differences to cause the vehicle to rise and be suspended in the atmosphere. Two ways are used to cause this difference in weight. The first method uses a closed envelope filled with a light gas. As you learned earlier, blimps use helium, and dirigibles used hydrogen, to generate lift. The vehicle is made to rise and lower very similar to a submarine. Inside the vehicle are air tanks or bags. The tanks are filled with outside air to make the aircraft heavier. Therefore, the aircraft descends. Removing the air from the tanks or bags makes the craft light, and the aircraft rises.

The second system uses circulating, warm air in an open-ended envelope to cause the craft (the hot air balloon) to ascend and descend. See **Figure 25-33.** The balloon has a burner suspended below its opening. See **Figure 25-34.** To cause the

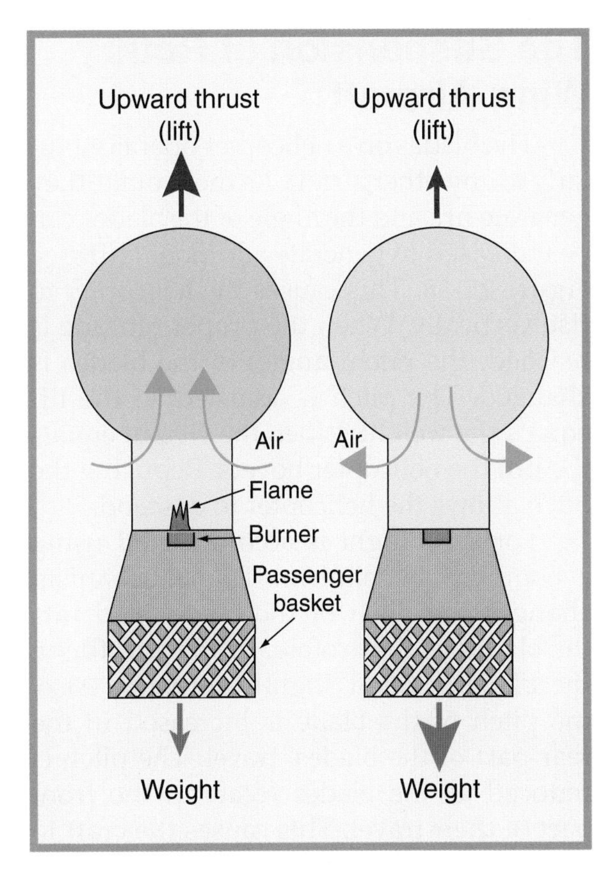

Figure 25-33. A hot air balloon uses warm air to produce lift.

©iStockphoto.com/Sage78

Figure 25-34. A close-up of a burner in a hot air balloon.

balloon to rise, the burner is ignited. The flame heats air, which rises into the balloon. The warm, rising air displaces colder, heavier air. The warmer air is less dense and makes the balloon lighter than the air it displaces in the atmosphere. Therefore, it rises until its displaced air is equivalent to its weight. As the balloon rises, it enters air that is less dense (lighter).

Also, the air in the balloon cools and becomes heavier. Therefore, the burner must continue to operate to keep the balloon flying. To descend to Earth, the operator turns the burner off. During flight, the operator uses the burner intermittently (turns it on and off) to keep the balloon at a constant height. Running the burner all the time might cause the balloon to rise too high. Not running the burner often enough allows the balloon to descend too low over the Earth.

The Suspension of Fixed-Wing Aircraft

As noted earlier, the most common aircraft is the fixed-wing airplane. An airplane has four major forces affecting its ability to fly. See **Figure 25-35.** These forces are the following:

- **Thrust.** This force causes the aircraft to move forward.
- **Lift.** This force holds or lifts the craft in the air.
- **Drag.** This force is the air-resistance force opposing the vehicle's forward motion.
- **Weight.** This force is the pull of gravity causing the craft to descend.

Critical for all flight is lift. Air flowing over the wing of the aircraft generates lift. The wing is shaped to form

Standards for Technological Literacy

18

Demonstrate

During class or lab, use an airfoil to demonstrate lift.

TechnoFact

The de Havilland Comet, the world's first big commercial jet airliner, was developed in Britain and first offered passenger service in 1952. This aircraft split apart in the sky on two separate occasions during 1954, killing everybody on board. These catastrophes led to key model alterations on these and later airplanes. The Boeing 707, the first American-built jetliner, first offered passenger service between the United States and Europe in 1958. The Douglas DC-8 and the Convair 880 jet airliners later offered passenger service.

Standards for Technological Literacy

18

Figure Discussion

Use Figure 25-35 to differentiate between lift, thrust, drag, and weight.

Extend

Describe how the rotors on a helicopter provide forward/reverse motion and lift.

Research

Have the class find out what the function of regional flight control centers is and where they are located.

TechnoFact

Large jets were initially developed in the United States. In 1969, the Lockheed C-5A Galaxy military vehicle started service in the U.S. Air Force. The Boeing 747, the world's earliest commercial jumbo jet, began service in 1970.

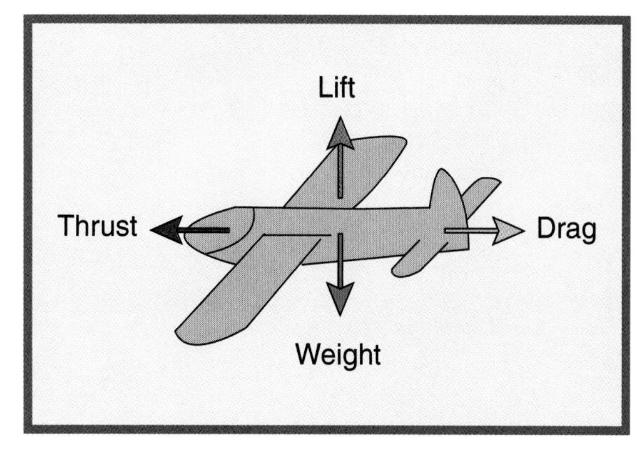

Figure 25-35. Four forces affect flight.

an ***airfoil***. See **Figure 25-36.** The wing separates the air into two streams. The airfoil shape causes the upper stream to move farther than the lower stream. Therefore, the upper stream speeds up. This increased speed causes a decrease in pressure. The high-pressure air below the wing forces the wing up. This gives the plane the required lift.

The greater the slope of the upper surface of the wing is, the greater the lift will be. Drag is also increased, however, because of the larger front profile of the wing. An airplane needs greater lift during takeoff and landing. Devices called *flaps* on the leading and trailing edges of the wings are extended. See **Figure 25-37.** The flaps increase the lift. They are also used to slow the aircraft during landings.

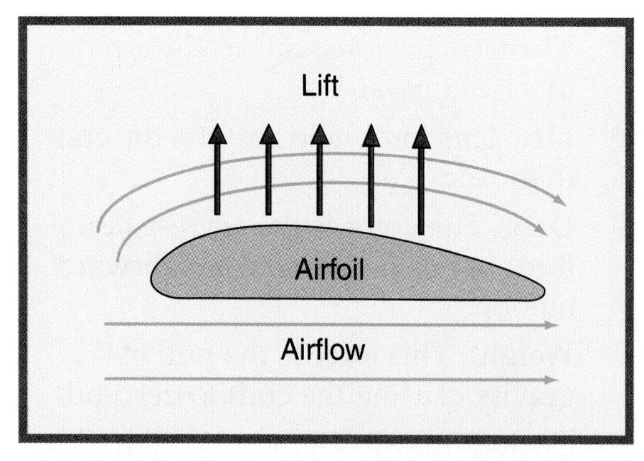

Figure 25-36. Air flows a greater distance over the top of an airfoil and produces lift.

The Suspension of Rotary-Wing Aircraft

The blades on a helicopter operate similarly to any other airfoil. As they rotate, they generate lift, and the angle of the blades can be increased to generate additional lift. See **Figure 25-38.** This causes the helicopter to rise vertically. When the proper altitude is reached, the pitch (angle) of the blades is changed. The pitch is adjusted so the lift equals the weight. When the weight equals the lift, the helicopter hovers. Reducing the pitch allows the helicopter to descend.

Forward flight is accomplished using a complex mechanism. This mechanism changes the pitch of the blades and tilts the blades as they rotate. In forward flight, the blade is tilted slightly forward. Also, the pitch of the blade is increased in the rear part of the blades' travel. The pitch is reduced as the blades rotate in the front part of their travel. This causes the craft to have more lift in the back than in the front. Therefore, the helicopter leans slightly forward. The combination of the lift and thrust of the rotors causes the helicopter to move forward.

Backward flight is accomplished by reversing the tilt. The rotors are pitched the most in the front part of their travel. This causes the craft to tilt slightly backward and travel in reverse. Helicopters can move left and right in the same way, by tilting the rotor and adjusting the pitch.

The Guidance of Air-Transportation Vehicles

Guidance for the pilot of a commercial aircraft comes in many forms. Ground personnel help the pilot bring the aircraft into the terminal safely. Control-tower personnel are in radio contact as the airplane taxies from the terminal and is cleared for takeoff. Once the airplane is in the air, regional air-traffic controllers take over. They monitor and direct the aircraft's

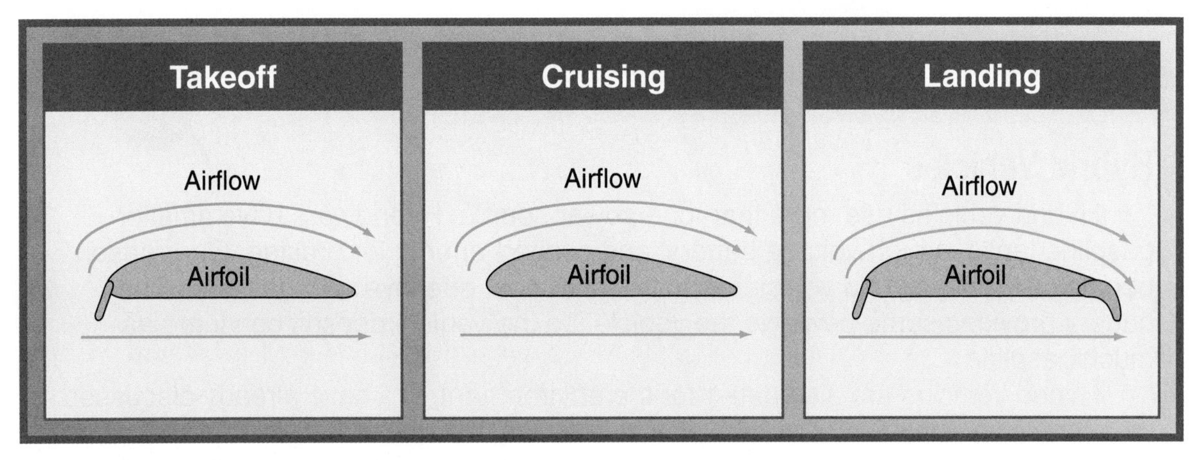

Figure 25-37. The airflow over the wing during takeoff, cruising, and landing. Devices called *flaps* are extended during takeoff and landing. The flaps increase the lift.

progress across their region and hand the craft off to controllers in adjoining regional centers as the vehicle leaves their region. See **Figure 25-39.** Operators of general aviation aircraft do not normally use regional air-traffic control centers. They use visual flight rules to govern their movements.

©iStockphoto.com/fotoVoyager, ©iStockphoto.com/Macatack

Figure 25-38. Mechanisms adjust the rotating blades or airfoils allowing the helicopter to adjust its lift and direction of flight.

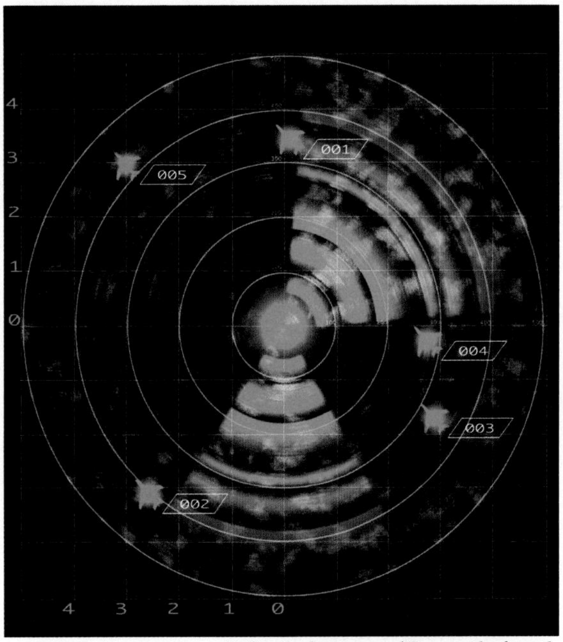

©iStockphoto.com/Georgethefourth

Figure 25-39. A radar screen helps air-traffic controllers to guide and direct aircraft as they travel along their assigned routes.

Standards for Technological Literacy

18

TechnoFact

The United States launched experimental production of rocket research planes for supersonic flight in about 1943. Charles Yeager piloted the Bell X-1 rocket plane in the first supersonic flight ever, in 1947. The North American X-15 rocket plane ascended to 314,750′ above Earth in 1962, setting a new altitude record. This certified Robert H. White, the pilot, as an astronaut.

Standards for Technological Literacy

5 18

Research

Have the students find out how an ILS (instrument landing system) works.

TechnoFact

A number of European inventors labored to build helicopters. The earliest practicable single-rotor helicopter was manufactured and flown in 1939, by Igor I. Sikorsky, a Russian-born aviator living in the United States.

Think Green

Hybrid Vehicles

Hybrid vehicles use more than one power source. Hybrid cars have smaller gasoline tanks, a high-voltage battery, and convert energy for braking. One reason people may drive hybrid vehicles is to be more cost effective with gasoline. The battery provides some power to the vehicle, so the vehicle doesn't consume as much gasoline.

Hybrid vehicles are also better for the environment. We have already discussed the increasing depletion of fossil fuels. Because hybrid vehicles consume less gasoline, the use of fossil fuels is reduced. Gasoline also does not burn cleanly, and conventional vehicles may emit large quantities of carbon dioxide. Because of the electric power source that works with the gasoline in a hybrid vehicle, the carbon dioxide emissions are reduced.

The Control of Air-Transportation Vehicles

Various instruments help pilots monitor and control the aircraft properly. See Figure 25-40. These instruments are called ***aviation electronics (avionics)***. This term includes all electronic instruments and systems that provide navigation and operating data to the pilot. The amount of avionics an airplane carries depends on the airplane's size and cost. Large passenger airliners have a wide variety of avionics. A small general aviation airplane has just basic equipment. Avionics include various functions and systems. These functions and systems are the following:

- **Communication systems.** Short-distance radio systems operate on frequencies between 118 and 135.975 MHz. These radios allow pilots to communicate with air-traffic controllers and other people on the ground.
- **Automatic pilot.** The automatic pilot is an electronic, computer-based system that monitors an aircraft's position and adjusts the control surfaces to keep the airplane on course.

Alaska Airlines

Figure 25-40. This passenger-jet cockpit contains a number of instruments.

- **ILS.** This system helps airplanes to land in bad weather. ILS uses a series of radio beams to help the pilot guide the airplane to the runway. A wide vertical beam helps the pilot stay on the center of the runway. A wide horizontal radio beam tells the pilot if the plane is above or below the correct approach path. Vertical beams serve as distance markers to tell pilots how far away they are from the runways. Pilots in aircraft without ILS equipment use approach lighting to guide the plane in for landing. See **Figure 25-41.**

- **Weather radar.** This system provides pilots with up-to-date weather conditions in front of the aircraft.
- **Navigation systems.** These systems guide the plane along its course and indicate airspeed.
- **Engine and flight instruments.** These instruments give pilots information about how the engine is operating. Flight instruments tell the direction (in three degrees of freedom) the airplane is flying.

Space-Transportation Vehicles

Space travel might be the transportation system of the future. Today, however, its role is limited. Space travel is restricted to conducting scientific experiments and placing communication, weather, and surveillance satellites into orbit.

Figure 25-41. Jets landing over runway approach lighting.

Types of Space Travel

Space travel has become a part of everyday human experience. Some space travel involves human passengers. Other voyages do not. Some space travel involves spacecraft orbiting Earth, while other missions probe outer space. Space travel, then, can be classified in two different ways, which we explore separately. These ways are the following:

- Unmanned or manned flight.
- Earth orbit or outer space travel.

Unmanned and Manned Flights

The first spaceflights were ***unmanned spaceflights***. These flights used rockets to place a payload into orbit. The term *payload* is usually used to describe the items for which a flight is made. Typically, the payload was either a scientific experiment or a communication satellite. See **Figure 25-42.** Today, unmanned space vehicles continue to launch these and other payloads.

Figure 25-42. Satellites are a typical payload for unmanned spaceflights.

Standards for Technological Literacy

18

Extend

Discuss the reasons for space travel.

TechnoFact

When the Soviet Union launched the first artificial satellite, Sputnik, into Earth orbit on October 4, 1957, the Space Age began. The Soviet Union also made the next key development of this era, with the first piloted spaceflight, on April 12, 1961. Yuri A. Gagarin was the pilot aboard the spacecraft Vostok I for this momentous flight.

Standards for Technological Literacy

18

Research

Have your students select a space capsule or satellite and develop a short report on its development and use.

Demonstrate

In class, draw a diagram for the class that shows orbits and their perigee and apogee.

TechnoFact

The earliest piloted journey to the moon was in 1968, when the U.S. Apollo 8 spaceship traveled around the moon ten times and returned to Earth. U.S. astronauts Neil A. Armstrong and Edwin "Buzz" Aldrin landed their Apollo 11 lunar module on the moon on July 20, 1969, and Armstrong became the first human to set foot on the moon.

A ***manned spaceflight*** carries human beings into space and returns them safely to the Earth. The first manned spaceflight was flown on April 12, 1961, when a Soviet cosmonaut named Yuri Gagarin made one orbit of the Earth. Manned spaceflight in the United States also started in 1961. Astronaut Alan Shepard completed a suborbital flight less than one month after Gagarin's flight. This was followed by John Glenn's first orbital flight in the Mercury Program. The Mercury Program was followed by the Gemini and Apollo Programs, which ended with the first man on the Moon. See **Figure 25-43.** All these programs used capsules attached to the nose of a rocket to place humans into orbit or into outer space.

Manned flights were then carried out by space shuttles. A shuttle would rocket into orbit, and at the end of the mission, it returned to Earth, gliding in for a landing similarly to an airplane. The next stage of manned flights will be carried out by Orion spacecraft. See **Figure 25-44.**

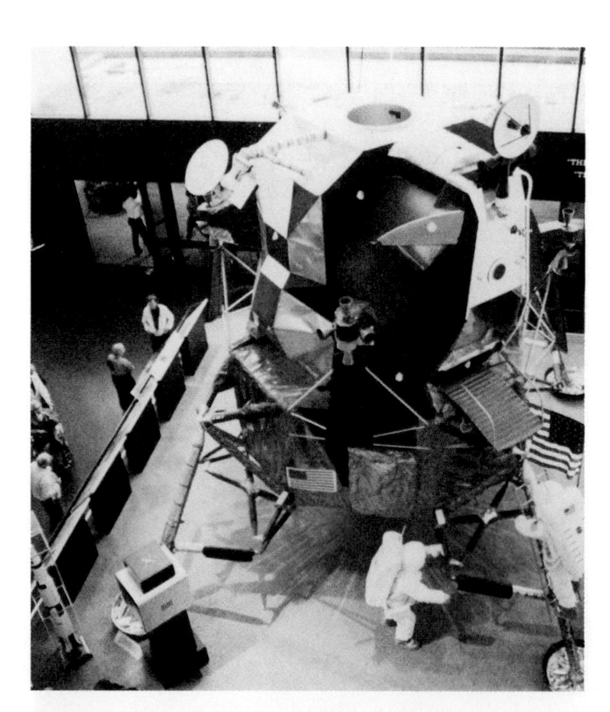

Figure 25-43. This model of the Moon lander is on display at the Smithsonian Institution.

NASA

Figure 25-44. An artist's representation of an Orion spacecraft.

Earth-Orbit and Outer Space Travel

The two types of space travel are Earth-orbit and outer space travel. Communication satellites and space shuttles use ***Earth-orbit travel.*** They travel at about 18,000 miles per hour (mph) in an elliptical orbit around the Earth. This orbit comes closest to Earth at a point called the ***perigee.*** The farthest distance away from Earth is called the ***apogee.***

Satellites can be placed into two types of orbits. One type allows the satellite to circle the Earth, viewing a complete path around the globe. This type of orbit can be used for such objectives as weather monitoring, geological and agricultural surveys, and military surveillance.

The second type of orbit is called ***geosynchronous orbit.*** In this type of orbit, the satellite travels the same speed the Earth is turning. Therefore, the satellite stays over a single point on the globe. This type of orbit is used for communication and weather satellites.

Outer space travel places the spacecraft in Earth orbit first. The speed of the spacecraft is then increased to more than 25,000 mph. At this speed, the spacecraft is thrown into a path causing it to move out of the Earth's gravitational field. The spacecraft can then travel to distant planets and out of the solar system. *Voyager 1* and *Voyager 2* are examples of this type of spacecraft. They moved out of Earth orbit and traveled past a number of planets. These spacecraft are now traveling out of the solar system into what is called *deep space.*

The Structure of Space-Transportation Vehicles

A space vehicle and its launch systems have three major parts. These parts are a rocket engine to place the vehicle or payload into space, an operating section, and a cargo and passenger compartment. As you have learned, early space programs integrated these components into a single launch vehicle, or rocket. The engine was at the rear. The operating controls were placed above and around the engine. The cargo or passenger capsule formed the nose of the rocket. The space shuttle is a newer type of space vehicle. This vehicle uses two distinct systems:

- Two solid-fuel rockets strapped to the external fuel tank.
- The shuttle orbiter.

The Rockets

The ***solid-fuel rockets*** lift the orbiter off the launchpad and give it initial acceleration. After they exhaust their fuel, they fall away. The solid-fuel rockets parachute into the ocean and are recovered to be used again. At this point, the three engines in the orbiter take over. They use the fuel in the external tank to place the shuttle into orbit. As the shuttle enters orbit, the external fuel tank falls away and burns up as it falls toward the Earth.

The Shuttle Orbiter

The shuttle is about the size of a small airliner (such as a Boeing 717)—122′ (37 m) long with a 78′ (24 m) wingspan. The shuttle's crew operates the shuttle from the flight deck at the front of the orbiter. The middle of the shuttle is a large cargo bay. This bay can carry nearly 30 tons (27 metric tons) into orbit and almost 15 tons (13.5 metric tons) on reentry.

The Propulsion of Space-Transportation Vehicles

Space transportation depends on rocket propulsion systems. The rocket is based on Newton's third law of motion, which we discussed earlier. Thus, rocket engines can be called *reaction engines.* The rocket applies this principle by doing the following:

- Burning fuel inside the engine.
- Allowing pressure from the exhaust gases to build up.
- Directing the pressurized gases out of an opening at one end of the engine.

This action produces motion in the opposite direction from the exiting gases. If you have ever blown up a balloon and allowed it to fly around as the air escapes, you have seen the principle of a rocket engine.

The fuel used in rockets is either liquid or solid. Early rockets were developed in China and India before 1800. The developers used gunpowder as a solid fuel. Robert Goddard designed and built the first modern liquid-fuel rocket in 1926.

Liquid-Fuel Rockets

Today's ***liquid-fuel rockets*** have two tanks. One tank contains the fuel, or propellant. The other contains oxygen—the oxidizer. The fuel and oxidizer are fed into the combustion chamber. There, they combine and burn to generate the thrust needed to propel the rocket and its payload into orbit.

Standards for Technological Literacy

18

Extend

Describe the space shuttle through pictures and diagrams.

TechnoFact

The U.S. space shuttle *Columbia* became the first reusable rocket ship to circle Earth, in 1981. It was also the first spaceship to land at a normal landing field.

Standards for Technological Literacy

18

Extend

Describe how rocket engines work.

TechnoFact

The first U.S. space station, Skylab, was unleashed into orbit in 1973. Three missions permitted astronauts to operate there until 1979, when it fell from its flight path. The Soviet space station, Mir, was launched in 1986. Several cosmonauts spent numerous months in outer space working on research on this space station. Russia destroyed Mir in March 2001, by directing it into the atmosphere. The United States, Brazil, Canada, Japan, Russia, and the European Space Agency (ESA) are partners in building the International Space Station.

Liquid-fuel rockets have several distinct advantages. First, the amount of fuel and oxidizer fed into the engine can control the amount of thrust. Second, a liquid-fuel rocket engine can be used intermittently. Finally, the rocket engine can be recovered and reused after the spaceflight.

Solid-Fuel Rockets

Solid-fuel rockets use a powder or a spongelike mixture of fuel and oxidizer. See **Figure 25-45.** Once the mixture is ignited, it burns without outside control. Therefore, the thrust cannot be changed. Also, the mixture burns completely. The mixture cannot be stopped from burning once it is started.

©iStockphoto.com/bmcent1

Figure 25-45. This model rocket is launched by a solid-fuel rocket engine.

The Suspension of Space-Transportation Vehicles

Spacecraft, similar to aircraft, stay aloft because the forces developed by the movement of the craft overcome the forces of gravity. In spacecraft, it is the velocity of the craft that counteracts the pull of gravity from the various solar bodies. These bodies include the Sun, moons, and planets.

The Guidance of Space-Transportation Vehicles

Spacecraft, similar to all other vehicles, require guidance. Sophisticated instruments measuring at least three parameters provide guidance. These parameters are the following:

- **Velocity.** This parameter is the speed at which the spacecraft is traveling.
- **Attitude.** This parameter is the orientation of the spacecraft in space.
- **Location.** This parameter is the position of the spacecraft in space.

The Control of Space-Transportation Vehicles

Similar to all other vehicles, spacecraft require control. The major control systems for a spacecraft are velocity control and attitude control. Attitude control keeps the spacecraft correctly oriented in space, while velocity control deals with the speed of the spacecraft.

Velocity control can be used to change the speed or the orbit of a spacecraft. This change can be affected in two ways. Bursts from a propulsion system within the spacecraft can be used to increase the speed of the craft. Also, the gravity from other solar bodies can be used to increase the speed of the spacecraft. This action is called *gravity assist*, or the *slingshot effect*.

Attitude-control systems consist of equipment that measures, reports, and changes the orientation of the spacecraft while the spacecraft is in flight. These systems respond to external forces so the spacecraft does not rotate wildly or move off course. Attitude-control systems contain sensors and actuators. The sensors determine the actual attitude of the vehicle. The actuators change the attitude.

There are many different types of attitude-control devices used in spacecraft. A common device is a thruster, which is a small propulsion device. Another attitude-control device is a momentum wheel. This device is a rotor that is spun in the opposite direction of the natural rotation of the spacecraft. The force of the momentum wheel cancels out the force making the spacecraft rotate. Another device is a solar sail. This device produces thrust as a reaction to reflecting light. Solar sails can be used for both small attitude and velocity adjustments.

Areas of Operation

Spacecraft and satellites operate in several regions. The lowest is called the ***troposphere***, which includes the first 6 miles (9.7 km) of space above the Earth. General aviation and commuter aircraft operate in this region.

Above this region is the ***stratosphere***. This region extends from 7 miles to 22 miles (11 km to 35 km) above the Earth. Commercial and military jet aircraft operate in the lower part of this region. The upper part of the stratosphere is called the ***ozone layer***. This layer absorbs much of the Sun's UV radiation. Evidence of damage to this layer has caused great concern about global warming and the health of the planet.

The next layer is the ***mesosphere***. This region extends from 22 miles to 50 miles (35 km to 80 km) above the Earth. The ***thermosphere*** lies just above the mesosphere. This region ranges from 50 miles to 62 miles (80 km to 99 km) above the Earth. Many satellites operate in this layer of the atmosphere. The last layer is called the ***exosphere*** and blends directly into outer space.

Standards for Technological Literacy

TechnoFact

It is possible to lower an amplifier into the Indian Ocean and transmit a low frequency sound that could be picked up by a U.S. listening post as a very faint signal three hours later. The signal could be broadcast as pure acoustic sound halfway around the globe.

Answers to Test Your Knowledge Questions

1. E. Control (Note: D. Guidance may also be accepted at the instructor's discretion.)
2. D. Guidance
3. A. Structure
4. C. Suspension
5. B. Propulsion
6. A. Structure
7. C. Suspension (Note: A. Structure may also be accepted at the instructor's discretion.)
8. E. Control (Note: D. Guidance may also be accepted at the instructor's discretion.)
9. Passenger or operator unit, cargo unit, and power unit
10. internal combustion
11. True
12. container ships
13. buoyancy
14. True
15. A fuselage is the body of the aircraft, in which cargo and passengers are carried and to which the wings and tail assembly are attached.
16. Turbojet, turbofan, and turboprop
17. Thrust, lift, drag, and weight
18. geosynchronous
19. False

Summary

Transportation involves all actions that move people and goods from one place to another. Today, four types of transportation exist: land, water, air, and space. All these systems use vehicles to move the people or cargo. Each of these vehicles has structural, propulsion, suspension, guidance, and control systems. The structure is the physical frame and covering. Structural systems provide spaces and protection for people, cargo, and other devices. The propulsion system uses energy to create power for motion. Suspension systems support the weight of the vehicles and cargo. Control systems affect the speed and direction of vehicles. Guidance systems provide information for control.

Test Your Knowledge

Write your answers on a separate piece of paper. Please do not write in this book.

Matching questions: For Questions 1 through 8, match each description on the left with the correct vehicular system on the right. (Note: Answers can be used more than once.)

1. ______ Involves degrees of freedom.
2. ______ A speedometer is one example.
3. ______ Includes the physical frame.
4. ______ The wing on an airplane is an example.
5. ______ A bicycle pedal is one example.
6. ______ Provides spaces for cargo.
7. ______ The hull of a ship is an example.
8. ______ Involves direction the vehicle moves.

A. Structure.
B. Propulsion.
C. Suspension.
D. Guidance.
E. Control.

9. What are the three structural units common to all land-transportation vehicles?
10. The most common engine in land-transportation vehicles is the ______ engine.
11. Most land-transportation vehicles are controlled manually by an operator. True or false?
12. Ships that carry cargo in sealed steel boxes are called ______.
13. Ships and barges float because of the scientific principle of ______.
14. Friction plays a role in the suspension system of a hydrofoil. True or false?

15. What is a fuselage?
16. List the three major types of jet engines.
17. What are the four main forces affecting a fixed-wing airplane's ability to fly?
18. A satellite traveling at the same speed the Earth is turning is said to be in ______ orbit.
19. Rocket propulsion systems are based on Newton's first law of motion. True or false?

STEM Applications

1. Using a rubber band for power, design a land vehicle that has all five vehicle systems. Analyze the vehicle. Explain the features of each system.
2. Build and fly either a model airplane or a model rocket.
3. Design a cargo container for a raw egg that can withstand the impact of a 15′ (4.6 m) fall.

Chapter Outline
Types of Transportation
Components of a Transportation System
Transporting People and Cargo
Maintaining Transportation Systems
Regulating Transportation Systems

This chapter covers the benchmark topics for the following Standards for Technological Literacy:

3 5 18

Learning Objectives

After studying this chapter, you will be able to do the following:

- Explain how the speed of transportation has increased over time.
- Compare personal and commercial transportation systems.
- Recall the common components of all transportation systems.
- Recall the major actions of transportation systems.
- Summarize the characteristics of maintenance and repair activities involved in transportation systems.
- Recall various regulatory agencies that have an effect on transportation systems.

Key Terms

commercial transportation	international transportation	regulation
domestic transportation	interstate commerce	route
fare	personal transportation	schedule
hub-and-spoke system		

Strategic Reading

As you read this chapter, outline the details of the different types of transportation systems, as well as their characteristics.

Transportation is an important part of everyday life. When we walk, we are using a common form of transportation. We might also use many different kinds of transportation technology. To improve our ability to move from one place to another, we use vehicles and systems.

Over time, technology has increased the speed of moving cargo and relocating people. See **Figure 26-1.** In early times, for example, we depended on our ability to walk. A human can walk at an average of 1 mph (1.6 km/h) over a great distance. Later, animals pulled wagons.

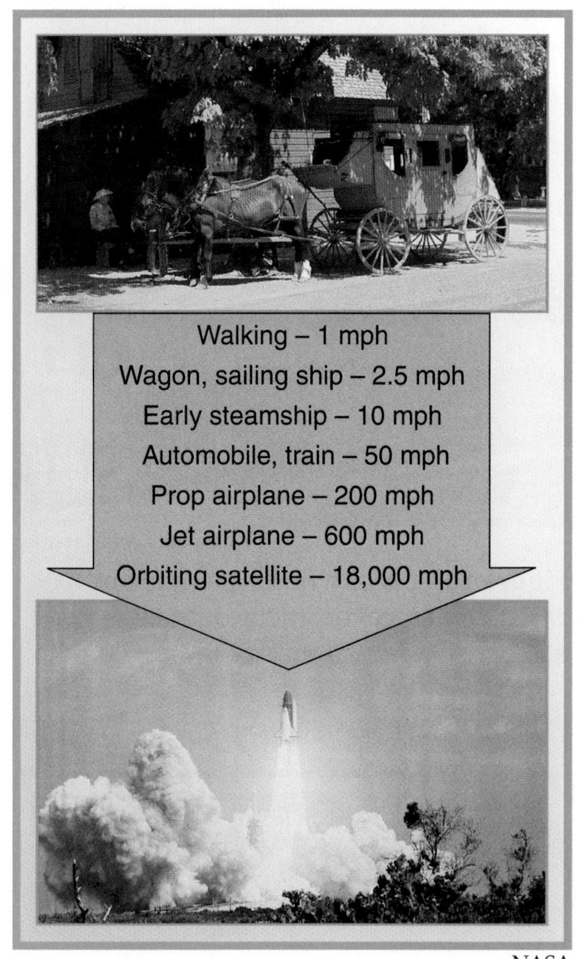

NASA

Figure 26-1. The speed of transportation has increased throughout history.

Sails captured wind to push boats. These advancements increased travel speed by nearly three times. The development of canals and steamships also made travel faster. The advent of the railroad multiplied this speed by 20-fold to 50 mph (80 km/h). Today, jet aircraft move people at about 600 mph (966 km/h). Satellites move at 18,000 mph (28,968 km/h). The advancements in the forms of travel and in the rates of speed have increased the kinds of operations involved in different transportation systems. In this chapter, we look at the types of transportation and various aspects of their successful operation.

Types of Transportation

Through the course of their lives, most people use two types of transportation. See **Figure 26-2.** These types are the following:

- Personal transportation.
- Commercial transportation.

Personal Transportation

Personal transportation is travel using a vehicle one person owns. The most common personal transportation vehicle

Standards for Technological Literacy

18

Research

Have the students compare and contrast the speed of many modes of transportation and the vehicles in them.

Reinforce

Describe operating transportation systems as a process technology (Chapter 5).

Figure Discussion

Use Figure 26-2 to elaborate on personal and commercial transportation systems.

TechnoFact

Subways are underground rail systems designed to transport individuals in cities quickly. They were one of the earliest types of public mass transportation developed. The oldest system is the London Underground, which went into service in 1863. In 1898, Boston established the first subway in the United States. There are almost one hundred subways in approximately sixty nations today.

Figure 26-2. The automobile is an example of personal transportation. The cruise ship is an example of a commercial vehicle.

Standards for Technological Literacy

18

Brainstorm

Ask your students to name and describe the advantages and disadvantages of traveling by (1) personal transportation and (2) commercial transportation.

Extend

Discuss vehicles, routes, schedules, and terminals as integral parts of a transportation system.

TechnoFact

Mass transportation is transportation that is available to all travelers, without the necessity of reservations. Its growth was directly linked to urbanization and industrialization. Nearly all major cities, including London, New York, Boston, and Paris, had fixed rail subway systems and trolley lines by 1900. Buses were widespread in these and other cities by the 1920s.

TechnoFact

There was a rising need for high-quality roads throughout the developing world by 1900. A number of the original roads developed in the United States were made to provide farmers admittance to the railroads that transported farm goods to markets.

is the automobile. Other typical personal vehicles are bicycles, motorcycles, mopeds, and skateboards.

Personal transportation vehicles are flexible. They allow us to travel where we want to go. Personal transportation (except for bicycles) is not always energy efficient, however. A car burns the same amount of fuel whether one or five people ride in it. We often drive around alone in our cars. The passenger seats are empty and, therefore, underutilized. Most personal transportation systems use public routes built using public funds that taxes and fees generate. Most often, these pathways are the streets, roads, and highways crisscrossing the nation.

Commercial Transportation

Commercial transportation includes all the enterprises that move people and goods for money. These are the land, water, and air carriers operating locally, nationally, and internationally. The carriers include taxi, bus, truck, rail, ship, and air-transport companies. Most are profit-centered enterprises that use private funds to finance their initial operations. People who expect to make a profit on their investments own these companies. Some commercial transportation companies, however, are government owned. These companies provide transportation services in which there is little chance for a private company to make a profit. For example, public-transit companies operate city and regional bus, rapid transit, and commuter rail services. See **Figure 26-3.**

©iStockphoto.com/prominx

Figure 26-3. A city bus system is an example of a government-operated transportation system.

Components of a Transportation System

All commercial transportation systems share some common elements. See **Figure 26-4.** They all have the following:

- A vehicle.
- A route the vehicle travels from the origin to the destination.
- An established schedule for the movement of people and goods.
- Terminals at the origin and destination points of the system.

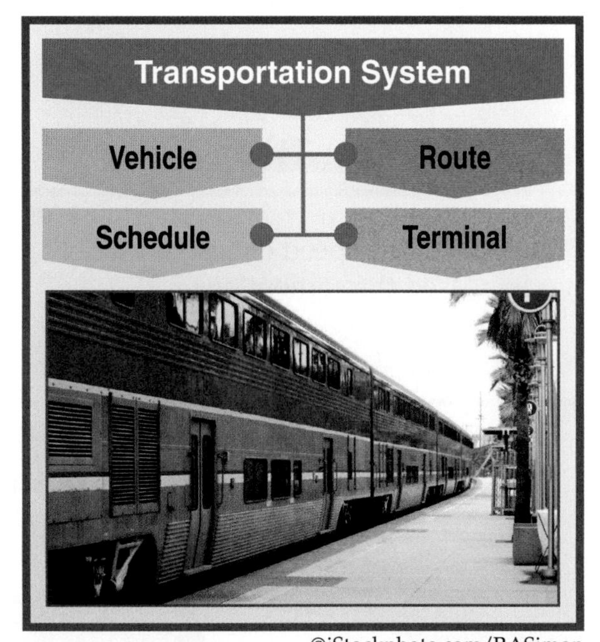

©iStockphoto.com/RASimon

Figure 26-4 All transportation systems have vehicles, routes, schedules, and terminals.

Vehicles

You studied vehicles in Chapter 25. Vehicles are powered carriers that support, move, and protect people and cargo. Also, all vehicles have structural, propulsion, suspension, guidance, and control systems.

Transportation Routes

All transportation vehicles move from a point of origin to a destination. The path a vehicle follows is called a ***route***. Personal transportation vehicles follow individual routes selected because of the destination and the purpose of the trip. For example, a person hurrying to get to work might follow the most direct route or the one having the shortest travel time. On the other hand, a person on vacation might take a less direct, but more scenic, route.

Commercial transportation vehicles often travel on specific routes. This means a vehicle follows one path from its origin to its destination. See **Figure 26-5.**

Typically, transportation systems are made up of many individual routes. These routes are designed to collect vehicle traffic. To understand this concept, consider the road system shown in **Figure 26-6.** These roads are typical of those found in the midwestern United States. In this part of the country, there are often roads on each section line. A section is a one-mile, square piece of land. The smallest of these paths gives access to individual homes and farms. These paths can be dirt, gravel, or paved roads. This series of local-access roads feeds into collector roads. Collector roads are larger pathways designed to carry more traffic. They feed into arterial roads. Arterial roads are high-speed highways connecting cities

Standards for Technological Literacy

18

Extend

Use a road map to show the progression of roads from local roads to interstate highways. Discuss the Federal highway numbering system.

TechnoFact

Germany started building a system of divided thoroughfares in 1934. These roads, called *autobahns*, had intersections and areas dedicated to service. The earliest American expressway was the Pennsylvania Turnpike, which had its first part open in 1940. California started the city thruway development in the same year, with the opening of the Arroyo Seco Parkway, connecting Pasadena and Los Angeles.

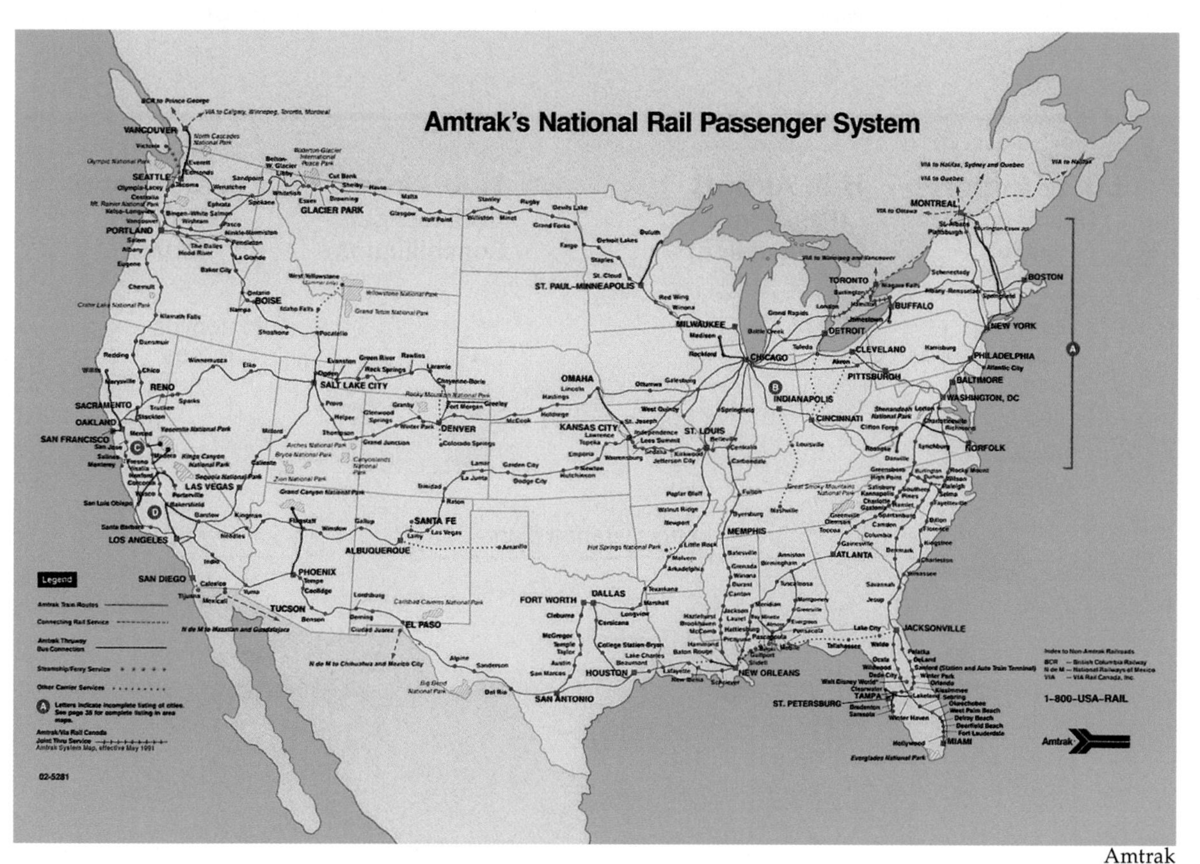

Amtrak

Figure 26-5. A partial route map for a major railway.

Standards for Technological Literacy

18

Research

Have your students use an airline route map to identify a hub and the local airports it connects.

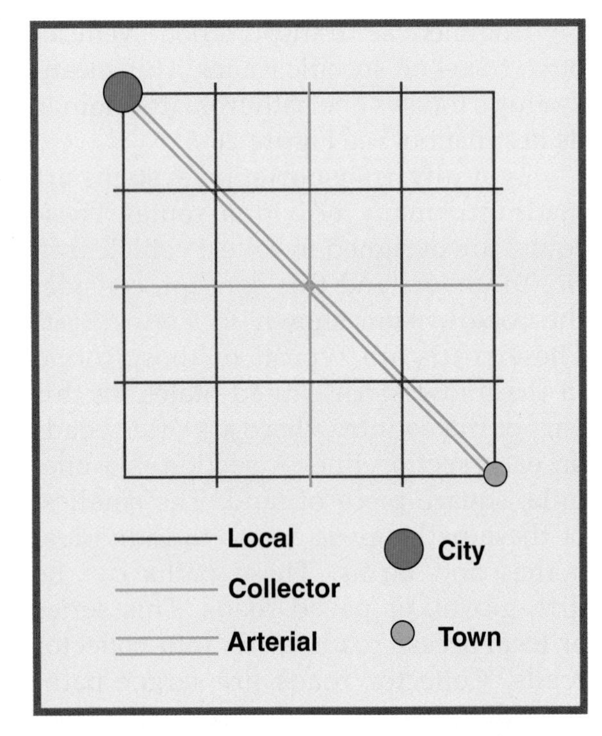

Figure 26-6. This map shows how the various parts of a highway transportation system feed into each other to provide local access and connections.

and towns. Some highways are limited-access highways. On this type of highway, vehicles can enter and leave only at interchange points. Thus, the access to some land along these highways is limited. Larger U.S. highways and the interstate highway system are good examples of arterial highways.

Many airlines use a route pattern called a ***hub-and-spoke system***. See **Figure 26-7.** These routes are less apparent because they are pathways through the sky. A hub-and-spoke system is made up of small local airports and large regional airports. The local airports are on the end of two-way routes radiating from the regional airport. Their routes appear to be spokes attached to a hub of a wheel.

The local airport serves as a collecting and dispersing point for the system. The passengers normally arrive at a local airport using personal cars, taxies, shuttle vans, buses, or trains. At scheduled times, flights leave the local airport and travel to

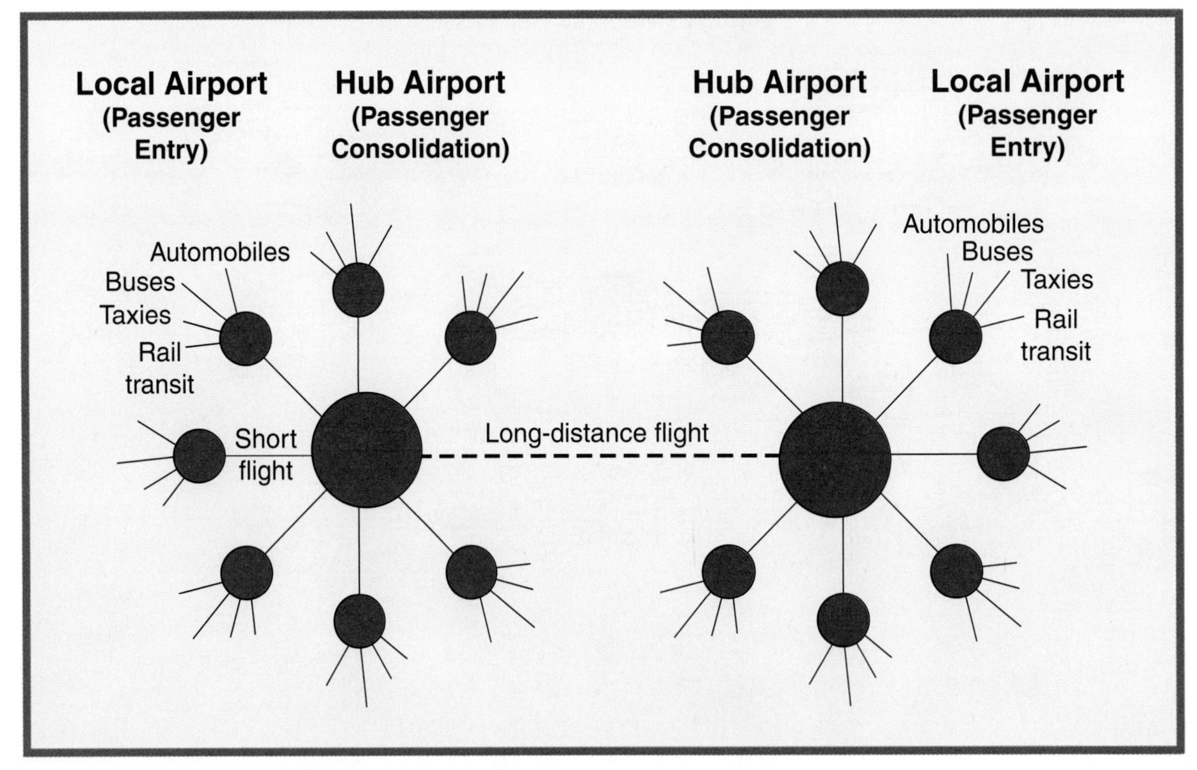

Figure 26-7. A model of a typical hub-and-spoke airline system.

the hub airport. Small- and medium-sized jet aircraft are used. In other cases, small commuter airlines fly these routes.

The passengers arriving at the hub airport can transfer to other flights traveling to their destinations. The simplest hub-and-spoke systems use one hub. For example, a wave of aircraft might arrive from points to the east and south of the hub. The same airplanes are then used for flights to the west and north. Later, the flow is reversed, with arrivals from the west and north and departures traveling to the east and south.

More complex systems tie two or more hubs together. Some passengers use one of the hub airports to travel within the region of the country it serves. Other passengers use both hubs. They fly into one hub and then take a flight connecting it to a second hub across the country. The final flight segment takes them from the second hub to their destinations. This system also allows the hub city to be served from a number of directions.

Often, the passenger who transfers at a hub never leaves the airport. For example, O'Hare International Airport in Chicago is a major hub terminal serving a major commercial and industrial city. Some passengers arriving at O'Hare stay in Chicago for personal or business reasons. Many more people who use the airport never step foot outside the airport. They simply arrive on one flight and leave on another.

Airlines are not the only transportation systems using hub-and-spoke-type systems. Trucking lines use small trucks to pick up and deliver freight. These vehicles work from a local terminal that collects and dispenses cargo for a city or local region. The freight might then move to a regional terminal. At this terminal, it is sorted for long-distance transfer on large cross-country trucks. At the other end, the freight is separated into smaller shipments. These shipments are delivered to local terminals and on to their final destinations.

Parcel services, such as Federal Express and the United Parcel Service, also gather and group shipments using the hub-and-spoke principle. Likewise, many railroads, interstate bus systems, and the postal service use this principle. See **Figure 26-8.**

Figure 26-8. The postal service uses a hub-and-spoke system—your mailbox, to the local post office, to a central sorting facility, to the destination post office, to the destination mailbox.

Transportation Schedules

People travel for various reasons. They might be going to attend a conference, visit other people, participate in a business meeting, or watch a sports event. For these people, the arrival time is very important. Therefore, they select a transportation mode by how long the trip takes and the expected arrival time.

This information is presented in transportation schedules. ***Schedules*** list the departure and arrival times for each trip. These schedules are available for all commercial transportation systems. The systems include freight hauling and passenger travel on bus, train, and airline systems.

Standards for Technological Literacy

18

TechnoFact

Primitive aircraft were lightweight and flew slowly. They only required basic airports, which typically had no designated landing strips. Airfields with heavier runways were constructed in the 1930s, when multiengine planes carrying up to seventy-five travelers were introduced. The introduction of commercial jet aircraft in 1959 required new, better airports with longer landing strips and bigger passenger terminals.

How long the trip takes and the expected arrival time are often presented on the Internet or available in printed form. Schedules for passenger travel are often displayed on television monitors or on schedule boards at the terminal. See **Figure 26-9.**

Standards for Technological Literacy

18

Demonstrate

During class or lab, show the class how to use an Internet travel service to show airline schedules from one city to another one.

TechnoFact

Airfields vary in size and layout, depending on their purposes. Military airports have 10,000′ to 15,000′ long landing strips and are used only by military airplanes. General aviation airports have 3000′ to 5000′ long runways and are used by small civilian planes. Commercial airports have 6000′ to 12,000′ long landing strips and are used by airlines.

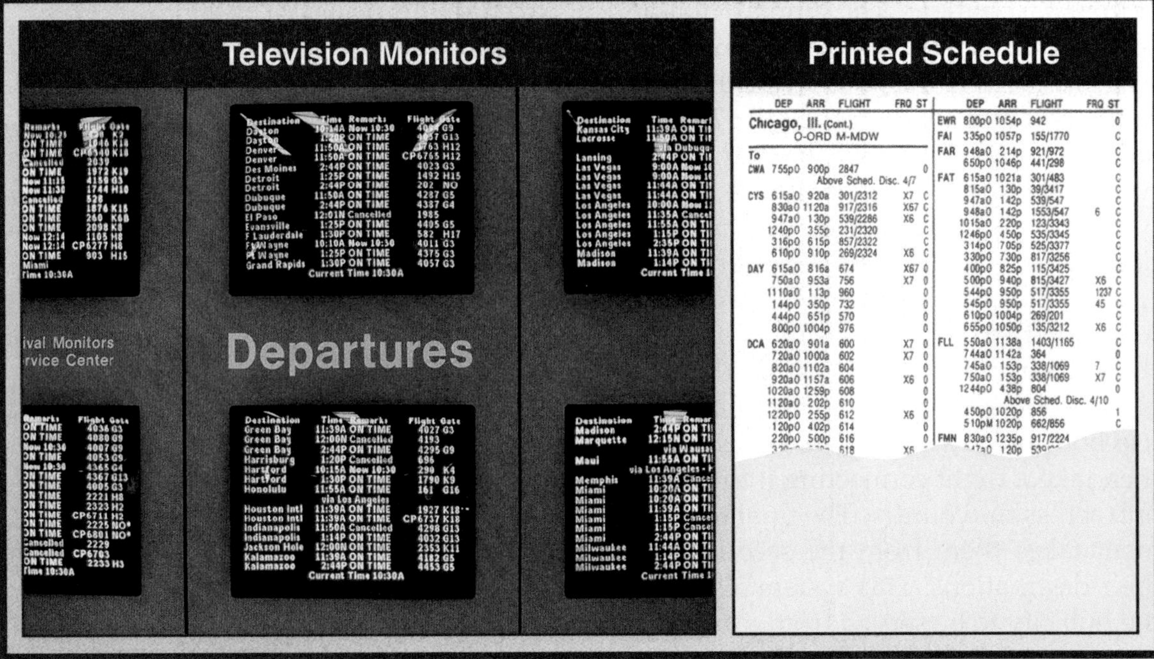

United Airlines, Bruce Kincheloe

Figure 26-9. A schedule might be displayed on television monitors in a terminal and contained in a printed schedule.

Career Corner

Bus Drivers

Bus drivers pick up and drop off passengers at bus stops, stations, and other locations on an established route. Drivers must maintain a set time schedule and operate vehicles safely. Local-transit drivers collect fares, answer riders' questions about schedules and routes, submit daily trip reports, and report mechanical problems. They might make several trips each day around the same city or suburban streets. The drivers usually have a five-day workweek. Saturdays and Sundays are considered regular workdays. Bus drivers must meet state and federal qualifications and hold a commercial driver's license (CDL). A CDL requires applicants to pass a written test on driving rules and regulations and demonstrate safe bus-driving skills.

Transportation Terminals

Terminals are important parts of commercial transportation systems. They are places where passengers and cargo are loaded onto and unloaded from vehicles. These structures are most often located at the origin and the destination of a transportation route.

Cargo Terminals

Cargo systems have terminals that provide storage for goods awaiting shipment. They also have methods to load cargo onto transportation vehicles. See **Figure 26-10.** Finally, they have spaces for vehicles to stop as they are loaded and unloaded. These places might be a dock for ships, an apron for aircraft, or a loading dock on a warehouse for trucks and railcars.

Passenger Terminals

Passenger terminals are more complex than cargo terminals. People need and demand more attention and comfort than cargo does. Many terminals resemble small cities. See **Figure 26-11.** They include at least six different areas:

- Passenger-arrival areas allowing vehicles from other modes to unload passengers. Also, parking lots and bus shuttle services are provided at many large terminals.
- Processing areas. In these areas, passengers can purchase or validate tickets and check in luggage.
- Passenger-movement systems allowing passengers to move from ticket counters to loading areas. Such systems include moving sidewalks, escalators, and automatic people movers.
- Passenger service areas providing food, gifts, and reading materials. Also, rest rooms and very important person (VIP) or frequent-traveler lounges are available.

Norfolk Southern

Figure 26-10. The water-transportation terminal includes warehouses; storage yards; and a dock for berthing, loading, and unloading ships. The rail terminal includes a rail yard for sorting and storing railcars, warehouses for product storage, and an area for loading and unloading.

- Passenger waiting areas. In these areas, people can sit while waiting to board their vehicle.
- Passenger boarding areas and mechanisms allowing people to board the vehicle.

Many functions in passenger terminals occur behind the scenes. Some workers sort baggage for departing travel. Others return baggage to arriving passengers. System managers direct the arrival,

Standards for Technological Literacy

18

Extend

Give several examples of the many areas in a modern airline terminal and discuss them.

Brainstorm

Ask the class what areas they would include in a school bus terminal at their school.

Standards for Technological Literacy

18

TechnoFact

Denver International Airport is the newest big airfield in the United States. The airport covers just about 34,000 acres and has five 12,000-feet landing strips. The passenger terminals and concourses contain 1,500,000 feet^2 and consist of roughly one hundred airline gates.

©iStockphoto.com/robas

Figure 26-11. Passenger terminals serve a number of purposes.

loading, unloading, and departure of vehicles. Operating crews have areas to prepare for trips and to rest between assignments.

Transporting People and Cargo

You have been introduced to transportation vehicles, routes, schedules, and terminals. These things exist for one reason only. This reason is to move people and cargo from one point to another. The task requires three major actions. See **Figure 26-12.** These actions are loading, moving, and unloading.

Loading

Loading involves placing cargo onto or allowing people to board a vehicle. This action entails checking bills of lading (shipping tickets) for cargo or passenger tickets. The cargo must then be loaded. In other cases, the people must be allowed to board the vehicle. During loading, cargo must be secured so it does not shift during the trip. Unwanted movement can cause damage to the freight.

People must be securely seated before a land trip or an air trip can begin. Air-transportation regulations, especially, require all passengers to be properly seated. The passengers must have their seat belts securely fastened before the plane can leave the terminal. Likewise, many states require seat belt use in automobiles. Bus- and rail-transportation systems do not normally have this requirement.

At the same time the people and cargo are loaded, workers complete a number of

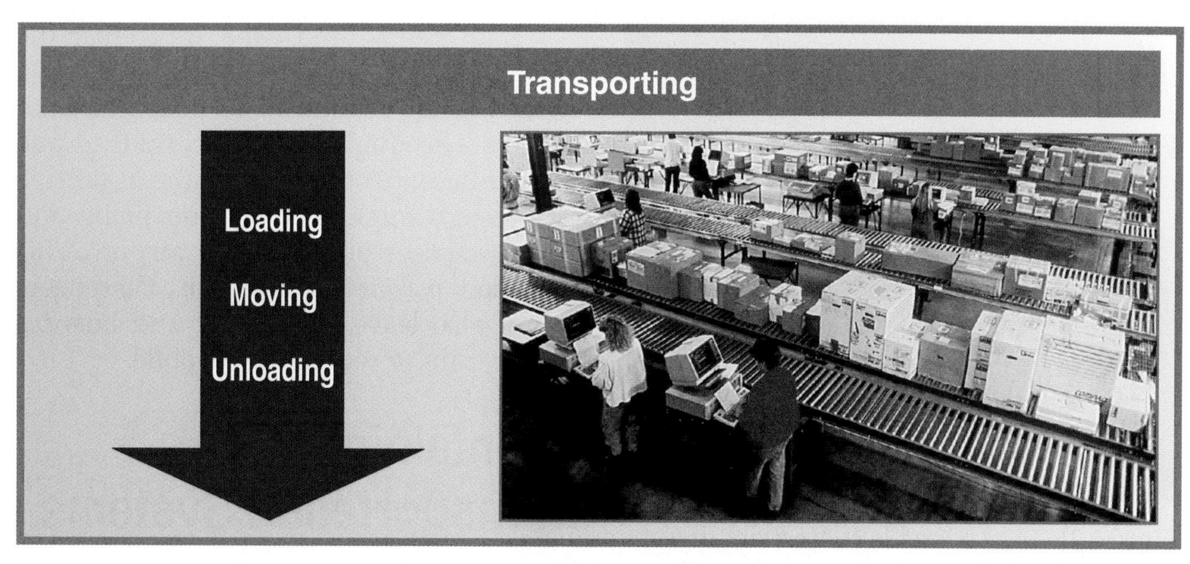

Figure 26-12. Transportation involves loading, moving, and unloading people and goods.

other tasks. See Figure 26-13. They check to be sure that vehicles receive scheduled service. These workers add fuel, check oil levels, deliver drinking water and food, and remove wastes. Also, they complete routine maintenance and minor repairs.

Moving

Moving people and cargo involves driving a vehicle, piloting a ship, or flying an airplane. Still, this is only part of the task. Moving also involves monitoring and controlling the progress of the vehicle. In air transportation, air-traffic controllers talk to the pilot of the aircraft. See Figure 26-14. In railroad systems, employees monitor the progress of a train and switch the train onto appropriate tracks.

Unloading

Unloading passengers and cargo is the opposite of loading them. The passengers are allowed to get off the vehicle. Cargo is untied and lifted out of the vehicle.

Standards for Technological Literacy

18

Figure Discussion

Explain that transportation, in it its simplest terms, involves loading, moving, and unloading people and cargo (Figure 26-12).

TechnoFact

Ports are vital support facilities for water transport systems. They include facilities to connect to other transportation means (rail lines, main roads, and pipelines), piers, cargo space, and freight packing and unpacking equipment.

United Airlines

Figure 26-13. During the loading phase of a trip, passengers and cargo are loaded. The vehicle is serviced. Routine maintenance is performed.

Standards for Technological Literacy

3 18

Research

Have the class find out how cargo is unloaded from ships.

TechnoFact

Malcolm McLean modified semitrailers so they could be moved on trains and freight vessels. This change allowed for quick loading and unloading of boats. The innovative system, called containerized shipping, has significantly altered seagoing delivery.

Figure 26-14. Air traffic controllers direct aircraft and provide information to pilots both in the air and on the ground.

Both people and cargo can enter the transportation system again by being loaded onto or allowed to board another vehicle. This might be a vehicle of the same or a different mode. For example, airline passengers might get off one plane and board another one at a hub airport. This is called *making a connection*. Passengers might also leave the airport and board a bus or train or rent an automobile.

Maintaining Transportation Systems

Transportation systems are made up of a group of technological devices and

STEM Connections: Mathematics

Relating Speed, Time, and Distance

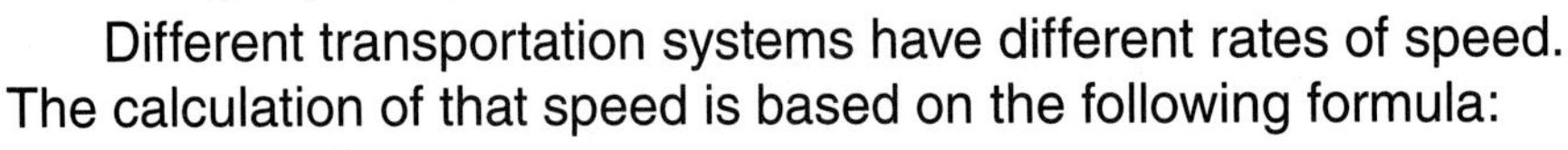

Different transportation systems have different rates of speed. The calculation of that speed is based on the following formula:

$$\text{speed} = \frac{\text{distance}}{\text{time}}$$

We, thus, know that, if we travel 10 miles (distance) in 1 hour (time), our rate of speed is 10 mph:

$$10 \text{ mph} = \frac{10 \text{ miles}}{1 \text{ hour}}$$

As we have learned in our mathematics courses, we can also determine the other parts of the formula by changing the equation. For example, we can change the equation as follows:

$$\text{time} = \frac{\text{distance}}{\text{speed}} \quad \text{and} \quad \text{distance} = \text{speed} \times \text{time}$$

©iStockphoto.com/onurdongel
©iStockphoto.com/Mlenny
©iStockphoto.com/gilleslougassi

Speed, time, and distance are important to all transportation activities.

Therefore, if you live 1 mile from school and you want to know how long it would take to get to school if you walk at a rate of 2 mph, you can calculate the time as follows:

$$\text{time} = \frac{1 \text{ mile (distance)}}{2 \text{ mph (speed)}} = 1/2 \text{ hour, or 30 minutes}$$

If you want to know how far you had traveled and knew you had gone at a rate of 2 mph for 1/2 hour, you calculate as follows:

$$\text{distance} = 2 \text{ mph (speed)} \times 1/2 \text{ hour (time)} = 1 \text{ mile}$$

How long would it take you to travel the 1 mile to school, if you ride a bicycle and travel at 10 mph?

structures. Each of these devices and structures requires maintenance. Typically, the roadways, rail lines, navigational structures, and terminals are maintained using construction knowledge. This knowledge is used to build the components and, therefore, is needed to maintain and repair them.

Structure Repairs

Roadway and railroad-track repair ensures that support structures meet the requirements of the system. See **Figure 26-15.** This kind of repair might involve patching holes and resurfacing roads or replacing rails and cross ties. Signals and signs must also be periodically checked, repaired, or replaced. Likewise, painted surfaces and stripes must be repainted.

Terminals also require repair of their functional parts. Their roofs need to be periodically repaired. Exterior surfaces will need painting or re-siding. Entrances might need new steps and sidewalks. All these repairs ensure that the building can fulfill its function. This function is to protect people and cargo.

Appearance is also important to passengers. They want and expect a clean, bright, and well-maintained terminal. Therefore, walls might be repainted. In other cases, wall coverings might be replaced. New carpeting and lighting fixtures might be installed. Seating might be repaired or replaced.

Figure 26-15. Transportation support structures must be kept in good repair. The first photo shows a railroad-track repair device. The second photo shows a waterway being dredged.

Vehicle Repairs

Everyone wants to travel in comfort and safety. This requires service and repair of vehicles and other equipment. See **Figure 26-16.** Maintenance is the routine

Alaska Airlines

Figure 26-16. Transportation vehicles must be maintained continuously and repaired when needed.

Standards for Technological Literacy

18

Extend

Use automotive service to differentiate between preventative maintenance and repair.

Brainstorm

Ask the class why preventative maintenance is better than repair.

TechnoFact

Three scientists developed a piece of equipment, called the catalytic converter, to decrease dangerous car emissions. This device changes 95 percent of automotive emissions to carbon dioxide and water vapor.

Standards for Technological Literacy

5 18

TechnoFact

From 1887 to 1995, the Interstate Commerce Commission was an autonomous organization of the U.S. government. It enforced federal regulations dealing with transport by ground and water across state lines. Congress deregulated the rail, trucking, and bus industries and eliminated the Interstate Commerce Commission in 1995.

tasks performed to keep a vehicle operating properly. Generally, when a vehicle (a bus, an airplane, or a train) reaches a terminal, it receives daily service. Workers check lubricants, adjust controls and gages, and perform other simple tasks. They also carry out more extensive servicing, such as engine overhauls, on a fixed schedule. Look at an automobile owner's manual. This manual details the servicing required for the vehicle on the basis of fixed times or mileage.

Repair involves diagnosing technical problems, replacing worn or damaged parts, and testing the repaired vehicle. Minor repairs and parts replacement might take place in the field (at the terminal). Extensive repairs are generally completed at a vehicle repair and servicing center.

Regulating Transportation Systems

Transportation is considered essential for the general welfare of a country and its people. Therefore, it is subject to ***regulation*** so all people are served. This regulation takes place on two levels. The first level is ***domestic transportation***. This type of transportation occurs within the geographic boundaries of one country. The other level is ***international transportation***. This transportation moves passengers and cargo between nations.

Domestic Transportation

In the United States, local, state, and federal agencies regulate domestic transportation. For example, cities often control the number of taxicabs and the area they serve. City or county governments often own and operate buses and rapid transit lines. Therefore, these governments regulate routes and the level of bus service. State agencies are responsible for building and maintaining the highway system. Their actions regulate the use of the highways, set speed limits, and control other operating conditions.

Much of the traffic in the United States crosses state lines. This type of traffic involves ***interstate commerce***. Interstate commerce means that business dealings extend across state lines.

Therefore, the federal government plays an active regulatory role over much

Think Green

Alternative Fuels

Hybrid vehicles use a type of alternative fuel—electricity. Alternative fuels are defined as any type of fuel that isn't conventional, such as coal or petroleum. There are many other types of alternative fuels currently in use or being studied for consumer use. Among these are ethanol, biomass, and compressed natural gas (CNG).

Ethanol is a type of bioalcohol. It can be produced from different major crops, such as sugar and corn. The production of this fuel does not cause fossil fuel sources to be depleted. Some gas stations have the option of using an ethanol-based fuel, called E85. Another alternative to fossil fuels is biomass. Biomass is created from live organisms, or organisms that were once alive, such as plants or sugar cane. Another alternative fuel that is gaining in popularity is compressed natural gas. For any compatible automobiles, the gas is stored in cylindrical tanks. Compressed natural gas burns more cleanly than conventional fuels.

of the transportation industry. A number of agencies do this. These agencies include the following:

- **The Department of Transportation.** This agency coordinates federal programs and policies to promote fast and safe transportation. Two of the most important parts of the department are the following:
  - **The Federal Aviation Administration (FAA).** This administration regulates airspace and air safety and provides navigational aids. The FAA licenses pilots and determines the airworthiness of aircraft. This administration also regulates the economics of air transportation.
  - **The Federal Highway Administration.** This administration sets highway-construction standards for federally financed roads and regulates the safety of interstate trucking.
- **The Federal Maritime Commission.** This agency regulates and promotes shipping on the oceans.

In addition, the National Transportation Safety Board promotes safe transportation and investigates serious accidents. The Environmental Protection Agency sets controls on auto emissions, traffic-control plans, and other transportation issues.

International Transportation

International transportation involves diplomacy. Diplomacy entails conducting negotiations between countries that end up with agreements. These agreements might establish transportation routes between two countries, set ***fares*** (the cost of tickets), and approve schedules. The actions often consider the needs of the passengers. They might also, however, attempt to protect one country's transportation companies.

Standards for Technological Literacy

18

TechnoFact

A dry dock is a construction in which a boat can be kept out of the water while maintenance is being performed on it. There are two kinds of dry docks: graving docks and floating docks. A graving dock is a big basin that has one side open to the port. After a craft enters, a gateway is closed, and the water is drained out of the basin to expose the entire ship. A floating dock is moved from location to location, as vessels require renovation. It includes compartments, which are flooded to sink the dock below the level of the boat. The craft is then situated over the dock, and the water is forced out of the compartments, permitting the dock to rise around the ship. The water in the dock is then pushed out to expose the vessel.

The action of loading a transportation system can be simple, as in a person loading bags into her car, or complex, as in loading shipping containers onto a ship.

Summary

Moving people and cargo can be completed by personal, privately owned vehicles. These vehicles serve only the owner, however, and cannot move large quantities of goods or large numbers of people. Therefore, commercial transportation systems have developed. These systems include vehicles, routes, schedules, and terminals. Through these systems, people and cargo are loaded, moved, and unloaded. The structures and vehicles involved in the system must be maintained and repaired because of the wear and tear involved in such movement. Various local, state, and federal agencies regulate each transportation system. Their goal is to encourage the development of an integrated transportation system that serves the individuals and society as a whole.

Test Your Knowledge

Write your answers on a separate piece of paper. Please do not write in this book.

1. Name one technological development and how it affected the speed of travel.
2. Name one difference between personal and commercial transportation.
3. List the four components of a transportation system.
4. What is a hub-and-spoke system?
5. Cargo terminals are more complex than passenger terminals. True or false?
6. What are the three major actions transportation systems perform?
7. Name one maintenance activity involved in transportation systems.
8. Give one example of how a city might regulate a taxicab company.
9. What is a major purpose of the Federal Highway Administration?
10. The Federal Maritime Commission regulates airspace. True or false?

STEM Applications

1. Obtain a rail schedule, a bus schedule, or an airline schedule. Use the schedule to plan a trip taking you from your city to a major destination. List the departure time, arrival time, and travel time for each segment of the trip and for the entire trip.
2. Obtain a map. Plot the routes for the trip you developed above.
3. Visit a local bus terminal or airport. Make a rough floor plan of the terminal. Indicate the various passenger areas.
4. Develop a drawing for a new terminal to replace the one you visited.

Answers to Test Your Knowledge Questions

1. Evaluate individually.
2. Evaluate individually.
3. A vehicle, a route the vehicle travels from the origin to the destination, an established schedule for the movement of people and goods, and terminals at the origin and destination points of the system
4. A hub-and-spoke system is made up of small local terminals and large regional terminals. The local terminals are on the ends of two-way routes that radiate from the regional terminal. The routes appear to be spokes attached to a hub of a wheel.
5. False
6. Loading, moving, and unloading
7. Evaluate individually.
8. Evaluate individually.
9. Sets highway construction standards for federally financed roads and regulates the safety of interstate trucking
10. False

TSA Modular Activity

This activity develops the skills used in TSA's System Control Technology event.

System Control Technology

Activity Overview

In this activity, you will build a computer-controlled mechanical model representing controls for an elevator serving two levels. You will also prepare a report explaining your design and listing directions for operation.

Materials

- A pencil.
- Paper.
- Two touch sensors.
- Two lights.
- Two motors.
- A computer hardware- and software-control system.

Background Information

- **General.** Use the sensors, lights, and motors to model an elevator serving two floors. The sensors represent the elevator call button. The motors represent the elevator doors. One motor represents the first-floor elevator doors. The other motor represents the second-floor elevator doors. A running motor represents an open elevator door. The lights represent the location of the elevator car. When the first-floor light is on, the elevator car is at that floor. When the second-floor light is on, the elevator is at that floor.
- **Control conditions.** Your control system must adhere to the following guidelines:
  - The elevator car can be at only one floor at any given time.
  - The elevator car must be at the floor before the doors can open.
  - The elevator car cannot leave the floor while the doors are open.
  - When a button is pushed, the elevator car is called to the floor. The doors open and close. The car moves to the other floor. The doors open and close.
  - The car remains at the current floor until called.
  - There must be a two-second delay between when the car arrives at the floor and when the doors open.

- The doors remain open for five seconds and then close automatically.
- If the elevator call button is pressed while the doors are open, the doors remain open for five seconds from the time when the button is pushed.
- The car must wait two seconds after the door closes before going to the other floor.

- **Report.** Your report must include the following:
  - A description of the solution.
  - Instructions for operation.
  - A printout of the control program.

Guidelines

- Create rough schematics of the design solution and control logic.
- Sketch the final design before constructing it.
- After constructing the model and writing the control program, test the model. Before you begin testing, create a list of the various conditions and situations that need to be tested.

Evaluation Criteria

Your project will be evaluated using the following criteria:

- The report.
- Model functionality, dependability, and ingenuity.
- Computer-program logic and functionality.

Section 7 Activities

Activity 7A

Design Problem

Background

Most transportation systems use a vehicle to move people and cargo to their destinations. The vehicle must have a propulsion system to cause it to move along its pathway.

Situation

You are employed as a designer for the Technology Kits Company. This company markets simple kits that are fun to assemble and teach basic principles of technology.

Challenge

You have been chosen to design a transportation-vehicle kit. You can do any or all of the following:

- Use air, land, or water as the medium of travel for the vehicle.
- Utilize a rubber band, mousetrap, or spring as the power source.
- Use any materials commonly available in a craft store, such as tongue depressors, balsa strips, wheels, and dowels.

Your supervisor wants a working model, a parts list for the kit, and a set of assembly directions for the purchaser.

Optional

1. Design a package for the product.
2. Create a catalog advertisement for the kit.

Activity 7B

Fabrication Problem

Background

The internal combustion engine is the most popular source of power for land vehicles. These engines use petroleum-based fuels, which, in the near future, will be in short supply. Internal combustion engines also produce emissions that pollute the environment. Developing alternative-fuel systems for land vehicles is a major challenge facing designers and engineers.

Challenge

Build a land vehicle that a prop can power. See **Figure 7B-1.** Develop an electric power system for the vehicle. Test the performance of the vehicle. Finally, change the number of blades on the prop.

Materials and Equipment

- One piece of 1 1/2″ × 1 1/2″ × 11 1/2″ wood (pine, fir, or spruce).

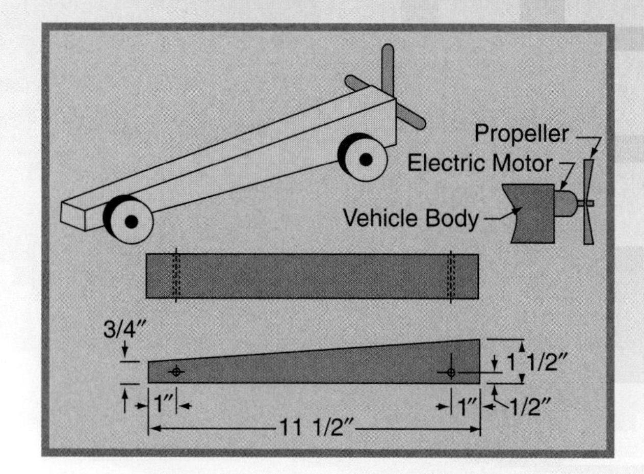

Figure 7B-1. The body of the vehicle.

- Four wheels (available from many hobby stores).
- Two pieces of 1/8″-diameter × 3″ welding rod (for use as the axles).
- Two pieces, 1 1/2″ long, of soda straw.
- One low-voltage electric motor (available from a hobby store or an electronics store).
- One battery matching the motor voltage.
- 12″ of electrical-hookup wire.
- One four-blade prop (available from a hobby store).
- One three-blade prop (available from a hobby store).

Procedure

All fabrication activities follow a set of procedures. Use the following procedures to build your prop-driven land vehicle.

Building the prototype

1. Obtain a block of wood 1 1/2″ × 1 1/2″ × 11 1/2″ long.
2. Lay out the block, including the following:
   - A taper from 3/4″ at one end to 1 1/2″ at the other end.
   - Axle holes 1″ from each end and 1/2″ above the bottom of the vehicle block.
3. Cut the block along the taper.

Safety
Follow all safety rules your teacher demonstrates!

4. Drill two holes for the axles.

Note
The diameter of the holes should be the same as the outside diameter of the soda-straw axle bearings.

5. Measure the diameter of the electric motor.
6. Drill a hole the same diameter of the electric motor into the tall end of the body.

Note
The depth of the hole should allow at least 1/2″ of the motor to extend out of the block. The diameter of the prop and the depth of the hole determine the vertical location of the hole. The motor should be as high in the vehicle as possible.

7. Drill a hole from the top of the vehicle into the motor hole.

Note
This hole allows the motor wires to be drawn out of the mounting hole.

8. Cut the axles to length.
9. Mount the axles and wheels onto the body.

Installing the propulsion system

1. Feed the motor wires into the motor hole and out the wire escape hole.
2. Press the motor into the body.
3. Attach one motor wire to the battery.
4. Mount the battery onto the vehicle.
5. Attach the three-blade prop to the motor.

Safety
The prop now rotates when the motor spins. Be sure to keep your hands out of the way.

Testing the vehicle

1. Attach the second wire to the motor.
2. Place the vehicle on the floor.
3. Test the distance it travels in a given period of time.
4. Disconnect one wire from the motor.
5. Change the prop to the four-blade model.
6. Repeat steps 1 through 4.

Optional

If the props are made of stamped metal, change the pitch (angle) of the blades. Do this by gently twisting each blade of the prop using your fingers or needle-nose pliers. Be sure to twist each blade in the same direction and an equal amount. Test the prop again to check the performance.

Section 8
Applying Technology: Using Energy

Tomorrow's Technology Today

Fuel Cells in Automobiles

The automobiles of the future will probably look much different, both inside and out, from what we see today. Engineers and others have developed a new automobile platform designed for use with a fuel cell engine (an electrochemical-cell engine) and a hydrogen-fuel tank. Many mechanical systems will be eliminated. These controls include the internal combustion engine, exhaust system, and brake and accelerator pedals. Electric motors in the hub of each wheel will propel the new chassis, making each vehicle all-wheel drive. All the controls will be located in the steering mechanism.

The fuel cell is a concept being explored for cars of the future. This cell is a system that uses stored hydrogen and oxygen from the air to create electrical energy. Fuel cells will most likely be widely seen in homes in the near future in products such as laptop computers, cellular telephones, vacuum cleaners, and hearing aids. Used in these ways, a fuel cell provides much longer life than a battery does. A fuel cell can be recharged quickly. Using fuel cells to power automobiles, however, is a much more difficult application. Recently, researchers have made major breakthroughs in creating a fuel cell with commercial prospects. Several automobile companies are now working on new vehicles with fuel cell engines.

We probably will not see hydrogen-powered vehicles mass-produced for a few years. Cars that fuel cells power are currently only demonstration models produced in very limited numbers. A few will be made available to the public to test their acceptance and performance. While researchers and automobile manufacturers work on developing fuel cell vehicles that run on pure hydrogen, gasoline-fed fuel cells will serve as a transitional technology. Some of these, already on the market, cut carbon dioxide emissions by 50% and can get up to 40 miles per gallon of fuel.

The advantages of the fuel cell engine are numerous. This engine has no moving parts. The fuel cell engine is a quiet, reliable source of power. This engine will lead to a cleaner environment. (The only emission is water vapor.) This engine will create less dependence on foreign oil. Another advantage is that the vehicle, while sitting in the garage, can generate enough electricity to power a home.

Some obstacles must be overcome first. A new fueling infrastructure (replacements for gas stations as we know them) will have to be installed across the country to distribute a usable fuel, such as methanol. Hopes are high for the emergence of this new technology. Several years will pass, however, before fuel cells replace the current technology presently on every street in America.

Fuel cells have many applications beyond powering automobiles. They are often used as power sources in remote locations, such as spacecraft and remote weather stations. Fuel cells are particularly useful in this type of application because they are lightweight and compact. Also, they have no moving parts, so they are very reliable.

Discussion Starters

Briefly discuss how fuel cells work and how reformers convert traditional fuels into hydrogen for the cells.

Group Activity

Have the class discuss whether they think, in time, fuel cells will completely replace the internal combustion engine. Ask some students to present their thoughts about the future role of the internal combustion engine in transportation.

Writing Assignment

Tell your students to imagine the global economic impact of widespread use of fuel cells. Have the students write a short paper discussing the effects that widespread fuel cell use would have on the global economy. Encourage the students to consider the ripple effect and look beyond the obvious.

Chapter Outline

- Types of Energy
- Energy, Work, and Power
- Forms of Energy
- Sources of Energy
- Effects of Energy Technology

This chapter covers the benchmark topics for the following Standards for Technological Literacy:

3 5 7 16

Learning Objectives

After studying this chapter, you will be able to do the following:

- Recall the definition of *energy*.
- Compare potential and kinetic energy.
- Compare energy, work, and power.
- Recall the six major forms of energy.
- Give examples of exhaustible, renewable, and inexhaustible energy sources.
- Explain how energy technology can make our lives better and how energy technology can cause damage.

Key Terms

biofuel	fusion	power
biogas	horsepower	radiant energy
biomass	inexhaustible energy resource	renewable energy resource
chemical energy	joule (J)	solar weather system
electrical energy	kilowatt-hour	thermal energy
exhaustible energy resource	kinetic energy	water cycle
fission	mechanical energy	watt
foot-pound (ft.-lb.)	potential energy	work

Strategic Reading

Before you read this chapter, consider your daily routine. Try to list the different types of energy associated with your actions. As you read, add to your list if necessary.

We use the word *energy* in different ways. For instance, we might say, "I don't have the energy to mow the lawn." We might worry about our energy dependence on foreign petroleum, and we might hear people talking about energy conservation. Not everyone using the word knows exactly what it means, however.

The word *energy* comes from the Greek word *energeia*, which means "work." As time passed, the word came to describe the force that makes things move. Today, *energy* is defined as "the ability to do work." This ability includes a broad spectrum of acts. Energy is used in simple human tasks, such as walking, running, and exercising. The ability to do work can be obtained from petroleum and then used to power a ship across the ocean. See **Figure 27-1.** Energy can be used to provide motion in vehicles and machines and produce heat and light. The ability to do work is fundamental to our communication technologies, and it is used in manufacturing products and constructing structures. Energy is everywhere and is used by all of us.

In this chapter, we look at energy from a variety of angles. We discuss the types of energy; the connection among energy, work, and power; and the forms of energy and their interrelationships. Also, we examine the various sources of energy and whether energy technology is helpful or harmful.

Types of Energy

Two types of energy exist. Energy can be associated either with a force doing the ***work*** or with a force that has the capability of doing work. ***Kinetic energy*** is energy involved in moving something. This energy is the energy in motion. A hammer striking a nail is an example of a technological act using kinetic energy. A sail capturing the wind to power a boat uses kinetic energy. Likewise, a river carrying a boat or turning a waterwheel is an example of kinetic energy.

Not all energy is being used at any given time. Some energy is stored for later use. Energy in this condition has the capability,

Standards for Technological Literacy

16

Reinforce

Describe energy as an input to technological systems.

Brainstorm

Ask your students how they use the word *energy* in everyday conversations.

TechnoFact

Throughout history, energy has been used to meet human needs. In the humanity's earliest days, muscle power and solar energy (sunlight) were the only forms of energy that could be harnessed. About 1.5 million years ago, people learned to create and control fire. With the introduction of this new energy source, people developed the ability to cook food and fire clay.

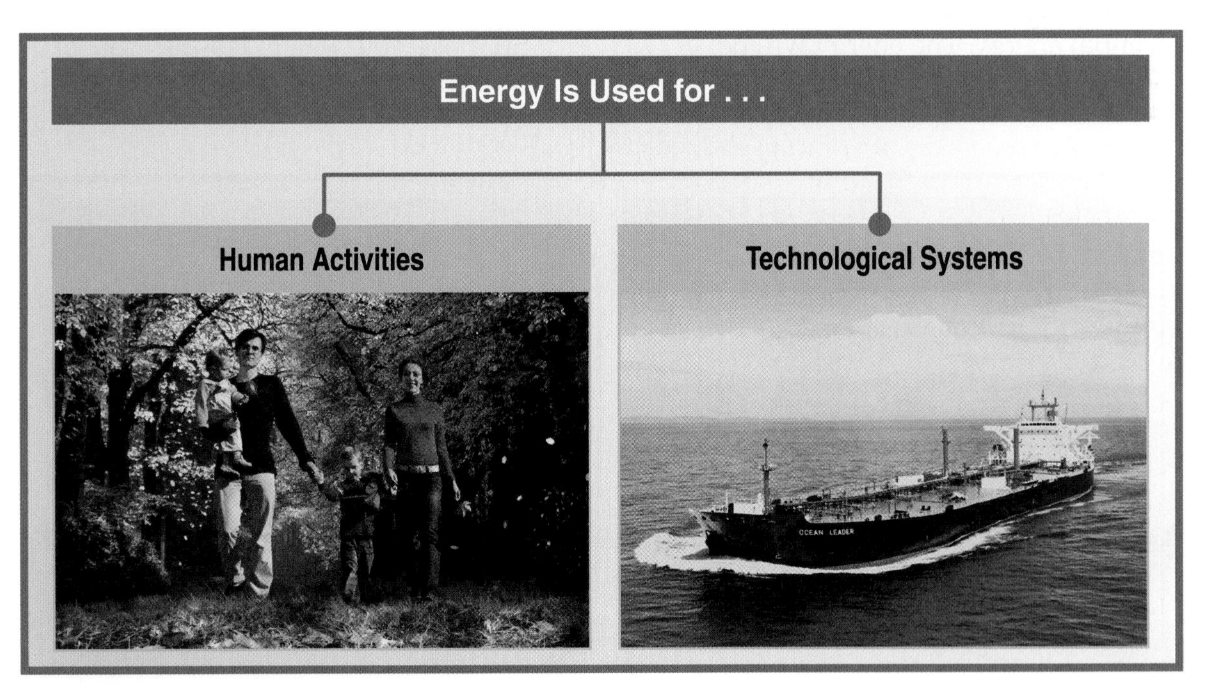

Figure 27-1. Energy is used in all actions, from walking to powering complex technological devices.

Standards for Technological Literacy

16

Demonstrate

Show the class how work can be calculated using a mathematical formula.

Figure Discussion

Use Figure 27-3 to differentiate between power and work.

TechnoFact

Wind and water currents were two of the earliest forms of energy to be harnessed by human innovation. The ancient Egyptians took advantage of wind energy by inventing sails to propel their boats. Also in ancient times, the waterwheel was developed to make use of the energy in a flowing river.

or potential, of doing work when it is needed. For example, water stored behind a hydroelectric dam possesses energy. The water releases this energy to turn a turbine when it flows through the power-generating plant. This stored energy is called ***potential energy***. See **Figure 27-2.** A flashlight battery and a gallon of gasoline are other examples of potential energy.

AMAX

Figure 27-2. The coal being mined has potential energy. The energy will be released when the coal is burned.

Energy, Work, and Power

It takes energy to do work. For example, you must eat well if you plan to run a marathon or hoe a garden. When you do one of these things, you might say, "I really worked hard!" What, however, does this mean? In scientific terms, work is applying a force that moves a mass a distance in the direction of the applied force. You might have lifted boards off the floor and placed them on a table. That was work. The boards have mass, and they were moved some distance.

Measuring Work

We measure work by multiplying the weight moved and the distance the weight was moved. See **Figure 27-3.** The result is a measurement called ***foot-pounds (ft.-lbs.)***. This figure tells you the amount of energy needed to move an object from one location to another. The amount of work completed can be measured with the following formula:

work (in ft.-lbs.) = force, or weight, (in lbs.) × distance (in ft.)

Suppose you weigh 140 lbs. You plan on walking across a 40′-wide room. To

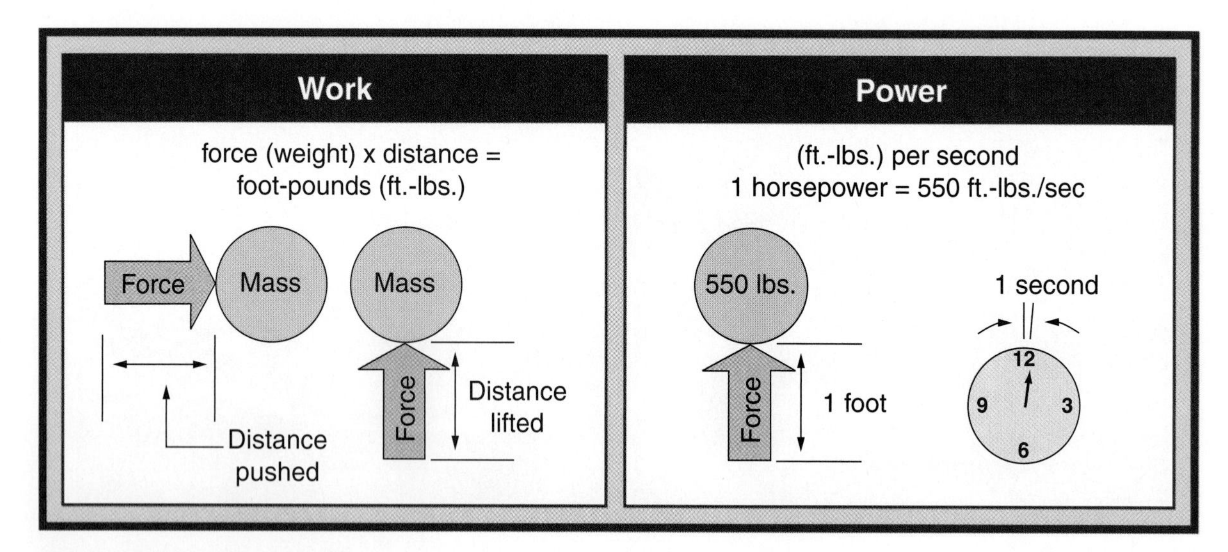

Figure 27-3. Work is done when a force moves a mass over a distance. Power is work done per unit of time.

complete the task, you need 5600 ft.-lbs. of energy (140 lbs. × 40′). Likewise, lifting a 20-lb. weight off the floor and placing it on top of a 36″-high table requires 60 ft.-lbs. of energy (20 lbs. × 3′ [36″]).

In the metric system, work is measured in newtons per meter, or ***joules (J)***. The force, or weight, is measured in newtons, and the distance is measured in meters. The metric work formula is the following:

work (in J) = force, or weight, (in newtons) × distance (in meters)

Measuring Power

Work is done in a context of time. Measuring the rate at which work is done gives you a term called ***power***. See **Figure 27-3.** Power can be calculated by dividing the work done by the time taken:

$$\text{power}\left(\text{in }\frac{\text{ft.-lbs.}}{\text{seconds}}\right) = \frac{\text{work done (in ft.-lbs.)}}{\text{time (in seconds)}}$$

The metric version is the following:

$$\text{power (in watts)} = \frac{\text{work done (in J)}}{\text{time (in seconds)}}$$

Two common power measurements are the horsepower and the kilowatt-hour. ***Horsepower*** is used to describe the power output of many mechanical systems. The force needed to move 550 lbs. 1′ (550 ft.-lbs.) in 1 second is 1 horsepower. The factor of time is important to power. A motor that lifts 550 lbs. in 1 minute can be smaller than one that lifts 550 lbs. the same distance in 1 second. Likewise, the engine that moves a car from 0 to 60 mph in 7 seconds must be more powerful than one that does the same job in 9 seconds.

The term *horsepower* is used in several different ways. The theoretical, or indicated horsepower, is the rated horsepower of an engine or a motor. This number suggests the maximum power that can be expected from the device under ideal operating conditions. Most often, this amount of power is not available from the device.

The brake horsepower is the power delivered at the rear of an engine operating under normal conditions. Drawbar horsepower is the power delivered to the hitch of tractors. Frictional horsepower is

Standards for Technological Literacy

16

Demonstrate

Show the class how to use the power formula (Watt's law) to calculate the power in an electrical circuit.

TechnoFact

Energy conversion is at the heart of many technological actions. In 1847, James Joule proved that mechanical energy could be converted into heat energy. Inspired by this discovery, later scientists learned that any one form of energy can be converted into any other form.

Career Corner

Power Plant Operators

Power plant operators control and monitor boilers, turbines, and generators in electrical power-generating plants. They monitor instruments to maintain voltage and regulate electricity flow from the plant. These operators start or stop generators and connect or disconnect them from circuits to meet changing power requirements.

To maintain round-the-clock operations, power plant operators usually work 8- or 12-hour shifts. Employers seek high school graduates with strong mathematics and science skills for entry-level power plant–operator positions. Most power plant operators receive extensive on-the-job and classroom instruction. Several years of training and experience are required to become a fully qualified operator.

Standards for Technological Literacy

16

TechnoFact

Wind energy, a form of mechanical energy, is a clean and inexhaustible resource for generating electrical power. However, because wind speeds are affected by a variety of conditions, such as time of year and weather patterns, wind power is considered an intermittent power source.

the power needed to overcome the internal friction of the technological device.

A ***watt*** is equal to 1 J of work per second. The work that 1000 watts complete in 1 hour is 1 ***kilowatt-hour***. In electrical apparatus, the resistance of the device determines the power consumed. The device's wattage rating is the product of the electrical current flowing through the device and the voltage drop across it. The formula can be expressed as follows:

P (power in watts) = I (current in amperes) × E (electromotive force in volts)

Forms of Energy

Energy is everywhere we look. The ability to do work is in the fires that burn coal and wood. Energy is in sunlight, wind, and moving water. In fact, we cannot exist without the aid of energy.

The hundreds of examples of energy can be grouped into six major forms. See **Figure 27-4.** These are the following:

- Mechanical.
- Radiant.
- Chemical.
- Thermal.
- Electrical.
- Nuclear.

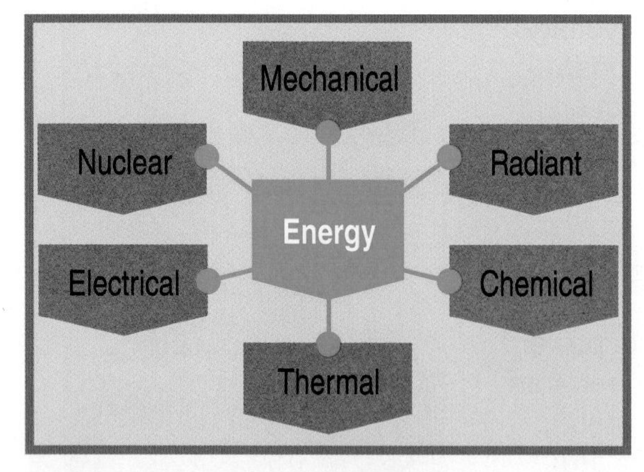

Figure 27-4. Energy can be grouped into six major forms.

These energy sources do work that ends up as motion, light, or heat. They are used to power manufacturing machines, light buildings, propel vehicles, and produce communication messages.

Mechanical Energy

Most of us are familiar with ***mechanical energy***. Often, it is produced by motion of technological devices. We associate machines with mechanical energy. This is correct, but it does not include all types of mechanical energy. Wind and moving water have motion and, thus, are also sources of mechanical energy. See **Figure 27-5.**

Figure 27-5. This waterfall is a natural source of mechanical energy.

Standards for Technological Literacy

3 7 16

Research

Have your class find out how the sun releases radiant energy.

TechnoFact

The sun is chiefly composed of hydrogen and helium. The sustained nuclear fusion in the sun causes it to emit enormous amounts of energy. Most of that energy reaches Earth in the forms of visible light and infrared rays, which we feel as heat.

Academic Connections: History

The Origin of Horsepower

After developing the first truly practical steam engine, James Watt needed to describe its output in a way that potential customers could understand. In those days, horses drove much of the machinery. Watt decided the best way to describe the output of an engine was in terms of the number of horses it could replace.

In 1782, Watt began working on the formula that would eventually lead to the unit of power known as *horsepower*. After consulting with experts who designed horse-driven machinery, Watt determined that an average horse can perform 22,000 ft.-lbs. of work per minute. Since the rotary-motion steam engine was new technology, potential customers were skeptical. Watt was afraid that, if factory or mine machinery failed because the engine used to power it was too weak, the public would blame the steam engine technology and not the miscalculation of the purchaser. Watt decided it would be better to understate the power of his engines.

This early steam tractor's power was rated in horsepower.

If a customer bought a 2-horsepower engine, Watt wanted that engine to be able to do the same amount of work as three actual horses. In order to accomplish this, Watt added 50% to his estimation of the amount of work a horse could do over a given time. This caused 1 horsepower to be equivalent to 33,000 ft.-lbs. of work per minute, which is the value still used today.

As technology improves, the amount of horsepower available to the average person increases. Research the horsepower rating for several types of automobiles. How many actual horses would be required to produce the same amount of power? (Remember, a horse can produce only 0.67 horsepower.)

Radiant Energy

Radiant energy is energy in the form of electromagnetic waves. You learned about these waves in Chapter 22. They extend from ELF radio waves (long waves) to gamma rays (short waves). Cool objects give off longer waves than hot objects do. Low-frequency waves contain less energy than high-frequency waves of the same amplitude do.

The main source of radiant energy is the Sun. Radiant energy is also emitted by objects heated with a flame or from a lightbulb in a lamp. Sometimes, radiant energy is called *light energy*. This is not completely correct because many waves with wavelengths longer or shorter than the wavelengths of light possess radiant energy. Examples of radiant energy include radio waves; microwaves; infrared, visible, and UV light; X rays; and gamma rays.

Standards for Technological Literacy

16

Extend

Describe how burning fuels (chemical reaction) releases energy.

Brainstorm

Ask your students how many ways they use electrical energy.

Research

Have your students investigate the advantages and disadvantages of nuclear energy.

Chemical Energy

Chemical energy is energy stored within a chemical substance. Typical sources of chemical energy are the fuels we use to power our technological machines. See **Figure 27-6.** The most common are petroleum, natural gas, and coal. Wood, grains (such as corn), and ***biomass*** (organic garbage) are less-frequently used sources of chemical energy. Chemical energy is released when a substance is put through a chemical reaction. This can be done by rapid oxidation (burning) or other chemical actions, such as digestion and reduction.

©iStockphoto.com/sasha_t

Figure 27-6. Energy to run this cordless drill is stored inside the drill's battery.

Thermal Energy

Thermal energy is another name for heat energy. See **Figure 27-7.** Thermal energy cannot be seen directly. You can see its effects, however, by watching the heated airwaves above a road on a very hot day. Thermal energy is usually felt. The energy strikes a surface, such as your skin, and elevates its temperature.

The internal movement of atoms in a substance creates thermal energy. These particles are always in motion. If the atoms move or vibrate rapidly, they give off heat, or thermal energy. The faster they move, the more heat they give off. Heat energy is widely used in technological devices. This energy provides the energy for our heating systems and some electrical generating plants.

©iStockphoto.com/DJClaassen

Figure 27-7. This log burning in a fireplace is giving off both radiant and thermal energy.

Electrical Energy

Electrical energy is associated with electrons moving along a conductor. This conductor can be a wire in a human-developed electrical system. Also, the conductor can be the air, as with lightning. Lightning is a natural source of electrical energy. Electrical energy is used as a basic source for other forms of energy. This energy is often converted into heat energy (for example, to warm buildings) and into light energy (for example, to illuminate our homes). See **Figure 27-8.**

Nuclear Energy

Nuclear energy is associated with the internal bonds of atoms. When atoms are split, they release vast quantities of energy. This process is called ***fission***. Likewise, combining two atoms into a new, larger atom releases large amounts of energy. This process is called ***fusion***.

Standards for Technological Literacy

5 16

TechnoFact

Heat energy comes from many sources. On a daily basis, humans harness heat energy from the sun, the ground, friction, chemical reactions, electricity, and nuclear reactions. Some of the sources are controllable; others are not.

Think Green

Renewable and Inexhaustible Energy

In this chapter, you have learned about the various types of energy sources. For some time now, people have been working toward alternatives to exhaustible energy sources. These sources have been steadily depleting for some time, and when they are gone, they will be gone forever. Looking ahead to this, the importance of renewable and inexhaustible energy sources is being stressed.

Renewable energy is a type of energy that can be grown, harvested, and then grown again. You have seen the use of this with alternative fuels, such as ethanol. Ethanol comes from crops such as sugar cane and corn. When the supply of ethanol has depleted, we can grow more resources to create more ethanol. Inexhaustible energy resources include the Sun and the wind. You have probably seen solar panels or wind turbines. Those devices harness energy. That energy is then converted to power. When that power is gone, those resources can be continually used to harness more energy.

Figure 27-8. This worker is using an electric arc to generate high heat to weld (melt) two steel parts together.

How the Forms of Energy Are Interrelated

All these forms of energy are related to one another. Radiant energy can be used to produce heat. If you have ever been sunburned, you have experienced this relationship. The radiant energy of the Sun heated your skin until your skin burned.

A fire causes fuel to undergo a chemical action. For example, coal can be changed into carbon dioxide and water. In the process of this chemical action, heat is given off. The mechanical motion of an electrical generator causes magnetic lines of force to cut across any nearby conductor. This process induces an electrical current in the conductor.

Sources of Energy

Energy is a basic input to all technological systems. All energy comes in one of three basic types of resources. See **Figure 27-9.** These are the following:

- Exhaustible energy resources.
- Renewable energy resources.
- Inexhaustible energy resources.

Standards for Technological Literacy

16

Figure Discussion

Use Figure 27-9 to elaborate on exhaustible, renewable, and inexhaustible energy resources.

Research

Have your students determine the depletion rate of major energy resources, such as coal, petroleum, and natural gas.

Research

Have your students find out what materials are used to make biogas.

TechnoFact

The term *energy supply* refers to energy resource reserves, or the total usable energy available for doing work.

Marathon, U.S. Department of Energy

Figure 27-9. The three types of energy resources are exhaustible, renewable, and inexhaustible. Pictured here are oil (exhaustible), wood (renewable), and wind (inexhaustible).

Exhaustible Energy Resources

Exhaustible energy resources are those materials that cannot be replaced. Once they are used up, we will no longer have that source. The most common exhaustible resources are petroleum, natural gas, and coal. These resources, called *fossil fuels*, originated from living matter. Millions of years ago, plant and animal matter was buried under the earth. Over time, this matter was subjected to pressure, and it decayed. This resulted in deposits of solid fuels (coal and peat), liquid fuels (petroleum), and gaseous fuels (natural gas). These deposits have been found in many locations on the Earth. Chapter 14 describes how the deposits are located and the fuels are extracted.

Uranium is another exhaustible energy source. This energy source is an element that developed when the solar system came into being. Uranium is a radioactive mineral used in nuclear power plants.

Renewable Energy Resources

Renewable energy resources are biological materials that can be grown and harvested. Human propagation, growing, and harvesting activities directly affect the supply of these resources. The most common renewable energy resources are wood and grains. They can be burned directly to generate thermal energy. Corn is often converted to alcohol (ethanol), which then can be used as a fuel. See **Figure 27-10.**

Organic matter, such as garbage, sewage, straw, animal waste, and other waste, can be an energy resource. This matter is often referred to as *biomass resources*. The prefix *bio-* means "having a biological, or living, origin." The resources can be traced back to plant or animal matter.

These organic materials can be burned directly as ***biofuels***. Also, these materials can be converted into methane, a highly flammable gas. This process generates a ***biogas***, which can replace some exhaustible fuel resources.

Inexhaustible Energy Resources

Inexhaustible energy resources are part of the ***solar weather system*** existing on Earth. This natural cycle starts with solar energy. About one-third of the solar energy reaching Earth's atmosphere is reflected back into space. The other two-thirds enter the atmosphere. The atmosphere absorbs much of this solar energy.

About one-fourth of this energy powers what can be called a ***water cycle***. See **Figure 27-11.** A small portion of Earth's water is in rivers and lakes. The majority of

©iStockphoto.com/photosbyjim

Figure 27-10. The corn in the foreground can be converted into ethanol at the plant in the background.

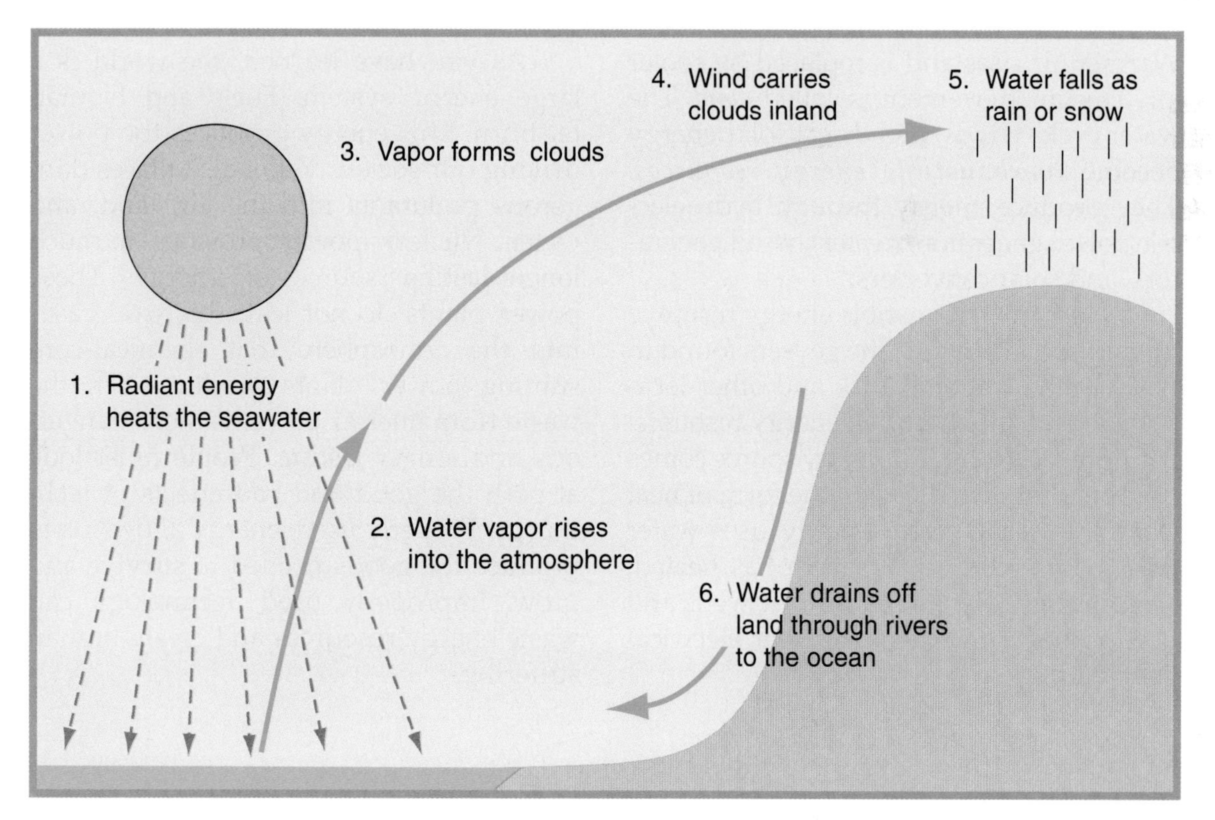

Figure 27-11. The Sun's radiant energy powers Earth's water cycle. The water cycle provides an inexhaustible energy source.

Standards for Technological Literacy

16

Brainstorm

Ask your students why the water cycle is important to energy conversion.

TechnoFact

Some of the current uses of inexhaustible energy sources include heating water with direct solar energy, generating electricity with solar cells, turning hydroelectric turbines with river currents, and turning generators with the wind.

Standards for Technological Literacy

16

Research

Have your students find out where major geothermal areas are located in the United States.

TechnoFact

As underground water passes hot, rocky layers in the Earth's crust, it heats up. Eventually the hot water erupts from the ground. Such geological features are known as hot springs, geysers, and geothermal vents. These geological features can be found in such diverse areas of the world as Yellowstone National Park, the North Island of New Zealand, and Iceland.

the water is in the oceans covering much of the globe. The solar energy causes the water in the oceans to heat and evaporate. The warm water vapor rises into the atmosphere and forms clouds. The clouds rise and are carried inland by the wind. As the clouds travel upward, the water vapor cools. This cooling effect condenses the water vapor into droplets. The droplets fall to Earth in the form of rain or snow. Much of the water runs off the land and collects in rivers. From there, the water flows into the ocean, where it begins the cycle once again. Not all the water follows this exact pattern.

Plants use some quantity of water in their respiratory cycles. Some water ends up in lakes, from which it evaporates into the air to join clouds. Other portions flow underground into rivers and oceans.

Solar energy also heats the land, but the heating effects on the oceans and the land are different. Temperature differences are created because different amounts of solar energy strike various areas of the globe. Warmer air rises and is replaced by cooler air. This air movement is called *wind*. The water cycle, winds, and direct solar energy become ***inexhaustible energy resources***. They produce energy through hydroelectric power generators (water), wind generators, and solar converters.

Another inexhaustible energy resource is geothermal energy. The geysers found in Yellowstone National Park and other locations are examples of this energy resource. See **Figure 27-12.** This energy source comes up from the earth (geo-) in the form of heat (thermal). Geothermal energy uses water that the hot core of the earth has heated. This energy is usually tapped by wells and used to heat buildings or power electrical generators.

Figure 27-12. Geothermal energy produces this geyser.

Effects of Energy Technology

As you have learned, the world is a large energy system. Fuels and biomatter burn. This energy produces the power driving our society. Yet, it also places dangerous pollutants into the air, land, and water. Nuclear power provides a much longer-lasting source of energy. These power plants do not leak the toxic gases into the atmosphere that chemical-consuming power plants produce. Yet, the waste from nuclear power is more dangerous and longer lasting. People must look at both the good and bad effects of technology. Properly used, energy allows us to produce the goods needed to survive and grow. Improperly used, technology can waste energy resources and create human suffering.

Radiant energy is the type of energy emitted by the Sun, by objects being heated by flame, or by a lightbulb. Radiant energy is also used to take X rays.

Answers to Test Your Knowledge Questions

1. Energy is the ability to do work.
2. Kinetic energy is a force that is doing work. Potential energy is a force that has the potential to do work.
3. B. Work
4. C. Power
5. B. Work
6. A. Energy
7. C. Power
8. A. Energy
9. B. Work
10. True
11. the Sun
12. False
13. fusion
14. True
15. B. Renewable
16. C. Inexhaustible
17. B. Renewable
18. C. Inexhaustible
19. A. Exhaustible
20. A. Exhaustible
21. C. Inexhaustible
22. B. Renewable
23. A. Exhaustible
24. Evaluate individually.
25. Evaluate individually.

Summary

Energy is a basic need for all technological activities. The ability to do work is the foundation for power generation and work. Energy takes the forms of mechanical, radiant, chemical, thermal, electrical, and nuclear energy. These energy forms are derived from exhaustible, renewable, and inexhaustible resources. Each form of energy has positive and negative impacts.

Test Your Knowledge

Write your answers on a separate piece of paper. Please do not write in this book.

1. Define *energy*.
2. What is the difference between kinetic and potential energy?

Matching questions: For Questions 3 through 9, match each description on the left with the correct term on the right. (Note: Answers can be used more than once.)

3. _____ Can be defined as "applying a force that moves a mass a distance in the direction of the applied force."
4. _____ One common measurement is the kilowatt-hour.
5. _____ Is measured by multiplying the weight moved and the distance moved.
6. _____ Can be defined as "the ability to do work."
7. _____ Can be defined as "the rate at which work is done."
8. _____ Is either in motion or stored.
9. _____ Is measured in newtons per meter or J, in the metric system.

A. Energy.
B. Work.
C. Power.

10. Wind is one source of mechanical energy. True or false?
11. The main source of radiant energy is _____.
12. Chemical energy is also known as *heat energy*. True or false?
13. Two processes associated with nuclear energy are fission and _____.
14. All forms of energy are interrelated. True or false?

Matching questions: For Questions 15 through 23, match each description on the left with the correct type of energy resource on the right. (Note: Answers can be used more than once.)

15. ______ Corn.
16. ______ Falling water.
17. ______ Wood.
18. ______ Wind.
19. ______ Natural gas.
20. ______ Petroleum.
21. ______ Sunshine.
22. ______ Biomass materials.
23. ______ Coal.

A. Exhaustible.
B. Renewable.
C. Inexhaustible.

24. Give one example of how energy technology can improve our lives.
25. Give one negative effect of energy technology.

STEM Applications

1. Construct a simple device to change wind energy into rotating mechanical motion.
2. Identify a renewable energy resource. Describe it in terms of the factors shown in the chart below. (Prepare a similar chart for your chosen resource.)

Energy resource:
Exhaustible resources it can replace:
Present location of the resource:
Advantages of using the resource:
Problems associated with producing the resource:
Problems associated with using the resource:

Chapter Outline
Inexhaustible-Energy Converters
Renewable-Energy Converters
Thermal-Energy Converters
Electrical-Energy Converters
Applying Energy to Do Work

This chapter covers the benchmark topics for the following Standards for Technological Literacy:

3 5 7 16

Learning Objectives

After studying this chapter, you will be able to do the following:

- Summarize the characteristics of devices that convert inexhaustible energy into mechanical motion.
- Compare passive and active solar-conversion systems.
- Summarize the main ways solar energy is converted into other forms of energy.
- Summarize the characteristics of a common geothermal energy–conversion system.
- Explain the operation of a common biomass converter.
- Summarize the characteristics of heat engines, in terms of energy conversion.
- Compare internal and external combustion engines.
- Summarize common ways to heat homes and buildings.
- Recall the major parts of an electric-energy generation-and-conversion system and their functions.
- Summarize the common energy-input systems for electric generation plants.
- Explain how energy is applied to do work.

Key Terms

active collector
anaerobic digestion
biochemical process
biomass resource
cam
conduction
convection
crank
direct active solar system
direct-gain solar system
external combustion engine
fermentation
fuel converter
gasification
gear
gear and rack
hydraulic system
hydroelectric generating plant
indirect active solar system
indirect-gain solar system
insolation
isolated solar system
liquefaction
liquidification
ocean mechanical energy–conversion system
ocean thermal energy–conversion (OTEC) system
passive collector
photovoltaic cell
pneumatic system

prime mover
pulleys and V belt
pyrolysis
radiation
rotary motion
solar converter
thermochemical conversion
waterwheel
windmill

Standards for Technological Literacy

16

Strategic Reading

Think about any energy converters you're familiar with. As you read this chapter, make an outline of the different types of converts and list any examples you can think of.

Science tells us that energy can be neither created nor destroyed. A great deal of human action, however, is devoted to converting energy from one form into another form. For example, we burn fuels to change water into steam. Steam contains energy in the form of heat. The steam might then be passed through devices used to warm rooms or dry lumber.

As you can see in this example, energy converters should be viewed as part of a larger system. An energy converter is a unique device. A converter has energy as its input and its output. Mechanical energy is the input to a turbine in a hydroelectric generator. Electricity (electrical energy) is the output. This same electricity can be the input to several other energy converters. Incandescent lamps convert electrical energy into light (radiant energy). Motors convert electrical energy into rotary motion (mechanical energy). Resistance heaters change electrical energy into heat (thermal energy).

Your body is an energy converter. This converter converts food (your fuel) into energy. Energy moves muscles, allowing you to walk, talk, and see. Likewise, an automobile engine is an energy converter. The engine converts the potential energy in gasoline into heat energy to produce mechanical motion. Energy converters power our factories, propel our transportation vehicles, heat and light our homes, and help produce our communication messages. See **Figure 28-1.**

©iStockphoto.com/Kativ, ©iStockphoto.com/onfilm

Figure 28-1. Energy is the foundation for technology. This foundation powers our machines and carries us across long distances.

Research
Have your students research the law of conservation of energy.

Reinforce
Discuss energy conversion as a technological system (Chapter 2) and energy as an input to the system (Chapter 4).

TechnoFact
Wood was the most important fuel throughout the majority of human history. So much timber was harvested for personal and industrial use that it started becoming scarce in the 18th century. Coal gradually replaced wood as the fuel of choice. Over the last century, petroleum, natural gas, and nuclear energy came into widespread use.

TechnoFact
During the 20th century, a growing population, improved standards of living, industrial expansion, the proliferation of energy-consuming inventions, and expanding uses for petroleum and natural gas caused fuel consumption to nearly double every 20 years.

Standards for Technological Literacy

7 16

Extend

Discuss the historical development and use of windmills.

TechnoFact

The ancient Egyptians may have been the first to use wind energy. As early as 5000 years ago, the Egyptians developed sails to move their boats up and down the Nile River. The Persians developed the windmill about 1300 years ago.

Humans have developed hundreds of energy converters to meet their needs. In this chapter, we explore four broad categories of energy-conversion systems:

- Inexhaustible-energy converters.
- Renewable-energy converters.
- Thermal-energy converters.
- Electrical-energy converters.

After our discussion of energy converters, we will examine how we apply energy to do work.

Inexhaustible-Energy Converters

The earliest energy-conversion technologies were designed to power simple devices. These devices fall into a category that mechanical engineers call ***prime movers***. A prime mover is any device that changes a natural source of energy into mechanical power.

Most early prime movers used wind power and waterpower. Wind power and waterpower are inexhaustible energy sources. Almost all societies used energy converters in transportation. Wind and flowing water helped move their boats.

Several uses for energy converters emerged, however. On land, windmills and waterwheels became important technological devices developed to harness the forces of wind and water. See **Figure 28-2.**

Figure 28-2. Windmills were early energy converters.

These two devices convert natural mechanical energy (flowing air or running water) into controlled mechanical energy. For example, they can produce the motion needed to power a water pump or an electric generator. Other important converters use solar, geothermal, and ocean energy. These converters can be used to produce energy needed to heat and light our homes or power other technological devices.

Wind-Energy Conversion

The Sun is the original source of most of the energy on Earth. This energy is stored in growing plants and animals and in decayed organic matter. Decayed organic matter includes peat, coal, natural gas, and petroleum.

Also, the Sun causes the winds to blow all over Earth. An unequal heating of Earth's surface produces these air currents. Each day, the Sun's rays heat the landmasses and water they touch. Not all areas are touched at the same time or with equal energy, however. Polar areas receive less solar energy than do areas near the equator. Areas under cloud cover receive less solar heating than do areas in direct sunlight.

The heat from the land and water warms the air above it. The warm air rises. Cooler air moves in to replace it. This movement produces air currents we call *wind*.

The air above the hot areas near the equator is always rising. The cooler polar air moves toward the equator. In addition, air above water heats and cools more rapidly than air over land. Thus, during the morning hours, the air above the water warms quickly and rises. Cool air from the land moves in to replace it. In the evenings, the air above the land stays warmer longer than the air above the water. The cooler sea air moves inland to replace the rising land air.

Sails

Early humans designed technological devices to use these air currents. An early use of wind power was the ship's sail. See **Figure 28-3.** This device was developed in Egypt around 12,000 years ago. Sails remained the primary power for ships until the development of the steam engine in the late 1700s.

©iStockphoto.com/rcp

Figure 28-3. Sails were one of the earliest technological devices to harness wind power.

Windmills and Turbines

Wind power, the principle of the sail, was adapted to land applications with the development of the ***windmill***. The windmill's first use was probably in the Middle East around 200 BC. See **Figure 28-4.** These mills were used for grinding grains into flour. This small start developed over the years, leading to today's windmills and turbines. The modern windmill is primarily used to pump water for livestock on large

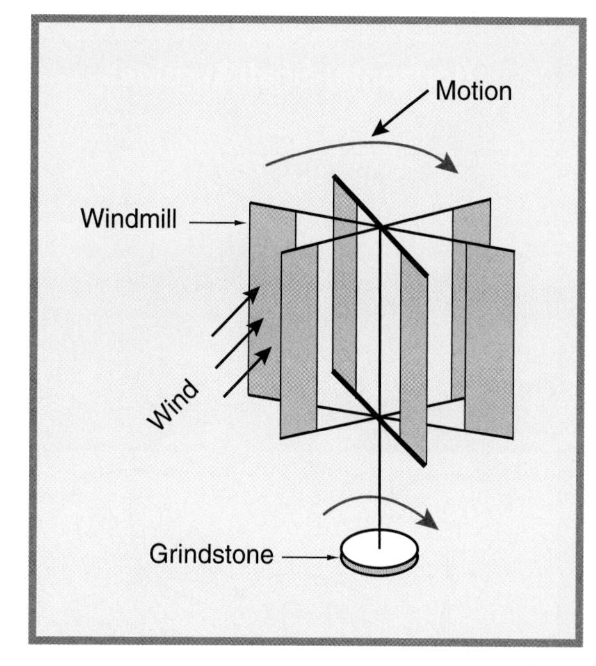

Figure 28-4. A design thought to be used in ancient windmills. The windmill harnesses the wind for ease in grinding.

western cattle ranches. The wind turbine is used to power electric generators.

Windmill and wind turbine designs can be grouped into two classes: horizontal axis and vertical axis. See **Figure 28-5.** The horizontal-axis design has one or more blades connected to a horizontal shaft. The wind flows over the blades, causing them to turn. To see this action, place a household cooling fan in front of a blast of air. The blades turn, even though the fan's power is off.

Vertical-shaft designs have blades arranged around the shaft. As the wind blows past the blades, torque (turning force) is generated, rotating the shaft. The most common vertical-shaft devices are the Darrieus and Savonius wind turbines.

Currently, a great deal of experimentation is being done to develop efficient wind turbines that can power electric generators. Large numbers of these devices are grouped together in wind farms located in various parts in the western United States. See **Figure 28-6.**

Standards for Technological Literacy

7 16

Research

Have your students find out how much wind is needed to operate a wind turbine.

Extend

Discuss where major wind farms (wind turbines generating electrical energy) are located.

TechnoFact

Most American windmills sit atop tall steel towers. The windmill itself is a wheel spoked with warped vanes (sails). A broad tail keeps the wheel facing the wind. In the past, such windmills were used to pump water in rural parts of the United States. Such a windmill would be capable of generating about one kilowatt of electricity.

Standards for Technological Literacy

7 16

Extend

Discuss how a turbine is used in a hydroelectric system.

Research

Have your students use the Internet to explore the operation of a waterwheel.

TechnoFact

On average, wind turbines installed on windy sites operate at about 25% to 35% of their full capacity. A power plant fueled with coal generally operates at 75% to 85% of its full capacity.

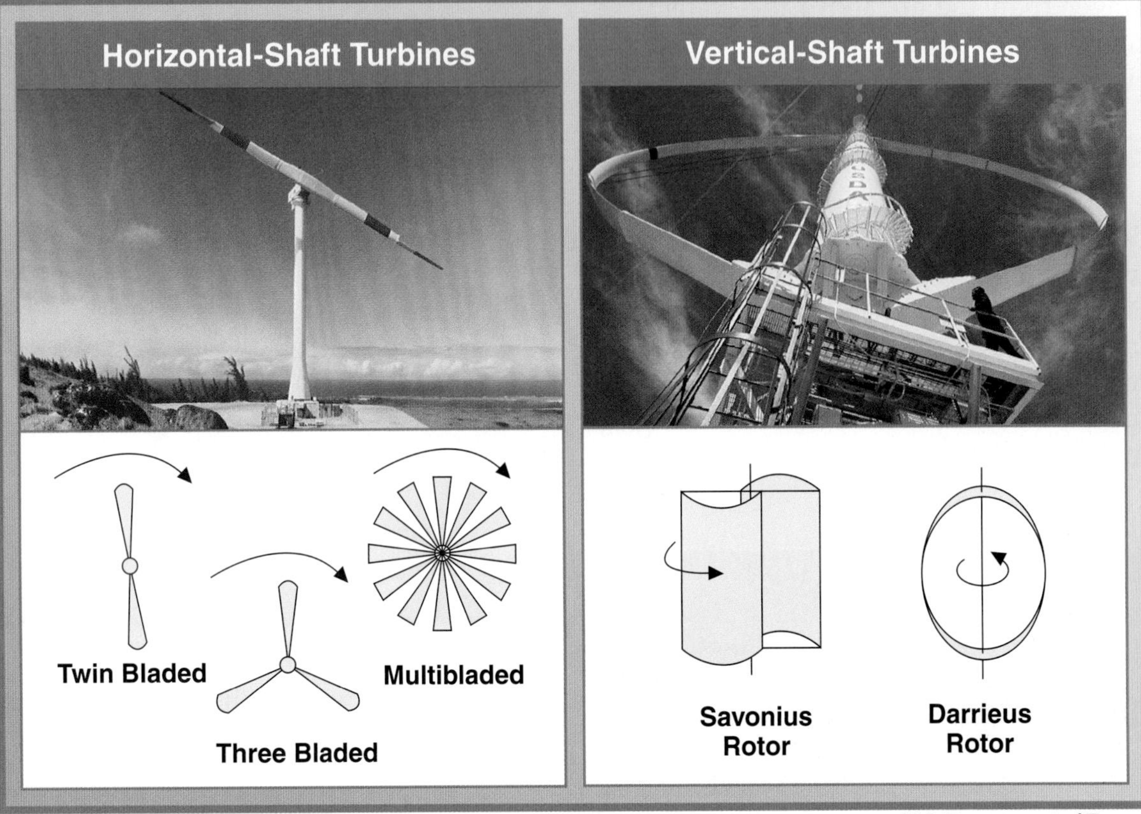

U.S. Department of Energy

Figure 28-5. The two classes of wind turbines. On the left are examples of wind turbines with their axes parallel with the airflow (horizontal shafts). On the right are examples with their axes at right angles to the wind (vertical shafts).

©iStockphoto.com/GregC

Figure 28-6. A windmill farm in California.

Water-Energy Conversion

A moving gas (air) powers windmills. Likewise, moving liquids can provide the energy to power technological devices. One of the earliest devices used to capture this energy was the ***waterwheel***. This device is essentially a series of paddles extending outward from a shaft. The flowing water drives the paddles, causing the wheel to rotate.

Waterwheels powered the first factories of the Industrial Revolution. The wheels were produced in two basic designs. These designs are undershot and overshot. See **Figure 28-7.** Water rushing under the undershot waterwheel powers this type of waterwheel. The overshot wheel is powered by water falling onto it from an overhead trough or pipe.

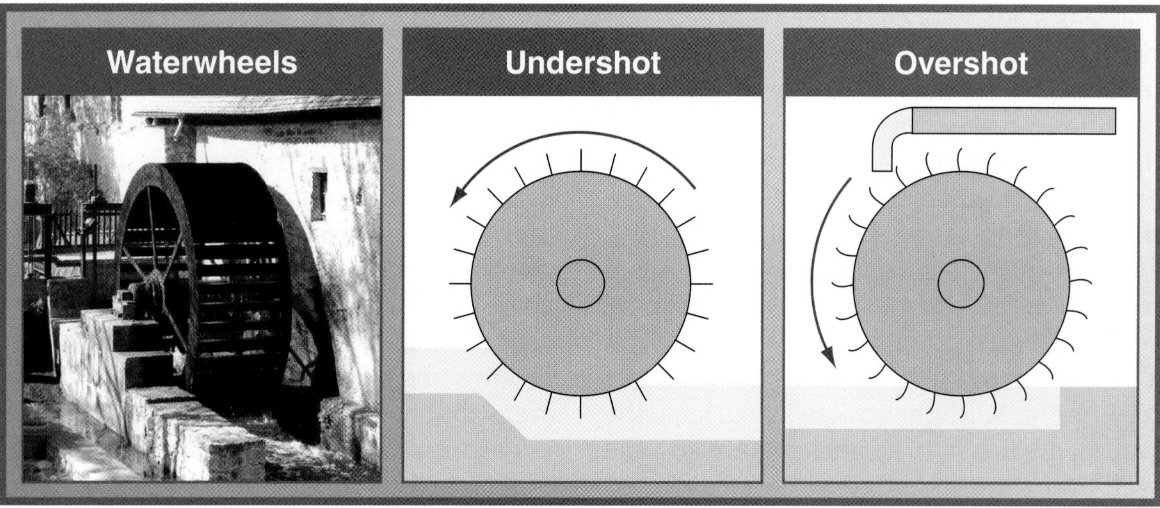

©iStockphoto.com/esemelwe

Figure 28-7. The two common types of waterwheels.

A modification of the waterwheel is the water turbine. The water turbine is a series of blades arranged around a shaft. As water passes through the turbine at a high speed, the blades spin the shaft. Water turbines are used widely to power electric generators in hydroelectric power plants.

The steam turbine is a similar device. This device uses steam (hot water vapor), however, to drive the turbine. Coal, oil, natural gas, and nuclear power plants use steam turbines to drive their generators.

Solar-Energy Conversion

A third inexhaustible-energy converter is the ***solar converter***. The solar converter uses the constant energy source of the Sun. The Sun generates 3.8×10^{20} MW (380 million million million megawatts) of power through internal nuclear fusion.

The need to conserve exhaustible energy resources has brought solar energy into consideration as a replacement resource. Most solar energy–conversion systems have two major parts. These parts are a collection system and a storage system.

Solar energy is very intense before it reaches Earth. The energy can produce temperatures approaching 10,800°F. As it reaches Earth's protective ozone layer, however, much of this energy is absorbed and heats our atmosphere. Water vapor in the air absorbs additional energy. Some solar energy reaches Earth and is available as an inexhaustible energy source. The amount of this available energy depends on the inclination (height above the horizon) of the Sun and the atmospheric conditions (cloud cover) over Earth. The term ***insolation*** is used to describe the solar energy available in a specific location at any given time. Insolation varies with the seasons and the weather. The maximum insolation on a clear, summer day is about 1000 megawatts (MW) per square kilometer (0.38 sq. mile).

Solar collectors depend on the principle that black surfaces absorb most of the solar energy that strikes them. This causes black surfaces to gather heat when they are exposed to sunlight. Typical solar collectors can be grouped into two categories. These categories are passive collectors and active collectors. Each is discussed in turn.

Passive Collectors

Passive collectors directly collect, store, and distribute the heat they convert from solar energy. Actually, an entire

Standards for Technological Literacy

5 7 16

Research

Have your students find out where, in the United States, the best areas for solar energy conversion are.

TechnoFact

When the waterwheel was invented about 2100 years ago, it was used to grind grain. Later, the wheel was adapted to perform a variety of mechanical actions. It was the power source that brought Europe out of the Dark Ages. It was the dominant source of industrial power until the steam engine was introduced in the 1700s.

Standards for Technological Literacy

5 16

Figure Discussion

Use Figure 28-8 to identify and describe the types of passive solar collectors.

Brainstorm

Ask the students why they would use solar energy to heat a home.

TechnoFact

A solar furnace concentrates sunlight to create intense heat. Some solar furnaces focus sunlight until it is 50,000 times more concentrated than that of sunlight. Such furnaces can generate temperatures up to 6300°F.

house is a solar collector in a passive solar-collection system. The building sits quietly in position, as the Sun heats it. This can be done in three ways. See **Figure 28-8.** A ***direct-gain solar system*** allows the radiant energy to enter the home through windows, heating inside surfaces.

An ***indirect-gain solar system*** uses a black concrete or masonry wall (Trombe wall) that has glass panels in front of it. The wall has openings at its bottom and top. As the sunlight strikes its surface, the wall heats up. In turn, the air between the wall and the glass panels becomes heated by energy radiating from the wall. The warm air rises and flows into the building through the openings at the top of the wall. This creates natural convection currents that draw cooler, heavier air into the openings at the bottom. This new air, in turn, is heated and rises. The Trombe wall also retains a great deal of heat. Consequently, after the Sun sets, the wall continues to radiate the heat and warms the air between the wall and the glass panels.

Adobe homes in the southwestern part of the United States use a similar solar-heating principle. The solar energy striking the adobe brick warms the surface. During the day, the energy slowly penetrates the thick wall. This penetration takes about 12 hours. In the evening, the heat finally reaches the inside of the dwelling and provides warmth during the cool nights. By morning, the wall is cool. The wall insulates the rooms from the daytime heat. The 12-hour lag between heating and cooling makes a very effective daytime cooling and nighttime heating system for areas that have hot days and cool nights.

An ***isolated solar system*** uses solar collectors or greenhouses separate from the house. These collectors are built below the level of the house. The heat generated in the collector can be channeled directly to heat the home. Additional heat is stored in a thermal mass (rock bed) for night and cloudy-day heating.

Active Collectors

Active collectors use pumps to circulate the water collecting, storing, and distributing the heat they convert from solar energy. These systems are used in many areas to provide hot water and room heat

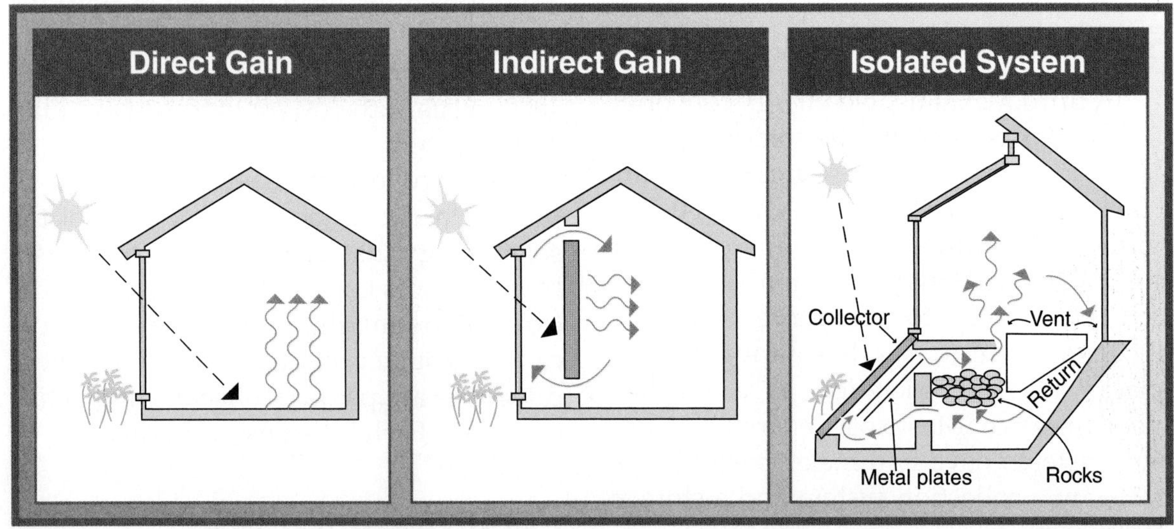

PPG Industries

Figure 28-8. Three types of passive solar systems can be used to heat a home. Direct gain is the simplest system. The indirect system uses a Trombe wall. One type of isolated system uses collectors separate from the house.

(space heating). Two major types of active-collector systems exist. These types are indirect and direct.

Indirect active-collector systems

The typical ***indirect active solar system*** has a series of collectors. Each collector has a black surface to absorb solar energy. Above or below this surface is a network of tubes, or pipes. Water is circulated through these channels. As the fluid passes a warm black surface, it absorbs heat. The warm water is then pumped to a heat exchanger. In the heat exchanger, it can heat water for domestic use or provide thermal energy for a heating system. See **Figure 28-9.**

Direct active-collector systems

A ***direct active solar system*** does not have a heat exchanger. The water circulated in the system is used as domestic hot water. This water flows directly to such areas as household faucets, washing machines, and showers.

A more recent use of active solar converters is in electric-power generation. The converters produce steam. The steam drives the turbines in the generation plant. See **Figure 28-10.**

Figure 28-10. This large bank of solar collectors collects solar energy to produce steam for an electrical generating plant located in the Mojave Desert in California.

Another type of solar converter is the ***photovoltaic cell*** (solar cell). The cell can be a small device powering a pocket calculator or part of an array of units providing electricity for such devices as satellites, solar-powered vehicles, and portable signs. See **Figure 28-11.**

Standards for Technological Literacy

5 16

Research

Have your students investigate how solar cells work.

TechnoFact

Alexandre Becquerel discovered the photovoltaic effect in 1839. His discovery led to the development of the first photovoltaic cell, which was made from selenium. The early versions of these cells only converted about 1% of the sunlight energy cast on them into electricity. In 1954, the first practical photovoltaic cell was developed by Bell Telephone Laboratories. This cell could convert about 6% of the solar energy into electricity.

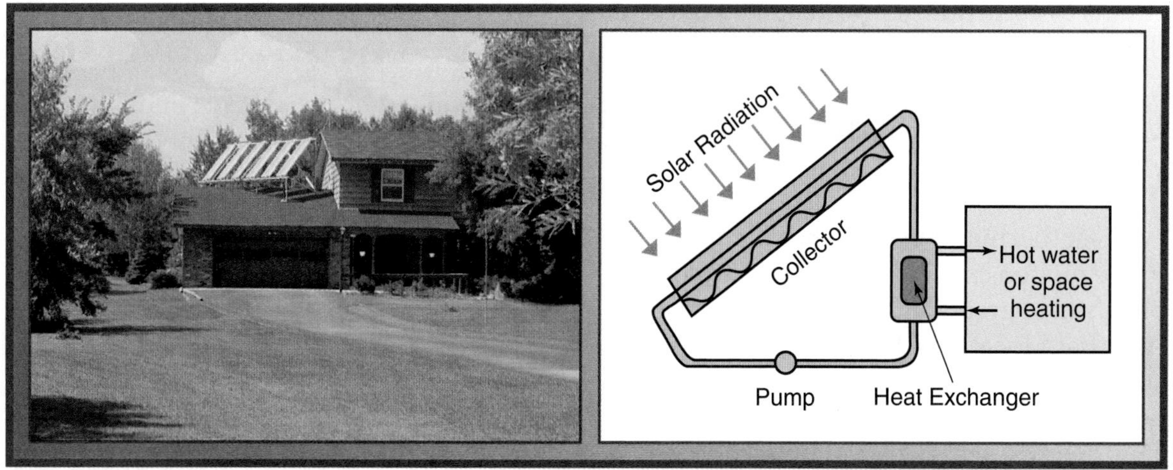

U.S. Department of Energy

Figure 28-9. The solar panels on the building power a system similar to the one in the diagram.

Standards for Technological Literacy

5 16

Extend

Describe how geothermal energy is used to produce electrical power in New Zealand or other locations.

TechnoFact

Geothermal engineers want to tap the Earth's internal heat. In some places, the earth's temperature rises as much as 270°F for every mile of depth. In an experimental program at Los Alamos, two holes were drilled to a depth of 2.5 miles. Cold water was pumped under high pressure into one hole. The water flowed through cracks in the hot rock bed and emerged from the other hole at a temperature of 465°F. Such superheated water could be used to drive a steam turbine.

©iStockphoto.com/kozmoat98

Figure 28-11. This portable highway warning sign is powered by a series of solar cells.

NPS Photos

Figure 28-12. Geothermal energy produces the Old Faithful geyser in Yellowstone National Park.

The cells are made of certain semiconductor materials, such as crystalline silicon. Small bundles of light energy, called *photons*, impact the cell. These photons strike the cell, causing electrons to dislodge from the silicon wafer. The electrons move in one direction across the wafer. A wire attached to the wafer provides a path for the electrons to enter an electrical circuit. A second wire allows electrons from the circuit to return to the wafer. This current flow supplies the power for each device.

Geothermal-Energy Conversion

Geothermal energy is heat originating in the molten core of Earth. This energy can be found at great depths all over the planet. At certain locations, however, it reaches the surface. The energy appears as volcanoes, hot springs, and geysers. See **Figure 28-12.** This energy is tapped in a number of ways. Electricity is readily produced using geothermal energy to form steam that drives a steam turbine–powered generator.

Other applications use geothermal energy for direct heating. Geothermal heat pumps use the constant 55°F temperature of groundwater to heat homes. Often, the water is pumped from a well extending into an underground aquifer. The water enters the heat pump. The pump removes heat from the water and transfers it to the dwelling. The cool water is returned to the aquifer through a second well.

Ocean-Energy Conversion

Ocean energy is an inexhaustible source that has only recently been considered a major source of energy for the coming generations. The oceans cover more than 70% of the globe. They contain two important sources of energy. These sources are thermal and mechanical (wave and tide motion).

Ocean Thermal Energy–Conversion (OTEC) Systems

Ocean thermal energy–conversion (OTEC) systems use the differences in temperature between the various depths of the ocean. The basic system has three steps. First, warm ocean water is used to evaporate a working fluid. Second, the vapors are fed into a turbine turning an electrical generator. Finally, cold ocean water is used to condense the vapors to complete the energy-conversion cycle. The process requires water with at least a 38°F difference. This occurs only at the equator.

Ocean Mechanical Energy–Conversion Systems

Ocean mechanical energy–conversion systems use the mechanical energy in the oceans to generate power. Two sources of mechanical energy are tapped. These sources are wave energy and tidal energy. See **Figure 28-13.**

Wave-energy conversion

Presently, several wave-generation devices showing commercial promise are being developed. The first is a mechanical surface follower. The surface follower is used as a navigational aid. This device is one of the simplest designs. The surface follower is a buoy floating in the water. Inside the device is a mechanism that uses the up-and-down movement of the buoy to ring a bell or blow a whistle.

The second system is a pressure-activated device. This is also a buoy used for navigational aid. The device uses the bobbing action the waves create to compress air in a cylinder. As the water rises, it compresses the air. When the buoy falls, the compressed air is released, powering a small generator. The resulting electricity can power a navigational light.

Tidal-energy conversion

Tidal-energy devices use the difference between the height of the ocean at high tide and the height at low tide to generate power. As the ocean rises, water is allowed to flow over a dam into a basin. As the tide recedes, the water flows back through turbines. This generates electricity. See **Figure 28-14.**

Figure 28-13. Ocean tides and waves contain tremendous energy that can be tapped to meet human needs.

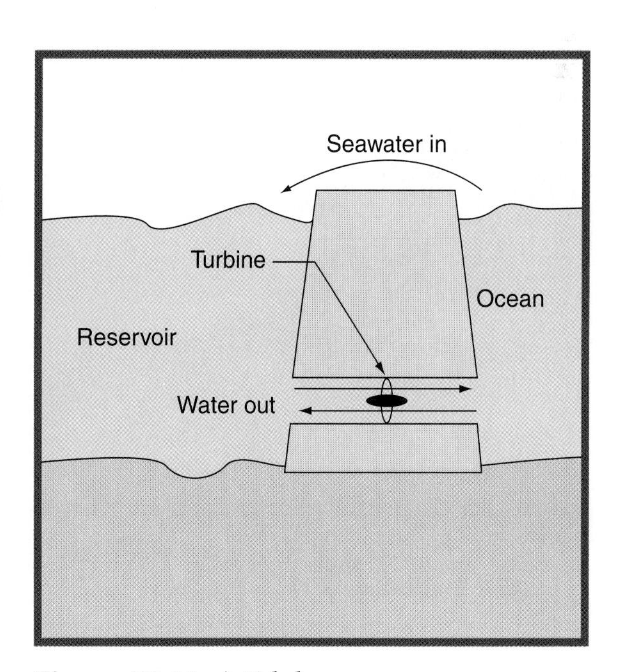

Figure 28-14. A tidal-power ocean energy–conversion system.

Standards for Technological Literacy

16

Research

Have your students investigate some of the research projects on ocean-energy conversion.

Brainstorm

Ask the class to name some advantages and disadvantages of ocean-energy conversion.

Standards for Technological Literacy

16

TechnoFact

The height of ocean waves is controlled by a number of variables including wind speed and duration.

A system in France creates power from water flow in both directions. The rising tide drives the turbine in one direction. Later, the falling tide powers the turbine in the other direction.

Renewable-Energy Converters

Our early ancestors depended heavily on renewable energy resources. They burned wood and cattle dung to heat their homes and cook their food. In 1850, these energy sources provided 90% of energy needs. In many parts of the world, these resources are still very important. In the United States today, however, they supply just a small percentage of our energy needs.

Biomass resources are one source of renewable energy being actively considered as an alternate energy supply. These resources are all the vegetable wastes and animal wastes generated through biological actions. Most biofuels come from three sources:

- **The forest-products industry.** This industry includes sawdust, bark, logging slash (waste), wood shavings, scrap lumber, and paper, for example. See **Figure 28-15.**
- **Agriculture and food processing.** This source includes corncobs, nutshells, fruit pits, grain hulls, sugarcane bagasse, and manure, for example.
- **Municipal waste.** This source includes sewage and solid waste (garbage).

Bioenergy conversion is completed using one of two basic processes. These processes are thermochemical conversion and biochemical conversion.

Thermochemical Conversion

Thermochemical conversion produces a chemical reaction by applying heat. The most common method is direct combustion. The biofuel is burned to produce heat for buildings or to produce steam to power electric generation plants. This system is widely used in the forest-products industry. Burning mill waste produces the steam that heats buildings, dries lumber, and operates the processing equipment.

©iStockphoto.com/LyaC

Figure 28-15. Wood from forest thinning and logging waste can provide material for biomass conversion.

A second thermochemical process is ***pyrolysis***. In this process, the material is heated in the absence of oxygen. The heat causes the biofuel to form liquids, solids, or gases. The solids are carbon and ash. The liquids are very similar to petroleum and require further processing. The gases are flammable hydrocarbons. All these materials can be directly or indirectly used as fuels. Other names used for this process are ***liquidification*** and ***gasification***.

A third process is ***liquefaction***. In this process, the biofuel is heated at moderate temperatures under high pressure. During heating, steam and carbon monoxide, or hydrogen and carbon monoxide, are present. A chemical action takes place, converting the material into an oil that has more oxygen than petroleum. This oil requires extensive refining to develop usable fuels.

Biochemical Conversion

Biochemical processes use chemical reactions that fungi, enzymes, or other microorganisms cause. The two common biochemical-conversion processes are anaerobic digestion and fermentation. ***Anaerobic digestion*** is a controlled decaying process that takes place without oxygen. The material used is agriculture waste, manure, algae, seaweed (kelp), municipal solid waste, and paper. The reaction produces methane (a flammable gas) as the biomaterials decay. See **Figure 28-16.**

©iStockphoto.com/LianeM

Figure 28-16. A biogas converter located next to a large dairy farm. The product is methane, a highly flammable gas.

Standards for Technological Literacy

16

Extend

Describe how methane gas is produced through anaerobic digestion or how ethanol is produced through fermentation.

TechnoFact

Biofuels is short for biomass fuels. Biomass fuels are liquid fuels that can be substituted for petroleum-based fuels. Common biofuels include biodiesel and ethanol. These fuels can be made from a wide variety of materials, including crops and vegetable waste. The material from which a biofuel is produced is known as the biomass feedstock.

Career Corner

Power-Line Installers

Power-line installers construct electrical power lines using construction equipment, such as trenchers and boring machines. They use digger derricks. These derricks are trucks equipped with augers and cranes that dig holes and set utility poles. Line installers string cable along the poles and towers or in underground trenches or conduit. Also, line installers maintain electrical, telecommunications, and cable television lines. They identify problems and repair or replace defective cables or equipment.

Line installers often encounter serious hazards and must follow safety procedures to reduce the risks of danger. They must have a high school diploma and are trained on the job. Many employers prefer a technical knowledge of electricity and electronics obtained through vocational or technical programs, community colleges, or the armed forces.

Standards for Technological Literacy

16

Reinforce

Review the discussion of the internal combustion engine presented in Chapter 7.

TechnoFact

In 1859, J. J. Étienne Lenoir developed the first practical gasoline-powered engine. It was a continuously operating, double acting engine with spark-ignition engine. In 1878, Nikolaus A. Otto built a successful four-stroke engine, which became known as the "Otto-cycle" engine.

Fermentation is a very old process using yeast (a living organism) to decompose the material. The yeast changes carbohydrates into ethyl alcohol (ethanol). Grain, particularly corn, is often used for this process. The ethanol can be directly burned or can be mixed with gasoline as an automobile fuel.

Thermal-Energy Converters

Heat and thermal energy have had important parts in history. The Industrial Revolution was greatly dependent on heat engines. Electrical motors have replaced these engines in most industrial applications. Transportation is one exception.

Applications of Thermal Energy

We still depend on thermal energy, however. The comfort of the home you live in depends on heating and cooling systems. Many industrial processes use heat to cook, cure, or dry materials and products. Here, we explore two major applications of thermal energy. These applications are heat engines and space heating.

Heat Engines

Today, most of our transportation systems are based on fossil fuel–powered engines. These technological devices burn fuel to produce heat. In turn, the heat is converted into mechanical energy. All heat engines can be classified as either internal combustion engines or external combustion engines.

These classifications are based on the location of the thermal energy source. Internal combustion engines burn the fuel within the engine. ***External combustion engines*** burn the fuel away from the engine.

Internal combustion engines

We discussed some features of internal combustion engines in earlier chapters of this book. For example, in Chapter 7, we learned that gasoline is widely used in land- and water-transportation vehicles. See **Figure 28-17.** Jet and rocket engines that were introduced in Chapter 25 are also internal combustion engines.

Figure 28-17. This internal combustion engine is used in Indy-class race cars.

To create power, all these engines use expanding gases that burning fuel produces. They change heat energy into mechanical motion. Let us review the common gasoline-powered internal combustion engine. You learned that the engine operates on a cycle that has four strokes. These strokes are intake, compression, power, and exhaust.

During the intake stroke, the piston moves downward to create a partial vacuum. A fuel-and-air mixture is then introduced into the cylinder. The compression stroke follows the intake stroke. During this stroke, the piston moves upward, compressing the fuel-air mixture into the small cavity at the top of the cylinder.

The power stroke then starts with an electrical spark that the spark plug produces. This action ignites the compressed fuel-air mixture. The fuel-air mixture expands. The resulting gases force the piston downward in a powerful movement.

The final stroke is the exhaust stroke. During this stroke, the piston moves upward to force the exhaust gases and water vapor from the cylinder. At the end of this stroke, the engine is ready to repeat the four-stroke cycle.

External combustion engines

Most external combustion engines are steam engines. The steam engine uses the principle that steam occupies more space than the water from which it came. In fact, 1 cubic centimeter (cc) of water produces 1700 cc of steam.

The operation of the steam engine is simple. See **Figure 28-18.** Water is heated in a boiler until it changes into steam. This high-pressure steam is introduced into a closed cylinder that has a free-moving piston in it. The steam forces the piston down. Next, cold water is introduced into the cylinder, condensing the steam. The resulting water takes up only 1/1700th as much space as the steam did, so a vacuum is formed in the cylinder. This causes the piston to be drawn up. At the top of the piston's stroke, a fresh supply of high-pressure steam is introduced into the cylinder. The engine repeats its cycle.

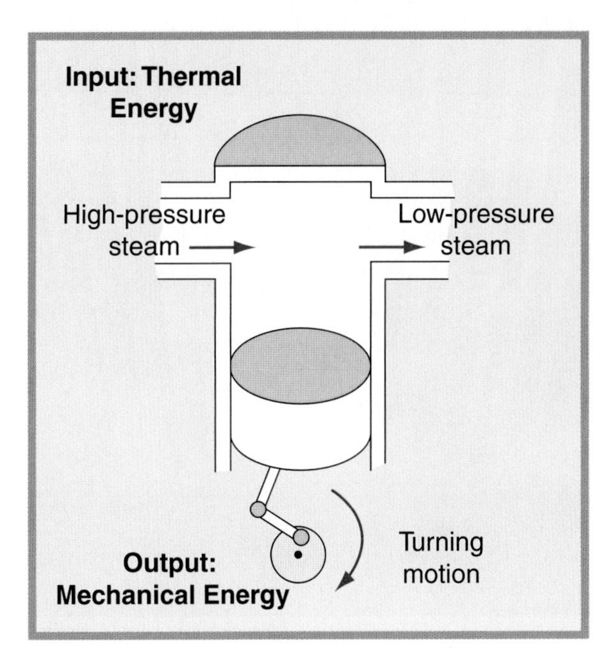

Figure 28-18. The operation of a simple steam engine.

A flywheel changes the reciprocating (up-and-down) motion of the engine to ***rotary motion***. This rotary motion can be used to power any number of technological devices. In past times, steam engines powered ships, locomotives, cars, and many machines in factories.

Space Heating

An important use of thermal energy is heating buildings and other enclosed spaces. Three basic types of heat transfer are used to heat space. See **Figure 28-19.** These are the following:

- Conduction.
- Convection.
- Radiation.

Conduction

Conduction is the movement of heat along a solid material or between two solid materials touching each other. This movement takes place without any flow of matter. The movement of energy is from the area with a higher temperature to the area with a lower temperature. Conduction heats a pan on an electric heating plate.

Convection

Convection is the transfer of heat between or within fluids (liquids or gases). This transfer of heat involves the actual movement of the substance. The process uses currents between colder areas and warmer areas within the material. Convection can occur through natural action or through the use of technological devices. The wind is an example of natural

Standards for Technological Literacy

16

Figure Discussion

Use Figure 28-18 to discuss how a steam engine operates.

Research

Have your students trace the historical development of the steam engine.

TechnoFact

In early steam engines, expanding steam pushed the piston up in the cylinder. When the piston was at the top of the cylinder, cold water was injected into the cylinder under the piston. This caused the steam to condense, creating a vacuum and drawing the piston back down in the cylinder. James Watt improved the steam engine by devising a valve system that allowed steam to drive the piston on both the upward and downward stroke.

Standards for Technological Literacy

16

Demonstrate

Use a small space heater to demonstrate convection, conduction, and radiation.

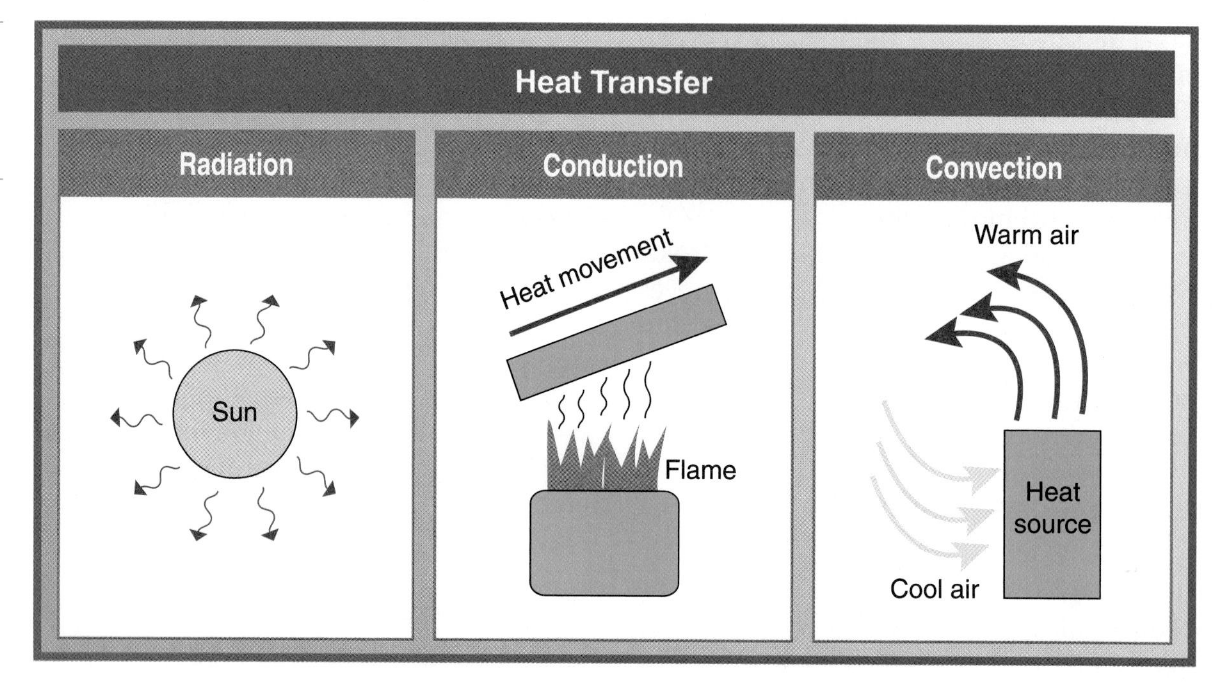

Figure 28-19. Examples of radiation, conduction, and convection. Radiation, conduction, and convection are the three important heating processes.

convection action. Forced hot–air heating systems are technological devices that warm homes with convection currents.

Radiation

Radiation is heat transfer using electromagnetic waves. The strength of the radiation is directly related to the temperature of the radiating medium. Hot objects radiate more heat than cooler objects do. The heat people feel on a bright, sunny day is from solar radiation. Also, if you bring your hand close to a hot metal bar, you can feel the heat radiate from it.

This heat transfer heats only the solid objects it strikes. Radiation does not heat the air it travels through. Radiant heaters in warehouses keep workers warm in a building. The air still feels cold. See **Figure 28-20.**

©iStockphoto.com/Acerebel

Figure 28-20. This radiant heater is used to heat an industrial space.

Heat Production

The conversion of energy into heat has been a goal of humans since before recorded history. People living in colder climates have always been challenged to heat their living spaces. A number of different methods are used today to produce thermal energy to heat materials and buildings. These include burning fuels, capturing heat from the surroundings, and converting electrical energy.

Fuel Conversion

Typical ***fuel converters*** include fossil fuel furnaces, wood-burning stoves, and fireplaces. A furnace has a firebox, a heat exchanger, and a means of heat distribution. See **Figure 28-21.** The fuel is burned in the firebox to generate thermal energy. Convection currents pass through the cells of the heat exchanger and raise its temperature. This thermal energy is transferred in the heat-distribution chamber to a heating medium (water or air). The medium is then passed over or through the heat exchanger.

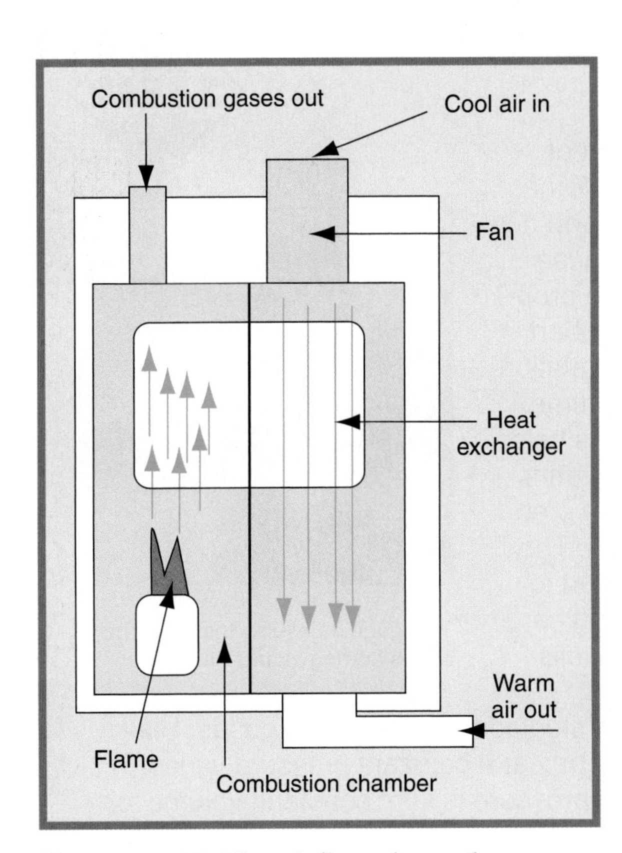

Figure 28-21. The airflow through a gas-fired, hot air furnace.

In some systems, water is heated or turned to steam. The fluid is then piped to radiators in various locations. These radiators use convection and radiation currents to heat the room. Other systems blow air through ducts to areas needing heat. Convection currents circulate the warm air within the enclosure.

Atmospheric Heat

The atmosphere has heat available, no matter how cold the day seems. The standard device used to capture this heat is called a *heat pump.* This pump is actually a refrigeration unit that can be run in two directions. In one direction, the pump removes heat from the room and releases it into the atmosphere. This is part of what an air-conditioning (cooling) unit does. When a heat pump is operated in the opposite direction, the pump takes heat from the outside air and releases the heat inside a building.

Heat pumps work on a simple principle. This principle is that, when a liquid vaporizes, it absorbs heat, and when a liquid is compressed, it releases heat. The system consists of a compressor, cooling or condenser coils, evaporator coils, and a refrigerant (volatile liquid). See **Figure 28-22.**

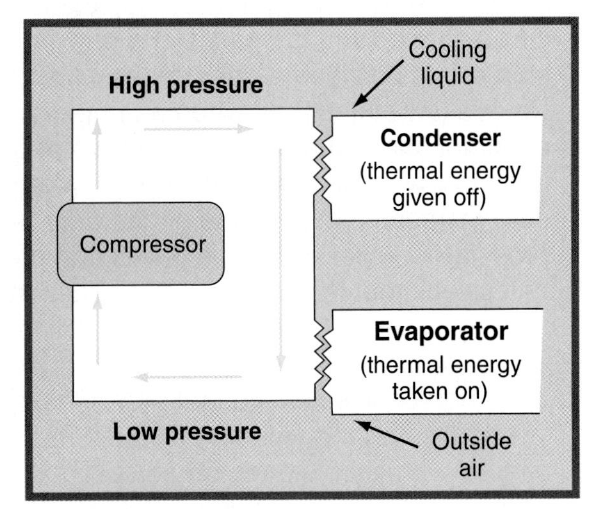

Figure 28-22. The operation of a simple heat pump.

Standards for Technological Literacy

16

TechnoFact

The ancient Romans developed clay and earthenware stoves. More than 1800 years ago, the Chinese were making stoves from cast iron. When settlers arrived in the New World, they brought with them five-plate stoves or the molds for making them. Benjamin Franklin developed the first heating stove in 1743.

Standards for Technological Literacy

3 16

Extend

Describe the type of electric heat available for homes.

In a heat pump, a heat-transfer medium, such as ammonia, is allowed to vaporize in the evaporator coils. The heat needed to complete this task is drawn from the material around the coils. This material might be air (in an atmospheric heat pump) or water from a well (in a groundwater heat pump).

The refrigerant gas is then compressed. This action causes the material to give off heat through the condenser coils. The heat can be used to warm air or water. The air or water is then transferred to the rooms needing heat.

The system can be reversed to produce cooling for air conditioning. The heat for the evaporation is drawn from within the building. The heat from compressing the gas is expelled into the outside atmosphere.

Electric Heat

We can use electricity in a variety of ways to heat a building. The heat pump described above is one way. Electricity powers the compressor drawing heat from the air. Another method using electricity is a furnace that uses an electric resistance heater.

One common method of heating uses electric resistance heaters in each room. These heaters have special wires that have a high resistance to electrical current. The wires become very hot when electricity passes through them. The hot wires warm the air around them. Convection currents transfer the heat to all parts of the room.

Another type of electric heating heats with radiation (radiant heating). This system uses high-resistance wires installed

STEM Connections: Science

Laws of Gases

As we note in this chapter, the expansion of gas that burning fuel produces powers the internal combustion engine. We were able to create and now are able to use this artifact successfully because of various scientific discoveries regarding the properties of gases. For example, the scientist Robert Boyle studied various gases and saw that a relationship exists between a gas's pressure and volume. Boyle's law states that, when a temperature is held constant, the volume of a fixed mass of gas varies inversely with the pressure. That is, when the pressure increases, the volume of gas decreases. The pressure doubles when the gas is compressed to half its volume. In the internal combustion engine, the pressure increases against the cylinder walls and piston.

©iStockphoto.com/wolv

This gage measures the pressure of gas flowing in an oil field.

Another scientist, Jacques Charles, also studied the behavior of gases. His law states that the volume of a fixed mass of gas at a constant pressure varies directly with absolute temperature. That is, if pressure is kept constant, volume and temperature are directly related. As the temperature increases, the volume increases. How does this law apply to the internal combustion engine?

in the ceiling. When electricity passes through them, they become warm. This warmth radiates into the room very similarly to the way heat radiates from a hot bar of steel. The electromagnetic waves emitted from the system warm objects in the room.

Electrical-Energy Converters

Life in the United States is closely linked with our electrical generation and distribution systems. Electricity is very important in our society. Without electricity, stores would close, food would spoil, and many homes would become dark and cold. Let us examine a typical electrical generation and distribution system.

Electricity Generation

Electricity generation uses the principles of electromechanical-energy conversion. In most commercial systems, water or steam is used to turn a turbine. A water-powered plant is called a ***hydroelectric generating plant.*** See **Figure 28-23.** This plant uses a dam to develop a water reservoir. The water is channeled through large pipes into the turbines in the generating plant.

A steam-powered electrical plant uses fossil fuels or nuclear energy to produce steam to drive the generator's turbines. Fossil fuel electrical plants burn coal, natural gas, or fuel oils to produce thermal energy. A nuclear plant uses atomic reactions to heat water in a primary system. The heated water is used to produce steam in a secondary system. The steam then drives the generator's turbine. See **Figure 28-24.** Keeping the loops separate prevents the water in the reactor from entering the steam turbines. This reduces the hazards for workers in the plant and for people living near the plant.

Steam and water turbines have a series of blades attached to a shaft. As the water

Standards for Technological Literacy

16

Extend

Discuss how a hydroelectric plant operates.

Research

Have your students identify and locate the major hydroelectric generation plants in the United States.

TechnoFact

The California Electric Light Company was the first company to produce and sell electric power to private customers. In 1882, the Edison Electric Illuminating Company began operating a steam-powered electric power plant on Pearl Street in New York City. In 1956, England built the first large-scale nuclear power plant. In 1966, the world's first tidal power plant was built in France.

U.S. Department of Energy

Figure 28-23. Hydroelectric power. Shasta Dam is in northern California. Note the water-delivery pipes that lead to the power plant in the lower-right portion of the left-hand picture. In the right-hand picture are the generators above the water turbines.

Standards for Technological Literacy

16

Demonstrate

In class or lab, show how an electric generator works.

Research

Have your students find out how a nuclear reactor in a power plant operates.

TechnoFact

The United States, Canada, Brazil, China, and Russia account for more than one-half of the world's hydroelectric power infrastructure. Among the world's nations, those in Europe and North America have the greatest development of their waterpower resources. The nations of Africa, Asia, and South America have the greatest potential for further development of hydroelectric power.

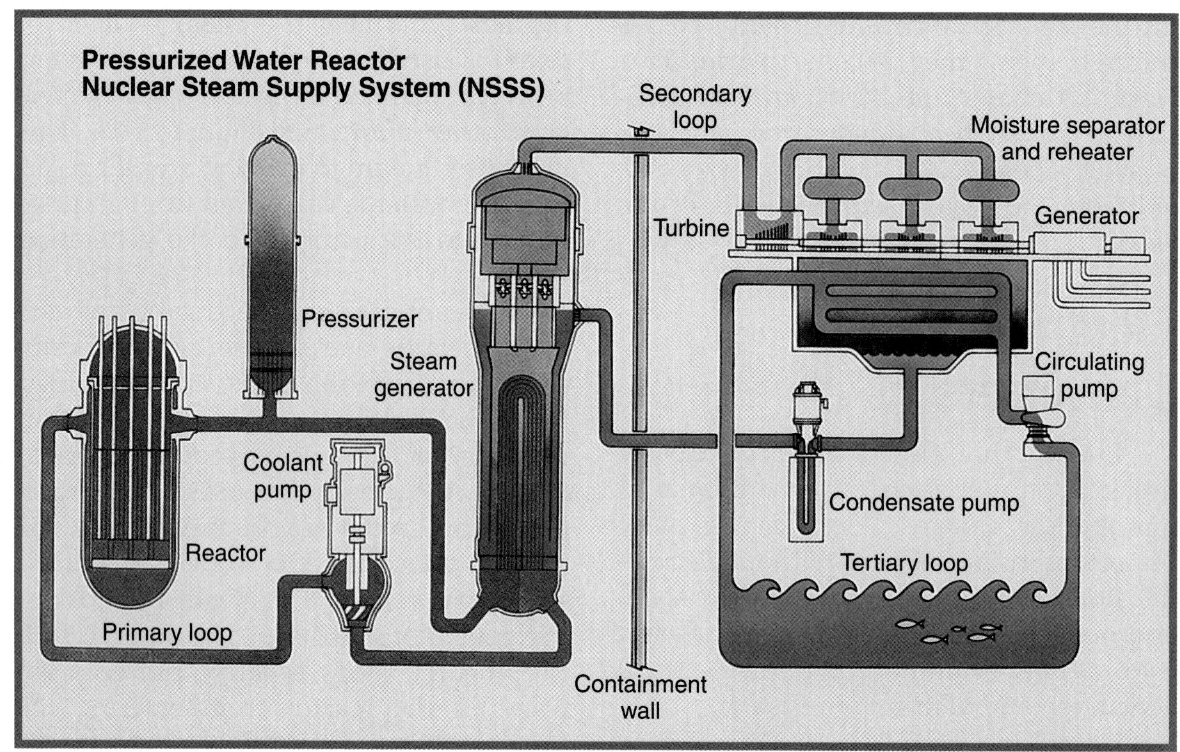

Westinghouse Electric Corp.

Figure 28-24. An NSSS. Notice how the water in the primary loop is used to heat the water in the secondary system. The contents of the two loops remain separate.

or steam strikes the turbine blades, the shaft turns. This shaft is attached to a generator. The generator changes the mechanical energy into electrical energy.

The electrical generator is the opposite of the electric motor described in Chapter 7. As we note in that chapter, two laws of physics are directly applied. The first law states that like poles of a magnet repel one another and unlike poles attract one another. The second law states that current flowing in a wire creates an electromagnetic field around the conductor.

Look at **Figure 28-25.** There are two magnets. Similar to a motor, the outside magnet of the generator is a stationary electromagnet called the *field magnet*. The inside magnet is a series of wires wound on a core. This part is called the *armature* and is able to rotate on its axis.

An electrical current is allowed to flow in the coils of the field magnet. This action

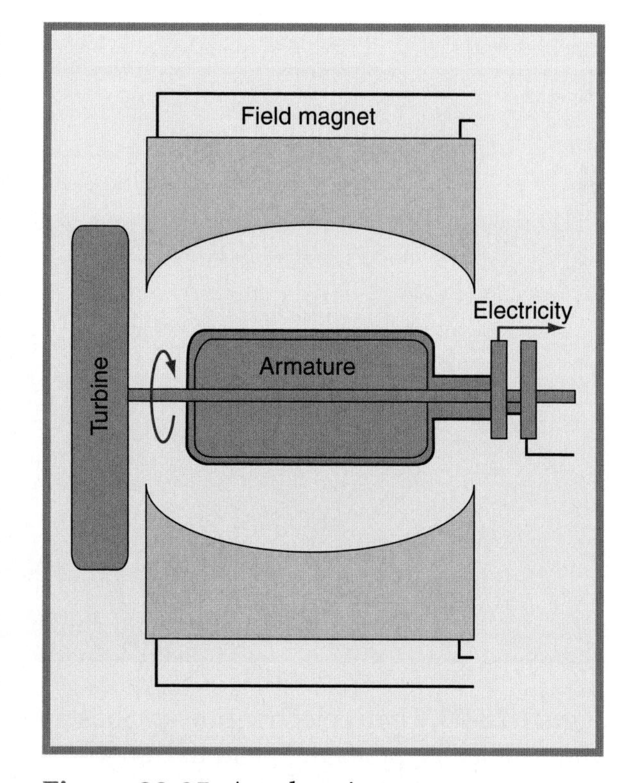

Figure 28-25. An electric generator.

produces an electromagnetic field around the field magnet that cuts through the armature. When the water or steam turbine spins the armature, the wires on the armature cut through the magnetic lines of force around the field magnet. This induces a current in the armature. The current is drawn off through commutators and fed into the distribution system.

Electricity Distribution

The electricity produced in the generating plant is passed through a step-up transformer. This transformer is called a *step-up transformer* because it steps up, or increases, the output voltage of the electrical current. The very high voltage reduces power losses in transmission.

Large transmission lines supported on tall steel towers usually carry the high-voltage electrical current to distant locations. See **Figure 28-26.** When the current reaches the area in which it will be used, the electricity flows through another transformer. The transformer reduces, or steps down, the voltage. This lower-voltage electrical current moves along the distribution lines.

Standards for Technological Literacy

16

Figure Discussion

Use Figure 28-26 to describe how electricity is delivered to a home. Discuss the voltage on various parts of the system.

TechnoFact

The term *power* can describe the source of energy used to operate machinery or to generate electricity (e.g. steam power). The term *power* is also used to describe the energy output of a machine or device.

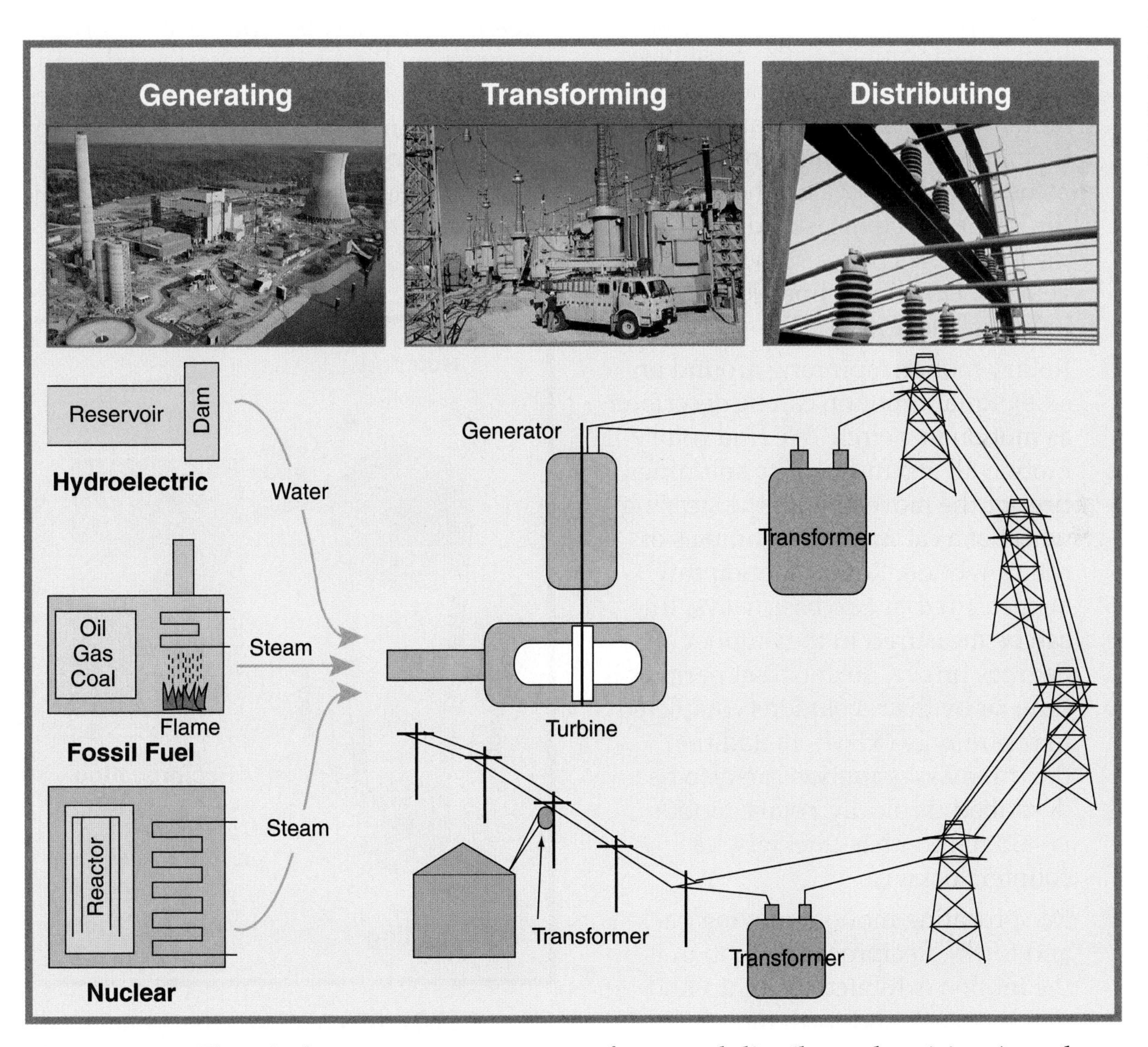

Figure 28-26. Electrical systems generate, transform, and distribute electricity. A nuclear power plant generates the electricity. A giant transformer station steps up and distributes electricity. High-voltage power lines take electric power over long distances.

Standards for Technological Literacy

16

Figure Discussion

Use Figure 28-27 to discuss the three types of motion.

Brainstorm

Ask the students to give examples of machines or devices that use linear, reciprocation, or rotary motion.

Just before it reaches its final destination, the electricity enters another step-down transformer. This transformer generally reduces the current to 110 and 220 volts for residential use. Some industrial applications use 440 or 880 volts.

Applying Energy to Do Work

Work involves moving a load. Therefore, motion is always present while work is being done. Three important types of motion need to be discussed:

- Linear motion (moving in a straight line). Linear motion is the most basic of all motion. Objects exhibiting linear motion will move in a straight line indefinitely. Linear motion can be measured in two ways. The speed and direction can be measured. These two measurements make up what is called velocity. Drawing a line along a ruler involves linear motion.
- Rotary motion (spinning around an axis). Rotary motion can be described as motion in a circle. We find rotary motion all around us. The spinning tires or the movement of the steering wheel of a car are both examples of rotary motion. Rotary motion can be measured in two basic ways. It can be measured in the number of degrees turned during a set period of time, or by the revolutions completed in one minute (RPM). In addition, the direction of motion can also be described. Typically, rotary motion is said to be either clockwise or counterclockwise.
- Reciprocating motion (moving back and forth). Reciprocating motion is the motion exhibited by an up-and-down or back-and-forth movement of an object. Reciprocating motion is usually repeated over and over again. The movement of a piston of an internal combustion engine is an example of reciprocating motion. There are two ways to measure reciprocating motion. Throw describes the distance between the two extremes of the motion. The period is the length of time required for each cycle.

To review the three types of motion, refer to Figure 7-12. The circular saw in the illustration uses rotating motion, the band saw uses linear motion, and the scroll saw uses reciprocating motion.

A main activity in energy conversion involves changing the type or direction of a load's motion. This action is called *power transmission*. Power transmission takes the energy a converter generates and changes it into motion. Look at the example shown in **Figure 28-27.** The figure shows the reciprocating motion of a piston in the cylinder of an internal combustion engine.

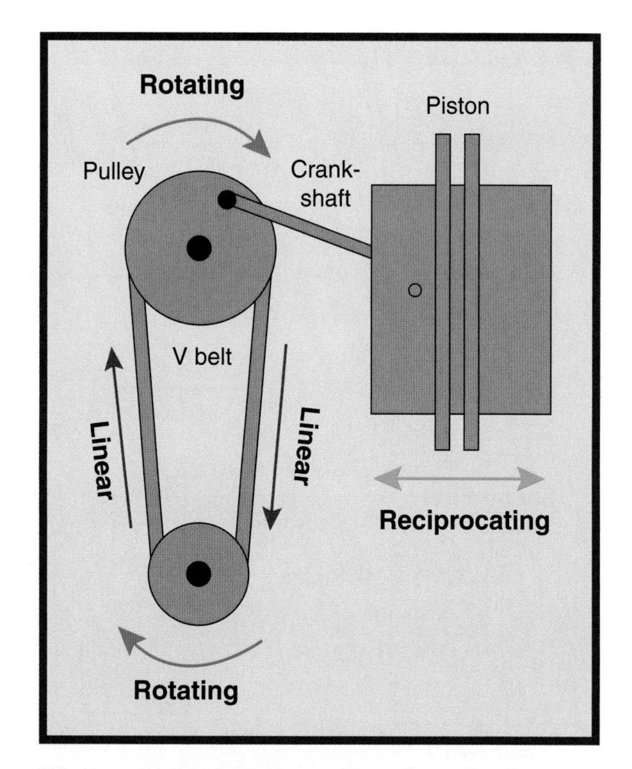

Figure 28-27. Mechanisms that change the type and direction of motion. This system can change rotating motion into reciprocating motion, or vice versa.

The crankshaft changes this motion to rotary motion. The end of the crankshaft is attached to a pulley. The pulley drives the V belt. The belt travels in a linear motion around a second pulley. The second pulley changes the linear movement back into rotary motion. Also, the two pulleys rotate in opposite directions. Applying motion to perform work often requires changing both the type of motion and the motion's direction. Two basic types of systems are used to change the type, direction, or speed of a force. These are mechanical-power and fluid-power (or fluidic) systems.

Mechanical-Power Systems

Mechanical systems use moving parts to transfer motion. See **Figure 28-28.** This is the oldest method of transferring energy. Various mechanical methods are used in technological devices. **Figure 28-29** shows six common techniques:

- **Levers.** This technique is a device that changes the direction or intensity of a linear force. A downward force can be applied to one end of a lever. This causes the lever arm to pivot on its fulcrum. The opposite end moves in an upward direction. Remember from Chapter 4 that the location of the fulcrum determines whether the device multiplies the amount of the output force or the distance it moves. Many door-handle mechanisms in automobiles transfer motion with levers.
- ***Cranks.*** A pivot pin near the outside edge of a wheel or disk changes reciprocating motion into rotating motion. The diameter of the swing of the crank determines whether the amount of the force or the distance of the force is multiplied. An internal combustion engine transfers power from the piston to the transmission using this type of drive.

American Electric Power

Figure 28-28. This crane uses a mechanical means to lift a load.

- ***Gears.*** Two or more wheels with teeth on their circumferences change the direction of a rotating force. The relative diameters of the input and output gears determine whether the system is a force multiplier (the output gear rotates faster) or a distance multiplier (the output gear turns over a greater area). If a smaller input gear is used, the unit increases the output force and reduces its speed. If a larger input gear is used, the unit decreases the output force and increases its speed. Some automobile transmissions use this type of power-transmission system.
- ***Cams.*** This technique is a pear-shaped disk with an off-center pivot point used to change rotating motion into reciprocating motion.

Standards for Technological Literacy

16

TechnoFact

One of the first energy converters to use rotary motion was the waterwheel. The first waterwheel was known as the Norse wheel and was developed in the Middle East. It had a vertical shaft, the top of which was connected to a millstone. The lower end of the shaft has a series of vanes, or paddles, that were driven by the flowing stream.

Standards for Technological Literacy

16

Figure Discussion

Use Figure 28-29 to discuss mechanical devices that change the type of motion in a device.

Brainstorm

Ask the students to give examples of the use of levers, cams, gears, cranks, gears and racks, and pulleys and V-belts in everyday devices.

TechnoFact

Hydraulic systems are widely used in automobiles. Most modern automobiles use hydraulic brake systems, which pressurize liquid (brake fluid) to increase the braking force applied to the pedal and transmit it to the wheel brakes.

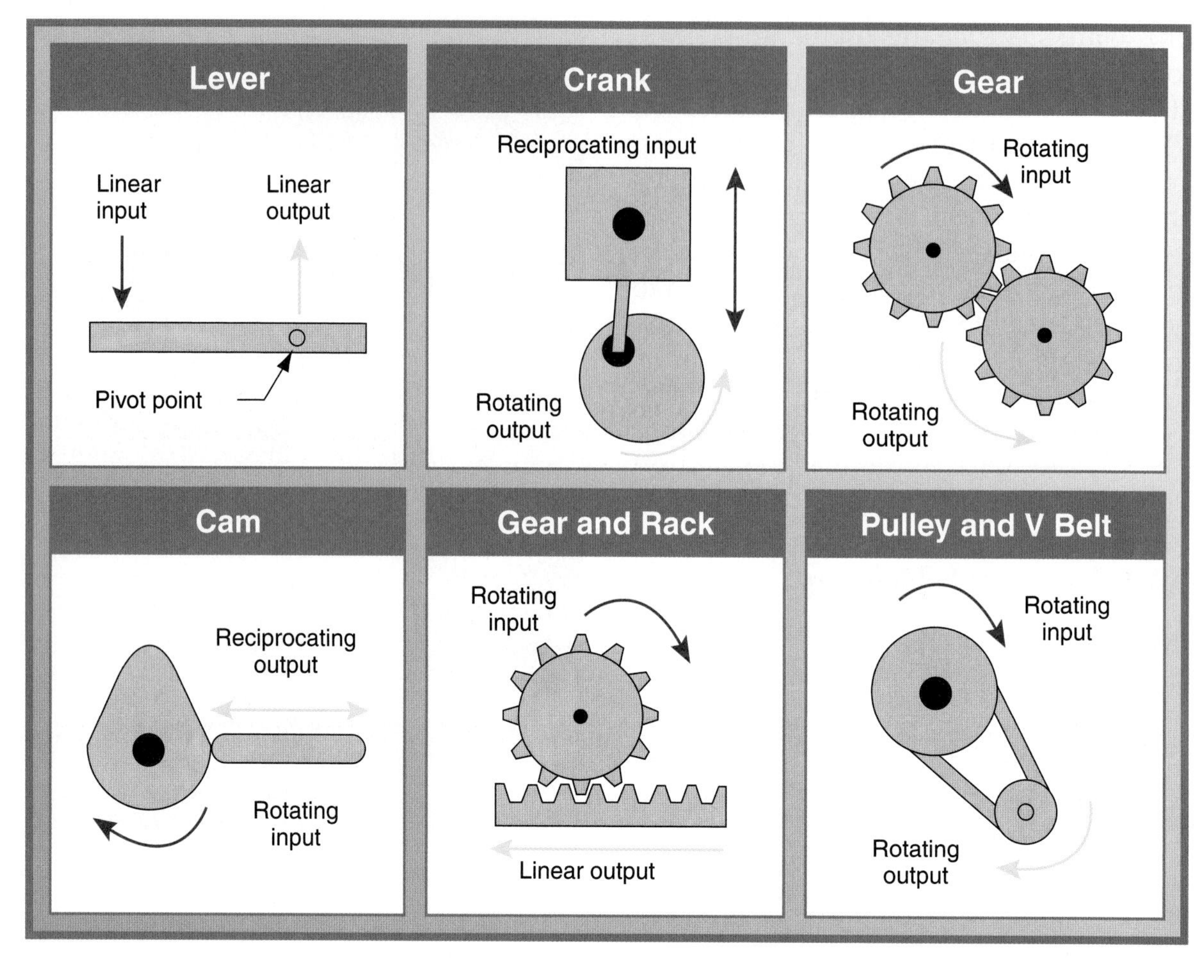

Figure 28-29. There are six important mechanical techniques that change the type, direction, or speed of a moving force.

A cam with a large lobe (an extended portion) creates longer strokes for the reciprocating member. The force is reduced, however. The valves in an internal combustion engine are opened using a cam system.

- ***Gears and racks.*** A rotating gear meshes with a bar that has gear teeth along its length (rack) and changes rotating motion into linear motion. As the gear turns, it slides the rack forward or backward. Rack-and-pinion (gear) steering for automobiles uses this system.
- ***Pulleys and V belts.*** Two pulleys with a V belt stretched between them change the speed or power of a motion. As one pulley turns, the V belt moves. This movement, in turn, rotates the second pulley. The two pulleys rotate in opposite directions. If the force is applied to a larger pulley, the smaller pulley turns faster. The smaller pulley, however, has less power. If the power is applied to the smaller-diameter pulley, the larger-diameter pulley turns slower. The larger-diameter pulley, however, turns over a greater area. Pulleys and V belts drive many machines. Under heavy loads, the V belt can slip as it drives the pulleys. The problem is overcome by using gears and a chain drive similar to the one used on bicycles.

Fluid-Power Systems

Fluid-power systems use either liquids or gases to transfer power from one place to another. Systems using air as the transfer medium are called ***pneumatic systems***. Liquids (usually oil) are used in ***hydraulic systems***. See **Figure 28-30.** Generally, these systems contain two cylinders with movable pistons, a pump, valves to control the flow, and piping to connect the components. **Figure 28-31** shows the three basic uses for hydraulic systems.

The view on the left shows a typical power-transfer system. This system increases neither the distance, nor the intensity, of the force. The power-transfer system simply moves the force from one location to another. This system changes the force's direction. The piston in the left cylinder forces the fluid downward. This action causes the fluid to flow into the right cylinder. The fluid's movement causes the right piston to move upward.

The center drawing shows a distance multiplier. The left piston is forced downward 1″. The motion displaces three cubic inches of fluid because the piston has an area of three square inches. This causes three cubic inches of fluid to flow into the right cylinder. The right piston's area is only one square inch. Therefore, the piston must move upward 3″ to accommodate the three cubic inches of fluid.

The drawing on the right shows a force-multiplying system. A force of 50 lbs. is applied to the left piston. This piston has an area of one square inch. Thus, the original force is 50 psi on the fluid beneath the piston. This force is transferred to the right piston. The right piston has a three-square-inch area. The 50-psi force from the piston on the left exerts a 150-psi upward force on the piston on the right.

Pneumatic systems operate in a similar way. The force and distance movement calculations are more difficult, however.

Figure 28-30. Hydraulic cylinders are used to activate the loading boom of a truck. A pneumatic cylinder is used in the nailer.

Standards for Technological Literacy

16

Research

Have your students investigate how an automotive brake system works.

Demonstrate

During lab or class, use two syringes and some plastic tubing to demonstrate hydraulic and pneumatic systems.

TechnoFact

When hydraulics were integrated into elevator designs, it allowed for taller buildings. Both hydraulic and steam-powered elevators were invented in the 1840s. These early elevators were less than ideal; hydraulic elevators were slow and the ropes used to hoist steam-powered elevators often snapped, sending the cars plummeting. Elisha Otis invented the safety brake for elevators in 1854.

TechnoFact

The first water clocks, called clepsydra, were invented more than 4000 years ago. They were very similar in function to an hourglass. The early water clock was no more than a container of water with a small opening in the bottom through which water seeped. As the water level dropped, it exposed incremental marks on the walls of the container. These marks indicated the time that had passed since the container was full.

Standards for Technological Literacy

16

Figure Discussion

Use Figure 28-31 to discuss using cylinders as direction changing and distance multipliers. Contrast this with using levers for the same tasks.

TechnoFact

A person living in a developing (third-world) country uses only 6%–7% of the energy used by a person in a developed nation. In third world countries, the energy is used primarily to satisfy basic human needs. Developed nations generate much more energy than they need to fulfill their basic needs. Much of this additional energy is applied to nonessential activities or is simply wasted.

Liquids do not compress, so nearly all the force is transferred from one cylinder to the other. A slight amount of the force is used to overcome the friction of the piston and the fluid in the pipes.

Air can be compressed, however. Therefore, some of the force in pneumatic systems is used in reducing the volume of the air in the system. The remainder of the force is applied to moving a load.

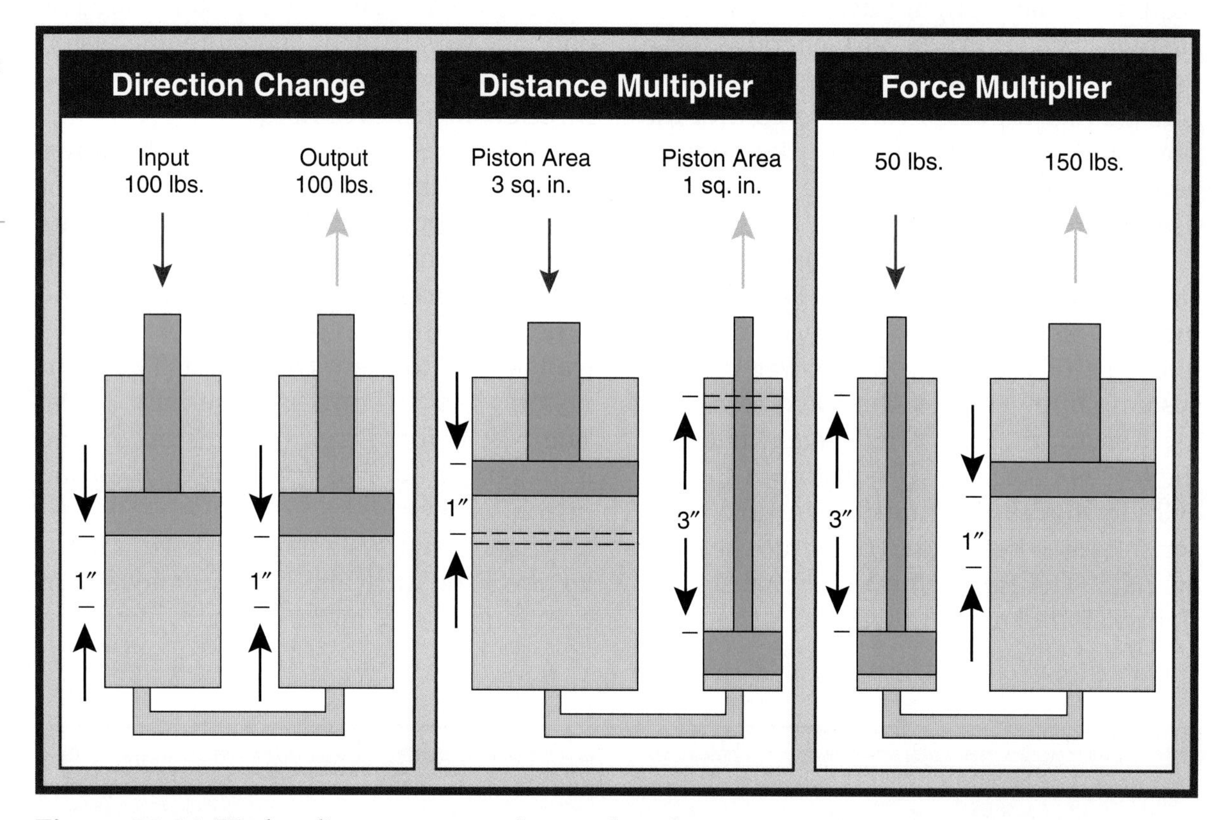

Figure 28-31. Hydraulic systems can be used in three ways. They can change the direction of a force. These systems can increase the distance of the force. They can also increase the strength of the force.

The biofuel ethanol is made by a biochemical process called fermentation. In this process, corn is decomposed and becomes ethyl alcohol, or ethanol, and may be added to gasoline as fuel.

Answers to Test Your Knowledge Questions

1. True
2. Evaluate individually. Common answers include windmills and sailboats.
3. Passive collectors directly collect, store, and distribute the heat they convert from solar energy. Active collectors use pumps to circulate the water that collects, stores, and distributes the heat they convert from solar energy.
4. Trombe wall
5. Evaluate individually.
6. Student answers may include either ocean thermal energy conversion systems or ocean mechanical energy conversion systems.
7. biochemical
8. heat engine
9. Evaluate individually.
10. Student answers may include by burning fuels, capturing heat from the surroundings, or converting electrical energy.
11. mechanical, electrical
12. True
13. fluid power

Summary

We spend much time and effort converting energy into different forms. The four major categories of energy-conversion systems are inexhaustible-energy converters, renewable-energy converters, thermal-energy converters, and electric-energy converters. Such devices as windmills, waterwheels, and solar collectors convert the inexhaustible energy sources of wind, water, and sunlight, respectively, into other forms. We use thermochemical and biochemical conversion systems to change renewable energy sources. Two major sources of thermal-energy conversion are fuel converters and heat pumps. Finally, electricity produced in a generating plant is passed through transformers that increase and decrease output voltage.

Energy is applied to do work. Motion is always present in work. One activity of energy conversion involves changing motion. This action is called *power transmission*. The two basic systems used to change motion are mechanical-power and fluid-power systems.

Test Your Knowledge

Write your answers on a separate piece of paper. Please do not write in this book.

1. An energy converter has energy as both its input and its output. True or false?
2. Name one device earlier people created that used wind energy to fulfill a need.
3. What is one difference between passive and active solar collectors?
4. A black concrete wall behind a glass wall that collects solar energy is called a(n) ______.
5. Name one geothermal energy–conversion system.
6. Name one of the two types of ocean energy–conversion systems.
7. Biomass conversion uses two processes. These processes are thermochemical and ______.
8. A device that burns fuel to produce heat that is converted into mechanical energy is called a(n) ______.
9. Name one difference between internal and external combustion engines.
10. Name one way thermal energy is used to heat buildings and homes.
11. A turbine is an electrical generation plant that converts ______ energy into ______ energy.
12. A steam-powered electrical plant might use fossil fuels to produce steam. True or false?
13. The two basic systems used to change motion are mechanical-power systems and ______ systems.

STEM Applications

1. Design and construct a passive solar device that heats the air in a shoe box. Prepare a sketch of the device and an explanation of how it works.
2. Select a major energy-conversion system, such as a coal-powered, electric generating plant. Explain the advantages and disadvantages of the conversion system. Next, list at least two energy converters that can be used to replace the system you selected. List their advantages and disadvantages.

Activity 8A

Design Problem

Background

All technological devices use energy. Often, this energy is converted from one form to another. For example, a motor changes electrical energy into mechanical energy. In addition, the motion an energy converter produces is often transformed. For example, a crankshaft changes the reciprocating motion of an engine's piston to rotating motion.

Situation

You are employed as a designer for a children's technology museum. The curator wants a series of models showing how humans change types of motion. She wants models demonstrating how the rotating motion a vertical waterwheel produces is changed into reciprocating motion for a colonial sawmill or pounding motion for a forge. See **Figure 8A-1.**

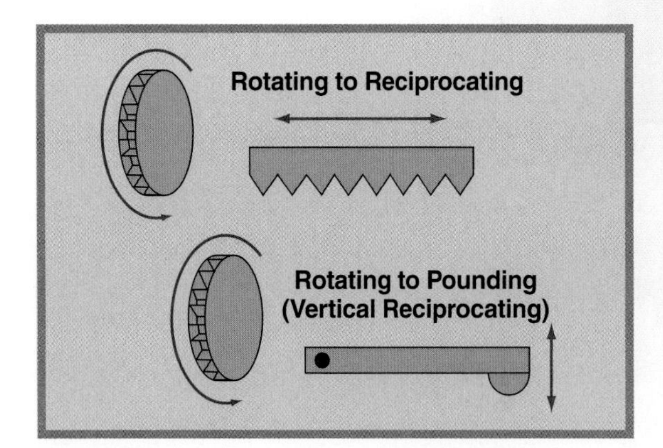

Figure 8A-1. Types of motion to be simulated.

Challenge

Select one of the two models listed above. Design and build a working model, using wood, cardboard, or other easily worked materials. See **Figure 8A-2.** This model converts vertical rotating motion into horizontal rotating motion.

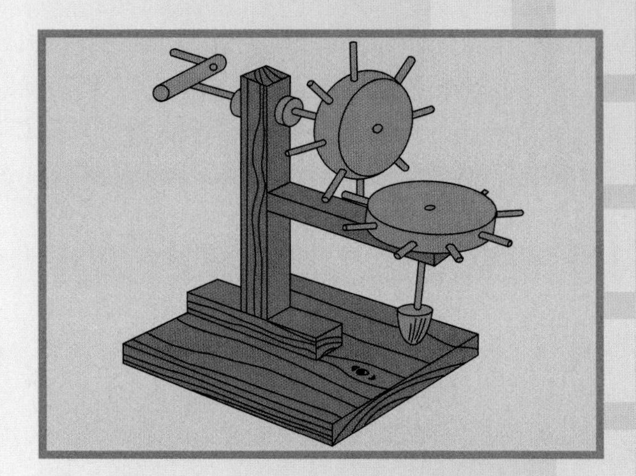

Figure 8A-2. A model changing vertical rotating motion into horizontal rotating motion.

Optional

Prepare a display, using your model, explaining your model's function and history.

Activity 8B

Fabrication Problem

Background

It is commonly accepted that deposits of fossil fuels (petroleum, natural gas, and coal) will be exhausted in the not-too-distant future. Therefore, finding ways to use inexhaustible energy sources is a major challenge for the future. An important inexhaustible energy source is wind.

Challenge

To understand wind power more fully, you are to build and test a working model of a wind-powered electricity-generation system. Use the drawings and following procedure to complete this challenge.

Materials and Equipment

- One 3/4″ × 5 1/2″ × 8″ wood base (pine, fir, or spruce).
- One 3/4″ × 18″ strip of thin-gage sheet metal.
- Two 1/2″ No. 6 sheet metal screws.
- One small electric motor.
- One multimeter.
- 24″ of electrical-hookup wire.
- One three-blade or four-blade prop.

Procedure

Building the prototype

1. Obtain a piece of wood with dimensions of 3/4″ × 5 1/2″ × 8″.
2. Cut a strip of sheet metal 3/4″ wide and 18″ long.

Safety
Follow all safety rules your teacher demonstrates!

3. Drill a 1/8″-diameter hole 1/2″ from each end of the strip of sheet metal.
4. Form the sheet metal strip around the motor. See **Figure 8B-1.** This forms the motor mount.

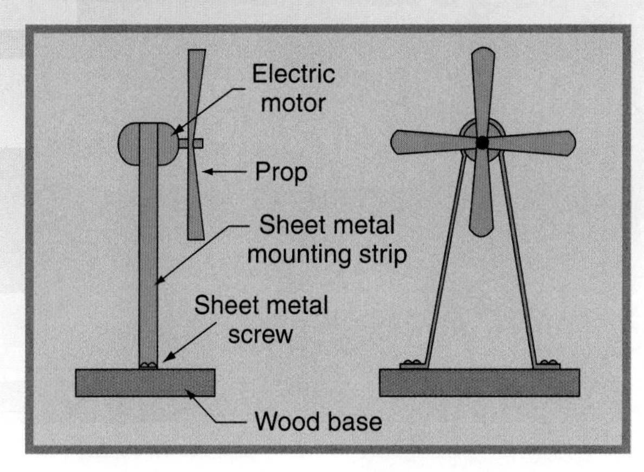

Figure 8B-1. A model windmill.

5. Bend the sheet metal strip to form 1″-long tabs that can be mounted to the wood base.
6. Use the two sheet metal screws to attach the motor mount to the base.
7. Attach the prop to the motor.
8. Press the motor into the loop of the motor mount.
9. Attach a piece of hookup wire to each wire coming out of the motor.
10. Solder your joints. Wrap them with electrical tape.
11. Set the multimeter to read current.
12. Attach the wires to the multimeter. See **Figure 8B-2.**

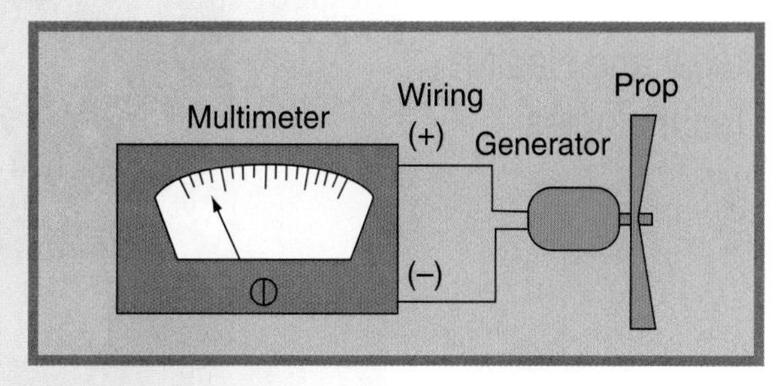

Figure 8B-2. A wiring schematic for testing the windmill operation.

Testing the device

1. Place the wind-powered generator in front of a window fan.
2. Turn the fan on to its low setting.
3. Observe the rotation of the prop.
4. Note and record the meter reading on the multimeter.
5. Repeat steps 3 and 4 using the fan's medium setting.
6. Repeat steps 3 and 4 for the fan's high setting.
7. Turn off the fan.

Optional

1. Change the prop to a model with a different number of blades or change the angle at which the wind is hitting the propeller to 30° off center.
2. Repeat steps 1 through 4 in the "Testing the device" section.
3. Write a report explaining your observations.

Section 9
Applying Technology: Meeting Needs through Biorelated Technologies

Tomorrow's Technology Today

Genetically Modified (GM) Foods

If you have ever spoken with a farmer, you are probably aware that farmers face many obstacles in producing the world's food supply. Besides dealing with insects and other pests, weeds, the weather, and rotting food, the world's population is increasing at a higher rate than our current food-production capabilities. This increase is placing additional burdens on our food producers. Fortunately, the latest research in genetic engineering suggests that we might soon have the answer to all these problems. Genetically modified (GM) foods might be able to put these concerns in the past and offer even more exciting benefits.

GM foods can be created by altering the genetic makeup of living organisms to produce organisms with desired characteristics. Specific genes can be inserted into a plant. In other cases, the genes from two or more different species can be combined to produce a transgenic organism. Using these techniques, scientists have already been successful in creating plants that produce human insulin, crops resistant to herbicides, and plants unaffected by insects and fungus. They have also successfully engineered vegetables that can remain on the vine longer to ripen, such as the Flavr Savr® tomato. This plant was the first commercially grown GM food to receive a license for human consumption. The tomatoes were approved for use in 1992 and were grown for a few years before being discontinued. They have less of a ripening gene. The tomatoes are firmer and tastier when they reach the grocery store.

Besides these benefits, GM foods have the potential to keep people healthier. For example, almost half of the world's population depends on eating rice daily for most of its caloric intake. This kind of diet can cause deficiencies in vitamin A. These deficiencies are already a serious health problem in many countries. Some GM foods can have extra genes inserted. These genes may increase the synthesis of beta-carotene, a producer of vitamin A. These types of GM foods are being developed for use in countries where the population has high levels of a vitamin A deficiency. These GM foods are still being developed and tested. If they are deemed acceptable, however, they might greatly improve the health of much of the world's population.

Further down the road, there are even more exciting possible applications for GM foods. Research is currently being done on ways to introduce other vitamins into foods. If this is successful, you might never need to take a multivitamin in pill form again. There also is the possibility that vaccines might someday be inserted into foods. This would allow you to eat a GM banana, instead of going to the doctor and getting a shot!

All this sounds promising. Great debate also exists, however, over the use of GM foods. Many scientists are concerned about the accidental creation of "superweeds," weeds resistant to herbicides, and inadvertent harm that might be done to other wildlife. Economists worry that GM foods might hurt traditional farmers and increase the gap between wealthy and poor countries. Many consumers are concerned about the safety of consuming artificially modified foods, especially when these foods are not labeled as such. All these concerns are being studied. The research and development, however, continue. Obviously, we need to consider the benefits and risks regarding the use of GM foods. There is no doubt, however, that this emerging technology holds exciting promise.

Discussion Starters

Briefly discuss the two primary methods of genetically modifying plants, the bacterial method and the particle gun method.

Group Activity

Divide the class into two groups. Have one group brainstorm possible benefits of genetically modified foods. Have the second group brainstorm the possible dangers of genetically modified foods.

Writing Assignment

Some countries, including the United States, do not require the labeling of genetically modified foods. Proponents of labeling suggest that it is the only way for people to be able to make an informed choice about the foods that they eat. Opponents of labeling frequently argue that it is unnecessary and would impede progress in a field of technology that offers nearly unlimited promise. Have your students write a persuasive letter either supporting or arguing against mandatory GM food labeling.

Chapter Outline
Types of Agriculture
Agriculture and Biotechnology

This chapter covers the benchmark topics for the following Standards for Technological Literacy:

3 5 7 15

Learning Objectives

After studying this chapter, you will be able to do the following:

- Recall the definition of *agriculture.*
- Compare science and technology in agriculture.
- Recall the major types of agriculture.
- Recall major crops grown on farms.
- Recall major equipment used in crop production.
- Explain how technology is used in hydroponics.
- Recall the types of livestock farms.
- Summarize the characteristics of the technology used in raising livestock.
- Explain how technology is used in aquaculture.
- Recall the definition of *biotechnology.*
- Summarize ways biotechnology can be used in agriculture.

Key Terms

agriculture
agricultural technology
animal husbandry
antibiotic
aquaculture
baler
barbed wire fence
berry
biotechnology
cable-wire fence
catalyst
combine
crop production
cultivator
drip irrigation
electric fence
fertilizer
flood irrigation
forage crop
fruit
furrow irrigation
gene splicing
genetic engineering
grain
grain drill
hydroponics
irrigation
nonfood crop
nut
pivot sprinkler
plow
rail fence
sprinkler irrigation
swather
tillage
vegetable
windrow
woven-wire fence

Strategic Reading

As you read this chapter, make a list of current agricultural trends, and relate them to how they were different before technological advances were made.

People have a number of basic needs and wants. Two of the primary needs are food and clothing. Agriculture directly addresses both of these needs. This human activity involves using science and technology to grow crops and raise livestock. ***Agriculture*** is people using materials, information, and machines to produce food and natural fibers. This activity takes place on farms and ranches around the world.

Modern farming uses both scientific and technological knowledge. Two important branches of the life sciences support agriculture. These areas of the life sciences are plant science and animal science. They are used in cross-pollinating plants to improve crops and in crossbreeding animals to improve livestock. Other sciences help farms manage their operations. For example, the science of weather allows farmers to plan planting and harvesting activities. The sciences describing the seasons allow farmers to select appropriate crops and livestock to raise. Knowledge of plant nutrients allows farmers to select appropriate fertilizers. These and other sciences have helped farmers become more efficient.

Technological advancements have also caused many massive changes in farming. See **Figure 29-1.** For example, they have caused changes in the size of farms and

Figure 29-1. Technological advancements have greatly changed agriculture.

Standards for Technological Literacy

15

Reinforce

Differentiate between scientific and technological knowledge (Chapter 1).

TechnoFact

For hundreds of thousands of years, people lived as simple hunters/gatherers. Their way of life consisted of catching fish, gathering wild plants, and hunting game. The discovery that some animals could be tamed and that plants grow from seeds 10,000 years ago gave way to the start of agriculture.

TechnoFact

The likely origin of animal domestication is the Middle East about 10,000 years ago. The first farming communities were established in modern day Jordan and Israel at that time.

Standards for Technological Literacy

15

Extend

Discuss the differences between agricultural technology and agricultural science.

Research

Have your students determine which crops and animal products are common in your state.

TechnoFact

Crop rotation is the oldest way of preserving soil for crops. By alternating the types of crops grown in a field every year, the minerals used by the previous year's crop are replaced. The crop rotation system has been gradually replaced by commercial fertilizers.

how farmers go about their work. New and modern machines and equipment allow for greater production. Fewer people are needed to grow more food on fewer acres. Technological advances have helped people work with greater ease and efficiency. These advancements can be attributed to ***agricultural technology***. This technology uses technical means (machines and equipment) to help plant, grow, and harvest crops and raise livestock.

Types of Agriculture

Agriculture involves people managing land, buildings and machinery, and crops and livestock. In doing this, people engage in two major types of agriculture. See **Figure 29-2.**

The first is ***crop production***, which grows plants for various uses. Crop production provides food for humans, feed for animals, and natural fibers for many applications. This type of agriculture produces trees and plants for ornamental use, such as landscaping. Crop production grows trees for lumber and paper production. This type of agriculture produces basic ingredients for medicines and health-care products. Crop production also provides materials for many industrial processes, such as textile weaving, plywood manufacture, and food processing.

The second type of agriculture is called ***animal husbandry***. This area involves breeding, raising, and training animals. These animals might be used for food and fiber for humans. In some cases, they are used to do physical work. This is especially true in developing countries. Many animals are also raised as hobbies or are used for riding and racing.

The United States uses more than 40% of its total land for farming activities. A total of about 350,000,000 acres is used for crop production. About half of this land is used for raising wheat and corn. Another billion acres is used for pastures, ranges, and forests.

Crop Production

Many different crops are raised on North American farms and ranches. Some of these crops were originally found in other areas of the world. For example, China and central Asia gave us lettuce, onions, peas, sugarcane, and soybeans. Rice, sugarcane,

©iStockphoto.com/nkurtzman, ©iStockphoto.com/LUGO

Figure 29-2. Agriculture involves growing crops and raising livestock.

bananas, and citrus fruits came from Asia. The Middle East, southern Europe, and North Africa gave us wheat, barley, and oats, along with alfalfa and sugar beets. Corn, beans, tomatoes, potatoes, peanuts, tobacco, and sunflowers came from North and South America.

A number of major crops are grown in North America. The most widely grown are forms of ***grain***. See **Figure 29-3.** Grain crops are members of the grass family that have large edible seeds. The commonly grown grains are wheat, rice, corn, barley, oats, rye, and sorghum. All these grains can be used in food products. Corn, barley, oats, and sorghum are widely used in animal feed.

Vegetables are another important farm crop. They have edible leaves, stems, roots, and seeds that provide important vitamins and minerals for the daily diet. Vegetables include root crops, such as beets, carrots, radishes, and potatoes. They also include leaf crops, such as lettuce, spinach, and celery. Other vegetables provide food from their fruit and seeds. This group includes sweet corn, peas, beans, melons, squash, and tomatoes.

Fruits and ***berries*** are grown in many parts of the country. See **Figure 29-4.** These plants are cultivated for their edible parts. The major fruit crops grown in temperate climates include apples, peaches, pears, plums, and cherries. Citrus fruits (oranges, lemons, limes, grapefruits, and tangerines), olives, and figs are grown in warmer climates. Tropical ***fruits*** include bananas, dates, and pineapples.

Nuts are grown in selected parts of the country. They are grown for their hard-shelled seeds. Walnuts, pecans, chestnuts, almonds, and filberts (hazelnuts) are grown in temperate climates. Palm oil nuts and coconuts are grown in more tropical areas. Peanuts and coconuts are the most important nut crops and are significant sources of food and oil.

Forage crops are grown for animal feed. See **Figure 29-5.** These plants include hay crops, such as alfalfa and clover. Grasses used for pasture and hay are also included in this group.

A number of ***nonfood crops*** are grown on farms. These plants include tobacco, cotton, and rubber. Nonfood crops also include nursery stock grown for landscape use and Christmas trees.

Standards for Technological Literacy

15

Research

Have your students select a grain or other crop and trace its historical origins.

Research

Have your students select a crop and find out how it is planted, grown, and harvested.

TechnoFact

Two branches of farming technology spread through Europe when it was first developed. While one form modified itself for temperate zones, the other was closer to the sea. The branch that moved through Greece developed more quickly than farming that was done in Italy.

U.S. Department of Agriculture

Figure 29-3. Grain, such as this corn, is grown widely for human and animal food.

U.S. Department of Agriculture

Figure 29-4. Berries and fruit are important crops grown on American farms.

Standards for Technological Literacy

15

Research

Have your students research when and where the tractor was developed.

TechnoFact

Wheat developed over thousands of centuries from a wild grain. Today, it is used to make most bread and pasta. Due to mutations that occurred about 10,000 years ago, a hybrid called emmer evolved. Shortly afterward, another mutation brought about a similar grain to wheat.

Case IH

Figure 29-5. Forage crops provide animal feed.

Technology in Crop Production

Crops are no different from any other living thing. They have set life cycles. Crops are "born" when seeds germinate, continue through growing cycles, and then mature. After a time, they die. To be of benefit to people and animals, the crops must be harvested before they spoil or shatter. Farming takes advantage of a plant's life cycle through four major processes. These processes are planting, growing, harvesting, and in some cases, storing.

In the past, farming was a very labor-intensive activity. Most of the population was involved in raising crops and animals. Technological advancements during the past 200 years, however, have changed all this in many countries. Now, just a small percentage of the population in developed countries is involved in agriculture. Farming has become very equipment intensive. To a large extent, machines and equipment have replaced human and animal labor. Farm equipment is used at all stages of crop production. This equipment can be classified into eight major groups:

- Power or pulling equipment.
- Tillage equipment.
- Planting equipment.
- Pest-control equipment.
- Irrigation equipment.
- Harvesting equipment.
- Transportation equipment.
- Storage equipment.

Career Corner

Agricultural Workers

Agricultural workers are essential to growing and harvesting grains, fruits, vegetables, nuts, fiber, trees, shrubs, and other crops. They generally work on farms, on ranches, or in nurseries. These workers do many different jobs requiring unique knowledge and skills they acquire through on-the-job training. Most farmworkers receive low pay and often must perform strenuous work outdoors in all kinds of weather.

Power or pulling equipment

People have a long history of replacing human power with other power sources. Over most of recorded history, people used animals to pull loads. The invention of the agriculture tractor in 1890 changed this, however. During the 1900s, this new power source replaced animal power on most farms. Today, the farm tractor provides the power to pull all types of farm equipment.

Farm tractors can be either wheel tractors or track machines. See **Figure 29-6.** Most agricultural tractors are the wheel type. Some wheel tractors have rear, power wheels. These tractors usually have smaller front wheels. Other wheel tractors have power to all wheels. These tractors generally have the same-size tires on all wheels. For additional traction, both types of wheel tractors might have dual drive wheels.

Track-type tractors are used for special purposes and are generally slower than wheel-type tractors. They are more suited for muddy fields. These tractors sink less and have less slippage than wheel-type tractors. Also, they compact the soil less and do not produce wheel ruts.

Most tractors are designed on a unit principle. The engine, transmission, and gearbox are a single unit. This structure provides a rigid backbone for the machine. The steering and drive wheels are attached to the basic unit. Implements are attached to it or pulled behind the tractor.

Tillage equipment

The soil must be prepared before crops can be planted. Residue from previous crops must be handled. The seedbed must be conditioned by breaking and pulverizing the soil. This process is called ***tillage***, or tilling the soil.

The most important piece of tilling equipment is the ***plow***. This piece of equipment performs the same task as the spade used in flower gardens. The plow breaks, raises, and turns the soil. This process loosens the ground and brings new soil into contact with the atmosphere.

Three major types of plows exist: the moldboard plow, the disc plow, and the chisel plow. The moldboard plow is made up of a frame and several plowshares. See **Figure 29-7.** When the plow is pulled through the earth, it cuts and rolls the soil.

Standards for Technological Literacy

15

Brainstorm

Have your students list gardening activities that could be considered tillage.

TechnoFact

The earliest tractors were driven by steam engines and were known as tractor engines. Modern tractors are driven by gasoline or diesel engines. Some of the early advancements in the tractor included the 1918 introduction of power takeoff systems, which transfer part of the engine's power to a shaft for driving implements, and the addition of rubber tires in 1932. More recent advances include four-wheel drive designs and enclosed, air-conditioned cabs.

Case IH, Deere and Co.

Figure 29-6. The two kinds of farm tractors are wheel and track tractors.

Standards for Technological Literacy

15

Research

Have your students research the common ingredients of fertilizer and the proportions at which they are normally used.

TechnoFact

The early plow consisted of an iron-tipped wooden wedge attached to a single handle. It was pulled by men or draft animals and dug into the soil, breaking it up but not turning it. During first half of the 18th century, the plow was made more efficient by adding a moldboard, a curved board that turned the soil over as it was plowed. In 1837, John Deere invented a steel plow that could easily turn the thick soil that covered much of the American Midwest.

Case IH

Figure 29-7. This moldboard plow is cutting and rolling the soil.

Disc plows have a frame that has several discs mounted on an axle. See **Figure 29-8.** The discs and axles are set at a steep angle to the direction of travel. When pulled, the discs turn and then cut and loosen the soil.

Chisel plows have a set of shaped chisels attached to a frame. When the plow is pulled through the earth, it breaks up the soil. The plow does not lift and turn the soil, however. Chisel plows are used in grain stubble and where the soil needs little tilling.

Planting equipment

Once the soil is prepared, the crop can be planted. Planting involves two actions that can be done separately or together. The first action involves applying fertilizer. ***Fertilizer*** is a liquid, powder, or pellet containing important chemicals. This liquid, powder, or pellet primarily delivers nitrogen, phosphorus, and potassium, which encourage and support plant growth. The fertilizer can be applied before, during, or after seeds are planted.

This liquid, powder, or pellet is applied with special equipment or along with a seed planter. In some cases, fertilizer is scattered (broadcast) on the ground before final tilling and planting. A machine with a series of knives can inject liquid and gaseous (anhydrous ammonia) fertilizers into the soil.

Deere and Co., U.S. Department of Agriculture

Figure 29-8. This disc plow is cutting and loosening the soil. Notice the close-up of the discs on the right.

The second action is planting the crop. This might involve putting seeds or starter plants into the fields. Grains, grasses, and many vegetables are started from seeds planted directly in soil. Some vegetables, such as tomatoes, cabbage, cauliflower, and broccoli, are started from seeds in nursery greenhouses.

The most widely used seed planter is the ***grain drill***. See **Figure 29-9.** This planter is pulled behind a tractor. As it moves across the field, it opens a narrow trench, drops seeds, and closes the trench.

Pest-control equipment

In nature, not all plants that sprout grow to maturity. Diseases and insects kill some of them. Neighboring plants crowd others out. Farm crops face many of the same dangers. Farmers can take action to reduce these dangers by using chemical and nonchemical pest-control measures. These techniques use special machines and devices.

Deere and Co.

Figure 29-9. A specially designed grain drill that plants seeds in soil that has not been tilled.

Chemical control techniques generally use a liquid spray to control weeds and insects. These sprays include herbicides to control weeds and pesticides to control diseases and insects. Ground equipment and aircraft can apply both of these materials.

Nonchemical pest control uses machines and other devices to control pests. A ***cultivator*** is the machine that usually performs this task. This machine uses a series of hoe-shaped blades pulled through the ground. See **Figure 29-10.** The blades break the crust and allow rain and irrigation water to enter the soil. They also cut off and pull out weeds. (A weed is any out-of-place plant.)

U.S. Department of Agriculture

Figure 29-10. This field of potatoes is being cultivated to remove weeds and loosen the soil.

Irrigation equipment

Rainfall is sufficient to raise crops in some parts of the world, but it is not sufficient in all parts. Many places are too dry, or the rain comes at the wrong time, for successful farming. Irrigation systems can be used to support agriculture in these dry or unpredictable climates.

Standards for Technological Literacy

15

Extend

Differentiate between a pesticide and an herbicide.

Research

Have your students research common herbicides and the types of weeds they control.

Standards for Technological Literacy

15

Research

Have your students research the historical development of irrigation.

Extend

Describe the areas in the United States where irrigation is widely used.

TechnoFact

To plant crops, ancient farmers simply scattered seed in all directions by hand, a technique known as broadcasting. Eventually, machines were developed to improve on and replace this inefficient planting method. Henry Blair, one of the first African-Americans to receive a patent, invented a corn seed planter in 1834 and a cotton planter in 1836. Today, machines are available for almost every crop. Some machines plant seeds while others place live seedlings.

Irrigation is artificial watering to maintain plant growth. Irrigation systems must have the following:

- A constant and dependable source of water.
- A series of canals and ditches to move the water from its source to various crops.
- A way to control the flow of the water in the ditches and canals.

The common sources of irrigation water are lakes, rivers, and underground aquifers. Various ways are used to control the water at its source. A dam can restrict water in a river to form a reservoir. See **Figure 29-11.** A dam at the outlet of a natural lake can control the level of the lake and divert water for irrigation. A well and pump can be used to obtain underground water. See **Figure 29-12.** Canals and pipes move the water from the source to farm fields. There, one of four irrigation methods can be used to water the crops.

If the land is level, ***flood irrigation*** can be used. These systems use a large quantity of water that advances across the fields. Ditches or pipes bring the water to one end of the field. The water is released from lateral ditches or pipes with holes along their lengths. As additional water is released, gravity causes the water to flow across the field. In some cases, another set of ditches or pipes carries off excess water.

In row crops, ***furrow irrigation*** can be used. This system uses small ditches, called *furrows,* which are created between the rows of plants. The furrows guide the water as it flows from one end of the field to the other. See **Figure 29-13.** Pipes or siphon tubes are often used to control the flow and allocate water to various sections of the field.

Another irrigation system is called ***sprinkler irrigation.*** See **Figure 29-14.** This approach produces artificial rain to water crops. Sprinkler irrigation is used on uneven ground or where the amount of

Figure 29-11. A dam, such as this one, can be used to form a reservoir. The water can be used for irrigation, to generate electricity, and for recreation.

Figure 29-12. This pump of a deep well supplies irrigation water for a sprinkler system.

U.S. Department of Agriculture

Figure 29-13. The furrows between these rows of potatoes are used to carry irrigation water.

Figure 29-14. This sprinkler-irrigation system is operating on a farm that raises nursery plants.

water applied must be closely controlled. This irrigation uses less water than flood-irrigation systems do.

In sprinkler irrigation, a pump is used to force water from the source into the main distribution lines. The pressurized water flows through the main lines to a series of lateral pipes. These pipes extend at right angles from the main line. Each lateral line has a series of small-diameter pipes sticking upward at even intervals. At the end of each standpipe is a sprinkler head. The water in the lateral lines enters the sprinkler heads, which spray water onto the land. Valves between the main and lateral lines shut off or control the water flow.

Many sprinkler systems have several straight sprinkler (lateral) lines, which are used to irrigate the fields. Each line applies water on a long, narrow band across the field. To irrigate the strip of land on each side of the sprinkler lines, the water is allowed to run for a set time. The water is then shut off from the lateral lines with a valve, and the lines are disconnected. They are moved by hand or rolled under power to the next position (set) along the main line. Here, they irrigate the next strip across the field. After a series of moves, the entire field is irrigated.

In some cases, many sprinkler lines are used to cover the entire field. When the lines are turned on, the entire field is irrigated at once. This is called a *solid-set sprinkler system*. The solid-set sprinkler system approach eliminates the need to move individual lines. This approach is used where the crop must be watered often or where frost protection must be provided using the sprinkler system.

Another type of sprinkler system is called a ***pivot sprinkler***. This system uses one long line. The line is attached at one end to a water source. This long line pivots around this point on large wheels that electric motors power. The line is constantly moving very slowly in a circle. Sprinkler or mist heads apply the water as the line pivots.

Standards for Technological Literacy

15

Figure Discussion

Diagram a sprinkler system like the one shown in Figure 29-14. Show the lines from the pump to the sprinkler head.

TechnoFact

Areas in China, Egypt, Mesopotamia, and India have been continuously irrigated for thousands of years. Native Americans of the Southwest were the first to practice irrigation in North America. Many modern, large-scale irrigation projects also incorporate flood control and hydroelectric power generation facilities.

Standards for Technological Literacy

3 7 15

Extend

Discuss why drip irrigation is the most efficient way to irrigate trees and landscape plants.

Extend

Have your students identify the land-grant university in your state and discuss its programs.

TechnoFact

In 1831, Cyrus McCormick invented the first popular reaper. In 1834, Hiram and John Pitts patented an improved threshing machine that separated straw and winnowed the chaff. A combined harvester and thresher, known as a combine, followed these inventions.

TechnoFact

Many breeds of livestock have been developed over the years. The important beef breeds are Hereford from England, Aberdeen Angus from Scotland, Brahman from the United States, and Simmental from Switzerland. Important dairy cattle breeds are Holstein-Friesian from Netherlands and Germany, Guernsey from Isle of Guernsey, Jersey from Isle of Jersey, Brown Swiss from Switzerland, Ayrshire from Scotland, and the Milking Shorthorn from England.

Drip irrigation is the fourth type of irrigation. The system delivers water slowly to the base of the plants. Drip systems use main lines to bring water near the plants. Individual tubes or emitters extend from the main lines to each plant. These tubes apply water, which soaks into the ground around the roots. This system ensures that each plant is properly watered, and it reduces the amount of water lost to evaporation. Drip systems are used in many orchards and vineyards (grape fields).

Harvesting equipment

To be of value, a mature crop must be harvested. Different harvesting machines exist. The most widely used is the grain ***combine***. See **Figure 29-15.** This machine can be used to harvest a wide range of grains and other seed crops.

Academic Connections: History

The Homestead Act and the Morrill Act

In this chapter and others, we learn how various agricultural and related biotechnologies have changed throughout history. By looking at history from a slightly different perspective, we can also see why these changes occurred. For example, two political events in 1862 forever changed the kinds of crop and livestock production developed in the United States. In May of 1862, Congress, in an effort to promote movement to the West, passed the Homestead Act. This act gave 160 acres of land free of charge to any adult citizen who would live on the land for five years and develop it. In July of 1862, Congress passed the Morrill Act. Under this act, the federal government gave each state 30,000 acres of land for each congressional representative that state had. The land was to be sold, with the resulting monies used to establish colleges of agriculture and the mechanical arts.

The Homestead Act of 1862 opened up vast quantities of land for farming and ranching.

Each act influenced agriculture in various ways. For example, because of the Homestead Act, thousands of people moved west. They discovered that the land in the West was ideal for raising livestock. These people also found that the dry conditions actually helped the growth of wheat. By the 1900s, the plains area of the United States had become one of the world's major sources of that crop. Moreover, because of the growth of these types of farming in the West and the resulting competition, farmers in the East started turning to dairy and truck farming.

Many colleges and agriculture departments within universities were established because of the Morrill Act. These colleges and departments had the research capabilities to help those in agriculture improve existing methods and products. Their pioneering research efforts continue to improve agricultural and related biotechnologies in the United Sates to this day.

Another event in the 1860s also greatly affected U.S. agriculture. This event was the completion of the transcontinental railroad at Provo, Utah, in 1869. Can you think of some reasons why this event made such an impact on agriculture?

Deere and Co.

Figure 29-15. This combine is harvesting a grain crop.

There are several stages in the operation of a combine. See **Figure 29-16.** First, a rotating reel pulls the grain into a cutter bar. There, the tops of the plants containing the grain or seeds are cut off and drawn to the center of the machine. The heads and straw move into the machine, where a revolving cylinder separates the grain from the heads. The grain and straw move onto straw walkers. These devices move the straw to the back of the machine. As the straw moves, the grain falls through holes in the walkers, onto a grain pan. There, blasts of air blow away chaff and other lightweight materials. An auger lifts the remaining grain from the grain pan into storage hoppers. The unwanted straw and waste materials are conveyed out the back of the machine and drop onto the ground. When the storage hopper is full, another auger unloads the grain onto a truck or wagon.

Standards for Technological Literacy

15

Research

Have your students trace the historical development of machines to harvest grain crops.

TechnoFact

The combine was developed in the 1830s but was not widely accepted for nearly 100 years. Self-propelled units were made available in the 1940s. Today's combines protect the operator in dust-free, air-conditioned cabs. The original combines were developed for cereal grains, but present models can also harvest corn, sorghum, grass seed, and soybeans.

TechnoFact

Originally, hand tools like the scythe were used to harvest forage crops. The introduction of the mowing machine, in 1847, made hand mowing a thing of the past. Today, farmers harvest forage crops with tractor-drawn or self-propelled mowers and swathers.

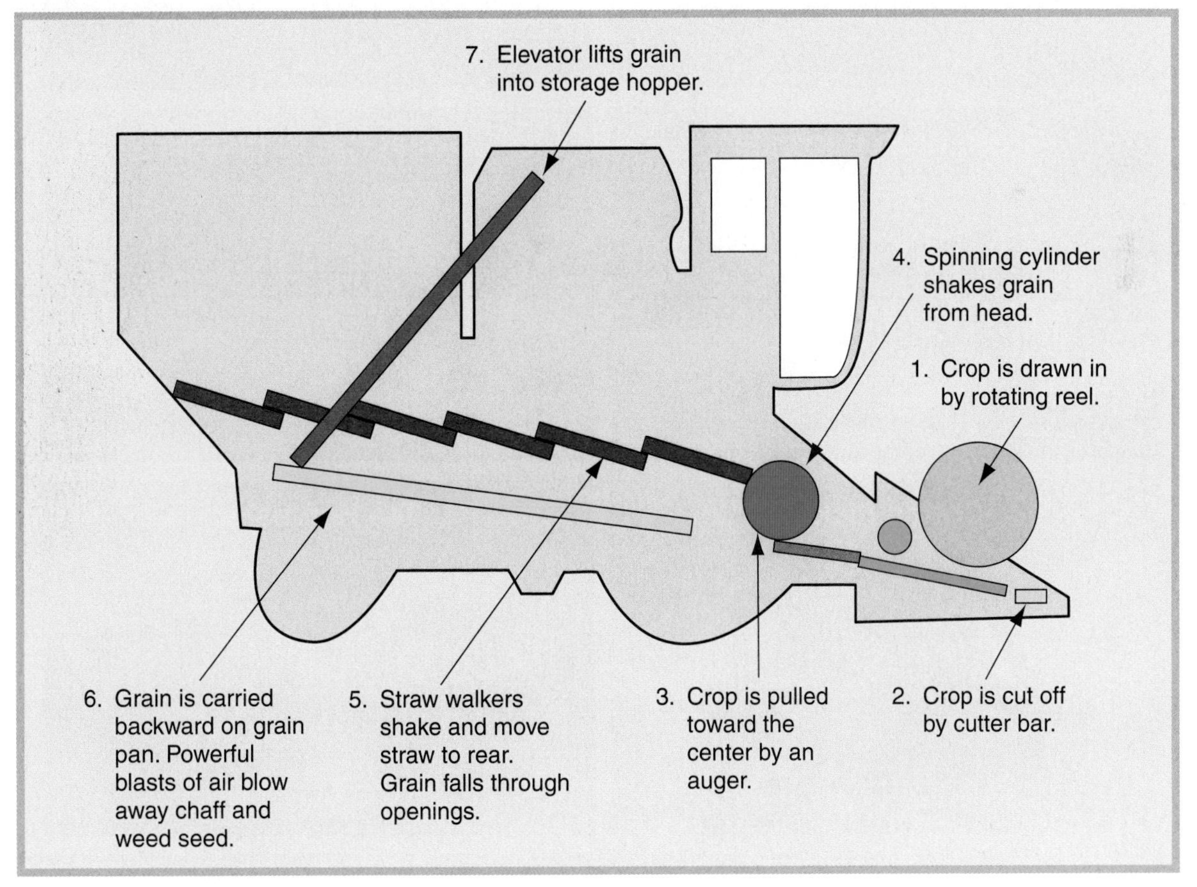

Figure 29-16. The operation of a typical combine.

Standards for Technological Literacy

15

Extend

Select one of the harvesters shown in Figure 29-17 and explain how it works.

Brainstorm

Ask your students why there are so many different harvesting machines.

Specialized harvesting machines have been developed for other crops. See **Figure 29-17.** Cotton is removed from the plants with a cotton picker. The picker strips the cotton bolls from the plant and deposits them in a bin.

Deere and Co., Case IH

Figure 29-17. Some special-purpose harvesting machines.

Potatoes and onions are dug from the ground with mechanical diggers or harvesters. The crop moves across conveyors, where people remove rocks and dirt clods. The remaining potatoes or onions are conveyed onto trucks. Peanuts are dug using special-purpose harvesters. Corn can be harvested with either a combine or a corn picker that strips the ears from the stalks.

Vegetables can be harvested by special -purpose machines or by hand. Special mechanical pickers are used to harvest green beans, sugar peas, and sweet corn. The vegetables are stripped from the plants in one pass. The crop is conveyed into a bin, which is later dumped into trucks for delivery to canning or freezing plants.

Fruits are often picked by hand and placed in boxes. Special machines might be used for some fruits and nuts, however. These machines shake the tree, causing the fruit or nuts to fall into raised catching frames. The crop is gathered from the frames and hauled from the orchards.

A series of special machines harvests grass and alfalfa hay. The crop must first be cut and laid on the ground to dry. In some cases, a mover might be used to cut the plants and let them fall on the ground. After the hay has dried for a day or more, a rake is used to gather it into ***windrows*** (bands of hay). In large-scale hay operations, however, a windrower or ***swather*** is used for these processes. See **Figure 29-18.** This machine cuts and windrows the hay in one pass over the field.

After the hay in the windrows has dried, it is usually baled. A machine called a ***baler*** is used to gather, compact, and contain the hay. Balers produce two basic types of bales: square and round. The most common is the square bale, which can weigh up to several hundred pounds. The standard bale, however, is about 4′ long and weighs 75 to 125 pounds. The bales are bound with either baling twine or wire, making the hay easy to handle and to store.

Deere and Co.

Figure 29-18. This windrower or swather is cutting and windrowing hay. Other swathers are self-propelled, eliminating the need for a tractor.

Transportation equipment

Crops grow in fields spread over a farm. The harvested crops must be moved from the fields to storage or processing plants. This movement can be of two types. First, the crop can be moved from the field to a central location or storage facilities on the farm.

Second, the crop can be moved on roads and highways from the farm to processing plants or commercial storage sites. In some cases, the crops are moved directly from the harvesting equipment to processing plants or central storage. See **Figure 29-19.**

©iStockphoto.com/ImagineGolf

Figure 29-19. This truck is used to haul the grain harvest to an elevator (a storage silo).

Standards for Technological Literacy

15

Extend

Discuss the way that hay is harvested and stored.

Research

Have your students identify the crops that are used for hay.

TechnoFact

The selective breeding of plants and livestock has a long history. The Romans were the first Europeans to employ selective breeding techniques, resulting in many new breeds of livestock. One example, the Holstein dairy cow, was developed more than 2000 years ago in what is now the Netherlands.

Standards for Technological Literacy

15

Extend

Discuss the storage requirements for different farm crops. Show pictures of the buildings used for storage.

TechnoFact

Grain elevators are tall buildings where grain is unloaded, cleaned, stored, and loaded into vehicles for distribution. There are two primary types of grain elevators, country elevators and terminal elevators. Country elevators receive grain from farmers, clean the grain, and then store it until it can be shipped. Terminal elevators are located at major grain markets and shipping hubs. Terminal elevators store grain awaiting processing or shipment to distant markets.

Often, crops are moved short distances with wagons and medium-duty trucks. See **Figure 29-20.** Longer hauls are done on heavy-duty (semi) trucks, railcars, barges, and ships.

Storage equipment

Many crops are stored on farms or in commercial locations before they are sent to processing plants. Most grain crops are stored in silos or buildings at grain elevators. See **Figure 29-21.** Here, the grain is dumped into a pit with sloping sides. At the bottom of the pit is a grain elevator. The most common type of elevator is a bucket elevator. See **Figure 29-22.** This elevator lifts the grain using a series of buckets or pans attached to a moving belt. The moving bucket digs into the pile of grain in the pit and moves the grain upward. When the grain reaches the top of the elevator, it is dumped in a shoot or pipe leading to the top of the silo.

Hay is stored in a dry location so rain and snow do not damage it. The most common hay-storage structure is a hay barn. This structure is simply a roof attached to long poles and does not have

©iStockphoto.com/CarolineSmith

Figure 29-20. This tractor with attached wagon is being used to move grapes from the field (vineyard) to storage.

Figure 29-21. A typical rural grain elevator.

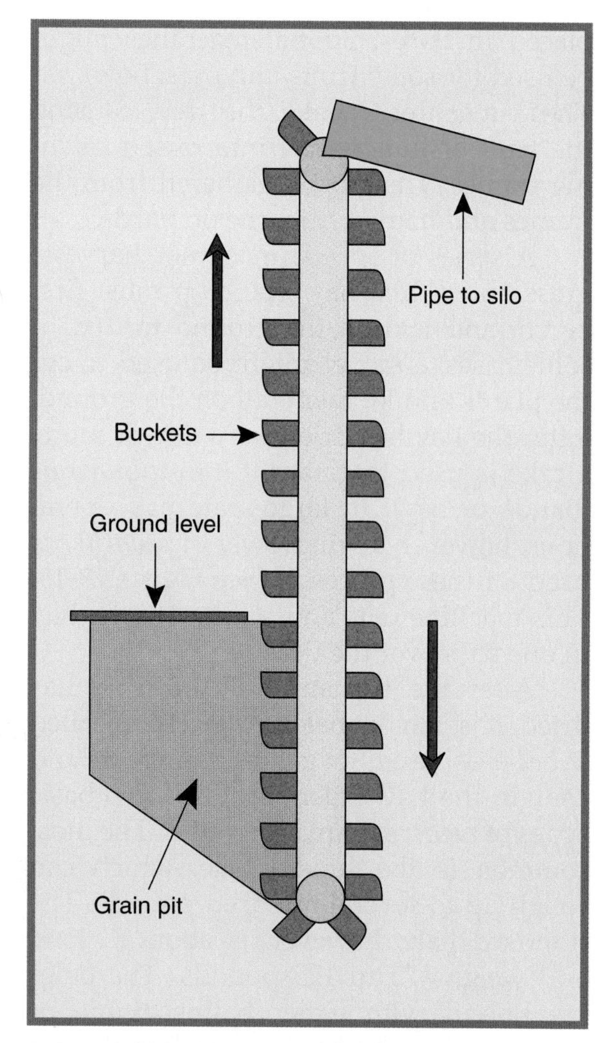

Figure 29-22. How a bucket elevator operates.

enclosed sides or ends. See **Figure 29-23.** In very wet climates, the sides might be enclosed to give added protection.

Many vegetables and fruits are stored in climate-controlled (cold storage) buildings. These insulated buildings use cooling equipment to maintain the crop at an appropriate temperature. The crops are transported from these storage sites to processing plants throughout the country and world, as demand requires.

Figure 29-23. This hay shed is protecting the hay from rain and snow.

A Special Type of Crop Production

A unique type of crop growing is hydroponics. ***Hydroponics*** can be defined as "the growing of plants in nutrient solutions without soil." See **Figure 29-24.** Hydroponic systems have two things in common:

- The nutrients are supplied in liquid solutions.
- Porous materials (peat, sand, gravel, or glass wool) drawing the nutrient solution from its source to the roots support the plants.

Various hydroponic systems exist. The most practical commercial method is

©iStockphoto.com/dstephens

Figure 29-24. Hydroponic lettuce being grown in a greenhouse.

subirrigation. In this system, the plants are grown in trays filled with a coarse material such as gravel or cinders. At specific times, the materials are flooded with nutrient solution. The solution is allowed to drain off after each flooding. Each flooding feeds the plants as they grow toward maturity.

Hydroponic methods can be used to grow plants in greenhouses. They can also be used in areas where the soil or climate is not suitable for the crop. For example, hydroponics is used to grow tomatoes for winter use.

Animal Husbandry

A major activity of agriculture is raising livestock. See **Figure 29-25.** This

Figure 29-25. Livestock raising is a major part of agriculture.

Standards for Technological Literacy

15

Research

Have your students research how hydroponic growing systems work, where they are used, and what crops can be grown with such systems.

TechnoFact

The terms *domesticated* and *tame* are very similar, but do have distinctions. Animals become domesticated after generations of breeding in captivity. Tamed animals are the offspring of wild animals and are made docile through conditioning.

includes the breeding and care of cattle, horses, swine (pigs), sheep, goats, horses, and poultry (chickens and turkeys). These animals are raised to provide meat, milk, or materials for clothing and for recreational purposes.

Standards for Technological Literacy

5 15

Extend

Describe the common buildings found on a dairy farm, a cattle feed lot, and a family farm.

TechnoFact

Confinement production (factory farm) is a method of mass-producing livestock products. Most poultry and eggs are produced by confinement techniques. In confinement operations, livestock are confined to small pens or cages, limiting their movement. Since the animals expend very little energy moving around, they produce more meat or other products.

Technology in Animal Husbandry

Most livestock are raised on single-purpose farms. These farms include cattle ranches; dairies; and swine, horse, and poultry farms. Livestock raising involves various technologies. These include the following:

- Constructing and maintaining livestock buildings.
- Constructing and maintaining fences and fencing to establish feedlots and pastures.
- Operating and maintaining buildings and machines used to feed animals.
- Constructing, operating, and maintaining animal-waste facilities.

Livestock buildings

Most livestock operations require specialized buildings. These building are used to house animals and feed-processing equipment. The types of buildings vary with the type of livestock being raised. Barns and sheds are built to protect animals from the weather. See **Figure 29-26.** Some facilities house animals while the animals' young are being born. Hog-factory buildings contain animals in pens, as the animals are raised for meat. Poultry houses contain chickens and turkeys raised

Figure 29-26. This barn protects dairy cattle from the weather.

Think Green

Sustainable Agriculture

You have already learned the purposes of sustainability. With conventional agriculture, the land may be used until the soil is no longer capable of growing or sustaining crops. Farmers using conventional methods of reusing the land while trying to sustain it with chemical fertilizers may need to find new land when the soil is no longer healthy. Sustainable agriculture finds ways of developing new methods of working with the land that are more environmentally friendly.

Sustainable agriculture is a system that uses the idea of crop rotation to keep the soil healthy without using chemical fertilizers. The soil becomes healthy if allowed to rest. Also, with sustainable agriculture, there is no factory farming. Factory farming takes an industrial approach to dealing with livestock and crop production, and it can pollute the soil, water, and air with various wastes. Sustainable agriculture doesn't produce various chemical byproducts, and, therefore, it leaves little to no impact on the environment.

for food. Laying houses or henhouses hold chickens raised for egg production. Stables house horses raised for pleasure riding and racing. The processes for constructing agricultural buildings are the same as are used for other buildings. These processes are discussed in Chapter 17.

Fences and fencing

Farmers and ranchers construct barriers to separate their land from that of neighbors. Also, barriers separating the farm into fields are needed. These barriers are called *fences*. Without fencing, livestock roam freely, damaging other crops.

Early barriers were made of rock, tree limbs, mud, and other natural materials. Later, smooth wire was adopted as a fencing material. Drawing hot iron through dies (blocks of steel with holes in them) produced it. The smooth wire did not always keep the animals in the enclosure, however. This led to the development of barbed wire, which is a smooth wire with pointed wire added along its length. See **Figure 29-27.** The technological advancement was so popular that more than 570 patents were issued for different designs for barbed wire.

Barbed wire was widely used because it was ideal for western conditions. Barbed wire fences divided and separated the vast prairies and plains. This led to range management, farming, and settlement.

Today, farmers and ranchers use several types of fences, including rail, barbed wire, woven wire, and electric. The type of fence selected depends on the livestock and crops being raised. Horses can run through a fence, causing injury to them. Cattle try to crawl over fences, whereas sheep try to crawl under them. Hogs attempt to root under a fence.

Rail fences are often used as border fences around farm buildings or homes. They are also popular on horse farms. Rail fences are made from posts set in the ground with boards or rails attached between the posts. The common materials used are treated wood, painted wood, vinyl-coated wooden boards, and polyvinyl chloride (PVC) plastic.

Barbed wire fences consist of wood or steel posts with strands of barbed wire attached to them. They usually have posts spaced 10′ to 12′ apart and use three to five strands of wire. ***Woven-wire fences*** use posts with a special wire product, called *woven wire*, attached to them. The woven

Figure 29-27. Barbed wire opened the West to ranching and farming.

Standards for Technological Literacy

15

Extend

Discuss the disputes and range wars (impact of technology) that resulted from the introduction of barbed wire to the western ranges.

TechnoFact

In 1868, Michael Kelly created the first practical barbed wire. The design consisted of two wires twisted together, one of which was strung with flat metal thorns. Joseph Glidden introduced the first commercially successful barbed wire in 1874.

Standards for Technological Literacy

15

Extend

Compare a dog dish with a feed trough.

Research

Have your students investigate the way silos are filled and emptied.

TechnoFact

A silo is a structure that stores the chopped plants and grains that are used for animal feed. The earliest form of silo was a pit covered with boards. In 1873, a square aboveground silo was designed. The design was a failure because the feed would not pack tightly into the corners of the structure. In 1882, Franklin King built the first round silo, which was airtight when it was tightly packed with feed.

wire consists of a number of horizontal smooth wires that vertical (stay) wires hold apart. The horizontal wires are arranged with narrow spacing at the bottom and wider spacing at the top. The height of the fence depends on the size and jumping ability of the contained animals. Typically the fences are 26″ to 48″ tall.

Cable-wire fences consist of 3/8″ steel-wire cables stretched from one anchor post to another. Heavy springs are attached to one end of each cable to absorb shock. The other end is rigidly attached to another anchor post. Each cable passes through holes in a number of line posts set between the two anchor (end or corner) posts. See **Figure 29-28.** ***Electric fences*** are temporary or permanent fences that use electrical charges (shocks) to contain animals in a field. This type of fence can be a separate fence or a strand of electric wire added to another type of fence.

Buildings and machines for feeding

Most livestock farms have livestock-feeding equipment and buildings. The buildings contain machines that grind and mix feed for the animals. As noted earlier, hay barns might be built to protect hay from the weather. Silos are used to contain and protect grain and hay (silage). See **Figure 29-29.** Feed troughs or bunkers are required so the animals can eat grain and hay. Water is provided using manufactured pumps and tanks.

Animal waste–disposal facilities

A major challenge for large-scale livestock production is animal-waste disposal. The waste must be controlled so it does not pollute streams, lakes, and underground water. Livestock farmers must plan to collect, store, treat, and apply animal waste to land properly. This requires them to identify sites for waste disposal and use appropriate land-application procedures.

U.S. Department of Agriculture

Figure 29-28. This cable-wire fence has a pipe on top to add strength.

Figure 29-29. These silos contain feed for a dairy.

All these actions require technological actions. They need a combination of structures and practices serving the animal-feeding operation. Typically, this requires collection both of animal wastes and of other kinds of wastes, including feed and litter. In many cases, a lagoon is built to contain these wastes. This lagoon is a confined body of wastewater holding animal and other waste. This waste is periodically removed from the lagoon and applied to land.

The type of equipment used to apply livestock waste to land depends on the type and consistency of the waste. Dry litter can be applied with a box (manure) spreader. Lagoon waste is often handled in two forms: wastewater and slurry. Wastewater is waste containing less than 2% solids. A slurry mixture is agitated sludge and wastewater. Many farmers apply wastewater using their regular irrigation systems, such as the ones discussed earlier in the chapter. Slurries require special pumping equipment and sprinklers that have large nozzles.

Another technique is to inject the waste into the soil using special application equipment. This equipment uses a tank to haul the waste. When the tank arrives in the field, the waste is inserted into the soil using knife applicators. See **Figure 29-30.** Animal manure can also be dried and sold to homeowners for lawn and garden fertilizer.

A Special Type of Animal Husbandry

A special kind of animal raising is called *aquaculture*. ***Aquaculture*** can be defined as "growing and harvesting aquatic (water) organisms in controlled conditions." This growing and harvesting involves raising fish, shellfish, and aquatic plants. Aquaculture is considered an agricultural activity. Many differences exist between aquaculture and traditional agriculture, however. Aquaculture mainly produces protein crops. Traditional farming focuses on starchy staple crops.

Aquaculture also has waste-disposal problems, similar to livestock raising. In traditional livestock operations, however, animal waste can be disposed of off-site. This waste can be spread on fields as fertilizer. In contrast, animal waste in aquaculture stays in the ponds. Therefore, careful management is required to maintain the water of the ponds.

Standards for Technological Literacy

Research

Have your students visit the photo gallery at the newsroom of the United States Department of Agriculture (USDA) Web site to find pictures of early homesteads.

Research

Have your students research where aquaculture is practiced in the United States and the types of products it produces.

Extend

Describe the technology (plant and equipment) used in aquaculture.

TechnoFact

Aquaculture began nearly 4000 years ago in China. Today, China grows more aquatic plants and raises more aquatic animals in controlled environments than any other country. The Chinese produce about 60% of the world's aquaculture products.

Applying Lagoon Waste

Inserting the Waste into the Soil

Figure 29-30. Lagoon waste is being applied with a special-purpose applicator. The applicator applies the waste by knifing it into the soil.

Standards for Technological Literacy

15

Extend

Differentiate between agriculture and biotechnology.

TechnoFact

Friedrich Miescher isolated a substance called nuclein (now known as DNA) in 1896. In the 1970s, technology was developed that allowed researchers to isolate and alter individual genes. Later in that decade, researchers used recombinant DNA techniques to create insulin- and interferon-producing bacteria. In 1982, insulin produced by this technique became the first genetically engineered drug approved for human use.

This growing and harvesting is also different from traditional fishing. Traditional fishing is called *capture fishing*. This fishing involves locating and capturing fish in nature (oceans, lakes, and rivers). In contrast, aquaculture requires human action in raising fish.

Aquaculture requires machines and equipment to perform a number of tasks. An environment must be constructed. Most aquaculture is done in ponds dug in the earth. These ponds usually have water inlets and outlets. They are stocked with young fish. Fertilizer is often added to the ponds. This promotes the growth of food supplies for the fish. The fish are fed a diet promoting their growth. When they have reached market size, they are harvested. In a complete harvest, the pond is drained. All the animals are removed from the pond. In a partial harvest, some of the animals are removed using a net. See **Figure 29-31.**

Figure 29-31. These workers are harvesting catfish at a fish farm.

Agriculture and Biotechnology

People have used scientific activities to improve plants and animals for hundreds of years. They have used selective breeding of livestock and cross-pollination of plants to create new or improved plants and animals. In recent history, technology has also been used for this goal. This technology is called ***biotechnology***. What, however, is biotechnology? *Biotechnology* can be defined as "using biological agents in processes to produce goods or services." The biological agents are generally microorganisms (very small living things), enzymes (a special group of proteins), or animal and plant cells. They are used as catalysts in the selected process. The word ***catalyst*** means they are used to cause a reaction. The catalyst, however, does not enter into the reaction itself.

The term *biotechnology* is fairly new. The practice, however, can be traced back into distant history. Evidence suggests that the Babylonians used biotechnology to brew beer as early as 6000 BC. As far back as 4000 BC, the Egyptians used biotechnology to produce bread. During World War I, scientists used an additive to change the output of a yeast-fermentation process. The result was glycerol, instead of ethanol. The glycerol was a basic input to explosives manufacturing. During World War II, scientists used the fermentation process to produce ***antibiotics*** (antibodies).

Today, biotechnology has a major impact on modern agriculture. Agricultural biotechnology is being used to create, improve, or modify plants, animals, and microorganisms. See **Figure 29-32.** Agricultural biotechnology is being used to produce new pest-resistant and chemical-tolerant crops. This is done through a process called ***genetic engineering***.

These new crops have helped combat diseases. For example, golden rice was developed using genetic engineering. This new rice provides infants in developing countries with beta-carotene to fight blindness.

©iStockphoto.com/chemicalbilly

Figure 29-32. This scientist is collecting tissue from a genetically modified plant.

Biotechnology is a major factor in increasing crop yields. This technology has helped produce more food on the same number of acres. For example, biotechnology has been used to produce soybeans resistant to certain herbicides. Also, it has been used to develop a cotton plant resistant to major pests.

This technology can be used to promote human health. The nutritional value of foods can be improved using genetic engineering. Genetic engineering is based on the fact that every living thing carries a genetic code (blueprint) determining precisely the traits it has. This code was linked to a major discovery called *recombinant DNA*. The structure of DNA is a double helix (spiral) structure. This spiral structure consists of a jigsawlike fit of biochemicals. The two strands have biochemical bonds between them.

The DNA molecule can be considered a set of plans for living organisms. This molecule carries the genetic code determining the traits of living organisms. Scientists can use enzymes to cut the DNA chain cleanly at any point. The enzyme selected determines where the chain is cut. Two desirable parts can then be spliced back together. This produces an organism with a new set of traits. The process is often called ***gene splicing***. See **Figure 29-33.**

This process allows scientists to engineer plants that have specific characteristics. For example, resistance to specific diseases can be engineered into the plant. This can reduce the need for pesticides to control insect damage to crops.

Gene splicing has received many headlines in newspapers and magazines. This activity is controversial. Some people think it makes life better. Others think we should not change the genetic structures of living things.

Standards for Technological Literacy

15

Figure Discussion

Use Figure 29-33 to discuss genetic engineering.

TechnoFact

In 1986, the first U.S. patent for a genetically engineered plant was issued for a strain of corn that was engineered for increased nutritional value.

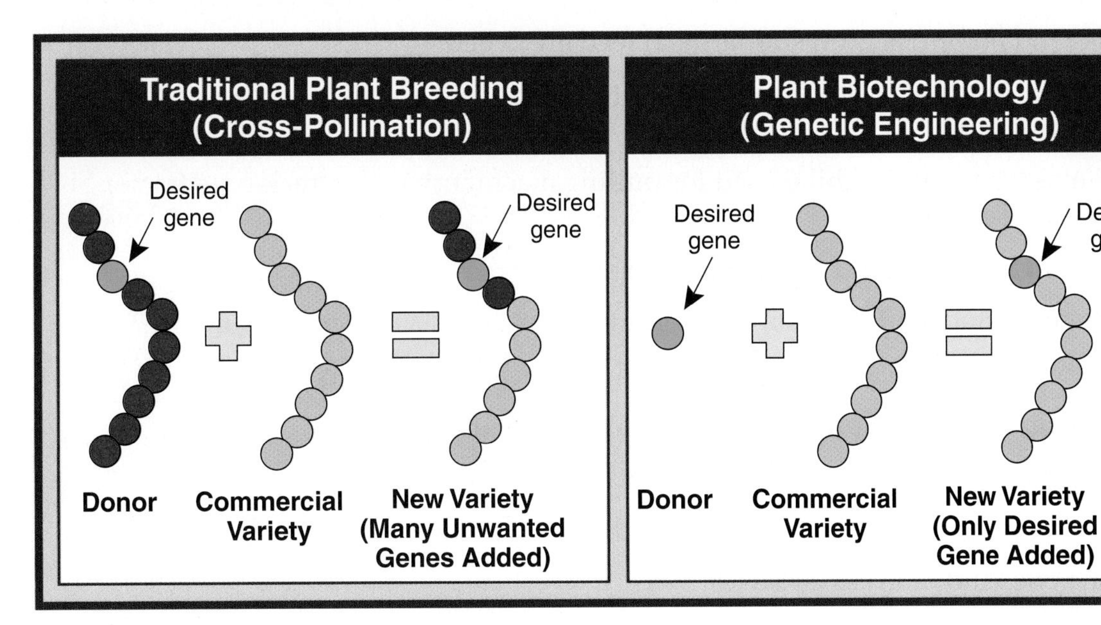

Figure 28-33. Genetic engineering can introduce a desired gene into an existing plant.

Answers to Test Your Knowledge Questions

1. Agriculture is people using materials, information, and machines to produce the food and natural fibers.
2. Evaluate individually.
3. Crop production and animal husbandry
4. grains
5. forage
6. storing
7. False
8. plow
9. True
10. True
11. sprinkler irrigation
12. Evaluate individually.
13. hydroponics
14. Student answers may include: cattle ranches, dairies, and swine, horse, and poultry farms. Additional answers may be accepted at the instructor's discretion.
15. Student answers may include: rail, barbed wire, woven wire, and electric.
16. Silos are used to contain and protect grain and hay (silage).
17. slurry
18. aquaculture
19. Biotechnology is using biological agents in processes to produce goods or services.
20. Evaluate individually.

Summary

Agriculture includes two major activities taking place on farms and ranches. These activities are growing crops and raising livestock. Crop production involves using machinery and equipment in planting, growing, harvesting, and storing plants and plant products. These tasks require various kinds of equipment: power or pulling equipment (tractors), tilling equipment, planting equipment, pest-control equipment, irrigation equipment, harvesting equipment, transportation equipment, and storage equipment.

Livestock operations use structures and equipment to feed and care for cattle, swine, sheep, goats, poultry, and horses. Such structures and equipment include livestock buildings, fences and fencing, buildings and machines for feedings, and facilities for animal waste. Agriculture also employs biotechnology. This type of technology applies biological organisms to production processes. We use biotechnology to produce new strains of crops and drugs.

Test Your Knowledge

Write your answers on a separate piece of paper. Please do not write in this book.

1. What is agriculture?
2. Name one difference between the role of science and the role of technology in agriculture.
3. What are the two branches of agriculture?
4. Members of the grass family that have edible seeds are called ______.
5. Grass and hay crops grown for animal feed are called ______.
6. The four agricultural processes that use technology are planting, growing, harvesting, and ______.
7. The most common machine used for pulling machinery is the track-type tractor. True or false?
8. The most important piece of tilling equipment is the ______.
9. A chemical applied to the soil to add nutrients is called *fertilizer.* True or false?
10. A cultivator is a piece of equipment used in nonchemical pest control. True or false?
11. The irrigation system producing artificial rain is called ______.
12. Name one type of harvesting equipment.
13. Growing plants in nutrient solutions without soil is called ______.
14. Name one kind of livestock farm.
15. Name one kind of fencing used to contain livestock.

16. For what purpose do farmers use silos?
17. The two kinds of lagoon waste are wastewater and ______.
18. The practice of growing and harvesting fish in controlled conditions is called ______.
19. Define *biotechnology.*
20. Name one form of genetic engineering.

STEM Applications

1. Farmers use many types of equipment during the growing season. Some equipment is used at different times throughout the season. Other equipment is used only part of the season. Using a form similar to the one below, list the names of some pieces of equipment (the plow or combine, for example) used at various times in the growing season.

Spring	Summer	Fall

2. Genetic engineering to modify different kinds of food crops has caused controversy. Research the topic of genetically modified products (such as corn) and make two lists: one of the positive effects (benefits) of genetically modified products and one of the negative effects or risks. Compare the lists and try to form an opinion on the subject.

This activity develops the skills used in TSA's Biotechnology Design event.

Agriculture and Biotechnology Design

Activity Overview

In this activity, you will research a contemporary problem, issue, or technology related to agriculture or biotechnology; prepare a report; and create a display. You will prepare an oral presentation incorporating a PowerPoint® presentation. Your report must be contained in a three-ring binder and consist of the following items:

- A cover page.
- The definition of a problem, an issue, or a technology.
- A report on the topic (4–10 pages).
- A printout of PowerPoint slides (three slides per page).
- A list of sources and references.

Materials

- A three-ring binder.
- Materials appropriate for a tabletop display. (These will vary greatly.)
- A computer with PowerPoint presentation software (or similar).

Background Information

- **Selection.** Before selecting the theme for your project, use brainstorming techniques to develop a list of possible themes. Some contemporary topics include the following:
  - Waste management.
  - DNA testing.
  - Soil-conservation techniques.
  - The Human Genome Project.
  - Genetically modified food.
  - Cloning.
  - Aquaculture.
  - Food-production techniques.
  - Irradiation.

- **Research.** Use a variety of sources to research your theme. Do not rely solely on information you find on the Internet. Use books and periodicals available at your local library. Research the historical developments of the topic. Did an individual develop the technology, or did a corporation develop the technology? What were some previous technologies that allowed this technology to become a reality? How did the public receive the technology, and was the response expected?
- **PowerPoint presentation.** When developing your PowerPoint presentation, consider the following design guidelines:
  - Develop a general slide design. Use it for all your slides.
  - Keep the design simple. Do not use more than two type fonts. Select easy-to-read type fonts. Be sure that the type size is large enough to be seen from the rear of the room in which you will be presenting.
  - Include a title on each slide.
  - Do not attempt to squeeze an abundance of information on a single slide. Create multiple slides instead.

Guidelines

- Research should focus on any cultural, social, economic, or political impacts. Both opportunities and risks should be addressed.
- The display can be no larger than 18′ deep × 3′ wide × 3′ high.
- If a source of electricity is desired, only dry cells or photovoltaic cells can be used.
- The oral presentation can be up to 10 minutes in length.

Evaluation Criteria

Your project will be evaluated using the following criteria:

- The content and accuracy of the report.
- The attractiveness and creativeness of the display.
- Communication skills and the presentation design of the oral presentation.

Chapter Outline
Primary Food Processing
Secondary Food Processing

This chapter covers the benchmark topics for the following Standards for Technological Literacy:

3 5 7 15

Learning Objectives

After studying this chapter, you will be able to do the following:

- Summarize the role of food in daily life.
- Recall the definition of *food-processing technology.*
- Compare primary and secondary food processing.
- Summarize ways used to process foods.
- Recall ways to preserve foods.
- Summarize the steps that might be used to develop a new food product.
- Summarize the activities involved in manufacturing a new food product.

Key Terms

aseptic packaging
can
chemical processing
cure
food-processing technology
freeze
irradiation
mechanical processing
milling
pasteurization
preservative
primary food-processing technology
refrigerate
secondary food-processing technology
smoke
thermal processing

Strategic Reading

Before you read this chapter, think about the different types of food you ate yesterday. As you read, see if you can determine the types of conversion, preservation, and processing your food underwent.

Humans have always needed an adequate food supply. Food is needed to sustain life and promote growth. Try to remember what you ate yesterday. See **Figure 30-1.** The food might have been some fruit, vegetables, and meat. You might have eaten several slices of bread and some cereal. All these substances are food.

Food is one of the three essentials of life. The others are clothing and shelter. Without a proper supply of food, human life is not possible. See **Figure 30-2.** Food sustains life by providing seven basic components:

- **Carbohydrates.** Carbohydrates provide the body with its basic fuel.
- **Proteins.** Proteins are amino acids that provide the building material cells need to maintain their structure and growth.
- **Fats.** Fats provide triglycerides, which are stored in muscle cells and fat cells and are burned as fuel.
- **Vitamins.** Vitamins are organic substances essential in the life of most animals and some plants. The human body needs many different vitamins including vitamin A, vitamin B, vitamin B_1 (thiamin), vitamin B_2 (riboflavin), vitamin B_3 (niacin), vitamin B_6, vitamin B_{12}, folic acid, vitamin C, vitamin D, vitamin E, vitamin K, pantothenic acid, and biotin.
- **Minerals.** Minerals are elements that the body needs to create specific molecules. Common minerals the body needs are calcium, chlorine, chromium, copper, fluorine, iodine, iron, magnesium, manganese, molybdenum, phosphorus, potassium, selenium, sodium, and zinc.
- **Fiber.** Fiber is the substance people eat that their bodies cannot digest. The three major fibers in food are cellulose, hemicellulose, and pectin.

Standards for Technological Literacy

Research

Have your students select a food product and look at its package to determine its nutritional value and ingredients.

TechnoFact

Early humans called hunter/gatherers probably did not store a large food supply. People spent most of their time searching for food in one area, looking for plants and animals they could catch. Food preservation was not likely used at this time, and when the food supply ran out, these people would relocate to another area.

TechnoFact

When prehistoric hunter/gatherers learned how to grow their own food, they were able to remain in the same area instead of moving around to find food. It was at this time that food preservation and food processing came about. With the discovery of fire and the invention of pottery, food preservation became more standard.

Figure 30-1. People eat many different types of food.

Figure 30-2. Food contains seven basic ingredients: carbohydrates, proteins, fats, vitamins, minerals, fiber, and water.

Standards for Technological Literacy

15

Extend

Show students the difference between primary and secondary processing. Use peanuts to peanut butter as the primary processing example. Use peanut butter and crackers to make a snack food product as the secondary processing example.

TechnoFact

Fermentation was an early food preservation method. Cheese was made in Asia by fermenting milk, and many societies discovered that various liquids could be converted into wine. Salt curing was another method used long ago. Salting was a process applied in ancient time to prevent the spoiling of such foods as meat and fish.

- **Water.** The human body is about 60% water. About 40 ounces of the water are lost each day and must be replaced. We replace this water through our consumption of moist foods and drinks.

To supply a reliable source of food, people have developed food-processing technology. ***Food-processing technology*** involves using knowledge, machines, and techniques to convert agricultural products into foods that have specific textures, appearances, and nutritional properties. These technological actions transform animal and vegetable materials into safe and edible food for humans. Also, food-processing technology includes processes used to make food more tasty and convenient to prepare. In addition, it involves actions used to extend the shelf lives of perishable foods.

Food-processing technology includes two basic types of processes. See **Figure 30-3.** These are the following:

- **Primary food processing.** These processes are technological actions that change raw agricultural materials into food commodities or ingredients.
- **Secondary food processing.** These processes are technological actions that convert food commodities and ingredients into edible products.

Primary food-processing technology produces the basic ingredients for food. For example, this type of technology changes wheat into flour. Primary food-processing technology changes animal carcasses into hamburger and transforms raw milk into pasteurized milk.

Secondary food-processing technology is used to make finished food products. For example, it converts flour and other ingredients into bread and changes hamburger and other ingredients into lasagna. Both of these types of processes are critical to our production of food. We now look at each one in more detail.

Primary Food Processing

Most agricultural products are not eaten directly from the fields in which they are grown. They are processed using technical means (machines) to change their form,

Wheat Foods Council

Figure 30-3. Agricultural products go through primary and secondary processing before becoming edible food.

Northwest Cherry Growers

Figure 30-4. Food processing can be as simple as receiving, washing, and grading produce.

appearance, or usefulness. Primary food processes include both material conversion (processing) and food preservation.

Material Conversion

Few of us eat products that come directly from farm fields or gardens. Most of our food is processed in one way or another. This processing might be as simple as washing, grading, and packing fresh fruits and vegetables. See **Figure 30-4.** In other cases, the produce might be preserved using one of several techniques. The fruits and vegetables might be canned or frozen for later use. They might be dried or cured. These and other preservation techniques are discussed later. All these actions constitute technology. They involve using machines to change the condition of the food.

Many food-processing techniques convert raw agricultural products into a different form. They might change the product so it has a different nature and appearance. For example, milk might be changed into cheese or butter. Corn and soybeans might be processed to produce cooking oils. Sugarcane and sugar beets might be refined into granulated sugar. These processes convert the original farm product into an entirely new ingredient.

Food can be processed in many ways. These ways can be grouped under the three headings of mechanical processing, thermal (heat) processing, and chemical processing. ***Mechanical processing*** uses machines to change the form of the food product physically. This processing might crush, slice, grind, or scrape the material to form a new ingredient. ***Thermal processing*** uses heat as the primary energy to convert a food. This processing might use the energy to melt, cook, blanch, or roast the material. ***Chemical processing*** uses energy to cause a basic chemical change in the food. This process might involve pickling, fermenting, coagulating, or other similar actions. A representative example of physical, thermal, and chemical processing follows.

Mechanical Processing: Flour Milling

Flour is finely ground grain, such as wheat, rye, corn, or rice. Wheat flour is the most commonly produced flour in the Western world. The composition of this flour depends on the type of wheat used and the milling processes employed. In a common milling process, four major steps are used. See **Figure 30-5.** These include the following:

1. **Grain receiving and cleaning.** Wheat is transported from farms to the

Standards for Technological Literacy

15

Demonstrate

Show mechanical processing by grinding some grain into flour using a hand process or a food processor.

Brainstorm

Ask students to give examples of food products (ingredients) that have been produced using mechanical, thermal, and chemical processes.

TechnoFact

Flour is a finely ground grain commonly found in baked bread. Usually, it is made from wheat, and it provides nourishment to people all over the world.

Standards for Technological Literacy

15

TechnoFact

Flour has been in use for over 12,000 years. Before early people settled into various areas, flour was made from wild grain. Once farming developed, wheat, barley, and rye were ground into flour with millstones.

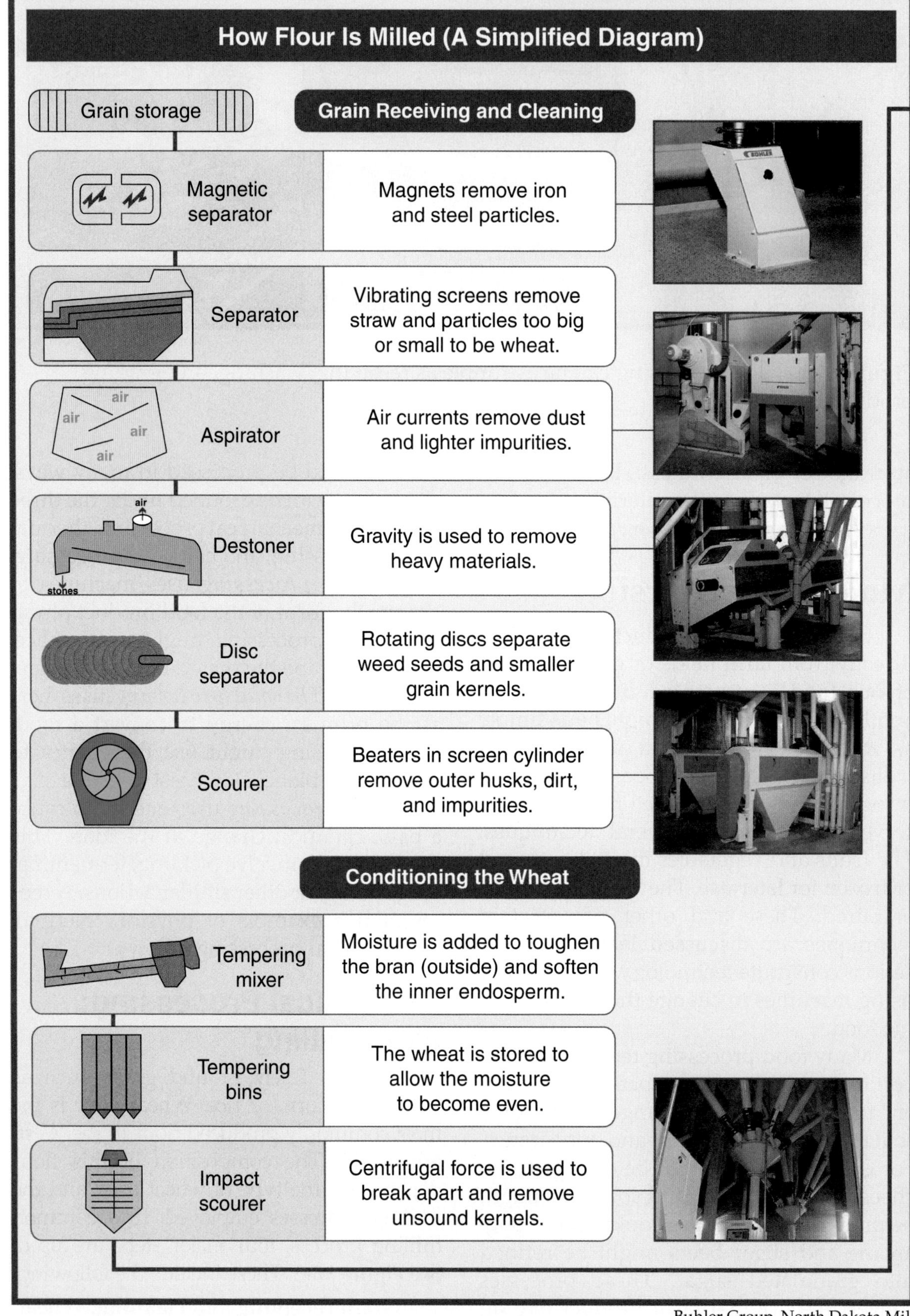

Buhler Group, North Dakota Mill

Figure 30-5. The flour-milling process involves four steps.

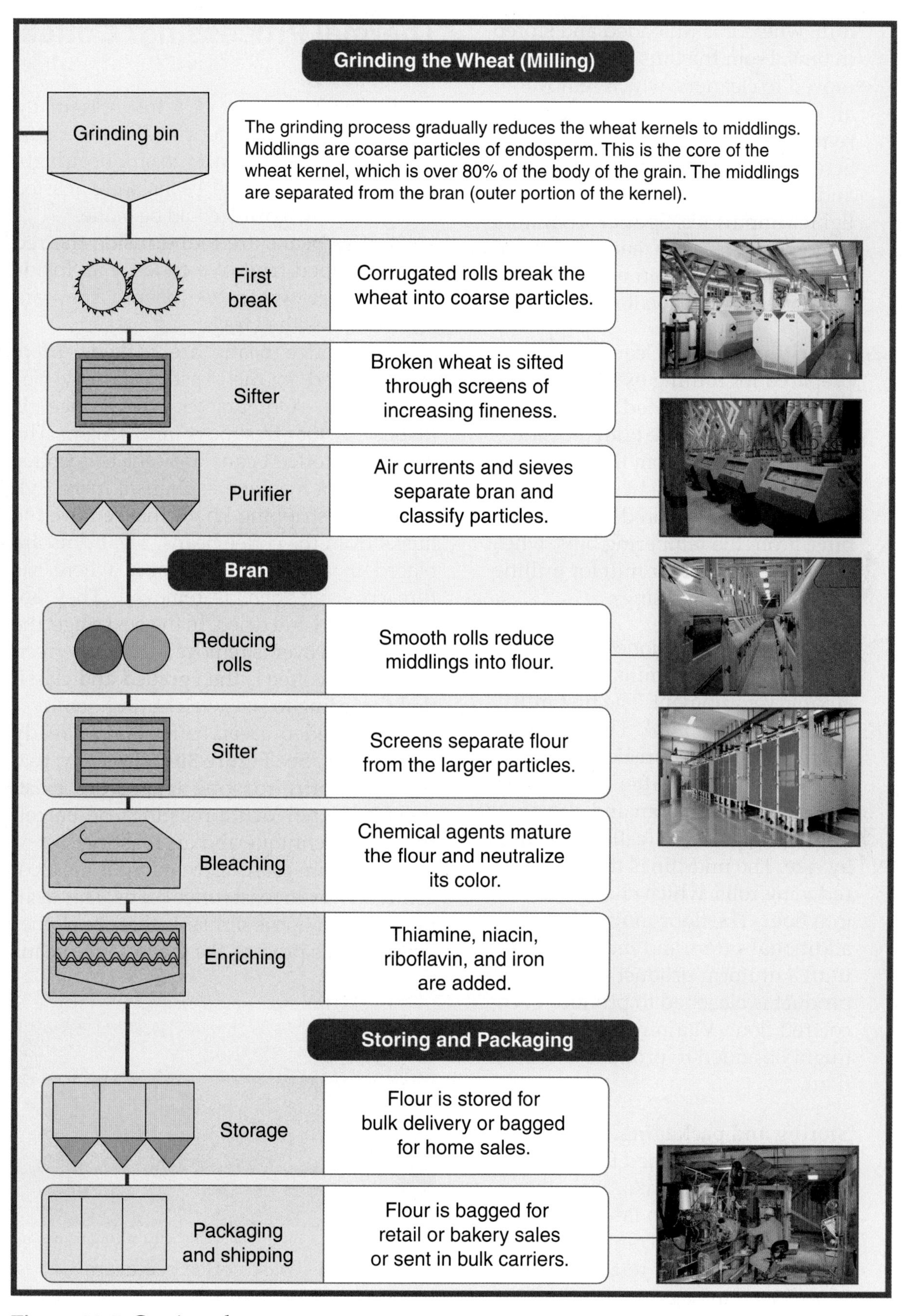

Figure 30-5. Continued.

Standards for Technological Literacy

15

TechnoFact

The progress of flour mills coincides somewhat with the development of steam-powered engines. James Watt, who patented the steam engine in the late 18th century, invented a steam-powered flour mill. Oliver Evans, who made progress with Watt's steam engine in the area of transportation, later built a more advanced automatic flour mill.

Standards for Technological Literacy

15

Research

Have students use the Internet to explore how coffee is ground.

TechnoFact

The effects of coffee beans were discovered in Ethiopia thousands of years ago. By the 12th century, the beans were ground into a drink. Over the centuries, coffee has become one of the largest world commodities.

mill, where it is unloaded and stored in bins. From the bins, the wheat is moved to cleaners, which remove all impurities. Magnetic separators remove iron and steel particles. Screens remove larger stones, sticks, and other materials. Air blasts remove lighter impurities. Special separators remove other grains (such as oats) and weed seeds. The clean wheat is stored in hopper bins to await ***milling***.

2. **Conditioning.** The clean wheat is prepared for milling by a conditioning process. Water is added to make it easier to separate the bran (outside hull of the grain) from the endosperm (flour portion) of the wheat kernel. The wet wheat is stored in tempering bins. From the tempering bins, wheat is moved to the flour mill for milling (grinding or processing).

3. **Milling.** The conditioned wheat is milled using roller mills, sifters, and auxiliary equipment. The first mill is called the *first break*. This mill uses corrugated rolls to break wheat into coarse particles. A sifter and purifier then separate the bran and classify the remaining particles (called *middlings*) by size. The middlings move to reducing rolls, which change them into flour. The flour moves through additional sifters and reducing rolls until a uniform product emerges. This product is bleached to produce even-colored flour. Vitamins and minerals might be added to produce enriched flour.

4. **Storing and packaging.** The milled, bleached, and enriched flour is conveyed to steel tanks. From there, it might be moved in tank cars to bakeries or other commercial customers. Flour for retail customers is packaged in bags and sent to grocery stores.

Thermal Processing: Coffee Roasting

Coffee is the fruit of a tree. Originally, coffee was a food, not a drink. People mixed coffee beans with animal fat to produce a high-energy food. Later, the beans were roasted and ground so a drink could be made.

Coffee beans are found inside the red cherry-shaped fruit of a coffee tree. Inside the fruit are two coffee beans. They are referred to as the *greens*.

The coffee beans are picked, dried, and shipped to the roasting factory. See **Figure 30-6.** On arrival, the coffee is inspected for freshness and color. The inspected coffee beans are sent to a coffee pulper. This machine contains a metal cylinder with stripping knobs that remove the husks from the coffee beans. The beans are placed in a fermentation tank, where the film covering them is removed. They are then washed and dried. In the next stage, the parchment cover is removed from the green beans. The coffee is then graded and classified according to size, weight, and quality.

The graded green coffee is now ready for roasting. See **Figure 30-7.** Basically, two types of coffee-roasting techniques exist: the traditional drum-roasting, or barrel-roasting, technique and air roasting.

The drum-roasting technique is the most common way to roast coffee beans. This technique's operation is similar to that of a clothes dryer. In this process, the green coffee beans

Figure 30-6. These coffee beans are being sun dried before being bagged and shipped to the United States.

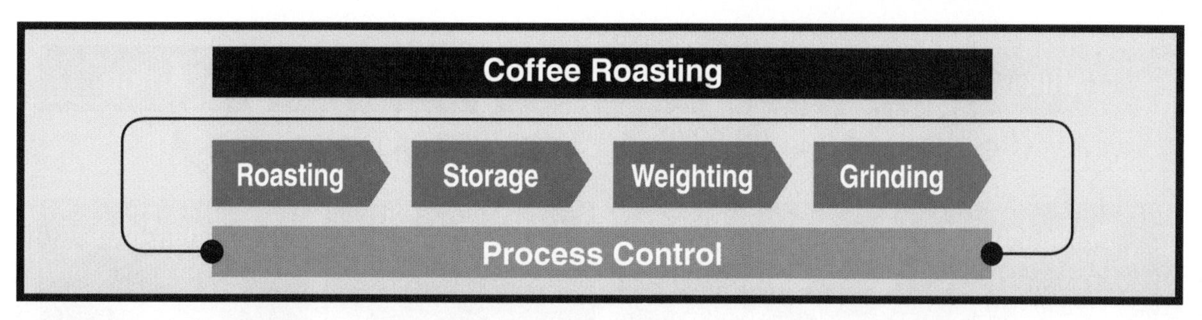

Figure 30-7. The coffee roasting-and-grinding process involves four steps.

are heated to a temperature of about 550°F in large rotating drums. The beans tumble in the cylinder and are roasted.

The air-roasting technique is similar to popping popcorn. The coffee beans are roasted at the desired temperature using hot air. They are constantly moving in the air blast, which allows them to roast evenly.

In both cases, the first stage of roasting turns the green beans to a yellowish color as they dry. They begin to smell similar to toast or popcorn. In the second step, called the *first crack*, the beans double in size. They pop very similarly to popcorn and become a light-brown color.

In the third step, the beans begin to brown as their oils start to emerge. A chemical reaction caused by the heat and oils produces the flavor and aroma of coffee. A second pop occurs several minutes later. At this point, the beans are fully roasted and can be moved to storage.

The roasted coffee moves out of storage and is weighed. This coffee then is ground using mechanical grinding machines. The grinding allows people to get the most flavor out of the bean when coffee drinks are prepared. The ground coffee is packaged in airtight bags or cans to help maintain its freshness.

Chemical Processing: Cheddar-Cheese Manufacturing

Cheese is a dairy product made from milk protein. This product is a coagulated, compressed, and ripened curd of milk that has been separated from the whey (moisture in milk). Cheese is made in hundreds of types and forms. See **Figure 30-8.** The following describes how cheddar cheese is made through a chemical conversion process:

1. **Receiving milk.** The cheese-making process starts with milk being delivered from local dairies. The milk moves from the farm to the cheese factory in refrigerated milk trucks. On arrival, it is tested for quality, flavor, and odor.

2. **Cooking.** The fresh milk is pasteurized to destroy any harmful bacteria. This milk is pumped into cooking vats. Special starter cultures (bacteria) are added to the warm milk and change a very small amount of the milk sugar into lactic acid. Coloring is added to produce a consistent cheese color. Finally, a substance to promote coagulation and firm the curd is added.

3. **Cutting the curd.** After about 30 minutes, the vat of milk sets up. In this action, a soft curd, consisting of casein and milk fat, is formed. Stainless steel knives cut the soft curd into 1/4″ pieces. The temperature in the cooking vat is then raised to about 100°F to drive out moisture and firm the curd. The liquid part in the vat is called *whey*.

Standards for Technological Literacy

15

TechnoFact

Transforming milk into a solid was practiced over 4000 years ago. Cheese is food that has been popular all over the world for centuries. Cheese-making was an international profession until the Industrial Revolution, when the first American cheese factory opened in New York.

Standards for Technological Literacy

15

Figure Discussion

Use Figure 30-8 to review the cheese-making process.

Research

Have students research and find some common types of cheese and document the differences in their manufacture.

TechnoFact

As a seller of cheese, James L. Kraft found that the food's perishability presented difficulty in his business. In order to preserve the natural cheese, Kraft experimented until he invented process cheese. The process of wrapping individual slices of cheese began in 1916 and is still a method of preservation used today.

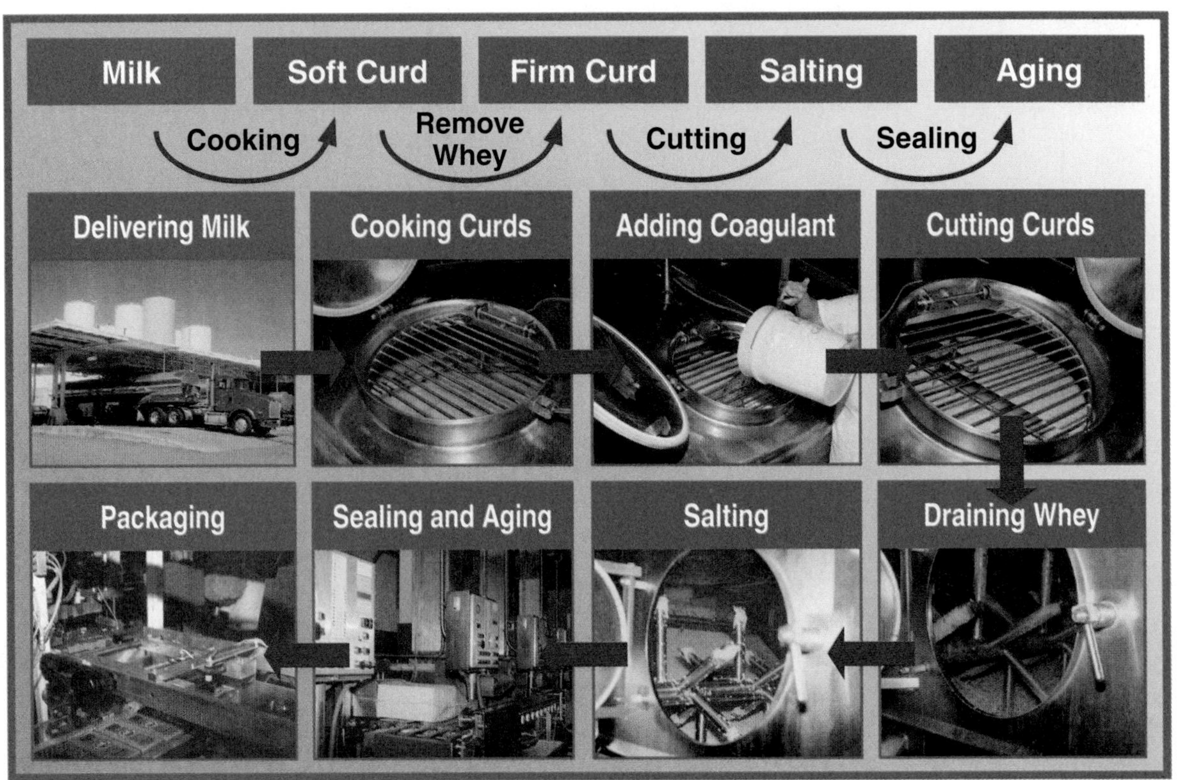

Tillamook County Creamery Association

Figure 30-8. In the cheese-making process, milk curds are cooked, separated from the whey, cut, and sealed.

4. **Making cheddar.** The curds and whey are pumped to the cheddar-making machine. The whey is removed, and the curd is matted on a wide belt inside the machine. During the cheddaring process, a chemical change occurs. The curd particles adhere to each other, giving them a stringy consistency. When the proper acidity is reached in the curd, the cheddar mat is forced through the curd mill, which chops the large slabs into small, 3″-long bits.
5. **Salting.** These curd chunks are passed through a salting chamber. In this machine, a thin layer of salt is applied to the surface of each curd. The salted curd is stirred, and it absorbs the salt. When the salt is completely absorbed, the curds are transferred to pressing towers, where a vacuum draws off the excess moisture in the cheese.
6. **Sealing and aging.** After 30 minutes in the pressing towers, large blocks of cheese are cut from the bases of the towers. These blocks are placed in plastic bags. A vacuum is drawn, and the package of cheese is sealed. The block is now contained in an airtight and moisture-proof bag. The sealed blocks are transported to a cooling room and held for about 24 hours at temperatures near 38°F. From there, they are palletized and placed in storage for aging and curing at about 42°F.
7. **Packaging.** After aging for a set number of days, the cheese blocks are removed from their aging bags. Each batch of cheese is tested and graded for quality. See Figure 30-9. The blocks are cut into specified sizes and packaged for various markets.

Wisconsin Milk Marketing Board

Figure 30-9. This inspector is checking cheese for unwanted odors.

Food Preservation

Food-preservation technologies are used to keep food from being unfit for human use. They reduce or eliminate the effects of bruising and insects. Preservation controls microorganisms such as bacteria, yeast, and molds. These techniques can destroy unwanted enzymes (proteins) present in raw foods. The enzymes cause chemical and physical changes that naturally occur after harvesting. They contribute to food spoilage.

These technologies also help eliminate moisture or temperature conditions that encourage the growth of microorganisms. Microorganisms can produce unwanted changes in the taste and appearance of the food. They can also cause food-borne illness.

Food processing can improve the nutritional value of foods. For example, high heat can destroy unwanted factors present in many foods. Prolonged boiling destroys the harmful lectins present in such foods as red kidney beans.

Some preservation techniques can reduce the quality of food. They can produce food that has lower nutritional value and poorer texture and flavor. A number of different technologies are used for food preservation. These technologies include drying, curing and smoking, canning, aseptic packaging, refrigeration and freezing, controlled-atmosphere storage, fermentation, pasteurization, irradiation, and preservatives.

Standards for Technological Literacy

15

Extend

Show some common dried, cured, smoked, and canned foods. Discuss how they were produced.

TechnoFact

Though dehydrated food became widespread during World War I, food drying has been one of the most common methods of food preservation for centuries. Even before modern methods of food drying were used, natural elements such as the sun and wind were applied.

Career Corner

Agriculture, Food & Natural Resources

Food-Processing Workers

There are many different jobs in food processing. Two of these are food-batch makers and food cooking–machine operators. Food-batch makers set up and operate equipment that mixes, blends, or cooks ingredients used in the manufacture of food products. Food cooking–machine operators operate cooking equipment to prepare food products. These food-machine operators usually are trained on the job. They learn to operate equipment by watching and helping other workers. Employment opportunities for these workers are expected to increase more slowly than in other fields. Many food-processing workers are members of the United Food and Commercial Workers International Union and earn considerably more than minimum wage.

Standards for Technological Literacy

15

Research

Have students use the Internet and other resources to explore how freeze-drying works.

Drying

Drying is the oldest method people have used to preserve food. This method removes water that microorganisms causing food spoilage need to grow. Drying includes three basic methods of food preservation. Sun drying allows foods to dry naturally in the sunlight. See **Figure 30-10.** Hot air drying exposes food to a blast of hot air. Freeze-drying uses a vacuum chamber to draw water out of frozen foods. In freeze-drying, water escapes from the food by a process in which ice changes from a solid directly to a vapor, without first becoming a liquid.

Dried foods keep well because the moisture content is so low that spoilage organisms cannot grow. Drying is used on a limited range of foods, including milk, eggs, instant coffee, and fruits. This method will never replace canning or freezing as the primary way to preserve foods. Canning and freezing retain the taste, appearance, and nutritive value of fresh food better than drying does.

Curing and Smoking

Curing and smoking are techniques used to preserve meat and fish. See **Figure 30-11.** ***Curing*** involves adding a combination of natural ingredients to the meat. These ingredients might be salt, sugar, spices, or vinegar. This process is used to produce products such as bacon, ham, and corned beef. ***Smoking*** is a process that adds flavor to meat and fish, while preserving them. This process involves slowly cooking the meat or fish over a low-heat wood fire.

These techniques preserve food by binding or removing water. This causes the water to be unavailable for the growth of microorganisms. Both curing and smoking produce a distinctive flavor and color in food. In many cases, these techniques eliminate the need for ***refrigeration***.

Canning

Canning is a preservation method that puts food into glass jars or metal cans. This method preserves food by heating it in a vacuum-sealed container. The process

©iStockphoto.com/ivanastar

Figure 30-10. These chili peppers have been sun dried in New Mexico.

©iStockphoto.com/davidf

Figure 30-11. This salami is being cured in warm, humid conditions, which encourage growth of the bacteria involved in the fermentation process.

removes oxygen from the container, kills microorganisms in the food, and destroys enzymes that can spoil the food.

During the process, the can is filled with food. See **Figure 30-12.** Air is pumped to form a vacuum, and the container is sealed. The food is heated and then cooled to prevent the food from becoming overcooked. Low-acid foods, such as meats and most vegetables, are heated to 240°F to 265°F, and foods with higher acid levels, such as fruits, are heated to about 212°F.

In many cases, the food is blanched before canning. In this process, the food is heated with steam or water for a short period of time. This heating inactivates enzymes that can change the food's color, flavor, or texture. Blanching reduces the volume of the vegetables by driving out gases.

Canning is used to preserve a wide variety of foods, including fruits, vegetables, jams and jellies, pickles, milk, meats, soups, sauces, and juices. Canned foods can be stored without refrigeration for a long time. Canning affects the color, texture, flavor, and nutrient content of foods because of the high temperatures used.

Standards for Technological Literacy

15

TechnoFact

In 18th century Italy, Lazzaro Spallanzani studied microbes and experimented with preventing them. It was at this time he created an early version of a canning process. Food that was sealed and heated could be preserved for many weeks.

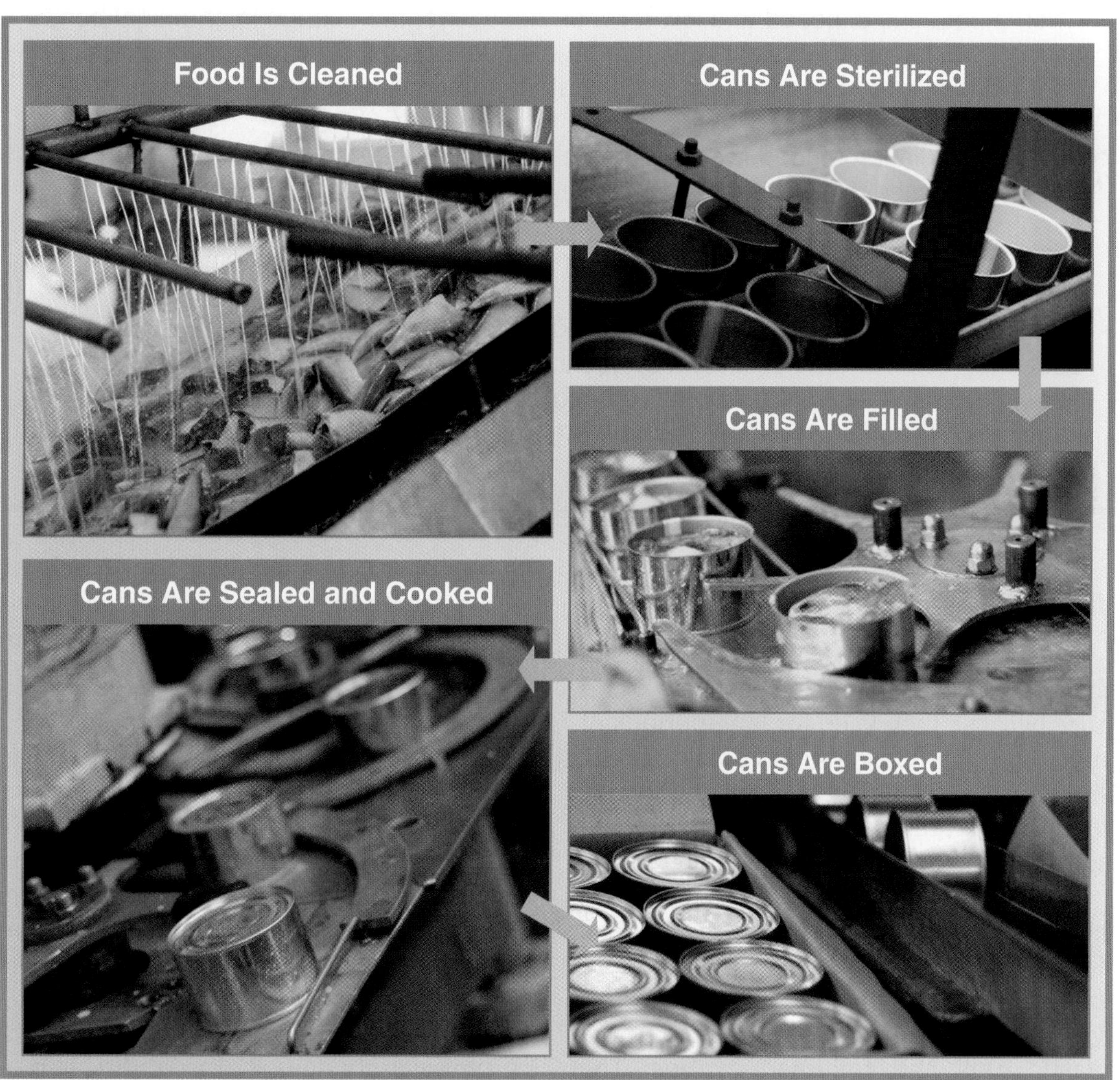

©iStockphoto.com/Anutik

Figure 30-12. The major steps in the canning process.

Standards for Technological Literacy

3 7 15

Extend

Discuss the aseptic packaging process, its development, and advantages.

Research

Have students research when and why freezing foods was developed.

TechnoFact

The canning process of food preservation became more common in the early 1800s, when Nicolas Appert experimented with the process. His book on canning described how packing food in sealed glass jars before heating them in boiling water helped prevent food spoilage.

STEM Connections: Science

Integrated STEM Curriculum

Irradiation

The question of whether or not we should consume irradiated food is hotly debated. Some groups fear that harmful chemicals might be formed in the process. Others argue that irradiated foods might cause cancer. Many scientists, however, believe that irradiation is both safe and beneficial.

To be able to make informed choices regarding this subject, we should know what irradiation of food involves. Irradiation itself is a form of radiant energy, which as we learned in earlier chapters, is energy in the form of electromagnetic waves. The electromagnetic waves used in irradiation are primarily gamma rays, UV rays, and X rays. Scientists have found that, when these rays penetrate food, chemical bonds are broken and the chemical composition of the food changes.

The change in chemical bonds can affect food in many ways. For example, when a potato is exposed to a low level of these rays, the change in chemical bonds causes the sprouting to stop. At higher levels of exposure, the changes in chemical bonds cause such changes as the elimination of insects and the destruction of harmful bacteria.

The irradiation of food is sometimes chosen over the cooking of food (thermal energy) because, even though chemical changes occur in both processes, fewer bonds are broken in irradiation. Therefore, the food is much fresher. Irradiation is also sometimes preferred to the addition of chemicals to destroy pests or extend storage life because of the possible hazards involved with certain chemicals.

Thus, with irradiation, we can keep food longer and in better condition. We still use other forms of food-preservation technologies, however, with particular foods. For example, we use salt to preserve bacon, and we use heat (pasteurization) to keep milk fresh. Can you think of some reasons for using these methods, as opposed to irradiation?

Aseptic Packaging

Aseptic packaging is commonly used for packaging aseptic, packaged milk and juice because they keep for long periods of time without refrigeration. This process, similar to canning, uses heat to sterilize food. Unlike canning, however, the package and food are sterilized separately.

The containers are sterilized with hydrogen peroxide, rather than with heat. This allows the use of lower-cost containers that heat sterilization would destroy. The most commonly used containers are plastic bags and foil-lined cartons. The food is sterilized more rapidly and at lower temperatures than in canning. This allows the food to have better flavor and to retain more nutrients.

Refrigeration and Freezing

Storage at low temperature slows many of the reactions causing foods to spoil. The two low-temperature preservation techniques are refrigeration and freezing. Refrigeration maintains foods at temperatures from 32°F to 40°F. *Freezing* keeps the foods at or below 32°F.

Refrigeration does not cause chemical or physical changes to food. Still, it preserves foods for a fairly short period of time.

Foods that should be refrigerated include the vast majority of milk, eggs, meats, fish, and some fruits and vegetables.

Freezing is used on a wide range of foods. See **Figure 30-13.** This process allows foods to be stored for longer periods because it greatly reduces enzyme activity and the growth of microbes. The ice crystals that form in the food disrupt the food's structure, however. When thawed, the food has a softer texture. Also, food deterioration is rapid after thawing because the organisms in the food attack cells that ice crystals injured.

Controlled-Atmosphere Storage

Fruits and vegetables can be stored in sealed environments where temperatures

Standards for Technological Literacy

15

Figure Discussion

Discuss Figure 30-13 as an example of producing a frozen food.

TechnoFact

While cold storage had been used by cutting ice from lakes, an ice-making machine made it possible to preserve foods with greater ease. While attempting to cool rooms in a hospital, John Gorrie created the first commercial ice-making machine.

Making and Freezing Commercial French Fries

Step 1: Getting Potatoes—The potatoes come directly from the producers or from large warehouses. The potatoes are checked for solids content, grade, and sugar content.

Step 2: Peeling the Potatoes—A batch of potatoes are put into a large, hot, pressurized tank. After a set amount of time, the pressure is quickly released, and the potato skin is said to "fly off". The potatoes are removed from the tank and sprayed with high-power water jets to remove any peel still clinging.

Step 3: Inspecting the Potatoes—The peeled potatoes pass an inspection line where sorters remove any defective potatoes.

Step 4: Cutting the Potatoes—The potatoes go through a pump that propels them at about 50 mph at some stationary blades, which chop the potatoes into strips.

Step 5: Inspecting the Strips—The small parts left from the outer edges of the potato are removed. The remaining strips are automatically inspected, and strips with black bits are removed. This is done at the rate of about 1,000 strips and chips each second!

Step 6: Blanching—The inspected strips are blanched on a moving conveyor chain that carries them through a large vat of hot water. This process removes excess sugars and gives the strips a consistent, uniform color.

Step 7: Drying—The blanched strips are partially dried as they are conveyed past blasts of hot air from both the top and bottom.

Step 8: Partial Frying—The strips are cooked for about a minute and a half in hot oil. The process is called *par fry*, or *partial fry*.

Step 9: The Deep Freeze—The strips enter blast freezing, where the French fries travel down on the wire conveyor surronded with air cooled to about –40° Fahrenheit.

Step 10: Packaging—After freezing, the product is bagged or boxed and shipped to customers.

J.R. Simplot

Figure 30-13. The freezing of French fries allows them to be available longer.

Standards for Technological Literacy

15

Extend

Discuss the historical development and use of fermentation in food processing.

Figure Discussion

Use Figure 30-15 to explain to students how milk is pasteurized.

Research

Have students research Louis Pasteur's contributions to top food processing.

TechnoFact

The first patent for practical refrigeration was given in 1834 to Jacob Perkins. His idea was that as a liquid changes to a gas, it absorbs heat. Another patent for mechanized refrigeration went to John Gorrie in 1851.

and humidity are controlled. In some cases, the composition of the air around the fruit is changed. Oxygen is often reduced, while the level of carbon dioxide is increased. This change reduces the chance of food spoilage. This controlled environment slows the reactions leading to decomposition and decay. Controlled atmospheres can extend the storage lives of fruits and vegetables by several months.

Fermentation

Fermentation uses microorganisms to break down complex organic compounds into simpler substances. Such compounds as alcohol and acids are produced in fermentation. These compounds act as preservatives that reduce further microbial growth.

In some cases, fermentation spoils the product. In other cases, it is desirable. In these cases, microorganisms are added to foods.

For example, in the manufacture of yogurt and cheese, bacteria convert a sugar found in milk (lactose) into lactic acid. Fermentation is also used to produce alcoholic beverages, cheese, yeast bread, soy sauce, and cucumber pickles. See **Figure 30-14.**

Figure 30-14. These grapes are being washed and readied for wine production.

Pasteurization

Pasteurization uses heat to kill harmful microorganisms. For example, milk is usually heated to 145°F for 30 minutes. See **Figure 30-15.** Pasteurization is commonly used for milk, fruit juices, beer, and wine.

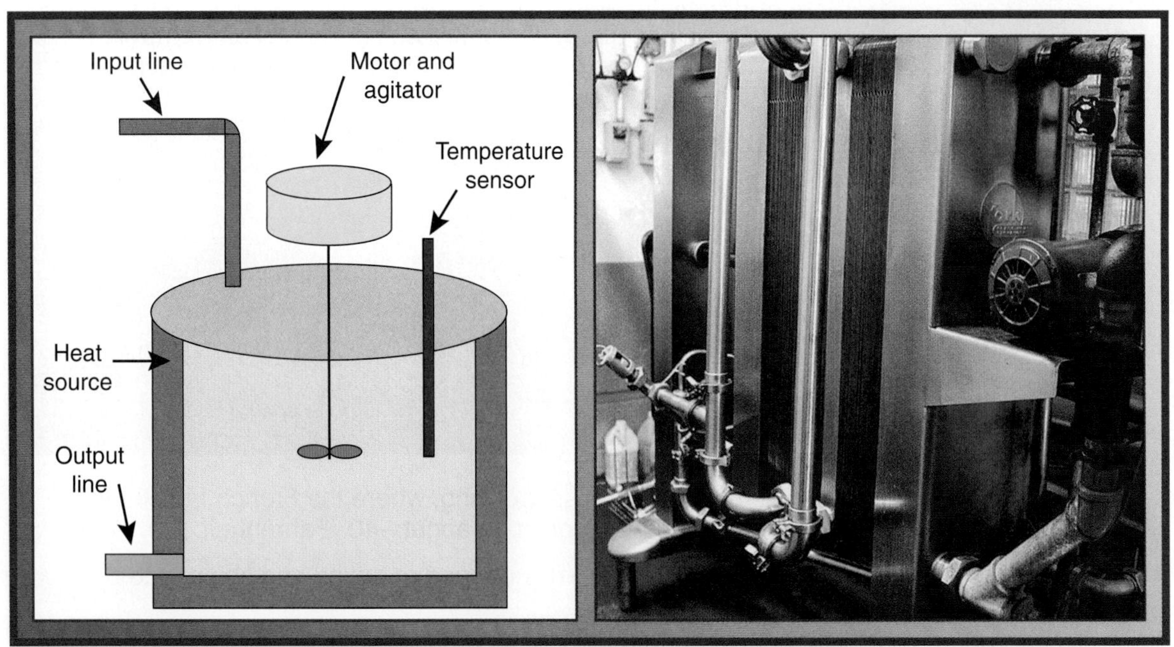

Wisconsin Milk Marketing Board

Figure 30-15. The operation of a batch pasteurizer.

In newer processes, this technique is called *ultrahigh-temperature (UHT) pasteurization*. UHT pasteurization uses higher temperatures and shorter heat times for foods in sterile packaging. The foods are heated to 280°F for two to four seconds. This rapid sterilization allows the food to have better flavor and retain more nutrients.

Irradiation

Irradiation uses gamma rays or X rays to kill most molds and bacteria that might be in the food. This technology is also used to delay the ripening of fruits and the sprouting of vegetables. The delay allows the produce to be stored for longer periods of time. In this process, the food passes through a chamber where it is exposed to high-energy rays. Irradiation involves little heating and does not change the taste, texture, or nutritive value of food.

Preservatives

Preservatives are chemicals added to food in small amounts. These preservatives act in two ways. First, they delay the spoilage of the food. Second, they ensure that the food retains its quality. The first method uses sugar (jams and jellies), vinegar (pickles and meats), and salt (hams and bacon). The second method uses acids, sulfur dioxide, and other agents to slow the growth of microorganisms in food.

Secondary Food Processing

Secondary food-processing technology converts the ingredients produced in primary food-processing activities into edible products. This technology involves combining and processing ingredients and food products to change their properties. There are literally thousands of end products of secondary food-processing activities. The shelves of a typical supermarket are evidence of this.

These products can be grouped in various ways. See **Figure 30-16.** One way is as follows:

- Starch products.
  - Bread.
  - Cakes and cookies.
  - Crackers.
  - Pasta.
- Dairy products.
  - Milk, buttermilk, and cream.
  - Butter.
  - Cheese, cottage cheese, and yogurt.
- Meat, poultry, and fish products.
  - Fresh meat, fish, and poultry.
  - Processed products (hot dogs and lunch meats, for example).
- Fruit and vegetable products.
  - Fruit preserves and fillings.
  - Vegetables dishes.

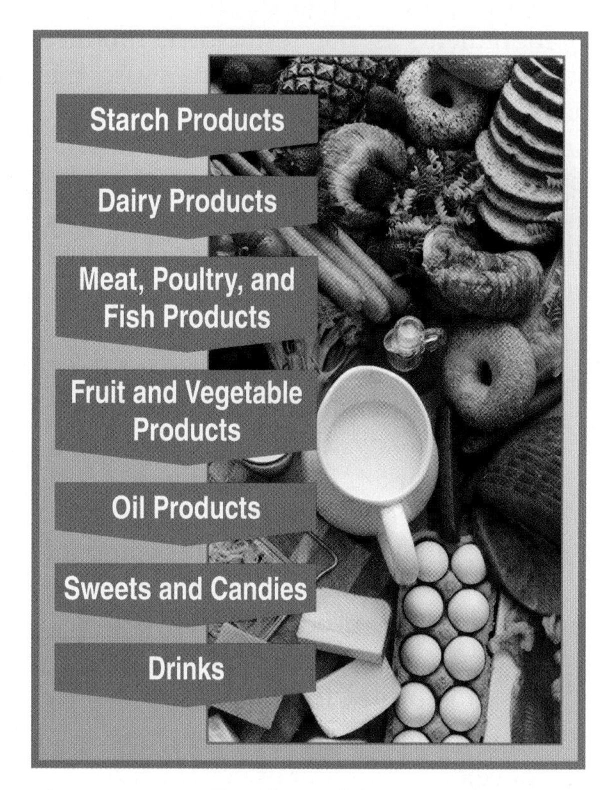

Figure 30-16. Food products can be grouped into seven major categories.

Standards for Technological Literacy

15

Extend

Discuss with students the typical preservatives added to food.

Brainstorm

Ask students if preservatives should be added to food. Also ask them what would happen if food processors stopped using them.

TechnoFact

Although freezing was an accepted method of food preservation by the early 20th century, it was found that the quality of the frozen food diminished after the process. In 1925, Clarence Birdseye discovered that rapid freezing would help keep food fresh for months afterward. His quick-freezing process led him to market quick-frozen foods until 1929, when the Postum Company purchased Birdseye's patents and marketed their own frozen foods.

Standards for Technological Literacy

Brainstorm
Ask students to name some good new food products.

Figure Discussion
Use Figure 30-17 to discuss the major steps in developing a new food product.

TechnoFact
Fermentation had been used in ancient times to create alcoholic beverages, such as wine and beer. At this time, vinegar also came from a stage in fermentation.

- Oil products.
  - Cooking oils.
  - Margarine and low-fat spreads.
  - Salad dressings.
- Sweets and candies.
- Drinks.

The creation of these products involves two steps. These steps are developing the food products and manufacturing them. We first examine each of these steps separately and then look at a specific example of secondary food processing.

Food-Product Development

Many products can be made using similar basic ingredients. For example, flour and water are the basic ingredients for bread, biscuits, cookies, and pasta. Other ingredients are added to create the uniqueness of each product.

Each product created from a series of ingredients is the result of a product-development process. Companies use this process to develop an array of new or modified food products. The process of product development involves a series of steps that moves the product from the original idea to the consumer. Some of these steps are shown in Figure 30-17. The steps are the following:

1. **New food product–idea identification.** A company identifies consumer trends and eating patterns. This company studies new products on the market. The company investigates advancements in food-preparation techniques. This company uses this information to brainstorm possible new products and line extensions (products building on existing products).
2. **Idea development.** The company develops a number of new recipes. This company specifies the ingredients and establishes their costs.
3. **Small-scale testing.** The company makes several versions of the product, usually using slightly different ingredients or processes. The products might be prototypes in the company's test kitchens. Company employees or small focus groups are asked to test the products and provide comments.
4. **Sensory evaluation.** The company asks trained personnel to comment on the appearance, taste, smell, and texture of the products.
5. **Product modification.** The company alters the composition and recipe of the product to incorporate the results of the initial test and sensory evaluations.
6. **Pilot plant production.** The company scales up the product to pilot plant production. This means the item is produced using small versions of the equipment that can be used in full-scale manufacture.

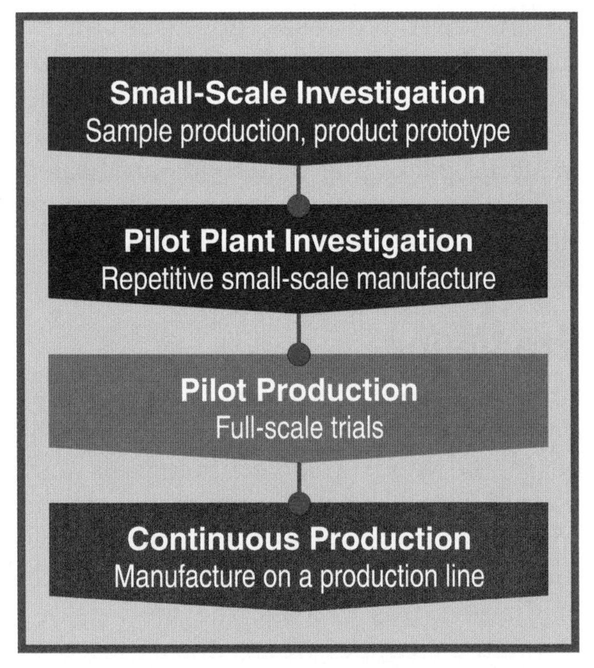

Figure 30-17. Companies follow a series of production steps when creating a new product.

7. **Consumer and sensory evaluation.** The company tests the output of the pilot production to determine its potential of being a profitable product.
8. **Product specification.** The company develops the final product specifications. This company develops the final ingredients and methods of production. These specifications are used in producing each batch of the product.
9. **Pilot production.** The company readies the production plant for the product. This company conducts production tests to determine the effectiveness of the production processes.
10. **Continuous or large-scale product production.** The company releases the product for full-scale production. The actual manufacture is often done in major processes, such as measuring, mixing, and cooking. The company carefully controls each of these processes to maintain product quality, promote food purity and safety, and reduce waste.
11. **Product packaging.** The company places the product in appropriate packages. This company places the individual packages in shipping containers ready for movement to the wholesaler or retailer. See **Figure 30-18.**
12. **Product promotion.** The company announces the availability of the product through advertising and product promotion.

Not all companies develop products using this procedure. For example, some companies move products directly from the "test kitchen" stage to trials on factory production equipment. The decision on how to proceed depends on competitive pressure, economic factors, the size of the production facility, and the type of product being developed.

Figure 30-18. This root beer is being bottled (packaged) as a final step in product manufacture.

Food-Product Manufacture

Making a consumer food product involves a number of manufacturing processes. See **Figure 30-19.** The procedure most foods go through includes some of the following:

1. **Storing.** This involves containing and protecting the condition of the raw materials.
2. **Cleaning.** This involves removing foreign matter from ingredients used in the product.

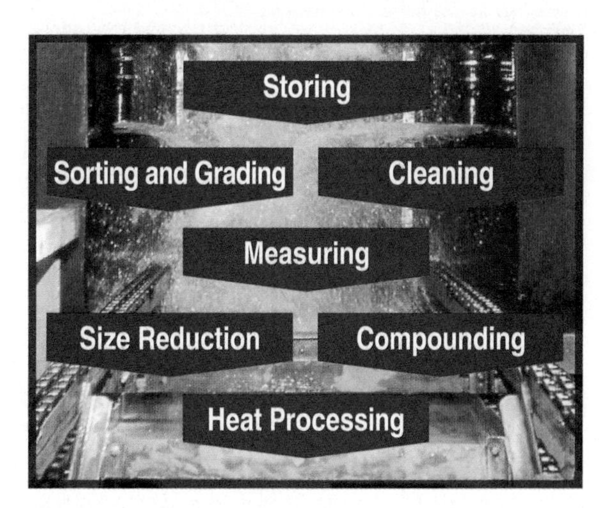

Figure 30-19. Various technological processes are used to produce nutritious foods.

Standards for Technological Literacy

15

Extend

Use soup-making to show the procedures that are employed to make a consumer product.

TechnoFact

Several scientists before the 19th century studied microbes and experimented with ways to destroy them. Louis Pasteur recognized that fermentation was caused by microbes. He found that heating wine helped destroy bacteria and did not affect the wine otherwise. This process, called pasteurization, is used in products such as milk before being sold.

Standards for Technological Literacy

5 15

TechnoFact

As it had done with many other processes, the mechanization of the Industrial Revolution provided a more efficient way to make cans. In 1876, machines were able to shape cans. Today's automation allows machines to make, fill, and seal cans.

3. **Sorting and grading.** This involves assessing the quality of the ingredients to be used in making the product. See **Figure 30-20.**
4. **Measuring.** This involves weighing and measuring the wet and dry ingredients to be used in the product.
5. **Size reduction.** This involves trimming, slicing, and crushing the basic ingredients.
6. **Compounding.** This involves mixing and combining the ingredients for the product.
7. **Heat processing.** This involves cooking and cooling the mixture to form the product.

These technological processes produce a nutritional and easy-to-use consumer product. Initially, food-processing technology produced a food unit such as bread, pasta, crackers, or stewed tomatoes. These products were combined at home to prepare dishes and complete meals. Now, however, people are demanding food that is easier and quicker to prepare. Today, food-processing technology provides both single unit–type foods and finished dishes. These finished dishes might be a meal entrée, such as lasagna, or a complete meal, such as that seen in a frozen dinner.

J.R. Simplot

Figure 30-20. These carrots are being sorted and graded before entering the manufacturing line.

An Example of Secondary Food Processing: Pasta Making

The goal of pasta production is to convert flour into an edible product. Pasta

Think Green

Local Organic Food

You have already learned that organic material comes from a living organism. Food can be organic by using alternative, organic approaches for pesticides and fertilizers. Conventional pesticides and fertilizers may taint the soil and water when they break down. They also may release toxins as they evaporate into the air. Organic food eliminates most of the chemicals used to help grow the food and to kill the insects it attracts.

Buying local food also has a more positive environmental impact than buying industry-made food. One reason for that is if the food is grown locally and not through industrial companies, there is less likelihood that the food is contained in plastic packaging. As you've already learned, plastic waste does not decompose. The primary benefit to the environment from buying locally is that the use of transportation resources is greatly reduced. Trucks transporting food over great distances use a large amount of fossil fuels and have a greater potential for emitting carbon dioxide into the atmosphere.

production includes blending the ingredients to form a dough, kneading and mixing the dough, forming the final product, and then drying it. See **Figure 30-21.**

Blending the Ingredients

The basic ingredients of pasta are durum wheat semolina and water. These ingredients are measured and added to a batch mixer. Additional ingredients, such as eggs, spinach, tomato powder, and flavorings, can be added at this time. They are mixed under vacuum to distribute moisture evenly throughout the flour. The semolina starts to absorb water and forms a crumb structure.

Kneading and Mixing the Dough

The lumps of semolina are kneaded (pressed together) to form a basic dough. This action forces the crumb structures together, which fuses the particles. The result is consistent dough.

Standards for Technological Literacy

15

Figure Discussion

Show an example of pasta and use Figure 30-21 to discuss its manufacture.

Dakota Growers Pasta Company

Figure 30-21. Making pasta involves several secondary food-processing techniques.

Standards for Technological Literacy

15

TechnoFact

The following are important dates associated with food preservation: In 1858, John Mason made home canning possible with his invention of the Mason jar. In 1861, Gail Borden's first evaporated milk factory opened. In 1905, the first patents for irradiation were issued to help destroy bacteria.

Forming the Product

An extruder is used to form the pasta shape. This machine uses the screw principle to push dough through a shaped die. These molds have round or oval holes through them to produce rods for products such as spaghetti. A steel rod can be placed in the center of each hole to form a tube-shaped product, such as macaroni. Grooves in the screw scoop up the dough. The rotating screw pushes the dough toward the die opening. As the material passes through the die, it is shaped and formed. Cutters cut the developing tube of product to length. Other pasta products, such as noodles, are formed in flat sheets. The shapes are stamped out of the sheets using shaped cutting dies.

Drying the Product

The cut products pass through long tunnel dryers to remove excess moisture. Inside the dryers, very hot, moist air removes the moisture. The product is cured.

Coffee beans are processed using thermal processing techniques. These beans have been roasted and are now cool and ready for storage.

Answers to Test Your Knowledge Questions

1. Evaluate individually.
2. Food-processing technology involves using knowledge, machines, and techniques to convert agricultural products into foods that have specific textures, appearances, and nutritional properties.
3. primary processes
4. secondary processes
5. Grain receiving and cleaning, conditioning, milling, and storage and packaging
6. drum or barrel
7. Student answers may include any of the following: receiving the milk, cooking, cutting the curd, making cheddar, salting, sealing and aging, and packaging.
8. I. Irradiation
9. C. Canning
10. A. Drying
11. F. Controlled atmosphere storage
12. J. Preservatives
13. B. Curing and smoking
14. E. Refrigeration and freezing
15. H. Pasteurization
16. D. Aseptic packaging
17. G. Fermentation
18. C. Canning
19. A. Drying
20. D. Aseptic packaging
21. True
22. product modification

Summary

Food-processing technology involves converting agricultural products into foods that have specific textures, appearances, and nutritional properties. This technology involves two basic types of processes: primary food processes and secondary food processes. Primary food processing involves two steps: material conversion and food preservation. Secondary food processing involves the development of the food products and their manufacture.

Test Your Knowledge

Write your answers on a separate piece of paper. Please do not write in this book.

1. Name one way in which food sustains life.
2. Define the term *food-processing technology.*
3. Food processes that make basic food ingredients are called ______.
4. Food processes that make finished food products are called ______.
5. What are the major steps in milling flour?
6. The two types of roasting used for coffee are air roasting and ______ roasting.
7. Name one step in the chemical conversion process of making cheddar cheese.

Matching questions: For Questions 8 through 20, match each description on the left with the correct food-preservation technology on the right. (Note: Answers can be used more than once.)

8. ______ Using X rays to kill mold and bacteria.
9. ______ Heating a food in a vacuum-sealed container.
10. ______ Removing water by placing a food in sunlight.
11. ______ Storing food in sealed areas with controlled environments.
12. ______ Using chemicals to preserve food.
13. ______ Treating meat with salt and adding flavor.
14. ______ Storing food at low temperatures.
15. ______ Using heat to kill microorganisms in milk and juices.
16. ______ Sterilizing a package and then placing sterilized food in it.
17. ______ Using microorganisms to break down the food's structure.
18. ______ Using blanching to drive out gases.
19. ______ Removing water by exposing food to hot air.
20. ______ Sterilizing the food containers with hydrogen peroxide.

A. Drying.
B. Curing and smoking.
C. Canning.
D. Aseptic packaging.
E. Refrigeration and freezing.
F. Controlled-atmosphere storage.
G. Fermentation.
H. Pasteurization.
I. Irradiation.
J. Preservatives.

21. Brainstorming is one of the first activities in developing a food product. True or false?
22. The step of altering the composition of a food product to incorporate the results of the initial test is called ______.
23. Producing a scaled-up food product using small versions of the equipment is called ______.
24. Listing the ingredients of a food and the methods used to produce it is called ______.
25. List the major manufacturing processes used in making a food product.

23. pilot production
24. product specification
25. Storing, cleaning, sorting and grading, measuring, size reduction, compounding, and heat processing

STEM Applications

1. Create a form similar to the one below. Take it to your neighborhood supermarket and find a specific product as an example of each type of preservation. Write the product name (for example, "Birdseye frozen baby peas") in the space provided.

Preservation Process	Product Example
Drying	
Curing or smoking	
Canning	
Aseptic packaging	
Refrigeration	
Freezing	
Fermentation	
Pasteurization	

2. Bake a batch of cookies as an example of secondary food processing. Make a step-by-step list as you convert flour and other ingredients into an edible finished product. Share your cookies with the class.

Chapter Outline
Technology and Wellness
Technology and Illness

This chapter covers the benchmark topics for the following Standards for Technological Literacy:

3 5 12 13 14 15

Learning Objectives

After studying this chapter, you will be able to do the following:

- Summarize the roles of science and technology in wellness and illness.
- Recall the four major factors involved in wellness.
- Summarize how technology is applied to exercise and sports.
- Recall the definition of *medicine.*
- Recall kinds of health-care professionals.
- Explain how technological devices are used in diagnosing illnesses and physical conditions.
- Explain how diagnostic imaging systems work.
- Explain how technological devices are used to treat illnesses and physical conditions.
- Explain how a new drug is developed.
- Summarize work done by biomedical engineers.
- Explain how technology is used in surgical procedures.

Key Terms

aerobic exercise
anaerobic exercise
computerized tomography (CT) scanner
dental hygienist
dentist
disease
drug
electrocardiograph (EKG) machine
emergency medicine
endoscope
ethical drug
illness
intervention radiology
invasive diagnostic equipment
magnetic resonance imaging (MRI)
medical technologist
medicine
noninvasive diagnostic equipment
nurse
nurse practitioner
pharmacist
physician
physician assistant
proprietary drug
prosthesis
radiation therapy
radiology
routine diagnostic equipment
surgery
ultrasound
wellness
X-ray machine

Strategic Reading

How has technology helped you when you were ill? As you read this chapter, list all the ways your health has benefited from technology.

Standards for Technological Literacy

14

Extend

Differentiate between illness and wellness, with emphasis on their impacts of quality of life.

Extend

Discuss with students, the relationship between wellness and preventative medicine.

TechnoFact

Divine causes were believed to be the reason for physical illness in prehistoric times. Priests treated sick people by attempting to chase out evil spirits or make peace with the gods. Physical conditions were not taken into account.

Throughout history, people have been concerned about living longer and better. They have sought ways to cure illnesses, repair damage to their bodies, and improve their health. Each of these challenges has led to the development of new technologies. People have created artifacts to care for, improve, and protect their health.

The search for healthy lives can be viewed from two perspectives. See **Figure 31-1.** The first is wellness, and the second is illness. ***Wellness*** is a state of physical well-being that is considered healthy. This state is generally achieved through proper diet and regular exercise.

Illness can be described as a state of poor health. A disease or sickness might cause illness. In the broadest sense, illness can be extended to include injuries caused by accidents.

Both science and technology play a role in wellness and illness. Science provides knowledge about the human body and health and describes the natural processes the body uses to maintain itself. Technology provides the tools and equipment needed to achieve wellness and treat illness and helps people maintain and restore the body's processes and functions.

Technology and Wellness

Wellness involves actions that keep the body healthy. This state contributes to physical fitness, which is a combination of good health and physical development. The objective of physical fitness is to maximize a person's health, strength, and endurance.

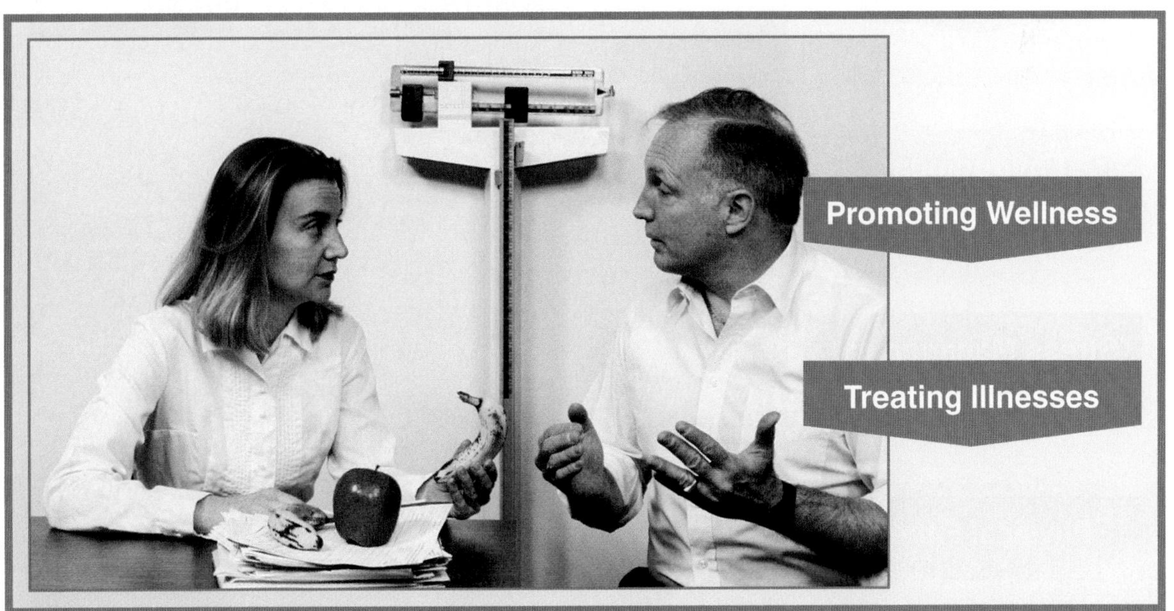

U.S. Department of Agriculture

Figure 31-1. People are living longer because of wellness programs and modern ways of treating illnesses.

Standards for Technological Literacy

14

Research

Ask students what people should do to remain well.

TechnoFact

Trephining is the earliest evidence of surgery. In this procedure, a hole was cut in a person's skull to release pressure in the head. While the physical conditions of headaches or other illnesses caused prehistoric people to develop this treatment 10,000 years ago, it was still believed that spirits were the cause of the illnesses.

The state of wellness can be considered preventive medicine and treatment that helps prevent people from becoming ill. Wellness often focuses on what people can do for themselves to maintain their well-being. This state considers at least four major factors. These factors are nutrition and diet, environment, stress management, and physical fitness. See **Figure 31-2.**

Wellness means people should be concerned with what they eat. Proper nutrition is important to physical fitness and wellness. The energy a person can expend depends on nutrition. If the diet is inadequate, the fitness level drops. Therefore, people should be conscious of the quantity and value of the food they consume. See **Figure 31-3.**

Figure 31-3. Wellness programs encourage people to select and eat proper foods.

The state of wellness also means individuals should be aware of the environment's effects on their body. For example, they should try to improve the quality of the air they breathe. They should control their exposure to direct sunlight. Technology has been applied to air and water purifiers and has been used to develop sunscreen lotions, sunlight-filtering clothing, and other effective products.

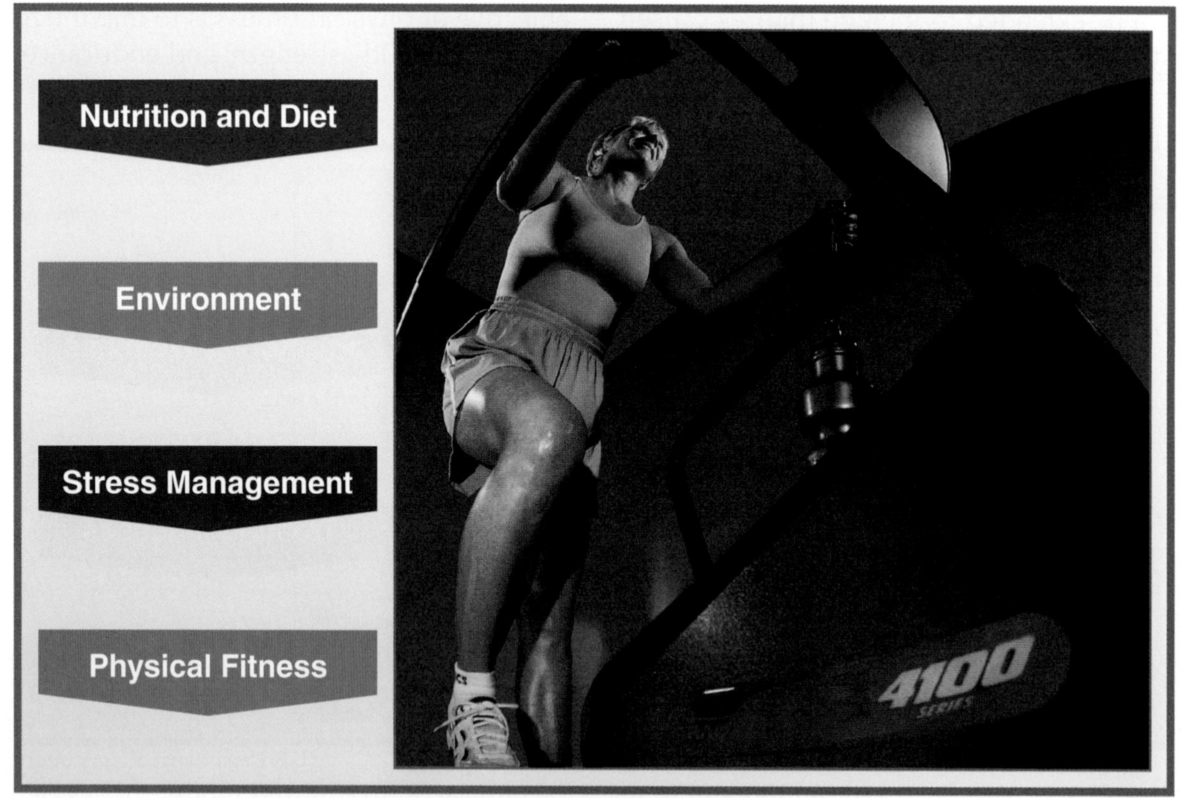

StarTrac

Figure 31-2. There are four parts to a wellness program.

The approach to wellness suggests that people should be aware of and control the emotional stress they encounter. Moreover, they should keep their bodies fit through activity and exercise. Technology has also been applied to stress management and physical fitness through two areas: exercise and sports.

Technology and Exercise

Exercise has been described as exerting the body or the mind. See **Figure 31-4.** We can have physical or mental exercise. Exercise is done for training or improving health. Two major types of exercise exist: anaerobic and aerobic.

Anaerobic exercise involves heavy work by a limited number of muscles. This type of exercise is maintained for short intervals of time. Examples of anaerobic exercise are weight lifting and sprinting. These activities are maintained only for short intervals. They increase strength and muscle mass, but they have limited benefits to cardiovascular health.

U.S. Department of Agriculture

Figure 31-4. Exercise prepares people for the demands of everyday life.

Aerobic exercise uses oxygen to keep large muscle groups moving continuously. The exercise must be maintained for at least 20 minutes. Aerobic exercise uses several major muscle groups throughout the body. This causes greater demands on the cardiovascular and respiratory systems to supply oxygen to the working muscles.

Standards for Technological Literacy

14

Brainstorm

Ask students why people should exercise.

Extend

Discuss with students why aerobic exercise is better that anaerobic exercise in a wellness program.

TechnoFact

The first organized methods of treating patients were developed by ancient Egypt. An Egyptian named Imhotep was an advisor to the Egyptian ruler at the time writing and recordkeeping were developed. Imhotep's abilities to heal made a great impression on his people. After his death, he was revered as the god of healing.

Career Corner

Biomedical Engineers

Biomedical engineers combine biology and medicine with engineering to develop devices and procedures that solve medical and health-related problems. They develop devices used in various medical procedures (such as the computers used to analyze blood and the laser systems used in corrective eye surgery), artificial organs, and imaging systems (such as magnetic resonance, ultrasound, and X-ray systems). Most biomedical engineers have a background in one of the basic engineering specialties, such as mechanical or electronics, and specialized biomedical training. The specialties within biomedicine include biomaterials, biomechanics, medical imaging, rehabilitation engineering, and orthopedic engineering. Unlike many other engineering specialties, a graduate degree is recommended or required for many entry-level jobs.

Standards for Technological Literacy

14

Research

Have students select an exercise machine and find out its specifications, uses, and construction.

TechnoFact

As some ancient Egyptian medicines are still used today, so are many medical treatments developed by ancient China. Acupuncture is one of the most common of these practices. It was created at a time when people still believed diseases were caused by spiritual forces; in this case, the imbalance of the yin and yang was seen as origins of illness.

Aerobic exercise includes walking, jogging, and swimming. This type of exercise can reduce the risk of heart disease and increase endurance.

Exercise can be done without special equipment. People can walk or jog to improve their health. Even these activities often require technology. Special shoes and clothing have been designed and produced to help in these activities. Also, many medical professionals and fitness experts suggest that people use exercise equipment to improve their health and well-being. This equipment includes treadmills, stationary bikes, stair climbers, rowers, and home gyms. See **Figure 31-5.**

Treadmills are moving belts that allow people to walk or jog in place. They provide an aerobic fitness workout. Stationary bikes are similar to bicycles, except they do not move. They allow people to obtain the benefits of bicycling without leaving home. In addition, on most of these bikes, the handlebars move and provide resistance. This feature provides an upper-body workout. Stair climbers allow people to obtain the benefits of climbing without having to have stairs available. These machines provide lower-body workouts.

Rowing machines simulate an actual rowboat with oars. They allow people to use their arms and legs without having a boat in the water. Home gyms are multistation exercise machines that allow people to work on many different muscle groups. These machines can provide a complete workout program.

Technology and Sports

Sports are another way to promote wellness and physical fitness. They are games, or contests, involving skill, physical strength, and endurance. Contests involving physical abilities can be traced back to prehistoric times. See **Figure 31-6.** Many different types of sports have developed throughout history. Medical proof of the benefits of physical exercise has reinforced the value of sports.

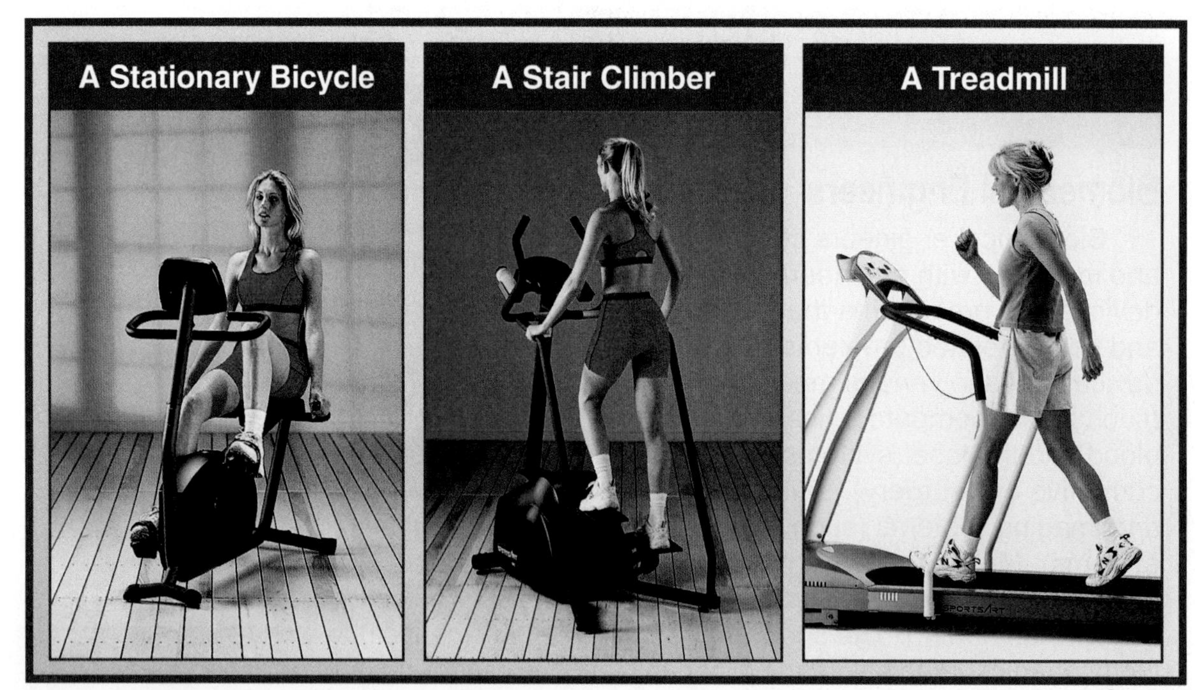

SportArt

Figure 31-5. Among the many types of exercise equipment are stationary bicycles, stair climbers, and treadmills.

©iStockphoto.com/Gala98

Figure 31-6. This amphitheater in historic Pompeii was used for sporting events and other activities in ancient times.

This discussion of sports and wellness is focusing on using sports for physical fitness. In this context, ordinary people use sports as a way to maintain health and well-being. We are not exploring sports as an economic activity here.

All sports require one or more types of technological products. These products include playing fields or venues, game equipment, and personal protection. Each game, sport, or contest uses specific technologies. For example, football's playing venue is different from tennis's. Likewise, the game equipment for football (the ball and goalposts, for example) is different from that for tennis (the ball, racket, and net, for example). Similarly, the personal clothing and protection equipment for football are different from those for tennis.

Technology and Playing Venues

Each sport requires playing courts, fields, or buildings. These venues are part of our built environment and are the result of construction technology. This technology is presented in Chapter 17.

Typically, a playing venue has a playing surface, goals, and constraints. See **Figure 31-7.** For example, golf uses grass-covered areas as a playing surface, holes for goals, and water and sand traps as constraints. Basketball uses a wood surface, a concrete surface, or an asphalt playing surface. This sport has a rim and backboard as a goal and painted lines as constraints. Skiing uses slopes covered with snow.

Technology is often used to improve the natural playing surfaces. For example, special grasses have been developed for golf courses. Fertilizers have been developed and manufactured to encourage the grass to grow. Special lawn-grooming and moving equipment has been developed to maintain the courses. Likewise, snow-grooming equipment prepares and renews the surfaces of ski runs. Snowmaking equipment has been developed to supplement

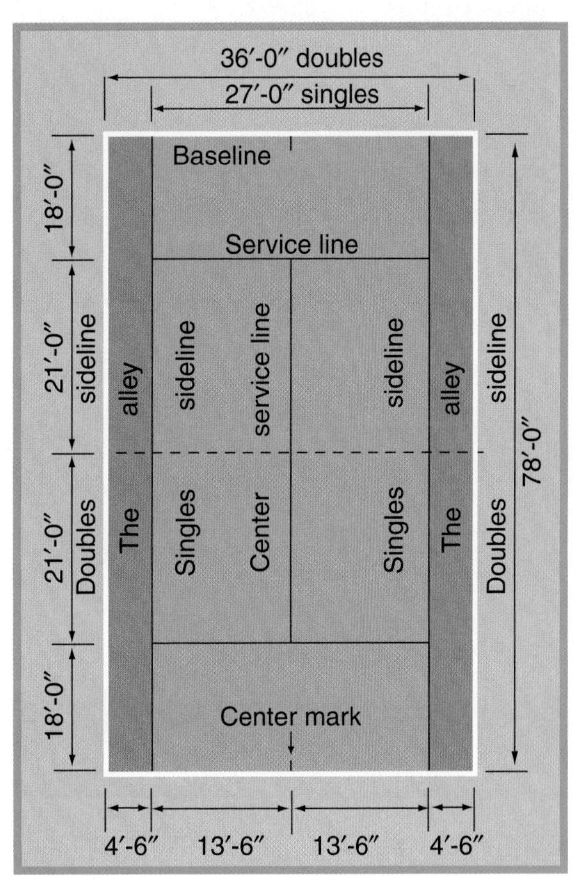

Figure 31-7. A tennis court.

Standards for Technological Literacy

14

Extend

Differentiate between sports played for competition and sports played for wellness.

Brainstorm

Ask students how competitive sports can work against wellness.

TechnoFact

Perhaps the first instance of believing in natural causes of illness was 2500 years ago in Greece. Hippocrates and his school took over with the development of diagnosis and intervention. Although Hippocratic doctors didn't perform surgery at this time, the methods of prevention and treatment are still the basis of today's medical treatments.

Standards for Technological Literacy

14

Research

Have students select a sport and determine the personal and game equipment required to play it.

Extend

Use the Internet to show how a piece of sports equipment is manufactured.

TechnoFact

After Hippocrates made his advancements in medicine, another Greek physician named Galen brought about his own medical developments. While Hippocrates didn't perform dissections, Galen used methods of surgery to discover how the human body worked. Though many of his theories were proved false in the 16th century, his works were considered authoritative.

natural snow for skiing. See **Figure 31-8.** This equipment makes snow by breaking water into small particles with compressed air. The water is cooled as it moves through cold air, forms small particles of ice, and is distributed as snow on a surface.

Technology and Game Equipment

As we noted earlier, each sport has its own playing equipment. Hockey has pucks and sticks. Baseball has bats and balls. Tennis has rackets and balls. These manufactured products are designed and produced to specifications developed by groups who establish rules for playing the sport.

Technology and Personal Protection

Players in many sports wear special clothing and protective gear. See **Figure 31-9.** The clothing might be made from fabrics that wick perspiration away from the body or shed rain. Special shoes

©iStockphoto.com/mollin

Figure 31-8. The photo shows a snowmaking machine in action. The drawing below it shows how the machine works.

Figure 31-9. The shoes and batting helmets are examples of special equipment designed for sports.

can be developed to provide foot support and absorb the shock that running on hard surfaces causes.

Some sports can lead to bodily injury or muscle damage. Participants in these sports wear protective gear. Baseball players wear batting helmets to protect their heads from wild pitches. Football players wear protective devices to protect their heads, necks, bodies, hands, and feet. The game equipment and protective equipment are technological products. They were designed for a specific purpose and have a customer base of amateur and professional athletes.

Technology and Illness

The second focus of health involves the area often called ***medicine***. Medicine involves diagnosing, treating, and preventing diseases and injuries. This area helps people live longer and lead more active lives. Medicine's goal is to reduce human suffering and physical disability. This goal involves health-care professionals working with patients who are ill or injured. Medicine's goal also includes searching for new drugs, treatments, and technology.

Disease is any change interfering with the appearance, structure, or function of the body. Treating disease and injury requires a number of different health-care professionals. See **Figure 31-10.** These professions include the following:

- ***Physicians.*** These are individuals who diagnose diseases and injuries. They administer appropriate treatment and advise patients on ways to stay healthy.
- ***Nurses.*** These are individuals who help physicians diagnose and treat illnesses and injuries. They assist physicians during examinations, treatment, and surgery. Nurses observe and record symptoms that patients exhibit. They administer medications and provide care in hospitals and nursing homes.

Physicians
Nurses
Nurse Practitioners
Physician Assistants
Medical Technologists
Dentists
Dental Hygienists
Pharmacists

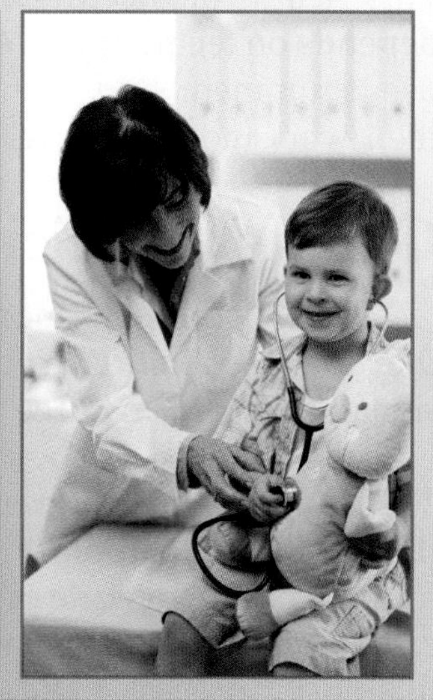

Figure 31-10. Health-care professionals work to cure illnesses and to help people stay healthy.

- ***Nurse practitioners.*** These are individuals who perform some of the basic duties physicians once provided. They diagnose and treat common illnesses and prescribe medication.
- ***Physician assistants.*** These are individuals who deliver basic health services under the supervision of a physician. They examine patients, order diagnostic tests and X-ray films, and prescribe drugs or other treatments.
- ***Medical technologists.*** These are individuals who gather and analyze specimens to assist physicians in diagnosis and treatment.
- ***Dentists.*** These are individuals who diagnose, treat, and help prevent diseases of the teeth and gums.
- ***Dental hygienists.*** These are individuals who assist dentists in surgery. They also clean teeth and advise patients on proper techniques to prevent tooth and gum disease.
- ***Pharmacists.*** These are individuals who dispense prescription drugs and advise people on their uses.

Standards for Technological Literacy

14 15

Research

Have your students select an occupation in the health care field and research the education requirements for it.

TechnoFact

The Islamic areas of southwest and central Asia developed new medical advancements during the Middle Ages. Rhazes identified and described contagious diseases, and Avicenna compiled a summary of medical knowledge into the *Canon of Medicine*.

Standards for Technological Literacy

3 14 15

Extend

Differentiate between medical science and medical technology.

Reinforce

Explain the relationship between medical diagnosing and product service diagnosing that was presented in Chapter 18.

TechnoFact

While the contributions of Islamic medical practitioners took place in the Middle Ages, medical schools and hospitals rose in Europe. One of the first prominent schools was established in Salerno, Italy in the 10th century, and it was a renowned medical education center for centuries.

This team of health-care professionals depends on technology in their work. They use equipment and techniques to make their work more effective.

Goals of Medicine

Many people seek medical care because they are ill or injured. Health-care professionals respond in three major ways. See **Figure 31-11.**

First, the medical personnel diagnose the illness or condition. Diagnosis is performed by conducting interviews, physical examinations, and medical tests. The diagnostic process tries to determine the nature or cause of the condition. Medical personnel then treat the illness or condition. Treatment involves applying medical procedures to cure diseases, heal injuries, or ease symptoms.

Prevention is the third common action health-care professionals take. One method of prevention has already been discussed. This method is wellness programs promoting exercise and proper diet.

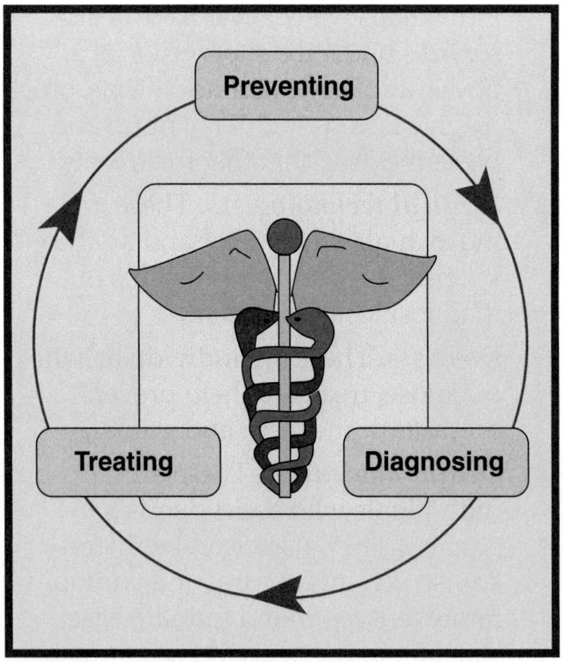

Preventing

Treating

Diagnosing

Figure 31-11. Health-care professionals have three major goals.

The other way is through immunization. See **Figure 31-12.** This is a process of exposing the body to small amounts of a bacterium, protein, or virus to cause antibodies to form. These antibodies are cells or proteins that circulate in the blood and attack foreign bacteria, viruses, or fungi. These foreign substances can cause diseases.

©iStockphoto.com/YvanDube

Figure 31-12. Immunizations (shots) help people from getting diseases.

Technology in Medicine

Many medical actions involve tools and equipment. Physicians and dentists use equipment to diagnose and treat illnesses and conditions. They use technology to extend the potential to deal with medical problems. The use of technology throughout medicine is different from *medical technology.* This term defines the narrow field of gathering and analyzing of specimens to assist physicians in diagnosing and treating illnesses.

Technology and Diagnosis

In the past, physicians depended on people to describe their symptoms. From these descriptions, the physicians planned treatment procedures. Physicians found, however, that descriptions are not always accurate and that many are hard to interpret. To deal with these problems, physicians and others have developed diagnostic equipment. Today, many different types of diagnostic devices exist. For this discussion, three major types are examined here:

- Routine diagnostic equipment.
- Noninvasive diagnostic equipment.
- Invasive diagnostic equipment.

Routine diagnostic equipment

Routine diagnostic equipment is used to gather general information about the patient. This equipment includes scales to determine the patient's weight and thermometers to determine body temperature. Routine diagnostic equipment also includes devices to do such things as measure the oxygen in the blood, listen to heart rhythm (stethoscopes), and measure blood pressure. See **Figure 31-13.** These items provide a baseline of general information.

Noninvasive diagnostic equipment

Noninvasive diagnostic equipment gathers information about the patient without entering the body. A typical example of this type of diagnostics is called ***radiology***. Radiology uses electromagnetic radiation (waves) and ultrasonics (high frequency sounds) to diagnose diseases and injuries. See **Figure 31-14.** Diagnostic radiology uses special equipment called *body scanners* or *body-imaging equipment*. They produce images (pictures) of the body without entering it.

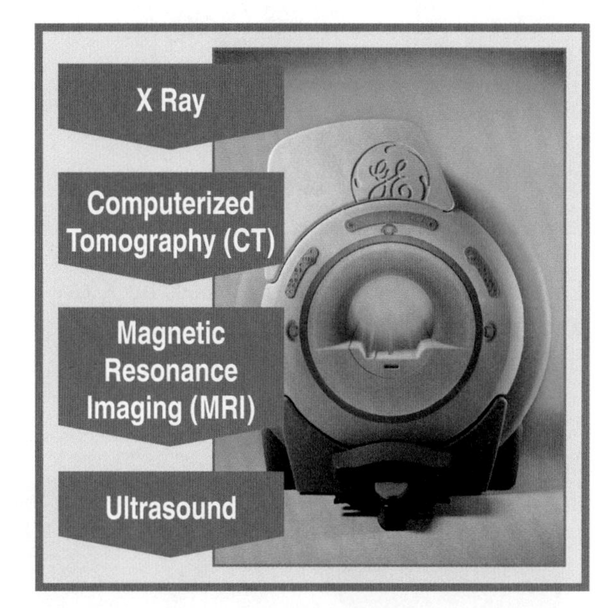

Figure 31-14. The four types of radiology are X ray, CT, MRI, and ultrasound.

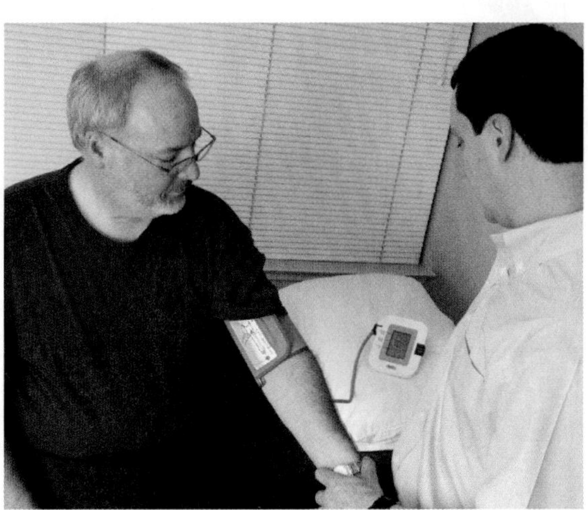

©iStockphoto.com/leezsnow

Figure 31-13. This physician is checking a patient's blood pressure using a diagnostic tool.

One of the most common diagnostic imaging machines is the ***X-ray machine***. An X-ray machine is essentially a camera that uses X rays instead of visible light to expose the film. X rays are electromagnetic waves that are so short they can pass through solid materials such as paper and human tissue. Denser materials, however, such as metals and human bones, absorb some or all of the waves. Thus, if you put a piece of film under your hand and then pass X rays through your hand, the skin and tissue let most of the X rays pass directly through. The film behind the skin and tissue is completely exposed. The bones, however, absorb most of the X rays. The film behind them is not exposed completely. When the

Standards for Technological Literacy

3 14 15

Research

Have your students research the historical development of X-rays and X-ray diagnosis and therapy.

TechnoFact

Like many other subjects of arts and sciences during the Renaissance, important advancements in the study of medicine occurred. Andreas Vesalius was a leader in the study of human anatomy, and in 1545, the first scientific textbook on the subject was published.

Standards for Technological Literacy

3 14 15

Figure Discussion

Use Figure 31-16 to explain to students how X-ray imaging works.

TechnoFact

In the 17th century, another aspect of Galenic theory was proved false after William Harvey concluded that the heart pumps blood throughout the body, not the liver. He discovered the heart is central to the human body, pumping blood out into the body through the arteries and back in through the veins.

film is developed, an image of the bones in the hand appears. Any fractures or joint deformities are shown. See **Figure 31-15.**

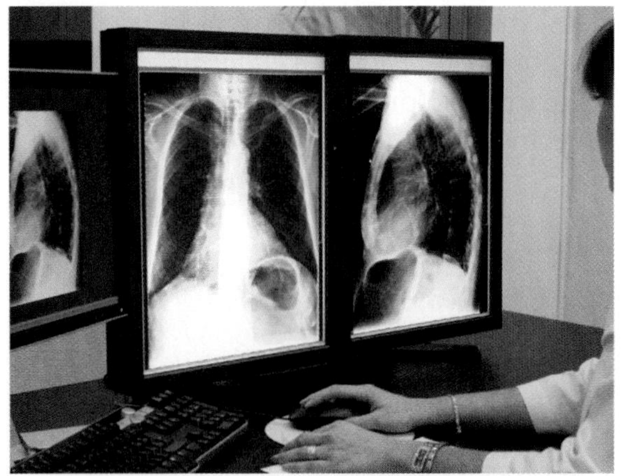

©iStockphoto.com/Adivin

Figure 31-15. A doctor analyzes the X ray to determine if a person needs medical treatment.

The first X-ray machines were used to detect fractures in bones and shadows on lungs. Later, substances that absorb X rays were introduced into the body. Various cavities in the body are filled with a material to aid the X-ray process. This material is more transparent or more opaque when X rays pass through than the surrounding tissue is. When the X rays are taken, it brings the particular organ more sharply into view. For example, a patient swallows barium sulfate so gastric ulcers can be located. In other cases, dyes are injected so the heart, kidneys, or gallbladder can be X-rayed. This allows people to produce radiographs, or X-ray photographs, of the desired area.

Figure 31-16 shows a diagram of an X-ray machine. This machine has an X-ray tube with a positive electrode (an anode) and a negative electrode (a cathode). The

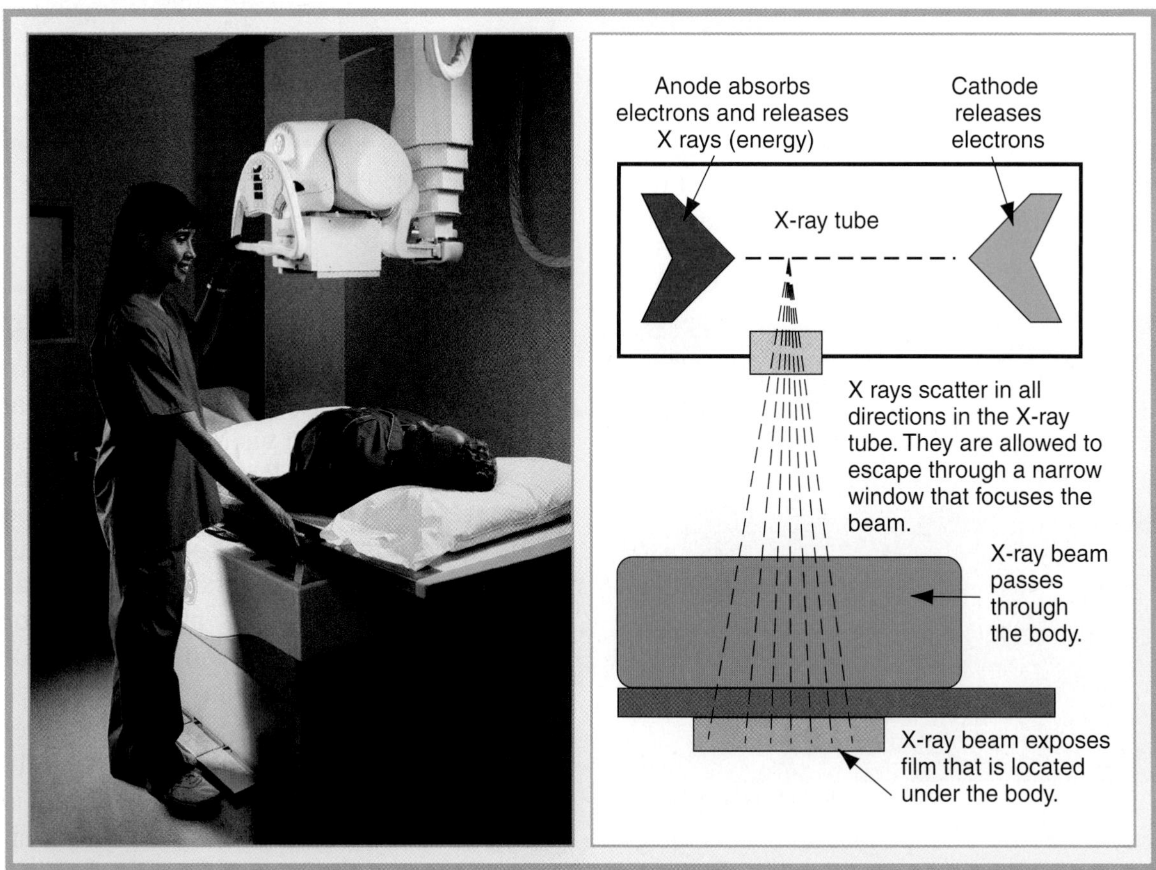

GE Medical Systems

Figure 31-16. How an X-ray machine works.

cathode is heated and gives off electrons. They travel toward the anode, where they hit and release energy (X rays). The X rays are concentrated into a beam that leaves the X-ray tube. The beam is directed at the desired location on the body. There, it passes through the body and exposes light-sensitive film. When the film is developed, an image of the section of the body is revealed. Some newer X-ray machines do not use film, but instead they produce digital images that can be viewed on computer systems.

The major disadvantage of X rays is that the images are two-dimensional. X rays are flat images taken of a three-dimensional object (the body). Therefore, depth is not shown. To deal with this shortcoming, people have developed ***computerized tomography (CT) scanners***. See **Figure 31-17.** A CT scanner produces images of any part of the body without using dyes. The scanner rotates around a patient's body and sends a thin X-ray beam at many different points. Crystals opposite the beam pick up and record the absorption rates of the bone and tissue. A computer processes the data into a cross-sectional image of the part of the body being scanned. This digital image can be viewed immediately or stored for later use.

X rays can also be dangerous to use. They can cause damage to body parts. To deal with the hazards associated with X rays, people have developed new imaging techniques. One of these is ***magnetic resonance imaging (MRI)***. See **Figure 31-18.** This technique can produce computer-developed, cross-sectional images of any part of the body very quickly. The images are developed using magnetic waves, rather than X rays. See **Figure 31-19.**

Standards for Technological Literacy

3 14 15

Extend

Compare and contrast X-ray, CT, and MRI systems.

TechnoFact

Though Edward Jenner is credited for creating the first vaccination in 1796, inoculations had been in practice for centuries. The further development of vaccines occurred in 1936, when Gerhard Domagk developed the treatment prontosil while in the fight against infection. Finally, penicillin became the first antibiotic in the 1940s as an effort by English and American scientists.

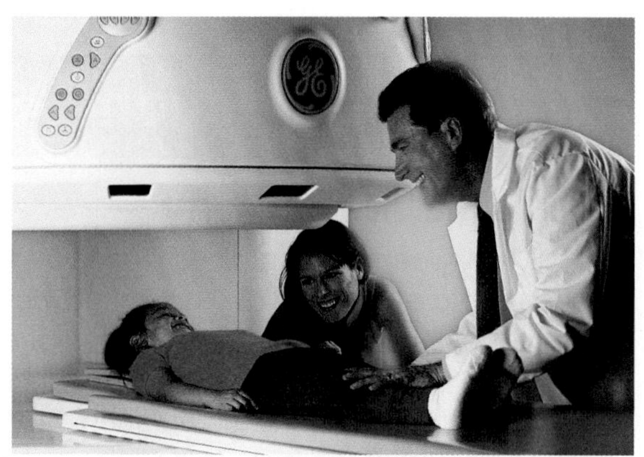

GE Medical Systems

Figure 31-18. An MRI unit uses magnetic waves, rather than X rays.

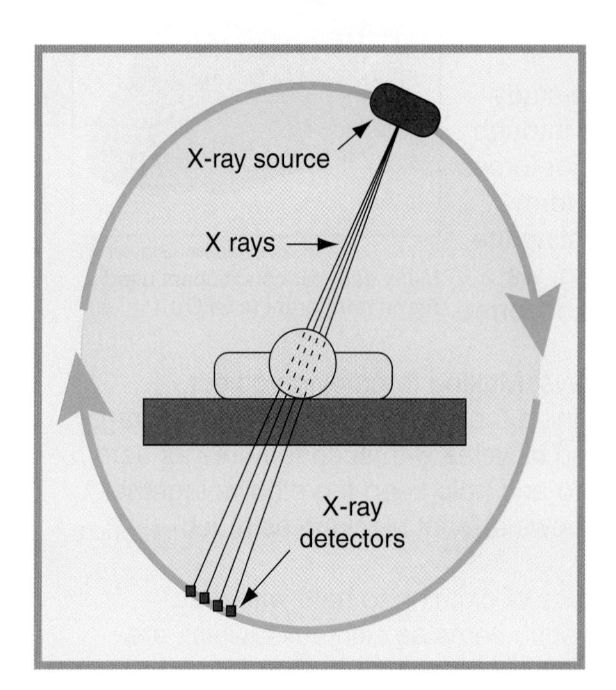

Figure 31-17. A CT scanner rotates around a person's body.

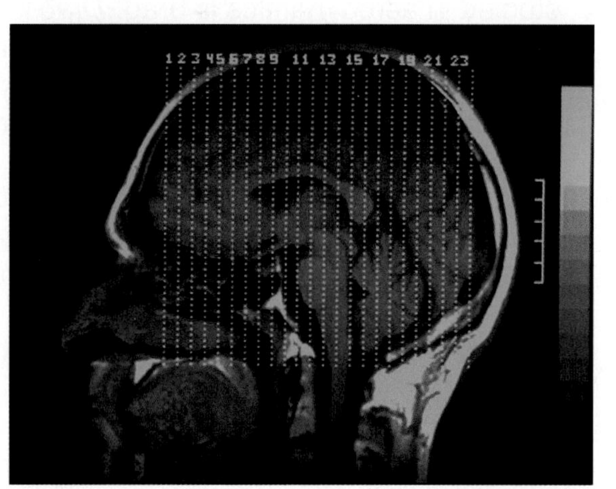

©iStockphoto.com/Seti

Figure 31-19. The result of an MRI of a patient's head.

Standards for Technological Literacy

3 14 15

TechnoFact

Until the late 19th century, bacteria created by operations were not taken into account. Therefore, anything treated in an operating room could become infected and cause a person to die. Joseph Lister was one of the first to work on destroying microorganisms that cause infection, and he began the promotion of sanitation in operating rooms by medical staff.

An MRI unit is a large tube surrounded by a circular magnet. The patient lies on his back on a bed, which is moved into the magnet. The body is positioned so the part to be scanned is in the exact center of the magnetic field. At this point, the scan can begin.

This unit exposes the patient to radio waves in a strong magnetic field. The magnetic field lines up the protons in the tissues. A beam of radio waves then spins these protons. As the protons spin, they produce signals that a receiver in the scanner picks up. The computer processes these signals to produce very sharp and detailed images of the body.

Another imaging technique is called ***ultrasound***. See **Figure 31-20.** This technique uses high frequency sound waves and their echoes to develop an image of the body. The ultrasound machine subjects the body to high frequency sound pulses using a probe. The sound waves travel into the body. There, they hit a boundary between tissues. This boundary can be between soft tissue and bone or between fluid and soft tissue. Some of the sound waves are reflected back to the probe. Others travel further, until they reach another boundary and get reflected. The probe picks up the reflected sound waves. A computer processes them to produce a

STEM Connections: Science

Integrated STEM Curriculum — Science, Technology, Engineering, Mathematics

Aerodynamics

Many people use a bicycle simply to get somewhere or to exercise. They do not, however, realize the role science plays in many areas of cycling. For example, the scientific study of the principles of aerodynamics has greatly affected such aspects of cycling as the design of the bicycle and the clothing a rider might wear for more effective cycling.

Generally speaking, aerodynamics is the study of the forces acting on an object as it moves through the air or other gaseous fluid. One of the major forces studied in aerodynamics is drag. *Drag* is the term used to describe the aerodynamic force resisting the forward movement of an object. When cycling, we encounter two kinds of drag: frictional drag and form drag. Both forces slow a cyclist.

©iStockphoto.com/SocjosensPG

Many early air conditioners used Freon refrigerant (a CFC).

Scientists counter these forces in two ways. Making the moving object smoother can lessen frictional drag. Streamlining the object can lessen form drag. To counter drag, then, designers have created bicycles with such features as aero bars. These handlebars lower the rider's torso and help keep the elbows together. Drag is decreased because the torso and elbows are not blocking as much airflow.

Designers have also created different kinds of clothing to help with drag. For example, many professional riders wear such items as skin suits, which are designed to reduce friction. Bicycles also have other features dealing with the principles of aerodynamics. Can you think of two of these?

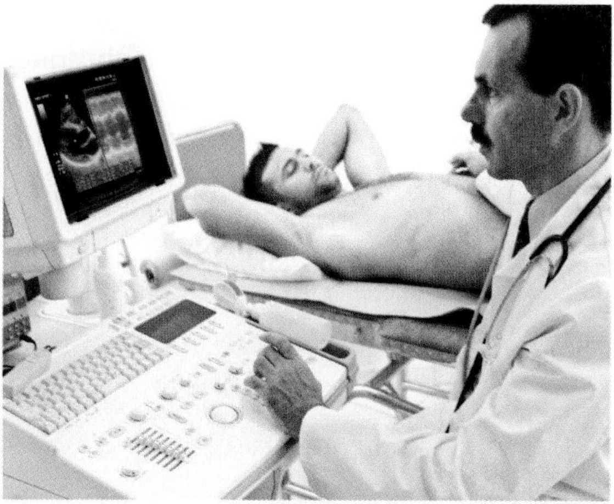

©iStockphoto.com/nyul

Figure 31-20. A cardiologist is using an ultrasound machine to scan the heart of a patient.

still image (photograph) or moving image (television). In a typical ultrasound procedure, millions of pulses are sent each second. The probe is moved along the surface of the body or angled to obtain different images.

Other technological devices are used in diagnosis besides imaging devices. A very important device is the ***electrocardiograph (EKG) machine***. See Figure 31-21. This device produces a visual record of the heart's electrical activity. As the heart works, it sends off very small electrical signals that can be detected on the skin. The EKG machine uses electrodes attached to the skin to capture the signals. The machine amplifies the signals, and it produces a graph of their values. Health-care professionals can read this graph to determine how the heart is functioning.

Another important diagnostic device is the ***endoscope***, which allows a physician to look inside the body. An endoscope is a narrow, flexible tube containing a number of fiber-optic fibers smaller in diameter than a human hair. The tube can be threaded through a natural opening, such as the throat, or through a small incision. Light sent through the fibers shines on an interior part of the body. This light is reflected back

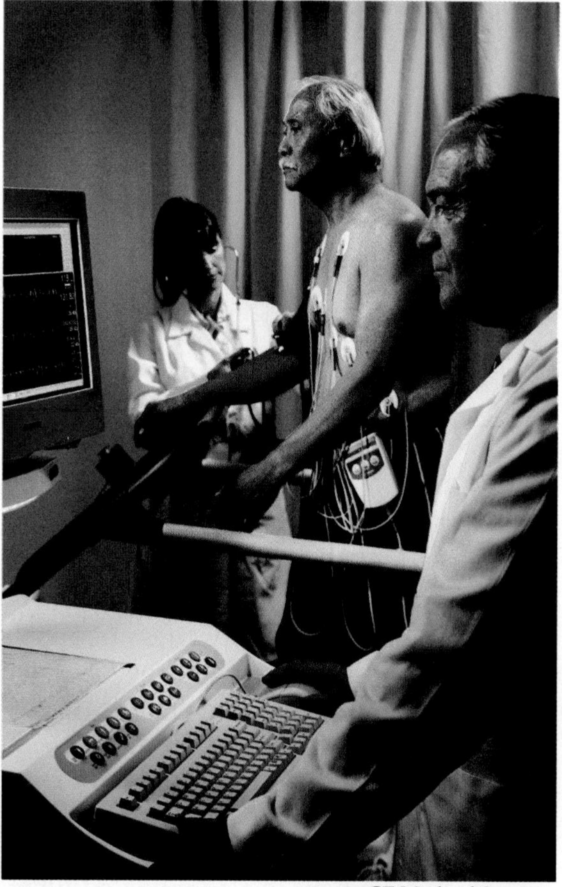

GE Medical Systems

Figure 31-21. This person is being diagnosed with the use of an EKG machine.

through the fibers to form a series of dots. Each fiber in the tube produces one dot. The dots form a picture of an internal organ or other part of the body. These examples are just a few of the many devices that have been designed and built to help diagnose illnesses and physical conditions. They show the dramatic use of technology to help reduce human suffering.

Invasive diagnostic equipment

Invasive diagnostic equipment is used when drawing and testing a blood sample. A blood test determines the chemical composition of a sample of blood. The sample is tested in a laboratory using a number of different technological procedures. See Figure 31-22. The tests can

Standards for Technological Literacy

3 14 15

Research

Ask students to research how an EKG machine works.

TechnoFact

After the War of 1859 in Europe, survivors moved to establish a coalition to transport wounded soldiers from the battlefield. The first American horse-drawn ambulance service was established in New York in 1869.

Standards for Technological Literacy

3 14 15

Extend

Differentiate between illegal and ethical drugs and between ethical and proprietary drugs.

TechnoFact

Although many physicians attempted to dull the pain of surgery with drinks and drugs, neither of these treatments was an effective anesthetic. It wasn't until the 1840s that two Americans began to use ether gas to put patients to sleep and keep them peaceful during surgery.

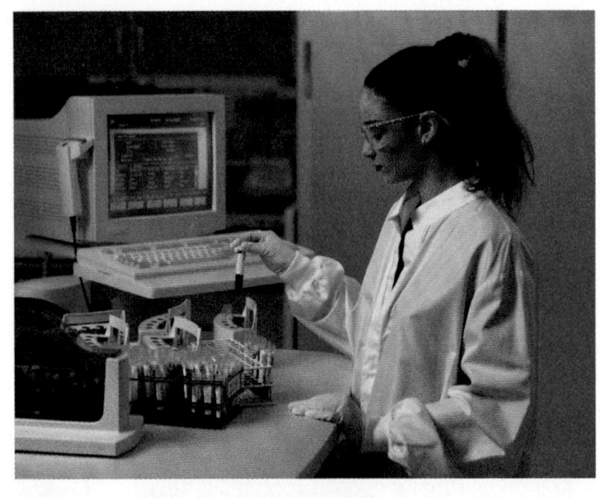

Figure 31-22. This technician is running tests of blood samples.

detect the presence of specific chemicals associated with a disease. They can also detect an imbalance in the chemical composition of the blood. These data provide health-care professionals with information needed to treat illnesses or physical conditions.

This equipment is also involved when taking tissue samples (biopsies) for laboratory examination. Many medical conditions, including cases of cancer, are diagnosed by removing a sample of tissue. A pathologist examines this tissue. This medical professional is a person who is trained to perform medical diagnoses by examining body tissues and fluids removed from the body.

Technology and Treatment

Treatment of illnesses and physical conditions can require drugs, specialized equipment, or both. Both of these approaches are the result of technology. They are the products of design and production actions.

Treatment with drugs and vaccines

Humans have always experimented with substances to treat pain and illness and to restore health. The development of new treatments and prevention techniques begins with the unmet medical needs of people. One common treatment method is drugs. See **Figure 31-23.**

Figure 31-23. Drugs are important tools in treating illness. They can be obtained either by prescription or over-the-counter.

A ***drug*** is a substance used to prevent, diagnose, or treat a disease. This substance can also be used to prolong the lives of patients with incurable conditions. Throughout history, drugs have improved the quality of life for people. For example, a vaccine, a special type of drug, is a substance administered to stimulate the immune system to produce antibodies against a disease. Vaccines have helped eliminate such diseases as smallpox and poliomyelitis (polio). Drugs can be classified in many ways. These classifications include the following:

- **Method of dispensing.** Drugs can be classified as prescription or over-the-counter (proprietary) drugs. Those that physicians prescribe and pharmacists dispense are known as ***ethical drugs. Proprietary drugs*** are those considered safe for unsupervised use by consumers. They are available to everyone through retail outlets.

Technology Explained

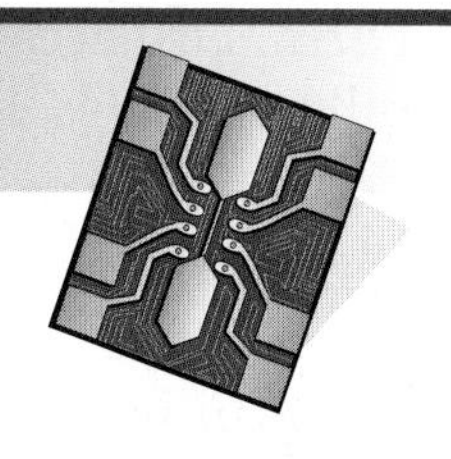

Dialysis Machine: a device that can do some of the functions of the human kidneys. This machine filters waste and fluid from the blood.

For any number of reasons, some people's kidneys stop functioning. They stop removing urea and certain salts from the blood. This condition is fatal, unless the person receives treatment. One treatment is called *hemodialysis*. This treatment is the process of removing waste and excess fluid from the blood. An artificial-kidney machine does this.

The word *dialysis* describes the movement of microscopic (very small) particles from one side of a semipermeable membrane to the other side of the membrane. Hemodialysis is a special type of dialysis. This type involves the blood (hemo-). *Hemodialysis* literally means "cleaning the blood."

In hemodialysis, the blood is pumped out of the patient's body to the artificial-kidney machine. See **Figure A.** In this dialysis machine, blood runs through tubes made of a semiporous membrane. Surrounding the tubes is a sterile liquid called the *dialysate solution*. This liquid is made up of water, sugars, and other components.

As the blood circulates, the red and white cells and other important components are too large to fit through the pores in the membranes. Impurities, such as urea and salt, however, pass through the membrane. See **Figure B.** The dialysate solution, which is discarded, carries them away. Tubes connected to the kidney machine return the cleaned blood to the bloodstream.

Dialysis is a treatment for people in the late stages of chronic kidney failure. Trained professionals generally perform the procedure. See **Figure C.** Dialysis normally takes three to five hours to complete. Typical patients receive three treatments a week. Dialysis allows these people to maintain many of their normal activities after the treatments.

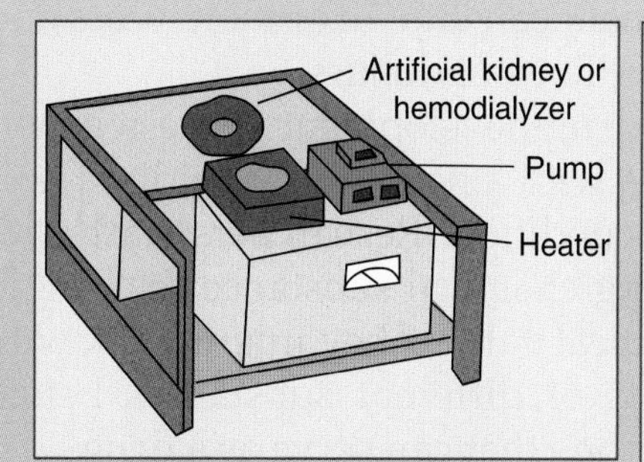

Figure A. The tools used for hemodialysis.

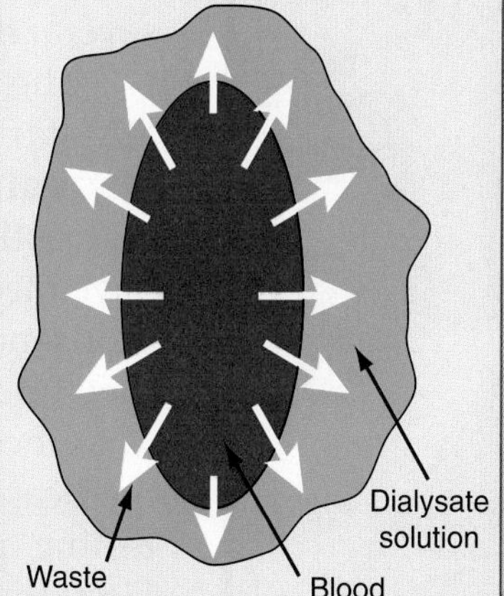

Figure B. Wastes pass from the blood, through the membrane of the tubing, into the dialysate solution.

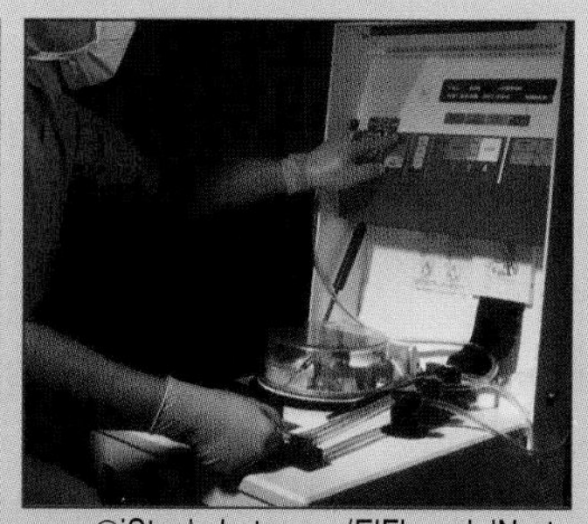

Figure C. A technician prepares a portable dialysis machine for use.

Standards for Technological Literacy

3 14 15

TechnoFact

Various scientists in the early 1900s, including F.G. Hopkins and Casimir Funk, performed experiments that allowed them to recognize the importance of vitamins in human diets. Their research showed that nutritional imbalances caused such diseases as beriberi and scurvy.

Standards for Technological Literacy

3 14 15

Research

Have students research the methods for developing new drugs.

TechnoFact

Chemotherapy is a treatment that was invented in the early 1900s. It wasn't commonly used as a treatment until 1948. Today, it is a process used to treat cancer in an attempt to destroy diseased cells.

- **Form.** Drugs can be classified as pills, capsules, liquids, or gases. See **Figure 31-24.**
- **Method of administering.** Drugs can be classified as injections (shots), oral drugs, inhalants, or absorbable drugs (patches).
- **Illness or condition treated.** Drugs can be classified as cancer drugs, measles vaccines, or high blood pressure drugs, for example.

Today, most drugs are the products of chemical laboratories. See **Figure 31-25.** They are called *synthetic drugs*, or *human-made drugs*. These drugs are developed because they can be controlled better than natural drugs can be controlled. Not all synthetic drugs are totally new. Some are developed by altering the structure of existing substances. These new drugs are called *analogs*. This means they have a structure similar to another compound, but they have a slightly different composition. These new drugs might be more effective than the original drugs. Also, they might cause fewer side effects.

A number of new drugs have been developed by using gene splicing or recombinant DNA. This approach joins the DNA of a selected human cell to the DNA of a second organism, such as a harmless bacterium. The new organism can produce the disease-fighting substance. This new substance is extracted from the bacterium and processed into a drug.

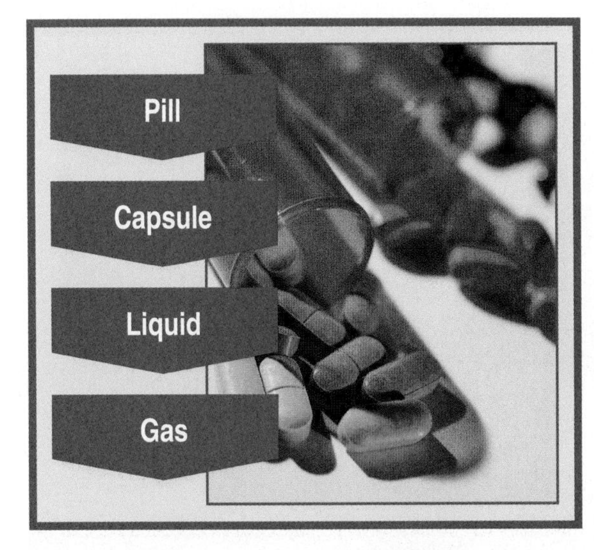

Figure 31-24. Drugs come in many different forms.

Figure 31-25. Most drugs are products of chemical laboratories.

The first drug produced using gene splicing was the hormone insulin. This drug was created in 1982 by inserting a human insulin gene into *E. coli* bacteria. The development of any new drug requires great amounts of time and money. This development can take 10 or more years and several millions of dollars.

The drug-development process generally starts with a need to treat a disease or physical condition. Researchers start with an existing chemical substance that might have medical value. They might work with thousands of different substances before they find one that can serve as a drug.

Once a new substance that might have medical value is discovered, an extensive testing program starts. See **Figure 31-26.** First, the drug is tested on small animals, such as rats and mice. If the tests show promising results, additional tests are conducted on larger animals, such as dogs and monkeys.

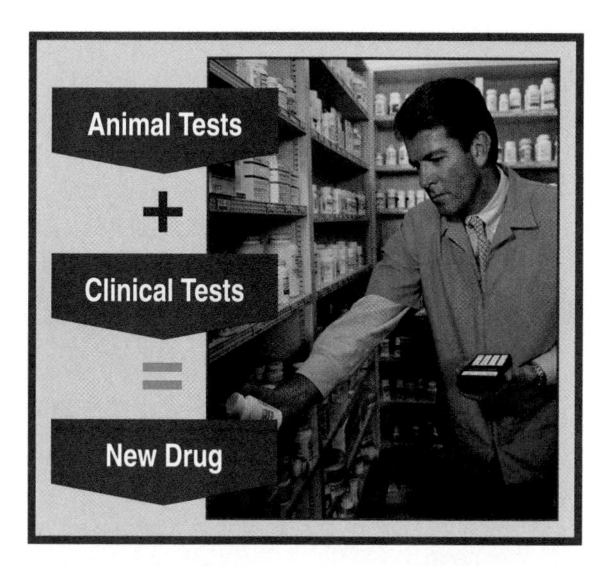

Figure 31-26. Before a drug is made available to treat illnesses, it must go through animal tests and three stages of clinical tests.

The tests are evaluated in terms of treating the disease and physical condition. Also, the drug must have a low level of toxicity (capability to poison a person). Drugs judged to be effective in animal tests are ready for the next level of testing. At this point, a request is made to the Food and Drug Administration (FDA) to conduct clinical tests.

If the FDA approves the request, the drug can be tested on humans. These tests are generally conducted in three phases. Each of these phases can take many months to complete. During the first phase, the drug is given to a small number of healthy individuals. These tests are designed to determine the drug's effect on people.

If the drug passes this test, it moves into the next phase. Here, it is given to a small number of people who have the disease or physical condition that the drug will treat. These individuals are divided into two groups. The first group is given the drug. The other group is given an inert substance, such as sugar, used in place of an active drug. This inactive compound is called a *placebo*. During the test, neither the group nor the researchers know who is receiving the drug or who is receiving the placebo. This research technique is called a *double-blind study*. A double-blind study keeps participants from influencing the results of the study.

Standards for Technological Literacy

5 14 15

Research

Ask students to research the Food and Drug Administration (FDA) and its role in the development of new drugs.

Extend

Discuss the historical development of surgery and anesthetics.

Think Green

Green Household Cleaners

One reason you clean your house is to disinfect, so you can stay healthy. Conventional household cleaners combine several types of components that may have adverse effects on the environment. Ammonia, which is often used as a cleaning agent, can taint the water supply. Bleach, commonly found in cleaners, can also contaminate water. In both of these cases, while the chemicals are not directly harmful to humans, they will kill fish and possibly other animals.

Without these toxins, household cleaners can still be effective. Several alternatives to these chemicals are available for purchase. Companies that produce these alternatives often try to be environmentally responsible from the product itself to the packaging it uses. They use carbonate- and plant-based materials (such as vegetable oils) rather than chemical cleaners. These materials are not harmful to the environment or to animals, and they biodegrade more easily. There are no VOCs emitted with green household cleaners. Some types of green cleaners are also easy to make yourself.

Standards for Technological Literacy

14 15

Research

Ask students to research the major departments in a hospital. Ask them how each of these departments contribute to treating ill or injured patients.

TechnoFact

Although many therapies were used in cases of poliomyelitis in the first half of the 20th century, it wasn't until the development of Salk's vaccine and Sabin's oral vaccine that new cases of polio were no longer an issue.

Drugs that pass the second phase move into a final phase of testing. Here, the drug is tested with a much larger group of people. The goal of these tests is to determine specific doses, side effects, interactions with other drugs, and other information. The data from these tests is used in drug labeling.

When the third phase is complete, the results of the tests are submitted to the FDA for approval. The agency must decide whether or not the drug is effective and safe. Also, it must weigh the drug's benefits against any risks that might be present. If the FDA determines that the drug meets its criteria, it approves the drug for use.

Treatment with medical equipment

Biomedical engineers develop a great deal of diagnostic and treatment equipment. See **Figure 31-27.** These professionals combine engineering with medicine to improve health care. They define and solve problems in biology and medicine. The work biomedical engineers do includes the following:

- Designing life-support equipment, such as cardiac pacemakers, defibrillators, and artificial kidneys.
- Designing artificial body parts, such as hearts, blood vessels, joints, arms, and legs.
- Designing computer systems to monitor patients while in surgery and intensive care.
- Designing sensors for the blood's chemistry.
- Designing instruments and devices for therapeutic uses, such as a laser system for eye surgery or a catheter to clear out a blocked blood vessel.
- Designing medical imaging and treatment systems.

Health-care professionals use these and other technological systems to treat injuries and diseases. In this section, we examine three major treatments: radiation therapy, surgery, and emergency medicine.

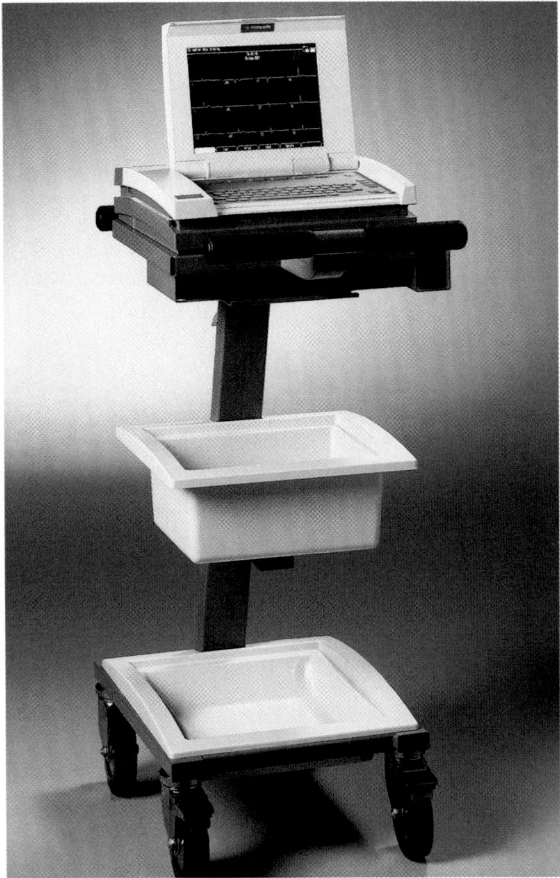

GE Medical Systems

Figure 31-27. This EKG machine is the result of many people's work, including biomedical engineers.

Obviously, these are only three among many treatment technologies.

Many types of cancers can be treated using ***radiation therapy***. This treatment is called *therapeutic radiology* (radiology for the treatment of disease or disorders). Therapeutic radiology uses high-energy radiation to treat the cancer cells. The technique works because the radiation destroys the cancer cells' ability to reproduce. Normal cells can recover from the effects of radiation better than cancer cells can.

Sometimes, radiation therapy is only part of a patient's treatment. Patients can be treated with radiation therapy and chemotherapy (chemicals or drugs). ***Surgery*** might follow this treatment.

Another use of radiology is called ***intervention radiology***. This technique uses images that radiology produces for nonsurgical treatment of ailments. These images allow the physician to guide catheters (hollow, flexible tubes), balloons, and other tiny instruments through blood vessels and organs. An example of this approach is balloon angioplasty, which uses a balloon to open blocked arteries.

Surgery is a common way to treat a disease. See **Figure 31-28.** Surgery treats diseases and injuries with procedures called *operations*. This treatment can be used to remove diseased organs, repair broken bones, and stop bleeding. Most surgery involves manually removing (cutting) diseased tissue and organs.

Newer technologies, however, are also being used for many types of surgery. High frequency sound waves (ultrasound) can be used to break up kidney stones. Lasers use a beam of light to vaporize or destroy tissue. An endoscope can be used with special devices to operate on a particular area of the body. In transplant surgery, organs removed from one person can be implanted into another person. Also, devices such as pacemakers can be implanted.

A practice closely related to surgery is prosthetics. A ***prosthesis*** is an artificial body part, such as an artificial heart or arm. These devices are developed through biomechanical engineering. This branch of engineering applies mechanical engineering principles and materials to surgery and prosthetics.

One of the oldest types of prosthesis is a denture. Dentures are removable, artificial teeth that replace full-mouth (upper or lower) teeth. A similar treatment for a few missing teeth is a tooth-supported bridge. These artificial teeth fill a space between existing teeth. Another type of artificial tooth is the dental implant. Dental implants are replacement teeth set into the gums. They replace one or more missing teeth.

Emergency medicine deals with unexpected illnesses and injuries. See **Figure 31-29.** In this area, for example, health-care professionals deal with people who are injured in automobile accidents and on jobs. They treat people who become ill very quickly or suffer heart attacks and strokes. An emergency room contains many technological devices allowing health-care professionals to diagnose and treat patients. Frequently, the job of these professionals is to determine what is wrong with the person and stabilize her condition. More traditional treatments, such as surgery or radiation treatment, for example, can then be administered.

Standards for Technological Literacy

14 15

TechnoFact

The first human heart transplant was completed in 1967 by Christiaan Barnard. Although the patient receiving the heart rejected it shortly afterward, the surgery was still considered successful by medical practitioners.

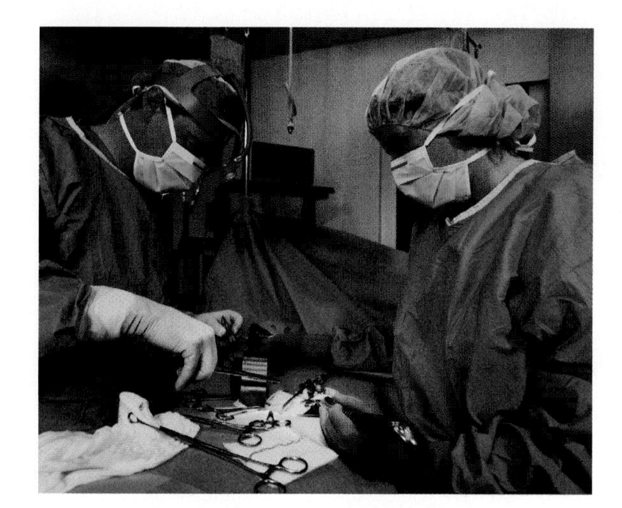

Figure 31-28. Surgery usually involves cutting diseased tissue and organs.

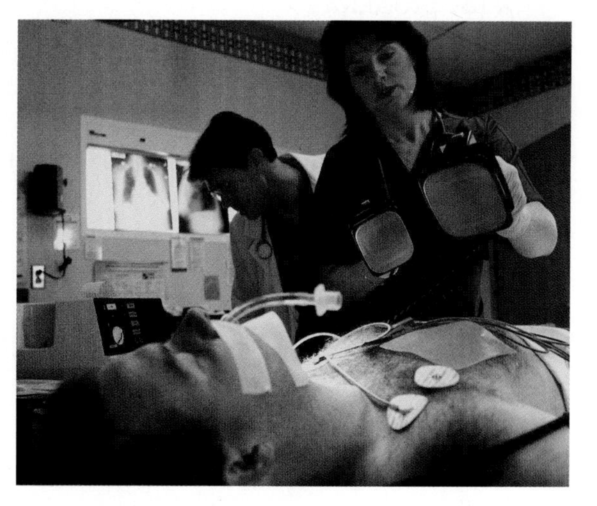

Figure 31-29. This patient is receiving treatment in an emergency room.

Summary

Technology in medicine involves both wellness and illness. Wellness programs are designed to help people maintain their health. Illness treatment involves diagnosing, treating, and preventing diseases and injuries. Diagnosis determines what is wrong with the person. Treatment attempts to restore the person's health. Prevention deals with promoting wellness programs and immunizations. In wellness programs and illness treatment, technology plays a vital role. Both technological equipment and technological knowledge aid health-care professionals in maintaining and restoring people's health.

Test Your Knowledge

Write your answers on a separate piece of paper. Please do not write in this book.

1. Name one difference between the roles of science and technology, in regard to wellness and illness.
2. List the four major factors involved in wellness.
3. Name one technological artifact used in exercise designed to improve a person's well-being.
4. Name one technological artifact used in sports.
5. Medicine involves diagnosing, treating, and ______ diseases and injuries.
6. Nurse practitioners can prescribe medication. True or false?
7. Give one example of routine diagnostic equipment a health-care professional uses.
8. Name one reason a physician might want a patient to have a CT scan instead of an X-ray procedure.
9. Briefly describe one step involved in ultrasound imaging.
10. What is an endoscope?
11. An EKG machine is an example of invasive diagnostic equipment. True or false?
12. What is a proprietary drug?
13. The second clinical phase of an FDA-approved drug involves a double-blind study. True or false?
14. Name one piece of equipment a biomedical engineer might design.
15. Give one example of a newer technological artifact used in surgical procedures.

Answers to Test Your Knowledge Questions

1. Evaluate individually.
2. Nutrition and diet, environment, stress management, and physical fitness
3. Evaluate individually.
4. Evaluate individually.
5. preventing
6. True
7. Student answers may include any of the following: scales, thermometers, devices to do such things as measure the oxygen in the blood, listen to heart rhythm (stethoscopes), and measure blood pressure.
8. A computerized tomography scan is three-dimensional, whereas a conventional X ray is two-dimensional.
9. The ultrasound machine subjects the body to high-frequency sound pulses using a probe. The sound waves travel into the body. There, they hit a boundary between tissues. This boundary could be between soft tissue and bone or between fluid and soft tissue. Some of the sound waves are reflected back to the probe. Others travel further until they reach another boundary and get reflected. The probe picks up the reflected sound waves. They are processed by a computer to produce a still (photograph) or moving (television) image.

STEM Applications

1. Make a list of all the foods (meals and snacks) you eat during a three-day period. Obtain a brochure or other information on healthy eating. (Your school nurse or a local drugstore should be able to supply the information.) Compare your eating with the dietary guidelines in the brochure. Are you eating a healthy diet, or are there some areas where you can improve?
2. For each of the sports listed, fill in the playing surface, goals, and constraints.

Sport	Playing Surface	Goals	Constraints
Auto racing			
Archery			
Baseball			
Billiards			
Downhill skiing			
Hockey			
Swimming			
Soccer			

10. An endoscope is a narrow flexible tube containing a number of fiber-optic fibers that allow a physician to look inside the body.
11. False
12. Proprietary drugs are those drugs considered safe for unsupervised use by consumers.
13. True
14. Student answers may include examples of: life support equipment, artificial body parts, computer systems to monitor patients, sensors for blood chemistry, instruments and devices for therapeutic uses, and medical imaging and treatment systems.
15. Student answers may include: ultrasound, lasers, or endoscopes. Other answers may be accepted at the instructor's discretion.

TSA Modular Activity

Standards for Technological Literacy

12 13

This activity develops the skills used in TSA's Biotechnology Design event.

Medical Technology

Activity Overview

In this activity, you will research a contemporary problem, issue, or technology related to medical technology; prepare a report; and create a display. You will prepare an oral presentation incorporating a PowerPoint presentation. Your report must be contained in a three-ring binder and consist of the following items:

- A cover page.
- A definition of a problem, an issue, or a technology.
- A report on the topic (4–10 pages).
- A printout of the PowerPoint slides (three slides per page).
- A list of sources and references.

Materials

- A three-ring binder.
- Materials appropriate for a tabletop display. (These will vary greatly.)
- A computer with PowerPoint presentation software (or similar).

Background Information

- **Selection.** Before selecting the theme for your project, use brainstorming techniques to develop a list of possible themes. Some contemporary topics include the following:
  - The U.S. health-care system.
  - Minimally invasive surgery.
  - Home health testing.
  - Exercise.
  - Immunizations.
  - Diagnostic equipment.
  - Prescription medications.
  - Prevention programs and techniques.

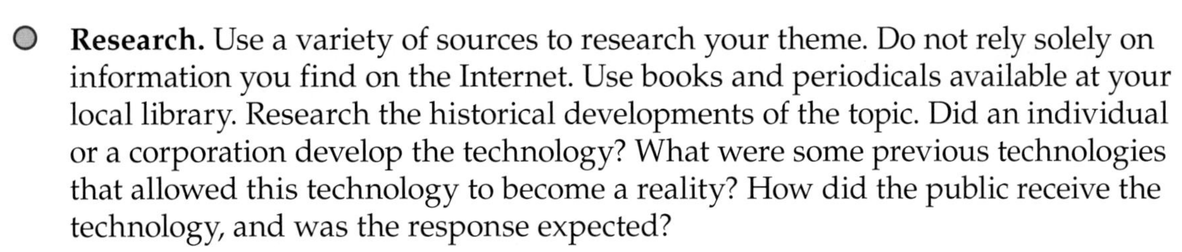

- **Research.** Use a variety of sources to research your theme. Do not rely solely on information you find on the Internet. Use books and periodicals available at your local library. Research the historical developments of the topic. Did an individual or a corporation develop the technology? What were some previous technologies that allowed this technology to become a reality? How did the public receive the technology, and was the response expected?
- **PowerPoint presentation.** When developing your PowerPoint presentation, consider the following design guidelines:
  - Develop a general slide design and use it for all your slides.
  - Keep the design simple. Do not use more than two type fonts. Select easy-to-read type fonts. Be sure the type size is large enough to be seen from the rear of the room in which you will be presenting.
  - Include a title on each slide.
  - Do not attempt to squeeze an abundance of information on a single slide. Create multiple slides instead.

Standards for Technological Literacy

12 13

Guidelines

- Research should focus on any cultural, social, economic, or political impacts. Both opportunities and risks should be addressed.
- The display can be no larger than 18″ deep × 3′ wide × 3′ high.
- If a source of electricity is desired, only dry cells or photovoltaic cells can be used.
- The oral presentation can be up to 10 minutes in length.

Evaluation Criteria

Your project will be evaluated using the following criteria:

- The content and accuracy of the report.
- The attractiveness and creativeness of the display.
- Communication skills and the presentation design of the oral presentation.

Section 9 Activities

Activity 9A

Design Problem

Background

Biomedical engineers develop a great deal of diagnostic and treatment equipment. They also develop artificial parts for human bodies and devices that allow people to do routine tasks.

Situation

You work with a volunteer agency that helps people with arthritis. Many of the clients of this agency have trouble bending over. They often complain that, when they drop something, it is very difficult to reach it and pick it up.

Challenge

Design a device that allows a person to pick up an object on the floor without bending over. The device should be easy to use and inexpensive to make.

Activity 9B

Fabrication Problem

Background

Drying is one of the oldest food-preservation methods. Originally, it was done by salting the food and then drying the food in the sunshine, in open rooms, or on stoves. In 1795, the first dehydrator was developed to dry (dehydrate) fruits and vegetables. Today, dried foods have become a multimillion-dollar industry.

Foods can be dehydrated using sunshine, a conventional oven, or an electric dehydrator.

- Solar drying is a type of sunshine drying in which solar energy is used to heat a specially designed unit that has adequate ventilation for removing moist air. This type of unit can develop temperatures up to 30° warmer than the outside temperature. The higher temperatures reduce the drying time.
- Oven drying is the easiest way to experiment with dehydration. This drying requires little initial investment. Oven drying produces darker, more brittle, and less flavorful foods, however, than foods a dehydrator dries.
- Electric dehydrators use trays to hold the food being processed. The units have a heat source and ventilation system. They produce a better product than those other drying methods produce.

Drying times in conventional ovens and dehydrators vary considerably, depending on the amount of food dried, the food's moisture content, the room temperature, and the humidity. Some foods require several hours, and others might take more than a day. It is important to control air temperature and circulation during the drying process.

Temperatures between 120°F and 140°F are recommended for drying fruits and vegetables. If the temperature is too low or the humidity is too high, the food dries too slowly to prevent microbial growth.

If the temperature is too high initially, a hard shell might develop on the outside, trapping moisture on the inside. If the temperature is too high at the end of the drying period, the food might scorch.

Challenge

You are a researcher for a food-processing company, and you want to develop a dried-fruit snack hikers can take with them on outings. Select a mixture of several fruits and vegetables. Experiment with drying them to develop a trail mix snack.

Materials and Equipment

- An electric dehydrator.
- One pound of fruit and vegetables.

Procedure

1. Carefully read the operating instructions for the electric dehydrator.
2. Select fresh, good-quality fruits and vegetables.
3. Trim away inedible and damaged portions.
4. Cut fruits and vegetables into halves, strips, or slices that will dry readily. The strips should be 1/8″ to 1/4″ thick.
5. If you are drying vegetables, they need to be blanched. Heat them in water to a temperature high enough to neutralize the natural enzymes.
6. Treat most fruits by dipping them in antioxidants, such as citric and ascorbic acid mixtures.
7. Preheat the dehydrator to 125°F (52°C).
8. Place a single layer of food on the tray.
9. Stack the trays in the dehydrator. Close the unit.
10. Gradually increase the temperature to 140°F (60°C). It takes 4 to 12 hours to dry fruits or vegetables in a dehydrator.
11. Examine the food often and turn trays frequently. Turn larger pieces, such as apricot halves, halfway through the drying time. Move the pieces from near the sides of the trays to the center.
12. After the drying process is complete, condition the food. This is done to equalize (evenly distribute) the moisture left in the food after drying. To condition a food, do the following:
    A. Allow the food to cool on trays.
    B. Pour the food into a large nonporous container until the container is about two-thirds full.
    C. Cover the container. Place it in a warm and dry place. Stir the contents at least once a day for 10 to 14 days.

Dried fruits can be eaten as snacks or soaked for one to two hours to rehydrate them. Most vegetables are refreshed with water before use. This can be done by soaking them in water for one to two hours, adding two cups of boiling water for each cup of food, or adding dried vegetables directly to soups or stews.

Section 10
Managing a Technological Enterprise

Tomorrow's Technology Today

Warm-Up Jackets

If you have ever taken part in sports such as track or basketball, you probably have worn a warm-up jacket over your uniform. "Warm-up jacket," of course, doesn't mean that the jacket actually warms your body. You generate the heat. The jacket merely traps it to keep you warm. Now, however, there is a jacket that actually does provide the heat to warm up the person wearing it. This makes it ideal for people who must be outdoors in very cold conditions, such as military personnel, mountain climbers, hunters, and construction workers.

The principle involved is similar to that used in the electric blanket you might have on your bed in the winter. An electric current causes wires in the fabric to give off heat. The electric blanket, however, uses fairly large wires and has to be plugged into a wall outlet to operate. That would not be very practical for use in a jacket, of course. The new jacket uses very thin wires that two small batteries heat. The wearer's movements are not restricted.

The breakthrough that made the new jacket possible was the development of microthin carbon fibers—electrical conductors thinner than human hairs. The microfibers can be woven right into the cloth of the jacket and are as soft and flexible as the fabric itself. They can even make a trip through the washing machine without harm. Two rechargeable batteries supply the power. Together, the batteries weigh less than one-half of a pound.

Heating wires are concentrated in the chest and back areas. Research has shown that heating the chest is the most effective method. Heating the chest warms the body's core. The heart and lungs are located in the core. The blood circulates through the core, carrying heat to all parts of the body. Heating the back provides an additional level of comfort. The batteries for one manufacturer's jacket can presently provide five hours of heating at the low setting, two-and-one-half hours at the medium setting, or one-and-one-half hours at the high setting. Improvements in battery technology will eventually make much longer periods possible.

To test the effectiveness of the system, a company that makes clothing and equipment for mountain climbers conducted a dramatic demonstration. The company testers outfitted a climber with a jacket, placed him in a deep crevasse on a glacier, and then buried him in snow. He was told to wait until the extreme cold caused his body to start shaking violently and then switch on the jacket. Within minutes, the jacket warmed his body enough to stop the shaking.

Several companies are marketing the jackets to consumers. Also, the technology is extending to other articles of clothing, such as gloves, socks, and vests. One future application being discussed might be the coolest of all—finding a way to use the technology to provide air-conditioning, as well as heating.

Discussion Starters

Electric clothing similar to the warm-up jacket in the Tomorrow's Technology Today feature use an electrically conductive textile sold under the brand name GorixTM. Some advantages of woven microfiber heating elements are increased reliability and even heat distribution.

Group Activity

Divide the class into groups and have each group brainstorm new products they could develop using electrically conductive textiles. Encourage the groups to be inventive. Have each group present their favorite idea to the class.

Writing Assignment

The developer of the warm-up jacket worked hand-in-hand with the U.S. Army to develop a heated jacket for soldiers. The Army has a long history of incorporating cutting edge technology into the basic personal gear of the common soldier. Have your students research a current product being developed at the U.S. Army's Soldier Systems Center, and write a paper describing the product.

Chapter 32 Organizing a Technological Enterprise

Chapter Outline
Technology and the Entrepreneur
Technology and Management
Risks and Rewards
Forming a Company

This chapter covers the benchmark topics for the following Standards for Technological Literacy:

Learning Objectives

After studying this chapter, you will be able to do the following:

- Compare entrepreneurship and intrapreneurship.
- Recall the definition of *management*.
- Recall the four functions of management.
- Recall the risks and rewards associated with being involved with a company.
- Recall the three main forms of business ownership.
- Recall problems that can occur in the partnership form of business ownership.
- Summarize why a corporation is said to be "similar to a person," in a legal sense.
- Recall the role of a board of directors in a corporation.
- Compare equity financing and debt financing.

Key Terms

articles of incorporation
authority
board of directors
bond
bylaw
chief executive officer (CEO)
corporate charter
dividend
inside director
intrapreneurship
limited liability
management
middle management
operating management
outside director
own
president
prime interest rate (prime)
private enterprise
proprietor
public enterprise
responsibility
reward
risk
share of stock
supervisor
top management
unlimited liability
vice president

Strategic Reading

Make a list of various businesses in your area. As you read this chapter, determine the type of ownership each business is.

Standards for Technological Literacy

1 2

Reinforce

Review the definition and scope of technology (Chapter 1).

Reinforce

Review management as a process (Chapter 5).

TechnoFact

Prior to the Industrial Revolution, the majority of Europeans lived in small towns. They were able to provide for themselves by growing food they needed and making their own clothes, furniture, and tools from the environment around them.

As you study this book, you are learning a great deal about technology. You are learning that technology consists of human-made devices designed to control and modify the natural environment. Technology is a system with goals, inputs, processes, and outputs. You are also learning that technology occurs in seven broad areas. These areas are communication, construction, manufacturing, agriculture, energy and power, medicine, and transportation.

©iStockphoto.com/egdigital

Figure 32-1. Complex systems and a large organization are needed to produce a product such as this airplane.

Technology and the Entrepreneur

Have you ever thought about where technology comes from? Technology is a product of the human mind, developed by people to serve people. At one time, most technology was developed, produced, and used by one person. In the modern world, however, organization is necessary. People use very complex systems to develop and produce technology. See **Figure 32-1.** At the base of many of these systems are entrepreneurs. See **Figure 32-2.** These are people with very special talents. Entrepreneurs look beyond present practices and products. They see new ways to meet human needs and wants. By focusing on what the customers value, entrepreneurs develop systems and products to meet desires and expectations. They might change the entire way something is being done.

Goodyear Tire and Rubber Co.

Figure 32-2. Entrepreneurs have the vision to recognize consumer wants and to devise ways to meet them with new products.

A good example of entrepreneurship is the McDonald's® restaurant chain. The first McDonald's restaurant was a small hamburger stand in southern California. The original owners had developed some innovative ways to make and sell their product.

Standards for Technological Literacy

2

Brainstorm

Ask your students to explain the following statement: "Technology is the purposeful product of managed human activity."

Research

Have your students identify and describe a present-day or historical entrepreneur.

TechnoFact

Though the Latin roots in the word manufacture mean "make by hand," the modern meaning has been altered by technology. It not only includes products made by hand, but also making products with the use of machines.

They did not, however, look beyond their local market. An outsider, Ray Kroc, saw greater possibilities. Under his leadership, the fast-food business was born. Kroc and his managers carefully studied the various jobs and developed special management techniques. They standardized the product, created effective training programs, and developed the chain into a worldwide organization. This is some of the work of the entrepreneur, to improve the use of the resources and create new products or markets.

The dictionary defines an *entrepreneur* as "any person taking the financial risks of starting a small business." This definition, however, leaves out the aspect of entrepreneurial spirit, the spirit of innovation. People starting another beauty shop, delicatessen, or bakery are taking financial ***risks*** and might become successful business operators. They are not, however, innovators. These people do not deal with change as an opportunity to produce a new product or service.

There also can be entrepreneurship within an existing company. Entrepreneurship involves an attitude and an approach and consists of searching for opportunities for change and responding to them. Large companies often encourage entrepreneurship within their organizations. In fact, a new term, ***intrapreneurship***, has evolved to describe this action. This term means "the application of entrepreneurial spirit and action within an existing company structure."

Technology and Management

Technology is purposeful. This system is developed to meet a problem or an opportunity. Identifying and responding to the need for change is only one part of developing technology. The production and use of technology must be managed. Therefore, technology is a product of managed human activity. See **Figure 32-3.** Actually, you or other people manage all your actions. This is not bad. ***Management*** is simply the act of planning, directing, and evaluating any activity. See **Figure 32-4.** This act can be

Figure 32-3. The actions of these workers must be managed to produce boats efficiently.

American Electric Power

Figure 32-4. Planning is an important part of managing a business. These managers are using a model of a boiler system as they plan a factory expansion.

as simple as managing personal expenditures or as complicated as managing an industrial complex. Management involves authority (the right to direct actions) and responsibility (an accountability for actions). Managers have the ***responsibility*** to make decisions that ensure the business is successful. Their ***authority*** might include hiring personnel, purchasing materials, developing products, and setting pay rates. Likewise, managers have the responsibility of protecting the rights of a company's owners, workers, and customers. This might include securing product patents, investing company funds wisely, providing a safe work environment, and producing a quality product.

The Functions of Management

To carry out their duties, company managers perform four important functions. See **Figure 32-5.** These functions are the following:

- **Planning.** This function includes setting long-term and short-term goals for the company or parts of it. Planning also involves selecting a course of action to meet those goals. Planning activities often result in an action plan, or plan of work. Such a plan lists what needs to be done, who will do it, and when it is to be done.
- **Organizing.** This involves structuring the company or workforce to address company goals. Typical activities include developing organizational charts, establishing chains of command, and determining the company's operating procedures. Organizing makes sure people, materials, and equipment are in place to meet the action plan.
- **Actuating.** Initiating the work related to the action plan is known as *actuating.* This function can include training employees, issuing work orders, providing a motivational work environment, or solving production problems. Actuating causes plans to take form. For example, products and structures are built. In other cases, services are provided.
- **Controlling.** Comparing results against the plan is the controlling function. Control actions ensure that resources are used properly and outputs meet stated standards. Typical terms applied to this area are *inventory control, production control, quality control,* and *process control.*

Figure 32-5. Managers plan, organize, actuate, and control company activities. This manager is performing the control function by gathering information.

The Authority and Responsibility of Management

Managerial functions are carried out through an organizational structure. See **Figure 32-6.** This structure typically begins

Standards for Technological Literacy

2

Extend

Discuss how management can be applied to any human activity.

Figure Discussion

Use a simple task, such as mowing a lawn, to explain the four activities of management, shown in Figure 32-5.

TechnoFact

Technology has helped create several new jobs over the last 200 years. For example, a manufacturing plant needs workers in the factory, but it is likely that the area surrounding the factory will be built up with consumer businesses, which provide more jobs outside the factory.

Standards for Technological Literacy

Brainstorm

Ask your students if you can have authority without responsibility. Have them explain their answers.

TechnoFact

The standard modern chain of command in corporate enterprises distributes power among several managers of various groups. Managers are employed to supervise and guide the work of individuals. This style of corporate bureaucracy directs the activities of a business.

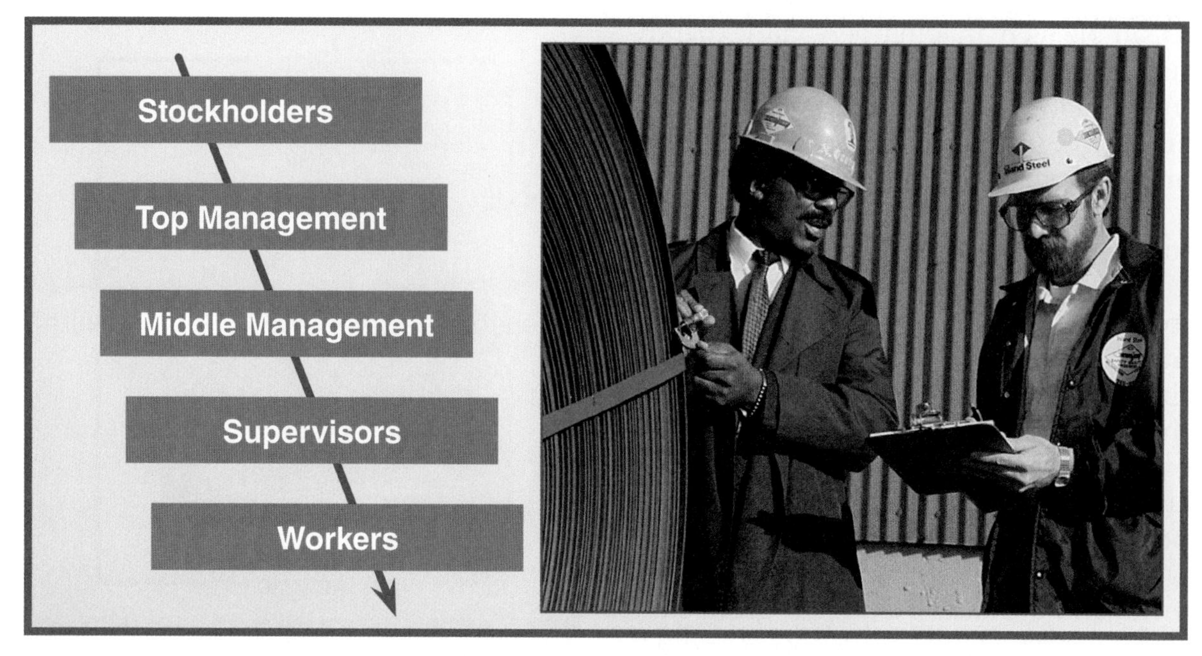

Figure 32-6. Managers have specific levels of authority and responsibility within a company.

with the ***owners*** of the business. The owners have ultimate control, or final authority, over company activities. The business is their company. They can hire or fire personnel, set policies, or close the business.

Top Management

In most larger companies, however, the owners are not the managers. They frequently have other jobs or interests. Also, they often do not have the skills needed to manage a large, complex business. Such owners delegate responsibility to full-time managers. In many companies, the top manager is the ***president*** or the ***chief executive officer (CEO)***. In very large companies, two different people usually hold these titles. In smaller companies, the same person might hold the two titles. In either case, this top level of management is responsible for the entire company's operation. People at this level have day-to-day control of the company.

Few people can manage a company by themselves. Therefore, the top managers employ other managers to assist them. The number of managers and levels of management vary with the size and type of the company. Larger companies generally have ***vice presidents***. Vice presidents report to the president or CEO. Each vice president is responsible for some segment of the company. The segment might be a functional area, such as marketing, finance, production, or engineering. In other cases, the vice president has a regional responsibility, such as foreign sales or West-Coast operations.

Middle Management

Most vice presidents have a scope of responsibilities. These responsibilities include a number of managers reporting to the vice presidents. Regional sales managers might report to a vice president for sales. Plant managers often report to vice presidents in charge of production. This level of management is often called ***middle management***. Middle management is below ***top management*** (the president and vice presidents). This management, however, is above ***operating management***.

Career Corner

Top Executives

Top executives set specific goals and objectives for a company and direct the overall operations of businesses. They formulate policies and strategies so these objectives are met. There is a range of titles for top executives. These titles include CEO, chief operating officer (COO), chair of the board, and president.

Executives typically have spacious offices and large support staffs. They might travel to national and international sites to monitor operations and meet with customers and staff members. Many top executives have a bachelor's degree or an advanced degree. A top executive must have excellent personal skills and an analytical mind. Executives must be able to communicate clearly, provide leadership, and make difficult decisions.

Standards for Technological Literacy

2 4

Extend

Discuss the roles and levels of management using a school organization chart.

TechnoFact

While the government is not allowed to restrict businesses in the free enterprise system, corporations may be involved in government enterprises. While these municipal corporations control school districts or the water supply, they also provide governments with community services.

Operating Management

The lowest levels of management directly oversee specific operations in the company. Managers at this level might be supervisors on the production floor, district sales managers, or human resources directors. They are the managers who are closest to the people who produce the company's products and services. These managers are often called ***supervisors*** or *operating management.*

The Coastal Corp.

Figure 32-7. Most technology is developed and produced by industrial companies. The operation of any company, large or small, involves both risks and rewards.

Risks and Rewards

As noted earlier, most technology is developed and produced through the planning of industrial companies. See **Figure 32-7.** The company designs, engineers, and produces the products, structures, transportation services, and communication media you depend on daily.

Everyone involved with a company has some basic elements in common. These elements are risks and rewards. Owners risk their money to finance the company. Banks also accept a level of risk. They and other lending institutions make loans to finance company growth. Employees risk missing other employment opportunities by working for the company. Consumers risk their money when they buy a product.

Standards for Technological Literacy

4

Research

Have your students visit the local Chamber of Commerce or its Web site to identify examples of proprietorships, partnerships, and corporations in your community.

TechnoFact

All activities engaged in by enterprises that produce goods and services are described as *businesses*. An individual enterprise may also be called a business.

In return, the risk takers expect a ***reward***. The owners want their investments to grow. They also expect periodic financial returns for the use of their money. Banks expect interest to be paid on their loans. Employees expect job promotions, pay raises, safe working conditions, and job security. Consumers expect performance and value in return for the money they spend on products and services.

Forming a Company

Companies are organized and operated under the laws and mores (accepted traditions and practices) of our society. There are several important features involved in the formation of a company. These features include the following:

- Selecting a type of ownership.
- Establishing the enterprise.
- Securing financing.

Selecting a Type of Ownership

Business enterprises can be divided into two different sectors. These sectors are public and private. ***Public enterprises*** are those enterprises that the government or a special form of corporation controls. They are generally those enterprises meeting two criteria:

- First, they are operated for the general welfare of the society.
- Second, they cannot or should not make a profit.

As an example, consider a police department. A police department is run similarly to a business. Police are commissioned to protect people and their property. They have managers (police chiefs and captains) and workers (patrol officers, traffic officers, and narcotics officers). If the police department had to show a profit and attract private investment, however, some aspects of law enforcement might get cut back. This might limit the department's market to the people who could pay for the service. The police would, thus, solve the crimes showing the most potential for profit. See **Figure 32-8.**

Another example of public ownership is road construction. We pay for it through taxes. Each segment of road, however, does not have to show a profit. If it did, we would not have many of our rural roads.

Individuals or groups of people own ***private enterprises***. This ownership can be through a direct means of investment or through an indirect means, such as a pension or an investment fund. Owners of private enterprises invest their money, take their risks, and hope to reap a profit. Within legal limits, the owners are free to select business activities, produce the products and services they choose, and divide the profits as they see fit.

Private enterprises can be either publicly held or privately held. A publicly held enterprise is one in which the public can purchase a portion of ownership in the form of ***shares of stock***. Individuals or a group of people own privately held enterprises. These enterprises, however, do not

©iStockphoto.com/slobo

Figure 32-8. Police and fire departments are examples of public enterprises.

offer stock for sale to members of the public. There are three main types of private-business ownership. See **Figure 32-9.**

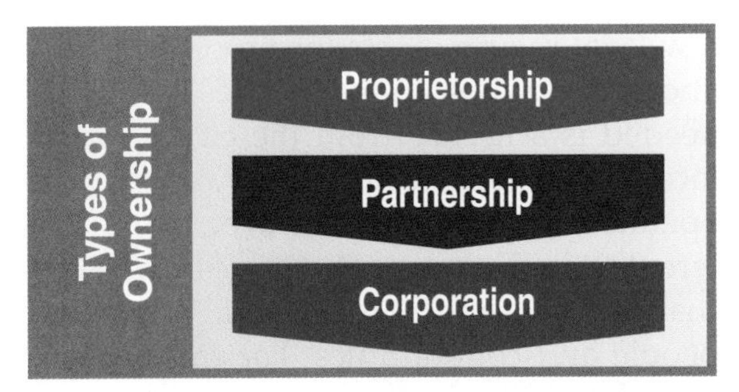

Figure 32-9. A private enterprise can be owned by one person (a proprietorship), a small number of people (a partnership), or many investors (a corporation).

Proprietorships

The first type is a ***proprietorship***. A proprietorship is a business with a single owner. See **Figure 32-10.** The owner has complete control of the company. He sets goals, manages activities, and has the right to all business profits. A proprietorship is a fairly easy type of ownership to form.

Proprietors might have difficulty raising money, especially to start a new business. The company's finances are typically limited to the owner's personal wealth or borrowing ability. Banks might be hesitant to loan a large quantity of money, however, to unproven businesses or individuals.

An additional problem of this type of business is limited knowledge. Many individuals do not have the skills and know-how needed to run all aspects of the business. This causes inefficiency in operating the company. Finally, the proprietor is responsible for all the debts the business incurs. She cannot separate business income and liabilities from her personal finances. This is a major disadvantage called ***unlimited liability***.

Proprietorships are the most common type of ownership in the United States.

©iStockphoto.com/jwilkinson

Figure 32-10. Many small retail and service businesses are sole proprietorships.

They are generally, however, small retail, service, and farming businesses. Thus, the dollar impact of this form of ownership on the economy is considerably less than the impact large corporations make.

Partnerships

A second form of private ownership is the partnership. Partnerships are businesses that two or more people own and operate. See **Figure 32-11.** Partnerships, thus, have more sources of money to finance the company. Also, the interests and abilities of the partners might complement each other. One partner might be strong in production. Another might have sales skills.

Having more than one owner active in a business can cause confusion, however. Employees might receive conflicting directions. The partners can disagree when making important business decisions. Also, a partnership, similar to a proprietorship, has unlimited liability. This is a particularly touchy problem, since one partner can commit the entire partnership to financial risk.

Standards for Technological Literacy

4

Figure Discussion

Use Figure 32-9 as a starting point to discuss the types of ownership and their advantages and disadvantages.

TechnoFact

Most states employ the Uniform Partnership Act, which protects and regulates partnerships, partnership agreements, and the sale of partnerships.

Standards for Technological Literacy

4

Brainstorm

Ask your students which type of ownership they would choose if they were starting a landscaping business. Ask the students to explain their decision.

TechnoFact

Until 1811, state legislatures were responsible for granting charters to corporations. However, corporate ownership became so popular that the power shifted to the Secretary of State.

©iStockphoto.com/LivingImages

Figure 32-11. Many small businesses, such as farms, are partnerships.

Corporations

The third form of ownership is the corporation. See **Figure 32-12.** A corporation is a business in which investors have purchased partial ownership in the form of shares of stock. Investors can be individuals, other companies, or groups (such as in a pension plan or an investment club). Legally, the corporation is similar to a person. This business can own property, sue, be sued, enter into contracts, and contribute to worthy causes.

Generally, the investor-owners of a corporation do not manage the business. They employ professional managers for this task. The owners invest their money and expect to receive a ***dividend*** (periodic payment from the company's profits) in return.

Since a corporation is a legal person, the company is responsible for the debts. (The company's owners are not responsible.) This feature is called ***limited liability.*** Limited liability means, if the company fails, an owner's loss is limited to the amount of money he has invested.

Establishing the Enterprise

Once the type of ownership is selected, the company must be established. There are few legal requirements for proprietorships and partnerships. In many cases, all that is needed is a license from the city or other local government where the business will operate. Corporations, however, are a different story. They are often large businesses that can have a serious financial impact on people and communities. For this reason, their formation is placed under state control. Each state establishes its own rules for forming a corporation. Most states require the following steps to be completed when forming a corporation. See **Figure 32-13.**

Filing Articles of Incorporation

A corporation, similar to a person, must be born. This process is begun by filing ***articles of incorporation*** with one of the

©iStockphoto.com/Freezingtime

Figure 32-12. Most large businesses are corporations.

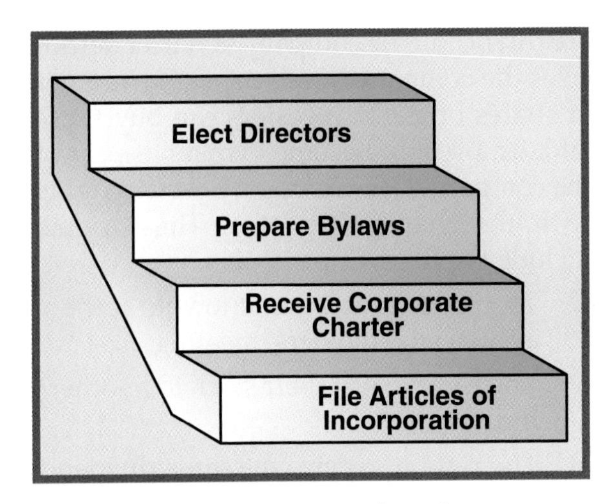

Figure 32-13. The steps in forming a corporation.

states. The articles of incorporation serve as an application for a ***corporate charter*** (a "birth certificate" for the corporation). The state usually asks for the company name, type of business the company plans to enter, location of the company offices, and type and value of any stock that will be issued.

Receiving a Corporate Charter

The articles of incorporation are filed with the appropriate state office. State officials then review the articles. They try to determine if the business will operate legal activities and provide customers with appropriate products or services. If they believe the business meets all state laws, a corporate charter is issued. This allows the company to conduct the specified business in the state. All the other states will recognize the corporate charter and allow the company to conduct business within their borders.

Approving Bylaws

An incorporated business must have a set of ***bylaws***. These are the general rules under which the company operates. A set of bylaws includes the information contained in the charter. This information is the name of the company, purpose of the business, and location of the corporate offices. In addition, the bylaws list the following:

- The corporate officers.
- The duties of, terms of office of, and method of selecting corporate officers.
- The number of directors, as well their duties and terms of office.
- The date, location, and frequency of the board of directors' meetings.
- The date and location of the annual stockholders' meeting.
- The types of proposals that can be presented at the annual stockholders' meeting.
- The procedure for changing the bylaws.

Electing a Board of Directors

The charter and bylaws allow the company to operate. The stockholders, however, want their investments to be wisely managed. This requires oversight and supervision. Many companies have hundreds or thousands of stockholders. Few stockholders can be, or wish to be, involved in managing the company. Therefore, a ***board of directors*** is elected to represent the interests of the stockholders. The directors are responsible for forming company policy and providing overall direction for the company.

A typical board of directors includes two groups of people. One group is made up of the top managers of the company. These individuals are known as ***inside directors***. Other directors are not involved in the day-to-day operation of the company. They are selected to provide a different view of the company's operation. Since these people are outside the managerial structure, they are called ***outside directors***.

Directors are elected using a voting system similar to our political system. The

Standards for Technological Literacy

4

TechnoFact

The rise of financial institutions accompanied the Industrial Revolution. The newly developed area of industry was supported by investments from individuals and institutions.

Standards for Technological Literacy

4

Reinforce

Differentiate between debt and equity financing.

Brainstorm

Ask your students which type of financing they would use if they were starting a new company. Have your students explain their answers.

Research

Have your students determine what city, county, or state projects in your area have been financed with bonds.

TechnoFact

A development of capitalism is free enterprise. While capitalism allows private individuals to own the means of production and profit through their investments in capital and employment of labor, free enterprise goes one step further. Government intervention in the economy in a free enterprise system is restricted, and a free market controls trade by determining supply and demand.

TechnoFact

Adam Smith's *The Wealth of Nations* defined the nature of capitalism as it existed in 1776. However, the importance of capitalism grew as landowners were replaced by merchants and industrialists as the Industrial Revolution progressed.

main difference is companies do not use a one-person and one-vote rule. Instead, they use a one-share and one-vote procedure. Each share of stock (equal portion of the total company) has a vote assigned to it. Stockholders each have as many votes as the shares of stock they own. Therefore, those who own a larger portion of the company and accept a larger risk have a greater say in forming company policy.

Securing Financing

There is more to starting a company than completing paperwork. The company needs money to operate. There are two basic methods of raising operating funds. These are equity financing and debt financing. See **Figure 32-14.**

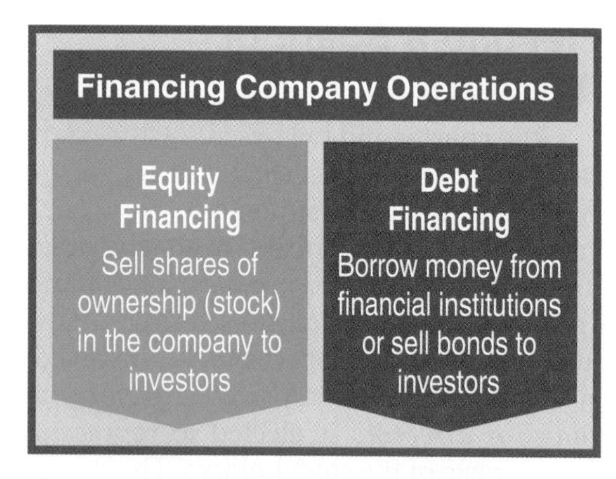

Figure 32-14. Two major methods of financing are available to build and expand a company.

Equity Financing

Equity financing involves selling portions of ownership in the company. This is an important way in which corporations are financed. The company's charter authorizes the company to sell a specific number of shares of stock. Investors can buy these and, as a result, become owners of part of the company. Owners receive certain rights with the shares they own. These rights include the following:

- The right to attend and to vote at the annual stockholders' meeting.
- The right to sell their stock to another individual.
- The right to receive the same dividend per share as other stockholders.
- The right to a portion of the company's assets (property and money) if the company is liquidated.

Debt Financing

Debt financing involves borrowing money from a financial institution or private investors. Banks and insurance companies loan corporations money to finance new buildings, equipment, and the company's daily operations. They charge interest for the use of their money. These banks and insurance companies charge their best (safest) customers a lower interest rate. This rate is called the ***prime interest rate (prime).*** Other borrowers pay a rate higher than the prime rate. This rate is often quoted in terms of the prime, such as "prime plus 2%."

Corporations can also sell debt securities called ***bonds.*** These securities are often sold in fairly large denominations, such as $5000. The bonds are usually long-term securities. This means they will be in force for 10 to 20 years. The company pays quarterly or yearly interest on the face value (original value) of the bonds. At maturity (the end of the bond's term), the company pays back the original investment.

Management is necessary in business. Various levels of management exist to differentiate among the different steps in developing technological solutions.

Answers to Test Your Knowledge Questions

1. intrapreneurship
2. B. authority
3. Actuating is initiating the work related to the action plan.
4. Risk: losing money invested in the company, Reward: financial return on investment.
5. An educational television station would be considered a public enterprise if it is not organized and operated to create a profit.
6. Proprietorship, partnership, and the corporation
7. B and C.
8. True
9. The corporation
10. D. File articles of incorporation
    B. Receive corporate charter
    A. Prepare bylaws
    C. Elect corporate directors
11. D. All of the above.
12. stock, bonds

Summary

People develop technology for people. At the very foundation of this action are entrepreneurs. They are people who can see possibilities and are willing to take risks. Often, the result of an entrepreneur's work is a company. A company is a business enterprise that is organized and managed to produce goods and services. A group of people called *managers* operates the company. They plan, organize, actuate, and control company activities. The managers set goals, structure systems, direct operations, and measure results. Without management, technological activities would become less efficient. With proper management, technology promises to serve our needs.

Test Your Knowledge

Write your answers on a separate piece of paper. Please do not write in this book.

1. Entrepreneurship within an existing company is referred to as ______.
2. Management involves ______ and responsibility.
   A. credibility
   B. authority
   C. discipline
   D. personality
3. Describe the management function called *actuating.*
4. What are the risks and rewards of company ownership?
5. When would an educational television station be considered a public enterprise?
6. List the three forms of business ownership.
7. Which of the following can be a problem in a partnership? (More than one answer can be used.)
   A. Multiple sources of money.
   B. Confusing lines of authority and directions for workers.
   C. One partner can commit all partners to financial risk.
   D. Partners can bring special talents to the company.
8. Similar to a person, a corporation can own property, enter into contracts, sue, and be sued. True or false?
9. Which form of business ownership provides the owners with limited liability?

10. Place the steps in forming a company, listed on the right, in their proper order:

_____ First step.
_____ Second step.
_____ Third step.
_____ Fourth step.

A. Prepare bylaws.
B. Receive corporate charter.
C. Elect corporate directors.
D. File articles of incorporation.

11. A corporation's board of directors is responsible for _____.
    A. forming company policy
    B. representing the interests of the stockholders
    C. providing overall direction for the company
    D. All of the above.
12. Equity financing involves the sale of _____. Debt financing involves the sale of _____.

STEM Applications

1. Assume your class is going to produce and sell popcorn at school basketball games. Develop an organization chart for the enterprise. Indicate who has the most authority and the chain of command from that person to all other members of the enterprise.
2. Call or visit the office of a local stockbroker. Find out if you can obtain brochures or other simple explanations of stocks and bonds and how they are traded. Use the information as the basis for a written report. You might want to invite the person to speak to your class.

Chapter Outline
Societal Institutions
Economic Enterprises
Industry
Areas of Industrial Activity
Industry-Consumer Product Cycle

This chapter covers the benchmark topics for the following Standards for Technological Literacy:

2 3 4 5 6 7 10 19

Learning Objectives

After studying this chapter, you will be able to do the following:

- Summarize the five major societal institutions.
- Recall the definition of an *economic enterprise*.
- Recall the primary definition of *industry* used in this book.
- Summarize the relationships among the five main managed areas of activity within a technological enterprise.
- Compare basic research and applied research.
- Recall the four common systems of manufacturing.
- Recall the meaning of the term *quality control*.
- Explain the channels of distribution used to move products from the producer to the consumer.
- Recall the major types of programs included in industrial relations.
- Summarize the different types of employee training.
- Summarize how expenses, income, and profit are related.
- Recall the steps in the industry-consumer product cycle.

Key Terms

accounting
advertising
applied research
apprenticeship training
architectural drawing
basic research
benefit
classroom training
commission
communication process
computer-aided design (CAD)
construction process
continuous manufacturing
custom manufacturing
distribution
economic activity
economic enterprise
economy
education
employee relations
employment
engineering
expense
family
grievance
income
industry

inspection
intermittent
manufacturing
just-in-time (JIT)
inventory-control system
labor agreement
labor relations
manufacturing process
market research
on-the-job training
package
politics and law
price
process development
product development
public relations
recruit
religion
retained earnings
salary
sales
screening
structure development
transportation process
wage

Standards for Technological Literacy

4 6

Extend

Give local examples of each of the five societal institutions.

TechnoFact

The five basic institutions are relevant to all cultures. In the same way, all cultures are related to each other. For example, many American cultural products are used around the world.

Strategic Reading

There are different kinds of hierarchies in technological enterprises. To better understand the hierarchies discussed in this chapter, make an outline of the main points as you read.

People generally live in a community, region, or nation with people who share common customs and laws. This arrangement is often called a *society*, which is made up of major parts called *institutions*. One major societal institution deals with the economic (goods and services) activities of society. Within this economic domain are organizations called *economic enterprises*, of which industry is one type. In this section, you will explore a number of topics dealing with technological enterprises and, more specifically, industry.

Societal Institutions

Over time, humans have developed a complex society to meet their wants and needs. Within this society are a series of institutions. Five basic institutions are listed in **Figure 33-1.** They are the following:

- ***Family.*** This institution provides the foundation for social and economic actions. Family is the basic unit within society.

Figure 33-1. These five basic institutions are important to our society.

Standards for Technological Literacy

4 6

Reinforce

Review technology and the economic structure (Chapter 3).

TechnoFact

Tony Blair, former Prime Minister of Great Britain, wrote that success comes from the ability to take an already existing product and transform it into something that exceeds customer needs and, therefore, surprises the user.

- ***Religion.*** This institution develops and communicates values and beliefs about life and appropriate ways of living.
- ***Education.*** This institution communicates information, ideas, and skills from one person to another and from one generation to another.
- ***Politics and law.*** This institution establishes and enforces society's rules of behavior and conduct.
- ***Economy.*** This institution designs, produces, and delivers the basic goods and services the society requires.

All these institutions use technology because they are concerned with efficient and appropriate action. They apply resources to meet human wants and needs. People in each institution use technical means to make their jobs more efficient.

Almost all technology, however, originates in the economic institution. For example, teachers work in the educational institution. They might use computers to make their teaching more efficient. The computer, however, is not a product of the educational institution. Computers are products of the economic institution. In the same way, politicians are part of the political and legal institution. They might use television and printed material to help win an election. Again, television, printing presses, and all associated communication devices are not developed in the political system. They are outputs of the economic institution.

Figure 33-2. There are four major types of economic enterprises. Only the industrial enterprises, however, actually develop and produce technology.

Economic Enterprises

Technology is directly associated with economic enterprises. ***Economic enterprises*** are organizations engaging in business efforts directed toward making a profit. See **Figure 33-2.** This ***economic activity*** includes all trade in goods and services paid for with money.

In a free enterprise economic system, the principles of supply and demand determine production and prices. The government has little impact on markets. Businesses are free to produce any products. As consumer demands change over time, some businesses are created. Some businesses close. To meet current demands, some businesses modify their production. In this system, business plays the important roles of owning production equipment and processes, determining product supply, and reacting to market demands.

Not all economic activity develops technology. Commercial trade includes the wholesale and retail merchants who form the link between producers and consumers. These traders do not change the products they distribute. They simply make these products easily available for people to buy.

Banks, insurance companies, and stockbrokerage firms provide financial services to people. They protect our wealth, buy and sell stocks and bonds, or insure our lives and possessions. Again, these companies use technology. They do not, however, develop it.

Similarly, many service businesses repair products and structures. They service and maintain technological devices. These businesses extend the useful lives of such devices. They do not, however, develop the devices.

Industry

Almost all of the design, development, and production of technology takes place within the type of economic enterprise called ***industry***. See **Figure 33-3.** The term *industry* can have several meanings. One definition groups together all businesses making similar products. Thus, we read about the steel industry or the electronics industry. In this book, however, a more restricted definition is used. *Industry* is "the area of economic activity that uses resources and systems to produce products, structures, and services with intent to make a profit."

Areas of Industrial Activity

In each industry, a number of actions take place. These actions are designed to capture, develop, produce, and market creative ideas. These technological activities form the link from the inventor or innovator to the customer. There are thousands of individual actions that cause a new product or service to take shape. They can be gathered, however, into five different areas of managed activity. See **Figure 33-4.** These areas are the following:

- **Research and development.** These activities might result in new or improved products and processes.
- **Production.** These activities develop methods for producing products or services and produce the desired outputs.
- **Marketing.** These activities encourage the flow of goods and services from the producer to the consumer.
- **Industrial relations.** These activities develop an efficient workforce and maintain positive relations with the workers and the public.
- **Financial affairs.** These activities obtain, account for, and disburse funds.

The first three activity areas are product or service centered. They directly contribute to the design, production, and delivery of the planned outputs. The other two areas are support areas. Financial affairs provides monetary support. Industrial relations contributes human or personnel support.

Standards for Technological Literacy

4 6

Figure Discussion

Discuss and give local examples of the four major types of economic enterprises listed in Figure 33-2. Encourage class participation.

Research

Have your students find at least three different definitions for the word *industry*.

Figure Discussion

Compare the definition of *industry* with that of *technology* given in Chapter 1.

TechnoFact

According to Danny Bone, a New Product Innovation Manager at Black & Decker Corporation, all innovation requires someone to maintain the momentum during times of uncertainty.

Figure 33-3. Industrial enterprises carry out the design and production of goods, such as these consumer products.

Standards for Technological Literacy

4 6

Figure Discussion

Use a local manufacturer as an example to discuss the five managed areas of activity presented in Figure 33-4.

Brainstorm

Ask your students why you would need all five managed areas of activity if you were to start a company to produce CD holders.

TechnoFact

According to an article in the *Guardian*, research that doesn't seem to have a practical application is still worth investing in. Curiosity-driven research, not economic motives, has been at the heart of almost every major breakthrough.

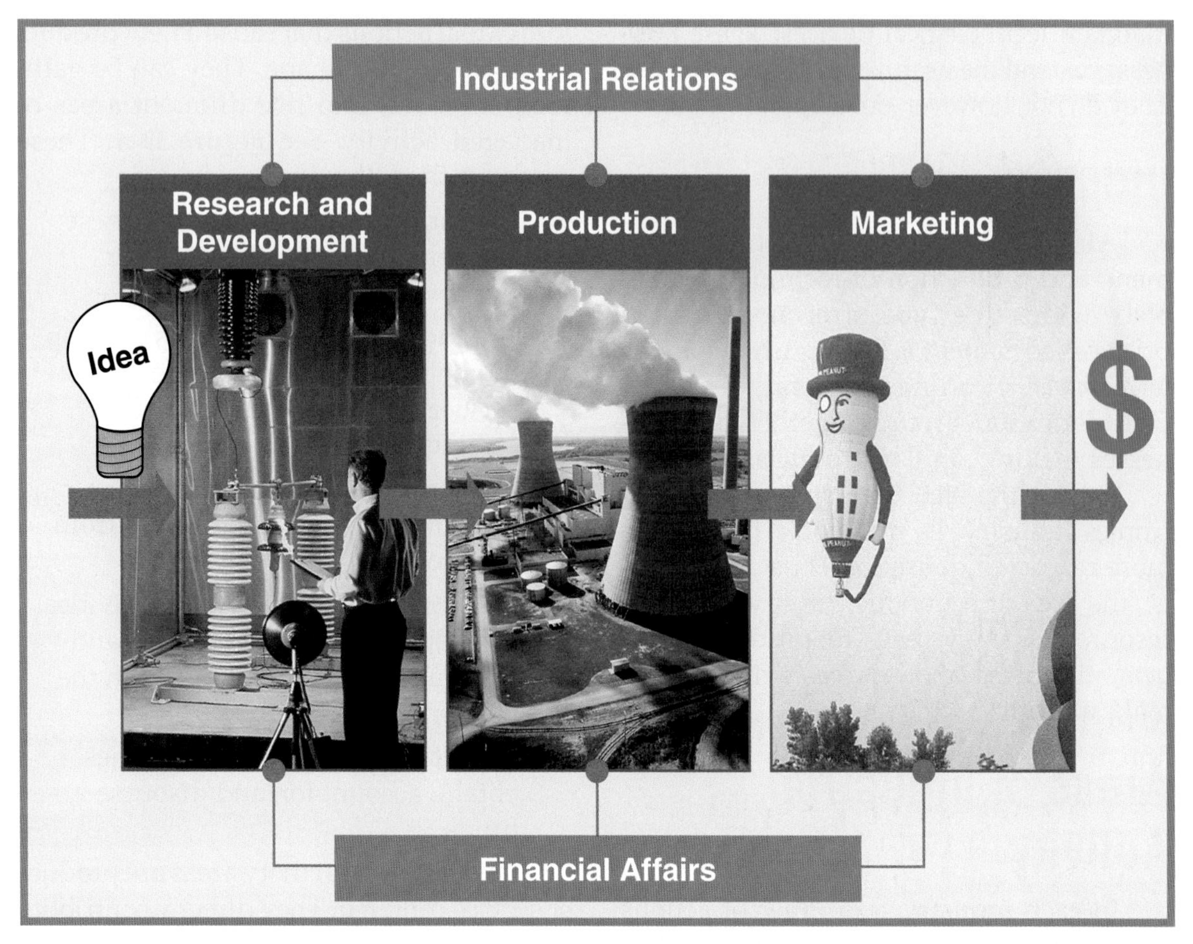

Figure 33-4. There are five major areas of managed activities in industry that change ideas into products, structures, or services. Notice how each area relates to the others.

Career Corner

Technical Illustrators

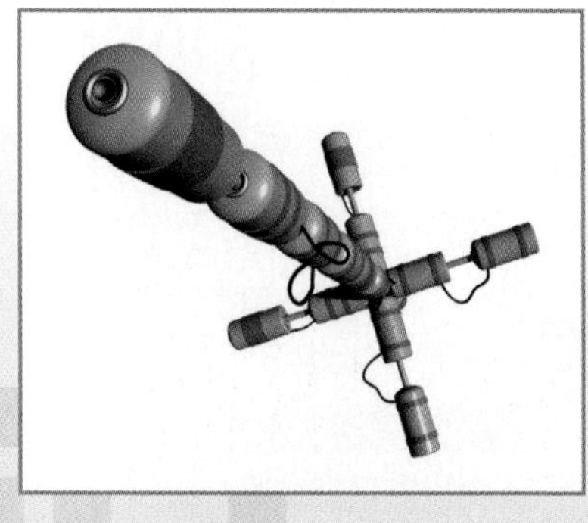

Technical illustrators create art to communicate ideas, thoughts, or feelings. They typically create pictures for publications, such as books, magazines, sales brochures, and advertisements, and for commercial products, such as textiles, wrapping paper, stationery, greeting cards, and calendars. Increasingly, illustrators prepare work directly on computers in a digital format.

Industrial enterprises, advertising agencies, publishing companies, and design firms employ illustrators. Many companies expect an applicant to have at least an associate's degree in technical illustration or a related field. Knowledge of computer graphics and skill in using visual-display software are important.

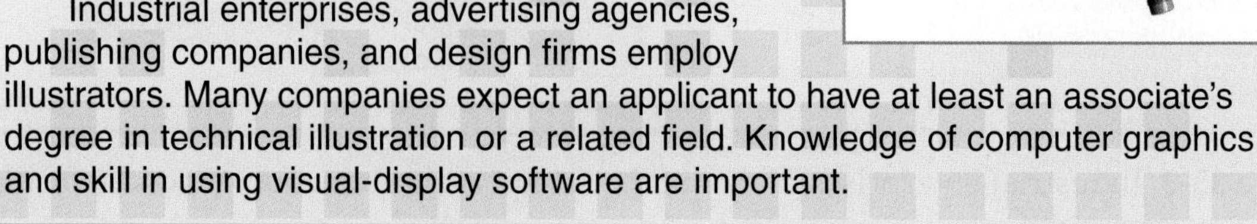

Research and Development

Research and development can be viewed as the "idea mill" of the enterprise. In this area, employees work with the true raw material of technology, human ideas. They convert what the mind envisions into physical products and services. These actions can be divided into three steps. The steps are research, development, and engineering.

Research

Research is the process of scientifically seeking and discovering knowledge. See **Figure 33-5.** Research explores the universe systematically and with purpose. This process also determines, to a large extent, what technology we will have in the future. Research determines the type of human-built world in which we will live.

AT&T

Figure 33-5. Research seeks to discover knowledge. Computers have proven to be a powerful aid for researchers.

There are two types of research. The first is ***basic research***. Basic research seeks knowledge for its own sake. We conduct basic research to enlarge the scope and depth of human understanding. People working in basic research are not concerned about creating new products. Their focus is on generating knowledge.

The second type of research is ***applied research***. This activity seeks to reach a commercial goal by selecting, applying, and adapting knowledge gathered during basic research. The focus of applied research is on tangible results such as products, structures, and technological systems.

Basic research and applied research complement one another. The former finds knowledge. The latter finds a use for this knowledge. For example, basic research might develop knowledge about the reactions of different materials to high temperatures. Applied research might then determine which material is appropriate for the reentry shield of a spacecraft.

Development

Development uses knowledge gained from research to derive specific answers to problems. This step converts knowledge into a physical form. The inputs for development are two-dimensional information, such as sketches, drawings, or reports. The outputs are models of three-dimensional artifacts, such as products or structures. This step takes place in two areas. See **Figure 33-6.**

Product or structure development

The first area is ***product development*** or ***structure development***. This area of development applies knowledge to design new or improved products, structures, and services. Development might result in a totally new product or structure or improve on an already-existing one. For example, the bicycle was originally a product from the 1800s. The 10-speed bicycle was later

Standards for Technological Literacy

4 10

Extend

Differentiate, with examples, between basic (pure) research and applied research.

TechnoFact

The role design plays in research and technological development is becoming more and more significant to businesses everywhere.

Standards for Technological Literacy

7 10

Reinforce

Review the types of drawings presented in Chapter 12.

TechnoFact

With the development of technology, the idea of good value has turned from low cost to design and quality manufacturing. Something of low cost may be outdated and wear out easily, while design and manufacturing dictate the worth of a product.

Boeing

Figure 33-6. The two types of development, product and process, were used in designing this aircraft.

developed from the standard bicycle. Likewise, the laser printer uses many processes originally developed for photocopiers.

Process development

The second type of development is ***process development***. This activity devises new or improved ways of completing tasks in manufacturing, construction, communication, agriculture, energy and power, medicine, or transportation. Process development might result in something totally new, such as fiber-optic communication. This development might also improve on existing processes, the way inert gas welding was developed. Inert gas welding was derived from standard arc welding.

Engineering

Developed products and structures must be built. Developed processes must be implemented. To do these things, people need information. Product and structure engineering is responsible for this activity. ***Engineering*** develops the specifications for products, structures, processes, and services. This is done through two basic activities. These activities are design interpretation and engineering testing.

Design interpretation conveys the information needed to produce the product or structure. This includes three main types of documents. These types are engineering and architectural drawings, bills of materials, and specification sheets.

Engineering drawings convey the characteristics of manufactured products. See **Figure 33-7.** A set of engineering drawings includes the following:

- Detail drawings conveying the sizes, shapes, and surface finishes of individual parts.
- Assembly drawings showing how parts go together to produce assemblies and finished products.
- Systems drawings showing the relationship of components in mechanical, fluidic (hydraulic and pneumatic), and electrical and electronic systems.

Architectural drawings are used to specify characteristics of buildings and other structures. See **Figure 33-8.** They include floor, plumbing, and electrical plans for a structure. The drawings also include elevations showing interior and exterior walls.

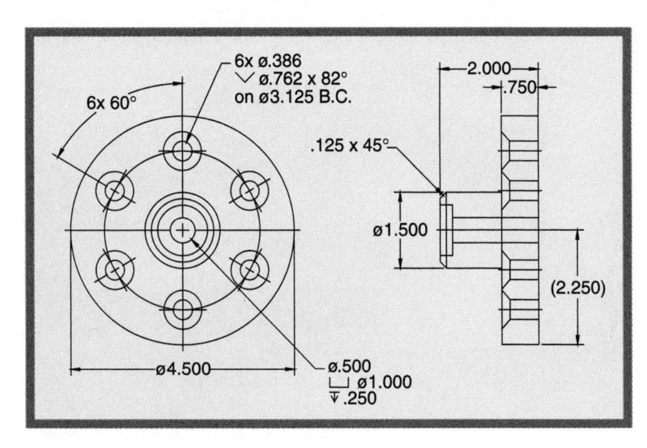

Figure 33-7. Engineering drawings provide detailed information about products or parts to be manufactured.

Exxon Corp.

Figure 33-8. These engineers are reviewing architectural drawings for an oil refinery under construction.

Drawings, however, do not convey all the information needed to build products and structures. The people who implement designs need to know the quantities, types, and sizes of the materials and hardware needed. This information is included on a bill of materials. Finally, data about material characteristics are contained on specification sheets. You might look back in Chapter 12 for a more extensive presentation of these three types of documents. Engineering testing is done during the design stage of a new or redesigned product or system. The goal of engineering testing is to ensure that the product or system meets design criteria and customer expectations.

Standards for Technological Literacy

3 10 19

TechnoFact

Over half of the firms listed in the Fortune 500 have disappeared since 1985. Technology was developed so rapidly by the rivals of these companies that their products quickly became obsolete.

Production

Research and development develop and specify ideas for products and structures. Production, then, must manufacture or

STEM Connections: Mathematics

Science Technology Integrated STEM Curriculum Engineering Mathematics

Calculating Bids

Members of the Rodriguez family want to add a 10′ × 10′ screened-in porch to their house. They ask Mr. Murphy, the sole proprietor of the A-Able Construction Company, to provide a bid for the work. Mr. Murphy decides he needs to charge enough to cover his costs and earn a 20% profit to finance new ventures. He calculates his bid as follows:

Rent equipment to clear the land and pour the foundation and footings.	$ 400.00
Purchase materials (cement, lumber, screening, shingles, felt paper, fasteners, and a door).	$2900.00
Hire two laborers for approximately one week (eight hours per day) at $25.00 per hour each.	$2000.00
Subtotal	$5300.00
Earn a 20% profit ($5300.00 × 20%).	$1060.00
Total	$6360.00

What would Mr. Murphy bid for the job if he found out that equipment rental is $550, instead of $400, and if he decided he needed to make a 25% profit?

Standards for Technological Literacy

19

Extend

Discuss the types of products produced with each type of manufacturing.

Extend

Discuss how building a custom-built home is like manufacturing a custom-designed product, such as a communication satellite.

TechnoFact

Today's technology strives to meet the needs and wants of people while preventing damage to the environment.

construct the physical item. There are a number of different systems used to produce products and structures. The four common manufacturing systems are the following:

- ***Custom manufacturing.*** This system involves producing a limited quantity of a product to a customer's specifications. Generally, the product is produced only once. The custom-manufacturing system requires highly skilled workers and has a low production rate. This makes it an expensive system to operate. Examples of custom-manufactured products are tailor-made clothing, some items of furniture, and the space shuttle.
- ***Intermittent manufacturing.*** With intermittent (job lot) manufacturing, a group of products is manufactured to the company's or a customer's specification. The parts move through the manufacturing sequence in a single batch. All parts are processed at each workstation before the batch moves to the next station. See **Figure 33-9.** Often, repeat orders for the product are expected. This manufacturing activity is relatively inexpensive. Considerable setup time is required, however, between batches of new products.
- ***Continuous manufacturing.*** In continuous manufacturing, a production line manufactures or assembles products continuously. The materials flow down a manufacturing line specifically designed to produce those products. See **Figure 33-10.** The parts flow from station to station in a steady stream. This type of manufacturing handles a high volume and has relatively low production costs. Continuous-manufacturing lines are fairly inflexible, however, and can be used for very few different products. Many are dedicated to a single product.

American Woodmark

Figure 33-9. These kitchen cabinets are being produced with an intermittent manufacturing system, or a job lot manufacturing system.

©iStockphoto.com/Kativ

Figure 33-10. The continuous-manufacturing process brought about rapid expansion in automobile production. Shown is a cutting-and-packaging line for cheese.

- **Flexible manufacturing.** Flexible manufacturing is a computer-based manufacturing system combining the advantages of intermittent manufacturing with the advantages of continuous manufacturing. Thus, it makes possible short runs with low unit-production cost. Machine setup and adjustment are computer controlled. This permits quick and relatively inexpensive product changeovers.

Similar production systems are used in the other technologies, such as construction. Housing can be built to an owner's specifications (custom-built). See **Figure 33-11.** Other dwellings are built in tracts with a common plan used to make a large number of buildings.

An analogy can be made in transportation. Driving an automobile is very similar to custom manufacturing. Automobiles are flexible. They are, however, relatively expensive. Rapid transit buses and trains are more similar to continuous manufacturing. They move people on set lines at lower costs.

Figure 33-11. Some homeowners choose to have their houses custom-built.

The actual production of products and structures can be divided into three important tasks. These tasks are planning to produce, producing, and maintaining quality of the product or structure. See **Figure 33-12.**

Planning

Planning determines the sequence of operations needed to complete a particular task. This task is the backbone of most production systems. Planning determines the needs for human, machine, and material resources. This task also assigns people and tasks to various workstations.

Hewlett-Packard

Figure 33-12. The three tasks of production. Shown are the planning and producing stages in the production of communication messages.

Standards for Technological Literacy

19

Research

Have your students research flexible manufacturing and computer-integrated manufacturing (CIM).

TechnoFact

With the idea of maintenance in mind, manufacturers like Ford and General Motors have been developing telematics systems, which would have the ability to notify both the customer and manufacturer when the customer's car needed service.

Standards for Technological Literacy

2 19

Extend

Discuss how energy-conversion processes and agricultural production are like manufacturing processes.

TechnoFact

Before the Industrial Revolution, the process of production consisted of domestic industry. People made things in their homes, and the merchants who had provided the raw materials collected the finished products, paid for the work done to produce the products, and found a suitable market for them.

Closely associated with planning and scheduling is production engineering. Production engineers design and install the system used to build the product or structure. They are concerned with the physical arrangement of the machines and workstations needed to produce the product.

Producing

Producing is the actual fabrication of the product. In manufacturing, it involves changing the forms of materials to add to the worth, or value, of the materials. These activities include locating and securing material resources, producing standard stock, and manufacturing the products.

Product ***manufacturing processes*** are used to change the sizes, shapes, combinations, and compositions of materials. These processes include using casting and molding, forming, and separating processes to size or shape materials. Conditioning processes change the internal properties of the material. Assembling processes put products together. Finishing processes protect or beautify the products' surfaces. ***Construction processes*** are used to produce buildings and heavy engineering structures. Typical construction processes include preparing the building site, setting foundations, erecting superstructures, enclosing and finishing structures, and installing utility systems.

Communication processes are used to produce graphic and electronic media. Generally, communication messages are designed, prepared for production, produced, and delivered. This is done through the processes of encoding, storing, transmitting, receiving, and decoding operations.

Transportation processes are used to move people and cargo. They are used in land, water, air, and space systems. Typical transportation production processes include loading, moving, and unloading vehicles.

Maintaining Quality

Throughout these processes, a standard of perfection is maintained. Customers want products, structures, and systems to meet their needs and desires. For this to happen, a process called *quality control* is used. Quality control includes all systems and programs that ensure the outputs of technological systems meet engineering standards and customer expectations.

Often, people think "quality" means smooth, shiny, and exactly sized. This is not always true. A smooth, shiny road makes a poor driving surface. Cars would have difficulty controlling and braking on such a surface. Likewise, holding the length of a nail to a tolerance of ±.001" is inappropriate. The cost of manufacturing to that tolerance is too high for the product. The important quality consideration in a nail is holding power, not exact length. Quality can be measured only when a person knows how the product or part is to be used. The product's function dictates quality standards.

Inspection

An important part of a quality control program is ***inspection***. The inspection process compares materials and products with set quality standards. There are three phases of an inspection program. See **Figure 33-13.** The first phase inspects materials and purchased parts as they enter production operations. The second phase inspects work during production. The final phase inspects the end product or structure.

Inspection can be done on every product or on a representative sample of the products. Expensive, complex, or critical components and products are subjected to 100% inspection. This means every part is inspected at least once. Products such as aircraft components and some medical devices are examples of outputs receiving 100% inspection.

Goodyear Tire and Rubber Co.

Figure 33-13. Quality control inspects materials entering the plant; work in progress; and finally, finished products.

Less expensive and less critical parts receive random inspections. A sample of the product is selected representing a typical production run. The sample size and the frequency of inspections are determined using statistics (mathematically based predictions). Inspection of representative samples is part of a program called *statistical quality control.*

The selected sample is inspected. If it passes, the entire run is accepted. If the sample fails to meet the quality standards, the entire run is rejected. Rejected production lots can be dealt with in various ways. The run can be sorted to remove rejects (parts failing to meet the standards). The whole run can be discarded. The run can also be reworked.

Random inspection is used whenever it is cost-effective. Often, the cost of 100% inspection outweighs the value of this type of inspection. For example, it would be expensive to use 100% inspection on roofing nails. Also, the user can discard the occasional defect that slips past random inspection without endangering the product or customers.

Production and Value

In addition to the aspects of planning, producing, and maintaining quality, production activities must deliver products at a reasonable cost. This challenge requires attention to both price and value. Delivering products at a reasonable cost also requires attention to cutting unnecessary costs out of production activities.

Price and value

In addition to quality, customers expect value. Price and value are two different things. ***Price*** is what someone must pay to buy or use the product or service. Initial prices are established by businesses and reflect market conditions. The customer determines value. Value is a measure of the functional worth the customer sees in the product. The customer expects the product or structure to deliver service and satisfaction equal to or greater than its cost. Answering the question, "Was the product worth what I paid for it?" can establish the product's value.

Standards for Technological Literacy

2 19

Demonstrate

Show how to use measuring tools to perform product inspections.

Brainstorm

Have your students select a product and then determine what features they would inspect.

Extend

Elaborate on the difference between price and value.

TechnoFact

A statistic that is applied to measure the contribution of manufacturing to a particular area is called value added by manufacturer. It shows the increase in value from raw materials to finished products.

Standards for Technological Literacy

19

TechnoFact

Time and motion are both related to human movement. Frank and Lillian Gilbreth developed the laws of human motion. The technique they used was called motion study, which differed from time study. Their development of scientific management is still used today in business.

Technology Explained

Wind Tunnel: a device used to test the aerodynamics of vehicles and structures under controlled conditions.

Wind resistance affects vehicles as they move people and cargo. This resistance has an impact on the operating efficiencies of the vehicles. In designing these vehicles, engineers often subject models to tests to maximize the efficiencies of the designs. One important test instrument in this quest is the wind tunnel.

Originally, this device was used solely to test airfoils during aircraft-design activities. The uses of wind tunnels, however, have been expanded. With current concern for fuel efficiency, wind tunnels are now used extensively to ensure that vehicles offer the least amount of wind resistance possible. Wind tunnels are also used to test the wind patterns over and around buildings and various other structures. See **Figure A.**

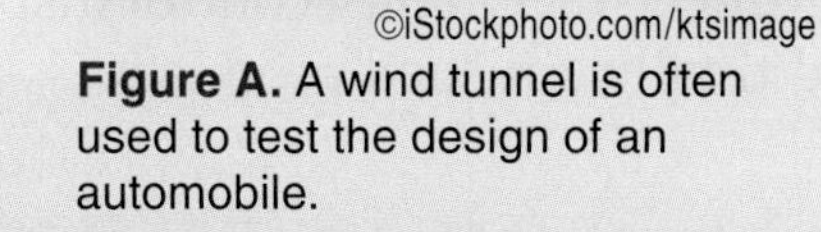

©iStockphoto.com/ktsimage

Figure A. A wind tunnel is often used to test the design of an automobile.

The wind tunnel is designed to pass high-speed air over a full-size or -scale model of a vehicle or structure. An important design requirement is that a smooth and uniform flow of air be produced in the tunnel. To accomplish this, a number of wind tunnel designs have been produced. One of these is shown below. See **Figure B.** Diagramed is a closed loop wind tunnel.

Most wind tunnels have fans or turbines developing airflow in large ducts. The diameter of the ducts increases as the air travels away from the fans. This reduces the airspeed, as well as frictional losses. The tunnels have mitered corners. This addition reduces the wind loss as the airflow changes directions. Also, in high-speed tunnel models, the air passes through cooling tubes to remove heat it gains while passing through the fans or turbines.

The air the fans produce has a swirling pattern. This airflow creates unreliable test results. Therefore, the airflow passes through an air-smoothing unit. This unit is a series of tubes removing the swirls and directing the air in a straight line.

As the air reaches the test chamber, the diameter of the tunnel shrinks rapidly. This causes the speed of the airflow to increase. The amount of the decrease in the tunnel's diameter controls the final airspeed in the test chamber.

Models in the chamber are carefully tested. Anemometers test the airspeed. Smoke might be introduced to visually observe the flow patterns of the air, as the air passes structures or vehicles. The vehicle itself might be attached to instruments to measure the lift and drag it develops as air passes.

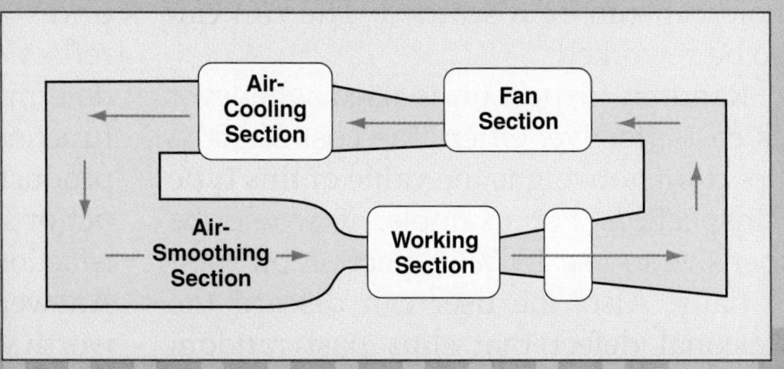

Figure B. The path the airflow follows in a common wind tunnel design.

Cost-cutting systems

A number of new production systems have been developed to reduce product cost and, in turn, increase product value. See **Figure 33-14.** One of these systems is ***computer-aided design (CAD).*** CAD reduces product design and engineering costs. ***Just-in-time (JIT) inventory-control systems*** schedule materials to arrive at manufacturing when the materials are needed. This reduces warehousing costs. Flexible manufacturing reduces machine-setup time. This type of manufacturing was discussed earlier.

Similar systems are used in the other technologies. Computer scheduling is used in construction to ensure that human and material resources are effectively used. Computer ticketing reduces transportation costs. Computer systems make layout and preparation of color illustrations for printed products more economical.

©iStockphoto.com/seraficus

Figure 33-14. This automatic control center in a newspaper-printing plant reduces costs by maintaining quality and reducing labor.

Marketing

Products, structures, and services are of little value to companies unless the companies can sell them to customers. The products and structures must be exchanged for money. This is the challenge for marketing personnel. Marketing efforts promote, sell, and distribute products, structures, and services. Specifically, marketing involves four important activities:

- ***Market research*** gathers information about the product's market. See **Figure 33-15.** This can include data about who will buy the product and where these people are located, in addition to these people's ages, genders, and marital statuses. Also, market research can measure the effectiveness of advertising campaigns, sales channels, or other marketing activities.
- ***Advertising*** includes the print and electronic messages promoting a company or its products. This activity can also present ideas to promote safety or public health. Advertising is designed to cause people to take

Figure 33-15. These people are participating in a quality-audit program. This program examines customers' reactions to new recipes.

Standards for Technological Literacy

2 6 19

Research

Have your students research what makes an advertisement effective.

TechnoFact

Consumers are made aware of new and existing products through advertising. Advertising is also responsible for linking a brand image to a particular product. While consumers are interested in a product's price and characteristics, they are also influenced by the image associated with a certain brand.

Standards for Technological Literacy

6 19

Brainstorm

Have your students list the ways a product is promoted and protected by its packaging.

action (buy a certain product) or think differently (buckle your seat belt while riding in an automobile). Closely related to advertising is ***packaging***. See **Figure 33-16.** This activity deals with designing, producing, and filling containers. The packages are designed to promote the product through colorful or interesting designs. Packaging also protects the product during shipment and display. Finally, the packages must include information that helps the customer select and use the product wisely.

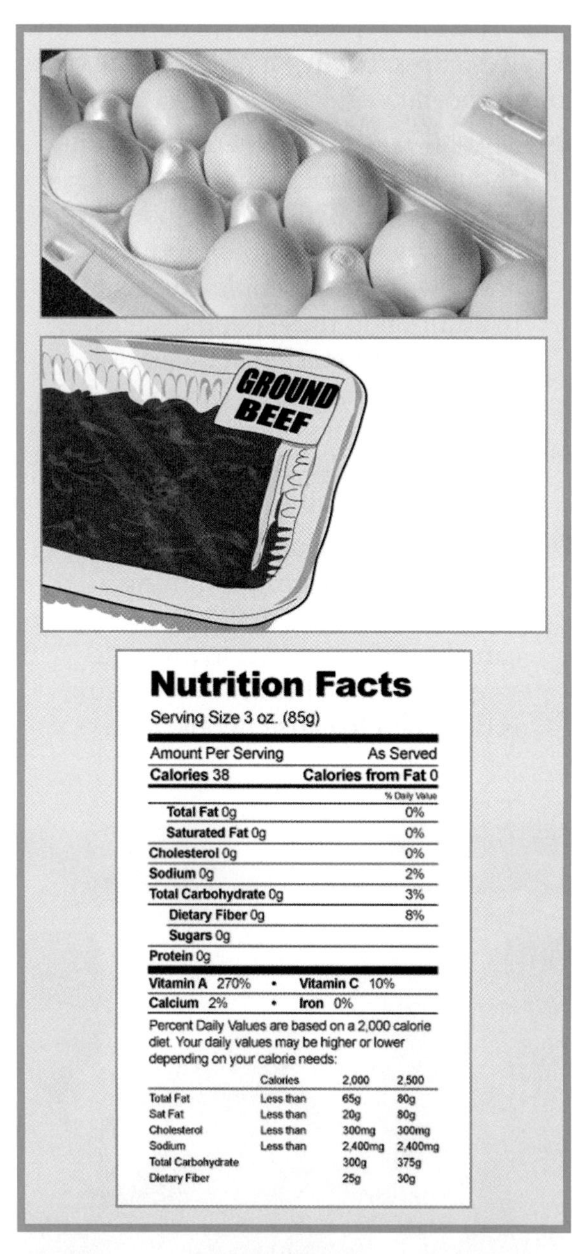

Figure 33-16. Packages are designed and produced to perform multiple functions. A package promotes the product, protects the product, and contains information to inform the customer about nutritional content and other topics.

- ***Sales*** is the activity involving the physical exchange of products for money. This activity includes sales planning. Sales planning develops selling methods and selects and trains sales personnel for their efforts. Sales also includes the act of selling. This act involves approaching customers, presenting the product, and closing the sale. This series of steps is all part of sales operations. The end results of sales operations are an order from the customer and income for the company.
- ***Distribution*** is physically moving the product from the producer to the consumer. This consumer might be another company or a retail customer. Consumer products follow at least three common channels. See **Figure 33-17.** The product might move from the producer directly to the consumer. This channel is called *direct sales.* Sales of homes, encyclopedias, cosmetics, vacuum cleaners, and transportation services often use this channel.

In another channel, the producer sells the products or service directly to a retailer. The retailer then makes the items available to the customer. Franchised businesses, such as new-automobile dealers and some restaurants, are examples of this distribution channel. This channel allows the producer to regulate the number of sales outlets and the quality of service those outlets provide.

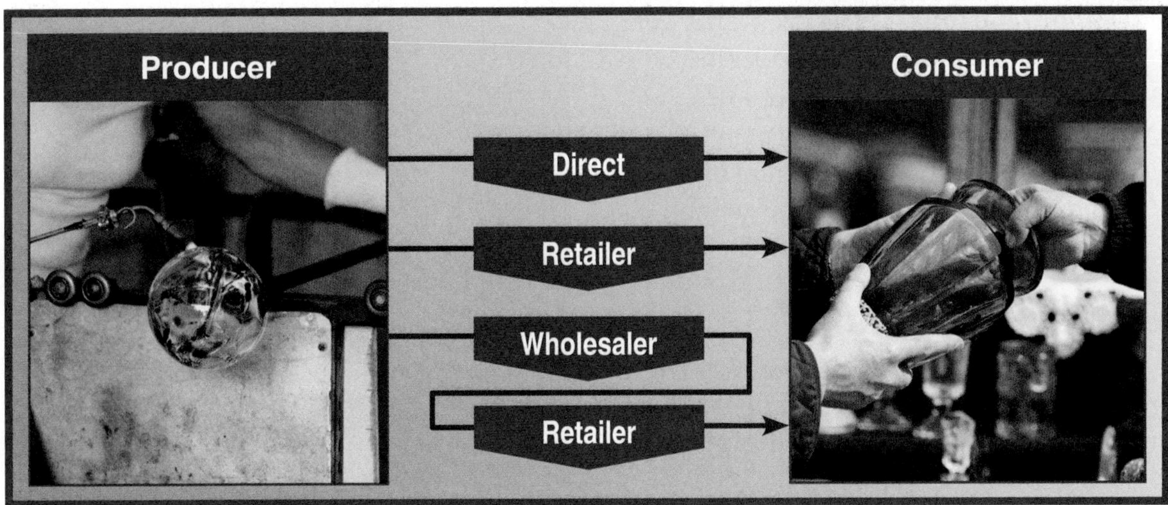

©iStockphoto.com/dscott, ©iStockphoto.com/E_B_E

Figure 33-17. There are three different paths moving products from producers to consumers.

A third channel has the producer selling products to wholesalers. Wholesalers buy and take possession of the products. They, in turn, sell their commodities to retailers. The retailers then sell the commodities to the customers. In this channel, producers have little control over the retailers who are selling their products.

As described earlier, research and development develop the product, structure, or service. Production produces the item. Marketing promotes and sells the item. These three activities, however, cannot stand alone. They require money and people to make everything work. These requirements are the responsibility of two managed support areas. The support areas are industrial relations and financial affairs.

Industrial Relations

You have learned that humans are the foundation for all technology. They create technology, develop it, produce it, and use it. Thus, people are fundamental to all company operations. They work for the company, buy the company's products, and pass laws regulating company operations.

Companies, therefore, are very concerned about their relationships with people. They nurture positive relationships with people by engaging in industrial relations (often called *human resources*) activities. These activities can be grouped under three main programs. See **Figure 33-18.** These programs are the following:

- ***Employee relations.*** These programs recruit, select, develop, and reward the company's employees.

Harris Corp.

Figure 33-18. Industrial relations personnel develop and administer three types of people-centered programs.

Standards for Technological Literacy

4 16 19

Figure Discussion

Use local examples to explain the three distribution channels shown in Figure 33-17.

TechnoFact

In the distribution process, products are packaged and moved from the producer to the consumer. Paper is one of the most common materials used in packaging products, and it is also a material commonly recycled. It is used in various boxes, cartons, and bags.

Standards for Technological Literacy

4

Research

Have your students select a job and use the U.S. Department of Labor Web site to determine its requirements.

TechnoFact

While part-time work is still not considered typical, jobs based on a 40-hour workweek have declined. The needs of employers and employees determine whether a business would require its workers to labor for shorter or longer hours.

- ***Labor relations.*** These programs deal with the employees' labor unions.
- ***Public relations.*** These programs communicate the company's policies and practices to governmental officials, community leaders, and the general population.

Employee Relations

All companies have employees. These employees do not magically appear, fully qualified to work. They are the result of managed employee-relations activities. These activities select, train, and reward the people producing products and managing operations.

Figure 33-19. This job applicant is being interviewed to gain information not gathered on the application form.

Selecting employees

The first action in the employee-relations process is called ***employment***. This task involves determining the company's need for qualified workers. Applicants are then acquired through a searching process called ***recruiting***. Recruiting can be done through newspaper advertisements, school recruiting visits, or employment agencies. Other applicants might come to the company seeking employment on their own.

Next, job applicants go through a screening process. ***Screening*** allows qualified people to be selected from the applicant pool. Generally, this selection process starts with an application form to gather personal and work-experience data. The promising applicants are interviewed to gather additional information. See **Figure 33-19.** Some jobs require special abilities and knowledge. In these cases, applicants might be given a test to find those who qualify.

Training employees

Successful applicants gain employment. Few new employees, however, are ready to begin work without training. Most need some basic training. Some employees might need special instruction. Basic information about the company and its rules and policies is provided to all workers. This is called *induction training* or *orientation.*

Special job skills can be provided through one of three programs. Simple skills are generally taught through ***on-the-job training***. In this method, experienced workers or managers train new workers at the new workers' workstations. More specialized skills might be developed in classroom training sessions. ***Classroom training*** involves qualified instructors providing information and demonstrating practices that each employee must learn.

Highly skilled workers are developed through ***apprenticeship training***. Apprentices receive a combination of on-the-job and classroom training over an extended period of time. Apprenticeships usually last from two to four years, depending on the skills to be learned.

In all three types of training, workplace safety is stressed. New employees are informed about company safety rules; shown how to work safely; and in many cases, tested on safe work practices. Safety training is vital to providing a safe workplace. Federal and state agencies provide companies with safety rules and regulations and are responsible for enforcing them. A principal source of these regulations is the Occupational Safety and Health Administration (OSHA).

Executives and professional employees also receive training. This training might be called *executive development, sales training,* or *managerial training.* Since the work these employees do does not involve making the products, most training sessions are given in classroom settings. See **Figure 33-20.** In many companies, the total training program is called *human resource development (HRD).*

Rewarding employees

People want to be recognized and rewarded for their work. Companies recognize and reward people in two ways. First, they pay employees a wage, salary, or commission as a direct reward for work accomplished. A ***wage*** is a set rate paid for each hour worked. A ***salary*** is payment based on a longer period of time, such as a week, month, or year. Wage earners are often called *hourly workers.* Hourly workers are usually the production workers who build products, erect buildings, print products, or provide transportation services.

Salaried employees are usually technical and managerial workers. They develop products, engineer facilities, maintain financial records, and direct the work of other people. These employees often have more formal education than hourly workers. Salaried employees are held accountable more in terms of the amount of work they do than the hours they work.

Some salespeople are paid in a different way. Instead of wages or a salary, they receive a ***commission*** for each sale they make. The commission is usually a percentage of the total dollar value of the goods sold.

The second type of reward is called a ***benefit.*** Benefits are the insurance plans, vacations, holidays, and other programs the company provides. These items cost the company money and, therefore, are a part of the total pay package for an employee. Some companies also make use of special rewards called *bonuses* or *incentives.* These rewards are typically awarded to employees for performance exceeding what is expected or for suggestions leading to improvements in efficiency, productivity, or workplace safety.

Standards for Technological Literacy

4

Brainstorm

Ask your students whether they would rather earn a wage or a salary. Have them explain their reasoning.

Research

Have your students use the Internet to find out what OSHA regulates and how it does it.

Quotes

"Faced with a deluge of undifferentiated information, people will need creative and analytical skills to assess that information and to discover the most effective ways of using it...these skills will be the toolbox of the new knowledge-driven society."
—David Patterson, in a report to the British Design Council.

Figure 33-20. These managers are receiving classroom training to improve communication skills.

Standards for Technological Literacy

4 5

Research

Have your students identify some typical benefits given hourly-paid and salaried workers.

Brainstorm

Ask your students why companies pay salespersons by commissions.

Research

Trace the historical development of labor unions.

Extend

Discuss the types of workers that are most likely to be unionized and how this is changing.

Extend

Discuss why public relations are important to a manufacturing company.

TechnoFact

Trade has been practiced since prehistoric times, when food and tools were exchanged. Trade is practiced today as the art of selling, but instead of trading goods for goods, it is more common to exchange goods for money.

TechnoFact

When wage earners, farmers, and merchants banded together in 1869, they formed the first union of workers: the Noble Order of the Knights of Labor. Although its growth continued through the end of the 19th century, a strike caused the union to disband in 1900.

Think Green

Sustainability Plan

Throughout this text, you have learned about the different types of environmental impacts technology can have and the ways individuals can change these impacts. In much the same way, companies can use greener alternatives to make a difference. Several businesses now include a sustainability plan in their business plans. A *sustainability plan* contains the guidelines and procedures used in order to help companies use, develop, and protect resources in a way that meets needs without harming the environment.

A company's sustainability plan may come from individual employee ideas, but the whole company will put these ideas into action. There are no set guidelines for all businesses to use in creating sustainability plans. The general goals typically focus on reducing the use of materials, energy, and water. Another common goal is to reduce the amount of pollution output by the company. The ways companies achieve these goals are similar to how individuals work to lessen their environmental impacts.

Labor Relations

In many larger companies, labor unions represent the employees. These companies require a labor-relations program. This program works on two levels. First, ***labor agreements***, called *contracts*, are negotiated between the company and the union. Labor agreements establish pay rates, hours, and working conditions for all employees the contract covers. The agreements cover a specific period of time, generally ranging from one to three years.

During the contract period, disputes often arise over the contract's interpretation. These disputes are called ***grievances***. This is the second level of work for labor relations. Labor-relations officials work with union representatives to settle the grievances.

Public Relations

Companies hire people, pay taxes, and have direct impacts on communities. Company managers form policies and have practices that they feel benefit the company. These practices often are subject to government regulation and might be affected by community pressures.

A company's public relations program is designed to gain acceptance for company operations and policies. The program informs governmental officials about the need for, and impact of, laws and regulations. Public relations also communicates with the community leaders so local actions do not hamper the company's legitimate interests. Finally, public relations communicates with the general public. This communication presents the company as a positive force in the community. See **Figure 33-21.** This image improves the company's ability to sell the company's products and to hire qualified workers.

Financial Affairs

Just as a company needs people, it also needs money. Companies must buy materials and equipment, pay wages and salaries, and rent or buy buildings. Taxes and insurance premiums must be paid. These

Figure 33-21. Community activities, such as entering a float in a parade, are part of a company's public relations program. The float is intended to help create a positive image of the company.

actions can be shown in a simple flowchart. See Figure 33-22. If you start at the upper-left corner and read down, you see the following: Management employs people to use machines to change the form of materials, which are paid for with money and produce products, which are converted in the marketplace into money.

The word *money* appears twice. Money pays for resources. These resources are people's time, materials, and machines. This money is called ***expenses***. Money is also the end result of sales—***income***. The difference between income and expenses is profit or loss. The goal of a company is to have more income than expenses. This makes a company profitable.

Profits must also be managed. They serve two important purposes. First, they can become ***retained earnings***. These earnings are profits the company holds and uses to enlarge its operations. Profits are

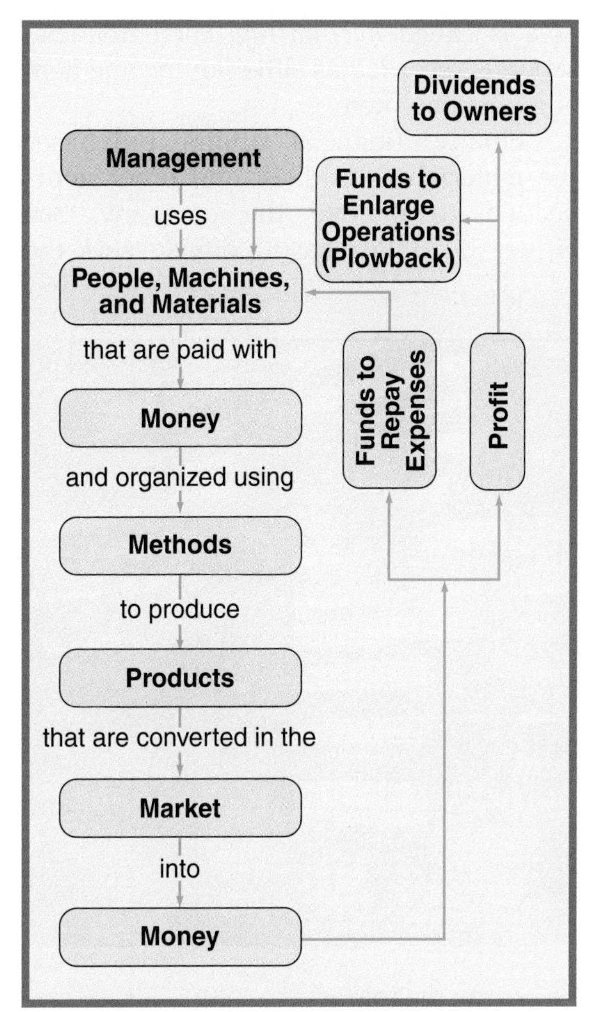

Figure 33-22. A simple flowchart showing how money cycles within company activities.

an important source of money for financing new products, plant expansions, and mergers.

Many companies pay out another portion of the profits as dividends. These dividends are quarterly or annual payments to the stockholders. They are the reward for investing in the company and sharing the risks of owning a business.

Managing the use of this money is the responsibility of financial-affairs employees. They raise money, pay for insurance, collect from customers, and pay taxes. These employees also keep records of the financial transactions of the company. This

Standards for Technological Literacy

Figure Discussion

Use Figure 33-22 to show how a manufacturing company adds value to resources and how it distributes money it gains from selling value-added products.

TechnoFact

Samuel Gompers helped organize and lead the first American federation of workers. Later, several AFL unions formed the Committee for Industrial Organization (CIO). When a disagreement caused the AFL to separate itself from the CIO in 1938, the CIO became a new union called the Congress of Industrial Organizations.

Standards for Technological Literacy

4 6

Figure Discussion

Use Figure 33-24 to discuss the Develop-Produce-Sell cycle and the Select-Use-Maintain-Discard cycle of a product.

Brainstorm

Have your students explore the social and environment responsibilities associated with manufacturing products and using products.

TechnoFact

Accounting provides financial information needed for an enterprise to plan for future activities and compare successes and failures. It is the process of gathering, summarizing, and communicating financial information.

TechnoFact

In the early 1900s, the Industrial Revolution came to the United States in what is called the second stage of the movement. Until that time, only Western Europe had benefited from the development of large manufacturing centers. Since the second stage, however, the United States has been the leading manufacturing nation.

area is called ***accounting***. Each financial action is recorded as either an income item or an expense item.

Finally, financial affairs purchases the materials, machines, and other items needed to operate the company. See **Figure 33-23.** Purchasing officers seek the "best" items, in relation to company needs. The term *best* takes into account price, quality, and delivery date.

Brush Wellman

Figure 33-23. One responsibility of financial-affairs personnel is purchasing materials the company needs.

The Industry-Consumer Product Cycle

We have been looking at company activities as a linear action. These activities are presented as starting with research and development. Production is the next step. Finally, marketing is the last step. This path is correct if you look at a single model or version of one product or structure. Our system, however, is much more complex. The economy is actually described as dynamic, or always changing. Products are developed, produced, and sold. Consumers select, use, maintain, and discard the products. See **Figure 33-24.** They communicate their satisfaction or dissatisfaction with current products. This might cause companies to redesign existing products or, in some cases, to develop new ones. These new products are sold. They in turn, are selected, used, maintained, and discarded. This cycle continues with a constant array of new products being developed and obsolete products disappearing.

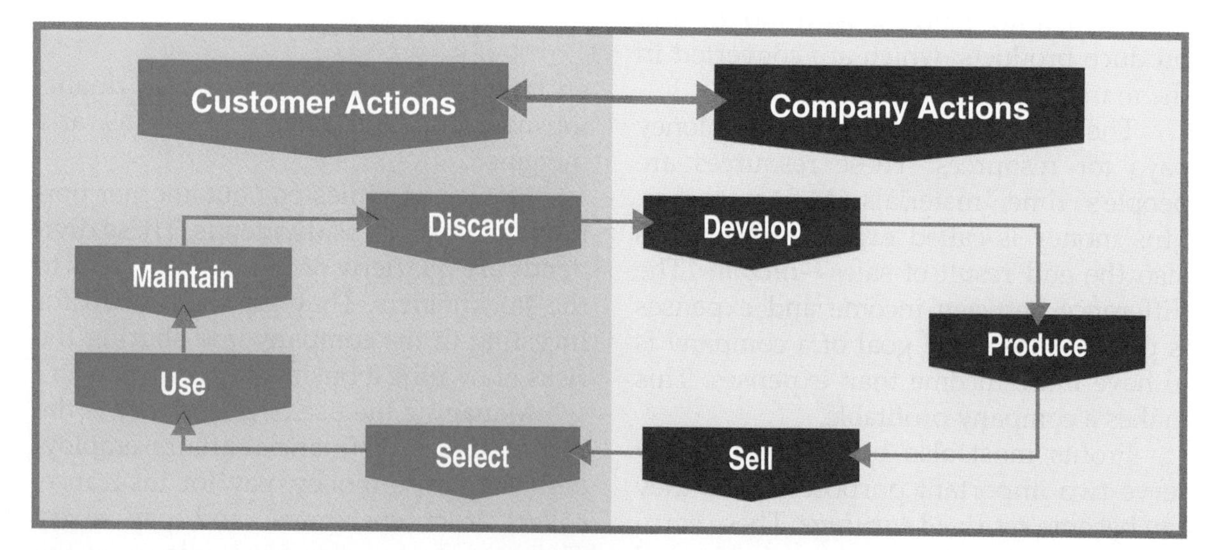

Figure 33-24. Customers' reactions to products cause companies to continually design, produce, and sell new items.

Quality control is an integral step in the production process. One method of quality control is the inspection. The inspection process is done in three phases to ensure the quality of the end product.

Answers to Test Your Knowledge Questions

1. B. Sports.
2. economic enterprises
3. Industry is the area of economic activity that uses resources and systems to produce products, structures, and services with intent to make a profit.
4. E. Financial affairs
5. B. Production
6. A. Research and development
7. C. Marketing
8. B. Production
9. D. Industrial relations
10. E. Financial affairs
11. applied research
12. Custom manufacturing, intermittent (job-lot) manufacturing, continuous manufacturing, and flexible manufacturing
13. materials and purchased parts, end product
14. False
15. Any order: Employee, Labor, Public.
16. True
17. Income
18. Consumers select, use, maintain, and discard the products.

Summary

Most technology is developed and produced by industrial enterprises. These enterprises create a steady flow of products, structures, and services through five main areas of activity. Research and development develop and specify the item. Production produces the item. Marketing promotes, sells, and distributes the item. Industrial relations and financial-affairs activities support these product- and structure-centered functions. Industrial relations recruit and develop the workforce. They also deal with labor unions and promote the company's image and policies. Financial affairs maintains financial records, pays the company's bills, and purchases material resources.

Test Your Knowledge

Write your answers on a separate piece of paper. Please do not write in this book.

1. Which of the following is *not* a societal institution?
   A. Religion.
   B. Sports.
   C. Family.
   D. Education.
2. Businesses directed toward making a profit are called _____.
3. In terms of technology, why is industry different from the other types of economic enterprises?

Matching questions: For Questions 4 through 10, match each responsibility on the left with the correct area of activity on the right. (Note: Answers can be used more than once.)

4. _____ Accounts for income and expenditures.
5. _____ Makes products or structures.
6. _____ Specifies characteristics of products.
7. _____ Promotes products.
8. _____ Responsible for quality control.
9. _____ Includes public relations.
10. _____ Purchases materials.

A. Research and development.
B. Production.
C. Marketing.
D. Industrial relations.
E. Financial affairs.

11. Research focused on developing products is called _____.
12. List the four types of manufacturing systems.
13. The three phases of an inspection program are _____, work in progress, and _____.

14. Franchised businesses typically purchase products and necessary services through a wholesaler. True or false?
15. The three major programs in a complete industrial relations program are the following:
    A. ______ relations.
    B. ______ relations.
    C. ______ relations.
16. Many skilled workers are trained through apprenticeship programs. True or false?
17. To show a profit, a company must have more ______ than expenses.
18. List the four steps on the consumer side of the industry-consumer product cycle.

STEM Applications

1. Select a simple product, such as a kite. Apply the principles of research and development, production, and marketing to design, produce, and advertise it.
2. Set up a production line for chocolate suckers, cookies, or another food product. Describe how you will plan for the product, produce the product, and maintain quality.

Chapter 34 Using and Assessing Technology

Chapter Outline
Using Technology
Assessing Technology

This chapter covers the benchmark topics for the following Standards for Technological Literacy:

12 13 20

Learning Objectives

After studying this chapter, you will be able to do the following:

- Summarize how using a technological product is different from using a technological service.
- Summarize how people can effectively select an appropriate product.
- Recall where people can obtain information about operating products.
- Summarize how products should be properly discarded.
- Summarize how people can effectively select an appropriate technological service.
- Summarize ways societies assess technological advancements.

Key Terms

owners' manual
technological assessment
technological product
technological service

Strategic Reading

As you read this chapter, think of how the technological products you use meet your needs. Do you spend enough time assessing your products' usefulness and effects?

People working in various industries design and make technological products and systems. This group of people is a part of a larger community or nation. They have one type of job. Other people in the community have different jobs. One job that all people share, however, is using these technological outputs. See **Figure 34-1.** Each of us uses the products of technological effort every day. As we use these products, we meet our needs and, in doing so, impact our lives and the world around us. The use and impacts of technology can be seen as two major technological actions:

- **Using technology.** This action is selecting and operating tools,

Hewlett-Packard

Figure 34-1. People with special skills and knowledge design and make products we need.

machines, and systems to modify or control the environment.

- **Assessing technology.** This action is measuring and reporting the impacts of technological use on people, society, and the environment.

Using Technology

People are constantly generating needs and wants. They might want to travel to a vacation spot. People might want to cook food faster. They might want to have better health or higher energy levels. These and thousands of other needs and wants can be met with technological devices and systems.

Technological outputs can be grouped into two categories. These categories are technological products and technological services. See **Figure 34-2.** ***Technological products*** are the artifacts people build. They can be a manufactured product, a constructed structure, or a communication medium (such as a book or CD recording). ***Technological services*** are outputs that we use, but do not own. These include transportation and communication services, such as airline travel and television programming.

Figure 34-2. Technological outputs can be products or services. These gondolas are providing a service by delivering goods. This church is a constructed product.

Standards for Technological Literacy

12 13

Brainstorm

Ask your students what technological products they have used today and where those products were produced.

Reinforce

Discuss using and assessing technology in relationship to the outputs and feedback of a technological system (Chapters 1 and 2).

Quotes

"Design is about ideas, and through ideas we get change. The world only moves forward by changing... everyone can be part of that."—Wayne Hemingway, founder of Red or Dead.

Standards for Technological Literacy

12 13

Figure Discussion

Use Figure 34-3 to differentiate between how people use technological products and technological services.

TechnoFact

The development of technology can be seen as the development of tools, techniques, and processes that people have made over time. Technology has reduced work required to meet basic needs, and it has also produced several helpful timesaving tools.

Using products and systems is important to our lives. This use can make life easier, more enjoyable, and more productive. Using products and systems can, however, also create frustration and negative impacts on people or the environment. This tension between good and bad, positive and negative, means that people should use technology correctly.

How do people know, however, that they are using technology properly? This knowledge requires them to understand and apply the steps in proper technological use. See **Figure 34-3.**

Using Technological Products

When we use technological products, each of us completes a series of tasks. We might do these tasks ourselves or have another person do them for us. The tasks might be seen as the "steps of life" for the product. These steps start when we obtain the product and end when the product is no longer of use to us. Simply put, the steps in using technological products are the following:

1. Selecting appropriate devices and systems.
2. Operating devices and systems properly.
3. Servicing devices and systems.
4. Repairing broken parts and systems.
5. Disposing of worn-out or obsolete devices and systems.

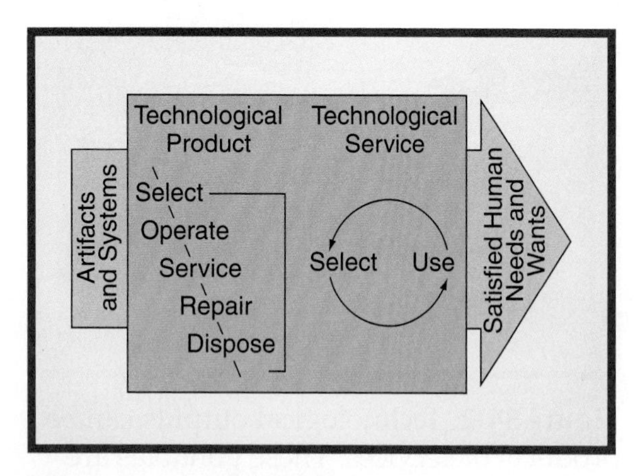

Figure 34-3. Five steps should be followed when using technological products.

When we use technological services, we do not use all these steps. We can only select and use the service. Most of us do not operate the airline or the television network. We do not have a product to service, repair, or discard.

Selecting Appropriate Devices and Systems

Look around you. You are surrounded with technological products. Someone selected each of these. They were not the only products that could have served their purposes. Consider electric lighting. People can choose from incandescent, halogen, and fluorescent lights. These lights can be of different lighting levels (brightnesses), such as 60 watts, 75 watts, or 100 watts. People can select ceiling, floor, or table lights.

Some people use trial and error to make product selections. They buy a product and try it. If it does not work well, they replace it with another product with different features. This is an inefficient and expensive approach, however, to selecting an appropriate product.

A better selection method is to deal with the challenge logically. See **Figure 34-4.** First, the person should determine the exact need being addressed. The need should then be described in terms of the product's operation, price, quality, and similar features. From these descriptions, a list of alternative products meeting the need can be made. The features of each product should be identified. A list of advantages and disadvantages (pros and cons) of each product should be developed from these features. For example, purchase price, operational features, and maintenance requirements for each product can be identified. The ease of use can be explored. The product's safety

Figure 34-4. Product selection follows a logical sequence.

can be examined. Likewise, the appearance and styling can be considered. From this list of features and pros and cons, the most appropriate product can be selected. The most appropriate product might not be the cheapest or best-operating product. This product should, however, be the one most closely meeting the overall need.

Operating Devices and Systems

Once a product has been selected and purchased, the new owner must learn how to operate the device. Some products need little or no training to be able to use them. See Figure 34-5. The owners already know all they need to know to operate the product. Few of us need to learn how to use a new pen or pencil.

Other products are replacements for an older model. Again, the owners know a lot about the product's operation. They need only to review the new product's operation and new features. For example, maybe you can effectively use a version of a word processing program. If you buy a new version, you need to learn how to use only the program's new features.

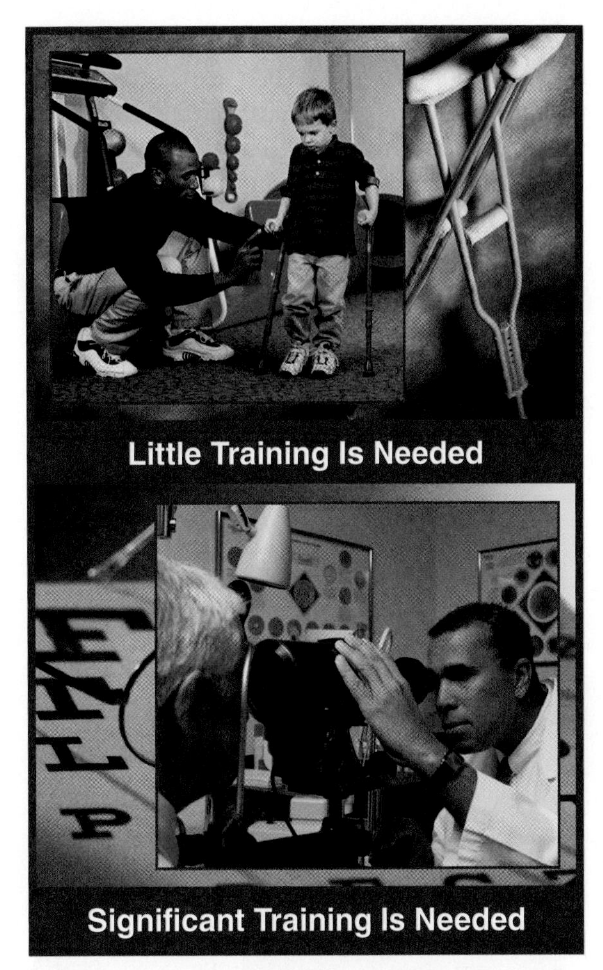

Figure 34-5. Most people need little instruction on how to use crutches. The doctor, however, needed training to use the eye-testing equipment.

Some products are new to the owners. These owners have never used the product or one similar to it. They then have a lot to learn. This includes learning how to do the following:

- Unpack and set up the product.
- Adjust the product for different operations.
- Correctly and safely operate the product.
- Care for and maintain the product.
- Obtain service and repair.

There are several ways to learn this information. One way is to carefully read

Standards for Technological Literacy

12 13

Research

Have your students determine what type of lightbulb will give the most light for the least amount of energy.

TechnoFact

Various technological devices have been used over time, from what we consider primitive to what is known as advanced. While some tools seem outdated, at the time of development, they were deemed appropriate.

Standards for Technological Literacy

12 13

Demonstrate

Use an owner's manual to show your students how to find operating and servicing requirements for a new product.

Research

Have your students use the Internet to find out how to make a simple home repair.

TechnoFact

Although repair is not always necessary, maintenance of some sort is often required for technological products. Learning how to service and repair products is frequently done by reading manuals that accompany the products.

the ***owners' manual***. See **Figure 34-6.** The new owner should read and study the information to obtain information about the five elements listed earlier. Only after developing this information should the device be used.

Another way to obtain this information is through training. This training can be from a person skilled in using the device. For example, a salesperson in a computer store can provide some basic information about setting up and using a simple computer system. New owners can also seek more formal instruction. They can attend a seminar or class that teaches about using the device.

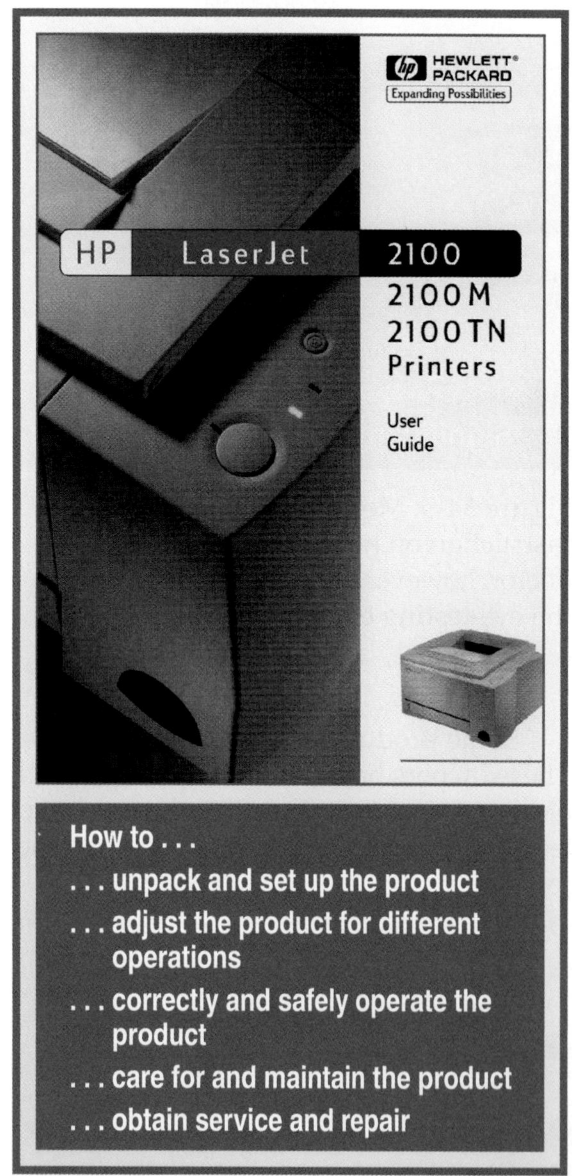

Hewlett-Packard

Figure 34-6. An owners' manual contains valuable information.

Servicing Devices and Systems

Many products need periodic maintenance or service during their lifetimes. Servicing is doing routine tasks that keep the product operating. This includes cleaning the product, oiling moving parts, making simple operating adjustments, replacing filters, and other similar tasks. See **Figure 34-7.** Most owners' manuals contain information needed to complete periodic servicing.

©iStockphoto.com/vladacanon, ©iStockphoto.com/skodonnell

Figure 34-7. Servicing involves simple tasks to keep equipment working, such as checking the oil level and adding oil as needed in an automobile.

Career Corner

Landscape Architects

Landscape architects design outside areas, such as residential yards, public parks, college campuses, shopping centers, and golf courses. They help plan the locations of buildings, roads, and walkways and determine the arrangement of plant materials, such as lawns, flowers, shrubs, and trees. These architects also work with environmental scientists and foresters to conserve or restore natural resources.

They create detailed plans for sites. These plans include soil slopes, vegetation, walkways, and landscaping details, such as fountains and statues. Often, these plans are prepared and presented using CAD systems and video simulation. Entry into the field requires a bachelor's or master's degree in landscape architecture. A successful landscape architect has creative vision, artistic talent, and good oral and written communication skills.

Standards for Technological Literacy

12 13

Extend

Discuss the difference between worn out, obsolete, and unwanted products.

TechnoFact

The destruction of the rain forests for reasons relating to technological development has decreased. Today, many people reduce the use of technology in order to better preserve the natural environment.

Repairing Devices and Systems

Occasionally, products stop functioning properly and require repair. Repairing devices involves replacing worn or broken parts. See **Figure 34-8.** Repairing also includes making major operating adjustments. The owner can do this type of work. Most often, however, trained mechanics and technicians do the work. For example, repairing an automobile transmission, a television set, or a furnace is beyond most people's ability or interest.

Figure 34-8. Repairing involves replacing broken and worn parts. This worker is replacing timbers in a mine.

Disposing of Devices and Systems

Few products last forever. Most products have useful lifetimes, after which they might wear out or become obsolete. When this happens, the owner should properly dispose of the product or material. See **Figure 34-9.** This can involve donating the product to a group that has use for it.

Standards for Technological Literacy

12 13

Brainstorm

Ask your students what criteria they would use to select a new mp3 player.

Research

Have your students research the terms reduce, recycle, and reuse as they apply to recycling.

TechnoFact

In response to the growing problem of pollution, individuals and industries alike have chosen an alternative to solid waste disposal. Recycling is a way to reuse certain parts of products that are no longer in use. Several businesses and communities have adopted recycling programs to help reduce waste.

Figure 34-9. These bins allow people to recycle wastepaper, bottles, and cans.

For example, many charities and schools have use for computer systems that are not state-of-the-art. Many businesses that replace operating systems with newer, faster models donate the older systems to such groups.

Worn or obsolete products can be sold or given to a recycling operation. This group, or company, dismantles the product to retrieve good parts or materials. For example, automobile "wrecking yards" recycle good parts by removing and selling them to people who are repairing similar models. They also sell scrap metals to steel mills and aluminum processors.

Some products cannot be reused or recycled. The product might not have any useful parts. Recycling the materials might be impossible or might not be cost-effective. For example, old dinnerware has little value. Most ceramic materials cannot be recycled. Likewise, broken concrete slabs have little use or value. In this case, the materials should be sent to a landfill.

Using Technological Services

Selecting and using a technological service requires the customer to take fewer actions than using technological products does. See **Figure 34-10.** We can only select and use the service. Most of us do not operate the airline or the television network. We do not have a product to service, repair, or discard.

Figure 34-10. Transportation services are selected, used, and evaluated.

Standards for Technological Literacy

12 13 20

TechnoFact

In technological assessment, goals and conditions must be recognized before an attempt to meet any goals is made. The Brundtland Commission defined sustainable energy as a development that "meets the needs of the present without compromising the ability of future generations to meet their own needs."

Technology Explained

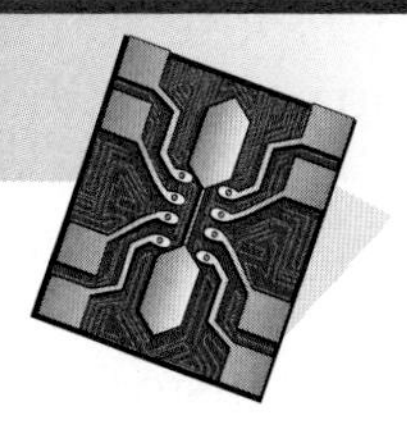

Earth-Sheltered Building: a structure built into the earth to take advantage of the insulation value and thermal properties of soil.

Energy-efficient buildings are important to many people. A number of different building techniques can be used to make buildings more efficient. One way is to use earth-sheltered construction. This method of construction partially or completely covers the dwelling with soil. The purpose is to use the soil's thermal qualities to keep the house at an even temperature. The temperature of soil a few feet below the surface does not vary much throughout the year.

Using the earth to shelter a dwelling is not a new idea. Native Americans in Arizona used earth-sheltered buildings in 1000 AD. This was 400 years before Columbus discovered America. Using protective cliff faces to shelter communities was common throughout the southwestern United States. Today, a number of these sites have been uncovered and restored. See **Figure A.**

Earth-sheltered construction requires careful planning. Special problems exist, such as moisture, loads from soil, and orientation for solar heat gain. Buildings can be partially earth-sheltered or totally covered. See **Figure B.** In the northern hemisphere, the building is normally sited so it faces south. This is so the house can absorb solar heat through large windows on the south face. The wall of windows helps make the front rooms light and airy. These are the daytime living rooms. Bathrooms and bedrooms can be placed toward the back of the structure. These rooms are normally used at night. At this time, artificial light is needed.

Figure A. This cliff dwelling at the Walnut Creek National Monument in Arizona uses earth sheltering.

This type of construction can be used for more than housing. Earth-sheltered construction is also appropriate for theaters, shopping malls, convention centers, and warehouses. The functions of these structures do not require windows or exterior views.

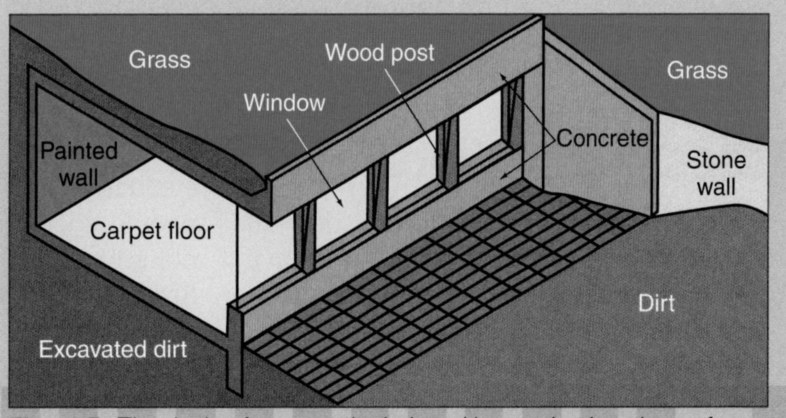

Figure B. The design for an earth-sheltered house that has the roof covered with earth. This house would be built using reinforced concrete.

Standards for Technological Literacy

12 13

Research

Have your students investigate how a federal agency, such as the Food and Drug Administration or the Environmental Protection Agency, uses assessment.

TechnoFact

The Office of Technology Assessment operated from 1972–1995. It reported to the United States Congress and gave object reports and analyses on development technologies and their possible effects.

Suppose you plan to travel across the country. You still need to establish your need (goal). This might be using the least amount of time to travel from point A to point B. Your need might, instead, be seeing as many historical sights as possible.

Once the need is defined, the possible ways to make the trip should be established. You can look at a road map and decide if you can make the trip in a car. Also, you can explore flight, bus, and train schedules, routes, and prices. Armed with this information, you can choose a method of travel.

After the trip is completed, you can evaluate the trip. Was it pleasant? Did you get where you were going on time? In short, you can decide if the service met your needs. If not, the next time you travel, you can select a different mode of travel or choose a different company to take you to your destination.

You can use the same "select-use-evaluate" model for any technological service. This model can be applied to a television program. You might want an entertainment program appropriate for all your family to watch at 7:00 PM on Tuesday. In this case, you can search the program listings, read the summaries, and select a program. You can then watch it. After you see the program, you can decide if it was entertaining and appropriate for your family. If not, you will select a different program the next Tuesday evening.

Assessing Technology

Technological systems were not all developed in the past few years. They have been with people for several centuries. These systems date back to before computer systems and space travel. One of the most complex technological systems is a city. This system arose during the Middle Ages and gathered people in one spot. Cities provided the workers for the Industrial Revolution. This provision launched a new way of working called *division of labor.* The task of making something was divided into a large number of jobs. Each job was assigned to a single worker. This model replaced the skilled worker who made a product from start to finish.

At the same time, agricultural productivity increased. Medical science was developed. Most people saw technological advancements as a positive force. Life became better as the years passed. People were making more money. Many could work fewer hours and have more time for family activities. New and improved products were appearing almost daily. Homes were becoming larger and more comfortable. Better health care was increasing life expectancies.

All was not good, however, in the country. World War I and the Great Depression caused some people to question technology's effects on people. This continued until the 1950s. A growing number of critics began to suggest that many products of technology had harmful aspects. The era of smog arrived in this country. See **Figure 34-11.** Automobile exhausts caused this smog. Industrial wastes were polluting lakes, rivers, and groundwater. Pesticides, such as DDT, were entering the food chain.

Several plans to deal with technological impacts have been developed. Individuals evaluate devices and systems according to their personal needs. This is, however, a personal view. A different type of evaluation is called *technology assessment.* This type of evaluation involves groups of people evaluating the impacts of technology on people, society, and the environment. Technology assessment places the task of controlling technology on the government and the courts. These agencies provide a way of assessing the effects of technological innovations on human life. They conduct studies and review those studies other groups do. In particular, they evaluate the economic, ethical, social, health, environmental, and political effects of various technological products and systems.

Figure 34-11. Early factories polluted the air.

One model of ***technological assessment*** suggests what groups need to do. See Figure 34-12. First, the underlying goal for the technology needs to be identified. This should clearly describe why the technology was developed or used. The goal should list the desired outcomes for any new or improved technology. For example, the goal might be a fuel-efficient way to transport people quickly and safely from city-center hotels to the local airport.

Second, a way to measure the success in meeting the goal needs to be developed. This requires criteria to clearly describe the proposed outcomes. Using the airport-transport example, several criteria can be developed. The system should be convenient to the major hotels in the downtown area. This transport should be fast and clean. The system should have a schedule that meshes with the major flights at the airport. This transport should produce at least 50% less emissions than auto travel produces. The system should be cost-effective. This transport should run on public right-of-ways.

Third, the assessment needs to monitor and measure the impact of the technological innovation. A method to gather data needs to be developed. The system should gather information about each criterion of success. Again, using the transport example, interviews with riders can tell people about the convenience, cleanliness, and speed of the service. Monitoring equipment can measure vehicle emissions. Financial records contain data about cost-effectiveness. The system's schedules and airline-flight schedules contain data about the interface between the two modes of travel.

Fourth, people must review the data and draw conclusions. They need to evaluate the results of the measuring and monitoring action. People need to measure the results against the criteria.

Finally, the assessment group must prepare a report. This report should present the problem and the criteria used to explore it, summarize the findings, and present the positive and negative aspects of the technology being assessed. Most importantly, it should recommend action. For example, the report might recommend that the technology be abandoned, altered, or maintained.

The findings of technological assessment can be applied to political and legal actions. Local, state, and national laws and regulations can be developed. Lawsuits can be filed when the actions are found to be against existing laws and regulations. People can support or boycott the product or service.

Standards for Technological Literacy

13

Figure Discussion

Use Figure 34-12 to discuss a technological assessment process.

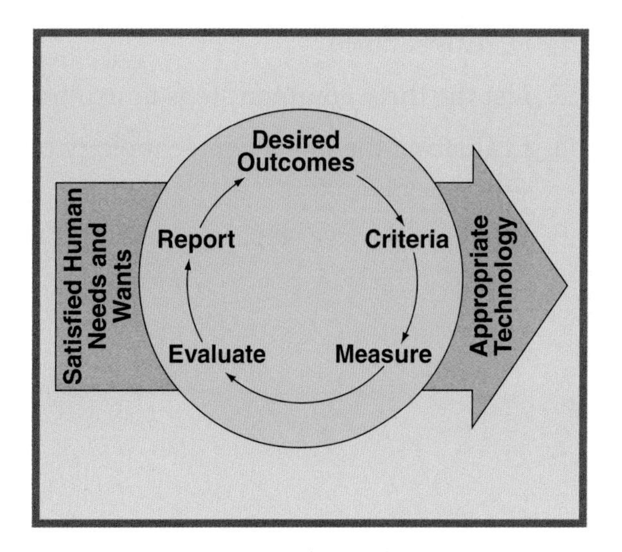

Figure 34-12. Technological assessment can be conducted in five steps.

Answers to Test Your Knowledge Questions

1. Selecting appropriate devices and systems, operating devices and systems properly, maintaining devices and systems, repairing broken parts and systems, and disposing of worn out or obsolete devices and systems
2. We cannot operate, service, repair, or dispose of a service; only select and use a service.
3. need
4. Owners' manual.
5. Any two of the following: donating, reusing, recycling, or taking them to a landfill.
6. C. Need.
7. Student answers should describe the select-use-evaluate model.
8. Assessment
9. First, the underlying goal for the technology needs to be identified. Second, a way to measure the success in meeting the goal needs to be developed. Third, the assessment needs to monitor and measure the impact of the technological innovation. Fourth, people must review the data and draw conclusions. Finally, the assessment group must prepare a report of its findings.

Summary

People select and purchase technological products and services almost daily. They use products and systems to meet their needs and wants. With the application of technology, each person is responsible for using technology correctly and wisely. Likewise, communities, states, and the nation should be concerned about the impacts of technological use on people, society, and the environment. These impacts can be measured through technological assessment.

Test Your Knowledge

Write your answers on a separate piece of paper. Please do not write in this book.

1. What are the common steps in using a technological product?
2. Why are there fewer steps needed to properly use a technological service than there are to use a technological product?
3. The first step in selecting an appropriate product is to determine the exact ______.
4. What is the primary source of information about using a specific product?
5. List two ways of disposing of unwanted products.
6. Which of the following is the best criterion for selecting a technological product?
   A. Want.
   B. Cost.
   C. Need.
   D. Appearance.
7. List the three common steps in using a technological service.
8. Evaluating the impact of technology on people, society, and the environment is called ______.
9. What are the five steps in assessing a technological product or system?

STEM Applications

Use the assessment model to deal with each of the following cases:

1. People of all walks of life are becoming more technologically literate. For example, people of all ages use telephone banking, automated teller machines (ATMs), the Internet, voice mail, e-mail, and fax machines. The Internet is becoming a vehicle for information access and shopping. Is this good or bad?
2. Solar-energy technology has failed to make the impact many people expected. This technology has been, for many years, similar to the related research-and-development area of fuel cells, a "promising technology." What are the barriers to solar technology?
3. Hybrid cars use a combination of a gasoline engine and an electric motor and generator for power. Many people believe we should shift quickly to the new vehicle. This vehicle has high fuel efficiency and low emissions. What government and industry action is needed to make this happen?

Activity 10A

Forming a Company

Background

Companies produce many technological devices and products. These enterprises are the products of human actions. They are formed and structured to efficiently use resources to produce artifacts meeting our wants and needs.

Situation

Students have mentioned they need a way to be informed about important sporting and social events at school. They also want a way to publicize happenings they feel are important. From these comments, you have concluded that an inexpensive, personalized calendar will meet their needs and earn you a profit. This type of calendar can be produced with limited finances by using new computer software. See Figure 10A-1.

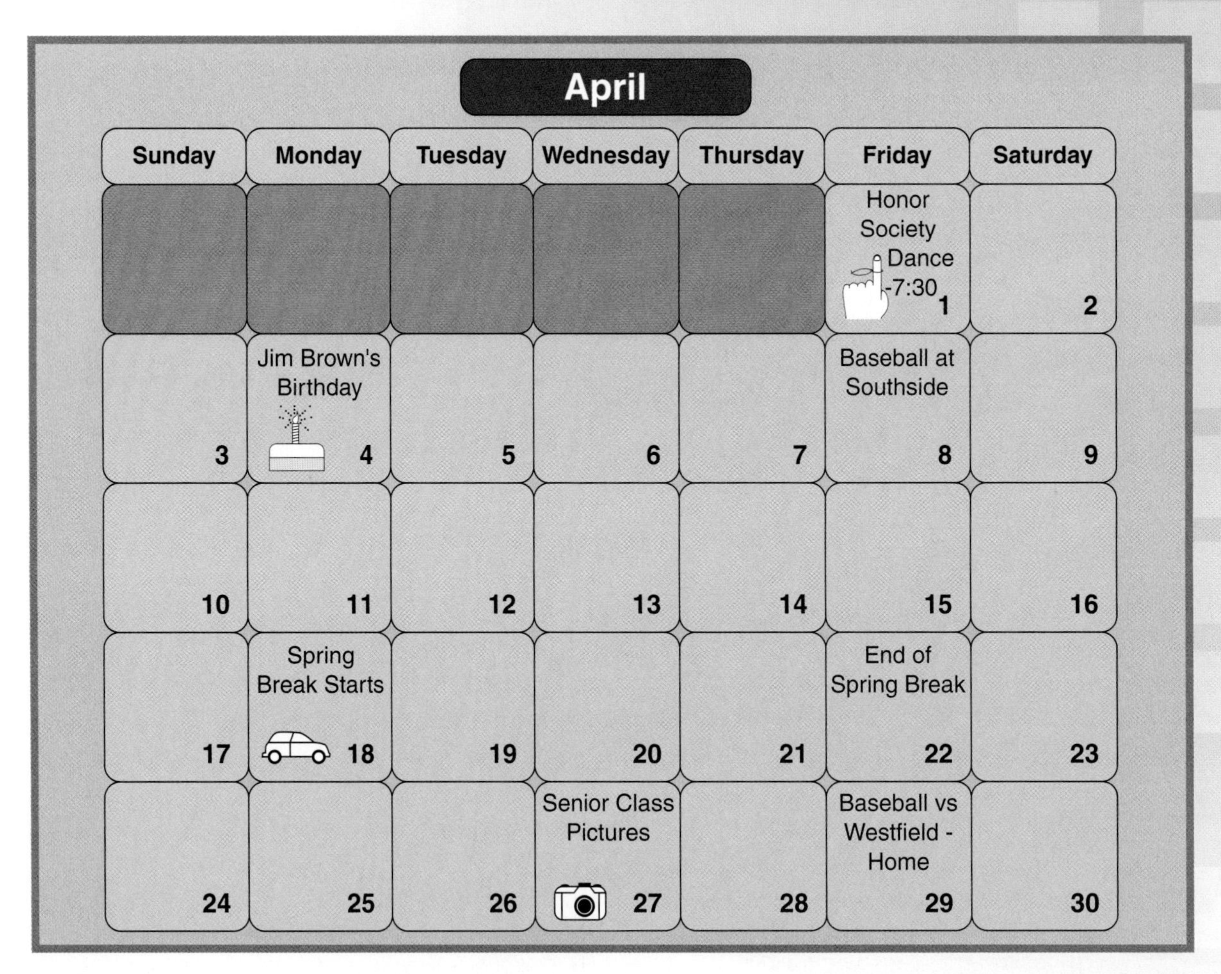

Figure 10A-1. A sample of a calendar.

Challenge

Organize a company to produce a nine-month calendar. The calendar should span the school year and have selected days personalized. Consider the tasks to be completed and the managerial structure needed to complete them. Be sure to recognize that there are production and marketing tasks. Also, there are two distinct phases of the company's operations. These phases might require two different organizations. One operation can finance the company, design the calendar, and sell calendar entries. The other operation can maintain financial records, produce the calendars, and sell the finished products. See **Figure 10A-2.**

Optional

1. Write a one-page job description for each job on the organization chart.
2. Develop a set of goals for each major department in the company.

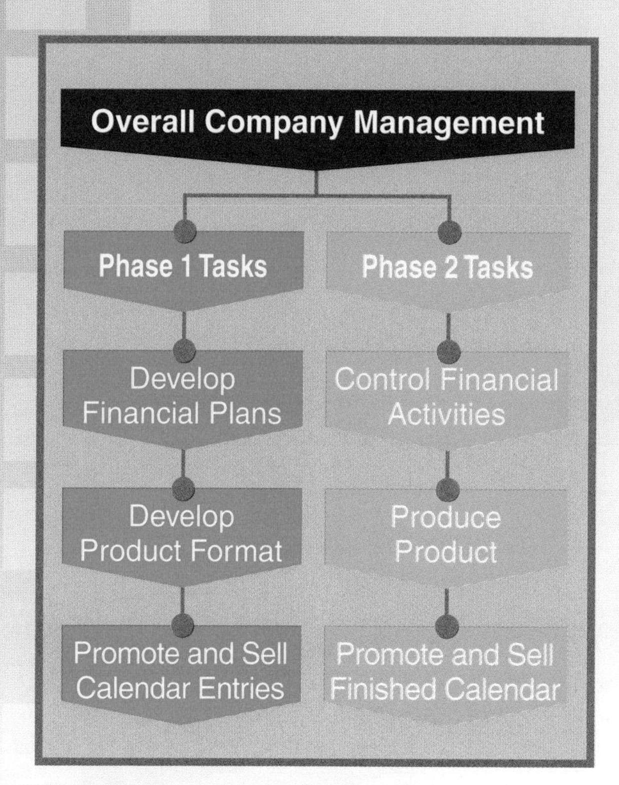

Figure 10A-2. Tasks for finance operations.

Activity 10B

Operating a Company

Background

Companies are series of independent tasks that have been integrated into functioning enterprises. Each task must be planned for and carried out with efficiency.

Challenge

Identify, schedule, and complete the several tasks required to produce and market a personalized calendar for your school.

Procedure

Design and development department

1. Obtain software that can be used to produce a personalized calendar.
2. Follow the instructions to produce a calendar for a single month. This will acquaint the department members with the operation of the software.
3. Establish the layout for the calendar.
4. Produce a common layout sheet for the marketing group to use in selling calendar entries.

Calendar-entry marketing department

1. Determine the selling price for a calendar entry.
2. Develop a calendar-entry order form.
3. Make posters to promote the sale of calendar entries.
4. Sell calendar entries.

Production department

1. Receive calendar-entry forms from the marketing department.
2. Enter data on the calendar layouts.
3. Print a proof of the calendar.
4. Submit the proof to marketing for approval.

5. Correct the calendar entries.
6. Print the master calendar.
7. Reproduce the calendar.

Calendar marketing department

1. Produce and distribute advertisements for the sale of calendars.
2. Select and train calendar salespeople.
3. Sell calendars.
4. Maintain sales records.

Finance department

1. Set budgets for company operations.
2. Sell stock. Maintain stockholder records.
3. Purchase materials and supplies.
4. Maintain all financial records.

Executive committee (president and vice presidents)

1. Set deadlines for important activities.
2. Monitor progress in completing tasks.
3. Set budgets.
4. Establish selling prices for calendar entries and finished calendars.
5. After the calendar copies have been sold, close the company. Liquidate the assets.

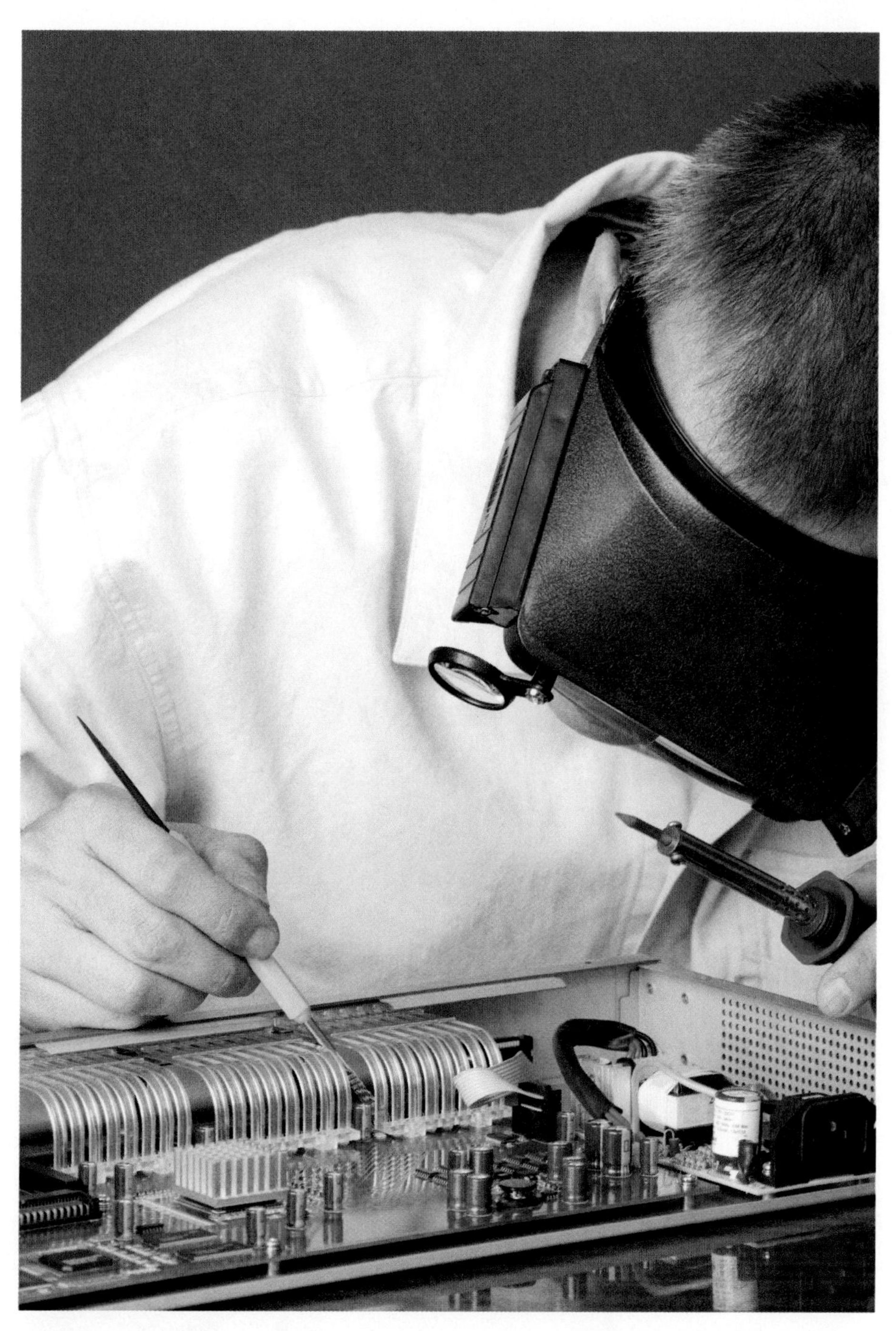

Part of using technology is repairing that technology. While owners can sometimes repair their own products or devices, trained technicians are sometimes needed to do the work.

Section 11
Technological Systems in Modern Society

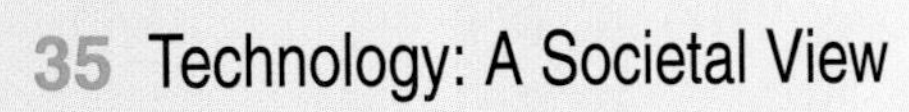

Tomorrow's Technology Today

Artificial Ecological Systems

You might have gone to your neighborhood greenhouse and seen hundreds of species of plants. However, the plants you saw were likely only the species that can grow and thrive in your local environment. In order to see an array of plants not native to your surroundings, you would have to find a type of artificial ecological system or closed ecological system; that is, a type of greenhouse that does not rely on anything from the outside to help its plants grow. Closed ecological systems are controlled environments.

Scientists have been attempting to study plants in closed ecological systems for decades. The BIOS-3 facilities were used to conduct various experiments until 1984. The Biosphere 2 structure has had various owners since 1991 and has been conducting research since it was constructed. It is the largest closed ecological system in the world. It has been used to research agriculture, recycling, and human health. Another artificial ecological system is the Eden Project, which is located in Cornwall, England and open to the public. The Eden Project is not considered a closed ecological system. It consists of two separate biomes and an open-air garden, which house over 1 million plants from around the world.

The multidomed greenhouses of the Eden Project are the Rainforest Biome and the Mediterranean Biome. The Rainforest Biome consists of several domes representing the environment of a tropical rain forest. It contains hundreds of trees and other plants from rain forests in Africa, Asia, South America, and Australia. Plants include rubber, cocoa, vanilla orchids, and bamboo. The second biome is another multidomed greenhouse. The Mediterranean Biome has plants from temperate rain forests in California, Southern Africa, and the Mediterranean. Here, you can find grapevines, olives, orange groves, and hundreds of colorful flowers. The Eden Project's main goal is to educate people about the natural world, especially about using natural resources efficiently so they will continue to be available in the future.

While the Eden Project is the only one of its kind here on Earth, other projects have begun in order to study plant life and to find ways of using resources more efficiently. MELiSSA, or Micro-Ecological Life Support System Alternative, uses similar research methods to other closed ecological systems. However, its purpose is to use the results for long-term space missions rather than to apply the results on Earth. The European Space Agency has been heading this project, which has been in work since 2000. The project is divided into various phases, which start with research on Earth and go through applying that research in structures in space.

These artificial ecological systems have been used for research for decades. The Eden Project and MELiSSA have taken the research some steps further. With these projects making a difference, you can expect to see similar projects around the world in the coming years.

Discussion Starters

The Eden Project is an experiment in the reclamation of industrial wasteland. The project's biomes are built in an abandoned clay pit. Instead of importing nutrient-rich soils from various parts of the world, the Eden Project chose to manipulate the existing soil to meet the various needs of the diverse plant life.

Group Activity

Divide the class into groups of four or five students. Tell the students their job is to design an indoor environment like the Eden Project's biomes. Have the students select a geographical region to reproduce in the enclosure. Ask the students to imagine what sort of technical challenges they would face, and brainstorm possible solutions to those problems. Remind the students that sometimes solutions to one problem can create new problems.

Chapter 35 Technology: A Societal View

This chapter covers the benchmark topics for the following Standards for Technological Literacy:

4 5 6 7

Learning Objectives

After studying this chapter, you will be able to do the following:

- Recall some destructive natural forces that affect human lives.
- Give some examples of how technology is used to help to control natural forces.
- Summarize the global impact of automobile use in North America.
- Differentiate between divergent thinking and convergent thinking as futuring tools.
- Summarize the four major types of futures considered in futures research.
- Recall the three forces causing a condition called an *environmental crisis.*
- Summarize the movement of the center of technical knowledge and power from east to west through human history.
- Recall several specific actions companies can take to remain competitive in a world economy.

Key Terms

aquifer
biological future
computer-integrated manufacturing (CIM)
finite
futures research
futuring
global impact
greenhouse effect
human-psyche future
inexhaustible resource
nonrenewable resource
pollution
renewable resource
social future
technological future

Strategic Reading

What do you think the term futuring means? As you read this chapter, determine the various ways futurists view technology's changing impacts on society.

You have been learning about technology in the first 34 chapters of this book and have been introduced to many concepts and a great deal of information. The key to all this learning is one principle. This principle is that people develop technology for people to use. See **Figure 35-1.** You have learned that technology is the product of human knowledge and ability. Technology is designed to help people modify and control the natural world. Also, technology is the sum total of all human-built systems and products. Technology is the human-built world. See **Figure 35-2.**

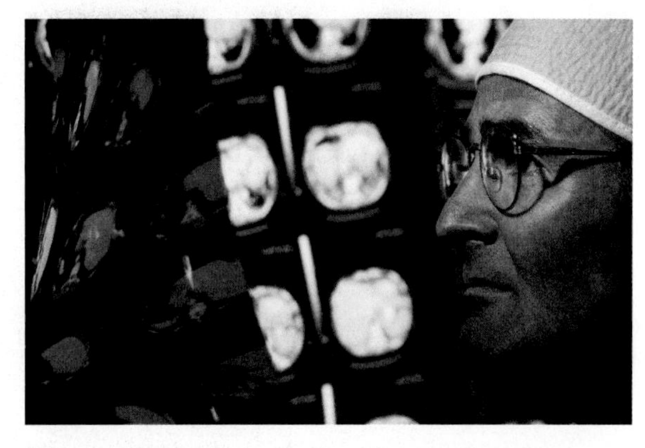

Figure 35-1. People develop technology, such as this medical imaging system, for people to use.

Standards for Technological Literacy

4

Reinforce

Review the societal contexts of technology (Chapter 3).

Extend

Discuss the concept "technology is developed by people for people to use."

Brainstorm

Ask your students how technology has changed their lives.

Research

Have your students select a task, such as moving cargo, harvesting crops, or healing illness, and research how technology has made it easier.

TechnoFact

Once technology is developed by people, they determine how the technology is used and how it will evolve.

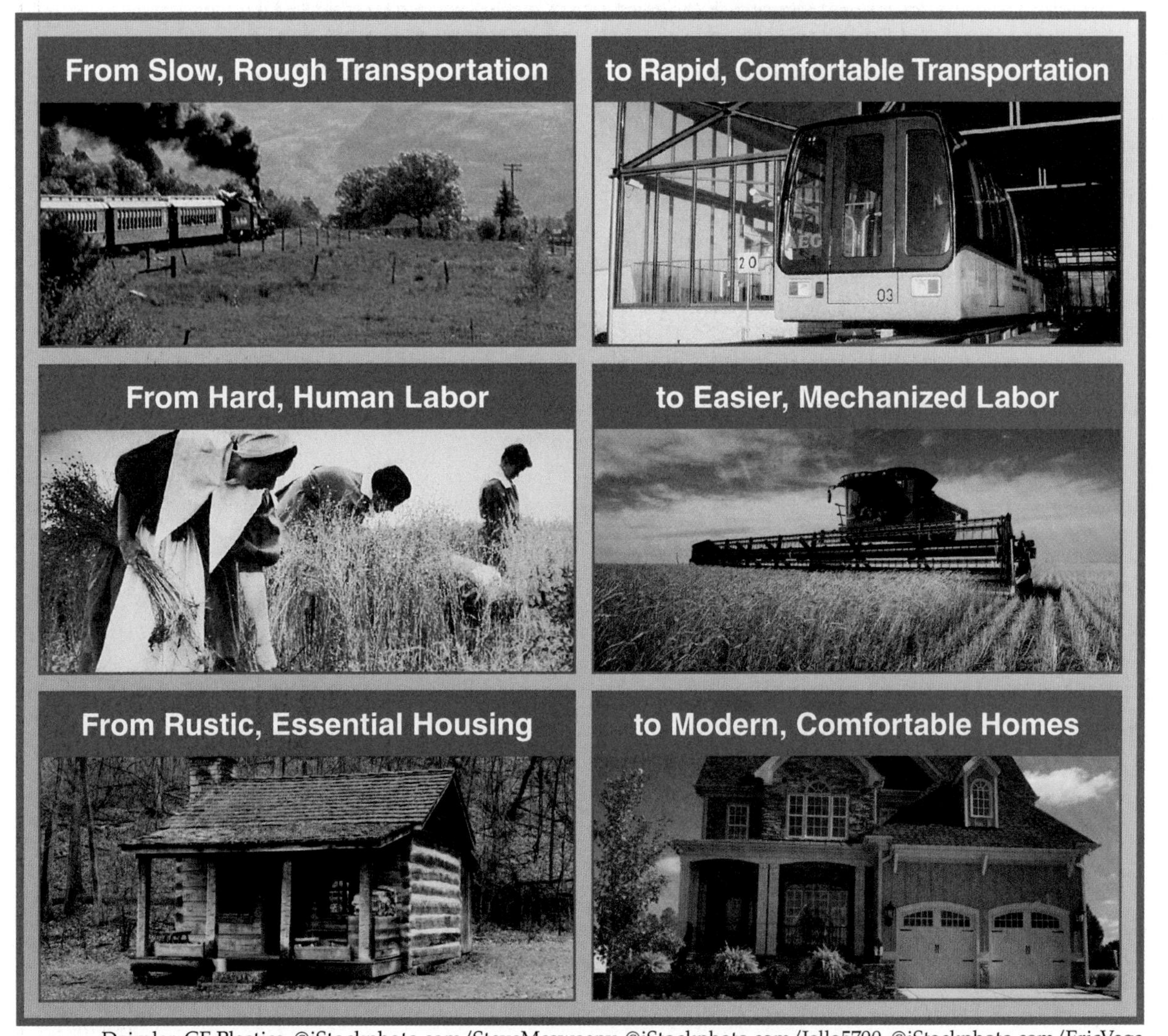

Daimler, GE Plastics, ©iStockphoto.com/SteveMcsweeny, ©iStockphoto.com/Jello5700, ©iStockphoto.com/EricVega

Figure 35-2. The application of technology has greatly changed our world and our way of life.

Standards for Technological Literacy

4 5

Extend

Discuss how a technological development implemented in one area of the world impacts all other areas.

Example

People from around the world know many English words and have contact with American cultural products, such as recordings, films, fast food, and brand-name clothing.

TechnoFact

Water is a natural resource whose usage has changed over time due to developing technology. Flooding has been reduced by technology because modern buildings have been built with the intention of withstanding these natural disasters. At the same time, dams can be used to store water.

Technology is neutral. By itself, technology does not affect people or the environment. How people use technology determines if it is good or bad, helpful or harmful. How we use technology can help or harm the world around us.

Technology and Natural Forces

Natural forces affect, and sometimes disrupt, human life. Hurricanes wreck ships and destroy coastal settlements. Tornadoes can level entire sections of towns and cities. Floods wash away homes and carry away vital topsoil, reducing the productivity of croplands. Fires burn buildings, crops, and forests. Earthquakes shake structures until the structures collapse. In all these natural events, people can be killed. Their possessions can be destroyed.

One of the earliest uses of technology was to harness natural forces. People started to design a human-built world that could reduce natural, destructive forces. Dams hold back floodwater and produce energy in hydroelectric plants. Fire is tamed to heat our homes and process industrial materials. See **Figure 35-3.** Geothermal power plants capture Earth's inner heat. Wind is used to generate electricity. Solar energy provides heat and electricity for businesses and homes.

Controlling and using natural forces, however, are not enough. Humans are starting to understand that having control over Earth's future carries with it serious responsibilities. Many people realize that humans must protect the environment and the plants and animals that live with us on this planet. We must live in harmony with the natural world.

Technology's Global Impacts

Responsible use of technology requires us to realize that we live in a world very different from that of our grandparents. The technology we develop and use today has impacts beyond our homes, cities, states, and countries. We must look at the impacts of our decisions on the entire world.

Figure 35-3. People use technology to harness and control the forces of nature. Fire can destroy a forest or can be harnessed for useful tasks, such as baking pastries.

For example, the attachment our society has to the automobile as its primary transportation vehicle has ***global impacts***. The citizens of North America represent a small percentage of the world's population. They use a large percentage, however, of the world's petroleum. Also, automobiles discharge great quantities of pollutants into the atmosphere.

Likewise, many North Americans are concerned about high birthrates in Third World countries. The impact of each new North American child on the world's resources, however, is many times greater than the impact of a child from the Third World. We must be concerned about population and resource use.

Technology and the Future

Since technology is a product of human activity, humans can control it. To do this, we must have an idea of the kind of future we want. Using a research technique called *futuring*, or ***futures research***, can develop this idea. See **Figure 35-4.** This process helps people select the best of many possible courses of action. ***Futuring*** emphasizes five distinct features:

- **Alternate avenues.** The futurist looks for many possible answers, rather than one answer.
- **Different futures.** Traditional planning sees the future as a refinement of the present. A futurist looks for an entirely new future.
- **Rational decision making.** Traditional planners rely heavily on statistical projections. They use mathematical formulas to help them predict events and impacts. A futurist uses logical thinking and considers the consequences when making decisions.
- **Designing the future.** Futurists are not concerned with improving present or past practices. They do not see the future as a variation of the present. Futurists focus on predicting a possible future that can be created.
- **Interrelationships.** Traditional planners use linear models that suggest one step leads to the next step. Futurists see alternatives, cross-impacts, and leaps forward.

Figure 35-4. People can use futuring to determine the type of future they want.

The Views of Futurists

Using a futures approach, exciting new technologies can be developed. Futurists must have a dual view. One view must be of a present challenge or problem. The other view must be of the world of the future. Their grandchildren will live in this world. Looking at both of these views requires a combination of short-term and long-term goals. The most important tool to develop these futures is the human mind. The human mind is the only tool capable of reasoning and making value judgments.

Standards for Technological Literacy

4 5 6 7

Demonstrate

Show your students how to use one or more futuring techniques.

TechnoFact

All technology systems yield desirable and undesirable outputs. By planning ahead, we are more able to reduce the undesirable outputs. One tool commonly used in preventing harmful effects is technology assessment. The negative outputs are identified in this process, and the government is in a position to develop regulations for reducing these effects or preventing them completely.

Standards for Technological Literacy

4 7

TechnoFact

Most technology developed today is intended for worldwide distribution and use. However, the less developed (Third World) nations have not benefited as significantly as some developed nations of Europe, Asia, and the Americas.

TechnoFact

The idea of technology is neither good nor bad. The people using the technology and their intentions determine how technology is considered. Many aspects, such as intelligence, imagination, and skill, shape the way technology is developed and used.

Career Corner

Public Relations

Public relations involves managing the public image of a company or an individual. The public relations personnel deal with all aspects of communicating an organization's message to the general public and specialized segments of the population. The field of public relations is growing, as it broadens into the area of international communications.

People entering this field must have effective writing skills and a background in media production. Many public relations personnel are university graduates with majors in journalism, mass communication, or business. Careers in public relations include publicity managers, press agents, lobbyists, corporate public-affairs specialists, and public-opinion researchers

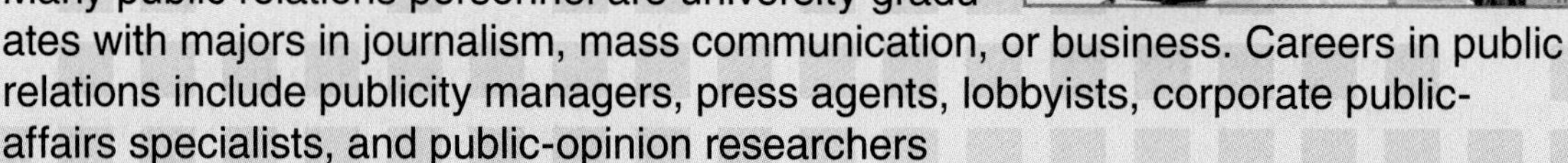

Types of Thinking in Futuring

Futuring requires two types of thinking. The first is divergent thinking. This type of thinking lets the mind soar. People are encouraged to explore all possible and, in many cases, impossible solutions. They look for interrelationships and connections. Out of this activity emerges a number of futures.

One possible future, however, must be selected as the one that will be created. This requires convergent thinking. The final solution must receive focus and attention. This solution's positive and negative impacts must be carefully analyzed.

Types of Futures

The analysis must be done with four types of futures in mind. The first is the ***social future***. This future suggests the types of relationships people want with each other. The second is the ***technological future***. This looks at the type of human-built world we desire. The third is the ***biological future***. This future deals with the types of plant and animal life we want. See **Figure 35-5**. The fourth is the ***human-psyche future***. This future deals with the mental condition of people. The human-psyche future stresses the spirit, rather than the mind—attitude,

FMC Corp.

Figure 35-5. The biological future might hold new developments in the use of plant and animal products, especially those from the sea. New technologies impact the environment, people, and society as a whole.

instead of physical condition. This future is concerned about how people will feel about life and themselves.

These futures are listed separately. They are actually, however, interconnected. For example, new technologies directly impact how people relate to one another. Television dramatically changed family life, recreation, and the entertainment industry. Technology also changes the natural environment. Acid rain that automobile emissions and coal-burning electric plants caused has destroyed forests in Canada and the eastern United States. Likewise, technology has changed how we view ourselves. Some people feel threatened by technology. Other people feel empowered by it.

Technology's Challenges and Promises

It is impossible to explore how each new technology has impacted our lives and how it will impact the future. We can explore some examples, however, to provide a foundation for personal study. Later, you can use these examples as you evaluate other technologies. Three widely discussed issues are energy use, environmental protection, and global economic competition. Let's look at these issues in terms of their challenges and promises.

Energy Use

The world as we know it would come to a grinding halt without energy. Almost everything in the human-built environment depends on energy. Therefore, the supply and use of energy resources are very important.

Nonrenewable Energy Resources

Every person uses energy resources. Many of these are exhaustible resources, or ***nonrenewable resources***. When we burn them all, there will be no more. The supply is said to be ***finite***. There is a limited quantity of the resource available. One of these resources is petroleum. This resource is still the fuel powering most transportation vehicles. Other exhaustible energy resources are coal and natural gas. The shrinking supply of these resources is a major concern, particularly in the case of petroleum. We are challenged to reduce our dependence on these resources.

Renewable Energy Resources

One alternative is to shift to ***renewable resources***. These resources have life cycles. They are the products of farming, forestry, and fishing. These resources are in limited supply, however, at any one time. For example, in many less developed countries, wood is the primary fuel for cooking. As the population grows, people must roam a greater distance to find the firewood they need. Thus, large regions are being stripped bare of trees. Also, using wood for fuel eliminates it as a source of building material.

Likewise, corn can be used to make ethanol. Ethanol can be used as a fuel. If we shift large quantities of corn from food production to fuel production, however, world hunger might be worsened.

Inexhaustible Energy Resources

Shifting from exhaustible resources to renewable resources might be a partial solution. Another solution is to make greater use of ***inexhaustible resources***. See **Figure 35-6**. The most common of these are solar energy, wind energy, and water energy. We can generate electricity using any or all of these three energy sources. This requires, however, a large expenditure of money and human energy. Generating electricity using these energy sources also

Standards for Technological Literacy

4 5 7

Research

Have your students investigate some new energy-saving devices and determine how much energy they can save.

TechnoFact

The use of electric means of transportation has been greatly reduced over the last 100 years. Even at the start of the 20th century, electric trains and buses were a standard of public transportation in most cities. Though most automobiles today run on different fuels, electric vehicles are still offered by some manufacturers.

Standards for Technological Literacy

4 5

Figure Discussion

Use Figure 35-6 to discuss exhaustible, renewable, and inexhaustible resources.

Research

Have your students visit the U.S. Department of Energy Web site to find and describe a new energy-saving device or vehicle.

Figure 35-6. The future requires a shift from exhaustible energy sources, such as petroleum, to renewable resources, such as wood, and inexhaustible sources, such as solar power.

covers large tracts of land with solar and wind generators. Additionally, it takes time to develop the technology to fully use these resources. For example, solar-powered automobiles are now an interesting experiment. See **Figure 35-7.** Practical vehicles of this type are years away.

Using Energy Efficiently

An immediate solution to our energy problems is to use energy more efficiently. This requires people to think about and change their lifestyles. Members of society have to ask themselves a number of very difficult questions. Should people drive to work alone in a personal car? Should people heat their homes to 75°F (24°C) in the winter

AC-Rochester

Figure 35-7. Solar-powered cars such as this one might be a practical transportation vehicle in the future.

and cool them to 65°F (18°C) in the summer? Should we make buildings more costly by using more insulation and installing double- and triple-glazed windows? Should people be strongly encouraged, through taxes or fees, to use public transportation instead of their cars? Should gas-guzzling cars be removed from the market?

Standards for Technological Literacy

4 5 7

TechnoFact

The main goal of the U.S. Department of Energy is to improve the safety, comfort, and energy usage of indoor environments. They conduct studies while developing technology to better meet the demands of energy efficiency for society.

Technology Explained

plasma display: a relatively lightweight, flat-panel display about 3″ to 6″ thick that is an alternative to cathode-ray tube (CRT)–based television sets.

CRT-based television sets were bulky, had few applications, and did not always present a clear picture. Gas-plasma displays are lightweight flat panels that are between 3″ and 6″ thick and use phosphorescent material contained between two surfaces. See **Figure A.** Light can be created in gas-plasma displays when an electric field passes through the gas into the layer of phosphor.

The phosphorescent material used in gas-plasma displays is actually millions of cells of gas. The phosphorescent material is then coated with red, green, and blue. See **Figure B.** When the transparent electrodes and address electrodes become electrically excited, they cause the cells to release a UV light. This light strikes the phosphor coating in a subpixel, causing the subpixel to release a light of corresponding color. By varying the pulses of current in a pixel area, the entire light spectrum of color can be duplicated.

While plasma displays have many advantages over traditional CRTs, there are some disadvantages. The panels are often heavy and are not economically manufactured in larger sizes. However, these displays can be found in other applications as well. They may be used to communicate information in billboards, or they may be used for computer monitors.

Figure A. Gas-plasma displays are made of a phosphorescent material contained between two thin surfaces.

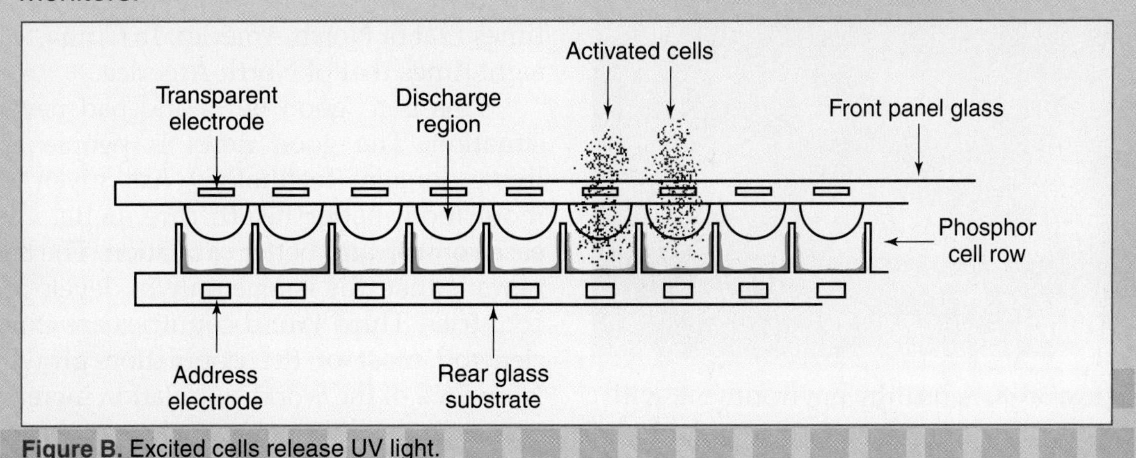

Figure B. Excited cells release UV light.

Standards for Technological Literacy

Research

Have your students research and plot the population growth of the major regions of the world, including the United States.

Brainstorm

Ask your students how technology has affected world population growth.

TechnoFact

The Environmental Protection Agency was established in 1970. The EPA works to control clean air, water, and land. Industrial factories are carefully monitored by the EPA to satisfy environmental regulations.

Environmental Protection

Open space, clean air, land for a home or farm, and safe drinking water were once viewed as birthrights for people in the United States and Canada. We simply expected them to be available. Today, we know better. We have discovered that unwise use of technology can threaten our quality of life.

This understanding was slow to come. These issues have been brought into sharper focus in recent years. People have participated in what can be called an *environmental revolution*. This revolution has resulted in three basic principles:

- The natural environment has a direct effect on the safety and health of people. See **Figure 35-8.**
- The long-term survival of any civilization is based on wisely managing natural resources.
- A healthy natural environment is essential for human life.

Figure 35-8. A healthy environment, with unpolluted air and water, is essential for a good quality of life.

Protecting the environment involves studying the relationship between the human population and the use of technology. This relationship directly impacts a number of environmental conditions. These include the climate and the supplies of food, water, energy, and material resources.

Many scientists say we are creating an environmental crisis. This means we must take action or the environment will be permanently damaged. Three important forces contributing to this crisis are overpopulation, resource depletion, and pollution.

Overpopulation

From the dawn of civilization to the end of the 1700s, the world population grew to about 1 billion people. During the 1800s, the population increased at a higher rate. By 1900, it reached a total of about 1.7 billion people. The world population grew even more rapidly during the 1900s. The U.S. Census Bureau reported that the world's population reached 3 billion by about 1960, 4 billion around 1974, 5 billion by about 1987, and 6 billion by 1999. This bureau predicts that the population will reach 7 billion in 2013, 8 billion in 2026, and 9 billion by 2043.

The world's population is not evenly distributed. The number of people per square mile of cultivated land varies greatly. In Europe, this density is three times that of North America. In south Asia, it is four times that of North America. In China, it is eight times that of North America.

This is a "good news and bad news" situation. The good news is people are living longer. Technology has given us more food, better health care, better disease control, and better sanitation. The bad news is that this is true only in developed countries. Third World countries are experiencing most of the population growth. Now, 96% of the world-population increase occurs in the developing regions of Africa, Asia, and Latin America. The economies of these countries cannot support rapidly growing populations. This leads to tremendous hardships. Many people in Third World countries go to sleep hungry each night. Millions of people starve to death each year. Still more die of disease.

As we continue to control diseases; use diplomacy, instead of war, to solve international conflicts; and improve health care, the population grows even faster. The increased population contributes to the other two problems. These problems are pollution and resource depletion. A concerted effort is needed to limit growth of the world's population.

Resource Depletion

Each living person places demands on the resources of the planet. Many material resources have finite supplies. Once used, they are gone forever. These resources include metal ores, petroleum, natural gas, sulfur, and gypsum.

As with energy resources, we can shift from using nonrenewable resources to employing renewable resources. These renewable resources are primarily the food and fiber that farming, forestry, and fishing produce. See **Figure 35-9.** The fertility and availability of land and the productivity of the oceans, however, limit the supply of these resources. If we take too much from the water and land, the water and land will be damaged.

Figure 35-9. Farming is one source of renewable material resources and food.

An example of "taking too much" is the intensive forestry practices used in many parts of the world. The logs, limbs, and bark of the tree are taken away to be processed into products. Nothing is left behind. This might seem wise. Leaving nothing behind, however, is not wise. The forest floor is left without the limbs and the bark. These parts of the tree normally break down and provide nutrients for other plants and trees. As a result of the bare forest floor, the soil becomes less fertile. Coupled with the acid rain falling on many forests, this produces trees that are not as healthy as before. The production of wood fiber is reduced.

Likewise, intensive farming practices deplete the soil. This makes heavy applications of commercial fertilizers necessary. Also, using prime farmland for housing forces us to cultivate less productive areas. We also strain precious water resources to irrigate crops growing in the desert and draw the water from rivers and wells. This, in turn, reduces the amount of water downstream and depletes aquifers. (***Aquifers*** are underground water-bearing layers of rock, sand, or gravel.)

A problem related to these issues is land use. Throughout the country, conflicts are raging between environmentalists (people who are concerned about leaving the environment unspoiled) and commercial interests. See **Figure 35-10.** An example of one controversy is the spotted owl habitat in the forests of Oregon and Washington. Commercial interests want to cut virgin timber and manage the forest for maximum fiber production. Environmentalists want most of these forests left untouched as a habitat for the owls.

Debate continues over how much wilderness is enough. Some people want large tracts of rangeland and forestland to remain

Standards for Technological Literacy

4 5

Extend

Discuss some of the political and social issues related to resource use and depletion.

TechnoFact

Natural resources are depended on by modern technology. There are some materials believed to be renewable resources, like iron and sand, but there are also exhaustible resources, such as hydrocarbon fuels. These fuel resources are likely to be depleted in the next 200 years.

Standards for Technological Literacy

4 5

Extend

Discuss how technological advancements can cause or reduce pollution.

Brainstorm

Have your students list the causes of pollution in your area. Ask the students to think of ways each type of pollution be reduced and controlled.

TechnoFact

Environmental pollution is sometimes attributed to the growing population. Increased human activity leads to increased waste. Early people were not as concerned with effects waste might have on the environment. Today, pollution has become a primary concern, and such developments as recycling are being used to reduce pollution's effects.

Figure 35-10. There is a continuing debate over land use in the United States. On one side are those who want land kept in a wilderness state. On the other side are people who want more land for lumber or agricultural production.

untouched. They want only hiking trails in these areas. Few people dispute the need to save some unspoiled areas. The debate is over the sizes and number of these areas. The difficulty with these debates is that there are no clear-cut answers. There are only opinions of what is right or wrong, good or bad.

As with energy, short-term solutions to resource depletion lie in better use of our material, land, and water resources. We must use our land, but we must not abuse it. Some of it should be set aside as nature preserves. Other areas should be considered multiuse land. These areas should combine recreational and commercial uses. Hunters, campers, ranchers, farmers, and loggers should all use the land. Finally, some tracts of land should be devoted to commercial, residential, and transportation uses.

We must reconsider how we use Earth's resources and learn to use all materials more efficiently. This means buying fewer items we really do not need. Buying products that will last becomes very important. Maintaining and repairing them, instead of discarding them, also become very important. Finally, when a product can no longer be used, the materials must be recycled.

Pollution

People living in some major cities rarely see clear skies. Haze reduces the visibility. Smog affects people's health. Water in many parts of the world is unsafe to drink. In some areas, the land is contaminated with hazardous waste. People cannot live on this land or travel over it. All these things are called ***pollution***. See **Figure 35-11.** Pollution is most often a product of human activity.

This product of human activity sometimes has been brought to our attention in dramatic ways. In 1969, the Cuyahoga River in northern Ohio literally caught fire. This river runs through the city of Cleveland. The Cuyahoga River had become severely polluted with debris and flammable liquids. The incident made many people more aware of the pollution issue. This fire led to the passage of new laws to protect the waterways and efforts to clean up streams, rivers, and lakes.

Figure 35-11. Pollution can affect the land, water, and air we depend on for life.

Pollution affects more than the air we breathe, the land we walk on, and the water we drink. Many scientists think it is changing the climate of our planet. In the spring of 1983, scientists observed a brown layer of air pollution over the Arctic region. Later, other scientists discovered a hole in the ozone layer over the Antarctic. This hole has grown larger over the years. Since that time, a controversy has raged. Some scientists say the hole lets more UV light into the atmosphere than before. They believe these rays, combined with increased levels of carbon dioxide and other gases, cause Earth to retain more heat. This causes global warming, with higher land and water temperatures. Scientists are concerned that a warmer climate will melt the polar ice caps and cause the oceans to rise. In turn, large coastal areas, including some major cities, might be flooded. Scientists call this problem the ***greenhouse effect***.

Other scientists feel that global warming is part of a natural cycle. They say warming and cooling have occurred a number of times over history. No one is sure what will happen. This warming, however, is an area of great concern. Changes have been called for in our technological actions. Certain chemicals in air conditioners, refrigerators, foam containers, and aerosol sprays are blamed for part of the problem. These chemicals are being removed from the market as replacements are developed.

A plan is needed to protect the atmosphere and the ozone layer. Failure to act might cause a significant change in the climate of our world. The greenhouse effect might turn lush farmland into desert, cause the extinction of some species of plants and animals, and create widespread human suffering.

Global Economic Competition

A third issue directly related to people and technology deals with the distribution of wealth and industrial power. Some historians suggest that technical knowledge and power have moved steadily around the world from east to west. At one time, China was a global power and the source of many innovations. This center of power moved across what is now India into the area around the Mediterranean. That area was the dominant economic center about 2000 years ago. Later, northern Europe became the economic leader, with the Industrial Revolution of the 1800s. The dominance of the United States followed this in the 1900s. Now, we are seeing the area called

Standards for Technological Literacy

4 5 6

Figure Discussion

Use Figure 35-11 to discuss types of land, water, and air pollution and their causes.

TechnoFact

While some experts believe the cause of pollution is the increase of population, others believe it is the further development of technology. There is a strong possibility the combination of these two factors is the true cause. Because technological advancements grew rapidly, the population grew as well, and the people didn't take the effects of these advancements into consideration. Poor sanitation and contaminated water led to epidemics and much pollution.

Standards for Technological Literacy

4 6

Research

Have your students research Japanese-style, or participatory, management.

TechnoFact

Technological developments have had a great impact on the way people work. A certain knowledge of technology is required in several jobs today, and many new careers have risen with these developments.

the *Pacific Rim* become more important. Japan, Korea, China, and other Far East countries are challenging the industries of North America and Europe. This represents nearly a full cycle of industrial development around the globe.

Today, we understand economics better than at any time in history. This allows countries on the back side of the economic wave to resist losing economic power. They can take a number of actions. These actions include the following:

- **Changing management styles.** Companies are redefining the roles of workers and managers. They are creating teams that design and produce products. See **Figure 35-12.** These teams include designers, engineers, production workers, quality control specialists, marketing people, and managers. Managers are not seen as bosses. They help others do their work better and with greater ease.
- **Increasing their use of computers.** Competitive companies are assigning routine work to computers. This means there are less manual labor and more computer-aided or computer-controlled work. See **Figure 35-13.** We now hear about an "alphabet" of computer actions:
  - CAD systems make drawings easy to produce, correct, and store.
  - ***Computer-integrated manufacturing (CIM)*** ties many manufacturing actions to the computer. Computer systems monitor machine control, quality control, parts movement, and an array of other operations.
  - JIT inventory control. This computer system monitors material orders so supplies arrive at the plant just before they are needed. Also, finished products are made only when they are needed.

Figure 35-12. People often work together in teams to design and produce better products.

Figure 35-13. In many business and industrial operations, computer applications have replaced manual methods.

- Many computer-controlled purchasing and warehousing operations. These systems include warehousing systems in which robot vehicles store and retrieve parts.
- **Producing world-class products.** Successful companies now make products meeting the needs of customers around the world. This means the products must function well, be fairly priced, and deliver excellent value to the customer. No longer can a local or national area provide a safe, protected market for a company. Political forces are causing the world to become one large market for all countries.
- **Using flexible, automated manufacturing systems.** Traditional manufacturing is based on long production runs. Semiskilled workers on a manufacturing line make a single product. Automated production lines called *flexible manufacturing systems* are replacing this type of production. See **Figure 35-14.** Such computer-controlled systems can produce a number of different products with simple tooling changes. Flexible manufacturing is cost-effective for small quantities of products.

Arvin Industries

Figure 35-14. Today, products are often produced on flexible, automated production lines. Computer control allows fast, simple changeover to produce different types of products.

Standards for Technological Literacy

4 6

Extend

Discuss what is meant by the phrase a world-class product.

TechnoFact

The first automated controls were built by James Watt as feedback controls to better regulate the speed of his steam engine. By the 20th century, automation in industry was a standard in such factories as automobile assembly plants.

TechnoFact

Sharing newer technologies with less developed countries is sometimes challenging. Without the full understanding of its possible effects on the country's social, political, and economic systems, the implementation of technology can sometimes cause problems.

Think Green

Shopping Bags

Throughout this text, we have discussed various ways of changing your environmental impact. Some of the changes to be made are group-oriented changes, while others can be done on an individual basis. The way you choose to transport products is a small but meaningful change you can make.

When you make your purchases at a store, you sometimes have the option of choosing to put your merchandise in paper bags or plastic bags. Paper bags are not always reusable, but you can recycle them when you have finished using them. If you throw them out, they can decompose naturally. If you choose plastic bags, you can typically save and reuse them. If they cannot be reused, however, they should be recycled. Recycling plastic bags uses less fossil fuels than producing new plastic bags. Plastic bags should not be thrown out. Another option you have is to invest in reusable bags and to carry at least one with you whenever you shop. Because there is no recycling or waste associated with reusable bags, using these bags has a lesser effect on the environment.

Answers to Test Your Knowledge Questions

1. True
2. Evaluate individually.
3. natural
4. The automobile.
5. Divergent thinking explores all possible and impossible solutions. Convergent thinking focuses attention on a single solution.
6. B. Economic.
7. technology
8. Overpopulation, resource depletion, and pollution
9. Industrial
10. CAD (computer-aided design), CIM (computer-integrated manufacturing), and just-in-time (JIT) inventory control.

Summary

Technology and human life cannot be separated. We use technology, depend on technology, and feel we must have technology. Technology lets us travel easily, lets us communicate quickly, lets us live in comfort, and has also caused us concern. Poor application of technology pollutes the air, water, and land around us. This application of technology can threaten our very lives.

The challenge facing all people is to determine the type of future we want. We must then develop the technology promoting that future. Likewise, we have to change our personal habits and preferences. Some technologies we now depend on might need to be modified or abandoned. Only then can future generations enjoy a good quality of life.

Test Your Knowledge

Write your answers on a separate piece of paper. Please do not write in this book.

1. Technology, by itself, does not affect people or the environment. True or false?
2. Name at least three destructive natural forces that impact humans.
3. Hydroelectric dams and wind farms are examples of harnessing _____ forces.
4. North America uses a large percentage of the world's petroleum production. What product of technology accounts for a major portion of that consumption?
5. Describe the two types of thinking used in the futuring process.
6. Which of the following is *not* one of the types of futures considered in futuring research?
   A. Social.
   B. Economic.
   C. Technological.
   D. Biological.
   E. Human-psyche.
7. Some people feel threatened by _____. Others feel empowered by it.
8. What are the three forces contributing to the current environmental crisis?
9. Northern Europe became the center of technical knowledge and power during the _____ Revolution of the 1800s.
10. Companies are adopting new, computer-based technologies, such as *CAD*, *CIM*, and *JIT*, to remain competitive in the global economy. What do these three abbreviations stand for?

STEM Applications

1. Choose an environmental problem. Gather information about it. Summarize your knowledge on a form similar to the one shown.

Problem:
Factors causing or contributing to the problem:
Possible solution #1:
Problems solution #1 might cause:
Possible solution #2:
Problems solution #2 might cause:

2. Prepare a drawing or a photomontage showing an ideal future in which you would like to live.

Chapter Outline

This chapter covers the benchmark topics for the following Standards for Technological Literacy:

4 5 6 7

Learning Objectives

After studying this chapter, you will be able to do the following:

- Summarize how jobs have evolved from the colonial period, through the Industrial Revolution, to the information age.
- Summarize the typical levels of technology-based jobs.
- Recall the factors that should be considered when selecting a job.
- Recall the three factors affecting job satisfaction.
- Summarize the ways individuals exercise control over technology.
- Summarize the three major concerns people have to which technology can contribute to a solution.
- Give examples of some technological activities introduced as fiction that have become actual processes.

Key Terms

activism
information skills
language and communication skills
lifestyle
microgravity
people skills
personal skills
political power
socioethical skills
thinking skills

Strategic Reading

As you read this chapter, think of different jobs in technology that appeal to you. What do you think the job requirements are? What kind of skills are necessary?

Your personal life is highly dependent on the technology people have developed. In the brief span of 50 years, life has changed dramatically for citizens of developed countries. We live in different housing, travel on different systems, have different products to purchase, and communicate in ways far different from in the past. To help investigate the changes affecting everyday life in the twenty-first century, let us look at five areas:

- Technology and lifestyle.
- Technology and employment.
- Technology and individual control.
- Technology and major concerns.
- Technology and new horizons.

Technology and Lifestyle

Each person lives in a specific way, or has a lifestyle. A ***lifestyle*** is what a person does with business and family life—their work, social, and recreational activities. See **Figure 36-1.**

Figure 36-1. What kind of lifestyle do you think this father and son enjoy?

Colonial Life and Technology

The lifestyle during America's colonial period was a harsh contrast to that of today. Housing was simple and modest. Most products were designed and produced to meet basic human needs. There were few decorative items available. Those items were mostly owned by the wealthy. Transportation systems included horses, animal-drawn wagons, and simple boats. The communication systems available were crude.

Most people lived on farms. A few people practiced basic crafts. These people were the carpenters, blacksmiths, and other tradespeople needed to produce the basic products the community required. The men and boys did most of the fieldwork on the farm and practiced the trades. Women and girls tended gardens and did household work. See **Figure 36-2.**

Figure 36-2. This woman is demonstrating the old craft of candle making, a common task in colonial life.

Standards for Technological Literacy

6 7

Brainstorm

Ask your students how technology plays a role in determining a person's lifestyle.

Reinforce

Review the evolution of technology (Chapters 1 and 2).

Research

Have your students determine what technologies the early settlers of the American colonies used.

TechnoFact

While people influence the development of technology, technology influences the change of people's lifestyles.

TechnoFact

The development of modern transportation, including the automobile and airplanes, altered the way and distance people traveled. It took people from only going a few miles from their birthplaces to traveling around the world.

Standards for Technological Literacy

6 7

Extend

Discuss how the Industrial Revolution changed the way products were made and the way people lived.

TechnoFact

The Industrial Revolution, which started in Britain in the late 18th century, created social and economic changes, altering the way people worked and lived. Before long, this shift spread throughout Europe and North America, causing these areas to become highly industrialized by the mid-19th century.

Everyone had to work long hours, six days a week, to raise small amounts of food. Most colonial families set aside the seventh day for church activities. People did not take vacations. They celebrated few holidays.

With the exception of slaves, however, most people were their own bosses. They owned their farms and stores. These people practiced their crafts as independent workers. Each person was an owner, a manager, and a worker, rolled into one.

The Industrial Revolution and Technology

The Industrial Revolution of the mid-1800s changed this lifestyle. A series of events took place. First, advanced technology was developed for the farm. This technology included the moldboard plow, the reaper, and the steam tractor. These and other devices made farmers more efficient. Fewer people could farm more land and produce more food to sustain the society. Second, the movement to the land west of the Appalachians allowed for larger and more efficient farms. During this period, the percentage of the workforce engaged in farming began dropping rapidly. At the start of the Industrial Revolution, more than 90% of the workforce were engaged in farming. Today, fewer than 3% of workers are employed on farms. Third, a large number of people from Europe immigrated to the United States during this time. These immigrants, plus the farmers who were no longer needed to till the soil, provided a vast labor supply.

This labor supply was a basic resource for the factory system that was then being developed. The local tradespeople working in their shops could no longer meet the demand for goods. Centralized manufacturing operations were replacing the functions of the tradespeople. See **Figure 36-3.** Such operations included several important factors:

- **Professional management.** This management established procedures, employed resources, and supervised work.
- **Division of labor.** This division assigned portions of the total job to individual workers. Each worker did only part of a total job, allowing the workers to quickly develop the specialized skills needed to do the assigned tasks.

Daimler

Figure 36-3. Many current manufacturing plants are based on the principles of the Industrial Revolution.

- **Continuous-manufacturing techniques.** These techniques increased production speed. Raw materials generally entered a production line at one end. Finished products left the line at the other end.
- **Material-handling devices.** These devices were used to move the products from workstation to workstation. Workers remained at their stations. The products moved to them.
- **Interchangeable parts.** These parts allowed the production of large quantities of uniform products.

Low wages and poor working conditions in the factories caused widespread worker unrest. Labor unions were formed to give the workers a voice in determining working conditions and pay rates. Bloody battles erupted between the workers and management. The government usually supported managerial positions, resisting attempts of the unions to deal with the issues.

These conflicts were finally settled with changes in governmental attitudes, new laws, and different management stances. This led to a strong industrial period for the country. Broad employment opportunities characterized the period. The workers enjoyed a high standard of living. The 40-hour, five-day workweek with a number of holidays and paid vacation time became fairly common.

The Information Age and Technology

The development of the computer changed the Industrial Revolution. During the industrial period, the company that could efficiently process the greatest amount of material was the most successful. This required major investment in large continuous-manufacturing plants. The huge automobile- and steel-manufacturing operations characterized these plants. These plants employed thousands of people and used millions of tons of materials.

The computer allowed the development of a new type of manufacturing, flexible manufacturing. This type of manufacturing can quickly and inexpensively respond to change. People with few skills are replaced with computer-controlled machines. See **Figure 36-4.** The workers who remain have more training and motivation to work. They accept change and responsibility more readily than the workers of the Industrial Revolution.

Also, management has changed greatly. Management is less distant from the workers. The entire workforce is seen as a team, with each person having an area of responsibility. See **Figure 36-5.** Managers might be

Cincinnati Milicron

Figure 36-4. This robot is placing cartons on a pallet. Semiskilled workers formerly did this task.

Standards for Technological Literacy

6 7

Brainstorm

Ask your students how the development of computer systems has changed people's lives.

Extend

Discuss the advantages and challenges created by the Information Age.

TechnoFact

Near the end of the 20th century, the Internet and World Wide Web became new methods of buying and selling things. Companies have altered their marketing tactics around this new medium of commerce.

Standards for Technological Literacy

6 7

TechnoFact

The first computers were built as mechanized mathematical devices. This computer age began in the 1940s, though further progress was made in the 1960s with the invention of the integrated circuit. Today's Information Age came from these early developments, and computers throughout the world are joined together on the Internet.

Air Products Co.

Figure 36-5. In today's industry, managers and workers work as a team. They cooperate to reach company goals.

responsible for setting goals and controlling money. Workers are responsible for producing products. Everyone is responsible, however, for work procedures and product quality.

These changes have given people a new lifestyle. Those people who are able to change and adjust to the demands of the new age can live very differently. They are better informed, work more with their brains than their muscles, travel greater distances, and have more control over their work. See **Figure 36-6.**

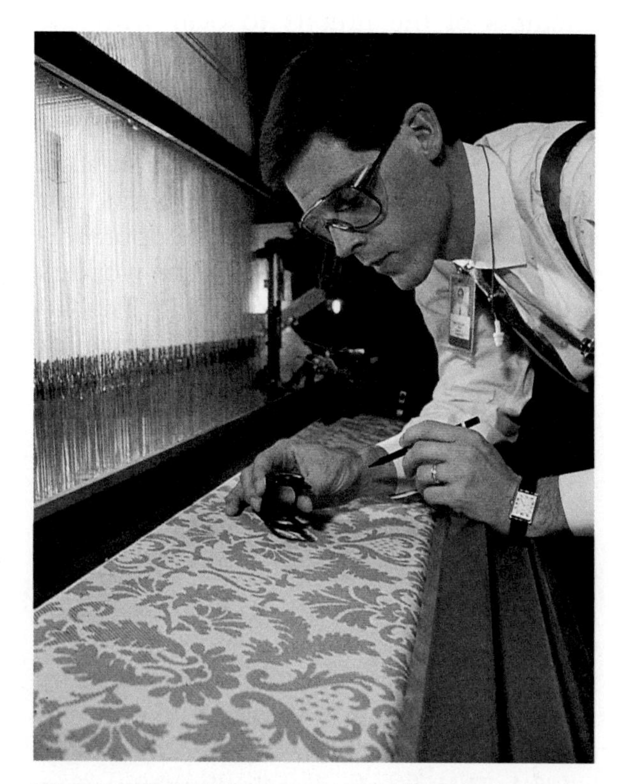

Figure 36-6. Workers in modern industry more often work with their brains than with their muscles.

Technology and Employment

Lifestyle and employment are closely connected. Most people need the money they earn through working to afford the type of life they want. Some general requirements for most jobs in the future can be identified. Fundamental to almost all jobs will be a high school education. Specialized technical training beyond that level quite often will be needed. See **Figure 36-7.** This might be technical training in career centers and community colleges or a university education. In addition, workers in the information age must be willing to do the following:

- Pursue additional education and training throughout their work lives.
- Accept job and career changes several times during their work lives.
- Work in teams and place team goals above personal ambitions.
- Exercise leadership and accept responsibility for their work.

Types of Technical Jobs

There is a wide variety of jobs requiring technical knowledge. They include five levels. See **Figure 36-8.** The first is

General Electric Co.

Figure 36-7. Specialized training beyond high school is required for many technical jobs.

Extend

Discuss the types of work that require technological knowledge and skills.

Figure Discussion

Use Figure 36-8 to differentiate among production workers, technicians, technologists, engineers, and managers.

TechnoFact

Engineering skills emerged with the developments during the Industrial Revolution. These skills were required as the technological advancements became more complex. Today, several types of engineering exist for the many types of technology available.

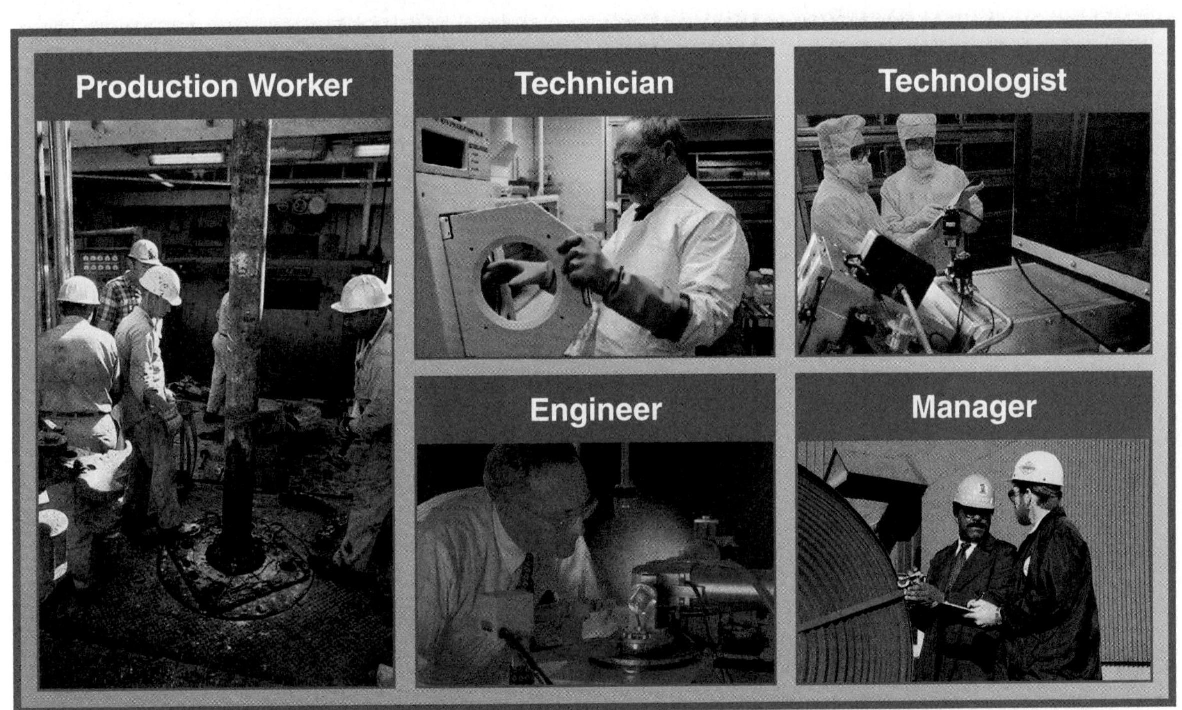

American Petroleum Institute, Goodyear Tire and Rubber Co., AT&T, Inland Steel Co.

Figure 36-8. People with technology backgrounds or an aptitude for technology can become production workers, technicians, technologists, engineers, or managers. Many people work in more than one of these areas during their careers.

Research
Have your students visit the U.S. Department of Labor Web site to find jobs that could give them the types of lifestyles they want.

Brainstorm
Ask your students what gives them satisfaction when they have a job to do.

the technically trained production worker. These workers include the people who process materials and make products in manufacturing companies, erect structures, and operate transportation vehicles. Technicians work closely with production workers. They, however, do more specialized jobs. Typically, technicians set up and repair equipment, service machinery, conduct product tests and laboratory experiments, work in dental and medical laboratories, and do quality control testing. Technologists are highly trained technical employees. They form the bridge between the engineers who design systems and the workers and technicians who must implement the systems. Many companies have a need for people with advanced technological knowledge. Engineers design products and structures, conduct research, and develop production processes and systems. Technically trained managers can set goals, plot courses of action, and motivate people to work together. They are people-oriented leaders who also have technical knowledge.

Selecting a Job

People should not just take the first job they can find. Whenever possible, a job should match your interests and abilities. When deciding on employment opportunities, you should consider at least three factors. These factors are lifestyle, job requirements, and job satisfaction.

Lifestyle

Each of us wants to live comfortably. A person's job has a direct impact on both life at work and life away from it. Some people like to travel and meet new people. An industrial-sales job can provide these aspects. A factory job that pays well and has good holiday and vacation benefits can also meet this need.

Job Requirements

A variety of job requirements affect each employee. In selecting employment, a person should consider job requirements from three important points. The first requirement considers your freedom to organize the tasks assigned to the job and the level of accountability that goes with the job. The second consideration is the balance of working with data, machines, and people. See **Figure 36-9.** The last factor is the level of education a job requires.

Job Satisfaction

Job satisfaction is a description of how happy a worker is with her job. Three

Goodyear Tire and Rubber Co.

Figure 36-9. In industry, or business, each job deals, in varying degrees, with machines, people, and data. Which areas do you prefer?

factors strongly affecting job satisfaction are values, recognition, and pay. A job should match the values of the person doing it. Some jobs allow for more visible recognition than others. The money you receive in exchange for your work is also important.

Job Skills

Each person seeking a job has a set of job skills. These are the activities a person does well. Each job has a set of skill requirements. For example, an engineer must be able to use mathematics, science, and technology to solve problems. An accountant possesses mathematics and accounting skills. A taxi driver needs to know the city's streets, major hotels, and tourist sites and how to drive a car. The challenge is for people to find jobs matching their own sets of skills.

In addition to specific job skills, all applicants need general job skills. General job skills are skills that can be used in many different jobs. They are skills that can be taken from one job to another. These skills are transferable skills. General job skills should be developed during grade school and high school. Most specific job skills are gained through additional education and training activities.

General job skills can be divided into six major groups. These groups are language and communication skills, thinking skills, information skills, socioethical skills, people skills, and personal skills. ***Language and communication skills*** are the abilities to read, write, and speak a language. These skills include the following:

- **Reading.** Reading skills involve the ability to locate text information and then identify and comprehend relevant details.
- **Writing.** Written-communication skills include the ability to write clearly and for a variety of audiences and purposes.
- **Speaking.** Oral-communication skills involve the ability to explain and present ideas in clear, concise ways to a variety of audiences.

Thinking skills involve the ability to use mental processes to address problems and issues. These skills include the following:

- **Problem solving.** Problem-solving skills include the ability to recognize a problem, create possible solutions, implement a selected solution, and evaluate the solution.
- **Creative thinking.** Creative-thinking skills involve the ability to use imagination to develop and combine ideas or information in new ways.
- **Decision making.** Decision-making skills include the ability to identify a goal and alternate courses of action, evaluate the advantages and disadvantages of each alternative, and then select the best action.
- **Visualizing.** Visualization skills involve the ability to envision a three-dimensional object that an engineering drawing or a schematic drawing presents.

Information skills are the abilities to locate, select, and use information. These skills include the following:

- **Information literacy.** Information-literacy skills include the ability to locate, evaluate, use, and cite information properly.
- **IT.** IT skills involve the ability to use computer systems to acquire, organize, and present information. See **Figure 36-10.**

Socioethical skills involve understanding the implications of actions on people, society, and the environment. These skills include the following:

- **Ethics.** Ethical skills include the ability to see the implications of actions on people, society, and the

Research

Have each student select three jobs and use the U.S. Department of Labor Web site to research them, in terms of their nature of work, working conditions, training requirements, and job outlook.

TechnoFact

While the major classes of jobs in early civilizations were priest, craftworker, merchant, and warrior, the changes in the late 18th century brought about separate divisions of labor including those who designed, built, and sold products.

TechnoFact

Careers have a great impact on a person's lifestyle, and they determine the area of work a person does. A career is the work and work-related activities performed in the jobs that a person has in his or her lifetime.

environment and then to act in legally and morally responsible ways.

- **Sociability.** Social skills involve the ability to show interest in, an understanding of, and respect for the feelings and actions of others.

People skills are the abilities needed to work with people in a cooperative way. These skills include the following:

- **Leadership.** Leadership skills include the ability to encourage and persuade people to accept and act on a common goal.
- **Teamwork.** Team skills involve the ability to work cooperatively with others to complete a project. See **Figure 36-11.**
- **Diversity.** Diversity skills include the ability to work with people from different ethnic, social, or educational backgrounds.

Figure 36-10. IT skills are required for most jobs.

Figure 36-11. Teamwork is an essential general job skill.

Career Corner

Business, Management & Administration

Technology Education Teachers

A teacher must have the ability to effectively organize and deliver material to students. Communication skills are necessary to facilitate comprehension and understanding, whether the subject is technology, mathematics, science, reading, or English. Effective teachers are good communicators. Technology education teachers must possess all the skills of a good teacher and know and understand how people have developed, produced, and used tools, materials, and machines to control and improve their lives. These teachers are developed in technology teacher education programs in a number of universities and are certified to teach by the state in which they work.

Personal skills involve the ability to grow and manage personal actions on a job. These skills include the following:

- **Self-management.** Self-management skills involve the ability to manage time, adapt to change, be aware of personal responsibilities, and maintain physical and mental health.
- **Self-learning.** Self-learning skills include the ability to recognize the need for new knowledge and seek ways to meet the demand.
- **Goal.** Goal skills involve the ability to set goals and work toward meeting them.

In addition to general and specific job skills, employers look for proper work attitudes. Employers look for cooperation, dependability, a good work ethic, and respect. This means the employees are expected to do the following:

- Cooperate with supervisors, other employees, and customers.
- Arrive at work on time and complete tasks in a timely manner.
- Put in an honest day's work for a day's pay.
- Show respect for others, the company, and themselves.

Today's workplace emphasizes equality—that is, the idea that all employees are to be treated alike. Harassment (an offensive and unwelcome action against another person) and discrimination (treating someone differently due to a personal characteristic such as age, sex, or race) are not tolerated. These negative behaviors often result in termination of employment.

Technology and Individual Control

Technology holds great promise and hidden dangers. People make the difference. See **Figure 36-12.** We often say, "They should control this," "They should do that," or "They should stop doing something else." The harsh reality is that "they" will never accomplish anything. Only when someone says "I," instead of "they," does anything meaningful get done.

The future lies in the hands of people who believe they can make a difference. Examples such as Thomas Edison, the Wright brothers, and George Washington Carver might come to mind. These individuals did not wait for a group to do something. They pursued their own visions of what was important and needed by society. This type of action requires people to understand technology. People must also comprehend the political and economic systems directing technology's development and implementation.

American Electric Power Co.

Figure 36-12. This quiet fishing lake was once a strip mine. After the coal was removed, people worked together to restore the area.

Extend

Discuss the power of the consumer in determining the types and uses of technology.

TechnoFact

Aptitudes, abilities, interests, and values should be considered to help determine a person's career field. Once a career goal is determined, selection of appropriate educational opportunities is significant.

Standards for Technological Literacy

4 5

Brainstorm

Ask your students how their buying habits shape product development and use.

TechnoFact

The goal of consumerism is to protect consumers from untruthful advertising and unsafe products. Governmental agencies monitor business practices in order to ensure consumers are better informed before making product and service selections.

Think Green

Compact Fluorescent Lamps (CFLs)

Conventional incandescent bulbs use a great deal of energy to give light, and they also waste energy by giving off heat. Compact fluorescent lamps, or CFLs, may be used to replace most incandescent bulbs. When used and disposed of properly, the use of CFLs rather than incandescent bulbs makes a great deal of difference to the environment.

Compact fluorescent lamps use gas and ballasts to produce light. There is no wasted energy in the form of heat because they do not use filaments. CFLs do not use as much energy to produce the same amount of light as incandescent bulbs. The wattage used is much lower with the same results. Because of the low energy usage, they also last much longer than incandescent bulbs. CFLs help the environment overall because they use less energy, and less electricity needs to be generated. The disposal of CFLs is important, however. You cannot just throw them out because of the mercury inside. This can contaminate the ground and water supply. Be sure to take your used CFLs to a recycling center to be disposed of properly.

The Role of the Consumer

Individuals control technology in a number of ways. The first is through the role of the consumer. This involves the following:

- Selecting proper products, structures, and services.
- Using products, structures, and services properly.
- Maintaining and servicing the products and structures they own.
- Properly disposing of worn or obsolete items.

Consumer action causes appropriate technology to be developed and helps inappropriate technology disappear. See **Figure 36-13.** When a product or service does not sell, production of that product or service stops.

U.S. Department of Energy

Figure 36-13. The types of products and structures you demand as a consumer make a difference. Solar heating, for example, is becoming increasingly popular. This solar-heated home saves energy without harm to the environment.

Political Power

An individual also has *political power.* All companies must operate under governmental regulations. These regulations include the Occupational Safety and Health Act, the Pure Food and Drug Act, the Environmental Protection Act, and the Clean Water Act. Few laws are passed solely because an elected official thinks they are important. Most of them come from people asking their elected representatives to deal with a problem or concern.

Activism

Finally, an individual can make a difference as an activist. ***Activists*** use public opinion to shape practices and societal values. Some people see this as a bad thing. They are disturbed by the actions and words of people whose views are different from their own. Actually, activists are simply people practicing their constitutional rights to freedom of speech and assembly. As a citizen, you have the ability to meet with others to discuss and promote your point of view on a subject.

Technology and Major Concerns

The human-built world concerns many people. People are worried about pollution, energy use, and unemployment. Technology is designed, produced, and used, however, by people. These issues can be resolved only through human action. Some areas of concern are at the forefront of our attention. These concerns include the following:

- **Nuclear power and nuclear-waste disposal.** Is nuclear-power generation an appropriate activity? How do we design and build safe and reliable nuclear power plants? Can the radioactive waste that nuclear power plants produce be disposed of safely?
- **Technological unemployment.** Should technology that causes unemployment be applied? Does a company have a responsibility to provide training and other benefits to workers who become unemployed through the adoption of new technologies? Should foreign products that cause technological unemployment be barred from domestic markets? What are the individual worker's responsibilities in seeking training to deal with changing job requirements?
- **Genetic engineering.** Is it right to change the genetic structure of living organisms? Who decides what genetic engineering activities are appropriate? Is it acceptable to alter the genetic structure of humans? How are religious and technological conflicts going to be dealt with?
- **Energy use.** How can we reduce our dependence on petroleum? Can society's reliance on the private automobile be changed? See **Figure 36-14.** What alternate energy sources and converters should be developed? Should mass transit

Figure 36-14. Greater use of mass transit will be needed in the future to minimize environmental damage and relieve traffic congestion in urban areas.

Standards for Technological Literacy

4 5

Extend

Discuss how technology is neutral but its application by people can create problems.

Brainstorm

Ask your students to list some problems that technology has created and solved for them.

Research

Have your students select a technological issue, such as genetic engineering, energy resource depletion, or pollution, and explore the issues related to it.

TechnoFact

Consumer protection began in the Middle Ages with standards set by workers themselves. It wasn't until later that consumers called for product protection legislation.

Standards for Technological Literacy

4 5

Brainstorm

Technology has been a major means for creating new physical and human environments. Ask your students if they think technology will also destroy the global civilization that human beings have created.

Research

Have your students select a new, cutting-edge technology and find out how it works and what its benefits will be.

TechnoFact

The growing problem of pollution has brought on various views on its origin. While some people see it as a connection to the increase in population, others believe industry and technology are at fault. Experts believe that controlling one of these two factors would help control pollution.

be financed with tax money? How can we control the environmental damage that burning fossil fuels causes? Should we place high taxes on gasoline?

- **Land use.** What are the rights of landowners? Should the desires of the majority overrule individual landowner rights? What responsibilities do governmental officials have for public lands? How do you balance environmental protection issues with economic issues? See **Figure 36-15.**
- **Pollution.** Should products polluting the environment be banned from manufacture and use, and should strict pollution controls apply equally to individuals and companies? What type of evidence is needed before a product can be banned? Should there be a pollution tax on fuels that damage the environment? How do you handle the economic and social impacts of banning products? Should we limit the use of wood as a fuel, and should the solid waste (garbage) from one state be allowed to enter another state?

Figure 36-15. Economic considerations must be balanced with concern for the environment as the population continues to grow.

This book does not attempt to provide answers to these or other technological problems. Remember, the right answer according to one person or group is often rejected by other people and other groups. The best we can hope for is an answer most people support.

Technology and New Horizons

Technology continues to be developed at a very rapid pace. See **Figure 36-16.** Many ideas that seem impossible now will be commonplace in the near future. This is not new. In the 1800s, Jules Verne wrote a fictional book called *20,000 Leagues Under the*

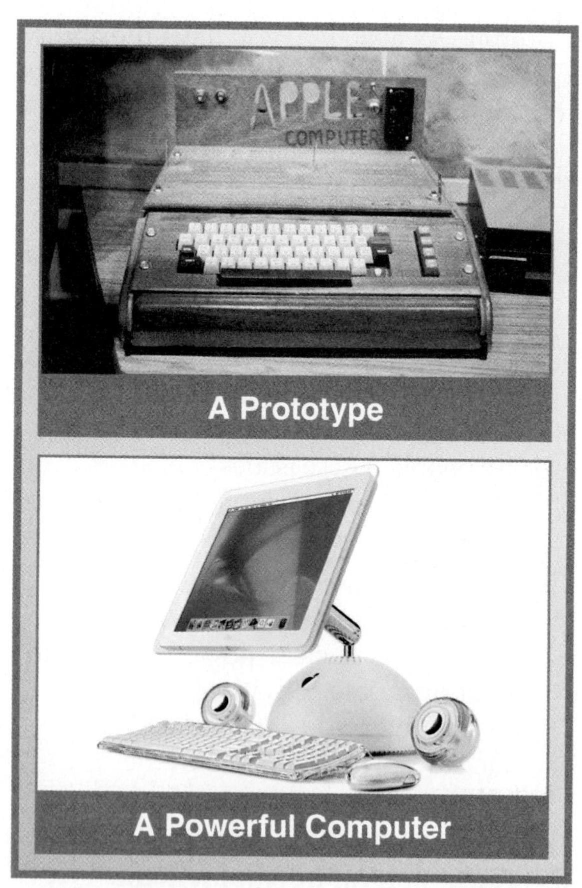

Apple Computer

Figure 36-16. In only about 30 years, the Apple® computer moved from a crude prototype to a polished, powerful computer.

Sea. This story deals with the then-impossible feat of traveling under the ocean in a submarine. In the 1940s and 1950s, people read a fictional comic strip in which the characters used the then-ridiculous means of rocket ships to travel in space. What is fiction today might become an everyday part of life for your children and grandchildren.

Many futuristic ideas have been proposed. One is mining the resources of outer space and our oceans. Will we have colonies on the Moon or other planets that extract and process precious mineral resources? Our oceans are the last of the vast resource beds on Earth. Will they be mined, or will concerns regarding pollution keep them off-limits?

Manufacturing in space has been proposed and tried on a small scale. See **Figure 36-17.** Experiments have shown that protein crystals can be grown in the ***microgravity*** (very low gravity) of space. Earth-grown crystals are often small and flawed. Crystals grown in space can be larger and more complex. Some experiments have dealt with growing zeolite crystals. These crystals might be used in portable kidney-dialysis machines and in cleanup efforts with radioactive waste. Earth's atmosphere and gravitational pull adversely affect some manufacturing processes. Will these processes be moved into space, and will we have space stations where manufacturing is routinely done?

NASA

Figure 36-17. In the future, we might build complex manufacturing systems in space.

A third future technological activity is commercial space travel. To date, space travel is government financed. Most of it is restricted to military and scientific missions or communication-satellite launching. A few private citizens have paid to have a ride into space. Will we routinely be traveling into space and back some day, and will future generations take vacation trips to the Moon or nearby planets?

These are only a few examples of possible technological advancements. Society has moved from the horse-drawn wagon to space travel in a single lifetime. What might be next?

Standards for Technological Literacy

4

Brainstorm

Ask your students what new technologies they would like to see developed.

TechnoFact

With the further development of communication technologies, such as satellites, digital broadcasting, and the Internet, instantaneous communication has become a standard. People are able to see events from all over the world with the click of a button.

TechnoFact

Computers and electronics are commonly connected with technology. Technology, like engineering, is much broader and is associated with several different areas.

TechnoFact

While science is more interested in discovering and explaining natural occurrences, technology's goal is producing products and systems. Science studies the universe through observation and the collection of data. Technology focuses on knowledge and skill, and though it may use scientific knowledge, an understanding of science is not required to work with technology.

Answers to Test Your Knowledge Questions

1. 90
2. Immigrants and out-of-work farmers
3. True
4. Production worker, technicians, technologists, engineers, and managers
5. Lifestyle, job requirements, and job satisfaction.
6. people
7. Values, recognition, and pay.
8. Through the role of consumer, through political power, and through activism
9. Evaluate individually.

Summary

The human-built world is as old as humanity. Technology helps us lead better lives by enhancing our abilities. This world provides better jobs and helps us evolve new opportunities. Technology also threatens us with pollution, a potential loss of personal worth, and possibly even nuclear disaster. How we design, produce, and use technology makes the difference. The choices are yours to make. You now know more about technology. Using technological knowledge, you can select products wisely, pursue appropriate employment, and better understand societal issues. You can be a better consumer, worker, and citizen.

Test Your Knowledge

Write your answers on a separate piece of paper. Please do not write in this book.

1. Before the Industrial Revolution of the mid-1800s, more than ______ % of the workforce were engaged in farming.
2. What two factors made a large labor supply available for factories during the Industrial Revolution?
3. Most jobs in the future will require specialized technical training beyond high school. True or false?
4. What are the five categories of technical jobs described in this chapter?
5. What are three key factors that should be considered when selecting a job?
6. When considering a job, you should look at whether it places emphasis on data, machines, or ______.
7. What are the three major factors that contribute to a person's satisfaction with a specific job?
8. List three ways an individual can exercise control over the type of technology that is developed.
9. List some examples of technology that were once found only in fiction, but are now part of everyday life.

STEM Applications

1. Select a technological idea that is now considered fiction. Write a description of how it will be used as an everyday item or process in the future.
2. Read a book about people who developed new technology. Write a report on their attitudes toward change and criticism, as well as their inventions or innovations.

Section 11 Activities

Activity 11A

Design Problem

Background

Technology, by itself, is neither good nor bad. How a technological object is used can create benefits or drawbacks. Each person should strive to use technology wisely to make the future better, protect the environment, and help people live in harmony with nature.

Challenge

Identify an important issue to people today that technology impacts. List the positive and negative factors related to the issue. Develop a plan to maximize the positive features and reduce the negative impacts.

Activity 11B

Production Problem

Background

Humans can control the use of technology. Many people, however, do not know about the impacts technology has on individuals and society.

Challenge

You are the communications director of a citizens group. Your group is concerned about public-policy issues. This group has determined that people in your community are not participating in the local recycling program. Design a 60-second public-service commercial for television. The commercial needs to explain the importance of recycling to your community. Identify the materials that can be recycled and the benefits of recycling.

Materials and Equipment

- Storyboard forms.
- Pencils.
- Felt-tip pens.
- A video camera.
- A video recorder.
- A video monitor.

Procedure

Designing the commercial

1. Select the theme of the commercial.
2. List the major points to be emphasized in the commercial.
3. Develop a storyboard for the commercial. Use photocopies of **Figure 11B-1.**
4. Write a script for the actors.
5. Develop a shot chart for the director and cameraperson to follow.

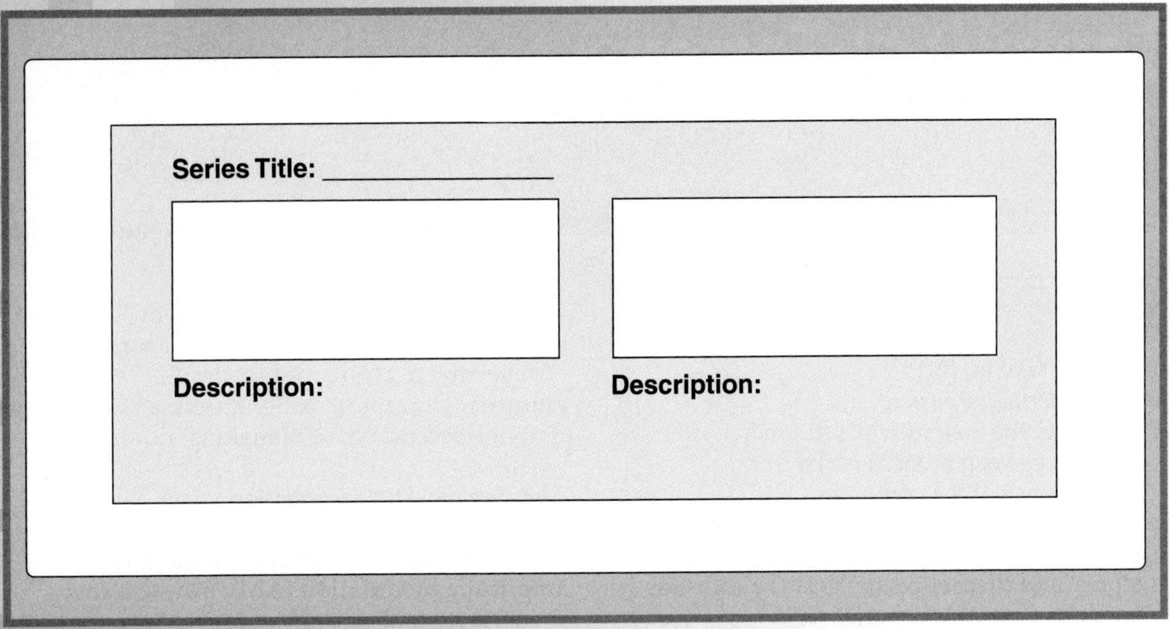

Series Title: ________________

Description:

Description:

Figure 11B-1. The storyboard layout.

Producing the commercial

1. Recruit and select actors for the commercial.
2. Present the script to the actors. Have them rehearse their parts.
3. Walk through the commercial with the director. Have the actors and camera operators block (plan) their movements.
4. Record the commercial.
5. Edit in any titles you need.
6. Present the commercial to an audience. Ask for their reactions.

Technical Terms

A

absorb: to take in. (21)
AC: See *alternating current.*
accounting: the area in which financial-affairs employees keep records of the financial transactions of the company. (33)
active collector: a solar collector that uses pumps to circulate the water that collects, stores, and distributes the heat the collector converts from solar energy. (28)
activism: using public opinion to shape practices and societal values. (36)
actuate: to initiate the work related to an action plan. (2)
adhesive bonding: a bonding technique that uses substances with high adhesive forces to hold parts together. (16)
adjusting device: a device that modifies a system to produce better outputs. (6)
adjustment: the step in the repair process in which misaligned parts are adjusted. (18)
advertising: an activity that includes the print and electronic messages promoting a company or the company's products. (33)
aerobic exercise: exercise that uses oxygen to keep large muscle groups moving continuously. (31)
agricultural and related biotechnology: the technology used in growing food and producing natural fibers. (3)
agricultural technology: a technology that uses machines and equipment to help plant, grow, and harvest crops and to raise livestock. (29)
agriculture: people using materials, information, and machines to produce food and natural fibers. (29)
AI: See *artificial intelligence.*
airfoil: an object designed to produce some directional motion when in movement relative to the air. (25)
air-transportation system: a system that uses airplanes and helicopters to lift passengers and cargo into the air so they can be moved from place to place. (24)
alter: to extend the useful life of a product. (18)
alternating current (AC): the flow of electrons in both directions along a conductor, reversing at regular intervals. (22)
alumina: aluminum oxide. Alumina is the input to the second phase of making aluminum. (15)
AM: See *amplitude modulation.*
amplitude: the height of a wave. Amplitude measures the strength of the wave. (21)
amplitude modulation (AM): a system that merges a message onto a carrier wave by changing the strength of the carrier signal. (22)
anaerobic digestion: a controlled decaying process that takes place without oxygen. (28)
anaerobic exercise: exercise involving heavy work from a limited number of muscles. (31)
analog signal: a continuous electrical signal that varies in amplitude or frequency to carry information or data. (22)
analysis: careful examination of the parts or structure of some object or event. (11)
analytical system: a system that mathematically or scientifically makes comparisons. (6)
animal husbandry: the type of agriculture that involves breeding, feeding, and training animals. (29)
anneal: to soften and remove internal stress in a part. (16)
antibiotic: a substance derived from microorganisms able to kill other microorganisms. (29)
apogee: the point in an orbit at which an object is farthest from Earth. (25)
appearance: a factor to consider before buying a product. (18)
applied research: a type of research that seeks to reach a commercial goal by selecting, using, and adapting knowledge gathered during basic research. (33)
apprenticeship training: a program through which highly skilled workers are developed. (33)
aquaculture: growing and harvesting water organisms in controlled conditions. (29)
aquifer: an underground water-bearing layer of rock, sand, or gravel. (35)

arbor: a spindle, or shaft, used to hold table saw blades and milling cutters. (7)
arch bridge: a bridge that uses curved members to support the deck. (17)
architectural drawing: a document used to specify characteristics of buildings and other structures. (33)
area: the size of the surface an object covers. (8)
armature: the inner magnet in an electric motor. This magnet is an electromagnet that can rotate. (7)
articles of incorporation: an application for a corporate charter. (32)
artifact: an object made by humans. (1)
artificial intelligence (AI): intelligence a manufactured device or system exhibits. (16)
aseptic packaging: a process that uses heat to separately sterilize a food and its package. (30)
assemble: to bring materials and parts together to make a finished product. (5)
assembling process: a process that connects parts together to make assemblies and products. (16)
assembly drawing: an engineering drawing that shows how parts fit together. (12)
audience assessment: actions that study audience needs and wants. This assessment is the first step in the process of communication design. (20)
audio message: sound. (22)
authority: the right to direct actions. (32)
automatic control system: a system that can monitor, compare, and adjust a system without human interference. (6)
automatic transmission: a transmission that uses valves to change hydraulic pressure so the transmission shifts its input and output ratios. (25)
aviation electronics (avionics): the instruments that help pilots monitor and properly control aircraft. (25)
avionics: See *aviation electronics.*

B

backbone: a fiber-optic communication line built by a large communications company to connect regions in its system. (23)
balance: the design principle of having the information on both sides of a centerline appear equal in visual weight. (20)
baler: a machine used to gather, compact, and contain hay. (29)
ballast: extra weight added to a locomotive to give the locomotive better traction. (25)
band saw: a saw that uses a blade made of a continuous strip, or band, of metal. (7)
barbed wire fence: a fence consisting of wood or steel posts with strands of barbed wire attached to them. (29)
basic research: a type of research that seeks knowledge for its own sake. (33)
bauxite: aluminum ore. (15)
beam bridge: a bridge that uses concrete or steel beams to support the deck. (17)
benefit: a type of reward a company provides. (33)
berry: a plant grown in many parts of the country and cultivated for its edible parts. (29)
billet: a long, square piece of steel. (15)
bill of materials: a document containing information regarding the materials and hardware needed to complete a project. (12)
biochemical process: a process that uses chemical reactions caused by fungi, enzymes, or other microorganisms. (28)
biofuel: an organic material that can be burned or converted into methane. (27)
biogas: a mixture of methane and carbon dioxide produced by the bacterial decomposition of organic wastes and used as a fuel. (27)
biological future: a type of future dealing with the sorts of plant and animal life we want. (35)
biomass: a type of resource having a living origin. (27)
biomass resource: vegetable and animal waste generated through biological actions. This resource is a source of renewable energy being actively considered as an alternate energy supply. (28)
biotechnology: using biological agents in processes to produce goods or services. (29)
blast furnace: a furnace commonly used in iron smelting. (15)
blimp: a lighter-than-air nonrigid airship. (25)
bloom: a short, rectangular piece of steel used to produce structural shapes and rails. (15)
blowout: a dangerous and wasteful occurrence in which oil surges out of a well. (14)
board of directors: a group of people elected to represent the interests of stockholders. (32)
body: the structure of a camera that holds other parts in place. (21)
bond: a debt security sold by corporations in large denominations. (32)
bonding: an assembling process that uses cohesive or adhesive forces to hold parts together. (16)
bonding agent: a material used to achieve bonding. (16)

bow thruster: a maneuvering prop mounted at a right angle to the keel on a large ship. (25)
brainstorming: seeking creative solutions to an identified problem. (10)
broadcast frequency: a frequency above audible sound. (22)
broadcast system: a system that sends radio waves through the air, carrying the signal from the sender to the receiver. (22)
Bronze Age: the stage in human history that took place after the Stone Age. (1)
browser: a software program that acts as an interface between a user and the WWW. (23)
buck: to remove the limbs and tops of trees. (14)
building: a structure erected to protect people, materials, and equipment from the outside environment. (5)
buoyancy: the upward force exerted on an object immersed in a fluid. (25)
buttress dam: a dam that uses its structure to hold back water. (17)
bylaw: a general rule under which an incorporated company operates. (32)

C

cable-wire fence: a fence consisting of 3/8″ steel-wire cables stretched from one anchor post to another. (29)
CAD: See *computer-aided design.*
CAD software: See *computer-aided design software.*
CAD system: See *computer-aided design system.*
cam: a pear-shaped disk with an off-center pivot point, used to change rotating motion into reciprocating motion. (28)
can: to put food into glass jars or metal cans. (30)
cant: the square, center section of a log. (15)
cantilever bridge: a bridge that uses trusses extending out similarly to arms. (17)
capacity: the amount of a substance an object can hold. (8)
carrier frequency: a frequency used by radio broadcast systems that radiates from the transmitter. (22)
casting: a process through which on-camera performers are employed. (22)
casting and molding: a method of shaping parts or products by pouring liquid material into a shaped cavity. (5)
casting and molding processes: processes in which a liquid material is poured into a cavity in a mold, where it solidifies into the proper shape and size. (16)
catalyst: a chemical agent used to cause a reaction. (29)
CD system: See *compact disc system.*
ceiling: the inside surface at the top of a room. (17)
ceiling joist: a beam, resting on an outside wall or an interior wall, that supports the weight of the ceiling. (17)
centerline: a line passing through the center of a hole. (12)
central processing unit (CPU): the heart of a computer. (23)
CEO: See *chief executive officer.*
ceramic mineral: a fine-grained mineral that is formable when wet and becomes hard when dried or fired. (14)
channel: a carrier. (19)
chart: a graphic model showing relationships among people, actions, or operations. (11)
chemical action: a change caused by adding chemicals. (16)
chemical conditioning: a type of conditioning process that uses chemical actions to change the properties of a material. (16)
chemical energy: energy stored within a chemical substance. (27)
chemical machining: a nontraditional machining process using chemical reactions to remove material from a workpiece. (16)
chemical process: a process that breaks down or builds up materials by changing their chemical compositions. (15)
chemical processing: using chemicals to change the form of materials. (30)
chief executive officer (CEO): the top manager in a company. (32)
chip removal: a separating process using a tool to cut away excess material in the form of small pieces, or chips. (16)
chop saw: a circular saw used to cut narrow strips of material to length. (7)
chuck: an attachment used to hold and rotate drills and router bits. (7)
CIM: See *computer-integrated manufacturing.*
circular saw: a saw that uses a blade in the shape of a disk with teeth arranged around the edge. (7)
civil engineering structure: a structure such as a bridge, an airport, or a highway that civil engineers design. (5)
civilized conditions: societies in which humans make tools, grow crops, engineer materials, and develop transportation systems. (1)
classification: the division of a problem into major segments. (10)
classroom training: a program in which specialized skills can be developed. (33)
clear-cutting: a logging method in which all trees, regardless of species or size, are removed from a plot of land. (14)

closed loop control: a type of control system that uses feedback. (6)
coal: a combustible solid composed mostly of carbon. (14)
cold bonding: a joining technique that uses extreme pressure to squeeze two parts to create a bond. (16)
combine: a machine that can be used to harvest a wide range of grains and other seed crops. (29)
commercial aviation: an industry that makes money by transporting people and cargo in airplanes. (25)
commercial building: a building used for business and government purposes. (17)
commercial ship: a large ship used for transporting people and cargo for a profit. (25)
commercial structure: a building used to conduct business. (5)
commercial transportation: enterprises that move people and goods for money. (26)
commission: a salesperson's pay, consisting of a percentage of the total dollar value of goods that person sells. (33)
communication and information technology: the technology used in processing data into information and communicating ideas and information. (3)
communication process: a process used to produce graphic and electronic media. (33)
communication technology: a system that uses technical means to transmit information or data from one place to another or from one person to another. (19)
compact disc (CD) system: an information-reproduction system using a laser to read optically encoded information, changing the optical signals into digital signals, and then recreating the information. (19)
composed: designed. (21)
composite material: a combination of natural and synthetic materials used to create items with other desirable properties. (4)
compound: to mix the parts of a fluid for casting. (16)
comprehensive layout: a layout for a final design. (20)
computer: an information-processing machine that has changed the way we handle information. This machine can store information, and its programs can be changed. (7)
computer-aided design (CAD): a computer-based system used to create, modify, and communicate a plan or product design. This system reduces product design and engineering costs. (33)
computer-aided design (CAD) software: software that can be used to produce engineering and architectural drawings. (12)
computer-aided design (CAD) system: a system using computers to create, change, test, and store drawings communicating design information. (12)
computer-controlled machining: a process that adds automatic control, provided by a computer program, to basic machines. (16)
computer-integrated manufacturing (CIM): the combination of a number of computer-based systems (such as CAD, CAM, and robotics) to improve manufacturing processes. (35)
computerized tomography (CT) scanner: a medical instrument that uses X rays and computers to create images of body organs. (31)
computer model: a model a computer produces that helps designers to test strengths of materials and structures and to observe the product during normal operation. (11)
conceptual model: a model showing a general view of the components and their relationships. (11)
condition: to alter and improve the internal structure of a material. This action changes the properties of the material. (5)
conditioning process: a process in which heat, chemicals, and mechanical forces are used to change the internal structure of a material. (16)
conduction: the movement of heat through a solid material or between two solid materials touching each other. (28)
conductor: a metal that conducts an electrical current. (22)
constraint: a limit on a design. (5)
construction: the process of using manufactured goods and industrial materials to build structures on a site. (13)
construction process: the type of activity producing structures using technological actions. (33)
construction technology: the technology used in building structures for housing, business, transportation, and energy transmission. (3)
consumer: a person who financially supports a technological system by spending money on products or services. (4)
consumer product: an output developed for end users in the product cycle. (13)
contact print: a print produced by a negative being placed directly on top of a piece of light-sensitive photographic paper. (21)
containership: a ship that carries quantities of goods sealed in large boxlike, metal containers. (25)

continuous manufacturing: a manufacturing system in which a production line assembles products continuously. (33)
contrast: the design principle used to emphasize portions of a message. (20)
control: the feedback loop that causes management and production activities to change through evaluation, feedback, and corrective action. (2)
control system: a vehicular system controlling the speed and direction of the vehicle. (25)
convection: the transfer of heat between or within fluids. (28)
convergent thinking: a type of thinking that seeks to narrow and focus ideas until the most feasible solution is found. (10)
conversion: a process that changes hydrocarbon molecules into different sizes, both smaller and larger. (15)
conversion and processing: the final step in agricultural practices that changes a food product into a foodstuff for human use. (5)
converted surface finish: a finish on the surface of a metal that has been chemically changed to protect the metal. (16)
copy: text. (20)
core: the center layer of plywood. (15)
corporate charter: an operating permit. (32)
corporate participation: the spirit of cooperation in developing and operating a technological system. (3)
corporation: a legal entity people form to own an operation. A corporation is a business in which investors have purchased partial ownership in the form of shares of stock. (4)
CPU: See *central processing unit.*
cracking: the process of breaking heavier hydrocarbons into smaller ones. (15)
crank: a pivot pin near the outside edge of a wheel or disk that changes reciprocating motion into rotating motion. (28)
creativity: the ability to see a need or a way of making life easier and design systems and products to meet the need or desire. (4)
criterion: a feature a product or system must have in order to meet the expectations of the customer. (5)
critical thinking: self-directed thinking that allows people to make judgments about what to believe or what to accept or reject about a claim. (9)
crop production: the type of agriculture growing large quantities of plants for food or other uses. (29)
cross band: a layer between the faces in plywood. (15)
CT scanner: See *computerized tomography scanner.*
cultivator: a machine used to control weeds. This machine uses a series of hoe-shaped blades pulled through the ground. (29)
cure: to add a combination of natural ingredients to meat. (30)
current technology: the range of techniques used to produce most products and services today. (3)
custom manufacturing: a manufacturing system that involves producing a limited quantity of a product to a customer's specifications. (33)
cutting motion: an action that causes material to be removed from a workpiece. (7)
cutting tool: a tool used in cutting actions. (7)
cycle: a complete set of motions needed to produce a surge of power. (7)
cylindrical grinder: a grinder that uses the lathe principle to machine a material. (7)

D

data: a raw, unorganized fact or figure. (4)
data-comparing device: a device that compares data gathered against a standard or other expectations. (6)
DC: See *direct current.*
debt financing: raising money by borrowing money from a financial institution or private investors. (4)
decode: to change coded information back into a recognizable form. (5)
dedicated access: a system in which a subscriber's computer is directly connected to the Internet at all times. (23)
degree of freedom: the limited number of ways or directions a vehicle can move. (25)
delayed output: a chemical that has accumulated over decades and is harmful to the environment. (6)
dental hygienist: an individual who cleans teeth and assists a dentist. (31)
dentist: an individual who diagnoses, treats, and helps prevent diseases of the teeth and gums. (31)
descriptive method: a method designers use to record observations of present conditions. (9)
design: a plan, or drawing, showing the appearance or operation of a technological device or system. To design is to create the appearance or operation of a technological device or system. (9)
desirable output: an output from agricultural and related biotechnology, communication and information technology, construction

technology, energy and power technology, manufacturing technology, medical technology, or transportation technology that benefits people. (6)

desktop publishing: a computer system that produces type and line-illustration layouts for printed messages. (20)

detail drawing: a drawing that communicates the designer's specifications and contains all the information needed to manufacture a particular part. (12)

detailed sketch: a sketch communicating the information needed to build a model of a product or structure. (10)

developing: the first step in developing film in which chemicals are used to alter light-sensitive crystals in an emulsion. (21)

development: the work technologists do by building products and structures to make lives better. (1)

diagnosis: the area of medicine that involves using knowledge, technological devices, and other means to determine the causes of abnormal body conditions. Diagnosis is performed by conducting interviews, physical examinations, and medical tests. (5)

diagram: a graphic model showing the relationships among components in a system. (11)

dial-up access: a system in which computers are connected to the Internet through a modem. (23)

dialysis machine: a device that can do some of the functions of the human kidneys. This machine filters waste and fluid from the blood. (31)

diaphragm: the aperture control of a camera. A diaphragm regulates the amount of light that can enter at any given moment. (21)

die: a forming tool made of hardened steel. (16)

digital signal: an electrical signal made up of discrete on-off pulses used to carry information or data. (22)

digital theater: a movie theater receiving and displaying digitally formatted movies. (21)

dimension line: a line between extension lines that has arrows pointing to the extension lines, indicating the range of a dimension. (12)

dip: to run stock through a vat of molten metal. (16)

direct active solar system: a type of active collector system that does not have a heat exchanger. (28)

direct current (DC): the flow of electrons in one direction through a conductor. (22)

direct-gain solar system: a system allowing radiant energy to enter a home through windows, heating inside surfaces. (28)

direction: the path a light wave travels. (21)

direction control: a type of control system that makes a vehicle change its direction. (25)

direct-reading measurement tool: a measurement tool an operator manipulates and reads. (8)

dirigible: a rigid airship with a metal frame covered with a skin of fabric. (25)

disease: any change interfering with the appearance, structure, or function of the body. (31)

distance: the separation between two points. Distance is also called *length*. (8)

distance multiplier: a simple machine that increases the amount of movement applied to the work at hand. (4)

distribution: physically moving a product from a producer to a consumer. (33)

divergent thinking: broad thinking that lets the mind soar and seeks to create as many different solutions as possible. (10)

dividend: a periodic payment to stockholders from a company's profits. (32)

domain name: the address of a site on the Internet. (23)

domestic transportation: transportation that takes place within the geographic boundaries of one country. (26)

drawing machine: a machine that pulls materials through die openings to form the materials. (16)

drift mining: a type of underground mining used when a coal vein extends to the surface of the earth. (14)

drilling: the process of obtaining materials by pumping them through holes drilled into the earth. (5)

drilling machine: a separating machine that produces or enlarges holes using a rotating cutter for the cutting motion. (7)

drip irrigation: a type of irrigation delivering water slowly to the base of plants. (29)

drug: a substance used to prevent, diagnose, or treat a disease. (31)

dry: to remove excess moisture from materials. (16)

dry-cargo ship: a ship used to haul both crated and bulk cargo. (25)

drywall: gypsum wallboard. Drywall is used as an interior wall covering. (17)

duplex system: a basic system used in mobile communication that uses two channels. (22)

dynamic process: a process that is constantly changing or causing change. (1)

E

Earth-orbit travel: a type of space travel represented by communication satellites and space shuttles. (25)
earth-sheltered building: a structure built into the earth to take advantage of the insulation value and thermal properties of soil. (34)
e-commerce: See *electronic commerce.*
economic activity: an activity including all trade in goods and services paid for with money. (33)
economic enterprise: an institution that designs, produces, and delivers the basic goods and services a society requires. (33)
economy: an institution that designs, produces, and delivers the basic goods and services a society requires. (33)
edger saw: a machine that has a number of blades on a shaft. (15)
EDM: See *electrical discharge machining.*
education: an institution that communicates information, ideas, and skills from one person to another and from one generation to another. (33)
edutainment: creating a situation in which people want to gain information. (19)
EKG machine: See *electrocardiograph machine.*
elastic range: the range between a material at rest and the material's yield point. (16)
electrical and electronic controllers: devices that control other devices to adjust the operation of machines. (6)
electrical discharge machining (EDM): a nontraditional process that uses electrical sparks to make a cavity in a piece of metal. (16)
electrical energy: energy associated with electrons moving along a conductor. (27)
electrical or electronic sensor: a type of monitoring device that can be used to determine the frequency of or changes in electric current or electromagnetic waves. (6)
electric fence: a fence that uses electrical charges to contain animals in a field. (29)
electrocardiograph (EKG) machine: a type of machine that produces a visual record of the heart's electrical activity. (31)
electrochemical process: a process that breaks down or builds up materials by changing their chemical compositions. (15)
electromechanical controller: an adjusting device that uses electromagnetic coils and forces to move control linkages and operate switches to adjust machines or other devices. (6)
electronic commerce (e-commerce): a type of commerce involving selling products and services over the Internet. (23)
electronic publishing: a complex system that can function as a typesetting and layout system. These systems produce and combine text and illustrations into one layout. (20)
electrostatic printing: a process for printing that uses a machine with a special drum. Electrostatic printing is also called *copying* or *photocopying.*
emergency medicine: an area dealing with unexpected illnesses and injuries. (31)
emerging technology: new technology that is not widely employed today, but might be commonly used in a later period of time. (3)
emotion: a feeling that can be communicated. (19)
employee relations: a program that recruits, selects, develops, and rewards a company's employees. (33)
employment: the task that involves determining a company's need for qualified workers. Employment is the first action in the employee-relations process. (33)
enamel: a varnish that has color pigment added. (16)
encode: to change a message into a format that can be transmitted. (5)
endoscope: a narrow, flexible tube containing a number of fiber-optic fibers that allows a physician to look inside the body. (31)
energy: the ability to do work. (2)
energy and power technology: the technology used in converting and applying energy to power devices and systems. (3)
energy conversion: the changing of one form of energy into another. (7)
energy-processing converter: a converter that processes energy in various ways. (7)
engineer: a person who conducts research and applies scientific and technological knowledge to the design and development of products, structures, and systems. (4)
engineering: an activity that develops the specifications for products, structures, processes, and services. (33)
engineering drawing: a document communicating the basic information needed to construct a manufactured product or structure. (12)
entertain: the goal of communication to amuse people as they participate in or observe events and performances. (19)
entrepreneur: a person with very special talents who looks beyond present practices and products and creates a business. (4)
equity financing: raising money by selling a portion of ownership in a company. (4)

ergonomics: the science of designing products and structures around the people who use them. Ergonomics is also called *human-factors analysis*. (11)
ethical drug: a drug physicians prescribe and pharmacists dispense. (31)
ethical information: information describing the values people have regarding devices and systems. (9)
evaporate: to extract minerals from the oceans using solar energy. (14)
exhaust: to entirely use up. (4)
exhaustible energy resource: a material that cannot be replaced. (27)
exhaustible material: a material that, once depleted, human action or nature cannot replace. (4)
exosphere: the last layer of space above the Earth that blends directly into outer space. (25)
expendable mold: a mold that is destroyed to remove the cast item. (16)
expense: money that pays for resources. (33)
experimental method: a method designers use to compare different conditions. (9)
exploded view: a drawing that shows the parts making up a product, as if the product is taken apart. (12)
extension line: a line indicating a point from which measurements are taken. (12)
external combustion engine: an engine powered by steam. (28)
external storage device: a device on which data can be stored outside the basic computer circuitry. (23)
extrusion: a process in which material is pushed through a hole in a die. (16)

F

face: an outside layer of plywood. (15)
family: an institution that provides a foundation for social and economic actions. (33)
fare: the cost of a ticket. (26)
fascia board: a type of board used to finish the ends of rafters and an overhang. (17)
fasten: to hold parts together. (16)
fax machine: a device that sends copies of documents over telephone lines or radio waves, using digital signals. (20)
FDM: See *frequency division multiplexing.*
feedback: the process of using information about the output of a system to regulate the inputs to the system. (2)
feed motion: an action that brings new material in contact with a cutting tool and allows the cutting action to be continuous. (7)
fell: to use a chain saw to cut down appropriate trees. (14)
fermentation: a technology that uses microorganisms to break down complex organic compounds into simpler substances. (28)
fertilizer: a liquid, powder, or pellet containing important chemicals that encourage and support plant growth. (29)
fiberglass: strands of glass used as the matrix for composite materials and insulation. (15)
fiber-optic cable: a strand of glass used to transmit voice, television, and computer data at high speeds. (15)
fiber optics: channeling messages, in the form of light, through glass fibers. (22)
field magnet: the stationary outer magnet used in electric motors. (7)
film speed: the measure of a photographic film's sensitivity to light. (21)
finance: the money and credit necessary for the economic system to operate. (2)
financial affairs: an activity that obtains, accounts for, and disburses the money and physical resources and maintains the financial records needed to manage a system. (3)
finish: to coat or modify the surfaces of parts and products to protect them or make them more appealing to consumers. (5)
finishing process: a secondary process that protects products and enhances their appearances. (16)
finite: having a limited quantity. (35)
firing: a thermal conditioning process used for ceramic products. (16)
first-class lever: a lever in which the fulcrum is between the load and the effort. (4)
fission: the process of splitting atoms to release vast quantities of energy. (27)
fixed-wing aircraft: passenger and cargo aircraft. (25)
fixing: the step in developing film that removes unexposed silver-halide crystals remaining in the emulsion. (21)
flame cutting: cutting material to size and shape using burning gases. (16)
flexible manufacturing: using a set of automatic, programmable machines to produce products in low-volume production runs. (7)
flexography: an adaptation of letterpress. Flexography uses a plastic or rubber image carrier. (20)
float glass: glass that is changed into sheets for windows and similar products by floating the molten glass on a bed of molten tin. (15)
flood irrigation: a system that uses a large quantity of water advancing across a field. (29)

floor joist: a beam that carries the weight of the floor. (17)
flow bonding: a method of joining materials using a metal alloy as a bonding agent. (16)
flow coating: a process that passes a product under a flowing stream of finishing material. (16)
fluidic controller: an adjusting device that uses fluids to adjust machines or other devices. (6)
fluid mining: a mining method in which hot water is pumped down a well into a mineral deposit. (14)
FM: See *frequency modulation.*
food-processing technology: the application of science and technology to select, process, and preserve food. (30)
foot-pound (ft.-lb.): a measurement of the amount of energy needed to move an object from one location to another. (27)
forage crop: a plant grown for animal feed. (29)
forced-air heating system: a type of system in which furnaces heat air as a conduction medium. (17)
force multiplier: a simple machine that increases the force applied to the work at hand. (4)
form: to squeeze or stretch materials into a desired shape. Forming also includes bending, shaping, stamping, and crushing. (5)
forming process: a process in which force applied by a die or roll is used to reshape materials. (16)
Forstner bit: a two-lipped woodcutter that produces flat-bottomed round holes. (7)
fossil fuel: an exhaustible resource that is a mixture of carbon and hydrogen. (14)
foundation: the base of a structure. (5)
fractional distillation: a process in which petroleum is pumped through tubes and heated until it becomes a series of hot liquids and vapors. (15)
fractionating tower: a tower used to separate the different liquids and vapors in petroleum. (15)
fracture point: the point at which a material cannot withstand any more force. (16)
freeze: to keep foods at or below 32°F. (30)
frequency: the number of cycles passing some point in one second. (22)
frequency division multiplexing (FDM): multiplexing that uses a separate frequency to transmit each message. (22)
frequency modulation (FM): a system that encodes a message on a carrier wave by changing its frequency. (22)
fruit: a plant grown in many areas of the country and cultivated for its edible parts. (29)
f-stop number: a number identifying the size of the opening in a camera's diaphragm. (21)
ft.-lb.: See *foot-pound.*
fuel converter: a device that converts fuel into energy. (28)
fulcrum: the support on a lever on which the lever arm rests and turns. (4)
function: a factor to consider before buying a product. (18)
furrow irrigation: a system that uses small ditches created between rows of plants. (29)
fuselage: the body of an aircraft containing the flight-crew, passenger, and cargo units. (25)
fusion: the process of combining two atoms into a new, larger atom to release large amounts of energy. (27)
fusion bonding: a bonding technique that uses heat or solvents to melt the edges of a joint. (16)
futures research: another term for *futuring.*
futuring: a research technique that helps people select the best of many possible courses of action. This technique is also called *futures research.*

G

galvanized steel: zinc-coated steel used for automobile parts and containers. (15)
gas: a material that easily disperses and expands to fill any space. (4)
gasification: a process in which a material is heated in the absence of oxygen. (28)
gear: a wheel with teeth on its circumference that changes the direction of a rotating force. (28)
gear and rack: a rotating gear that meshes with a bar that has gear teeth along its length. A gear and rack changes rotating motion into linear motion. (28)
gem: a stone cut, polished, and prized for its beauty and hardness. (14)
general aviation: travel for pleasure or business in an aircraft owned by a person or business. (25)
gene splicing: the process of producing an organism with a new set of traits. (29)
genetic engineering: a process producing new pest-resistant and chemical-tolerant crops that help combat diseases. (29)
genetic material: an organic material that has a life cycle and can be regenerated. These materials are obtained during the normal life cycles of plants or animals. (4)
geometry dimension: a dimension indicating the shapes of features and the angles at which surfaces meet. (12)

geosynchronous orbit: a type of orbit in which a satellite travels the same speed the Earth is turning. (25)
germinate: to give birth. (14)
glass: a material produced using thermal processes by solidifying molten silica in an amorphous state. (15)
global impact: an effect the action of a small percentage of the world's population has on the world as a whole. (35)
goal: the reason, or purpose, for a system. (2)
grain: a widely grown crop that has large edible seeds. This crop is a member of the grass family. (29)
grain drill: a seed planter pulled behind a tractor. (29)
graph: a graphic model allowing designers to organize and plot data. Graphs display numerical information that can be used to design products and assess testing results. (11)
graphic communication: a communications process in which messages are visual and have two dimensions. (5)
graphic model: a model used to explore ideas for components and systems. (11)
gravity dam: a dam in which the lakeside is vertical, while the other side slopes outward. (17)
gravure: a type of process that prints finely detailed items. (20)
greenhouse effect: the problem in which UV rays, combined with increased levels of carbon dioxide and other gases, cause Earth to retain more heat. (35)
grievance: a dispute that arises over a contract's interpretation. (33)
grinding machine: a machine that uses bonded abrasives to cut material. (7)
growth: a major step in agricultural practices that involves providing feed and water for animals or cultivating and watering crops. (5)
guidance: the system that gathers and displays information so a vehicle can be kept on course. (5)
guidance system: a vehicular system that provides information. (25)

H

hammer: a device that delivers force to complete a forming action. Hammers drop or drive a ram down with a quick action. (16)
hand tool: a simple, handheld artifact requiring human-muscle power, air, or electric power to make it work. (4)
hardening: a process used to increase the hardness of a material. (16)
hardwired system: a system that sends its signals through a physical channel. (22)
hardwood lumber: a type of lumber produced from trees that lose their leaves at the end of each growing season. (15)
harmony: the design principle achieved by blending the parts of a design to create a pleasing message. (20)
harvest: to remove edible parts of plants from trees and stocks and butcher animals to produce meat and other products for consumption. Harvesting is the process of gathering genetic materials from the earth or bodies of water at the proper stage of their life cycles. (5)
header: a part of a building framework that carries the weight from the roof and ceiling across the door and window openings. Headers are held up by trimmer studs. (17)
head rig: a very large band saw that cuts narrow slabs from a log. (15)
heat pump: a unit used in climate control that works as a cooling and heating system by capturing heat in the atmosphere. (17)
heat-treat: to use heat to change the properties of a material. (16)
heavy engineering structure: a structure, such as a bridge, a highway, or an airport, that helps our economy function effectively. (5)
hertz: the unit of measurement for the number of cycles per second. (22)
hidden line: a light, dotted line used to show details hidden in one or more of the views in a drawing. (12)
high tech: See *high technology.*
high technology (high tech): new technology that is not in wide use today, but might become common in time. This technology is also called *emerging technology.* (3)
historical information: information about devices and systems that were developed to solve problems similar to the current problem to be solved. (9)
historical method: a method designers use to gather information from existing records. (9)
horsepower: a measurement used to describe the power output of mechanical systems. (27)
hot water heating: an indirect climate-control system that uses water to carry heat. (17)
hovercraft: a special type of boat suspended on a cushion of air. (25)
HTML tag: See *hypertext markup language tag.*
hub-and-spoke system: a route pattern airlines use. (26)
human information: information affecting the acceptance and use of a device or system. (9)

humanities: a type of knowledge describing the relationships between and among groups of people. (1)
human-psyche future: a type of future dealing with the mental condition of people. (35)
human-to-human communication: a type of communication used to inform, persuade, and entertain other people. (19)
human-to-machine communication: a type of communication system that starts, changes, or ends a machine's operations. (19)
hybrid vehicle: a vehicle combining two or more sources of power. (5)
hydraulic system: a system that uses a liquid as the transfer medium. (28)
hydroelectric generating plant: a water-powered plant that uses a dam to develop a water reservoir. (28)
hydrofoil: a type of special boat that has a normal hull and a set of underwater wings. (25)
hydroponics: the growing of plants in nutrient solutions without soil. (29)
hyperlink: a connection between one Web page and another page or Web site through a URL. (23)
hypertext markup language (HTML) tag: a type of coding that tells a receiving computer how a page should look. (23)

I

IC: See *integrated circuit.*
idea: a mental image of what a person thinks something should be. (19)
ideation: a process in which designers create many possible answers by letting their minds create solutions. (5)
illness: a state of poor health. (31)
illustration: a picture or symbol that adds interest and clarity to printed communication. (20)
illustration preparation: an activity required for useful communication. Illustration preparation includes sizing and converting line art and photographs. (20)
image carrier: a printing block. (20)
immediate output: a product or service that has been designed and produced for immediate use. (6)
inclined plane: an application of the principle that it is easier to move up a slope than a vertical surface. (4)
income: money that is the end result of sales or employment. (33)
indirect active solar system: a system that has a series of collectors that absorb solar energy. (28)
indirect-gain solar system: a system that uses a black concrete wall with glass panels in front of it. (28)
indirect-reading measurement tool: a system bringing sensors and computers together to automate measurement. (8)
induction: the process in which magnetic lines of force increase and decrease in strength, causing electrons to flow in an adjacent wire. (22)
industrial building: a building that houses the machines that make products. (17)
industrial material: a material that is an input to secondary manufacturing activities. (13)
industrial product: an item companies use in conducting their businesses. (13)
industrial relations: the activities that develop and manage programs ensuring an efficient workforce and positive relations among the company, the workers, and the public. (3)
Industrial Revolution: the historical period from 1750 to 1850, when tremendous changes in technology occurred. (1)
industrial structure: a building housing machines that make products or used to store raw materials or finished products. (5)
industry: an economic enterprise that uses resources and systems to produce products, structures, and services with intent to make a profit. (33)
inexhaustible: unable to be entirely used up or consumed. (4)
inexhaustible energy resource: a part of the solar weather system existing on Earth. (27)
inexhaustible resource: a resource that is incapable of being used up. (35)
inform: the goal of communication to provide information about people, events, or relationships. (19)
information: facts and figures, called *data,* that have been sorted and arranged for human use. Information is vital to taking an active part in society. (2)
information age: occurring after the Industrial Revolution, this time period places most importance on information processing and cooperative working relations between production workers and managers. (1)
information processing: gathering, storing, manipulating, and retrieving information that can be found in books and photographs and on tape and film. (7)
information skills: the abilities to locate, select, and use information. (36)
infotainment: providing information in an entertaining way. (19)

ink-jet printing: a printing process in which a computer generates a printed message made up of tiny ink dots. (20)

inland-waterway transportation: transportation on rivers, on lakes, and along coastal waterways. (25)

inorganic material: a material that does not come from living organisms. (4)

input: a material that flows into a system and is consumed or processed by the system. (2)

input device: a device that allows an operator to enter data into a computer's operating system. (23)

input unit: a device used to enter data into a system. (7)

inside director: a top manager of a company who serves on the company's board of directors. (32)

insolation: the solar energy available in a specific location at any given time. (28)

inspection: the part of a quality control program that compares materials and products with set quality standards. (33)

integrated circuit (IC): a piece of semiconducting material, in which a large number of electronic components are formed. (6)

intended output: a product or service designed and produced with a specific goal in mind. (6)

interference: anything impairing the accurate communication of a message. (19)

intermittent manufacturing: a manufacturing system in which a group of products is manufactured to a company's or customer's specification. (33)

intermodal shipping: cargo traveling on two or more modes of transport before reaching its destination. (24)

internal combustion engine: a common power source in land vehicles, in which fuel is burned inside the engine to convert energy from one form to another. (25)

international transportation: a level of transportation that moves passengers and cargo between nations. (26)

Internet: an interconnected network of computers sharing information. (19)

Internet access: the way a computer is connected to the Internet. (23)

Internet Protocol (IP) address: the identifying number assigned to each computer connected to the Internet. (23)

Internet service provider (ISP): a company that forms a network into which home computers are connected. (23)

interstate commerce: business dealings that extend across state lines. (26)

intervention radiology: a technique using images that radiology produces for nonsurgical treatment of ailments. (31)

intrapreneur: a person who applies entrepreneurship skills within a company. (32)

invasive diagnostic equipment: a type of diagnostic device used when drawing and testing a blood sample. (31)

IP address: See *Internet Protocol address.*

Iron Age: the historical period beginning around 1200 BC. (1)

irradiation: a technology that uses gamma rays or X rays to kill most molds and bacteria that might be in food. (30)

irrigation: artificial watering to maintain plant growth. (29)

isolated solar system: a system that uses solar collectors separate from the house. (28)

isometric sketch: a sketch in which the angles that the lines in the upper-right corner form are equal. (10)

ISP: See *Internet service provider.*

J

J: See *joule.*

jet engine: an engine that powers business and commercial aircraft. (25)

JIT inventory-control system: See *just-in-time inventory-control system.*

joint: a place where parts meet. (16)

joule (J): one newton per meter. (27)

judgmental system: a system that uses human opinions and values to enter into the control process. (6)

just-in-time (JIT) inventory-control system: a type of inventory-control system that schedules materials to arrive at manufacturing sites when needed. (33)

K

KHz: See *kilohertz.*

kilohertz (KHz): a basic unit of measurement in telecommunication. A KHz equals 1000 cycles per second. (22)

kilowatt-hour: the work that 1000 watts complete in one hour. (27)

kinetic energy: energy involved in moving something. (27)

knowledge: information learned and applied to a task. (4)

L

labor agreement: a contract negotiated between a company and a union to establish pay rates, hours, and working conditions for all employees the contract covers. (33)
labor relations: programs that deal with employees' labor unions. (33)
lacquer: a solvent-based, synthetic coating that dries through solvent evaporation. (16)
LAN: See *local area network.*
landscape: to help prevent erosion and improve the appearance of a site. (17)
land-transportation system: a transportation system that moves people and goods on the surface of the earth from place to place. (24)
language and communication skills: the abilities to read, write, and speak a language. (36)
laser machining: a nontraditional process that uses the intense light a laser generates to cut material. (16)
lathe: a machine that produced a cutting motion by rotating the workpiece. (7)
layout: the stage in which a message is put together. Layout is the physical act of designing a message. (20)
legal information: information about the laws and regulations controlling the installation and operation of a device or system. (9)
lehr: an annealing oven. (15)
length: the separation between two points. Length is also called *distance.* (8)
lens: the part of a camera that focuses light on the film. (21)
letterpress: a type of relief printing that uses metal plates or type as the image carrier. (20)
lever: a simple machine that multiplies the force applied to it. This machine changes the direction of a linear force. (4)
lever arm: a rod or bar on a lever that rests and turns on the fulcrum. (4)
lifestyle: what a person does with business and family life. (36)
lift: the force holding or lifting a craft in the air. Lift is critical for all flight. (25)
lighter-than-air vehicle: an air vehicle that uses either a light gas or hot air to produce lift. (25)
limited liability: the feature of a corporation limiting an owner's loss, if the company fails, to the amount of money the owner has invested. (32)
linear motion: a cutting and feed motion in which the cutter or work moves in one direction along a straight line. (7)
link: a connection between one Web page and another page or Web site through a URL. (23)
liquefaction: a process in which a biofuel is heated at moderate temperatures under high pressure. (28)
liquid: a visible, fluid material that does not normally hold its size and shape. (4)
liquid-fuel rocket: a rocket with two tanks. (25)
liquidification: a process in which a material is heated in the absence of oxygen. (28)
lithographic printing: a method of printing that uses a flat-surface image carrier. Lithographic printing is also called *offset lithography.* (20)
local area network (LAN): a system used in a single building or site to connect several personal computers, or workstations, to a central server. (23)
location dimension: a dimension indicating the positions of features on an object. (12)
locomotive: in rail systems, the vehicle in which the power and operator units are placed. (25)
log: to cut down, trim, and haul off timber. (14)
lumber: wood used to make frameworks in residences or other types of structures. (15)
lumber-core plywood: a type of core used for plywood made from pieces of solid lumber that have been glued to form a sheet. (15)

M

machine: an artifact that amplifies the speed, amount, or direction of a force. Machines transmit or change the application of power, force, or motion. (2)
machine-to-human communication: a type of communication system used to display machine operating conditions. (19)
machine-to-machine communication: the type of communication computer-controlled operations use. (19)
machine tool: a machine used to make other machines. (7)
maglev train: See *magnetic levitation train.*
magnetic (electromagnetic) sensor: a monitoring device that can be used to determine whether or not changes are occurring in the amount of current flowing in a circuit. (6)
magnetic levitation (maglev) train: a transportation vehicle using magnetism to suspend and propel itself along a guideway. (25)
magnetic resonance imaging (MRI): an imaging technique that can produce computer-developed, cross-sectional images of any part of the body very quickly. (31)
maintenance: a type of program that strives to keep products or structures in good condition and in good working order. (18)

management: the act of planning, directing, and evaluating any activity. (32)
management process: the action people use to ensure that production processes operate efficiently and appropriately. These processes are also designed and used to guide and direct the design, development, production, and marketing of the technological device, service, structure, or system. (2)
manager: a person who organizes and directs the work of others in a business by setting goals, structuring tasks to be completed, assigning work, and monitoring results. (4)
manned spaceflight: a spaceflight carrying human beings into space and returning them safely to Earth. (25)
manual control system: a system requiring humans to adjust the process. (6)
manufacture: to change raw materials into useful products for public use. (13)
manufactured home: a special type of building mostly built in a factory. (17)
manufacturing process: a process used to change the sizes, shapes, combinations, and compositions of materials. (33)
manufacturing technology: the technology used in converting materials into industrial and consumer products. (3)
maritime shipping: water transportation on oceans and large inland lakes. (25)
marketing: the stage of promoting, selling, and delivering a product, structure, or service. (3)
market research: an activity that gathers information about a product's market. (33)
mass: the quantity of matter present in an object. (8)
mated die: a die that has the desired shape machined into one or both halves of the die set. (16)
material: a natural, synthetic, or composite substance from which artifacts are made. (2)
material processing: changing the form of materials using tools and machines. (7)
mathematical model: a model showing relationships in terms of formulas. (11)
maturity: the period of time in the life cycles of plants and animals when growth slows down as the plants and animals reach older age. (14)
measurement: the practice of comparing the qualities of an object to a standard. (8)
mechanic: a skilled worker in a service operation. (4)
mechanical conditioning: a type of conditioning process that uses mechanical forces to change the internal structure of a material. (16)
mechanical controller: an adjusting device that uses cams, levers, and other types of linkages to adjust machines or other devices. (6)
mechanical energy: energy produced by motion of technological devices. (27)
mechanical fastening: an assembling process that uses mechanical forces to hold parts together. (16)
mechanical process: a process that uses mechanical forces to change the form of natural resources. (15)
mechanical processing: a way food can be processed using machines to change the form of the food product physically. (30)
mechanical sensor: a monitoring device that can be used to determine the position of components, force applied, or movement of parts. (6)
mechanical transmission: a transmission that has a clutch between the engine and the transmission. (25)
mechanization: replacing human labor with machines that people or other machines operate or control. (35)
medical technologist: an individual who gathers and analyzes specimens to assist physicians in diagnosis and treatment. (31)
medical technology: the technology used in maintaining health and curing illnesses. (3)
medicine: an area that involves diagnosing, treating, and preventing diseases and injuries. (31)
megahertz (MHz): a basic unit of measurement in telecommunication. A MHz equals 1 million cycles per second. (22)
memory: a major part of a computer in which data and operating instructions are stored. (23)
memory unit: the section of the computer holding information and instructions. (7)
merchant ship: a cargo-carrying ship. (25)
mesosphere: the layer extending from 22 miles to 50 miles (35 km to 80 km) above Earth. (25)
metric system: a measurement standard based on a unit of length called a *meter.* (8)
MHz: See *megahertz.*
microgravity: very low gravity. (36)
micrometer: a measuring device used to establish precise diameters. (8)
Middle Ages: the historical period beginning around 400 AD. This period is known for its various upheavals, as tribes continually fought each other for territory, but technology still progressed. (1)
middle management: the level of management below the president and vice presidents of a company, but above operating management. (32)

military ship: a vessel owned by a government that provides for the defense of the country. (25)
milling: grinding or processing. (30)
milling machine: a separating machine that uses a rotating cutter for the cutting motion. (16)
mineral: any substance with a specific chemical composition that occurs naturally. (14)
mining: the process of obtaining materials from the earth through shafts or pits. (5)
mock-up: an appearance model designed to show people how a product or structure will look. (11)
model: to simulate expected conditions to test design ideas. (11)
monitoring device: a device that gathers data about the action being controlled. (6)
motion picture: a series of transparencies, shot over a span of time, that create the illusion of motion. (21)
MRI: See *magnetic resonance imaging.*
mud: a mixture of water, clay, and chemicals. (14)
multiple-point tool: a cutting device on which a series of single-point tools is arranged. (7)
multiplex: to increase the capacity of a waveguide. (22)
multiplex system: a basic system used in mobile communication that uses multiple frequencies to accommodate different types of units. (22)
multiview method: a drawing method that places one or more views of the object in one drawing. (12)

N

NAP: See *network access point.*
natural gas: a combustible gas occurring in porous rock. (14)
natural material: a material occurring naturally on Earth. (4)
negative film: film that produces a reverse image of the photographed scene. (21)
network: a connection through which computers can communicate with other computers. (23)
network access point (NAP): a point allowing a customer of one company to connect with a customer of another company. (23)
noise: unwanted sounds or signals that become mixed in with desired information. (19)
nonfood crop: a plant grown on farms (such as cotton) that are not for human consumption. (29)
noninvasive diagnostic equipment: a type of diagnostic device that gathers information about a patient without entering the body. (31)
nonmetallic mineral: a substance that does not have metallic qualities. (14)
nonrenewable resource: an exhaustible energy resource. (35)
nontraditional machining process: a process that uses electrical, sound, chemical, and light energy to size and shape materials. (16)
nuclear energy: the energy holding atoms together, which is released by using controlled reactions. (15)
nurse: an individual who helps physicians diagnose and treat illnesses and injuries. (31)
nurse practitioner: an individual who performs some of the basic duties physicians once provided. (31)
nut: a crop grown in selected parts of the world for its hard-shelled seeds. (29)

O

object line: a solid, dark line in a drawing that outlines an object and the object's major details. (12)
oblique sketch: a pictorial sketch showing the front view, as if a person is looking directly at the front. (10)
obsolete technology: technology that can no longer efficiently meet human needs for products and services. (3)
ocean mechanical energy–conversion system: a system that uses the mechanical energy in the oceans to generate power. (28)
ocean thermal energy–conversion (OTEC) system: a system that uses the differences in temperature between the various depths of the ocean to generate power. (28)
offset lithography: a method of printing that uses a flat-surface image carrier. (20)
one-view drawing: a drawing used to show the layout of flat, sheet metal parts. (12)
on-the-job training: a program through which an experienced worker teaches simple skills at a workstation. (33)
open die: a simple die consisting of two flat die halves. (16)
open loop control: a type of control system in which output information is not used to adjust the process. (6)
open-pit mining: a type of mining used when a coal vein is not very deep underground. (14)
operating management: the managers closest to the people producing a company's products and services. (32)
optical sensor: a type of monitoring device that can be used to determine the level of light or changes in the intensity of light. (6)
ore: a mineral that has a metal chemically combined with other elements. (14)

organic material: a material that comes from a living organism. (4)

organize: to divide tasks into major segments and structure a workforce so goals can be met and resources can be assigned to complete each task. (2)

orthographic assembly drawing: an assembly drawing that uses a single view to show the mating of parts. (12)

orthographic projection: the projection of a single view of an object onto a drawing surface in which the lines of projection are perpendicular to the drawing surface. (12)

OTEC system: See *ocean thermal energy–conversion system.*

outboard motor: a type of power source attached to the stern of a boat. (25)

output: a result, good or bad, of the operation of any system. (2)

output device: a device on which data can be viewed. (23)

output unit: a device used to display and record the results of the processing unit's actions. (7)

outside director: a person outside the managerial structure who is selected to serve on a company's board of directors. This director provides a different view of the company's operation. (32)

own: to have final authority over company activities. (32)

owners' manual: a manual containing information needed to complete periodic servicing of a device. (34)

ozone layer: the upper part of the stratosphere. (25)

P

package: to design, produce, and fill containers. (33)

page: a text file someone creates to share information or ideas. (23)

pagination system: a complex and expensive computer system that allows the operator to merge text and illustrations very accurately. (20)

paint: a coating that dries through polymerization. (16)

panchromatic film: a type of black-and-white film that reacts to all colors of visible light and records them as shades of gray. (21)

particleboard-core plywood: plywood that has a core made of particleboard. (15)

partnership: a form of private ownership in which two or more people own and operate a business. (4)

passenger ship: a vessel carrying people. (25)

passive collector: a solar collector that directly collects, stores, and distributes the heat it converts from solar energy. (28)

pasteup: a sheet that looks exactly as the finished message will look. (20)

pasteurization: a technology that uses heat to kill harmful microorganisms. (30)

pathway: a structure along which vehicles travel. (5)

people skills: the abilities needed to work with people in a cooperative way. (36)

perigee: the point at which an orbit comes closest to Earth. (25)

permanent mold: a mold that withstands repeated use. (16)

personal skills: the abilities to grow and manage personal actions on a job. (36)

personal transportation: travel using a vehicle one person owns. (26)

perspective sketch: a sketch showing an object as the human eye or a camera sees the object. (10)

persuade: the goal of communication to convince people to act in a certain way. (19)

petroleum: an oily, flammable, nonuniform mixture of a large number of different solid and liquid hydrocarbons. (14)

pharmacist: an individual who dispenses prescription drugs and advises people on their uses. (31)

photograph: a picture made on light-sensitive material using a camera. (21)

photographic communication: the process of using photographs to communicate an idea or information. (19)

photography: the act of producing a photograph. (21)

photovoltaic cell: a converter that generates an electrical current when light strikes it. (28)

physical model: a three-dimensional representation of reality. (11)

physician: an individual who diagnoses diseases and injuries. (31)

physician assistant: an individual who delivers basic health services under the supervision of a physician. (31)

pickle: to dip a material in a solvent to remove unwanted materials. Pickling is also called *chemical cleaning.*

pictorial assembly drawing: an assembly drawing that shows an assembly using oblique, isometric, or perspective views. (12)

pig iron: the basic input for steelmaking that results from thermal and chemical actions. (15)

pile foundation: a type of foundation used on wet, marshy, or sandy soils. (17)

pivot sprinkler: a type of sprinkler system that uses one long line attached at one end to a water source. (29)

place utility: a value provided by the movement of people and cargo from one place to another. (24)

plan: to set goals and develop courses of action for a company or parts of the company to reach the goals. (2)

planing machine: a machine tool that produces flat surfaces. These machines move a workpiece back and forth under a tool to generate a cutting motion. (7)

planning: the establishing of goals and objectives for a project. (5)

plastic range: the range in which a material can be stretched, compressed, or bent. (16)

plating: an electrolytic process. (16)

PLC: See *programmable logic controller.*

pleasure craft: a vessel that a private citizen owns for recreation. (25)

plow: a piece of tilling equipment that breaks, raises, and turns soil. (29)

pneumatic system: a system that uses air as the transfer medium. (28)

point of interest: the place to which your eye is drawn. (21)

point of presence (POP): a connection point that lets local users access a company's network. (23)

political power: the ability of a person or an organization to gain the attention and cooperation of elected officials. (36)

politics and law: an institution that establishes and enforces society's rules of behavior and conduct. (33)

pollution: a product of human activity that diminishes air or water quality. (35)

polymerization: a conversion process that causes small hydrocarbon molecules to join together. (15)

POP: See *point of presence.*

positive transparency: a slide. (21)

potable water: water safe for drinking. (17)

potential energy: stored energy. (27)

potential field: an area that has never produced oil or gas. (14)

power: the rate at which work is done. (27)

power-generation system: a system that uses an engine as an energy converter. (25)

power-transmission system: a system that controls and directs the power of an engine to do work. (25)

precision measurement: a type of measurement used when exact size is critical to the function of a device. (8)

preservative: a chemical added to food in small amounts to delay spoilage and ensure that the food retains its quality. (30)

president: the top manager in a company. (32)

press: a device in which force is delivered to complete a forming action. Presses slowly close die halves by lowering a ram to produce a squeezing action. (16)

press fit: a fit in which friction between parts causes them to remain together. (16)

pressure bonding: a bonding technique that applies heat and pressure to a bond area. (16)

prevention: the area of medicine that involves using knowledge, technological devices, and other means to help people maintain healthy bodies. (5)

preventive maintenance: maintenance designed to prevent breakdowns. (18)

price: what someone must pay to buy or use a product or service. (33)

primary food-processing technology: a type of process that produces the basic ingredients for food. (30)

primary process: a step in which material resources are converted into industrial materials. (5)

prime: See *prime interest rate.*

prime interest rate (prime): the low interest rate banks and insurance companies charge their safest customers. (32)

prime mover: a device that changes a natural source of energy into mechanical power. (28)

primitive conditions: conditions determined by nature. (1)

printed graphic communication: a mass-communication system that uses technology to communicate through a printed medium. (19)

printing: a communication system that places the images of written words on a material. (20)

private enterprise: an enterprise that individuals or groups of people own. (32)

problem-solving and design process: the procedure used to develop technology attempting to satisfy people's technological needs and wants. (2)

process: the steps needed to complete a series of identifiable tasks within a system. (2)

process development: a type of development that devises new or improved ways of completing tasks in manufacturing, construction, communication, or transportation. (33)

processing unit: the part of the computer, also called a *CPU* or *microprocessor,* that manipulates the data. (7)

product development: an area of development that applies knowledge to design new or improved products, structures, and services. (33)
production: the stage of developing and operating systems for producing a product, structure, or service. (3)
production process: the action completed to perform the function of a technological system. (2)
production worker: a person who processes materials, builds structures, operates transportation vehicles, services products, or produces and delivers communication products. (4)
profit: the amount of money left over after all the expenses of a business have been paid. (1)
program: the instructions a computer uses to process data and produce output. (7)
programmable logic controller (PLC): a device that uses a microprocessor to control machines or processes. (16)
projection printing: a basic technique used to produce prints by shining light through a negative onto a piece of photographic paper. (21)
propagation: a step in agricultural practices that allows a biological organism to reproduce. (5)
proportion: the design principle dealing with the relative sizes of the parts of a design. (20)
proprietary drug: a drug considered safe for unsupervised use by consumers. (31)
proprietor: the single owner of a business. (32)
propulsion: the system in a vehicle that generates motion through energy conversion and transmission. (5)
propulsion system: a vehicular system that produces a force to propel the vehicle from its starting point to its destination. (25)
prosthesis: an artificial body part developed through biomechanical engineering. (31)
prototype: a working model of a system, an assembly, or a product built to test the operation, maintenance, and safety of the item. (11)
proven reserve: a producing oil or gas field. (14)
public enterprise: an enterprise that the government or a special form of corporation controls. (32)
public relations: a program that communicates a company's policies and practices to governmental officials, community leaders, and the general population. (33)
pulley: a grooved wheel attached to an axle that can be used to change the direction of a force, multiply force, or multiply distance. (4)
pulleys and V belt: a technique in which two pulleys with a V belt stretched between them change the speed or power of a motion. (28)
pyrolysis: a thermochemical process in which materials are heated in the absence of oxygen. (28)

Q

quality control: a process including all the systems and programs that ensure the outputs of technological systems will meet engineering standards and customer expectations. (8)

R

radar: a system that can detect, locate, and determine the speed of distant objects with the use of radio waves. (11)
radial saw: a circular saw that moves a rotating blade across a workpiece. (7)
radiant energy: energy in the form of electromagnetic waves. (27)
radiation: heat transfer by using electromagnetic waves. (28)
radiation therapy: a treatment using high-energy radiation to treat cancer cells. (31)
radiology: methods using electromagnetic waves and ultrasonics to diagnose diseases and injuries. (31)
radio wave: a series of frequencies within the electromagnetic spectrum extending from around 30 hertz to 300 gigahertz. (22)
rafter: an angled board extending from the top plate of an exterior wall to the ridge of the roof. (17)
rail fence: a fence used as a border fence around a farm building or home. (29)
RAM: See *random-access memory.*
random-access memory (RAM): memory that both the computer and the user can read or change. (23)
read-only memory (ROM): memory that can be read by the computer, but cannot be changed. (23)
receive: to recognize and accept information. (5)
receiver: the end of the communication channel that gathers and decodes a message. (19)
reciprocating motion: a back-and-forth movement. (7)
recruit: to acquire job applicants through a searching process. (33)
recycle: to process used or abandoned materials to create new materials or products. (18)
refine: to remove impurities or unwanted elements from a material being processed. (15)

refined sketch: a sketch merging ideas from two or more rough sketches. (10)
reflect: to prevent passage of and cause to change direction. (21)
refrigerate: to maintain foods at temperatures from 32°F to 40°F. (30)
regulation: an order that an executive authority issues. (26)
reinforced concrete: concrete that has wire mesh or steel bars embedded into it to increase its tensile strength. (17)
relief printing: a printing process that uses an image on a raised surface. (20)
religion: an institution that develops and communicates values and beliefs about life and appropriate ways of living. (33)
Renaissance: the historical period that began in the early 1300s in Italy and lasted until 1600. This period is known for new ideas in art, literature, history, and political science, but technological developments, such as the calculator and the telescope, also occurred. (1)
renewable: capable of being used up, but replaceable with the normal life cycle of the energy source. (4)
renewable energy resource: a biological material that can be grown and harvested. (27)
renewable resource: a resource that has a life cycle. (35)
repair: the process of putting a broken, damaged, or defective product or building back into good working order. (13)
replacement: the step in the repair process in which worn or broken parts are replaced. (18)
resaw: a machine that uses a group of evenly spaced circular or scroll saw–type blades to cut many boards at once. (15)
research: the work scientists do by gathering information to try to explain why something exists or happens in a certain way. (1)
research and development: the stage of designing, developing, and specifying the characteristics of a product, structure, or service. These activities might result in new or improved products and processes. (3)
residential building: a structure in which people live. (17)
residential structure: a place where people live. (5)
responsibility: accountability for actions. (32)
retained earnings: profits held by a company and used to enlarge its operations. (33)
retrieve: to bring back information. (5)
reward: something received for some service or attainment. (32)
rhythm: the design principle dealing with the flow of a communication. (20)
risk: the possibility of loss. (32)
robot: a mechanical device that can perform tasks automatically or with varying degrees of direct human control. (16)
rolling machine: a machine that uses two rolls rotating in opposing directions to form a material. (16)
ROM: See *read-only memory.*
rotary motion: a cutting and feed motion in which the work or the tool rotates. (28)
rotary-wing aircraft: an aircraft that develops lift by spinning an airfoil. (25)
rotating motion: a motion that uses round cutters or spins a workpiece around an axis. (7)
rough sketch: a drawing that shows only basic ideas of the size, shape, and appearance of a product. These drawings integrate and refine ideas generated in the thumbnails. (10)
route: the path a vehicle follows. (26)
router: a specialized computer that determines how to send information from one computer to another. (23)
routine diagnostic equipment: a type of device used to gather general information about a patient. (31)
rudder: a large flat plate at the stern of a ship, used to guide the vessel. (25)
rule: a strip of metal, wood, or plastic with measuring marks on its face. A rule is the most common linear measurement device. (8)

S

salary: payment based on work over a period of time, such as a month or year. (33)
sales: the activity involving the physical exchange of products for money. (33)
sawing machine: a machine that uses blades with teeth to cut materials to desired sizes and shapes. (7)
schedule: a list of departure and arrival times for a trip. (26)
science: knowledge of the natural world. (1)
scientific information: information about natural laws and principles that must be considered in developing a solution. (9)
scientist: a person who generally develops a basic knowledge of physics, materials science, geology, or chemistry to help create products and processes. (4)
scope: panorama. (21)
screening: a type of process that allows qualified people to be selected from an applicant pool. (33)

screen printing: a printing process that uses stencils with openings the shape of the message. (20)
screw: an inclined plane wrapped around a shaft. (4)
script: a document identifying characters, developing a situation, and communicating a story. (22)
scroll saw: a saw with a blade that is a strip of metal with teeth on one edge. (7)
search engine: a special site on the Internet operating on the principle of key words. (23)
secondary food-processing technology: a type of process used to make finished food products. (30)
secondary manufacturing process: the action used to change industrial materials into products. (16)
secondary process: a manufacturing process that changes industrial materials into industrial equipment and consumer products. (5)
second-class lever: a lever in which the load is between the effort and the fulcrum. (4)
seed-tree cutting: a logging method in which all trees, regardless of species, are removed from a large area, except three or four per acre. (14)
seismographic study: an accurate way to explore for petroleum and natural gas using shock waves. (14)
selective cutting: a logging method in which mature trees of a desired species are selected and cut from a plot of land. (14)
separate: to use tools to shear or machine away unwanted material. (5)
separating process: a process in which tools or machines are used to remove excess material to make an object of the correct size and shape. (16)
separation: a process used in petroleum refineries that breaks petroleum into major hydrocarbon groups. (15)
server: a special computer used to store programs and data for a network. (23)
service: routine tasks that keep a product operating. (5)
shaft mining: a type of underground mining that requires a vertical shaft to reach a coal deposit. (14)
shaped die: a die used to form plastic objects. (16)
shaping machine: a metalworking machine tool that produces flat surfaces. These machines move a single-point tool back and forth over a workpiece to produce a cutting motion. (7)
share of stock: a portion of the ownership of a company. (32)
shear: to use opposing edges of blades, knives, or dies to fracture unwanted material away from a workpiece. (16)
sheathing: a covering. (17)
shutter: the device that opens and closes to permit light into or prevent light from entering a camera. (21)
silk screening: a printing process that uses stencils mounted on silk fabric. Silk screening is also called *screen printing*. (20)
sill: a wood piece attached to the top of a foundation. (17)
simplex system: a basic system used in mobile communication that uses the same channel for both base and mobile transmissions. (22)
simulation: the process of imitating expected conditions to test design ideas. (11)
single-point tool: a simple cutting device with a cutting edge on the end or along the edge of a rod, bar, or strip. (7)
site preparation: the step in a construction project that involves removing existing buildings, structures, brush, and trees that will interfere with locating the new structure. (5)
size dimension: a dimension indicating the size and major features of an object. (12)
skelp: a strip of steel used to form pipes. (15)
slab: a wide, flat piece of steel. (15)
slab foundation: a type of foundation used for structures erected on soft soils. (17)
slide: a single transparency designed to be viewed independently. (21)
slope mining: a type of underground mining used when a coal vein is not too deep under the ground. (14)
smart house: a house that allows computers to control appliances and energy use. (1)
smelt: to extract metals from their ores using heat. (15)
smoke: to add flavor to meat and fish, while preserving them. (30)
social future: a type of future suggesting the types of relationships people want with each other. (35)
socioethical skills: the skills involving the understanding of the implications of actions on people, society, and the environment. (36)
soffit: the underside of a building's roof edge. (17)
softwood lumber: a type of lumber produced from needle-bearing trees. This lumber is used for construction, for shipping containers and crates, and for railroad ties. (15)

solar collector: a high-temperature device heated by concentrated solar energy. (2)
solar converter: an inexhaustible-energy converter that uses the constant energy source of the Sun. (28)
solar weather system: the natural cycle starting with solar energy. (27)
sole plate: the strip at the bottom of a framed wall. (17)
sole proprietor: one person who owns a business or an operation. (4)
solid: a material that holds its size and shape and can support loads. (4)
solid-fuel rocket: a rocket that uses a powder or spongelike mixture of fuel and oxidizer. (25)
solid model: a complex computer model that takes into account both the surface and the interior substance of an object. (11)
space-transportation system: a mode of transportation that uses manned and unmanned flights to explore the universe. (24)
spade bit: a flat cutter on the end of a shaft, used to drill holes. (7)
specification sheet: a document communicating the properties a material must possess for a specific application. (12)
speed control: a type of control system that makes a vehicle go faster through acceleration or slower by braking or coasting. (25)
spray: to use air to carry fine particles of finishing materials to the surface of a product. (16)
spread foundation: a type of foundation used on rock and in hard soils. (17)
sprinkler irrigation: an irrigation system producing artificial rain to water crops. (29)
square: an angle-measuring device that has a blade at a right angle to the head. (8)
standard measurement: a type of measurement in which the exact size of a part is not critical to the function of a product. (8)
standard view: a drawing that shows a product in one piece, as the product is after it is assembled. (12)
steel: an alloy of iron and carbon used for frameworks in industrial and commercial buildings. (15)
Stone Age: the earliest period in history. (1)
stop: to end the chemical action of a developer. (21)
stop bath: an acidic solution that neutralizes developers. (21)
store: to retain information for later use. (5)
stratosphere: the region above the troposphere. (25)
stroke: the movement of a piston from one end of a cylinder to another. (7)
structural system: a vehicular system that helps people arrive safely and in comfort and protects cargo. (25)
structure: the system that provides spaces for devices in vehicles. (5)
structure development: an area of development that applies knowledge to design new or improved structures. (33)
stud: an upright nailed to a sole plate. (17)
subfloor: a base, usually made from plywood or particleboard, that goes under tile, carpeting, or other flooring materials. (17)
submersible: a type of special boat that can travel on the surface of water or underwater. (25)
substrate: a material on which a printing is applied. (20)
superstructure: the framework of a building or tower constructed on a foundation. A superstructure also includes the pipes for pipelines, surfaces for roads and airport runways, and tracks for railroads. (5)
supervisor: a manager closest to the people producing a company's products and services. (32)
support staff: a nonmanagerial worker who carries out such tasks as keeping financial records, maintaining sales documents, and developing personnel systems. (4)
support system: the external operations and facilities that maintain transportation systems. (5)
surface grinder: a grinding machine that works on the metal-planer principle. (7)
surface mining: a type of coal mining used when the coal vein is not very deep underground. (14)
surface model: a three-dimensional computer model that is a wire frame with a sheet. (11)
surgery: a way to treat diseases and injuries with operations. (31)
suspension: the subsystem that maintains a vehicle on a pathway. (5)
suspension bridge: a bridge that uses cables to carry loads. (17)
suspension system: a vehicular system that produces proper support for the weight of the vehicle and cargo. (25)
swather: a machine that cuts and windrows hay in one pass over a field. (29)
synergism: the concept that solutions one or more individuals in a group propose often cause other members of the group to think of more ideas. (10)

synthetic material: a human-made material. (4)

system: a group of parts working together in a predictable way, designed to achieve a goal. (2)

systems drawing: a drawing used to show how parts in a system relate to each other and work together. (12)

T

table saw: a circular saw that uses a linear feed of a material. (7)

tail assembly: a structure that provides steering capability for an aircraft. (25)

tanker: a large vessel used to move liquids across oceans. (25)

tapping: the final step in the process of making steel in a basic oxygen furnace. (15)

TDM: See *time division multiplexing.*

technical data sheet: a document communicating the specifications for manufactured products. (12)

technical graphic communication: a type of system that prepares and reproduces engineering drawings and technical illustrations. (19)

technician: a skilled worker in a laboratory or product-testing facility. These workers work closely with production workers, but they do more specialized jobs. (4)

technological assessment: an evaluation involving groups of people evaluating the impacts of technology on people, society, and the environment. (34)

technological future: a type of future looking at the type of human-built world we desire. (35)

technological information: facts about how human-designed devices and systems operate. (9)

technologically literate: having the understanding and ability to direct new technology. (1)

technological opportunity: a nonproblem condition that can be improved with technology. (9)

technological problem: a problem that can affect individuals and groups of people. These problems can be solved with devices or systems. (9)

technological product: an artifact people built. (34)

technological service: an output we use, but do not own. (34)

technology: humans using objects to change the natural and human-made environments. (1)

telecommunication: a communication process that depends on electromagnetic waves to carry a message over a distance. (5)

telecommunications technology: the transmission of information over a distance for the purpose of communication. (5)

temper: to relieve internal stress in a part. (16)

temperature: the measurement of how hot or cold a material is. (8)

terminal: a structure where transportation activities begin and end. Terminals house passenger and cargo storage and loading facilities. (5)

testing: the step in the repair process in which a repaired product is checked to ensure it works properly. (18)

thermal conditioning: a conditioning process using heat. (16)

thermal energy: heat energy. (27)

thermal process: the type of process that uses heat to melt and reform natural resources. (15)

thermal processing: a food-processing method that uses heat as the primary energy to convert the food. (30)

thermal sensor: a monitoring device that can be used to determine changes in temperature. (6)

thermochemical conversion: a basic process that produces a chemical reaction by applying heat. (28)

thermosphere: the region lying just above the mesosphere. (25)

thinking skills: the skills involving the ability to use mental processes to address problems and issues. (36)

third-class lever: a lever in which the effort is placed between the load and the fulcrum. (4)

three-view drawing: a multiview drawing used to show the sizes and shapes of rectangular and complex parts. (12)

thumbnail sketch: a sketch that allows a graphic designer to experiment with various arrangements of copy and illustrations. (20)

tillage: the process of breaking and pulverizing soil to condition a seedbed. (29)

timber cruising: a process in which foresters measure the diameters and heights of trees to find stands of trees that can be economically harvested. (14)

time: the measurement of how long an event lasts. Time is a key resource in developing and operating technological systems. (2)

time division multiplexing (TDM): multiplexing that divides time into very brief segments. (22)

tin plate: tin-coated steel used to make food cans. (15)

tolerance: a number indicating the amount of deviation allowed in a dimension. (12)

tool: an artifact humans use to expand their capabilities. (2)
top management: presidents and vice presidents. (32)
top plate: a double ribbon of 2 × 4s. (17)
tractor: the part of a long-distance truck where the power and operator units are placed. (25)
trailer: a cargo unit attached to a tractor. (25)
transducer: a technological device that changes energy of one form into energy of another form. (22)
transmit: to send a message from one person or place to another. (5)
transportation: all activities that relocate humans or their possessions. (24)
transportation process: a process used to move people and cargo. (33)
transportation technology: the technology used in moving people and cargo from one place to another. (3)
treat: to add or remove chemicals to change the properties of petroleum products. (15)
treatment: the area of medicine that involves using knowledge and technological devices and applying medical procedures to fight diseases, heal injuries, or ease symptoms. (5)
trim saw: a machine that has a series of spaced blades. (15)
troposphere: the lowest region of space above Earth in which spacecraft and satellites operate. (25)
truss: a triangle-shaped structure that includes both the rafter and ceiling joist in one unit. (17)
truss bridge: a bridge that uses small parts arranged in triangles to support the deck. (17)
turbofan engine: an engine used in most commercial aircraft in use today. This engine operates at lower speeds than a turbojet engine does. (25)
turbojet engine: a type of jet engine developed during World War II. (25)
turboprop engine: a variation of the jet engine that operates efficiently at low speeds. (25)
turning machine: a separating machine that rotates a workpiece against a single-point tool to produce a cutting motion. (7)
twist drill: a shaft of steel with points on the ends to produce chips. (7)
two-view drawing: a multiview drawing used to show the size and shape of cylindrical parts. (12)
typesetting: an activity that produces the words of a message. (20)

U

UHF station: See *ultrahigh frequency station.*
ultrahigh frequency (UHF) station: a television station broadcasting within a range of 300 MHz and 3 GHz (3000 MHz). (22)
ultrasound: an imaging technique that uses high frequency sound waves and their echoes to develop an image of the body. (31)
underground mining: a type of mining that requires shafts in the earth to reach the coal deposits. (14)
undesirable output: an output that is not wanted. (6)
uniform resource locator (URL): a code that is a method of naming documents or sites on the WWW. (23)
unintended output: an output that was not considered when the system was designed. (6)
unlimited liability: a disadvantage in which proprietors cannot separate business income and liabilities from their personal finances. (32)
unmanned spaceflight: a spaceflight that uses rockets to place payloads into orbit. (25)
URL: See *uniform resource locator.*
U.S. customary system: a type of system used as a measurement standard in the United States today. (8)
utility: a system of a structure that provides water, electricity, heat, cooling, or communications. (5)

V

value: a measure of the functional worth a customer sees in a product. Value is a factor to consider before buying a product. (18)
variety: the design principle that makes a message unique and interesting. (20)
varnish: a clear finish made from a mixture of oil, resin, solvent, and a drying agent. (16)
vegetable: an important farm crop. Vegetables have edible leaves, stems, roots, and seeds that provide important vitamins and minerals for the daily diet. (29)
vehicle: a technological artifact designed to carry people and cargo on a pathway. (24)
vehicular system: an onboard technical system that makes a vehicle work. (5)
veneer: a thin sheet of wood sliced, sawed, or peeled from a log. (15)
veneer-core plywood: the most common type of core used for plywood. (15)

very high frequency (VHF) station: a television station broadcasting within a range of 30 MHz and 300 MHz. (22)
VHF station: See *very high frequency station.*
vice president: a manager who reports to the president or CEO of a company. (32)
video recording: a recording that contains both visual and audible components. (22)
virtual reality: a computer interface allowing a user to interact with three-dimensional, computer-generated images. (23)
volume: the amount of space an object occupies or encloses. (8)

W

wage: a set rate paid for each hour worked. (33)
WAN: See *wide area network.*
wastewater: the water from sinks, showers, tubs, toilets, and washing machines. This water is drained away by part of the plumbing system. (17)
water cycle: a cycle powered partially by solar energy. (27)
water-transportation system: a system that uses water to support a vehicle. (24)
waterwheel: a series of paddles extending into flowing water, which produces a rotating mechanical motion. (28)
watt: one J of work per second. (27)
wavelength: the distance from the beginning to the end of one wave cycle. (21)
wedge: a device used to split and separate materials and to grip parts. (4)
weight: the force of the Earth's pull on a mass. (8)
wellness: a state of physical well-being. (31)
what-if scenario: an outrageous proposal that can lead to solutions after its good and bad points have been investigated. (10)
wheel and axle: a shaft attached to a disk. (4)
wide area network (WAN): a type of computer network covering large geographic areas. (23)
windmill: a wind-driven wheel producing a rotating mechanical motion. (28)
windrow: a band of hay. (29)
wind tunnel: a device used to test the aerodynamics of vehicles and structures under controlled conditions. (33)
wing: the part of an aircraft separating the air into two streams, providing lift. (25)
wire-frame model: a three-dimensional computer model developed by connecting all the edges of an object. (11)
work: applying a force that moves a mass a distance in the direction of the applied force. (27)
World Wide Web (WWW): a computer-based network of information resources. (23)
woven-wire fence: a fence that uses posts with a special wire product attached to them. (29)
WWW: See *World Wide Web.*

X

X-ray machine: a diagnostic imaging machine that essentially is a camera. (31)

Y

yard: to gather logs in a central location. (14)
yield point: the point at which a material does not return to its original shape after being stretched. (16)

Index

A

B

C

D

E

F

G

H

J

K

L

M

Q

R

S

T

U

V

W